BEGINNING ALGEBRA

edition 5

BEGINNING ALGEBRA

James Streeter
Late Professor of Mathematics
Clackamas Community College
Donald Hutchison
Clackamas Community College
Barry Bergman
Clackamas Community College
Louis Hoelzle
Bucks County Community College

Boston Burr Ridge, IL Dubuque, IA Madison, WI
New York San Francisco St. Louis
Bangkok Bogotá Caracas Lisbon London Madrid Mexico City
Milan New Delhi Seoul Singapore Sydney Taipei Toronto

McGraw-Hill Higher Education
*A Division of The **McGraw-Hill** Companies*

BEGINNING ALGEBRA, FIFTH EDITION

Published by McGraw-Hill, an imprint of The McGraw-Hill Companies, Inc., 1221 Avenue of the Americas, New York, NY 10020.

Some ancillaries, including electronic and print components, may not be available to customers outside the United States.

 This book is printed on recycled, acid-free paper containing 10% postconsumer waste.

1 2 3 4 5 6 7 8 9 0 VNH/VNH 0 9 8 7 6 5 4 3 2 1 0

ISBN 0–07–231693–4
ISBN 0–07–237719–4 (AIE)

Vice president and editor-in-chief: *Kevin T. Kane*
Publisher: *JP Lenney*
Senior sponsoring editor: *William K. Barter*
Developmental editor: *Burrston House*
Editorial coordinator: *Beatrice Wikander*
Marketing manager: *Mary K. Kittell*
Project manager: *Susan J. Brusch*
Media technology lead producer: *Steve Metz*
Production supervisor: *Enboge Chong*
Designer: *K. Wayne Harms*
Interior designer: *Sheilah Barrett*
Cover designer: *Jamie O'Neal*
Cover credits: Train Photo: *Paul Chesley/National Geographic Image*
Senior photo research coordinator: *Carrie K. Burger*
Supplement coordinator: *Brenda A. Ernzen*
Compositor: *Interactive Composition Corporation*
Typeface: *10/12 NewTimes Roman*
Printer: *Von Hoffmann Press, Inc.*

Photo Credits

Chapter Openers: 0: © PhotoDisc website; 1: © R. Lord/The Image Works; 2: Alan Levenson/Tony Stone Images; 3, 4, 5: © PhotoDisc website; 6: © Memorable Moments/PhotoDisc, Inc.; 7: © Hank Morgan/Photo Researchers, Inc.; 8: © Charles Thatcher/Tony Stone Images; 9: © Small Business/Corbis CD; 10: © Kathy McLaughlin/The Image Works

Interior: Page 184, p. 196: © Lifestyles Today/PhotoDisc; p. 198: © Business & Industry/PhotoDisc; p. 266: © Spacescapes/PhotoDisc; p. 268: © Earth in Focus/PhotoDisc; p. 383: © Banking & Finance/PhotoDisc; p. 450: © Outdoor Celebrations and Lifestyles/PhotoDisc; p. 456: © American Vignette/Corbis CD; p. 490: © People and Lifestyles/PhotoDisc; p. 540: © American Vignette/Corbis CD; p. 563: © Far Eastern Business & Culture/PhotoDisc; p. 595: © Banking & Finance/PhotoDisc; p. 655: © Manufacturing & Industry/Corbis CD; p. 678: © Home & Family/Corbis CD; p. 729: © Nature, Wildlife, and Environment, 2/PhotoDisc; p. 738: © Homes and Gardens/PhotoDisc; p. 775: © Homes and Gardens/PhotoDisc; p. 801: © Sporting Goods/PhotoDisc

www.mhhe.com

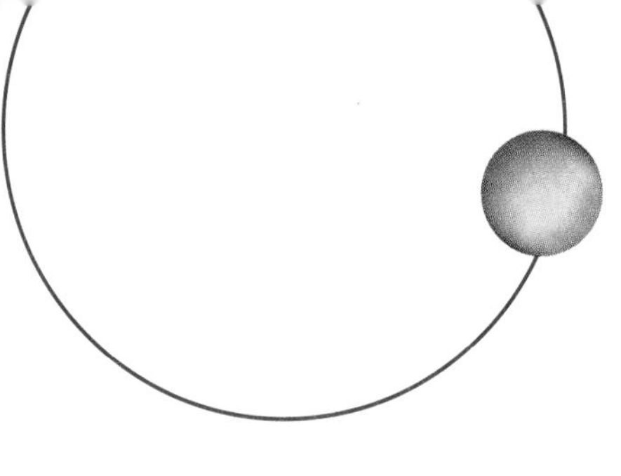

About the Authors

Don Hutchison spent his first 10 years of teaching working with disadvantaged students. He taught in an inner city elementary school and an inner city high school. Don also worked with physically and mentally challenged children in state agencies in New York and Oregon.

In 1982, Don completed his graduate work in mathematics. He was then hired by Jim Streeter to teach at Clackamas Community College. Through Jim's tutelage, Don developed a fascination with the relationship between teaching and writing mathematics. He has come to believe that his best writing is a result of his classroom experience, and his best teaching is a result of the thinking involved in manuscript preparation.

Don is also active in several professional organizations. He was a member of the ACM committee that undertook the writing of computer curriculum for the 2-year college. From 1989 to 1994 he was chair of the AMATYC Technology in Mathematics Education committee. He was president of ORMATYC from 1996 to 1998.

Barry Bergman has enjoyed teaching mathematics to a wide variety of students over the years. He began in the field of adult basic education, and moved into the teaching of high school mathematics in 1977. He taught at that level for 11 years, at which point he served for a year as a K-12 mathematics specialist for his county. This work allowed him to help promote the emerging NCTM standards in his region.

In 1990 Barry began the present portion of his career, having been hired to teach at Clackamas Community College. He maintains a strong interest in the appropriate use of technology in the learning of mathematics.

Throughout the past 23 years, Barry has played an active role in professional organizations. As a member of OCTM, he contributed several articles and activities to the group's journal. Recently, he served as an officer of ORMATYC for 4 years, and participated on an AMATYC committee to provide feedback and reactions to NCTM's revision of the standards.

Louis Hoelzle has been teaching at Bucks County Community College for 30 years. In 1989, Lou became chair of the Mathematics Department at Bucks County Community College. He has taught the entire range of courses from arithmetic to calculus, giving him an excellent view of the current and future developmental needs of students.

Over the past 36 years, Lou has also taught physics courses at 4-year colleges, which has enabled him to have the perspective of the practical applications of mathematics. Lou has always focused his writing on the student.

Lou is also active in several professional organizations. He has served on the Placement and Assessment committee and the Grants committee for AMATYC. He was president of PSMATYC from 1997 to 1999.

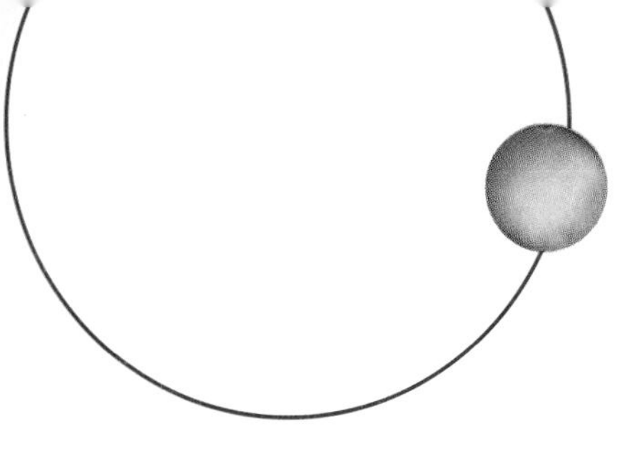

Dedication

This text is dedicated to the teachers who have inspired us with their compassion, energy, and ingenuity. In particular, we wish to acknowledge the many contributions of Marj Enneking, Sam Ensor, and Linda Foreman, who taught us how to teach.

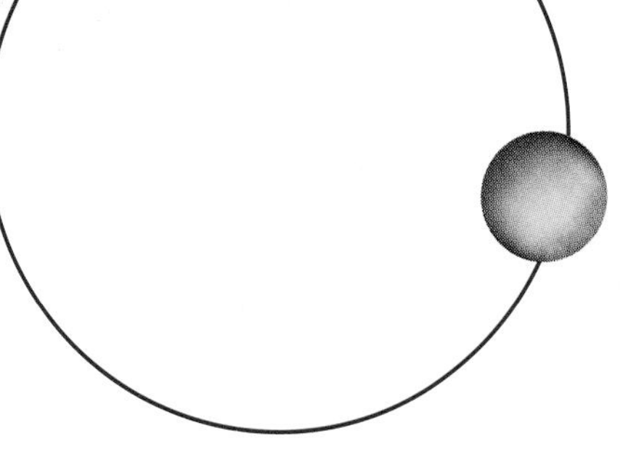

Table of Contents

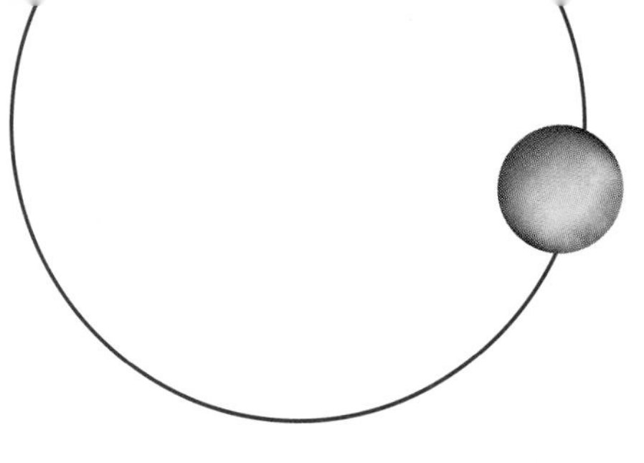

Preface

STATEMENT OF PHILOSOPHY

We believe that the key to learning mathematics, at any level, is active participation. When students are active participants in the learning process, they have the opportunity to construct their own mathematical ideas and make connections to previously studied material. Such participation leads to understanding, success, and confidence. We developed this text with that philosophy in mind and integrated many features throughout the book to reflect that philosophy. The *Check Yourself* exercises are designed to keep the students active and involved with every page of exposition. The calculator references involve students actively in the development of mathematical ideas. Almost every exercise set has application problems, challenging exercises, writing exercises, and/or collaborative exercises. Each exercise is designed to awaken interest and insight within students. Not all of the exercises will be appropriate for every student, but each one provides another opportunity for both the instructor and the student. Our hope is that every student who uses this text will be a better mathematical thinker as a result.

FEATURES OF THE FIFTH EDITION

In the previous four editions of *Beginning Algebra,* we have been fortunate enough to have great instructors to work with at Clackamas Community College and Bucks County Community College. Coupled with the excellent reviews provided by McGraw-Hill, we have been able to put together a text that reflected the ideas of good teachers and the needs of a variety of students.

In this edition, McGraw-Hill, in conjunction with Burrston House, gave us even more. We were given the opportunity to participate in discussion groups with faculty in seven different states. These groups, a combination of users of this text and at least nine other texts, inspired us with tough questions, creative solutions, and helpful suggestions. It was a testament to the value of group work. Many of those discussion participants are mentioned in the acknowledgements that follow this preface. All of them deserve thanks.

For those who have used previous editions of the text, we will describe the specific changes from the fourth edition.

1. A Pre-Test has been added for each chapter.

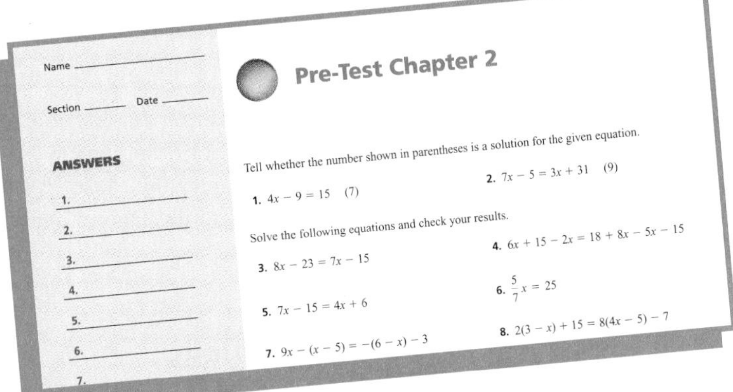

Name

Section Date

ANSWERS

1.
2.
3.
4.
5.
6.
7.

Pre-Test Chapter 2

Tell whether the number shown in parentheses is a solution for the given equation.

1. $4x - 9 = 15$ (7)
2. $7x - 5 = 3x + 31$ (9)

Solve the following equations and check your results.

3. $8x - 23 = 7x - 15$
4. $6x + 15 - 2x = 18 + 8x - 5x - 15$
5. $7x - 15 = 4x + 6$
6. $\frac{5}{7}x = 25$
7. $9x - (x - 5) = -(6 - x) - 3$
8. $2(3 - x) + 15 = 8(4x - 5) - 7$

2. An arithmetic review has been added (Chapter 0).
3. The treatment of signed numbers has been condensed from a chapter to three sections (1.2, 1.3, and 1.4).

4. Each of the first five chapters opens with a suggestion on "Overcoming Math Anxiety." These suggestions are designed to be timely and useful. They are the same suggestions most of us make in class, but sometimes those words are given extra weight when students see them in print.

Overcoming Math Anxiety

Throughout this text, we will present you with a series of class-tested techniques that are designed to improve your performance in this math class.

Hint #1 Become familiar with your text book.

Perform each of the following tasks.

1. Use the Table of Contents to find the title of Section 5.1.
2. Use the index to find the earliest reference to the term *mean.* (By the way, this term has nothing to do with the personality of either your instructor or the text book author!)
3. Find the answer to the first Check Yourself exercise in Section 0.1.
4. Find the answers to the Self-Test for Chapter 1.
5. Find the answers to the odd-numbered exercises in Section 0.1.
6. In the margin notes for Section 0.1, find the definition for the term *relatively prime.*

Now you know where some of the most important features of the text are. When you have a moment of confusion, think about using one of these features to help you clear up that confusion.

5. The treatment of graphing has been expanded to two chapters (Chapters 6 and 7).
6. The material on algebraic fractions has been reordered. Addition and subtraction are discussed before multiplication and division.
7. The material on complex fractions has been included in the treatment of division of fractions rather than in a separate section.
8. A new section on percent applications has been added (2.6).
9. A new section on direct variation has been added (6.5).
10. The material on probability has been deleted.
11. A new section on systems of linear inequalities has been added (8.4).
12. The material on exponents has been condensed into one section (3.1) prior to the treatment of polynomials.

Each of the features of this edition was scrutinized by the discussion groups. Almost all were modified in some way. Every supplement to this text was thoroughly discussed. More than ever, we are confident that the entire learning package is of value to students of beginning algebra. We will describe each of the features of that package.

PEDAGOGICAL FEATURES

Application Areas

Each chapter opens with a real-world vignette that showcases an example of how mathematics is used in a wide variety of jobs and professions. Problem sets for each section then

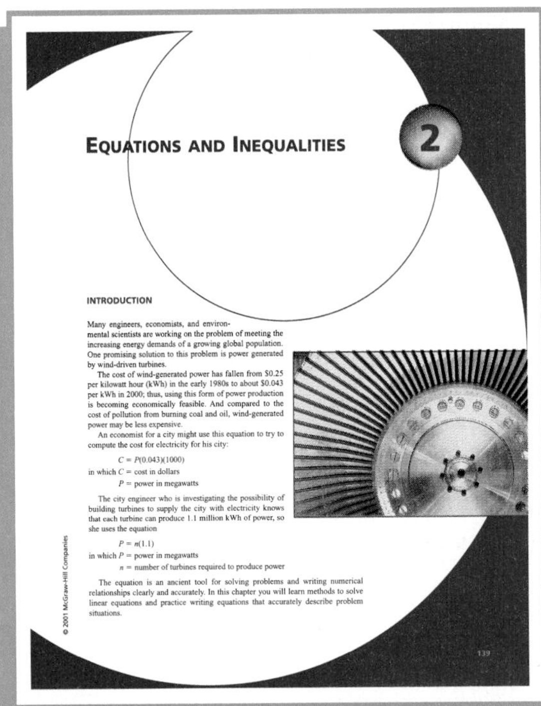
EQUATIONS AND INEQUALITIES 2

INTRODUCTION

Many engineers, economists, and environmental scientists are working on the problem of meeting the increasing energy demands of a growing global population. One promising solution to this problem is power generated by wind-driven turbines.

The cost of wind-generated power has fallen from \$0.25 per kilowatt hour (kWh) in the early 1980s to about \$0.043 per kWh in 2000; thus, using this form of power production is becoming economically feasible. And compared to the cost of pollution from burning coal and oil, wind-generated power may be less expensive.

An economist for a city might use this equation to try to compute the cost for electricity for his city:

$$C = P(0.043)(1000)$$

in which C = cost in dollars
P = power in megawatts

The city engineer who is investigating the possibility of building turbines to supply the city with electricity knows that each turbine can produce 1.1 million kWh of power, so she uses the equation

$$P = n(1.1)$$

in which P = power in megawatts
n = number of turbines required to produce power

The equation is an ancient tool for solving problems and writing numerical relationships clearly and accurately. In this chapter you will learn methods to solve linear equations and practice writing equations that accurately describe problem situations.

© 2001 McGraw-Hill Companies

139

feature one or more modeling/word problems that relate to the chapter-opening vignette. The application areas and the chapter each area appears in are:

Application Area	Chapter
Archeology	**0.** An Arithmetic Review
Anthropology	**1.** The Language of Algebra
Environmental Science	**2.** Equations and Inequalities
Postal Work	**3.** Polynomials
Encoding	**4.** Factoring
Dietary Science	**5.** Algebraic Fractions
Pediatric Medicine	**6.** An Introduction to Graphing
Pharmacology	**7.** Graphing and Inequalities
Electrical Engineering	**8.** Systems of Linear Equations
Civil Engineering	**9.** Exponents and Radicals
Pyrotechnics	**10.** Quadratic Equations

4.1 **OBJECTIVES**

1. Remove the greatest common factor (GCF)
2. Remove a binomial GCF

Section Objectives

Objectives for each section are clearly identified.

REMEMBER: The x coordinate gives the *horizontal* distance from the y axis. The y coordinate gives the *vertical* distance from the x axis.

(3, 2) means $x = 3$ and $y = 2$. (2, 3) means $x = 2$ and $y = 3$. (3, 2) and (2, 3) are entirely different. That's why we call them *ordered pairs*.

Marginal Notes and Caution Icons

Marginal notes are provided throughout and are designed to help the students focus on important topics and techniques. Caution icons point out potential trouble spots.

Check Yourself Exercises

These exercises have been the hallmark of the text; they are designed to actively involve students throughout the learning process. Each example is followed by an exercise that encourages students to solve a problem similar to the one just presented. Answers are provided at the end of the section for immediate feedback.

Example 5

Choosing an Appropriate Method for Solving a System

Select the most appropriate meth…

(a) $5x + 3y = 9$
$2x - 7y = 8$

CHECK YOURSELF 5

Select the most appropriate method for solving each of the following systems.

(a) $2x + 5y = 3$
$8x - 5y = -13$

(b) $4x - 3y = 2$
$y = 3x - 4$

Comprehensive Exercise Sets and Challenge Exercises

Complete exercise sets are at the end of each section as well as after the summary at the end of each chapter. These exercises were designed to reinforce basic skills and develop critical thinking and communication abilities. Exercise sets include writing and word problems, collaborative and group exercises, and challenge exercises, all denoted by distinctive icons. The calculator icon points out examples and exercises that illustrate when the calculator can best be used for further understanding of the concept at hand.

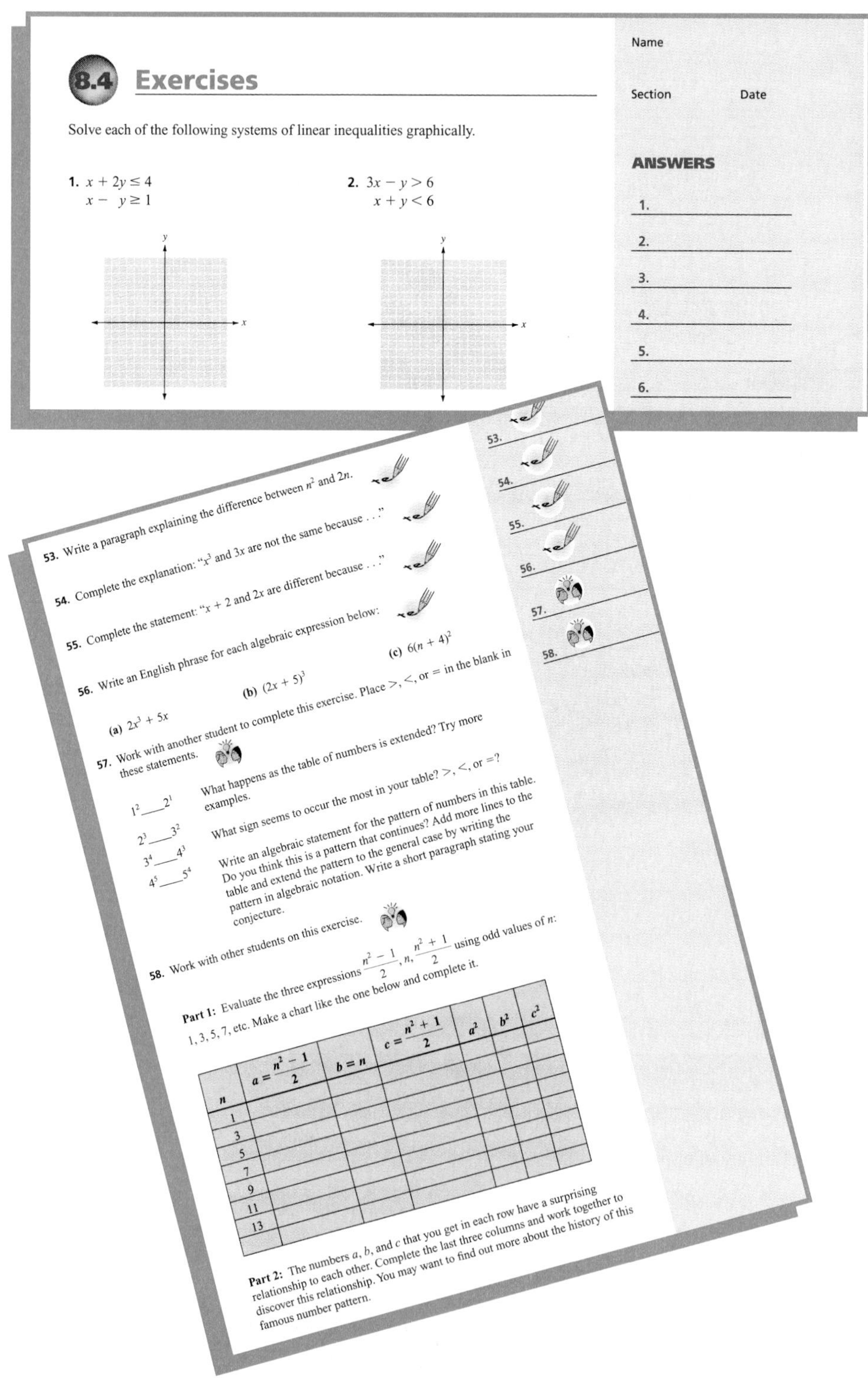

8.4 Exercises

Name

Section Date

Solve each of the following systems of linear inequalities graphically.

1. $x + 2y \le 4$
$x - y \ge 1$

2. $3x - y > 6$
$x + y < 6$

ANSWERS

1.
2.
3.
4.
5.
6.

53. Write a paragraph explaining the difference between n^2 and $2n$.

54. Complete the explanation: "x^3 and $3x$ are not the same because . . ."

55. Complete the statement: "$x + 2$ and $2x$ are different because . . ."

56. Write an English phrase for each algebraic expression below:

(a) $2x^3 + 5x$ (b) $(2x + 5)^3$ (c) $6(n + 4)^2$

57. Work with another student to complete this exercise. Place >, <, or = in the blank in these statements.

1^2 __ 2^1

2^3 __ 3^2

3^4 __ 4^3

4^5 __ 5^4

What happens as the table of numbers is extended? Try more examples.

What sign seems to occur the most in your table? >, <, or =?

Write an algebraic statement for the pattern of numbers in this table. Do you think this is a pattern that continues? Add more lines to the table and extend the pattern to the general case by writing the pattern in algebraic notation. Write a short paragraph stating your conjecture.

58. Work with other students on this exercise.

Part 1: Evaluate the three expressions $\frac{n^2 - 1}{2}$, n, $\frac{n^2 + 1}{2}$ using odd values of n: 1, 3, 5, 7, etc. Make a chart like the one below and complete it.

n	$a = \frac{n^2 - 1}{2}$	$b = n$	$c = \frac{n^2 + 1}{2}$	a^2	b^2	c^2
1						
3						
5						
7						
9						
11						
13						

Part 2: The numbers a, b, and c that you get in each row have a surprising relationship to each other. Complete the last three columns and work together to discover this relationship. You may want to find out more about the history of this famous number pattern.

53.
54.
55.
56.
57.
58.

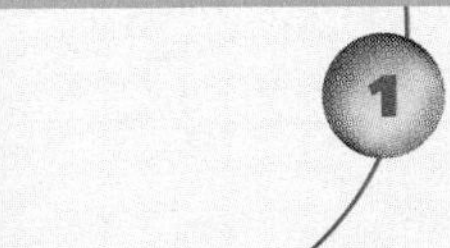

Getting Ready for Section 5.3 [Section 0.2]

(a) $\frac{3}{4} + \frac{1}{2}$ (b) $\frac{5}{6} - \frac{2}{3}$

(c) $\frac{7}{10} - \frac{3}{5}$ (d) $\frac{5}{8} + \frac{3}{4}$

(e) $\frac{5}{6} + \frac{3}{8}$ (f) $\frac{7}{8} - \frac{3}{5}$

(g) $\frac{9}{10} - \frac{2}{15}$ (h) $\frac{5}{12} + \frac{7}{18}$

Getting Ready Exercises

These exercises draw on problems from previous sections of the text and are designed to help students review concepts that will be applied in the following section. This preview helps students make important connections with upcoming material.

1 Summary

DEFINITION/PROCEDURE	EXAMPLE	REFERENCE
From Arithmetic to Algebra		**Section 1.1**
Addition $x + y$ means the **sum** of x **and** y or x **plus** y. Some other words indicating addition are "more than" and "increased by."	The sum of x and 5 is $x + 5$. 7 more than a is $a + 7$. b increased by 3 is $b + 3$.	p. 54
Subtraction $x - y$ means the **difference** of x **and** y or x **minus** y. Some other words indicating subtraction are "less than" and "decreased by."	The difference of x and 3 is $x - 3$.	
Multiplication $x \cdot y$ $(x)(y)$ These all mean the *product* of x an xy		

Summary and Summary Exercises

These comprehensive sections give students an opportunity to practice and review important concepts at the end of each chapter. Answers are provided with section references to aid in summarizing the material effectively.

Summary Exercises

This exercise set is provided to give you practice with each of the objectives of the chapter. Each exercise is keyed to the appropriate chapter section. The answers are provided in the *Instructor's Manual*. Your instructor will give you guidelines on how to best use these exercises.

[1.1] Write, using symbols.

1. 5 more than y

2. c decreased by 10

3. The product of 8 and a

4. The quotient when y is divided by 3

5. 5 times the product of m and n

6. The product of a and 5 less than a

7. 3 more than the product of 17 and x

8. The quotient when a plus 2 is divided by a minus 2

Identify which are expressions and which are not.

9. $4(x + 3)$

10. $7 \div \cdot 8$

11. $y + 5 = 9$

12. $11 + 2(3x - 9)$

[1.2] Identify the property that is illustrated by each of the following statements.

13. $5 + (7 + 12) = (5 + 7) + 12$

14. $2(8 + 3) = 2 \cdot 8 + 2 \cdot 3$

15. $4 \cdot (5 \cdot 3) = (4 \cdot 5) \cdot 3$

16. $4 \cdot 7 = 7 \cdot 4$

Verify that each of the following statements is true by evaluating each side of the equation separately and comparing the results.

Cumulative Tests

These tests help students build on what was previously covered and give them more opportunity to build skills necessary in preparing for midterm and final exams. Answers are at the back of the book.

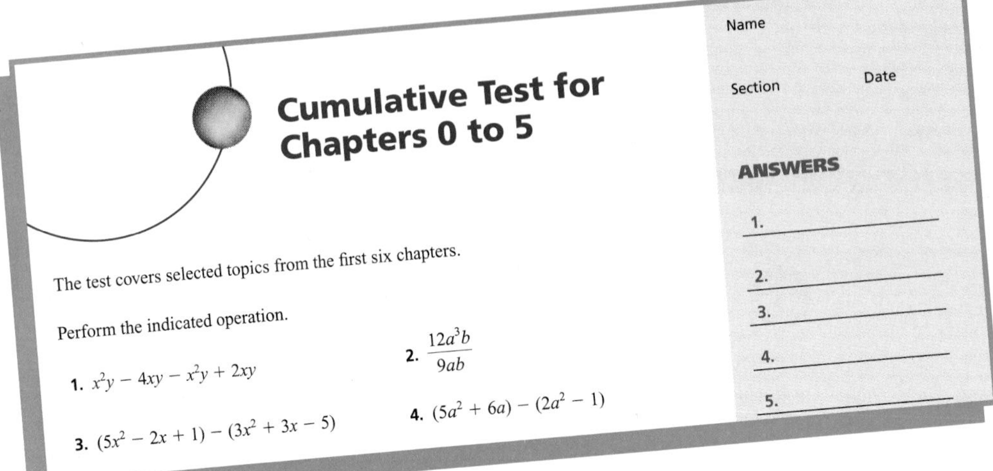

Self-Tests

Each chapter ends with a self-test to give students confidence and guidance in preparing for in-class tests. Answers are at the back of the book.

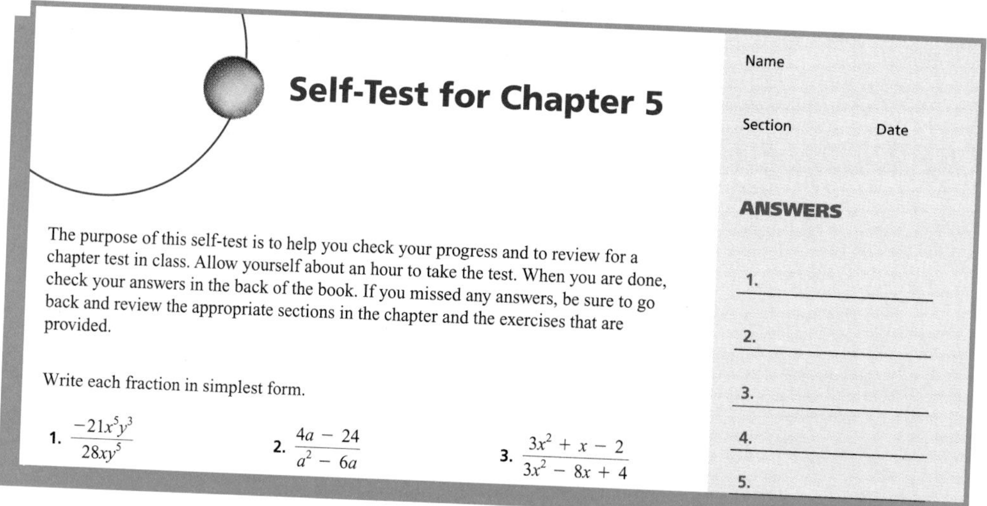

SUPPLEMENTS

A comprehensive set of ancillary materials for both the student and the instructor is available with this text.

Annotated Instructor's Edition

This ancillary includes answers to all exercises and tests. These answers are printed in a second color for ease of use by the instructor and are located on the appropriate pages throughout the text.

Instructor's Solutions Manual

The manual provides worked-out solutions to all the exercises in the text.

Print Test Bank

The print test bank contains (1) a diagnostic pretest for each chapter; (2) five forms of multiple-choice chapter tests; (3) two forms of multiple-choice cumulative tests; and (4) four forms of multiple-choice final tests. Answers to all these tests are provided.

Print and Computerized Testing

The testing materials provide an array of formats that allow the instructor to create tests using both algorithmically generated test questions and those from a standard testbank. This testing system enables the instructor to choose questions either manually or randomly by section, question type, difficulty level, and other criteria. Testing is available for IBM, IBM-compatible, and Macintosh computers.

Student's Solutions Manual

The manual provides worked-out solutions to the odd-numbered exercises in the text.

Streeter Video Series

The video series gives students additional reinforcement of the topics presented in the book. The videos were developed especially for the Streeter pedagogy, and features are tied directly to the main text's individual chapters and section objectives. The videos feature an effective combination of learning techniques, including personal instruction, state-of-the-art graphics, and real-world applications. Students are encouraged to work examples on their own and check their results with those provided.

Streeter Smart Tutorial CD-ROM

This interactive CD-ROM is a self-paced tutorial specifically linked to the text and reinforces topics through unlimited opportunities to review concepts and practice problem solving. The CD-ROM contains chapter- and section-specific tutorials, multiple choice questions with feedback, as well as algorithmically generated questions. It requires virtually no computer training on the part of students and supports IBM and Macintosh computers.

In addition, a number of other technology and Web-based ancillaries are under development; they will support the ever-changing technology needs in developmental mathematics. For further information about these or any supplements, please contact your local McGraw-Hill sales representative.

Web Site and Online Learning Center

Web-based, interactive learning for your students. Student resources include learning objectives, chapter reviews, PowerPoint slides, exercises, online quizzing, and web links.

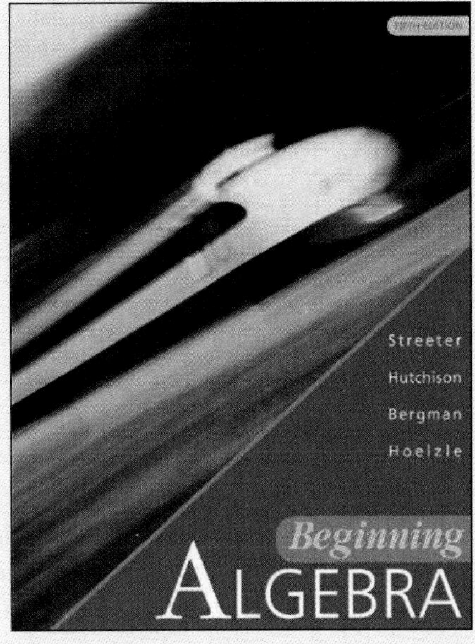

MCGRAW-HILL IS PROUD TO OFFER AN EXCITING NEW SUITE OF MULTIMEDIA PRODUCTS AND SERVICES CALLED COURSE SOLUTIONS.

Designed specifically to help you with your individual course needs, Course Solutions will assist you in integrating your syllabus with our premier titles and state-of-the-art new media tools that support them.

AT THE HEART OF COURSE SOLUTIONS YOU'LL FIND:

- Fully integrated multimedia
- A full-scale Online Learning Center
- A Course Integration Guide

AS WELL AS THESE UNPARALLELED SERVICES:

- McGraw-Hill Learning Architecture
- McGraw-Hill Course Consultant Service
- Visual Resource Library (VRL) Image Licensing
- McGraw-Hill Student Tutorial Service
- McGraw-Hill Instructor Syllabus Service
- PageOut Lite
- PageOut: The Course Web Site Development Center
- Other Delivery Options

COURSE SOLUTIONS truly has the solutions to your every teaching need. Read on to learn how we can specifically help you with your classroom challenges.

SPECIAL ATTENTION

to your specific needs.

These "perks" are all part of the extra service delivered through McGraw-Hill's Course Solutions:

McGRAW-HILL LEARNING ARCHITECTURE

Each McGraw-Hill *Online Learning Center* is ready to be ported into our *McGraw-Hill Learning Architecture*—a full course management software system for Local Area Networks and Distance Learning Classes. Developed in conjunction with Top Class software, *McGraw-Hill Learning Architecture* is a powerful course management system available upon special request.

McGRAW-HILL COURSE CONSULTANT SERVICE

In addition to the *Course Integration Guide,* instructors using Course Solutions textbooks can access a special curriculum-based *Course Consultant Service* via a web-based threaded discussion list within each *Online Learning Center.* A McGraw-Hill Course Solutions Consultant will personally help you—as a text adopter—integrate this text and media into your course to fit your specific needs. This content-based service is offered in addition to our usual software support services.

VISUAL RESOURCE LIBRARY (VRL) IMAGE LICENSING

Most of our Course Solutions titles are accompanied by a *Visual Resource Library (VRL) CD-ROM,* which features text figures in electronic format. Previously, use of these images was restricted to in-class presentation only. Now, McGraw-Hill will license adopters the right to use appropriate VRL image files—FREE OF CHARGE—for placement on their local Web site! Some restrictions apply. Consult your McGraw-Hill sales representative for more details.

McGRAW-HILL INSTRUCTOR SYLLABUS SERVICE

For *new* adopters of Course Solutions textbooks, McGraw-Hill will help correlate all text, supplement, and appropriate materials and services to your course syllabus. Simply call your McGraw-Hill sales representative for assistance.

PAGEOUT LITE

Free to Course Solutions textbook adopters, *PageOut Lite* is perfect for instructors who want to create their own Web site. In just a few minutes, even novices can turn their syllabus into a Web site using *PageOut Lite.*

PAGEOUT: THE COURSE WEB SITE DEVELOPMENT CENTER

For those that want the benefits of *PageOut Lite's* no-hassle approach to site development, but with even more features, we offer *PageOut: The Course Web Site Development Center.*

PageOut shares many of *PageOut Lite's* features, but also enables you to create links that will take your students to your original material, other Web site addresses, and to *McGraw-Hill Online Learning Center* content. This means you can assign *Online Learning Center* content within your syllabus-based Web site. *PageOut's* gradebook function will tell you when each student has taken a quiz or worked through an exercise, automatically recording their scores for you. *PageOut* also features a discussion board list where you and your students can exchange questions and post announcements, as well as an area for students to build personal Web pages.

OTHER DELIVERY OPTIONS

Online Learning Centers are also compatible with a number of full-service online course delivery systems or outside educational service providers. For a current list of compatible delivery systems, contact your McGraw-Hill sales representative.

And for your students...

McGRAW-HILL STUDENT TUTORIAL SERVICE

Within each *Online Learning Center* resides a FREE *Student Tutorial Service.* This web-based "homework hotline"—available via a threaded discussion list—features guaranteed, 24-hour response time on weekdays.

www.mhhe.com/streeter

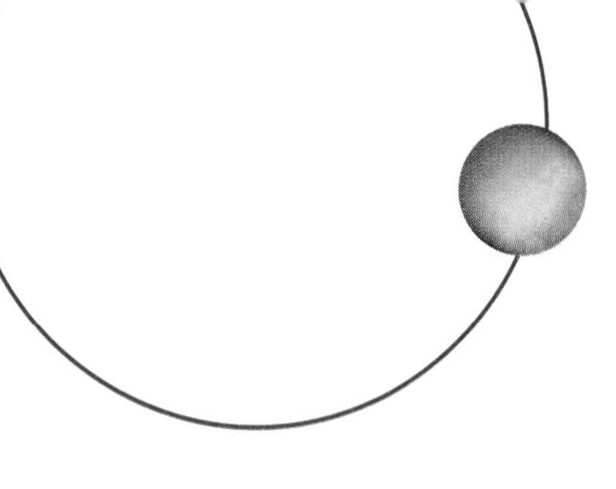

Acknowledgments

Those familiar with the publishing process will attest that change is inevitable. The same is true at McGraw-Hill. The difference there is that change invariably seems to lead to something positive. We have been most fortunate to work with our editors Bill Barter, Erin Brown, and Bea Wikander. All three manage that fine line by being both demanding managers and supportive coworkers. We have appreciated their talents, energy, and time. As always, we encourage prospective authors to talk with the staff at McGraw-Hill. It will be a valuable use of your time.

We would like to thank the many reviewers who reviewed and improved this text:

Sharon Abramson, Nassau Community College (NY)
Patricia Allaire, Queensborough Community College (NY)
Sharon Berrian, Northwest Shoals Community College (AL)
Matthews Chakkanakuzh, Palomar Community College (CA)
Alan Chutsky, Queensborough Community College (NY)
John Davidson, Southern State Community College (OH)
Katherine D'Orazio, Cumberland County Community College (NJ)
Bill Dunn, Las Positas College (CA)
Ellen Freedman, Camden County Community College (NJ)
Kelly Jackson, Camden County Community College (NJ)
Karen Jensen, Southeastern Community College (IA)
Ginny Licata, Camden County Community College (NJ)
S. Maheshwari, William Paterson University (NJ)
Laurie McManus, St. Louis Community College-Meramec (MO)
Diane Metzger, Rend Lake College (IL)
Wayne Miller, Lee College (TX)
Jeff Mock, Diablo Valley College (CA)
Ellen Musen, Brookdale Community College (NJ)
Larry Newman, Holyoke Community College (MA)
Lilia Orlova, Nassau Community College (NY)
Betty Pate, St. Petersburg Junior College (FL)
Kathryn Pletsch, Antelope Valley College (CA)
Larry Pontaski, Pueblo Community College (CO)
Donna Russo, Quincy College (MA)
Bruce Sisko, Belleville Area College (IL)
Barbara Jane Sparks, Camden County Community College (NJ)
Peter Speier, Prince George's Community College (MD)
Sharon Testone, Onandaga Community College (NY)
Patricia Wake, San Jacinto College (TX)

We would also like to thank those who contributed to the development of the fourth edition:

Richard Butterworth, Massasoit Community College (MA)
Charles Clare, Diablo Valley College (CA)
John DeCoursey, Vincennes University (IN)
David Donaldson, Northwest State Community College (OH)
Cheryl Groff, Florida Community College, South Campus
Mel Hamburger, Laramie County Community College (WY)
Paul Wayne Lee, Saint Philip's College (TX)
Carl Mancuso, William Paterson College (NJ)
Robert Mooney, Salem State College (MA)
Larry Pontaski, Pueblo Community College (CO)
Donna Russo, Quincy College (MA)

Randy Sowell, Central Virginia Community College
Gene Steinmeyer, Doane College (NE)
Mary Thurow, Minneapolis Community College
Annette Wiesner, University of Wisconsin-Parkside

We are especially grateful to the following focus group participants, who offered many valuable suggestions for the fifth edition:

Mary Kay Abbey, Montgomery College (MD)
Cynthia Albee, Tarrant County College (TX)
Sabah Alquaddoomi, Pasadena City College
Dimos Arsenidis, California State University Long Beach
Deidre Baker, College of Alameda (CA)
Jerry Bartolomeo, Nova Southeastern University (FL)
Palma Benko, Passaic County Community College (NJ)
Connie Bish, Las Positas College (CA)
Barbara Britton, Malcolm X (IL)
Brenda D. Brown, Prince George's Community College (MD)
Eleanor Browne, Richland College (TX)
Linda Burton, Miami Dade Community College-Homestead
Bob Caldwell, El Camino College (CA)
Donna Carlson, College of Lake County (IL)
Jim Castro, California State University Northridge
Yong S. Colen, Monroe College (NY)
Jorge Cossio, Miami-Dade Community College-Kendall
Charles Dietz, Charles County Community College (MD)
C. Toland Draper, College of Alameda (CA)
C. Wayne Ehler, Anne Arundel Community College (MD)
Gerald Floyd, Malcolm X (IL)
Dorothy Fujimura, California State University Hayward
Judy Godwin, Collin County Community College (TX)
David E. Gustafson, Tarrant County College (TX)
Garry Hart, California State University Dominguez Hills
Laxman Hegde, Frostburg State University (MD)
Celeste Hernandez, Richland College (TX)
Nancy Johnson, Broward Community College (FL)
Rosamma Joseph, Oakton Community College (IL)
Rosemary M. Karr, Collin County Community College (TX)
Joseph Kazimir, East Los Angeles College
Surinder K. Khurana, Fullerton College (CA)
Serge Kuznetsov, Kennedy-King College (IL)
Shirley Lathrop, Truman College (IL)
William Lepowsky, Laney College (CA)
Greg Liano, Brookdale Community College (NJ)
Sandy Lynn, Foothill College (CA)
Alice Madson, Kankakee Community College (IL)
Dorothy S. Marshall, Edison Community College
Bob Martin, Tarrant County College, NE Campus (TX)
Marilyn Massey-Moss, Collin County Community College (TX)
Victor Mastrovincenzo, Hudson County Community College (NJ)
William M. Mays, Gloucester County College (NJ)
Shyla McGill, Columbia College (IL)
Janet McLaughlin, S.C, Montclair State University (NJ)
Constance McNair, Broward Community College (FL)
Michael J. Morse, East Los Angeles College
Ann C. Mugavero, College of Staten Island

Shoeleh Mutameni, Morton College (IL)
Pat Newell, Edison Community College (FL)
Louise Olshan, County College of Morris (NJ)
Michael N. Payne, College of Alameda (CA)
Gary Piercy, Moraine Valley Community College (IL)
Virginia Puckett, Miami-Dade Community College-North
Linda Retterath, Mission College (CA)
Beth Rinehart, Anne Arundel Community College (MD)
David Ross, College of Alameda (CA)
Radha Sankaran, Passaic County Community College (NJ)
David P. Schaefer, College of Lake County (IL)
Richard Semmlen, Northern Virginia Community College
Luz V. Shin, Los Angeles Valley College
Alexis Thurman, County College of Morris (NJ)
Anthony Valenti, Nova University (FL)
Frissell Walker, College of Alameda (CA)
Anne Walsh, Monroe College (NJ)
Martin Weissman, Essex County College (NJ)
Pam Zener, Governors State University (IL)

We thank all of the students whom we have taught, talked to, questioned, and tested. This text was created for them. We also thank our community college compatriots. Professionals such as Betsy Farber, Alice Hayden, Susan Hopkirk, and Mark Yannotta are constantly providing us with both intentional and inadvertent guidance in our writing projects.

Donald Hutchison
Barry Bergman
Louis Hoelzle

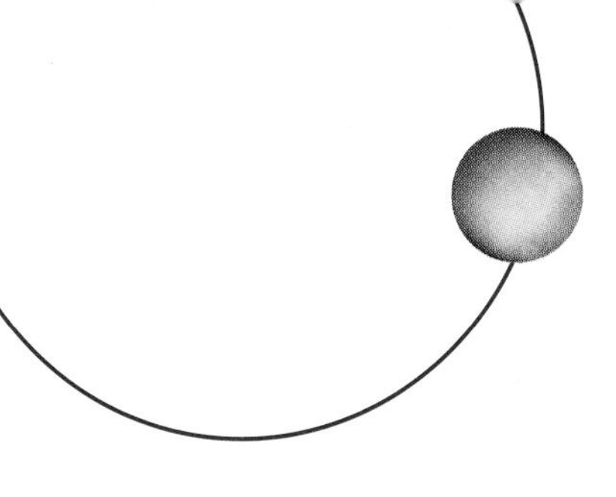

To the Student

You are about to begin a course in algebra. We made every attempt to provide a text that will help you understand what algebra is about and how to effectively use it. We made no assumptions about your previous experience with algebra. Your progress through the course will depend on the amount of time and effort you devote to the course and your previous background in math. There are some specific features in this book that will aid you in your studies. Here are some suggestions about how to use this book. (Keep in mind that a review of *all* the chapter and summary material will further enhance your ability to grasp later topics and to move more effectively through the text.)

1. If you are in a lecture class, make sure that you take the time to read the appropriate text section *before* your instructor's lecture on the subject. Then take careful notes on the examples that your instructor presents during class.
2. After class, work through similar examples in the text, making sure that you understand each of the steps shown. Examples are followed in the text by *Check Yourself* exercises. Algebra is best learned by being involved in the process, and that is the purpose of these exercises. Always have a pencil and paper at hand, and work out the problems presented and check your results immediately. If you have difficulty, go back and carefully review the previous exercises. Make sure you understand what you are doing and why. The best test of whether you do understand a concept lies in your ability to explain that concept to one of your classmates. Try working together.
3. At the end of each chapter section you will find a set of exercises. Work these carefully to check your progress on the section you have just finished. You will find the solutions for the odd-numbered exercises following the problem set. If you have difficulties with any of the exercises, review the appropriate parts of the chapter section. If your questions are not completely cleared up, by all means do not become discouraged. Ask your instructor or an available tutor for further assistance. A word of caution: Work the exercises on a regular (preferably daily) basis. Again, learning algebra requires becoming involved. As is the case with learning any skill, the main ingredient is practice.
4. When you complete a chapter, review by using the *Summary.* You will find all the important terms and definitions in this section, along with examples illustrating all the techniques developed in the chapter. Following the summary are *Summary Exercises* for further practice. The exercises are keyed to chapter sections, so you will know where to turn if you are still having problems.
5. When you finish with the *Summary Exercises,* try the *Self-Test* that appears at the end of each chapter. This test will give you an actual practice test to work as you review for in-class testing. Again, answers with section references are provided.
6. Finally, an important element of success in studying algebra is the process of regular review. We provide a series of *Cumulative Tests* throughout the textbook, beginning at the end of Chapter 2. These tests will help you review not only the concepts of the chapter that you have just completed but those of previous chapters. Use these tests in preparation for any midterm or final exams. If it appears that you have forgotten some concepts that are being tested, don't worry. Go back and review the sections where the idea was initially explained, or the appropriate chapter summary. That is the purpose of the cumulative tests.

We hope that you will find our suggestions helpful as you work through this material, and we wish you the best of luck in the course.

Donald Hutchison
Barry Bergman
Louis Hoelzle

AN ARITHMETIC REVIEW

0

Cultures from all over the world have developed number systems and ways to record patterns in their natural surroundings. The Mayans in Central America had one of the most sophisticated number systems in the world in the twelfth century A.D. The Chinese numbering and recording system dates from around 1200 B.C.E. The oldest evidence of numerical record is in Africa, where a bone notched in numerical patterns and dating from about 35,000 B.C.E. was found in the Lebembo Mountains near modern-day Swaziland in southern Africa.

The roots of algebra developed among the Babylonians 4000 years ago in an area now part of the country of Iraq. The Babylonians developed ways to record useful numerical relationships so that they were easy to remember, easy to record, and helpful in solving problems. Archeologists have found many tables, such as one giving successive powers of a given number, $9^2, 9^3, 9^4, \ldots, 9^n$. The tables include instructions for solving problems in engineering, economics, city planning, and agriculture. The writing is on clay tablets. Some of the formulas developed by the Babylonians are still in use today.

You are about to embark on an exciting and useful endeavor: learning to use algebra to help you solve problems. It will take some time and effort, but do not be discouraged. Everyone can master this topic—people just like you have used it for many centuries! Today algebra is even more useful than in the past because it is used in nearly every field of human endeavor.

Name ____________________

Section __________ Date __________

ANSWERS

1. 1, 2, 3, 6, 7, 14, 21, 42
2. Prime: 2, 3, 7, 17, 23; composite: 6, 9, 18, 21
3. $2 \times 2 \times 3 \times 5$
4. $2 \times 5 \times 5 \times 7$
5. 4
6. 6
7. $\frac{5}{4}$
8. $\frac{3}{2}$
9. $\frac{19}{12}$
10. $\frac{7}{18}$
11. 3.767
12. 22.8404
13. 6
14. 24
15. 4
16. 8
17. 13
18. 2
19. See exercise
20. −4, −2, −1, 0, 1, 5
21. Max: 7; Min: −5
22. 5
23. 6
24. 6
25. 6
26. 6
27. 16
28. −23

Pre-Test Chapter 0

This pre-test will point out any difficulties you may be having with basic arithmetic. Do all the problems, then check your answers with those in the back of the book.

1. List all the factors of 42.

2. For the group of numbers 2, 3, 6, 7, 9, 17, 18, 21, and 23, list the prime and composite numbers.

Write the prime factorizations for each of the following numbers.

3. 60

4. 350

Find the greatest common factor (GCF) for each of the following groups of numbers.

5. 12 and 32

6. 24, 36, and 42

Perform the indicated operations.

7. $\frac{3}{5} \cdot \frac{25}{12}$

8. $\frac{6}{7} \div \frac{12}{21}$

9. $\frac{5}{6} + \frac{3}{4}$

10. $\frac{17}{18} - \frac{5}{9}$

11. $8.123 - 4.356$

12. $7.16 \cdot 3.19$

Evaluate the following expressions.

13. $21 - 3 \cdot 5$

14. $(16 - 12) \cdot 6$

15. $8 - 2^2$

16. $3 \cdot 4 - 2^2$

17. $(18 \div 9) \cdot 2 + 3^2$

18. $(15 - 12 + 5) \div 2^2$

Represent the integers on the number line shown.

19. 6, −8, 4, −2, 10

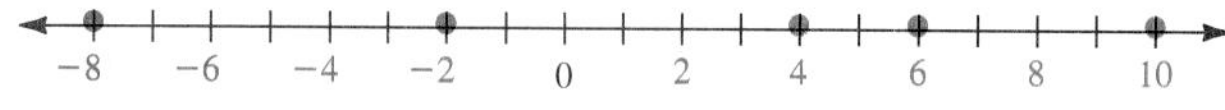

20. Place the following data set in ascending order: 5, −2, −4, 0, −1, 1.

21. Determine the maximum and minimum of the following data set: −4, 1, −5, 7, 3, 2.

Evaluate:

22. $|-5|$

23. $|6|$

24. $|11 - 5|$

25. $|-11| - |5|$

26. $|4 + 5| - |6 - 3|$

Find the opposite of each of the following.

27. −16

28. 23

0.1 Prime Factorization

0.1 OBJECTIVES

1. Find the factors of a natural number
2. Determine whether a number is prime, composite, or neither
3. Find the prime factorization for a number
4. Find the GCF for two or more numbers

Overcoming Math Anxiety

Throughout this text, we will present you with a series of class-tested techniques that are designed to improve your performance in this math class.

Hint #1 Become familiar with your text book.

Perform each of the following tasks.

1. Use the Table of Contents to find the title of Section 5.1.
2. Use the index to find the earliest reference to the term *mean.* (By the way, this term has nothing to do with the personality of either your instructor or the text book author!)
3. Find the answer to the first Check Yourself exercise in Section 0.1.
4. Find the answers to the Self-Test for Chapter 1.
5. Find the answers to the odd-numbered exercises in Section 0.1.
6. In the margin notes for Section 0.1, find the definition for the term *relatively prime.*

Now you know where some of the most important features of the text are. When you have a moment of confusion, think about using one of these features to help you clear up that confusion.

How would you arrange the following list of objects: cow, dog, daisy, fox, lily, sunflower, cat, tulip?

Although there are many ways to arrange the objects, most people would break them into two groups, the animals and the flowers. In mathematics, we call a group of things that have something in common a *set.*

Definitions: Set

A **set** is a collection of objects.

We generally use braces to enclose the elements of a set.

{cow, dog, fox, cat} or {daisy, lily, sunflower, tulip}

Of course, in mathematics many (but not all) of the sets we are interested in are sets of numbers.

The numbers used to count things—1, 2, 3, 4, 5, and so on—are called the **natural** (or **counting**) **numbers.** The **whole numbers** consist of the natural numbers and

zero—0, 1, 2, 3, 4, 5, and so on. They can be represented on a number line like the one shown. Zero (0) is considered the origin.

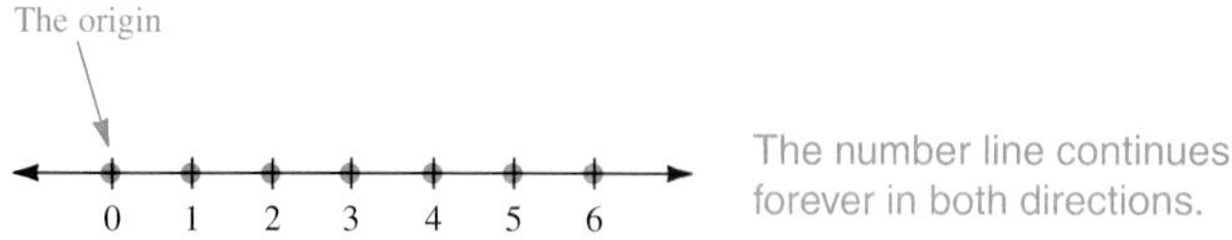

Any whole number can be written as a product of two whole numbers. For example, we say that $3 \cdot 4 = 12$. We call 3 and 4 **factors** of 12.

NOTE The centered dot represents multiplication.

Definitions: Factor

A **factor** of a whole number is another whole number that will *divide exactly* into that number. This means that the division will have a remainder of 0.

NOTE 2 and 5 can also be called *divisors* of 10. They divide 10 exactly.

Example 1

Finding Factors

List all factors of 18.

$3 \cdot 6 = 18$ Because $3 \cdot 6 = 18$, 3 and 6 are factors (or divisors) of 18.

$2 \cdot 9 = 18$ 2 and 9 are also factors of 18.

$1 \cdot 18 = 18$ 1 and 18 are factors of 18.

1, 2, 3, 6, 9, and 18 are all the factors of 18.

NOTE This is a complete list of the factors. There are no other whole numbers that divide 18 exactly. Note that the factors of 18, except for 18 itself, are *smaller* than 18.

CHECK YOURSELF 1*

List all the factors of 24.

Listing factors leads us to an important classification of whole numbers. Any whole number larger than 1 is either a *prime* or a *composite* number. Let's look at the following definitions.

NOTE A whole number greater than 1 will always have itself and 1 as factors. Sometimes these will be the *only* factors. For instance, 1 and 3 are the only factors of 3.

Definitions: Prime Number

A **prime number** is any whole number greater than 1 that has only 1 and itself as factors.

NOTE How large can a prime number be? There is no largest prime number. To date, the largest *known* prime is $2^{6972593} - 1$. This is a number with 2,098,960 digits, if you are curious. Of course, a computer had to be used to verify that a number of this size is prime. By the time you read this, someone may very well have found an even larger prime number.

As examples, 2, 3, 5, and 7 are prime numbers. Their only factors are 1 and themselves.

To check whether a number is prime, one approach is simply to divide the smaller primes, 2, 3, 5, 7, and so on, into the given number. If no factors other than 1 and the given number are found, the number is prime.

*Check Yourself answers appear at the end of each section throughout the book.

Here is the method known as the ***sieve of Eratosthenes*** for identifying prime numbers.

1. Write down a series of counting numbers, starting with the number 2. In the example below, we stop at 50.
2. Start at the number 2. Delete every second number after the 2.
3. Move to the number 3. Delete every third number after 3 (some numbers will be deleted twice).
4. Continue this process, deleting every fourth number after 4, every fifth number after 5, and so on.
5. When you have finished, the undeleted numbers are the prime numbers.

	2	3	~~4~~	5	~~6~~	7	~~8~~	~~9~~	~~10~~
11	~~12~~	13	~~14~~	~~15~~	~~16~~	17	~~18~~	19	~~20~~
~~21~~	~~22~~	23	~~24~~	~~25~~	~~26~~	~~27~~	~~28~~	29	~~30~~
31	~~32~~	~~33~~	~~34~~	~~35~~	~~36~~	37	~~38~~	~~39~~	~~40~~
41	~~42~~	43	~~44~~	~~45~~	~~46~~	47	~~48~~	~~49~~	~~50~~

The prime numbers less than 50 are 2, 3, 5, 7, 11, 13, 17, 19, 23, 29, 31, 37, 41, 43, and 47.

Example 2

Identifying Prime Numbers

Which of the following numbers are prime?

17 is a prime number. 1 and 17 are the only factors.

29 is a prime number. 1 and 29 are the only factors.

33 is *not* prime. 1, 3, 11, and 33 are all factors of 33.

Note: For two-digit numbers, if the number is *not* a prime, it will have one or more of the numbers 2, 3, 5, or 7 as factors.

CHECK YOURSELF 2

Which of the following numbers are prime numbers?

2, 6, 9, 11, 15, 19, 23, 35, 41

We can now define a second class of whole numbers.

NOTE This definition tells us that a composite number *does* have factors other than 1 and itself.

Definitions: Composite Number

A **composite number** is any whole number greater than 1 that is not prime.

Example 3

Identifying Composite Numbers

Which of the following numbers are composite?

18 is a composite number. 1, 2, 3, 6, 9, and 18 are all factors of 18.

23 is not a composite number. 1 and 23 are the only factors. This means that 23 is a *prime number.*

25 is a composite number. 1, 5, and 25 are factors.

38 is a composite number. 1, 2, 19, and 38 are factors.

CHECK YOURSELF 3

Which of the following numbers are composite numbers?

2, 6, 10, 13, 16, 17, 22, 27, 31, 35

By the definitions of prime and composite numbers:

Rules and Properties: 0 and 1

The whole numbers 0 and 1 are neither prime nor composite.

To **factor a number** means to write the number as a product of its whole-number factors.

Example 4

Factoring a Composite Number

Factor the number 10.

$10 = 2 \cdot 5$

The order in which you write the factors does not matter, so $10 = 5 \cdot 2$ would also be correct.

Of course, $10 = 10 \cdot 1$ is also a correct statement. However, in this section we are interested in factors other than 1 and the given number.

Factor the number 21.

$21 = 3 \cdot 7$

CHECK YOURSELF 4

Factor 35.

In writing composite numbers as a product of factors, there may be several different possible factorizations.

Example 5

Factoring a Composite Number

Find three ways to factor 72.

NOTE There have to be at least two different factorizations, because a composite number has factors other than 1 and itself.

$$72 = 8 \cdot 9 \quad (1)$$
$$= 6 \cdot 12 \quad (2)$$
$$= 3 \cdot 24 \quad (3)$$

CHECK YOURSELF 5

Find three ways to factor 42.

We now want to write composite numbers as a product of their **prime factors.** Look again at the first factored line of Example 5. The process of factoring can be continued until all the factors are prime numbers.

Example 6

Factoring a Composite Number

NOTE This is often called a **factor tree.**

$$72 = 8 \cdot 9$$
$$= 2 \cdot 4 \cdot 3 \cdot 3$$ 4 is still not prime, and so we continue by factoring 4.
$$= 2 \cdot 2 \cdot 2 \cdot 3 \cdot 3$$ 72 is now written as a product of prime factors.

When we write 72 as $2 \cdot 2 \cdot 2 \cdot 3 \cdot 3$, no further factorization is possible. This is called the *prime factorization* of 72.

NOTE Finding the prime factorization of a number will be important in our later work in adding fractions.

Now, what if we start with the second factored line from the same example, $72 = 6 \cdot 12$?

$$72 = 6 \cdot 12$$ Continue to factor 6 and 12.
$$= 2 \cdot 3 \cdot 3 \cdot 4$$ Continue again to factor 4. Other choices for the factors of 12 are possible. As we shall see, the end result will be the same.
$$= 2 \cdot 3 \cdot 3 \cdot 2 \cdot 2$$

No matter which pair of factors you start with, you will find the same prime factorization. In this case, there are three factors of 2 and two factors of 3. The order in which we write the factors does not matter.

CHECK YOURSELF 6

We could also write

$72 = 2 \cdot 36$

Continue the factorization.

Rules and Properties: The Fundamental Theorem of Arithmetic

There is exactly one prime factorization for any composite number.

The method of the previous example will always work. However, an easier method for factoring composite numbers exists. This method is particularly useful when numbers get large, in which case factoring with a number tree becomes unwieldy.

NOTE The prime factorization is then the product of all the prime divisors and the final quotient.

Rules and Properties: Factoring by Division

To find the prime factorization of a number, divide the number by a series of primes until the final quotient is a prime number.

Example 7

Finding Prime Factors

To write 60 as a product of prime factors, divide 2 into 60 for a quotient of 30. Continue to divide by 2 again for the quotient 15. Because 2 won't divide exactly into 15, we try 3. Because the quotient 5 is prime, we are done.

$$2\overset{30}{\overline{)60}} \longrightarrow 2\overset{15}{\overline{)30}} \longrightarrow 3\overset{5}{\overline{)15}} \quad \text{Prime}$$

Our factors are the prime divisors and the final quotient. We have

$60 = 2 \cdot 2 \cdot 3 \cdot 5$

CHECK YOURSELF 7

Complete the process to find the prime factorization of 90.

$$2\overset{45}{\overline{)90}} \longrightarrow ?\overset{?}{\overline{)45}}$$

Remember to continue until the final quotient is prime.

Writing composite numbers in their completely factored form can be simplified if we use a format called **continued division.**

Example 8

Finding Prime Factors Using Continued Division

Use the continued-division method to divide 60 by a series of prime numbers.

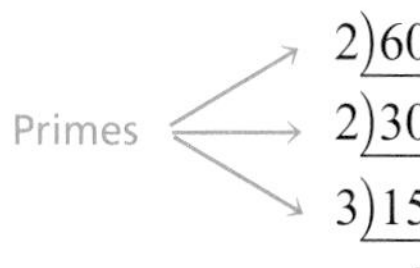

NOTE In each short division, we write the quotient *below* rather than above the dividend. This is just a convenience for the next division.

To write the factorization of 60, we include each divisor used and the final prime quotient. In our example, we have

$60 = 2 \cdot 2 \cdot 3 \cdot 5$

CHECK YOURSELF 8

Find the prime factorization of 234.

We know that a factor or a divisor of a whole number divides that number exactly.

The factors or divisors of 20 are

1, 2, 4, 5, 10, 20

NOTE Again the factors of 20, other than 20 itself, are less than 20.

Each of these numbers divides 20 exactly, that is, with no remainder.

Our work in the rest of this section involves common factors or divisors. A **common factor** or **divisor** for two numbers is any factor that divides both the numbers exactly.

Example 9

Finding Common Factors

Look at the numbers 20 and 30. Is there a common factor for the two numbers?

First, we list the factors. Then we circle the ones that appear in both lists.

Factors

20: (1), (2), 4, (5), (10), 20

30: (1), (2), 3, (5), 6, (10), 15, 30

We see that 1, 2, 5, and 10 are common factors of 20 and 30. Each of these numbers divides both 20 and 30 exactly.

Our later work with fractions will require that we find the greatest common factor (GCF) of a group of numbers.

Definition: Greatest Common Factor

The **greatest common factor** (GCF) of a group of numbers is the *largest* number that will divide each of the given numbers exactly.

In the first part of Example 9, the common factors of the numbers 20 and 30 were listed as

1, 2, 5, 10 Common factors of 20 and 30

The GCF of the two numbers is then 10, because 10 is the *largest* of the four common factors.

CHECK YOURSELF 9

List the factors of 30 *and* 36, *and then find the GCF.*

The method of Example 9 will also work in finding the GCF of a group of more than two numbers.

Example 10

Finding the GCF by Listing Factors

Find the GCF of 24, 30, and 36. We list the factors of each of the three numbers.

NOTE Looking at the three lists, we see that 1, 2, 3, and 6 are common factors.

24: (1), (2), (3), 4, (6), 8, 12, 24

30: (1), (2), (3), 5, (6), 10, 15, 30

36: (1), (2), (3), 4, (6), 9, 12, 18, 36

6 is the GCF of 24, 30, and 36.

CHECK YOURSELF 10

Find the GCF of 16, 24, *and* 32.

The process shown in Example 10 is very time-consuming when larger numbers are involved. A better approach to the problem of finding the GCF of a group of numbers uses the prime factorization of each number. Let's outline the process.

Finding the GCF

NOTE If there are no common prime factors, the GCF is 1.

Step 1 Write the prime factorization for each of the numbers in the group.
Step 2 Locate the prime factors that are *common* to all the numbers.
Step 3 The GCF will be the *product* of all the common prime factors.

Example 11

Finding the GCF

Find the GCF of 20 and 30.

Step 1 Write the prime factorizations of 20 and 30.

$20 = 2 \cdot 2 \cdot 5$

$30 = 2 \cdot 3 \cdot 5$

Step 2 Find the prime factors common to each number.

$20 = \textcircled{2} \cdot 2 \cdot \textcircled{5}$

$30 = \textcircled{2} \cdot 3 \cdot \textcircled{5}$

2 and 5 are the common prime factors.

Step 3 Form the product of the common prime factors.

$2 \cdot 5 = 10$

10 is the greatest common factor.

CHECK YOURSELF 11

Find the GCF of 30 *and* 36.

To find the GCF of a group of more than two numbers, we use the same process.

Example 12

Finding the GCF

Find the GCF of 24, 30, and 36.

$24 = \textcircled{2} \cdot 2 \cdot 2 \cdot \textcircled{3}$

$30 = \textcircled{2} \cdot \textcircled{3} \cdot 5$

$36 = \textcircled{2} \cdot 2 \cdot \textcircled{3} \cdot 3$

2 and 3 are the prime factors common to *all three numbers.*

$2 \cdot 3 = 6$ is the GCF.

CHECK YOURSELF 12

Find the GCF of 15, 30, *and* 45.

Example 13

Finding the GCF

NOTE If two numbers, such as 15 and 28, have no common factor other than 1, they are called **relatively prime.**

Find the GCF of 15 and 28.

$15 = 3 \cdot 5$

$28 = 2 \cdot 2 \cdot 7$

There are no common prime factors listed. But remember that 1 is a factor of every whole number.

The greatest common factor of 15 and 28 is 1.

CHECK YOURSELF 13

Find the greatest common factor of 30 *and* 49.

CHECK YOURSELF ANSWERS

1. 1, 2, 3, 4, 6, 8, 12, and 24. **2.** 2, 11, 19, 23, and 41 are prime numbers.
3. 6, 10, 16, 22, 27, and 35 are composite numbers. **4.** $5 \cdot 7$
5. $2 \cdot 21, 3 \cdot 14, 6 \cdot 7$ **6.** $2 \cdot 2 \cdot 2 \cdot 3 \cdot 3$
7. $2\overline{)90}$ = 45 → $3\overline{)45}$ = 15 → $3\overline{)15}$ = 5 **8.** $2 \cdot 3 \cdot 3 \cdot 13$

$90 = 2 \cdot 3 \cdot 3 \cdot 5$

9. 30: (1), (2), (3), 5, (6), 10, 15, 30

36: (1), (2), (3), 4, (6), 9, 12, 18, 36

6 is the GCF.

10. 16: (1), (2), (4), (8), 16

24: (1), (2), 3, (4), 6, (8), 12, 24

32: (1), (2), (4), (8), 16, 32

The GCF is 8.

11. $30 = (2) \cdot (3) \cdot 5$

$36 = (2) \cdot 2 \cdot (3) \cdot 3$

The GCF is $2 \cdot 3 = 6$.

12. 15 **13.** GCF is 1; 30 and 49 are relatively prime.

Name ______________________

Section __________ Date __________

0.1 Exercises

List the factors of each of the following numbers.

1. 4 **2.** 6

3. 10 **4.** 12

5. 15 **6.** 21

7. 24 **8.** 32

9. 64 **10.** 66

11. 11 **12.** 37

Use the following list of numbers for Exercises 13 and 14.

0, 1, 15, 19, 23, 31, 49, 55, 59, 87, 91, 97, 103, 105

13. Which of the given numbers are prime?

14. Which of the given numbers are composite?

15. List all the prime numbers between 30 and 50.

16. List all the prime numbers between 55 and 75.

Find the prime factorization of each number.

17. 18 **18.** 22

19. 30 **20.** 35

21. 51 **22.** 42

23. 63 **24.** 94

ANSWERS

1. 1, 2, 4
2. 1, 2, 3, 6
3. 1, 2, 5, 10
4. 1, 2, 3, 4, 6, 12
5. 1, 3, 5, 15
6. 1, 3, 7, 21
7. 1, 2, 3, 4, 6, 8, 12, 24
8. 1, 2, 4, 8, 16, 32
9. 1, 2, 4, 8, 16, 32, 64
10. 1, 2, 3, 6, 11, 22, 33, 66
11. 1, 11
12. 1, 37
13. 19, 23, 31, 59, 97, 103
14. 15, 49, 55, 87, 91, 105
15. 31, 37, 41, 43, 47
16. 59, 61, 67, 71, 73
17. $2 \cdot 3 \cdot 3$
18. $2 \cdot 11$
19. $2 \cdot 3 \cdot 5$
20. $5 \cdot 7$
21. $3 \cdot 17$
22. $2 \cdot 3 \cdot 7$
23. $3 \cdot 3 \cdot 7$
24. $2 \cdot 47$

ANSWERS

25. $2 \cdot 5 \cdot 7$
26. $2 \cdot 3 \cdot 3 \cdot 5$
27. $2 \cdot 3 \cdot 11$
28. $2 \cdot 2 \cdot 5 \cdot 5$
29. $2 \cdot 5 \cdot 13$
30. $2 \cdot 2 \cdot 2 \cdot 11$
31. $3 \cdot 3 \cdot 5 \cdot 7$
32. $2 \cdot 2 \cdot 2 \cdot 2 \cdot 5 \cdot 5$
33. $3 \cdot 3 \cdot 5 \cdot 5$
34. $2 \cdot 2 \cdot 3 \cdot 11$
35. $3 \cdot 3 \cdot 3 \cdot 7$
36. $2 \cdot 3 \cdot 5 \cdot 11$
37. 4, 6
38. 3, 5
39. 5, 6
40. 4, 7
41. 2
42. 3
43. 5
44. 2
45. 3
46. 11
47. 1
48. 14
49. 6
50. 1
51. 18
52. 12

25. 70 **26.** 90

27. 66 **28.** 100

29. 130 **30.** 88

31. 315 **32.** 400

33. 225 **34.** 132

35. 189 **36.** 330

In later mathematics courses, you often will want to find factors of a number with a given sum or difference. The following problems use this technique.

37. Find two factors of 24 with a sum of 10.

38. Find two factors of 15 with a difference of 2.

39. Find two factors of 30 with a difference of 1.

40. Find two factors of 28 with a sum of 11.

Find the GCF for each of the following groups of numbers.

41. 4 and 6 **42.** 6 and 9

43. 10 and 15 **44.** 12 and 14

45. 21 and 24 **46.** 22 and 33

47. 20 and 21 **48.** 28 and 42

49. 18 and 24 **50.** 35 and 36

51. 18 and 54 **52.** 12 and 48

53. 36 and 48

54. 36 and 54

55. 84 and 105

56. 70 and 105

57. 45, 60, and 75

58. 36, 54, and 180

59. 12, 36, and 60

60. 15, 45, and 90

61. 105, 140, and 175

62. 32, 80, and 112

63. 25, 75, and 150

64. 36, 72, and 144

65. Prime numbers that differ by two are called *twin primes.* Examples are 3 and 5, 5 and 7, and so on. Find one pair of twin primes between 85 and 105.

66. The following questions refer to "twin primes" (see Exercise 65).

(a) Search for, and make a list of several pairs of twin primes, in which the primes are greater than 3.
(b) What do you notice about each number that lies *between* a pair of twin primes?
(c) Write an explanation for your observation in part (b).

67. Obtain (or imagine that you have) a quantity of square tiles. Six tiles can be arranged in the shape of a rectangle in two different ways:

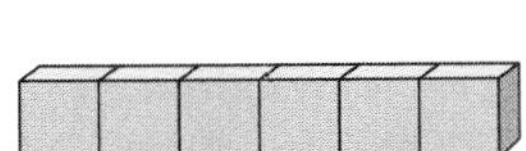

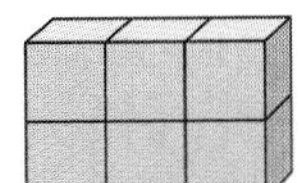

(a) Record the dimensions of the rectangles shown above.
(b) If you use seven tiles, how many different rectangles can you form?
(c) If you use ten tiles, how many different rectangles can you form?
(d) What kind of number (of tiles) permits *only one* arrangement into a rectangle? *More than* one arrangement?

68. The number 10 has four factors: 1, 2, 5, and 10. We can say that 10 has an even number of factors. Investigate several numbers to determine which numbers have an *even number* of factors and which numbers have an *odd number* of factors.

ANSWERS

53. 12
54. 18
55. 21
56. 35
57. 15
58. 18
59. 12
60. 15
61. 35
62. 16
63. 25
64. 36

65.

66.

67.

68.

69.

70

71

69. A natural number is said to be perfect if it is equal to the sum of its divisors.

(a) Show that 28 is a perfect number.

(b) Identify another perfect number less than 28.

70. Find the smallest natural number that is divisible by all of the following: 2, 3, 4, 6, 8, 9.

71. Suppose that a school has 1000 lockers and that they are all closed. A person passes through, opening every other locker, beginning with locker #2. Then another person passes through, changing every third locker (closing it if it is open, opening it if it is closed), starting with locker #3. Yet another person passes through, changing every fourth locker, beginning with locker #4. This process continues until 1000 people pass through.

(a) At the end of this process, which locker numbers are closed?

(b) Write an explanation for your answer to part (a). (Hint: It may help to attempt Exercise 68 first.)

Answers

We provide the answers for the odd numbered problems at the end of each Exercise set.

1. 1, 2, and 4 **3.** 1, 2, 5, and 10 **5.** 1, 3, 5, and 15
7. 1, 2, 3, 4, 6, 8, 12, and 24 **9.** 1, 2, 4, 8, 16, 32, and 64 **11.** 1 and 11
13. 19, 23, 31, 59, 97, 103 **15.** 31, 37, 41, 43, 47 **17.** $2 \cdot 3 \cdot 3$
19. $2 \cdot 3 \cdot 5$ **21.** $3 \cdot 17$ **23.** $3 \cdot 3 \cdot 7$ **25.** $2 \cdot 5 \cdot 7$ **27.** $2 \cdot 3 \cdot 11$
29. $2\overline{)130}$, $5\overline{)65}$, 13; $130 = 2 \cdot 5 \cdot 13$ **31.** $3 \cdot 3 \cdot 5 \cdot 7$ **33.** $3 \cdot 3 \cdot 5 \cdot 5$

35. $3\overline{)189}$, $3\overline{)63}$, $3\overline{)21}$, 7; $189 = 3 \cdot 3 \cdot 3 \cdot 7$ **37.** 4, 6 **39.** 5, 6 **41.** 2 **43.** 5 **45.** 3

47. 1 **49.** 6 **51.** 18 **53.** 12 **55.** 21 **57.** 15 **59.** 12
61. 35 **63.** 25
65. **67.** **69.**  **71.** 

0.2 Fractions

0.2 OBJECTIVES

1. Simplify a fraction
2. Multiply or divide two fractions
3. Add or subtract two fractions

This section provides a review of the basic operations, addition, subtraction, division, and multiplication, on fractions.

As mentioned in Section 0.1, the numbers used for counting are called the **natural numbers.** If we include zero in this group of numbers, we then call them the **whole numbers.**

The **numbers of ordinary arithmetic** consist of all the whole numbers and all fractions, whether they are proper fractions such as $\frac{1}{2}$ and $\frac{2}{3}$ or improper fractions such as $\frac{7}{2}$ or $\frac{19}{5}$.

Every number of ordinary arithmetic can be written in fraction form $\frac{a}{b}$.

The number 1 has many different fractional forms. Any fraction in which the numerator and denominator are the same (and not zero) is another name for the number one.

$$1 = \frac{2}{2} \qquad 1 = \frac{12}{12} \qquad 1 = \frac{257}{257}$$

Because these fractions are just different names for the same quantity, they are called **equivalent fractions.**

To write equivalent fractions, we use the **Fundamental Principle of Fractions (FPF).**

Rules and Properties: The Fundamental Principle of Fractions

$\frac{a}{b} = \frac{a \cdot c}{b \cdot c}$ or $\frac{a \cdot c}{b \cdot c} = \frac{a}{b}$, in which neither b nor c can equal zero.

Example 1

Rewriting Fractions

Write three fractional representations for each number.

NOTE Each representation is a numeral, or name for the number. Each number has many names.

(a) $\frac{2}{3}$

We use the fundamental principle to multiply the numerator and denominator by the same number.

NOTE In each case, we have used the Fundamental Principle of Fractions with c equal to a different number.

$$\frac{2}{3} = \frac{2 \cdot 2}{3 \cdot 2} = \frac{4}{6}$$

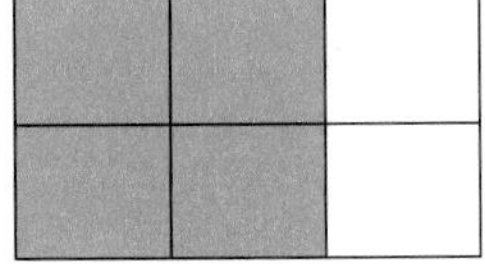

$\frac{4}{6}$

$$\frac{2}{3} = \frac{2 \cdot 3}{3 \cdot 3} = \frac{6}{9}$$

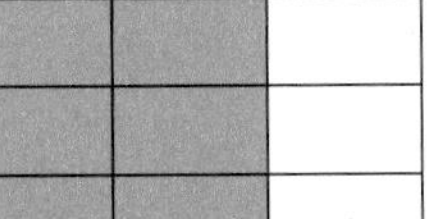

$\frac{6}{9}$

$$\frac{2}{3} = \frac{2 \cdot 10}{3 \cdot 10} = \frac{20}{30}$$

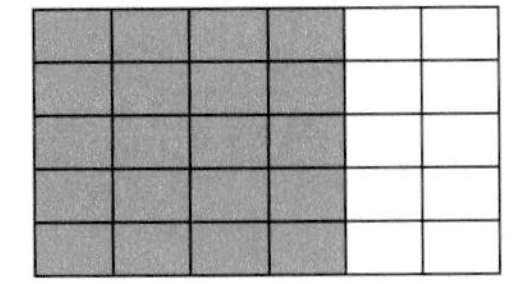

$\frac{20}{30}$

(b) 5

$$5 = \frac{5 \cdot 2}{1 \cdot 2} = \frac{10}{2}$$

$$5 = \frac{5 \cdot 3}{1 \cdot 3} = \frac{15}{3}$$

$$5 = \frac{5 \cdot 100}{1 \cdot 100} = \frac{500}{100}$$

CHECK YOURSELF 1

Write three fractional representations for each number.

(a) $\frac{5}{8}$ **(b)** $\frac{4}{3}$ **(c)** 3

The simplest fractional representation for a number has the smallest numerator and denominator. Fractions written in this form are said to be **simplified.**

Example 2

Simplifying Fractions

Simplify each fraction.

(a) $\frac{22}{55}$ **(b)** $\frac{35}{45}$ **(c)** $\frac{24}{36}$

In each case, we first find the prime factors for the numerator and for the denominator.

(a) $\frac{22}{55} = \frac{2 \cdot 11}{5 \cdot 11}$

We then use the fundamental principle.

$$\frac{22}{55} = \frac{2 \cdot 11}{5 \cdot 11} = \frac{2}{5}$$

(b) $\frac{35}{45} = \frac{7 \cdot 5}{3 \cdot 3 \cdot 5} = \frac{7 \cdot 5}{9 \cdot 5}$

Using the fundamental principle to remove the common factor of 5 yields

$$\frac{35}{45} = \frac{7}{9}$$

(c) $\frac{24}{36} = \frac{2 \cdot 2 \cdot 2 \cdot 3}{2 \cdot 2 \cdot 3 \cdot 3}$

Removing the common factor $2 \cdot 2 \cdot 3$ yields

$$\frac{2}{3}$$

CHECK YOURSELF 2

Simplify each fraction.

(a) $\frac{21}{33}$ **(b)** $\frac{15}{30}$ **(c)** $\frac{12}{54}$

Rules and Properties: Multiplication of Fractions

NOTE This is how two fractions, under the operation of multiplication, become one fraction.

$$\frac{a}{b} \cdot \frac{c}{d} = \frac{a \cdot c}{b \cdot d}$$

When multiplying two fractions, rewrite them in factored form, and then simplify before multiplying.

Example 3

Multiplying Fractions

Find the product of the two fractions.

NOTE A product is the result from multiplication.

$$\frac{9}{2} \cdot \frac{4}{3}$$

$$\begin{aligned} \frac{9}{2} \cdot \frac{4}{3} &= \frac{9 \cdot 4}{2 \cdot 3} \\ &= \frac{3 \cdot 3 \cdot 2 \cdot 2}{2 \cdot 3} \\ &= \frac{3 \cdot 2}{1} \\ &= \frac{6}{1} \quad \text{The denominator of one is not necessary.} \\ &= 6 \end{aligned}$$

CHECK YOURSELF 3

Multiply and simplify each pair of fractions.

(a) $\frac{3}{5} \cdot \frac{10}{7}$ **(b)** $\frac{12}{5} \cdot \frac{10}{6}$

Rules and Properties: Division of Fractions

NOTE This is how two fractions, under the operation of division, become one fraction.

$$\frac{a}{b} \div \frac{c}{d} = \frac{a}{b} \cdot \frac{d}{c} = \frac{a \cdot d}{b \cdot c}$$

To divide two fractions, the divisor is inverted, then the fractions are multiplied.

Example 4

Dividing Fractions

Find the quotient of the two fractions.

NOTE A quotient is the result from division.

$$\frac{7}{3} \div \frac{5}{6}$$

$$\begin{aligned} \frac{7}{3} \div \frac{5}{6} &= \frac{7}{3} \cdot \frac{6}{5} = \frac{7 \cdot 6}{3 \cdot 5} \\ &= \frac{7 \cdot 2 \cdot 3}{3 \cdot 5} = \frac{7 \cdot 2}{5} \\ &= \frac{14}{5} \end{aligned}$$

CHECK YOURSELF 4

Find the quotient of the two fractions

$$\frac{9}{2} \div \frac{3}{5}$$

Rules and Properties: Addition of Fractions

NOTE This is how two fractions with the same denominator, under the operation of addition, become one fraction.

$$\frac{a}{b} + \frac{c}{b} = \frac{a + c}{b}$$

When adding two fractions, find the **least common denominator (LCD)** first. The least common denominator is the smallest number that both denominators evenly divide. If you have forgotten how to find the LCD, you might want to review the process from your arithmetic book. After rewriting the fractions with this denominator, add the numerators, then simplify the result.

Example 5

Adding Fractions

Find the sum of the two fractions.

NOTE A sum is the result from addition.

$$\frac{5}{8} + \frac{7}{12}$$

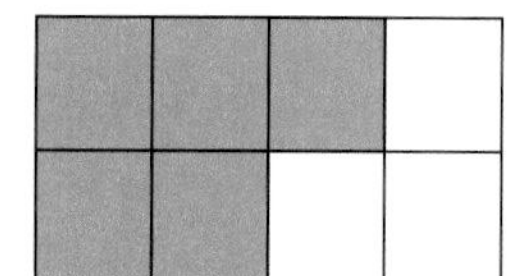

+

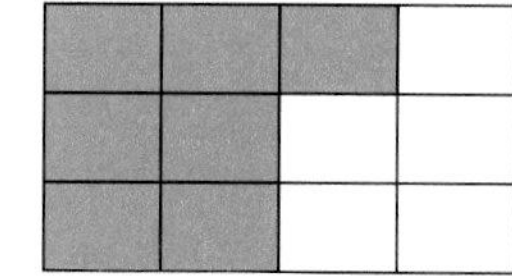

The LCD of 8 and 12 is 24. Each fraction should be rewritten as a fraction with that denominator.

$$\frac{5}{8} = \frac{15}{24}$$ Multiply the numerator and denominator by 3.

$$\frac{7}{12} = \frac{14}{24}$$ Multiply the numerator and denominator by 2.

$$\frac{5}{8} + \frac{7}{12} = \frac{15}{24} + \frac{14}{24} = \frac{29}{24}$$ This fraction cannot be simplified.

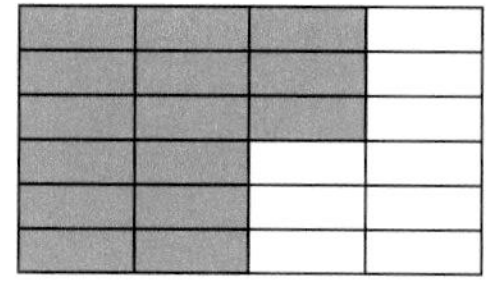

+

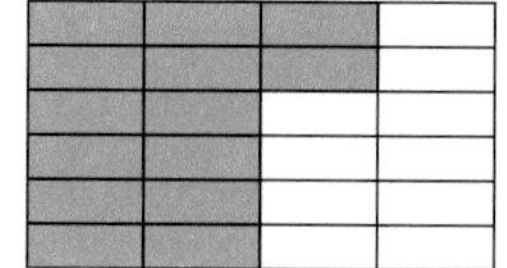

CHECK YOURSELF 5

Find the sum for each pair of fractions.

(a) $\frac{4}{5} + \frac{7}{9}$ **(b)** $\frac{5}{6} + \frac{4}{15}$

Rules and Properties: Subtraction of Fractions

NOTE This is how two fractions with like denominators become one fraction under the operation of subtraction.

$$\frac{a}{b} - \frac{c}{b} = \frac{a - c}{b}$$

Subtracting fractions is treated exactly like adding them, except the numerator becomes the difference of the two numerators.

Example 6

Subtracting Fractions

Find the difference.

NOTE The difference is the result from subtraction.

$$\frac{7}{9} - \frac{1}{6}$$

The LCD is 18. We rewrite the fractions with that denominator.

$$\frac{7}{9} = \frac{14}{18}$$

$$\frac{1}{6} = \frac{3}{18}$$

$$\frac{7}{9} - \frac{1}{6} = \frac{14}{18} - \frac{3}{18} = \frac{11}{18}$$

This fraction cannot be simplified.

CHECK YOURSELF 6

Find the difference $\frac{11}{12} - \frac{5}{8}$.

Fractions with denominator 10 (or 100, 1000, etc.) can be written in **decimal form.** Example 7 demonstrates the addition or subtraction of decimal fractions.

Example 7

Adding or Subtracting Two Decimals

Perform the indicated operation.

(a) Add 2.356 and 15.6

Aligning the decimal points, we get

$$\begin{array}{r} 2.356 \\ +15.600 \\ \hline 17.956 \end{array}$$

Although the zeros are not necessary, they ensure proper alignment.

(b) Subtract 3.84 from 8.1

Again, we align the decimal points

$$\begin{array}{r} 8.10 \\ -3.84 \\ \hline 4.26 \end{array}$$

When subtracting, always add zeros so that the right columns line up.

CHECK YOURSELF 7

Perform the indicated operation.

(a) 34.76 + 2.419 **(b)** 71.82 − 8.197

Example 8 illustrates the multiplication of two decimal fractions.

Example 8

Multiplying Decimal Fractions

Multiply 4.6 and 3.27

$$\begin{array}{r} 4.6 \\ \times\ 3.27 \\ \hline 322 \\ 920 \\ 13800 \\ \hline 15.042 \end{array}$$

It is not necessary to align decimals being multiplied. Note that the two factors have a total of three digits to the right of the decimal point.

The decimal point of the product is moved three digits to the left.

CHECK YOURSELF 8

Multiply 5.8 *and* 9.62.

CHECK YOURSELF ANSWERS

1. Answers will vary. **2.** (a) $\frac{7}{11}$; (b) $\frac{1}{2}$; (c) $\frac{2}{9}$ **3.** (a) $\frac{6}{7}$; (b) 4

4. $\frac{15}{2}$ **5.** (a) $\frac{71}{45}$; (b) $\frac{11}{10}$ **6.** $\frac{7}{24}$ **7.** (a) 37.179; (b) 63.623 **8.** 55.796

Name ______________

Section ________ Date ________

0.2 Exercises

In Exercises 1 to 12, write three fractional representations for each number.

1. $\frac{3}{7}$
2. $\frac{2}{5}$
3. $\frac{4}{9}$
4. $\frac{7}{8}$
5. $\frac{5}{6}$
6. $\frac{11}{13}$
7. $\frac{10}{17}$
8. $\frac{3}{7}$
9. $\frac{9}{16}$
10. $\frac{6}{11}$
11. $\frac{7}{9}$
12. $\frac{15}{16}$

Write each fraction in simplest form.

13. $\frac{8}{12}$
14. $\frac{12}{15}$
15. $\frac{10}{14}$
16. $\frac{15}{50}$
17. $\frac{12}{18}$
18. $\frac{28}{35}$
19. $\frac{35}{40}$
20. $\frac{21}{24}$
21. $\frac{11}{44}$
22. $\frac{10}{25}$
23. $\frac{12}{36}$
24. $\frac{18}{48}$
25. $\frac{24}{27}$
26. $\frac{30}{50}$
27. $\frac{32}{40}$
28. $\frac{17}{51}$
29. $\frac{75}{105}$
30. $\frac{62}{93}$
31. $\frac{48}{60}$
32. $\frac{48}{66}$
33. $\frac{105}{135}$

ANSWERS

1. $\frac{6}{14}, \frac{9}{21}, \frac{12}{28}$
2. $\frac{4}{10}, \frac{6}{15}, \frac{8}{20}$
3. $\frac{8}{18}, \frac{16}{36}, \frac{40}{90}$
4. $\frac{14}{16}, \frac{35}{40}, \frac{70}{80}$
5. $\frac{10}{12}, \frac{15}{18}, \frac{50}{60}$
6. $\frac{22}{26}, \frac{55}{65}, \frac{110}{130}$
7. $\frac{20}{34}, \frac{30}{51}, \frac{100}{170}$
8. $\frac{12}{28}, \frac{18}{42}, \frac{30}{70}$
9. $\frac{18}{32}, \frac{27}{48}, \frac{90}{160}$
10. $\frac{12}{22}, \frac{18}{33}, \frac{24}{44}$
11. $\frac{14}{18}, \frac{35}{45}, \frac{140}{180}$
12. $\frac{30}{32}, \frac{45}{48}, \frac{150}{160}$
13. $\frac{2}{3}$
14. $\frac{4}{5}$
15. $\frac{5}{7}$
16. $\frac{3}{10}$
17. $\frac{2}{3}$
18. $\frac{4}{5}$
19. $\frac{7}{8}$
20. $\frac{7}{8}$
21. $\frac{1}{4}$
22. $\frac{2}{5}$
23. $\frac{1}{3}$
24. $\frac{3}{8}$
25. $\frac{8}{9}$
26. $\frac{3}{5}$
27. $\frac{4}{5}$
28. $\frac{1}{3}$
29. $\frac{5}{7}$
30. $\frac{2}{3}$
31. $\frac{4}{5}$
32. $\frac{8}{11}$
33. $\frac{7}{9}$

34. $\frac{3}{7}$ 35. $\frac{15}{44}$

36. $\frac{10}{63}$ 37. $\frac{21}{20}$

38. $\frac{16}{15}$

39. $\frac{3}{7}$

40. $\frac{8}{11}$

41. $\frac{8}{39}$

42. $\frac{10}{33}$

43. $\frac{7}{33}$

44. $\frac{7}{15}$

45. $\frac{1}{6}$

46. $\frac{2}{15}$ 47. $\frac{4}{15}$

48. $\frac{6}{5}$ 49. $\frac{8}{15}$

50. $\frac{5}{6}$ 51. $\frac{2}{3}$

52. $\frac{55}{72}$ 53. $\frac{63}{50}$

54. $\frac{40}{33}$ 55. $\frac{4}{3}$

56. $\frac{2}{3}$ 57. $\frac{4}{15}$

58. $\frac{5}{12}$

59. $\frac{13}{20}$ 60. $\frac{29}{30}$

61. $\frac{13}{15}$ 62. $\frac{53}{60}$

63. $\frac{19}{24}$ 64. $\frac{31}{72}$

34. $\frac{54}{126}$ **35.** $\frac{15}{44}$ **36.** $\frac{10}{63}$

Multiply. Be sure to simplify each product.

37. $\frac{3}{4} \cdot \frac{7}{5}$ **38.** $\frac{2}{3} \cdot \frac{8}{5}$ **39.** $\frac{3}{5} \cdot \frac{5}{7}$

40. $\frac{6}{11} \cdot \frac{8}{6}$ **41.** $\frac{6}{13} \cdot \frac{4}{9}$ **42.** $\frac{5}{9} \cdot \frac{6}{11}$

43. $\frac{3}{11} \cdot \frac{7}{9}$ **44.** $\frac{7}{9} \cdot \frac{3}{5}$ **45.** $\frac{3}{10} \cdot \frac{5}{9}$

Divide. Write each result in simplest form.

46. $\frac{5}{21} \div \frac{25}{14}$ **47.** $\frac{1}{5} \div \frac{3}{4}$ **48.** $\frac{2}{5} \div \frac{1}{3}$

49. $\frac{2}{5} \div \frac{3}{4}$ **50.** $\frac{5}{8} \div \frac{3}{4}$ **51.** $\frac{8}{9} \div \frac{4}{3}$

52. $\frac{5}{9} \div \frac{8}{11}$ **53.** $\frac{7}{10} \div \frac{5}{9}$ **54.** $\frac{8}{9} \div \frac{11}{15}$

55. $\frac{8}{15} \div \frac{2}{5}$ **56.** $\frac{5}{27} \div \frac{15}{54}$

57. $\frac{5}{27} \div \frac{25}{36}$ **58.** $\frac{9}{28} \div \frac{27}{35}$

Add.

59. $\frac{2}{5} + \frac{1}{4}$ **60.** $\frac{2}{3} + \frac{3}{10}$ **61.** $\frac{2}{5} + \frac{7}{15}$

62. $\frac{3}{10} + \frac{7}{12}$ **63.** $\frac{3}{8} + \frac{5}{12}$ **64.** $\frac{5}{36} + \frac{7}{24}$

65. $\frac{2}{15} + \frac{9}{20}$

66. $\frac{9}{14} + \frac{10}{21}$

67. $\frac{7}{15} + \frac{13}{18}$

68. $\frac{12}{25} + \frac{19}{30}$

69. $\frac{1}{2} + \frac{1}{4} + \frac{1}{8}$

70. $\frac{1}{3} + \frac{1}{5} + \frac{1}{10}$

Subtract.

71. $\frac{8}{9} - \frac{3}{9}$

72. $\frac{9}{10} - \frac{6}{10}$

73. $\frac{5}{8} - \frac{1}{8}$

74. $\frac{11}{12} - \frac{7}{12}$

75. $\frac{7}{8} - \frac{2}{3}$

76. $\frac{5}{6} - \frac{3}{5}$

77. $\frac{11}{18} - \frac{2}{9}$

78. $\frac{5}{6} - \frac{1}{4}$

79. $\frac{5}{8} - \frac{1}{6}$

80. $\frac{13}{18} - \frac{5}{12}$

81. $\frac{8}{21} - \frac{1}{14}$

82. $\frac{13}{18} - \frac{7}{15}$

Perform the indicated operations.

83. $7.1562 + 14.78$

84. $6.2358 + 3.14$

85. $11.12 + 8.3792$

86. $6.924 + 5.2$

87. $9.20 - 2.85$

88. $17.345 - 11.12$

89. $18.234 - 13.64$

90. $21.983 - 9.395$

91. $3.21 \cdot 2.1$

92. $15.6 \cdot 7.123$

93. $6.29 \cdot 9.13$

94. $8.245 \cdot 3.1$

ANSWERS

65. $\frac{7}{12}$

66. $\frac{47}{42}$

67. $\frac{107}{90}$

68. $\frac{167}{150}$

69. $\frac{7}{8}$

70. $\frac{19}{30}$

71. $\frac{5}{9}$ 72. $\frac{3}{10}$

73. $\frac{1}{2}$ 74. $\frac{1}{3}$

75. $\frac{5}{24}$ 76. $\frac{7}{30}$

77. $\frac{7}{18}$ 78. $\frac{7}{12}$

79. $\frac{11}{24}$ 80. $\frac{11}{36}$

81. $\frac{13}{42}$ 82. $\frac{23}{90}$

83. 21.9362

84. 9.3758

85. 19.4992

86. 12.124

87. 6.35

88. 6.225

89. 4.594

90. 12.588

91. 6.741

92. 39.8888

93. 57.4277

94. 25.5595

Answers

1. $\frac{6}{14}, \frac{9}{21}, \frac{12}{28}$ **3.** $\frac{8}{18}, \frac{16}{36}, \frac{40}{90}$ **5.** $\frac{10}{12}, \frac{15}{18}, \frac{50}{60}$ **7.** $\frac{20}{34}, \frac{30}{51}, \frac{100}{170}$

9. $\frac{18}{32}, \frac{27}{48}, \frac{90}{160}$ **11.** $\frac{14}{18}, \frac{35}{45}, \frac{140}{180}$ **13.** $\frac{2}{3}$ **15.** $\frac{5}{7}$ **17.** $\frac{2}{3}$ **19.** $\frac{7}{8}$

21. $\frac{1}{4}$ **23.** $\frac{1}{3}$ **25.** $\frac{8}{9}$ **27.** $\frac{4}{5}$ **29.** $\frac{5}{7}$ **31.** $\frac{4}{5}$ **33.** $\frac{7}{9}$

35. $\frac{15}{44}$ **37.** $\frac{21}{20}$ **39.** $\frac{3}{7}$ **41.** $\frac{8}{39}$ **43.** $\frac{7}{33}$ **45.** $\frac{1}{6}$ **47.** $\frac{4}{15}$

49. $\frac{8}{15}$ **51.** $\frac{2}{3}$ **53.** $\frac{63}{50}$ **55.** $\frac{4}{3}$ **57.** $\frac{4}{15}$ **59.** $\frac{13}{20}$ **61.** $\frac{13}{15}$

63. $\frac{19}{24}$ **65.** $\frac{7}{12}$ **67.** $\frac{107}{90}$ **69.** $\frac{7}{8}$ **71.** $\frac{5}{9}$ **73.** $\frac{1}{2}$ **75.** $\frac{5}{24}$

77. $\frac{7}{18}$ **79.** $\frac{11}{24}$ **81.** $\frac{13}{42}$ **83.** 21.9362 **85.** 19.4992 **87.** 6.35

89. 4.594 **91.** 6.741 **93.** 57.4277 **95.** $\frac{13}{12}$ yd **97.** $160

99. 75 mi **101.** 60 in. **103.** $\frac{5}{12}$ **105.** $12\frac{1}{2}$ yd^2 **107.** 2520 mi

109. 66 in. **111.** $42\frac{9}{64}$ in.^{3} **113.**

65. $\frac{2}{15} + \frac{9}{20}$

66. $\frac{9}{14} + \frac{10}{21}$

67. $\frac{7}{15} + \frac{13}{18}$

68. $\frac{12}{25} + \frac{19}{30}$

69. $\frac{1}{2} + \frac{1}{4} + \frac{1}{8}$

70. $\frac{1}{3} + \frac{1}{5} + \frac{1}{10}$

Subtract.

71. $\frac{8}{9} - \frac{3}{9}$

72. $\frac{9}{10} - \frac{6}{10}$

73. $\frac{5}{8} - \frac{1}{8}$

74. $\frac{11}{12} - \frac{7}{12}$

75. $\frac{7}{8} - \frac{2}{3}$

76. $\frac{5}{6} - \frac{3}{5}$

77. $\frac{11}{18} - \frac{2}{9}$

78. $\frac{5}{6} - \frac{1}{4}$

79. $\frac{5}{8} - \frac{1}{6}$

80. $\frac{13}{18} - \frac{5}{12}$

81. $\frac{8}{21} - \frac{1}{14}$

82. $\frac{13}{18} - \frac{7}{15}$

Perform the indicated operations.

83. $7.1562 + 14.78$

84. $6.2358 + 3.14$

85. $11.12 + 8.3792$

86. $6.924 + 5.2$

87. $9.20 - 2.85$

88. $17.345 - 11.12$

89. $18.234 - 13.64$

90. $21.983 - 9.395$

91. $3.21 \cdot 2.1$

92. $15.6 \cdot 7.123$

93. $6.29 \cdot 9.13$

94. $8.245 \cdot 3.1$

ANSWERS

65. $\frac{7}{12}$

66. $\frac{47}{42}$

67. $\frac{107}{90}$

68. $\frac{167}{150}$

69. $\frac{7}{8}$

70. $\frac{19}{30}$

71. $\frac{5}{9}$

72. $\frac{3}{10}$

73. $\frac{1}{2}$

74. $\frac{1}{3}$

75. $\frac{5}{24}$

76. $\frac{7}{30}$

77. $\frac{7}{18}$

78. $\frac{7}{12}$

79. $\frac{11}{24}$

80. $\frac{11}{36}$

81. $\frac{13}{42}$

82. $\frac{23}{90}$

83. 21.9362

84. 9.3758

85. 19.4992

86. 12.124

87. 6.35

88. 6.225

89. 4.594

90. 12.588

91. 6.741

92. 39.8888

93. 57.4277

94. 25.5595

ANSWERS

95. $\frac{13}{12}$ yd

96. $\frac{20}{3}$ mi

97. $160

98. $\frac{1}{2}$

99. 75 mi

100. $67.50

101. 60 in.

102. $700, $1050

103. $\frac{5}{12}$

104. $\frac{7}{15}$

95. Sewing. Roseann is making shirts for her three children. One shirt requires $\frac{1}{2}$ yard of material, a second shirt requires $\frac{1}{3}$ yard of material, and the third shirt requires $\frac{1}{4}$ yard of material. How much material is required for all three shirts?

96. Hiking. Jose rode his trail bike for 10 miles. Two-thirds of the distance was over a mountain trail. How long is the mountain trail?

97. Salary. You make $240 a day on a job. What will you receive for working $\frac{2}{3}$ of a day?

98. Surveys. A survey has found that $\frac{3}{4}$ of the people in a city own pets. Of those who own pets, $\frac{2}{3}$ have cats. What fraction of those surveyed own cats?

Solve the following applications.

99. Map scales. The scale on a map is 1 inch (in.) = 200 miles (mi). What actual distance, in miles, does $\frac{3}{8}$ in. represent?

100. Salary. You make $90 a day on a job. What will you receive for working $\frac{3}{4}$ of a day?

101. Size. A lumberyard has a stack of 80 sheets of plywood. If each sheet is $\frac{3}{4}$ in. thick, how high will the stack be?

102. Family budget. A family uses $\frac{2}{5}$ of its monthly income for housing and utilities on average. If the family's monthly income is $1750, what is spent for housing and utilities? What amount remains?

103. Elections. Of the eligible voters in an election, $\frac{3}{4}$ were registered. Of those registered, $\frac{5}{9}$ actually voted. What fraction of those people who were eligible voted?

104. Surveys. A survey has found that $\frac{7}{10}$ of the people in a city own pets. Of those who own pets, $\frac{2}{3}$ have dogs. What fraction of those surveyed own dogs?

105. Area. A kitchen has dimensions $3\frac{1}{3}$ by $3\frac{3}{4}$ yards (yd). How many square yards (yd^2) of linoleum must be bought to cover the floor?

106. Distance. If you drive at an average speed of 52 miles per hour (mi/h) for $1\frac{3}{4}$ h, how far will you travel?

107. Distance. A jet flew at an average speed of 540 mi/h on a $4\frac{2}{3}$-h flight. What was the distance flown?

108. Area. A piece of land that has $11\frac{2}{3}$ acres is being subdivided for home lots. It is estimated that $\frac{2}{7}$ of the area will be used for roads. What amount remains to be used for lots?

109. Circumference. To find the approximate circumference or distance around a circle, we multiply its diameter by $\frac{22}{7}$. What is the circumference of a circle with a diameter of 21 in.?

110. Area. The length of a rectangle is $\frac{6}{7}$ yd, and its width is $\frac{21}{26}$ yd. What is its area in square yards?

111. Volume. Find the volume of a box that measures $2\frac{1}{4}$ in. by $3\frac{7}{8}$ in. by $4\frac{5}{6}$ in.

112. Topsoil. Nico wishes to purchase mulch to cover his garden. The garden measures $7\frac{7}{8}$ feet (ft) by $10\frac{1}{8}$ ft. He wants the mulch to be $\frac{1}{3}$ ft deep. How much mulch should Nico order if he must order a whole number of cubic feet?

113. Every fraction (rational number) has a corresponding decimal form that either terminates or repeats. For example, $\frac{5}{16} = 0.3125$ (the decimal form terminates), and $\frac{4}{11} = 0.363636........$ (the decimal form repeats). Investigate a number of fractions to determine which ones terminate and which ones repeat. (Hint: you can focus on the denominator; study the prime factorizations of several denominators.)

114. Complete the following sums:

$$\frac{1}{2} + \frac{1}{4} =$$

$$\frac{1}{2} + \frac{1}{4} + \frac{1}{8} =$$

$$\frac{1}{2} + \frac{1}{4} + \frac{1}{8} + \frac{1}{16} =$$

Based on these, predict the sum:

$$\frac{1}{2} + \frac{1}{4} + \frac{1}{8} + \frac{1}{16} + \frac{1}{32} + \frac{1}{64} + \frac{1}{128}$$

ANSWERS

105. $12\frac{1}{2}$ yd^2

106. 91 mi

107. 2520 mi

108. $8\frac{1}{3}$ acres

109. 66 in.

110. $\frac{9}{13}$ yd^2

111. $42\frac{9}{64}$ in.^{3}

112. 27 ft^3

113.

114. $\frac{3}{4}, \frac{7}{8}, \frac{15}{16}, \frac{127}{128}$

Answers

1. $\frac{6}{14}, \frac{9}{21}, \frac{12}{28}$ **3.** $\frac{8}{18}, \frac{16}{36}, \frac{40}{90}$ **5.** $\frac{10}{12}, \frac{15}{18}, \frac{50}{60}$ **7.** $\frac{20}{34}, \frac{30}{51}, \frac{100}{170}$

9. $\frac{18}{32}, \frac{27}{48}, \frac{90}{160}$ **11.** $\frac{14}{18}, \frac{35}{45}, \frac{140}{180}$ **13.** $\frac{2}{3}$ **15.** $\frac{5}{7}$ **17.** $\frac{2}{3}$ **19.** $\frac{7}{8}$

21. $\frac{1}{4}$ **23.** $\frac{1}{3}$ **25.** $\frac{8}{9}$ **27.** $\frac{4}{5}$ **29.** $\frac{5}{7}$ **31.** $\frac{4}{5}$ **33.** $\frac{7}{9}$

35. $\frac{15}{44}$ **37.** $\frac{21}{20}$ **39.** $\frac{3}{7}$ **41.** $\frac{8}{39}$ **43.** $\frac{7}{33}$ **45.** $\frac{1}{6}$ **47.** $\frac{4}{15}$

49. $\frac{8}{15}$ **51.** $\frac{2}{3}$ **53.** $\frac{63}{50}$ **55.** $\frac{4}{3}$ **57.** $\frac{4}{15}$ **59.** $\frac{13}{20}$ **61.** $\frac{13}{15}$

63. $\frac{19}{24}$ **65.** $\frac{7}{12}$ **67.** $\frac{107}{90}$ **69.** $\frac{7}{8}$ **71.** $\frac{5}{9}$ **73.** $\frac{1}{2}$ **75.** $\frac{5}{24}$

77. $\frac{7}{18}$ **79.** $\frac{11}{24}$ **81.** $\frac{13}{42}$ **83.** 21.9362 **85.** 19.4992 **87.** 6.35

89. 4.594 **91.** 6.741 **93.** 57.4277 **95.** $\frac{13}{12}$ yd **97.** \$160

99. 75 mi **101.** 60 in. **103.** $\frac{5}{12}$ **105.** $12\frac{1}{2}$ yd^2 **107.** 2520 mi

109. 66 in. **111.** $42\frac{9}{64}$ in.^{3} **113.**

0.3 Exponents and the Order of Operations

0.3 OBJECTIVES

1. Write a product of factors in exponential form
2. Evaluate an expression involving several operations

Often in mathematics we define symbols that allow us to write a mathematical statement in a more compact or "shorthand" form. This is an idea that you have encountered before. For example, the repeated addition:

$5 + 5 + 5$

can be rewritten as

$3 \cdot 5$

NOTE
$5 + 5 + 5 = 15$
and
$3 \cdot 5 = 15$

Thus multiplication is shorthand for repeated addition.

In algebra, we frequently have a number or variable that is repeated as a factor in an expression several times. For instance, we might have

$5 \cdot 5 \cdot 5$

NOTE A factor is a number or a variable that is being multiplied by another number or variable.

To abbreviate this product, we write

$5 \cdot 5 \cdot 5 = 5^3$

This is called **exponential notation** or **exponential form.** The exponent or power, here 3, indicates the number of times that the factor or base, here 5, appears in a product.

CAUTION

Be careful: 5^3 is *not* the same as $5 \cdot 3$. Notice that $5^3 = 5 \cdot 5 \cdot 5 = 125$ and $5 \cdot 3 = 15$.

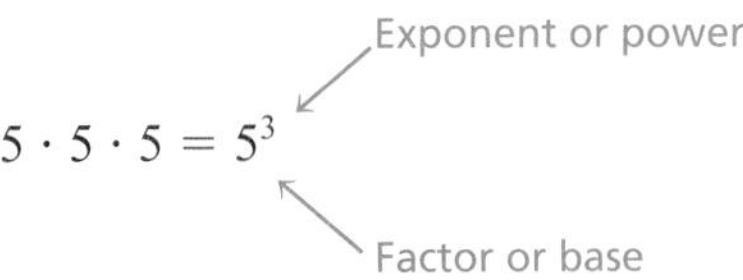

This is read "5 to the third power" or "5 cubed."

Example 1

Writing in Exponential Form

Write $3 \cdot 3 \cdot 3 \cdot 3$, using exponential form. The number 3 appears 4 times in the product, so

Four factors of 3

$3 \cdot 3 \cdot 3 \cdot 3 = 3^4$

This is read "3 to the fourth power."

CHECK YOURSELF 1

Rewrite each using exponential form.

(a) $4 \cdot 4 \cdot 4 \cdot 4 \cdot 4 \cdot 4$ **(b)** $7 \cdot 7 \cdot 7 \cdot 7$

To evaluate an arithmetic expression, you need to know the order in which the operations are done. To see why, simplify the expression $5 + 2 \cdot 3$.

CAUTION

Only one of these results can be correct.

Method 1	or	*Method 2*
$\underbrace{5 + 2} \cdot 3$		$5 + \underbrace{2 \cdot 3}$
Add first.		Multiply first.
$= 7 \cdot 3$		$= 5 + 6$
$= 21$		$= 11$

Because we get different answers depending on how we do the problem, the language of mathematics would not be clear if there were no agreement on which method is correct. The following rules tell us the order in which operations should be done.

NOTE Parentheses and brackets are both grouping symbols. Later we will see that fraction bars and radicals are also grouping symbols.

Step by Step: The Order of Operations

Step 1 Evaluate all expressions inside grouping symbols first.
Step 2 Evaluate all expressions involving exponents.
Step 3 Do any multiplication or division in order, working from left to right.
Step 4 Do any addition or subtraction in order, working from left to right.

Example 2

Evaluating Expressions

Evaluate $5 + 2 \cdot 3$.

There are no parentheses or exponents, so start with step 3: First multiply and then add.

$5 + 2 \cdot 3$ — Multiply first.

$= 5 + 6$ — Then add.

$= 11$

NOTE Method 2 shown above is the correct one.

CHECK YOURSELF 2

Evaluate the following expressions.

(a) $20 - 3 \cdot 4$ **(b)** $9 + 6 \div 3$

When there are no parentheses, evaluate the exponents first.

Example 3

Evaluating Expressions

Evaluate $5 \cdot 3^2$.

$5 \cdot 3^2 = 5 \cdot 9$ — Evaluate the power first.

$= 45$

CHECK YOURSELF 3

Evaluate $4 \cdot 2^4$.

© 2001 McGraw-Hill Companies

Both scientific and graphing calculators correctly interpret the order of operations. This is demonstrated in Example 4.

Example 4

Using a Calculator to Evaluate Expressions

Use your scientific or graphing calculator to evaluate each expression. Round the answer to the nearest tenth.

(a) $24.3 + 6.2 \cdot 3.53$

When evaluating expressions by hand, you must consider the order of operations. In this case, the multiplication must be done first, then the addition. With a calculator, you need only enter the expression correctly. The calculator is programmed to follow the order of operations.

Entering 24.3 [+] 6.2 [×] 3.53 [ENTER]

yields the evaluation 46.186. Rounding to the nearest tenth, we have 46.2.

NOTE With most graphing calculators, the final command is [ENTER]. With most other scientific calculators; the key is marked [=].

(b) $2.45^3 - 49 \div 8000 + 12.2 \cdot 1.3$

Some calculators use the carat (^) to designate powers. Others use the symbol x^y (or y^x).

Entering 2.45 [^] 3 [−] 49 [÷] 8000 [+] 12.2 [×] 1.3 [ENTER]

or 2.45 [y^x] 3 [−] 49 [÷] 8000 [+] 12.2 [×] 1.3 [=]

yields the evaluation 30.56. Rounding to the nearest tenth, we have 30.6.

CHECK YOURSELF 4

Use your scientific or graphing calculator to evaluate each expression.

(a) $67.89 - 4.7 \cdot 12.7$ **(b)** $4.3 \cdot 55.5 - 3.75^3 + 8007 \div 1600$

Operations inside grouping symbols are done first.

Example 5

Evaluating Expressions

Evaluate $(5 + 2) \cdot 3$.

Do the operation inside the parentheses as the first step.

$(5 + 2) \cdot 3 = 7 \cdot 3 = 21$

Add.

CHECK YOURSELF 5

Evaluate $4 \cdot (9 - 3)$.

The principle is the same when more than two "levels" of operations are involved.

Example 6

(a) Evaluate $4 \cdot (2 + 3)^3$.

Add inside the parentheses first.

$4 \cdot (2 + 3)^3 = 4 \cdot (5)^3$ — Evaluate the power.

$= 4 \cdot 125$ — Multiply.

$= 500$

(b) Evaluate $5 \cdot (7 - 3)^2 - 10$.

Evaluate the expression inside the parentheses.

$5 \cdot (7 - 3)^2 - 10 = 5(4)^2 - 10$ — Evaluate the power.

$= 5 \cdot 16 - 10$ — Multiply.

$= 80 - 10 = 70$ — Subtract.

CHECK YOURSELF 6

Evaluate.

(a) $4 \cdot 3^3 - 8 \cdot 11$ **(b)** $12 + 4 \cdot (2 + 3)^2$

CHECK YOURSELF ANSWERS

1. (a) 4^6; **(b)** 7^4 **2. (a)** 8; **(b)** 11 **3.** 64 **4. (a)** 8.2; **(b)** 190.92 **5.** 24
6. (a) 20; **(b)** 112

Name ____________

Section ________ Date ________

0.3 Exercises

Write each expression in exponential form.

1. $7 \cdot 7 \cdot 7 \cdot 7$

2. $2 \cdot 2 \cdot 2 \cdot 2 \cdot 2 \cdot 2$

3. $6 \cdot 6 \cdot 6 \cdot 6 \cdot 6$

4. $4 \cdot 4 \cdot 4 \cdot 4 \cdot 4 \cdot 4 \cdot 4$

5. $8 \cdot 8 \cdot 8 \cdot 8 \cdot 8 \cdot 8 \cdot 8 \cdot 8 \cdot 8 \cdot 8$

6. $10 \cdot 10 \cdot 10$

7. $15 \cdot 15 \cdot 15 \cdot 15 \cdot 15 \cdot 15$

8. $31 \cdot 31 \cdot 31 \cdot 31 \cdot 31 \cdot 31 \cdot 31 \cdot 31 \cdot 31 \cdot 31$

Evaluate each of the following expressions.

9. $7 + 2 \cdot 6$

10. $10 - 4 \cdot 2$

11. $(7 + 2) \cdot 6$

12. $(10 - 4) \cdot 2$

13. $12 - 8 \div 4$

14. $10 + 20 \div 5$

15. $(12 - 8) \div 4$

16. $(10 + 20) \div 5$

17. $8 \cdot 7 + 2 \cdot 2$

18. $48 \div 8 - 4 \div 2$

19. $8 \cdot (7 + 2) \cdot 2$

20. $48 \div (8 - 4) \div 2$

21. $3 \cdot 5^2$

22. $5 \cdot 2^3$

23. $(3 \cdot 5)^2$

24. $(5 \cdot 2)^3$

25. $4 \cdot 3^2 - 2$

26. $3 \cdot 2^4 - 8$

ANSWERS

1. 7^4
2. 2^6
3. 6^5
4. 4^7
5. 8^{10}
6. 10^3
7. 15^6
8. 31^{10}
9. 19
10. 2
11. 54
12. 12
13. 10
14. 14
15. 1
16. 6
17. 60
18. 4
19. 144
20. 6
21. 75
22. 40
23. 225
24. 1000
25. 34
26. 40

ANSWERS

27. 21
28. 8
29. 36
30. 2
31. 40
32. 195
33. 256
34. 48
35. 196
36. 400
37. 147
38. 40
39. 21
40. 12
41. 75
42. 80
43. 96
44. 89
45. 1.2
46. 1.5
47. 7.8
48. 5.4
49. 2^5
50. 9^3
51. $36 \div (4 + 2) - 4$
52. $48 \div (3 \cdot 2) - 2 \cdot 3$
53. $(6 + 9) \div 3 + (16 - 4) \cdot 2$
54. $(5 - 3) \cdot 2 + 8 \cdot (5 - 2)$

27. $7 \cdot (2^3 - 5)$

28. $4 \cdot (3^2 - 7)$

29. $3 \cdot 2^4 - 6 \cdot 2$

30. $4 \cdot 2^3 - 5 \cdot 6$

31. $(2 \cdot 4)^2 - 8 \cdot 3$

32. $(3 \cdot 2)^3 - 7 \cdot 3$

33. $4 \cdot (2 + 6)^2$

34. $3 \cdot (8 - 4)^2$

35. $(4 \cdot 2 + 6)^2$

36. $(3 \cdot 8 - 4)^2$

37. $3 \cdot (4 + 3)^2$

38. $5 \cdot (4 - 2)^3$

39. $3 \cdot 4 + 3^2$

40. $5 \cdot 4 - 2^3$

41. $4 \cdot (2 + 3)^2 - 25$

42. $8 + 2 \cdot (3 + 3)^2$

43. $(4 \cdot 2 + 3)^2 - 25$

44. $8 + (2 \cdot 3 + 3)^2$

Evaluate using your calculator. Round your answer to the nearest tenth.

45. $(1.2)^3 \div 2.0736 \cdot 2.4 + 1.6935 - 2.4896$

46. $(5.21 \cdot 3.14 - 6.2154) \div 5.12 - .45625$

47. $1.23 \cdot 3.169 - 2.05194 + (5.128 \cdot 3.15 - 10.1742)$

48. $4.56 + (2.34)^4 \div 4.7896 \cdot 6.93 \div 27.5625 - 3.1269 + (1.56)^2$

49. Population doubling. Over the last 2000 years, the Earth's population has doubled approximately 5 times. Write this last factor in exponential form.

50. Volume of a cube. The volume of a cube with each edge of length 9 in. is given by $9 \cdot 9 \cdot 9$. Write the volume using exponential notation.

Insert grouping symbols in the proper place so that the given value of the expression is obtained.

51. $36 \div 4 + 2 - 4; 2$

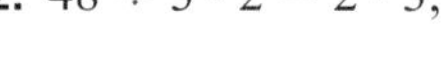

52. $48 \div 3 \cdot 2 - 2 \cdot 3; 2$

53. $6 + 9 \div 3 + 16 - 4 \cdot 2; 29$

54. $5 - 3 \cdot 2 + 8 \cdot 5 - 2; 28$

Answers

1. 7^4 **3.** 6^5 **5.** 8^{10} **7.** 15^6 **9.** 19 **11.** 54 **13.** 10
15. 1 **17.** 60 **19.** 144 **21.** 75 **23.** 225 **25.** 34 **27.** 21
29. 36 **31.** 40 **33.** 256 **35.** 196 **37.** 147 **39.** 21 **41.** 75
43. 96 **45.** 1.2 **47.** 7.8 **49.** 2^5 **51.** $36 \div (4 + 2) - 4$
53. $(6 + 9) \div 3 + (16 - 4) \cdot 2$

Positive and Negative Integers

OBJECTIVES

1. Represent integers on a number line
2. Order signed numbers
3. Evaluate numerical expressions involving absolute value

When numbers are used to represent physical quantities (altitudes, temperatures, and amounts of money are examples), it may be necessary to distinguish between *positive* and *negative* quantities. It is convenient to represent these quantities with plus (+) or minus (−) signs. For instance,

The altitude of Mount Whitney is 14,495 feet (ft) *above* sea level (+14,495).

Mount Whitney

The altitude of Death Valley is 282 ft *below* sea level (−282).

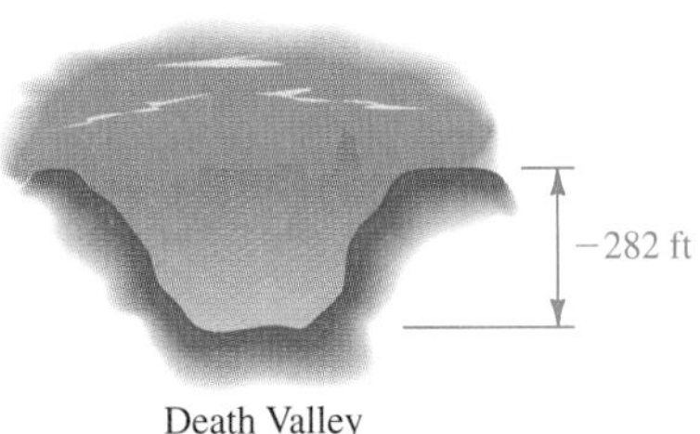

Death Valley

The temperature in Chicago is 10° *below* zero (−10°).

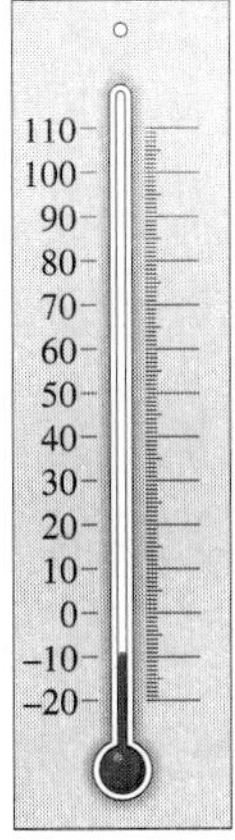

An account could show a *gain* of \$100 (+100), or a *loss* of \$100 (−100).

\$20
-\$25
-\$95
-\$100

These numbers suggest the need to extend the whole numbers to include both positive numbers (like +100) and negative numbers (like −282).

To represent the negative numbers, we extend the number line to the *left* of zero and name equally spaced points.

Numbers used to name points to the right of zero are positive numbers. They are written with a positive (+) sign or with no sign at all.

+6 and 9 are positive numbers

Numbers used to name points to the left of zero are negative numbers. They are always written with a negative (−) sign.

−3 and −20 are negative numbers

Read "negative 3."

Positive and negative numbers considered together are **signed numbers.**

Here is the number line extended to include both positive and negative numbers.

NOTE 0 is not considered a signed number.

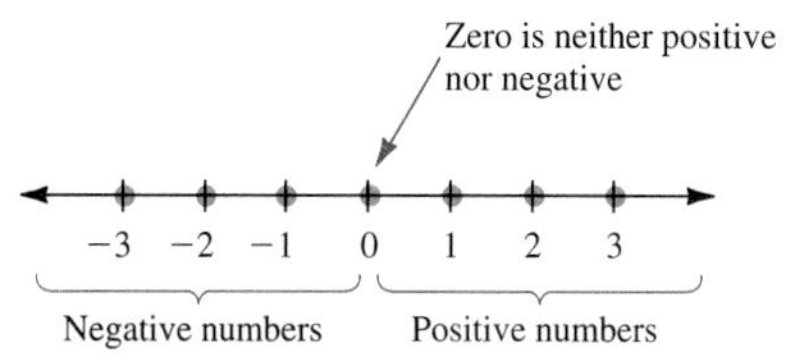

The numbers used to name the points shown on the number line above are called the **integers.** The integers consist of the natural numbers, their negatives, and the number 0. We can represent the set of integers by

NOTE The dots are called *ellipses* and indicate that the pattern continues.

$\{\ldots, -3, -2, -1, 0, 1, 2, 3, \ldots\}$

Example 1

Representing Integers on the Number Line

Represent the following integers on the number line shown.

−3, −12, 8, 15, −7

−12 −7 −3 8 15

−15 −10 −5 0 5 10 15

CHECK YOURSELF 1

Represent the following integers on a number line.

−1, −9, 4, −11, 8, 20

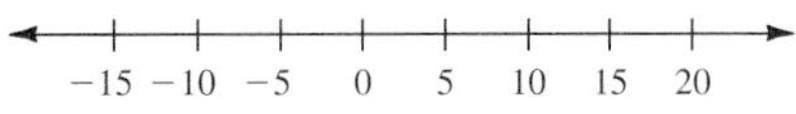

The set of numbers on the number line is *ordered.* The numbers get smaller moving to the left on the number line and larger moving to the right.

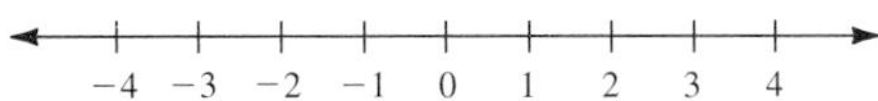

When a set of numbers is written from smallest to largest, the numbers are said to be in *ascending order.*

Example 2

Ordering Signed Numbers

Place each set of numbers in ascending order.

(a) 9, −5, −8, 3, 7

From smallest to largest, the numbers are

−8, −5, 3, 7, 9

Note that this is the order in which the numbers appear on a number line as we move from left to right.

(b) 3, −2, 18, −20, −13

From smallest to largest, the numbers are

−20, −13, −2, 3, 18

CHECK YOURSELF 2

Place each set of numbers in ascending order.

(a) 12, −13, 15, 2, −8, −3 **(b)** 3, 6, −9, −3, 8

The least and greatest numbers in a set are called the **extreme values.** The least element is called the **minimum** and the greatest element is called the **maximum.**

Example 3

Labeling Extreme Values

For each set of numbers, determine the minimum and maximum values.

(a) 9, −5, −8, 3, 7

From our previous ordering of these numbers, we see that −8, the least element, is the minimum, and 9, the greatest element, is the maximum.

(b) 3, −2, 18, −20, −13

−20 is the minimum and 18 is the maximum.

CHECK YOURSELF 3

For each set of numbers, determine the minimum and maximum values.

(a) 12, −13, 15, 2, −8, −3 **(b)** 3, 6, −9, −3, 8

Integers are not the only kind of signed numbers. Decimals and fractions can also be thought of as signed numbers.

Example 4

Identifying Signed Numbers that are Integers

Which of the following signed numbers are also integers?

(a) 145 is an integer.

(b) −28 is an integer.

(c) 0.35 is not an integer.

(d) $-\frac{2}{3}$ is not an integer.

CHECK YOURSELF 4

Which of the following signed numbers are also integers?

−23 1054 −0.23 0 −500 $-\frac{4}{5}$

Sometimes we refer to the negative of a number as its "opposite." But what is the opposite of the opposite of a number? It is the number itself. The next example illustrates.

Example 5

Find the Opposite for Each Number

(a) 5 The opposite of 5 is -5.
(b) -9 The opposite of -9 is 9.

CHECK YOURSELF 5

Find the opposite for each number.

(a) 17 **(b)** -12

An important idea for our work in this chapter is the **absolute value** of a number. This represents the distance of the point named by the number from the origin on the number line.

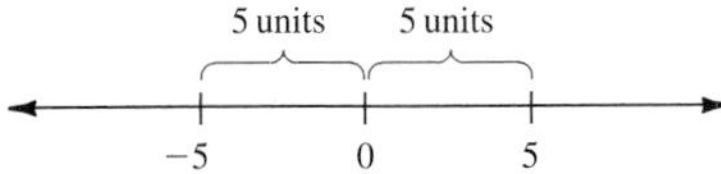

The absolute value of 5 is 5. The absolute value of -5 is also 5.

The **absolute value** of a positive number or zero is itself. The absolute value of a negative number is its opposite.

In symbols we write

$|5| = 5$ and $|-5| = 5$

Read "the absolute value of 5." Read "the absolute value of negative 5."

The absolute value of a number does *not* depend on whether the number is to the right or to the left of the origin, but on its *distance* from the origin.

Example 6

Simplifying Absolute Value Expressions

(a) $|7| = 7$
(b) $|-7| = 7$
(c) $-|-7| = -7$

This is the *negative,* or opposite, of the absolute value of negative 7.

(d) $|-10| + |10| = 10 + 10 = 20$

Absolute value bars serve as another set of grouping symbols, so do the operation *inside* first.

(e) $|8 - 3| = |5| = 5$
(f) $|8| - |3| = 8 - 3 = 5$

Here, evaluate the absolute values, then subtract.

CHECK YOURSELF 6

Evaluate.

(a) $|8|$ **(b)** $|-8|$ **(c)** $-|-8|$
(d) $|-9| + |4|$ **(e)** $|9 - 4|$ **(f)** $|9| - |4|$

CHECK YOURSELF ANSWERS

1.

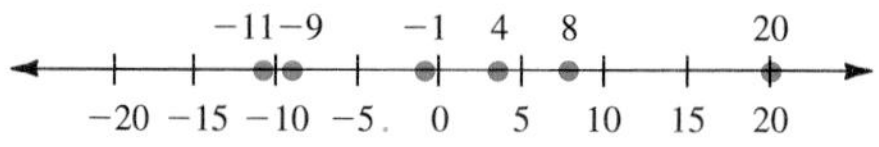

2. (a) $-13, -8, -3, 2, 12, 15$ **(b)** $-9, -3, 3, 6, 8$

3. (a) minimum is -13; maximum is 15 **(b)** minimum is -9; maximum is 8

4. -23, 1054, 0, and -500

5. (a) -17; **(b)** 12

6. (a) 8; **(b)** 8; **(c)** -8; **(d)** 13; **(e)** 5; **(f)** 5

Name ____________________

Section ________ Date ________

Exercises

Represent each quantity with a signed number.

1. An altitude of 400 feet (ft) above sea level
2. An altitude of 80 ft below sea level
3. A loss of $200
4. A profit of $400
5. A decrease in population of 25,000
6. An increase in population of 12,500

Represent the integers on the number lines shown.

7. 5, −15, 18, −8, 3

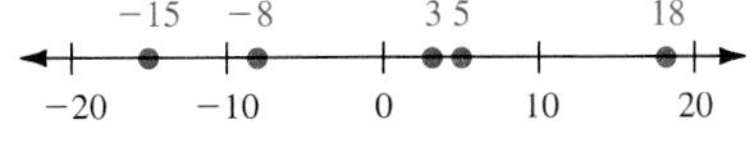

8. −18, 4, −5, 13, 9

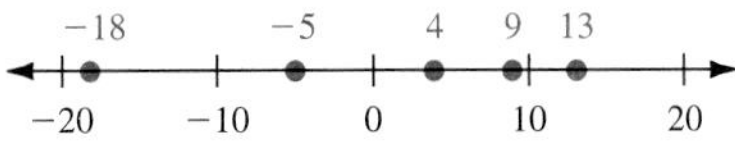

Which numbers in the following sets are integers?

9. $\left\{5, -\frac{2}{9}, 175, -234, -0.64\right\}$

10. $\left\{-45, 0.35, \frac{3}{5}, 700, -26\right\}$

Place each of the following sets in ascending order.

11. 3, −5, 2, 0, −7, −1, 8
12. −2, 7, 1, −8, 6, −1, 0
13. 9, −2, −11, 4, −6, 1, 5
14. 23, −18, −5, −11, −15, 14, 20
15. −6, 7, −7, 6, −3, 3
16. 12, −13, 14, −14, 15, −15

For each set, determine the maximum and minimum values.

17. 5, −6, 0, 10, −3, 15, 1, 8
18. 9, −1, 3, 11, −4, 2, 5, −2
19. 21, −15, 0, 7, −9, 16, −3, 11
20. −22, 0, 22, −31, 18, −5, 3
21. 3, 0, 1, −2, 5, 4, −1
22. 2, 7, −3, 5, −10, −5

Find the opposite of each number.

23. 15
24. 18

ANSWERS

1. +400
2. −80
3. −200
4. +400
5. −25,000
6. +12,500
7. See exercise
8. See exercise
9. 5, 175, −234
10. −45, 700, −26
11. −7, −5, −1, 0, 2, 3, 8
12. −8, −2, −1, 0, 1, 6, 7
13. −11, −6, −2, 1, 4, 5, 9
14. −18, −15, −11, −5, 14, 20, 23
15. −7, −6, −3, 3, 6, 7
16. −15, −14, −13, 12, 14, 15
17. Max: 15; Min: −6
18. Max: 11; Min: −4
19. Max: 21; Min: −15
20. Max: 22; Min: −31
21. Max: 5; Min: −2
22. Max: 7; Min: −10
23. −15
24. −18

ANSWERS

25. −11
26. −34
27. 19
28. 5
29. 7
30. 54
31. 17
32. 28
33. 10
34. 7
35. −3
36. −5
37. −8
38. −13
39. 5
40. 7
41. 18
42. 22
43. 0
44. 0
45. 7
46. 8
47. 7
48. 8
49. 11
50. 11
51. 1
52. 5

25. 11 **26.** 34

27. -19 **28.** -5

29. -7 **30.** -54

Evaluate.

31. $|17|$ **32.** $|28|$

33. $|-10|$ **34.** $|-7|$

35. $-|3|$ **36.** $-|5|$

37. $-|-8|$ **38.** $-|-13|$

39. $|-2| + |3|$ **40.** $|4| + |-3|$

41. $|-9| + |9|$ **42.** $|11| + |-11|$

43. $|4| - |-4|$ **44.** $|5| - |-5|$

45. $|15| - |8|$ **46.** $|11| - |3|$

47. $|15 - 8|$ **48.** $|11 - 3|$

49. $|-9| + |2|$ **50.** $|-7| + |4|$

51. $|-8| - |-7|$ **52.** $|-9| - |-4|$

Label each statement as true or false.

53. All whole numbers are integers.

54. All nonzero integers are signed numbers.

55. All integers are whole numbers.

56. All signed numbers are integers.

57. All negative integers are whole numbers.

58. Zero is neither positive nor negative.

Place absolute value bars in the proper location on the left side of the expression so that the equation is true.

59. $6 + (-2) = 4$

60. $8 + (-3) = 5$

61. $6 + (-2) = 8$

62. $8 + (-3) = 11$

Represent each quantity with a signed number.

63. Soil erosion. The erosion of 5 centimeters (cm) of topsoil from an Iowa corn field.

64. Soil formation. The formation of 2.5 cm of new topsoil on the African savanna.

65. Checking accounts. The withdrawal of $50 from a checking account.

66. Saving accounts. The deposit of $200 in a savings account.

67. Temperature. The temperature change pictured.

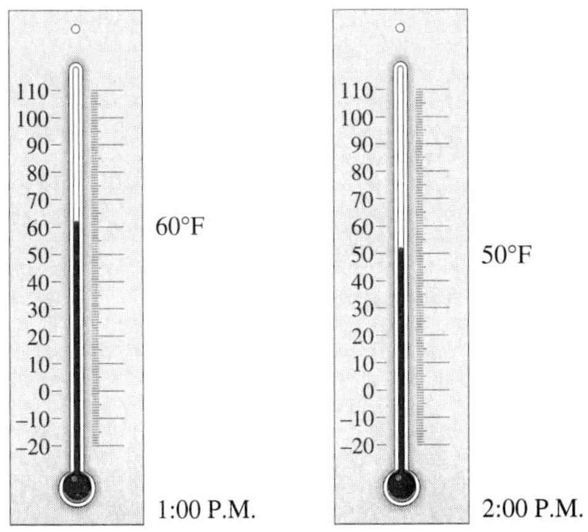

68. Stocks. An increase of 75 points in the Dow-Jones average.

ANSWERS

53. True
54. True
55. False
56. False
57. False
58. True
59. $|6 + (-2)| = 4$
60. $|8 + (-3)| = 5$
61. $|6| + |-2| = 8$
62. $|8| + |(-3)| = 11$
63. −5
64. +2.5
65. −50
66. +200
67. −10°
68. +75

ANSWERS

69. −8

70. +25,000

71. +90,000,000

72. −60,000,000

73. −6; 8; 8; −2

74. −8; 9; 9; 3

75. −2; 6; 6; 0

76. −9; −9; −9; 0

77.

69. Baseball. An eight-game losing streak by the local baseball team.

70. Population. An increase of 25,000 in the population of the city.

71. Positive trade balance. A country exported \$90,000,000 more than it imported, creating a positive trade balance.

72. Negative trade balance. A country exported \$60,000,000 less than it imported, creating a negative trade balance.

For each collection of numbers given in exercises 73 to 76, answer the following:

(a) Which number is smallest?
(b) Which number lies farthest from the origin?
(c) Which number has the largest absolute value?
(d) Which number has the smallest absolute value?

73. $-6, 3, 8, 7, -2$

74. $-8, 3, -5, 4, 9$

75. $-2, 6, -1, 0, 2, 5$

76. $-9, 0, -2, 3, 6$

77. Simplify each of the following:

$-(-7) \qquad -(-(-7)) \qquad -(-(-(-7)))$

Based on your answers, generalize your results.

Answers

1. 400 or (+400) **3.** −200 **5.** −25,000

7. 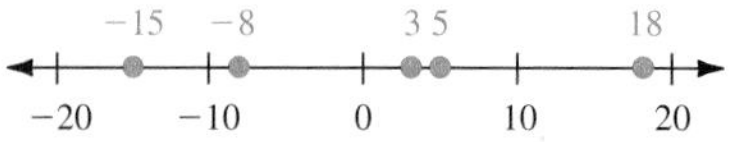 Number line from −20 to 20 (ticks at −20, −10, 0, 10, 20) with points marked at −15, −8, 3, 5, 18. **9.** 5, 175, −234

11. −7, −5, −1, 0, 2, 3, 8 **13.** −11, −6, −2, 1, 4, 5, 9
15. −7, −6, −3, 3, 6, 7 **17.** Max: 15; Min: −6 **19.** Max: 21, Min: −15
21. Max: 5; Min: −2 **23.** −15 **25.** −11 **27.** 19 **29.** 7 **31.** 17
33. 10 **35.** −3 **37.** −8 **39.** 5 **41.** 18 **43.** 0 **45.** 7
47. 7 **49.** 11 **51.** 1 **53.** True **55.** False **57.** False
59. $|6 + (-2)| = 4$ **61.** $|6| + |-2| = 8$ **63.** −5 **65.** −50
67. −10°F **69.** −8 **71.** +90,000,000
73. −6; 8; 8; −2 **75.** −2; 6; 6; 0 **77.**

Summary

Definition/Procedure	Example	Reference
Prime Factorization		**Section 0.1**
Factor A **factor** of a whole number is another whole number that will divide exactly into that number, leaving a remainder of zero.	The factors of 12 are 1, 2, 3, 4, 6, and 12.	**p. 4**
Prime Number Any whole number greater than 1 that has only 1 and itself as factors.	7, 13, 29, and 73 are prime numbers.	**p. 4**
Composite Number Any whole number greater than 1 that is not prime.	8, 15, 42, and 65 are composite numbers.	**p. 5**
Zero and 1 0 and 1 are not classified as prime or composite numbers.		**p. 6**
Greatest Common Factor (GCF) The GCF is the *largest* number that is a factor of each of a group of numbers.		**p. 9**
To find the GCF 1. Write the prime factorization for each of the numbers in the group. 2. Locate the prime factors that are common to all the numbers. 3. The greatest common factor (GCF) will be the product of all of the common prime factors. If there are no common prime factors, the GCF is 1.	To find the GCF of 24, 30, and 36: $24 = ② \cdot 2 \cdot 2 \cdot ③$ $30 = ② \cdot ③ \cdot 5$ $36 = ② \cdot 2 \cdot ③ \cdot 3$ The GCF is $2 \cdot 3 = 6$.	**p. 10**
Fractions		**Section 0.2**
The Fundamental Principle of Fractions $\frac{a}{b} = \frac{a \cdot c}{b \cdot c}$ in which neither b nor c can equal zero.	$\frac{2}{3} = \frac{2 \cdot 3}{3 \cdot 3}$	**p. 17**
Multiplying Fractions 1. Multiply numerator by numerator. This gives the numerator of the product. 2. Multiply denominator by denominator. This gives the denominator of the product. 3. Simplify the resulting fraction if possible. In multiplying fractions it is usually easiest to divide by any common factors in the numerator and denominator *before* multiplying.	$\frac{5}{8} \cdot \frac{3}{7} = \frac{5 \cdot 3}{8 \cdot 7} = \frac{15}{56}$ $\frac{5}{9} \cdot \frac{3}{10} = \frac{\overset{1}{\cancel{5}} \cdot \overset{1}{\cancel{3}}}{\underset{3}{\cancel{9}} \cdot \underset{2}{\cancel{10}}} = \frac{1}{6}$	**p. 19**
Dividing Fractions Invert the divisor and multiply.	$\frac{3}{7} \div \frac{4}{5} = \frac{3}{7} \cdot \frac{5}{4} = \frac{15}{28}$	**p. 19**

Continued

DEFINITION/PROCEDURE	EXAMPLE	REFERENCE
Fractions		**Section 0.2**
To Add or Subtract Fractions with Different Denominators 1. Find the LCD of the fractions. 2. Change each fraction to an equivalent fraction with the LCD as a common denominator. 3. Add or subtract the resulting like fractions as before.	$\frac{3}{4} + \frac{7}{10} = \frac{15}{20} + \frac{14}{20} = \frac{29}{20}$ $\frac{8}{9} - \frac{5}{6} = \frac{16}{18} - \frac{15}{18} = \frac{1}{18}$	pp. 20, 21
Exponents and the Order of Operations		**Section 0.3**
Using Exponents **Base** The number that is raised to a power. **Exponent** The exponent is written to the right and above the base. The exponent tells the number of times the base is to be used as a factor.	Exponent $5^3 = \underbrace{5 \cdot 5 \cdot 5}_{\text{Three factors}} = 125$ Base This is read "5 to the third power" or "5 cubed."	p. 29
The Order of Operations *Mixed Operations* in an expression should be done in the following order: 1. Do any operations inside parentheses. 2. Evaluate any powers. 3. Do all multiplication and division in order from left to right. 4. Do all addition and subtraction in order from left to right.	$4 \cdot (2 + 3)^2 - 7$ $= 4 \cdot 5^2 - 7$ $= 4 \cdot 25 - 7$ $= 100 - 7$ $= 93$	p. 30
Positive and Negative Integers		**Section 0.4**
Positive Numbers Numbers used to name points to the right of the origin on the number line. *Negative Numbers* Numbers used to name points to the left of the origin on the number line. *Signed Numbers* The positive and negative numbers. *Integers* The natural (or counting) numbers, their negatives, and zero. The integers are $\{\ldots, -3, -2, -1, 0, 1, 2, 3, \ldots\}$	The origin −3 −2 −1 0 1 2 3 Negative numbers Positive numbers	p. 37
Absolute Value The distance (on the number line) between the point named by a signed number and the origin. The absolute value of x is written $\|x\|$.	$\|7\| = 7$ $\|-10\| = 10$	p. 39

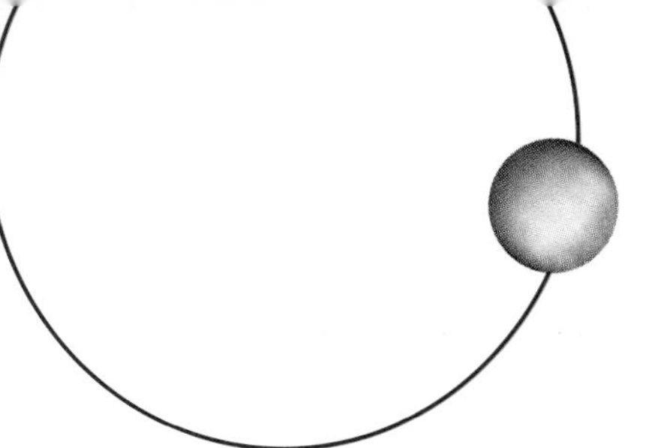

Summary Exercises

This exercise set will give you practice with each of the objectives of the chapter. Each exercise is keyed to the appropriate chapter section. The answers are provided in the *Instructor's Manual.* Your instructor will give you guidelines on how to best use these exercises in your instructional setting.

[0.1] In Exercises 1 to 4, list all the factors of the given numbers.

1. 52 **1, 2, 4, 13, 26, 52**

2. 41 **1, 41**

3. 76 **1, 2, 4, 19, 38, 76**

4. 315 **1, 3, 5, 7, 9, 15, 21, 35, 45, 63, 105, 315**

In Exercise 5, use the group of numbers 2, 5, 7, 11, 14, 17, 21, 23, 27, 39, and 43.

5. List the prime numbers; then list the composite numbers. **Prime: 2, 5, 7, 11, 17, 23, 43; Composite: 14, 21, 27, 39**

In Exercises 6 to 9, find the prime factorization for the given numbers.

6. 48 $2^4 \cdot 3$

7. 420 $2^2 \cdot 3 \cdot 5 \cdot 7$

8. 60 $2^2 \cdot 3 \cdot 5$

9. 180 $2^2 \cdot 3^2 \cdot 5$

In Exercises 10 to 13, find the greatest common factor (GCF).

10. 15 and 20 **5**

11. 30 and 31 **1**

12. 72 and 180 **36**

13. 240 and 900 **60**

[0.2] In Exercises 14 to 16, write three fractional representations for each number.

14. $\frac{5}{7}$ $\frac{10}{14}, \frac{15}{21}, \frac{20}{28}$

15. $\frac{3}{11}$ $\frac{6}{22}, \frac{9}{33}, \frac{12}{44}$

16. $\frac{4}{9}$ $\frac{8}{18}, \frac{12}{27}, \frac{16}{36}$

17. Write the fraction $\frac{24}{64}$ in simplest form. $\frac{3}{8}$

[0.2] In Exercises 18 to 28, perform the indicated operations.

18. $\frac{7}{15} \cdot \frac{5}{21}$ $\frac{1}{9}$

19. $\frac{10}{27} \cdot \frac{9}{20}$ $\frac{1}{6}$

20. $\frac{5}{12} \div \frac{5}{8}$ $\frac{2}{3}$

21. $\frac{7}{15} \div \frac{14}{25}$ $\frac{5}{6}$

22. $\frac{5}{6} + \frac{11}{18}$ $\frac{13}{9}$

23. $\frac{5}{18} + \frac{7}{12}$ $\frac{31}{36}$

24. $\frac{11}{18} - \frac{2}{9}$ $\frac{7}{18}$

25. $\frac{11}{27} - \frac{5}{18}$ $\frac{7}{54}$

26. $5.123 + 6.4$ **11.523**

27. $10.127 - 5.49$ **4.637**

28. $5.26 \cdot 3.796$ **19.96696**

[0.3] Evaluate each of the following expressions.

29. $18 - 3 \cdot 5$ **3**

30. $(18 - 3) \cdot 5$ **75**

31. $5 \cdot 4^2$ **80**

32. $(5 \cdot 4)^2$ **400**

33. $5 \cdot 3^2 - 4$ **41**

34. $5 \cdot (3^2 - 4)$ **25**

35. $5 \cdot (4 - 2)^2$ **20**

36. $5 \cdot 4 - 2^2$ **16**

37. $(5 \cdot 4 - 2)^2$ **324**

38. $3 \cdot (5 - 2)^2$ **27**

39. $3 \cdot 5 - 2^2$ **11**

40. $(3 \cdot 5 - 2)^2$ **169**

41. $8 \div 4 \cdot 2$ **4**

42. $19 - 14 + 2 \cdot 5$ **15**

43. $36 + 4 \cdot 2 - 7 \cdot 6$ **2**

[0.4] Represent the integers on the number line shown.

44. $6, -18, -3, 2, 15, -9$

−18 −9 −3 2 6 15

−20 −10 0 10 20

Place each of the following sets in ascending order.

45. $4, -3, 6, -7, 0, 1, -2$ **−7, −3, −2, 0, 1, 4, 6**

46. $-7, 8, -8, 1, 2, -3, 3, 0, 7$ **−8, −7, −3, 0, 1, 2, 3, 7, 8**

For each data set, determine the maximum and minimum.

47. $4, -2, 5, 1, -6, 3, -4$ **Max: 5 Min: −6**

48. $-4, 2, 5, -9, 8, 1, -6$ **Max: 8 Min: −9**

Find the opposite of each number.

49. 17 **−17**

50. -63 **63**

Evaluate.

51. $|9|$ **9**

52. $|-9|$ **9**

53. $-|9|$ **−9**

54. $-|-9|$ **−9**

55. $|12 - 8|$ **4**

56. $|8| + |-12|$ **20**

57. $-|8 + 12|$ **−20**

58. $|-18| - |-12|$ **6**

59. $|-7| - |-3|$ **4**

60. $|-9| + |-5|$ **14**

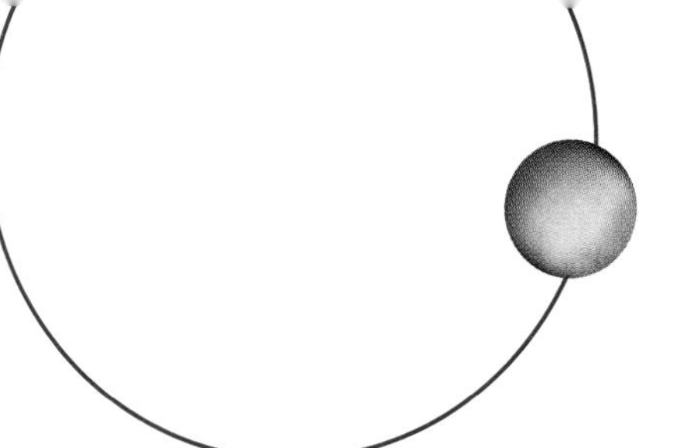

Self-Test for Chapter 0

Name ________________

Section ________ Date ________

The purpose of this self-test is to help you check your progress and to review for a chapter test in class. Allow yourself about an hour to take the test. When you are done, check your answers in the back of the book. If you missed any problems, go back and review the appropriate sections in the chapter and the exercises provided.

1. Which of the numbers 5, 9, 13, 17, 22, 27, 31, and 45 are prime numbers? Which are composite numbers?

2. Find the prime factorization for 264.

In Exercises 3 and 4, find the greatest common factor (GCF) for the given numbers.

3. 36 and 84

4. 16, 24, and 72

Perform the indicated operations.

5. $\frac{8}{21} \cdot \frac{3}{4}$

6. $\frac{7}{12} \div \frac{28}{36}$

7. $\frac{3}{4} + \frac{5}{6}$

8. $\frac{8}{21} - \frac{2}{7}$

9. $3.25 + 4.125$

10. $16.234 - 12.35$

11. $7.29 \cdot 3.15$

12. $6.10 \cdot 13.1$

Write, using exponents.

13. $4 \cdot 4 \cdot 4 \cdot 4$

14. $9 \cdot 9 \cdot 9 \cdot 9 \cdot 9$

Evaluate the following expressions.

15. $23 - 4 \cdot 5$

16. $4 \cdot 5^2 - 35$

17. $4 \cdot (2 + 4)^2$

18. $16 \cdot 2 - 5^2$

19. $(3 \cdot 2 - 4)^3$

20. $8 - 3 \cdot 2 + 5$

ANSWERS

1. Prime: 5, 13, 17, 31
   Composite: 9, 22, 27, 45
2. $2 \cdot 2 \cdot 2 \cdot 3 \cdot 11$
3. 12
4. 8
5. $\frac{2}{7}$
6. $\frac{3}{4}$
7. $\frac{19}{12}$
8. $\frac{2}{21}$
9. 7.375
10. 3.884
11. 22.9635
12. 79.91
13. 4^4
14. 9^5
15. 3
16. 65
17. 144
18. 7
19. 8
20. 7

ANSWERS

21. See exercise
22. −6, −3, −2, 0, 2, 4, 5
23. Max: 6; Min: −5
24. 7
25. 7
26. 11
27. 11
28. −19
29. −40
30. 19

Represent the integers on the number line shown.

21. 5, −12, 4, −7, 18, −17

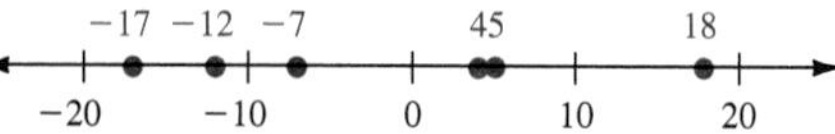

22. Place the following data set in ascending order: 4, −3, −6, 5, 0, 2, −2

23. Determine the maximum and minimum of the following data set: 3, 2, −5, 6, 1, −2

Evaluate.

24. $|7|$

25. $|-7|$

26. $|18 - 7|$

27. $|18| - |-7|$

28. $-|24 - 5|$

Find the opposite of each of the following.

29. 40

30. −19

1

THE LANGUAGE OF ALGEBRA

INTRODUCTION

Anthropologists and archeologists investigate modern human cultures and societies as well as cultures that existed so long ago that their characteristics must be inferred from objects found buried in lost cities or villages. When some interesting object is found, such as the Babylonian tablets mentioned in Chapter 0, often the first questions that arise are "How old is this? When did this culture flourish?" With methods such as carbon dating, it has been established that large, organized cultures existed around 3000 B.C.E. in Egypt, 2800 B.C.E. in India, no later than 1500 B.C.E. in China, and around 1000 B.C.E. in the Americas.

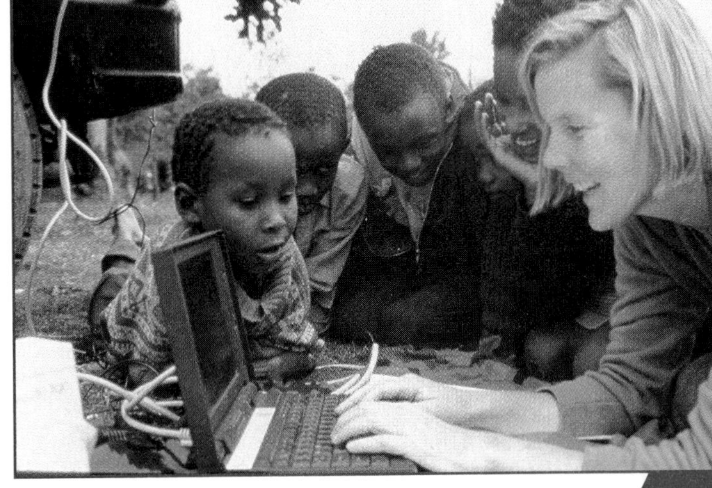

How long ago was 1500 B.C.E.? Which is older, an object from 3000 B.C.E. or an object from 500 A.D.*? Using the Christian notation for dates, we have to count A.D. years and B.C.E. years differently. An object from 500 A.D. is 2000 − 500 years old, or about 1500 years old. But an object from 3000 B.C.E. is 2000 + 3000 years old, or about 5000 years old. Why subtract in the first case but add in the other? Because of the way years are counted before the Christian era (B.C.E.) and after the birth of Christ (A.D.), the B.C.E. dates must be considered as *negative* numbers.

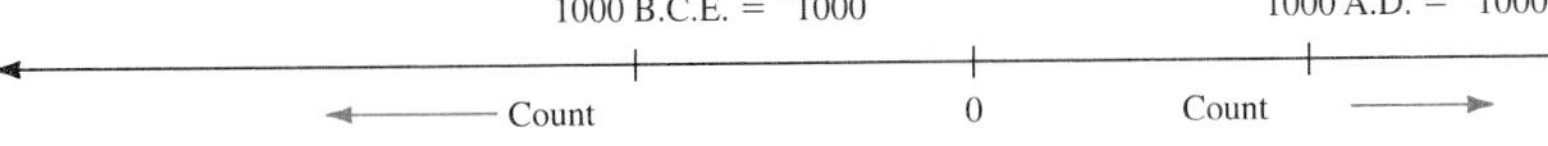

Very early on, the Chinese accepted the idea that a number could be negative; they used red calculating rods for positive numbers and black for negative numbers. Hindu mathematicians in India worked out the arithmetic of negative numbers as long ago as 400 A.D., but western mathematicians did not recognize this idea until the sixteenth century. It would be difficult today to think of measuring things such as temperature, altitude, and money without using negative numbers.

*A.D. stands for the Latin *Anno Domini,* which means "in the year of the Lord."

Name ______________________

Section __________ Date __________

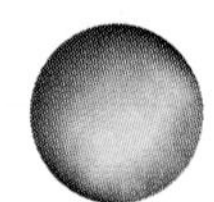

Pre-Test Chapter 1

ANSWERS

1. $x - 8$
2. $\frac{w}{17x}$
3. No
4. Yes
5. Commutative Prop. of Mult.
6. Distributive Property
7. Associative Property of Addition
8. -10
9. -1
10. -5
11. -1
12. -3
13. -19
14. 12
15. 0
16. 21
17. 14
18. $\frac{1}{2}$
19. 7
20. -3
21. 55
22. -7
23. $8w^2t$
24. $-a^2 + 4a + 3$
25. $12xy^2$

Write each of the following using symbols.

1. 8 less than x

2. the quotient when w is divided by the product of x and 17

Identify which are expressions and which are not.

3. $7x - 5 = 11$

4. $3x - 2(x + 1)$

Identify the property that is illustrated by the following statements.

5. $8 \cdot 9 = 9 \cdot 8$

6. $3(4 + 2) = 3 \cdot 4 + 3 \cdot 2$

7. $9 + (1 + 7) = (9 + 1) + 7$

Perform the indicated operations.

8. $-7 + (-3)$ **9.** $8 + (-9)$ **10.** $(-3) + (-2)$

11. $-\frac{7}{4} + \frac{3}{4}$ **12.** $8 - 11$ **13.** $-8 - 11$

14. $9 - (-3)$ **15.** $6 + (-6)$ **16.** $(-7)(-3)$

17. $(3.5)(4)$ **18.** $(3)\left(\frac{1}{6}\right)$ **19.** $\frac{-27 + 6}{-3}$

Evaluate the following expressions,

20. $5 - 4^2 \cdot 3 \div 6$ **21.** $(45 - 3 \cdot 5) + 5^2$

22. If $x = -2$, $y = 7$, and $w = -4$, evaluate the expression $\frac{x^2y}{w}$.

Combine like terms.

23. $5w^2t + 3w^2t$ **24.** $4a^2 - 3a + 5 + 7a - 2 - 5a^2$

Divide.

25. $\frac{96x^3y^5}{8x^2y^3}$

1.1 From Arithmetic to Algebra

1.1 OBJECTIVES

1. Represent addition, subtraction, multiplication, and division by using the symbols of algebra
2. Identify algebraic expressions

Overcoming Math Anxiety

Throughout this text, we will present you with a series of class-tested techniques that are designed to improve your performance in this math class.

Hint #2 Become familiar with your syllabus.

In the first class meeting, your instructor probably handed out a class syllabus. If you haven't done so already, you need to incorporate important information into your calendar and address book.

1. Write all important dates in your calendar. This includes homework due dates, quiz dates, test dates, and the date and time of the final exam. Never allow yourself to be surprised by any deadline!
2. Write your instructor's name, contact number, and office number in your address book. Also include the office hours. Make it a point to see your instructor early in the term. Although this is not the only person who can help clear up your confusion, it is the most important person.
3. Make note of other resources that are made available to you. This includes CDs, video tapes, web pages, and tutoring.

Given all of these resources, it is important that you never let confusion or frustration mount. If you can't "get it" from the text, try another resource. All of the resources are there specifically for you, so take advantage of them!

In arithmetic, you learned how to do calculations with numbers by using the basic operations of addition, subtraction, multiplication, and division.

In algebra, you will still use numbers and the same four operations. However, you will also use letters to represent numbers. Letters such as x, y, L, or W are called **variables** when they represent numerical values.

Here we see two rectangles whose lengths and widths are labeled with numbers.

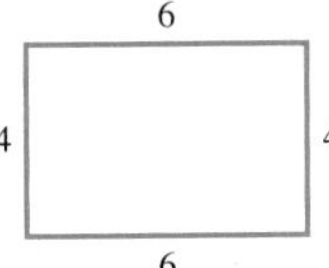

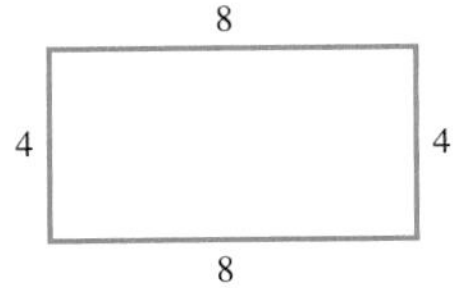

If we need to represent the length and width of *any* rectangle, we can use the variables L and W.

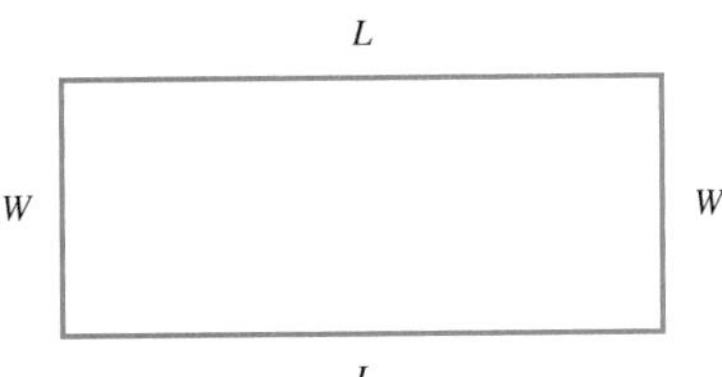

NOTE In arithmetic:
+ denotes addition
− denotes subtraction
× denotes multiplication
÷ denotes division.

You are familiar with the four symbols (+, −, ×, ÷) used to indicate the fundamental operations of arithmetic.

Let's look at how these operations are indicated in algebra. We begin by looking at addition.

Definitions: Addition

$x + y$ means the *sum* of x and y or x *plus* y.

Example 1

Writing Expressions That Indicate Addition

(a) *The sum of a and* 3 is written as $a + 3$.
(b) *L plus W* is written as $L + W$.
(c) 5 *more than m* is written as $m + 5$.
(d) *x increased by* 7 is written as $x + 7$.

CHECK YOURSELF 1

Write, using symbols.

(a) The sum of y and 4
(b) a plus b
(c) 3 more than x
(d) n increased by 6

Let's look at how subtraction is indicated in algebra.

Definitions: Subtraction

$x - y$ means the *difference* of x and y or x *minus* y.

Example 2

Writing Expressions That Indicate Subtraction

(a) *r minus s* is written as $r - s$.
(b) *The difference of m and* 5 is written as $m - 5$.
(c) *x decreased by* 8 is written as $x - 8$.
(d) 4 *less than a* is written as $a - 4$.

CHECK YOURSELF 2

Write, using symbols.

(a) w minus z
(b) The difference of a and 7
(c) y decreased by 3
(d) 5 less than b

You have seen that the operations of addition and subtraction are written exactly the same way in algebra as in arithmetic. This is not true in multiplication because the sign × looks like the letter x. So in algebra we use other symbols to show multiplication to avoid any confusion. Here are some ways to write multiplication.

NOTE x and y are called the **factors** of the product xy.

Definitions: Multiplication

A centered dot	$x \cdot y$	These all indicate the *product* of x and y or x times y.
Parentheses	$(x)(y)$	
Writing the letters next to each other	xy	

NOTE You can place letters next to each other or numbers and letters next to each other to show multiplication. But you *cannot* place numbers side by side to show multiplication: 37 means the number "thirty-seven," not 3 times 7.

Example 3

Writing Expressions That Indicate Multiplication

(a) The product of 5 and a is written as $5 \cdot a$, $(5)(a)$, or $5a$. The last expression, $5a$, is the shortest and the most common way of writing the product.

(b) 3 times 7 can be written as $3 \cdot 7$ or $(3)(7)$.

(c) Twice z is written as $2z$.

(d) The product of 2, s, and t is written as $2st$.

(e) 4 more than the product of 6 and x is written as $6x + 4$.

CHECK YOURSELF 3

Write, using symbols.

(a) m times n

(b) The product of h and b

(c) The product of 8 and 9

(d) The product of 5, w, and y

(e) 3 more than the product of 8 and a

Before we move on to division, let's look at how we can combine the symbols we have learned so far.

NOTE Not every collection of symbols is an expression.

Definitions: Expression

An **expression** is a meaningful collection of numbers, variables, and signs of operation.

Example 4

Identifying Expressions

(a) $2m + 3$ is an expression. It means that we multiply 2 and m, then add 3.

(b) $x + \cdot + 3$ is not an expression. The three operations in a row have no meaning.

(c) $y = 2x - 1$ is not an expression. The equals sign is not an operation sign.

(d) $3a + 5b - 4c$ is an expression. Its meaning is clear.

CHECK YOURSELF 4

Identify which are expressions and which are not.

(a) $7 - \cdot x$ **(b)** $6 + y = 9$

(c) $a + b - c$ **(d)** $3x - 5yz$

To write more complicated products in algebra, we need some "punctuation marks." Parentheses ( ) mean that an expression is to be thought of as a single quantity. Brackets [ ] are used in exactly the same way as parentheses in algebra. Look at the following example showing the use of these signs of grouping.

Example 5

Expressions with More Than One Operation

(a) 3 times the sum of a and b is written as

NOTE This can be read as "3 times the quantity a plus b."

$$3(a + b)$$

The sum of a and b is a single quantity, so it is enclosed in parentheses.

NOTE No parentheses are needed here because the 3 multiplies *only* the a.

(b) The sum of 3 times a and b is written as $3a + b$.

(c) 2 times the difference of m and n is written as $2(m - n)$.

(d) The product of s plus t and s minus t is written as $(s + t)(s - t)$.

(e) The product of b and 3 less than b is written as $b(b - 3)$.

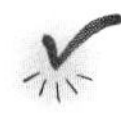

CHECK YOURSELF 5

Write, using symbols.

(a) Twice the sum of p and q

(b) The sum of twice p and q

(c) The product of a and the quantity $b - c$

(d) The product of x plus 2 and x minus 2

(e) The product of x and 4 more than x

NOTE In algebra the fraction form is usually used.

Now let's look at the operation of division. In arithmetic, you use the division sign ÷, the long division symbol $\overline{)\quad}$, and the fraction notation. For example, to indicate the quotient when 9 is divided by 3, you could write

$$9 \div 3 \quad \text{or} \quad 3\overline{)9} \quad \text{or} \quad \frac{9}{3}$$

Definitions: Division

$\frac{x}{y}$ means *x divided by y* or *the quotient of x and y.*

Example 6

Writing Expressions That Indicate Division

(a) m divided by 3 is written as $\frac{m}{3}$.

(b) The quotient of a plus b divided by 5 is written as $\frac{a + b}{5}$.

(c) The sum p plus q divided by the difference p minus q is written as $\frac{p + q}{p - q}$.

CHECK YOURSELF 6

Write, using symbols.

(a) r divided by s
(b) The quotient when x minus y is divided by 7
(c) The difference a minus 2 divided by the sum a plus 2

Notice that we can use many different letters to represent variables. In Example 6 the letters m, a, b, p, and q represented different variables. We often choose a letter that reminds us of what it represents, for example, L for *length* or W for *width*.

Example 7

Writing Geometric Expressions

(a) *Length* times *width* is written $L \cdot W$.

(b) One-half of *altitude* times *base* is written $\frac{1}{2} a \cdot b$.

(c) *Length* times *width* times *height* is written $L \cdot W \cdot H$.

(d) Pi (π) times *diameter* is written πd.

CHECK YOURSELF 7

Write each geometric expression, using symbols.

(a) Two times *length* plus two times *width* **(b)** Two times pi (π) times *radius*

CHECK YOURSELF ANSWERS

1. **(a)** $y + 4$; **(b)** $a + b$; **(c)** $x + 3$; **(d)** $n + 6$ **2.** **(a)** $w - z$; **(b)** $a - 7$; **(c)** $y - 3$; **(d)** $b - 5$ **3.** **(a)** mn; **(b)** hb; **(c)** $8 \cdot 9$ or $(8)(9)$; **(d)** $5wy$; **(e)** $8a + 3$
4. **(a)** Not an expression; **(b)** not an expression; **(c)** an expression; **(d)** an expression
5. **(a)** $2(p + q)$; **(b)** $2p + q$; **(c)** $a(b - c)$; **(d)** $(x + 2)(x - 2)$; **(e)** $x(x + 4)$
6. **(a)** $\frac{r}{s}$; **(b)** $\frac{x - y}{7}$; **(c)** $\frac{a - 2}{a + 2}$ **7.** **(a)** $2L + 2W$; **(b)** $2\pi r$

Name ______________

Section ________ Date ________

1.1 Exercises

Write each of the following phrases, using symbols.

1. The sum of c and d
2. a plus 7
3. w plus z
4. The sum of m and n
5. x increased by 2
6. 3 more than b
7. 10 more than y
8. m increased by 4
9. a minus b
10. 5 less than s
11. b decreased by 7
12. r minus 3
13. 6 less than r
14. x decreased by 3
15. w times z
16. The product of 3 and c
17. The product of 5 and t
18. 8 times a
19. The product of 8, m, and n
20. The product of 7, r, and s
21. The product of 3 and the quantity p plus q
22. The product of 5 and the sum of a and b
23. Twice the sum of x and y
24. 3 times the sum of m and n
25. The sum of twice x and y

ANSWERS

1. $c + d$
2. $a + 7$
3. $w + z$
4. $m + n$
5. $x + 2$
6. $b + 3$
7. $y + 10$
8. $m + 4$
9. $a - b$
10. $s - 5$
11. $b - 7$
12. $r - 3$
13. $r - 6$
14. $x - 3$
15. wz
16. $3c$
17. $5t$
18. $8a$
19. $8mn$
20. $7rs$
21. $3(p + q)$
22. $5(a + b)$
23. $2(x + y)$
24. $3(m + n)$
25. $2x + y$

ANSWERS

26. $3m + n$
27. $2(x - y)$
28. $3(c - d)$
29. $(a + b)(a - b)$
30. $(x + y)(x - y)$
31. $m(m - 3)$
32. $a(a + 7)$
33. $\frac{x}{5}$
34. $\frac{b}{8}$
35. $\frac{a + b}{7}$
36. $\frac{x - y}{9}$
37. $\frac{p - q}{4}$
38. $\frac{a + 5}{9}$
39. $\frac{a + 3}{a - 3}$
40. $\frac{m - n}{m + n}$
41. $x + 5$
42. $x + 8$
43. $x - 7$
44. $x - 10$
45. $9x$
46. $2x$
47. $3x + 6$
48. $5x - 10$
49. $2(x + 5)$
50. $3(x - 4)$
51. $(x + 2)(x - 2)$

26. The sum of 3 times m and n

27. Twice the difference of x and y

28. 3 times the difference of c and d

29. The quantity a plus b times the quantity a minus b

30. The product of x plus y and x minus y

31. The product of m and 3 less than m

32. The product of a and 7 more than a

33. x divided by 5

34. The quotient when b is divided by 8

35. The quotient of a plus b, divided by 7

36. The difference x minus y, divided by 9

37. The difference of p and q, divided by 4

38. The sum of a and 5, divided by 9

39. The sum of a and 3, divided by the difference of a and 3

40. The difference of m and n, divided by the sum of m and n

Write each of the following phrases, using symbols. Use the variable x to represent the number in each case.

41. 5 more than a number

42. A number increased by 8

43. 7 less than a number

44. A number decreased by 10

45. 9 times a number

46. Twice a number

47. 6 more than 3 times a number

48. 5 times a number, decreased by 10

49. Twice the sum of a number and 5

50. 3 times the difference of a number and 4

51. The product of 2 more than a number and 2 less than that same number

52. The product of 5 less than a number and 5 more than that same number

53. The quotient of a number and 7

54. A number divided by 3

55. The sum of a number and 5, divided by 8

56. The quotient when 7 less than a number is divided by 3

57. 6 more than a number divided by 6 less than that same number

58. The quotient when 3 less than a number is divided by 3 more than that same number

Write each of the following geometric expressions using symbols.

59. Four times the length of a side (s)

60. $\frac{4}{3}$ times π times the cube of the radius (r)

61. The radius (r) squared times the height (h) times π

62. Twice the length (L) plus twice the width (W)

63. One-half the product of the height (h) and the sum of two unequal sides (b_1 and b_2)

64. Six times the length of a side (s) squared

Identify which are expressions and which are not.

65. $2(x + 5)$

66. $4 + (x - 3)$

67. $4 + \div m$

68. $6 + a = 7$

69. $2b = 6$

70. $x(y + 3)$

71. $2a + 5b$

72. $4x + \cdot 7$

73. **Population growth.** The Earth's population has doubled in the last 40 years. If we let x represent the Earth's population 40 years ago, what is the population today?

74. **Species extinction.** It is estimated that the Earth is losing 4000 species of plants and animals every year. If S represents the number of species living last year, how many species are on Earth this year?

75. **Interest.** The simple interest (I) earned when a principal (P) is invested at a rate (r) for a time (t) is calculated by multiplying the principal times the rate times the time. Write a formula for the interest earned.

ANSWERS

52. $(x - 5)(x + 5)$

53. $\frac{x}{7}$

54. $\frac{x}{3}$

55. $\frac{x + 5}{8}$

56. $\frac{x - 7}{3}$

57. $\frac{x + 6}{x - 6}$

58. $\frac{x - 3}{x + 3}$

59. $4s$

60. $\frac{4}{3}\pi r^3$

61. $\pi r^2 h$

62. $2L + 2W$

63. $\frac{1}{2}h(b_1 + b_2)$

64. $6s^2$

65. Expression

66. Expression

67. Not an expression

68. Not an expression

69. Not an expression

70. Expression

71. Expression

72. Not an expression

73. $2x$

74. $S - 4000$

75. $I = Prt$

76. $KE = \frac{1}{2}mv^2$

77.

a. 1

b. 5

c. 8

d. 2

e. 12

f. 12

76. Kinetic energy. The kinetic energy of a particle of mass m is found by taking one-half of the product of the mass and the square of the velocity (v). Write a formula for the kinetic energy of a particle.

77. Rewrite the following algebraic expressions in English phrases. Exchange papers with another student to edit your writing. Be sure the meaning in English is the same as in algebra. These expressions are not complete sentences, so your English does not have to be in complete sentences. Here is an example.

Algebra: $2(x - 1)$

English: We could write "One less than a number is doubled." Or we might write "A number is diminished by one and then multiplied by two."

(a) $n + 3$ **(b)** $\frac{x + 2}{5}$ **(c)** $3(5 + a)$ **(d)** $3 - 4n$ **(e)** $\frac{x + 6}{x - 1}$

*Getting Ready for Section 1.2 [Section 0.3]

Evaluate the following:

(a) $8 - (5 + 2)$ (b) $(8 - 5) + 2$ (c) $16 \div 4 \cdot 2$

(d) $16 \div (4 \cdot 2)$ (e) $6 \cdot 2$ (f) $2 \cdot 6$

Answers

1. $c + d$ **3.** $w + z$ **5.** $x + 2$ **7.** $y + 10$ **9.** $a - b$ **11.** $b - 7$ **13.** $r - 6$ **15.** wz **17.** $5t$ **19.** $8mn$ **21.** $3(p + q)$ **23.** $2(x + y)$ **25.** $2x + y$ **27.** $2(x - y)$ **29.** $(a + b)(a - b)$ **31.** $m(m - 3)$ **33.** $\frac{x}{5}$ **35.** $\frac{a + b}{7}$ **37.** $\frac{p - q}{4}$ **39.** $\frac{a + 3}{a - 3}$ **41.** $x + 5$ **43.** $x - 7$ **45.** $9x$ **47.** $3x + 6$ **49.** $2(x + 5)$ **51.** $(x + 2)(x - 2)$ **53.** $\frac{x}{7}$ **55.** $\frac{x + 5}{8}$ **57.** $\frac{x + 6}{x - 6}$ **59.** $4s$ **61.** $\pi r^2 h$ **63.** $\frac{1}{2}h(b_1 + b_2)$ **65.** Expression **67.** Not an expression **69.** Not an expression **71.** Expression **73.** $2x$ **75.** $I = Prt$ **77.** **a.** 1 **b.** 5 **c.** 8 **d.** 2 **e.** 12 **f.** 12

*Exercises headed "Getting Ready for . . ." are designed to help you prepare for material in the next section of the text. If you have any difficulty with these exercises, please review the section referred to in brackets.

Properties of Signed Numbers

OBJECTIVES

1. Recognize applications of the commutative property
2. Recognize applications of the associative property
3. Recognize applications of the distributive property

All that we do in algebra is based on the rules for the operations introduced in Section 1.1. We call these rules **properties of the real numbers.** In this section we consider those properties that we will use in the remainder of this chapter.

The **commutative properties** tell us that we can add or multiply in any order.

Rules and Properties: The Commutative Properties

If a and b are any numbers,

1. $a + b = b + a$ Commutative property of addition
2. $a \cdot b = b \cdot a$ Commutative property of multiplication

NOTE All integers, decimals, and fractions that we see in this course are real numbers.

Example 1

Identifying the Commutative Properties

(a) $5 + 9 = 9 + 5$ and $x + 7 = 7 + x$

These are applications of the commutative property of addition.

(b) $5 \cdot 9 = 9 \cdot 5$

This is an application of the commutative property of multiplication.

CHECK YOURSELF 1

Identify the property being applied.

(a) $7 + 3 = 3 + 7$ **(b)** $7 \cdot 3 = 3 \cdot 7$

(c) $a + 4 = 4 + a$ **(d)** $x \cdot 2 = 2 \cdot x$

We also want to be able to change the grouping in simplifying expressions. This is possible because of the **associative properties.** Numbers or variables can be grouped in any manner to find a sum or a product.

Rules and Properties: The Associative Properties

If a, b, and c are any numbers,

1. $a + (b + c) = (a + b) + c$ Associative property of addition
2. $a \cdot (b \cdot c) = (a \cdot b) \cdot c$ Associative property of multiplication

Example 2

Demonstrating the Associative Properties

(a) Show that $2 + (3 + 8) = (2 + 3) + 8$.

NOTE Remember, as we saw in Section 0.3, we always do the operation in the parentheses first.

$2 + \underbrace{(3 + 8)}_{\text{Add first.}}$ $\qquad$ $\underbrace{(2 + 3)}_{\text{Add first.}} + 8$

$= 2 + 11$ $\qquad$ $= 5 + 8$

$= 13$ $\qquad$ $= 13$

So

$$2 + (3 + 8) = (2 + 3) + 8$$

(b) Show that $\frac{1}{3} \cdot (6 \cdot 5) = \left(\frac{1}{3} \cdot 6\right) \cdot 5$.

$\frac{1}{3} \cdot \underbrace{(6 \cdot 5)}_{\text{Multiply first.}}$ $\qquad$ $\underbrace{\left(\frac{1}{3} \cdot 6\right)}_{\text{Multiply first.}} \cdot 5$

$= \frac{1}{3} \cdot (30)$ $\qquad$ $= (2) \cdot 5$

$= 10$ $\qquad$ $= 10$

So

$$\frac{1}{3} \cdot (6 \cdot 5) = \left(\frac{1}{3} \cdot 6\right) \cdot 5$$

CHECK YOURSELF 2

Show that the following statements are true.

(a) $3 + (4 + 7) = (3 + 4) + 7$

(b) $3 \cdot (4 \cdot 7) = (3 \cdot 4) \cdot 7$

(c) $\left(\frac{1}{5} \cdot 10\right) \cdot 4 = \frac{1}{5} \cdot (10 \cdot 4)$

The **distributive property** involves addition and multiplication together. We can illustrate this property with an application.

REMEMBER: The area of a rectangle is the product of its length and width:

$A = L \cdot W$

Suppose that we want to find the total of the two areas shown in the following figure.

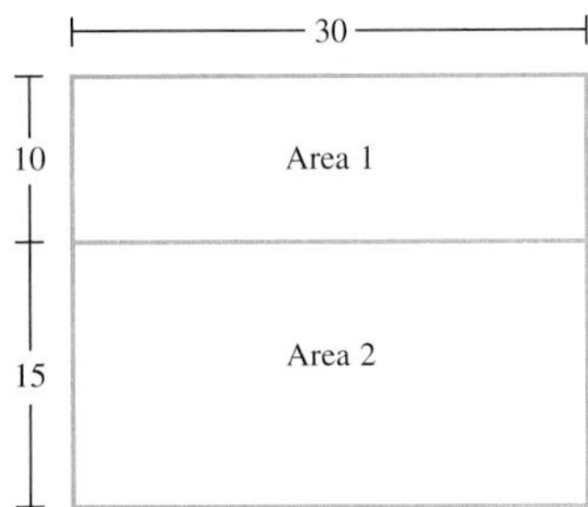

We can find the total area by multiplying the length by the overall width, which is found by adding the two widths.	[or]	We can find the total area as a sum of the two areas.

Length · Overall Width: $30 \cdot (10 + 15)$

$= 30 \cdot 25$

$= 750$

(Area 1) Length · Width + (Area 2) Length · Width: $30 \cdot 10 + 30 \cdot 15$

$= 300 + 450$

$= 750$

So

$30 \cdot (10 + 15) = 30 \cdot 10 + 30 \cdot 15$

This leads us to the following property.

NOTE Notice the pattern.

$a(b + c) = a \cdot b + a \cdot c$

We "distributed" the multiplication "over" the addition.

Rules and Properties: The Distributive Property

If a, b, and c are any numbers,

$a(b + c) = a \cdot b + a \cdot c$ and $(b + c)a = b \cdot a + c \cdot a$

Example 3

Using the Distributive Property

Use the distributive property to simplify (remove the parentheses in) the following.

NOTE $5(3 + 4) = 5 \cdot 7 = 35$

or

$5 \cdot 3 + 5 \cdot 4 = 15 + 20 = 35$

(a) $5(3 + 4)$

$5(3 + 4) = 5 \cdot 3 + 5 \cdot 4$

$= 15 + 20 = 35$

NOTE Because the variables are different, $8x + 8y$ cannot be simplified further.

(b) $8(x + y)$

$8(x + y) = 8x + 8y$

(c) $2(3x + 5)$

$2(3x + 5) = 2 \cdot 3x + 2 \cdot 5$

$= 6x + 10$

NOTE It is also true that

$\frac{1}{3}(9 + 12) = \frac{1}{3}(21) = 7$

(d) $\frac{1}{3}(9 + 12) = \frac{1}{3} \cdot 9 + \frac{1}{3} \cdot 12$

$= 3 + 4 = 7$

CHECK YOURSELF 3

Use the distributive property to simplify (remove the parentheses).

(a) $4(6 + 7)$ **(b)** $9(m + n)$

(c) $3(5a + 7)$ **(d)** $\frac{1}{5}(10 + 15)$

Example 4 requires that you identify which property is being demonstrated. Look for patterns that will help you remember each of the properties.

Example 4

Identifying Properties

Name the property demonstrated.

(a) $3(x + 2) = 3x + 3 \cdot 2$ demonstrates the distributive property.
(b) $2 + (3 + 5) = (2 + 3) + 5$ demonstrates the associative property of addition.
(c) $3 \cdot 5 = 5 \cdot 3$ demonstrates the commutative property of multiplication.

CHECK YOURSELF 4

Name the property demonstrated.

(a) $2 \cdot (3 \cdot 5) = (2 \cdot 3) \cdot 5$

(b) $4(a + b) = 4a + 4b$

(c) $x + 8 = 8 + x$

CHECK YOURSELF ANSWERS

1. **(a)** Commutative property of addition; **(b)** commutative property of multiplication; **(c)** commutative property of addition; **(d)** commutative property of multiplication
2. **(a)** $3 + (4 + 7) = 3 + 11 = 14$ **(b)** $3 \cdot (4 \cdot 7) = 3 \cdot 28 = 84$
   $(3 + 4) + 7 = 7 + 7 = 14$ $(3 \cdot 4) \cdot 7 = 12 \cdot 7 = 84$

   (c) $\left(\frac{1}{5} \cdot 10\right) \cdot 4 = 2 \cdot 4 = 8$

   $\frac{1}{5} \cdot (10 \cdot 4) = \frac{1}{5} \cdot 40 = 8$
3. **(a)** $4 \cdot 6 + 4 \cdot 7 = 24 + 28 = 52$; **(b)** $9m + 9n$; **(c)** $15a + 21$;
   (d) $\frac{1}{5} \cdot 10 + \frac{1}{5} \cdot 15 = 2 + 3 = 5$
4. **(a)** Associative property of multiplication; **(b)** distributive property; **(c)** commutative property of addition

Name ______________________

Section ________ Date ________

1.2 Exercises

Identify the property that is illustrated by each of the following statements.

1. $5 + 9 = 9 + 5$

2. $6 + 3 = 3 + 6$

3. $2 \cdot (3 \cdot 5) = (2 \cdot 3) \cdot 5$

4. $3 \cdot (5 \cdot 6) = (3 \cdot 5) \cdot 6$

5. $10 \cdot 5 = 5 \cdot 10$

6. $8 \cdot 4 = 4 \cdot 8$

7. $8 + 12 = 12 + 8$

8. $6 + 2 = 2 + 6$

9. $(5 \cdot 7) \cdot 2 = 5 \cdot (7 \cdot 2)$

10. $(8 \cdot 9) \cdot 2 = 8 \cdot (9 \cdot 2)$

11. $9 \cdot 8 = 8 \cdot 9$

12. $6 \cdot 4 = 4 \cdot 6$

13. $2(3 + 5) = 2 \cdot 3 + 2 \cdot 5$

14. $5 \cdot (4 + 6) = 5 \cdot 4 + 5 \cdot 6$

15. $5 + (7 + 8) = (5 + 7) + 8$

16. $8 + (2 + 9) = (8 + 2) + 9$

17. $(10 + 5) + 9 = 10 + (5 + 9)$

18. $(5 + 5) + 3 = 5 + (5 + 3)$

19. $7 \cdot (3 + 8) = 7 \cdot 3 + 7 \cdot 8$

20. $5 \cdot (6 + 8) = 5 \cdot 6 + 5 \cdot 8$

Verify that each of the following statements is true by evaluating each side of the equation separately and comparing the results.

21. $7 \cdot (3 + 4) = 7 \cdot 3 + 7 \cdot 4$

22. $4 \cdot (5 + 1) = 4 \cdot 5 + 4 \cdot 1$

23. $2 + (9 + 8) = (2 + 9) + 8$

24. $6 + (15 + 3) = (6 + 15) + 3$

25. $5 \cdot (6 \cdot 3) = (5 \cdot 6) \cdot 3$

26. $2 \cdot (9 \cdot 10) = (2 \cdot 9) \cdot 10$

ANSWERS

1. Comm. prop. of add.
2. Comm. prop. of add.
3. Assoc. prop. of mult.
4. Assoc. prop. of mult.
5. Comm. prop. of mult.
6. Comm. prop. of mult.
7. Comm. prop. of add.
8. Comm. prop. of add.
9. Assoc. prop. of mult.
10. Assoc. prop. of mult.
11. Comm. prop. of mult.
12. Comm. prop. of mult.
13. Distributive prop.
14. Distributive prop.
15. Assoc. prop. of add.
16. Assoc. prop. of add.
17. Assoc. prop. of add.
18. Assoc. prop. of add.
19. Distributive prop.
20. Distributive prop.
21. $49 = 49$
22. $24 = 24$
23. $19 = 19$
24. $24 = 24$
25. $90 = 90$
26. $180 = 180$

ANSWERS

27. 50 = 50
28. 36 = 36
29. 23 = 23
30. 27 = 27
31. 56 = 56
32. 90 = 90
33. 4 = 4
34. 5 = 5
35. $\frac{7}{6} = \frac{7}{6}$
36. $\frac{15}{8} = \frac{15}{8}$
37. 10.3 = 10.3
38. 13.4 = 13.4
39. 8 = 8
40. 3 = 3
41. $\frac{2}{3} = \frac{2}{3}$
42. 2 = 2
43. 50 = 50
44. 42 = 42
45. 16
46. 50
47. $3x + 15$
48. $5y + 40$
49. $4w + 4v$
50. $7c + 7d$
51. $6x + 10$
52. $21a + 12$
53. 8
54. 10

27. $5 \cdot (2 + 8) = 5 \cdot 2 + 5 \cdot 8$

28. $3 \cdot (10 + 2) = 3 \cdot 10 + 3 \cdot 2$

29. $(3 + 12) + 8 = 3 + (12 + 8)$

30. $(8 + 12) + 7 = 8 + (12 + 7)$

31. $(4 \cdot 7) \cdot 2 = 4 \cdot (7 \cdot 2)$

32. $(6 \cdot 5) \cdot 3 = 6 \cdot (5 \cdot 3)$

33. $\frac{1}{2} \cdot (2 + 6) = \frac{1}{2} \cdot 2 + \frac{1}{2} \cdot 6$

34. $\frac{1}{3} \cdot (6 + 9) = \frac{1}{3} \cdot 6 + \frac{1}{3} \cdot 9$

35. $\left(\frac{2}{3} + \frac{1}{6}\right) + \frac{1}{3} = \frac{2}{3} + \left(\frac{1}{6} + \frac{1}{3}\right)$

36. $\frac{3}{4} + \left(\frac{5}{8} + \frac{1}{2}\right) = \left(\frac{3}{4} + \frac{5}{8}\right) + \frac{1}{2}$

37. $(2.3 + 3.9) + 4.1 = 2.3 + (3.9 + 4.1)$

38. $(1.7 + 4.1) + 7.6 = 1.7 + (4.1 + 7.6)$

39. $\frac{1}{2} \cdot (2 \cdot 8) = \left(\frac{1}{2} \cdot 2\right) \cdot 8$

40. $\frac{1}{5} \cdot (5 \cdot 3) = \left(\frac{1}{5} \cdot 5\right) \cdot 3$

41. $\left(\frac{3}{5} \cdot \frac{5}{6}\right) \cdot \frac{4}{3} = \frac{3}{5} \cdot \left(\frac{5}{6} \cdot \frac{4}{3}\right)$

42. $\frac{4}{7} \cdot \left(\frac{21}{16} \cdot \frac{8}{3}\right) = \left(\frac{4}{7} \cdot \frac{21}{16}\right) \cdot \frac{8}{3}$

43. $2.5 \cdot (4 \cdot 5) = (2.5 \cdot 4) \cdot 5$

44. $4.2 \cdot (5 \cdot 2) = (4.2 \cdot 5) \cdot 2$

Use the distributive property to remove the parentheses in each of the following expressions. Then simplify your result where possible.

45. $2(3 + 5)$

46. $5(4 + 6)$

47. $3(x + 5)$

48. $5(y + 8)$

49. $4(w + v)$

50. $7(c + d)$

51. $2(3x + 5)$

52. $3(7a + 4)$

53. $\frac{1}{3} \cdot (15 + 9)$

54. $\frac{1}{6} \cdot (36 + 24)$

Use the properties of addition and multiplication to complete each of the following statements.

55. $5 + 7 = \quad + 5$

56. $(5 + 3) + 4 = 5 + (\quad + 4)$

57. $(8)(3) = (3)(\quad)$

58. $8(3 + 4) = 8 \cdot 3 + \quad \cdot 4$

59. $7(2 + 5) = 7 \cdot \quad + 7 \cdot 5$

60. $4 \cdot (2 \cdot 4) = (\quad \cdot 2) \cdot 4$

Use the indicated property to write an expression that is equivalent to each of the following expressions.

61. $3 + 7$ (commutative property of addition)

62. $2(3 + 4)$ (distributive property)

63. $5 \cdot (3 \cdot 2)$ (associative property of multiplication)

64. $(3 + 5) + 2$ (associative property of addition)

65. $2 \cdot 4 + 2 \cdot 5$ (distributive property)

66. $7 \cdot 9$ (commutative property of multiplication)

Evaluate each of the following pairs of expressions. Then answer the given question.

67. $8 - 5$ and $5 - 8$
Do you think subtraction is commutative?

68. $12 \div 3$ and $3 \div 12$
Do you think division is commutative?

69. $(12 - 8) - 4$ and $12 - (8 - 4)$
Do you think subtraction is associative?

70. $(48 \div 16) \div 4$ and $48 \div (16 \div 4)$
Do you think division is associative?

71. $3(6 - 2)$ and $3 \cdot 6 - 3 \cdot 2$
Do you think multiplication is distributive over subtraction?

72. $\frac{1}{2}(16 - 10)$ and $\frac{1}{2} \cdot 16 - \frac{1}{2} \cdot 10$
Do you think multiplication is distributive over subtraction?

In Exercises 73 and 74, complete the statement using

(a) the distributive property,
(b) the commutative property of addition,
(c) the commutative property of multiplication.

ANSWERS

55. 7
56. 3
57. 8
58. 8
59. 2
60. 4
61. $7 + 3$
62. $2 \cdot 3 + 2 \cdot 4$
63. $(5 \cdot 3) \cdot 2$
64. $3 + (5 + 2)$
65. $2 \cdot (4 + 5)$
66. $9 \cdot 7$
67. No
68. No
69. No
70. No
71. Yes
72. Yes

(a) $5 \cdot 3 + 5 \cdot 4$
(b) $5 \cdot (4 + 3)$
73. (c) $(3 + 4) \cdot 5$

(a) $6 \cdot 5 + 6 \cdot 4$
(b) $6 \cdot (4 + 5)$
74. (c) $(5 + 4) \cdot 6$

75. Assoc. prop. of add.

76. Comm. prop. of add.

77. Comm. prop. of add.

78. Comm. prop. of mult.

a. 20

b. 21

c. 24

d. 1

e. $\frac{13}{15}$

f. $\frac{2}{9}$

73. $5 \cdot (3 + 4) =$ **74.** $6 \cdot (5 + 4) =$

In Exercises 75 to 78, identify the property that is used.

75. $5 + (6 + 7) = (5 + 6) + 7$ **76.** $5 + (6 + 7) = 5 + (7 + 6)$

77. $4 \cdot (3 + 2) = 4 \cdot (2 + 3)$ **78.** $4 \cdot (3 + 2) = (3 + 2) \cdot 4$

Getting Ready for Section 1.3 [Section 1.2]

Find each sum.

(a) $3 + (8 + 9)$ (b) $6 + (12 + 3)$
(c) $(3 + 8) + (9 + 4)$ (d) $15 - 11 - (2 + 1)$
(e) $\frac{3}{5} + \frac{4}{15}$ (f) $\frac{12}{27} - \frac{2}{9}$

Answers

1. Commutative property of addition **3.** Associative property of multiplication **5.** Commutative property of multiplication **7.** Commutative property of addition **9.** Associative property of multiplication **11.** Commutative property of multiplication **13.** Distributive property **15.** Associative property of addition **17.** Associative property of addition **19.** Distributive property **21.** $49 = 49$ **23.** $19 = 19$ **25.** $90 = 90$ **27.** $50 = 50$ **29.** $23 = 23$ **31.** $56 = 56$ **33.** $4 = 4$ **35.** $\frac{7}{6} = \frac{7}{6}$ **37.** $10.3 = 10.3$ **39.** $8 = 8$ **41.** $\frac{2}{3} = \frac{2}{3}$ **43.** $50 = 50$ **45.** 16 **47.** $3x + 15$ **49.** $4w + 4v$ **51.** $6x + 10$ **53.** 8 **55.** 7 **57.** 8 **59.** 2 **61.** $7 + 3$ **63.** $(5 \cdot 3) \cdot 2$ **65.** $2 \cdot (4 + 5)$ **67.** No **69.** No **71.** Yes **73.** **(a)** $5 \cdot 3 + 5 \cdot 4$ **(b)** $5 \cdot (4 + 3)$ **(c)** $(3 + 4) \cdot 5$ **75.** Associative property of addition **77.** Commutative property of addition **a.** 20 **b.** 21 **c.** 24 **d.** 1 **e.** $\frac{13}{15}$ **f.** $\frac{2}{9}$

1.3 Adding and Subtracting Signed Numbers

1.3 OBJECTIVES

1. Find the median of a set of signed numbers
2. Find the difference of two signed numbers
3. Find the range of a set of signed numbers

In Section 0.4 we introduced the idea of signed numbers. Now we will examine the four arithmetic operations (addition, subtraction, multiplication, and division) and see how those operations are performed when signed numbers are involved. We start by considering addition.

An application may help. As before, let's represent a gain of money as a positive number and a loss as a negative number.

If you gain \$3 and then gain \$4, the result is a gain of \$7:

$$3 + 4 = 7$$

If you lose \$3 and then lose \$4, the result is a loss of \$7:

$$-3 + (-4) = -7$$

If you gain \$3 and then lose \$4, the result is a loss of \$1:

$$3 + (-4) = -1$$

If you lose \$3 and then gain \$4, the result is a gain of \$1:

$$-3 + 4 = 1$$

The number line can be used to illustrate the addition of signed numbers. Starting at the origin, we move to the *right* for positive numbers and to the *left* for negative numbers.

Example 1

Adding Signed Numbers

(a) Add $(-3) + (-4)$.

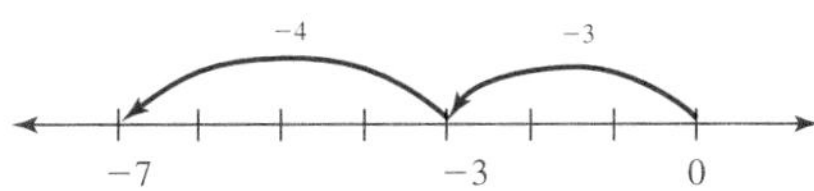

Start at the origin and move 3 units to the left. Then move 4 more units to the left to find the sum. From the number line we see that the sum is

$$(-3) + (-4) = -7$$

(b) Add $\left(-\frac{3}{2}\right) + \left(-\frac{1}{2}\right)$.

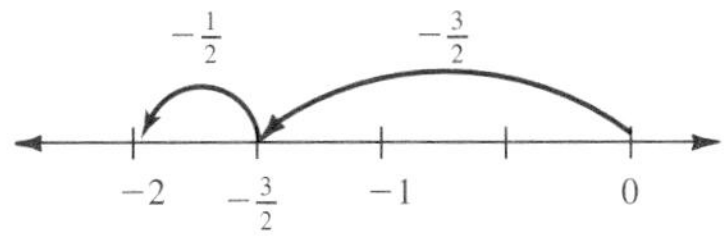

As before, we start at the origin. From that point move $\frac{3}{2}$ units left. Then move another $\frac{1}{2}$ unit left to find the sum. In this case

$$\left(-\frac{3}{2}\right) + \left(-\frac{1}{2}\right) = -2$$

CHECK YOURSELF 1

Add.

(a) $(-4) + (-5)$

(b) $(-3) + (-7)$

(c) $(-5) + (-15)$

(d) $\left(-\frac{5}{2}\right) + \left(-\frac{3}{2}\right)$

You have probably noticed some helpful patterns in the previous examples. These patterns will allow you to do the work mentally without having to use the number line. Look at the following rule.

Rules and Properties: Adding Signed Numbers Case 1: Same Sign

If two numbers have the same sign, add their absolute values. Give the sum the sign of the original numbers.

NOTE This means that the sum of two positive numbers is positive and the sum of two negative numbers is negative. We first encountered absolute values in Section 0.4.

Let's again use the number line to illustrate the addition of two numbers. This time the numbers will have *different* signs.

Example 2

Adding Signed Numbers

(a) Add $3 + (-6)$.

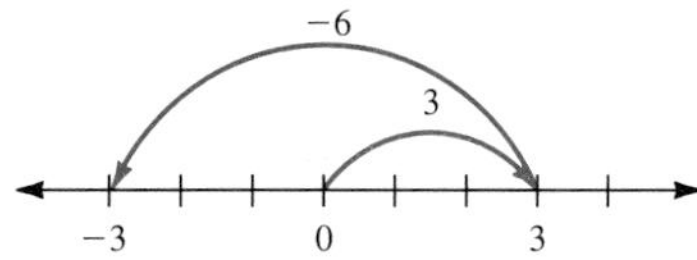

First move 3 units to the right of the origin. Then move 6 units to the left.

$$3 + (-6) = -3$$

(b) Add $-4 + 7$.

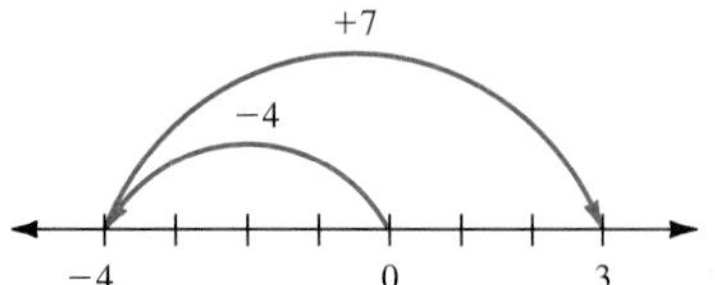

This time move 4 units to the left of the origin as the first step. Then move 7 units to the right.

$-4 + 7 = 3$

CHECK YOURSELF 2

Add.

(a) $7 + (-5)$ **(b)** $4 + (-8)$ **(c)** $-\frac{1}{3} + \frac{16}{3}$ **(d)** $-7 + 3$

You have no doubt noticed that, in adding a positive number and a negative number, sometimes the sum is positive and sometimes it is negative. This depends on which of the numbers has the larger absolute value. This leads us to the second part of our addition rule.

Rules and Properties: Adding Signed Numbers Case 2: Different Signs

If two numbers have different signs, subtract their absolute values, the smaller from the larger. Give the result the sign of the number with the larger absolute value.

NOTE Again, we first encountered absolute values in Section 0.4.

Example 3

Adding Signed Numbers

(a) $7 + (-19) = -12$

Because the two numbers have different signs, subtract the absolute values $(19 - 7 = 12)$. The sum has the sign $(-)$ of the number with the larger absolute value, -19.

(b) $-\frac{13}{2} + \frac{7}{2} = -3$

Subtract the absolute values $\left(\frac{13}{2} - \frac{7}{2} = \frac{6}{2} = 3\right)$. The sum has the sign $(-)$ of the number with the larger absolute value, $-\frac{13}{2}$.

NOTE Remember, **signed numbers** can be fractions and decimals as well as integers.

(c) $-8.2 + 4.5 = -3.7$

Subtract the absolute values $(8.2 - 4.5 = 3.7)$. The sum has the sign $(-)$ of the number with the larger absolute value, -8.2.

CHECK YOURSELF 3

Add mentally.

(a) $5 + (-14)$ **(b)** $-7 + (-8)$ **(c)** $-8 + 15$

(d) $7 + (-8)$ **(e)** $-\frac{2}{3} + \left(-\frac{7}{3}\right)$ **(f)** $5.3 + (-2.3)$

In Section 1.2 we discussed the commutative, associative, and distributive properties. There are two other properties of addition that we should mention. First, the sum of any number and 0 is always that number. In symbols,

Rules and Properties: Additive Identity Property

For any number a,

$a + 0 = 0 + a = a$

NOTE No number loses its identity after addition with 0. Zero is called the **additive identity.**

Example 4

Adding Signed Numbers

Add.

(a) $9 + 0 = 9$

(b) $0 + \left(-\frac{5}{4}\right) = -\frac{5}{4}$

(c) $(-25) + 0 = -25$

CHECK YOURSELF 4

Add.

(a) $8 + 0$ **(b)** $0 + \left(-\frac{8}{3}\right)$ **(c)** $(-36) + 0$

NOTE The opposite of a number is also called the **additive inverse** of that number.

Recall that every number has an *opposite.* It corresponds to a point the same distance from the origin as the given number, but in the opposite direction.

NOTE 3 and −3 are opposites.

The opposite of 9 is −9.
The opposite of −15 is 15.

Our second property states that the sum of any number and its opposite is 0.

NOTE Here $-a$ represents the opposite of the number a. The sum of any number and its opposite, or additive inverse, is 0.

Rules and Properties: Additive Inverse Property

For any number a, there exists a number $-a$ such that

$$a + (-a) = (-a) + a = 0$$

Example 5

Adding Signed Numbers

(a) $9 + (-9) = 0$

(b) $-15 + 15 = 0$

(c) $(-2.3) + 2.3 = 0$

(d) $\frac{4}{5} + \left(-\frac{4}{5}\right) = 0$

CHECK YOURSELF 5

Add.

(a) $(-17) + 17$

(b) $12 + (-12)$

(c) $\frac{1}{3} + \left(-\frac{1}{3}\right)$

(d) $(-1.6) + 1.6$

In Section 0.4 we saw that the least and greatest elements of a set were called the minimum and maximum. The middle value of an ordered set is called the **median.** The median is sometimes used to represent an *average* of the set of numbers.

Example 6

Finding the Median

Find the median for each set of numbers.

(a) $9, -5, -8, 3, 7$

First, rewrite the set in ascending order.

$-8, -5, 3, 7, 9$

The median is then the element that has just as many numbers to its right as it has to its left. In this set, 3 is the median, because there are two numbers that are larger (7 and 9) and two numbers that are smaller (-8 and -5).

(b) $3, -2, 18, -20, -13$

First, rewrite the set in ascending order.

$-20, -13, -2, 3, 18$

The median is then the element that is exactly in the middle. The median for this set is -2.

CHECK YOURSELF 6

Find the median for each set of numbers.

(a) $-3, 2, 7, -6, -1$ **(b)** $5, 1, -10, 2, -20$

In the previous example, each set had an odd number of elements. If we had an even number of elements, there would be no single middle number.

To find the median from a set with an even number of elements, add the two middle numbers and divide their sum by 2.

Example 7

Finding the Median

Find the median for each set of numbers.

(a) $-3, 3, -8, 4, -1, -7, 5, 9$

First, rewrite the set in ascending order.

$-8, -7, -3, -1, 3, 4, 5, 9$

Add the middle two numbers (-1 and 3), then divide their sum by 2.

$$\frac{(-1) + (3)}{2} = \frac{2}{2} = 1$$

The median is 1.

(b) $8, 3, -2, 4, -5, -7$

Rewrite the set in ascending order.

$-7, -5, -2, 3, 4, 8$

The median is one-half the sum of the middle two numbers.

$$\frac{-2 + 3}{2} = \frac{1}{2} = 0.5$$

CHECK YOURSELF 7

Find the median for each set of numbers.

(a) $2, -5, 15, 8, 3, -4$
(b) $8, 3, 6, -8, 9, -7$

To begin our discussion of subtraction when signed numbers are involved, we can look back at a problem using natural numbers. Of course, we know that

$$8 - 5 = 3 \tag{1}$$

From our work in adding signed numbers, we know that it is also true that

$$8 + (-5) = 3 \tag{2}$$

Comparing equations (1) and (2), we see that the results are the same. This leads us to an important pattern. Any subtraction problem can be written as a problem in addition. Subtracting 5 is the same as adding the opposite of 5, or -5. We can write this fact as follows:

$$8 - 5 = 8 + (-5) = 3$$

This leads us to the following rule for subtracting signed numbers.

Rules and Properties: Subtracting Signed Numbers

1. Rewrite the subtraction problem as an addition problem by
   a. Changing the minus sign to a plus sign.
   b. Replacing the number being subtracted with its opposite.
2. Add the resulting signed numbers as before.
   In symbols,

   $a - b = a + (-b)$

NOTE This is the *definition* of subtraction.

Example 8 illustrates the use of this definition while subtracting.

Example 8

Subtracting Signed Numbers

Subtraction *Addition*

Change the subtraction symbol (−) to an addition symbol (+).

(a) $15 - 7 = 15 + (-7)$

Replace 7 with its opposite, −7.

$= 8$

(b) $9 - 12 = 9 + (-12) = -3$

(c) $-6 - 7 = -6 + (-7) = -13$

(d) $-\frac{3}{5} - \frac{7}{5} = -\frac{3}{5} + \left(-\frac{7}{5}\right) = -\frac{10}{5} = -2$

(e) $2.1 - 3.4 = 2.1 + (-3.4) = -1.3$

(f) Subtract 5 from -2. We write the statement as $-2 - 5$ and proceed as before:

$-2 - 5 = -2 + (-5) = -7$

CHECK YOURSELF 8

Subtract.

(a) $18 - 7$ **(b)** $5 - 13$ **(c)** $-7 - 9$

(d) $-\frac{5}{6} - \frac{7}{6}$ **(e)** $-2 - 7$ **(f)** $5.6 - 7.8$

The subtraction rule is used in the same way when the number being subtracted is negative. Change the subtraction to addition. Replace the negative number being subtracted with its opposite, which is positive. Example 9 will illustrate this principle.

Example 9

Subtracting Signed Numbers

Subtraction *Addition*

Change the subtraction to an addition.

(a) $5 - (-2) = 5 + (+2) = 5 + 2 = 7$

Replace −2 with its opposite, +2 or 2.

(b) $7 - (-8) = 7 + (+8) = 7 + 8 = 15$

(c) $-9 - (-5) = -9 + 5 = -4$

(d) $-12.7 - (-3.7) = -12.7 + 3.7 = -9$

(e) $-\frac{3}{4} - \left(-\frac{7}{4}\right) = -\frac{3}{4} + \left(+\frac{7}{4}\right) = \frac{4}{4} = 1$

(f) Subtract −4 from −5. We write

$-5 - (-4) = -5 + 4 = -1$

CHECK YOURSELF 9

Subtract.

(a) $8 - (-2)$ **(b)** $3 - (-10)$ **(c)** $-7 - (-2)$

(d) $-9.8 - (-5.8)$ **(e)** $7 - (-7)$

Given a set of numbers, the **range** is the difference between the maximum and the minimum.

Example 10

Finding the Range

Find the range for each set of numbers.

(a) 5, −2, −7, 9, 3

Rewrite the set in ascending order. The maximum is 9, the minimum is −7. The range is the difference.

$$9 - (-7) = 9 + 7 = 16$$

The range is 16.

(b) 3, 8, −17, 12, −2

Rewrite the set in ascending order. The maximum is 12. The minimum is −17. The range is $12 - (-17) = 29$.

CHECK YOURSELF 10

Find the range for each set of numbers.

(a) 2, −4, 7, −3, −1 **(b)** −3, 4, −7, 5, 9, −4

Your scientific calculator can be used to do arithmetic with signed numbers. Before we look at an example, there are some keys on your calculator with which you should become familiar.

There are two similar keys you must find on the calculator. The first is used for subtraction ([−]) and is usually found in the right column of calculator keys. The second will "change the sign" of a number. It is usually a [+/−] and is found on the bottom row.

NOTE Some graphing calculators have a negative sign [(−)] that acts to change the sign of a number.

We will use these keys in our next example.

Example 11

Subtracting Signed Numbers

NOTE If you have a graphing calculator, the key sequence will be

[(−)] 12.43 [−] 3.516 [ENTER]

Using your calculator, find the difference.

(a) −12.43 − 3.516

Enter the 12.43 and push the [+/−] to make it negative. Then push [−] 3.516 [=]. The result should be −15.946.

(b) 23.56 − (−4.7)

The key sequence is

23.56 [−] 4.7 [+/−] [=]

The answer should be 28.26.

CHECK YOURSELF 11

Use your calculator to find the difference.

(a) $-13.46 - 5.71$ **(b)** $-3.575 - (-6.825)$

CHECK YOURSELF ANSWERS

1. **(a)** -9; **(b)** -10; **(c)** -20; **(d)** -4 **2.** **(a)** 2; **(b)** -4; **(c)** 5; **(d)** -4
3. **(a)** -9; **(b)** -15; **(c)** 7; **(d)** -1; **(e)** -3; **(f)** 3 **4.** **(a)** 8; **(b)** $-\frac{8}{3}$; **(c)** -36
5. **(a)** 0; **(b)** 0; **(c)** 0; **(d)** 0 **6.** **(a)** -1; **(b)** 1 **7.** **(a)** 2.5; **(b)** 4.5
8. **(a)** 11; **(b)** -8; **(c)** -16; **(d)** -2; **(e)** -9; **(f)** -2.2
9. **(a)** 10; **(b)** 13; **(c)** -5; **(d)** -4; **(e)** 14 **10.** **(a)** $7 - (-4) = 11$;
(b) $9 - (-7) = 16$ **11.** **(a)** -19.17; **(b)** 3.25

Name ______________________

Section __________ Date __________

1.3 Exercises

Add.

1. $3 + 6$

2. $5 + 9$

3. $11 + 5$

4. $8 + 7$

5. $\frac{3}{4} + \frac{5}{4}$

6. $\frac{7}{3} + \frac{8}{3}$

7. $\frac{1}{2} + \frac{4}{5}$

8. $\frac{2}{3} + \frac{5}{9}$

9. $(-2) + (-3)$

10. $(-1) + (-9)$

11. $\left(-\frac{3}{5}\right) + \left(-\frac{7}{5}\right)$

12. $\left(-\frac{3}{5}\right) + \frac{12}{5}$

13. $\left(-\frac{1}{2}\right) + \left(-\frac{3}{8}\right)$

14. $\left(-\frac{4}{7}\right) + \left(-\frac{3}{14}\right)$

15. $(-1.6) + (-2.3)$

16. $(-3.5) + (-2.6)$

17. $9 + (-3)$

18. $10 + (-4)$

19. $\frac{3}{4} + \left(-\frac{1}{2}\right)$

20. $\frac{2}{3} + \left(-\frac{1}{6}\right)$

21. $\left(-\frac{4}{5}\right) + \frac{9}{20}$

22. $\left(-\frac{11}{6}\right) + \frac{5}{12}$

ANSWERS

1. 9
2. 14
3. 16
4. 15
5. 2
6. 5
7. $\frac{13}{10}$
8. $\frac{11}{9}$
9. -5
10. -10
11. -2
12. $\frac{9}{5}$
13. $-\frac{7}{8}$
14. $-\frac{11}{14}$
15. -3.9
16. -6.1
17. 6
18. 6
19. $\frac{1}{4}$
20. $\frac{1}{2}$
21. $-\frac{7}{20}$
22. $-\frac{17}{12}$

Answers

1. 9 **3.** 16 **5.** 2 **7.** $\frac{13}{10}$ **9.** -5 **11.** -2 **13.** $-\frac{7}{8}$

15. -3.9 **17.** 6 **19.** $\frac{1}{4}$ **21.** $-\frac{7}{20}$ **23.** 2 **25.** 4 **27.** -9

29. 0 **31.** 0 **33.** -1 **35.** -2 **37.** -3 **39.** -6 **41.** 8

43. 37 **45.** 1 **47.** 2.5 **49.** -2 **51.** -21 **53.** -2 **55.** -3.8

57. -8 **59.** -23 **61.** $-\frac{11}{10}$ **63.** -8.1 **65.** 16 **67.** 19

69. $\frac{9}{4}$ **71.** $\frac{17}{14}$ **73.** 14 **75.** 20.6 **77.** -12 **79.** 8 **81.** 2

83. -7 **85.** 3.2 **87.** -9.491 **89.** -1.0155 **91.** 2.3522

93. 5 **95.** 7 **97.** 22 **99.** 13 **101.** 10 **103.** 11 **105.** \$128

107. 23 yards **109.** 70° **111.** \$95.50 **113.** 1175 **115.** 24

117. **119.** 14°F **121.** **a.** -4 **b.** 15 **c.** 27

d. -30 **e.** -25 **f.** -32

Name ____________________

Section ________ Date ________

1.3 Exercises

Add.

1. $3 + 6$

2. $5 + 9$

3. $11 + 5$

4. $8 + 7$

5. $\frac{3}{4} + \frac{5}{4}$

6. $\frac{7}{3} + \frac{8}{3}$

7. $\frac{1}{2} + \frac{4}{5}$

8. $\frac{2}{3} + \frac{5}{9}$

9. $(-2) + (-3)$

10. $(-1) + (-9)$

11. $\left(-\frac{3}{5}\right) + \left(-\frac{7}{5}\right)$

12. $\left(-\frac{3}{5}\right) + \frac{12}{5}$

13. $\left(-\frac{1}{2}\right) + \left(-\frac{3}{8}\right)$

14. $\left(-\frac{4}{7}\right) + \left(-\frac{3}{14}\right)$

15. $(-1.6) + (-2.3)$

16. $(-3.5) + (-2.6)$

17. $9 + (-3)$

18. $10 + (-4)$

19. $\frac{3}{4} + \left(-\frac{1}{2}\right)$

20. $\frac{2}{3} + \left(-\frac{1}{6}\right)$

21. $\left(-\frac{4}{5}\right) + \frac{9}{20}$

22. $\left(-\frac{11}{6}\right) + \frac{5}{12}$

ANSWERS

1. 9
2. 14
3. 16
4. 15
5. 2
6. 5
7. $\frac{13}{10}$
8. $\frac{11}{9}$
9. -5
10. -10
11. -2
12. $\frac{9}{5}$
13. $-\frac{7}{8}$
14. $-\frac{11}{14}$
15. -3.9
16. -6.1
17. 6
18. 6
19. $\frac{1}{4}$
20. $\frac{1}{2}$
21. $-\frac{7}{20}$
22. $-\frac{17}{12}$

ANSWERS

23. 2
24. 4
25. 4
26. 2.3
27. −9
28. −15
29. 0
30. 0
31. 0
32. 0
33. −1
34. −1
35. −2
36. −3
37. −3
38. −1
39. −6
40. −11
41. 8
42. 14
43. 37
44. 47
45. 1
46. 1
47. 2.5
48. 7.2
49. −2
50. −5

23. $-11.4 + 13.4$

24. $-5.2 + 9.2$

25. $-3.6 + 7.6$

26. $-2.6 + 4.9$

27. $-9 + 0$

28. $-15 + 0$

29. $7 + (-7)$

30. $12 + (-12)$

31. $-4.5 + 4.5$

32. $\left(-\frac{2}{3}\right) + \frac{2}{3}$

33. $7 + (-9) + (-5) + 6$

34. $(-4) + 6 + (-3) + 0$

35. $7 + (-3) + 5 + (-11)$

36. $-\frac{6}{5} + \left(-\frac{13}{5}\right) + \frac{4}{5}$

37. $-\frac{3}{2} + \left(-\frac{7}{4}\right) + \frac{1}{4}$

38. $\frac{1}{3} + \left(-\frac{5}{6}\right) + \left(-\frac{1}{2}\right)$

39. $2.3 + (-5.4) + (-2.9)$

40. $-5.4 + (-2.1) + (-3.5)$

Subtract.

41. $21 - 13$

42. $36 - 22$

43. $82 - 45$

44. $103 - 56$

45. $\frac{15}{7} - \frac{8}{7}$

46. $\frac{17}{8} - \frac{9}{8}$

47. $7.9 - 5.4$

48. $11.7 - 4.5$

49. $8 - 10$

50. $14 - 19$

51. $24 - 45$

52. $136 - 352$

53. $\frac{7}{6} - \frac{19}{6}$

54. $\frac{5}{9} - \frac{32}{9}$

55. $7.8 - 11.6$

56. $14.3 - 25.5$

57. $-5 - 3$

58. $-15 - 8$

59. $-9 - 14$

60. $-8 - 12$

61. $-\frac{2}{5} - \frac{7}{10}$

62. $-\frac{5}{9} - \frac{7}{18}$

63. $-3.4 - 4.7$

64. $-8.1 - 7.6$

65. $5 - (-11)$

66. $7 - (-5)$

67. $7 - (-12)$

68. $3 - (-10)$

69. $\frac{3}{4} - \left(-\frac{3}{2}\right)$

70. $\frac{5}{6} - \left(-\frac{7}{6}\right)$

71. $\frac{6}{7} - \left(-\frac{5}{14}\right)$

72. $\frac{11}{16} - \left(-\frac{7}{8}\right)$

73. $8.3 - (-5.7)$

74. $6.5 - (-4.3)$

75. $8.9 - (-11.7)$

76. $14.5 - (-24.6)$

77. $-36 - (-24)$

78. $-28 - (-11)$

ANSWERS

51. -21
52. -216
53. -2
54. -3
55. -3.8
56. -11.2
57. -8
58. -23
59. -23
60. -20
61. $-\frac{11}{10}$
62. $-\frac{17}{18}$
63. -8.1
64. -15.7
65. 16
66. 12
67. 19
68. 13
69. $\frac{9}{4}$
70. 2
71. $\frac{17}{14}$
72. $\frac{25}{16}$
73. 14
74. 10.8
75. 20.6
76. 39.1
77. -12
78. -17

ANSWERS

79. 8
80. 5
81. 2
82. $\frac{1}{8}$
83. −7
84. −3
85. 3.2
86. 4.2
87. −9.491
88. 9.5194
89. −1.0155
90. −8.2516
91. 2.3522
92. −1.3467
93. 5
94. 6
95. 7
96. 32
97. 22
98. 23
99. 13
100. 18
101. 10
102. 10
103. 11
104. 16
105. $128
106. $225

79. $-19 - (-27)$

80. $-11 - (-16)$

81. $\left(-\frac{3}{4}\right) - \left(-\frac{11}{4}\right)$

82. $-\frac{1}{2} - \left(-\frac{5}{8}\right)$

83. $-12.7 - (-5.7)$

84. $-5.6 - (-2.6)$

85. $-6.9 - (-10.1)$

86. $-3.4 - (-7.6)$

Use your calculator to evaluate each expression.

87. $-4.1967 - 5.2943$

88. $5.3297 - (-4.1897)$

89. $-4.1623 - (-3.1468)$

90. $(-3.6829) - 4.5687$

91. $-6.3267 + 8.6789$

92. $-6.6712 + 5.3245$

Find the median for each of the following sets.

93. 1, 3, 5, 7, 9

94. 2, 4, 6, 8, 10

95. 8, 7, 2, 25, 5, 13, 3

96. 53, 23, 34, 21, 32, 30, 32

Determine the range for each of the following sets.

97. 2, 7, 9, 15, 24

98. 4, 8, 11, 15, 27

99. −4, −3, 2, 7, 9

100. −7, −2, 1, 8, 11

101. $\frac{7}{8}, 2, -\frac{1}{2}, -8, \frac{3}{4}$

102. $3, \frac{5}{6}, -7, -\frac{1}{3}, \frac{2}{3}$

103. 3, 2, −5, 6, −3

104. 1, −9, 7, −2, 3

Solve the following problems.

105. Checking account. Amir has $100 in his checking account. He writes a check for $23 and makes a deposit of $51. What is his new balance?

106. Checking account. Olga has $250 in her checking account. She deposits $52 and then writes a check for $77. What is her new balance?

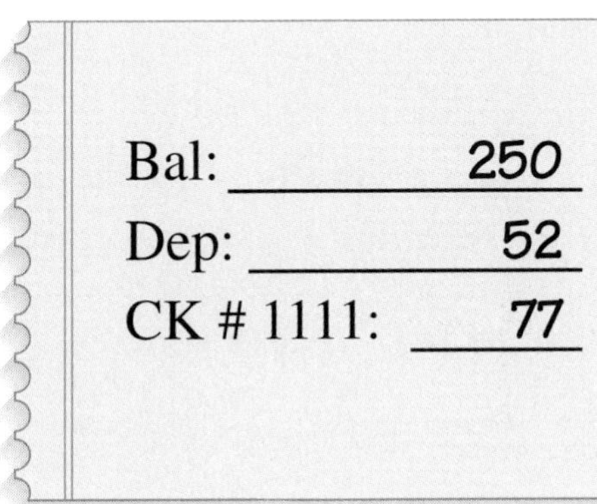

107. Football yardage. On four consecutive running plays, Ricky Watters of the Seattle Seahawks gained 23 yards, lost 5 yards, gained 15 yards, and lost 10 yards. What was his net yardage change for the series of plays?

108. VISA balance. Ramon owes \$780 on his VISA account. He returns three items costing \$43.10, \$36.80, and \$125.00 and receives credit on his account. Next, he makes a payment of \$400. He then makes a purchase of \$82.75. How much does Tom still owe?

109. Temperature. The temperature at noon on a June day was 82°. It fell by 12° in the next 4 hours. What was the temperature at 4:00 P.M.?

110. Mountain climbing. Chia is standing at a point 6000 feet (ft) above sea level. She descends to a point 725 ft lower. What is her distance above sea level?

111. Checking account. Omar's checking account was overdrawn by \$72. He wrote another check for \$23.50. How much was his checking account overdrawn after writing the check?

112. Personal finance. Angelo owed his sister \$15. He later borrowed another \$10. What positive or negative number represents his current financial condition?

113. Education. A local community college had a decrease in enrollment of 750 students in the fall of 1999. In the spring of 2000, there was another decrease of 425 students. What was the total decrease in enrollment for both semesters?

114. Temperature. At 7 A.M., the temperature was −15°F. By 1 P.M., the temperature had increased by 18°F. What was the temperature at 1 P.M.?

115. Education. Ezra's scores on five tests taken in a mathematics class were 87, 71, 95, 81, and 90. What was the range of his scores?

ANSWERS

107. 23 yards

108. \$257.85

109. 70°

110. 5275 ft

111. \$95.50

112. −25

113. 1175

114. 3°F

115. 24

ANSWERS

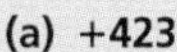

(a) +423
(b) −1878
(c) +753
116. (d) −47

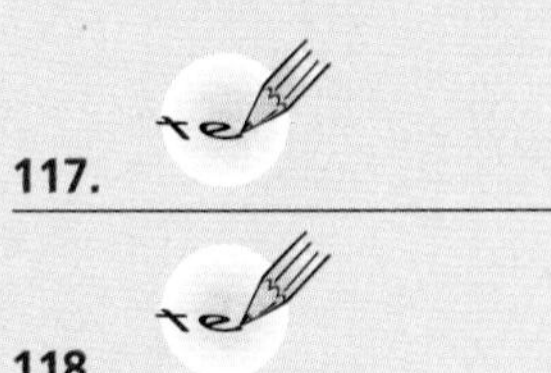

117.

118.

116. The bar chart shown represents the total yards gained by the leading passer in the NFL from 1993 to 1997. Use a signed number to represent the change in total passing yards from one year to the next.

(a) from 1993 to 1994 **(b)** from 1994 to 1995
(c) from 1995 to 1996 **(d)** from 1996 to 1997

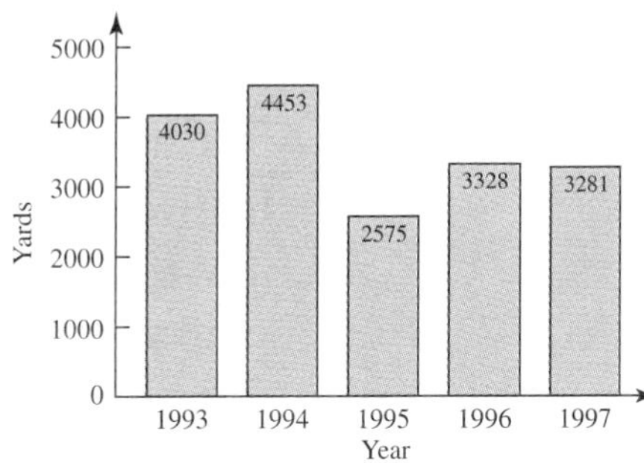

117. In this chapter, it is stated that "Every number has an opposite." The opposite of 9 is −9. This corresponds to the idea of an opposite in English. In English an opposite is often expressed by a prefix, for example, *un-* or *ir-*.

(a) Write the opposite of these words: unmentionable, uninteresting, irredeemable, irregular, uncomfortable.

(b) What is the meaning of these expressions: not uninteresting, not irredeemable, not irregular, not unmentionable?

(c) Think of other prefixes that *negate* or change the meaning of a word to its *opposite.* Make a list of words formed with these prefixes, and write a sentence with three of the words you found. Make a sentence with two words and phrases from each of the lists above. Look up the meaning of the word *irregardless.*

What is the value of $-[-(-5)]$? What is the value of $-(-6)$? How does this relate to the above examples? Write a short description about this relationship.

118. The temperature on the plains of North Dakota can change rapidly, falling or rising many degrees in the course of an hour. Here are some temperature changes during each day over a week.

Day	Mon.	Tues.	Wed.	Thurs.	Fri.	Sat.	Sun.
Temp. Change from 10 A.M. to 3 P.M.	+13°	+20°	−18°	+10°	−25°	−5°	+15°

Write a short speech for the TV weather reporter that summarizes the daily temperature change. Use the median as you characterize the average daily midday change.

119. Science. The daily average temperatures in degrees Fahrenheit for a week in February were $-1, 3, 5, -2, 4, 12$, and 10. What was the range of temperatures for that week?

120. How long ago was the year 1250 B.C.E.? What year was 3300 years ago? Make a number line and locate the following events, cultures, and objects on it. How long ago was each item in the list? Which two events are the closest to each other? You may want to learn more about some of the cultures in the list and the mathematics and science developed by that culture.

Inca culture in Peru—1400 A.D.
The *Ahmes Papyrus,* a mathematical text from Egypt—1650 B.C.E.
Babylonian arithmetic develops the use of a zero symbol—300 B.C.E.
First Olympic Games—776 B.C.E.
Pythagoras of Greece dies—580 B.C.E.
Mayans in Central America independently develop use of zero—500 A.D.
The *Chou Pei,* a mathematics classic from China—1000 B.C.E.
The *Aryabhatiya,* a mathematics work from India—499 A.D.
Trigonometry arrives in Europe via the Arabs and India—1464 A.D.
Arabs receive algebra from Greek, Hindu, and Babylonian sources and develop it into a new systematic form—850 A.D.
Development of calculus in Europe—1670 A.D.
Rise of abstract algebra—1860 A.D.
Growing importance of probability and development of statistics—1902 A.D.

121. Complete the following statement: "$3 - (-7)$ is the same as ____ because . . . " Write a problem that might be answered by doing this subtraction.

122. Explain the difference between the two phrases: "a number subtracted from 5" and "a number less than 5." Use algebra and English to explain the meaning of these phrases. Write other ways to express subtraction in English. Which ones are confusing?

Getting Ready for Section 1.4 [Section 0.3]

Add.

(a) $(-1) + (-1) + (-1) + (-1)$
(b) $3 + 3 + 3 + 3 + 3$
(c) $9 + 9 + 9$
(d) $(-10) + (-10) + (-10)$
(e) $(-5) + (-5) + (-5) + (-5) + (-5)$
(f) $(-8) + (-8) + (-8) + (-8)$

ANSWERS

119. 14°F

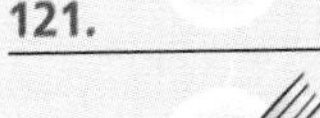

120.

121.

122.

a. −4

b. 15

c. 27

d. −30

e. −25

f. −32

Answers

1. 9 **3.** 16 **5.** 2 **7.** $\frac{13}{10}$ **9.** −5 **11.** −2 **13.** $-\frac{7}{8}$

15. −3.9 **17.** 6 **19.** $\frac{1}{4}$ **21.** $-\frac{7}{20}$ **23.** 2 **25.** 4 **27.** −9

29. 0 **31.** 0 **33.** −1 **35.** −2 **37.** −3 **39.** −6 **41.** 8

43. 37 **45.** 1 **47.** 2.5 **49.** −2 **51.** −21 **53.** −2 **55.** −3.8

57. −8 **59.** −23 **61.** $-\frac{11}{10}$ **63.** −8.1 **65.** 16 **67.** 19

69. $\frac{9}{4}$ **71.** $\frac{17}{14}$ **73.** 14 **75.** 20.6 **77.** −12 **79.** 8 **81.** 2

83. −7 **85.** 3.2 **87.** −9.491 **89.** −1.0155 **91.** 2.3522

93. 5 **95.** 7 **97.** 22 **99.** 13 **101.** 10 **103.** 11 **105.** $128

107. 23 yards **109.** 70° **111.** $95.50 **113.** 1175 **115.** 24

117. **119.** 14°F **121.** **a.** −4 **b.** 15 **c.** 27

d. −30 **e.** −25 **f.** −32

 Multiplying and Dividing Signed Numbers

1.4 OBJECTIVES

1. Find the product of two signed numbers
2. Find the quotient of two signed numbers

When you first considered multiplication in arithmetic, it was thought of as repeated addition. Let's see what our work with the addition of signed numbers can tell us about multiplication when signed numbers are involved. For example,

$$3 \cdot 4 = \underbrace{4 + 4 + 4} = 12$$

We interpret multiplication as repeated addition to find the product, 12.

Now, consider the product $(3)(-4)$:

$$(3)(-4) = (-4) + (-4) + (-4) = -12$$

Looking at this product suggests the first portion of our rule for multiplying signed numbers. The product of a positive number and a negative number is negative.

Rules and Properties: Multiplying Signed Numbers Case 1: Different Signs

The product of two numbers with different signs is negative.

To use this rule in multiplying two numbers with different signs, multiply their absolute values and attach a negative sign.

Example 1

Multiplying Signed Numbers

Multiply.

(a) $(5)(-6) = -30$

The product is negative.

(b) $(-10)(10) = -100$

(c) $(8)(-12) = -96$

NOTE Multiply together numerators and then denominators and reduce.

(d) $\left(-\frac{3}{4}\right)\left(\frac{2}{5}\right) = -\frac{3}{10}$

 CHECK YOURSELF 1

Multiply.

(a) $(-7)(5)$ **(b)** $(-12)(9)$ **(c)** $(-15)(8)$ **(d)** $\left(-\frac{4}{7}\right)\left(\frac{14}{5}\right)$

The product of two negative numbers is harder to visualize. The following pattern may help you see how we can determine the sign of the product.

$$(3)(-2) = -6$$

$$(2)(-2) = -4$$

NOTE This number is decreasing by 1.

$$(1)(-2) = -2$$

Do you see that the product is *increasing* by 2 each time?

$$(0)(-2) = 0$$

$$(-1)(-2) = 2$$

NOTE $(-1)(-2)$ is the opposite of -2.

What should the product $(-2)(-2)$ be? Continuing the pattern shown, we see that

$(-2)(-2) = 4$

This suggests that the product of two negative numbers is positive. That is the case. We can extend our multiplication rule.

NOTE If you would like a more detailed explanation, see the discussion at the end of this section.

Rules and Properties: Multiplying Signed Numbers Case 2: Same Sign

The product of two numbers with the same sign is positive.

Example 2

Multiplying Signed Numbers

Multiply.

(a) $9 \cdot 7 = 63$ The product of two positive numbers (same sign, +) is positive.

(b) $(-8)(-5) = 40$ The product of two negative numbers (same sign, −) is positive.

(c) $\left(-\frac{1}{2}\right)\left(-\frac{1}{3}\right) = \frac{1}{6}$

CHECK YOURSELF 2

Multiply.

(a) $10 \cdot 12$ **(b)** $(-8)(-9)$ **(c)** $\left(-\frac{2}{3}\right)\left(-\frac{6}{7}\right)$

Two numbers, 0 and 1, have special properties in multiplication.

Rules and Properties: Multiplicative Identity Property

The product of 1 and any number is that number. In symbols,

$a \cdot 1 = 1 \cdot a = a$

NOTE The number 1 is called the **multiplicative identity** for this reason.

Rules and Properties: Multiplicative Property of Zero

The product of 0 and any number is 0. In symbols,

$a \cdot 0 = 0 \cdot a = 0$

Example 3

Multiplying Signed Numbers

Find each product.

(a) $(1)(-7) = -7$

(b) $(15)(1) = 15$

(c) $(-7)(0) = 0$

(d) $0 \cdot 12 = 0$

(e) $\left(-\frac{4}{5}\right)(0) = 0$

CHECK YOURSELF 3

Multiply.

(a) $(-10)(1)$ **(b)** $(0)(-17)$ **(c)** $\left(\frac{5}{7}\right)(1)$ **(d)** $(0)\left(\frac{3}{4}\right)$

Before we continue, consider the following equivalent fractions:

$$-\frac{1}{a} = \frac{-1}{a} = \frac{1}{-a}$$

Any of these forms can occur in the course of simplifying an expression. The first form is generally preferred.

To complete our discussion of the properties of multiplication, we state the following.

Rules and Properties: Multiplicative Inverse Property

For any number a, where $a \neq 0$, there is a number $\frac{1}{a}$ such that

$a \cdot \frac{1}{a} = 1$

NOTE $\frac{1}{a}$ is called the **multiplicative inverse,** or the **reciprocal,** of a. The product of any nonzero number and its reciprocal is 1.

Example 4 illustrates this property.

Example 4

Multiplying Signed Numbers

(a) $3 \cdot \frac{1}{3} = 1$ The reciprocal of 3 is $\frac{1}{3}$.

(b) $-5\left(-\frac{1}{5}\right) = 1$ The reciprocal of -5 is $\frac{1}{-5}$ or $-\frac{1}{5}$.

(c) $\frac{2}{3} \cdot \frac{3}{2} = 1$ The reciprocal of $\frac{2}{3}$ is $\frac{1}{\frac{2}{3}}$, or $\frac{3}{2}$.

CHECK YOURSELF 4

Find the multiplicative inverse (or the reciprocal) of each of the following numbers.

(a) 6 **(b)** -4 **(c)** $\frac{1}{4}$ **(d)** $-\frac{3}{5}$

You know from your work in arithmetic that multiplication and division are related operations. We can use that fact, and our work of the last section, to determine rules for the division of signed numbers. Every division problem can be stated as an equivalent multiplication problem. For instance,

$\frac{15}{5} = 3$ because $15 = 5 \cdot 3$

$\frac{-24}{6} = -4$ because $-24 = (6)(-4)$

$\frac{-30}{-5} = 6$ because $-30 = (-5)(6)$

The examples above illustrate that because the two operations are related, the rule of signs that we stated in the last section for multiplication is also true for division.

Rules and Properties: Dividing Signed Numbers

1. The quotient of two numbers with different signs is negative.
2. The quotient of two numbers with the same sign is positive.

Again, the rule is easy to use. To divide two signed numbers, divide their absolute values. Then attach the proper sign according to the rule above.

Example 5

Dividing Signed Numbers

Divide.

(a) Positive → 28, Positive → 7: $\frac{28}{7} = 4$ ← Positive

(b) Negative → −36, Negative → −4: $\frac{-36}{-4} = 9$ ← Positive

(c) Negative → −42, Positive → 7: $\frac{-42}{7} = -6$ ← Negative

(d) Positive → 75, Negative → −3: $\frac{75}{-3} = -25$ ← Negative

(e) Positive → 15.2, Negative → −3.8: $\frac{15.2}{-3.8} = -4$ ← Negative

CHECK YOURSELF 5

Divide.

(a) $\dfrac{-55}{11}$ (b) $\dfrac{80}{20}$ (c) $\dfrac{-48}{-8}$ (d) $\dfrac{144}{-12}$ (e) $\dfrac{-13.5}{-2.7}$

You should be very careful when 0 is involved in a division problem. Remember that 0 divided by any nonzero number is just 0. Recall that

$$\frac{0}{-7} = 0 \qquad \text{because} \qquad 0 = (-7)(0)$$

However, if zero is the *divisor,* we have a special problem. Consider

$$\frac{9}{0} = ?$$

This means that $9 = 0 \cdot ?$.

Can 0 times a number ever be 9? No, so there is no solution.

Because $\frac{9}{0}$ cannot be replaced by any number, we agree that *division by 0 is not allowed.*

We say that

Rules and Properties: Division by Zero

Division by 0 is undefined.

Example 6

Dividing Signed Numbers

Divide, if possible.

(a) $\dfrac{7}{0}$ is undefined.

(b) $\dfrac{-9}{0}$ is undefined.

(c) $\dfrac{0}{5} = 0$

(d) $\dfrac{0}{-8} = 0$

Note: The expression $\frac{0}{0}$ is called an **indeterminate form.** You will learn more about this in later mathematics classes.

CHECK YOURSELF 6

Divide if possible.

(a) $\dfrac{0}{3}$ (b) $\dfrac{5}{0}$ (c) $\dfrac{-7}{0}$ (d) $\dfrac{0}{-9}$

The fraction bar serves as a *grouping symbol.* This means that all operations in the numerator and denominator should be performed separately. Then the division is done as the last step. Example 7 illustrates this property.

Example 7

Dividing Signed Numbers

Evaluate each expression.

(a) $\dfrac{(-6)(-7)}{3} = \dfrac{42}{3} = 14$ Multiply in the numerator, then divide.

(b) $\dfrac{3 + (-12)}{3} = \dfrac{-9}{3} = -3$ Add in the numerator, then divide.

(c) $\dfrac{-4 + (2)(-6)}{-6 - 2} = \dfrac{-4 + (-12)}{-6 - 2}$ Multiply in the numerator. Then add in the numerator and subtract in the denominator.

$= \dfrac{-16}{-8} = 2$ Divide as the last step.

CHECK YOURSELF 7

Evaluate each expression.

(a) $\dfrac{-4 + (-8)}{6}$ **(b)** $\dfrac{3 - (2)(-6)}{-5}$ **(c)** $\dfrac{(-2)(-4) - (-6)(-5)}{(-4)(11)}$

Evaluating fractions with a calculator poses a special problem. Example 8 illustrates this problem.

Example 8

Using a Calculator to Divide

Use your scientific calculator to evaluate each fraction.

(a) $\dfrac{4}{2 - 3}$

As you can see, the correct answer should be -4. To get this answer with your calculator, you must place the denominator in parentheses. The key stroke sequence will be

4 [÷] [(] 2 [−] 3 [)] [=]

(b) $\dfrac{-7 - 7}{3 - 10}$

In this problem, the correct answer is 2. This can be found on your calculator by placing the numerator in parentheses and then placing the denominator in parentheses. The key stroke sequence will be

[(] 7 [+/−] [−] 7 [)] [÷] [(] 3 [−] 10 [)] [=]

When evaluating a fraction with a calculator, it is safest to use parentheses in both the numerator and the denominator.

CHECK YOURSELF 8

Evaluate using your calculator.

(a) $\dfrac{-8}{5 - 7}$ **(b)** $\dfrac{-3 - 2}{-13 + 23}$

Example 9

Multiplying Signed Numbers

Evaluate each expression.

(a) $7(-9 + 12)$ Evaluate inside the parentheses first.

$=7(3) = 21$

(b) $(-8)(-7) - 40$ Multiply first, then subtract.

$= 56 - 40$

$= 16$

(c) $(-5)^2 - 3$ Evaluate the power first.

$= (-5)(-5) - 3$ Note that $(-5)^2 = (-5)(-5) = 25$

$= 25 - 3$

$= 22$

(d) $-5^2 - 3$ Note that $-5^2 = -25$. The power applies *only* to the 5.

$= -25 - 3$

$= -28$

CHECK YOURSELF 9

Evaluate each expression.

(a) $8(-9 + 7)$ **(b)** $(-3)(-5) + 7$

(c) $(-4)^2 - (-4)$ **(d)** $-4^2 - (-4)$

NOTE Here is a more detailed explanation of why the product of two negative numbers is positive.

Rules and Properties: The Product of Two Negative Numbers

From our earlier work, we know that the sum of a number and its opposite is 0:

$5 + (-5) = 0$

Multiply both sides of the equation by -3:

$(-3)[5 + (-5)] = (-3)(0)$

Because the product of 0 and any number is 0, on the right we have 0.

$(-3)[5 + (-5)] = 0$

We use the distributive property on the left.

$(-3)(5) + (-3)(-5) = 0$

We know that $(-3)(5) = -15$, so the equation becomes

$-15 + (-3)(-5) = 0$

We now have a statement of the form

$-15 + \square = 0$

in which $\square$ is the value of $(-3)(-5)$. We also know that $\square$ is the number that must be added to -15 to get 0, so $\square$ is the opposite of -15, or 15. This means that

$(-3)(-5) = 15$ The product is positive!

It doesn't matter what numbers we use in this argument. The resulting product of two negative numbers will always be positive.

CHECK YOURSELF ANSWERS

1. **(a)** -35; **(b)** -108; **(c)** -120; **(d)** $-\frac{8}{5}$ **2.** **(a)** 120; **(b)** 72; **(c)** $\frac{4}{7}$
3. **(a)** -10; **(b)** 0; **(c)** $\frac{5}{7}$; **(d)** 0 **4.** **(a)** $\frac{1}{6}$; **(b)** $-\frac{1}{4}$; **(c)** 4; **(d)** $-\frac{5}{3}$
5. **(a)** -5; **(b)** 4; **(c)** 6; **(d)** -12; **(e)** 5 **6.** **(a)** 0; **(b)** undefined; **(c)** undefined; **(d)** 0
7. **(a)** -2; **(b)** -3; **(c)** $\frac{1}{2}$ **8.** **(a)** 4; **(b)** -0.5 **9.** **(a)** -16; **(b)** 22; **(c)** 20; **(d)** -12

Exercises

Name ______________________

Section ________ Date ________

Multiply.

1. $4 \cdot 10$

2. $3 \cdot 14$

3. $(5)(-12)$

4. $(10)(-2)$

5. $(-8)(9)$

6. $(-12)(3)$

7. $(4)\left(-\frac{3}{2}\right)$

8. $(9)\left(-\frac{2}{3}\right)$

9. $\left(-\frac{1}{4}\right)(8)$

10. $\left(-\frac{3}{2}\right)(4)$

11. $(3.25)(-4)$

12. $(5.4)(-5)$

13. $(-8)(-7)$

14. $(-9)(-8)$

15. $(-5)(-12)$

16. $(-7)(-3)$

17. $(-9)\left(-\frac{2}{3}\right)$

18. $(-6)\left(-\frac{3}{2}\right)$

19. $(-1.25)(-12)$

20. $(-1.5)(-20)$

21. $(0)(-18)$

22. $(-17)(0)$

23. $(15)(0)$

24. $(0)(25)$

25. $\left(-\frac{11}{12}\right)(0)$

26. $\left(-\frac{8}{9}\right)(0)$

ANSWERS

1. 40
2. 42
3. −60
4. −20
5. −72
6. −36
7. −6
8. −6
9. −2
10. −6
11. −13
12. −27
13. 56
14. 72
15. 60
16. 21
17. 6
18. 9
19. 15
20. 30
21. 0
22. 0
23. 0
24. 0
25. 0
26. 0

ANSWERS

27. 0
28. 0
29. 1
30. 1
31. −1
32. −1
33. 5
34. 5
35. 8
36. −3
37. −10
38. 4
39. −13
40. −8
41. 25
42. −4
43. 0
44. 5
45. 9
46. Undefined
47. 12
48. −10
49. Undefined
50. 0
51. −17
52. 27
53. 9
54. −25
55. −6
56. 7

27. $(-3.57)(0)$

28. $(-2.37)(0)$

29. $\left(-\frac{3}{2}\right)\left(-\frac{2}{3}\right)$

30. $\left(-\frac{4}{5}\right)\left(-\frac{5}{4}\right)$

31. $\left(\frac{4}{7}\right)\left(-\frac{7}{4}\right)$

32. $\left(\frac{8}{9}\right)\left(-\frac{9}{8}\right)$

Divide.

33. $\frac{-20}{-4}$

34. $\frac{70}{14}$

35. $\frac{48}{6}$

36. $\frac{-24}{8}$

37. $\frac{50}{-5}$

38. $\frac{-32}{-8}$

39. $\frac{-52}{4}$

40. $\frac{56}{-7}$

41. $\frac{-75}{-3}$

42. $\frac{-60}{15}$

43. $\frac{0}{-8}$

44. $\frac{-125}{-25}$

45. $\frac{-9}{-1}$

46. $\frac{-10}{0}$

47. $\frac{-96}{-8}$

48. $\frac{-20}{2}$

49. $\frac{18}{0}$

50. $\frac{0}{8}$

51. $\frac{-17}{1}$

52. $\frac{-27}{-1}$

53. $\frac{-144}{-16}$

54. $\frac{-150}{6}$

55. $\frac{-29.4}{4.9}$

56. $\frac{-25.9}{-3.7}$

57. $\frac{-8}{32}$

58. $\frac{-6}{-30}$

59. $\frac{24}{-16}$

60. $\frac{-25}{10}$

61. $\frac{-28}{-42}$

62. $\frac{-125}{-75}$

Perform the indicated operations.

63. $\frac{(-6)(-3)}{2}$

64. $\frac{(-9)(5)}{-3}$

65. $\frac{(-8)(2)}{-4}$

66. $\frac{(7)(-8)}{-14}$

67. $\frac{24}{-4 - 8}$

68. $\frac{36}{-7 + 3}$

69. $\frac{-12 - 12}{-3}$

70. $\frac{-14 - 4}{-6}$

71. $\frac{55 - 19}{-12 - 6}$

72. $\frac{-11 - 7}{-14 + 8}$

73. $\frac{7 - 5}{2 - 2}$

74. $\frac{10 - 6}{4 - 4}$

Do the indicated operations. Remember the rules for the order of operations.

75. $5(7 - 2)$

76. $7(8 - 5)$

77. $2(5 - 8)$

78. $6(14 - 16)$

79. $-3(9 - 7)$

80. $-6(12 - 9)$

81. $-3(-2 - 5)$

82. $-2(-7 - 3)$

83. $(-2)(3) - 5$

84. $(-6)(8) - 27$

ANSWERS

57. $-\frac{1}{4}$

58. $\frac{1}{5}$

59. $-\frac{3}{2}$

60. $-\frac{5}{2}$

61. $\frac{2}{3}$

62. $\frac{5}{3}$

63. 9

64. 15

65. 4

66. 4

67. −2

68. −9

69. 8

70. 3

71. −2

72. 3

73. Undefined

74. Undefined

75. 25

76. 21

77. −6

78. −12

79. −6

80. −18

81. 21

82. 20

83. −11

84. −75

ANSWERS

85. −33
86. 16
87. −2
88. −4
89. −1
90. −4
91. −22
92. −11
93. −1
94. −5
95. 32
96. 16
97. 43
98. 14
99. −40
100. −28
101. 6
102. −43
103. 39
104. 20
105. 27
106. 48
107. −89
108. −45
109. −89
110. −117
111. 253 points
112. −$540

85. $4(-7) - 5$

86. $(-3)(-9) - 11$

87. $(-5)(-2) - 12$

88. $(-7)(-3) - 25$

89. $(3)(-7) + 20$

90. $(2)(-6) + 8$

91. $-4 + (-3)(6)$

92. $-5 + (-2)(3)$

93. $7 - (-4)(-2)$

94. $9 - (-2)(-7)$

95. $(-7)^2 - 17$

96. $(-6)^2 - 20$

97. $(-5)^2 + 18$

98. $(-2)^2 + 10$

99. $-6^2 - 4$

100. $-5^2 - 3$

101. $(-4)^2 - (-2)(-5)$

102. $(-3)^3 - (-8)(-2)$

103. $(-8)^2 - 5^2$

104. $(-6)^2 - 4^2$

105. $(-6)^2 - (-3)^2$

106. $(-8)^2 - (-4)^2$

107. $-8^2 - 5^2$

108. $-6^2 - 3^2$

109. $-8^2 - (-5)^2$

110. $-9^2 - (-6)^2$

111. Basketball. You score 23 points a game for 11 straight games. What is the total number of points that you scored?

112. Gambling. In Atlantic City, Nick played the slot machines for 12 hours. He lost $45 an hour. Use signed numbers to represent the change in Nick's financial status at the end of the 12 hours.

113. Stocks. Suppose you own 35 shares of stock. If the price increases $1.25 per share, how much money have you made?

114. Checking account. Your bank charges a flat service charge of $3.50 per month on your checking account. You have had the account for 3 years. How much have you paid in service charges?

115. Temperature. The temperature is $-6°F$ at 5:00 in the evening. If the temperature drops 2°F every hour, what is the temperature at 1:00 A.M.?

116. Dieting. A woman lost 42 pounds (lb). If she lost 3 lb each week, how long has she been dieting?

117. Mowing lawns. Patrick worked all day mowing lawns and was paid $9 per hour. If he had $125 at the end of a 9-hour day, how much did he have before he started working?

118. Unit pricing. A 4.5-lb can of food costs $8.91. What is the cost per pound?

119. Investment. Suppose that you and your two brothers bought equal shares of an investment for a total of $20,000 and sold it later for $16,232. How much did each person lose?

120. Temperature. Suppose that the temperature outside is dropping at a constant rate. At noon, the temperature is 70°F and it drops to 58°F at 5:00 P.M. How much did the temperature change each hour?

121. Test tube count. A chemist has 84 ounces (oz) of a solution. He pours the solution into test tubes. Each test tube holds $\frac{2}{3}$ oz. How many test tubes can he fill?

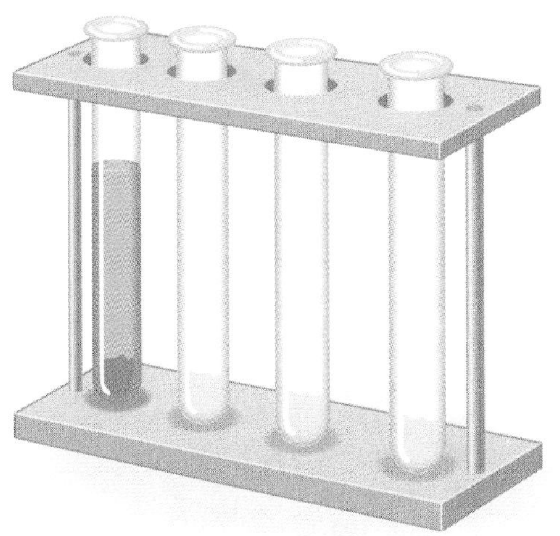

ANSWERS

113. $43.75

114. $126

115. −22°F

116. 14 weeks

117. $44

118. $1.98

119. $1256

120. 2.4°F

121. 126 test tubes

122. −7

123. 4

124. 5

125. −2

126.

a. 10

b. 2

c. 1

d. 1

e. −4

f. 9

Use your calculator to evaluate each expression.

122. $\dfrac{7}{4 - 5}$

123. $\dfrac{-8}{-4 + 2}$

124. $\dfrac{-6 - 9}{-4 + 1}$

125. $\dfrac{-10 + 4}{-7 + 10}$

126. Some animal ecologists in Minnesota are planning to reintroduce a group of animals into a wilderness area. The animals, a mammal on the endangered species list, will be released into an area where they once prospered and where there is an abundant food supply. But, the animals will face predators. The ecologists expect the number of mammals to grow about 25 percent each year but that 30 of the animals will die from attacks by predators and hunters.

The ecologists need to decide how many animals they should release to establish a stable population. Work with other students to try several beginning populations and follow the numbers through 8 years. Is there a number of animals that will lead to a stable population? Write a letter to the editor of your local newspaper explaining how to decide what number of animals to release. Include a formula for the number of animals next year based on the number this year. Begin by filling out this table to track the number of animals living each year after the release:

No. Initially Released	Year 1	2	3	4	5	6	7	8
20	+___−___ =_______							
100	+___−___ =_______							
200	+___−___ =_______							

Getting Ready for Section 1.5 [Sections 1.3 and 1.4]

(a) $\dfrac{6 \cdot 2 + 8}{5 - 3}$ (b) $\dfrac{4 \cdot 5 - 8}{8 - 4 \div 2}$ (c) $\dfrac{-8 + 3 \cdot 2}{-12 \div 6}$

(d) $\dfrac{-3^2 - (-4 - 1)}{-2 \cdot 2}$ (e) $8 \div 4 - 3 \cdot 2$ (f) $6^2 - 18 \div 2 \cdot 3$

Answers

1. 40 **3.** −60 **5.** −72 **7.** −6 **9.** −2 **11.** −13 **13.** 56 **15.** 60 **17.** 6 **19.** 15 **21.** 0 **23.** 0 **25.** 0 **27.** 0 **29.** 1 **31.** −1 **33.** 5 **35.** 8 **37.** −10 **39.** −13 **41.** 25 **43.** 0 **45.** 9 **47.** 12 **49.** Undefined **51.** −17 **53.** 9 **55.** −6 **57.** $-\dfrac{1}{4}$ **59.** $-\dfrac{3}{2}$ **61.** $\dfrac{2}{3}$ **63.** 9 **65.** 4 **67.** −2 **69.** 8 **71.** −2 **73.** Undefined **75.** 25 **77.** −6 **79.** −6 **81.** 21 **83.** −11 **85.** −33 **87.** −2 **89.** −1 **91.** −22 **93.** −1 **95.** 32 **97.** 43 **99.** −40 **101.** 6 **103.** 39 **105.** 27 **107.** −89 **109.** −89 **111.** 253 points **113.** \$43.75 **115.** −22°F **117.** \$44 **119.** \$1256 **121.** 126 **123.** 4 **125.** −2 **a.** 10 **b.** 2 **c.** 1 **d.** 1 **e.** −4 **f.** 9

1.5 Evaluating Algebraic Expressions

1.5 OBJECTIVES

1. Evaluate algebraic expressions given any signed number value for the variables
2. Use a calculator to evaluate algebraic expressions
3. Find the sum of a set of signed numbers
4. Interpret summation notation

In applying algebra to problem solving, you will often want to find the value of an algebraic expression when you know certain values for the letters (or variables) in the expression. Finding the value of an expression is called *evaluating the expression* and uses the following steps.

Step by Step: To Evaluate an Algebraic Expression

Step 1 Replace each variable by the given number value.

Step 2 Do the necessary arithmetic operations, following the rules for order of operations.

Example 1

Evaluating Algebraic Expressions

Suppose that $a = 5$ and $b = 7$.

(a) To evaluate $a + b$, we replace a with 5 and b with 7.

$a + b = 5 + 7 = 12$

(b) To evaluate $3ab$, we again replace a with 5 and b with 7.

$3ab = 3 \cdot 5 \cdot 7 = 105$

CHECK YOURSELF 1

If $x = 6$ and $y = 7$, evaluate.

(a) $y - x$ **(b)** $5xy$

We are now ready to evaluate algebraic expressions that require following the rules for the order of operations.

Example 2

Evaluating Algebraic Expressions

Evaluate the following expressions if $a = 2$, $b = 3$, $c = 4$, and $d = 5$.

CAUTION

This is different from
$(3c)^2 = (3 \cdot 4)^2$
$= 12^2 = 144$

(a) $5a + 7b = 5 \cdot 2 + 7 \cdot 3$ Multiply first.

$= 10 + 21 = 31$ Then add.

(b) $3c^2 = 3 \cdot 4^2$ Evaluate the power.

$= 3 \cdot 16 = 48$ Then multiply.

(c) $7(c + d) = 7(4 + 5)$ Add inside the parentheses.

$= 7 \cdot 9 = 63$

(d) $5a^4 - 2d^2 = 5 \cdot 2^4 - 2 \cdot 5^2$ Evaluate the powers.

$= 5 \cdot 16 - 2 \cdot 25$ Multiply.

$= 80 - 50 = 30$ Subtract.

CHECK YOURSELF 2

If $x = 3$, $y = 2$, $z = 4$, and $w = 5$, evaluate the following expressions.

(a) $4x^2 + 2$ **(b)** $5(z + w)$ **(c)** $7(z^2 - y^2)$

To evaluate algebraic expressions when a fraction bar is used, do the following: Start by doing all the work in the numerator, then do the work in the denominator. Divide the numerator by the denominator as the last step.

Example 3

Evaluating Algebraic Expressions

If $p = 2$, $q = 3$, and $r = 4$, evaluate:

(a) $\dfrac{8p}{r}$

NOTE As we mentioned in Section 1.4, the fraction bar is a grouping symbol, like parentheses. Work first in the numerator and then in the denominator.

Replace p with 2 and r with 4.

$\dfrac{8p}{r} = \dfrac{8 \cdot 2}{4} = \dfrac{16}{4} = 4$ Divide as the last step.

(b) $\dfrac{7q + r}{p + q} = \dfrac{7 \cdot 3 + 4}{2 + 3}$ Now evaluate the top and bottom separately.

$= \dfrac{21 + 4}{2 + 3} = \dfrac{25}{5} = 5$

CHECK YOURSELF 3

Evaluate the following if $c = 5$, $d = 8$, and $e = 3$.

(a) $\dfrac{6c}{e}$ **(b)** $\dfrac{4d + e}{c}$ **(c)** $\dfrac{10d - e}{d + e}$

Example 4 shows how a scientific calculator can be used to evaluate algebraic expressions.

Example 4

Using a Calculator to Evaluate Expressions

Use a scientific calculator to evaluate the following expressions.

(a) $\frac{4x + y}{z}$ if $x = 2$, $y = 1$, and $z = 3$

Replace x with 2, y with 1, and z with 3:

$$\frac{4x + y}{z} = \frac{4 \cdot 2 + 1}{3}$$

Now, use the following keystrokes:

[(] 4 [×] 2 [+] 1 [)] [÷] 3 [=]

The display will read 3.

(b) $\frac{7x - y}{3z - x}$ if $x = 2$, $y = 6$, and $z = 2$

$$\frac{7x - y}{3z - x} = \frac{7 \cdot 2 - 6}{3 \cdot 2 - 2}$$

Use the following keystrokes:

[(] 7 [×] 2 [−] 6 [)] [÷] [(] 3 [×] 2 − 2 [)] [=]

The display will read 2.

CHECK YOURSELF 4

Use a scientific calculator to evaluate the following if $x = 2$, $y = 6$, and $z = 5$.

(a) $\frac{2x + y}{z}$ **(b)** $\frac{4y - 2z}{x}$

Example 5

Evaluating Expressions

Evaluate $5a + 4b$ if $a = -2$ and $b = 3$.

Replace *a* with −2 and *b* with 3.

$$5a + 4b = 5(-2) + 4(3)$$
$$= -10 + 12$$
$$= 2$$

NOTE Remember the rules for the order of operations. Multiply first, then add.

CHECK YOURSELF 5

Evaluate $3x + 5y$ if $x = -2$ and $y = -5$.

We follow the same rules no matter how many variables are in the expression.

Example 6

Evaluating Expressions

Evaluate the following expressions if $a = -4$, $b = 2$, $c = -5$, and $d = 6$.

This becomes $-(-20)$, or $+20$.

(a) $7a - 4c = 7(-4) - 4(-5)$

$= -28 + 20$

$= -8$

Evaluate the power first, then multiply by 7.

(b) $7c^2 = 7(-5)^2 = 7 \cdot 25$

$= 175$

CAUTION

When a squared variable is replaced by a negative number, square the negative.

$(-5)^2 = (-5)(-5) = 25$

The exponent applies to -5!

$-5^2 = -(5 \cdot 5) = -25$

The exponent applies only to 5!

(c) $b^2 - 4ac = 2^2 - 4(-4)(-5)$

$= 4 - 4(-4)(-5)$

$= 4 - 80$

$= -76$

Add inside the parentheses first.

(d) $b(a + d) = 2(-4 + 6)$

$= 2(2)$

$= 4$

CHECK YOURSELF 6

Evaluate if $p = -4$, $q = 3$, *and* $r = -2$.

(a) $5p - 3r$ **(b)** $2p^2 + q$ **(c)** $p(q + r)$
(d) $-q^2$ **(e)** $(-q)^2$

If an expression involves a fraction, remember that the fraction bar is a grouping symbol. This means that you should do the required operations first in the numerator and then the denominator. Divide as the last step.

Example 7

Evaluating Expressions

Evaluate the following expressions if $x = 4$, $y = -5$, $z = 2$, and $w = -3$.

(a) $\frac{z - 2y}{x} = \frac{2 - 2(-5)}{4} = \frac{2 + 10}{4}$

$= \frac{12}{4} = 3$

(b) $\frac{3x - w}{2x + w} = \frac{3(4) - (-3)}{2(4) + (-3)} = \frac{12 + 3}{8 + (-3)}$

$= \frac{15}{5} = 3$

CHECK YOURSELF 7

Evaluate if $m = -6$, $n = 4$, and $p = -3$.

(a) $\frac{m + 3n}{p}$ (b) $\frac{4m + n}{m + 4n}$

When an expression is evaluated by a calculator, the same order of operations that we introduced in Section 0.3 is followed.

	Algebraic Notation	Calculator Notation
Addition	$6 + 2$	6 [+] 2
Subtraction	$4 - 8$	4 [−] 8
Multiplication	$(3)(-5)$	3 [×] [(−)] 5 or 3 [×] 5 [+/−]
Division	$\frac{8}{6}$	8 [÷] 6
Exponential	3^4	3^4 or 3 [y^x] 4

In many applications, you will need to find the sum of a set of numbers that you are working with. In mathematics, the shorthand symbol for "sum of" is the Greek letter Σ (capital sigma, the "S" of the Greek alphabet). The expression Σx, in which x refers to all the numbers in a given set, means the sum of all the numbers in that set.

Example 8

Summing a Set

Find Σx for the following set of numbers:

$-2, -6, 3, 5, -4$

$$\Sigma x = -2 + (-6) + 3 + 5 + (-4)$$
$$= (-8) + 3 + 5 + (-4)$$
$$= (-8) + 8 + (-4)$$
$$= -4$$

CHECK YOURSELF 8

Find Σx for each set of numbers.

(a) $-3, 4, -7, -9, 8$ **(b)** $-2, 6, -5, -3, 4, 7$

CHECK YOURSELF ANSWERS

1. **(a)** 1; **(b)** 210 **2.** **(a)** 38; **(b)** 45; **(c)** 84 **3.** **(a)** 10; **(b)** 7; **(c)** 7
4. **(a)** 2; **(b)** 7 **5.** -31 **6.** **(a)** -14; **(b)** 35; **(c)** -4; **(d)** -9; **(e)** 9
7. **(a)** -2; **(b)** -2 **8.** **(a)** -7; **(b)** 7

Name ______________

Section ________ Date ________

1.5 Exercises

Evaluate each of the expressions if $a = -2$, $b = 5$, $c = -4$, and $d = 6$.

1. $3c - 2b$

2. $4c - 2b$

3. $8b + 2c$

4. $7a - 2c$

5. $-b^2 + b$

6. $(-b)^2 + b$

7. $3a^2$

8. $6c^2$

9. $c^2 - 2d$

10. $3a^2 + 4c$

11. $2a^2 + 3b^2$

12. $4b^2 - 2c^2$

13. $2(a + b)$

14. $5(b - c)$

15. $4(2a - d)$

16. $6(3c - d)$

17. $a(b + 3c)$

18. $c(3a - d)$

19. $\dfrac{6d}{c}$

20. $\dfrac{8b}{5c}$

21. $\dfrac{3d + 2c}{b}$

22. $\dfrac{2b + 3d}{2a}$

ANSWERS

1. −22
2. −26
3. 32
4. −6
5. −20
6. 30
7. 12
8. 96
9. 4
10. −4
11. 83
12. 68
13. 6
14. 45
15. −40
16. −108
17. 14
18. 48
19. −9
20. −2
21. 2
22. −7

ANSWERS

23. 2
24. −2
25. 11
26. 12
27. 1
28. 4
29. 11
30. 12
31. 91
32. −72
33. 1
34. −216
35. 91
36. −72
37. 29
38. 32
39. 9
40. 64
41. 16
42. 81
43. −15.3
44. −11.4
45. −11.5
46. 15.3

23. $\dfrac{2b - 3a}{c + 2d}$

24. $\dfrac{3d - 2b}{5a + d}$

25. $d^2 - b^2$

26. $c^2 - a^2$

27. $(d - b)^2$

28. $(c - a)^2$

29. $(d - b)(d + b)$

30. $(c - a)(c + a)$

31. $d^3 - b^3$

32. $c^3 + a^3$

33. $(d - b)^3$

34. $(c + a)^3$

35. $(d - b)(d^2 + db + b^2)$

36. $(c + a)(c^2 - ac + a^2)$

37. $b^2 + a^2$

38. $d^2 - a^2$

39. $(b + a)^2$

40. $(d - a)^2$

41. $a^2 + 2ad + d^2$

42. $b^2 - 2bc + c^2$

Use your calculator to evaluate each expression if $x = -2.34$, $y = -3.14$, and $z = 4.12$. Round your answer to the nearest tenth.

43. $x + yz$

44. $y - 2z$

45. $x^2 - z^2$

46. $x^2 + y^2$

47. $\dfrac{xy}{z - x}$

48. $\dfrac{y^2}{zy}$

49. $\dfrac{2x + y}{2x + z}$

50. $\dfrac{x^2y^2}{xz}$

For the following data sets, evaluate Σx.

51. 1, 2, 3, 7, 8, 9, 11

52. 2, 4, 5, 6, 10, 11, 12

53. −5, −3, −1, 2, 3, 4, 8

54. −4, −2, −1, 5, 7, 8, 10

55. 3, 2, −1, −4, −3, 8, 6

56. 3, −4, 2, −1, 2, −7, 9

57. $-\frac{1}{2}, -\frac{3}{4}, 2, 3, \frac{1}{4}, \frac{3}{2}, -1$

58. $-\frac{1}{3}, -\frac{5}{3}, -1, 1, 3, \frac{2}{3}, \frac{5}{3}$

59. −2.5, −3.2, 2.6, −1, 2, 4, −3

60. −2.4, −3.1, −1.7, 3, 1, 2, 5

In each of the following problems, decide if the given values make the statement true or false.

61. $x - 7 = 2y + 5$; $x = 22$, $y = 5$

62. $3(x - y) = 6$; $x = 5$, $y = -3$

63. $2(x + y) = 2x + y$; $x = -4$, $y = -2$

64. $x^2 - y^2 = x - y$; $x = 4$, $y = -3$

65. **Electrical resistance.** The formula for the total resistance in a parallel circuit is given by the formula $R_T = R_1R_2/(R_1 + R_2)$. Find the total resistance if $R_1 = 6$ ohms (Ω) and $R_2 = 10\ \Omega$.

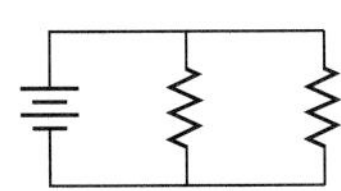

66. **Area.** The formula for the area of a triangle is given by $A = \frac{1}{2}ab$. Find the area of a triangle if $a = 4$ centimeters (cm) and $b = 8$ cm.

ANSWERS

47. 1.1
48. −0.8
49. 14.0
50. −5.6
51. 41
52. 50
53. 8
54. 23
55. 11
56. 4
57. $\frac{9}{2}$
58. $\frac{10}{3}$
59. −1.1
60. 3.8
61. True
62. False
63. False
64. True
65. 3.75 Ω
66. 16 cm^2

ANSWERS

67. 30 in.

68. \$1440

69. \$1875

70. 5%

71. 14°F

72. 28.26 m^2

73.

67. Perimeter. The perimeter of a rectangle of length L and width W is given by the formula $P = 2L + 2W$. Find the perimeter when $L = 10$ inches (in.) and $W = 5$ in.

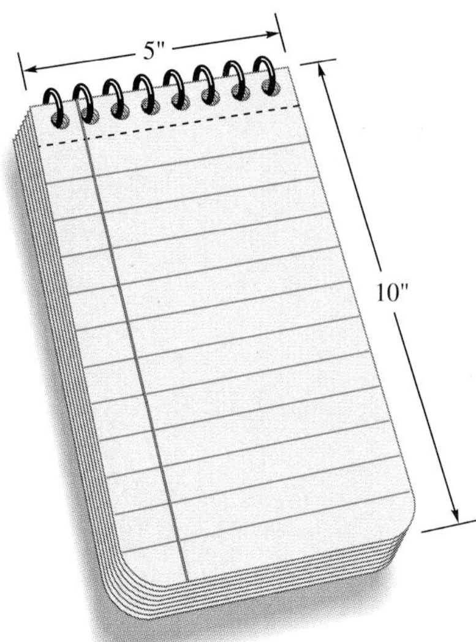

68. Simple interest. The simple interest I on a principal of P dollars at interest rate r for time t, in years, is given by $I = Prt$. Find the simple interest on a principal of \$6000 at 8 percent for 3 years. (**Note:** 8% = 0.08)

69. Simple interest. Use the simple interest formula to find the principal if the total interest earned was \$150 and the rate of interest was 4% for 2 years.

70. Simple interest. Use the simple interest formula to find the rate of interest if \$10,000 earns \$1500 interest in 3 years.

71. Temperature conversion. The formula that relates Celsius and Fahrenheit temperature is $F = \frac{9}{5}C + 32$. If the temperature of the day is -10°C, what is the Fahrenheit temperature?

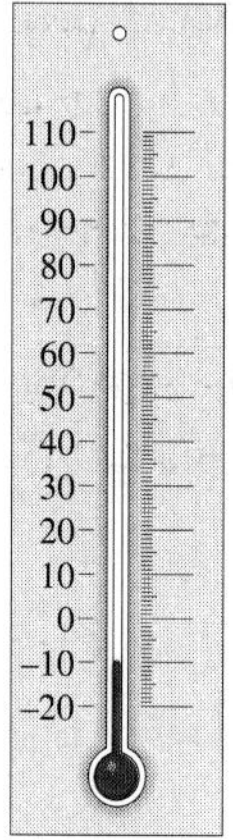

72. Geometry. If the area of a circle whose radius is r is given by $A = \pi r^2$, where $\pi = 3.14$, find the area when $r = 3$ meters (m).

73. Write an English interpretation of each of the following algebraic expressions.

(a) $(2x^2 - y)^3$ **(b)** $3n - \frac{n-1}{2}$ **(c)** $(2n + 3)(n - 4)$

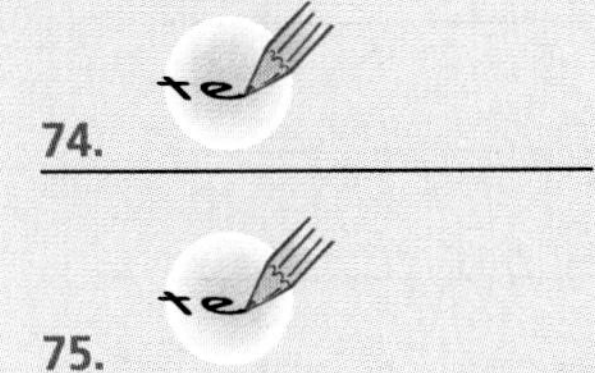

74. ______

75. ______

74. Is $a^n + b^n = (a + b)^n$? Try a few numbers and decide if you think this is true for all numbers, for some numbers, or never true. Write an explanation of your findings and give examples.

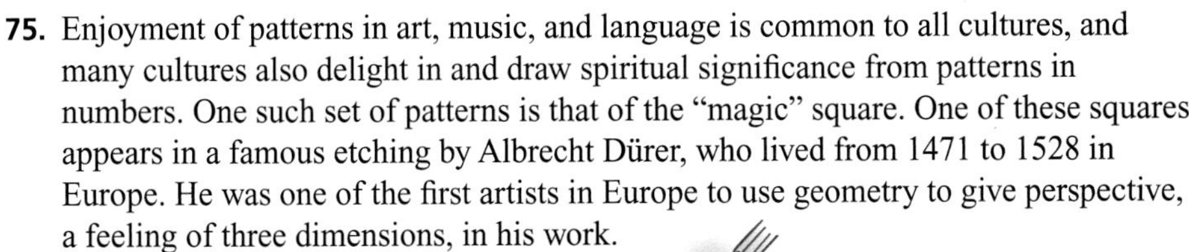

75. Enjoyment of patterns in art, music, and language is common to all cultures, and many cultures also delight in and draw spiritual significance from patterns in numbers. One such set of patterns is that of the "magic" square. One of these squares appears in a famous etching by Albrecht Dürer, who lived from 1471 to 1528 in Europe. He was one of the first artists in Europe to use geometry to give perspective, a feeling of three dimensions, in his work.

The magic square in his work is this one:

16	3	2	13
5	10	11	8
9	6	7	12
4	15	15	1

Why is this square "magic"? It is magic because every row, every column, and both diagonals add to the same number. In this square there are sixteen spaces for the numbers 1 through 16.

Part 1: What number does each row and column add to?

Write the square that you obtain by adding -17 to each number. Is this still a magic square? If so, what number does each column and row add to? If you add 5 to each number in the original magic square, do you still have a magic square? You have been studying the operations of addition, multiplication, subtraction, and division with integers and with rational numbers. What operations can you perform on this magic square and still have a magic square? Try to find something that will not work. Use algebra to help you decide what will work and what won't. Write a description of your work and explain your conclusions.

Part 2: Here is the oldest published magic square. It is from China, about 250 B.C.E. Legend has it that it was brought from the River Lo by a turtle to the Emperor Yii, who was a hydraulic engineer.

4	9	2
3	5	7
8	1	6

Check to make sure that this is a magic square. Work together to decide what operation might be done to every number in the magic square to make the sum of each row, column, and diagonal the *opposite* of what it is now. What would you do to every number to cause the sum of each row, column, and diagonal to equal zero?

ANSWERS

a. 12

b. 0

c. 26

d. 17

e. 8

f. 64

Getting Ready for Section 1.6 [Sections 1.3 and 1.4]

(a) $(8 + 9) - 5$

(b) $15 - 4 - 11$

(c) $5(4 + 3) - 9$

(d) $-3(5 - 7) + 11$

(e) $-6(-9 + 7) - 4$

(f) $8 - 7(-2 - 6)$

Answers

1. -22 **3.** 32 **5.** -20 **7.** 12 **9.** 4 **11.** 83 **13.** 6 **15.** -40 **17.** 14 **19.** -9 **21.** 2 **23.** 2 **25.** 11 **27.** 1 **29.** 11 **31.** 91 **33.** 1 **35.** 91 **37.** 29 **39.** 9 **41.** 16 **43.** -15.3 **45.** -11.5 **47.** 1.1 **49.** 14.0 **51.** 41 **53.** 8 **55.** 11 **57.** $\frac{9}{2}$ **59.** -1.1 **61.** True **63.** False **65.** 3.75 Ω **67.** 30 in. **69.** $1875 **71.** 14°F **73.** **75.**

a. 12 **b.** 0 **c.** 26 **d.** 17 **e.** 8 **f.** 64

Adding and Subtracting Terms

OBJECTIVES

1. Identify terms and like terms
2. Combine like terms
3. Add algebraic expressions
4. Subtract algebraic expressions

To find the perimeter of (or the distance around) a rectangle, we add 2 times the length and 2 times the width. In the language of algebra, this can be written as

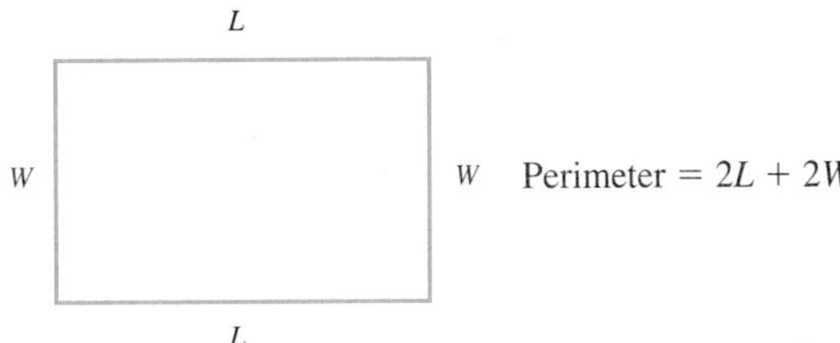

Perimeter $= 2L + 2W$

We call $2L + 2W$ an **algebraic expression,** or more simply an **expression.** Recall from Section 1.1 that an expression allows us to write a mathematical idea in symbols. It can be thought of as a meaningful collection of letters, numbers, and operation signs.

Some expressions are

$5x^2$ $\qquad$ $3a + 2b$ $\qquad$ $4x^3 + (-2y) + 1$

In algebraic expressions, the addition and subtraction signs break the expressions into smaller parts called *terms.*

Definitions: Term

A **term** is a number, or the product of a number and one or more variables, raised to a power.

In an expression, each sign (+ or −) is a part of the term that follows the sign.

Example 1

Identifying Terms

(a) $5x^2$ has one term.

(b) $\underbrace{3a}_{\text{Term}} + \underbrace{2b}_{\text{Term}}$ has two terms: $3a$ and $2b$.

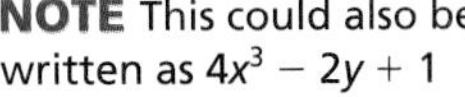

NOTE This could also be written as $4x^3 - 2y + 1$

(c) $\underbrace{4x^3}_{\text{Term}} + \underbrace{(-2y)}_{\text{Term}} + \underbrace{1}_{\text{Term}}$ has three terms: $4x^3$, $-2y$, and 1.

CHECK YOURSELF 1

List the terms of each expression.

(a) $2b^4$ $\qquad$ **(b)** $5m + 3n$ $\qquad$ **(c)** $2s^2 - 3t - 6$

Note that a term in an expression may have any number of factors. For instance, $5xy$ is a term. It has factors of 5, x, and y. The number factor of a term is called the **numerical coefficient.** So for the term $5xy$, the numerical coefficient is 5.

Example 2

Identifying the Numerical Coefficient

(a) $4a$ has the numerical coefficient 4.

(b) $6a^3b^4c^2$ has the numerical coefficient 6.

(c) $-7m^2n^3$ has the numerical coefficient -7.

(d) Because $1 \cdot x = x$, the numerical coefficient of x is understood to be 1.

CHECK YOURSELF 2

Give the numerical coefficient for each of the following terms.

(a) $8a^2b$ **(b)** $-5m^3n^4$ **(c)** y

If terms contain exactly the *same letters* (or variables) raised to the *same powers*, they are called **like terms.**

Example 3

Identifying Like Terms

(a) The following are like terms.

$6a$ and $7a$

$5b^2$ and b^2

$10x^2y^3z$ and $-6x^2y^3z$

$-3m^2$ and m^2

Each pair of terms has the same letters, with each letter raised to the same power—the numerical coefficients can be any number.

(b) The following are *not* like terms.

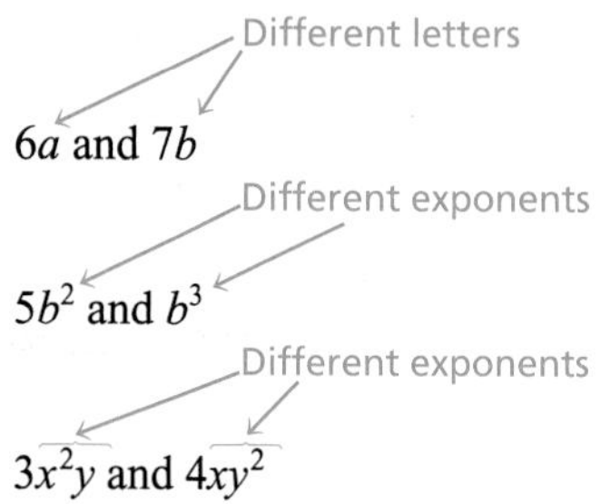

CHECK YOURSELF 3

Circle the like terms.

$5a^2b$ ab^2 a^2b $-3a^2$ $4ab$ $3b^2$ $-7a^2b$

Like terms of an expression can always be combined into a single term. Look at the following:

$$\overbrace{x + x}^{2x} + \overbrace{x + x + x + x + x}^{5x} = \overbrace{x + x + x + x + x + x + x}^{7x}$$

Rather than having to write out all those x's, try

NOTE Here we use the distributive property from Section 1.2.

$2x + 5x = (2 + 5)x = 7x$

In the same way,

NOTE You don't have to write all this out—just do it mentally!

$9b + 6b = (9 + 6)b = 15b$

and $10a + (-4a) = (10 + (-4))a = 6a$

This leads us to the following rule.

Step by Step: To Combine Like Terms

To combine like terms, use the following steps.

Step 1 Add or subtract the numerical coefficients.
Step 2 Attach the common variables.

Example 4

Combining Like Terms

Combine like terms.*

(a) $8m + 5m = (8 + 5)m = 13m$
(b) $5pq^3 - 4pq^3 = 5pq^3 + (-4pq^3) = 1pq^3 = pq^3$
(c) $7a^3b^2 - 7a^3b^2 = 7a^3b^2 + (-7a^3b^2) = 0a^3b^2 = 0$

NOTE Remember that when any factor is multiplied by 0, the product is 0.

CHECK YOURSELF 4

Combine like terms.

(a) $6b + 8b$
(b) $12x^2 - 3x^2$
(c) $8xy^3 - 7xy^3$
(d) $9a^2b^4 - 9a^2b^4$

Let's look at some expressions involving more than two terms. The idea is just the same.

Example 5

Combining Like Terms

NOTE The distributive property can be used over any number of like terms.

Combine like terms.

(a) $5ab - 2ab + 3ab$

$= 5ab + (-2ab) + 3ab$

$= (5 + (-2) + 3)ab = 6ab$

*When an example requires simplification of an expression, that expression will be screened. The simplification will then follow the equals sign.

Only like terms can be combined.

(b) $8x - 2x \quad + 5y$

$(8 + (-2))\,x + 5y$

$= 6x \quad + 5y$

Like terms   Like terms

NOTE With practice you won't be writing out these steps, but doing it mentally.

(c) $5m + 8n \quad + 4m - 3n$

$= (5m + 4m) + (8n + (-3n))$

$= \quad 9m \quad + \quad 5n$

Here we have used the associative and commutative properties.

(d) $4x^2 + 2x - 3x^2 + x$

$= (4x^2 + (-3x^2)) + (2x + x)$

$= x^2 + 3x$

As these examples illustrate, combining like terms often means changing the grouping and the order in which the terms are written. Again all this is possible because of the properties of addition that we introduced in Section 1.2.

CHECK YOURSELF 5

Combine like terms.

(a) $4m^2 - 3m^2 + 8m^2$ **(b)** $9ab + 3a - 5ab$ **(c)** $4p + 7q + 5p - 3q$

As you have seen in arithmetic, subtraction can be performed directly. As this is the form used for most of mathematics, we will use that form throughout this text. Just remember, by using negative numbers, you can always rewrite a subtraction problem as an addition problem.

Example 6

Combining Like Terms

Combine the like terms.

(a) $2xy - 3xy + 5xy$

$= (2 - 3 + 5)xy$

$= 4xy$

(b) $5a - 2b + 7b - 8a$

$= (5a - 8a) + (-2b + 7b)$

$= -3a + 5b$

CHECK YOURSELF 6

Combine like terms.

(a) $4ab + 5ab - 3ab - 7ab$ **(b)** $2x - 7y - 8x - y$

CHECK YOURSELF ANSWERS

1. **(a)** $2b^4$; **(b)** $5m$, $3n$; **(c)** $2s^2$, $-3t$, -6 **2.** **(a)** 8; **(b)** -5; **(c)** 1

3. The like terms are $5a^2b$, a^2b, and $-7a^2b$ **4.** **(a)** $14b$; **(b)** $9x^2$; **(c)** xy^3; **(d)** 0

5. **(a)** $9m^2$; **(b)** $4ab + 3a$; **(c)** $9p + 4q$ **6.** **(a)** $-ab$; **(b)** $-6x - 8y$

Name ____________

Section ________ Date ________

1.6 Exercises

List the terms of the following expressions.

1. $5a + 2$
2. $7a - 4b$
3. $4x^3$
4. $3x^2$
5. $3x^2 + 3x - 7$
6. $2a^3 - a^2 + a$

Circle the like terms in the following groups of terms.

7. $5ab, 3b, 3a, 4ab$
8. $9m^2, 8mn, 5m^2, 7m$
9. $4xy^2, 2x^2y, 5x^2, -3x^2y, 5y, 6x^2y$
10. $8a^2b, 4a^2, 3ab^2, -5a^2b, 3ab, 5a^2b$

Combine the like terms.

11. $3m + 7m$
12. $6a^2 + 8a^2$
13. $7b^3 + 10b^3$
14. $7rs + 13rs$
15. $21xyz + 7xyz$
16. $4mn^2 + 15mn^2$
17. $9z^2 - 3z^2$
18. $7m - 6m$
19. $5a^3 - 5a^3$
20. $13xy - 9xy$
21. $19n^2 - 18n^2$
22. $7cd - 7cd$
23. $21p^2q - 6p^2q$
24. $17r^3s^2 - 8r^3s^2$
25. $10x^2 - 7x^2 + 3x^2$
26. $13uv + 5uv - 12uv$

ANSWERS

1. 5a, 2
2. 7a, −4b
3. $4x^3$
4. $3x^2$
5. $3x^2, 3x, -7$
6. $2a^3, -a^2, a$
7. 5ab, 4ab
8. $9m^2, 5m^2$
9. $2x^2y, -3x^2y, 6x^2y$
10. $8a^2b, -5a^2b, 5a^2b$
11. 10m
12. $14a^2$
13. $17b^3$
14. 20rs
15. 28xyz
16. $19mn^2$
17. $6z^2$
18. m
19. 0
20. 4xy
21. n^2
22. 0
23. $15p^2q$
24. $9r^3s^2$
25. $6x^2$
26. 6uv

ANSWERS

a. 4^3

b. 6^6

c. 3^5

d. $(-2)^3$

e. $(-8)^4$

f. 9^8

Getting Ready for Section 1.7 [Section 0.3]

Write the following using exponential notation.

(a) $4 \cdot 4 \cdot 4$

(b) $6 \cdot 6 \cdot 6 \cdot 6 \cdot 6 \cdot 6$

(c) $3 \cdot 3 \cdot 3 \cdot 3 \cdot 3$

(d) $(-2) \cdot (-2) \cdot (-2)$

(e) $(-8) \cdot (-8) \cdot (-8) \cdot (-8)$

(f) $9 \cdot 9 \cdot 9 \cdot 9 \cdot 9 \cdot 9 \cdot 9 \cdot 9$

Answers

1. $5a, 2$ **3.** $4x^3$ **5.** $3x^2, 3x, -7$ **7.** $5ab, 4ab$ **9.** $2x^2y, -3x^2y, 6x^2y$ **11.** $10m$ **13.** $17b^3$ **15.** $28xyz$ **17.** $6z^2$ **19.** 0 **21.** n^2 **23.** $15p^2q$ **25.** $6x^2$ **27.** $2a + 4b$ **29.** $3x + y$ **31.** $2a + 10b + 1$ **33.** $2m + 3$ **35.** $2x + 7$ **37.** $7a + 10$ **39.** $13a^4$ **41.** $3a^3$ **43.** $7x$ **45.** $11mn^2$ **47.** $6x + 8$ **49.** $42a - 10$ **51.** $6s + 12$

53. **55.**  **57.** **a.** 4^3 **b.** 6^6 **c.** 3^5

d. $(-2)^3$ **e.** $(-8)^4$ **f.** 9^8

Name ______________________

Section ________ Date ________

1.6 Exercises

List the terms of the following expressions.

1. $5a + 2$

2. $7a - 4b$

3. $4x^3$

4. $3x^2$

5. $3x^2 + 3x - 7$

6. $2a^3 - a^2 + a$

Circle the like terms in the following groups of terms.

7. $5ab, 3b, 3a, 4ab$

8. $9m^2, 8mn, 5m^2, 7m$

9. $4xy^2, 2x^2y, 5x^2, -3x^2y, 5y, 6x^2y$

10. $8a^2b, 4a^2, 3ab^2, -5a^2b, 3ab, 5a^2b$

Combine the like terms.

11. $3m + 7m$

12. $6a^2 + 8a^2$

13. $7b^3 + 10b^3$

14. $7rs + 13rs$

15. $21xyz + 7xyz$

16. $4mn^2 + 15mn^2$

17. $9z^2 - 3z^2$

18. $7m - 6m$

19. $5a^3 - 5a^3$

20. $13xy - 9xy$

21. $19n^2 - 18n^2$

22. $7cd - 7cd$

23. $21p^2q - 6p^2q$

24. $17r^3s^2 - 8r^3s^2$

25. $10x^2 - 7x^2 + 3x^2$

26. $13uv + 5uv - 12uv$

ANSWERS

1. $5a, 2$
2. $7a, -4b$
3. $4x^3$
4. $3x^2$
5. $3x^2, 3x, -7$
6. $2a^3, -a^2, a$
7. $5ab, 4ab$
8. $9m^2, 5m^2$
9. $2x^2y, -3x^2y, 6x^2y$
10. $8a^2b, -5a^2b, 5a^2b$
11. $10m$
12. $14a^2$
13. $17b^3$
14. $20rs$
15. $28xyz$
16. $19mn^2$
17. $6z^2$
18. m
19. 0
20. $4xy$
21. n^2
22. 0
23. $15p^2q$
24. $9r^3s^2$
25. $6x^2$
26. $6uv$

ANSWERS

27. $2a + 4b$
28. $11m^2 - 3m$
29. $3x + y$
30. $13a^2 + 2a$
31. $2a + 10b + 1$
32. $9p^2 + 7p + 2$
33. $2m + 3$
34. $a - 2$
35. $2x + 7$
36. $2y + 4$
37. $7a + 10$
38. $3m + 15$
39. $13a^4$
40. $21p^2$
41. $3a^3$
42. $13m^3$
43. $7x$
44. $4ab$
45. $11mn^2$
46. $14x^2y$
47. $6x + 8$
48. $12z + 6$
49. $42a - 10$
50. $3w - 21$
51. $6s + 12$
52. $9p + 4$

27. $9a - 7a + 4b$

28. $5m^2 - 3m + 6m^2$

29. $7x + 5y - 4x - 4y$

30. $6a^2 + 11a + 7a^2 - 9a$

31. $4a + 7b + 3 - 2a + 3b - 2$

32. $5p^2 + 2p + 8 + 4p^2 + 5p - 6$

33. $\frac{2}{3}m + 3 + \frac{4}{3}m$

34. $\frac{1}{5}a - 2 + \frac{4}{5}a$

35. $\frac{13}{5}x + 2 - \frac{3}{5}x + 5$

36. $\frac{17}{12}y + 7 + \frac{7}{12}y - 3$

37. $2.3a + 7 + 4.7a + 3$

38. $5.8m + 4 - 2.8m + 11$

Perform the indicated operations.

39. Find the sum of $5a^4$ and $8a^4$.

40. Find the sum of $9p^2$ and $12p^2$.

41. Subtract $12a^3$ from $15a^3$.

42. Subtract $5m^3$ from $18m^3$.

43. Subtract $4x$ from the sum of $8x$ and $3x$.

44. Subtract $8ab$ from the sum of $7ab$ and $5ab$.

45. Subtract $3mn^2$ from the sum of $9mn^2$ and $5mn^2$.

46. Subtract $4x^2y$ from the sum of $6x^2y$ and $12x^2y$.

Use the distributive property to remove the parentheses in each expression. Then simplify by combining like terms.

47. $2(3x + 2) + 4$

48. $3(4z + 5) - 9$

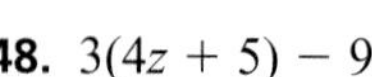

49. $5(6a - 2) + 12a$

50. $7(4w - 3) - 25w$

51. $4s + 2(s + 4) + 4$

52. $5p + 4(p + 3) - 8$

53. Write a paragraph explaining the difference between n^2 and $2n$. 

54. Complete the explanation: "x^3 and $3x$ are not the same because . . ." 

55. Complete the statement: "$x + 2$ and $2x$ are different because . . ." 

56. Write an English phrase for each algebraic expression below: 

(a) $2x^3 + 5x$ **(b)** $(2x + 5)^3$ **(c)** $6(n + 4)^2$

57. Work with another student to complete this exercise. Place >, <, or = in the blank in these statements.

1^2____2^1 What happens as the table of numbers is extended? Try more examples.

2^3____3^2

3^4____4^3 What sign seems to occur the most in your table? >, <, or =?

4^5____5^4 Write an algebraic statement for the pattern of numbers in this table. Do you think this is a pattern that continues? Add more lines to the table and extend the pattern to the general case by writing the pattern in algebraic notation. Write a short paragraph stating your conjecture.

58. Work with other students on this exercise. 

Part 1: Evaluate the three expressions $\frac{n^2 - 1}{2}$, n, $\frac{n^2 + 1}{2}$ using odd values of n: 1, 3, 5, 7, etc. Make a chart like the one below and complete it.

n	$a = \frac{n^2 - 1}{2}$	$b = n$	$c = \frac{n^2 + 1}{2}$	a^2	b^2	c^2
1						
3						
5						
7						
9						
11						
13						

Part 2: The numbers a, b, and c that you get in each row have a surprising relationship to each other. Complete the last three columns and work together to discover this relationship. You may want to find out more about the history of this famous number pattern.

ANSWERS

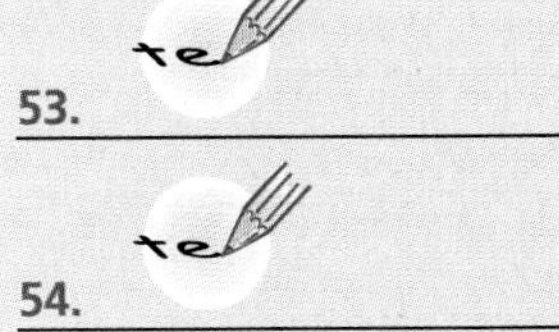

53. ________

54. ________

55. ________

56. ________

57. ________

58. ________

ANSWERS

a. 4^3

b. 6^6

c. 3^5

d. $(-2)^3$

e. $(-8)^4$

f. 9^8

Getting Ready for Section 1.7 [Section 0.3]

Write the following using exponential notation.

(a) $4 \cdot 4 \cdot 4$

(b) $6 \cdot 6 \cdot 6 \cdot 6 \cdot 6 \cdot 6$

(c) $3 \cdot 3 \cdot 3 \cdot 3 \cdot 3$

(d) $(-2) \cdot (-2) \cdot (-2)$

(e) $(-8) \cdot (-8) \cdot (-8) \cdot (-8)$

(f) $9 \cdot 9 \cdot 9 \cdot 9 \cdot 9 \cdot 9 \cdot 9 \cdot 9$

Answers

1. $5a, 2$ **3.** $4x^3$ **5.** $3x^2, 3x, -7$ **7.** $5ab, 4ab$ **9.** $2x^2y, -3x^2y, 6x^2y$ **11.** $10m$ **13.** $17b^3$ **15.** $28xyz$ **17.** $6z^2$ **19.** 0 **21.** n^2 **23.** $15p^2q$ **25.** $6x^2$ **27.** $2a + 4b$ **29.** $3x + y$ **31.** $2a + 10b + 1$ **33.** $2m + 3$ **35.** $2x + 7$ **37.** $7a + 10$ **39.** $13a^4$ **41.** $3a^3$ **43.** $7x$ **45.** $11mn^2$ **47.** $6x + 8$ **49.** $42a - 10$ **51.** $6s + 12$

53. **55.** **57.** **a.** 4^3 **b.** 6^6 **c.** 3^5 **d.** $(-2)^3$ **e.** $(-8)^4$ **f.** 9^8

Multiplying and Dividing Terms

OBJECTIVES

1. Find the product of two algebraic terms
2. Find the quotient of two algebraic terms

In Section 0.3, we introduced exponential notation. Remember that the exponent tells us how many times the base is to be used as a factor.

Exponent

$$2^5 = 2 \cdot 2 \cdot 2 \cdot 2 \cdot 2 = 32$$

Base — The fifth power of 2

NOTE In general,

$$x^m = \underbrace{x \cdot x \cdot \cdots \cdot x}_{m \text{ factors}}$$

where m is a natural number. **Natural numbers** are the numbers we use for counting: 1, 2, 3, and so on.

The notation can also be used when you are working with letters or variables.

$$x^4 = \underbrace{x \cdot x \cdot x \cdot x}_{4 \text{ factors}}$$

Now look at the product $x^2 \cdot x^3$.

$$x^2 \cdot x^3 = (x \cdot x)(x \cdot x \cdot x) = x \cdot x \cdot x \cdot x \cdot x = x^5$$

2 factors + 3 factors = 5 factors

So

$$x^2 \cdot x^3 = x^{2+3} = x^5$$

NOTE Note that the exponent of x^5 is the *sum* of the exponents in x^2 and x^3.

This leads us to the following property of exponents.

Rules and Properties: Property 1 of Exponents

For any positive integers m and n and any real number a,

$$a^m \cdot a^n = a^{m+n}$$

In words, to multiply expressions with the same base, keep the base and add the exponents.

Example 1

Using the First Property of Exponents

(a) $a^5 \cdot a^7 = a^{5+7} = a^{12}$

(b) $x \cdot x^8 = x^1 \cdot x^8 = x^{1+8} = x^9$ $x = x^1$

(c) $3^2 \cdot 3^4 = 3^{2+4} = 3^6$

(d) $y^2 \cdot y^3 \cdot y^5 = y^{2+3+5} = y^{10}$

(e) $x^3 \cdot y^4$ *cannot* be simplified. The bases are not the same.

CAUTION

The product is *not* 9^6. The base does not change.

CHECK YOURSELF 1

Multiply.

(a) $b^6 \cdot b^8$ **(b)** $y^7 \cdot y$ **(c)** $2^3 \cdot 2^4$ **(d)** $a^2 \cdot a^4 \cdot a^3$

Suppose that numerical coefficients (other than 1) are involved in a product. To find the product, multiply the coefficients and then use the first property of exponents to combine the variables.

NOTE Note that although we have several factors, this is still a single term.

$$
\begin{aligned}
2x^3 \cdot 3x^5 &= (2 \cdot 3)(x^3 \cdot x^5) && \text{Multiply the coefficients.} \\
&= 6x^{3+5} && \text{Add the exponents.} \\
&= 6x^8
\end{aligned}
$$

You may have noticed that we have again changed the order and grouping. This method uses the commutative and associative properties of Section 1.2.

Example 2

Using the First Property of Exponents

Multiply.

NOTE Again we have written out all the steps. You can do the multiplication mentally with practice.

(a) $5a^4 \cdot 7a^6 = (5 \cdot 7)(a^4 \cdot a^6) = 35a^{10}$

(b) $y^2 \cdot 3y^3 \cdot 6y^4 = (1 \cdot 3 \cdot 6)(y^2 \cdot y^3 \cdot y^4) = 18y^9$

(c) $2x^2y^3 \cdot 3x^5y^2 = (2 \cdot 3)(x^2 \cdot x^5)(y^3 \cdot y^2) = 6x^7y^5$

CHECK YOURSELF 2

Multiply.

(a) $4x^3 \cdot 7x^5$ **(b)** $3a^2 \cdot 2a^4 \cdot 2a^5$ **(c)** $3m^2n^4 \cdot 5m^3n$

What about dividing expressions when exponents are involved? For instance, what if we want to divide x^5 by x^2? We can use the following approach to division:

$$\frac{x^5}{x^2} = \frac{\overbrace{x \cdot x \cdot x \cdot x \cdot x}^{5 \text{ factors}}}{\underbrace{x \cdot x}_{2 \text{ factors}}} = \frac{x \cdot x \cdot x \cdot x \cdot x}{x \cdot x}$$

We can divide by 2 factors of x.

$$= \overbrace{x \cdot x \cdot x}^{3 \text{ factors}} = x^3$$

So

NOTE Note that the exponent of x^3 is the *difference* of the exponents in x^5 and x^2.

$$\frac{x^5}{x^2} = x^{5-2} = x^3$$

This leads us to a second property of exponents.

Rules and Properties: Property 2 of Exponents

For any positive integers *m* and *n*, where *m* is greater than *n*, and any real number *a*, where *a* is not equal to zero,

$$\frac{a^m}{a^n} = a^{m-n}$$

In words, to divide expressions with the same base, keep the base and subtract the exponents.

Example 3

Using the Second Property of Exponents

Divide the following.

(a) $\frac{y^7}{y^3} = y^{7-3} = y^4$

(b) $\frac{m^6}{m} = \frac{m^6}{m^1} = m^{6-1} = m^5$

(c) $\frac{a^3b^5}{a^2b^2} = a^{3-2} \cdot b^{5-2} = ab^3$

Apply the second property to each variable separately.

CHECK YOURSELF 3

Divide.

(a) $\frac{m^9}{m^6}$ (b) $\frac{a^8}{a}$ (c) $\frac{a^3b^5}{a^2}$ (d) $\frac{r^5s^6}{r^3s^2}$

If numerical coefficients are involved, just divide the coefficients and then use the second law of exponents to divide the variables. Look at Example 4.

Example 4

Using the Second Property of Exponents

Divide the following.

Subtract the exponents.

(a) $\frac{6x^5}{3x^2} = 2x^{5-2} = 2x^3$

6 divided by 3

20 divided by 5

(b) $\frac{20a^7b^5}{5a^3b^4} = 4a^{7-3} \cdot b^{5-4}$

Again apply the second property to each variable separately.

$= 4a^4b$

CHECK YOURSELF 4

Divide.

(a) $\frac{4x^3}{2x}$ (b) $\frac{20a^6}{5a^2}$ (c) $\frac{24x^5y^3}{4x^2y^2}$

CHECK YOURSELF ANSWERS

1. **(a)** b^{14}; **(b)** y^8; **(c)** 2^7; **(d)** a^9 **2.** **(a)** $28x^8$; **(b)** $12a^{11}$; **(c)** $15m^5n^5$
3. **(a)** m^3; **(b)** a^7; **(c)** ab^5; **(d)** r^2s^4 **4.** **(a)** $2x^2$; **(b)** $4a^4$; **(c)** $6x^3y$

Name ______________________

Section ________ Date ________

1.7 Exercises

Multiply.

1. $x^5 \cdot x^7$
2. $b^2 \cdot b^4$
3. $5 \cdot 5^5$
4. $y^6 \cdot y^4$
5. $a^9 \cdot a$
6. $3^4 \cdot 3^5$
7. $z^{10} \cdot z^3$
8. $x^7 \cdot x$
9. $p^5 \cdot p^7$
10. $s^6 \cdot s^9$
11. $x^3y \cdot x^2y^4$
12. $m^2n^3 \cdot mn^4$
13. $w^5 \cdot w^2 \cdot w$
14. $x^5 \cdot x^4 \cdot x^6$
15. $m^3 \cdot m^2 \cdot m^4$
16. $r^3 \cdot r \cdot r^5$
17. $a^3b \cdot a^2b^2 \cdot ab^3$
18. $w^2z^3 \cdot wz \cdot w^3z^4$
19. $p^2q \cdot p^3q^5 \cdot pq^4$
20. $c^3d \cdot c^4d^2 \cdot cd^5$
21. $3a^6 \cdot 2a^3$
22. $5s^6 \cdot s^4$
23. $x^2 \cdot 3x^5$
24. $2m^4 \cdot 6m^7$

ANSWERS

1. x^{12}
2. b^6
3. 5^6
4. y^{10}
5. a^{10}
6. 3^9
7. z^{13}
8. x^8
9. p^{12}
10. s^{15}
11. x^5y^5
12. m^3n^7
13. w^8
14. x^{15}
15. m^9
16. r^9
17. a^6b^6
18. w^6z^8
19. p^6q^{10}
20. c^8d^8
21. $6a^9$
22. $5s^{10}$
23. $3x^7$
24. $12m^{11}$

ANSWERS

25. $20m^4n^5$
26. $42x^3y^9$
27. $54x^4y^6$
28. $50a^4b^5$
29. $6a^{12}$
30. $24x^{10}$
31. $24c^8d^5$
32. $15p^6q^6$
33. $30m^{10}$
34. $12a^{13}$
35. $30r^7s^5$
36. $36a^6b^6$
37. a^3
38. m^6
39. y^6
40. b^5
41. p^5
42. s^6
43. x^3y
44. s^2t^2
45. $2m^2$
46. $2x^4$
47. $4a^3$
48. $5x$
49. $2m^2n$
50. $5a^4b$
51. $4w^2z^4$
52. $6p^2q^6$

25. $5m^3n^2 \cdot 4mn^3$

26. $7x^2y^5 \cdot 6xy^4$

27. $6x^3y \cdot 9xy^5$

28. $5a^3b \cdot 10ab^4$

29. $2a^2 \cdot a^3 \cdot 3a^7$

30. $4x^5 \cdot 2x^3 \cdot 3x^2$

31. $3c^2d \cdot 4cd^3 \cdot 2c^5d$

32. $5p^2q \cdot p^3q^2 \cdot 3pq^3$

33. $5m^2 \cdot m^3 \cdot 2m \cdot 3m^4$

34. $3a^3 \cdot 2a \cdot a^4 \cdot 2a^5$

35. $2r^3s \cdot rs^2 \cdot 3r^2s \cdot 5rs$

36. $6a^2b \cdot ab \cdot 3ab^3 \cdot 2a^2b$

Divide.

37. $\frac{a^9}{a^6}$

38. $\frac{m^8}{m^2}$

39. $\frac{y^{10}}{y^4}$

40. $\frac{b^9}{b^4}$

41. $\frac{p^{15}}{p^{10}}$

42. $\frac{s^{18}}{s^{12}}$

43. $\frac{x^5y^3}{x^2y^2}$

44. $\frac{s^5t^4}{s^3t^2}$

45. $\frac{6m^3}{3m}$

46. $\frac{8x^5}{4x}$

47. $\frac{24a^7}{6a^4}$

48. $\frac{25x^9}{5x^8}$

49. $\frac{26m^8n}{13m^6}$

50. $\frac{30a^4b^5}{6b^4}$

51. $\frac{28w^3z^5}{7wz}$

52. $\frac{48p^6q^7}{8p^4q}$

53. $\dfrac{18x^3y^4z^5}{9xy^2z^2}$

54. $\dfrac{25a^5b^4c^3}{5a^4bc^2}$

Simplify each of the following expressions where possible.

55. $2a^3b \cdot 3a^2b$

56. $2xy^3 \cdot 3xy^2$

57. $2a^3b + 3a^2b$

58. $2xy^3 + 3xy^2$

59. $2x^2y^3 \cdot 3x^2y^3$

60. $5a^3b^2 \cdot 10a^3b^2$

61. $2x^2y^3 + 3x^2y^3$

62. $5a^3b^2 + 10a^3b^2$

63. $\dfrac{8a^2b \cdot 6a^2b}{2ab}$

64. $\dfrac{6x^2y^3 \cdot 9x^2y^3}{3x^2y^2}$

65. $\dfrac{8a^2b + 6a^2b}{2ab}$

66. $\dfrac{6x^2y^3 + 9x^2y^3}{3x^2y^2}$

67. Complete the following statements: 

(a) a^n is negative when ____________ because ____________.

(b) a^n is positive when ____________ because ____________.

(give all possibilities)

68. "Earn Big Bucks!" reads an ad for a job. "You will be paid 1 cent for the first day and 2 cents for the second day, 4 cents for the third day, 8 cents for the fourth day, and so on, doubling each day. Apply now!" What kind of deal is this—where is the big money offered in the headline? The fine print at the bottom of the ad says: "Highly qualified people may be paid $1,000,000 for the first 28 working days if they choose." Well, *that* does sound like big bucks! Work with other students to decide which method of payment is better and how much better. You may want to make a table and try to find a formula for the first offer.

69. An oil spill from a tanker in pristine Prince Williams Sound in Alaska begins in a circular shape only 2 ft across. The area of the circle is $A = \pi r^2$. Make a table to decide what happens to the area if the diameter is doubling each hour. How large will the spill be in 24 h?

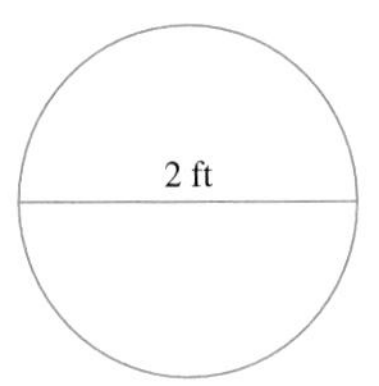

ANSWERS

53. $2x^2y^2z^3$

54. $5ab^3c$

55. $6a^5b^2$

56. $6x^2y^5$

57. Cannot simplify

58. Cannot simplify

59. $6x^4y^6$

60. $50a^6b^4$

61. $5x^2y^3$

62. $15a^3b^2$

63. $24a^3b$

64. $18x^2y^4$

65. $7a$

66. $5y$

67. ____________

68. ____________

69. ____________

Answers

1. x^{12} **3.** 5^6 **5.** a^{10} **7.** z^{13} **9.** p^{12} **11.** x^5y^5 **13.** w^8
15. m^9 **17.** a^6b^6 **19.** p^6q^{10} **21.** $6a^9$ **23.** $3x^7$ **25.** $20m^4n^5$
27. $54x^4y^6$ **29.** $6a^{12}$ **31.** $24c^8d^5$ **33.** $30m^{10}$ **35.** $30r^7s^5$ **37.** a^3
39. y^6 **41.** p^5 **43.** x^3y **45.** $2m^2$ **47.** $4a^3$ **49.** $2m^2n$
51. $4w^2z^4$ **53.** $2x^2y^2z^3$ **55.** $6a^5b^2$ **57.** Cannot simplify **59.** $6x^4y^6$
61. $5x^2y^3$ **63.** $24a^3b$ **65.** $7a$ **67.** **69.**

Summary

DEFINITION/PROCEDURE	EXAMPLE	REFERENCE
From Arithmetic to Algebra		**Section 1.1**
Addition $x + y$ means the **sum** of x **and** y or x **plus** y. Some other words indicating addition are "more than" and "increased by."	The sum of x and 5 is $x + 5$. 7 more than a is $a + 7$. b increased by 3 is $b + 3$.	p. 54
Subtraction $x - y$ means the **difference** of x **and** y or x **minus** y. Some other words indicating subtraction are "less than" and "decreased by."	The difference of x and 3 is $x - 3$. 5 less than p is $p - 5$. a decreased by 4 is $a - 4$.	p. 54
Multiplication $\left.\begin{array}{l} x \cdot y \\ (x)(y) \\ xy \end{array}\right\}$ These all mean the *product* of x and y or x *times* y.	The product of m and n is mn. The product of 2 and the sum of a and b is $2(a + b)$.	p. 55
Division $\frac{x}{y}$ means x *divided by* y or the *quotient* when x is divided by y.	n divided by 5 is $\frac{n}{5}$. The sum of a and b, divided by 3, is $\frac{a + b}{3}$.	p. 57
The Properties of Signed Numbers		**Section 1.2**
The Commutative Properties If a and b are any numbers, **1.** $a + b = b + a$ **2.** $a \cdot b = b \cdot a$	$3 + 8 = 8 + 3$ $2 \cdot 5 = 5 \cdot 2$	p. 63
The Associative Properties If a, b, and c are any numbers, **1.** $a + (b + c) = (a + b) + c$ **2.** $a \cdot (b \cdot c) = (a \cdot b) \cdot c$	$3 + (7 + 12) = (3 + 7) + 12$ $2 \cdot (5 \cdot 12) = (2 \cdot 5) \cdot 12$	p. 63
The Distributive Property If a, b, and c are any numbers, $a(b + c) = a \cdot b + a \cdot c$	$6 \cdot (8 + 15) = 6 \cdot 8 + 6 \cdot 15$	p. 65
Adding and Subtracting Signed Numbers		**Section 1.3**
Adding Signed Numbers **1.** If two numbers have the same sign, add their absolute values. Give the sum the sign of the original numbers.	$9 + 7 = 16$ $(-9) + (-7) = -16$	p. 72
2. If two numbers have different signs, subtract their absolute values, the smaller from the larger. Give the sum the sign of the number with the larger absolute value.	$15 + (-10) = 5$ $(-12) + 9 = -3$	p. 73

Continued

DEFINITION/PROCEDURE	EXAMPLE	REFERENCE
Adding and Subtracting Signed Numbers		**Section 1.3**
Subtracting Signed Numbers **1.** Rewrite the subtraction problem as an addition problem by **a.** Changing the subtraction symbol to an addition symbol **b.** Replacing the number being subtracted with its opposite **2.** Add the resulting signed numbers as before.	$16 - 8 = 16 + (-8) = 8$ $8 - 15 = 8 + (-15) = -7$ $-9 - (-7) = -9 + 7 = -2$	**p. 77**
Multiplying and Dividing Signed Numbers		**Section 1.4**
Multiplying Signed Numbers Multiply the absolute values of the two numbers. **1.** If the numbers have different signs, the product is negative. **2.** If the numbers have the same sign, the product is positive.	$5(-7) = -35$ $(-10)(9) = -90$ $8 \cdot 7 = 56$ $(-9)(-8) = 72$	**p. 89** **p. 90**
Dividing Signed Numbers Divide the absolute values of the two numbers. **1.** If the numbers have different signs, the quotient is negative. **2.** If the numbers have the same sign, the quotient is positive.	$\frac{-32}{4} = -8$ $\frac{75}{-5} = -15$ $\frac{20}{5} = 4$ $\frac{-18}{-9} = 2$	**p. 92**
Evaluating Algebraic Expressions		**Section 1.5**
Algebraic Expressions An expression that contains numbers and letters (called *variables*).		**p. 103**
Evaluating Algebraic Expressions To evaluate an algebraic expression: **1.** Replace each variable or letter with its number value. **2.** Do the necessary arithmetic, following the rules for the order of operations.	Evaluate $2x + 3y$ if $x = 5$ and $y = -2$. $2x + 3y$ $= 2 \cdot 5 + (3)(-2)$ $= 10 - 6 = 4$	**p. 103**
Adding and Subtracting Terms		**Section 1.6**
Term A number or the product of a number and one or more variables.		**p. 115**
Combining Like Terms To combine like terms: **1.** Add or subtract the coefficients (the numbers multiplying the variables). **2.** Attach the common variable.	$5x + 2x = 7x$ $5 + 2$ $8a - 5a = 3a$ $8 - 5$	**p. 117**
Multiplying and Dividing Terms		**Section 1.7**
Property 1 of Exponents $a^m \cdot a^n = a^{m+n}$	$2^7 \cdot 2^3 = 2^{7+3} = 2^{10}$	**p. 123**
Property 2 of Exponents $\frac{a^m}{a^n} = a^{m-n}$	$\frac{3^7}{3^3} = 3^{7-3} = 3^4$	**p.125**

Summary Exercises

This exercise set is provided to give you practice with each of the objectives of the chapter. Each exercise is keyed to the appropriate chapter section. The answers are provided in the *Instructor's Manual.* Your instructor will give you guidelines on how to best use these exercises.

[1.1] Write, using symbols.

1. 5 more than y $y + 5$

2. c decreased by 10 $c - 10$

3. The product of 8 and a $8a$

4. The quotient when y is divided by 3 $\frac{y}{3}$

5. 5 times the product of m and n $5mn$

6. The product of a and 5 less than a $a(a - 5)$

7. 3 more than the product of 17 and x $17x + 3$

8. The quotient when a plus 2 is divided by a minus 2 $\frac{a + 2}{a - 2}$

Identify which are expressions and which are not.

9. $4(x + 3)$ Yes

10. $7 \div \cdot 8$ No

11. $y + 5 = 9$ No

12. $11 + 2(3x - 9)$ Yes

[1.2] Identify the property that is illustrated by each of the following statements.

13. $5 + (7 + 12) = (5 + 7) + 12$ Associative property of addition

14. $2(8 + 3) = 2 \cdot 8 + 2 \cdot 3$ Distributive property

15. $4 \cdot (5 \cdot 3) = (4 \cdot 5) \cdot 3$ Associative property of multiplication

16. $4 \cdot 7 = 7 \cdot 4$ Commutative property of multiplication

Verify that each of the following statements is true by evaluating each side of the equation separately and comparing the results.

17. $8(5 + 4) = 8 \cdot 5 + 8 \cdot 4$ $72 = 72$

18. $2(3 + 7) = 2 \cdot 3 + 2 \cdot 7$ $20 = 20$

19. $(7 + 9) + 4 = 7 + (9 + 4)$ $20 = 20$

20. $(2 + 3) + 6 = 2 + (3 + 6)$ $11 = 11$

21. $(8 \cdot 2) \cdot 5 = 8(2 \cdot 5)$ $80 = 80$

22. $(3 \cdot 7) \cdot 2 = 3 \cdot (7 \cdot 2)$ $42 = 42$

Use the distributive law to remove parentheses.

23. $3(7 + 4)$ $3 \cdot 7 + 3 \cdot 4$

24. $4(2 + 6)$ $4 \cdot 2 + 4 \cdot 6$

25. $4(w + v)$ $4w + 4v$

26. $6(x + y)$ $6x + 6y$

27. $3(5a + 2)$ $3 \cdot 5a + 3 \cdot 2$

28. $2(4x^2 + 3x)$ $2 \cdot 4x^2 + 2 \cdot 3x$

[1.3] Add.

29. $-3 + (-8)$ -11

30. $10 + (-4)$ 6

31. $6 + (-6)$ 0

32. $-16 + (-16)$ -32

33. $-18 + 0$ -18

34. $\frac{3}{8} + \left(-\frac{11}{8}\right)$ -1

35. $5.7 + (-9.7)$ -4

36. $-18 + 7 + (-3)$ -14

Subtract.

37. $8 - 13$ −5

38. $-7 - 10$ −17

39. $10 - (-7)$ 17

40. $-5 - (-1)$ −4

41. $-9 - (-9)$ 0

42. $0 - (-2)$ 2

43. $-\frac{5}{4} - \left(-\frac{17}{4}\right)$ 3

44. $7.9 - (-8.1)$ 16

Find the median for each of the following sets.

45. 2, 4, 9, 10, 15 9

46. −7, −3, 2, 4, 5 2

47. −3, −8, 4, 1, 6 1

48. 6, −3, 2, −5, 1 1

49. 2, 4, 1, 8, 6, 7 5

50. −3, −1, −5, 3, 4, 1 0

Determine the range for each of the following sets.

51. 3, 5, 1, 8, 9 8

52. −4, −5, 6, 4, 2, 1 11

53. −5, 2, −1, 3, 8 13

54. 7, 3, 5, 3, −4 11

[1.4] Multiply.

55. $(10)(-7)$ −70

56. $(-8)(-5)$ 40

57. $(-3)(-15)$ 45

58. $(1)(-15)$ −15

59. $(0)(-8)$ 0

60. $\left(\frac{2}{3}\right)\left(-\frac{3}{2}\right)$ −1

61. $(-4)\left(\frac{3}{8}\right)$ $-\frac{3}{2}$

62. $\left(-\frac{5}{4}\right)(-1)$ $\frac{5}{4}$

Divide.

63. $\frac{80}{16}$ 5

64. $\frac{-63}{7}$ −9

65. $\frac{-81}{-9}$ 9

66. $\frac{0}{-5}$ 0

67. $\frac{32}{-8}$ −4

68. $\frac{-7}{0}$ Undefined

Perform the indicated operations.

69. $\frac{-8 + 6}{-8 - (-10)}$ −1

70. $\frac{-6 - 1}{5 - (-2)}$ −1

71. $\frac{25 - 4}{-5 - (-2)}$ −7

Evaluate each of the following expressions.

72. $18 - 3 \cdot 5$ 3

73. $(18 - 3) \cdot 5$ 75

74. $5 \cdot 4^2$ 80

75. $(5 \cdot 4)^2$ 400

76. $5 \cdot 3^2 - 4$ 41

77. $5(3^2 - 4)$ 25

78. $5(4 - 2)^2$ 20

79. $5 \cdot 4 - 2^2$ 16

80. $(5 \cdot 4 - 2)^2$ 324

81. $3(5 - 2)^2$ 27

82. $3 \cdot 5 - 2^2$ 11

83. $(3 \cdot 5 - 2)^2$ 169

[1.5] Evaluate the expressions if $x = -3$, $y = 6$, $z = -4$, and $w = 2$.

84. $3x + w$ **−7**

85. $5y - 4z$ **46**

86. $x + y - 3z$ **15**

87. $5z^2$ **80**

88. $3x^2 - 2w^2$ **19**

89. $3x^3$ **−81**

90. $5(x^2 - w^2)$ **25**

91. $\frac{6z}{2w}$ **−6**

92. $\frac{2x - 4z}{y - z}$ **1**

93. $\frac{3x - y}{w - x}$ **−3**

94. $\frac{x(y^2 - z^2)}{(y + z)(y - z)}$ **−3**

95. $\frac{y(x - w)^2}{x^2 - 2xw + w^2}$ **6**

[1.6] List the terms of the expressions.

96. $4a^3 - 3a^2$ **$4a^3, -3a^2$**

97. $5x^2 - 7x + 3$ **$5x^2, -7x, 3$**

Circle like terms.

98. $5m^2, -3m, -4m^2, 5m^3, m^2$ **$5m^2, -4m^2, m^2$**

99. $4ab^2, 3b^2, -5a, ab^2, 7a^2, -3ab^2, 4a^2b$ **$4ab^2, ab^2, -3ab^2$**

Combine like terms.

100. $5c + 7c$ **$12c$**

101. $2x + 5x$ **$7x$**

102. $4a - 2a$ **$2a$**

103. $6c - 3c$ **$3c$**

104. $9xy - 6xy$ **$3xy$**

105. $5ab^2 + 2ab^2$ **$7ab^2$**

106. $7a + 3b + 12a - 2b$ **$19a + b$**

107. $6x - 2x + 5y - 3x$ **$x + 5y$**

108. $5x^3 + 17x^2 - 2x^3 - 8x^2$ **$3x^3 + 9x^2$**

109. $3a^3 + 5a^2 + 4a - 2a^3 - 3a^2 - a$ **$a^3 + 2a^2 + 3a$**

110. Subtract $4a^3$ from the sum of $2a^3$ and $12a^3$. **$10a^3$**

111. Subtract the sum of $3x^2$ and $5x^2$ from $15x^2$. **$7x^2$**

[1.7] Divide.

112. $\frac{x^{10}}{x^3}$ **x^7**

113. $\frac{a^5}{a^4}$ **a**

114. $\frac{x^2 \cdot x^3}{x^4}$ **x**

115. $\frac{m^2 \cdot m^3 \cdot m^4}{m^5}$ **m^4**

116. $\frac{18p^7}{9p^5}$ **$2p^2$**

117. $\frac{24x^{17}}{8x^{13}}$ **$3x^4$**

118. $\frac{30m^7n^5}{6m^2n^3}$ **$5m^5n^2$**

119. $\frac{108x^9y^4}{9xy^4}$ **$12x^8$**

120. $\frac{48p^5q^3}{6p^3q}$ **$8p^2q^2$**

121. $\frac{52a^5b^3c^5}{13a^4c}$ **$4ab^3c^4$**

122. $(4x^3)(5x^4)$ **$20x^7$**

123. $(3x)^2(4xy)$ **$36x^3y$**

124. $(8x^2y^3)(3x^3y^2)$ **$24x^5y^5$**

125. $(-2x^3y^3)(-5xy)$ **$10x^4y^4$**

126. $(6x^4)(2x^2y)$ **$12x^6y$**

Write the algebraic expression that answers the question.

[1.1–1.7]

127. Carpentry. If x feet (ft) are cut off the end of a board that is 23 ft long, how much is left? $23 - x$

128. Money. Joan has 25 nickels and dimes in her pocket. If x of these are dimes, how many of the coins are nickels? $25 - x$

129. Age. Sam is 5 years older than Angela. If Angela is x years old now, how old is Sam? $x + 5$

130. Money. Margaret has \$5 more than twice as much money as Gerry. Write an expression for the amount of money that Margaret has. $2x + 5$

131. Geometry. The length of a rectangle is 4 meters (m) more than the width. Write an expression for the length of the rectangle. $x + 4$

132. Number problem. A number is 7 less than 6 times the number n. Write an expression for the number. $6n - 7$

133. Carpentry. A 25-ft plank is cut into two pieces. Write expressions for the length of each piece. $x, 25 - x$

134. Money. Bernie has x dimes and q quarters in his pocket. Write an expression for the amount of money that Bernie has in his pocket. $.10x + .25q$

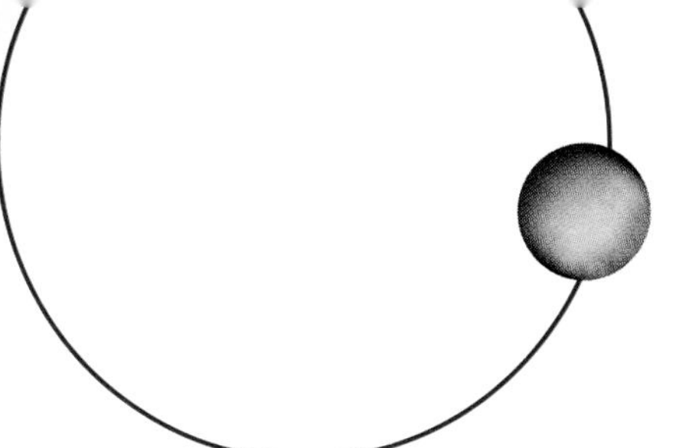

Self-Test for Chapter 1

Name ______________________

Section __________ Date __________

ANSWERS

1. $a - 5$
2. $6m$
3. $4(m + n)$
4. $\frac{a + b}{3}$
5. Commutative property of multiplication
6. Distributive property
7. Associative property of addition
8. 21
9. $20x + 12$
10. Not an Expression
11. Expression
12. −13
13. −3
14. −21
15. 1
16. −6
17. −24
18. 9
19. 0
20. 3
21. 1
22. −40
23. 63

The purpose of this self-test is to help you check your progress and to review for a chapter test in class. Allow yourself about an hour to take the test. When you are done, check your answers in the back of the book. If you missed any problems, be sure to go back and review the appropriate sections in the chapter and the exercises that are provided.

Write, using symbols.

1. 5 less than a

2. The product of 6 and m

3. 4 times the sum of m and n

4. The quotient when the sum of a and b is divided by 3

Identify the property that is illustrated by each of the following statements.

5. $6 \cdot 7 = 7 \cdot 6$

6. $2(6 + 7) = 2 \cdot 6 + 2 \cdot 7$

7. $4 + (3 + 7) = (4 + 3) + 7$

Use the distributive property to remove parentheses. Then simplify your result.

8. $3(5 + 2)$

9. $4(5x + 3)$

Identify which are expressions and which are not.

10. $5x + 6 = 4$

11. $4 + (6 + x)$

Add.

12. $-8 + (-5)$

13. $6 + (-9)$

14. $(-9) + (-12)$

15. $-\frac{5}{3} + \frac{8}{3}$

Subtract.

16. $9 - 15$

17. $-9 - 15$

18. $5 - (-4)$

19. $-7 - (-7)$

Find the median of each of the following sets.

20. 2, −4, 5, −7, 8, 3, 10

21. −4, 6, −1, −9, 3, 7, −6, 11

Multiply.

22. $(-8)(5)$

23. $(-9)(-7)$

ANSWERS

24. −27
25. −24
26. 14
27. −25
28. 3
29. −5
30. Undefined
31. 3
32. 65
33. 144
34. −9
35. −4
36. 15*a*
37. 19*x* + 5*y*
38. $8a^2$
39. a^{14}
40. $15x^3y^7$
41. $2x^3$
42. $4ab^3$
43. x^9
44. 2*x* − 8
45. 2*w* + 4

24. $(4.5)(-6)$ **25.** $(6)(-4)$

26. Determine the range for the following set: 4, −1, 6, 3, −6, 2, 8, 5

Evaluate each expression.

27. $\frac{75}{-3}$ **28.** $\frac{-36 + 9}{-9}$

29. $\frac{(-15)(-3)}{-9}$ **30.** $\frac{9}{0}$

Evaluate the following expressions.

31. $23 - 4 \cdot 5$ **32.** $4 \cdot 5^2 - 35$

33. $4(2 + 4)^2$ **34.** $16 \div (-4) + (-5)$

35. If $x = 2, y = -1$, and $z = 3$, evaluate the expression $\frac{9x^2y}{3z}$.

Combine like terms.

36. $8a + 7a$ **37.** $10x + 8y + 9x - 3y$

38. Subtract $9a^2$ from the sum of $12a^2$ and $5a^2$.

Multiply.

39. $a^5 \cdot a^9$ **40.** $3x^2y^3 \cdot 5xy^4$

Divide.

41. $\frac{4x^5}{2x^2}$ **42.** $\frac{20a^3b^5}{5a^2b^2}$

43. $\frac{x^{10} \cdot x^5}{x^6}$

44. Tom is 8 years younger than twice Moira's age. Write an expression for Tom's age. Let *x* represent Moira's age.

45. The length of a rectangle is 4 more than twice the width. Write an expression for the length of the rectangle.

EQUATIONS AND INEQUALITIES

2

INTRODUCTION

Many engineers, economists, and environmental scientists are working on the problem of meeting the increasing energy demands of a growing global population. One promising solution to this problem is power generated by wind-driven turbines.

The cost of wind-generated power has fallen from \$0.25 per kilowatt hour (kWh) in the early 1980s to about \$0.043 per kWh in 2000; thus, using this form of power production is becoming economically feasible. And compared to the cost of pollution from burning coal and oil, wind-generated power may be less expensive.

An economist for a city might use this equation to try to compute the cost for electricity for his city:

$$C = P(0.043)(1000)$$

in which C = cost in dollars

P = power in megawatts

The city engineer who is investigating the possibility of building turbines to supply the city with electricity knows that each turbine can produce 1.1 million kWh of power, so she uses the equation

$$P = n(1.1)$$

in which P = power in megawatts

n = number of turbines required to produce power

The equation is an ancient tool for solving problems and writing numerical relationships clearly and accurately. In this chapter you will learn methods to solve linear equations and practice writing equations that accurately describe problem situations.

NOTE An equation such as

$x + 3 = 5$

is called a **conditional equation** because it can be either true or false depending on the value given to the variable.

Just as the balance scale may be in balance or out of balance, an equation may be either true or false. For instance, $3 + 4 = 7$ is true because both sides name the same number. What about an equation such as $x + 3 = 5$ that has a letter or variable on one side? Any number can replace x in the equation. However, only one number will make this equation a true statement.

$$\text{If } x = \begin{matrix} 1 \\ 2 \\ 3 \end{matrix} \begin{cases} 1 + 3 = 5 \text{ is false} \\ 2 + 3 = 5 \text{ is true} \\ 3 + 3 = 5 \text{ is false} \end{cases}$$

The number 2 is called the **solution** (or *root*) of the equation $x + 3 = 5$ because substituting 2 for x gives a true statement.

Definitions: Solution

A **solution** for an equation is any value for the variable that makes the equation a true statement.

Example 1

Verifying a Solution

(a) Is 3 a solution for the equation $2x + 4 = 10$?

To find out, replace x with 3 and evaluate $2x + 4$ on the left.

Left side		*Right side*
$2 \cdot 3 + 4$	$\stackrel{?}{=}$	10
$6 + 4$	$\stackrel{?}{=}$	10
10	$=$	10

Because $10 = 10$ is a true statement, 3 is a solution of the equation.

(b) Is 5 a solution of the equation $3x - 2 = 2x + 1$?

To find out, replace x with 5 and evaluate each side separately.

NOTE Remember the rules for the order of operation. Multiply first; then add or subtract.

Left side		*Right side*
$3 \cdot 5 - 2$	$\stackrel{?}{=}$	$2 \cdot 5 + 1$
$15 - 2$	$\stackrel{?}{=}$	$10 + 1$
13	$\neq$	11

Because the two sides do not name the same number, we do not have a true statement, and 5 is not a solution.

CHECK YOURSELF 1

For the equation

$2x - 1 = x + 5$

(a) Is 4 a solution? **(b)** Is 6 a solution?

EQUATIONS AND INEQUALITIES

2

INTRODUCTION

Many engineers, economists, and environmental scientists are working on the problem of meeting the increasing energy demands of a growing global population. One promising solution to this problem is power generated by wind-driven turbines.

The cost of wind-generated power has fallen from \$0.25 per kilowatt hour (kWh) in the early 1980s to about \$0.043 per kWh in 2000; thus, using this form of power production is becoming economically feasible. And compared to the cost of pollution from burning coal and oil, wind-generated power may be less expensive.

An economist for a city might use this equation to try to compute the cost for electricity for his city:

$$C = P(0.043)(1000)$$

in which C = cost in dollars

P = power in megawatts

The city engineer who is investigating the possibility of building turbines to supply the city with electricity knows that each turbine can produce 1.1 million kWh of power, so she uses the equation

$$P = n(1.1)$$

in which P = power in megawatts

n = number of turbines required to produce power

The equation is an ancient tool for solving problems and writing numerical relationships clearly and accurately. In this chapter you will learn methods to solve linear equations and practice writing equations that accurately describe problem situations.

Name ____________________

Section ________ Date ________

Pre-Test Chapter 2

ANSWERS

1. No
2. Yes
3. 8
4. −12
5. 7
6. 35
7. −2
8. 2
9. $W = \frac{P - 2L}{2}$ or $\frac{P}{2} - L$
10. $y = \frac{5x - 14}{3}$ or $\frac{5}{3}x - \frac{14}{3}$
11. $4x + 5 = 17$
12. $4(y + 6) = 10y + 6$
13. $x \le 15$
14. $x \le -1$
15. 6
16. 15, 17
17. 4 cm × 13 cm
18. $540
19. $13,125
20. 15%

Tell whether the number shown in parentheses is a solution for the given equation.

1. $4x - 9 = 15 \quad (7)$

2. $7x - 5 = 3x + 31 \quad (9)$

Solve the following equations and check your results.

3. $8x - 23 = 7x - 15$

4. $6x + 15 - 2x = 18 + 8x - 5x - 15$

5. $7x - 15 = 4x + 6$

6. $\frac{5}{7}x = 25$

7. $9x - (x - 5) = -(6 - x) - 3$

8. $2(3 - x) + 15 = 8(4x - 5) - 7$

Solve for the indicated variable.

9. $p = 2L + 2W$ for W

10. $5x - 3y = 14$ for y

Translate each statement into an algebraic equation.

11. 5 more than 4 times a number is 17.

12. 4 times the sum of a number and 6 is 6 more than 10 times the number.

Solve the following inequalities.

13. $x - 7 \le 8$

14. $6 - 2x \ge 9 + x$

Solve the following word problems.

15. 7 times a number decreased by 5 is 37. Find the number.

16. The sum of 2 consecutive odd integers is 32. Find the integers.

17. The perimeter of a rectangle is 34 cm. If the length is 1 cm more than 3 times the width, what are the dimensions of the rectangle?

18. A state sales tax rate is 2.5%. If the tax on a purchase is $13.50, what was the amount of the purchase?

19. A house sells for $250,000 and the rate of commission is 5.25%. How much will the salesperson make for the sale?

20. A stereo system is marked down from $790 to $671.50. What was the rate of discount?

Solving Equations by the Addition Property

OBJECTIVES

1. Determine whether a given number is a solution for an equation
2. Use the addition property to solve an equation

Overcoming Math Anxiety

Throughout this text, we will present you with a series of class-tested techniques that are designed to improve your performance in this math class.

Hint #3 Don't Procrastinate!

1. Do your math homework while you're still fresh. If you wait until too late at night, your tired mind will have much more difficulty understanding the concepts.
2. Do your homework the day it is assigned. The more recent the explanation is, the easier it is to recall.
3. When you've finished your homework, try reading the next section through one time. This will give you a sense of direction when you next hear the material. This works whether you are in a lecture or lab setting.

Remember that, in a typical math class, you are expected to do two or three hours of homework for each weekly class hour. This means two or three hours per night. Schedule the time and stay to your schedule.

In this chapter you will begin working with one of the most important tools of mathematics, the equation. The ability to recognize and solve various types of equations is probably the most useful algebraic skill you will learn. We will continue to build upon the methods of this chapter throughout the remainder of the text. To start, let's describe what we mean by an *equation.*

Definitions: Equation

An **equation** is a mathematical statement that two expressions are equal.

Some examples are $3 + 4 = 7$, $x + 3 = 5$, $P = 2L + 2W$.

As you can see, an equals sign (=) separates the two equal expressions. These expressions are usually called the *left side* and the *right side* of the equation.

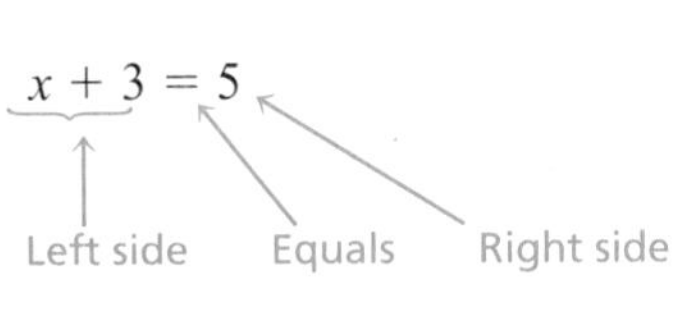

NOTE An equation such as

$x + 3 = 5$

is called a **conditional equation** because it can be either true or false depending on the value given to the variable.

Just as the balance scale may be in balance or out of balance, an equation may be either true or false. For instance, $3 + 4 = 7$ is true because both sides name the same number. What about an equation such as $x + 3 = 5$ that has a letter or variable on one side? Any number can replace x in the equation. However, only one number will make this equation a true statement.

$$\text{If } x = \begin{matrix} 1 \\ 2 \\ 3 \end{matrix} \begin{cases} 1 + 3 = 5 \text{ is false} \\ 2 + 3 = 5 \text{ is true} \\ 3 + 3 = 5 \text{ is false} \end{cases}$$

The number 2 is called the **solution** (or *root*) of the equation $x + 3 = 5$ because substituting 2 for x gives a true statement.

Definitions: Solution

A **solution** for an equation is any value for the variable that makes the equation a true statement.

Example 1

Verifying a Solution

(a) Is 3 a solution for the equation $2x + 4 = 10$?

To find out, replace x with 3 and evaluate $2x + 4$ on the left.

Left side		*Right side*
$2 \cdot 3 + 4$	$\stackrel{?}{=}$	10
$6 + 4$	$\stackrel{?}{=}$	10
10	=	10

Because $10 = 10$ is a true statement, 3 is a solution of the equation.

(b) Is 5 a solution of the equation $3x - 2 = 2x + 1$?

To find out, replace x with 5 and evaluate each side separately.

NOTE Remember the rules for the order of operation. Multiply first; then add or subtract.

Left side		*Right side*
$3 \cdot 5 - 2$	$\stackrel{?}{=}$	$2 \cdot 5 + 1$
$15 - 2$	$\stackrel{?}{=}$	$10 + 1$
13	$\neq$	11

Because the two sides do not name the same number, we do not have a true statement, and 5 is not a solution.

CHECK YOURSELF 1

For the equation

$2x - 1 = x + 5$

(a) Is 4 a solution?

(b) Is 6 a solution?

You may be wondering whether an equation can have more than one solution. It certainly can. For instance,

NOTE This is an example of a **quadratic equation.** We will consider methods of solution in Chapter 4 and then again in Chapter 10.

$$x^2 = 9$$

has two solutions. They are 3 and -3 because

$$3^2 = 9 \quad \text{and} \quad (-3)^2 = 9$$

In this chapter, however, we will always work with *linear equations in one variable.* These are equations that can be put into the form

$$ax + b = 0$$

in which the variable is x, a and b are any numbers, and a is not equal to 0. In a linear equation, the variable can appear only to the first power. No other power (x^2, x^3, etc.) can appear. Linear equations are also called **first-degree equations.** The degree of an equation in one variable is the highest power to which the variable appears.

Rules and Properties: Linear Equations

Linear equations in one variable are equations that can be written in the form

$$ax + b = 0 \qquad a \neq 0$$

Every such equation will have exactly one solution.

Example 2

Identifying Expressions and Equations

Label each of the following as an expression, a linear equation, or an equation that is not linear.

(a) $4x + 5$ is an expression

(b) $2x + 8 = 0$ is a linear equation

(c) $3x^2 - 9 = 0$ is an equation that is not linear

(d) $5x = 15$ is a linear equation

CHECK YOURSELF 2

Label each as an expression, a linear equation, or an equation that is not linear.

(a) $2x^2 = 8$ **(b)** $2x - 3 = 0$ **(c)** $5x - 10$ **(d)** $2x + 1 = 7$

It is not difficult to find the solution for an equation such as $x + 3 = 8$ by guessing the answer to the question "What plus 3 is 8?" Here the answer to the question is 5, and that is also the solution for the equation. But for more complicated equations you are going to need something more than guesswork. A better method is to transform the given equation to an *equivalent equation* whose solution can be found by inspection. Let's make a definition.

Definitions: Equivalent Equations

Equations that have the same solution are called **equivalent equations.**

The following are all equivalent equations:

$2x + 3 = 5 \qquad 2x = 2 \qquad \text{and} \qquad x = 1$

They all have the same solution, 1. We say that a linear equation is *solved* when it is transformed to an equivalent equation of the form

NOTE In some cases we'll write the equation in the form

$\square = x$

The number will be our solution when the equation has the variable isolated on the left or on the right.

$x = \square$

The variable is alone on the left side.

The right side is some number, the solution.

The addition property of equality is the first property you will need to transform an equation to an equivalent form.

REMEMBER An equation is a statement that the two sides are equal. Adding the same quantity to both sides does not change the equality or "balance."

Rules and Properties: The Addition Property of Equality

If $\quad a = b$

then $\quad a + c = b + c$

In words, adding the same quantity to both sides of an equation gives an equivalent equation.

Recall that we said that a true equation was like a scale in balance.

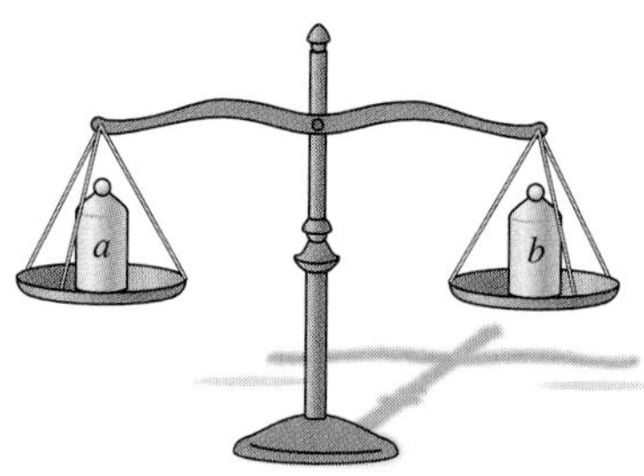

The addition property is equivalent to adding the same weight to both sides of the scale. It will remain in balance.

Example 3

Using the Addition Property to Solve an Equation

Solve

$x - 3 = 9$

Remember that our goal is to isolate x on one side of the equation. Because 3 is being subtracted from x, we can add 3 to remove it. We must use the addition property to add 3 to both sides of the equation.

NOTE To check, replace x with 12 in the original equation:

$x - 3 \stackrel{?}{=} 9$

$12 - 3 \stackrel{?}{=} 9$

$9 = 9$

Because we have a true statement, 12 is the solution.

$$\begin{array}{rcr} x - 3 & = & 9 \\ +3 & & +3 \\ \hline x & = & 12 \end{array}$$

Adding 3 "undoes" the subtraction and leaves x alone on the left.

Because 12 is the solution for the equivalent equation $x = 12$, it is the solution for our original equation.

CHECK YOURSELF 3

Solve and check.

$x - 5 = 4$

The addition property also allows us to add a negative number to both sides of an equation. This is really the same as subtracting the same quantity from both sides.

Example 4

Using the Addition Property to Solve an Equation

Solve

$x + 5 = 9$

NOTE Recall our comment that we could write an equation in the equivalent forms $x = \square$ or $\square = x$, in which $\square$ represents some number. Suppose we have an equation like

$12 = x + 7$

Adding -7 will isolate x *on the right:*

$$\begin{array}{rcr} 12 & = & x + 7 \\ -7 & & -7 \\ \hline 5 & = & x \end{array}$$

and the solution is 5.

In this case, 5 is *added* to x on the left. We can use the addition property to add a -5 to both sides. Because $5 + (-5) = 0$, this will "undo" the addition and leave the variable x alone on one side of the equation.

$$\begin{array}{rcr} x + 5 & = & 9 \\ -5 & & -5 \\ \hline x & = & 4 \end{array}$$

The solution is 4. To check, replace x with 4:

$4 + 5 = 9$ (True)

CHECK YOURSELF 4

Solve and check.

$x + 6 = 13$

What if the equation has a variable term on both sides? You will have to use the addition property to add or subtract a term involving the variable to get the desired result.

Example 5

Using the Addition Property to Solve an Equation

Solve

$5x = 4x + 7$

We will start by adding $-4x$ to both sides of the equation. Do you see why? Remember that an equation is solved when we have an equivalent equation of the form $x = \square$.

NOTE Recall that adding $-4x$ is identical to subtracting $4x$.

$$\begin{array}{rrl} 5x = & 4x + 7 \\ -4x & -4x \\ \hline x = & 7 \end{array}$$

Adding $-4x$ to both sides *removes* $4x$ from the right.

To check: Because 7 is a solution for the equivalent equation $x = 7$, it should be a solution for the original equation. To find out, replace x with 7:

$$5 \cdot 7 \stackrel{?}{=} 4 \cdot 7 + 7$$
$$35 \stackrel{?}{=} 28 + 7$$
$$35 = 35 \quad \text{(True)}$$

CHECK YOURSELF 5

Solve and check.

$7x = 6x + 3$

You may have to apply the addition property more than once to solve an equation. Look at Example 6.

Example 6

Using the Addition Property to Solve an Equation

Solve

$7x - 8 = 6x$

We want all variables on *one* side of the equation. If we choose the left, we add $-6x$ to both sides of the equation. This will remove $6x$ from the right:

$$\begin{array}{rr} 7x - 8 = & 6x \\ -6x \qquad & -6x \\ \hline x - 8 = & 0 \end{array}$$

We want the variable alone, so we add 8 to both sides. This isolates x on the left.

$$\begin{array}{rr} x - 8 = & 0 \\ +8 & +8 \\ \hline x \quad = & 8 \end{array}$$

The solution is 8. We'll leave it to you to check this result.

CHECK YOURSELF 6

Solve and check.

$9x + 3 = 8x$

Often an equation will have more than one variable term *and* more than one number. You will have to apply the addition property twice in solving these equations.

Example 7

Using the Addition Property to Solve an Equation

Solve

$5x - 7 = 4x + 3$

We would like the variable terms on the left, so we start by adding $-4x$ to remove the $4x$ term from the right side of the equation:

$$\begin{array}{rcr} 5x - 7 &=& 4x + 3 \\ -4x & & -4x \\ \hline x - 7 &=& 3 \end{array}$$

Now, to isolate the variable, we add 7 to both sides.

$$\begin{array}{rcr} x - 7 &=& 3 \\ +7 & & +7 \\ \hline x &=& 10 \end{array}$$

NOTE You could just as easily have added 7 to both sides and *then* added $-4x$. The result would be the same. In fact, some students prefer to combine the two steps.

The solution is 10. To check, replace x with 10 in the original equation:

$$5 \cdot 10 - 7 \stackrel{?}{=} 4 \cdot 10 + 3$$
$$43 = 43 \quad \text{(True)}$$

CHECK YOURSELF 7

Solve and check.

(a) $4x - 5 = 3x + 2$ **(b)** $6x + 2 = 5x - 4$

NOTE Remember, by *simplify* we mean to combine all like terms.

In solving an equation, you should always simplify each side as much as possible before using the addition property.

Example 8

Combining Like Terms and Solving the Equation

Solve

Like terms ↙ ↘ Like terms ↙ ↘

$5 + 8x - 2 = 2x - 3 + 5x$

Because like terms appear on each side of the equation, we start by combining the numbers on the left (5 and -2). Then we combine the like terms ($2x$ and $5x$) on the right. We have

$$3 + 8x = 7x - 3$$

Now we can apply the addition property, as before:

$$\begin{array}{rcrl} 3 + 8x &=& 7x - 3 & \\ -7x &=& -7x & \text{Add } -7x. \\ \hline 3 + x &=& -3 & \\ -3 & & -3 & \text{Add } -3. \\ \hline x &=& -6 & \text{Isolate } x. \end{array}$$

The solution is -6. To check, always return to the original equation. That will catch any possible errors in simplifying. Replacing x with -6 gives

$$5 + 8(-6) - 2 \stackrel{?}{=} 2(-6) - 3 + 5(-6)$$
$$5 - 48 - 2 \stackrel{?}{=} -12 - 3 - 30$$
$$-45 = -45 \quad \text{(True)}$$

CHECK YOURSELF 8

Solve and check.

(a) $3 + 6x + 4 = 8x - 3 - 3x$ **(b)** $5x + 21 + 3x = 20 + 7x - 2$

We may have to apply some of the properties discussed in Section 1.2 in solving equations. Example 9 illustrates the use of the distributive property to clear an equation of parentheses.

Example 9

Using the Distributive Property and Solving Equations

Solve

NOTE $2(3x + 4)$
$= 2(3x) + 2(4)$
$= 6x + 8$

$$2(3x + 4) = 5x - 6$$

Applying the distributive property on the left, we have

$$6x + 8 = 5x - 6$$

We can then proceed as before:

$$\begin{array}{rcrl} 6x + 8 &=& 5x - 6 & \\ -5x & & -5x & \text{Add } -5x. \\ \hline x + 8 &=& -6 & \\ -8 & & -8 & \text{Add } -8. \\ \hline x &=& -14 & \end{array}$$

NOTE Remember that $x = -14$ and $-14 = x$ are equivalent equations.

The solution is -14. We will leave the checking of this result to the reader.

Remember: Always return to the original equation to check.

CHECK YOURSELF 9

Solve and check each of the following equations.

(a) $4(5x - 2) = 19x + 4$

(b) $3(5x + 1) = 2(7x - 3) - 4$

Given an expression such as

$-2(x - 5)$

the distributive property can be used to create the equivalent expression.

$-2x + 10$

The distribution of a negative number is used in Example 10.

Example 10

Distributing a Negative Number

Solve each of the following equations.

(a) $-2(x - 5) = -3x + 2$

$$\begin{aligned} -2x + 10 &= -3x + 2 && \text{Distribute the } -2.\\ +3x \quad &\quad\ \ +3x && \text{Add } 3x.\\ \hline x + 10 &= 2 \\ -10 &= -10 && \text{Add } -10.\\ \hline x &= -8 \end{aligned}$$

(b) $-3(3x + 5) = -5(2x - 2)$

$$\begin{aligned} -9x - 15 &= -5(2x - 2) && \text{Distribute the } -3.\\ -9x - 15 &= -10x + 10 && \text{Distribute the } -5.\\ +10x \quad &\quad\ \ +10x && \text{Add } 10x.\\ \hline x - 15 &= 10 \\ +15 &= +15 && \text{Add } 15.\\ \hline x &= 25 \end{aligned}$$

CHECK YOURSELF 10

Solve each of the following.

(a) $-2(x - 3) = -x + 5$

(b) $-4(2x - 1) = -3(3x + 2)$

When parentheses are preceded only by a negative, or by the minus sign, we say that we have a silent negative one. Example 11 illustrates this case.

Example 11

Distributing the Silent Negative One

Solve

$-(2x + 3) = -3x + 7$

$-1(2x + 3) = -3x + 7$

$(-1)(2x) + (-1)(3) = -3x + 7$

$$\begin{aligned} -2x - 3 &= -3x + 7 \\ +3x \quad &\quad\ +3x \qquad \text{Add } 3x. \\ \hline x - 3 &= 7 \\ +3 &\quad +3 \qquad \text{Add 3.} \\ \hline x &= 10 \end{aligned}$$

CHECK YOURSELF 11

Solve $-(3x + 2) = -2x - 6$.

CHECK YOURSELF ANSWERS

1. (a) 4 is not a solution; **(b)** 6 is a solution
2. (a) Nonlinear equation; **(b)** linear equation; **(c)** expression; **(d)** linear equation
3. 9 **4.** 7 **5.** 3 **6.** −3 **7. (a)** 7; **(b)** −6 **8. (a)** −10; **(b)** −3
9. (a) 12; **(b)** −13 **10. (a)** 1; **(b)** −10 **11.** 4

Name ____________

Section ______ Date ______

2.1 Exercises

Is the number shown in parentheses a solution for the given equation?

1. $x + 4 = 9$ (5)

2. $x + 2 = 11$ (8)

3. $x - 15 = 6$ (-21)

4. $x - 11 = 5$ (16)

5. $5 - x = 2$ (4)

6. $10 - x = 7$ (3)

7. $4 - x = 6$ (-2)

8. $5 - x = 6$ (-3)

9. $3x + 4 = 13$ (8)

10. $5x + 6 = 31$ (5)

11. $4x - 5 = 7$ (2)

12. $2x - 5 = 1$ (3)

13. $5 - 2x = 7$ (-1)

14. $4 - 5x = 9$ (-2)

15. $4x - 5 = 2x + 3$ (4)

16. $5x + 4 = 2x + 10$ (4)

17. $x + 3 + 2x = 5 + x + 8$ (5)

18. $5x - 3 + 2x = 3 + x - 12$ (-2)

19. $\frac{3}{4}x = 20$ (18)

20. $\frac{3}{5}x = 24$ (40)

21. $\frac{3}{5}x + 5 = 11$ (10)

22. $\frac{2}{3}x + 8 = -12$ (-6)

Label each of the following as an expression or a linear equation.

23. $2x + 1 = 9$

24. $7x + 14$

25. $2x - 8$

26. $5x - 3 = 12$

ANSWERS

1. Yes
2. No
3. No
4. Yes
5. No
6. Yes
7. Yes
8. No
9. No
10. Yes
11. No
12. Yes
13. Yes
14. No
15. Yes
16. No
17. Yes
18. No
19. No
20. Yes
21. Yes
22. No
23. Linear equation
24. Expression
25. Expression
26. Linear equation

ANSWERS

27. Expression
28. Linear equation
29. Linear equation
30. Expression
31. 2
32. 10
33. 11
34. 4
35. −2
36. −3
37. −7
38. 1
39. 6
40. −7
41. 4
42. −8
43. −10
44. 5
45. −3
46. 6
47. 4
48. 7
49. 2
50. 3
51. $\frac{7}{10}$
52. $\frac{33}{8}$
53. $\frac{5}{2}$
54. $\frac{19}{6}$

27. $7x + 2x + 8 - 3$

28. $x + 5 = 13$

29. $2x - 8 = 3$

30. $12x - 5x + 2 + 5$

Solve and check the following equations.

31. $x + 9 = 11$

32. $x - 4 = 6$

33. $x - 8 = 3$

34. $x + 11 = 15$

35. $x - 8 = -10$

36. $x + 5 = 2$

37. $x + 4 = -3$

38. $x - 5 = -4$

39. $11 = x + 5$

40. $x + 7 = 0$

41. $4x = 3x + 4$

42. $7x = 6x - 8$

43. $11x = 10x - 10$

44. $9x = 8x + 5$

45. $6x + 3 = 5x$

46. $12x - 6 = 11x$

47. $8x - 4 = 7x$

48. $9x - 7 = 8x$

49. $2x + 3 = x + 5$

50. $3x - 2 = 2x + 1$

51. $4x - \frac{3}{5} = 3x + \frac{1}{10}$

52. $5\left(x - \frac{3}{4}\right) = 4x + \frac{3}{8}$

53. $\frac{7}{8}(x - 2) = \frac{3}{4} - \frac{1}{8}x$

54. $\frac{5}{6}(3x - 2) = \frac{3}{2}(x + 1)$

55. $3x - 0.54 = 2(x - 0.15)$

56. $7x + 0.125 = 6x - 0.289$

57. $6x + 3(x - 0.2789) = 4(2x + 0.3912)$

58. $9x - 2(3x - 0.124) = 2x + 0.965$

59. $3x - 5 + 2x - 7 + x = 5x + 2$

60. $5x + 8 + 3x - x + 5 = 6x - 3$

61. $5x - (0.345 - x) = 5x + 0.8713$

62. $-3(0.234 - x) = 2(x + 0.974)$

63. $3(7x + 2) = 5(4x + 1) + 17$

64. $5(5x + 3) = 3(8x - 2) + 4$

65. $\frac{5}{4}x - 1 = \frac{1}{4}x + 7$

66. $\frac{7}{5}x + 3 = \frac{2}{5}x - 8$

67. $\frac{9}{2}x - \frac{3}{4} = \frac{7}{2}x + \frac{5}{4}$

68. $\frac{11}{3}x + \frac{1}{6} = \frac{8}{3}x + \frac{19}{6}$

69. Which of the following is equivalent to the equation $5x - 7 = 4x - 12$?

a. $9x = 19$ **b.** $9x - 7 = -12$ **c.** $x = -18$ **d.** $x - 7 = -12$

70. Which of the following is equivalent to the equation $12x - 6 = 8x + 14$?

a. $4x - 6 = 14$ **b.** $x = 20$ **c.** $20x = 20$ **d.** $4x = 8$

71. Which of the following is equivalent to the equation $7x + 5 = 12x - 10$?

a. $5x = -15$ **b.** $7x - 5 = 12x$ **c.** $-5 = 5x$ **d.** $7x + 15 = 12x$

True or false?

72. Every linear equation with one variable has exactly one solution.

73. Isolating the variable on the right side of the equation will result in a negative solution.

74. An algebraic equation is a complete sentence. It has a subject, a verb, and a predicate. For example, $x + 2 = 5$ can be written in English as "Two more than a number is five." Or, "A number added to two is five." Write an English version of the following equations. Be sure you write complete sentences and that the sentences express the same idea as the equations. Exchange sentences with another student, and see if your interpretation of each other's sentences result in the same equation.

(a) $2x - 5 = x + 1$ **(b)** $2(x + 2) = 14$

(c) $n + 5 = \frac{n}{2} - 6$ **(d)** $7 - 3a = 5 + a$

75. Complete the following explanation in your own words: "The difference between $3(x - 1) + 4 - 2x$ and $3(x - 1) + 4 = 2x$ is . . . ."

ANSWERS

55. 0.24

56. −0.414

57. 2.4015

58. 0.717

59. 14

60. −16

61. 1.2163

62. 2.65

63. 16

64. −17

65. 8

66. −11

67. 2

68. 3

69. d

70. a

71. d

72. True

73. False

74.

75.

76.

a. 1

b. 1

c. 1

d. 1

e. 1

f. 1

g. 1

h. 1

76. "Surprising Results!" Work with other students to try this experiment. Each person should do the following six steps mentally, not telling anyone else what their calculations are:

(a) Think of a number.
(b) Add 7.
(c) Multiply by 3.
(d) Add 3 more than the original number.
(e) Divide by 4.
(f) Subtract the original number.

What number do you end up with? Compare your answer with everyone else's. Does everyone have the same answer? Make sure that everyone followed the directions accurately. How do you explain the results? Algebra makes the explanation clear. Work together to do the problem again, using a variable for the number. Make up another series of computations that give "surprising results."

Getting Ready for Section 2.2 [Section 1.4]

Multiply.

(a) $\left(\frac{1}{3}\right)(3)$

(b) $(-6)\left(-\frac{1}{6}\right)$

(c) $(7)\left(\frac{1}{7}\right)$

(d) $\left(-\frac{1}{4}\right)(-4)$

(e) $\left(\frac{3}{5}\right)\left(\frac{5}{3}\right)$

(f) $\left(\frac{7}{8}\right)\left(\frac{8}{7}\right)$

(g) $\left(-\frac{4}{7}\right)\left(-\frac{7}{4}\right)$

(h) $\left(-\frac{6}{11}\right)\left(-\frac{11}{6}\right)$

Answers

1. Yes **3.** No **5.** No **7.** Yes **9.** No **11.** No **13.** Yes **15.** Yes **17.** Yes **19.** No **21.** Yes **23.** Linear equation **25.** Expression **27.** Expression **29.** Linear equation **31.** 2 **33.** 11 **35.** -2 **37.** -7 **39.** 6 **41.** 4 **43.** -10 **45.** -3 **47.** 4 **49.** 2 **51.** $\frac{7}{10}$ **53.** $\frac{5}{2}$ **55.** 0.24 **57.** 2.4015 **59.** 14 **61.** 1.2163 **63.** 16 **65.** 8 **67.** 2 **69.** d **71.** d **73.** False **75.**  **a.** 1 **b.** 1 **c.** 1 **d.** 1 **e.** 1 **f.** 1 **g.** 1

h. 1

2.2 Solving Equations by the Multiplication Property

OBJECTIVES

1. Determine whether a given number is a solution for an equation
2. Use the multiplication property to solve equations
3. Find the mean for a given set

Let's look at a different type of equation. For instance, what if we want to solve an equation like the following?

$6x = 18$

Using the addition property of the last section won't help. We will need a second property for solving equations.

NOTE Again, as long as you do the *same* thing to *both* sides of the equation, the "balance" is maintained.

Rules and Properties: The Multiplication Property of Equality

If $a = b$ then $ac = bc$ where $c \neq 0$

In words, multiplying both sides of an equation by the same nonzero number gives an equivalent equation.

NOTE Do you see why the number cannot be 0? Multiplying by 0 gives $0 = 0$. We have lost the variable!

Again, we return to the image of the balance scale. We start with the assumption that a and b have the same weight.

The multiplication property tells us that the scale will be in balance as long as we have the same number of "a weights" as we have of "b weights."

Let's work through some examples, using this second rule.

Example 1

Solving Equations by Using the Multiplication Property

Solve

$6x = 18$

Here the variable x is multiplied by 6. So we apply the multiplication property and multiply both sides by $\frac{1}{6}$. Keep in mind that we want an equation of the form

NOTE

$$\frac{1}{6}(6x) = \left(\frac{1}{6} \cdot 6\right)x$$
$$= 1 \cdot x, \text{ or } x$$

We then have x alone on the left, which is what we want.

$x = \square$

$$\frac{1}{6}(6x) = \left(\frac{1}{6}\right)18$$

We can now simplify.

$1 \cdot x = 3$ or $x = 3$

The solution is 3. To check, replace x with 3:

$6 \cdot 3 \stackrel{?}{=} 18$

$18 = 18$ (True)

CHECK YOURSELF 1

Solve and check.

$8x = 32$

In Example 1 we solved the equation by multiplying both sides by the reciprocal of the coefficient of the variable.

Example 2 illustrates a slightly different approach to solving an equation by using the multiplication property.

Example 2

Solving Equations by Using the Multiplication Property

Solve

$5x = -35$

NOTE Because division is defined in terms of multiplication, we can also divide both sides of an equation by the same nonzero number.

The variable x is multiplied by 5. We *divide* both sides by 5 to "undo" that multiplication:

$$\frac{5x}{5} = \frac{-35}{5}$$

$x = -7$ Note that the right side reduces to −7. Be careful with the rules for signs.

We will leave it to you to check the solution.

CHECK YOURSELF 2

Solve and check.

$7x = -42$

Example 3

Solving Equations by Using the Multiplication Property

Solve

$-9x = 54$

In this case, x is multiplied by -9, so we divide both sides by -9 to isolate x on the left:

$$\frac{-9x}{-9} = \frac{54}{-9}$$

$$x = -6$$

The solution is -6. To check:

$$(-9)(-6) \stackrel{?}{=} 54$$

$$54 = 54 \qquad \text{(True)}$$

CHECK YOURSELF 3

Solve and check.

$-10x = -60$

Example 4 illustrates the use of the multiplication property when fractions appear in an equation.

Example 4

Solving Equations by Using the Multiplication Property

(a) Solve

$$\frac{x}{3} = 6$$

Here x is *divided* by 3. We will use multiplication to isolate x.

$$3\left(\frac{x}{3}\right) = 3 \cdot 6$$

$$x = 18$$

This leaves x alone on the left because $3\left(\frac{x}{3}\right) = \frac{3}{1} \cdot \frac{x}{3} = \frac{x}{1} = x$

To check:

$$\frac{18}{3} \stackrel{?}{=} 6$$

$$6 = 6 \qquad \text{(True)}$$

(b) Solve

$$\frac{x}{5} = -9$$

$$5\left(\frac{x}{5}\right) = 5(-9)$$ Because x is divided by 5, multiply both sides by 5

$$x = -45$$

The solution is -45. To check, we replace x with -45:

$$\frac{-45}{5} \stackrel{?}{=} -9$$

$$-9 = -9 \quad \text{(True)}$$

The solution is verified.

CHECK YOURSELF 4

Solve and check.

(a) $\frac{x}{7} = 3$ **(b)** $\frac{x}{4} = -8$

When the variable is multiplied by a fraction that has a numerator other than 1, there are two approaches to finding the solution.

Example 5

Solving Equations by Using Reciprocals

Solve

$$\frac{3}{5}x = 9$$

One approach is to multiply by 5 as the first step.

$$5\left(\frac{3}{5}x\right) = 5 \cdot 9$$

$$3x = 45$$

Now we divide by 3.

$$\frac{3x}{3} = \frac{45}{3}$$

$$x = 15$$

To check:

$$\frac{3}{5} \cdot 15 \stackrel{?}{=} 9$$

$$9 = 9 \quad \text{(True)}$$

A second approach combines the multiplication and division steps and is generally a bit more efficient. We multiply by $\frac{5}{3}$.

NOTE Recall that $\frac{5}{3}$ is the *reciprocal* of $\frac{3}{5}$, and the product of a number and its reciprocal is just 1! So

$$\left(\frac{5}{3}\right)\left(\frac{3}{5}\right) = 1$$

$$\frac{5}{3}\left(\frac{3}{5}x\right) = \frac{5}{3} \cdot 9$$

$$x = \frac{5}{\cancel{3}_1} \cdot \frac{\cancel{9}^3}{1} = 15$$

So $x = 15$, as before.

CHECK YOURSELF 5

Solve and check.

$$\frac{2}{3}x = 18$$

You may sometimes have to simplify an equation before applying the methods of this section. Example 6 illustrates this property.

Example 6

Combining Like Terms and Solving Equations

Solve and check:

$$3x + 5x = 40$$

Using the distributive property, we can combine the like terms on the left to write

$$8x = 40$$

We can now proceed as before.

$$\frac{8x}{8} = \frac{40}{8} \qquad \text{Divide by 8.}$$

$$x = 5$$

The solution is 5. To check, we return to the original equation. Substituting 5 for x yields

$$3 \cdot 5 + 5 \cdot 5 \stackrel{?}{=} 40$$

$$15 + 25 \stackrel{?}{=} 40$$

$$40 = 40 \qquad \text{(True)}$$

The solution is verified.

CHECK YOURSELF 6

Solve and check.

$7x + 4x = -66$

An **average** is a value that is representative of a set of numbers. One kind of average is the *mean.*

Definitions: Mean

The **mean** of a set is the sum of the set divided by the number of elements in the set. The mean is written as $\bar{x}$ (sometimes called "x-bar"). In mathematical symbols, we say

$$\bar{x} = \frac{\Sigma x}{n}$$

←The sum of the set
←The number of elements in the set

Example 7

Finding the Mean

Find the mean for each set of numbers.

(a) 2, −3, 5, 4, 7

We begin by finding Σx.

$\Sigma x = 2 + (-3) + 5 + 4 + 7 = 15$

Next we find n.

$n = 5$ Remember that n is the number of elements in the set.

Finally, we substitute our numbers into the equation.

$$\bar{x} = \frac{\Sigma x}{n} = \frac{15}{5} = 3$$

The mean of the set is 3.

(b) −4, 7, 9, −3, 6, −2, −3, 8

First find Σx.

$\Sigma x = (-4) + 7 + 9 + (-3) + 6 + (-2) + (-3) + 8 = 18$

Next find n.

$n = 8$

Substitute these numbers into the equation

$$\bar{x} = \frac{\Sigma x}{n} = \frac{18}{8} = \frac{9}{4} \text{ (or 2.25)}$$

The mean of this set is $\frac{9}{4}$ or 2.25

CHECK YOURSELF 7

Find the mean for each set of numbers.

(a) 5, −2, 6, 3, −2 **(b)** 6, −2, 3, 8, 5, −6, 1, −3

Example 8

Finding the Mean

During a week in February the low temperature in Fargo, North Dakota, was recorded each day. The results are presented in the following table. Find both the median and the mean for the set of numbers.

M	T	W	Th	F	Sa	Su
−11	−17	−15	−18	−20	−2	20

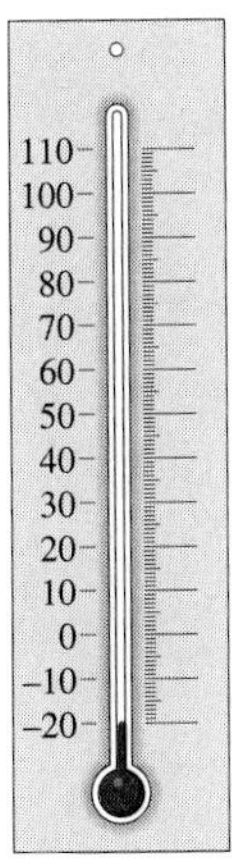

NOTE You can review the discussion of the median in Section 1.3.

To find the median we place the numbers in ascending order:

−20 −18 −17 −15 −11 −2 20

The median is the middle value, so the median is −15 degrees.

To find the mean, we first find Σx.

$$\Sigma x = (-11) + (-17) + (-15) + (-18) + (-20) + (-2) + 20 = -63$$

Then, given that $n = 7$, we use the equation for the mean.

$$\bar{x} = \frac{\Sigma x}{n} = \frac{-63}{7} = -9$$

The mean is −9.

Which average was more appropriate? There is really no "right" answer to that question. In this case, the median would probably be preferred by most statisticians. It yields a temperature that was actually the low temperature on Wednesday of that week, so it is more representative of the set of low temperatures.

CHECK YOURSELF 8

The low temperatures in Anchorage, Alaska, for one week in January are given in the following table. Compute both the median and the mean low temperature for that week.

M	T	W	Th	F	Sa	Su
6	−10	−12	−22	−28	−26	−27

CHECK YOURSELF ANSWERS

1. 4 **2.** -6 **3.** 6 **4.** **(a)** 21; **(b)** -32 **5.** 27 **6.** -6
7. **(a)** 2; **(b)** 1.5 **8.** mean $= -17$, median $= -22$

Name ____________

Section ________ Date ________

2.2 Exercises

Solve for x and check your result.

1. $5x = 20$
2. $6x = 30$
3. $9x = 54$
4. $6x = -42$
5. $63 = 9x$
6. $66 = 6x$
7. $4x = -16$
8. $-3x = 27$
9. $-9x = 72$
10. $10x = -100$
11. $6x = -54$
12. $-7x = 49$
13. $-4x = -12$
14. $52 = -4x$
15. $-42 = 6x$
16. $-7x = -35$
17. $-6x = -54$
18. $-4x = -24$
19. $\frac{x}{2} = 4$
20. $\frac{x}{3} = 2$
21. $\frac{x}{5} = 3$
22. $\frac{x}{8} = 5$
23. $6 = \frac{x}{7}$
24. $6 = \frac{x}{3}$
25. $\frac{x}{5} = -4$
26. $\frac{x}{7} = -5$
27. $-\frac{x}{3} = 8$
28. $-\frac{x}{4} = -3$
29. $\frac{2}{3}x = 0.9$
30. $\frac{4}{5}x = 8$
31. $\frac{3}{4}x = -15$
32. $\frac{3}{5}x = 10 - \frac{6}{5}$
33. $-\frac{5}{6}x = -15$
34. $5x + 4x = 36$
35. $16x - 9x = -16.1$
36. $4x - 2x + 7x = 36$

ANSWERS

1. 4
2. 5
3. 6
4. −7
5. 7
6. 11
7. −4
8. −9
9. −8
10. −10
11. −9
12. −7
13. 3
14. −13
15. −7
16. 5
17. 9
18. 6
19. 8
20. 6
21. 15
22. 40
23. 42
24. 18
25. −20
26. −35
27. −24
28. 12
29. 1.35
30. 10
31. −20
32. $14\frac{2}{3}$
33. 18
34. 4
35. −2.3
36. 4

37. 4

38. −3

39. −0.78

40. 3.79

41. 3

42. −2

43. Mean 4 Median 4

44. Mean 8 Median 8

45. Mean 3 Median 3

46. Mean 2 Median $\frac{5}{2}$

47. Mean 2 Median $\frac{9}{4}$

48. Mean 2 Median $\frac{2}{3}$

49. Mean 17.5 oz Median 17 oz

50. Mean $157.80 Median $156

a. $2x - 6$

b. $3a + 12$

c. $10b + 5$

d. $9p - 12$

e. $21x - 28$

f. $-20x - 16$

g. $-12x + 9$

h. $-15y + 10$

Once again, certain equations involving decimal fractions can be solved by the methods of this section. For instance, to solve $2.3x = 6.9$ we simply use our multiplication property to divide both sides of the equation by 2.3. This will isolate x on the left as desired. Use this idea to solve each of the following equations for x.

37. $3.2x = 12.8$

38. $5.1x = -15.3$

39. $-4.5x = 3.51$

40. $-8.2x = -31.078$

41. $1.3x + 2.8x = 12.3$

42. $2.7x + 5.4x = -16.2$

Find the median and the mean of each data set.

43. 2, 3, 4, 5, 6

44. 1, 3, 8, 10, 18

45. −3, −1, 2, 4, 6, 10

46. −5, −2, 1, 4, 6, 8

47. $-\frac{3}{2}, -1, 2, \frac{5}{2}, 3, 7$

48. $-\frac{4}{3}, -\frac{1}{3}, \frac{2}{3}, 5, 6$

49. Average weight. Kareem bought four bags of candy. The weights of the bags were 16 ounces (oz), 21 oz, 18 oz, and 15 oz. Find the median and the mean weight of the bags of candy.

50. Average savings. Jose has savings accounts for each of his five children. They contain $215, $156, $318, $75, and $25. Find the median and the mean amount of money per account.

Getting Ready for Section 2.3 [Section 1.2]

Use the distributive property to remove the parentheses in the following expressions.

(a) $2(x - 3)$ (b) $3(a + 4)$ (c) $5(2b + 1)$ (d) $3(3p - 4)$
(e) $7(3x - 4)$ (f) $-4(5x + 4)$ (g) $-3(4x - 3)$ (h) $-5(3y - 2)$

Answers

1. 4 **3.** 6 **5.** 7 **7.** −4 **9.** −8 **11.** −9 **13.** 3 **15.** −7 **17.** 9 **19.** 8 **21.** 15 **23.** 42 **25.** −20 **27.** −24 **29.** 1.35 **31.** −20 **33.** 18 **35.** −2.3 **37.** 4 **39.** −0.78 **41.** 3 **43.** 4 **45.** 3 **47.** 2 **49.** Mean: 17.5, Median: 17 oz. **a.** $2x - 6$ **b.** $3a + 12$ **c.** $10b + 5$ **d.** $9p - 12$ **e.** $21x - 28$ **f.** $-20x - 16$ **g.** $-12x + 9$ **h.** $-15y + 10$

Combining the Rules to Solve Equations

OBJECTIVES

1. Combine the addition and multiplication properties to solve an equation
2. Use the order of operations when solving an equation
3. Recognize identities
4. Recognize equations with no solutions

In all our examples thus far, either the addition property or the multiplication property was used in solving an equation. Often, finding a solution will require the use of both properties.

Example 1

Solving Equations

(a) Solve

$4x - 5 = 7$

Here x is *multiplied* by 4. The result, $4x$, then has 5 subtracted from it (or -5 added to it) on the left side of the equation. These two operations mean that both properties must be applied in solving the equation.

Because the variable term is already on the left, we start by adding 5 to both sides:

$$\begin{array}{rcl} 4x - 5 &=& 7 \\ +5 && +5 \\ \hline 4x &=& 12 \end{array}$$

We now divide both sides by 4:

$$\frac{4x}{4} = \frac{12}{4}$$

$$x = 3$$

The solution is 3. To check, replace x with 3 in the original equation. Be careful to follow the rules for the order of operations.

$$4 \cdot 3 - 5 \stackrel{?}{=} 7$$

$$12 - 5 \stackrel{?}{=} 7$$

$$7 = 7 \quad \text{(True)}$$

(b) Solve

$$\begin{array}{rcl} 3x + 8 &=& -4 \\ -8 && -8 \\ \hline 3x &=& -12 \end{array}$$

Add -8 to both sides.

Now divide both sides by 3 to isolate x on the left.

$$\frac{3x}{3} = \frac{-12}{3}$$

$$x = -4$$

The solution is -4. We'll leave the check of this result to you.

CHECK YOURSELF 1

Solve and check.

(a) $6x + 9 = -15$ **(b)** $5x - 8 = 7$

The variable may appear in any position in an equation. Just apply the rules carefully as you try to write an equivalent equation, and you will find the solution. Example 2 illustrates this property.

Example 2

Solving Equations

Solve

$$\begin{array}{rcr} 3 - 2x &=& 9 \\ -3 & & -3 \\ \hline -2x &=& 6 \end{array}$$

First add -3 to both sides.

NOTE $\frac{-2}{-2} = 1$, so we divide by -2 to isolate x on the left.

Now divide both sides by -2. This will leave x alone on the left.

$$\frac{-2x}{-2} = \frac{6}{-2}$$

$$x = -3$$

The solution is -3. We'll leave it to you to check this result.

CHECK YOURSELF 2

Solve and check.

$10 - 3x = 1$

You may also have to combine multiplication with addition or subtraction to solve an equation. Consider Example 3.

Example 3

Solving Equations

(a) Solve

$$\frac{x}{5} - 3 = 4$$

To get the x term alone, we first add 3 to both sides.

$$\begin{array}{rcr} \frac{x}{5} - 3 &=& 4 \\ +3 & & +3 \\ \hline \frac{x}{5} &=& 7 \end{array}$$

Now, to undo the division multiply both sides of the equation by 5.

$$5\left(\frac{x}{5}\right) = 5 \cdot 7$$

$$x = 35$$

The solution is 35. Just return to the original equation to check the result.

$$\frac{35}{5} - 3 \stackrel{?}{=} 4$$

$$7 - 3 \stackrel{?}{=} 4$$

$$4 = 4 \quad \text{(True)}$$

(b) Solve

$$\begin{aligned} \frac{2}{3}x + 5 &= 13 \\ -5 \quad &\ -5 \\ \hline \frac{2}{3}x \quad &= 8 \end{aligned}$$

First add −5 to both sides.

Now multiply both sides by $\frac{3}{2}$, the reciprocal of $\frac{2}{3}$.

$$\left(\frac{3}{2}\right)\left(\frac{2}{3}x\right) = \left(\frac{3}{2}\right)8$$

or

$$x = 12$$

The solution is 12. We'll leave it to you to check this result.

CHECK YOURSELF 3

Solve and check.

(a) $\frac{x}{6} + 5 = 3$ **(b)** $\frac{3}{4}x - 8 = 10$

In Section 2.1, you learned how to solve certain equations when the variable appeared on both sides. Example 4 will show you how to extend that work by using the multiplication property of equality.

Example 4

Solving an Equation

Solve

$$6x - 4 = 3x - 2$$

First add 4 to both sides. This will undo the subtraction on the left.

$$\begin{aligned} 6x - 4 &= 3x - 2 \\ +4 \quad &\quad +4 \\ \hline 6x \quad &= 3x + 2 \end{aligned}$$

Now add $-3x$ so that the terms in x will be on the left only.

$$\begin{aligned} 6x &= 3x + 2 \\ -3x \quad &\ -3x \\ \hline 3x &= \quad 2 \end{aligned}$$

Finally divide by 3.

$$\frac{3x}{3} = \frac{2}{3}$$

$$x = \frac{2}{3}$$

Check:

$$6\left(\frac{2}{3}\right) - 4 \stackrel{?}{=} 3\left(\frac{2}{3}\right) - 2$$

$$4 - 4 \stackrel{?}{=} 2 - 2$$

$$0 = 0 \quad \text{(True)}$$

As you know, the basic idea is to use our two properties to form an equivalent equation with the x isolated. Here we added 4 and then subtracted $3x$. You can do these steps in either order. Try it for yourself the other way. In either case, the multiplication property is then used as the *last step* in finding the solution.

CHECK YOURSELF 4

Solve and check.

$7x - 5 = 3x + 5$

Let's look at two approaches to solving equations in which the coefficient on the right side is greater than the coefficient on the left side.

Example 5

Solving an Equation (Two Methods)

Solve $4x - 8 = 7x + 7$.

Method 1

$$\begin{array}{rcl} 4x - 8 &=& 7x + 7 \\ +8 & & +8 \\ \hline 4x &=& 7x + 15 \\ -7x & & -7x \\ \hline -3x &=& 15 \end{array}$$

Adding 8 will leave the x term alone on the left.

Adding $-7x$ will get the variable terms on the left.

$$\frac{-3x}{-3} = \frac{15}{-3}$$

Dividing by -3 will isolate x on the left.

$$x = -5$$

We'll let you check this result.

To avoid a negative coefficient (in Example 5, -3), some students prefer a different approach.

This time we'll work toward having the number on the *left* and the x term on the *right*, or

$\square = x$

Method 2

NOTE It is usually easier to isolate the variable term on the side that will result in a positive coefficient.

$$\begin{array}{rcl} 4x - 8 &=& 7x + 7 \\ -7 && -7 \\ \hline 4x - 15 &=& 7x \\ -4x && -4x \\ \hline -15 &=& 3x \end{array}$$

Add −7.

Add −4*x* to get the variables on the right.

$$\frac{-15}{3} = \frac{3x}{3}$$ Dividing by 3 to isolate *x* on the right.

$$-5 = x$$

Because $-5 = x$ and $x = -5$ are equivalent equations, it really makes no difference; the solution is still −5! You can use whichever approach you prefer.

CHECK YOURSELF 5

Solve $5x + 3 = 9x - 21$ *by finding equivalent equations of the form* $x = \square$ *and* $\square = x$ *to compare the two methods of finding the solution.*

It may also be necessary to remove grouping symbols in solving an equation.

Example 6

Solving Equations That Contain Parentheses

Solve and check.

NOTE

$5(x - 3)$
$= 5\,(x + (-3))$
$= 5x + 5\,(-3)$
$= 5x + (-15)$
$= 5x - 15$

$$\begin{array}{rcl} 5(x - 3) - 2x &=& x + 7 \\ 5x - 15 - 2x &=& x + 7 \\ 3x - 15 &=& x + 7 \\ +15 && +15 \\ \hline 3x &=& x + 22 \\ -x && -x \\ \hline 2x &=& 22 \\ x &=& 11 \end{array}$$

First, apply the distributive property.
Combine like terms.
Add 15.
Add −*x*.
Divide by 2.

The solution is 11. To check, substitute 11 for *x* in the original equation. Again note the use of our rules for the order of operations.

$5(11 - 3) - 2 \cdot 11 \stackrel{?}{=} 11 + 7$ Simplify terms in parentheses.

$5 \cdot 8 - 2 \cdot 11 \stackrel{?}{=} 11 + 7$ Multiply.

$40 - 22 \stackrel{?}{=} 11 + 7$ Add and subtract.

$18 = 18$ A true statement.

CHECK YOURSELF 6

Solve and check.

$$7(x + 5) - 3x = x - 7$$

An equation that is true for any value of *x* is called an **identity**.

NOTE Here $\square$ means an expression containing all the numbers or letters *other* than *h*.

We now have the height *h* in terms of the area *A* and the base *b*. This is called **solving the equation for *h*** and means that we are rewriting the formula as an equivalent equation of the form

$$h = \square$$

CHECK YOURSELF 1

Solve $V = \frac{1}{3}Bh$ *for h.*

You have already learned the methods needed to solve most literal equations or formulas for some specified variable. As Example 1 illustrates, the rules of Sections 2.1 and 2.2 are applied in exactly the same way as they were applied to equations with one variable.

You may have to apply both the addition and the multiplication properties when solving a formula for a specified variable. Example 2 illustrates this property.

Example 2

Solving a Literal Equation

NOTE This is a linear equation in two variables. You will see this again in Chapter 6.

Solve $y = mx + b$ for x.

Remember that we want to end up with x alone on one side of the equation. Let's start by subtracting b from both sides to undo the addition on the right.

$$\begin{array}{r} y = mx + b \\ \underline{-b \qquad\quad -b} \\ y - b = mx \end{array}$$

If we now divide both sides by m, then x will be alone on the right-hand side.

$$\frac{y-b}{m} = \frac{mx}{m}$$

$$\frac{y-b}{m} = x$$

or

$$x = \frac{y-b}{m}$$

CHECK YOURSELF 2

Solve $v = v_0 + gt$ *for t.*

Let's summarize the steps illustrated by our examples.

Step by Step: Solving a Formula or Literal Equation

Step 1 If necessary, multiply both sides of the equation by the same term to clear it of fractions.

Step 2 Add or subtract the same term on both sides of the equation so that all terms involving the variable that you are solving for are on one side of the equation and all other terms are on the other side.

Step 3 Divide both sides of the equation by the coefficient of the variable that you are solving for.

Method 2

NOTE It is usually easier to isolate the variable term on the side that will result in a positive coefficient.

$$\begin{aligned} 4x - 8 &= 7x + 7 \\ -7 \quad & \quad -7 \end{aligned}$$ Add −7.

$$\begin{aligned} 4x - 15 &= 7x \\ -4x \quad & \quad -4x \end{aligned}$$ Add −4x to get the variables on the right.

$$-15 = 3x$$

$$\frac{-15}{3} = \frac{3x}{3}$$ Dividing by 3 to isolate x on the right.

$$-5 = x$$

Because $-5 = x$ and $x = -5$ are equivalent equations, it really makes no difference; the solution is still -5! You can use whichever approach you prefer.

CHECK YOURSELF 5

Solve $5x + 3 = 9x - 21$ by finding equivalent equations of the form $x = \square$ and $\square = x$ to compare the two methods of finding the solution.

It may also be necessary to remove grouping symbols in solving an equation.

Example 6

Solving Equations That Contain Parentheses

Solve and check.

NOTE
$5(x - 3)$
$= 5\,(x + (-3))$
$= 5x + 5\,(-3)$
$= 5x + (-15)$
$= 5x - 15$

$$5(x - 3) - 2x = x + 7$$ First, apply the distributive property.

$$5x - 15 - 2x = x + 7$$ Combine like terms.

$$\begin{aligned} 3x - 15 &= x + 7 \\ +15 \quad & \quad +15 \end{aligned}$$ Add 15.

$$\begin{aligned} 3x &= x + 22 \\ -x \quad & -x \end{aligned}$$ Add −x.

$$2x = 22$$ Divide by 2.

$$x = 11$$

The solution is 11. To check, substitute 11 for x in the original equation. Again note the use of our rules for the order of operations.

$$5(11 - 3) - 2 \cdot 11 \stackrel{?}{=} 11 + 7$$ Simplify terms in parentheses.

$$5 \cdot 8 - 2 \cdot 11 \stackrel{?}{=} 11 + 7$$ Multiply.

$$40 - 22 \stackrel{?}{=} 11 + 7$$ Add and subtract.

$$18 = 18$$ A true statement.

CHECK YOURSELF 6

Solve and check.

$$7(x + 5) - 3x = x - 7$$

An equation that is true for any value of x is called an **identity.**

Example 7

Solving an Equation

Solve the equation $2(x - 3) = 2x - 6$

$$
\begin{aligned}
2(x - 3) &= 2x - 6 \\
2x - 6 &= 2x - 6 \\
-2x \quad &\quad -2x \\
\hline
-6 &= -6
\end{aligned}
$$

NOTE We could ask the question "For what values of x does $-6 = -6$?"

The statement $-6 = -6$ is true for any value of x. The original equation is an identity.

CHECK YOURSELF 7

Solve the equation $3(x - 4) - 2x = x - 12$

There are also equations for which there are no solutions.

Example 8

Solving an Equation

Solve the equation $3(2x - 5) - 4x = 2x + 1$

$$
\begin{aligned}
3(2x - 5) - 4x &= 2x + 1 \\
6x - 15 \quad - 4x &= 2x + 1 \\
2x \quad - 15 &= 2x + 1 \\
-2x \qquad &\quad -2x \\
\hline
-15 &= 1
\end{aligned}
$$

NOTE We could ask the question "For what values of x does $-15 = 1$?"

These two numbers are never equal. The original equation has no solutions.

CHECK YOURSELF 8

Solve the equation $2(x - 5) + x = 3x - 3$

NOTE Such an outline of steps is sometimes called an **algorithm** for the process.

Step by Step: Solving Linear Equations

Step 1 Use the distributive property to remove any grouping symbols. Then simplify by combining like terms on each side of the equation.

Step 2 Add or subtract the same term on each side of the equation until the variable term is on one side and a number is on the other.

Step 3 Multiply or divide both sides of the equation by the same nonzero number so that the variable is alone on one side of the equation. If no variable remains, determine whether the original equation is an identity or whether it has no solutions.

Step 4 Check the solution in the original equation.

CHECK YOURSELF ANSWERS

1. **(a)** -4; **(b)** 3 **2.** 3 **3.** **(a)** -12; **(b)** 24 **4.** $\frac{5}{2}$ **5.** 6 **6.** -14

7. The equation is an identity, x is any real number. **8.** There are no solutions.

Name ____________

Section ________ Date ________

2.3 Exercises

Solve for x and check your result.

1. $2x + 1 = 9$

2. $3x - 1 = 17$

3. $3x - 2 = 7$

4. $5x + 3 = 23$

5. $4x + 7 = 35$

6. $7x - 8 = 13$

7. $2x + 9 = 5$

8. $6x + 25 = -5$

9. $4 - 7x = 18$

10. $8 - 5x = -7$

11. $3 - 4x = -9$

12. $5 - 4x = 25$

13. $\frac{x}{2} + 1 = 5$

14. $\frac{x}{3} - 2 = 3$

15. $\frac{x}{4} - 5 = 3$

16. $\frac{x}{5} + 3 = 8$

17. $\frac{2}{3}x + 5 = 17$

18. $\frac{3}{4}x - 5 = 4$

19. $\frac{4}{5}x - 3 = 13$

20. $\frac{5}{7}x + 4 = 14$

21. $5x = 2x + 9$

22. $7x = 18 - 2x$

23. $3x = 10 - 2x$

24. $11x = 7x + 20$

ANSWERS

1. 4
2. 6
3. 3
4. 4
5. 7
6. 3
7. −2
8. −5
9. −2
10. 3
11. 3
12. −5
13. 8
14. 15
15. 32
16. 25
17. 18
18. 12
19. 20
20. 14
21. 3
22. 2
23. 2
24. 5

ANSWERS

25. 6
26. 5
27. −2
28. −8
29. $-\frac{10}{3}$
30. $-\frac{5}{2}$
31. −4
32. 3
33. 6
34. 7
35. 4
36. 2
37. 5
38. −5
39. −4
40. 11
41. $\frac{5}{3}$
42. $-\frac{3}{2}$
43. 4
44. −5
45. $-\frac{3}{5}$
46. $\frac{2}{3}$
47. 9
48. 8
49. $-\frac{13}{2}$
50. −4

25. $9x + 2 = 3x + 38$

26. $8x - 3 = 4x + 17$

27. $4x - 8 = x - 14$

28. $6x - 5 = 3x - 29$

29. $5x + 7 = 2x - 3$

30. $9x + 7 = 5x - 3$

31. $7x - 3 = 9x + 5$

32. $5x - 2 = 8x - 11$

33. $5x + 4 = 7x - 8$

34. $2x + 23 = 6x - 5$

35. $2x - 3 + 5x = 7 + 4x + 2$

36. $8x - 7 - 2x = 2 + 4x - 5$

37. $6x + 7 - 4x = 8 + 7x - 26$

38. $7x - 2 - 3x = 5 + 8x + 13$

39. $9x - 2 + 7x + 13 = 10x - 13$

40. $5x + 3 + 6x - 11 = 8x + 25$

41. $8x - 7 + 5x - 10 = 10x - 12$

42. $10x - 9 + 2x - 3 = 8x - 18$

43. $7(2x - 1) - 5x = x + 25$

44. $9(3x + 2) - 10x = 12x - 7$

45. $3x + 2(4x - 3) = 6x - 9$

46. $7x + 3(2x + 5) = 10x + 17$

47. $\frac{8}{3}x - 3 = \frac{2}{3}x + 15$

48. $\frac{12}{5}x + 7 = 31 - \frac{3}{5}x$

49. $\frac{2}{5}x - 5 = \frac{12}{5}x + 8$

50. $\frac{3}{7}x - 5 = \frac{24}{7}x + 7$

51. $5.3x - 7 = 2.3x + 5$

52. $9.8x + 2 = 3.8x + 20$

53. $4(x + 5) = 4x + 20$

54. $-3(2x - 4) - 12 = -6x$

55. $5(x + 1) - 4x = x - 5$

56. $-4(2x - 3) = -8x + 5$

57. $6x - 4x + 1 = 12 + 2x - 11$

58. $-2x + 5x - 9 = 3(x - 4) - 5$

59. $-4(x + 2) - 11 = 2(-2x - 3) - 13$

60. $4(-x - 2) + 5 = -2(2x + 7)$

61. Create an equation of the form $ax + b = c$ that has 2 as a solution.

62. Create an equation of the form $ax + b = c$ that has 7 as a solution.

63. The equation $3x = 3x + 5$ has no solution, whereas the equation $7x + 8 = 8$ has zero as a solution. Explain the difference between a solution of zero and no solution.

64. Construct an equation for which every real number is a solution.

In exercises 65 to 68, find the length of each side of the figure for the given perimeter.

65. (Triangle with sides x, $2x - 2$, $x + 2$)

$P = 24$ in.

66. (Rectangle with sides $3x - 4$, x)

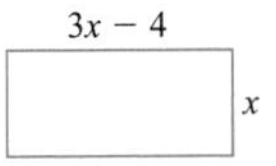

$P = 32$ cm

67. 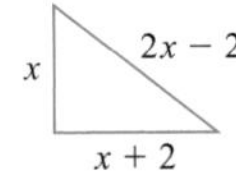

$P = 90$ in.

68. 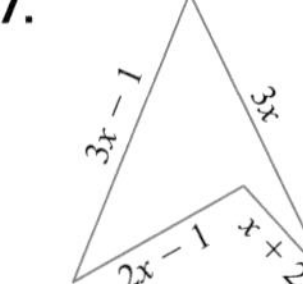

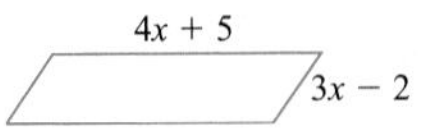

$P = 34$ cm

ANSWERS

51. 4
52. 3
53. Identity
54. Identity
55. No solution
56. No solution
57. Identity
58. No solution
59. Identity
60. No solution
61. $6x + 5 = 17$
62. $3x + 11 = 32$
63. 
64. 
65. 6 in., 8 in., 10 in.
66. 5 cm, 11 cm
67. 12 in., 19 in., 29 in., 30 in.
68. 4 cm, 13 cm

ANSWERS

a. b

b. x

c. y

d. b

e. m

f. b

g. t

h. y

Getting Ready for Section 2.4 [Section 1.7]

Divide.

(a) $\frac{3b}{3}$ (b) $\frac{5x}{5}$

(c) $\frac{4xy}{4x}$ (d) $\frac{6a^2b}{6a^2}$

(e) $\frac{7mn^2}{7n^2}$ (f) $\frac{\pi ab}{\pi a}$

(g) $\frac{srt}{sr}$ (h) $\frac{x^2yz}{x^2z}$

Answers

1. 4 **3.** 3 **5.** 7 **7.** -2 **9.** -2 **11.** 3 **13.** 8 **15.** 32 **17.** 18 **19.** 20 **21.** 3 **23.** 2 **25.** 6 **27.** -2 **29.** $-\frac{10}{3}$ **31.** -4 **33.** 6 **35.** 4 **37.** 5 **39.** -4 **41.** $\frac{5}{3}$ **43.** 4 **45.** $-\frac{3}{5}$ **47.** 9 **49.** $-\frac{13}{2}$ **51.** 4 **53.** Identity **55.** No solution **57.** Identity **59.** Identity **61.** $6x + 5 = 17$ **63.** (writing icon) **65.** 6 in., 8 in., 10 in. **67.** 12 in., 19 in., 29 in., 30 in.

a. b **b.** x **c.** y **d.** b **e.** m **f.** b **g.** t **h.** y

Formulas and Problem Solving

OBJECTIVES

1. Solve a literal equation for one of its variables
2. Translate a word statement to an equation
3. Use an equation to solve an application

Formulas are extremely useful tools in any field in which mathematics is applied. Formulas are simply equations that express a relationship between more than one letter or variable. You are no doubt familiar with all kinds of formulas, such as

$A = \frac{1}{2}bh$ The area of a triangle

$I = Prt$ Interest

$V = \pi r^2 h$ The volume of a cylinder

Actually a formula is also called a **literal equation** because it involves several letters or variables. For instance, our first formula or literal equation, $A = \frac{1}{2}bh$, involves the three letters A (for area), b (for base), and h (for height).

Unfortunately, formulas are not always given in the form needed to solve a particular problem. Then algebra is needed to change the formula to a more useful equivalent equation, which is solved for a particular letter or variable. The steps used in the process are very similar to those you used in solving linear equations. Let's consider an example.

Example 1

Solving a Literal Equation Involving a Triangle

Suppose that we know the area A and the base b of a triangle and want to find its height h.

We are given

$$A = \frac{1}{2}bh$$

Our job is to find an equivalent equation with h, the unknown, by itself on one side. We call $\frac{1}{2}b$ the **coefficient** of h. We can remove the two *factors* of that coefficient, $\frac{1}{2}$ and b, separately.

$$2A = 2\left(\frac{1}{2}bh\right)$$ Multiply both sides by 2 to clear the equation of fractions.

or

$$2A = bh$$

$$\frac{2A}{b} = \frac{bh}{b}$$ Divide by b to isolate h.

$$\frac{2A}{b} = h$$

or

$$h = \frac{2A}{b}$$ Reverse the sides to write h on the left.

NOTE

$2\left(\frac{1}{2}bh\right) = \left(2 \cdot \frac{1}{2}\right)(bh)$
$= 1 \cdot bh$
$= bh$

We now have the height h in terms of the area A and the base b. This is called **solving the equation for h** and means that we are rewriting the formula as an equivalent equation of the form

NOTE Here $\square$ means an expression containing all the numbers or letters *other* than h.

$$h = \square$$

CHECK YOURSELF 1

Solve $V = \frac{1}{3}Bh$ for h.

You have already learned the methods needed to solve most literal equations or formulas for some specified variable. As Example 1 illustrates, the rules of Sections 2.1 and 2.2 are applied in exactly the same way as they were applied to equations with one variable.

You may have to apply both the addition and the multiplication properties when solving a formula for a specified variable. Example 2 illustrates this property.

Example 2

Solving a Literal Equation

NOTE This is a linear equation in two variables. You will see this again in Chapter 6.

Solve $y = mx + b$ for x.

Remember that we want to end up with x alone on one side of the equation. Let's start by subtracting b from both sides to undo the addition on the right.

$$\begin{aligned} y &= mx + b \\ -b &\quad\;\; -b \\ \hline y - b &= mx \end{aligned}$$

If we now divide both sides by m, then x will be alone on the right-hand side.

$$\frac{y - b}{m} = \frac{mx}{m}$$

$$\frac{y - b}{m} = x$$

or

$$x = \frac{y - b}{m}$$

CHECK YOURSELF 2

Solve $v = v_0 + gt$ for t.

Let's summarize the steps illustrated by our examples.

Step by Step: Solving a Formula or Literal Equation

Step 1 If necessary, multiply both sides of the equation by the same term to clear it of fractions.

Step 2 Add or subtract the same term on both sides of the equation so that all terms involving the variable that you are solving for are on one side of the equation and all other terms are on the other side.

Step 3 Divide both sides of the equation by the coefficient of the variable that you are solving for.

Let's look at one more example, using the above steps.

Example 3

Solving a Literal Equation Involving Money

NOTE This is a formula for the *amount* of money in an account after interest has been earned.

Solve $A = P + Prt$ for r.

$$\begin{array}{rcl} A & = & P + Prt \\ -P & & -P \\ \hline A - P & = & Prt \end{array}$$

Adding $-P$ to both sides will leave the term involving r alone on the right.

$$\frac{A - P}{Pt} = \frac{Prt}{Pt}$$

Dividing both sides by Pt will isolate r on the right.

$$\frac{A - P}{Pt} = r$$

or

$$r = \frac{A - P}{Pt}$$

CHECK YOURSELF 3

Solve $2x + 3y = 6$ for y.

Now let's look at an application of solving a literal equation.

Example 4

Solving a Literal Equation Involving Money

Suppose that the amount in an account, 3 years after a principal of \$5000 was invested, is \$6050. What was the interest rate?

From our previous example,

$$A = P + Prt \tag{1}$$

in which A is the amount in the account, P is the principal, r is the interest rate, and t is the time that the money has been invested. By the result of Example 3 we have

$$r = \frac{A - P}{Pt} \tag{2}$$

NOTE Do you see the advantage of having our equation solved for the desired variable?

and we can substitute the known values in equation (2):

$$r = \frac{6050 - 5000}{(5000)(3)}$$

$$= \frac{1050}{15{,}000} = 0.07 = 7\%$$

The interest rate is 7 percent.

CHECK YOURSELF 4

Suppose that the amount in an account, 4 years after a principal of \$3000 *was invested, is* \$3720. *What was the interest rate?*

The main reason for learning how to set up and solve algebraic equations is so that we can use them to solve word problems. In fact, algebraic equations were *invented* to make solving word problems much easier. The first word problems that we know about are over 4000 years old. They were literally "written in stone," on Babylonian tablets, about 500 years before the first algebraic equation made its appearance.

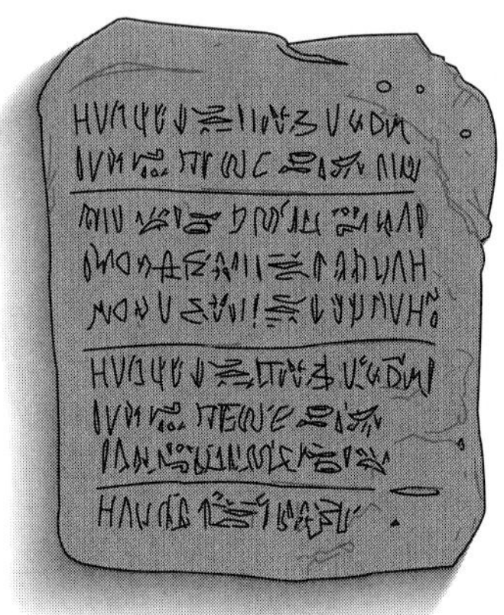

Before algebra, people solved word problems primarily by **substitution,** which is a method of finding unknown numbers by using trial and error in a logical way. Example 5 shows how to solve a word problem using substitution.

Example 5

Solving a Word Problem by Substitution

The sum of two consecutive integers is 37. Find the two integers.

If the two integers were 20 and 21, their sum would be 41. Because that's more than 37, the integers must be smaller. If the integers were 15 and 16, the sum would be 31. More trials yield that the sum of 18 and 19 is 37.

CHECK YOURSELF 5

The sum of two consecutive integers is 91. *Find the two integers.*

Most word problems are not so easily solved by substitution. For more complicated word problems, a five-step procedure is used. Using this step-by-step approach will, with

practice, allow you to organize your work. Organization is the key to solving word problems. Here are the five steps.

Step by Step: To Solve Word Problems

Step 1 Read the problem carefully. Then reread it to decide what you are asked to find.
Step 2 Choose a letter to represent one of the unknowns in the problem. Then represent all other unknowns of the problem with expressions that use the same letter.
Step 3 Translate the problem to the language of algebra to form an equation.
Step 4 Solve the equation and answer the question of the original problem.
Step 5 Check your solution by returning to the original problem.

NOTE We discussed these translations in Section 1.1. You might find it helpful to review that section before going on.

The third step is usually the hardest part. We must translate words to the language of algebra. Before we look at a complete example, the following table may help you review that translation step.

Translating Words to Algebra

Words	Algebra
The sum of x and y	$x + y$
3 plus a	$3 + a$ or $a + 3$
5 more than m	$m + 5$
b increased by 7	$b + 7$
The difference of x and y	$x - y$
4 less than a	$a - 4$
s decreased by 8	$s - 8$
The product of x and y	$x \cdot y$ or xy
5 times a	$5 \cdot a$ or $5a$
Twice m	$2m$
The quotient of x and y	$\frac{x}{y}$
a divided by 6	$\frac{a}{6}$
One-half of b	$\frac{b}{2}$ or $\frac{1}{2}b$

Now let's look at some typical examples of translating phrases to algebra.

Example 6

Translating Statements

Translate each statement to an algebraic expression.

(a) The sum of a and 2 times b $\quad a + 2b$

Sum — 2 times b

(b) 5 times m increased by 1 $\quad 5m + 1$

5 times m — Increased by 1

(c) 5 less than 3 times x $\quad 3x - 5$

3 times x — 5 less than

(d) The product of x and y, divided by 3 $\quad \frac{xy}{3}$

The product of x and y — Divided by 3

CHECK YOURSELF 6

Translate to algebra.

(a) 2 more than twice x

(b) 4 less than 5 times n

(c) The product of twice a and b

(d) The sum of s and t, divided by 5

Now let's work through a complete example. Although this problem could be solved by substitution, it is presented here to help you practice the five-step approach.

Example 7

Solving an Application

The sum of a number and 5 is 17. What is the number?

Step 1 *Read carefully.* You must find the unknown number.

Step 2 *Choose letters or variables.* Let x represent the unknown number. There are no other unknowns.

Step 3 *Translate.*

The sum of

$$x + 5 = 17$$

is

NOTE Always return to the *original problem* to check your result and *not* to the equation of step 3. This will prevent possible errors!

Step 4 *Solve.*

$$\begin{aligned} x + 5 &= 17 \\ -5 \quad & \;\; -5 \qquad \text{Add } -5. \\ \hline x &= 12 \end{aligned}$$

So the number is 12.

Step 5 *Check.* Is the sum of 12 and 5 equal to 17? Yes ($12 + 5 = 17$). We have checked our solution.

CHECK YOURSELF 7

The sum of a number and 8 *is* 35. *What is the number?*

Definitions: Consecutive Integers

Consecutive integers are integers that follow one another, like 10, 11, and 12. To represent them in algebra:

If x is an integer, then $x + 1$ is the next consecutive integer, $x + 2$ is the next, and so on.

We'll need this idea in Example 8.

Example 8

Solving an Application

REMEMBER THE STEPS! Read the problem carefully. What do you need to find?

Assign letters to the unknown or unknowns.

Write an equation.

The sum of two consecutive integers is 41. What are the two integers?

Step 1 We want to find the two consecutive integers.

Step 2 Let x be the first integer. Then $x + 1$ must be the next.

Step 3

The first integer — The second integer

$$\underbrace{x}_{\text{The first integer}} \;\underset{\text{The sum}}{+}\; \overbrace{x + 1}^{\text{The second integer}} \underset{\text{Is}}{=} 41$$

NOTE Solve the equation.

Step 4

$$x + x + 1 = 41$$
$$2x + 1 = 41$$
$$2x = 40$$
$$x = 20$$

The first integer (x) is 20, and the next integer ($x + 1$) is 21.

NOTE Check.

Step 5 The sum of the two integers 20 and 21 is 41.

CHECK YOURSELF 8

The sum of three consecutive integers is 51. *What are the three integers?*

Sometimes algebra is used to reconstruct missing information. Example 9 does just that with some election information.

Example 9

Solving an Application

There were 55 more yes votes than no votes on an election measure. If 735 votes were cast in all, how many yes votes were there? How many no votes?

NOTE What do you need to find?

Step 1 We want to find the number of yes votes and the number of no votes.

NOTE Assign letters to the unknowns.

Step 2 Let x be the number of no votes. Then

$$\underbrace{x + 55}_{\text{55 more than } x}$$

is the number of yes votes.

NOTE Write an equation.

Step 3

$$\underset{\text{No votes}}{x} + \underbrace{x + 55}_{\text{Yes votes}} = 735$$

NOTE Solve the equation.

Step 4

$$\begin{aligned} x + x + 55 &= 735 \\ 2x + 55 &= 735 \\ 2x &= 680 \\ x &= 340 \\ \text{No votes } (x) &= 340 \\ \text{Yes votes } (x + 55) &= 395 \end{aligned}$$

NOTE Check.

Step 5 Thus 340 no votes plus 395 yes votes equals 735 total votes. The solution checks.

CHECK YOURSELF 9

Francine earns $120 per month more than Rob. If they earn a total of $2680 per month, what are their monthly salaries?

Similar methods will allow you to solve a variety of word problems. Example 10 includes three unknown quantities but uses the same basic solution steps.

NOTE There are other choices for x, but choosing the smallest quantity will usually give the easiest equation to write and solve.

Example 10

Solving an Application

Juan worked twice as many hours as Jerry. Marcia worked 3 more hours than Jerry. If they worked a total of 31 hours, find out how many hours each worked.

Step 1 We want to find the hours each worked, so there are three unknowns.

Step 2 Let x be the hours that Jerry worked.

Twice Jerry's hours

Then $2x$ is Juan's hours worked

3 more hours than Jerry worked

and $x + 3$ is Marcia's hours.

Step 3

Jerry Juan Marcia

$$x + 2x + x + 3 = 31$$

Sum of their hours

Step 4

$$x + 2x + x + 3 = 31$$
$$4x + 3 = 31$$
$$4x = 28$$
$$x = 7$$

Jerry's hours $(x) = 7$

Juan's hours $(2x) = 14$

Marcia's hours $(x + 3) = 10$

Step 5 The sum of their hours $(7 + 14 + 10)$ is 31, and the solution is verified.

CHECK YOURSELF 10

Lucy jogged twice as many miles (mi) as Paul but 3 less than Isaac. If the three ran a total of 23 mi, how far did each person run?

CHECK YOURSELF ANSWERS

1. $h = \frac{3V}{B}$ **2.** $t = \frac{v - v_0}{g}$ **3.** $y = \frac{6 - 2x}{3}$ or $y = -\frac{2}{3}x + 2$

4. 6 percent **5.** 45 and 46 **6. (a)** $2x + 2$; **(b)** $5n - 4$; **(c)** $2ab$; **(d)** $\frac{s + t}{5}$

7. The equation is $x + 8 = 35$. The number is 27.

8. The equation is $x + x + 1 + x + 2 = 51$. The integers are 16, 17, and 18.

9. The equation is $x + x + 120 = 2680$. Rob's salary is \$1280, and Francine's is \$1400. **10.** Paul: 4 mi; Lucy: 8 mi; Isaac: 11 mi

Name ____________

Section ________ Date ________

2.4 Exercises

Solve each literal equation for the indicated variable.

1. $p = 4s$ (for s) Perimeter of a square

2. $V = Bh$ (for B) Volume of a prism

3. $E = IR$ (for R) Voltage in an electric circuit

4. $I = Prt$ (for r) Simple interest

5. $V = LWH$ (for H) Volume of a rectangular solid

6. $V = \pi r^2 h$ (for h) Volume of a cylinder

7. $A + B + C = 180$ (for B) Measure of angles in a triangle

8. $P = I^2R$ (for R) Power in an electric circuit

9. $ax + b = 0$ (for x) Linear equation in one variable

10. $y = mx + b$ (for m) Slope-intercept form for a line

11. $s = \frac{1}{2}gt^2$ (for g) Distance

12. $K = \frac{1}{2}mv^2$ (for m) Energy

13. $x + 5y = 15$ (for y) Linear equation

14. $2x + 3y = 6$ (for x) Linear equation

15. $P = 2L + 2W$ (for L) Perimeter of a rectangle

16. $ax + by = c$ (for y) Linear equation in two variables

17. $V = \frac{KT}{P}$ (for T) Volume of a gas

18. $V = \frac{1}{3}\pi r^2 h$ (for h) Volume of a cone

ANSWERS

1. $\frac{p}{4}$
2. $\frac{V}{h}$
3. $\frac{E}{I}$
4. $\frac{I}{Pt}$
5. $\frac{V}{LW}$
6. $\frac{V}{\pi r^2}$
7. $180 - A - C$
8. $\frac{P}{I^2}$
9. $-\frac{b}{a}$
10. $\frac{y - b}{x}$
11. $\frac{2s}{t^2}$
12. $\frac{2K}{v^2}$
13. $\frac{15 - x}{5}$ or $-\frac{1}{5}x + 3$
14. $\frac{6 - 3y}{2}$ or $3 - \frac{3}{2}y$
15. $\frac{P - 2W}{2}$ or $\frac{P}{2} - W$
16. $\frac{c - ax}{b}$
17. $\frac{PV}{K}$
18. $\frac{3V}{\pi r^2}$

ANSWERS

19. $2x - a$

20. $C - nD$

21. $\frac{5}{9}(F - 32)$ or $\frac{5(F - 32)}{9}$

22. $\frac{A - P}{Pr}$ or $\frac{A}{Pr} - \frac{1}{r}$

23. $\frac{S - 2\pi r^2}{2\pi r}$ or $\frac{S}{2\pi r} - r$

24. $\frac{2A - hB}{h}$ or $\frac{2A}{h} - B$

25. 3 cm

26. 3 in.

27. 5 percent

28. 18 ft

29. 25°C

30. 8 m

19. $x = \frac{a + b}{2}$ (for b) Mean of two numbers

20. $D = \frac{C - s}{n}$ (for s) Depreciation

21. $F = \frac{9}{5}C + 32$ (for C) Celsius/Fahrenheit

22. $A = P + Prt$ (for t) Amount at simple interest

23. $S = 2\pi r^2 + 2\pi rh$ (for h) Total surface area of a cylinder

24. $A = \frac{1}{2}h(B + b)$ (for b) Area of a trapezoid

25. **Height of a solid.** A rectangular solid has a base with length 8 centimeters (cm) and width 5 cm. If the volume of the solid is 120 cm^3, find the height of the solid. (See Exercise 5.)

26. **Height of a cylinder.** A cylinder has a radius of 4 inches (in.). If the volume of the cylinder is 48π in.^3, what is the height of the cylinder? (See Exercise 6.)

27. **Interest rate.** A principal of \$3000 was invested in a savings account for 3 years. If the interest earned for the period was \$450, what was the interest rate? (See Exercise 4.)

28. **Length of a rectangle.** If the perimeter of a rectangle is 60 feet (ft) and the width is 12 ft, find its length. (See Exercise 15.)

29. **Temperature conversion.** The high temperature in New York for a particular day was reported at 77°F. How would the same temperature have been given in degrees Celsius? (See Exercise 21.)

30. **Garden length.** Rose's garden is in the shape of a trapezoid. If the height of the trapezoid is 16 meters (m), one base is 20 m, and the area is 224 m^2, find the length of the other base. (See Exercise 24.)

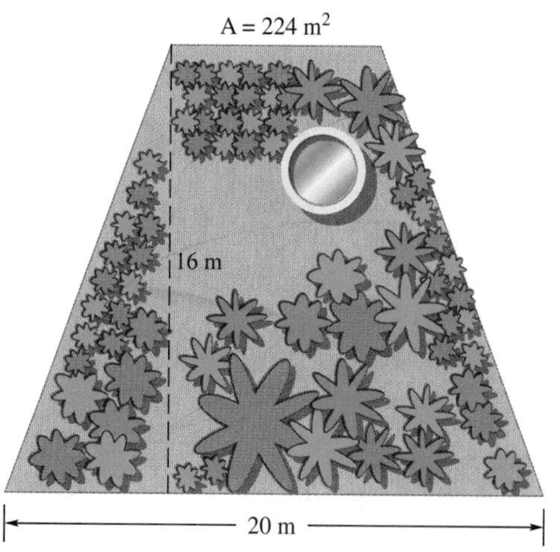

Translate each statement to an algebraic equation. Let x represent the number in each case.

31. 3 more than a number is 7.

32. 5 less than a number is 12.

33. 7 less than 3 times a number is twice that same number.

34. 4 more than 5 times a number is 6 times that same number.

35. 2 times the sum of a number and 5 is 18 more than that same number.

36. 3 times the sum of a number and 7 is 4 times that same number.

37. 3 more than twice a number is 7.

38. 5 less than 3 times a number is 25.

39. 7 less than 4 times a number is 41.

40. 10 more than twice a number is 44.

41. 5 more than two-thirds of a number is 21.

42. 3 less than three-fourths of a number is 24.

43. 3 times a number is 12 more than that number.

44. 5 times a number is 8 less than that number.

Solve the following word problems. Be sure to label the unknowns and to show the equation you use for the solution.

45. Number problem. The sum of a number and 7 is 33. What is the number?

46. Number problem. The sum of a number and 15 is 22. What is the number?

47. Number problem. The sum of a number and -15 is 7. What is the number?

48. Number problem. The sum of a number and -8 is 17. What is the number?

49. Number of votes cast. In an election, the winning candidate has 1840 votes. If the total number of votes cast was 3260, how many votes did the losing candidate receive?

ANSWERS

31. $x + 3 = 7$

32. $x - 5 = 12$

33. $3x - 7 = 2x$

34. $5x + 4 = 6x$

35. $2(x + 5) = x + 18$

36. $3(x + 7) = 4x$

37. $2x + 3 = 7$

38. $3x - 5 = 25$

39. $4x - 7 = 41$

40. $2x + 10 = 44$

41. $\frac{2}{3}x + 5 = 21$

42. $\frac{3}{4}x - 3 = 24$

43. $3x = x + 12$

44. $5x = x - 8$

45. 26

46. 7

47. 22

48. 25

49. 1420

Answers

1. $\frac{p}{4}$ **3.** $\frac{E}{I}$ **5.** $\frac{V}{LW}$ **7.** $180 - A - C$ **9.** $-\frac{b}{a}$ **11.** $\frac{2s}{t^2}$
13. $\frac{15 - x}{5}$ or $-\frac{1}{5}x + 3$ **15.** $\frac{P - 2W}{2}$ **17.** $\frac{PV}{K}$ **19.** $2x - a$
21. $\frac{5}{9}(F - 32)$ or $\frac{5(F - 32)}{9}$ **23.** $\frac{S - 2\pi r^2}{2\pi r}$ or $\frac{S}{2\pi r} - r$ **25.** 3 cm
27. 5% **29.** 25°C **31.** $x + 3 = 7$ **33.** $3x - 7 = 2x$
35. $2(x + 5) = x + 18$ **37.** $2x + 3 = 7$ **39.** $4x - 7 = 41$
41. $\frac{2}{3}x + 5 = 21$ **43.** $3x = x + 12$ **45.** 26 **47.** 22 **49.** 1420
51. 13 **53.** 18 **55.** 35, 36 **57.** 20, 21, 22 **59.** 32, 34 **61.** 25, 27
63. 33, 35, 37 **65.** 20, 21, 22, 23 **67.** 12, 13 **69.** 1710, 1550
71. Washer, $360; dryer, $290 **73.** 9 years old **75.** 29 years old **77.** $820
79. 8 A.M.:23; 10 A.M.:46, 12 P.M.:30 **81.** **83.** **a.** 15
b. 14 **c.** 3 **d.** −16 **e.** −6 **f.** 3600 **g.** −7 **h.** 40

Translate each statement to an algebraic equation. Let x represent the number in each case.

31. 3 more than a number is 7.

32. 5 less than a number is 12.

33. 7 less than 3 times a number is twice that same number.

34. 4 more than 5 times a number is 6 times that same number.

35. 2 times the sum of a number and 5 is 18 more than that same number.

36. 3 times the sum of a number and 7 is 4 times that same number.

37. 3 more than twice a number is 7.

38. 5 less than 3 times a number is 25.

39. 7 less than 4 times a number is 41.

40. 10 more than twice a number is 44.

41. 5 more than two-thirds of a number is 21.

42. 3 less than three-fourths of a number is 24.

43. 3 times a number is 12 more than that number.

44. 5 times a number is 8 less than that number.

Solve the following word problems. Be sure to label the unknowns and to show the equation you use for the solution.

45. Number problem. The sum of a number and 7 is 33. What is the number?

46. Number problem. The sum of a number and 15 is 22. What is the number?

47. Number problem. The sum of a number and -15 is 7. What is the number?

48. Number problem. The sum of a number and -8 is 17. What is the number?

49. Number of votes cast. In an election, the winning candidate has 1840 votes. If the total number of votes cast was 3260, how many votes did the losing candidate receive?

ANSWERS

31. $x + 3 = 7$

32. $x - 5 = 12$

33. $3x - 7 = 2x$

34. $5x + 4 = 6x$

35. $2(x + 5) = x + 18$

36. $3(x + 7) = 4x$

37. $2x + 3 = 7$

38. $3x - 5 = 25$

39. $4x - 7 = 41$

40. $2x + 10 = 44$

41. $\frac{2}{3}x + 5 = 21$

42. $\frac{3}{4}x - 3 = 24$

43. $3x = x + 12$

44. $5x = x - 8$

45. 26

46. 7

47. 22

48. 25

49. 1420

ANSWERS

50. $1360
51. 13
52. 14
53. 18
54. 16
55. 35, 36
56. 72, 73
57. 20, 21, 22
58. 30, 31, 32
59. 32, 34
60. 42, 44
61. 25, 27
62. 43, 45
63. 33, 35, 37
64. 40, 42, 44
65. 20, 21, 22, 23

50. Monthly earnings. Mike and Stefanie work at the same company and make a total of $2760 per month. If Stefanie makes $1400 per month, how much does Mike earn every month?

51. Number addition. The sum of twice a number and 7 is 33. What is the number?

52. Number addition. 3 times a number, increased by 8, is 50. Find the number.

53. Number subtraction. 5 times a number, minus 12, is 78. Find the number.

54. Number subtraction. 4 times a number, decreased by 20, is 44. What is the number?

55. Consecutive integers. The sum of two consecutive integers is 71. Find the two integers.

56. Consecutive integers. The sum of two consecutive integers is 145. Find the two integers.

57. Consecutive integers. The sum of three consecutive integers is 63. What are the three integers?

58. Consecutive integers. If the sum of three consecutive integers is 93, find the three integers.

59. Even integers. The sum of two consecutive even integers is 66. What are the two integers? (*Hint:* Consecutive even integers such as 10, 12, and 14 can be represented by x, $x + 2$, $x + 4$, and so on.)

60. Even integers. If the sum of two consecutive even integers is 86, find the two integers.

61. Odd integers. If the sum of two consecutive odd integers is 52, what are the two integers? (*Hint:* Consecutive odd integers such as 21, 23, and 25 can be represented by x, $x + 2$, $x + 4$, and so on.)

62. Odd integers. The sum of two consecutive odd integers is 88. Find the two integers.

63. Odd integers. The sum of three consecutive odd integers is 105. What are the three integers?

64. Even integers. The sum of three consecutive even integers is 126. What are the three integers?

65. Consecutive integers. The sum of four consecutive integers is 86. What are the four integers?

66. **Consecutive integers.** The sum of four consecutive integers is 62. What are the four integers?

67. **Consecutive integers.** 4 times an integer is 9 more than 3 times the next consecutive integer. What are the two integers?

68. **Consecutive integers.** 4 times an integer is 30 less than 5 times the next consecutive even integer. Find the two integers.

69. **Election votes.** In an election, the winning candidate had 160 more votes than the loser. If the total number of votes cast was 3260, how many votes did each candidate receive?

70. **Monthly salaries.** Jody earns $140 more per month than Frank. If their monthly salaries total $2760, what amount does each earn?

71. **Appliance costs.** A washer-dryer combination costs $650. If the washer costs $70 more than the dryer, what does each appliance cost?

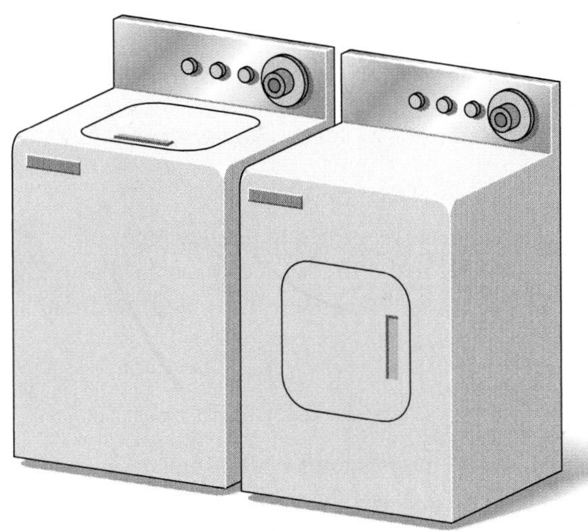

72. **Length of materials.** Yuri has a board that is 98 inches (in.) long. He wishes to cut the board into two pieces so that one piece will be 10 in. longer than the other. What should be the length of each piece?

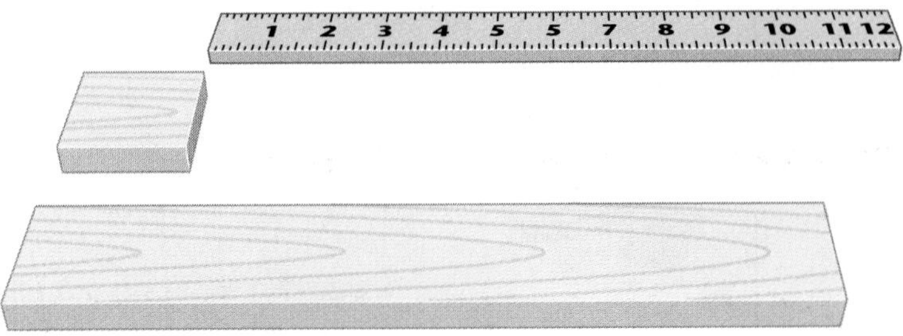

ANSWERS

66. 14, 15, 16, 17

67. 12, 13

68. 20, 22

69. 1710, 1550

70. Jody, $1450; Frank, $1310

71. Washer, $360; dryer, $290

72. 44 in., 54 in.

ANSWERS

73. 9 years old

74. Diane, 18 years old; Dan, 9 years old

75. 29 years old

76. 32 years old

77. $820

78. Tom, $18,000; Rachel, $30,000

79. 8 A.M.: 23, 10 A.M.: 46, 12 P.M.: 30

80. September: 30 gal, October: 42 gal, November: 60 gal

73. Age. Yan Ling is 1 year less than twice as old as his sister. If the sum of their ages is 14 years, how old is Yan Ling?

74. Age. Diane is twice as old as her brother Dan. If the sum of their ages is 27 years, how old are Diane and her brother?

75. Age. Maritza is 3 years less than 4 times as old as her daughter. If the sum of their ages is 37, how old is Maritza?

76. Age. Mrs. Jackson is 2 years more than 3 times as old as her son. If the difference between their ages is 22 years, how old is Mrs. Jackson?

77. Airfare costs. On her vacation in Europe, Jovita's expenses for food and lodging were $60 less than twice as much as her airfare. If she spent $2400 in all, what was her airfare?

78. Earnings. Rachel earns $6000 less than twice as much as Tom. If their two incomes total $48,000, how much does each earn?

79. Number of students. There are 99 students registered in three sections of algebra. There are twice as many students in the 10 A.M. section as the 8 A.M. section and 7 more students at 12 P.M. than at 8 A.M. How many students are in each section?

80. Gallons of fuel. The Randolphs used 12 more gallons (gal) of fuel oil in October than in September and twice as much oil in November as in September. If they used 132 gal for the 3 months, how much was used during each month?

81. "I make \$2.50 an hour more in my new job." If x = the amount I used to make per hour and y = the amount I now make, which equation(s) below say the same thing as the statement above? Explain your choice(s) by translating the equation into English and comparing with the original statement.

(a) $x + y = 2.50$

(b) $x - y = 2.50$

(c) $x + 2.50 = y$

(d) $2.50 + y = x$

(e) $y - x = 2.50$

(f) $2.50 - x = y$

82. "The river rose 4 feet above flood stage last night." If a = the river's height at flood stage, b = the river's height now (the morning after), which equations below say the same thing as the statement? Explain your choices by translating the equations into English and comparing the meaning with the original statement.

(a) $a + b = 4$

(b) $b - 4 = a$

(c) $a - 4 = b$

(d) $a + 4 = b$

(e) $b + 4 = b$

(f) $b - a = 4$

83. Maxine lives in Pittsburgh, Pennsylvania, and pays 8.33 cents per kilowatt hour (kWh) for electricity. During the 6 months of cold winter weather, her household uses about 1500 kWh of electric power per month. During the two hottest summer months, the usage is also high because the family uses electricity to run an air conditioner. During these summer months, the usage is 1200 kWh per month; the rest of the year, usage averages 900 kWh per month.

(a) Write an expression for the total yearly electric bill.

(b) Maxine is considering spending \$2000 for more insulation for her home so that it is less expensive to heat and to cool. The insulation company claims that "with proper installation the insulation will reduce your heating and cooling bills by 25 percent." If Maxine invests the money in insulation, how long will it take her to get her money back in saving on her electric bill? Write to her about what information she needs to answer this question. Give her your opinion about how long it will take to save \$2000 on heating and cooling bills, and explain your reasoning. What is your advice to Maxine?

Getting Ready for Section 2.5 [Section 1.4]

Perform the indicated operations.

(a) $4 \cdot (8 - 6) + 7$

(b) $3 \cdot (8 - 5) + 10 \div 2$

(c) $4(6 - 8) \div 4 + 5$

(d) $8(7 - 3 \cdot 4) + 12(5 - 3)$

(e) $-2(16 \div 4 \cdot 2) - 5(3 - 5)$

(f) $8 \cdot (4 \cdot 5 - 2)(10 \div 2 \cdot 5)$

(g) $-7(13 - 4 \cdot 2) \div (25 - 10 \cdot 2)$

(h) $2(-25 + 5 \cdot 3) \cdot (4 - 3 \cdot 2)$

ANSWERS

81.

82.

83.

a. 15

b. 14

c. 3

d. −16

e. −6

f. 3600

g. −7

h. 40

Answers

1. $\frac{p}{4}$ **3.** $\frac{E}{I}$ **5.** $\frac{V}{LW}$ **7.** $180 - A - C$ **9.** $-\frac{b}{a}$ **11.** $\frac{2s}{t^2}$

13. $\frac{15 - x}{5}$ or $-\frac{1}{5}x + 3$ **15.** $\frac{P - 2W}{2}$ **17.** $\frac{PV}{K}$ **19.** $2x - a$

21. $\frac{5}{9}(F - 32)$ or $\frac{5(F - 32)}{9}$ **23.** $\frac{S - 2\pi r^2}{2\pi r}$ or $\frac{S}{2\pi r} - r$ **25.** 3 cm

27. 5% **29.** 25°C **31.** $x + 3 = 7$ **33.** $3x - 7 = 2x$

35. $2(x + 5) = x + 18$ **37.** $2x + 3 = 7$ **39.** $4x - 7 = 41$

41. $\frac{2}{3}x + 5 = 21$ **43.** $3x = x + 12$ **45.** 26 **47.** 22 **49.** 1420

51. 13 **53.** 18 **55.** 35, 36 **57.** 20, 21, 22 **59.** 32, 34 **61.** 25, 27

63. 33, 35, 37 **65.** 20, 21, 22, 23 **67.** 12, 13 **69.** 1710, 1550

71. Washer, \$360; dryer, \$290 **73.** 9 years old **75.** 29 years old **77.** \$820

79. 8 A.M.:23; 10 A.M.:46, 12 P.M.:30 **81.** **83.** **a.** 15

b. 14 **c.** 3 **d.** −16 **e.** −6 **f.** 3600 **g.** −7 **h.** 40

2.5 Applications of Linear Equations

OBJECTIVES

1. Solve linear equations when signs of grouping are present
2. Solve applications involving numbers
3. Solve geometry problems
4. Solve mixture problems
5. Solve motion problems

In Section 1.2, we looked at the distributive property. That property is used when solving equations in which parentheses are involved.

Let's start by reviewing an example similar to those we considered earlier. We will then solve other equations involving grouping symbols.

Example 1

Solving Equations with Parentheses

Solve for x:

$5(2x - 1) = 25$

First, multiply on the left to remove the parentheses, then solve as before.

NOTE Again, returning to the *original equation* will catch any possible errors in the removal of the parentheses.

Left side	Right side
$5(2 \cdot 3 - 1) \stackrel{?}{=} 25$	
$5(6 - 1) \stackrel{?}{=} 25$	
$5 \cdot 5 \stackrel{?}{=} 25$	
$25 = 25$ (True)	

$$
\begin{aligned}
10x - 5 &= 25 \\
+5 \quad &\;\; +5 \qquad \text{Add 5.} \\
\hline
10x &= 30 \\
\frac{10x}{10} &= \frac{30}{10} \qquad \text{Divide by 10.} \\
x &= 3
\end{aligned}
$$

The answer is 3. To check, return to the *original equation.* Substitute 3 for x. Then evaluate the left and right sides separately.

CHECK YOURSELF 1

Solve for x.

$8(3x + 5) = 16$

NOTE Given an expression such as $a - (b + c)$ you could rewrite it as $a + (-(b + c))$.

Be especially careful if a negative sign precedes a grouping symbol. The sign of each term inside the grouping symbol must be changed.

Example 2

Solving Equations with Parentheses

Solve $8 - (3x + 1) = -8$.

First, remove the parentheses. The original equation then becomes

NOTE Remember,

$-(3x + 1) = -3x - 1$

↑ ↑

Change *both* signs.

$$
\begin{aligned}
8 - 3x - 1 &= -8 \\
-3x + 7 &= -8 \qquad \text{Combine like terms.} \\
-7 \quad &\;\; -7 \qquad \text{Add } -7 \text{ to each side.} \\
\hline
-3x &= -15 \\
x &= 5 \qquad \text{Divide by } -3.
\end{aligned}
$$

The solution is 5. You should verify this result.

CHECK YOURSELF 2

Solve for x.

$7 - (4x - 3) = 22$

Example 3 illustrates the solution process when more than one grouping symbol is involved in an equation.

Example 3

Solving Equations with Parentheses

Solve $2(3x - 1) - 3(x + 5) = 4$.

$2(3x - 1) - 3(x + 5) = 4$ Use the distributive property to remove the parentheses.

$6x - 2 - 3x - 15 = 4$ Combine like terms on the left.

$3x - 17 = 4$ Add 17.

$3x = 21$ Divide by 3.

$x = 7$

The solution is 7.

To check, return to the original equation to replace x with 7.

NOTE Notice how the rules for the order of operations are applied.

$2(3 \cdot 7 - 1) - 3(7 + 5) \stackrel{?}{=} 4$

$2(21 - 1) - 3(7 + 5) \stackrel{?}{=} 4$

$2 \cdot 20 - 3 \cdot 12 \stackrel{?}{=} 4$

$40 - 36 \stackrel{?}{=} 4$

$4 = 4$ (True)

The solution is verified.

CHECK YOURSELF 3

Solve for x.

$5(2x + 4) = 7 - 3(1 - 2x)$

Many applications lead to equations involving parentheses. That means the methods of Examples 2 and 3 will have to be applied during the solution process. Before we look at examples, you should review the five-step process for solving word problems found in Section 2.4.

These steps are illustrated in Example 4.

Example 4

Solving Applications Using Parentheses

One number is 5 more than a second number. If 3 times the smaller number plus 4 times the larger is 104, find the two numbers.

Step 1 What are you asked to find? You must find the two numbers.

Step 2 Represent the unknowns. Let x be the smaller number. Then

NOTE "5 more than" x

$x + 5$

is the larger number.

Step 3 Write an equation.

$$3x + 4(x + 5) = 104$$

3 times the smaller — Plus — 4 times the larger

NOTE Note that the parentheses are *essential* in writing the correct equation.

Step 4 Solve the equation.

$$3x + 4(x + 5) = 104$$
$$3x + 4x + 20 = 104$$
$$7x + 20 = 104$$
$$7x = 84$$
$$x = 12$$

The smaller number (x) is 12, and the larger number ($x + 5$) is 17.

Step 5 Check the solution: 12 is the smaller number, and 17 is the larger number.

$3 \cdot 12 + 4 \cdot 17 = 104$ (True)

CHECK YOURSELF 4

One number is 4 more than another. If 6 times the smaller minus 4 times the larger is 4, what are the two numbers?

The solutions for many problems from geometry will also yield equations involving parentheses. Consider Example 5.

Example 5

Solving a Geometry Application

NOTE Whenever you are working on an application involving geometric figures, you should draw a sketch of the problem, including the labels assigned in step 2.

Length $3x - 1$

Width x

The length of a rectangle is 1 centimeter (cm) less than 3 times the width. If the perimeter is 54 cm, find the dimensions of the rectangle.

Step 1 You want to find the dimensions (the width and length).

Step 2 Let x be the width.

Then $3x - 1$ is the length.

3 times the width — 1 less than

Step 3 To write an equation, we'll use this formula for the perimeter of a rectangle:

$$P = 2W + 2L$$

So

$$2x + 2(3x - 1) = 54$$

Twice the width; Twice the length; Perimeter

Step 4 Solve the equation.

$$2x + 2(3x - 1) = 54$$
$$2x + 6x - 2 = 54$$
$$8x = 56$$
$$x = 7$$

NOTE Be sure to return to the original statement of the problem when checking your result.

The width x is 7 cm, and the length, $3x - 1$, is 20 cm. We leave step 5, the check, to you.

CHECK YOURSELF 5

The length of a rectangle is 5 *inches (in.) more than twice the width. If the perimeter of the rectangle is* 76 *in., what are the dimensions of the rectangle?*

You will also often use parentheses in solving *mixture problems.* Mixture problems involve combining things that have a different value, rate, or strength. Look at Example 6.

Example 6

Solving a Mixture Problem

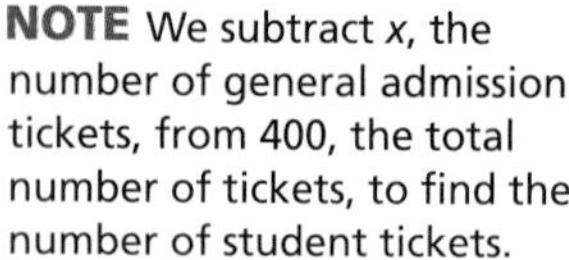

Four hundred tickets were sold for a school play. General admission tickets were \$4, and student tickets were \$3. If the total ticket sales were \$1350, how many of each type of ticket were sold?

Step 1 You want to find the number of each type of ticket sold.

Step 2 Let x be the number of general admission tickets.

Then $400 - x$ student tickets were sold.

400 tickets were sold in all.

NOTE We subtract x, the number of general admission tickets, from 400, the total number of tickets, to find the number of student tickets.

Step 3 The sales value for each kind of ticket is found by multiplying the price of the ticket by the number sold.

General admission tickets:	$4x$	\$4 for each of the x tickets
Student tickets:	$3(400 - x)$	\$3 for each of the $400 - x$ tickets

So to form an equation, we have

$$4x + 3(400 - x) = 1350$$

Value of general admission tickets; Value of student tickets; Total value

Step 4 Solve the equation.

$$4x + 3(400 - x) = 1350$$

$$4x + 1200 - 3x = 1350$$

$$x + 1200 = 1350$$

$$x = 150$$

So 150 general admission and 250 student tickets were sold. We leave the check to you.

CHECK YOURSELF 6

Beth bought 35¢ stamps and 15¢ stamps at the post office. If she purchased 60 stamps at a cost of $17, how many of each kind did she buy?

The next group of applications we will look at in this section involves *motion problems*. They involve a distance traveled, a rate or speed, and time. To solve motion problems, we need a relationship among these three quantities.

Suppose you travel at a rate of 50 miles per hour (mi/h) on a highway for 6 hours (h). How far (what distance) will you have gone? To find the distance, you multiply:

NOTE Be careful to make your units consistent. If a rate is given in *miles per hour,* then the time must be given in *hours* and the distance in *miles.*

(50 mi/h)(6 h) = 300 mi

↑ Speed or rate ↑ Time ↑ Distance

Definitions: Relationship for Motion Problems

In general, if r is a rate, t is the time, and d is the distance traveled,

$d = r \cdot t$

This is the key relationship, and it will be used in all motion problems. Let's see how it is applied in Example 7.

Example 7

Solving a Motion Problem

On Friday morning Ricardo drove from his house to the beach in 4 h. In coming back on Sunday afternoon, heavy traffic slowed his speed by 10 mi/h, and the trip took 5 h. What was his average speed (rate) in each direction?

Step 1 We want the speed or rate in each direction.

Step 2 Let x be Ricardo's speed to the beach. Then $x - 10$ is his return speed.

It is always a good idea to sketch the given information in a motion problem. Here we would have

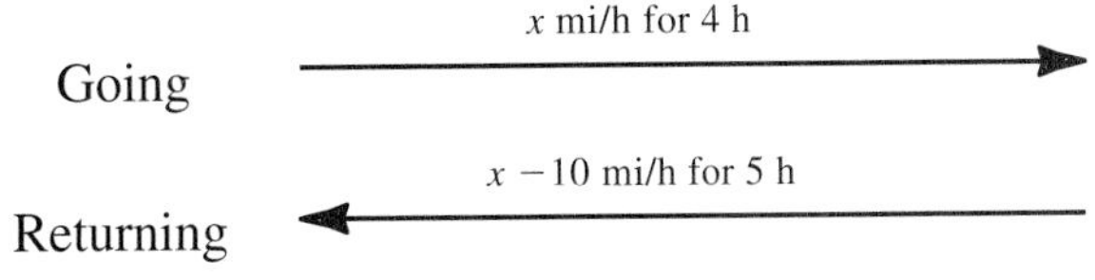

Step 3 Because we know that the distance is the same each way, we can write an equation, using the fact that the product of the rate and the time each way must be the same. So

NOTE Distance (going) = distance (returning)

or

Time · rate (going) = time · rate (returning)

$$\underbrace{4x}_{\substack{\uparrow \\ \text{Time} \cdot \text{rate} \\ \text{(going)}}} = \underbrace{5(x - 10)}_{\substack{\uparrow \\ \text{Time} \cdot \text{rate} \\ \text{(returning)}}}$$

A chart can help summarize the given information. We begin by filling in the information given in the problem.

	Distance	Rate	Time
Going		x	4
Returning		$x - 10$	5

Now we fill in the missing information. Here we use the fact that $d = rt$ to complete the chart.

	Distance	Rate	Time
Going	$4x$	x	4
Returning	$5(x - 10)$	$x - 10$	5

From here we set the two distances equal to each other and solve as before.

Step 4 Solve.

$$4x = 5(x - 10)$$

$$4x = 5x - 50$$

$$-x = -50$$

$$x = 50 \text{ mi/h}$$

NOTE x was his rate going, $x - 10$ his rate returning.

So Ricardo's rate going to the beach was 50 mi/h, and his rate returning was 40 mi/h.

Step 5 To check, you should verify that the product of the time and the rate is the same in each direction.

CHECK YOURSELF 7

A plane made a flight (with the wind) between two towns in 2 h. Returning against the wind, the plane's speed was 60 mi/h slower, and the flight took 3 h. What was the plane's speed in each direction?

Example 8 illustrates another way of using the distance relationship.

Example 8

Solving a Motion Problem

Katy leaves Las Vegas for Los Angeles at 10 A.M., driving at 50 mi/h. At 11 A.M. Jensen leaves Los Angeles for Las Vegas, driving at 55 mi/h along the same route. If the cities are 260 mi apart, at what time will they meet?

Step 1 Let's find the time that Katy travels until they meet.

Step 2 Let x be Katy's time.

Then $x - 1$ is Jensen's time.

Jensen left 1 h later!

Again, you should draw a sketch of the given information.

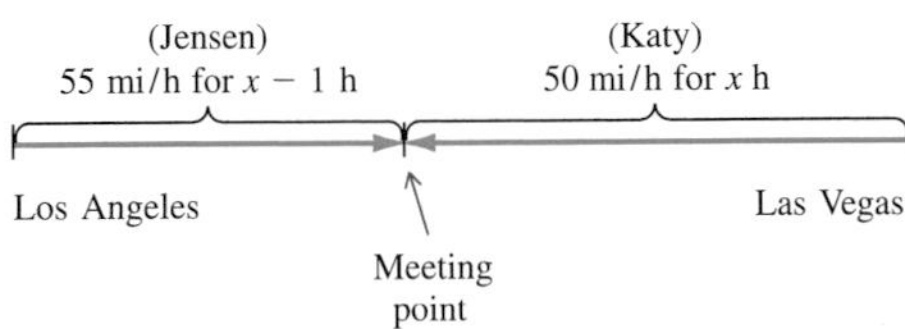

Step 3 To write an equation, we will again need the relationship $d = rt$. From this equation, we can write

Katy's distance $= 50x$

Jensen's distance $= 55(x - 1)$

As before, we can use a table to solve.

	Distance	Rate	Time
Katy	$50x$	50	x
Jensen	$55(x - 1)$	55	$x - 1$

From the original problem, the sum of those distances is 260 mi, so

$$50x + 55(x - 1) = 260$$

Step 4

$$50x + 55(x - 1) = 260$$

$$50x + 55x - 55 = 260$$

$$105x - 55 = 260$$

$$105x = 315$$

$$x = 3 \text{ h}$$

NOTE Be sure to answer the question asked in the problem.

Finally, because Katy left at 10 A.M., the two will meet at 1 P.M. We leave the check of this result to you.

CHECK YOURSELF 8

At noon a jogger leaves one point, running at 8 mi/h. One hour later a bicyclist leaves the same point, traveling at 20 mi/h in the opposite direction. At what time will they be 36 mi apart?

CHECK YOURSELF ANSWERS

1. -1 **2.** -3 **3.** -4 **4.** The numbers are 10 and 14. **5.** The width is 11 in.; the length is 27 in. **6.** 40 at 35¢, and 20 at 15¢ **7.** 180 mi/h with the wind and 120 mi/h against the wind **8.** At 2 P.M.

Name ______________________

Section ________ Date ________

2.5 Exercises

Solve each of the following equations for x, and check your results.

1. $3(x - 5) = 6$

2. $2(x + 3) = -6$

3. $5(2x + 3) = 35$

4. $4(3x - 5) = 88$

5. $7(5x + 8) = -84$

6. $6(3x + 2) = -60$

7. $10 - (x - 2) = 15$

8. $12 - (x + 3) = 3$

9. $5 - (2x + 1) = 12$

10. $9 - (3x - 2) = 2$

11. $7 - (3x - 5) = 13$

12. $5 - (4x + 3) = 4$

13. $5x = 3(x - 6)$

14. $5x = 2(x + 12)$

15. $7(2x - 3) = 20x$

16. $4(3x + 5) = 18x$

17. $6(6 - x) = 3x$

18. $5(8 - x) = 3x$

19. $2(2x - 1) = 3(x + 1)$

20. $3(3x - 1) = 4(2x + 1)$

21. $5(4x + 2) = 6(3x + 4)$

22. $4(6x - 1) = 7(3x + 2)$

23. $9(8x - 1) = 5(4x + 6)$

24. $7(3x + 11) = 9(3 - 6x)$

ANSWERS

1. 7
2. -6
3. 2
4. 9
5. -4
6. -4
7. -3
8. 6
9. -4
10. 3
11. $-\frac{1}{3}$
12. $-\frac{1}{2}$
13. -9
14. 8
15. $-\frac{7}{2}$
16. $\frac{10}{3}$
17. 4
18. 5
19. 5
20. 7
21. 7
22. 6
23. $\frac{3}{4}$
24. $-\frac{2}{3}$

ANSWERS

25. 2
26. 5
27. 3
28. −2
29. 5
30. −2
31. −2
32. 2
33. All real numbers
34. All real numbers
35. 10, 18
36. 13, 16
37. 8, 15
38. 5, 15
39. 5, 6
40. 9, 11
41. 12 in., 25 in.
42. 7 cm, 16 cm
43. 6 m, 22 m

25. $-4(2x - 1) + 3(3x + 1) = 9$

26. $7(3x + 4) = 8(2x + 5) + 13$

27. $5(2x - 1) - 3(x - 4) = 4(x + 4)$

28. $2(x - 3) - 3(x + 5) = 3(x - 2) - 7$

29. $3(3 - 4x) + 30 = 5x - 2(6x - 7)$

30. $3x - 5(3x - 7) = 2(x + 9) + 45$

31. $-2x + [3x - (-2x + 5)] = -(15 + 2x)$

32. $-3x + [5x - (-x + 4)] = -2(x - 3)$

33. $3x^2 - 2(x^2 + 2) = x^2 - 4$

34. $5x^2 - [2(2x^2 + 3)] - 3 = x^2 - 9$

Solve the following word problems. Be sure to show the equation you use for the solution.

35. Number problem. One number is 8 more than another. If the sum of the smaller number and twice the larger number is 46, find the two numbers.

36. Number problem. One number is 3 less than another. If 4 times the smaller number minus 3 times the larger number is 4, find the two numbers.

37. Number problem. One number is 7 less than another. If 4 times the smaller number plus 2 times the larger number is 62, find the two numbers.

38. Number problem. One number is 10 more than another. If the sum of twice the smaller number and 3 times the larger number is 55, find the two numbers.

39. Consecutive integers. Find two consecutive integers such that the sum of twice the first integer and 3 times the second integer is 28. (*Hint:* If x represents the first integer, $x + 1$ represents the next consecutive integer.)

40. Consecutive integers. Find two consecutive odd integers such that 3 times the first integer is 5 more than twice the second. (*Hint:* If x represents the first integer, $x + 2$ represents the next consecutive odd integer.)

41. Dimensions of a rectangle. The length of a rectangle is 1 inch (in.) more than twice its width. If the perimeter of the rectangle is 74 in., find the dimensions of the rectangle.

42. Dimensions of a rectangle. The length of a rectangle is 5 centimeters (cm) less than 3 times its width. If the perimeter of the rectangle is 46 cm, find the dimensions of the rectangle.

43. Garden size. The length of a rectangular garden is 4 meters (m) more than 3 times its width. The perimeter of the garden is 56 m. What are the dimensions of the garden?

44. Size of a playing field. The length of a rectangular playing field is 5 feet (ft) less than twice its width. If the perimeter of the playing field is 230 ft, find the length and width of the field.

45. Isosceles triangle. The base of an isosceles triangle is 3 cm less than the length of the equal sides. If the perimeter of the triangle is 36 cm, find the length of each of the sides.

46. Isosceles triangle. The length of one of the equal legs of an isosceles triangle is 3 in. less than twice the length of the base. If the perimeter is 29 in., find the length of each of the sides.

47. Ticket sales. Tickets for a play cost $8 for the main floor and $6 in the balcony. If the total receipts from 500 tickets were $3600, how many of each type of ticket were sold?

48. Ticket sales. Tickets for a basketball tournament were $6 for students and $9 for nonstudents. Total sales were $10,500, and 250 more student tickets were sold than nonstudent tickets. How many of each type of ticket were sold?

49. Number of stamps. Maria bought 80 stamps at the post office in 33¢ and 25¢ denominations. If she paid $24 for the stamps, how many of each denomination did she buy?

50. Money denominations. A bank teller had a total of 125 $10 bills and $20 bills to start the day. If the value of the bills was $1650, how many of each denomination did he have?

ANSWERS

44. 75 ft, 40 ft

45. Legs, 13 cm; base, 10 cm

46. Legs, 11 in.; base, 7 in.

47. 200 $6 tickets, 300 $8 tickets

48. 850 student, 600 nonstudent

49. 30 25¢ stamps, 50 33¢ stamps

50. 85 $10 bills, 40 $20 bills

ANSWERS

51. 60 coach, 40 berth, and 20 sleeping room

52. 500 box seats, 600 grandstand, and 1000 bleachers

53. 40 mi/h, 30 mi/h

54. 20 mi/h, 25 mi/h

55. 6 P.M.

56. 4:30 P.M.

57. 2 P.M.

58. 1 P.M.

59. 3 P.M.

51. Ticket sales. Tickets for a train excursion were \$120 for a sleeping room, \$80 for a berth, and \$50 for a coach seat. The total ticket sales were \$8600. If there were 20 more berth tickets sold than sleeping room tickets and 3 times as many coach tickets as sleeping room tickets, how many of each type of ticket were sold?

52. Baseball tickets. Admission for a college baseball game is \$6 for box seats, \$5 for the grandstand, and \$3 for the bleachers. The total receipts for one evening were \$9000. There were 100 more grandstand tickets sold than box seat tickets. Twice as many bleacher tickets were sold as box seat tickets. How many tickets of each type were sold?

53. Driving speed. Patrick drove 3 hours (h) to attend a meeting. On the return trip, his speed was 10 miles per hour (mi/h) less and the trip took 4 h. What was his speed each way?

54. Bicycle speed. A bicyclist rode into the country for 5 h. In returning, her speed was 5 mi/h faster and the trip took 4 h. What was her speed each way?

55. Driving speed. A car leaves a city and goes north at a rate of 50 mi/h at 2 P.M. One hour later a second car leaves, traveling south at a rate of 40 mi/h. At what time will the two cars be 320 mi apart?

56. Bus distance. A bus leaves a station at 1 P.M., traveling west at an average rate of 44 mi/h. One hour later a second bus leaves the same station, traveling east at a rate of 48 mi/h. At what time will the two buses be 274 mi apart?

57. Traveling time. At 8:00 A.M., Catherine leaves on a trip at 45 mi/h. One hour later, Max decides to join her and leaves along the same route, traveling at 54 mi/h. When will Max catch up with Catherine?

58. Bicycling time. Martina leaves home at 9 A.M., bicycling at a rate of 24 mi/h. Two hours later, John leaves, driving at the rate of 48 mi/h. At what time will John catch up with Martina?

59. Traveling time. Mika leaves Boston for Baltimore at 10:00 A.M., traveling at 45 mi/h. One hour later, Hiroko leaves Baltimore for Boston on the same route, traveling at 50 mi/h. If the two cities are 425 mi apart, when will Mika and Hiroko meet?

60. Traveling time. A train leaves town A for town B, traveling at 35 mi/h. At the same time, a second train leaves town B for town A at 45 mi/h. If the two towns are 320 mi apart, how long will it take for the two trains to meet?

61. Tree inventory. There are 500 Douglas fir and hemlock trees in a section of forest bought by Hoodoo Logging Co. The company paid an average of $250 for each Douglas fir and $300 for each hemlock. If the company paid $132,000 for the trees, how many of each kind did the company buy?

62. Tree inventory. There are 850 Douglas fir and ponderosa pine trees in a section of forest bought by Sawz Logging Co. The company paid an average of $300 for each Douglas fir and $225 for each ponderosa pine. If the company paid $217,500 for the trees, how many of each kind did the company buy?

63. There is a universally agreed on "order of operations" used to simplify expressions. Explain how the order of operations is used in solving equations. Be sure to use complete sentences.

64. A common mistake when solving equations is the following:

The equation: $2(x - 2) = x + 3$
First step in solving: $2x - 2 = x + 3$

Write a clear explanation of what error has been made. What could be done to avoid this error?

65. Another very common mistake is in the equation below:

The equation: $6x - (x + 3) = 5 + 2x$
First step in solving: $6x - x + 3 = 5 + 2x$

Write a clear explanation of what error has been made and what could be done to avoid the mistake.

66. Write an algebraic equation for the English statement "Subtract 5 from the sum of x and 7 times 3 and the result is 20." Compare your equation with other students. Did you all write the same equation? Are all the equations correct even though they don't look alike? Do all the equations have the same solution? What is wrong? The English statement is *ambiguous*. Write another English statement that leads correctly to more than one algebraic equation. Exchange with another student and see if they think the statement is ambiguous. Notice that the algebra is *not* ambiguous!

ANSWERS

60. 4 h

61. 360 Douglas fir, 140 hemlock

62. 350 Douglas fir, 500 ponderosa pine

63.

64.

65.

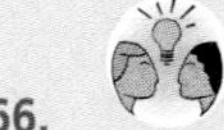

66.

ANSWERS

a. 30

b. 6000

c. 3

d. 12,800

e. 10.35

f. 6.9

Getting Ready for Section 2.6 [Section 1.4]

Evaluate the following.

(a) $\frac{270 \cdot 10}{90}$ (b) $\frac{660 \cdot 100}{11}$ (c) $\frac{120 \cdot 100}{4000}$

(d) $\frac{320 \cdot 100}{2.5}$ (e) $\frac{23 \cdot 4.5}{10}$ (f) $\frac{46 \cdot 15}{100}$

Answers

1. 7 **3.** 2 **5.** -4 **7.** -3 **9.** -4 **11.** $-\frac{1}{3}$ **13.** -9
15. $-\frac{7}{2}$ **17.** 4 **19.** 5 **21.** 7 **23.** $\frac{3}{4}$ **25.** 2 **27.** 3
29. 5 **31.** -2 **33.** All real numbers **35.** 10, 18 **37.** 8, 15
39. 5, 6 **41.** 12 in., 25 in. **43.** 6 m, 22 m **45.** Legs, 13 cm; base, 10 cm
47. 200 \$6 tickets, 300 \$8 tickets **49.** 30 25¢ stamps, 50 33¢ stamps
51. 60 coach, 40 berth, and 20 sleeping room **53.** 40 mi/h, 30 mi/h **55.** 6 P.M.
57. 2 P.M. **59.** 3 P.M. **61.** 360 Douglas fir, 140 hemlock
63. **65.** **a.** 30 **b.** 6000 **c.** 3 **d.** 12,800
e. 10.35 **f.** 6.9

2.6 Solving Percent Applications

2.6 OBJECTIVES

1. Solve a percent application for the amount
2. Solve a percent application for the rate
3. Solve a percent application for the base

The word **percent** is Latin for "for each hundred." Look at the following drawing.

NOTE The rectangle is divided into 100 parts, 25 of them are shaded.

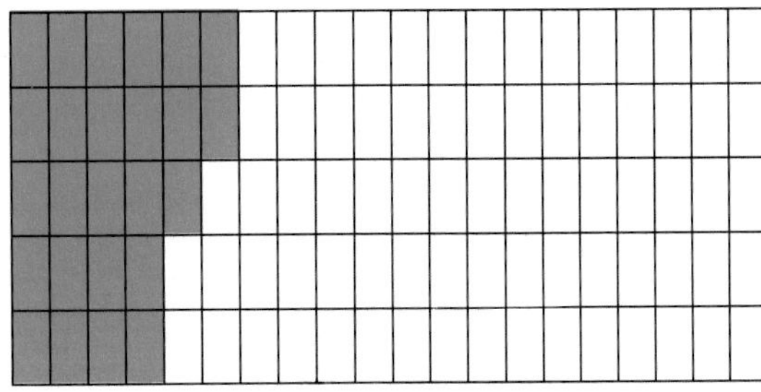

How can we describe how much of the rectangle is shaded? We can say that

1. $\frac{1}{4}$ of the rectangle is shaded,
2. 0.25 of the rectangle is shaded, or
3. 25% of the rectangle is shaded.

Example 1

Converting Rational Numbers to Percents

Convert each rational number to a percent.

(a) $0.36 = 36\%$ To convert a decimal number to a percent, we move the decimal point two places to the right.

(b) $1.575 = 157.5\%$ Again, we move the decimal point two places to the right.

(c) $\frac{2}{5} = 0.4 = 40\%$ When converting a fraction to a percent, rewrite the fraction as a decimal and move the decimal point two places to the right.

CHECK YOURSELF 1

Convert each rational number to a percent.

(a) 0.45 **(b)** 2.1 **(c)** $\frac{1}{8}$

There are many practical applications that use percents. Every complete percent statement has three parts that need to be identified. Let's look at some definitions that will help with that process.

Definitions: Base, Amount, and Rate

The **base** is the whole in a statement. It is the standard used for comparison.

The **amount** is the part of the whole that is being compared to the base.

The **rate** is the ratio of the amount to the base. It is written as a percent.

Example 2

Identifying Base, Amount, and Rate.

Identify the indicated quantity in each statement.

(a) The rate in the statement "50% of 480 is 240." The rate is 50%.

(b) The amount in the statement "20% of 400 is 80." The amount is 80.

(c) The base in the statement "125% of 200 is 250." The base is 200. Note that the base is almost always preceded by the word *of.*

CHECK YOURSELF 2

Identify the indicated quantity in each statement.

(a) The rate in the statement "40% of 80 is 32."

(b) The amount in the statement "150% of 300 is 450."

(c) The base in the statement "5% of 800 is 40."

Given an application of percents, it is frequently easiest to rewrite the statement in the form

R% of B is A

to identify the three parts.

Example 3

Identifying Rate, Base, and Amount

Identify the rate, base, and amount in the following statement.

Delia borrows $10,000 for 1 year at 11% interest. How much interest will she pay?

We can rewrite the statement as "11% of $10,000 is what amount?"

11% is the rate, $10,000 is the base, and the amount is unknown (at least at the moment).

CHECK YOURSELF 3

Identify the rate, base, and amount in the following statement.

Melina earned $140 *last year from a certificate of deposit that paid* 6%. *How much did she invest?*

Definitions: Percent Proportion

To solve such problems as those in the previous example, we use the **percent proportion.**

$$\frac{\textit{Amount}}{\textit{Base}} = \frac{\textit{Rate}}{100}$$

This proportion could also be written as the equation

$100 \cdot \text{amount} = \text{rate} \cdot \text{base}$ or $100A = R \cdot B$

Example 4

Solving a Problem Involving an Unknown Amount

Find the interest you must pay if you borrow \$2000 for 1 year with an interest rate of $9\frac{1}{2}\%$.

The base (the principal) is \$2000, the rate is $9\frac{1}{2}\%$, and we want to find the interest (the amount). Using the percent proportion gives

REMEMBER

$9\frac{1}{2}\% = 9.5\%$

$$\frac{A}{2000} = \frac{9.5}{100}$$

so

$$100A = 9.5 \cdot 2000$$

or

$$A = \frac{19{,}000}{100} = 190$$

The interest (amount) is \$190.

CHECK YOURSELF 4

You invest \$5000 for 1 year at $8\frac{1}{2}\%$. *How much interest will you earn?*

Let's look at an application that requires finding the rate.

Example 5

Solving a Problem Involving an Unknown Rate

You borrow \$2000 from a bank for 1 year and are charged \$150 interest. What is the interest rate?

The base is the amount of the loan (the principal). The amount is the interest paid. To find the interest rate, we again use the percent proportion.

$$\frac{150}{2000} = \frac{R}{100}$$

Then

$$100 \cdot 150 = R \cdot 2000$$

$$R = \frac{15{,}000}{2000} = 7.5$$

The interest rate is 7.5%.

CHECK YOURSELF 5

Xian borrowed \$3200 *and was charged* \$352 *in interest for* 1 *year. What was the interest rate?*

Now let's look at an application that requires finding the base.

Example 6

Solving a Problem Involving an Unknown Base

Ms. Hobson agrees to pay 11% interest on a loan for her new automobile. She is charged \$550 interest on a loan for 1 year. How much did she borrow?

The rate is 11%. The amount, or interest, is \$550. We want to find the base, which is the principal, or the size of the loan. To solve the problem, we have

$$\frac{550}{B} = \frac{11}{100}$$

$$100 \cdot 550 = 11B$$

$$B = \frac{55{,}000}{11} = 5000$$

She borrowed \$5000.

CHECK YOURSELF 6

Sue pays \$210 interest for a 1-year loan at 10.5%. What was the size of her loan?

Percents are used in too many ways for us to list. Look at the variety in the following examples, which illustrate some additional situations in which you will find percents.

Another common application of percents involves tax rates.

Example 7

Solving a Percent Problem

A state taxes sales at 5.5%. How much sales tax will you pay on a purchase of \$48?

The tax you pay is the amount (the part of the whole). Here the base is the purchase price, \$48, and the rate is the tax rate, 5.5%.

NOTE In an application involving taxes, the tax paid is always the amount.

$$\frac{A}{48} = \frac{5.5}{100} \quad \text{or} \quad 100A = 5.5 \cdot 48$$

Now

NOTE $48 \cdot 5.5 = 264$

$$A = \frac{264}{100} = 2.64$$

The sales tax paid is \$2.64.

CHECK YOURSELF 7

Suppose that a state has a sales tax rate of $6\frac{1}{2}$%. *If you buy a used car for* \$1200, *how much sales tax must you pay?*

Percents are also used to deal with store markups or discounts. Consider Example 8.

Example 8

Solving a Percent Problem

A store marks up items to make a 30% profit. If an item cost \$2.50 from the supplier, what will the selling price be?

The base is the cost of the item, \$2.50, and the rate is 30%. In the percent proportion, the markup is the amount in this application.

$$\frac{A}{2.50} = \frac{30}{100} \quad \text{or} \quad 100A = 30 \cdot 2.50$$

Then

$$A = \frac{75}{100} = 0.75$$

The markup is \$0.75. Finally we have

NOTE
Selling price = original cost + markup

$$\text{Selling price} = \$2.50 + \$0.75 = \$3.25$$

Add the cost and the markup to find the selling price.

CHECK YOURSELF 8

A store wants to discount (or mark down) an item by 25% for a sale. If the original price of the item was \$45, find the sale price. [Hint: Find the discount (the amount the item will be marked down), and subtract that from the original price.]

Our final examples illustrate increases and decreases stated in terms of percents.

Example 9

Solving a Percent Problem

The population of a town increased 15% in a 3-year period. If the original population was 12,000, what was the population at the end of the period?

First we find the increase in the population. That increase is the amount in the problem.

$$\frac{A}{12{,}000} = \frac{15}{100} \quad \text{so} \quad 100A = 15 \cdot 12{,}000$$

$$A = \frac{180{,}000}{100}$$

$$= 1800$$

To find the population at the end of the period, we add

$$12{,}000 + 1800 = 13{,}800$$

Original population (12,000) — Increase (1800) — New population (13,800)

CHECK YOURSELF 9

A school's enrollment decreased by 8% from a given year to the next. If the enrollment was 550 students the first year, how many students were enrolled the second year?

Example 10

Solving a Percent Problem

Enrollment at a school increased from 800 to 888 students from a given year to the next. What was the rate of increase?

First we must subtract to find the actual increase.

Increase: $888 - 800 = 88$ students

Now to find the rate, we have

NOTE We use the *original* enrollment, 800, as our base.

$$\frac{88}{800} = \frac{R}{100} \quad \text{so} \quad 100 \cdot 88 = R \cdot 800$$

$$R = \frac{8800}{800} = 11$$

The enrollment has increased at a rate of 11%.

CHECK YOURSELF 10

Car sales at a dealership decreased from 350 units one year to 322 units the next. What was the rate of decrease?

Example 11

Solving a Percent Problem

A company hired 18 new employees in 1 year. If this was a 15% increase, how many employees did the company have before the increase?

The rate is 15%. The amount is 18, the number of new employees. The base in this problem is the number of employees *before the increase.* So

$$\frac{18}{B} = \frac{15}{100}$$

$$100 \cdot 18 = 15B \quad \text{or} \quad B = \frac{1800}{15} = 120$$

The company had 120 employees before the increase.

CHECK YOURSELF 11

A school had 54 new students in one term. If this was a 12% increase over the previous term, how many students were there before the increase?

CHECK YOURSELF ANSWERS

1. **(a)** 45%; **(b)** 210%; **(c)** 12.5% **2.** **(a)** 40%; **(b)** 450; **(c)** 800
3. Rate = 6%; base unknown; amount = \$140 **4.** \$425 **5.** 11% **6.** \$2000
7. \$78 **8.** \$33.75 **9.** 506 **10.** 8% **11.** 450

2.6 Exercises

Name ______________

Section ________ Date ________

Convert each rational number to a percent.

1. 0.23 **2.** 0.31 **3.** 2.5 **4.** 1.8 **5.** $\frac{3}{8}$ **6.** $\frac{7}{16}$

In Exercises 7 to 12, identify the indicated quantity in each statement.

7. The rate in the statement "23% of 400 is 92."

8. The base in the statement "40% of 600 is 240."

9. The amount in the statement "200 is 40% of 500."

10. The rate in the statement "480 is 60% of 800."

11. The base in the statement "16% of 350 is 56."

12. The amount in the statement "150 is 75% of 200."

Identify the rate, base, and amount in the following applications. *Do not solve* the applications at this point.

13. Commission. Jan has a 5% (R%) commission rate on all her sales. If she sells $40,000 (B) worth of merchandise in 1 month, what commission (A) will she earn?

14. Salary. 22% (R%) of Shirley's monthly salary is deducted for withholding. If those deductions total $209 (A), what is her salary (B)?

15. Chemistry. In a chemistry class of 30 (B) students, 5 (A) received a grade of A. What percent (R%) of the students received A's?

16. Mixtures. A can of mixed nuts contains 80% (R%) peanuts. If the can holds 16 (B) ounces (oz), how many ounces of peanuts (A) does it contain?

17. Selling price. The sale tax rate in a state is 5.5% (R%). If you pay a tax of $3.30 (A) on an item that you purchase, what is its selling price (B)?

18. Manufacturing. In a shipment of 750 (B) parts, 75 (A) were found to be defective. What percent (R%) of the parts were faulty?

19. Enrollments. A college had 9000 (B) students at the start of a school year. If there is an enrollment increase of 6% (R%) by the beginning of the next year, how many additional students (A) were there?

20. Investments. Paul invested $5000 (B) in a time deposit. What interest will he earn (A) for 1 year if the interest rate is 6.5% (R%)?

ANSWERS

1. 23%
2. 31%
3. 250%
4. 180%
5. 37.5%
6. 43.75%
7. 23%
8. 600
9. 200
10. 60%
11. 350
12. 150
13. See exercise
14. See exercise
15. See exercise
16. See exercise
17. See exercise
18. See exercise
19. See exercise
20. See exercise

ANSWERS

21. $408

22. 54 mL

23. $143

24. $5100

25. 5%

26. 11%

27. 1.5%

28. 13%

Solve each of the following applications.

21. Interest. What interest will you pay on a $3400 loan for 1 year if the interest rate is 12%?

22. Chemistry. A chemist has 300 milliliters (mL) of solution that is 18% acid. How many milliliters of acid are in the solution?

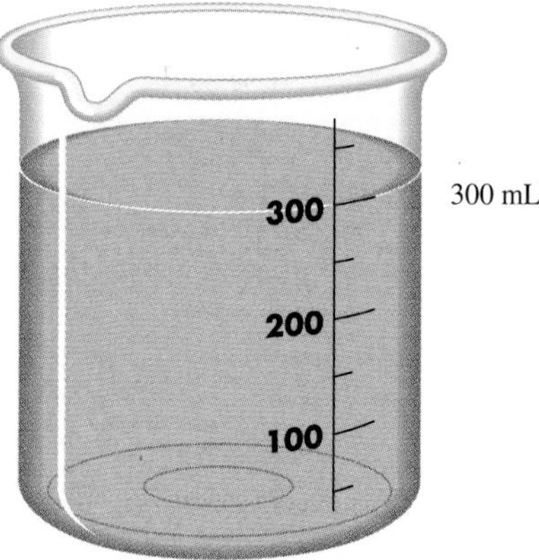

23. Payroll deductions. Roberto has 26% of his pay withheld for deductions. If he earns $550 per week, what amount is withheld?

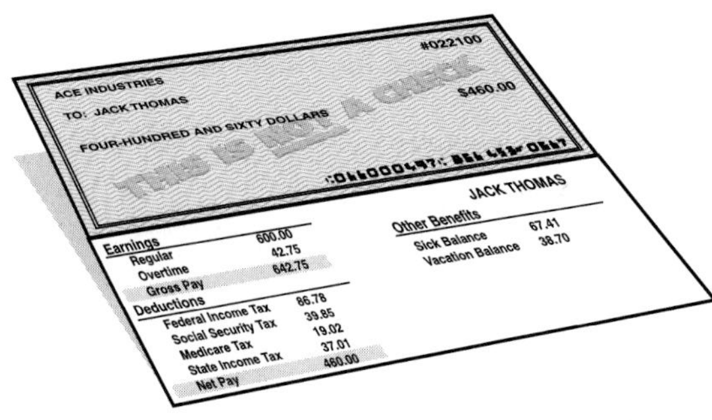

24. Commission. A real estate agent's commission rate is 6%. What will be the amount of the commission on the sale of an $85,000 home?

25. Commission. If a salesman is paid a $140 commission on the sale of a $2800 sailboat, what is his commission rate?

26. Interest. Ms. Jordan has been given a loan of $2500 for 1 year. If the interest charged is $275, what is the interest rate on the loan?

27. Interest. Joan was charged $18 interest for 1 month on a $1200 credit card balance. What was the monthly interest rate?

28. Chemistry. There is 117 grams (g) of acid in 900 g of a solution of acid and water. What percent of the solution is acid?

29. **Test scores.** On a test, Alice had 80% of the problems right. If she had 20 problems correct, how many questions were on the test?

30. **Sales tax.** A state sales tax rate is 3.5%. If the tax on a purchase is \$7, what was the amount of the purchase?

31. **Loan amount.** Patty pays \$525 interest for a 1-year loan at 10.5%. What was the amount of her loan?

32. **Commission.** A saleswoman is working on a 5% commission basis. If she wants to make \$1800 in 1 month, how much must she sell?

33. **Sales tax.** A state sales tax is levied at a rate of 6.4%. How much tax would one pay on a purchase of \$260?

34. **Down payment.** Betty must make a $9\frac{1}{2}$% down payment on the purchase of a \$2000 motorcycle. How much must she pay down?

35. **Commission.** If a house sells for \$125,000 and the commission rate is $6\frac{1}{2}$%, how much will the salesperson make for the sale?

36. **Test scores.** Marla needs 70% on a final test to receive a C for a course. If the exam has 120 questions, how many questions must she answer correctly?

37. **Unemployment.** A study has shown that 102 of the 1200 people in the workforce of a small town are unemployed. What is the town's unemployment rate?

38. **Surveys.** A survey of 400 people found that 66 were left-handed. What percent of those surveyed were left-handed?

39. **Dropout rate.** Of 60 people who start a training program, 45 complete the course. What is the dropout rate?

ANSWERS

29. 25 questions
30. \$200
31. \$5000
32. \$36,000
33. \$16.64
34. \$190
35. \$8125
36. 84 questions
37. 8.5%
38. 16.5%
39. 25%

ANSWERS

40. 16%

41. 1200 people

42. 3600 students

43. $950

44. $2400

45. $732

46. 963

47. $156,250

48. $12,960

40. **Manufacturing.** In a shipment of 250 parts, 40 are found to be defective. What percent of the parts are faulty?

41. **Surveys.** In a recent survey, 65% of those responding were in favor of a freeway improvement project. If 780 people were in favor of the project, how many people responded to the survey?

42. **Enrollments.** A college finds that 42% of the students taking a foreign language are enrolled in Spanish. If 1512 students are taking Spanish, how many foreign language students are there?

43. **Salary.** 22% of Samuel's monthly salary is deducted for withholding. If those deductions total $209, what is his salary?

44. **Budgets.** The Townsends budget 36% of their monthly income for food. If they spend $864 on food, what is their monthly income?

45. **Markup.** An appliance dealer marks up refrigerators 22% (based on cost). If the cost of one model was $600, what will its selling price be?

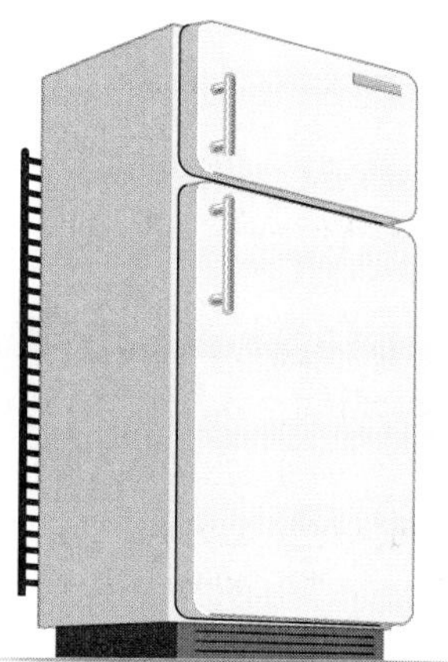

46. **Enrollments.** A school had 900 students at the start of a school year. If there is an enrollment increase of 7% by the beginning of the next year, what is the new enrollment?

47. **Land value.** A home lot purchased for $125,000 increased in value by 25% over 3 years. What was the lot's value at the end of the period?

48. **Depreciation.** New cars depreciate an average of 28% in their first year of use. What will an $18,000 car be worth after 1 year?

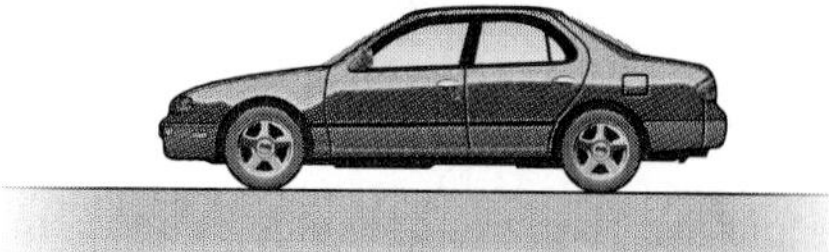

49. Enrollment. A school's enrollment was up from 950 students in 1 year to 1064 students in the next. What was the rate of increase?

50. Salary. Under a new contract, the salary for a position increases from $31,000 to $33,635. What rate of increase does this represent?

51. Markdown. A stereo system is marked down from $450 to $382.50. What is the discount rate?

52. Business. The electricity costs of a business decrease from $12,000 one year to $10,920 the next. What is the rate of decrease?

53. Price changes. The price of a new van has increased $4830, which amounts to a 14% increase. What was the price of the van before the increase?

54. Markdown. A television set is marked down $75, to be placed on sale. If this is a 12.5% decrease from the original price, what was the selling price before the sale?

55. Workforce. A company had 66 fewer employees in July 1999 than in July 1998. If this represents a 5.5% decrease, how many employees did the company have in July 1998?

56. Salary. Carlotta received a monthly raise of $162.50. If this represented a 6.5% increase, what was her monthly salary before the raise?

57. Retail sales. A pair of shorts is advertised for $48.75 and as being 25% off the original price. What was the original price?

ANSWERS

49. 12%

50. 8.5%

51. 15%

52. 9%

53. $34,500

54. $600

55. 1200 employees

56. $2500

57. $65

ANSWERS

58. $56.80

59. 76%

60. 144%

61. ≈15%

62. ≈20%

63. $4494.40

64. $3434.70

65. $4630.50

66. $5955.08

58. Tipping. If the total bill at a restaurant, including a 15% tip is $65.32, what was the cost of the meal alone?

The following chart shows U.S. trade with Mexico from 1992–1997. Use this information for exercises 59–62.

U.S. Trade With Mexico, 1992-97
Source: Office of Trade and Economic Analysis, U.S. Dept. of Commerce
(millions of dollars)

MEXICO

Year	Exports	Imports	Trade Balance[1]
1992	$40,592	$35,211	$5,381
1993	41,581	39,917	1,664
1994[2]	50,844	49,494	1,350
1995	46,292	61,685	−15,393
1996	56,792	74,297	−17,506
1997	71,388	85,938	−14,549

(1) Totals may not add due to rounding
(2) NAFTA provisions began to take effect: Jan. 1, 1994

59. What is the rate of increase (to the nearest whole percent) of exports from 1992 to 1997?

60. What is the rate of increase (to the nearest whole percent) of imports from 1992 to 1997?

61. By what percent did exports exceed imports in 1992?

62. By what percent did imports exceed exports in 1997?

Many percent problems involve calculating what is known as **compound interest.**

Suppose that you invest $1000 at 5% in a savings account for 1 year. For year 1, the interest is 5% of $1000, or 0.05 × $1000 = $50. At the end of year 1, you will have $1050 in the account.

$$\underset{\text{Start}}{\$1000} \xrightarrow{\text{At } 5\%} \underset{\text{Year 1}}{\$1050}$$

Now if you leave that amount in the account for a second year, the interest will be calculated on the original principal, $1000, plus the first year's interest, $50. This is called *compound interest.*

For year 2, the interest is 5% of $1050, or 0.05 × $1050 = $52.50. At the end of year 2, you will have $1102.50 in the account.

$$\underset{\text{Start}}{\$1000} \xrightarrow{\text{At } 5\%} \underset{\text{Year 1}}{\$1050} \xrightarrow{\text{At } 5\%} \underset{\text{Year 2}}{\$1102.50}$$

In Exercises 63 to 66, assume the interest is compounded annually (at the end of each year), and find the amount in an account with the given interest rate and principal.

63. $4000, 6%, 2 years

64. $3000, 7%, 2 years

65. $4000, 5%, 3 years

66. $5000, 6%, 3 years

67. Automobiles. In 1990, there were an estimated 145.0 million passenger cars registered in the United States. The total number of vehicles registered in the United States for 1990 was estimated at 194.5 million. What percent of the vehicles registered were passenger cars?

68. Gasoline. Gasoline accounts for 85% of the motor fuel consumed in the United States every day. If 8882 thousand barrels (bbl) of motor fuel are consumed each day, how much gasoline is consumed each day in the United States?

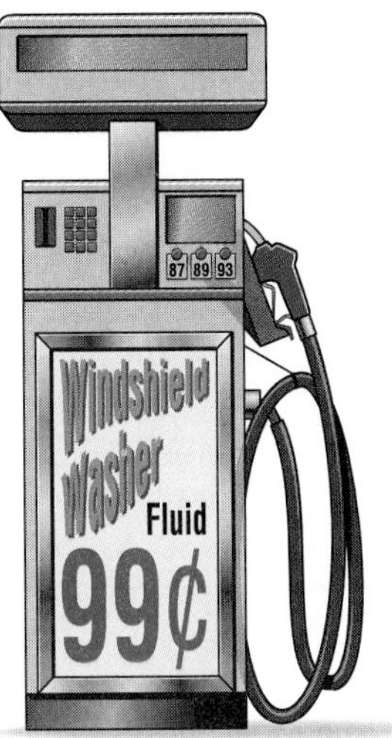

69. Petroleum. In 1999, transportation accounted for 63% of U.S. petroleum consumption. Assuming that same rate applies now, and 10.85 million bbl of petroleum are used each day for transportation in the United States, what is the total daily petroleum consumption by all sources in the United States?

70. Pollution. Each year, 540 million metric tons (t) of carbon dioxide are added to the atmosphere by the United States. Burning gasoline and other transportation fuels is responsible for 35% of the carbon dioxide emissions in the United States. How much carbon dioxide is emitted each year by the burning of transportation fuels in the United States?

71. The progress of the local Lions club is shown below. What percent of the goal has been achieved so far?

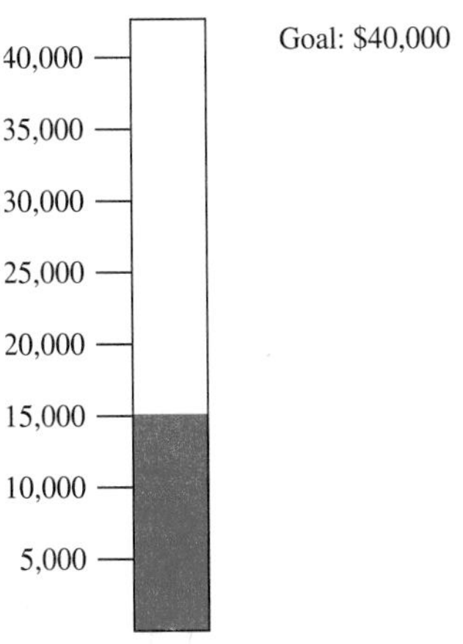

ANSWERS

67. ≈74.6%

68. ≈7550 thousand bbl

69. ≈17.22 million bbl

70. 189 million t

71. 37.5%

72. 25%

73. 75%

74. 37.5%

75. 50%

76. ≈66.67%

a. See exercise

b. See exercise

c. See exercise

d. See exercise

e. See exercise

f. See exercise

g. See exercise

h. See exercise

In Exercises 72 to 76, use the following number line.

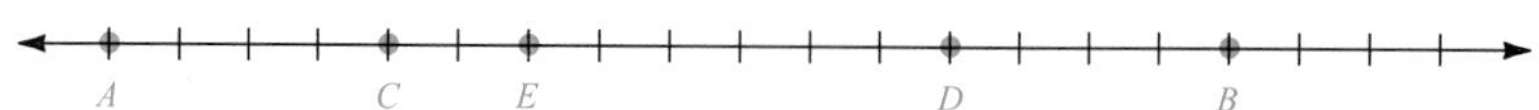

72. Length AC is what percent of length AB?

73. Length AD is what percent of AB?

74. Length AE is what percent of AB?

75. Length AE is what percent of AD?

76. Length AC is what percent of AE?

Getting Ready for Section 2.7 [Section 1.2]

Locate each of the following numbers on the number line.

(a) 4 (b) -5 (c) -3 (d) 2

(e) $-\frac{7}{2}$ (f) $\frac{2}{3}$ (g) 2.5 (h) -1.1

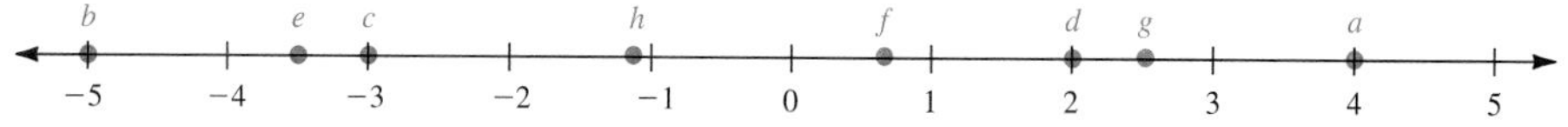

Answers

1. 23% **3.** 250% **5.** 37.5% **7.** 23% **9.** 200 **11.** 350 **13.** R: 5%; A: Commission; B: $40,000 **15.** A: 5; R: percent; B: 30 **17.** A: $3.30; R: 5.5%; B: selling price **19.** A: students; B: 9000; R: 6% **21.** $408 **23.** $143 **25.** 5% **27.** 1.5% **29.** 25 questions **31.** $5000 **33.** $16.64 **35.** $8125 **37.** 8.5% **39.** 25% **41.** 1200 people **43.** $950 **45.** $732 **47.** $156,250 **49.** 12% **51.** 15% **53.** $34,500 **55.** 1200 employees **57.** $65 **59.** 76% **61.** ≈15% **63.** $4494.40 **65.** $4630.50 **67.** ≈74.6% **69.** ≈17.22 million bbl **71.** 37.5% **73.** 75% **75.** 50%

a.–h.

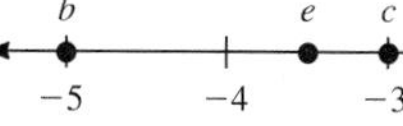

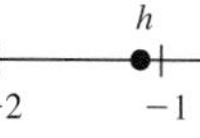

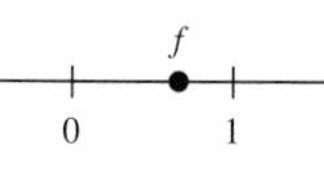

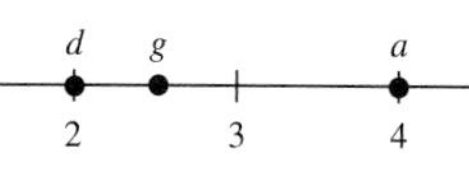

2.7 Inequalities—An Introduction

2.7 OBJECTIVES

1. Use the notation of inequalities
2. Graph the solution set of an inequality
3. Solve an inequality and graph the solution set

As pointed out in the introduction to this chapter, an equation is just a statement that two expressions are equal. In algebra, an **inequality** is a statement that one expression is less than or greater than another. Four new symbols are used in writing inequalities. The use of two of them is illustrated in Example 1.

Example 1

Reading the Inequality Symbol

NOTE To help you remember, the "arrowhead" always points toward the smaller quantity.

$5 < 8$ is an inequality read "5 is less than 8."

$9 > 6$ is an inequality read "9 is greater than 6."

CHECK YOURSELF 1

Fill in the blanks, using the symbols $<$ and $>$.

(a) 12 ______ 8 **(b)** 20 ______ 25

Like an equation, an inequality can be represented by a balance scale. Note that, in each case, the inequality arrow points to the side that is "lighter."

$2x < 4x - 3$

NOTE The $2x$ side is less than the $4x - 3$ side, so it is "lighter."

$5x - 6 > 9$

Just as was the case with equations, inequalities that involve variables may be either true or false depending on the value that we give to the variable. For instance, consider the inequality

$x < 6$

$$\text{If } x = \begin{cases} 3 & 3 < 6 \text{ is true} \\ 5 & 5 < 6 \text{ is true} \\ -10 & -10 < 6 \text{ is true} \\ 8 & 8 < 6 \text{ is false} \end{cases}$$

Therefore 3, 5, and -10 are some *solutions* for the inequality $x < 6$; they make the inequality a true statement. You should see that 8 is *not* a solution. We call the set of all solutions the **solution set** for the inequality. Of course, there are many possible solutions.

Because there are so many solutions (an infinite number, in fact), we certainly do not want to try to list them all! A convenient way to show the solution set of an inequality is with the use of a number line.

Example 2

Solving Inequalities

To graph the solution set for the inequality $x < 6$, we want to include all real numbers that are "less than" 6. This means all numbers *to the left* of 6 on the number line. We then start at 6 and draw an arrow extending left, as shown:

NOTE The colored arrow indicates the direction of the *solution.*

Note: The **open circle** at 6 means that we do not include 6 in the solution set (6 is not less than itself). The colored arrow shows all the numbers in the solution set, with the arrowhead indicating that the solution set continues indefinitely to the left.

CHECK YOURSELF 2

Graph the solution set of $x < -2$.

Two other symbols are used in writing inequalities. They are used with inequalities such as

$x \geq 5$ and $x \leq 2$

Here $x \geq 5$ is really a combination of the two statements $x > 5$ and $x = 5$. It is read "x is greater than or equal to 5." The solution set includes 5 in this case.

The inequality $x \leq 2$ combines the statements $x < 2$ and $x = 2$. It is read "x is less than or equal to 2."

Example 3

Graphing Inequalities

NOTE Here the filled-in circle means that we want to include 5 in the solution set. This is often called a **closed** circle.

The solution set for $x \geq 5$ is graphed as follows.

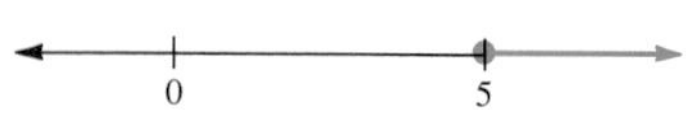

CHECK YOURSELF 3

Graph the solution sets.

(a) $x \le -4$ **(b)** $x \ge 3$

You have learned how to graph the solution sets of some simple inequalities, such as $x < 8$ or $x \ge 10$. Now we will look at more complicated inequalities, such as

$$2x - 3 < x + 4$$

This is called a **linear inequality in one variable.** Only one variable is involved in the inequality, and it appears only to the first power. Fortunately, the methods used to solve this type of inequality are very similar to those we used earlier in this chapter to solve linear equations in one variable. Here is our first property for inequalities.

Rules and Properties: The Addition Property of Inequality

If $a < b$ then $a + c < b + c$

In words, adding the same quantity to both sides of an inequality gives an **equivalent inequality.**

NOTE Equivalent inequalities have exactly the same solution sets.

Again, we can use the idea of a balance scale to see the significance of this property. If we add the same weight to both sides of an unbalanced scale, it stays unbalanced.

Example 4

Solving Inequalities

NOTE The inequality is solved when an equivalent inequality has the form

$x < \square$ or $x > \square$

Solve and graph the solution set for $x - 8 < 7$.

To solve $x - 8 < 7$, add 8 to both sides of the inequality by the addition property.

$$\begin{array}{rcl} x - 8 & < & 7 \\ +8 & & +8 \\ \hline x & < & 15 \end{array}$$

(The inequality is solved)

The graph of the solution set is 

CHECK YOURSELF 4

Solve and graph the solution set for

$x - 9 > -3$

Example 5

Solving Inequalities

Solve and graph the solution set for $4x - 2 \geq 3x + 5$.

First, we add $-3x$ to both sides of the inequality.

NOTE We added $-3x$ and then added 2 to both sides. If these steps are done in the other order, the resulting inequality will be the same.

$$\begin{array}{rcr} 4x - 2 & \geq & 3x + 5 \\ -3x & & -3x \\ \hline x - 2 & \geq & 5 \\ +2 & & +2 \\ \hline x & \geq & 7 \end{array}$$

Now we add 2 to both sides.

The graph of the solution set is

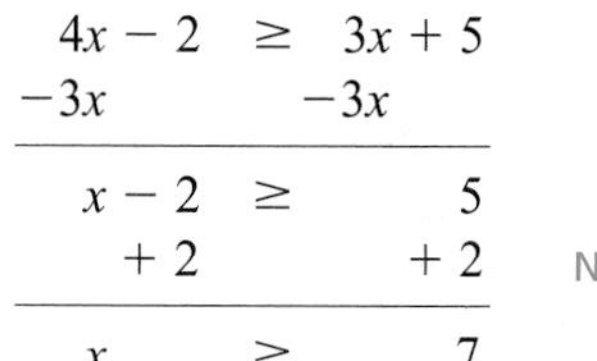

CHECK YOURSELF 5

Solve and graph the solution set.

$7x - 8 \leq 6x + 2$

You will also need a rule for multiplying on both sides of an inequality. Here you'll have to be a bit careful. There is a difference between the multiplication property for inequalities and that for equations. Look at the following:

$2 < 7$ (A true inequality)

Let's multiply both sides by 3.

$$2 < 7$$
$$3 \cdot 2 < 3 \cdot 7$$
$$6 < 21 \quad \text{(A true inequality)}$$

Now we multiply both sides by -3.

$$2 < 7$$
$$(-3)(2) < (-3)(7)$$
$$-6 < -21 \quad \text{(\textit{Not} a true inequality)}$$

Let's try something different.

$$2 < 7$$
$$(-3)(2) > (-3)(7)$$
$$-6 > -21$$

Change the "sense" of the inequality: $<$ becomes $>$. (This is now a true inequality.)

This suggests that multiplying both sides of an inequality by a negative number changes the "sense" of the inequality.

We can state the following general property.

Rules and Properties: The Multiplication Property of Inequality

If $a < b$ then $ac < bc$ when $c > 0$
and $ac > bc$ when $c < 0$

In words, multiplying both sides of an inequality by the same *positive* number gives an equivalent inequality.

When both sides of an inequality are multiplied by the same *negative* number, it is necessary to *reverse the sense* of the inequality to give an equivalent inequality.

Example 6

Solving and Graphing Inequalities

(a) Solve and graph the solution set for $5x < 30$.

Multiplying both sides of the inequality by $\frac{1}{5}$ gives

$$\frac{1}{5}(5x) < \frac{1}{5}(30)$$

Simplifying, we have

$$x < 6$$

The graph of the solution set is

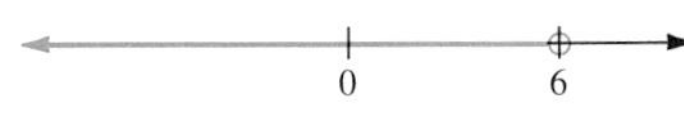

(b) Solve and graph the solution set for $-4x \geq 28$.

In this case we want to multiply both sides of the inequality by $-\frac{1}{4}$ to leave x alone on the left.

$$\left(-\frac{1}{4}\right)(-4x) \leq \left(-\frac{1}{4}\right)(28)$$

Reverse the sense of the inequality because you are multiplying by a negative number!

or $\quad x \leq -7$

The graph of the solution set is

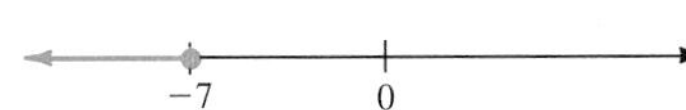

CHECK YOURSELF 6

Solve and graph the solution sets:

(a) $7x > 35$ **(b)** $-8x \leq 48$

Example 7 illustrates the use of the multiplication property when fractions are involved in an inequality.

Example 7

Solving and Graphing Inequalities

(a) Solve and graph the solution set for

$$\frac{x}{4} > 3$$

Here we multiply both sides of the inequality by 4. This will isolate x on the left.

$$4\left(\frac{x}{4}\right) > 4(3)$$

$$x > 12$$

The graph of the solution set is

0 12

(b) Solve and graph the solution set for

$$-\frac{x}{6} \geq -3$$

In this case, we multiply both sides of the inequality by -6:

NOTE Note that we reverse the sense of the inequality because we are multiplying by a negative number.

$$(-6)\left(-\frac{x}{6}\right) \leq (-6)(-3)$$

$$x \leq 18$$

The graph of the solution set is

0 18

CHECK YOURSELF 7

Solve and graph the solution sets for the following inequalities.

(a) $\frac{x}{5} \leq 4$ **(b)** $-\frac{x}{3} < -7$

Example 8

Solving and Graphing Inequalities

(a) Solve and graph the solution set for $5x - 3 < 2x$.

First, add 3 to both sides to undo the subtraction on the left.

$$\begin{array}{rcl} 5x - 3 & < & 2x \\ +3 & & +3 \\ \hline 5x & < & 2x + 3 \end{array}$$

Add 3 to both sides to undo the subtraction.

Now add $-2x$, so that only the number remains on the right.

$$\begin{array}{rcr} 5x & < & 2x + 3 \\ +(-2x) & & +(-2x) \\ \hline 3x & < & 3 \end{array}$$

Add $-2x$ to isolate the number on the right.

NOTE Note that the multiplication property also allows us to divide both sides by a nonzero number.

Next *divide* both sides by 3.

$$\frac{3x}{3} < \frac{3}{3}$$

$$x < 1$$

The graph of the solution set is

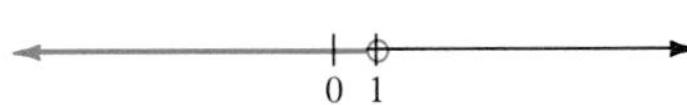

(b) Solve and graph the solution set for $2 - 5x < 7$.

$$\begin{array}{rcl} 2 - 5x & < & 7 \\ -2 & & -2 \\ \hline -5x & < & 5 \end{array} \quad \text{Add } -2.$$

$$\frac{-5x}{-5} > \frac{5}{-5} \quad \text{Divide by } -5. \text{ Be sure to reverse the sense of the inequality.}$$

or $\quad x > -1$

The graph is

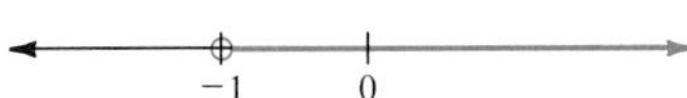

CHECK YOURSELF 8

Solve and graph the solution sets.

(a) $4x + 9 \geq x$ **(b)** $5 - 6x < 41$

As with equations, we will collect all variable terms on one side and all constant terms on the other.

Example 9

Solving and Graphing Inequalities

Solve and graph the solution set for $5x - 5 \geq 3x + 4$.

$$\begin{array}{rcl} 5x - 5 & \geq & 3x + 4 \\ + 5 & & + 5 \\ \hline 5x & \geq & 3x + 9 \\ -3x & & -3x \\ \hline 2x & \geq & 9 \end{array} \quad \begin{array}{l} \text{Add } 5. \\ \\ \text{Add } -3x. \end{array}$$

$$\frac{2x}{2} \geq \frac{9}{2} \quad \text{Divide by } 2.$$

$$x \geq \frac{9}{2}$$

The graph of the solution set is 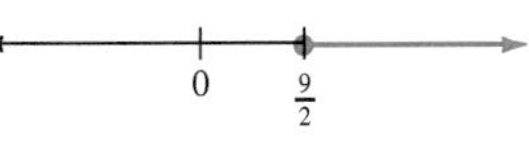

CHECK YOURSELF 9

Solve and graph the solution set for

$8x + 3 < 4x - 13$

Be especially careful when negative coefficients occur in the solution process.

Example 10

Solving and Graphing Inequalities

Solve and graph the solution set for $2x + 4 < 5x - 2$.

$$\begin{array}{rcl} 2x + 4 & < & 5x - 2 \\ -4 & & -4 \\ \hline 2x & < & 5x - 6 \\ -5x & & -5x \\ \hline -3x & < & -6 \end{array}$$

Add −4.

Add −5x.

$$\frac{-3x}{-3} > \frac{-6}{-3}$$

Divide by −3, and reverse the sense of the inequality.

$$x > 2$$

The graph of the solution set is 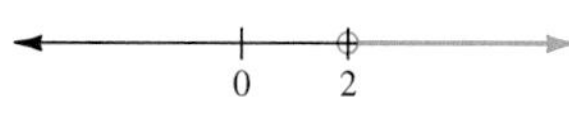

CHECK YOURSELF 10

Solve and graph the solution set.

$5x + 12 \geq 10x - 8$

The solution of inequalities may also require the use of the distributive property.

Example 11

Solving and Graphing Inequalities

Solve and graph the solution set for

$5(x - 2) \geq -8$

Applying the distributive property on the left yields

$5x - 10 \geq -8$

Solving as before yields

$$\begin{array}{rcl} 5x - 10 & \geq & -8 \\ +10 & & +10 \\ \hline 5x & \geq & 2 \end{array}$$

Add 10.

or $x \geq \frac{2}{5}$ Divide by 5.

The graph of the solution set is 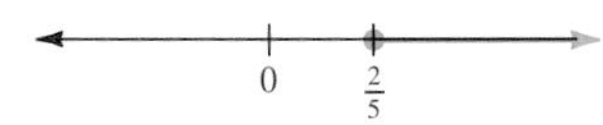

CHECK YOURSELF 11

Solve and graph the solution set.

$4(x + 3) < 9$

Some applications are solved by using an inequality instead of an equation. Example 12 illustrates such an application.

Example 12

Solving an Application with Inequalities

Mohammed needs a mean score of 92 or higher on four tests to get an A. So far his scores are 94, 89, and 88. What score on the fourth test will get him an A?

Name: Mohammed

88

2 x 3 = 6	5 x 4 = 20
1 + 5 = 6	3 x 4 = 12
2 x 5 = 10	5 x 2 = 10
4 + 5 = 9	5 + 4 = 9
15 - 2 = 13	15 - 4 = 11
4 x 3 = 12	8 x 3 = 22
3 + 6 = 9	6 + 3 = 9
9 + 4 = 13	5 + 6 = 11
3 + 9 = 11	6 + 9 = 15
1 x 2 = 2	2 x 1 = 2
13 - 4 = 9	13 - 3 = 10
5 + 6 = 11	9 + 4 = 12

8 x 4 = 32

NOTE What do you need to find?

Step 1 We are looking for the score that will, when combined with the other scores, give Mohammed an A.

NOTE Assign a letter to the unknown.

Step 2 Let x represent a fourth-test score that will get him an A.

NOTE Write an inequality.

Step 3 The inequality will have the mean on the left side, which must be greater than or equal to the 92 on the right.

$$\frac{94 + 89 + 88 + x}{4} \geq 92$$

NOTE Solve the inequality.

Step 4 First, multiply both sides by 4:

$$94 + 89 + 88 + x \geq 368$$

Then add the test scores:

$$183 + 88 + x \geq 368$$

$$271 + x \geq 368$$

Subtracting 271 from both sides,

$$x \geq 97$$

Step 5 To check the solution, we find the mean of the four test scores, 94, 89, 88, and 97.

$$\frac{94 + 89 + 88 + 97}{4} = \frac{368}{4} = 92$$

CHECK YOURSELF 12

Felicia needs a mean score of at least 75 *on five tests to get a passing grade in her health class. On her first four tests she has scores of* 68, 79, 71, *and* 70. *What score on the fifth test will give her a passing grade?*

The following outline (or algorithm) summarizes our work in this section.

Step by Step: Solving Linear Inequalities

Step 1 Remove any grouping symbols and combine any like terms appearing on either side of the inequality.

Step 2 Apply the addition property to write an equivalent inequality with the variable term on one side of the inequality and the number on the other.

Step 3 Apply the multiplication property to write an equivalent inequality with the variable isolated on one side of the inequality. Be sure to reverse the sense of the inequality if you multiply or divide by a negative number. The set of solutions derived in step 3 can then be graphed on a number line.

CHECK YOURSELF ANSWERS

1. **(a)** >; **(b)** <

2. 

3. **(a)** 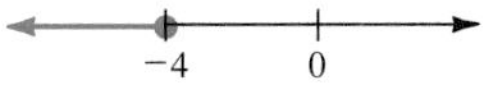

; **(b)** 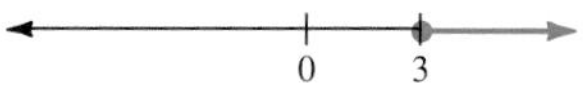

4. $x > 6$ 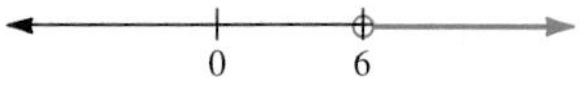

5. $x \le 10$ 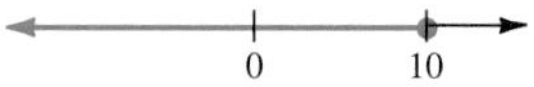

6. **(a)** $x > 5$ 

; **(b)** $x \ge -6$ 

7. **(a)** $x \le 20$ 

; **(b)** $x > 21$ 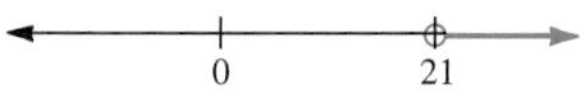

8. **(a)** $x \ge -3$ 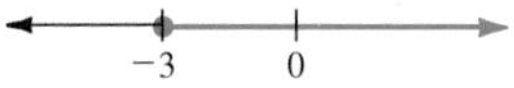

; **(b)** $x > -6$ 

9. $x < -4$ 

10. $x \le 4$ 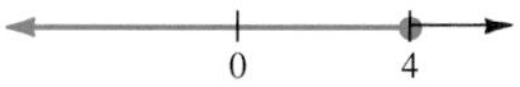

11. $x < -\frac{3}{4}$ 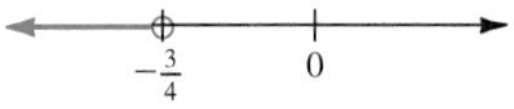

12. 87 or greater

2.7 Exercises

Complete the statements, using the symbol < or >.

1. 5 __________ 10

2. 9 __________ 8

3. 7 __________ −2

4. 0 __________ −5

5. 0 __________ 4

6. −10 __________ −5

7. −2 __________ −5

8. −4 __________ −11

Write each inequality in words.

9. $x < 3$

10. $x \leq -5$

11. $x \geq -4$

12. $x < -2$

13. $-5 \leq x$

14. $2 < x$

Graph the solution set of each of the following inequalities.

15. $x > 2$

16. $x < -3$

17. $x < 9$

18. $x > 4$

19. $x > 1$

20. $x < -2$

21. $x < 8$

22. $x > 3$

23. $x > -5$

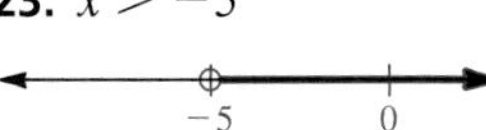

24. $x < -4$

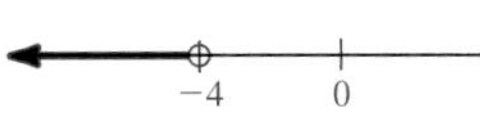

ANSWERS

1. <
2. >
3. >
4. >
5. <
6. <
7. >
8. >
9. *x* is less than 3
10. *x* is less than or equal to −5
11. *x* is greater than or equal to −4
12. *x* is less than −2
13. −5 is less than or equal to *x*
14. 2 is less than *x*
15. See exercise
16. See exercise
17. See exercise
18. See exercise
19. See exercise
20. See exercise
21. See exercise
22. See exercise
23. See exercise
24. See exercise

ANSWERS

25. See exercise
26. See exercise
27. See exercise
28. See exercise
29. $x < 13$
30. $x \le -1$
31. $x \ge 2$
32. $x > -3$
33. $x < 7$
34. $x \ge -4$
35. $x \le 8$
36. $x > -2$
37. $x \ge 8$
38. $x \le -8$
39. $x < -9$
40. $x > 10$
41. $x \le 3$
42. $x > 4$
43. $x > -7$
44. $x \le -3$
45. $x \le -3$
46. $x > -5$
47. $x > 6$
48. $x \le 4$

25. $x \ge 9$

26. $x \ge 0$

27. $x < 0$

28. $x \le -3$

Solve and graph the solution set of each of the following inequalities.

29. $x - 7 < 6$

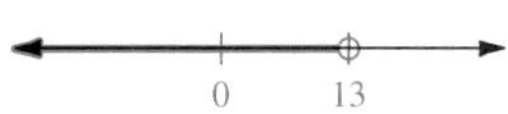

30. $x + 5 \le 4$

31. $x + 8 \ge 10$

32. $x - 11 > -14$

33. $5x < 4x + 7$

34. $3x \ge 2x - 4$

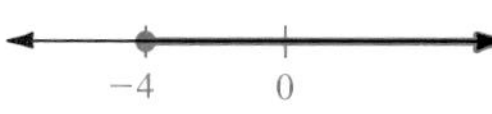

35. $6x - 8 \le 5x$

36. $3x + 2 > 2x$

37. $4x - 3 \ge 3x + 5$

38. $5x + 2 \le 4x - 6$

39. $7x + 5 < 6x - 4$

40. $8x - 7 > 7x + 3$

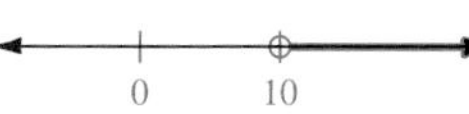

41. $3x \le 9$

42. $5x > 20$

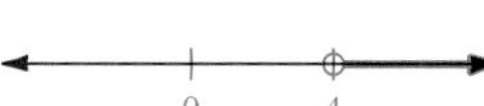

43. $5x > -35$

44. $7x \le -21$

45. $-6x \ge 18$

46. $-9x < 45$

47. $-10x < -60$

48. $-12x \ge -48$

49. $\frac{x}{4} > 5$

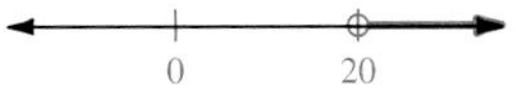

50. $\frac{x}{3} \le -3$

51. $-\frac{x}{2} \ge -3$

52. $-\frac{x}{5} < 4$

53. $\frac{2x}{3} < 6$

54. $\frac{3x}{4} \ge -9$

55. $5x > 3x + 8$

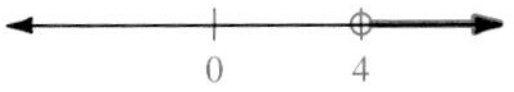

56. $4x \le x - 9$

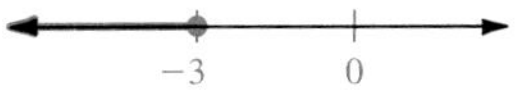

57. $5x - 2 > 3x$

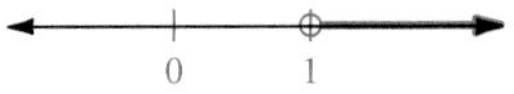

58. $7x + 3 \ge 2x$

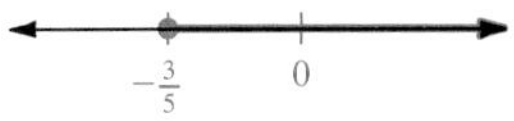

59. $3 - 2x > 5$

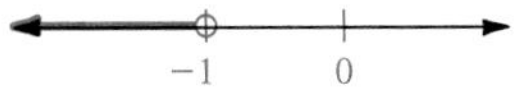

60. $5 - 3x \le 17$

61. $2x \ge 5x + 18$

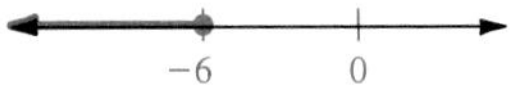

62. $3x < 7x - 28$

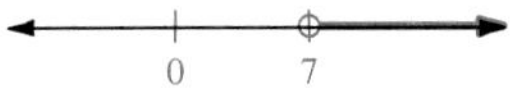

63. $5x - 3 \le 3x + 15$

64. $8x + 7 > 5x + 34$

65. $9x + 7 > 2x - 28$

66. $10x - 5 \le 8x - 25$

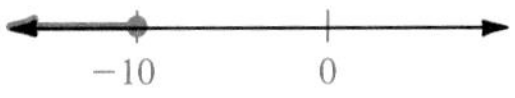

67. $7x - 5 < 3x + 2$

68. $5x - 2 \ge 2x - 7$

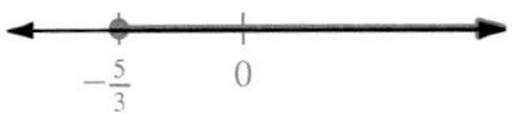

69. $5x + 7 > 8x - 17$

70. $4x - 3 \le 9x + 27$

71. $3x - 2 \le 5x + 3$

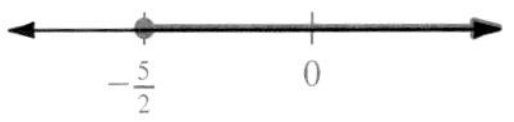

72. $2x + 3 > 8x - 2$

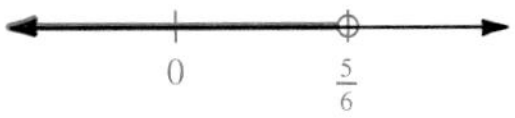

ANSWERS

49. $x > 20$

50. $x \le -9$

51. $x \le 6$

52. $x > -20$

53. $x < 9$

54. $x \ge -12$

55. $x > 4$

56. $x \le -3$

57. $x > 1$

58. $x \ge -\frac{3}{5}$

59. $x < -1$

60. $x \ge -4$

61. $x \le -6$

62. $x > 7$

63. $x \le 9$

64. $x > 9$

65. $x > -5$

66. $x \le -10$

67. $x < \frac{7}{4}$

68. $x \ge -\frac{5}{3}$

69. $x < 8$

70. $x \ge -6$

71. $x \ge -\frac{5}{2}$

72. $x < \frac{5}{6}$

ANSWERS

73. $x \le \frac{3}{2}$

74. $x > \frac{4}{3}$

75. $x < -\frac{2}{3}$

76. $x \ge \frac{5}{4}$

77. $x + 5 > 3$

78. $x - 3 \le 5$

79. $2x - 4 \le 7$

80. $x + 10 > -2$

81. $4x - 15 > x$

82. $2x + 28 \le 6x$

83. a

84. f

85. c

86. d

87. b

88. e

89. $P < 1000$

73. $4(x + 7) \le 2x + 31$

74. $6(x - 5) > 3x - 26$

75. $2(x - 7) > 5x - 12$

76. $3(x + 4) \le 7x + 7$

Translate the following statements into inequalities. Let x represent the number in each case.

77. 5 more than a number is greater than 3.

78. 3 less than a number is less than or equal to 5.

79. 4 less than twice a number is less than or equal to 7.

80. 10 more than a number is greater than negative 2.

81. 4 times a number, decreased by 15, is greater than that number.

82. 2 times a number, increased by 28, is less than or equal to 6 times that number.

Match each inequality on the right with a statement on the left.

83. x is nonnegative	**a.** $x \ge 0$
84. x is negative	**b.** $x \ge 5$
85. x is no more than 5	**c.** $x \le 5$
86. x is positive	**d.** $x > 0$
87. x is at least 5	**e.** $x < 5$
88. x is less than 5	**f.** $x < 0$

89. Panda population. There are fewer than 1000 wild giant pandas left in the bamboo forests of China. Write an inequality expressing this relationship.

$P < 1000$

90. **Forestry.** Let C represent the amount of Canadian forest and M represent the amount of Mexican forest. Write an inequality showing the relationship of the forests of Mexico and Canada if Canada contains at least 9 times as much forest as Mexico.

$C \geq 9M$

91. **Test scores.** To pass a course with a grade of B or better, Liza must have an average of 80 or more. Her grades on three tests are 72, 81, and 79. Write an inequality representing the score that Liza must get on the fourth test to obtain a B average or better for the course.

$\frac{72+81+79+x}{4} \geq 80$

92. **Test scores.** Sam must have an average of 70 or more in his summer course to obtain a grade of C. His first three test grades were 75, 63, and 68. Write an inequality representing the score that Sam must get in the last test to get a C grade.

$\frac{75+63+68+x}{4} \geq 70$

93. **Commission.** Juanita is a salesperson for a manufacturing company. She may choose to receive \$500 or 5 percent commission on her sales as payment for her work. How much does she need to sell to make the 5 percent offer a better deal?

$.05x > 500$

94. **Telephone costs.** The cost for a long distance telephone call is \$0.36 for the first minute and \$0.21 for each additional minute or portion thereof. The total cost of the call cannot exceed \$3. Write an inequality representing the number of minutes a person could talk without exceeding \$3.

$.36(1) + .21(x-1) \leq 3$

95. **Geometry.** The perimeter of a rectangle is to be no greater than 250 centimeters (cm) and the length must be 105 cm. Find the maximum width of the rectangle.

$105 + 2x \leq 250$

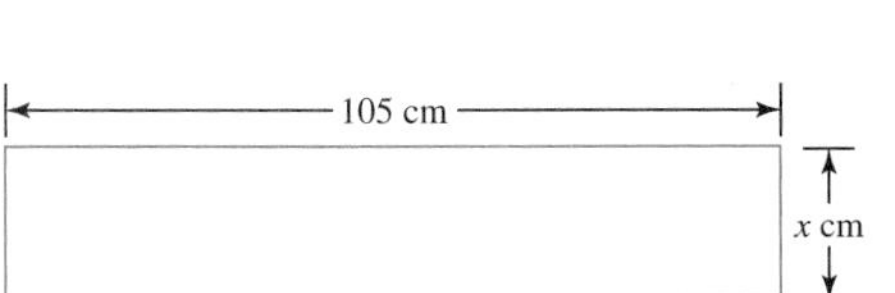

96. **Recreation.** Sarah bowled 136 and 189 in her first two games. What must she bowl in her third game to have an average of at least 170?

$\frac{136+189+x}{3} \geq 170$

97. You are the office manager for a small company. You need to acquire a new copier for the office. You find a suitable one that leases for \$250 a month from the copy machine company. It costs 2.5¢ per copy to run the machine. You purchase paper for \$3.50 a ream (500 sheets). If your copying budget is no more than \$950 per month, is this machine a good choice? Write a brief recommendation to the Purchasing Department. Use equations and inequalities to explain your recommendation.

ANSWERS

90. $C \geq 9M$

91. $x \geq 88$

92. $x \geq 74$

93. $>$\$10,000

94. $.36 + .21(t - 1) \leq 3$

95. 20 cm

96. ≥ 185

97.

98. ______

98. Your aunt calls to ask your help in making a decision about buying a new refrigerator. She says that she found two that seem to fit her needs, and both are supposed to last at least 14 years, according to *Consumer Reports.* The initial cost for one refrigerator is \$712, but it only uses 88 kilowatt-hours (kWh) per month. The other refrigerator costs \$519 and uses an estimated 100 kWh/per month. You do not know the price of electricity per kilowatt-hour where your aunt lives, so you will have to decide what in cents per kilowatt-hour will make the first refrigerator cheaper to run for its 14 years of expected usefulness. Write your aunt a letter explaining what you did to calculate this cost, and tell her to make her decision based on how the kilowatt-hour rate she has to pay in her area compares with your estimation.

Answers

1. $5 < 10$ **3.** $7 > -2$ **5.** $0 < 4$ **7.** $-2 > -5$ **9.** x is less than 3
11. x is greater than or equal to -4 **13.** -5 is less than or equal to x
15. 0 2 **17.** 0 9
19. 0 1 **21.** 0 8
23. −5 0 **25.** 0 9
27. 0 **29.** $x < 13$ 0 13
31. $x \geq 2$ 0 2 **33.** $x < 7$ 0 7
35. $x \leq 8$ 0 8 **37.** $x \geq 8$ 0 8
39. $x < -9$ −9 0 **41.** $x \leq 3$ 0 3
43. $x > -7$ −7 0 **45.** $x \leq -3$ −3 0
47. $x > 6$ 0 6 **49.** $x > 20$ 0 20
51. $x \leq 6$ 0 6 **53.** $x < 9$ 0 9
55. $x > 4$ 0 4 **57.** $x > 1$ 0 1
59. $x < -1$ −1 0 **61.** $x \leq -6$ −6 0
63. $x \leq 9$ 0 9 **65.** $x > -5$ −5 0
67. $x < \frac{7}{4}$ 0 $\frac{7}{4}$ **69.** $x < 8$ 0 8
71. $x \geq -\frac{5}{2}$ $-\frac{5}{2}$ 0 **73.** $x \leq \frac{3}{2}$ 0 $\frac{3}{2}$
75. $x < -\frac{2}{3}$ $-\frac{2}{3}$ 0 **77.** $x + 5 > 3$ **79.** $2x - 4 \leq 7$
81. $4x - 15 > x$ **83.** a **85.** c **87.** b **89.** $P < 1000$ **91.** $x \geq 88$
93. >\$10,000 **95.** 20 cm **97.**

2 Summary

DEFINITION/PROCEDURE	EXAMPLE	REFERENCE
Solving Equations by the Addition Property		**Section 2.1**
Equation A statement that two expressions are equal	$2x - 3 = 5$ is an equation	**p. 141**
Solution A value for a variable that makes an equation a true statement	4 is a solution for the above equation because $2(4) - 3 = 5$	**p. 142**
Equivalent Equations Equations that have exactly the same solutions	$2x - 3 = 5$ and $x = 4$ are equivalent equations	**p. 144**
The Addition Property of Equality If $a = b$ then $a + c = b + c$	If $2x - 3 = 7$ then $2x - 3 + 3 = 7 + 3$	**p. 144**
Solving Equations by the Multiplication Property		**Section 2.2**
The Multiplication Property of Equality If $a = b$ then $a \cdot c = b \cdot c$	If $\frac{1}{2}x = 7$ then $2\left(\frac{1}{2}x\right) = 2(7)$	**p. 155**
The mean of a set is the sum of that set divided by the number of things in the set.	Given the set $-2, -1, 6, 9$ The mean is $12/4 = 3$	**p. 160**
Combining the Rules to Solve Equations		**Section 2.3**
Solving Linear Equations The steps of solving a linear equation are as follows: **1.** Use the distributive property to remove any grouping symbols. Then simplify by combining like terms. **2.** Add or subtract the same term on both sides of the equation until the variable term is on one side and a number is on the other. **3.** Multiply or divide both sides of the equation by the same nonzero number so that the variable is alone on one side of the equation. **4.** Check the solution in the original equation.	Solve: $3(x - 2) + 4x = 3x + 14$ $3x - 6 + 4x = 3x + 14$ $7x - 6 = 3x + 14$ $+6 \quad +6$ $7x = 3x + 20$ $-3x \quad -3x$ $4x = 20$ $\frac{4x}{4} = \frac{20}{4}$ $x = 5$	**p. 170**
Formulas and Problem Solving		**Section 2.4**
Literal Equation An equation that involves more than one letter or variable.	$a = \frac{2b + c}{3}$	**p. 175**
Solving Literal Equations **1.** Multiply both sides of the equation by the same term to clear it of fractions. **2.** Add or subtract the same term on both sides of the equation so that all terms containing the variable you are solving for are on one side. **3.** Divide both sides by the coefficient of the variable that you are solving for.	Solve for b: $a = \frac{2b + c}{3}$ $3a = \left(\frac{2b + c}{3}\right)3$ $3a = 2b + c$ $3a - c = 2b$ $\frac{3a - c}{2} = b$	**p. 176**

Continued

DEFINITION/PROCEDURE	EXAMPLE	REFERENCE
Solving Percent Applications		**Section 2.6**
The base is the whole in a statement.	14 is 25% of 56 56 is the base	**p. 207**
The amount is the part being compared to the base.	14 is the amount	**p. 207**
The rate is the ratio of the amount to the base.	25% is the rate	**p. 207**
The proportion used for solving most percent applications is $\frac{A}{B} = \frac{R}{100}$	$\frac{14}{56} = \frac{25}{100}$	**p. 208**
Inequalities—An Introduction		**Section 2.7**
Inequality A statement that one quantity is less than (or greater than) another. Four symbols are used: $a < b$ — a is less than b $a > b$ — a is greater than b $a \leqslant b$ — a is less than or equal to b $a \geqslant b$ — a is greater than or equal to b	$-4 < -1$ $x^2 + 1 \geqslant x + 1$	**p. 221**
Graphing Inequalities To graph $x < a$, we use an open circle and an arrow pointing left. The heavy arrow indicates all numbers less than (or to the left of) a. The open circle means a is not included.	Graph $x < 3$ 0 3	**p. 222**
To graph $x \geq b$, we use a closed circle and an arrow pointing right. The closed circle means that in this case b is included.	Graph $x \geq -1$ −1 0	**p. 223**
Solving Inequalities An inequality is "solved" when it is in the form $x < \square$ or $x > \square$. Proceed as in solving equations by using the following properties. **1.** If $a < b$, then $a + c < b + c$. Adding (or subtracting) the same quantity to both sides of an inequality gives an equivalent inequality. **2.** If $a < b$, then $ac < bc$ when $c > 0$ and $ac > bc$ when $c < 0$. Multiplying both sides of an inequality by the same *positive number* gives an equivalent inequality. When both sides of an inequality are multiplied by the same *negative number, you must reverse the sense* of the inequality to give an equivalent inequality.	$2x - 3 > 5x + 6$ $\underline{\quad + 3 \qquad + 3}$ $2x > 5x + 9$ $\underline{-5x \qquad -5x}$ $-3x > 9$ $\frac{-3x}{-3} < \frac{9}{-3}$ $x < -3$ −3 0	**p. 225**

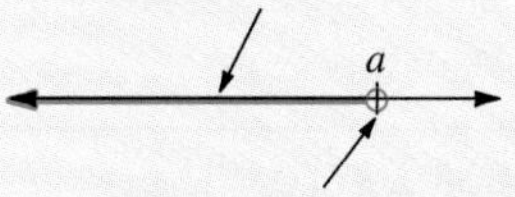

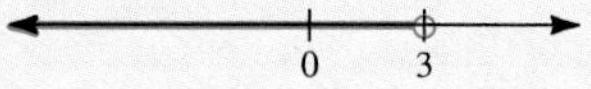

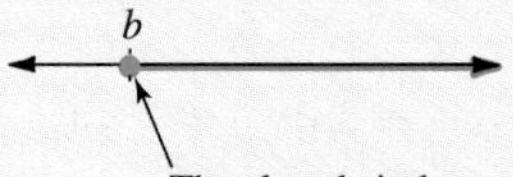

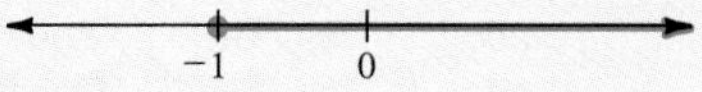

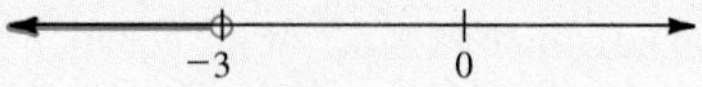

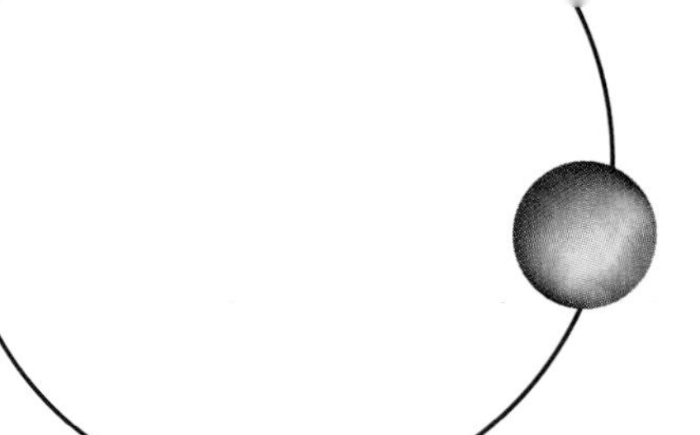

Summary Exercises

This summary exercise set is provided to give you practice with each of the objectives of the chapter. Each exercise is keyed to the appropriate chapter section. The answers are provided in the *Instructor's Manual.* Your instructor will give you guidelines on how to best use these exercises in your instructional setting.

[2.1] Tell whether the number shown in parentheses is a solution for the given equation.

1. $7x + 2 = 16$ (2) **Yes**

2. $5x - 8 = 3x + 2$ (4) **No**

3. $7x - 2 = 2x + 8$ (2) **Yes**

4. $4x + 3 = 2x - 11$ (−7) **Yes**

5. $x + 5 + 3x = 2 + x + 23$ (6) **No**

6. $\frac{2}{3}x - 2 = 10$ (21) **No**

[2.1–2.3] Solve the following equations and check your results.

7. $x + 5 = 7$ **2**

8. $x - 9 = 3$ **12**

9. $5x = 4x - 5$ **−5**

10. $3x - 9 = 2x$ **9**

11. $5x - 3 = 4x + 2$ **5**

12. $9x + 2 = 8x - 7$ **−9**

13. $7x - 5 = 6x - 4$ **1**

14. $3 + 4x - 1 = x - 7 + 2x$ **−9**

15. $4(2x + 3) = 7x + 5$ **−7**

16. $5(5x - 3) = 6(4x + 1)$ **21**

17. $5x = 35$ **7**

18. $7x = -28$ **−4**

19. $-6x = 24$ **−4**

20. $-9x = -63$ **7**

21. $\frac{x}{4} = 8$ **32**

22. $-\frac{x}{5} = -3$ **15**

23. $\frac{2}{3}x = 18$ **27**

24. $\frac{3}{4}x = 24$ **32**

25. $5x - 3 = 12$ **3**

26. $4x + 3 = -13$ **−4**

27. $7x + 8 = 3x$ **−2**

28. $3 - 5x = -17$ **4**

29. $3x - 7 = x$ $\frac{7}{2}$

30. $2 - 4x = 5$ $-\frac{3}{4}$

31. $\frac{x}{3} - 5 = 1$ **18**

32. $\frac{3}{4}x - 2 = 7$ **12**

33. $6x - 5 = 3x + 13$ **6**

34. $3x + 7 = x - 9$ **−8**

35. $7x + 4 = 2x + 6$ $\frac{2}{5}$

36. $9x - 8 = 7x - 3$ $\frac{5}{2}$

37. $2x + 7 = 4x - 5$ **6**

38. $3x - 15 = 7x - 10$ $-\frac{5}{4}$

39. $\frac{10}{3}x - 5 = \frac{4}{3}x + 7$ **6**

40. $\frac{11}{4}x - 15 = 5 - \frac{5}{4}x$ **5**

41. $3.7x + 8 = 1.7x + 16$ **4**

42. $5.4x - 3 = 8.4x + 9$ **−4**

43. $3x - 2 + 5x = 7 + 2x + 21$ **5**

44. $8x + 3 - 2x + 5 = 3 - 4x$ $-\frac{1}{2}$

45. $5(3x - 1) - 6x = 3x - 2$ $\frac{1}{2}$

[2.4] Solve for the indicated variable.

46. $V = LWH$ (for L) $\frac{V}{WH}$

47. $P = 2L + 2W$ (for L) $\frac{P - 2W}{2}$ or $\frac{P}{2} - W$

48. $ax + by = c$ (for y) $\frac{c - ax}{b}$

49. $A = \frac{1}{2}bh$ (for h) $\frac{2A}{b}$

50. $A = P + Prt$ (for t) $\frac{A - P}{Pr}$

51. $m = \frac{n - p}{q}$ (for n) $mq + p$

[2.4–2.6] Solve the following word problems. Be sure to label the unknowns and to show the equation you used.

52. The sum of 3 times a number and 7 is 25. What is the number? **6**

53. 5 times a number, decreased by 8, is 32. Find the number. **8**

54. If the sum of two consecutive integers is 85, find the two integers. **42, 43**

55. The sum of three consecutive odd integers is 57. What are the three integers? **17, 19, 21**

56. Rafael earns \$35 more per week than Andrew. If their weekly salaries total \$715, what amount does each earn?
Rafael: \$375, Andrew: \$340

57. Larry is 2 years older than Susan, and Nathan is twice as old as Susan. If the sum of their ages is 30 years, find each of their ages. **Susan: 7 years, Larry: 9 years, Nathan: 14 years**

58. **Commission.** Joan works on a 4% commission basis. She sold $45,000 in merchandise during 1 month. What was the amount of her commission? **$1800**

59. **Discount rate.** David buys a dishwasher that is marked down $77 from its original price of $350. What is the discount rate? **22%**

60. **Chemistry.** A chemist prepares a 400-milliliter (400-mL) acid-water solution. If the solution contains 30 mL of acid, what percent of the solution is acid? **7.5%**

61. **Price increase.** The price of a new compact car has increased $819 over the previous year. If this amounts to a 4.5% increase, what was the price of the car before the increase? **$18,200**

62. **Markdown.** A store advertises, "Buy the red-tagged items at 25% off their listed price." If you buy a coat marked $136, what will you pay for the coat during the sale? **$102**

63. **Salary.** Tom has 6% of his salary deducted for a retirement plan. If that deduction is $168, what is his monthly salary? **$2800**

64. **Enrollment.** A college finds that 35% of its science students take biology. If there are 252 biology students, how many science students are there altogether? **720 students**

65. **Increase rate.** A company finds that its advertising costs increased from $72,000 to $76,680 in 1 year. What was the rate of increase? **6.5%**

66. **Interest.** A savings bank offers 5.25% on 1-year time deposits. If you place $3000 in an account, how much will you have at the end of the year? **$3157.50**

67. **Salary.** Maria's company offers her a 4% pay raise. This will amount to a $126 per month increase in her salary. What is her monthly salary before and after the raise? **$3150 before, $3276 after**

68. **Computers.** A computer has 8 gigabytes (GB) of storage space. Arlene is going to add 16 GB of storage space. By what percent will the available storage space be increased? **200%**

69. **Computers.** A virus scanning program is checking every file for viruses. It has completed 30% of the files in 150 seconds (s). How long should it take to check all the files? **500 s**

70. **Tipping.** If the total bill at a restaurant for 10 people is $572.89, including an 18% tip, what was the cost of the food? **$485.50**

71. **Sales.** A pair of running shoes is advertised at 30% off the original price for $80.15. What was the original price? **$114.50**

[2.7] Solve and graph the solution sets for the following inequalities.

72. $x - 4 \le 7$ $\quad x \le 11$

73. $x + 3 > -2$ $\quad x > -5$

74. $5x > 4x - 3$ $\quad x > -3$

75. $4x \ge -12$ $\quad x \ge -3$

76. $-12x < 36$ $\quad x > -3$

77. $-\frac{x}{5} \ge 3$ $\quad x \le -15$

78. $2x \le 8x - 3$ $\quad x \ge \frac{1}{2}$

79. $2x + 3 \ge 9$ $\quad x \ge 3$

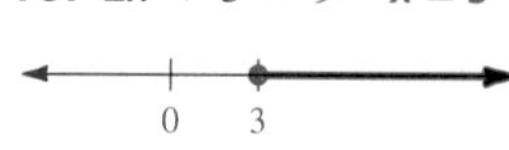

80. $4 - 3x > 8$ $\quad x < -\frac{4}{3}$

81. $5x - 2 \le 4x + 5$ $\quad x \le 7$

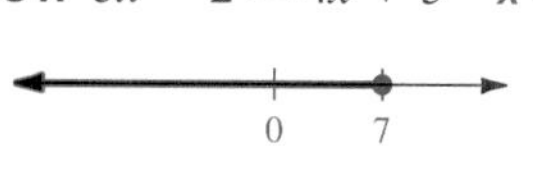

82. $7x + 13 \ge 3x + 19$ $\quad x \ge \frac{3}{2}$

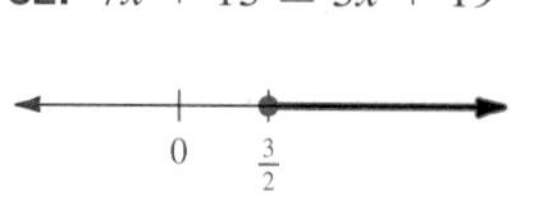

83. $4x - 2 < 7x + 16$ $\quad x > -6$

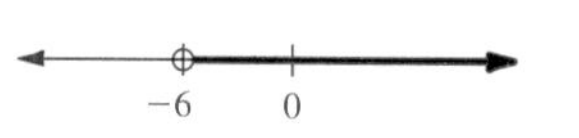

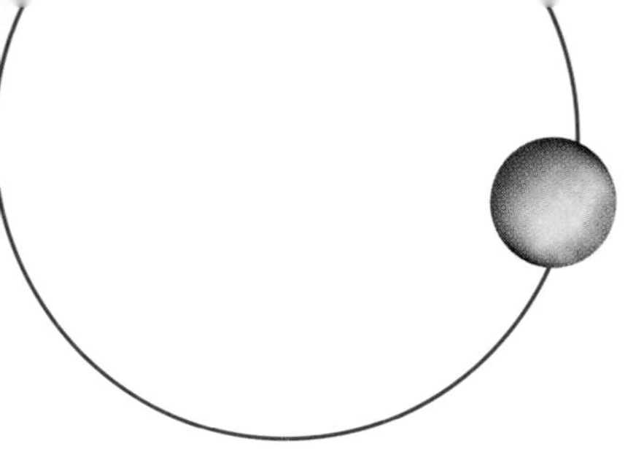

Self-Test for Chapter 2

Name ____________________

Section ________ Date ________

The purpose of this self-test is to help you check your progress and to review for a chapter test in class. Allow yourself about an hour to take the test. When you are done, check your answers in the back of the book. If you missed any answers, be sure to go back and review the appropriate sections in the chapter and the exercises that are provided.

Tell whether the number shown in parentheses is a solution for the given equation.

1. $7x - 3 = 25 \quad (5)$

2. $8x - 3 = 5x + 9 \quad (4)$

Solve the following equations and check your results.

3. $x - 7 = 4$

4. $7x - 12 = 6x$

5. $9x - 2 = 8x + 5$

Solve the following equations and check your results.

6. $7x = 49$

7. $\frac{1}{4}x = -3$

8. $\frac{4}{5}x = 20$

Solve the following equations and check your results.

9. $7x - 5 = 16$

10. $10 - 3x = -2$

11. $7x - 3 = 4x - 5$

12. $2x - 7 = 5x + 8$

Solve for the indicated variable.

13. $C = 2\pi r$ (for r)

14. $V = \frac{1}{3}Bh$ (for h)

15. $3x + 2y = 6$ (for y)

Solve and graph the solution sets for the following inequalities.

16. $x - 5 \le 9$

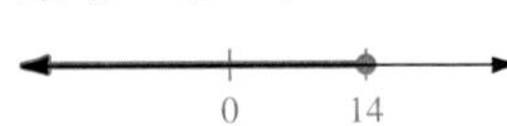

17. $5 - 3x > 17$

-4 0

18. $5x + 13 \ge 2x + 17$

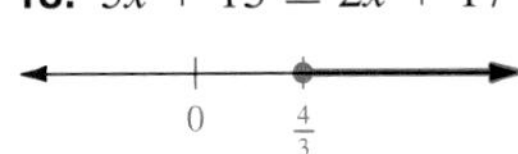

19. $2x - 3 < 7x + 2$

-1 0

ANSWERS

1. No
2. Yes
3. 11
4. 12
5. 7
6. 7
7. −12
8. 25
9. 3
10. 4
11. $-\frac{2}{3}$
12. −5
13. $\frac{C}{2\pi}$
14. $\frac{3V}{B}$
15. $\frac{6 - 3x}{2}$
16. $x \le 14$
17. $x < -4$
18. $x \ge \frac{4}{3}$
19. $x > -1$

ANSWERS

20. 7

21. 21, 22, 23

22. Juwan, 6; Jan, 12; Rick, 17

23. 10 in., 21 in.

24. 5%

25. $35,000

Solve the following word problems. Be sure to show the equation you used for the solution.

20. 5 times a number, decreased by 7, is 28. What is the number?

21. The sum of three consecutive integers is 66. Find the three integers.

22. Jan is twice as old as Juwan, and Rick is 5 years older than Jan. If the sum of their ages is 35 years, find each of their ages.

23. The perimeter of a rectangle is 62 inches (in.). If the length of the rectangle is 1 in. more than twice its width, what are the dimensions of the rectangle?

24. **Commission.** Mrs. Moore made a $450 commission on the sale of a $9000 pickup truck. What was her commission rate?

25. **Salary.** Cynthia makes a 5% commission on all her sales. She earned $1750 in commissions during 1 month. What were her gross sales for the month?

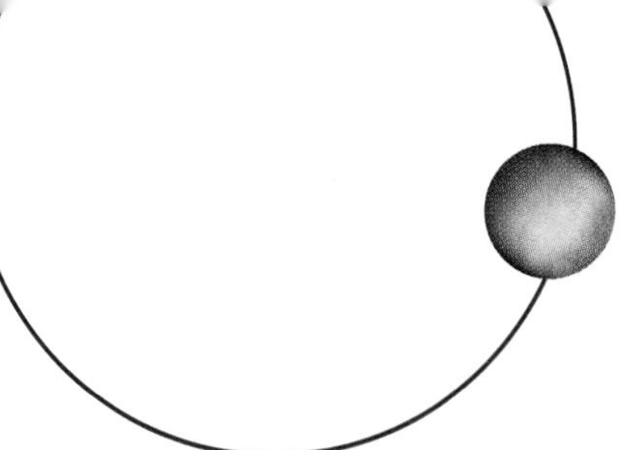

Cumulative Test Chapters 1 and 2

Name ______________________

Section __________ Date __________

This test covers selected topics from Chapters 1 and 2.

Perform the indicated operations.

1. $8 + (-4)$
2. $-7 + (-5)$
3. $6 - (-2)$
4. $-4 - (-7)$
5. $(-6)(3)$
6. $(-11)(-4)$
7. $20 \div (-4)$
8. $(-50) \div (-5)$
9. $0 \div (-26)$
10. $15 \div 0$

Evaluate the expressions if $x = 5$, $y = 2$, $z = -3$, and $w = -4$.

11. $2xy$
12. $2x + 7z$
13. $3z^2$
14. $4(x + 3w)$
15. $\frac{2w}{y}$
16. $\frac{2x - w}{2y - z}$

Simplify each of the following expressions.

17. $14x^2y - 11x^2y$
18. $2x^3(3x - 5y)$
19. $\frac{x^2y - 2xy^2 + 3xy}{xy}$
20. $10x^2 + 5x + 2x^2 - 2x$

Solve the following equations and check your results.

21. $9x - 5 = 8x$
22. $-\frac{3}{4}x = 18$
23. $6x - 8 = 2x - 3$
24. $2x + 3 = 7x + 5$
25. $\frac{4}{3}x - 6 = 4 - \frac{2}{3}x$

ANSWERS

1. 4
2. -12
3. 8
4. 3
5. -18
6. 44
7. -5
8. 10
9. 0
10. Undefined
11. 20
12. -11
13. 27
14. -28
15. -4
16. 2
17. $3x^2y$
18. $6x^4 - 10x^3y$
19. $x - 2y + 3$
20. $12x^2 + 3x$
21. 5
22. -24
23. $\frac{5}{4}$
24. $-\frac{2}{5}$
25. 5

26. $\frac{I}{Pt}$

27. $\frac{2A}{b}$

28. $\frac{c - ax}{b}$

29. $x < 3$

30. $x \leq -\frac{3}{2}$

31. $x > 4$

32. $x \geq \frac{4}{3}$

33. 13

34. 42, 43

35. 7

36. $420

37. 5 cm, 17 cm

38. 8 in., 13 in., 16 in.

39. 2.5%

40. 7.5%

Solve the following equations for the indicated variable.

26. $I = Prt$ (for r)

27. $A = \frac{1}{2}bh$ (for h)

28. $ax + by = c$ (for y)

Solve and graph the solution sets for the following inequalities.

29. $3x - 5 < 4$

30. $7 - 2x \geq 10$

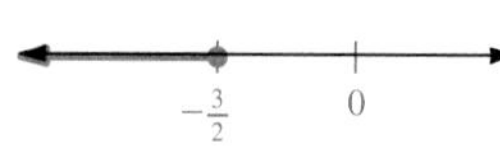

31. $7x - 2 > 4x + 10$

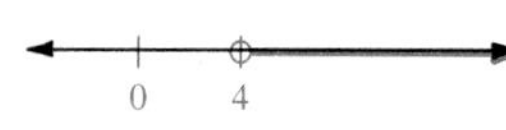

32. $2x + 5 \leq 8x - 3$

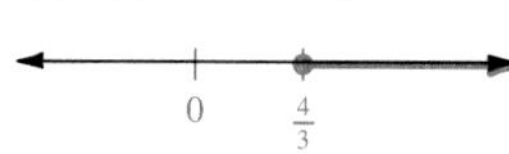

Solve the following word problems. Be sure to show the equation used for the solution.

33. If 4 times a number decreased by 7 is 45, find that number.

34. The sum of two consecutive integers is 85. What are those two integers?

35. If 3 times an integer is 12 more than the next consecutive odd integer, what is that integer?

36. Michelle earns $120 more per week than Dmitri. If their weekly salaries total $720, how much does Michelle earn?

37. The length of a rectangle is 2 centimeters (cm) more than 3 times its width. If the perimeter of the rectangle is 44 cm, what are the dimensions of the rectangle?

38. One side of a triangle is 5 inches (in.) longer than the shortest side. The third side is twice the length of the shortest side. If the triangle perimeter is 37 in., find the length of each leg.

39. Jesse paid $1562.50 in state income tax last year. If his salary was $62,500 what was the rate of tax?

40. A car is marked down from $31,500 to $29,137.50. What was the discount rate?

3

POLYNOMIALS

INTRODUCTION

The U.S. Post Office limits the size of rectangular boxes it will accept for mailing. The regulations state that "length plus girth cannot exceed 108 inches." "Girth" means the distance around a cross section; in this case, this measurement is $2h + 2w$. Using the polynomial $l + 2w + 2h$ to describe the measurement required by the Post Office, the regulations say that $l + 2w + 2h \leq 108$ inches.

The volume of a rectangular box is expressed by another polynomial: $V = lwh$

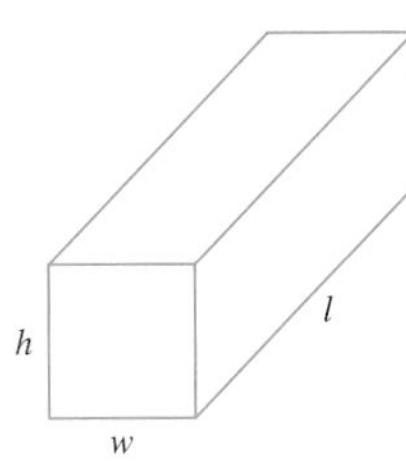

A company that wishes to produce boxes for use by postal patrons must use these formulas and do a statistical survey about the shapes that are useful to the most customers. The surface area, expressed by another polynomial expression, $2lw + 2wh + 2lh$, is also used so each box can be manufactured with the least amount of material, to help lower costs.

Name ______________________

Section ________ Date ________

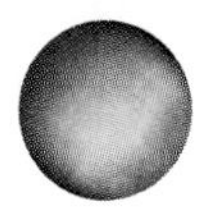

Pre-Test Chapter 3

ANSWERS

1. x^{12}
2. $8x^5y^7$
3. $3x^3y$
4. $4x^6y^8$
5. x^{16}
6. $\frac{2y^3}{x^5}$
7. Binomial
8. Trinomial
9. $2x^2 - 2x - 2$
10. $9x^2 - 11x + 6$
11. $12x^3y^3 - 6x^2y^2 + 21x^2y^4$
12. $6x^2 - 11x - 10$
13. $x^2 - 4y^2$
14. $16m^2 + 40m + 25$
15. $3x^3 - 14x^2y + 17xy^2 - 6y^3$
16. $9x^3 - 30x^2y + 25xy^2$
17. $4y - 5x^2y^3$
18. $x + 2$
19. $x - 3$
20. $3x - 5 - \frac{5}{x + 4}$

Simplify each of the following expressions. Write your answers with positive exponents only.

1. x^5x^7

2. $(2x^3y^2)(4x^2y^5)$

3. $\frac{9x^5y^2}{3x^2y}$

4. $(2x^3y^4)^2(x^{-4}y^{-6})^0$

5. $\frac{(x^{-2})^{-4}}{x^{-8}}$

6. $2x^{-5}y^3$

Classify each of the following polynomials as a monomial, binomial, or trinomial.

7. $6x^2 - 7x$

8. $-4x^3 + 5x - 9$

Add.

9. $4x^2 - 7x + 5$ and $-2x^2 + 5x - 7$

Subtract.

10. $-2x^2 + 3x - 1$ from $7x^2 - 8x + 5$

Multiply.

11. $3xy(4x^2y^2 - 2xy + 7xy^3)$

12. $(3x + 2)(2x - 5)$

13. $(x + 2y)(x - 2y)$

14. $(4m + 5)^2$

15. $(3x - 2y)(x^2 - 4xy + 3y^2)$

16. $x(3x - 5y)^2$

Divide.

17. $\frac{28x^2y^3 - 35x^4y^5}{7x^2y^2}$

18. $\frac{x^2 - x - 6}{x - 3}$

19. $\frac{x^3 - 2x - 3x^2 + 6}{x^2 - 2}$

20. $\frac{3x^2 + 7x - 25}{x + 4}$

Exponents and Polynomials

OBJECTIVES

1. Recognize the five properties of exponents
2. Use the properties to simplify expressions
3. Identify types of polynomials
4. Find the degree of a polynomial
5. Write a polynomial in descending exponent form
6. Evaluate a polynomial

Overcoming Math Anxiety

Hint #4 Preparing for a Test

Preparation for a test really begins on the first day of class. Everything you have done in class and at home has been part of that preparation. However, there are a few things that you should focus on in the last few days before a scheduled test.

1. Plan your test preparation to end at least 24 hours before the test. The last 24 hours is too late, and besides, you will need some rest before the test.
2. Go over your homework and class notes with pencil and paper in hand. Write down all of the problem types, formulas, and definitions that you think might give you trouble on the test.
3. The day before the test, take the page(s) of notes from step 2, and transfer the most important ideas to a 3×5 card.
4. Just before the test, review the information on the card. You will be surprised at how much you remember about each concept.
5. Understand that, if you have been successful at completing your homework assignments, you can be successful on the test. This is an obstacle for many students, but it is an obstacle that can be overcome. Truly anxious students are often surprised that they scored as well as they did on a test. They tend to attribute this to blind luck. It is not. It is the first sign that you really do "get it." Enjoy the success.

In Chapter 0, we introduced the idea of exponents. Recall that the exponent notation indicates repeated multiplication and that the exponent tells us how many times the base is to be used as a factor.

Exponent

$$3^5 = \underbrace{3 \cdot 3 \cdot 3 \cdot 3 \cdot 3}_{\text{5 factors}} = 243$$

Base

Now, we will look at the properties of exponents.

The first property is used when multiplying two values with the same base.

Rules and Properties: Property 1 of Exponents

For any real number a and positive integers m and n,

$a^m \cdot a^n = a^{m+n}$

For example,

$2^5 \cdot 2^7 = 2^{12}$

The second property is used when dividing two values with the same base.

Rules and Properties: Property 2 of Exponents

For any real number a and positive integers m and n, with $m > n$,

$a^m/a^n = a^{m-n}$

For example,

$2^{12}/2^7 = 2^5$

Consider the following:

NOTE Notice that this means that the base, x^2, is used as a factor 4 times.

$(x^2)^4 = x^2 \cdot x^2 \cdot x^2 \cdot x^2 = x^8$

This leads us to our third property for exponents.

Rules and Properties: Property 3 of Exponents

For any real number a and positive integers m and n,

$(a^m)^n = a^{m \cdot n}$

For example,

$(2^3)^2 = 2^{3 \cdot 2} = 2^6$

The use of this new property is illustrated in Example 1.

Example 1

Using the Third Property of Exponents

CAUTION

Be careful! Be sure to distinguish between the correct use of Property 1 and Property 3.

$(x^4)^5 = x^{4 \cdot 5} = x^{20}$

but

$x^4 \cdot x^5 = x^{4+5} = x^9$

Simplify each expression.

(a) $(x^4)^5 = x^{4 \cdot 5} = x^{20}$ Multiply the exponents.

(b) $(2^3)^4 = 2^{3 \cdot 4} = 2^{12}$

CHECK YOURSELF 1

Simplify each expression.

(a) $(m^5)^6$ **(b)** $(m^5)(m^6)$ **(c)** $(3^2)^4$ **(d)** $(3^2)(3^4)$

Suppose we now have a product raised to a power. Consider an expression such as

NOTE Here the base is $3x$.

$(3x)^4$

We know that

NOTE Here we have applied the commutative and associative properties.

$$(3x)^4 = (3x)(3x)(3x)(3x)$$
$$= (3 \cdot 3 \cdot 3 \cdot 3)(x \cdot x \cdot x \cdot x)$$
$$= 3^4 \cdot x^4 = 81x^4$$

Note that the power, here 4, has been applied to each factor, 3 and x. In general, we have

Rules and Properties: Property 4 of Exponents

For any real numbers a and b and positive integer m,

$(ab)^m = a^m b^m$

For example,

$(3x)^3 = 3^3 \cdot x^3 = 27x^3$

The use of this property is shown in Example 2.

Example 2

Using the Fourth Property of Exponents

NOTE Notice that $(2x)^5$ and $2x^5$ are entirely different expressions. For $(2x)^5$, the base is $2x$, so we raise each factor to the fifth power. For $2x^5$, the base is x, and so the exponent applies only to x.

Simplify each expression.

(a) $(2x)^5 = 2^5 \cdot x^5 = 32x^5$

(b) $(3ab)^4 = 3^4 \cdot a^4 \cdot b^4 = 81a^4b^4$

(c) $5(2r)^3 = 5 \cdot 2^3 \cdot r^3 = 40r^3$

CHECK YOURSELF 2

Simplify each expression.

(a) $(3y)^4$ **(b)** $(2mn)^6$ **(c)** $3(4x)^2$ **(d)** $5x^3$

We may have to use more than one of our properties in simplifying an expression involving exponents. Consider Example 3.

Example 3

Using the Properties of Exponents

NOTE To help you understand each step of the simplification, we refer to the property being applied. Make a list of the properties now to help you as you work through the remainder of this and the next section.

Simplify each expression.

(a) $(r^4s^3)^3 = (r^4)^3 \cdot (s^3)^3$ Property 4

$= r^{12}s^9$ Property 3

(b) $(3x^2)^2 \cdot (2x^3)^3$

$= 3^2(x^2)^2 \cdot 2^3 \cdot (x^3)^3$ Property 4

$= 9x^4 \cdot 8x^9$ Property 3

$= 72x^{13}$ Multiply the coefficients and apply Property 1.

(c) $\dfrac{(a^3)^5}{a^4} = \dfrac{a^{15}}{a^4}$ Property 3

$= a^{11}$ Property 2

CHECK YOURSELF 3

Simplify each expression.

(a) $(m^5n^2)^3$ (b) $(2p)^4(4p^2)^2$ (c) $\dfrac{(s^4)^3}{s^5}$

We have one final exponent property to develop. Suppose we have a quotient raised to a power. Consider the following:

$$\left(\frac{x}{3}\right)^3 = \frac{x}{3} \cdot \frac{x}{3} \cdot \frac{x}{3} = \frac{x \cdot x \cdot x}{3 \cdot 3 \cdot 3} = \frac{x^3}{3^3}$$

Note that the power, here 3, has been applied to the numerator x and to the denominator 3. This gives us our fifth property of exponents.

Rules and Properties: Property 5 of Exponents

For any real numbers a and b, when b is not equal to 0, and positive integer m,

$$\left(\frac{a}{b}\right)^m = \frac{a^m}{b^m}$$

For example,

$$\left(\frac{2}{5}\right)^3 = \frac{2^3}{5^3} = \frac{8}{125}$$

Example 4 illustrates the use of this property. Again note that the other properties may also have to be applied in simplifying an expression.

Example 4

Using the Fifth Property of Exponents

Simplify each expression.

(a) $\left(\dfrac{3}{4}\right)^3 = \dfrac{3^3}{4^3} = \dfrac{27}{64}$ Property 5

(b) $\left(\dfrac{x^3}{y^2}\right)^4 = \dfrac{(x^3)^4}{(y^2)^4}$ Property 5

$= \dfrac{x^{12}}{y^8}$ Property 3

(c) $\left(\dfrac{r^2s^3}{t^4}\right)^2 = \dfrac{(r^2s^3)^2}{(t^4)^2}$ Property 5

$= \dfrac{(r^2)^2(s^3)^2}{(t^4)^2}$ Property 4

$= \dfrac{r^4s^6}{t^8}$ Property 3

CHECK YOURSELF 4

Simplify each expression.

(a) $\left(\dfrac{2}{3}\right)^4$ **(b)** $\left(\dfrac{m^3}{n^4}\right)^5$ **(c)** $\left(\dfrac{a^2b^3}{c^5}\right)^2$

The following table summarizes the five properties of exponents that were discussed in this section:

General Form	Example
1. $a^m a^n = a^{m+n}$	$x^2 \cdot x^3 = x^5$
2. $\dfrac{a^m}{a^n} = a^{m-n} \quad (m > n)$	$\dfrac{5^7}{5^3} = 5^4$
3. $(a^m)^n = a^{mn}$	$(z^5)^4 = z^{20}$
4. $(ab)^m = a^m b^m$	$(4x)^3 = 4^3x^3 = 64x^3$
5. $\left(\dfrac{a}{b}\right)^m = \dfrac{a^m}{b^m}$	$\left(\dfrac{2}{3}\right)^6 = \dfrac{2^6}{3^6} = \dfrac{64}{729}$

Our work in this chapter deals with the most common kind of algebraic expression, a *polynomial*. To define a polynomial, let's recall our earlier definition of the word *term*.

Definitions: Term

A **term** is a number or the product of a number and one or more variables.

For example, x^5, $3x$, $-4xy^2$, and 8 are terms. A **polynomial** consists of one or more terms in which the only allowable exponents are the whole numbers, 0, 1, 2, 3, . . . and so on. These terms are connected by addition or subtraction signs.

NOTE In a polynomial, terms are separated by + and − signs.

Definitions: Numerical Coefficient

In each term of a polynomial, the number is called the **numerical coefficient,** or more simply the **coefficient,** of that term.

Example 5

Identifying Polynomials

(a) $x + 3$ is a polynomial. The terms are x and 3. The coefficients are 1 and 3.

(b) $3x^2 - 2x + 5$, or $3x^2 + (-2x) + 5$, is also a polynomial. Its terms are $3x^2$, $-2x$, and 5. The coefficients are 3, −2, and 5.

(c) $5x^3 + 2 - \dfrac{3}{x}$ is *not* a polynomial because of the division by x in the third term.

CHECK YOURSELF 5

Which of the following are polynomials?

(a) $5x^2$ **(b)** $3y^3 - 2y + \frac{5}{y}$ **(c)** $4x^2 - 2x + 3$

Certain polynomials are given special names because of the number of terms that they have.

NOTE The prefix *mono-* means 1. The prefix *bi-* means 2. The prefix *tri-* means 3. There are no special names for polynomials with four or more terms.

Definitions: Monomial, Binomial, and Trinomial

A polynomial with one term is called a **monomial.**
A polynomial with two terms is called a **binomial.**
A polynomial with three terms is called a **trinomial.**

Example 6

Identifying Types of Polynomials

(a) $3x^2y$ is a monomial. It has one term.

(b) $2x^3 + 5x$ is a binomial. It has two terms, $2x^3$ and $5x$.

(c) $5x^2 - 4x + 3$, or $5x^2 + (-4x) + 3$, is a trinomial. Its three terms are $5x^2$, $-4x$, and 3.

CHECK YOURSELF 6

Classify each of these as a monomial, binomial, or trinomial.

(a) $5x^4 - 2x^3$ **(b)** $4x^7$ **(c)** $2x^2 + 5x - 3$

NOTE Remember, in a polynomial the allowable exponents are the whole numbers 0, 1, 2, 3, and so on. The degree will be a whole number.

We also classify polynomials by their *degree.* The **degree** of a polynomial that has only one variable is the highest power appearing in any one term.

Example 7

Classifying Polynomials by Their Degree

The highest power

(a) $5x^3 - 3x^2 + 4x$ has degree 3.

The highest power

(b) $4x - 5x^4 + 3x^3 + 2$ has degree 4.

(c) $8x$ has degree 1. (Because $8x = 8x^1$)

(d) 7 has degree 0.

NOTE We will see in the next section that $x^0 = 1$.

Note: Polynomials can have more than one variable, such as $4x^2y^3 + 5xy^2$. The degree is then the sum of the highest powers in any single term (here 2 + 3, or 5). In general, we will be working with polynomials in a single variable, such as x.

CHECK YOURSELF 7

Find the degree of each polynomial.

(a) $6x^5 - 3x^3 - 2$ **(b)** $5x$ **(c)** $3x^3 + 2x^6 - 1$ **(d)** 9

Working with polynomials is much easier if you get used to writing them in **descending-exponent form** (sometimes called *descending-power form*). This simply means that the term with the highest exponent is written first, then the term with the next highest exponent, and so on.

Example 8

Writing Polynomials in Descending Order

The exponents get smaller from left to right.

(a) $5x^7 - 3x^4 + 2x^2$ is in descending-exponent form.

(b) $4x^4 + 5x^6 - 3x^5$ is *not* in descending-exponent form. The polynomial should be written as

$$5x^6 - 3x^5 + 4x^4$$

Notice that the degree of the polynomial is the power of the *first,* or *leading,* term once the polynomial is arranged in descending-exponent form.

CHECK YOURSELF 8

Write the following polynomials in descending-exponent form.

(a) $5x^4 - 4x^5 + 7$ **(b)** $4x^3 + 9x^4 + 6x^8$

A polynomial can represent any number. Its value depends on the value given to the variable.

Example 9

Evaluating Polynomials

Given the polynomial

$$3x^3 - 2x^2 - 4x + 1$$

(a) Find the value of the polynomial when $x = 2$.

Substituting 2 for x, we have

$$3(2)^3 - 2(2)^2 - 4(2) + 1$$
$$= 3(8) - 2(4) - 4(2) + 1$$
$$= 24 - 8 - 8 + 1$$
$$= 9$$

NOTE Again note how the rules for the order of operations are applied. See Section 0.3 for a review.

CAUTION

Be particularly careful when dealing with powers of negative numbers!

(b) Find the value of the polynomial when $x = -2$.

Now we substitute -2 for x.

$$3(-2)^3 - 2(-2)^2 - 4(-2) + 1$$

$$= 3(-8) - 2(4) - 4(-2) + 1$$

$$= -24 - 8 + 8 + 1$$

$$= -23$$

CHECK YOURSELF 9

Find the value of the polynomial

$4x^3 - 3x^2 + 2x - 1$

When

(a) $x = 3$ **(b)** $x = -3$

CHECK YOURSELF ANSWERS

1. **(a)** m^{30}; **(b)** m^{11}; **(c)** 3^8; **(d)** 3^6 **2.** **(a)** $81y^4$; **(b)** $64m^6n^6$; **(c)** $48x^2$; **(d)** $5x^3$
3. **(a)** $m^{15}n^6$; **(b)** $256p^8$; **(c)** s^7 **4.** **(a)** $\frac{16}{81}$; **(b)** $\frac{m^{15}}{n^{20}}$; **(c)** $\frac{a^4b^6}{c^{10}}$
5. **(a)** and **(c)** are polynomials. **6.** **(a)** Binomial; **(b)** monomial; **(c)** trinomial
7. **(a)** 5; **(b)** 1; **(c)** 6; **(d)** 0 **8.** **(a)** $-4x^5 + 5x^4 + 7$; **(b)** $6x^8 + 9x^4 + 4x^3$
9. **(a)** 86; **(b)** -142

Name ______________________

Section ________ Date ________

3.1 Exercises

Use Property 3 of exponents to simplify each of the following expressions.

1. $(x^2)^3$
2. $(a^5)^3$
3. $(m^4)^4$
4. $(p^7)^2$
5. $(2^4)^2$
6. $(3^3)^2$
7. $(5^3)^5$
8. $(7^2)^4$

Use the five properties of exponents to simplify each of the following expressions.

9. $(3x)^3$
10. $(4m)^2$
11. $(2xy)^4$
12. $(5pq)^3$
13. $5(3ab)^3$
14. $4(2rs)^4$
15. $\left(\frac{3}{4}\right)^2$
16. $\left(\frac{2}{3}\right)^3$
17. $\left(\frac{x}{5}\right)^3$
18. $\left(\frac{a}{2}\right)^5$
19. $(2x^2)^4$
20. $(3y^2)^5$
21. $(a^8b^6)^2$
22. $(p^3q^4)^2$
23. $(4x^2y)^3$
24. $(4m^4n^4)^2$
25. $(3m^2)^4(m^3)^2$
26. $(y^4)^3(4y^3)^2$
27. $\frac{(x^4)^3}{x^2}$
28. $\frac{(m^5)^3}{m^6}$
29. $\frac{(s^3)^2(s^2)^3}{(s^5)^2}$
30. $\frac{(y^5)^3(y^3)^2}{(y^4)^4}$
31. $\left(\frac{m^3}{n^2}\right)^3$
32. $\left(\frac{a^4}{b^3}\right)^4$
33. $\left(\frac{a^3b^2}{c^4}\right)^2$
34. $\left(\frac{x^5y^2}{z^4}\right)^3$

Which of the following expressions are polynomials?

35. $7x^3$
36. $5x^3 - \frac{3}{x}$
37. $4x^4y^2 - 3x^3y$
38. 7
39. -7
40. $4x^3 + x$
41. $\frac{3 + x}{x^2}$
42. $5a^2 - 2a + 7$

ANSWERS

1. x^6
2. a^{15}
3. m^{16}
4. p^{14}
5. 2^8
6. 3^6
7. 5^{15}
8. 7^8
9. $27x^3$
10. $16m^2$
11. $16x^4y^4$
12. $125p^3q^3$
13. $135a^3b^3$
14. $64r^4s^4$
15. $\frac{9}{16}$
16. $\frac{8}{27}$
17. $\frac{x^3}{125}$
18. $\frac{a^5}{32}$
19. $16x^8$
20. $243y^{10}$
21. $a^{16}b^{12}$
22. p^6q^8
23. $64x^6y^3$
24. $16m^8n^8$
25. $81m^{14}$
26. $16y^{18}$
27. x^{10}
28. m^9
29. s^2
30. y^5
31. $\frac{m^9}{n^6}$
32. $\frac{a^{16}}{b^{12}}$
33. $\frac{a^6b^4}{c^8}$
34. $\frac{x^{15}y^6}{z^{12}}$
35. Polynomial
36. Not a polynomial
37. Polynomial
38. Polynomial
39. Polynomial
40. Polynomial
41. Not a polynomial
42. Polynomial

ANSWERS

43. $2x^2, -3x; 2, -3$
44. $5x^3, x; 5, 1$
45. $4x^3, -3x, 2; 4, -3, 2$
46. $7x^2; 7$
47. Binomial
48. Monomial
49. Trinomial
50. Trinomial
51. Not classified
52. Not a polynomial
53. Monomial
54. Not classified
55. Not a polynomial
56. Binomial
57. $4x^5 - 3x^2; 5$
58. $3x^3 + 5x^2 + 4; 3$
59. $-5x^9 + 7x^7 + 4x^3; 9$
60. $x + 2; 1$
61. $4x; 1$
62. $x^{17} - 3x^4; 17$
63. $x^6 - 3x^5 + 5x^2 - 7; 6$
64. $5; 0$
65. $7, -5$
66. $5, -15$
67. $4, -4$
68. $34, 34$
69. $62, 30$
70. $-1, 19$
71. $0, 0$
72. $0, 0$

For each of the following polynomials, list the terms and the coefficients.

43. $2x^2 - 3x$

44. $5x^3 + x$

45. $4x^3 - 3x + 2$

46. $7x^2$

Classify each of the following as a monomial, binomial, or trinomial where possible.

47. $7x^3 - 3x^2$

48. $4x^7$

49. $7y^2 + 4y + 5$

50. $2x^2 + 3xy + y^2$

51. $2x^4 - 3x^2 + 5x - 2$

52. $x^4 + \frac{5}{x} + 7$

53. $6y^8$

54. $4x^4 - 2x^2 + 5x - 7$

55. $x^5 - \frac{3}{x^2}$

56. $4x^2 - 9$

Arrange in descending-exponent form if necessary, and give the degree of each polynomial.

57. $4x^5 - 3x^2$

58. $5x^2 + 3x^3 + 4$

59. $7x^7 - 5x^9 + 4x^3$

60. $2 + x$

61. $4x$

62. $x^{17} - 3x^4$

63. $5x^2 - 3x^5 + x^6 - 7$

64. 5

Find the values of each of the following polynomials for the given values of the variable.

65. $6x + 1, x = 1$ and $x = -1$

66. $5x - 5, x = 2$ and $x = -2$

67. $x^3 - 2x, x = 2$ and $x = -2$

68. $3x^2 + 7, x = 3$ and $x = -3$

69. $3x^2 + 4x - 2, x = 4$ and $x = -4$

70. $2x^2 - 5x + 1, x = 2$ and $x = -2$

71. $-x^2 - 2x + 3, x = 1$ and $x = -3$

72. $-x^2 - 5x - 6, x = -3$ and $x = -2$

Indicate whether each of the following statements is always true, sometimes true, or never true.

73. A monomial is a polynomial.

74. A binomial is a trinomial.

75. The degree of a trinomial is 3.

76. A trinomial has three terms.

77. A polynomial has four or more terms.

78. A binomial must have two coefficients.

Solve the following problems.

79. Write x^{12} as a power of x^2.

80. Write y^{15} as a power of y^3.

81. Write a^{16} as a power of a^2.

82. Write m^{20} as a power of m^5.

83. Write each of the following as a power of 8. (Remember that $8 = 2^3$.)

$2^{12}, 2^{18}, (2^5)^3, (2^7)^6$

84. Write each of the following as a power of 9.

$3^8, 3^{14}, (3^5)^8, (3^4)^7$

85. What expression raised to the third power is $-8x^6y^9z^{15}$?

86. What expression raised to the fourth power is $81x^{12}y^8z^{16}$?

The formula $(1 + R)^Y = G$ gives us useful information about the growth of a population. Here R is the rate of growth expressed as a decimal, y is the time in years, and G is the growth factor. If a country has a 2 percent growth rate for 35 years, then it will double its population:

$(1.02)^{35} \approx 2$

87. **a.** With this growth rate, how many doublings will occur in 105 years? How much larger will the country's population be?

b. The less developed countries of the world had an average growth rate of 2 percent in 1986. If their total population was 3.8 billion, what will their population be in 105 years if this rate remains unchanged?

88. The United States has a growth rate of 0.7 percent. What will be its growth factor after 35 years?

89. Write an explanation of why $(x^3)(x^4)$ is *not* x^{12}.

90. Your algebra study partners are confused. "Why isn't $x^2 \cdot x^3 = 2x^5$?", they ask you. Write an explanation that will convince them.

ANSWERS

73. Always

74. Never

75. Sometimes

76. Always

77. Sometimes

78. Always

79. $(x^2)^6$

80. $(y^3)^5$

81. $(a^2)^8$

82. $(m^5)^4$

83. $8^4, 8^6, 8^5, 8^{14}$

84. $9^4, 9^7, 9^{20}, 9^{14}$

85. $-2x^2y^3z^5$

86. $3x^3y^2z^4$

87. a. Three doublings, 8 times as large
b. 30.4 billion

88. $(1.007)^{35} \approx 1.28$, so 28 percent

89.

90.

ANSWERS

91.	4
92.	2
93.	11
94.	11
95.	14
96.	21
97.	5
98.	3
99.	10
100.	10
101.	7
102.	5
103.	−2
104.	1
105.	$3y + 20$, \$170
106.	$20s + 150$, \$290

Capital italic letters such as *P* or *Q* are often used to name polynomials. For example, we might write $P(x) = 3x^3 - 5x^2 + 2$ in which $P(x)$ is read "*P* of *x*." The notation permits a convenient shorthand. We write $P(2)$, read "*P* of 2," to indicate the value of the polynomial when $x = 2$. Here

$$\begin{aligned} P(2) &= 3(2)^3 - 5(2)^2 + 2 \\ &= 3 \cdot 8 - 5 \cdot 4 + 2 \\ &= 6 \end{aligned}$$

Use the information above in the following problems.

If $P(x) = x^3 - 2x^2 + 5$ and $Q(x) = 2x^2 + 3$, find:

91. $P(1)$ **92.** $P(-1)$ **93.** $Q(2)$ **94.** $Q(-2)$

95. $P(3)$ **96.** $Q(-3)$ **97.** $P(0)$

98. $Q(0)$ **99.** $P(2) + Q(-1)$ **100.** $P(-2) + Q(3)$

101. $P(3) - Q(-3) \div Q(0)$ **102.** $Q(-2) \div Q(2) \cdot P(0)$

103. $|Q(4)| - |P(4)|$ **104.** $\dfrac{P(-1) + Q(0)}{P(0)}$

105. Cost of typing. The cost, in dollars, of typing a term paper is given as 3 times the number of pages plus 20. Use *y* as the number of pages to be typed and write a polynomial to describe this cost. Find the cost of typing a 50-page paper.

106. Manufacturing. The cost, in dollars, of making suits is described as 20 times the number of suits plus 150. Use *s* as the number of suits and write a polynomial to describe this cost. Find the cost of making seven suits.

107. Revenue. The revenue, in dollars, when x pairs of shoes are sold is given by $3x^2 - 95$. Find the revenue when 12 pairs of shoes are sold. What is the average revenue per pair of shoes?

108. Manufacturing. The cost in dollars of manufacturing w wing nuts is given by the expression $0.07w + 13.3$. Find the cost when 375 wing nuts are made. What is the average cost to manufacture one wing nut?

109. Suppose that when you were born, a rich uncle put \$500 in the bank for you. He never deposited money again, but the bank paid 5 percent interest on the money every year on your birthday. How much money was in the bank after 1 year? After 2 years? After 1 year (as you know), the amount is $\$500 + 500(0.05)$, which can be written as $\$500(1 + 0.05)$ because of the distributive property. $1 + 0.05 = 1.05$, so after 1 year the amount in the bank was $500(1.05)$. After 2 years, this amount was again multiplied by 1.05. How much is in the bank today? Complete the following chart.

Birthday	Computation	Amount
0 (Day of Birth)		\$500
1	\$500(1.05)	
2	\$500(1.05)(1.05)	
3	\$500(1.05)(1.05)(1.05)	
4	$\$500(1.05)^4$	
5	$\$500(1.05)^5$	
6		
7		
8		

Write a formula for the amount in the bank on your nth birthday. About how many years does it take for the money to double? How many years for it to double again? Can you see any connection between this and the rules for exponents? Explain why you think there may or may not be a connection.

110. Work with another student to correctly complete the statements:

(a) $\dfrac{m^3}{n^3} < 1$ when . . .

$\dfrac{m^3}{n^3} > 1$ when . . .

$\dfrac{m^3}{n^3} = 1$ when . . .

$\dfrac{m^3}{n^3} < 0$ (is negative) when . . .

$\dfrac{m^3}{n^3} = 0$ when . . .

(b) $\dfrac{a^x}{a^y} > 1$ when . . .

$\dfrac{a^x}{a^y} = 1$ when . . .

$\dfrac{a^x}{a^y} < 1$ when . . .

$\dfrac{a^x}{a^y} = 0$ when . . .

$\dfrac{a^x}{a^y} < 0$ when . . .

ANSWERS

107. \$337, \$28.08

108. \$39.55, \$0.11

109.

110.

ANSWERS

a. $\frac{1}{m^2}$

b. $\frac{1}{x^3}$

c. $\frac{1}{a^6}$

d. $\frac{1}{y^4}$

e. 1

f. 1

g. 1

h. 1

Getting Ready for Section 3.2 [Section 1.7]

Reduce each of the following fractions to simplest form.

(a) $\frac{m^3}{m^5}$ (b) $\frac{x^7}{x^{10}}$ (c) $\frac{a^3}{a^9}$ (d) $\frac{y^4}{y^8}$

(e) $\frac{x^3}{x^3}$ (f) $\frac{b^5}{b^5}$ (g) $\frac{s^7}{s^7}$ (h) $\frac{r^{10}}{r^{10}}$

Answers

1. x^6 **3.** m^{16} **5.** 2^8 **7.** 5^{15} **9.** $27x^3$ **11.** $16x^4y^4$ **13.** $135a^3b^3$
15. $\frac{9}{16}$ **17.** $\frac{x^3}{125}$ **19.** $16x^8$ **21.** $a^{16}b^{12}$ **23.** $64x^6y^3$ **25.** $81m^{14}$
27. x^{10} **29.** s^2 **31.** $\frac{m^9}{n^6}$ **33.** $\frac{a^6b^4}{c^8}$ **35.** Polynomial
37. Polynomial **39.** Polynomial **41.** Not a polynomial
43. $2x^2, -3x$; 2, -3 **45.** $4x^3, -3x$, 2; 4, -3, 2 **47.** Binomial
49. Trinomial **51.** Not classified **53.** Monomial **55.** Not a polynomial
57. $4x^5 - 3x^2$; 5 **59.** $-5x^9 + 7x^7 + 4x^3$; 9 **61.** $4x$; 1
63. $x^6 - 3x^5 + 5x^2 - 7$; 6 **65.** 7, -5 **67.** 4, -4 **69.** 62, 30
71. 0, 0 **73.** Always **75.** Sometimes **77.** Sometimes **79.** $(x^2)^6$
81. $(a^2)^8$ **83.** $8^4, 8^6, 8^5, 8^{14}$ **85.** $-2x^2y^3z^5$
87. **(a)** Three doublings, 8 times as large; **(b)** 30.4 billion **89.** 

91. 4 **93.** 11 **95.** 14 **97.** 5 **99.** 10 **101.** 7 **103.** -2
105. $3y + 20$, \$170 **107.** \$337, \$28.08 **109.** **a.** $\frac{1}{m^2}$ **b.** $\frac{1}{x^3}$
c. $\frac{1}{a^6}$ **d.** $\frac{1}{y^4}$ **e.** 1 **f.** 1 **g.** 1 **h.** 1

3.2 Negative Exponents and Scientific Notation

3.2 OBJECTIVES

1. Evaluate expressions involving zero or a negative exponent
2. Simplify expressions involving zero or a negative exponent
3. Write a decimal number in scientific notation
4. Solve an application of scientific notation

In Section 3.1, we discussed exponents.

We now want to extend our exponent notation to include 0 and negative integers as exponents.

First, what do we do with x^0? It will help to look at a problem that gives us x^0 as a result. What if the numerator and denominator of a fraction have the same base raised to the same power and we extend our division rule? For example,

$$\frac{a^5}{a^5} = a^{5-5} = a^0 \tag{1}$$

NOTE By Property 2, $\frac{a^m}{a^n} = a^{m-n}$ when $m > n$. Here m and n are both 5 so $m = n$.

But from our experience with fractions we know that

$$\frac{a^5}{a^5} = 1 \tag{2}$$

By comparing equations (1) and (2), it seems reasonable to make the following definition:

NOTE As was the case with $\frac{0}{0}$, 0^0 will be discussed in a later course.

Definitions: Zero Power

For any number a, $a \neq 0$,

$a^0 = 1$

In words, any expression, except 0, raised to the 0 power is 1.

Example 1 illustrates the use of this definition.

Example 1

Raising Expressions to the Zero Power

CAUTION

In part (d) the 0 exponent applies only to the x and *not* to the factor 6, because the base is x.

Evaluate. Assume all variables are nonzero.

(a) $5^0 = 1$

(b) $27^0 = 1$

(c) $(x^2y)^0 = 1 \quad$ if $x \neq 0$ and $y \neq 0$

(d) $6x^0 = 6 \cdot 1 = 6 \quad$ if $x \neq 0$

CHECK YOURSELF 1

Evaluate. Assume all variables are nonzero.

(a) 7^0 **(b)** $(-8)^0$ **(c)** $(xy^3)^0$ **(d)** $3x^0$

The second property of exponents allows us to define a negative exponent. Suppose that the exponent in the denominator is *greater than* the exponent in the numerator. Consider the expression $\frac{x^2}{x^5}$.

Our previous work with fractions tells us that

NOTE Divide the numerator and denominator by the two common factors of x.

$$\frac{x^2}{x^5} = \frac{x \cdot x}{x \cdot x \cdot x \cdot x \cdot x} = \frac{1}{x^3} \qquad (1)$$

However, if we extend the second property to let n be greater than m, we have

REMEMBER: $\frac{a^m}{a^n} = a^{m-n}$

$$\frac{x^2}{x^5} = x^{2-5} = x^{-3} \qquad (2)$$

Now, by comparing equations (1) and (2), it seems reasonable to define x^{-3} as $\frac{1}{x^3}$.

In general, we have this result:

Definitions: Negative Powers

NOTE John Wallis (1616–1703), an English mathematician, was the first to fully discuss the meaning of 0 and negative exponents.

For any number a, $a \neq 0$, and any positive integer n,

$$a^{-n} = \frac{1}{a^n}$$

Example 2

Rewriting Expressions That Contain Negative Exponents

Rewrite each expression, using only positive exponents.

Negative exponent in numerator

(a) $x^{-4} = \frac{1}{x^4}$

Positive exponent in denominator

(b) $m^{-7} = \frac{1}{m^7}$

(c) $3^{-2} = \frac{1}{3^2}$ or $\frac{1}{9}$

(d) $10^{-3} = \frac{1}{10^3}$ or $\frac{1}{1000}$

CAUTION

(e) $2x^{-3} = 2 \cdot \frac{1}{x^3} = \frac{2}{x^3}$

The -3 exponent applies only to x, because x is the base.

(f) $\frac{a^5}{a^9} = a^{5-9} = a^{-4} = \frac{1}{a^4}$

(g) $-4x^{-5} = -4 \cdot \frac{1}{x^5} = -\frac{4}{x^5}$

CHECK YOURSELF 2

Write, using only positive exponents.

(a) a^{-10} **(b)** 4^{-3} **(c)** $3x^{-2}$ **(d)** $\frac{x^5}{x^8}$

We will now allow negative integers as exponents in our first property for exponents. Consider Example 3.

Example 3

Simplifying Expressions Containing Exponents

NOTE $a^m \cdot a^n = a^{m+n}$ for *any* integers m and n. So add the exponents.

Simplify (write an equivalent expression that uses only positive exponents).

(a) $x^5x^{-2} = x^{5+(-2)} = x^3$

Note: An alternative approach would be

NOTE By definition $x^{-2} = \frac{1}{x^2}$

$$x^5x^{-2} = x^5 \cdot \frac{1}{x^2} = \frac{x^5}{x^2} = x^3$$

(b) $a^7a^{-5} = a^{7+(-5)} = a^2$

(c) $y^5y^{-9} = y^{5+(-9)} = y^{-4} = \frac{1}{y^4}$

CHECK YOURSELF 3

Simplify (write an equivalent expression that uses only positive exponents).

(a) x^7x^{-2} **(b)** b^3b^{-8}

Example 4 shows that all the properties of exponents introduced in the last section can be extended to expressions with negative exponents.

Example 4

Simplifying Expressions Containing Exponents

Simplify each expression.

(a) $\frac{m^{-3}}{m^4} = m^{-3-4}$ Property 2

$= m^{-7} = \frac{1}{m^7}$

(b) $\frac{a^{-2}b^6}{a^5b^{-4}} = a^{-2-5}b^{6-(-4)}$ Apply Property 2 to each variable.

$= a^{-7}b^{10} = \frac{b^{10}}{a^7}$

NOTE This could also be done by using Property 4 first, so

$(2x^4)^{-3} = 2^{-3} \cdot (x^4)^{-3} = 2^{-3}x^{-12}$

$= \dfrac{1}{2^3x^{12}}$

$= \dfrac{1}{8x^{12}}$

(c) $(2x^4)^{-3} = \dfrac{1}{(2x^4)^3}$ Definition of the negative exponent

$= \dfrac{1}{2^3(x^4)^3}$ Property 4

$= \dfrac{1}{8x^{12}}$ Property 3

(d) $\dfrac{(y^{-2})^4}{(y^3)^{-2}} = \dfrac{y^{-8}}{y^{-6}}$ Property 3

$= y^{-8-(-6)}$ Property 2

$= y^{-2} = \dfrac{1}{y^2}$

CHECK YOURSELF 4

Simplify each expression.

(a) $\dfrac{x^5}{x^{-3}}$ **(b)** $\dfrac{m^3n^{-5}}{m^{-2}n^3}$ **(c)** $(3a^3)^{-4}$ **(d)** $\dfrac{(r^3)^{-2}}{(r^{-4})^2}$

Let us now take a look at an important use of exponents, scientific notation.

We begin the discussion with a calculator exercise. On most calculators, if you multiply 2.3 times 1000, the display will read

2300

Multiply by 1000 a second time. Now you will see

2300000.

Multiplying by 1000 a third time will result in the display

2.3 09 or 2.3 E09

NOTE This must equal 2,300,000,000.

And multiplying by 1000 again yields

2.3 12 or 2.3 E12

NOTE Consider the following table:

$2.3 = 2.3 \times 10^0$
$23 = 2.3 \times 10^1$
$230 = 2.3 \times 10^2$
$2300 = 2.3 \times 10^3$
$23{,}000 = 2.3 \times 10^4$
$230{,}000 = 2.3 \times 10^5$

Can you see what is happening? This is the way calculators display very large numbers. The number on the left is always between 1 and 10, and the number on the right indicates the number of places the decimal point must be moved to the right to put the answer in standard (or decimal) form.

This notation is used frequently in science. It is not uncommon in scientific applications of algebra to find yourself working with very large or very small numbers. Even in the time of Archimedes (287–212 B.C.E.), the study of such numbers was not unusual. Archimedes estimated that the universe was 23,000,000,000,000,000 m in diameter, which is the approximate distance light travels in $2\frac{1}{2}$ years. By comparison, Polaris (the North Star) is actually 680 light-years from the earth. Example 6 will discuss the idea of light-years.

In scientific notation, Archimedes's estimate for the diameter of the universe would be

2.3×10^{16} m

In general, we can define scientific notation as follows.

Definitions: Scientific Notation

Any number written in the form

$a \times 10^n$

in which $1 \le a < 10$ and n is an integer, is written in scientific notation.

Example 5

Using Scientific Notation

Write each of the following numbers in scientific notation.

NOTE Notice the pattern for writing a number in scientific notation.

(a) $120{,}000. = 1.2 \times 10^5$
5 places — The power is 5.

(b) $88{,}000{,}000. = 8.8 \times 10^7$
7 places — The power is 7.

NOTE The exponent on 10 shows the *number of places* we must move the decimal point. A positive exponent tells us to move right, and a negative exponent indicates to move left.

(c) $520{,}000{,}000. = 5.2 \times 10^8$
8 places

(d) $4{,}000{,}000{,}000. = 4 \times 10^9$
9 places

(e) $0.0005 = 5 \times 10^{-4}$
4 places — If the decimal point is to be moved to the left, the exponent will be negative.

NOTE To convert back to standard or decimal form, the process is simply reversed.

(f) $0.0000000081 = 8.1 \times 10^{-9}$
9 places

CHECK YOURSELF 5

Write in scientific notation.

(a) 212,000,000,000,000,000 **(b)** 0.00079
(c) 5,600,000 **(d)** 0.0000007

Example 6

An Application of Scientific Notation

(a) Light travels at a speed of 3.05×10^8 meters per second (m/s). There are approximately 3.15×10^7 s in a year. How far does light travel in a year?

We multiply the distance traveled in 1 s by the number of seconds in a year. This yields

$$\begin{aligned}(3.05 \times 10^8)(3.15 \times 10^7) &= (3.05 \cdot 3.15)(10^8 \cdot 10^7) \\ &= 9.6075 \times 10^{15}\end{aligned}$$

Multiply the coefficients, and add the exponents.

NOTE Notice that $9.6075 \times 10^{15} \approx 10 \times 10^{15} = 10^{16}$

For our purposes we round the distance light travels in 1 year to 10^{16} m. This unit is called a **light-year,** and it is used to measure astronomical distances.

(b) The distance from earth to the star Spica (in Virgo) is 2.2×10^{18} m. How many light-years is Spica from earth?

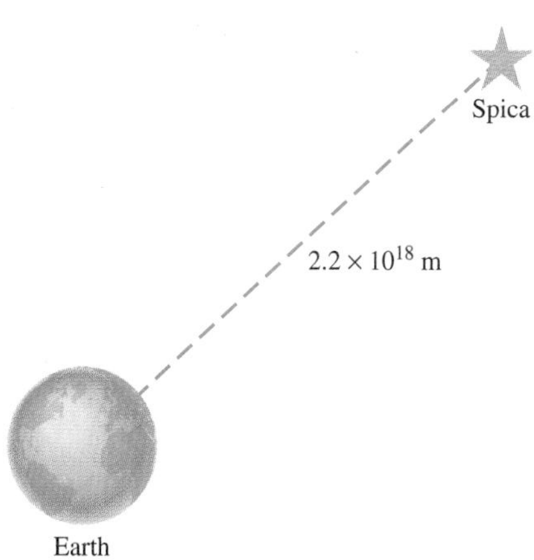

NOTE We divide the distance (in meters) by the number of meters in 1 light-year.

$$\begin{aligned}\frac{2.2 \times 10^{18}}{10^{16}} &= 2.2 \times 10^{18-16} \\ &= 2.2 \times 10^2 = 220 \text{ light-years}\end{aligned}$$

CHECK YOURSELF 6

The farthest object that can be seen with the unaided eye is the Andromeda galaxy. This galaxy is 2.3×10^{22} m from earth. What is this distance in light-years?

CHECK YOURSELF ANSWERS

1. (a) 1; (b) 1; (c) 1; (d) 3 2. (a) $\frac{1}{a^{10}}$; (b) $\frac{1}{4^3}$ or $\frac{1}{64}$; (c) $\frac{3}{x^2}$; (d) $\frac{1}{x^3}$
3. (a) x^5; (b) $\frac{1}{b^5}$ 4. (a) x^8; (b) $\frac{m^5}{n^8}$; (c) $\frac{1}{81a^{12}}$; (d) r^2
5. (a) 2.12×10^{17}; (b) 7.9×10^{-4}; (c) 5.6×10^6; (d) 7×10^{-7}
6. 2,300,000 light-years

Name ______________

Section ________ Date ________

3.2 Exercises

Evaluate (assume the variables are nonzero).

1. 4^0

2. $(-7)^0$

3. $(-29)^0$

4. 75^0

5. $(x^3y^2)^0$

6. $7m^0$

7. $11x^0$

8. $(2a^3b^7)^0$

9. $(-3p^6q^8)^0$

10. $-7x^0$

Write each of the following expressions using positive exponents; simplify when possible.

11. b^{-8}

12. p^{-12}

13. 3^{-4}

14. 2^{-5}

15. 5^{-2}

16. 4^{-3}

17. 10^{-4}

18. 10^{-5}

19. $5x^{-1}$

20. $3a^{-2}$

21. $(5x)^{-1}$

22. $(3a)^{-2}$

23. $-2x^{-5}$

24. $3x^{-4}$

25. $(-2x)^{-5}$

26. $(3x)^{-4}$

Use Properties 1 and 2 to simplify each of the following expressions. Write your answers with positive exponents only.

27. a^5a^3

28. m^5m^7

29. x^8x^{-2}

30. $a^{12}a^{-8}$

31. b^7b^{-11}

32. y^5y^{-12}

33. x^0x^5

34. $r^{-3}r^0$

35. $\dfrac{a^8}{a^5}$

ANSWERS

1. 1
2. 1
3. 1
4. 1
5. 1
6. 7
7. 11
8. 1
9. 1
10. −7
11. $\frac{1}{b^8}$
12. $\frac{1}{p^{12}}$
13. $\frac{1}{81}$
14. $\frac{1}{32}$
15. $\frac{1}{25}$
16. $\frac{1}{64}$
17. $\frac{1}{10{,}000}$
18. $\frac{1}{100{,}000}$
19. $\frac{5}{x}$
20. $\frac{3}{a^2}$
21. $\frac{1}{5x}$
22. $\frac{1}{9a^2}$
23. $-\frac{2}{x^5}$
24. $\frac{3}{x^4}$
25. $-\frac{1}{32x^5}$
26. $\frac{1}{81x^4}$
27. a^8
28. m^{12}
29. x^6
30. a^4
31. $\frac{1}{b^4}$
32. $\frac{1}{y^7}$
33. x^5
34. $\frac{1}{r^3}$
35. a^3

ANSWERS

36. m^5

37. $\frac{1}{x^2}$

38. $\frac{1}{a^7}$

39. $\frac{1}{r^8}$

40. x^8

41. x

42. $\frac{1}{p^3}$

43. $\frac{m^9}{n^8}$

44. $\frac{q}{p^7}$

45. $\frac{16}{a^{12}}$

46. $\frac{1}{27x^6}$

47. $\frac{x^4}{y^6}$

48. $\frac{b^9}{a^{15}}$

49. $\frac{1}{r^2}$

50. $\frac{1}{y^6}$

51. $\frac{1}{x}$

52. $\frac{1}{m^4}$

53. $\frac{1}{a^{11}}$

54. 1

55. 9.3×10^7 mi

56. 2.1×10^{-5} m

57. 1.3×10^{11} cm

58. 6.02×10^{23} molecules

59. 28

36. $\frac{m^9}{m^4}$ **37.** $\frac{x^7}{x^9}$ **38.** $\frac{a^3}{a^{10}}$

39. $\frac{r^{-3}}{r^5}$ **40.** $\frac{x^3}{x^{-5}}$ **41.** $\frac{x^{-4}}{x^{-5}}$

42. $\frac{p^{-6}}{p^{-3}}$

Simplify each of the following expressions. Write your answers with positive exponents only.

43. $\frac{m^5n^{-3}}{m^{-4}n^5}$ **44.** $\frac{p^{-3}q^{-2}}{p^4q^{-3}}$ **45.** $(2a^{-3})^4$

46. $(3x^2)^{-3}$ **47.** $(x^{-2}y^3)^{-2}$ **48.** $(a^5b^{-3})^{-3}$

49. $\frac{(r^{-2})^3}{r^{-4}}$ **50.** $\frac{(y^3)^{-4}}{y^{-6}}$ **51.** $\frac{(x^{-3})^3}{(x^4)^{-2}}$

52. $\frac{(m^4)^{-3}}{(m^{-2})^4}$ **53.** $\frac{(a^{-3})^2(a^4)}{(a^{-3})^{-3}}$ **54.** $\frac{(x^2)^{-3}(x^{-2})}{(x^2)^{-4}}$

In exercises 55 to 58, express each number in scientific notation.

55. The distance from the earth to the sun: 93,000,000 mi.

56. The diameter of a grain of sand: 0.000021 m.

57. The diameter of the sun: 130,000,000,000 cm.

58. The number of molecules in 22.4 L of a gas: 602,000,000,000,000,000,000,000 (Avogadro's number).

59. The mass of the sun is approximately 1.98×10^{30} kg. If this were written in standard or decimal form, how many 0s would follow the digit 8?

60. Archimedes estimated the universe to be 2.3×10^{19} millimeters (mm) in diameter. If this number were written in standard or decimal form, how many 0s would follow the digit 3?

In exercises 61 to 64, write each expression in standard notation.

61. 8×10^{-3}
62. 7.5×10^{-6}
63. 2.8×10^{-5}
64. 5.21×10^{-4}

In exercises 65 to 68, write each of the following in scientific notation.

65. 0.0005
66. 0.000003
67. 0.00037
68. 0.000051

In exercises 69 to 72, compute the expressions using scientific notation, and write your answer in that form.

69. $(4 \times 10^{-3})(2 \times 10^{-5})$
70. $(1.5 \times 10^{-6})(4 \times 10^{2})$

71. $\dfrac{9 \times 10^3}{3 \times 10^{-2}}$
72. $\dfrac{7.5 \times 10^{-4}}{1.5 \times 10^{2}}$

In exercises 73 to 78, perform the indicated calculations. Write your result in scientific notation.

73. $(2 \times 10^5)(4 \times 10^4)$
74. $(2.5 \times 10^7)(3 \times 10^5)$
75. $\dfrac{6 \times 10^9}{3 \times 10^7}$

76. $\dfrac{4.5 \times 10^{12}}{1.5 \times 10^7}$
77. $\dfrac{(3.3 \times 10^{15})(6 \times 10^{15})}{(1.1 \times 10^8)(3 \times 10^6)}$
78. $\dfrac{(6 \times 10^{12})(3.2 \times 10^8)}{(1.6 \times 10^7)(3 \times 10^2)}$

In 1975 the population of Earth was approximately 4 billion and doubling every 35 years. The formula for the population P in year Y for this doubling rate is

$$P \text{ (in billions)} = 4 \times 2^{(Y-1975)/35}$$

79. What was the approximate population of Earth in 1960?

80. What will Earth's population be in 2025?

The United States population in 1990 was approximately 250 million, and the average growth rate for the past 30 years gives a doubling time of 66 years. The above formula for the United States then becomes

$$P \text{ (in millions)} = 250 \times 2^{(Y-1990)/66}$$

81. What was the approximate population of the United States in 1960?

82. What will be the population of the United States in 2025 if this growth rate continues?

ANSWERS

60. 18
61. 0.008
62. 0.0000075
63. 0.000028
64. 0.000521
65. 5×10^{-4}
66. 3×10^{-6}
67. 3.7×10^{-4}
68. 5.1×10^{-5}
69. 8×10^{-8}
70. 6×10^{-4}
71. 3×10^{5}
72. 5×10^{-6}
73. 8×10^{9}
74. 7.5×10^{12}
75. 2×10^{2}
76. 3×10^{5}
77. 6×10^{16}
78. 4×10^{11}
79. 2.97 billion
80. 10.77 billion
81. 182 million
82. 361 million

CHECK YOURSELF 1

Add $6x^2 + 2x$ *and* $4x^2 - 7x$.

The same technique is used to find the sum of two trinomials.

Example 2

Adding Polynomials Using the Horizontal Method

Add $4a^2 - 7a + 5$ and $3a^2 + 3a - 4$.

Write the sum.

$(4a^2 - 7a + 5) + (3a^2 + 3a - 4)$

$= 4a^2 - 7a + 5 + 3a^2 + 3a - 4 = 7a^2 - 4a + 1$

REMEMBER Only the like terms are combined in the sum.

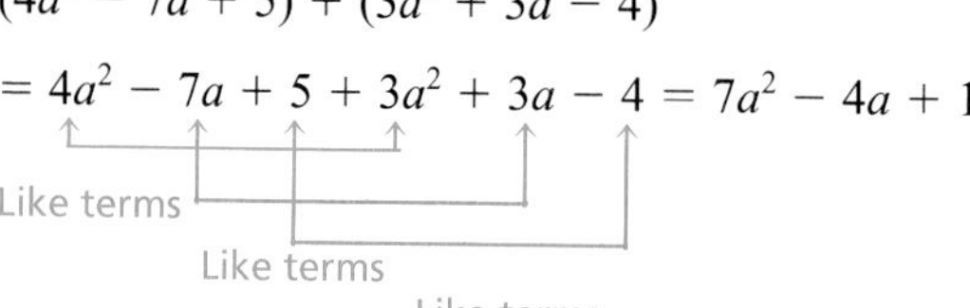

CHECK YOURSELF 2

Add $5y^2 - 3y + 7$ *and* $3y^2 - 5y - 7$.

Example 3

Adding Polynomials Using the Horizontal Method

Add $2x^2 + 7x$ and $4x - 6$.

Write the sum.

$(2x^2 + 7x) + (4x - 6)$

$= 2x^2 + 7x + 4x - 6$

These are the only like terms; $2x^2$ and -6 cannot be combined.

$= 2x^2 + 11x - 6$

CHECK YOURSELF 3

Add $5m^2 + 8$ *and* $8m^2 - 3m$.

As we mentioned in Section 3.1 writing polynomials in descending-exponent form usually makes the work easier. Look at Example 4.

Example 4

Adding Polynomials Using the Horizontal Method

Add $3x - 2x^2 + 7$ and $5 + 4x^2 - 3x$.

ANSWERS

60. 18
61. 0.008
62. 0.0000075
63. 0.000028
64. 0.000521
65. 5×10^{-4}
66. 3×10^{-6}
67. 3.7×10^{-4}
68. 5.1×10^{-5}
69. 8×10^{-8}
70. 6×10^{-4}
71. 3×10^{5}
72. 5×10^{-6}
73. 8×10^{9}
74. 7.5×10^{12}
75. 2×10^{2}
76. 3×10^{5}
77. 6×10^{16}
78. 4×10^{11}
79. 2.97 billion
80. 10.77 billion
81. 182 million
82. 361 million

60. Archimedes estimated the universe to be 2.3×10^{19} millimeters (mm) in diameter. If this number were written in standard or decimal form, how many 0s would follow the digit 3?

In exercises 61 to 64, write each expression in standard notation.

61. 8×10^{-3} **62.** 7.5×10^{-6} **63.** 2.8×10^{-5} **64.** 5.21×10^{-4}

In exercises 65 to 68, write each of the following in scientific notation.

65. 0.0005 **66.** 0.000003 **67.** 0.00037 **68.** 0.000051

In exercises 69 to 72, compute the expressions using scientific notation, and write your answer in that form.

69. $(4 \times 10^{-3})(2 \times 10^{-5})$ **70.** $(1.5 \times 10^{-6})(4 \times 10^{2})$

71. $\dfrac{9 \times 10^{3}}{3 \times 10^{-2}}$ **72.** $\dfrac{7.5 \times 10^{-4}}{1.5 \times 10^{2}}$

In exercises 73 to 78, perform the indicated calculations. Write your result in scientific notation.

73. $(2 \times 10^{5})(4 \times 10^{4})$ **74.** $(2.5 \times 10^{7})(3 \times 10^{5})$ **75.** $\dfrac{6 \times 10^{9}}{3 \times 10^{7}}$

76. $\dfrac{4.5 \times 10^{12}}{1.5 \times 10^{7}}$ **77.** $\dfrac{(3.3 \times 10^{15})(6 \times 10^{15})}{(1.1 \times 10^{8})(3 \times 10^{6})}$ **78.** $\dfrac{(6 \times 10^{12})(3.2 \times 10^{8})}{(1.6 \times 10^{7})(3 \times 10^{2})}$

In 1975 the population of Earth was approximately 4 billion and doubling every 35 years. The formula for the population P in year Y for this doubling rate is

$$P \text{ (in billions)} = 4 \times 2^{(Y-1975)/35}$$

79. What was the approximate population of Earth in 1960?

80. What will Earth's population be in 2025?

The United States population in 1990 was approximately 250 million, and the average growth rate for the past 30 years gives a doubling time of 66 years. The above formula for the United States then becomes

$$P \text{ (in millions)} = 250 \times 2^{(Y-1990)/66}$$

81. What was the approximate population of the United States in 1960?

82. What will be the population of the United States in 2025 if this growth rate continues?

ANSWERS

83. 66 years

84. 210 years

85. 1.55×10^{23} L
2.58×10^{13} L

86. 5.27×10^{15} L

87. 8.32×10^{14} L

a. $15m$

b. $4x$

c. $9m^2 - 8m$

d. x^2

e. $20c^3$

f. $17s^3$

g. $10c^2 - 6c$

h. $13r^3 - 7r^2$

83. Megrez, the nearest of the Big Dipper stars, is 6.6×10^{17} m from Earth. Approximately how long does it take light, traveling at 10^{16} m/year, to travel from Megrez to Earth?

84. Alkaid, the most distant star in the Big Dipper, is 2.1×10^{18} m from Earth. Approximately how long does it take light to travel from Alkaid to Earth?

85. The number of liters (L) of water on Earth is 15,500 followed by 19 zeros. Write this number in scientific notation. Then use the number of liters of water on Earth to find out how much water is available for each person on Earth. The population of Earth is 6 billion.

86. If there are 6×10^9 people on Earth and there is enough freshwater to provide each person with 8.79×10^5 L, how much freshwater is on Earth?

87. The United States uses an average of 2.6×10^6 L of water per person each year. The United States has 3.2×10^8 people. How many liters of water does the United States use each year?

Getting Ready for Section 3.3 [Section 1.6]

Combine like terms where possible.

(a) $8m + 7m$
(b) $9x - 5x$
(c) $9m^2 - 8m$
(d) $8x^2 - 7x^2$
(e) $5c^3 + 15c^3$
(f) $9s^3 + 8s^3$
(g) $8c^2 - 6c + 2c^2$
(h) $8r^3 - 7r^2 + 5r^3$

Answers

1. 1 **3.** 1 **5.** 1 **7.** 11 **9.** 1 **11.** $\frac{1}{b^8}$ **13.** $\frac{1}{81}$ **15.** $\frac{1}{25}$ **17.** $\frac{1}{10,000}$ **19.** $\frac{5}{x}$ **21.** $\frac{1}{5x}$ **23.** $-\frac{2}{x^5}$ **25.** $-\frac{1}{32x^5}$ **27.** a^8 **29.** x^6 **31.** $\frac{1}{b^4}$ **33.** x^5 **35.** a^3 **37.** $\frac{1}{x^2}$ **39.** $\frac{1}{r^8}$ **41.** x **43.** $\frac{m^9}{n^8}$ **45.** $\frac{16}{a^{12}}$ **47.** $\frac{x^4}{y^6}$ **49.** $\frac{1}{r^2}$ **51.** $\frac{1}{x}$ **53.** $\frac{1}{a^{11}}$ **55.** 9.3×10^7 mi **57.** 1.3×10^{11} cm **59.** 28 **61.** 0.008 **63.** 0.000028 **65.** 5×10^{-4} **67.** 3.7×10^{-4} **69.** 8×10^{-8} **71.** 3×10^5 **73.** 8×10^9 **75.** 2×10^2 **77.** 6×10^{16} **79.** 2.97 billion **81.** 182 million **83.** 66 years **85.** 1.55×10^{23} L; 2.58×10^{13} L **87.** 8.32×10^{14} L **a.** $15m$ **b.** $4x$ **c.** $9m^2 - 8m$ **d.** x^2 **e.** $20c^3$ **f.** $17s^3$ **g.** $10c^2 - 6c$ **h.** $13r^3 - 7r^2$

Adding and Subtracting Polynomials

OBJECTIVES

1. Add two polynomials
2. Subtract two polynomials

Addition is always a matter of combining like quantities (two apples plus three apples, four books plus five books, and so on). If you keep that basic idea in mind, adding polynomials will be easy. It is just a matter of combining like terms. Suppose that you want to add

$5x^2 + 3x + 4$ and $4x^2 + 5x - 6$

Parentheses are sometimes used in adding, so for the sum of these polynomials, we can write

NOTE The plus sign between the parentheses indicates the addition.

$(5x^2 + 3x + 4) + (4x^2 + 5x - 6)$

Now what about the parentheses? You can use the following rule.

Rules and Properties: Removing Signs of Grouping Case 1

If a plus sign (+) or nothing at all appears in front of parentheses, just remove the parentheses. No other changes are necessary.

Now let's return to the addition.

NOTE Just remove the parentheses. No other changes are necessary.

$(5x^2 + 3x + 4) + (4x^2 + 5x - 6)$

$= 5x^2 + 3x + 4 + 4x^2 + 5x - 6$

Like terms   Like terms   Like terms

NOTE Note the use of the associative and commutative properties in reordering and regrouping.

Collect like terms. (*Remember:* Like terms have the same variables raised to the same power).

$= (5x^2 + 4x^2) + (3x + 5x) + (4 - 6)$

Combine like terms for the result:

NOTE Here we use the distributive property. For example,

$5x^2 + 4x^2 = (5 + 4)x^2$
$= 9x^2$

$= 9x^2 + 8x - 2$

As should be clear, much of this work can be done mentally. You can then write the sum directly by locating like terms and combining. Example 1 illustrates this property.

Example 1

Combining Like Terms

NOTE We call this the "horizontal method" because the entire problem is written on one line.
$3 + 4 = 7$ is the horizontal method.

$$\begin{array}{r} 3 \\ +\ 4 \\ \hline 7 \end{array}$$

is the vertical method.

Add $3x - 5$ and $2x + 3$.

Write the sum.

$(3x - 5) + (2x + 3)$

$= 3x - 5 + 2x + 3 = 5x - 2$

Like terms   Like terms

CHECK YOURSELF 1

Add $6x^2 + 2x$ *and* $4x^2 - 7x$.

The same technique is used to find the sum of two trinomials.

Example 2

Adding Polynomials Using the Horizontal Method

Add $4a^2 - 7a + 5$ and $3a^2 + 3a - 4$.

Write the sum.

$(4a^2 - 7a + 5) + (3a^2 + 3a - 4)$

REMEMBER Only the like terms are combined in the sum.

$= 4a^2 - 7a + 5 + 3a^2 + 3a - 4 = 7a^2 - 4a + 1$

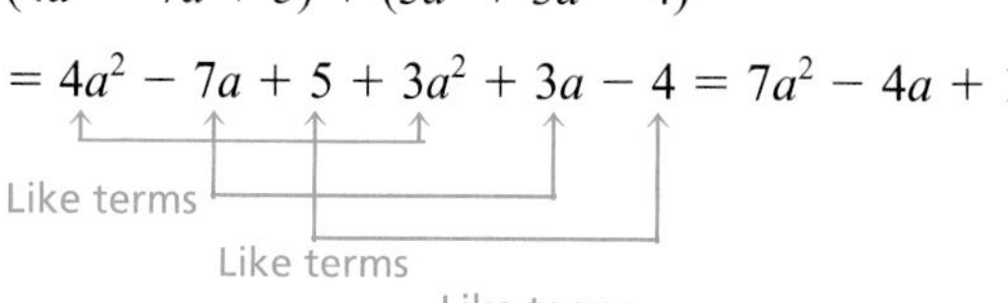

CHECK YOURSELF 2

Add $5y^2 - 3y + 7$ *and* $3y^2 - 5y - 7$.

Example 3

Adding Polynomials Using the Horizontal Method

Add $2x^2 + 7x$ and $4x - 6$.

Write the sum.

$(2x^2 + 7x) + (4x - 6)$

$= 2x^2 + 7x + 4x - 6$

These are the only like terms; $2x^2$ and -6 cannot be combined.

$= 2x^2 + 11x - 6$

CHECK YOURSELF 3

Add $5m^2 + 8$ *and* $8m^2 - 3m$.

As we mentioned in Section 3.1 writing polynomials in descending-exponent form usually makes the work easier. Look at Example 4.

Example 4

Adding Polynomials Using the Horizontal Method

Add $3x - 2x^2 + 7$ and $5 + 4x^2 - 3x$.

Write the polynomials in descending-exponent form, then add.

$(-2x^2 + 3x + 7) + (4x^2 - 3x + 5)$

$= 2x^2 + 12$

CHECK YOURSELF 4

Add $8 - 5x^2 + 4x$ *and* $7x - 8 + 8x^2$.

Subtracting polynomials requires another rule for removing signs of grouping.

Rules and Properties: Removing Signs of Grouping Case 2

If a minus sign (−) appears in front of a set of parentheses, the parentheses can be removed by changing the sign of each term inside the parentheses.

The use of this rule is illustrated in Example 5.

Example 5

Removing Parentheses

In each of the following, remove the parentheses.

(a) $-(2x + 3y) = -2x - 3y$ Change each sign to remove the parentheses.

NOTE This uses the distributive property, because

$-(2x + 3y) = (-1)(2x + 3y)$
$= -2x - 3y$

(b) $m - (5n - 3p) = m \underbrace{- 5n + 3p}$
Sign changes.

(c) $2x - (-3y + z) = 2x \underbrace{+ 3y - z}$
Sign changes.

CHECK YOURSELF 5

(a) $-(3m + 5n)$ **(b)** $-(5w - 7z)$ **(c)** $3r - (2s - 5t)$ **(d)** $5a - (-3b - 2c)$

Subtracting polynomials is now a matter of using the previous rule to remove the parentheses and then combining the like terms. Consider Example 6.

Example 6

Subtracting Polynomials Using the Horizontal Method

(a) Subtract $5x - 3$ from $8x + 2$.

Write

NOTE The expression following "from" is written first in the problem.

$(8x + 2) - (5x - 3)$

$= 8x + 2 \underbrace{- 5x + 3}$ Recall that subtracting $5x$ is the same as adding $-5x$.
Sign changes.

$= 3x + 5$

(b) Subtract $4x^2 - 8x + 3$ from $8x^2 + 5x - 3$.

Write

$(8x^2 + 5x - 3) - (4x^2 - 8x + 3)$

$= 8x^2 + 5x - 3 \underbrace{- 4x^2 + 8x - 3}_{\text{Sign changes.}}$

$= 4x^2 + 13x - 6$

CHECK YOURSELF 6

(a) Subtract $7x + 3$ from $10x - 7$.
(b) Subtract $5x^2 - 3x + 2$ from $8x^2 - 3x - 6$.

Again, writing all polynomials in descending-exponent form will make locating and combining like terms much easier. Look at Example 7.

Example 7

Subtracting Polynomials Using the Horizontal Method

(a) Subtract $4x^2 - 3x^3 + 5x$ from $8x^3 - 7x + 2x^2$.

Write

$(8x^3 + 2x^2 - 7x) - (-3x^3 + 4x^2 + 5x)$

$= 8x^3 + 2x^2 - 7x \underbrace{+ 3x^3 - 4x^2 - 5x}_{\text{Sign changes.}}$

$= 11x^3 - 2x^2 - 12x$

(b) Subtract $8x - 5$ from $-5x + 3x^2$.

Write

$(3x^2 - 5x) - (8x - 5)$

$= 3x^2 \underbrace{- 5x - 8x} + 5$

Only the like terms can be combined.

$= 3x^2 - 13x + 5$

CHECK YOURSELF 7

(a) Subtract $7x - 3x^2 + 5$ from $5 - 3x + 4x^2$.
(b) Subtract $3a - 2$ from $5a + 4a^2$.

If you think back to addition and subtraction in arithmetic, you'll remember that the work was arranged vertically. That is, the numbers being added or subtracted were placed under one another so that each column represented the same place value. This meant that in adding or subtracting columns you were always dealing with "like quantities."

It is also possible to use a vertical method for adding or subtracting polynomials. First rewrite the polynomials in descending-exponent form, then arrange them one under another, so that each column contains like terms. Then add or subtract in each column.

Example 8

Adding Using the Vertical Method

Add $2x^2 - 5x$, $3x^2 + 2$, and $6x - 3$.

Like terms

$$\begin{array}{r} 2x^2 - 5x \phantom{{}+2} \\ 3x^2 \phantom{{}-5x} + 2 \\ 6x - 3 \\ \hline 5x^2 + x - 1 \end{array}$$

CHECK YOURSELF 8

Add $3x^2 + 5$, $x^2 - 4x$, *and* $6x + 7$.

The following example illustrates subtraction by the vertical method.

Example 9

Subtracting Using the Vertical Method

(a) Subtract $5x - 3$ from $8x - 7$.

Write

$$\begin{array}{r} 8x - 7 \\ (-)\ 5x - 3 \\ \hline 3x - 4 \end{array}$$

To subtract, change each sign of $5x - 3$ to get $-5x + 3$, then add.

$$\begin{array}{r} 8x - 7 \\ = -5x + 3 \\ \hline 3x - 4 \end{array}$$

(b) Subtract $5x^2 - 3x + 4$ from $8x^2 + 5x - 3$.

Write

$$\begin{array}{r} 8x^2 + 5x - 3 \\ (-)\ 5x^2 - 3x + 4 \\ \hline 3x^2 + 8x - 7 \end{array}$$

To subtract, change each sign of $5x^2 - 3x + 4$ to get $-5x^2 + 3x - 4$, then add.

$$\begin{array}{r} 8x^2 + 5x - 3 \\ = -5x^2 + 3x - 4 \\ \hline 3x^2 + 8x - 7 \end{array}$$

Subtracting using the vertical method takes some practice. Take time to study the method carefully. You'll be using it in long division in Section. 3.6.

CHECK YOURSELF 9

Subtract, using the vertical method.

(a) $4x^2 - 3x$ from $8x^2 + 2x$ **(b)** $8x^2 + 4x - 3$ from $9x^2 - 5x + 7$

CHECK YOURSELF ANSWERS

1. $10x^2 - 5x$ **2.** $8y^2 - 8y$ **3.** $13m^2 - 3m + 8$ **3.** $3x^2 + 11x$
5. **(a)** $-3m - 5n$; **(b)** $-5w + 7z$; **(c)** $3r - 2s + 5t$; **(d)** $5a + 3b + 2c$
6. **(a)** $3x - 10$; **(b)** $3x^2 - 8$ **7.** **(a)** $7x^2 - 10x$; **(b)** $4a^2 + 2a + 2$
8. $4x^2 + 2x + 12$ **9.** **(a)** $4x^2 + 5x$; **(b)** $x^2 - 9x + 10$

Name ______________

Section ______ Date ______

3.3 Exercises

Add.

1. $6a - 5$ and $3a + 9$

2. $9x + 3$ and $3x - 4$

3. $8b^2 - 11b$ and $5b^2 - 7b$

4. $2m^2 + 3m$ and $6m^2 - 8m$

5. $3x^2 - 2x$ and $-5x^2 + 2x$

6. $3p^2 + 5p$ and $-7p^2 - 5p$

7. $2x^2 + 5x - 3$ and $3x^2 - 7x + 4$

8. $4d^2 - 8d + 7$ and $5d^2 - 6d - 9$

9. $2b^2 + 8$ and $5b + 8$

10. $4x - 3$ and $3x^2 - 9x$

11. $8y^3 - 5y^2$ and $5y^2 - 2y$

12. $9x^4 - 2x^2$ and $2x^2 + 3$

13. $2a^2 - 4a^3$ and $3a^3 + 2a^2$

14. $9m^3 - 2m$ and $-6m - 4m^3$

15. $4x^2 - 2 + 7x$ and
$5 - 8x - 6x^2$

16. $5b^3 - 8b + 2b^2$ and
$3b^2 - 7b^3 + 5b$

Remove the parentheses in each of the following expressions, and simplify when possible.

17. $-(2a + 3b)$

18. $-(7x - 4y)$

19. $5a - (2b - 3c)$

20. $7x - (4y + 3z)$

21. $9r - (3r + 5s)$

22. $10m - (3m - 2n)$

23. $5p - (-3p + 2q)$

5p + 3p - 2q
8p - 2q

24. $8d - (-7c - 2d)$

8d + 7c + 2d

ANSWERS

1. $9a + 4$
2. $12x - 1$
3. $13b^2 - 18b$
4. $8m^2 - 5m$
5. $-2x^2$
6. $-4p^2$
7. $5x^2 - 2x + 1$
8. $9d^2 - 14d - 2$
9. $2b^2 + 5b + 16$
10. $3x^2 - 5x - 3$
11. $8y^3 - 2y$
12. $9x^4 + 3$
13. $-a^3 + 4a^2$
14. $5m^3 - 8m$
15. $-2x^2 - x + 3$
16. $-2b^3 + 5b^2 - 3b$
17. $-2a - 3b$
18. $-7x + 4y$
19. $5a - 2b + 3c$
20. $7x - 4y - 3z$
21. $6r - 5s$
22. $7m + 2n$
23. $8p - 2q$
24. $7c + 10d$

ANSWERS

25. $x - 7$
26. $2x + 7$
27. $m^2 - 3m$
28. $2a^2 - 5a$
29. $-2y^2$
30. $-2n^2$
31. $2x^2 - x + 1$
32. $2x^2 - 6x - 7$
33. $8a^2 - 12a - 7$
34. $x^3 - x^2 - 5x$
35. $-6b^2 + 8b$
36. $6y^2 - 9y$
37. $2x^2 + 12$
38. $-x^2 - 3x$
39. $6b - 1$
40. $6m - 3$
41. $10x - 9$
42. $-x^2 + 5$
43. $2x^2 + 5x - 12$
44. $6a + 2$
45. $-6y^2 - 8y$
46. $-2r^3 - 3r^2$
47. $6w^2 - 2w + 2$
48. $5x^2 + 2x + 3$
49. $9x^2 - x$
50. $8x^2 - 4x - 10$

Subtract.

25. $x + 4$ from $2x - 3$

26. $x - 2$ from $3x + 5$

27. $3m^2 - 2m$ from $4m^2 - 5m$

28. $9a^2 - 5a$ from $11a^2 - 10a$

29. $6y^2 + 5y$ from $4y^2 + 5y$

30. $9n^2 - 4n$ from $7n^2 - 4n$

31. $x^2 - 4x - 3$ from $3x^2 - 5x - 2$

32. $3x^2 - 2x + 4$ from $5x^2 - 8x - 3$

33. $3a + 7$ from $8a^2 - 9a$

34. $3x^3 + x^2$ from $4x^3 - 5x$

35. $4b^2 - 3b$ from $5b - 2b^2$

36. $7y - 3y^2$ from $3y^2 - 2y$

37. $x^2 - 5 - 8x$ from $3x^2 - 8x + 7$

38. $4x - 2x^2 + 4x^3$ from $4x^3 + x - 3x^2$

Perform the indicated operations.

39. Subtract $3b + 2$ from the sum of $4b - 2$ and $5b + 3$.

40. Subtract $5m - 7$ from the sum of $2m - 8$ and $9m - 2$. 11m − 10 −5m + 7 6m − 3

41. Subtract $3x^2 + 2x - 1$ from the sum of $x^2 + 5x - 2$ and $2x^2 + 7x - 8$.

42. Subtract $4x^2 - 5x - 3$ from the sum of $x^2 - 3x - 7$ and $2x^2 - 2x + 9$.

43. Subtract $2x^2 - 3x$ from the sum of $4x^2 - 5$ and $2x - 7$.

44. Subtract $5a^2 - 3a$ from the sum of $3a - 3$ and $5a^2 + 5$.

45. Subtract the sum of $3y^2 - 3y$ and $5y^2 + 3y$ from $2y^2 - 8y$.

46. Subtract the sum of $7r^3 - 4r^2$ and $-3r^3 + 4r^2$ from $2r^3 + 3r^2$.

Add, using the vertical method.

47. $2w^2 + 7$, $3w - 5$, and $4w^2 - 5w$

48. $3x^2 - 4x - 2$, $6x - 3$, and $2x^2 + 8$

49. $3x^2 + 3x - 4$, $4x^2 - 3x - 3$, and $2x^2 - x + 7$

50. $5x^2 + 2x - 4$, $x^2 - 2x - 3$, and $2x^2 - 4x - 3$

Subtract, using the vertical method.

51. $3a^2 - 2a$ from $5a^2 + 3a$

52. $6r^3 + 4r^2$ from $4r^3 - 2r^2$

53. $5x^2 - 6x + 7$ from $8x^2 - 5x + 7$

54. $8x^2 - 4x + 2$ from $9x^2 - 8x + 6$

55. $5x^2 - 3x$ from $8x^2 - 9$

56. $7x^2 + 6x$ from $9x^2 - 3$

Perform the indicated operations.

57. $[(9x^2 - 3x + 5) - (3x^2 + 2x - 1)] - (x^2 - 2x - 3)$

58. $[(5x^2 + 2x - 3) - (-2x^2 + x - 2)] - (2x^2 + 3x - 5)$

Find values for a, b, c, and d so that the following equations are true.

59. $3ax^4 - 5x^3 + x^2 - cx + 2 = 9x^4 - bx^3 + x^2 - 2d$

60. $(4ax^3 - 3bx^2 - 10) - 3(x^3 + 4x^2 - cx - d) = x^2 - 6x + 8$

61. Geometry. A rectangle has sides of $8x + 9$ and $6x - 7$. Find the polynomial that represents its perimeter.

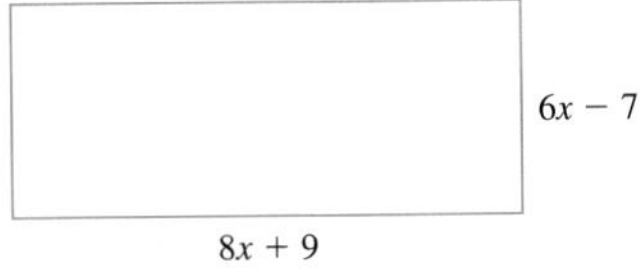

62. Geometry. A triangle has sides $3x + 7$, $4x - 9$, and $5x + 6$. Find the polynomial that represents its perimeter.

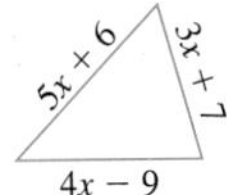

63. Business. The cost of producing x units of an item is $C = 150 + 25x$. The revenue for selling x units is $R = 90x - x^2$. The profit is given by the revenue minus the cost. Find the polynomial that represents profit.

64. Business. The revenue for selling y units is $R = 3y^2 - 2y + 5$ and the cost of producing y units is $C = y^2 + y - 3$. Find the polynomial that represents profit.

ANSWERS

51. $2a^2 + 5a$

52. $-2r^3 - 6r^2$

53. $3x^2 + x$

54. $x^2 - 4x + 4$

55. $3x^2 + 3x - 9$

56. $2x^2 - 6x - 3$

57. $5x^2 - 3x + 9$

58. $5x^2 - 2x + 4$

59. $a = 3, b = 5, c = 0, d = -1$

60. $a = \frac{3}{4}, b = -\frac{13}{3}, c = -2, d = 6$

61. $28x + 4$

62. $12x + 4$

63. $-x^2 + 65x - 150$

64. $2y^2 - 3y + 8$

ANSWERS

a. x^{12}

b. y^{20}

c. $2a^7$

d. $3m^7$

e. $12r^6$

f. $30w^5$

g. $-16x^9$

h. $30a^6$

Getting Ready for Section 3.4 [Section 1.7]

Multiply.

(a) $x^5 \cdot x^7$ (b) $y^8 \cdot y^{12}$ (c) $2a^3 \cdot d^4$

(d) $3m^5 \cdot m^2$ (e) $4r^5 \cdot 3r$ (f) $6w^2 \cdot 5w^3$

(g) $(-2x^2)(8x^7)$ (h) $(-10a)(-3a^5)$

Answers

1. $9a + 4$ **3.** $13b^2 - 18b$ **5.** $-2x^2$ **7.** $5x^2 - 2x + 1$ **9.** $2b^3 + 5b + 16$ **11.** $8y^3 - 2y$ **13.** $-a^3 + 4a^2$ **15.** $-2x^2 - x + 3$ **17.** $-2a - 3b$ **19.** $5a - 2b + 3c$ **21.** $6r - 5s$ **23.** $8p - 2q$ **25.** $x - 7$ **27.** $m^2 - 3m$ **29.** $-2y^2$ **31.** $2x^2 - x + 1$ **33.** $8a^2 - 12a - 7$ **35.** $-6b^2 + 8b$ **37.** $2x^2 + 12$ **39.** $6b - 1$ **41.** $10x - 9$ **43.** $2x^2 + 5x - 12$ **45.** $-6y^2 - 8y$ **47.** $6w^2 - 2w + 2$ **49.** $9x^2 - x$ **51.** $2a^2 + 5a$ **53.** $3x^2 + x$ **55.** $3x^2 + 3x - 9$ **57.** $5x^2 - 3x + 9$ **59.** $a = 3, b = 5, c = 0, d = -1$ **61.** $28x + 4$ **63.** $-x^2 + 65x - 150$ **a.** x^{12} **b.** y^{20} **c.** $2a^7$ **d.** $3m^7$ **e.** $12r^6$ **f.** $30w^5$ **g.** $-16x^9$ **h.** $30a^6$

3.4 Multiplying Polynomials

3.4 OBJECTIVES

1. Find the product of a monomial and a polynomial
2. Find the product of two polynomials

You have already had some experience in multiplying polynomials. In Section 1.7 we stated the first property of exponents and used that property to find the product of two monomial terms. Let's review briefly.

Step by Step: To Find the Product of Monomials

Step 1 Multiply the coefficients.

Step 2 Use the first property of exponents to combine the variables.

NOTE The first property of exponents:

$x^m \cdot x^n = x^{m+n}$

Example 1

Multiplying Monomials

Multiply $3x^2y$ and $2x^3y^5$.

NOTE Once again we have used the commutative and associative properties to rewrite the problem.

Write

$(3x^2y)(2x^3y^5)$

$= \underbrace{(3 \cdot 2)}\underbrace{(x^2 \cdot x^3)}\underbrace{(y \cdot y^5)}$

Multiply the coefficients. Add the exponents.

$= 6x^5y^6$

CHECK YOURSELF 1

Multiply.

(a) $(5a^2b)(3a^2b^4)$ **(b)** $(-3xy)(4x^3y^5)$

NOTE You might want to review Section 1.2 before going on.

Our next task is to find the product of a monomial and a polynomial. Here we use the distributive property, which we introduced in Section 1.2. That property leads us to the following rule for multiplication.

Rules and Properties: To Multiply a Polynomial by a Monomial

Use the distributive property to multiply each term of the polynomial by the monomial.

NOTE Distributive property:

$a(b + c) = ab + ac$

Example 2

Multiplying a Monomial and a Binomial

(a) Multiply $2x + 3$ by x.

Write

NOTE With practice you will do this step mentally.

$x(2x + 3)$

$= x \cdot 2x + x \cdot 3$

$= 2x^2 + 3x$

Multiply x by $2x$ and then by 3, the terms of the polynomial. That is, "distribute" the multiplication over the sum.

(b) Multiply $2a^3 + 4a$ by $3a^2$.

Write

$3a^2(2a^3 + 4a)$

$= 3a^2 \cdot 2a^3 + 3a^2 \cdot 4a = 6a^5 + 12a^3$

CHECK YOURSELF 2

Multiply.

(a) $2y(y^2 + 3y)$

(b) $3w^2(2w^3 + 5w)$

The patterns of Example 2 extend to *any* number of terms.

Example 3

Multiplying a Monomial and a Polynomial

Multiply the following.

(a) $3x(4x^3 + 5x^2 + 2)$

$= 3x \cdot 4x^3 + 3x \cdot 5x^2 + 3x \cdot 2 = 12x^4 + 15x^3 + 6x$

NOTE Again we have shown all the steps of the process. With practice you can write the product directly, and you should try to do so.

(b) $5y^2(2y^3 - 4)$

$= 5y^2 \cdot 2y^3 - 5y^2 \cdot 4 = 10y^5 - 20y^2$

(c) $-5c(4c^2 - 8c)$

$= (-5c)(4c^2) - (-5c)(8c) = -20c^3 + 40c^2$

(d) $3c^2d^2(7cd^2 - 5c^2d^3)$

$= 3c^2d^2 \cdot 7cd^2 - 3c^2d^2 \cdot 5c^2d^3 = 21c^3d^4 - 15c^4d^5$

CHECK YOURSELF 3

Multiply.

(a) $3(5a^2 + 2a + 7)$

(b) $4x^2(8x^3 - 6)$

(c) $-5m(8m^2 - 5m)$

(d) $9a^2b(3a^3b - 6a^2b^4)$

Example 4

Multiplying Binomials

(a) Multiply $x + 2$ by $x + 3$.

NOTE Note that this ensures that each term, x and 2, of the first binomial is multiplied by each term, x and 3, of the second binomial.

We can think of $x + 2$ as a single quantity and apply the distributive property.

$(x + 2)(x + 3)$ Multiply $x + 2$ by x and then by 3.

$= (x + 2)x + (x + 2)\,3$

$= x \cdot x + 2 \cdot x + x \cdot 3 + 2 \cdot 3$

$= x^2 + 2x + 3x + 6$

$= x^2 + 5x + 6$

(b) Multiply $a - 3$ by $a - 4$. (Think of $a - 3$ as a single quantity and distribute.)

$(a - 3)(a - 4)$

$= (a - 3)a - (a - 3)(4)$

$= a \cdot a - 3 \cdot a - [(a \cdot 4) - (3 \cdot 4)]$

$= a^2 - 3a - (4a - 12)$ Note that the parentheses are needed here because a *minus sign* precedes the binomial.

$= a^2 - 3a - 4a + 2$

$= a^2 - 7a + 12$

CHECK YOURSELF 4

Multiply.

(a) $(x + 4)(x + 5)$ **(b)** $(y + 5)(y - 6)$

Fortunately, there is a pattern to this kind of multiplication that allows you to write the product of the two binomials directly without going through all these steps. We call it the **FOIL method** of multiplying. The reason for this name will be clear as we look at the process in more detail.

To multiply $(x + 2)(x + 3)$:

1. $(x + 2)(x + 3)$ — $x \cdot x$ Find the product of the *first* terms of the factors.

NOTE Remember this by F!

2. $(x + 2)(x + 3)$ — $x \cdot 3$ Find the product of the *outer* terms.

NOTE Remember this by O!

3. $(x + 2)(x + 3)$ — $2 \cdot x$ Find the product of the *inner* terms.

NOTE Remember this by I!

4. $(x + 2)(x + 3)$ — $2 \cdot 3$ Find the product of the *last* terms.

NOTE Remember this by L!

Combining the four steps, we have

NOTE Of course these are the same four terms found in Example 4*a*.

$(x + 2)(x + 3)$

$= x^2 + 3x + 2x + 6$

$= x^2 + 5x + 6$

NOTE It's called FOIL to give you an easy way of remembering the steps: *F*irst, *O*uter, *I*nner, and *L*ast.

With practice, the FOIL method will let you write the products quickly and easily. Consider Example 5, which illustrates this approach.

Example 5

Using the FOIL Method

Find the following products, using the FOIL method.

F: $x \cdot x$ L: $4 \cdot 5$

(a) $(x + 4)(x + 5)$

I: $4x$ O: $5x$

$= x^2 + 5x + 4x + 20$

F O I L

$= x^2 + 9x + 20$

NOTE When possible, you should combine the outer and inner products mentally and write just the final product.

F: $x \cdot x$ L: $(-7)(3)$

(b) $(x - 7)(x + 3)$

I: $-7x$ O: $3x$

Combine the outer and inner products as $-4x$.

$= x^2 - 4x - 21$

CHECK YOURSELF 5

Multiply.

(a) $(x + 6)(x + 7)$ **(b)** $(x + 3)(x - 5)$ **(c)** $(x - 2)(x - 8)$

Using the FOIL method, you can also find the product of binomials with coefficients other than 1 or with more than one variable.

Example 6

Using the FOIL Method

Find the following products, using the FOIL method.

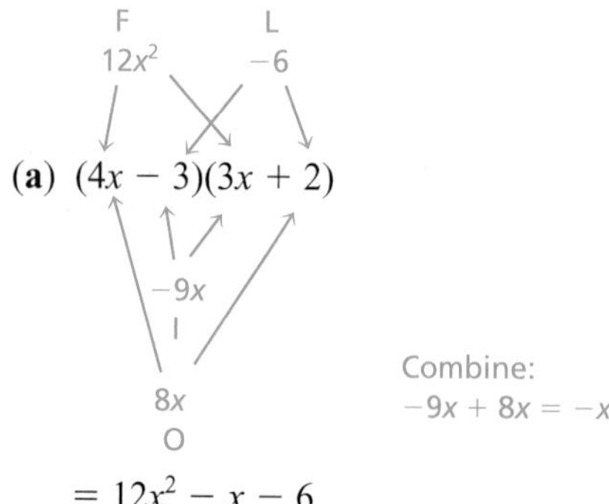

$= 12x^2 - x - 6$

$6x^2$ $35y^2$

(b) $(3x - 5y)(2x - 7y)$

$-10xy$

$-21xy$

Combine:
$-10xy - 21xy = -31xy$

$= 6x^2 - 31xy + 35y^2$

The following rule summarizes our work in multiplying binomials.

Step by Step: To Multiply Two Binomials

Step 1 Find the first term of the product of the binomials by multiplying the first terms of the binomials (F).

Step 2 Find the middle term of the product as the sum of the outer and inner products (O + I).

Step 3 Find the last term of the product by multiplying the last terms of the binomials (L).

CHECK YOURSELF 6

Multiply.

(a) $(5x + 2)(3x - 7)$ **(b)** $(4a - 3b)(5a - 4b)$ **(c)** $(3m + 5n)(2m + 3n)$

Sometimes, especially with larger polynomials, it is easier to use the vertical method to find their product. This is the same method you originally learned when multiplying two large integers.

Example 7

Multiplying Using the Vertical Method

Use the vertical method to find the product of $(3x + 2)(4x - 1)$.

First, we rewrite the multiplication in vertical form.

$$\begin{array}{l} 3x + 2 \\ \underline{4x + (-1)} \end{array}$$

Multiplying the quantity $3x + 2$ by -1 yields

$$\begin{array}{r} 3x + 2 \\ \underline{4x + (-1)} \\ -3x + (-2) \end{array}$$

Note that we maintained the columns of the original binomial when we found the product. We will continue with those columns as we multiply by the $4x$ term.

$$\begin{array}{r} 3x + 2 \\ \underline{4x + (-1)} \\ -3x + (-2) \\ \underline{12x^2 + 8x \qquad\quad} \\ 12x^2 + 5x + (-2) \end{array}$$

We could write the product as $(3x + 2)(4x - 1) = 12x^2 + 5x - 2$.

CHECK YOURSELF 7

Use the vertical method to find the product of $(5x - 3)(2x + 1)$.

We'll use the vertical method again in our next example. This time, we will multiply a binomial and a trinomial. Note that the FOIL method can never work for anything but the product of two binomials.

Example 8

Using the Vertical Method

Multiply $x^2 - 5x + 8$ by $x + 3$.

Step 1

$$\begin{array}{r} x^2 - 5x + 8 \\ x + 3 \\ \hline 3x^2 - 15x + 24 \end{array}$$

Multiply each term of $x^2 - 5x + 8$ by 3.

Step 2

$$\begin{array}{rrr} & x^2 - 5x + 8 \\ & x + 3 \\ \hline & 3x^2 - 15x + 24 \\ x^3 - 5x^2 + 8x & \\ \hline \end{array}$$

Now multiply each term by x.

Note that this line is shifted over so that like terms are in the same columns.

NOTE Using this vertical method ensures that each term of one factor multiplies each term of the other. That's why it works!

Step 3

$$\begin{array}{r} x^2 - 5x + 8 \\ x + 3 \\ \hline 3x^2 - 15x + 24 \\ x^3 - 5x^2 + 8x \quad\;\; \\ \hline x^3 - 2x^2 - 7x + 24 \end{array}$$

Now add to combine like terms to write the product.

CHECK YOURSELF 8

Multiply $2x^2 - 5x + 3$ by $3x + 4$.

CHECK YOURSELF ANSWERS

1. **(a)** $15a^4b^5$; **(b)** $-12x^4y^6$ **2.** **(a)** $2y^3 + 6y^2$; **(b)** $6w^5 + 15w^3$
3. **(a)** $15a^2 + 6a + 21$; **(b)** $32x^5 - 24x^2$; **(c)** $-40m^3 + 25m^2$; **(d)** $27a^5b^2 - 54a^4b^5$
4. **(a)** $x^2 + 9x + 20$; **(b)** $y^2 - y - 30$
5. **(a)** $x^2 + 13x + 42$; **(b)** $x^2 - 2x - 15$; **(c)** $x^2 - 10x + 16$
6. **(a)** $15x^2 - 29x - 14$; **(b)** $20a^2 - 31ab + 12b^2$; **(c)** $6m^2 + 19mn + 15n^2$
7. $10x^2 - x - 3$ **8.** $6x^3 - 7x^2 - 11x + 12$

Name ______________________

Section ________ Date ________

3.4 Exercises

Multiply.

1. $(5x^2)(3x^3)$

2. $(7a^5)(4a^6)$

3. $(-2b^2)(14b^8)$

4. $(14y^4)(-4y^6)$

5. $(-10p^6)(-4p^7)$

6. $(-6m^8)(9m^7)$

7. $(4m^5)(-3m)$

8. $(-5r^7)(-3r)$

9. $(4x^3y^2)(8x^2y)$

10. $(-3r^4s^2)(-7r^2s^5)$

11. $(-3m^5n^2)(2m^4n)$

12. $(7a^3b^5)(-6a^4b)$

13. $5(2x + 6)$

14. $4(7b - 5)$

15. $3a(4a + 5)$

16. $5x(2x - 7)$

17. $3s^2(4s^2 - 7s)$

18. $9a^2(3a^3 + 5a)$

19. $2x(4x^2 - 2x + 1)$

20. $5m(4m^3 - 3m^2 + 2)$

21. $3xy(2x^2y + xy^2 + 5xy)$

22. $5ab^2(ab - 3a + 5b)$

23. $6m^2n(3m^2n - 2mn + mn^2)$

24. $8pq^2(2pq - 3p + 5q)$

ANSWERS

1. $15x^5$
2. $28a^{11}$
3. $-28b^{10}$
4. $-56y^{10}$
5. $40p^{13}$
6. $-54m^{15}$
7. $-12m^6$
8. $15r^8$
9. $32x^5y^3$
10. $21r^6s^7$
11. $-6m^9n^3$
12. $-42a^7b^6$
13. $10x + 30$
14. $28b - 20$
15. $12a^2 + 15a$
16. $10x^2 - 35x$
17. $12s^4 - 21s^3$
18. $27a^5 + 45a^3$
19. $8x^3 - 4x^2 + 2x$
20. $20m^4 - 15m^3 + 10m$
21. $6x^3y^2 + 3x^2y^3 + 15x^2y^2$
22. $5a^2b^3 - 15a^2b^2 + 25ab^3$
23. $18m^4n^2 - 12m^3n^2 + 6m^3n^3$
24. $16p^2q^3 - 24p^2q^2 + 40pq^3$

ANSWERS

25. $x^2 + 5x + 6$
26. $a^2 - 10a + 21$
27. $m^2 - 14m + 45$
28. $b^2 + 12b + 35$
29. $p^2 - p - 56$
30. $x^2 - x - 90$
31. $w^2 + 30w + 200$
32. $s^2 - 20s + 96$
33. $3x^2 - 29x + 40$
34. $4w^2 + 13w - 35$
35. $6x^2 - x - 12$
36. $15a^2 + 38a + 7$
37. $12a^2 - 31ab + 9b^2$
38. $21s^2 + 47st - 24t^2$
39. $21p^2 - 13pq - 20q^2$
40. $10x^2 - 13xy + 4y^2$
41. $6x^2 + 23xy + 20y^2$
42. $16x^2 - 8xy - 15y^2$
43. $x^2 + 10x + 25$
44. $y^2 + 16y + 64$
45. $y^2 - 18y + 81$
46. $4a^2 + 12a + 9$
47. $36m^2 + 12mn + n^2$
48. $49b^2 - 14bc + c^2$
49. $a^2 - 25$
50. $x^2 - 49$
51. $x^2 - 4y^2$
52. $49x^2 - y^2$

Multiply.

25. $(x + 3)(x + 2)$

26. $(a - 3)(a - 7)$

27. $(m - 5)(m - 9)$

28. $(b + 7)(b + 5)$

29. $(p - 8)(p + 7)$

30. $(x - 10)(x + 9)$

31. $(w + 10)(w + 20)$

32. $(s - 12)(s - 8)$

33. $(3x - 5)(x - 8)$

34. $(w + 5)(4w - 7)$

35. $(2x - 3)(3x + 4)$

36. $(5a + 1)(3a + 7)$

37. $(3a - b)(4a - 9b)$

38. $(7s - 3t)(3s + 8t)$

39. $(3p - 4q)(7p + 5q)$

40. $(5x - 4y)(2x - y)$

41. $(2x + 5y)(3x + 4y)$

42. $(4x - 5y)(4x + 3y)$

43. $(x + 5)^2$

44. $(y + 8)^2$

45. $(y - 9)^2$

46. $(2a + 3)^2$

47. $(6m + n)^2$

48. $(7b - c)^2$

49. $(a - 5)(a + 5)$

50. $(x - 7)(x + 7)$

51. $(x - 2y)(x + 2y)$

52. $(7x + y)(7x - y)$

53. $(5s + 3t)(5s - 3t)$

54. $(9c - 4d)(9c + 4d)$

Multiply, using the vertical method.

55. $(x + 2)(3x + 5)$

56. $(a - 3)(2a + 7)$

57. $(2m - 5)(3m + 7)$

58. $(5p + 3)(4p + 1)$

59. $(3x + 4y)(5x - 2y)$

60. $(7a - 2b)(2a + 4b)$

61. $(a^2 + 3ab - b^2)(a^2 - 5ab + b^2)$

62. $(m^2 - 5mn + 3n^2)(m^2 + 4mn - 2n^2)$

63. $(x - 2y)(x^2 + 2xy + 4y^2)$

64. $(m + 3n)(m^2 - 3mn + 9n^2)$

65. $(3a + 4b)(9a^2 - 12ab + 16b^2)$

66. $(2r - 3s)(4r^2 + 6rs + 9s^2)$

Multiply.

67. $2x(3x - 2)(4x + 1)$

68. $3x(2x + 1)(2x - 1)$

69. $5a(4a - 3)(4a + 3)$

70. $6m(3m - 2)(3m - 7)$

71. $3s(5s - 2)(4s - 1)$

72. $7w(2w - 3)(2w + 3)$

73. $(x - 2)(x + 1)(x - 3)$

74. $(y + 3)(y - 2)(y - 4)$

75. $(a - 1)^3$

76. $(x + 1)^3$

ANSWERS

53. $25s^2 - 9t^2$
54. $81c^2 - 16d^2$
55. $3x^2 + 11x + 10$
56. $2a^2 + a - 21$
57. $6m^2 - m - 35$
58. $20p^2 + 17p + 3$
59. $15x^2 + 14xy - 8y^2$
60. $14a^2 + 24ab - 8b^2$
61. $a^4 - 2a^3b - 15a^2b^2 + 8ab^3 - b^4$
62. $m^4 - m^3n - 19m^2n^2 + 22mn^3 - 6n^4$
63. $x^3 - 8y^3$
64. $m^3 + 27n^3$
65. $27a^3 + 64b^3$
66. $8r^3 - 27s^3$
67. $24x^3 - 10x^2 - 4x$
68. $12x^3 - 3x$
69. $80a^3 - 45a$
70. $54m^3 - 162m^2 + 84m$
71. $60s^3 - 39s^2 + 6s$
72. $28w^3 - 63w$
73. $x^3 - 4x^2 + x + 6$
74. $y^3 - 3y^2 - 10y + 24$
75. $a^3 - 3a^2 + 3a - 1$
76. $x^3 + 3x^2 + 3x + 1$

ANSWERS

77. $\frac{x^2}{3} + \frac{11x}{45} - \frac{4}{15}$

78. $\frac{x^2}{4} + \frac{29x}{80} - \frac{9}{20}$

79. $x^2 - y^2 + 4y - 4$

80. $x^2 - y^2 + 6y - 9$

81. False

82. False

83. True

84. True

85. $6x^2 - 11x - 35$ cm^2

86. $3y^2 + \frac{5}{2}y - \frac{21}{2}$ in.^{2}

87. $2x^2 - 10x$

88. $2x^3 - 100x$

Multiply the following.

77. $\left(\frac{x}{2} + \frac{2}{3}\right)\left(\frac{2x}{3} - \frac{2}{5}\right)$

78. $\left(\frac{x}{3} + \frac{3}{4}\right)\left(\frac{3x}{4} - \frac{3}{5}\right)$

79. $[x + (y - 2)][x - (y - 2)]$

80. $[x + (3 - y)][x - (3 - y)]$

Label the following as true or false.

81. $(x + y)^2 = x^2 + y^2$

82. $(x - y)^2 = x^2 - y^2$

83. $(x + y)^2 = x^2 + 2xy + y^2$

84. $(x - y)^2 = x^2 - 2xy + y^2$

85. Length. The length of a rectangle is given by $3x + 5$ centimeters (cm) and the width is given by $2x - 7$ cm. Express the area of the rectangle in terms of x.

86. Area. The base of a triangle measures $3y + 7$ inches (in.) and the height is $2y - 3$ in. Express the area of the triangle in terms of y.

87. Revenue. The price of an item is given by $p = 2x - 10$. If the revenue generated is found by multiplying the number of items (x) sold by the price of an item, find the polynomial which represents the revenue.

88. Revenue. The price of an item is given by $p = 2x^2 - 100$. Find the polynomial that represents the revenue generated from the sale of x items.

89. Work with another student to complete this table and write the polynomial. A paper box is to be made from a piece of cardboard 20 inches (in.) wide and 30 in. long. The box will be formed by cutting squares out of each of the four corners and folding up the sides to make a box.

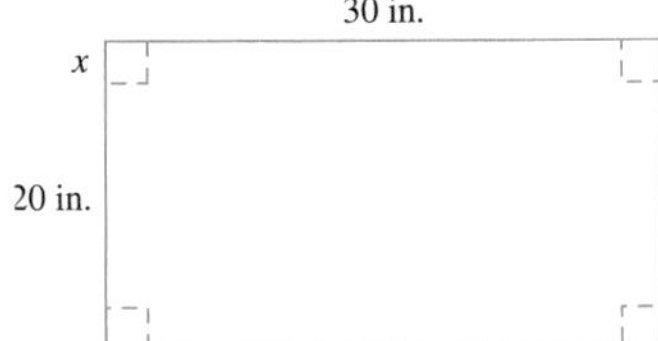

If x is the dimension of the side of the square cut out of the corner, when the sides are folded up, the box will be x inches tall. You should use a piece of paper to try this to see how the box will be made. Complete the following chart.

Length of Side of Corner Square	Length of Box	Width of Box	Depth of Box	Volume of Box
1 in.				
2 in.				
3 in.				
n in.				

Write a general formula for the width, length, and height of the box and a general formula for the *volume* of the box, and simplify it by multiplying. The variable will be the height, the side of the square cut out of the corners. What is the highest power of the variable in the polynomial you have written for the volume ______?

90. (a) Multiply $(x - 1)(x + 1)$

(b) Multiply $(x - 1)(x^2 + x + 1)$

(c) Multiply $(x - 1)(x^3 + x^2 + x + 1)$

(d) Based on your results to (a), (b), and (c), find the product $(x - 1)(x^{29} + x^{28} + \cdots + x + 1)$.

Getting Ready for Section 3.5 [Section 1.4]

Simplify.

(a) $(3a)(3a)$

(b) $(3a)^2$

(c) $(5x)(5x)$

(d) $(5x)^2$

(e) $(-2w)(-2w)$

(f) $(-2w)^2$

(g) $(-4r)(-4r)$

(h) $(-4r)^2$

ANSWERS

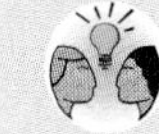

89. ______

90. (a) $x^2 - 1$

(b) $x^3 - 1$

(c) $x^4 - 1$

(d) $x^{30} - 1$

a. $9a^2$

b. $9a^2$

c. $25x^2$

d. $25x^2$

e. $4w^2$

f. $4w^2$

g. $16r^2$

h. $16r^2$

Answers

1. $15x^5$ **3.** $-28b^{10}$ **5.** $40p^{13}$ **7.** $-12m^6$ **9.** $32x^5y^3$ **11.** $-6m^9n^3$
13. $10x + 30$ **15.** $12a^2 + 15a$ **17.** $12s^4 - 21s^3$ **19.** $8x^3 - 4x^2 + 2x$
21. $6x^3y^2 + 3x^2y^3 + 15x^2y^2$ **23.** $18m^4n^2 - 12m^3n^2 + 6m^3n^3$ **25.** $x^2 + 5x + 6$
27. $m^2 - 14m + 45$ **29.** $p^2 - p - 56$ **31.** $w^2 + 30w + 200$
33. $3x^2 - 29x + 40$ **35.** $6x^2 - x - 12$ **37.** $12a^2 - 31ab + 9b^2$
39. $21p^2 - 13pq - 20q^2$ **41.** $6x^2 + 23xy + 20y^2$ **43.** $x^2 + 10x + 25$
45. $y^2 - 18y + 81$ **47.** $36m^2 + 12mn + n^2$ **49.** $a^2 - 25$ **51.** $x^2 - 4y^2$
53. $25s^2 - 9t^2$ **55.** $3x^2 + 11x + 10$ **57.** $6m^2 - m - 35$
59. $15x^2 + 14xy - 8y^2$ **61.** $a^4 - 2a^3b - 15a^2b^2 + 8ab^3 - b^4$
63. $x^3 - 8y^3$ **65.** $27a^3 + 64b^3$ **67.** $24x^3 - 10x^2 - 4x$
69. $80a^3 - 45a$ **71.** $60s^3 - 39s^2 + 6s$ **73.** $x^3 - 4x^2 + x + 6$
75. $a^3 - 3a^2 + 3a - 1$ **77.** $\frac{x^2}{3} + \frac{11x}{45} - \frac{4}{15}$ **79.** $x^2 - y^2 + 4y - 4$
81. False **83.** True **85.** $6x^2 - 11x - 35\text{cm}^2$ **87.** $2x^2 - 10x$
89. **a.** $9a^2$ **b.** $9a^2$ **c.** $25x^2$ **d.** $25x^2$ **e.** $4w^2$ **f.** $4w^2$
g. $16r^2$ **h.** $16r^2$

Special Products

OBJECTIVES

1. Square a binomial
2. Find the product of two binomials that differ only in their sign

Certain products occur frequently enough in algebra that it is worth learning special formulas for dealing with them. First, let's look at the **square of a binomial,** which is the product of two equal binomial factors.

$$(x + y)^2 = (x + y)(x + y)$$
$$= x^2 + 2xy + y^2$$

$$(x - y)^2 = (x - y)(x - y)$$
$$= x^2 - 2xy + y^2$$

The patterns above lead us to the following rule.

Step by Step: To Square a Binomial

Step 1 Find the first term of the square by squaring the first term of the binomial.

Step 2 Find the middle term of the square as twice the product of the two terms of the binomial.

Step 3 Find the last term of the square by squaring the last term of the binomial.

Example 1

Squaring a Binomial

CAUTION

A very common mistake in squaring binomials is to forget the middle term.

(a) $(x + 3)^2 = x^2 + 2 \cdot x \cdot 3 + 3^2$

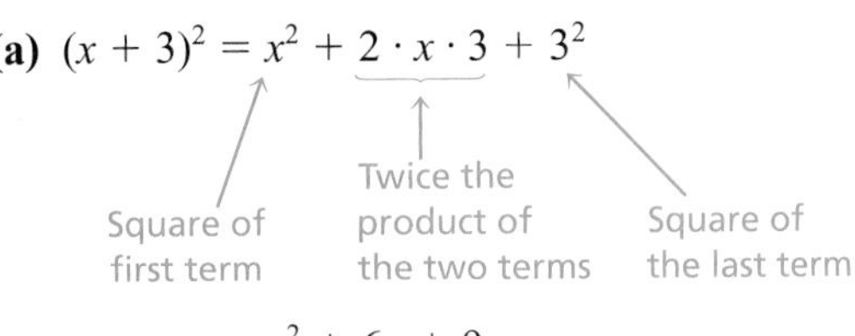

$$= x^2 + 6x + 9$$

(b) $(3a + 4b)^2 = (3a)^2 + 2(3a)(4b) + (4b)^2$

$$= 9a^2 + 24ab + 16b^2$$

(c) $(y - 5)^2 = y^2 + 2 \cdot y \cdot (-5) + (-5)^2$

$$= y^2 - 10y + 25$$

(d) $(5c - 3d)^2 = (5c)^2 + 2(5c)(-3d) + (-3d)^2$

$$= 25c^2 - 30cd + 9d^2$$

Again we have shown all the steps. With practice you can write just the square.

CHECK YOURSELF 1

Multiply.

(a) $(2x + 1)^2$ **(b)** $(4x - 3y)^2$

Example 2

Squaring a Binomial

Find $(y + 4)^2$.

NOTE You should see that $(2 + 3)^2 \neq 2^2 + 3^2$ because $5^2 \neq 4 + 9$

$(y + 4)^2$ is *not* equal to $y^2 + 4^2$ or $y^2 + 16$

The correct square is

$$(y + 4)^2 = y^2 + 8y + 16$$

The middle term is twice the product of y and 4.

CHECK YOURSELF 2

Multiply.

(a) $(x + 5)^2$ **(b)** $(3a + 2)^2$ **(c)** $(y - 7)^2$ **(d)** $(5x - 2y)^2$

A second special product will be very important in the next chapter, which deals with factoring. Suppose the form of a product is

$$(x + y)(x - y)$$

The two terms differ only in sign.

Let's see what happens when we multiply.

$$(x + y)(x - y)$$

$$= x^2 \underbrace{- xy + xy}_{= 0} - y^2$$

$$= x^2 - y^2$$

Because the middle term becomes 0, we have the following rule.

Rules and Properties: Special Product

The product of two binomials that differ only in the sign between the terms is the square of the first term minus the square of the second term.

Let's look at the application of this rule in Example 3.

Example 3

Multiplying Polynomials

Multiply each pair of binomials.

(a) $(x + 5)(x - 5) = x^2 - 5^2$

Square of the first term; Square of the second term

$= x^2 - 25$

NOTE
$(2y)^2 = (2y)(2y)$
$= 4y^2$

(b) $(x + 2y)(x - 2y) = x^2 - (2y)^2$

Square of the first term; Square of the second term

$= x^2 - 4y^2$

(c) $(3m + n)(3m - n) = 9m^2 - n^2$

(d) $(4a - 3b)(4a + 3b) = 16a^2 - 9b^2$

CHECK YOURSELF 3

Find the products.

(a) $(a - 6)(a + 6)$ **(b)** $(x - 3y)(x + 3y)$
(c) $(5n + 2p)(5n - 2p)$ **(d)** $(7b - 3c)(7b + 3c)$

When finding the product of three or more factors, it is useful to first look for the pattern in which two binomials differ only in their sign. Finding this product first will make it easier to find the product of all the factors.

Example 4

Multiplying Polynomials

(a) $x(x - 3)(x + 3)$ These binomials differ only in the sign.

$= x(x^2 - 9)$
$= x^3 - 9x$

(b) $(x + 1)(x - 5)(x + 5)$ — These binomials differ only in the sign.

$= (x + 1)(x^2 - 25)$ — With two binomials, use the FOIL method.

$= x^3 + x^2 - 25x - 25$

(c) $(2x - 1)(x + 3)(2x + 1)$ — These two binomials differ only in the sign of the second term. We can use the commutative property to rearrange the terms.

$= (x + 3)(2x - 1)(2x + 1)$

$= (x + 3)(4x^2 - 1)$

$= 4x^3 + 12x^2 - x - 3$

CHECK YOURSELF 4

Multiply.

(a) $3x(x - 5)(x + 5)$ **(b)** $(x - 4)(2x + 3)(2x - 3)$

(c) $(x - 7)(3x - 1)(x + 7)$

CHECK YOURSELF ANSWERS

1. **(a)** $4x^2 + 4x + 1$; **(b)** $16x^2 - 24xy + 9y^2$
2. **(a)** $x^2 + 10x + 25$; **(b)** $9a^2 + 12a + 4$; **(c)** $y^2 - 14y + 49$; **(d)** $25x^2 - 20xy + 4y^2$
3. **(a)** $a^2 - 36$; **(b)** $x^2 - 9y^2$; **(c)** $25n^2 - 4p^2$; **(d)** $49b^2 - 9c^2$
4. **(a)** $3x^3 - 75x$; **(b)** $4x^3 - 16x^2 - 9x + 36$; **(c)** $3x^3 - x^2 - 147x + 49$

Name ______________

Section ________ Date ________

3.5 Exercises

Find each of the following squares.

1. $(x + 5)^2$

2. $(y + 9)^2$

3. $(w - 6)^2$

4. $(a - 8)^2$

5. $(z + 12)^2$

6. $(p - 20)^2$

7. $(2a - 1)^2$

8. $(3x - 2)^2$

9. $(6m + 1)^2$

10. $(7b - 2)^2$

11. $(3x - y)^2$

12. $(5m + n)^2$

13. $(2r + 5s)^2$

14. $(3a - 4b)^2$

15. $(8a - 9b)^2$

16. $(7p + 6q)^2$

17. $\left(x + \frac{1}{2}\right)^2$

18. $\left(w - \frac{1}{4}\right)^2$

Find each of the following products.

19. $(x - 6)(x + 6)$

20. $(y + 8)(y - 8)$

21. $(m + 12)(m - 12)$

22. $(w - 10)(w + 10)$

23. $\left(x - \frac{1}{2}\right)\left(x + \frac{1}{2}\right)$

24. $\left(x + \frac{2}{3}\right)\left(x - \frac{2}{3}\right)$

ANSWERS

1. $x^2 + 10x + 25$
2. $y^2 + 18y + 81$
3. $w^2 - 12w + 36$
4. $a^2 - 16a + 64$
5. $z^2 + 24z + 144$
6. $p^2 - 40p + 400$
7. $4a^2 - 4a + 1$
8. $9x^2 - 12x + 4$
9. $36m^2 + 12m + 1$
10. $49b^2 - 28b + 4$
11. $9x^2 - 6xy + y^2$
12. $25m^2 + 10mn + n^2$
13. $4r^2 + 20rs + 25s^2$
14. $9a^2 - 24ab + 16b^2$
15. $64a^2 - 144ab + 81b^2$
16. $49p^2 + 84pq + 36q^2$
17. $x^2 + x + \frac{1}{4}$
18. $w^2 - \frac{1}{2}w + \frac{1}{16}$
19. $x^2 - 36$
20. $y^2 - 64$
21. $m^2 - 144$
22. $w^2 - 100$
23. $x^2 - \frac{1}{4}$
24. $x^2 - \frac{4}{9}$

ANSWERS

25. $p^2 - 0.16$
26. $m^2 - 0.36$
27. $a^2 - 9b^2$
28. $p^2 - 16q^2$
29. $16r^2 - s^2$
30. $49x^2 - y^2$
31. $64w^2 - 25z^2$
32. $49c^2 - 4d^2$
33. $25x^2 - 81y^2$
34. $36s^2 - 25t^2$
35. $x^3 - 4x$
36. $a^3 - 25a$
37. $2s^3 - 18r^2s$
38. $20w^3 - 5wz^2$
39. $5r^3 + 30r^2 + 45r$
40. $3x^3 - 12x^2 + 12x$
41. $x^2 - 36$
42. $x^2 + 10x + 25$
43. $x^2 - 8x + 16$
44. $x^2 - 25$
45. 2499
46. 891
47. 884
48. 9996
49. 3575
50. 3584

25. $(p - 0.4)(p + 0.4)$

26. $(m - 0.6)(m + 0.6)$

27. $(a - 3b)(a + 3b)$

28. $(p + 4q)(p - 4q)$

29. $(4r - s)(4r + s)$

30. $(7x - y)(7x + y)$

31. $(8w + 5z)(8w - 5z)$

32. $(7c + 2d)(7c - 2d)$

33. $(5x - 9y)(5x + 9y)$

34. $(6s - 5t)(6s + 5t)$

35. $x(x - 2)(x + 2)$

36. $a(a + 5)(a - 5)$

37. $2s(s - 3r)(s + 3r)$

38. $5w(2w - z)(2w + z)$

39. $5r(r + 3)^2$

40. $3x(x - 2)^2$

For each of the following problems, let x represent the number, then write an expression for the product.

41. The product of 6 more than a number and 6 less than that number

42. The square of 5 more than a number

43. The square of 4 less than a number

44. The product of 5 less than a number and 5 more than that number

Note that $(28)(32) = (30 - 2)(30 + 2) = 900 - 4 = 896$. Use this pattern to find each of the following products.

45. (49)(51)

46. (27)(33)

47. (34)(26)

48. (98)(102)

49. (55)(65)

50. (64)(56)

51. **Tree planting.** Suppose an orchard is planted with trees in straight rows. If there are $5x - 4$ rows with $5x - 4$ trees in each row, how many trees are there in the orchard?

52. **Area of a square.** A square has sides of length $3x - 2$ centimeters (cm). Express the area of the square as a polynomial.

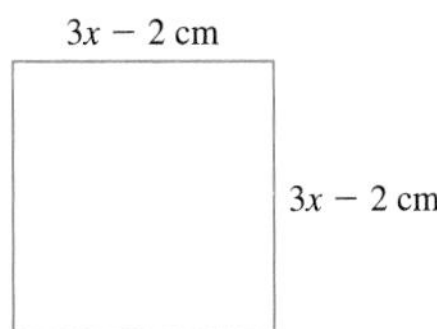

53. Complete the following statement: $(a + b)^2$ is not equal to $a^2 + b^2$ because. . . . But, wait! Isn't $(a + b)^2$ *sometimes* equal to $a^2 + b^2$? What do you think?

54. Is $(a + b)^3$ ever equal to $a^3 + b^3$? Explain.

55. In the following figures, identify the length, width, and area of the square:

a b
a
b

Length = ________

Width = ________

Area = ________

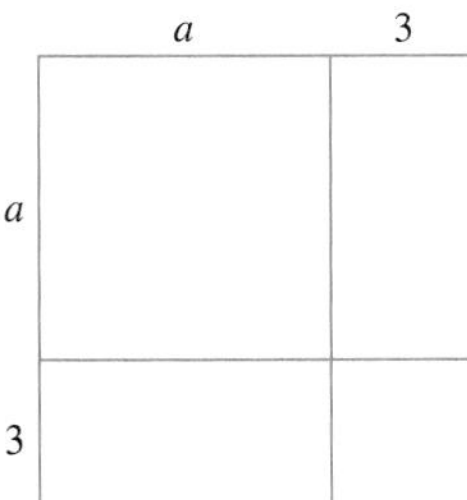

Length = ________

Width = ________

Area = ________

x
x x^2 $2x$
$2x$

Length = ________

Width = ________

Area = ________

ANSWERS

51. $25x^2 - 40x + 16$

52. $(9x^2 - 12x + 4)$ cm^2

53. ________

54. ________

55. ________

ANSWERS

56.

a. x

b. a^2

c. $3p$

d. $2m^2$

e. 4

f. $2x$

g. $3r^2s$

h. $7c^3d^3$

56. The square below is x units on a side. The area is ________.

Draw a picture of what happens when the sides are doubled. The area is ________.

Continue the picture to show what happens when the sides are tripled. The area is ________.

If the sides are quadrupled, the area is ________.

In general, if the sides are multiplied by n, the area is ________.

If each side is increased by 3, the area is increased by ________.

If each side is decreased by 2, the area is decreased by ________.

In general, if each side is increased by n, the area is increased by ________, and if each side is decreased by n, the area is decreased by ________.

Getting Ready for Section 3.6 [Section 1.7]

Divide.

(a) $\frac{2x^2}{2x}$ (b) $\frac{3a^3}{3a}$ (c) $\frac{6p^3}{2p^2}$

(d) $\frac{10m^4}{5m^2}$ (e) $\frac{20a^3}{5a^3}$ (f) $\frac{6x^2y}{3xy}$

(g) $\frac{12r^3s^2}{4rs}$ (h) $\frac{49c^4d^6}{7cd^3}$

Answers

1. $x^2 + 10x + 25$ **3.** $w^2 - 12w + 36$ **5.** $z^2 + 24z + 144$ **7.** $4a^2 - 4a + 1$
9. $36m^2 + 12m + 1$ **11.** $9x^2 - 6xy + y^2$ **13.** $4r^2 + 20rs + 25s^2$
15. $64a^2 - 144ab + 81b^2$ **17.** $x^2 + x + \frac{1}{4}$ **19.** $x^2 - 36$ **21.** $m^2 - 144$
23. $x^2 - \frac{1}{4}$ **25.** $p^2 - 0.16$ **27.** $a^2 - 9b^2$ **29.** $16r^2 - s^2$
31. $64w^2 - 25z^2$ **33.** $25x^2 - 81y^2$ **35.** $x^3 - 4x$ **37.** $2s^3 - 18r^2s$
39. $5r^3 + 30r^2 + 45r$ **41.** $x^2 - 36$ **43.** $x^2 - 8x + 16$ **45.** 2499
47. 884 **49.** 3575 **51.** $25x^2 - 40x + 16$
53. **55.** **a.** x **b.** a^2 **c.** $3p$ **d.** $2m^2$ **e.** 4
f. $2x$ **g.** $3r^2s$ **h.** $7c^3d^3$

3.6 Dividing Polynomials

3.6 OBJECTIVES

1. Find the quotient when a polynomial is divided by a monomial
2. Find the quotient of two polynomials

In Section 1.7, we introduced the second property of exponents, which was used to divide one monomial by another monomial. Let's review that process.

Step by Step: To Divide a Monomial by a Monomial

Step 1 Divide the coefficients.

Step 2 Use the second property of exponents to combine the variables.

NOTE The second property says: If x is not zero,

$$\frac{x^m}{x^n} = x^{m-n}$$

Example 1

Dividing Monomials

(a) $\frac{8x^4}{2x^2} = 4x^{4-2}$

Divide: $\frac{8}{2} = 4$

Subtract the exponents.

$= 4x^2$

(b) $\frac{45a^5b^3}{9a^2b} = 5a^3b^2$

CHECK YOURSELF 1

Divide.

(a) $\frac{16a^5}{8a^3}$ (b) $\frac{28m^4n^3}{7m^3n}$

Now let's look at how this can be extended to divide any polynomial by a monomial. For example, to divide $12a^3 + 8a^2$ by $4a$, proceed as follows:

NOTE Technically, this step depends on the distributive property and the definition of division.

$$\frac{12a^3 + 8a^2}{4a} = \frac{12a^3}{4a} + \frac{8a^2}{4a}$$

Divide each term in the numerator by the denominator, $4a$.

Now do each division.

$= 3a^2 + 2a$

The work above leads us to the following rule.

Step by Step: To Divide a Polynomial by a Monomial

1. Divide each term of the polynomial by the monomial.
2. Simplify the results.

Example 2

Dividing by Monomials

Divide each term by 2.

(a) $\frac{4a^2 + 8}{2} = \frac{4a^2}{2} + \frac{8}{2}$

$= 2a^2 + 4$

Divide each term by 6*y*.

(b) $\frac{24y^3 + (-18y^2)}{6y} = \frac{24y^3}{6y} + \frac{-18y^2}{6y}$

$= 4y^2 - 3y$

Remember the rules for signs in division.

(c) $\frac{15x^2 + 10x}{-5x} = \frac{15x^2}{-5x} + \frac{10x}{-5x}$

$= -3x - 2$

NOTE With practice you can write just the quotient.

(d) $\frac{14x^4 + 28x^3 - 21x^2}{7x^2} = \frac{14x^4}{7x^2} + \frac{28x^3}{7x^2} - \frac{21x^2}{7x^2}$

$= 2x^2 + 4x - 3$

(e) $\frac{9a^3b^4 - 6a^2b^3 + 12ab^4}{3ab} = \frac{9a^3b^4}{3ab} - \frac{6a^2b^3}{3ab} + \frac{12ab^4}{3ab}$

$= 3a^2b^3 - 2ab^2 + 4b^3$

CHECK YOURSELF 2

Divide.

(a) $\frac{20y^3 - 15y^2}{5y}$

(b) $\frac{8a^3 - 12a^2 + 4a}{-4a}$

(c) $\frac{16m^4n^3 - 12m^3n^2 + 8mn}{4mn}$

We are now ready to look at dividing one polynomial by another polynomial (with more than one term). The process is very much like long division in arithmetic, as Example 3 illustrates.

Example 3

Dividing by Binomials

Divide $x^2 + 7x + 10$ by $x + 2$.

NOTE The first term in the dividend, x^2, is divided by the first term in the divisor, x.

Step 1

$$\begin{array}{r} x \\ x + 2\overline{)x^2 + 7x + 10} \end{array}$$

Divide x^2 by x to get x.

Step 2

$$\begin{array}{r} x \phantom{{}+7x+10} \\ x + 2\overline{)x^2 + 7x + 10} \\ \underline{x^2 + 2x} \phantom{{}+10} \end{array}$$

Multiply the divisor, $x + 2$, by x.

REMEMBER To subtract $x^2 + 2x$, mentally change each sign to $-x^2 - 2x$, and add. Take your time and be careful here. It's where most errors are made.

Step 3

$$\begin{array}{r} x \phantom{{}+7x+10} \\ x + 2\overline{)x^2 + 7x + 10} \\ \underline{x^2 + 2x} \phantom{{}+10} \\ 5x + 10 \end{array}$$

Subtract and bring down 10.

Step 4

$$\begin{array}{r} x + 5 \\ x + 2\overline{)x^2 + 7x + 10} \\ \underline{x^2 + 2x} \phantom{{}+10} \\ 5x + 10 \end{array}$$

Divide $5x$ by x to get 5.

NOTE Notice that we repeat the process until the degree of the remainder is less than that of the divisor or until there is no remainder.

Step 5

$$\begin{array}{r} x + 5 \\ x + 2\overline{)x^2 + 7x + 10} \\ \underline{x^2 + 2x} \phantom{{}+10} \\ 5x + 10 \\ \underline{5x + 10} \\ 0 \end{array}$$

Multiply $x + 2$ by 5 and then subtract.

The quotient is $x + 5$.

CHECK YOURSELF 3

Divide $x^2 + 9x + 20$ by $x + 4$.

In Example 3, we showed all the steps separately to help you see the process. In practice, the work can be shortened.

Example 4

Dividing by Binomials

Divide $x^2 + x - 12$ by $x - 3$.

NOTE You might want to write out a problem like 408 ÷ 17, to compare the steps.

$$
\begin{array}{r}
x + 4 \\
x - 3\overline{)x^2 + x - 12} \\
\underline{x^2 - 3x} \\
4x - 12 \\
\underline{4x - 12} \\
0
\end{array}
$$

Step 1 Divide x^2 by x to get x, the first term of the quotient.
Step 2 Multiply $x - 3$ by x.
Step 3 Subtract and bring down −12. Remember to mentally change the signs to $-x^2 + 3x$ and add.
Step 4 Divide $4x$ by x to get 4, the second term of the quotient.
Step 5 Multiply $x - 3$ by 4 and subtract.

The quotient is $x + 4$.

CHECK YOURSELF 4

Divide.

$(x^2 + 2x - 24) \div (x - 4)$

You may have a remainder in algebraic long division just as in arithmetic. Consider Example 5.

Example 5

Dividing by Binomials

Divide $4x^2 - 8x + 11$ by $2x - 3$.

$$
\begin{array}{r}
2x - 1 \\
2x - 3\overline{)4x^2 - 8x + 11} \\
\underline{4x^2 - 6x} \\
-2x + 11 \\
\underline{-2x + 3} \\
8
\end{array}
$$

Quotient

Divisor

Remainder

This result can be written as

$$\frac{4x^2 - 8x + 11}{2x - 3}$$

$$= 2x - 1 + \frac{8}{2x - 3}$$

Remainder

Divisor

Quotient

CHECK YOURSELF 5

Divide.

$(6x^2 - 7x + 15) \div (3x - 5)$

The division process shown in our previous examples can be extended to dividends of a higher degree. The steps involved in the division process are exactly the same, as Example 6 illustrates.

Example 6

Dividing by Binomials

Divide $6x^3 + x^2 - 4x - 5$ by $3x - 1$.

$$
\begin{array}{r}
2x^2 + x - 1 \\
3x - 1 \overline{)6x^3 + x^2 - 4x - 5} \\
\underline{6x^3 - 2x^2} \qquad\qquad \\
3x^2 - 4x \qquad \\
\underline{3x^2 - x} \qquad \\
-3x - 5 \\
\underline{-3x + 1} \\
-6
\end{array}
$$

The result can be written as

$$\frac{6x^3 + x^2 - 4x - 5}{3x - 1} = 2x^2 + x - 1 + \frac{-6}{3x - 1}$$

CHECK YOURSELF 6

Divide $4x^3 - 2x^2 + 2x + 15$ *by* $2x + 3$.

Suppose that the dividend is "missing" a term in some power of the variable. You can use 0 as the coefficient for the missing term. Consider Example 7.

Example 7

Dividing by Binomials

Divide $x^3 - 2x^2 + 5$ by $x + 3$.

$$
\begin{array}{r}
x^2 - 5x + 15 \\
x + 3 \overline{)x^3 - 2x^2 + 0x + 5} \\
\underline{x^3 + 3x^2} \qquad\qquad \\
-5x^2 + 0x \qquad \\
\underline{-5x^2 - 15x} \qquad \\
15x + 5 \\
\underline{15x + 45} \\
-40
\end{array}
$$

Write $0x$ for the "missing" term in x.

This result can be written as

$$\frac{x^3 - 2x^2 + 5}{x + 3} = x^2 - 5x + 15 + \frac{-40}{x + 3}$$

CHECK YOURSELF 7

Divide.

$(4x^3 + x + 10) \div (2x - 1)$

You should always arrange the terms of the divisor and dividend in descending-exponent form before starting the long division process, as illustrated in Example 8.

Example 8

Dividing by Binomials

Divide $5x^2 - x + x^3 - 5$ by $-1 + x^2$.

Write the divisor as $x^2 - 1$ and the dividend as $x^3 + 5x^2 - x - 5$.

$$\begin{array}{r} x + 5 \\ x^2 - 1 \overline{) x^3 + 5x^2 - x - 5} \\ \underline{x^3 - x } \\ 5x^2 - 5 \\ \underline{5x^2 - 5} \\ 0 \end{array}$$

Write $x^3 - x$, the product of x and $x^2 - 1$, so that like terms fall in the same columns.

CHECK YOURSELF 8

Divide:

$(5x^2 + 10 + 2x^3 + 4x) \div (2 + x^2)$

CHECK YOURSELF ANSWERS

1. **(a)** $2a^2$; **(b)** $4mn^2$ **2.** **(a)** $4y^2 - 3y$; **(b)** $-2a^2 + 3a - 1$; **(c)** $4m^3n^2 - 3m^2n + 2$

3. $x + 5$ **4.** $x + 6$ **5.** $2x + 1 + \dfrac{20}{3x - 5}$ **6.** $2x^2 - 4x + 7 + \dfrac{-6}{2x + 3}$

7. $2x^2 + x + 1 + \dfrac{11}{2x - 1}$ **8.** $2x + 5$

Name ______________________

Section __________ Date __________

3.6 Exercises

Divide.

1. $\frac{18x^6}{9x^2}$

2. $\frac{20a^7}{5a^5}$

3. $\frac{35m^3n^2}{7mn^2}$

4. $\frac{42x^5y^2}{6x^3y}$

5. $\frac{3a + 6}{3}$

6. $\frac{4x - 8}{4}$

7. $\frac{9b^2 - 12}{3}$

8. $\frac{10m^2 + 5m}{5}$

9. $\frac{16a^3 - 24a^2}{4a}$

10. $\frac{9x^3 + 12x^2}{3x}$

11. $\frac{12m^2 + 6m}{-3m}$

12. $\frac{20b^3 - 25b^2}{-5b}$

13. $\frac{18a^4 + 12a^3 - 6a^2}{6a}$

14. $\frac{21x^5 - 28x^4 + 14x^3}{7x}$

15. $\frac{20x^4y^2 - 15x^2y^3 + 10x^3y}{5x^2y}$

16. $\frac{16m^3n^3 + 24m^2n^2 - 40mn^3}{8mn^2}$

Perform the indicated divisions.

17. $\frac{x^2 + 5x + 6}{x + 2}$

18. $\frac{x^2 + 8x + 15}{x + 3}$

19. $\frac{x^2 - x - 20}{x + 4}$

20. $\frac{x^2 - 2x - 35}{x + 5}$

21. $\frac{2x^2 + 5x - 3}{2x - 1}$

22. $\frac{3x^2 + 20x - 32}{3x - 4}$

23. $\frac{2x^2 - 3x - 5}{x - 3}$

24. $\frac{3x^2 + 17x - 12}{x + 6}$

ANSWERS

1. $2x^4$
2. $4a^2$
3. $5m^2$
4. $7x^2y$
5. $a + 2$
6. $x - 2$
7. $3b^2 - 4$
8. $2m^2 + m$
9. $4a^2 - 6a$
10. $3x^2 + 4x$
11. $-4m - 2$
12. $-4b^2 + 5b$
13. $3a^3 + 2a^2 - a$
14. $3x^4 - 4x^3 + 2x^2$
15. $4x^2y - 3y^2 + 2x$
16. $2m^2n + 3m - 5n$
17. $x + 3$
18. $x + 5$
19. $x - 5$
20. $x - 7$
21. $x + 3$
22. $x + 8$
23. $2x + 3 + \frac{4}{x - 3}$
24. $3x - 1 + \frac{-6}{x + 6}$

ANSWERS

25. $4x + 2 + \frac{-5}{x - 5}$

26. $3x + 6 + \frac{16}{x - 8}$

27. $2x + 3 + \frac{5}{3x - 5}$

28. $2x - 4 + \frac{3}{2x + 7}$

29. $x^2 - x - 2$

30. $x^2 + x + 7$

31. $x^2 + 2x + 3 + \frac{8}{4x - 1}$

32. $x^2 - 2x + 3 + \frac{1}{2x + 1}$

33. $x^2 + x + 2 + \frac{9}{x - 2}$

34. $x^2 - 3x + 13 + \frac{-42}{x + 3}$

35. $5x^2 + 2x + 1 + \frac{2}{5x - 2}$

36. $2x^2 - 2x + 1 + \frac{-1}{4x + 1}$

37. $x^2 + 4x + 5 + \frac{2}{x - 2}$

38. $2x^2 - 7x + 10 + \frac{-8}{x + 4}$

39. $x^3 + x^2 + x + 1$

40. $x^3 - 2x^2 + 5x - 10 + \frac{4}{x + 2}$

41. $x - 3$

42. $x + 2$

43. $x^2 - 1 + \frac{1}{x^2 + 3}$

44. $x^2 + 3 + \frac{1}{x^2 - 2}$

45. $y^2 - y + 1$

46. $y^2 + 2y + 4$

47. $x^2 + 1$

48. $x^3 + 1$

25. $\frac{4x^2 - 18x - 15}{x - 5}$

26. $\frac{3x^2 - 18x - 32}{x - 8}$

27. $\frac{6x^2 - x - 10}{3x - 5}$

28. $\frac{4x^2 + 6x - 25}{2x + 7}$

29. $\frac{x^3 + x^2 - 4x - 4}{x + 2}$

30. $\frac{x^3 - 2x^2 + 4x - 21}{x - 3}$

31. $\frac{4x^3 + 7x^2 + 10x + 5}{4x - 1}$

32. $\frac{2x^3 - 3x^2 + 4x + 4}{2x + 1}$

33. $\frac{x^3 - x^2 + 5}{x - 2}$

34. $\frac{x^3 + 4x - 3}{x + 3}$

35. $\frac{25x^3 + x}{5x - 2}$

36. $\frac{8x^3 - 6x^2 + 2x}{4x + 1}$

37. $\frac{2x^2 - 8 - 3x + x^3}{x - 2}$

38. $\frac{x^2 - 18x + 2x^3 + 32}{x + 4}$

39. $\frac{x^4 - 1}{x - 1}$

40. $\frac{x^4 + x^2 - 16}{x + 2}$

41. $\frac{x^3 - 3x^2 - x + 3}{x^2 - 1}$

42. $\frac{x^3 + 2x^2 + 3x + 6}{x^2 + 3}$

43. $\frac{x^4 + 2x^2 - 2}{x^2 + 3}$

44. $\frac{x^4 + x^2 - 5}{x^2 - 2}$

45. $\frac{y^3 + 1}{y + 1}$

46. $\frac{y^3 - 8}{y - 2}$

47. $\frac{x^4 - 1}{x^2 - 1}$

48. $\frac{x^6 - 1}{x^3 - 1}$

49. Find the value of c so that $\frac{y^2 - y + c}{y + 1} = y - 2$

50. Find the value of c so that $\frac{x^3 + x^2 + x + c}{x^2 + 1} = x + 1$

51. Write a summary of your work with polynomials. Explain how a polynomial is recognized, and explain the rules for the arithmetic of polynomials—how to add, subtract, multiply, and divide. What parts of this chapter do you feel you understand very well, and what part(s) do you still have questions about, or feel unsure of? Exchange papers with another student and compare your questions.

52. A funny (and useful) thing about division of polynomials: To find out about this funny thing, do this division. Compare your answer with another student's.

$(x - 2)\overline{)2x^2 + 3x - 5}$ Is there a remainder?

Now, evaluate the polynomial $2x^2 + 3x - 5$ when $x = 2$. Is this value the same as the remainder?

Try $(x + 3)\overline{)5x^2 - 2x + 1}$ Is there a remainder?

Evaluate the polynomial $5x^2 - 2x + 1$ when $x = -3$. Is this value the same as the remainder?

What happens when there is no remainder?

Try $(x - 6)\overline{)3x^3 + 14x^2 - 23x + 6}$ Is the remainder zero?

Evaluate the polynomial $3x^3 + 14x - 23x + 6$ when $x = 6$. Is this value zero? Write a description of the patterns you see. When does the pattern hold? Make up several more examples, and test your conjecture.

53. (a) Divide $\frac{x^2 - 1}{x - 1}$ (b) Divide $\frac{x^3 - 1}{x - 1}$ (c) Divide $\frac{x^4 - 1}{x - 1}$

(d) Based on your results to (a), (b), and (c), predict $\frac{x^{50} - 1}{x - 1}$

54. (a) Divide $\frac{x^2 + x + 1}{x - 1}$ (b) Divide $\frac{x^3 + x^2 + x + 1}{x - 1}$

(c) Divide $\frac{x^4 + x^3 + x^2 + x + 1}{x - 1}$

(d) Based on your results to (a), (b), and (c), predict $\frac{x^{10} + x^9 + x^8 + \cdots + x + 1}{x - 1}$

ANSWERS

49. $c = -2$

50. $c = 1$

51.

52.

53. (a) $x + 1$

(b) $x^2 + x + 1$

(c) $x^3 + x^2 + x + 1$

(d) $x^{49} + x^{48} + \cdots + x + 1$

54. (a) $x + 2 + \frac{3}{x - 1}$

(b) $x^2 + 2x + 3 + \frac{4}{x - 1}$

(c) $x^3 + 2x^2 + 3x + 4 + \frac{5}{x - 1}$

(d) $x^9 + 2x^8 + 3x^7 + \cdots + 9x + 10 + \frac{11}{x - 1}$

Answers

1. $2x^4$ **3.** $5m^2$ **5.** $a + 2$ **7.** $3b^2 - 4$ **9.** $4a^2 - 6a$ **11.** $-4m - 2$
13. $3a^3 + 2a^2 - a$ **15.** $4x^2y - 3y^2 + 2x$ **17.** $x + 3$ **19.** $x - 5$
21. $x + 3$ **23.** $2x + 3 + \frac{4}{x - 3}$ **25.** $4x + 2 + \frac{-5}{x - 5}$
27. $2x + 3 + \frac{5}{3x - 5}$ **29.** $x^2 - x - 2$ **31.** $x^2 + 2x + 3 + \frac{8}{4x - 1}$
33. $x^2 + x + 2 + \frac{9}{x - 2}$ **35.** $5x^2 + 2x + 1 + \frac{2}{5x - 2}$
37. $x^2 + 4x + 5 + \frac{2}{x - 2}$ **39.** $x^3 + x^2 + x + 1$ **41.** $x - 3$
43. $x^2 - 1 + \frac{1}{x^2 + 3}$ **45.** $y^2 - y + 1$ **47.** $x^2 + 1$ **49.** $c = -2$
51. **53.** **(a)** $x + 1$; **(b)** $x^2 + x + 1$; **(c)** $x^3 + x^2 + x + 1$;
(d) $x^{49} + x^{48} + \cdots + x + 1$

3 Summary

DEFINITION/PROCEDURE	EXAMPLE	REFERENCE
Exponents and Polynomials		**Section 3.1**
Properties of Exponents		
1. $a^m \cdot a^n = a^{m+n}$	$3^3 \cdot 3^4 = 3^7$	**p. 247**
2. $\frac{a^m}{a^n} = a^{m-n}$	$\frac{10^6}{10^2} = 10^4$	**p. 248**
3. $(a^m)^n = a^{mn}$	$(2^3)^5 = 2^{15}$	**p. 248**
4. $(ab)^m = a^m b^m$	$(3x)^2 = 9x^2$	**p. 249**
5. $\left(\frac{a}{b}\right)^m = \frac{a^m}{b^m}$	$\left(\frac{2}{3}\right)^3 = \frac{8}{27}$	**p. 250**
Term A number, or the product of a number and variables. *Polynomial* An algebraic expression made up of terms in which the exponents are whole numbers. These terms are connected by plus or minus signs. Each sign (+ or −) is attached to the term following that sign.	$4x^3 - 3x^2 + 5x$ is a polynomial. The terms of $4x^3 - 3x^2 + 5x$ are $4x^3$, $-3x^2$, and $5x$.	**p. 251**
Coefficient In each term of a polynomial, the number is called the *numerical coefficient* or, more simply, the *coefficient* of that term.	The coefficients of $4x^3 - 3x^2$ are 4 and −3.	**p. 251**
Types of Polynomials A polynomial can be classified according to the number of terms it has. A *mono*mial has one term. A *bi*nomial has two terms. A *tri*nomial has three terms.	$2x^3$ is a monomial. $3x^2 - 7x$ is a binomial. $5x^5 - 5x^3 + 2$ is a trinomial.	**p. 252**
Degree The highest power of the variable appearing in any one term.	The degree of $4x^5 - 5x^3 + 3x$ is 5.	**p. 252**
Descending-Exponent Form The form of a polynomial when it is written with the highest-degree term first, the next highest-degree term second, and so on.	$4x^5 - 5x^3 + 3x$ is written in descending-exponent form.	**p. 253**
Negative Exponents and Scientific Notation		**Section 3.2**
The Zero Power Any expression taken to the power zero equals one.	$3^0 = 1$ $(5x)^0 = 1$	**p. 261**
Negative Powers An expression taken to a negative power equals its reciprocal taken to the absolute value of its power.	$\left(\frac{x}{3}\right)^{-4} = \left(\frac{3}{x}\right)^4 = \frac{3^4}{x^4}$	**p. 262**

Continued

DEFINITION/PROCEDURE	EXAMPLE	REFERENCE
Negative Exponents and Scientific Notation		**Section 3.2**
Scientific Notation Any number written in the form $a \times 10^n$ in which $1 \le a < 10$ and n is an integer, is written in scientific notation.	6.2×10^{23}	p. 265
Adding and Subtracting Polynomials		**Section 3.3**
Removing Signs of Grouping 1. If a plus sign (+) or no sign at all appears in front of parentheses, just remove the parentheses. No other changes are necessary.	$3x + (2x - 3)$ $= 3x + 2x - 3$	p. 271
2. If a minus sign (−) appears in front of parentheses, the parentheses can be removed by changing the sign of each term inside the parentheses.	$2x - (x - 4)$ $= 2x - x + 4$	p. 273
Adding Polynomials Remove the signs of grouping. Then collect and combine any like terms.	$(2x + 3) + (3x - 5)$ $= 2x + 3 + 3x - 5 = 5x - 2$	p. 272
Subtracting Polynomials Remove the signs of grouping by changing the sign of each term in the polynomial being subtracted. Then combine any like terms.	$(3x^2 + 2x) - (2x^2 + 3x - 1)$ $= 3x^2 + 2x - 2x^2 - 3x + 1$ Sign changes $3x^2 - 2x^2 + 2x - 3x + 1$ $x^2 - x + 1$	p. 273
Multiplying Polynomials		**Section 3.4**
To Multiply a Polynomial by a Monomial Multiply each term of the polynomial by the monomial, and add the results.	$3x(2x + 3)$ $= 3x \cdot 2x + 3x \cdot 3$ $= 6x^2 + 9x$	p. 281
To Multiply a Binomial by a Binomial Use the FOIL method: F O I L $(a + b)(c + d) = a \cdot c + a \cdot d + b \cdot c + b \cdot d$	$(2x - 3)(3x + 5)$ $= 6x^2 + 10x - 9x - 15$ F O I L $= 6x^2 + x - 15$	p. 283
To Multiply a Polynomial by a Polynomial Arrange the polynomials vertically. Multiply each term of the upper polynomial by each term of the lower polynomial, and add the results.	$\begin{array}{r} x^2 - 3x + 5 \\ 2x - 3 \\ \hline -3x^2 + 9x - 15 \\ 2x^3 - 6x^2 + 10x \qquad \\ \hline 2x^3 - 9x^2 + 19x - 15 \end{array}$	p. 285
Special Products		**Section 3.5**
The Square of a Binomial $(a + b)^2 = a^2 + 2ab + b^2$	$(2x - 5)^2$ $= 4x^2 + 2 \cdot 2x \cdot (-5) + 25$ $= 4x^2 - 20x + 25$	p. 293
The Product of Binomials That Differ Only in Sign Subtract the square of the second term from the square of the first term. $(a + b)(a - b) = a^2 - b^2$	$(2x - 5y)(2x + 5y)$ $= (2x)^2 - (5y)^2$ $= 4x^2 - 25y^2$	p. 294

Summary Exercises

This summary exercise set is provided to give you practice with each of the objectives of the chapter. Each exercise is keyed to the appropriate chapter section. The answers are provided in the *Instructor's Manual.* Your instructor will give you guidelines on how to best use these exercises in your instructional setting.

[3.1] Simplify each of the following expressions.

1. $\frac{x^{10}}{x^3}$ x^7

2. $\frac{a^5}{a^4}$ a

3. $\frac{x^2 \cdot x^3}{x^4}$ x

4. $\frac{m^2 \cdot m^3 \cdot m^4}{m^5}$ m^4

5. $\frac{18p^7}{9p^5}$ $2p^2$

6. $\frac{24x^{17}}{8x^{13}}$ $3x^4$

7. $\frac{30m^7n^5}{6m^2n^3}$ $5m^5n^2$

8. $\frac{108x^9y^4}{9xy^4}$ $12x^8$

9. $\frac{48p^5q^3}{6p^3q}$ $8p^2q^2$

10. $\frac{52a^5b^3c^5}{13a^4c}$ $4ab^3c^4$

11. $(2ab)^2$ $4a^2b^2$

12. $(p^2q^3)^3$ p^6q^9

13. $(2x^2y^2)^3(3x^3y)^2$ $72x^{12}y^8$

14. $\left(\frac{p^2q^3}{t^4}\right)^2$ $\frac{p^4q^6}{t^8}$

15. $\frac{(x^5)^2}{(x^3)^3}$ x

16. $(4w^2t)^2(3wt^2)^3$ $432\,w^7t^8$

17. $(y^3)^2(3y^2)^3$ $27y^{12}$

18. $\left(\frac{4x^4}{3y}\right)^2$ $\frac{16x^8}{9y^2}$

[3.1] Find the value of each of the following polynomials for the given value of the variable.

19. $5x + 1; x = -1$ -4

20. $2x^2 + 7x - 5; x = 2$ 17

21. $-x^2 + 3x - 1; x = 6$ -19

22. $4x^2 + 5x + 7; x = -4$ 51

Classify each of the following polynomials as a monomial, binomial, or trinomial, where possible.

23. $5x^3 - 2x^2$ **Binomial**

24. $7x^5$ **Monomial**

25. $4x^5 - 8x^3 + 5$ **Trinomial**

26. $x^3 + 2x^2 - 5x + 3$ **Not classified**

27. $9a^2 - 18a^2$ **Binomial**

[3.1] Arrange in descending-exponent form, if necessary, and give the degree of each polynomial.

28. $5x^5 + 3x^2$ $5x^5 + 3x^2, 5$

29. $9x$ $9x, 1$

30. $6x^2 + 4x^4 + 6$ $4x^4 + 6x^2 + 6, 4$

31. $5 + x$ $x + 5, 1$

32. -8 $-8, 0$

33. $9x^4 - 3x + 7x^6$ $7x^6 + 9x^4 - 3x, 6$

2

[3.2] Evaluate each of the following expressions.

34. 4^0 1

35. $(3a)^0$ 1

36. $6x^0$ 6

37. $(3a^4b)^0$ 1

Write, using positive exponents.

38. x^{-5} $\frac{1}{x^5}$

39. 3^{-3} $\frac{1}{3^3}$

40. 10^{-4} $\frac{1}{10^4}$

41. $4x^{-4}$ $\frac{4}{x^4}$

42. $\frac{x^6}{x^8}$ $\frac{1}{x^2}$

43. m^7m^{-9} $\frac{1}{m^2}$

44. $\frac{a^{-4}}{a^{-9}}$ a^5

45. $\frac{x^2y^{-3}}{x^{-3}y^2}$ $\frac{x^5}{y^5}$

46. $(3m^{-3})^2$ $\frac{9}{m^6}$

47. $\frac{(a^4)^{-3}}{(a^{-2})^{-3}}$ $\frac{1}{a^{18}}$

[3.2] In Exercises 48 to 50, express each number in scientific notation.

48. The average distance from the Earth to the sun is 150,000,000,000 meters. 1.5×10^{11} m

49. A bat emits a sound with a frequency of 51,000 cycles per second. 5.1×10^4 cps

50. The diameter of a grain of salt is 0.000062 meters. 6.2×10^{-5} m

[3.2] In Exercises 51 to 54, compute the expression using scientific notation and express your answers in that form.

51. $(2.3 \times 10^{-3})(1.4 \times 10^{12})$ 3.22×10^9

52. $(4.8 \times 10^{-10})(6.5 \times 10^{34})$ 3.12×10^{25}

53. $\frac{(8 \times 10^{23})}{(4 \times 10^6)}$ 2×10^{17}

54. $\frac{(5.4 \times 10^{-12})}{(4.5 \times 10^{16})}$ 1.2×10^{-28}

[3.3] Add.

55. $9a^2 - 5a$ and $12a^2 + 3a$ $21a^2 - 2a$

56. $5x^2 + 3x - 5$ and $4x^2 - 6x - 2$ $9x^2 - 3x - 7$

57. $5y^3 - 3y^2$ and $4y + 3y^2$ $5y^3 + 4y$

[3.3] Subtract.

58. $4x^2 - 3x$ from $8x^2 + 5x$ $4x^2 + 8x$

59. $2x^2 - 5x - 7$ from $7x^2 - 2x + 3$ $5x^2 + 3x + 10$

60. $5x^2 + 3$ from $9x^2 - 4x$ $4x^2 - 4x - 3$

© 2001 McGraw-Hill Companies

[3.3] Perform the indicated operations.

61. Subtract $5x - 3$ from the sum of $9x + 2$ and $-3x - 7$. $x - 2$

62. Subtract $5a^2 - 3a$ from the sum of $5a^2 + 2$ and $7a - 7$. $10a - 5$

63. Subtract the sum of $16w^2 - 3w$ and $8w + 2$ from $7w^2 - 5w + 2$. $-9w^2 - 10w$

[3.3] Add, using the vertical method.

64. $x^2 + 5x - 3$ and $2x^2 + 4x - 3$ $3x^2 + 9x - 6$

65. $9b^2 - 7$ and $8b + 5$ $9b^2 + 8b - 2$

66. $x^2 + 7$, $3x - 2$, and $4x^2 - 8x$ $5x^2 - 5x + 5$

[3.3] Subtract, using the vertical method.

67. $5x^2 - 3x + 2$ from $7x^2 - 5x - 7$ $2x^2 - 2x - 9$

68. $8m - 7$ from $9m^2 - 7$ $9m^2 - 8m$

[3.4] Multiply.

69. $(5a^3)(a^2)$ $5a^5$

70. $(2x^2)(3x^5)$ $6x^7$

71. $(-9p^3)(-6p^2)$ $54p^5$

72. $(3a^2b^3)(-7a^3b^4)$ $-21a^5b^7$

73. $5(3x - 8)$ $15x - 40$

74. $4a(3a + 7)$ $12a^2 + 28a$

75. $(-5rs)(2r^2s - 5rs)$ $-10r^3s^2 + 25r^2s^2$

76. $7mn(3m^2n - 2mn^2 + 5mn)$ $21m^3n^2 - 14m^2n^3 + 35m^2n^2$

77. $(x + 5)(x + 4)$ $x^2 + 9x + 20$

78. $(w - 9)(w - 10)$ $w^2 - 19w + 90$

79. $(a - 7b)(a + 7b)$ $a^2 - 49b^2$

80. $(p - 3q)^2$ $p^2 - 6pq + 9q^2$

81. $(a + 4b)(a + 3b)$ $a^2 + 7ab + 12b^2$

82. $(b - 8)(2b + 3)$ $2b^2 - 13b - 24$

83. $(3x - 5y)(2x - 3y)$ $6x^2 - 19xy + 15y^2$

84. $(5r + 7s)(3r - 9s)$ $15r^2 - 24rs - 63s^2$

85. $(y + 2)(y^2 - 2y + 3)$ $y^3 - y + 6$

86. $(b + 3)(b^2 - 5b - 7)$ $b^3 - 2b^2 - 22b - 21$

87. $(x - 2)(x^2 + 2x + 4)$ $x^3 - 8$

88. $(m^2 - 3)(m^2 + 7)$ $m^4 + 4m^2 - 21$

89. $2x(x + 5)(x - 6)$ $2x^3 - 2x^2 - 60x$

90. $a(2a - 5b)(2a - 7b)$ $4a^3 - 24a^2b + 35ab^2$

[3.5] Find the following products.

91. $(x + 7)^2$ $x^2 + 14x + 49$

92. $(a - 8)^2$ $a^2 - 16a + 64$

93. $(2w - 5)^2$ $4w^2 - 20w + 25$

94. $(3p + 4)^2$ $9p^2 + 24p + 16$

95. $(a + 7b)^2$ $a^2 + 14ab + 49b^2$

96. $(8x - 3y)^2$ $64x^2 - 48xy + 9y^2$

97. $(x - 5)(x + 5)$ $x^2 - 25$

98. $(y + 9)(y - 9)$ $y^2 - 81$

99. $(2m + 3)(2m - 3)$ $4m^2 - 9$

100. $(3r - 7)(3r + 7)$ $9r^2 - 49$

101. $(5r - 2s)(5r + 2s)$ $25r^2 - 4s^2$

102. $(7a + 3b)(7a - 3b)$ $49a^2 - 9b^2$

103. $2x(x - 5)^2$ $2x^3 - 20x^2 + 50x$

104. $3c(c + 5d)(c - 5d)$ $3c^3 - 75cd^2$

[3.6] Divide.

105. $\frac{9a^5}{3a^2}$ $3a^3$

106. $\frac{24m^4n^2}{6m^2n}$ $4m^2n$

107. $\frac{15a - 10}{5}$ $3a - 2$

108. $\frac{32a^3 + 24a}{8a}$ $4a^2 + 3$

109. $\frac{9r^2s^3 - 18r^3s^2}{-3rs^2}$ $-3rs + 6r^2$

110. $\frac{35x^3y^2 - 21x^2y^3 + 14x^3y}{7x^2y}$ $5xy - 3y^2 + 2x$

[3.6] Perform the indicated long division.

111. $\frac{x^2 - 2x - 15}{x + 3}$ $x - 5$

112. $\frac{2x^2 + 9x - 35}{2x - 5}$ $x + 7$

113. $\frac{x^2 - 8x + 17}{x - 5}$ $x - 3 + \frac{2}{x - 5}$

114. $\frac{6x^2 - x - 10}{3x + 4}$ $2x - 3 + \frac{2}{3x + 4}$

115. $\frac{6x^3 + 14x^2 - 2x - 6}{6x + 2}$ $x^2 + 2x - 1 + \frac{-4}{6x + 2}$

116. $\frac{4x^3 + x + 3}{2x - 1}$ $2x^2 + x + 1 + \frac{4}{2x - 1}$

117. $\frac{3x^2 + x^3 + 5 + 4x}{x + 2}$ $x^2 + x + 2 + \frac{1}{x + 2}$

118. $\frac{2x^4 - 2x^2 - 10}{x^2 - 3}$ $2x^2 + 4 + \frac{2}{x^2 - 3}$

4

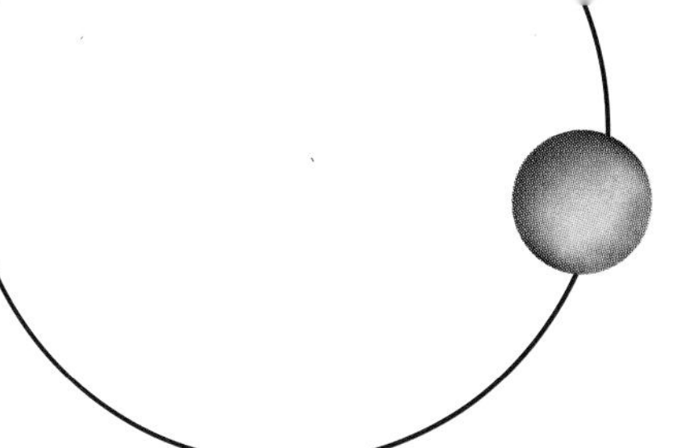

Self-Test for Chapter 3

Name ______________

Section ________ Date ________

The purpose of this self-test is to help you check your progress and to review for a chapter test in class. Allow yourself about an hour to take the test. When you are done, check your answers in the back of the book. If you missed any answers, be sure to go back and review the appropriate sections in the chapter and do the exercises that are provided.

Simplify each of the following expressions.

1. $a^5 \cdot a^9$

2. $3x^2y^3 \cdot 5xy^4$

3. $\dfrac{4x^5}{2x^2}$

4. $\dfrac{20a^3b^5}{5a^2b^2}$

5. $(3x^2y)^3$

6. $\left(\dfrac{2w^2}{3t^3}\right)^2$

7. $(2x^3y^2)^4(x^2y^3)^3$

8. Find the value of the polynomial $y = -3x^2 - 5x + 8$ if $x = -2$

Classify each of the following polynomials as a monomial, binomial, or trinomial.

9. $6x^2 + 7x$

10. $5x^2 + 8x - 8$

Arrange in descending-exponent form, and give the coefficients and degree of the polynomial.

11. $-3x^2 + 8x^4 - 7$

Evaluate (assume the variables are nonzero).

12. 8^0

13. $6x^0$

Rewrite, using positive exponents.

14. y^{-5}

15. $3b^{-7}$

16. y^4y^{-8}

17. $\dfrac{p^{-5}}{p^5}$

Add.

18. $3x^2 - 7x + 2$ and $7x^2 - 5x - 9$

19. $7a^2 - 3a$ and $7a^3 + 4a^2$

Subtract.

20. $5x^2 - 2x + 5$ from $8x^2 + 9x - 7$

21. $2b^2 + 5$ from $3b^2 - 7b$

22. $5a^2 + a$ from the sum of $3a^2 - 5a$ and $9a^2 - 4a$

ANSWERS

1. a^{14}
2. $15x^3y^7$
3. $2x^3$
4. $4ab^3$
5. $27x^6y^3$
6. $\dfrac{4w^4}{9t^6}$
7. $16x^{18}y^{17}$
8. 6
9. Binomial
10. Trinomial
11. $8x^4 - 3x^2 - 7$; 8, − 3, −7; 4
12. 1
13. 6
14. $\dfrac{1}{y^5}$
15. $\dfrac{3}{b^7}$
16. $\dfrac{1}{y^4}$
17. $\dfrac{1}{p^{10}}$
18. $10x^2 - 12x - 7$
19. $7a^3 + 11a^2 - 3a$
20. $3x^2 + 11x - 12$
21. $b^2 - 7b - 5$
22. $7a^2 - 10a$

ANSWERS

23. $4x^2 + 5x - 6$

24. $2x^2 - 7x + 5$

25. $15a^3b^2 - 10a^2b^2 + 20a^2b^3$

26. $3x^2 + x - 14$

27. $a^2 - 49b^2$

28. $8x^2 - 14xy - 15y^2$

29. $12x^3 + 11x^2y - 5xy^2$

30. $9m^2 + 12mn + 4n^2$

31. $2x^3 + 7x^2y - xy^2 - 2y^3$

32. $2x^2 - 3y$

33. $4c^2 - 6 + 9cd$

34. $x - 6$

35. $x + 2 + \dfrac{10}{2x - 3}$

36. $2x^2 - 3x + 2 + \dfrac{7}{3x + 1}$

37. $x^2 - 4x + 5 - \dfrac{4}{x - 1}$

38. 1.68×10^{20}

39. 3.12×10^{-10}

40. 5.2×10^{19}

Add, using the vertical method.

23. $x^2 + 3$, $5x - 7$, and $3x^2 - 2$

Subtract, using the vertical method.

24. $3x^2 - 5$ from $5x^2 - 7x$

Multiply.

25. $5ab(3a^2b - 2ab + 4ab^2)$

26. $(x - 2)(3x + 7)$

27. $(a - 7b)(a + 7b)$

28. $(4x + 3y)(2x - 5y)$

29. $x(3x - y)(4x + 5y)$

30. $(3m + 2n)^2$

31. $(2x + y)(x^2 + 3xy - 2y^2)$

Divide.

32. $\dfrac{14x^3y - 21xy^2}{7xy}$

33. $\dfrac{20c^3d - 30cd + 45c^2d^2}{5cd}$

34. $(x^2 - 2x - 24) \div (x + 4)$

35. $(2x^2 + x + 4) \div (2x - 3)$

36. $(6x^3 - 7x^2 + 3x + 9) \div (3x + 1)$

37. $(x^3 - 5x^2 + 9x - 9) \div (x - 1)$

Compute and answer using scientific notation.

38. $(2.1 \times 10^7)(8 \times 10^{12})$

39. $(6 \times 10^{-23})(5.2 \times 10^{12})$

40. $\dfrac{7.28 \times 10^3}{1.4 \times 10^{-16}}$

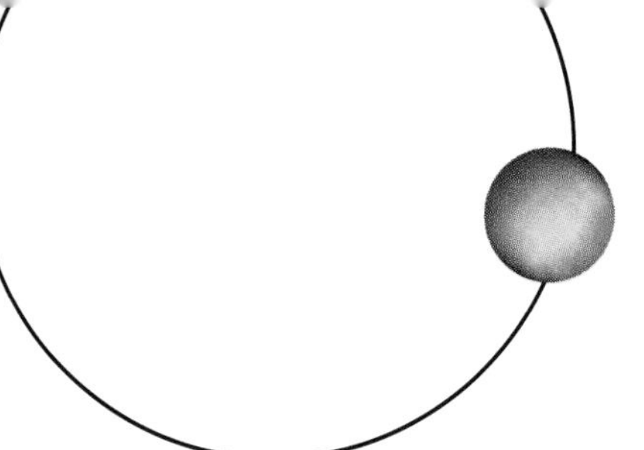

Cumulative Test Chapters 1 to 3

Name ____________

Section ________ Date ________

Perform the indicated operations.

1. $8 - (-9)$ **2.** $-26 + 32$ **3.** $(-25)(-6)$ **4.** $(-48) \div (-12)$

Evaluate the expressions if $x = -2$, $y = 5$, and $z = -2$

5. $-5(-3y - 2z)$ **6.** $\dfrac{3x - 4y}{2z + 5y}$

Use the properties of exponents to simplify each of the following expressions.

7. $(3x^2)^2 (x^3)^4$ **8.** $\left(\dfrac{x^5}{y^3}\right)^2$ **9.** $(2x^3y)^3$

10. $7y^0$ **11.** $(3x^4y^5)^0$

Simplify each expression using positive exponents only.

12. x^{-4} **13.** $3x^{-2}$ **14.** x^5x^{-9} **15.** $\dfrac{x^{-3}}{y^3}$

Simplify each of the following expressions.

16. $21x^5y - 17x^5y$ **17.** $(3x^2 + 4x - 5) - (2x^2 - 3x - 5)$

18. $(-4x^3 + 5x^2 - 7) - (-8x^3 + 2x^2 - 9)$ **19.** $(x + 3)(x - 5)$

20. $(x + y)^2$ **21.** $(3x - 4y)^2$

22. $\dfrac{x^2 + 2x - 8}{x - 2}$ **23.** $x(x + y)(x - y)$

ANSWERS

1. 17
2. 6
3. 150
4. 4
5. 55
6. $-\dfrac{26}{21}$
7. $9x^{16}$
8. $\dfrac{x^{10}}{y^6}$
9. $8x^9y^3$
10. 7
11. 1
12. $\dfrac{1}{x^4}$
13. $\dfrac{3}{x^2}$
14. $\dfrac{1}{x^4}$
15. $\dfrac{1}{x^3y^3}$
16. $4x^5y$
17. $x^2 + 7x$
18. $4x^3 + 3x^2 + 2$
19. $x^2 - 2x - 15$
20. $x^2 + 2xy + y^2$
21. $9x^2 - 24xy + 16y^2$
22. $x + 4$
23. $x^3 - xy^2$

ANSWERS

24. −2
25. −2
26. 84
27. 1
28. 2A − b
29. $x \geq -2$
30. $x < -22$
31. Sam: $510
Larry: $250
32. 37, 39
33. $2120
34. $645

Solve the following equations.

24. $7x - 4 = 3x - 12$

25. $4(2 - x) - 3x = 2(x + 13)$

26. $\frac{3}{4}x - 2 = 5 + \frac{2}{3}x$

27. $6(x - 1) - 3(1 - x) = 0$

28. Solve the equation $A = \frac{1}{2}(b + B)$ for B.

Solve the following inequalities.

29. $-5x - 7 \leq 3x + 9$

30. $-3(x + 5) > -2x + 7$

Solve the following problems.

31. Sam made $10 more than twice what Larry earned in one month. If together they earned $760, how much did each earn that month?

32. The sum of two consecutive odd integers is 76. Find the two integers.

33. Two-fifths of a woman's income each month goes to taxes. If she pays $848 in taxes each month, what is her monthly income?

34. The retail selling price of a sofa is $806.25. What is the cost to the dealer if she sells at 25% markup on the cost?

FACTORING

4

INTRODUCTION

Developing secret codes is big business because of the widespread use of computers and the Internet. Corporations all over the world sell encryption systems that are supposed to keep data secure and safe.

In 1977, three professors from the Massachusetts Institute of Technology developed an encryption system they call RSA, a name derived from the first letters of their last names. They offered a $100 reward to anyone who could break their security code, which was based on a number that has *129 digits.* They called the code RSA-129. For the code to be broken, the 129-digit number must be factored into two prime numbers; that is, two prime numbers must be found that when multiplied together give the 129-digit number. The three professors predicted that it would take *40 quadrillion* years to find the two numbers.

In April 1994, a research scientist, three computer hobbyists, and more than 600 volunteers from the Internet, using 1600 computers, found the two numbers after 8 months of work and won the $100.

A data security company says that people who are using their system are safe because as yet no truly efficient algorithm for finding prime factors of massive numbers has been found, although one may someday exist. This company, hoping to test its encrypting system, now sponsors contests challenging people to factor more very large numbers into two prime numbers. RSA-150 up to RSA-500 are being worked on now.

Software companies are waging a legal battle against the U.S. government because the government does not allow any codes to be used for which it does not have the key. The software firms claim that this prohibition is costing them about $60 billion in lost sales because many companies will not buy an encryption system knowing they can be monitored by the U.S. government.

Name ______________________

Section ________ Date ________

Pre-Test Chapter 4

ANSWERS

1. $5(3c + 7)$
2. $4q^3(2q - 5)$
3. $6(x^2 - 2x + 4)$
4. $7cd(c^2d - 3 + 2d^2)$
5. $(b - 3)(b + 5)$
6. $(x + 4)(x + 6)$
7. $(x - 9)(x - 5)$
8. $(a + 3b)(a + 4b)$
9. $(3y - 4)(y + 3)$
10. $(5w + 3)(w + 4)$
11. $(3x + 7y)(2x - 3y)$
12. $x(2x + 3)(x - 5)$
13. $(b + 7)(b - 7)$
14. $(6p + q)(6p - q)$
15. $(3x - 2y)^2$
16. $3x(3y - 4x)(3y + 4x)$
17. 4, 7
18. −2, 7
19. $\frac{3}{5}, -2$
20. 0, 2

Factor each of the following polynomials.

1. $15c + 35$

2. $8q^4 - 20q^3$

3. $6x^2 - 12x + 24$

4. $7c^3d^2 - 21cd + 14cd^3$

Factor each of the following polynomials completely.

5. $b^2 + 2b - 15$

6. $x^2 + 10x + 24$

7. $x^2 - 14x + 45$

8. $a^2 + 7ab + 12b^2$

Factor each of the following polynomials completely.

9. $3y^2 + 5y - 12$

10. $5w^2 + 23w + 12$

11. $6x^2 + 5xy - 21y^2$

12. $2x^3 - 7x^2 - 15x$

Factor each of the following polynomials completely.

13. $b^2 - 49$

14. $36p^2 - q^2$

15. $9x^2 - 12xy + 4y^2$

16. $27xy^2 - 48x^3$

Solve each of the following equations for x.

17. $x^2 - 11x + 28 = 0$

18. $x^2 - 5x = 14$

19. $5x^2 + 7x - 6 = 0$

20. $9p^2 - 18p = 0$

4.1 An Introduction to Factoring

OBJECTIVES

1. Remove the greatest common factor (GCF)
2. Remove a binomial GCF

Overcoming Math Anxiety

Hint #5

Working Together

How many of your classmates do you know? Whether you are by nature gregarious or shy, you have much to gain by getting to know your classmates.

1. It is important to have someone to call when you have missed class or if you are unclear on an assignment.
2. Working with another person is almost always beneficial to both people. If you don't understand something, it helps to have someone to ask about it. If you do understand something, nothing will cement that understanding more than explaining the idea to another person.
3. Sometimes we need to commiserate. If an assignment is particularly frustrating, it is reassuring to find that it is also frustrating for other students.
4. Have you ever thought you had the right answer, but it doesn't match the answer in the text? Frequently the answers are equivalent, but that's not always easy to see. A different perspective can help you see that. Occasionally there is an error in a textbook (here we are talking about *other* textbooks). In such cases it is wonderfully reassuring to find that someone else has the same answer you do.

In Chapter 3 you were given factors and asked to find a product. We are now going to reverse the process. You will be given a polynomial and asked to find its factors. This is called **factoring.**

Let's start with an example from arithmetic. To *multiply* $5 \cdot 7$, you write

$$5 \cdot 7 = 35$$

To *factor* 35, you would write

$$35 = 5 \cdot 7$$

Factoring is the *reverse* of multiplication.

Now let's look at factoring in algebra. You have used the distributive property as

$$a(b + c) = ab + ac$$

For instance,

NOTE 3 and $x + 5$ are the factors of $3x + 15$.

$$3(x + 5) = 3x + 15$$

To use the distributive property in factoring, we apply that property in the opposite fashion, as

$$ab + ac = a(b + c)$$

The property lets us remove the common monomial factor a from the terms of $ab + ac$. To use this in factoring, the first step is to see whether each term of the polynomial has a common monomial factor. In our earlier example,

$$3x + 15 = 3 \cdot x + 3 \cdot 5$$

Common factor (the 3 in each term)

So, by the distributive property,

$3x + 15 = 3(x + 5)$ The original terms are each divided by the greatest common factor to determine the terms in parentheses.

NOTE Again, factoring is the reverse of multiplication.

To check this, multiply $3(x + 5)$.

NOTE This diagram relates the idea of multiplication and factoring.

Multiplying →

$3(x + 5) = 3x + 15$

← Factoring

The first step in factoring is to identify the *greatest common factor* (GCF) of a set of terms. This is the monomial with the largest common numerical coefficient and the largest power common to any variables.

NOTE In fact, we will see that factoring out the GCF is the *first* method to try in any of the factoring problems we will discuss.

Definitions: Greatest Common Factor

The **greatest common factor (GCF)** of a polynomial is the monomial with the highest degree and the largest numerical coefficient that is a factor of each term of the polynomial.

Example 1

Finding the GCF

Find the GCF for each set of terms.

(a) 9 and 12 The largest number that is a factor of both is 3.

(b) 10, 25, 150 The GCF is 5.

(c) x^4 and x^7 The largest power that divides both terms is x^4.

(d) $12a^3$ and $18a^2$ The GCF is $6a^2$.

CHECK YOURSELF 1

Find the GCF for each set of terms.

(a) 14, 24 **(b)** 9, 27, 81 **(c)** a^9, a^5 **(d)** $10x^5$, $35x^4$

Step by Step: To Factor a Monomial from a Polynomial

NOTE Checking your answer is always important and perhaps is never easier than after you have factored.

Step 1 Find the *greatest common factor* (GCF) for all the terms.
Step 2 Factor the GCF from each term, then apply the distributive property.
Step 3 Mentally check your factoring by multiplication.

Example 2

Finding the GCF of a Binomial

(a) Factor $8x^2 + 12x$.

The largest common numerical factor of 8 and 12 is 4, and x is the variable factor with the largest power. So $4x$ is the GCF. Write

$$8x^2 + 12x = \underbrace{4x}_{\text{GCF}} \cdot 2x + \underbrace{4x}_{\text{GCF}} \cdot 3$$

NOTE It is always a good idea to check your answer by multiplying to make sure that you get the original polynomial. Try it here. Multiply $4x$ by $2x + 3$.

Now, by the distributive property, we have

$$8x^2 + 12x = 4x(2x + 3)$$

(b) Factor $6a^4 - 18a^2$.

The GCF in this case is $6a^2$. Write

$$6a^4 + (-18a^2) = \underbrace{6a^2}_{\text{GCF}} \cdot a^2 + \underbrace{6a^2}_{\text{GCF}} \cdot (-3)$$

NOTE It is also true that $6a^4 + (-18a^2) = 3a(2a^3 + (-6a))$. However, this is *not completely factored.* Do you see why? You want to find the common monomial factor with the *largest possible* coefficient and the *largest* exponent, in this case $6a^2$.

Again, using the distributive property yields

$$6a^4 - 18a^2 = 6a^2(a^2 - 3)$$

You should check this by multiplying.

CHECK YOURSELF 2

Factor each of the following polynomials.

(a) $5x + 20$ **(b)** $6x^2 - 24x$ **(c)** $10a^3 - 15a^2$

The process is exactly the same for polynomials with more than two terms. Consider Example 3.

Example 3

Finding the GCF of a Polynomial

(a) Factor $5x^2 - 10x + 15$.

NOTE The GCF is 5.

$$5x^2 - 10x + 15 = \underbrace{5}_{\text{GCF}} \cdot x^2 - \underbrace{5}_{\text{GCF}} \cdot 2x + \underbrace{5}_{\text{GCF}} \cdot 3$$

$$= 5(x^2 - 2x + 3)$$

(b) Factor $6ab + 9ab^2 - 15a^2$.

NOTE The GCF is $3a$.

$$6ab + 9ab^2 - 15a^2 = \underbrace{3a}_{\text{GCF}} \cdot 2b + \underbrace{3a}_{\text{GCF}} \cdot 3b^2 - \underbrace{3a}_{\text{GCF}} \cdot 5a$$

$$= 3a(2b + 3b^2 - 5a)$$

(c) Factor $4a^4 + 12a^3 - 20a^2$.

NOTE The GCF is $4a^2$.

$$4a^4 + 12a^3 - 20a^2 = \underbrace{4a^2}_{\text{GCF}} \cdot a^2 + \underbrace{4a^2}_{\text{GCF}} \cdot 3a - \underbrace{4a^2}_{\text{GCF}} \cdot 5$$

$$= 4a^2(a^2 + 3a - 5)$$

NOTE In each of these examples, you will want to check the result by multiplying the factors.

(d) Factor $6a^2b + 9ab^2 + 3ab$.

Mentally note that 3, a, and b are factors of each term, so

$$6a^2b + 9ab^2 + 3ab = 3ab(2a + 3b + 1)$$

CHECK YOURSELF 3

Factor each of the following polynomials.

(a) $8b^2 + 16b - 32$ **(b)** $4xy - 8x^2y + 12x^3$

(c) $7x^4 - 14x^3 + 21x^2$ **(d)** $5x^2y^2 - 10xy^2 + 15x^2y$

We can have two or more terms that have a binomial factor in common, as is the case in Example 4.

Example 4

Finding a Common Factor

(a) Factor $3x(x + y) + 2(x + y)$.

We see that *the binomial* $x + y$ is a common factor and can be removed.

NOTE Because of the commutative property, the factors can be written in either order.

$$3x(x + y) + 2(x + y)$$
$$= (x + y) \cdot 3x + (x + y) \cdot 2$$
$$= (x + y)(3x + 2)$$

(b) Factor $3x^2(x - y) + 6x(x - y) + 9(x - y)$.

We note that here the GCF is $3(x - y)$. Factoring as before, we have

$$3(x - y)(x^2 + 2x + 3)$$

CHECK YOURSELF 4

Completely factor each of the polynomials.

(a) $7a(a - 2b) + 3(a - 2b)$ **(b)** $4x^2(x + y) - 8x(x + y) - 16(x + y)$

CHECK YOURSELF ANSWERS

1. (a) 2; **(b)** 9; **(c)** a^5; **(d)** $5x^4$ **2. (a)** $5(x + 4)$; **(b)** $6x(x - 4)$; **(c)** $5a^2(2a - 3)$
3. (a) $8(b^2 + 2b - 4)$; **(b)** $4x(y - 2xy + 3x^2)$; **(c)** $7x^2(x^2 - 2x + 3)$;
(d) $5xy(xy - 2y + 3x)$ **4. (a)** $(a - 2b)(7a + 3)$; **(b)** $4(x + y)(x^2 - 2x - 4)$

Name ______

4.1 Exercises

Section ______ Date ______

Find the greatest common factor for each of the following sets of terms.

1. 10, 12

2. 15, 35

3. 16, 32, 88

4. 55, 33, 132

5. x^2, x^5

6. y^7, y^9

7. a^3, a^6, a^9

8. b^4, b^6, b^8

9. $5x^4, 10x^5$

10. $8y^9, 24y^3$

11. $8a^4, 6a^6, 10a^{10}$

12. $9b^3, 6b^5, 12b^4$

13. $9x^2y, 12xy^2, 15x^2y^2$

14. $12a^3b^2, 18a^2b^3, 6a^4b^4$

15. $15ab^3, 10a^2bc, 25b^2c^3$

16. $9x^2, 3xy^3, 6y^3$

17. $15a^2bc^2, 9ab^2c^2, 6a^2b^2c^2$

18. $18x^3y^2z^3, 27x^4y^2z^3, 81xy^2z$

19. $(x + y)^2, (x + y)^3$

20. $12(a + b)^4, 4(a + b)^3$

Factor each of the following polynomials.

21. $8a + 4$

22. $5x - 15$

23. $24m - 32n$

24. $7p - 21q$

25. $12m^2 + 8m$

26. $24n^2 - 32n$

ANSWERS

1. 2
2. 5
3. 8
4. 11
5. x^2
6. y^7
7. a^3
8. b^4
9. $5x^4$
10. $8y^3$
11. $2a^4$
12. $3b^3$
13. $3xy$
14. $6a^2b^2$
15. $5b$
16. 3
17. $3abc^2$
18. $9xy^2z$
19. $(x + y)^2$
20. $4(a + b)^3$
21. $4(2a + 1)$
22. $5(x - 3)$
23. $8(3m - 4n)$
24. $7(p - 3q)$
25. $4m(3m + 2)$
26. $8n(3n - 4)$

ANSWERS

27. $5s(2s + 1)$
28. $6y(2y - 1)$
29. $12x(x + 2)$
30. $14b(b - 2)$
31. $5a^2(3a - 5)$
32. $12b^2(3b^2 + 2)$
33. $6pq(1 + 3p)$
34. $8ab(1 - 3b)$
35. $7mn(m^2 - 3n^2)$
36. $9pq(4pq - 1)$
37. $6(x^2 - 3x + 5)$
38. $7(a^2 + 3a - 6)$
39. $3a(a^2 + 2a - 4)$
40. $5x(x^2 - 3x + 5)$
41. $3m(2 + 3n - 5n^2)$
42. $2s(2 + 3t - 7t^2)$
43. $5xy(2x + 3 - y)$
44. $3ab(b + 2 - 5a)$
45. $5r^2s^2(2r + 5 - 3s)$
46. $7x^2y(4y^2 - 5y + 6x)$
47. $3a(3a^4 - 5a^3 + 7a^2 - 9)$
48. $8p^2(p^4 - 5p^2 + 3p - 2)$
49. $5mn(3m^2n - 4m + 7n^2 - 2)$
50. $7ab^2(2b^2 + 3ab - 5a^2 + 4)$
51. $(x - 2)(x + 3)$
52. $(y + 5)(y - 3)$

27. $10s^2 + 5s$

28. $12y^2 - 6y$

29. $12x^2 + 24x$

30. $14b^2 - 28b$

31. $15a^3 - 25a^2$

32. $36b^4 + 24b^2$

33. $6pq + 18p^2q$

34. $8ab - 24ab^2$

35. $7m^3n - 21mn^3$

36. $36p^2q^2 - 9pq$

37. $6x^2 - 18x + 30$

38. $7a^2 + 21a - 42$

39. $3a^3 + 6a^2 - 12a$

40. $5x^3 - 15x^2 + 25x$

41. $6m + 9mn - 15mn^2$

42. $4s + 6st - 14st^2$

43. $10x^2y + 15xy - 5xy^2$

44. $3ab^2 + 6ab - 15a^2b$

45. $10r^3s^2 + 25r^2s^2 - 15r^2s^3$

46. $28x^2y^3 - 35x^2y^2 + 42x^3y$

47. $9a^5 - 15a^4 + 21a^3 - 27a$

48. $8p^6 - 40p^4 + 24p^3 - 16p^2$

49. $15m^3n^2 - 20m^2n + 35mn^3 - 10mn$

50. $14ab^4 + 21a^2b^3 - 35a^3b^2 + 28ab^2$

51. $x(x - 2) + 3(x - 2)$

52. $y(y + 5) - 3(y + 5)$

53. The GCF of $2x - 6$ is 2. The GCF of $5x + 10$ is 5. Find the greatest common factor of the product $(2x - 6)(5x + 10)$.

54. The GCF of $3z + 12$ is 3. The GCF of $4z + 8$ is 4. Find the GCF of the product $(3z + 12)(4z + 8)$.

55. The GCF of $2x^3 - 4x$ is $2x$. The GCF of $3x + 6$ is 3. Find the GCF of the product $(2x^3 - 4x)(3x + 6)$.

56. State, in a sentence, the rule that the previous three exercises illustrated.

Find the GCF for each product.

57. $(2a + 8)(3a - 6)$

58. $(5b - 10)(2b + 4)$

59. $(2x^2 + 5x)(7x - 14)$

60. $(6y^2 - 3y)(y + 7)$

61. **Area of a rectangle.** The area of a rectangle with width t is given by $33t - t^2$. Factor the expression and determine the length of the rectangle in terms of t.

62. **Area of a rectangle.** The area of a rectangle of length x is given by $3x^2 + 5x$. Find the width of the rectangle.

63. For centuries, mathematicians have found factoring numbers into prime factors a fascinating subject. A prime number is a number that cannot be written as a product of any numbers but 1 and itself. The list of primes begins with 2 because 1 is not considered a prime number and then goes on: 3, 5, 7, 11, . . . What are the first 10 primes? What are the primes less than 100? If you list the numbers from 1 to 100 and then cross out all numbers that are multiples of 2, 3, 5, and 7, what is left? Are all the numbers not crossed out prime? Write a paragraph to explain why this might be so. You might want to investigate the sieve of Eratosthenes, a system from 230 B.C.E. for finding prime numbers.

64. If we made a list of all the prime numbers, what number would be at the end of the list? Because there are an infinite number of prime numbers, there is no "largest prime number." But is there some formula that will give us all the primes? Here are some formulas proposed over the centuries:

$n^2 + n + 17 \qquad 2n^2 + 29 \qquad n^2 - n + 11$

In all these expressions, $n = +1, 2, 3, 4, \ldots$, that is, a positive integer beginning with 1. Investigate these expressions with a partner. Do the expressions give prime numbers when they are evaluated for these values of n? Do the expressions give *every* prime in the range of resulting numbers? Can you put in *any* positive number for n?

65. How are primes used in coding messages and for security? Work together to decode the messages. The messages are coded using this code: After the numbers are factored into prime factors, the power of 2 gives the number of the letter in the alphabet. This code would be easy for a code breaker to figure out, but you might make up code that would be more difficult to break.

a. 1310720, 229376, 1572864, 1760, 460, 2097152, 336

b. 786432, 143, 4608, 278528, 1344, 98304, 1835008, 352, 4718592, 5242880

c. Code a message using this rule. Exchange your message with a partner to decode it.

ANSWERS

53. 10

54. 12

55. 6x

56. The GCF for the product of two factors is the product of their GCFs.

57. 6

58. 10

59. 7x

60. 3y

61. t(33 − t); 33 − t

62. 3x + 5

63.

64.

65.

a. $a^2 + 3a - 4$

b. $x^2 + 2x - 3$

c. $x^2 - 6x + 9$

d. $y^2 - 8y - 33$

e. $x^2 + 12x + 35$

f. $y^2 - 12y - 13$

Getting Ready for Section 4.2 [Section 3.4]

Multiply.

(a) $(a - 1)(a + 4)$

(b) $(x - 1)(x + 3)$

(c) $(x - 3)(x - 3)$

(d) $(y - 11)(y + 3)$

(e) $(x + 5)(x + 7)$

(f) $(y + 1)(y - 13)$

Answers

1. 2 **3.** 8 **5.** x^2 **7.** a^3 **9.** $5x^4$ **11.** $2a^4$ **13.** $3xy$ **15.** $5b$ **17.** $3abc^2$ **19.** $(x + y)^2$ **21.** $4(2a + 1)$ **23.** $8(3m - 4n)$ **25.** $4m(3m + 2)$ **27.** $5s(2s + 1)$ **29.** $12x(x + 2)$ **31.** $5a^2(3a - 5)$ **33.** $6pq(1 + 3p)$ **35.** $7mn(m^2 - 3n^2)$ **37.** $6(x^2 - 3x + 5)$ **39.** $3a(a^2 + 2a - 4)$ **41.** $3m(2 + 3n - 5n^2)$ **43.** $5xy(2x + 3 - y)$ **45.** $5r^2s^2(2r + 5 - 3s)$ **47.** $3a(3a^4 - 5a^3 + 7a^2 - 9)$ **49.** $5mn(3m^2n - 4m + 7n^2 - 2)$ **51.** $(x - 2)(x + 3)$ **53.** 10 **55.** $6x$ **57.** 6 **59.** $7x$ **61.** $t(33 - t)$; $33 - t$ **63.** **65.**

a. $a^2 + 3a - 4$ **b.** $x^2 + 2x - 3$ **c.** $x^2 - 6x + 9$ **d.** $y^2 - 8y - 33$ **e.** $x^2 + 12x + 35$ **f.** $y^2 - 12y - 13$

Factoring Trinomials of the Form $x^2 + bx + c$

OBJECTIVES

1. Factor a trinomial of the form $x^2 + bx + c$
2. Factor a trinomial containing a common factor

NOTE The process used to factor here is frequently called the *trial-and-error method.* You'll see the reason for the name as you work through this section.

You learned how to find the product of any two binomials by using the FOIL method in Section 3.4. Because factoring is the reverse of multiplication, we now want to use that pattern to find the factors of certain trinomials.

Recall that to multiply two binomials, we have

$$(x + 2)(x + 3) = x^2 + 5x + 6$$

The product of the first terms ($x \cdot x$).

The sum of the products of the outer and inner terms ($3x$ and $2x$).

The product of the last terms ($2 \cdot 3$).

Not every trinomial can be written as the product of two binomials.

Suppose now that you are given $x^2 + 5x + 6$ and want to find its factors. First, you know that the factors of a trinomial may be two binomials. So write

$$x^2 + 5x + 6 = (\qquad)(\qquad)$$

Because the first term of the trinomial is x^2, the first terms of the binomial factors must be x and x. We now have

$$x^2 + 5x + 6 = (x \qquad)(x \qquad)$$

The product of the last terms must be 6. Because 6 is positive, the factors must have *like* signs. Here are the possibilities:

$$\begin{aligned} 6 &= 1 \cdot 6 \\ &= 2 \cdot 3 \\ &= (-1)(-6) \\ &= (-2)(-3) \end{aligned}$$

This means that the possible factors of the trinomial are

$(x + 1)(x + 6)$

$(x + 2)(x + 3)$

$(x - 1)(x - 6)$

$(x - 2)(x - 3)$

How do we tell which is the correct pair? From the FOIL pattern we know that the sum of the outer and inner products must equal the middle term of the trinomial, in this case $5x$. This is the crucial step!

Possible Factors	Middle Terms
$(x + 1)(x + 6)$	$7x$
$(x + 2)(x + 3)$	$5x$
$(x - 1)(x - 6)$	$-7x$
$(x - 2)(x - 3)$	$-5x$

The correct middle term!

So we know that the correct factorization is

$$x^2 + 5x + 6 = (x + 2)(x + 3)$$

Are there any clues so far that will make this process quicker? Yes, there is an important one that you may have spotted. We started with a trinomial that had a positive middle term and a positive last term. The negative pairs of factors for 6 led to negative middle terms. So you don't need to bother with the negative factors if the middle term and the last term of the trinomial are both positive.

Example 1

Factoring a Trinomial

(a) Factor $x^2 + 9x + 8$.

Because the middle term and the last term of the trinomial are both positive, consider only the positive factors of 8, that is, $8 = 1 \cdot 8$ or $8 = 2 \cdot 4$.

NOTE If you are wondering why we didn't list $(x + 8)(x + 1)$ as a possibility, remember that multiplication is commutative. The order doesn't matter!

Possible Factors	Middle Terms
$(x + 1)(x + 8)$	$9x$
$(x + 2)(x + 4)$	$6x$

Because the first pair gives the correct middle term,

$$x^2 + 9x + 8 = (x + 1)(x + 8)$$

(b) Factor $x^2 + 12x + 20$.

NOTE The factors for 20 are

$$20 = 1 \cdot 20$$
$$= 2 \cdot 10$$
$$= 4 \cdot 5$$

Possible Factors	Middle Terms
$(x + 1)(x + 20)$	$21x$
$(x + 2)(x + 10)$	$12x$
$(x + 4)(x + 5)$	$9x$

So

$$x^2 + 12x + 20 = (x + 2)(x + 10)$$

CHECK YOURSELF 1

Factor.

(a) $x^2 + 6x + 5$ **(b)** $x^2 + 10x + 16$

Let's look at some examples in which the middle term of the trinomial is negative but the first and last terms are still positive. Consider

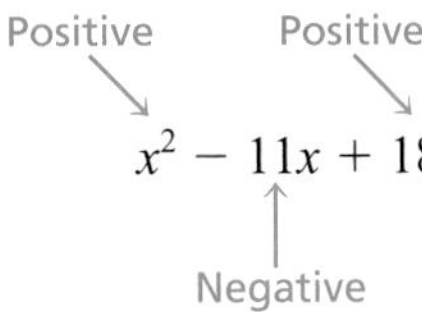

Because we want a negative middle term $(-11x)$, we use *two negative factors* for 18. Recall that the product of two negative numbers is positive.

Example 2

Factoring a Trinomial

(a) Factor $x^2 - 11x + 18$.

NOTE The negative factors of 18 are

$18 = (-1)(-18)$
$= (-2)(-9)$
$= (-3)(-6)$

Possible Factors	Middle Terms
$(x - 1)(x - 18)$	$-19x$
$(x - 2)(x - 9)$	$-11x$
$(x - 3)(x - 6)$	$-9x$

So

$$x^2 - 11x + 18 = (x - 2)(x - 9)$$

(b) Factor $x^2 - 13x + 12$.

NOTE The negative factors of 12 are

$12 = (-1)(-12)$
$= (-2)(-6)$
$= (-3)(-4)$

Possible Factors	Middle Terms
$(x - 1)(x - 12)$	$-13x$
$(x - 2)(x - 6)$	$-8x$
$(x - 3)(x - 4)$	$-7x$

So

$$x^2 - 13x + 12 = (x - 1)(x - 12)$$

A few more clues: We have listed all the possible factors in the above examples. It really isn't necessary. Just work until you find the right pair. Also, with practice much of this work can be done mentally.

CHECK YOURSELF 2

Factor.

(a) $x^2 - 10x + 9$ **(b)** $x^2 - 10x + 21$

Let's look now at the process of factoring a trinomial whose last term is negative. For instance, to factor $x^2 + 2x - 15$, we can start as before:

$$x^2 + 2x - 15 = (x \quad ?)(x \quad ?)$$

Note that the product of the last terms must be negative (-15 here). So we must choose factors that have different signs.

What are our choices for the factors of -15?

$$\begin{aligned} -15 &= (1)(-15) \\ &= (-1)(15) \\ &= (3)(-5) \\ &= (-3)(5) \end{aligned}$$

This means that the possible factors and the resulting middle terms are

Possible Factors	Middle Terms
$(x + 1)(x - 15)$	$-14x$
$(x - 1)(x + 15)$	$14x$
$(x + 3)(x - 5)$	$-2x$
$(x - 3)(x + 5)$	$2x$

NOTE Another clue: Some students prefer to look at the list of numerical factors rather than looking at the actual algebraic factors. Here you want the pair whose sum is 2, the coefficient of the middle term of the trinomial. That pair is -3 and 5, which leads us to the correct factors.

So $x^2 + 2x - 15 = (x - 3)(x + 5)$.

Let's work through some examples in which the constant term is negative.

Example 3

Factoring a Trinomial

(a) Factor $x^2 - 5x - 6$.

First, list the factors of -6. Of course, one factor will be positive, and one will be negative.

NOTE You may be able to pick the factors directly from this list. You want the pair whose sum is -5 (the coefficient of the middle term).

$$\begin{aligned} -6 &= (1)(-6) \\ &= (-1)(6) \\ &= (2)(-3) \\ &= (-2)(3) \end{aligned}$$

For the trinomial, then, we have

Possible Factors	Middle Terms
$(x + 1)(x - 6)$	$-5x$
$(x - 1)(x + 6)$	$5x$
$(x + 2)(x - 3)$	$-x$
$(x - 2)(x + 3)$	x

So $x^2 - 5x - 6 = (x + 1)(x - 6)$.

(b) Factor $x^2 + 8xy - 9y^2$.

The process is similar if two variables are involved in the trinomial you are to factor. Start with

$$x^2 + 8xy - 9y^2 = (x \quad ?)(x \quad ?).$$

The product of the last terms must be $-9y^2$.

$$-9y^2 = (-y)(9y)$$
$$= (y)(-9y)$$
$$= (3y)(-3y)$$

Possible Factors	Middle Terms
$(x - y)(x + 9y)$	$8xy$
$(x + y)(x - 9y)$	$-8xy$
$(x + 3y)(x - 3y)$	0

So $x^2 + 8xy - 9y^2 = (x - y)(x + 9y)$.

CHECK YOURSELF 3

Factor.

(a) $x^2 + 7x - 30$ **(b)** $x^2 - 3xy - 10y^2$

As was pointed out in the last section, any time that we have a common factor, that factor should be removed *before* we try any other factoring technique. Consider the following example.

Example 4

Factoring a Trinomial

(a) Factor $3x^2 - 21x + 18$.

$3x^2 - 21x + 18 = 3(x^2 - 7x + 6)$ Remove the common factor of 3.

We now factor the remaining trinomial. For $x^2 - 7x + 6$:

Possible Factors	Middle Terms
$(x - 2)(x - 3)$	$-5x$
$(x - 1)(x - 6)$	$-7x$

The correct middle term

CAUTION

A common mistake is to forget to write the 3 that was factored out as the first step.

So $3x^2 - 21x + 18 = 3(x - 1)(x - 6)$.

(b) Factor $2x^3 + 16x^2 - 40x$.

$2x^3 + 16x^2 - 40x = 2x(x^2 + 8x - 20)$ Remove the common factor of $2x$.

To factor the remaining trinomial, which is $x^2 + 8x - 20$, we have

Possible Factors	Middle Terms	
$(x - 4)(x + 5)$	x	
$(x - 5)(x + 4)$	$-x$	
$(x - 10)(x + 2)$	$-8x$	
$(x - 2)(x + 10)$	$8x$	The correct middle term

NOTE Once we have found the desired middle term, there is no need to continue.

So $2x^3 + 16x^2 - 40x = 2x(x - 2)(x + 10)$.

CHECK YOURSELF 4

Factor.

(a) $3x^2 - 3x - 36$ **(b)** $4x^3 + 24x^2 + 32x$

One further comment: Have you wondered if all trinomials are factorable? Look at the trinomial

$x^2 + 2x + 6$

The only possible factors are $(x + 1)(x + 6)$ and $(x + 2)(x + 3)$. Neither pair is correct (you should check the middle terms), and so this trinomial does not have factors with integer coefficients. Of course, there are many others.

CHECK YOURSELF ANSWERS

1. (a) $(x + 1)(x + 5)$; **(b)** $(x + 2)(x + 8)$ **2. (a)** $(x - 9)(x - 1)$; **(b)** $(x - 3)(x - 7)$
3. (a) $(x + 10)(x - 3)$; **(b)** $(x + 2y)(x - 5y)$
4. (a) $3(x - 4)(x + 3)$; **(b)** $4x(x + 2)(x + 4)$

Name ______________________

Section ________ Date ________

4.2 Exercises

Complete each of the following statements.

1. $x^2 - 8x + 15 = (x - 3)(\quad)$

2. $y^2 - 3y - 18 = (y - 6)(\quad)$

3. $m^2 + 8m + 12 = (m + 2)(\quad)$

4. $x^2 - 10x + 24 = (x - 6)(\quad)$

5. $p^2 - 8p - 20 = (p + 2)(\quad)$

6. $a^2 + 9a - 36 = (a + 12)(\quad)$

7. $x^2 - 16x + 64 = (x - 8)(\quad)$

8. $w^2 - 12w - 45 = (w + 3)(\quad)$

9. $x^2 - 7xy + 10y^2 = (x - 2y)(\quad)$

10. $a^2 + 18ab + 81b^2 = (a + 9b)(\quad)$

Factor each of the following trinomials.

11. $x^2 + 8x + 15$

12. $x^2 - 11x + 24$

13. $x^2 - 11x + 28$

14. $y^2 - y - 20$

15. $s^2 + 13s + 30$

16. $b^2 + 14b + 33$

17. $a^2 - 2a - 48$

18. $x^2 - 17x + 60$

19. $x^2 - 8x + 7$

20. $x^2 + 7x - 18$

21. $m^2 + 3m - 28$

22. $a^2 + 10a + 25$

ANSWERS

1. $x - 5$
2. $y + 3$
3. $m + 6$
4. $x - 4$
5. $p - 10$
6. $a - 3$
7. $x - 8$
8. $w - 15$
9. $x - 5y$
10. $a + 9b$
11. $(x + 3)(x + 5)$
12. $(x - 3)(x - 8)$
13. $(x - 4)(x - 7)$
14. $(y - 5)(y + 4)$
15. $(s + 3)(s + 10)$
16. $(b + 3)(b + 11)$
17. $(a - 8)(a + 6)$
18. $(x - 5)(x - 12)$
19. $(x - 1)(x - 7)$
20. $(x - 2)(x + 9)$
21. $(m + 7)(m - 4)$
22. $(a + 5)(a + 5)$

Answers

1. $x - 5$ **3.** $m + 6$ **5.** $p - 10$ **7.** $x - 8$ **9.** $x - 5y$
11. $(x + 3)(x + 5)$ **13.** $(x - 4)(x - 7)$ **15.** $(s + 10)(s + 3)$
17. $(a - 8)(a + 6)$ **19.** $(x - 1)(x - 7)$ **21.** $(m + 7)(m - 4)$
23. $(x - 10)(x + 4)$ **25.** $(x - 7)(x - 7)$ **27.** $(p - 12)(p + 2)$
29. $(x + 11)(x - 6)$ **31.** $(c + 4)(c + 15)$ **33.** $(n + 10)(n - 5)$
35. $(x + 2y)(x + 5y)$ **37.** $(a - 7b)(a + 6b)$ **39.** $(x - 5y)(x - 8y)$
41. $(b + 3a)(b + 3a)$ **43.** $(x - 4y)(x + 2y)$ **45.** $(5m + n)(5m + n)$
47. $3(a - 7)(a + 6)$ **49.** $r(r + 9)(r - 2)$ **51.** $2x(x - 12)(x + 2)$
53. $y(x - 12y)(x + 3y)$ **55.** $m(m - 5n)(m - 24n)$ **57.** 6 or 9
59. 8, 10, or 17 **61.** 4 **63.** 2 **65.** 3, 8, 15, 24, . . . **a.** $4x^2 + 4x - 3$
b. $3a^2 + 11a - 4$ **c.** $2x^2 - 11x + 12$ **d.** $2w^2 - 7w - 22$
e. $2y^2 + 19y + 45$ **f.** $2x^2 - 23x - 12$ **g.** $2p^2 + 23p + 45$
h. $6a^2 + 2a - 20$

Name ______

Section ______ Date ______

4.2 Exercises

Complete each of the following statements.

1. $x^2 - 8x + 15 = (x - 3)(\quad)$

2. $y^2 - 3y - 18 = (y - 6)(\quad)$

3. $m^2 + 8m + 12 = (m + 2)(\quad)$

4. $x^2 - 10x + 24 = (x - 6)(\quad)$

5. $p^2 - 8p - 20 = (p + 2)(\quad)$

6. $a^2 + 9a - 36 = (a + 12)(\quad)$

7. $x^2 - 16x + 64 = (x - 8)(\quad)$

8. $w^2 - 12w - 45 = (w + 3)(\quad)$

9. $x^2 - 7xy + 10y^2 = (x - 2y)(\quad)$

10. $a^2 + 18ab + 81b^2 = (a + 9b)(\quad)$

Factor each of the following trinomials.

11. $x^2 + 8x + 15$

12. $x^2 - 11x + 24$

13. $x^2 - 11x + 28$

14. $y^2 - y - 20$

15. $s^2 + 13s + 30$

16. $b^2 + 14b + 33$

17. $a^2 - 2a - 48$

18. $x^2 - 17x + 60$

19. $x^2 - 8x + 7$

20. $x^2 + 7x - 18$

21. $m^2 + 3m - 28$

22. $a^2 + 10a + 25$

ANSWERS

1. $x - 5$
2. $y + 3$
3. $m + 6$
4. $x - 4$
5. $p - 10$
6. $a - 3$
7. $x - 8$
8. $w - 15$
9. $x - 5y$
10. $a + 9b$
11. $(x + 3)(x + 5)$
12. $(x - 3)(x - 8)$
13. $(x - 4)(x - 7)$
14. $(y - 5)(y + 4)$
15. $(s + 3)(s + 10)$
16. $(b + 3)(b + 11)$
17. $(a - 8)(a + 6)$
18. $(x - 5)(x - 12)$
19. $(x - 1)(x - 7)$
20. $(x - 2)(x + 9)$
21. $(m + 7)(m - 4)$
22. $(a + 5)(a + 5)$

ANSWERS

23. $(x + 4)(x - 10)$
24. $(x - 1)(x - 10)$
25. $(x - 7)(x - 7)$
26. $(s - 8)(s + 4)$
27. $(p - 12)(p + 2)$
28. $(x + 4)(x - 15)$
29. $(x + 11)(x - 6)$
30. $(a - 8)(a + 10)$
31. $(c + 4)(c + 15)$
32. $(t + 6)(t - 10)$
33. $(n - 5)(n + 10)$
34. $(x - 7)(x - 9)$
35. $(x + 2y)(x + 5y)$
36. $(x - 2y)(x - 6y)$
37. $(a + 6b)(a - 7b)$
38. $(m - 4n)(m - 4n)$
39. $(x - 5y)(x - 8y)$
40. $(r + 3s)(r - 12s)$
41. $(b + 3a)(b + 3a)$
42. $(x - 2y)(x + 5y)$
43. $(x + 2y)(x - 4y)$
44. $(u - 5v)(u + 11v)$
45. $(n + 5m)(n + 5m)$
46. $(n - 8m)(n - 8m)$

23. $x^2 - 6x - 40$

24. $x^2 - 11x + 10$

25. $x^2 - 14x + 49$

26. $s^2 - 4s - 32$

27. $p^2 - 10p - 24$

28. $x^2 - 11x - 60$

29. $x^2 + 5x - 66$

30. $a^2 + 2a - 80$

31. $c^2 + 19c + 60$

32. $t^2 - 4t - 60$

33. $n^2 + 5n - 50$

34. $x^2 - 16x + 63$

35. $x^2 + 7xy + 10y^2$

36. $x^2 - 8xy + 12y^2$

37. $a^2 - ab - 42b^2$

38. $m^2 - 8mn + 16n^2$

39. $x^2 - 13xy + 40y^2$

40. $r^2 - 9rs - 36s^2$

41. $b^2 + 6ab + 9a^2$

42. $x^2 + 3xy - 10y^2$

43. $x^2 - 2xy - 8y^2$

44. $u^2 + 6uv - 55v^2$

45. $25m^2 + 10mn + n^2$

46. $64m^2 - 16mn + n^2$

Factor each of the following trinomials completely. Factor out the greatest common factor first.

47. $3a^2 - 3a - 126$

48. $2c^2 + 2c - 60$

49. $r^3 + 7r^2 - 18r$

50. $m^3 + 5m^2 - 14m$

51. $2x^3 - 20x^2 - 48x$

52. $3p^3 + 48p^2 - 108p$

53. $x^2y - 9xy^2 - 36y^3$

54. $4s^4 - 20s^3t - 96s^2t^2$

55. $m^3 - 29m^2n + 120mn^2$

56. $2a^3 - 52a^2b + 96ab^2$

Find a positive value for k for which each of the following can be factored.

57. $x^2 + kx + 8$

58. $x^2 + kx + 9$

59. $x^2 - kx + 16$

60. $x^2 - kx + 17$

61. $x^2 - kx - 5$

62. $x^2 - kx - 7$

63. $x^2 + 3x + k$

64. $x^2 + 5x + k$

65. $x^2 + 2x - k$

66. $x^2 + x - k$

Getting Ready for Section 4.3 [Section 3.3]

Multiply.

(a) $(2x - 1)(2x + 3)$

(b) $(3a - 1)(a + 4)$

(c) $(x - 4)(2x - 3)$

(d) $(2w - 11)(w + 2)$

(e) $(y + 5)(2y + 9)$

(f) $(2x + 1)(x - 12)$

(g) $(p + 9)(2p + 5)$

(h) $(3a - 5)(2a + 4)$

ANSWERS

47. $3(a + 6)(a - 7)$

48. $2(c - 5)(c + 6)$

49. $r(r - 2)(r + 9)$

50. $m(m - 2)(m + 7)$

51. $2x(x - 12)(x + 2)$

52. $3p(p - 2)(p + 18)$

53. $y(x + 3y)(x - 12y)$

54. $4s^2(s + 3t)(s - 8t)$

55. $m(m - 5n)(m - 24n)$

56. $2a(a - 2b)(a - 24b)$

57. 6 or 9

58. 6 or 10

59. 8, 10, or 17

60. 18

61. 4

62. 6

63. 2

64. 4 or 6

65. 3, 8, 15, 24, . . .

66. 2, 6, 12, 20, . . .

a. $4x^2 + 4x - 3$

b. $3a^2 + 11a - 4$

c. $2x^2 - 11x + 12$

d. $2w^2 - 7w - 22$

e. $2y^2 + 19y + 45$

f. $2x^2 - 23x - 12$

g. $2p^2 + 23p + 45$

h. $6a^2 + 2a - 20$

Answers

1. $x - 5$ **3.** $m + 6$ **5.** $p - 10$ **7.** $x - 8$ **9.** $x - 5y$
11. $(x + 3)(x + 5)$ **13.** $(x - 4)(x - 7)$ **15.** $(s + 10)(s + 3)$
17. $(a - 8)(a + 6)$ **19.** $(x - 1)(x - 7)$ **21.** $(m + 7)(m - 4)$
23. $(x - 10)(x + 4)$ **25.** $(x - 7)(x - 7)$ **27.** $(p - 12)(p + 2)$
29. $(x + 11)(x - 6)$ **31.** $(c + 4)(c + 15)$ **33.** $(n + 10)(n - 5)$
35. $(x + 2y)(x + 5y)$ **37.** $(a - 7b)(a + 6b)$ **39.** $(x - 5y)(x - 8y)$
41. $(b + 3a)(b + 3a)$ **43.** $(x - 4y)(x + 2y)$ **45.** $(5m + n)(5m + n)$
47. $3(a - 7)(a + 6)$ **49.** $r(r + 9)(r - 2)$ **51.** $2x(x - 12)(x + 2)$
53. $y(x - 12y)(x + 3y)$ **55.** $m(m - 5n)(m - 24n)$ **57.** 6 or 9
59. 8, 10, or 17 **61.** 4 **63.** 2 **65.** 3, 8, 15, 24, . . . **a.** $4x^2 + 4x - 3$
b. $3a^2 + 11a - 4$ **c.** $2x^2 - 11x + 12$ **d.** $2w^2 - 7w - 22$
e. $2y^2 + 19y + 45$ **f.** $2x^2 - 23x - 12$ **g.** $2p^2 + 23p + 45$
h. $6a^2 + 2a - 20$

Factoring Trinomials of the Form $ax^2 + bx + c$

OBJECTIVES

1. Factor a trinomial of the form $ax^2 + bx + c$
2. Completely factor a trinomial

Factoring trinomials is more time-consuming when the coefficient of the first term is not 1. Look at the following multiplication.

$$(5x + 2)(2x + 3) = 10x^2 + 19x + 6$$

Factors of $10x^2$ Factors of 6

Do you see the additional problem? We must consider all possible factors of the first coefficient (10 in the example) as well as those of the third term (6 in our example).

There is no easy way out! You need to form all possible combinations of factors and then check the middle term until the proper pair is found. If this seems a bit like guesswork, you're almost right. In fact some call this process factoring by *trial and error.*

We can simplify the work a bit by reviewing the sign patterns found in Section 4.2.

NOTE Any time the leading coefficient is negative, factor out a negative one from the trinomial. This will leave one of these cases.

Rules and Properties: Sign Patterns for Factoring Trinomials

1. If all terms of a trinomial are positive, the signs between the terms in the binomial factors are both plus signs.
2. If the third term of the trinomial is positive and the middle term is negative, the signs between the terms in the binomial factors are both minus signs.
3. If the third term of the trinomial is negative, the signs between the terms in the binomial factors are opposite (one is + and one is −).

Example 1

Factoring a Trinomial

Factor $3x^2 + 14x + 15$.

First, list the possible factors of 3, the coefficient of the first term.

$$3 = 1 \cdot 3$$

Now list the factors of 15, the last term.

$$15 = 1 \cdot 15$$
$$= 3 \cdot 5$$

Because the signs of the trinomial are all positive, we know any factors will have the form

$$(\_x + \_)(\_x + \_)$$

The product of the last terms must be 15.

The product of the numbers in the first blanks must be 3.

So the following are the possible factors and the corresponding middle terms:

Possible Factors	Middle Terms
$(x + 1)(3x + 15)$	$18x$
$(x + 15)(3x + 1)$	$46x$
$(3x + 3)(x + 5)$	$18x$
$(3x + 5)(x + 3)$	$14x$

The correct middle term

NOTE Take the time to multiply the binomial factors. This habit will ensure that you have an expression equivalent to the original problem.

So

$3x^2 + 14x + 15 = (3x + 5)(x + 3)$

CHECK YOURSELF 1

Factor.

(a) $5x^2 + 14x + 8$ **(b)** $3x^2 + 20x + 12$

Example 2

Factoring a Trinomial

Factor $4x^2 - 11x + 6$.

Because only the middle term is negative, we know the factors have the form

$(\_x - \_)(\_x - \_)$

Both signs are negative.

Now look at the factors of the first coefficient and the last term.

$4 = 1 \cdot 4 \qquad 6 = 1 \cdot 6$

$\;\;= 2 \cdot 2 \qquad\;\; = 2 \cdot 3$

This gives us the possible factors:

Possible Factors	Middle Terms
$(x - 1)(4x - 6)$	$-10x$
$(x - 6)(4x - 1)$	$-25x$
$(x - 2)(4x - 3)$	$-11x$

The correct middle term

NOTE Again, at least mentally, check your work by multiplying the factors.

Note that, in this example, we *stopped* as soon as the correct pair of factors was found. So

$4x^2 - 11x + 6 = (x - 2)(4x - 3)$

CHECK YOURSELF 2

Factor.

(a) $2x^2 - 9x + 9$ **(b)** $6x^2 - 17x + 10$

Let's factor a trinomial whose last term is negative.

Example 3

Factoring a Trinomial

Factor $5x^2 + 6x - 8$.

Because the last term is negative, the factors have the form

$(\_x + \_)(\_x - \_)$

Consider the factors of the first coefficient and the last term.

$5 = 1 \cdot 5 \qquad 8 = 1 \cdot 8$

$\qquad\qquad\quad = 2 \cdot 4$

The possible factors are then

Possible Factors	Middle Terms
$(x + 1)(5x - 8)$	$-3x$
$(x + 8)(5x - 1)$	$39x$
$(5x + 1)(x - 8)$	$-39x$
$(5x + 8)(x - 1)$	$3x$
$(x + 2)(5x - 4)$	$6x$

Again we stop as soon as the correct pair of factors is found.

$5x^2 + 6x - 8 = (x + 2)(5x - 4)$

CHECK YOURSELF 3

Factor $4x^2 + 5x - 6$.

The same process is used to factor a trinomial with more than one variable.

Example 4

Factoring a Trinomial

Factor $6x^2 + 7xy - 10y^2$.

The form of the factors must be

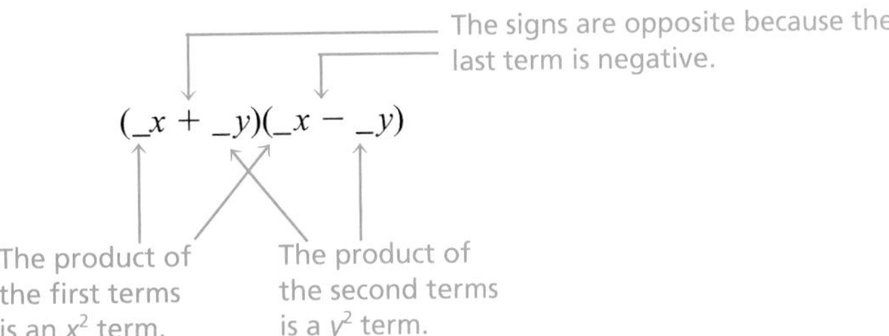

Again look at the factors of the first and last coefficients.

$6 = 1 \cdot 6 \qquad 10 = 1 \cdot 10$

$\;\; = 2 \cdot 3 \qquad\quad = 2 \cdot 5$

NOTE Be certain that you have a pattern that matches up every possible pair of coefficients.

Possible Factors	Middle Terms
$(x + y)(6x - 10y)$	$-4xy$
$(x + 10y)(6x - y)$	$59xy$
$(6x + y)(x - 10y)$	$-59xy$
$(6x + 10y)(x - y)$	$4xy$
$(x + 2y)(6x - 5y)$	$7xy$

Once more, we stop as soon as the correct factors are found.

$$6x^2 + 7xy - 10y^2 = (x + 2y)(6x - 5y)$$

CHECK YOURSELF 4

Factor $15x^2 - 4xy - 4y^2$.

The next example illustrates a special kind of trinomial called a *perfect square trinomial.*

Example 5

Factoring a Trinomial

Factor $9x^2 + 12xy + 4y^2$.

Because all terms are positive, the form of the factors must be

$$(\_x + \_y)(\_x + \_y)$$

Consider the factors of the first and last coefficients.

$$9 = 9 \cdot 1 \qquad 4 = 4 \cdot 1$$
$$= 3 \cdot 3 \qquad = 2 \cdot 2$$

Possible Factors	Middle Terms
$(x + y)(9x + 4y)$	$13xy$
$(x + 4y)(9x + y)$	$37xy$
$(3x + 2y)(3x + 2y)$	$12xy$

So

NOTE Perfect square trinomials can be factored by using previous methods. Recognizing the special pattern simply saves time.

$$9x^2 + 12xy + 4y^2 = (3x + 2y)(3x + 2y)$$
$$= (3x + 2y)^2$$

Square of $3x$ $2(3x)(2y)$ Square of $2y$

This trinomial is the result of squaring a binomial, thus the special name of perfect square trinomial.

CHECK YOURSELF 5

Factor.

(a) $4x^2 + 28x + 49$ **(b)** $16x^2 - 40xy + 25y^2$

Before we look at our next example, let's review one important point from Section 4.2. Recall that when you factor trinomials, you should not forget to look for a common factor as the first step. If there is a common factor, remove it and factor the remaining trinomial as before.

Example 6

Factoring a Trinomial

Factor $18x^2 - 18x + 4$.

First look for a common factor in all three terms. Here that factor is 2, so write

$$18x^2 - 18x + 4 = 2(9x^2 - 9x + 2)$$

By our earlier methods, we can factor the remaining trinomial as

NOTE If you don't see why this is true, you need to use your pencil to work it out before you move on!

$$9x^2 - 9x + 2 = (3x - 1)(3x - 2)$$

So

$$18x^2 - 18x + 4 = 2(3x - 1)(3x - 2)$$

Don't forget the 2 that was factored out!

CHECK YOURSELF 6

Factor $16x^2 + 44x - 12$.

Let's look at an example in which the common factor includes a variable.

Example 7

Factoring a Trinomial

Factor

$$6x^3 + 10x^2 - 4x$$

The common factor is $2x$.

So

$$6x^3 + 10x^2 - 4x = 2x(3x^2 + 5x - 2)$$

Because

$$3x^2 + 5x - 2 = (3x - 1)(x + 2)$$

we have

NOTE Remember to include the monomial factor.

$$6x^3 + 10x^2 - 4x = 2x(3x - 1)(x + 2)$$

CHECK YOURSELF 7

Factor $6x^3 - 27x^2 + 30x$.

You have now had a chance to work with a variety of factoring techniques. Your success in factoring polynomials depends on your ability to recognize when to use which technique. Here are some guidelines to help you apply the factoring methods you have studied in this chapter.

Step by Step: Factoring Polynomials

Step 1 Look for a greatest common factor other than 1. If such a factor exists, factor out the GCF.

Step 2 If the polynomial that remains is a *trinomial*, try to factor the trinomial by the trial-and-error methods of Sections 4.2 and 4.3.

The following example illustrates the use of this strategy.

Example 8

Factoring a Trinomial

(a) Factor $5m^2n + 20n$.

First, we see that the GCF is $5n$. Removing that factor gives

$$5m^2n + 20n = 5n(m^2 + 4)$$

(b) Factor $3x^3 - 24x^2 + 48x$.

First, we see that the GCF is $3x$. Factoring out $3x$ yields

$$\begin{aligned} 3x^3 - 24x^2 + 48x &= 3x(x^2 - 8x + 16) \\ &= 3x(x - 4)(x - 4) \end{aligned}$$

(c) Factor $8r^2s + 20rs^2 - 12s^3$.

First, the GCF is $4s$, and we can write the original polynomial as

$$8r^2s + 20rs^2 - 12s^3 = 4s(2r^2 + 5rs - 3s^2)$$

Because the remaining polynomial is a trinomial, we can use the trial-and-error method to complete the factoring as

$$8r^2s + 20rs^2 - 12s^3 = 4s(2r - s)(r + 3s)$$

CHECK YOURSELF 8

Factor the following polynomials.

(a) $8a^3 + 32a^2b + 32ab^2$ **(b)** $7x^3 + 7x^2y - 42xy^2$ **(c)** $5m^4 + 15m^3 + 5m^2$

CHECK YOURSELF ANSWERS

1. (a) $(5x + 4)(x + 2)$; **(b)** $(3x + 2)(x + 6)$ **2. (a)** $(2x - 3)(x - 3)$; **(b)** $(6x - 5)(x - 2)$ **3.** $(4x - 3)(x + 2)$ **4.** $(3x - 2y)(5x + 2y)$ **5. (a)** $(2x + 7)^2$; **(b)** $(4x - 5y)^2$ **6.** $4(4x - 1)(x + 3)$ **7.** $3x(2x - 5)(x - 2)$ **8. (a)** $8a(a + 2b)(a + 2b)$; **(b)** $7x(x + 3y)(x - 2y)$; **(c)** $5m^2(m^2 + 3m + 1)$

Name ____________

Section ________ Date ________

4.3 Exercises

Complete each of the following statements.

1. $4x^2 - 4x - 3 = (2x + 1)(\quad)$

2. $3w^2 + 11w - 4 = (w + 4)(\quad)$

3. $6a^2 + 13a + 6 = (2a + 3)(\quad)$

4. $25y^2 - 10y + 1 = (5y - 1)(\quad)$

5. $15x^2 - 16x + 4 = (3x - 2)(\quad)$

6. $6m^2 + 5m - 4 = (3m + 4)(\quad)$

7. $16a^2 + 8ab + b^2 = (4a + b)(\quad)$

8. $6x^2 + 5xy - 4y^2 = (3x + 4y)(\quad)$

9. $4m^2 + 5mn - 6n^2 = (m + 2n)(\quad)$

10. $10p^2 - pq - 3q^2 = (5p - 3q)(\quad)$

Factor each of the following polynomials.

11. $3x^2 + 7x + 2$

12. $5y^2 + 8y + 3$

13. $2w^2 + 13w + 15$

14. $3x^2 - 16x + 21$

15. $5x^2 - 16x + 3$

16. $2a^2 + 7a + 5$

17. $4x^2 - 12x + 5$

18. $2x^2 + 11x + 12$

19. $3x^2 - 5x - 2$

20. $4m^2 - 23m + 15$

21. $4p^2 + 19p - 5$

22. $5x^2 - 36x + 7$

23. $6x^2 + 19x + 10$

24. $6x^2 - 7x - 3$

ANSWERS

1. $2x - 3$
2. $3w - 1$
3. $3a + 2$
4. $5y - 1$
5. $5x - 2$
6. $2m - 1$
7. $4a + b$
8. $2x - y$
9. $4m - 3n$
10. $2p + q$
11. $(3x + 1)(x + 2)$
12. $(5y + 3)(y + 1)$
13. $(2w + 3)(w + 5)$
14. $(3x - 7)(x - 3)$
15. $(5x - 1)(x - 3)$
16. $(2a + 5)(a + 1)$
17. $(2x - 5)(2x - 1)$
18. $(2x + 3)(x + 4)$
19. $(3x + 1)(x - 2)$
20. $(4m - 3)(m - 5)$
21. $(4p - 1)(p + 5)$
22. $(5x - 1)(x - 7)$
23. $(3x + 2)(2x + 5)$
24. $(3x + 1)(2x - 3)$

ANSWERS

25. $(5x - 3)(3x + 2)$
26. $(4w + 1)(3w + 4)$
27. $(6m - 5)(m + 5)$
28. $(4x + 3)(2x - 3)$
29. $(3x - 2)(3x - 2)$
30. $(4x - 3)(5x - 2)$
31. $(6x + 5)(2x - 3)$
32. $(4a + 5)(4a + 5)$
33. $(3y - 2)(y + 3)$
34. $(4x - 3)(3x + 5)$
35. $(8x + 5)(x - 4)$
36. $(8v - 9)(3v + 4)$
37. $(2x + y)(x + y)$
38. $(3x - 2y)(x - y)$
39. $(5a + 2b)(a - 2b)$
40. $(5x - 3y)(x + 2y)$
41. $(9x - 5y)(x + y)$
42. $(4x + 3y)(4x + 5y)$
43. $(3m - 4n)(2m - 3n)$
44. $(5x + 3y)(3x - 2y)$
45. $(12a - 5b)(3a + b)$
46. $(3q + r)(q - 6r)$
47. $(x + 2y)(x + 2y)$
48. $(5b - 8c)(5b - 8c)$

25. $15x^2 + x - 6$

26. $12w^2 + 19w + 4$

27. $6m^2 + 25m - 25$

28. $8x^2 - 6x - 9$

29. $9x^2 - 12x + 4$

30. $20x^2 - 23x + 6$

31. $12x^2 - 8x - 15$

32. $16a^2 + 40a + 25$

33. $3y^2 + 7y - 6$

34. $12x^2 + 11x - 15$

35. $8x^2 - 27x - 20$

36. $24v^2 + 5v - 36$

37. $2x^2 + 3xy + y^2$

38. $3x^2 - 5xy + 2y^2$

39. $5a^2 - 8ab - 4b^2$

40. $5x^2 + 7xy - 6y^2$

41. $9x^2 + 4xy - 5y^2$

42. $16x^2 + 32xy + 15y^2$

43. $6m^2 - 17mn + 12n^2$

44. $15x^2 - xy - 6y^2$

45. $36a^2 - 3ab - 5b^2$

46. $3q^2 - 17qr - 6r^2$

47. $x^2 + 4xy + 4y^2$

48. $25b^2 - 80bc + 64c^2$

Factor each of the following polynomials completely.

49. $20x^2 - 20x - 15$

50. $24x^2 - 18x - 6$

51. $8m^2 + 12m + 4$

52. $14x^2 - 20x + 6$

53. $15r^2 - 21rs + 6s^2$

54. $10x^2 + 5xy - 30y^2$

55. $2x^3 - 2x^2 - 4x$

56. $2y^3 + y^2 - 3y$

57. $2y^4 + 5y^3 + 3y^2$

58. $4z^3 - 18z^2 - 10z$

59. $36a^3 - 66a^2 + 18a$

60. $20n^4 - 22n^3 - 12n^2$

61. $9p^2 + 30pq + 21q^2$

62. $12x^2 + 2xy - 24y^2$

Factor each of the following polynomials completely.

63. $10(x + y)^2 - 11(x + y) - 6$

64. $8(a - b)^2 + 14(a - b) - 15$

65. $5(x - 1)^2 - 15(x - 1) - 350$

66. $3(x + 1)^2 - 6(x + 1) - 45$

67. $15 + 29x - 48x^2$

68. $12 + 4a - 21a^2$

69. $-6x^2 + 19x - 15$

70. $-3s^2 - 10s + 8$

ANSWERS

49. $5(2x - 3)(2x + 1)$

50. $3(4x + 1)(2x - 2)$

51. $4(2m + 1)(m + 1)$

52. $2(7x - 3)(x - 1)$

53. $3(5r - 2s)(r - s)$

54. $5(2x - 3y)(x + 2y)$

55. $2x(x - 2)(x + 1)$

56. $y(2y + 3)(y - 1)$

57. $y^2(2y + 3)(y + 1)$

58. $2z(2z + 1)(z - 5)$

59. $6a(3a - 1)(2a - 3)$

60. $2n^2(5n + 2)(2n - 3)$

61. $3(p + q)(3p + 7q)$

62. $2(3x - 4y)(2x + 3y)$

63. $(5x + 5y + 2)(2x + 2y - 3)$

64. $(4a - 4b - 3)(2a - 2b + 5)$

65. $5(x - 11)(x + 6)$

66. $3(x - 4)(x + 4)$

67. $(1 + 3x)(15 - 16x)$

68. $(6 - 7a)(2 + 3a)$

69. $(3x - 5)(-2x + 3)$

70. $(-3s + 2)(s + 4)$

ANSWERS

a. $x^2 - 1$

b. $a^2 - 49$

c. $x^2 - y^2$

d. $4x^2 - 25$

e. $9a^2 - b^2$

f. $25a^2 - 16b^2$

Getting Ready for Section 4.4 [Section 3.5]

Multiply.

(a) $(x - 1)(x + 1)$
(b) $(a + 7)(a - 7)$
(c) $(x - y)(x + y)$
(d) $(2x - 5)(2x + 5)$
(e) $(3a - b)(3a + b)$
(f) $(5a - 4b)(5a + 4b)$

Answers

1. $2x - 3$ **3.** $3a + 2$ **5.** $5x - 2$ **7.** $4a + b$ **9.** $4m - 3n$
11. $(3x + 1)(x + 2)$ **13.** $(2w + 3)(w + 5)$ **15.** $(5x - 1)(x - 3)$
17. $(2x - 5)(2x - 1)$ **19.** $(3x + 1)(x - 2)$ **21.** $(4p - 1)(p + 5)$
23. $(3x + 2)(2x + 5)$ **25.** $(5x - 3)(3x + 2)$ **27.** $(6m - 5)(m + 5)$
29. $(3x - 2)(3x - 2)$ **31.** $(6x + 5)(2x - 3)$ **33.** $(3y - 2)(y + 3)$
35. $(8x + 5)(x - 4)$ **37.** $(2x + y)(x + y)$ **39.** $(5a + 2b)(a - 2b)$
41. $(9x - 5y)(x + y)$ **43.** $(3m - 4n)(2m - 3n)$ **45.** $(12a - 5b)(3a + b)$
47. $(x + 2y)^2$ **49.** $5(2x - 3)(2x + 1)$ **51.** $4(2m + 1)(m + 1)$
53. $3(5r - 2s)(r - s)$ **55.** $2x(x - 2)(x + 1)$ **57.** $y^2(2y + 3)(y + 1)$
59. $6a(3a - 1)(2a - 3)$ **61.** $3(p + q)(3p + 7q)$
63. $(5x + 5y + 2)(2x + 2y - 3)$ **65.** $5(x - 11)(x + 6)$
67. $(1 + 3x)(15 - 16x)$ **69.** $(3x - 5)(-2x + 3)$ **a.** $x^2 - 1$
b. $a^2 - 49$ **c.** $x^2 - y^2$ **d.** $4x^2 - 25$ **e.** $9a^2 - b^2$ **f.** $25a^2 - 16b^2$

4.4 Difference of Squares and Perfect Square Trinomials

OBJECTIVES

1. Factor a binomial that is the difference of two squares
2. Factor a perfect square trinomial

In Section 3.5, we introduced some special products. Recall the following formula for the product of a sum and difference of two terms:

$$(a + b)(a - b) = a^2 - b^2 \qquad (1)$$

This also means that a binomial of the form $a^2 - b^2$, called a **difference of two squares,** has as its factors $a + b$ and $a - b$.

To use this idea for factoring, we can write

$$a^2 - b^2 = (a + b)(a - b) \qquad (2)$$

A **perfect square** term has a coefficient that is a square (1, 4, 9, 16, 25, 36, etc.), and any variables have exponents that are multiples of 2 (x^2, y^4, z^6, etc.).

Example 1

Identifying Perfect Square Terms

For each of the following, decide whether it is a perfect square term. If it is, find the expression that was squared (called the *root*).

(a) $36x$
(b) $24x^6$
(c) $9x^4$
(d) $64x^6$
(e) $16x^9$

Only parts c and d are perfect square terms.

$9x^4 = (3x^2)^2$

$64x^6 = (8x^3)^2$

CHECK YOURSELF 1

For each of the following, decide whether it is a perfect square term. If it is, find the expression that was squared.

(a) $36x^{12}$ **(b)** $4x^6$
(c) $9x^7$ **(d)** $25x^8$
(e) $16x^{25}$

We will now use equation 2 above to factor the difference between two perfect square terms.

Example 2

Factoring the Difference of Two Squares

Factor $x^2 - 16$.

Think $x^2 - 4^2$

NOTE You could also write $(x - 4)(x + 4)$. The order doesn't matter because multiplication is commutative.

Because $x^2 - 16$ is a difference of squares, we have

$$x^2 - 16 = (x + 4)(x - 4)$$

CHECK YOURSELF 2

Factor $m^2 - 49$.

Any time an expression is a difference of two squares, it can be factored.

Example 3

Factoring the Difference of Two Squares

Factor $4a^2 - 9$.

Think $(2a)^2 - 3^2$

So

$$4a^2 - 9 = (2a)^2 - (3)^2$$
$$= (2a + 3)(2a - 3)$$

CHECK YOURSELF 3

Factor $9b^2 - 25$.

The process for factoring a difference of squares does not change when more than one variable is involved.

Example 4

Factoring the Difference of Two Squares

NOTE Think $(5a)^2 - (4b^2)^2$

Factor $25a^2 - 16b^4$.

$$25a^2 - 16b^4 = (5a + 4b^2)(5a - 4b^2)$$

CHECK YOURSELF 4

Factor $49c^4 - 9d^2$.

We will now consider an example that combines common-term factoring with difference-of-squares factoring. Note that the common factor is always removed as the *first step.*

Example 5

Removing the GCF First

NOTE Step 1
Remove the GCF.
Step 2
Factor the remaining binomial.

Factor $32x^2y - 18y^3$.

Note that $2y$ is a common factor, so

$$32x^2y - 18y^3 = 2y\underbrace{(16x^2 - 9y^2)}_{\text{Difference of squares}}$$

$$= 2y(4x + 3y)(4x - 3y)$$

CHECK YOURSELF 5

Factor $50a^3 - 8ab^2$.

CAUTION

Note that this is different from the sum of two squares (like $x^2 + y^2$), which never has integer factors.

Recall the following multiplication pattern.

$$(a + b)^2 = a^2 + 2ab + b^2$$

For example,

$$(x + 2)^2 = x^2 + 4x + 4$$

$$(x + 5)^2 = x^2 + 10x + 25$$

$$(2x + 1)^2 = 4x^2 + 4x + 1$$

Recognizing this pattern can simplify the process of factoring perfect square trinomials.

Example 6

Factoring a Perfect Square Trinomial

Factor the trinomial $4x^2 + 12xy + 9y^2$.

Note that this is a perfect square trinomial in which

$a = 2x$ and $b = 3y$.

In factored form, we have

$$4x^2 + 12xy + 9y^2 = (2x + 3y)^2$$

CHECK YOURSELF 6

Factor the trinomial $16u^2 + 24uv + 9v^2$.

Recognizing the same pattern can simplify the process of factoring perfect square trinomials in which the second term is negative.

Example 7

Factoring a Perfect Square Trinomial

Factor the trinomial $25x^2 - 10xy + y^2$.

This is also a perfect square trinomial, in which

$a = 5x$ and $b = -y$.

In factored form, we have

$$25x^2 - 10xy + y^2 = (5x + (-y))^2 = (5x - y)^2$$

CHECK YOURSELF 7

Factor the trinomial $4u^2 - 12uv + 9v^2$.

CHECK YOURSELF ANSWERS

1. (a) $(6x^6)^2$; **(b)** $(2x^3)^2$; **(d)** $(5x^4)^2$ **2.** $(m + 7)(m - 7)$ **3.** $(3b + 5)(3b - 5)$
4. $(7c^2 + 3d)(7c^2 - 3d)$ **5.** $2a(5a + 2b)(5a - 2b)$ **6.** $(4u + 3v)^2$
7. $(2u - 3v)^2$

Name ____________

Section ______ Date ______

4.4 Exercises

For each of the following binomials, state whether the binomial is a difference of squares.

1. $3x^2 + 2y^2$

2. $5x^2 - 7y^2$

3. $16a^2 - 25b^2$

4. $9n^2 - 16m^2$

5. $16r^2 + 4$

6. $p^2 - 45$

7. $16a^2 - 12b^3$

8. $9a^2b^2 - 16c^2d^2$

9. $a^2b^2 - 25$

10. $4a^3 - b^3$

Factor the following binomials.

11. $m^2 - n^2$

12. $r^2 - 9$

13. $x^2 - 49$

14. $c^2 - d^2$

15. $49 - y^2$

16. $81 - b^2$

17. $9b^2 - 16$

18. $36 - x^2$

19. $16w^2 - 49$

20. $4x^2 - 25$

21. $4s^2 - 9r^2$

22. $64y^2 - x^2$

23. $9w^2 - 49z^2$

24. $25x^2 - 81y^2$

ANSWERS

1. No
2. No
3. Yes
4. Yes
5. No
6. No
7. No
8. Yes
9. Yes
10. No
11. $(m + n)(m - n)$
12. $(r + 3)(r - 3)$
13. $(x + 7)(x - 7)$
14. $(c + d)(c - d)$
15. $(7 + y)(7 - y)$
16. $(9 + b)(9 - b)$
17. $(3b + 4)(3b - 4)$
18. $(6 + x)(6 - x)$
19. $(4w + 7)(4w - 7)$
20. $(2x + 5)(2x - 5)$
21. $(2s + 3r)(2s - 3r)$
22. $(8y + x)(8y - x)$
23. $(3w + 7z)(3w - 7z)$
24. $(5x + 9y)(5x - 9y)$

ANSWERS

25. $(4a + 7b)(4a - 7b)$
26. $(8m + 3n)(8m - 3n)$
27. $(x^2 + 6)(x^2 - 6)$
28. $(y^3 + 7)(y^3 - 7)$
29. $(xy + 4)(xy - 4)$
30. $(mn + 8)(mn - 8)$
31. $(5 + ab)(5 - ab)$
32. $(7 + wz)(7 - wz)$
33. $(r^2 + 2s)(r^2 - 2s)$
34. $(p + 3q^2)(p - 3q^2)$
35. $(9a + 10b^3)(9a - 10b^3)$
36. $(8x^2 + 5y^2)(8x^2 - 5y^2)$
37. $2x(3x + y)(3x - y)$
38. $2b(5a + b)(5a - b)$
39. $3mn(2m + 5n)(2m - 5n)$
40. $7p^2(3p + q)(3p - q)$
41. $3b^2(4a + 3b)(4a - 3b)$
42. $5w^3(2w + 3z^2)(2w - 3z^2)$
43. Yes; $(x - 7)^2$
44. No
45. No
46. Yes; $(x + 5)^2$
47. Yes; $(x - 9)^2$
48. No
49. $(x + 2)^2$
50. $(x + 3)^2$
51. $(x - 5)^2$
52. $(x - 4)^2$

25. $16a^2 - 49b^2$

26. $64m^2 - 9n^2$

27. $x^4 - 36$

28. $y^6 - 49$

29. $x^2y^2 - 16$

30. $m^2n^2 - 64$

31. $25 - a^2b^2$

32. $49 - w^2z^2$

33. $r^4 - 4s^2$

34. $p^2 - 9q^4$

35. $81a^2 - 100b^6$

36. $64x^4 - 25y^4$

37. $18x^3 - 2xy^2$

38. $50a^2b - 2b^3$

39. $12m^3n - 75mn^3$

40. $63p^4 - 7p^2q^2$

41. $48a^2b^2 - 27b^4$

42. $20w^5 - 45w^3z^4$

Determine whether each of the following trinomials is a perfect square. If it is, factor the trinomial.

43. $x^2 - 14x + 49$

44. $x^2 + 9x + 16$

45. $x^2 - 18x - 81$

46. $x^2 + 10x + 25$

47. $x^2 - 18x + 81$

48. $x^2 - 24x + 48$

Factor the following trinomials.

49. $x^2 + 4x + 4$

50. $x^2 + 6x + 9$

51. $x^2 - 10x + 25$

52. $x^2 - 8x + 16$

53. $4x^2 + 12xy + 9y^2$

54. $16x^2 + 40xy + 25y^2$

55. $9x^2 - 24xy + 16y^2$

56. $9w^2 - 30wv + 25v^2$

57. $y^3 - 10y^2 + 25y$

58. $12b^3 - 12b^2 + 3b$

Factor each expression.

59. $x^2(x + y) - y^2(x + y)$

60. $a^2(b - c) - 16b^2(b - c)$

61. $2m^2(m - 2n) - 18n^2(m - 2n)$

62. $3a^3(2a + b) - 27ab^2(2a + b)$

63. Find a value for k so that $kx^2 - 25$ will have the factors $2x + 5$ and $2x - 5$.

64. Find a value for k so that $9m^2 - kn^2$ will have the factors $3m + 7n$ and $3m - 7n$.

65. Find a value for k so that $2x^3 - kxy^2$ will have the factors $2x$, $x - 3y$, and $x + 3y$.

66. Find a value for k so that $20a^3b - kab^3$ will have the factors $5ab$, $2a - 3b$, and $2a + 3b$.

67. Complete the following statement in complete sentences: "To factor a number you . . . ."

68. Complete this statement: To factor an algebraic expression into prime factors means . . . .

Getting Ready for Section 4.5 [Section 4.1]

Factor.

(a) $2x(3x + 2) - 5(3x + 2)$

(b) $3y(y - 4) + 5(y - 4)$

(c) $3x(x + 2y) + y(x + 2y)$

(d) $5x(2x - y) - 3(2x - y)$

(e) $4x(2x - 5y) - 3y(2x - 5y)$

ANSWERS

53. $(2x + 3y)^2$

54. $(4x + 5y)^2$

55. $(3x - 4y)^2$

56. $(3w - 5v)^2$

57. $y(y - 5)^2$

58. $3b(2b - 1)^2$

59. $(x + y)^2(x - y)$

60. $(b - c)(a + 4b)(a - 4b)$

61. $2(m - 2n)(m + 3n)(m - 3n)$

62. $3a(2a + b)(a + 3b)(a - 3b)$

63. 4

64. 49

65. 18

66. 45

67.

68.

a. $(3x + 2)(2x - 5)$

b. $(y - 4)(3y + 5)$

c. $(x + 2y)(3x + y)$

d. $(2x - y)(5x - 3)$

e. $(2x - 5y)(4x - 3y)$

Answers

1. No **3.** Yes **5.** No **7.** No **9.** Yes **11.** $(m + n)(m - n)$
13. $(x + 7)(x - 7)$ **15.** $(7 + y)(7 - y)$ **17.** $(3b + 4)(3b - 4)$
19. $(4w + 7)(4w - 7)$ **21.** $(2s + 3r)(2s - 3r)$ **23.** $(3w + 7z)(3w - 7z)$
25. $(4a + 7b)(4a - 7b)$ **27.** $(x^2 + 6)(x^2 - 6)$ **29.** $(xy + 4)(xy - 4)$
31. $(5 + ab)(5 - ab)$ **33.** $(r^2 + 2s)(r^2 - 2s)$ **35.** $(9a + 10b^3)(9a - 10b^3)$
37. $2x(3x + y)(3x - y)$ **39.** $3mn(2m + 5n)(2m - 5n)$
41. $3b^2(4a + 3b)(4a - 3b)$ **43.** Yes; $(x - 7)^2$ **45.** No **47.** Yes; $(x - 9)^2$
49. $(x + 2)^2$ **51.** $(x - 5)^2$ **53.** $(2x + 3y)^2$ **55.** $(3x - 4y)^2$
57. $y(y - 5)^2$ **59.** $(x + y)^2(x - y)$ **61.** $2(m - 2n)(m + 3n)(m - 3n)$
63. 4 **65.** 18 **67.** **a.** $(3x + 2)(2x - 5)$

b. $(y - 4)(3y + 5)$ **c.** $(x + 2y)(3x + y)$ **d.** $(2x - y)(5x - 3)$
e. $(2x - 5y)(4x - 3y)$

Factoring by Grouping

OBJECTIVES

1. Factor a polynomial by grouping terms
2. Rewrite a polynomial so that it can be factored by the method of grouping terms

Some polynomials can be factored by grouping the terms and finding common factors within each group. Such a process is called factoring by grouping, and will be explored in this section.

Recall that in Section 4.1, we looked at the expression

$3x(x + y) + 2(x + y)$

and found that we could factor out the common binomial, $(x + y)$, giving us

$(x + y)(3x + 2)$

That technique will be used in the first example.

Example 1

Factoring by Grouping Terms

Suppose we want to factor the polynomial

$ax - ay + bx - by$

As you can see, the polynomial has no common factors. However, look at what happens if we separate the polynomial into *two groups of two terms.*

NOTE Note that our example has *four* terms. That is the clue for trying the factoring by grouping method.

$$ax - ay + bx - by$$
$$= \underbrace{ax - ay}_{(1)} + \underbrace{bx - by}_{(2)}$$

Now *each* group has a common factor, and we can write the polynomial as

$a(x - y) + b(x - y)$

In this form, we can see that $x - y$ is the GCF. Factoring out $x - y$, we get

$a(x - y) + b(x - y) = (x - y)(a + b)$

CHECK YOURSELF 1

Use the factoring by grouping method.

$x^2 - 2xy + 3x - 6y$

Be particularly careful of your treatment of algebraic signs when applying the factoring by grouping method. Consider Example 2.

Example 2

Factoring by Grouping Terms

Factor $2x^3 - 3x^2 - 6x + 9$.

We group the polynomial as follows.

$\underbrace{2x^3 - 3x^2}_{(1)} \underbrace{- 6x + 9}_{(2)}$ Remove the common factor of −3 from the second two terms.

NOTE Notice that $9 = (-3)(-3)$.

$= x^2(2x - 3) - 3(2x - 3)$

$= (2x - 3)(x^2 - 3)$

CHECK YOURSELF 2

Factor by grouping.

$3y^3 + 2y^2 - 6y - 4$

It may also be necessary to change the order of the terms as they are grouped. Look at Example 3.

Example 3

Factoring by Grouping Terms

Factor $x^2 - 6yz + 2xy - 3xz$.

Grouping the terms as before, we have

$\underbrace{x^2 - 6yz}_{(1)} + \underbrace{2xy - 3xz}_{(2)}$

Do you see that we have accomplished nothing because there are no common factors in the first group?

We can, however, rearrange the terms to write the original polynomial as

$\underbrace{x^2 + 2xy}_{(1)} \underbrace{- 3xz - 6yz}_{(2)}$

$= x(x + 2y) - 3z(x + 2y)$ We can now remove the common factor of $x + 2y$ in group (1) and group (2).

$= (x + 2y)(x - 3z)$

Note: It is often true that the grouping can be done in more than one way. The factored form will be the same.

CHECK YOURSELF 3

We can write the polynomial of Example 3 as

$x^2 - 3xz + 2xy - 6yz$

Factor, and verify that the factored form is the same in either case.

CHECK YOURSELF ANSWERS

1. $(x - 2y)(x + 3)$ **2.** $(3y + 2)(y^2 - 2)$ **3.** $(x - 3z)(x + 2y)$

Name ______________________

Section ________ Date ________

4.5 Exercises

Factor each polynomial by grouping the first two terms and the last two terms.

1. $x^3 - 4x^2 + 3x - 12$

2. $x^3 - 6x^2 + 2x - 12$

3. $a^3 - 3a^2 + 5a - 15$

4. $6x^3 - 2x^2 + 9x - 3$

5. $10x^3 + 5x^2 - 2x - 1$

6. $x^5 + x^3 - 2x^2 - 2$

7. $x^4 - 2x^3 + 3x - 6$

8. $x^3 - 4x^2 + 2x - 8$

Factor each polynomial completely by removing any common factors, and then factor by grouping. Do not combine like terms.

9. $3x - 6 + xy - 2y$

10. $2x - 10 + xy - 5y$

11. $ab - ac + b^2 - bc$

12. $ax + 2a + bx + 2b$

13. $3x^2 - 2xy + 3x - 2y$

14. $xy - 5y^2 - x + 5y$

15. $5s^2 + 15st - 2st - 6t^2$

$5s(s+3t) - 2t(s+3t)$

$(s+3t)(5s-2t)$

16. $3a^3 + 3ab^2 + 2a^2b + 2b^3$

17. $3x^3 + 6x^2y - x^2y - 2xy^2$

18. $2p^4 + 3p^3q - 2p^3q - 3p^2q^2$

19. $x^4 + 5x^3 - 2x^2 - 10x$

20. $x^4y - 2x^3y + x^4 - 2x^3$

21. $2x^3 - 2x^2 + 3x^2 - 3x$

22. $3b^4 - 3b^3c + 2b^3c - 2b^2c^2$

ANSWERS

1. $(x - 4)(x^2 + 3)$
2. $(x - 6)(x^2 + 2)$
3. $(a - 3)(a^2 + 5)$
4. $(2x^2 + 3)(3x - 1)$
5. $(5x^2 - 1)(2x + 1)$
6. $(x^2 + 1)(x^3 - 2)$
7. $(x - 2)(x^3 + 3)$
8. $(x - 4)(x^2 + 2)$
9. $(x - 2)(3 + y)$
10. $(x - 5)(2 + y)$
11. $(b - c)(a + b)$
12. $(x + 2)(a + b)$
13. $(3x - 2y)(x + 1)$
14. $(x - 5y)(y - 1)$
15. $(s + 3t)(5s - 2t)$
16. $(a^2 + b^2)(3a + 2b)$
17. $x(x + 2y)(3x - y)$
18. $p^2(2p + 3q)(p - q)$
19. $x(x + 5)(x^2 - 2)$
20. $x^3(x - 2)(y + 1)$
21. $x(x - 1)(2x + 3)$
22. $b^2(3b + 2c)(b - c)$

ANSWERS

a. $4x^2 + 4x - 3$

b. $3a^2 + 11a - 4$

c. $2x^2 - 11x + 12$

d. $2w^2 - 7w - 22$

e. $2y^2 + 19y + 45$

f. $2x^2 - 23x - 12$

g. $2p^2 + 23p + 45$

h. $6a^2 + 2a - 20$

Getting Ready for Section 4.6 [Section 3.4]

Multiply.

(a) $(2x - 1)(2x + 3)$
(b) $(3a - 1)(a + 4)$
(c) $(x - 4)(2x - 3)$
(d) $(2w - 11)(w + 2)$
(e) $(y + 5)(2y + 9)$
(f) $(2x + 1)(x - 12)$
(g) $(p + 9)(2p + 5)$
(h) $(3a - 5)(2a + 4)$

Answers

1. $(x - 4)(x^2 + 3)$ **3.** $(a - 3)(a^2 + 5)$ **5.** $(5x^2 - 1)(2x + 1)$ **7.** $(x - 2)(x^3 + 3)$ **9.** $(x - 2)(3 + y)$ **11.** $(b - c)(a + b)$ **13.** $(3x - 2y)(x + 1)$ **15.** $(s + 3t)(5s - 2t)$ **17.** $x(x + 2y)(3x - y)$ **19.** $x(x + 5)(x^2 - 2)$ **21.** $x(x - 1)(2x + 3)$ **a.** $4x^2 + 4x - 3$ **b.** $3a^2 + 11a - 4$ **c.** $2x^2 - 11x + 12$ **d.** $2w^2 - 7w - 22$ **e.** $2y^2 + 19y + 45$ **f.** $2x^2 - 23x - 12$ **g.** $2p^2 + 23p + 45$ **h.** $6a^2 + 2a - 20$

Using the *ac* Method to Factor

OBJECTIVES

1. Use the *ac* test to determine factorability
2. Use the results of the *ac* test
3. Completely factor a trinomial

In Sections 4.2 and 4.3 we used the trial-and-error method to factor trinomials. We also learned that not all trinomials can be factored. In this section we will look at the same kinds of trinomials, but in a slightly different context. We first determine whether a trinomial is factorable. We then use the results of that analysis to factor the trinomial.

Some students prefer the trial-and-error method for factoring because it is generally faster and more intuitive. Other students prefer the method of this section (called the *ac* method) because it yields the answer in a systematic way. We will let you determine which method you prefer.

We will begin by looking at some factored trinomials.

Example 1

Matching Trinomials and Their Factors

Determine which of the following are true statements.

(a) $x^2 - 2x - 8 = (x - 4)(x + 2)$

This is a true statement. Using the FOIL method, we see that

$$\begin{aligned}(x - 4)(x + 2) &= x^2 + 2x - 4x - 8\\ &= x^2 - 2x - 8\end{aligned}$$

(b) $x^2 + 6x + 5 = (x + 2)(x + 3)$

This is not a true statement.

$(x + 2)(x + 3) = x^2 + 3x + 2x + 6 = x^2 + 5x + 6$

(c) $x^2 + 5x - 14 = (x - 2)(x + 7)$

This is true: $(x - 2)(x + 7) = x^2 + 7x - 2x - 14 = x^2 + 5x - 14$

(d) $x^2 - 8x - 15 = (x - 5)(x - 3)$

This is false: $(x - 5)(x - 3) = x^2 - 3x - 5x + 15 = x^2 - 8x + 15$

CHECK YOURSELF 1

Determine which of the following are true statements.

(a) $2x^2 - 2x - 3 = (2x - 3)(x + 1)$
(b) $3x^2 + 11x - 4 = (3x - 1)(x + 4)$
(c) $2x^2 - 7x + 3 = (x - 3)(2x - 1)$

The first step in learning to factor a trinomial is to identify its coefficients. So that we are consistent, we first write the trinomial in standard $ax^2 + bx + c$ form, then label the three coefficients as a, b, and c.

Example 2

Identifying the Coefficients of $ax^2 + bx + c$

First, when necessary, rewrite the trinomial in $ax^2 + bx + c$ form. Then give the values for a, b, and c, in which a is the coefficient of the x^2 term, b is the coefficient of the x term, and c is the constant.

(a) $x^2 - 3x - 18$

$a = 1 \qquad b = -3 \qquad c = -18$

NOTE Notice that the negative sign is attached to the coefficients.

(b) $x^2 - 24x + 23$

$a = 1 \qquad b = -24 \qquad c = 23$

(c) $x^2 + 8 - 11x$

First rewrite the trinomial in descending order:

$x^2 - 11x + 8$

$a = 1 \qquad b = -11 \qquad c = 8$

CHECK YOURSELF 2

First, when necessary, rewrite the trinomials in $ax^2 + bx + c$ form. Then label a, b, and c, in which a is the coefficient of the x^2 term, b is the coefficient of the x term, and c is the constant.

(a) $x^2 + 5x - 14$ **(b)** $x^2 - 18x + 17$ **(c)** $x - 6 + 2x^2$

Not all trinomials can be factored. To discover if a trinomial is factorable, we try the ***ac* test.**

Definitions: The *ac* Test

A trinomial of the form $ax^2 + bx + c$ is factorable if (and only if) there are two integers, m and n, such that

$ac = mn \qquad \text{and} \qquad b = m + n$

In Example 3 we will look for m and n to determine whether each trinomial is factorable.

Example 3

Using the *ac* Test

Use the *ac* test to determine which of the following trinomials can be factored. Find the values of m and n for each trinomial that can be factored.

(a) $x^2 - 3x - 18$

First, we find the values of a, b, and c, so that we can find ac.

$a = 1 \qquad b = -3 \qquad c = -18$

$ac = 1(-18) = -18$ and $b = -3$

Then, we look for two numbers, m and n, such that $mn = ac$, and $m + n = b$. In this case, that means

$mn = -18$ and $m + n = -3$

We now look at all pairs of integers with a product of -18. We then look at the sum of each pair of integers, looking for a sum of -3.

mn	$m + n$
$1(-18) = -18$	$1 + (-18) = -17$
$2(-9) = -18$	$2 + (-9) = -7$
$3(-6) = -18$	$3 + (-6) = -3$
$6(-3) = -18$	
$9(-2) = -18$	
$18(-1) = -18$	

We need look no further than 3 and −6.

3 and -6 are the two integers with a product of ac and a sum of b. We can say that

$m = 3$ and $n = -6$

NOTE We could have chosen $m = -6$ and $n = 3$ as well.

Because we found values for m and n, we know that $x^2 - 3x - 18$ is factorable.

(b) $x^2 - 24x + 23$

We find that

$a = 1 \qquad b = -24 \qquad c = 23$

$ac = 1(23) = 23$ and $b = -24$

So

$mn = 23$ and $m + n = -24$

We now calculate integer pairs, looking for two numbers with a product of 23 and a sum of -24.

mn	$m + n$
$1(23) = 23$	$1 + 23 = 24$
$-1(-23) = 23$	$-1 + (-23) = -24$

$m = -1$ and $n = -23$

So, $x^2 - 24x + 23$ is factorable.

(c) $x^2 - 11x + 8$

We find that $a = 1$, $b = -11$, and $c = 8$. Therefore, $ac = 8$ and $b = -11$. Thus $mn = 8$ and $m + n = -11$. We calculate integer pairs:

mn	$m + n$
$1(8) = 8$	$1 + 8 = 9$
$2(4) = 8$	$2 + 4 = 6$
$-1(-8) = 8$	$-1 + (-8) = -9$
$-2(-4) = 8$	$-2 + (-4) = -6$

There are no other pairs of integers with a product of 8, and none of these pairs has a sum of -11. The trinomial $x^2 - 11x + 8$ is not factorable.

(d) $2x^2 + 7x - 15$

We find that $a = 2$, $b = 7$, and $c = -15$. Therefore, $ac = 2(-15) = -30$ and $b = 7$. Thus $mn = -30$ and $m + n = 7$. We calculate integer pairs:

mn	$m + n$
$1(-30) = -30$	$1 + (-30) = -29$
$2(-15) = -30$	$2 + (-15) = -13$
$3(-10) = -30$	$3 + (-10) = -7$
$5(-6) = -30$	$5 + (-6) = -1$
$6(-5) = -30$	$6 + (-5) = 1$
$10(-3) = -30$	$10 + (-3) = 7$

There is no need to go any further. We see that 10 and -3 have a product of -30 and a sum of 7, so

$m = 10$ and $n = -3$

Therefore, $2x^2 + 7x - 15$ is factorable.

It is not always necessary to evaluate all the products and sums to determine whether a trinomial is factorable. You may have noticed patterns and shortcuts that make it easier to find m and n. By all means, use them to help you find m and n. This is essential in mathematical thinking. You are taught a mathematical process that will always work for solving a problem. Such a process is called an **algorithm.** It is very easy to teach a computer to use an algorithm. It is very difficult (some would say impossible) for a computer to have insight. Shortcuts that you discover are *insights.* They may be the most important part of your mathematical education.

CHECK YOURSELF 3

Use the ac test to determine which of the following trinomials can be factored. Find the values of m and n for each trinomial that can be factored.

(a) $x^2 - 7x + 12$ **(b)** $x^2 + 5x - 14$

(c) $3x^2 - 6x + 7$ **(d)** $2x^2 + x - 6$

So far we have used the results of the *ac* test only to determine whether a trinomial is factorable. The results can also be used to help factor the trinomial.

Example 4

Using the Results of the *ac* Test to Factor

Rewrite the middle term as the sum of two terms, then factor by grouping.

(a) $x^2 - 3x - 18$

We find that $a = 1$, $b = -3$, and $c = -18$, so $ac = -18$ and $b = -3$. We are looking for two numbers, m and n, where $mn = -18$ and $m + n = -3$. In Example 3, part a, we looked at every pair of integers whose product (mn) was -18, to find a pair that had a sum ($m + n$) of -3. We found the two integers to be 3 and -6, because $3(-6) = -18$ and $3 + (-6) = -3$, so $m = 3$ and $n = -6$. We now use that result to rewrite the middle term as the sum of $3x$ and $-6x$.

$$x^2 + 3x - 6x - 18$$

We then factor by grouping:

$$\begin{aligned} x^2 + 3x - 6x - 18 &= x(x + 3) - 6(x + 3) \\ &= (x + 3)(x - 6) \end{aligned}$$

(b) $x^2 - 24x + 23$

We use the results from Example 3, part b, in which we found $m = -1$ and $n = -23$, to rewrite the middle term of the equation.

$$x^2 - 24x + 23 = x^2 - x - 23x + 23$$

Then we factor by grouping:

$$\begin{aligned} x^2 - x - 23x + 23 &= (x^2 - x) - (23x - 23) \\ &= x(x - 1) - 23(x - 1) \\ &= (x - 1)(x - 23) \end{aligned}$$

(c) $2x^2 + 7x - 15$

From Example 3, part d, we know that this trinomial is factorable, and $m = 10$ and $n = -3$. We use that result to rewrite the middle term of the trinomial.

$$\begin{aligned} 2x^2 + 7x - 15 &= 2x^2 + 10x - 3x - 15 \\ &= (2x^2 + 10x) - (3x + 15) \\ &= 2x(x + 5) - 3(x + 5) \\ &= (x + 5)(2x - 3) \end{aligned}$$

Careful readers will note that we did not ask you to factor Example 3, part c, $x^2 - 11x + 8$. Recall that, by the *ac* method, we determined that this trinomial was not factorable.

CHECK YOURSELF 4

Use the results of Check Yourself 3 to rewrite the middle term as the sum of two terms, then factor by grouping.

(a) $x^2 - 7x + 12$ **(b)** $x^2 + 5x - 14$ **(c)** $2x^2 + x - 6$

Let's look at some examples that require us to first find m and n, then factor the trinomial.

Example 5

Rewriting Middle Terms to Factor

Rewrite the middle term as the sum of two terms, then factor by grouping.

(a) $2x^2 - 13x - 7$

We find that $a = 2$, $b = -13$, and $c = -7$, so $mn = ac = -14$ and $m + n = b = -13$. Therefore,

mn	$m + n$
$1(-14) = -14$	$1 + (-14) = -13$

So, $m = 1$ and $n = -14$. We rewrite the middle term of the trinomial as follows:

$$\begin{aligned} 2x^2 - 13x - 7 &= 2x^2 + x - 14x - 7 \\ &= (2x^2 + x) - (14x + 7) \\ &= x(2x + 1) - 7(2x + 1) \\ &= (2x + 1)(x - 7) \end{aligned}$$

(b) $6x^2 - 5x - 6$

We find that $a = 6$, $b = -5$, and $c = -6$, so $mn = ac = -36$ and $m + n = b = -5$.

mn	$m + n$
$1(-36) = -36$	$1 + (-36) = -35$
$2(-18) = -36$	$2 + (-18) = -16$
$3(-12) = -36$	$3 + (-12) = -9$
$4(-9) = -36$	$4 + (-9) = -5$

So, $m = 4$ and $n = -9$. We rewrite the middle term of the trinomial:

$$\begin{aligned} 6x^2 - 5x - 6 &= 6x^2 + 4x - 9x - 6 \\ &= (6x^2 + 4x) - (9x + 6) \\ &= 2x(3x + 2) - 3(3x + 2) \\ &= (3x + 2)(2x - 3) \end{aligned}$$

CHECK YOURSELF 5

Rewrite the middle term as the sum of two terms, then factor by grouping.

(a) $2x^2 - 7x - 15$

(b) $6x^2 - 5x - 4$

Be certain to check trinomials and binomial factors for any common monomial factor. (There is no common factor in the binomial unless it is also a common factor in the original trinomial.) Example 6 shows the removal of monomial factors.

Example 6

Removing Common Factors

Completely factor the trinomial.

$3x^2 + 12x - 15$

We could first remove the common factor of 3:

$3x^2 + 12x - 15 = 3(x^2 + 4x - 5)$

Finding m and n for the trinomial $x^2 + 4x - 5$ yields $mn = -5$ and $m + n = 4$.

mn	$m + n$
$1(-5) = -5$	$1 + (-5) = -4$
$5(-1) = -5$	$-1 + (5) = 4$

So, $m = 5$ and $n = -1$. This gives us

$$\begin{aligned} 3x^2 + 12x - 15 &= 3(x^2 + 4x - 5) \\ &= 3(x^2 + 5x - x - 5) \\ &= 3[(x^2 + 5x) - (x + 5)] \\ &= 3[x(x + 5) - (x + 5)] \\ &= 3[(x + 5)(x - 1)] \\ &= 3(x + 5)(x - 1) \end{aligned}$$

CHECK YOURSELF 6

Completely factor the trinomial.

$6x^3 + 3x^2 - 18x$

CHECK YOURSELF ANSWERS

1. (a) False; **(b)** true; **(c)** true **2. (a)** $a = 1, b = 5, c = -14$; **(b)** $a = 1, b = -18, c = 17$; **(c)** $a = 2, b = 1, c = -6$
3. (a) Factorable, $m = -3, n = -4$; **(b)** factorable, $m = 7, n = -2$; **(c)** not factorable; **(d)** factorable, $m = 4, n = -3$
4. (a) $x^2 - 3x - 4x + 12 = (x - 3)(x - 4)$; **(b)** $x^2 + 7x - 2x - 14 = (x + 7)(x - 2)$; **(c)** $2x^2 + 4x - 3x - 6 = (2x - 3)(x + 2)$
5. (a) $2x^2 - 10x + 3x - 15 = (2x + 3)(x - 5)$; **(b)** $6x^2 - 8x + 3x - 4 = (3x - 4)(2x + 1)$ **6.** $3x(2x - 3)(x + 2)$

Not all possible product pairs need to be tried to find m and n. A look at the sign pattern of the trinomial will eliminate many of the possibilities. Assuming the leading coefficient is positive, there are four possible sign patterns.

Pattern	Example	Conclusion
1. b and c are both positive.	$2x^2 + 13x + 15$	m and n must both be positive.
2. b is negative and c is positive.	$x^2 - 7x + 12$	m and n must both be negative.
3. b is positive and c is negative.	$x^2 + 3x - 10$	m and n are of opposite signs. (The value with the larger absolute value is positive.)
4. b is negative and c is negative.	$x^2 - 3x - 10$	m and n are of opposite signs. (The value with the larger absolute value is negative.)

Name ______________________

Section __________ Date __________

4.6 Exercises

State whether each of the following is true or false.

1. $x^2 + 2x - 3 = (x + 3)(x - 1)$

2. $y^2 - 3y - 18 = (y - 6)(y + 3)$

3. $x^2 - 10x - 24 = (x - 6)(x + 4)$

4. $a^2 + 9a - 36 = (a - 12)(a + 4)$

5. $x^2 - 16x + 64 = (x - 8)(x - 8)$

6. $w^2 - 12w - 45 = (w - 9)(w - 5)$

7. $25y^2 - 10y + 1 = (5y - 1)(5y + 1)$

8. $6x^2 + 5xy - 4y^2 = (6x - 2y)(x + 2y)$

9. $10p^2 - pq - 3q^2 = (5p - 3q)(2p + q)$

10. $6a^2 + 13a + 6 = (2a + 3)(3a + 2)$

For each of the following trinomials, label a, b, and c.

11. $x^2 + 4x - 9$

12. $x^2 + 5x + 11$

13. $x^2 - 3x + 8$

14. $x^2 + 7x - 15$

15. $3x^2 + 5x - 8$

16. $2x^2 + 7x - 9$

17. $4x^2 + 8x + 11$

18. $5x^2 + 7x - 9$

19. $-3x^2 + 5x - 10$

20. $-7x^2 + 9x - 18$

ANSWERS

1. True
2. True
3. False
4. False
5. True
6. False
7. False
8. False
9. True
10. True
11. $a = 1, b = 4, c = -9$
12. $a = 1, b = 5, c = 11$
13. $a = 1, b = -3, c = 8$
14. $a = 1, b = 7, c = -15$
15. $a = 3, b = 5, c = -8$
16. $a = 2, b = 7, c = -9$
17. $a = 4, b = 8, c = 11$
18. $a = 5, b = 7, c = -9$
19. $a = -3, b = 5, c = -10$
20. $a = -7, b = 9, c = -18$

ANSWERS

21. Factorable; 3, −2
22. Factorable; 5, −3
23. Not factorable
24. Not factorable
25. Factorable; −3, −2
26. Not factorable
27. Factorable; 6, −1
28. Factorable; −15, 1
29. Factorable; −15, −4
30. Not factorable
31. $x^2 + 2x + 4x + 8$; $(x + 2)(x + 4)$
32. $x^2 + 5x - 2x - 10$; $(x - 2)(x + 5)$
33. $x^2 - 5x - 4x + 20$; $(x - 5)(x - 4)$
34. $x^2 - 5x - 3x + 15$; $(x - 5)(x - 3)$
35. $x^2 - 9x + 7x - 63$; $(x - 9)(x + 7)$
36. $x^2 + 11x - 5x - 55$; $(x + 11)(x - 5)$
37. $(x + 3)(x + 5)$
38. $(x - 8)(x - 3)$
39. $(x - 4)(x - 7)$
40. $(y - 5)(y + 4)$
41. $(s + 10)(s + 3)$
42. $(b + 3)(b + 11)$
43. $(a - 8)(a + 6)$
44. $(x - 12)(x - 5)$

Use the *ac* test to determine which of the following trinomials can be factored. Find the values of m and n for each trinomial that can be factored.

21. $x^2 + x - 6$

22. $x^2 + 2x - 15$

23. $x^2 + x + 2$

24. $x^2 - 3x + 7$

25. $x^2 - 5x + 6$

26. $x^2 - x + 2$

27. $2x^2 + 5x - 3$

28. $3x^2 - 14x - 5$

29. $6x^2 - 19x + 10$

30. $4x^2 + 5x + 6$

Rewrite the middle term as the sum of two terms and then factor by grouping.

31. $x^2 + 6x + 8$

32. $x^2 + 3x - 10$

33. $x^2 - 9x + 20$

34. $x^2 - 8x + 15$

35. $x^2 - 2x - 63$

36. $x^2 + 6x - 55$

Rewrite the middle term as the sum of two terms and then factor completely.

37. $x^2 + 8x + 15$

38. $x^2 - 11x + 24$

39. $x^2 - 11x + 28$

40. $y^2 - y - 20$

41. $s^2 + 13s + 30$

42. $b^2 + 14b + 33$

43. $a^2 - 2a - 48$

44. $x^2 - 17x + 60$

45. $x^2 - 8x + 7$

46. $x^2 + 7x - 18$

47. $x^2 - 6x - 40$

48. $x^2 - 11x + 10$

49. $x^2 - 14x + 49$

50. $s^2 - 4s - 32$

51. $p^2 - 10p - 24$

52. $x^2 - 11x - 60$

53. $x^2 + 5x - 66$

54. $a^2 + 2a - 80$

55. $c^2 + 19c + 60$

56. $t^2 - 4t - 60$

57. $n^2 + 5n - 50$

58. $x^2 - 16x + 63$

59. $x^2 + 7xy + 10y^2$

60. $x^2 - 8xy + 12y^2$

61. $a^2 - ab - 42b^2$

62. $m^2 - 8mn + 16n^2$

63. $x^2 - 13xy + 40y^2$

64. $r^2 - 9rs - 36s^2$

65. $6x^2 + 19x + 10$

66. $6x^2 - 7x - 3$

67. $15x^2 + x - 6$

68. $12w^2 + 19w + 4$

69. $6m^2 + 25m - 25$

70. $8x^2 - 6x - 9$

71. $9x^2 - 12x + 4$

72. $20x^2 - 23x + 6$

ANSWERS

45. $(x - 1)(x - 7)$
46. $(x + 9)(x - 2)$
47. $(x - 10)(x + 4)$
48. $(x - 1)(x - 10)$
49. $(x - 7)(x - 7)$
50. $(s - 8)(s + 4)$
51. $(p - 12)(p + 2)$
52. $(x - 15)(x + 4)$
53. $(x + 11)(x - 6)$
54. $(a + 10)(a - 8)$
55. $(c + 4)(c + 15)$
56. $(t - 10)(t + 6)$
57. $(n + 10)(n - 5)$
58. $(x - 9)(x - 7)$
59. $(x + 2y)(x + 5y)$
60. $(x - 6y)(x - 2y)$
61. $(a - 7b)(a + 6b)$
62. $(m - 4n)(m - 4n)$
63. $(x - 5y)(x - 8y)$
64. $(r - 12s)(r + 3s)$
65. $(3x + 2)(2x + 5)$
66. $(2x - 3)(3x + 1)$
67. $(5x - 3)(3x + 2)$
68. $(4w + 1)(3w + 4)$
69. $(6m - 5)(m + 5)$
70. $(4x + 3)(2x - 3)$
71. $(3x - 2)(3x - 2)$
72. $(5x - 2)(4x - 3)$

Answers

1. True **3.** False **5.** True **7.** False **9.** True
11. $a = 1, b = 4, c = -9$ **13.** $a = 1, b = -3, c = 8$
15. $a = 3, b = 5, c = -8$ **17.** $a = 4, b = 8, c = 11$
19. $a = -3, b = 5, c = -10$ **21.** Factorable; 3, −2
23. Not factorable **25.** Factorable; −3, −2 **27.** Factorable; 6, −1
29. Factorable; −15, −4 **31.** $x^2 + 2x + 4x + 8; (x + 2)(x + 4)$
33. $x^2 - 5x - 4x + 20; (x - 5)(x - 4)$ **35.** $x^2 - 9x + 7x - 63; (x - 9)(x + 7)$
37. $(x + 3)(x + 5)$ **39.** $(x - 4)(x - 7)$ **41.** $(s + 10)(s + 3)$
43. $(a - 8)(a + 6)$ **45.** $(x - 1)(x - 7)$ **47.** $(x - 10)(x + 4)$
49. $(x - 7)(x - 7)$ **51.** $(p - 12)(p + 2)$ **53.** $(x + 11)(x - 6)$
55. $(c + 4)(c + 15)$ **57.** $(n + 10)(n - 5)$ **59.** $(x + 2y)(x + 5y)$
61. $(a - 7b)(a + 6b)$ **63.** $(x - 5y)(x - 8y)$ **65.** $(3x + 2)(2x + 5)$
67. $(5x - 3)(3x + 2)$ **69.** $(6m - 5)(m + 5)$ **71.** $(3x - 2)(3x - 2)$
73. $(6x + 5)(2x - 3)$ **75.** $(3y - 2)(y + 3)$ **77.** $(8x + 5)(x - 4)$
79. $(2x + y)(x + y)$ **81.** $(5a + 2b)(a - 2b)$ **83.** $(9x - 5y)(x + y)$
85. $(3m - 4n)(2m - 3n)$ **87.** $(12a - 5b)(3a + b)$
89. $(x + 2y)^2$ **91.** $5(2x - 3)(2x + 1)$ **93.** $4(2m + 1)(m + 1)$
95. $3(5r - 2s)(r - s)$ **97.** $2x(x - 2)(x + 1)$ **99.** $y^2(2y + 3)(y + 1)$
101. $6a(3a - 1)(2a - 3)$ **103.** $3(p + q)(3p + 7q)$ **105.** 6 or 9
107. 8 or 10 or 17 **109.** 4 **111.** 2 **113.** 3, 8, 15, 24, . . .

a. $x = 5$ **b.** $x = \frac{1}{2}$ **c.** $x = -\frac{2}{3}$ **d.** $x = -4$ **e.** $x = 7$ **f.** $x = \frac{9}{4}$

45. $x^2 - 8x + 7$

46. $x^2 + 7x - 18$

47. $x^2 - 6x - 40$

48. $x^2 - 11x + 10$

49. $x^2 - 14x + 49$

50. $s^2 - 4s - 32$

51. $p^2 - 10p - 24$

52. $x^2 - 11x - 60$

53. $x^2 + 5x - 66$

54. $a^2 + 2a - 80$

55. $c^2 + 19c + 60$

56. $t^2 - 4t - 60$

57. $n^2 + 5n - 50$

58. $x^2 - 16x + 63$

59. $x^2 + 7xy + 10y^2$

60. $x^2 - 8xy + 12y^2$

61. $a^2 - ab - 42b^2$

62. $m^2 - 8mn + 16n^2$

63. $x^2 - 13xy + 40y^2$

64. $r^2 - 9rs - 36s^2$

65. $6x^2 + 19x + 10$

66. $6x^2 - 7x - 3$

67. $15x^2 + x - 6$

68. $12w^2 + 19w + 4$

69. $6m^2 + 25m - 25$

70. $8x^2 - 6x - 9$

71. $9x^2 - 12x + 4$

72. $20x^2 - 23x + 6$

ANSWERS

45. $(x - 1)(x - 7)$
46. $(x + 9)(x - 2)$
47. $(x - 10)(x + 4)$
48. $(x - 1)(x - 10)$
49. $(x - 7)(x - 7)$
50. $(s - 8)(s + 4)$
51. $(p - 12)(p + 2)$
52. $(x - 15)(x + 4)$
53. $(x + 11)(x - 6)$
54. $(a + 10)(a - 8)$
55. $(c + 4)(c + 15)$
56. $(t - 10)(t + 6)$
57. $(n + 10)(n - 5)$
58. $(x - 9)(x - 7)$
59. $(x + 2y)(x + 5y)$
60. $(x - 6y)(x - 2y)$
61. $(a - 7b)(a + 6b)$
62. $(m - 4n)(m - 4n)$
63. $(x - 5y)(x - 8y)$
64. $(r - 12s)(r + 3s)$
65. $(3x + 2)(2x + 5)$
66. $(2x - 3)(3x + 1)$
67. $(5x - 3)(3x + 2)$
68. $(4w + 1)(3w + 4)$
69. $(6m - 5)(m + 5)$
70. $(4x + 3)(2x - 3)$
71. $(3x - 2)(3x - 2)$
72. $(5x - 2)(4x - 3)$

ANSWERS

73. $(6x + 5)(2x - 3)$
74. $(4a + 5)(4a + 5)$
75. $(3y - 2)(y + 3)$
76. $(3x + 5)(4x - 3)$
77. $(8x + 5)(x - 4)$
78. $(8v - 9)(3v + 4)$
79. $(2x + y)(x + y)$
80. $(3x - 2y)(x - y)$
81. $(5a + 2b)(a - 2b)$
82. $(5x - 3y)(x + 2y)$
83. $(9x - 5y)(x + y)$
84. $(4x + 3y)(4x + 5y)$
85. $(3m - 4n)(2m - 3n)$
86. $(5x + 3y)(3x - 2y)$
87. $(12a - 5b)(3a + b)$
88. $(3q + r)(q - 6r)$
89. $(x + 2y)^2$
90. $(5b - 8c)^2$
91. $5(2x - 3)(2x + 1)$
92. $6(4x + 1)(x - 1)$
93. $4(2m + 1)(m + 1)$
94. $2(7x - 3)(x - 1)$

73. $12x^2 - 8x - 15$

74. $16a^2 + 40a + 25$

75. $3y^2 + 7y - 6$

76. $12x^2 + 11x - 15$

77. $8x^2 - 27x - 20$

78. $24v^2 + 5v - 36$

79. $2x^2 + 3xy + y^2$

80. $3x^2 - 5xy + 2y^2$

81. $5a^2 - 8ab - 4b^2$

82. $5x^2 + 7xy - 6y^2$

83. $9x^2 + 4xy - 5y^2$

84. $16x^2 + 32xy + 15y^2$

85. $6m^2 - 17mn + 12n^2$

86. $15x^2 - xy - 6y^2$

87. $36a^2 - 3ab - 5b^2$

88. $3q^2 - 17qr - 6r^2$

89. $x^2 + 4xy + 4y^2$

90. $25b^2 - 80bc + 64c^2$

91. $20x^2 - 20x - 15$

92. $24x^2 - 18x - 6$

93. $8m^2 + 12m + 4$

94. $14x^2 - 20x + 6$

95. $15r^2 - 21rs + 6s^2$

96. $10x^2 + 5xy - 30y^2$

97. $2x^3 - 2x^2 - 4x$

98. $2y^3 + y^2 - 3y$

99. $2y^4 + 5y^3 + 3y^2$

100. $4z^3 - 18z^2 - 10z$

101. $36a^3 - 66a^2 + 18a$

102. $20n^4 - 22n^3 - 12n^2$

103. $9p^2 + 30pq + 21q^2$

104. $12x^2 + 2xy - 24y^2$

Find a positive value for k for which each of the following can be factored.

105. $x^2 + kx + 8$

106. $x^2 + kx + 9$

107. $x^2 - kx + 16$

108. $x^2 - kx + 17$

109. $x^2 - kx - 5$

110. $x^2 - kx - 7$

111. $x^2 + 3x + k$

112. $x^2 + 5x + k$

113. $x^2 + 2x - k$

114. $x^2 + x - k$

Getting Ready for Section 4.7 [Section 2.3]

Solve.

(a) $x - 5 = 0$ (b) $2x - 1 = 0$ (c) $3x + 2 = 0$

(d) $x + 4 = 0$ (e) $7 - x = 0$ (f) $9 - 4x = 0$

ANSWERS

95. $3(5r - 2s)(r - s)$

96. $5(2x - 3y)(x + 2y)$

97. $2x(x - 2)(x + 1)$

98. $y(2y + 3)(y - 1)$

99. $y^2(2y + 3)(y + 1)$

100. $2z(2z + 1)(z - 5)$

101. $6a(3a - 1)(2a - 3)$

102. $2n^2(2n - 3)(5n + 2)$

103. $3(p + q)(3p + 7q)$

104. $2(2x + 3y)(3x - 4y)$

105. 6 or 9

106. 6 or 10

107. 8 or 10 or 17

108. 18

109. 4

110. 6

111. 2

112. 4 or 6

113. 3, 8, 15, 24, . . .

114. 2, 6, 12, 20, . . .

a. $x = 5$

b. $x = \frac{1}{2}$

c. $x = -\frac{2}{3}$

d. $x = -4$

e. $x = 7$

f. $x = \frac{9}{4}$

Answers

1. True **3.** False **5.** True **7.** False **9.** True
11. $a = 1, b = 4, c = -9$ **13.** $a = 1, b = -3, c = 8$
15. $a = 3, b = 5, c = -8$ **17.** $a = 4, b = 8, c = 11$
19. $a = -3, b = 5, c = -10$ **21.** Factorable; $3, -2$
23. Not factorable **25.** Factorable; $-3, -2$ **27.** Factorable; $6, -1$
29. Factorable; $-15, -4$ **31.** $x^2 + 2x + 4x + 8; (x + 2)(x + 4)$
33. $x^2 - 5x - 4x + 20; (x - 5)(x - 4)$ **35.** $x^2 - 9x + 7x - 63; (x - 9)(x + 7)$
37. $(x + 3)(x + 5)$ **39.** $(x - 4)(x - 7)$ **41.** $(s + 10)(s + 3)$
43. $(a - 8)(a + 6)$ **45.** $(x - 1)(x - 7)$ **47.** $(x - 10)(x + 4)$
49. $(x - 7)(x - 7)$ **51.** $(p - 12)(p + 2)$ **53.** $(x + 11)(x - 6)$
55. $(c + 4)(c + 15)$ **57.** $(n + 10)(n - 5)$ **59.** $(x + 2y)(x + 5y)$
61. $(a - 7b)(a + 6b)$ **63.** $(x - 5y)(x - 8y)$ **65.** $(3x + 2)(2x + 5)$
67. $(5x - 3)(3x + 2)$ **69.** $(6m - 5)(m + 5)$ **71.** $(3x - 2)(3x - 2)$
73. $(6x + 5)(2x - 3)$ **75.** $(3y - 2)(y + 3)$ **77.** $(8x + 5)(x - 4)$
79. $(2x + y)(x + y)$ **81.** $(5a + 2b)(a - 2b)$ **83.** $(9x - 5y)(x + y)$
85. $(3m - 4n)(2m - 3n)$ **87.** $(12a - 5b)(3a + b)$
89. $(x + 2y)^2$ **91.** $5(2x - 3)(2x + 1)$ **93.** $4(2m + 1)(m + 1)$
95. $3(5r - 2s)(r - s)$ **97.** $2x(x - 2)(x + 1)$ **99.** $y^2(2y + 3)(y + 1)$
101. $6a(3a - 1)(2a - 3)$ **103.** $3(p + q)(3p + 7q)$ **105.** 6 or 9
107. 8 or 10 or 17 **109.** 4 **111.** 2 **113.** 3, 8, 15, 24, . . .

a. $x = 5$ **b.** $x = \frac{1}{2}$ **c.** $x = -\frac{2}{3}$ **d.** $x = -4$ **e.** $x = 7$ **f.** $x = \frac{9}{4}$

Solving Quadratic Equations by Factoring

OBJECTIVE

1. Solve quadratic equations by factoring

The factoring techniques you have learned provide us with tools for solving equations that can be written in the form

$$ax^2 + bx + c = 0 \qquad a \neq 0$$

This is a quadratic equation in one variable, here x. You can recognize such a quadratic equation by the fact that the highest power of the variable x is the second power.

in which a, b, and c are constants.

An equation written in the form $ax^2 + bx + c = 0$ is called a **quadratic equation in standard form.** Using factoring to solve quadratic equations requires the **zero-product principle,** which says that if the product of two factors is 0, then one or both of the factors must be equal to 0. In symbols:

Definitions: Zero-Product Principle

If $a \cdot b = 0$, then $a = 0$ or $b = 0$ or $a = b = 0$.

Let's see how the principle is applied to solving quadratic equations.

Example 1

Solving Equations by Factoring

Solve.

$x^2 - 3x - 18 = 0$

NOTE To use the zero-product principle, 0 must be on one side of the equation.

Factoring on the left, we have

$(x - 6)(x + 3) = 0$

By the zero-product principle, we know that one or both of the factors must be zero. We can then write

$x - 6 = 0 \qquad \text{or} \qquad x + 3 = 0$

Solving each equation gives

$x = 6 \qquad \text{or} \qquad x = -3$

The two solutions are 6 and -3.

Quadratic equations can be checked in the same way as linear equations were checked: by substitution. For instance, if $x = 6$, we have

$$6^2 - 3 \cdot 6 - 18 \stackrel{?}{=} 0$$

$$36 - 18 - 18 \stackrel{?}{=} 0$$

$$0 = 0$$

which is a true statement. We leave it to you to check the solution -3.

CHECK YOURSELF 1

Solve $x^2 - 9x + 20 = 0$.

Other factoring techniques are also used in solving quadratic equations. Example 2 illustrates this concept.

Example 2

Solving Equations by Factoring

(a) Solve $x^2 - 5x = 0$.

Again, factor the left side of the equation and apply the zero-product principle.

$x(x - 5) = 0$

Now

$x = 0 \quad \text{or} \quad x - 5 = 0$

$x = 5$

The two solutions are 0 and 5.

CAUTION

A *common mistake* is to forget the statement $x = 0$ when you are solving equations of this type. Be sure to include the *two statements* obtained.

(b) Solve $x^2 - 9 = 0$.

Factoring yields

$(x + 3)(x - 3) = 0$

$x + 3 = 0 \quad \text{or} \quad x - 3 = 0$

$x = -3 \qquad x = 3$

The solutions may be written as $x = \pm 3$.

NOTE The symbol $\pm$ is read "plus or minus."

CHECK YOURSELF 2

Solve by factoring.

(a) $x^2 + 8x = 0$

(b) $x^2 - 16 = 0$

Example 3 illustrates a crucial point. Our solution technique depends on the zero-product principle, which means that the product of factors *must be equal to 0.* The importance of this is shown now.

CAUTION

Consider the equation

$x(2x - 1) = 3$

NOTE Students are sometimes tempted to write

$x = 3 \quad \text{or} \quad 2x - 1 = 3$

This is *not correct.* Instead, subtract 3 from both sides of the equation *as the first step* to write

$x^2 - 2x - 3 = 0$

in standard form. Only *now* can you factor and proceed as before.

Example 3

Solving Equations by Factoring

Solve $2x^2 - x = 3$.

The first step in the solution is to write the equation in standard form (that is, when one side of the equation is 0). So start by adding -3 to both sides of the equation.
Then,

$2x^2 - x - 3 = 0$ Make sure all terms are on one side of the equation. The other side will be 0.

You can now factor and solve by using the zero-product principle.

$$(2x - 3)(x + 1) = 0$$

$$2x - 3 = 0 \quad \text{or} \quad x + 1 = 0$$

$$2x = 3 \qquad\qquad x = -1$$

$$x = \frac{3}{2}$$

The solutions are $\frac{3}{2}$ and -1.

CHECK YOURSELF 3

Solve $3x^2 = 5x + 2$.

In all the previous examples, the quadratic equations had two distinct real number solutions. That may not always be the case, as we shall see.

Example 4

Solving Equations by Factoring

Solve $x^2 - 6x + 9 = 0$.

Factoring, we have

$$(x - 3)(x - 3) = 0$$

and

$$x - 3 = 0 \quad \text{or} \quad x - 3 = 0$$

$$x = 3 \qquad\qquad x = 3$$

The solution is 3.

A quadratic (or second-degree) equation always has *two* solutions. When an equation such as this one has two solutions that are the same number, we call 3 the **repeated** (or **double**) **solution** of the equation.

Although a quadratic equation will always have two solutions, they may not always be real numbers. You will learn more about this in a later course.

CHECK YOURSELF 4

Solve $x^2 + 6x + 9 = 0$.

Always examine the quadratic member of an equation for common factors. It will make your work much easier, as Example 5 illustrates.

Example 5

Solving Equations by Factoring

Solve $3x^2 - 3x - 60 = 0$.

First, note the common factor 3 in the quadratic member of the equation. Factoring out the 3, we have

$$3(x^2 - x - 20) = 0$$

Now divide both sides of the equation by 3.

NOTE Notice the advantage of dividing both members by 3. The coefficients in the quadratic member become smaller, and that member is much easier to factor.

$$\frac{3(x^2 - x - 20)}{3} = \frac{0}{3}$$

or

$$x^2 - x - 20 = 0$$

We can now factor and solve as before.

$$(x - 5)(x + 4) = 0$$

$$x - 5 = 0 \quad \text{or} \quad x + 4 = 0$$

$$x = 5 \qquad\qquad x = -4$$

CHECK YOURSELF 5

Solve $2x^2 - 10x - 48 = 0$.

CHECK YOURSELF ANSWERS

1. 4, 5 **2.** **(a)** 0, −8; **(b)** 4, −4 **3.** $-\frac{1}{3}$, 2 **4.** −3 **5.** −3, 8

Name ______________________

Section ________ Date ________

4.7 Exercises

Solve each of the following quadratic equations.

1. $(x - 3)(x - 4) = 0$

2. $(x - 7)(x + 1) = 0$

3. $(3x + 1)(x - 6) = 0$

4. $(5x - 4)(x - 6) = 0$

5. $x^2 - 2x - 3 = 0$

6. $x^2 + 5x + 4 = 0$

7. $x^2 - 7x + 6 = 0$

8. $x^2 + 3x - 10 = 0$

9. $x^2 + 8x + 15 = 0$

10. $x^2 - 3x - 18 = 0$

11. $x^2 + 4x - 21 = 0$

12. $x^2 - 12x + 32 = 0$

13. $x^2 - 4x = 12$

14. $x^2 + 8x = -15$

15. $x^2 + 5x = 14$

16. $x^2 = 11x - 24$

17. $2x^2 + 5x - 3 = 0$

18. $3x^2 + 7x + 2 = 0$

19. $4x^2 - 24x + 35 = 0$

20. $6x^2 + 11x - 10 = 0$

21. $4x^2 + 11x = -6$

22. $5x^2 + 2x = 3$

ANSWERS

1. 3, 4
2. −1, 7
3. $-\frac{1}{3}, 6$
4. $\frac{4}{5}, 6$
5. −1, 3
6. −4, −1
7. 1, 6
8. −5, 2
9. −3, −5
10. −3, 6
11. −7, 3
12. 4, 8
13. −2, 6
14. −3, −5
15. −7, 2
16. 3, 8
17. $-3, \frac{1}{2}$
18. $-\frac{1}{3}, -2$
19. $\frac{5}{2}, \frac{7}{2}$
20. $-\frac{5}{2}, \frac{2}{3}$
21. $-\frac{3}{4}, -2$
22. $-1, \frac{3}{5}$

ANSWERS

23. $-3, \frac{2}{5}$
24. $-\frac{3}{4}, 4$
25. 0, 2
26. 0, −5
27. 0, −8
28. 0, 7
29. 0, 3
30. 0, −5
31. −5, 5
32. −7, 7
33. −9, 9
34. −8, 8
35. −3, 3
36. −5, 5
37. −5, −3
38. −2, 3
39. $-5, \frac{1}{3}$
40. $-2, \frac{1}{5}$
41. 4, −5
42. 7, −4
43. 11, 12 or −12, −11

23. $5x^2 + 13x = 6$

24. $4x^2 = 13x + 12$

25. $x^2 - 2x = 0$

26. $x^2 + 5x = 0$

27. $x^2 = -8x$

28. $x^2 = 7x$

29. $5x^2 - 15x = 0$

30. $4x^2 + 20x = 0$

31. $x^2 - 25 = 0$

32. $x^2 = 49$

33. $x^2 = 81$

34. $x^2 = 64$

35. $2x^2 - 18 = 0$

36. $3x^2 - 75 = 0$

37. $3x^2 + 24x + 45 = 0$

38. $4x^2 - 4x = 24$

39. $6x^2 + 28x = 10$

40. $15x^2 + 27x = 6$

41. $(x + 3)(x - 2) = 14$

42. $(x - 5)(x + 2) = 18$

Solve the following problems.

43. Consecutive integers. The product of two consecutive integers is 132. Find the two integers.

$x(x+1) = 132$
$x^2 + x - 132 = 0$
$(x+12)(x-11) = 0$
$-12, 11$

11, 12
OR
−12, −11

44. Consecutive integers. If the product of two consecutive positive even integers is 120, find the two integers.

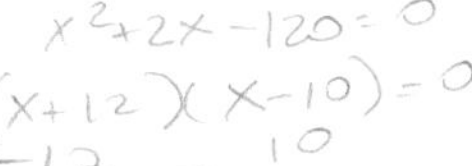

45. Integers. The sum of an integer and its square is 72. What is the integer?

46. Integers. The square of an integer is 56 more than the integer. Find the integer.

47. Geometry. If the sides of a square are increased by 3 in., the area is increased by 39 in.^{2}. What were the dimensions of the original square?

48. Geometry. If the sides of a square are decreased by 2 cm, the area is decreased by 36 cm^2. What were the dimensions of the original square?

49. Business. The profit on a small appliance is given by $P = x^2 - 3x - 60$, in which x is the number of appliances sold per day. How many appliances were sold on a day when there was a \$20 loss?

50. Business. The relationship between the number x of calculators that a company can sell per month and the price of each calculator p is given by $x = 1700 - 100p$. Find the price at which a calculator should be sold to produce a monthly revenue of \$7000. (*Hint:* Revenue $= xp$.)

51. Write a short comparison that explains the difference between $ax^2 + bx + c$ and $ax^2 + bx + c = 0$.

52. When solving quadratic equations, some people try to solve an equation in the manner shown below, but this doesn't work! Write a paragraph to explain what is

ANSWERS

44. 10, 12

45. −9 or 8

46. −7 or 8

47. 5 in. by 5 in.

48. 10 cm by 10 cm

49. 8

50. \$7 or \$10

51.

52. ______________________

wrong with this approach. 

$$2x^2 + 7x + 3 = 52$$

$$(2x + 1)(x + 3) = 52$$

$$2x + 1 = 52 \quad \text{or} \quad x + 3 = 52$$

$$x = \frac{51}{2} \quad \text{or} \quad x = 49$$

Answers

1. 3, 4 **3.** $-\frac{1}{3}, 6$ **5.** $-1, 3$ **7.** 1, 6 **9.** $-3, -5$ **11.** $-7, 3$
13. $-2, 6$ **15.** $-7, 2$ **17.** $-3, \frac{1}{2}$ **19.** $\frac{5}{2}, \frac{7}{2}$ **21.** $-\frac{3}{4}, -2$
23. $-3, \frac{2}{5}$ **25.** 0, 2 **27.** $0, -8$ **29.** 0, 3 **31.** $-5, 5$ **33.** $-9, 9$
35. $-3, 3$ **37.** $-5, -3$ **39.** $-5, \frac{1}{3}$ **41.** $4, -5$
43. 11, 12 or $-12, -11$ **45.** -9 or 8 **47.** 5 in. by 5 in. **49.** 8
51.

Summary

DEFINITION/PROCEDURE	EXAMPLE	REFERENCE
An Introduction to Factoring		**Section 4.1**
Common Monomial Factor A single term that is a factor of every term of the polynomial. The greatest common factor (GCF) of a polynomial is the common monomial factor that has the largest possible numerical coefficient and the largest possible exponents.	$4x^2$ is the greatest common monomial factor of $8x^4 - 12x^3 + 16x^2$.	**p. 324**
Factoring a Monomial from a Polynomial **1.** Determine the GCF for all terms. **2.** Factor the GCF from each term, then apply the distributive law in the form $ab + ac = a(b + c)$ (↑ The greatest common factor) **3.** Mentally check by multiplication.	$8x^4 - 12x^3 + 16x^2$ $= 4x^2(2x^2 - 3x + 4)$	**p. 324**
Factoring Trinomials		**Sections 4.2 and 4.3**
Trial and Error To factor a trinomial, find the appropriate sign pattern, then find integer values that yield the appropriate coefficients for the trinomial.	$x^2 - 5x - 24$ $= (x - \ )(x + \ )$ $= (x - 8)(x + 3)$	**p. 331**
Difference of Squares and Perfect Square Trinomials		**Section 4.4**
Factoring a Difference of Squares Use the following form: $a^2 - b^2 = (a + b)(a - b)$	To factor: $16x^2 - 25y^2$: Think: $(4x)^2 - (5y)^2$ so $16x^2 - 25y^2$ $= (4x + 5y)(4x - 5y)$	**p. 351**
Factoring a Perfect Square Trinomial Use the following form: $a^2 + 2ab + b^2 = (a + b)^2$	$4x^2 + 12xy + 9y^2$ $= (2x)^2 + 2(2x)(3y) + (3y)^2$ $= (2x + 3y)^2$	**p. 353**
Factoring by Grouping		**Section 4.5**
When there are four terms of a polynomial, factor the first pair and factor the last pair. If these two pairs have a common binomial factor, factor that out. The result will be the product of two binomials.	$4x^2 - 6x + 10x - 15$ $= 2x(2x - 3) + 5(2x - 3)$ $= (2x - 3)(2x + 5)$	**p. 359**

Continued

DEFINITION/PROCEDURE	EXAMPLE	REFERENCE
Using the *ac* Method to Factor		**Section 4.6**
Factoring Trinomials To factor a trinomial, first use the *ac* test to determine factorability. If the trinomial is factorable, the *ac* test will yield two terms (which have as their sum the middle term) that allow the factoring to be completed by using the grouping method.	$x^2 + 3x - 28$ $ac = -28; b = 3$ $mn = -28; m + n = 3$ $m = 7, n = -4$ $x^2 + 7x - 4x - 28$ $= x(x + 7) - 4(x + 7)$ $= (x + 7)(x - 4)$	**p. 363**
Solving Quadratic Equations by Factoring		**Section 4.7**
1. Add or subtract the necessary terms on both sides of the equation so that the equation is in standard form (set equal to 0). **2.** Factor the quadratic expression. **3.** Set each factor equal to 0. **4.** Solve the resulting equations to find the solutions. **5.** Check each solution by substituting in the original equation.	To solve $x^2 + 7x = 30$ $x^2 + 7x - 30 = 0$ $(x + 10)(x - 3) = 0$ $x + 10 = 0$ or $x - 3 = 0$ $x = -10$ and $x = 3$ are solutions.	**p. 377**

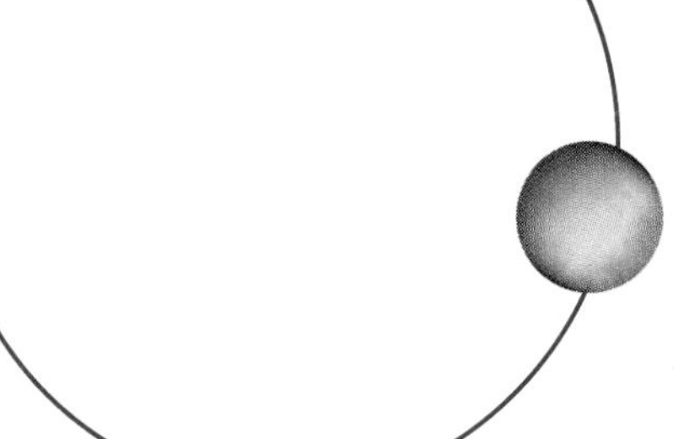

Summary Exercises

This summary exercise set is provided to give you practice with each of the objectives of the chapter. Each exercise is keyed to the appropriate chapter section. The answers are provided in the *Instructor's Manual.* Your instructor will give you guidelines on how to best use these exercises in your instructional setting.

[4.1] Factor each of the following polynomials.

1. $18a + 24$ $6(3a + 4)$

2. $9m^2 - 21m$ $3m(3m - 7)$

3. $24s^2t - 16s^2$ $8s^2(3t - 2)$

4. $18a^2b + 36ab^2$ $18ab(a + 2b)$

5. $35s^3 - 28s^2$ $7s^2(5s - 4)$

6. $3x^3 - 6x^2 + 15x$ $3x(x^2 - 2x + 5)$

7. $18m^2n^2 - 27m^2n + 45m^2n^3$ $9m^2n(2n - 3 + 5n^2)$

8. $121x^8y^3 + 77x^6y^3$ $11x^6y^3(11x^2 + 7)$

9. $8a^2b + 24ab - 16ab^2$ $8ab(a + 3 - 2b)$

10. $3x^2y - 6xy^3 + 9x^3y - 12xy^2$ $3xy(x - 2y^2 + 3x^2 - 4y)$

11. $x(2x - y) + y(2x - y)$ $(2x - y)(x + y)$

12. $5(w - 3z) - w(w - 3z)$ $(w - 3z)(5 - w)$

[4.2] Factor each of the following trinomials completely.

13. $x^2 + 9x + 20$ $(x + 4)(x + 5)$

14. $x^2 - 10x + 24$ $(x - 4)(x - 6)$

15. $a^2 - a - 12$ $(a - 4)(a + 3)$

16. $w^2 - 13w + 40$ $(w - 8)(w - 5)$

17 $x^2 + 12x + 36$ $(x + 6)(x + 6)$

18. $r^2 - 9r - 36$ $(r - 12)(r + 3)$

19. $b^2 - 4bc - 21c^2$ $(b - 7c)(b + 3c)$

20. $m^2n + 4mn - 32n$ $n(m + 8)(m - 4)$

21. $m^3 + 2m^2 - 35m$ $m(m + 7)(m - 5)$

22. $2x^2 - 2x - 40$ $2(x - 5)(x + 4)$

23. $3y^3 - 48y^2 + 189y$ $3y(y - 7)(y - 9)$

24. $3b^3 - 15b^2 - 42b$ $3b(b - 7)(b + 2)$

[4.3] Factor each of the following trinomials completely.

25. $3x^2 + 8x + 5$ $(3x + 5)(x + 1)$

26. $5w^2 + 13w - 6$ $(5w - 2)(w + 3)$

27. $2b^2 - 9b + 9$ $(2b - 3)(b - 3)$

28. $8x^2 + 2x - 3$ $(4x + 3)(2x - 1)$

29. $10x^2 - 11x + 3$ $(5x - 3)(2x - 1)$

30. $4a^2 + 7a - 15$ $(4a - 5)(a + 3)$

31. $9y^2 - 3yz - 20z^2$ $(3y - 5z)(3y + 4z)$

32. $8x^2 + 14xy - 15y^2$ $(2x + 5y)(4x - 3y)$

33. $8x^3 - 36x^2 - 20x$ $4x(2x + 1)(x - 5)$

34. $9x^2 - 15x - 6$ $3(3x + 1)(x - 2)$

35. $6x^3 - 3x^2 - 9x$ $3x(2x - 3)(x + 1)$

36. $5w^2 - 25wz + 30z^2$ $5(w - 2z)(w - 3z)$

[4.4] Factor each of the following completely.

37. $p^2 - 49$ $(p + 7)(p - 7)$

38. $25a^2 - 16$ $(5a + 4)(5a - 4)$

39. $m^2 - 9n^2$ $(m + 3n)(m - 3n)$

40. $16r^2 - 49s^2$ $(4r + 7s)(4r - 7s)$

41. $25 - z^2$ $(5 + z)(5 - z)$

42. $a^4 - 16b^2$ $(a^2 + 4b)(a^2 - 4b)$

43. $25a^2 - 36b^2$ $(5a + 6b)(5a - 6b)$

44. $x^6 - 4y^2$ $(x^3 + 2y)(x^3 - 2y)$

45. $3w^3 - 12wz^2$ $3w(w + 2z)(w - 2z)$

46. $16a^4 - 49b^2$ $(4a^2 + 7b)(4a^2 - 7b)$

47. $2m^2 - 72n^4$ $2(m + 6n^2)(m - 6n^2)$

48. $3w^3z - 12wz^3$ $3wz(w + 2z)(w + 2z)$

49. $x^2 + 8x + 16$ $(x + 4)^2$

50. $x^2 - 18x + 81$ $(x - 9)^2$

51. $4x^2 + 12x + 9$ $(2x + 3)^2$

52. $9x^2 - 12x + 4$ $(3x - 2)^2$

53 $16x^3 + 40x^2 + 25x$ $x(4x + 5)^2$

54. $4x^3 - 4x^2 + x$ $x(2x - 1)^2$

[4.5] Factor the following polynomials completely.

55 $x^2 - 4x + 5x - 20$ $(x - 4)(x + 5)$

56. $x^2 + 7x - 2x - 14$ $(x + 7)(x - 2)$

57 $6x^2 + 4x - 15x - 10$ $(3x + 2)(2x - 5)$

58. $12x^2 - 9x - 28x + 21$ $(4x - 3)(3x - 7)$

59 $6x^3 + 9x^2 - 4x^2 - 6x$ $x(2x + 3)(3x - 2)$

60. $3x^4 + 6x^3 + 5x^3 + 10x^2$ $x^2(x + 2)(3x + 5)$

[4.7] Solve each of the following quadratic equations.

61. $(x - 1)(2x + 3) = 0$ $1, -\frac{3}{2}$

62. $x^2 - 5x + 6 = 0$ $2, 3$

63. $x^2 - 10x = 0$ $0, 10$

64. $x^2 = 144$ $-12, 12$

65. $x^2 - 2x = 15$ $-3, 5$

66. $3x^2 - 5x - 2 = 0$ $-\frac{1}{3}, 2$

67. $4x^2 - 13x + 10 = 0$ $2, \frac{5}{4}$

68. $2x^2 - 3x = 5$ $-1, \frac{5}{2}$

69. $3x^2 - 9x = 0$ $0, 3$

70. $x^2 - 25 = 0$ $-5, 5$

71. $2x^2 - 32 = 0$ $-4, 4$

72. $2x^2 - x - 3 = 0$ $-1, \frac{3}{2}$

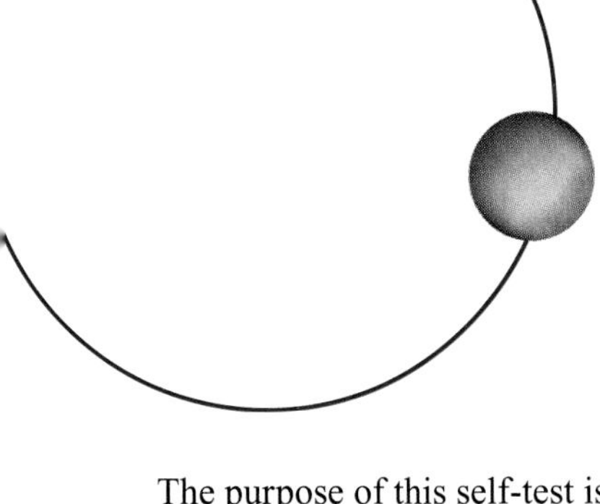

Self-Test for Chapter 4

Name ______________________

Section __________ Date __________

The purpose of this self-test is to help you check your progress and to review for a chapter test in class. Allow yourself about an hour to take the test. When you are done, check your answers in the back of the book. If you missed any answers, be sure to go back and review the appropriate sections in the chapter and the exercises that are provided.

Factor each of the following polynomials.

1. $12b + 18$

2. $9p^3 - 12p^2$

3. $5x^2 - 10x + 20$

4. $6a^2b - 18ab + 12ab^2$

Factor each of the following polynomials completely.

5. $a^2 - 25$

6. $64m^2 - n^2$

7. $49x^2 - 16y^2$

8. $32a^2b - 50b^3$

Factor each of the following polynomials completely.

9. $a^2 - 5a - 14$

10. $b^2 + 8b + 15$

11. $x^2 - 11x + 28$

12. $y^2 + 12yz + 20z^2$

13. $x^2 + 2x - 5x - 10$

14. $6x^2 + 2x - 9x - 3$

Factor each of the following polynomials completely.

15. $2x^2 + 15x - 8$

16. $3w^2 + 10w + 7$

17. $8x^2 - 2xy - 3y^2$

18. $6x^3 + 3x^2 - 30x$

ANSWERS

1. $6(2b + 3)$
2. $3p^2(3p - 4)$
3. $5(x^2 - 2x + 4)$
4. $6ab(a - 3 + 2b)$
5. $(a + 5)(a - 5)$
6. $(8m + n)(8m - n)$
7. $(7x + 4y)(7x - 4y)$
8. $2b(4a + 5b)(4a - 5b)$
9. $(a - 7)(a + 2)$
10. $(b + 3)(b + 5)$
11. $(x - 4)(x - 7)$
12. $(y + 10z)(y + 2z)$
13. $(x + 2)(x - 5)$
14. $(2x - 3)(3x + 1)$
15. $(2x - 1)(x + 8)$
16. $(3w + 7)(w + 1)$
17. $(4x - 3y)(2x + y)$
18. $3x(2x + 5)(x - 2)$

ANSWERS

19. 3, 5

20. −1, 4

21. $-1, \frac{2}{3}$

22. 0, 3

23. 0, 4

24. −3, 8

25. 7

Solve each of the following equations for x.

19. $x^2 - 8x + 15 = 0$

20. $x^2 - 3x = 4$

21. $3x^2 + x - 2 = 0$

22. $4x^2 - 12x = 0$

23. $x(x - 4) = 0$

24. $(x - 3)(x - 2) = 30$

25. $x^2 - 14x = -49$

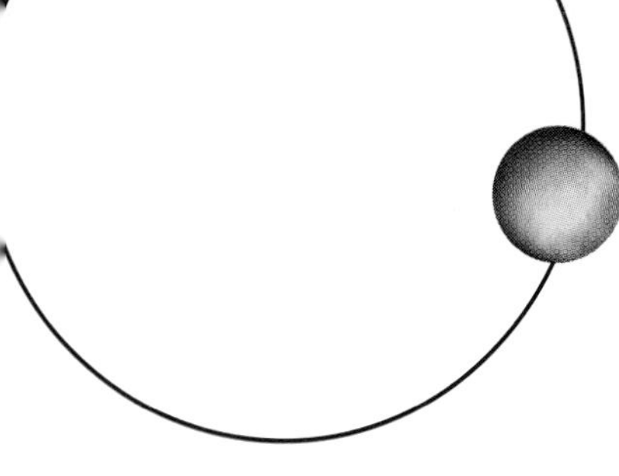

Cumulative Test Chapters 0 to 4

Name ____________

Section ______ Date ______

Perform the indicated operations.

1. $7 - (-10)$

2. $(-34) \div (17)$

Perform each of the indicated operations.

3. $(7x^2 + 5x - 4) + (2x^2 - 6x - 1)$

4. $(3a^2 - 2a) - (7a^2 + 5)$

5. Subtract $4b^2 - 3b$ from the sum of $6b^2 + 5b$ and $4b^2 - 3$.

6. $3rs(5r^2s - 4rs + 6rs^2)$

7. $(2a - b)(3a^2 - ab + b^2)$

8. $\dfrac{7xy^3 - 21x^2y^2 + 14x^3y}{-7xy}$

9. $\dfrac{3a^2 - 10a - 8}{a - 4}$

10. $\dfrac{2x^3 - 8x + 5}{2x + 4}$

Solve the following equation for x.

11. $2 - 4(3x + 1) = 8 - 7x$

Solve the following inequality.

12. $4(x - 7) \le -(x - 5)$

Solve the following equation for the indicated variable.

13. $S = \dfrac{n}{2}(a + t)$ for t

Simplify the following expressions.

14. x^6x^{11}

15. $(3x^2y^3)(2x^3y^4)$

16. $(3x^2y^3)^2(-4x^3y^2)^0$

17. $\dfrac{16x^2y^5}{4xy^3}$

18. $(3x^2)^3(2x)^2$

Factor each of the following polynomials completely.

19. $36w^5 - 48w^4$

20. $5x^2y - 15xy + 10xy^2$

ANSWERS

1. 17
2. -2
3. $9x^2 - x - 5$
4. $-4a^2 - 2a - 5$
5. $6b^2 + 8b - 3$
6. $15r^3s^2 - 12r^2s^2 + 18r^2s^3$
7. $6a^3 - 5a^2b + 3ab^2 - b^3$
8. $-y^2 + 3xy - 2x^2$
9. $3a + 2$
10. $x^2 - 2x + \dfrac{5}{2x + 4}$
11. $x = -2$
12. $x \le \dfrac{33}{5}$
13. $t = \dfrac{2S - na}{n}$
14. x^{17}
15. $6x^5y^7$
16. $9x^4y^6$
17. $4xy^2$
18. $108x^8$
19. $12w^4(3w - 4)$
20. $5xy(x - 3 + 2y)$

ANSWERS

21. $(5x + 3y)^2$

22. $4p(p + 6q)(p - 6q)$

23. $(a + 3)(a + 1)$

24. $2w(w^2 - 2w - 12)$

25. $(3x + 2y)(x + 3y)$

26. 3, 4

27. −4, 4

28. $\frac{2}{3}, -1$

29. 6

30. 5 in. by 21 in.

Factor each of the following polynomials completely.

21. $25x^2 + 30xy + 9y^2$

22. $4p^3 - 144pq^2$

Factor each of the following trinomials completely.

23. $a^2 + 4a + 3$

24. $2w^3 - 4w^2 - 24w$

25. $3x^2 + 11xy + 6y^2$

Solve each of the following equations.

26. $a^2 - 7a + 12 = 0$

27. $3w^2 - 48 = 0$

28. $15x^2 + 5x = 10$

Solve the following problems.

29. Twice the square of a positive integer is 12 more than 10 times that integer. What is the integer?

30. The length of a rectangle is 1 in. more than 4 times its width. If the area of the rectangle is 105 in.^{2}, find the dimensions of the rectangle.

ALGEBRAIC FRACTIONS

5

INTRODUCTION

In the United States, disorders of the heart and circulatory system kill more people than all other causes combined. The major risk factors for heart disease are smoking, high blood pressure, obesity, cholesterol over 240 g/dL, and a family history of heart problems. Although nothing can be done about family history, everyone can affect the first four risk factors by diet and exercise.

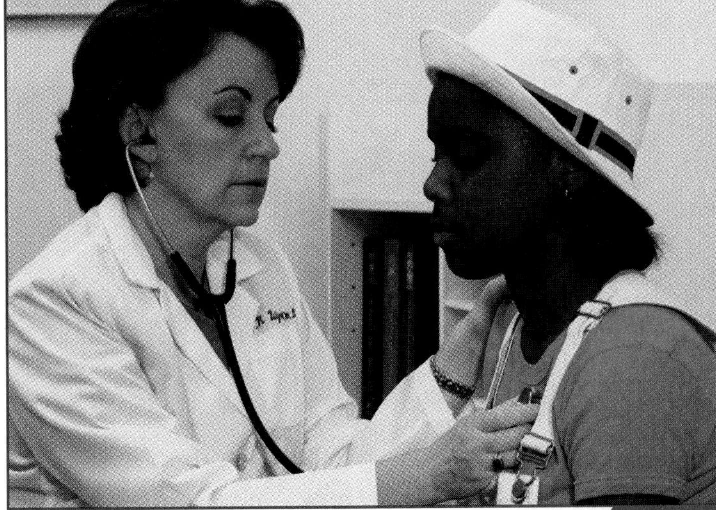

One quick way to check your risk of heart problems is to compare your waist and hip measurements. Measure around your waist at the navel and around your hips at the largest point. These measures may be in inches or centimeters. Use the ratio w/h to assess your risk. For women, $w/h \geq 0.8$ indicates an increased health risk, and for men, $w/h \geq 0.95$ is the indicator of an increased risk.

The American Medical Association sponsored a study using Body Mass Index, or BMI, which used height and weight measurements:

$$\text{BMI} = \frac{705w}{h^2}$$

in which w = weight in pounds
h = height in inches

This study concluded that people with BMI ≤ 21 had the lowest rates of heart disease, and that an increase of only 2 points in the BMI dramatically raises the risk of heart problems.

Medical professionals and researchers continue to disagree about how accurate these indicators are because each is a statistical average. One issue is how well the measures relate to the percentage of total body fat. A person may have a relatively low percentage of body fat and be in excellent health but have a BMI over 21 because of a very muscular build or large bone structure.

Name ______________________

Section ________ Date ________

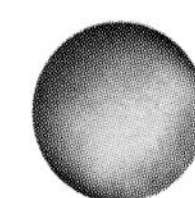

Pre-Test Chapter 5

ANSWERS

1. $\frac{-3b^6}{5a^2}$
2. $\frac{x+4}{2}$
3. $\frac{x-1}{2x}$
4. $\frac{13a}{6}$
5. 5
6. $x+6$
7. $\frac{5w-6}{2w^2}$
8. $\frac{3b+3}{b(b-3)}$
9. $\frac{-11}{6(x-1)}$
10. $\frac{10}{x-5}$
11. $\frac{a}{2b^2}$
12. $\frac{x+3}{2x}$
13. $2b^2$
14. $\frac{x+y}{4x}$
15. $\frac{3x}{2}$
16. $\frac{y}{2y+x}$
17. 3
18. −2, 5
19. 40
20. No solution
21. .5
22. 7
23. 5, 20
24. 48mi/hr going 40mi/hr returning
25. 15 ft, 40 ft

Write each fraction in simplest form.

1. $\frac{-15a^4b^7}{25a^6b}$ 2. $\frac{x^2-16}{2x-8}$ 3. $\frac{3x^2-2x-1}{6x^2+2x}$

Add or subtract as indicated.

4. $\frac{7a}{12}+\frac{19a}{12}$ 5. $\frac{5x}{x+1}+\frac{5}{x+1}$ 6. $\frac{x^2}{x-6}-\frac{36}{x-6}$ 7. $\frac{5}{2w}-\frac{3}{w^2}$

8. $\frac{4}{b-3}-\frac{1}{b}$ 9. $\frac{2}{3x-3}-\frac{5}{2x-2}$ 10. $\frac{4x}{x^2-8x+15}+\frac{6}{x-3}$

Multiply or divide as indicated.

11. $\frac{-4a^2}{6ab^3}\cdot\frac{3ab^2}{-4ab}$ 12. $\frac{x^2+5x+4}{2x^2+2x}\cdot\frac{x^2-x-12}{x^2-16}$

13. $\frac{8b^4}{5bc}\div\frac{12b^2c^2}{15bc^3}$ 14. $\frac{x^2y+2xy^2}{x^2-4y^2}\div\frac{4x^2y}{x^2-xy-2y^2}$

Simplify the complex fractions.

15. $\dfrac{\frac{x^3}{16}}{\frac{x^2}{24}}$ 16. $\dfrac{2-\frac{x}{y}}{4-\frac{x^2}{y^2}}$

What values for x, if any, must be excluded in the following algebraic fractions?

17. $\frac{5}{x-3}$ 18. $\frac{4}{x^2-3x-10}$

Solve the following equations for x.

19. $\frac{x}{4}-\frac{x}{5}=2$ 20. $\frac{x}{x-2}+1=\frac{x+4}{x-2}$

21. $\frac{7}{x}-\frac{1}{x-3}=\frac{9}{x^2-3x}$ 22. $\frac{x-3}{8}=\frac{x-2}{10}$

Solve the following applications.

23. One number is 4 times another. If the sum of their reciprocals is $\frac{1}{4}$, find the two numbers.

24. Mark drove 240 mi to visit Sandra. Returning by a shorter route, he found that the trip was only 200 mi, but traffic slowed his speed by 8 mi/h. If the two trips took exactly the same time, what was his rate each way?

25. A 55-ft cable is to be cut into two pieces whose lengths have the ratio 3 to 8. Find the lengths of the two pieces.

5.1 Simplifying Algebraic Fractions

5.1 OBJECTIVES

1. Find the GCF for two monomials and simplify a fraction
2. Find the GCF for two polynomials and simplify a fraction

Much of our work with algebraic fractions will be similar to your work in arithmetic. For instance, in algebra, as in arithmetic, many fractions name the same number. You will remember from Chapter 0 that

$$\frac{1}{4} = \frac{1 \cdot 2}{4 \cdot 2} = \frac{2}{8}$$

or

$$\frac{1}{4} = \frac{1 \cdot 3}{4 \cdot 3} = \frac{3}{12}$$

So $\frac{1}{4}, \frac{2}{8},$ and $\frac{3}{12}$ all name the same number. They are called **equivalent fractions.** These examples illustrate what is called the **Fundamental Principle of Fractions.** In algebra it becomes

Rules and Properties: Fundamental Principle of Algebraic Fractions

For polynomials P, Q, and R,

$$\frac{P}{Q} = \frac{PR}{QR} \qquad \text{when } Q \neq 0 \text{ and } R \neq 0$$

This principle allows us to multiply or divide the numerator and denominator of a fraction by the same nonzero polynomial. The result will be an expression that is equivalent to the original one.

Our objective in this section is to simplify algebraic fractions by using the fundamental principle. In algebra, as in arithmetic, to write a fraction in simplest form, you divide the numerator and denominator of the fraction by their greatest common factor (GCF). The numerator and denominator of the resulting fraction will have no common factors other than 1, and the fraction is then in **simplest form.** The following rule summarizes this procedure.

NOTE Notice that step 2 uses the Fundamental Principle of Fractions. The GCF is R in the rule above.

Step by Step: To Write Algebraic Fractions in Simplest Form

Step 1 Factor the numerator and denominator.

Step 2 Divide the numerator and denominator by the greatest common factor (GCF). The resulting fraction will be in lowest terms.

Example 1

Writing Fractions in Simplest Form

(a) Write $\frac{18}{30}$ in simplest form.

NOTE This is the same as dividing both the numerator and denominator of $\frac{18}{30}$ by 6.

$$\frac{18}{30} = \frac{2 \cdot 3 \cdot 3}{2 \cdot 3 \cdot 5} = \frac{\overset{1}{\cancel{2}} \cdot \overset{1}{\cancel{3}} \cdot 3}{\underset{1}{\cancel{2}} \cdot \underset{1}{\cancel{3}} \cdot 5} = \frac{3}{5}$$

Divide by the GCF. The slash lines indicate that we have divided the numerator and denominator by 2 and by 3.

(b) Write $\frac{4x^3}{6x}$ in simplest form.

$$\frac{4x^3}{6x} = \frac{\overset{1}{\cancel{2}} \cdot 2 \cdot \overset{1}{\cancel{x}} \cdot x \cdot x}{\underset{1}{\cancel{2}} \cdot 3 \cdot \underset{1}{\cancel{x}}} = \frac{2x^2}{3}$$

(c) Write $\frac{15x^3y^2}{20xy^4}$ in simplest form.

$$\frac{15x^3y^2}{20xy^4} = \frac{3 \cdot \overset{1}{\cancel{5}} \cdot \overset{1}{\cancel{x}} \cdot x \cdot x \cdot \overset{1}{\cancel{y}} \cdot \overset{1}{\cancel{y}}}{2 \cdot 2 \cdot \underset{1}{\cancel{5}} \cdot \underset{1}{\cancel{x}} \cdot \underset{1}{\cancel{y}} \cdot \underset{1}{\cancel{y}} \cdot y \cdot y} = \frac{3x^2}{4y^2}$$

(d) Write $\frac{3a^2b}{9a^3b^2}$ in simplest form.

$$\frac{3a^2b}{9a^3b^2} = \frac{\overset{1}{\cancel{3}} \cdot \overset{1}{\cancel{a}} \cdot \overset{1}{\cancel{a}} \cdot \overset{1}{\cancel{b}}}{\underset{1}{\cancel{3}} \cdot 3 \cdot \underset{1}{\cancel{a}} \cdot \underset{1}{\cancel{a}} \cdot a \cdot \underset{1}{\cancel{b}} \cdot b} = \frac{1}{3ab}$$

(e) Write $\frac{10a^5b^4}{2a^2b^3}$ in simplest form.

$$\frac{10a^5b^4}{2a^2b^3} = \frac{5 \cdot \overset{1}{\cancel{2}} \cdot \overset{1}{\cancel{a}} \cdot \overset{1}{\cancel{a}} \cdot a \cdot a \cdot a \cdot \overset{1}{\cancel{b}} \cdot \overset{1}{\cancel{b}} \cdot \overset{1}{\cancel{b}} \cdot b}{\underset{1}{\cancel{2}} \cdot \underset{1}{\cancel{a}} \cdot \underset{1}{\cancel{a}} \cdot \underset{1}{\cancel{b}} \cdot \underset{1}{\cancel{b}} \cdot \underset{1}{\cancel{b}}} = \frac{5a^3b}{1} = 5a^3b$$

NOTE Most of the methods of this chapter build on our factoring work of the last chapter.

CHECK YOURSELF 1

Write each fraction in simplest form.

(a) $\frac{30}{66}$ **(b)** $\frac{5x^4}{15x}$ **(c)** $\frac{12xy^4}{18x^3y^2}$ **(d)** $\frac{5m^2n}{10m^3n^3}$ **(e)** $\frac{12a^4b^6}{2a^3b^4}$

In simplifying arithmetic fractions, common factors are generally easy to recognize. With algebraic fractions, the factoring techniques you studied in Chapter 4 will have to be used as the *first step* in determining those factors.

Example 2

Writing Fractions in Simplest Form

Write each fraction in simplest form.

(a) $\frac{2x-4}{x^2-4} = \frac{2(x-2)}{(x+2)(x-2)}$ Factor the numerator and denominator.

$= \frac{2(\overset{1}{\cancel{x-2}})}{(x+2)(\underset{1}{\cancel{x-2}})}$ Divide by the GCF $x - 2$. The slash lines indicate that we have divided by that common factor.

$= \frac{2}{x+2}$

(b) $\frac{3x^2-3}{x^2-2x-3} = \frac{3(x-1)(\overset{1}{\cancel{x+1}})}{(x-3)(\underset{1}{\cancel{x+1}})}$

$= \frac{3(x-1)}{x-3}$

(c) $\frac{2x^2+x-6}{2x^2-x-3} = \frac{(x+2)(\overset{1}{\cancel{2x-3}})}{(x+1)(\underset{1}{\cancel{2x-3}})}$

$= \frac{x+2}{x+1}$

 **CAUTION**

Pick any value, other than 0, for x and substitute. You will quickly see that

$\frac{x+2}{x+1} \neq \frac{2}{1}$

Be Careful! The expression $\frac{x+2}{x+1}$ is already in simplest form. Students are often tempted to divide as follows:

$\frac{\cancel{x}+2}{\cancel{x}+1}$ is *not equal* to $\frac{2}{1}$

The x's are *terms* in the numerator and denominator. They *cannot* be divided out. Only *factors* can be divided. The fraction

$\frac{x+2}{x+1}$

is in its simplest form.

CHECK YOURSELF 2

Write each fraction in simplest form.

(a) $\frac{5x-15}{x^2-9}$ **(b)** $\frac{a^2-5a+6}{3a^2-6a}$

(c) $\frac{3x^2+14x-5}{3x^2+2x-1}$ **(d)** $\frac{5p-15}{p^2-4}$

Remember the rules for signs in division. The quotient of a positive number and a negative number is always negative. Thus there are three equivalent ways to write such a quotient. For instance,

$$\frac{-2}{3} = \frac{2}{-3} = -\frac{2}{3}$$

NOTE $\frac{-2}{3}$, with the negative sign in the numerator, is the most common way to write the quotient.

The quotient of two positive numbers or two negative numbers is always positive. For example,

$$\frac{-2}{-3} = \frac{2}{3}$$

Example 3

Writing Fractions in Simplest Form

Write each fraction in simplest form.

NOTE In part (a), the final quotient is written in the most common way with the minus sign in the numerator.

(a) $\dfrac{6x^2}{-3xy} = \dfrac{2 \cdot \overset{1}{\cancel{3}} \cdot \overset{1}{\cancel{x}} \cdot x}{(-1) \cdot \underset{1}{\cancel{3}} \cdot \underset{1}{\cancel{x}} \cdot y} = \dfrac{2x}{-y} = \dfrac{-2x}{y}$

(b) $\dfrac{-5a^2b}{-10b^2} = \dfrac{\overset{1}{\cancel{(-1)}} \cdot \overset{1}{\cancel{5}} \cdot a \cdot a \cdot \overset{1}{\cancel{b}}}{\underset{1}{\cancel{(-1)}} \cdot 2 \cdot \underset{1}{\cancel{5}} \cdot \underset{1}{\cancel{b}} \cdot b} = \dfrac{a^2}{2b}$

CHECK YOURSELF 3

Write each fraction in simplest form.

(a) $\dfrac{8x^3y}{-4xy^2}$ **(b)** $\dfrac{-16a^4b^2}{-12a^2b^5}$

It is sometimes necessary to factor out a monomial before simplifying the fraction.

Example 4

Writing Fractions in Simplest Form

Write each fraction in simplest form.

(a) $\dfrac{6x^2 + 2x}{2x^2 + 12x} = \dfrac{2x(3x + 1)}{2x(x + 6)} = \dfrac{3x + 1}{x + 6}$

(b) $\dfrac{x^2 - 4}{x^2 + 6x + 8} = \dfrac{(x + 2)(x - 2)}{(x + 2)(x + 4)} = \dfrac{x - 2}{x + 4}$

CHECK YOURSELF 4

Simplify each fraction.

(a) $\dfrac{3x^3 - 6x^2}{9x^4 - 3x^2}$ **(b)** $\dfrac{x^2 - 9}{x^2 - 12x + 27}$

Reducing certain algebraic fractions will be easier with the following result. First, verify for yourself that

$5 - 8 = -(8 - 5)$

In general, it is true that

$a - b = -(b - a)$

or, by dividing both sides of the equation by $b - a$,

$$\frac{a - b}{b - a} = \frac{-(b - a)}{b - a}$$

So dividing by $b - a$ on the right, we have

NOTE Remember that *a* and *b* cannot be divided out because they are not factors.

$$\frac{a - b}{b - a} = -1$$

Let's look at some applications of that result in Example 5.

Example 5

Writing Fractions in Simplest Form

Write each fraction in simplest form.

(a) $\dfrac{2x - 4}{4 - x^2} = \dfrac{2(x - 2)}{(2 + x)(2 - x)}$ This is equal to −1.

$$= \frac{2(-1)}{2 + x} = \frac{-2}{2 + x}$$

(b) $\dfrac{9 - x^2}{x^2 + 2x - 15} = \dfrac{(3 + x)(3 - x)}{(x + 5)(x - 3)}$ This is equal to −1.

$$= \frac{(3 + x)(-1)}{x + 5}$$

$$= \frac{-x - 3}{x + 5}$$

CHECK YOURSELF 5

Write each fraction in simplest form.

(a) $\dfrac{3x - 9}{9 - x^2}$ **(b)** $\dfrac{x^2 - 6x - 27}{81 - x^2}$

CHECK YOURSELF ANSWERS

1. (a) $\dfrac{5}{11}$; **(b)** $\dfrac{x^3}{3}$; **(c)** $\dfrac{2y^2}{3x^2}$; **(d)** $\dfrac{1}{2mn^2}$; **(e)** $6ab^2$ **2. (a)** $\dfrac{5}{x + 3}$; **(b)** $\dfrac{a - 3}{3a}$; **(c)** $\dfrac{x + 5}{x + 1}$; **(d)** $\dfrac{5(p - 3)}{(p + 2)(p - 2)}$ **3. (a)** $\dfrac{-2x^2}{y}$; **(b)** $\dfrac{4a^2}{3b^3}$ **4. (a)** $\dfrac{x - 2}{3x^2 - 1}$; **(b)** $\dfrac{x + 3}{x - 9}$ **5. (a)** $\dfrac{-3}{x + 3}$; **(b)** $\dfrac{-x - 3}{x + 9}$

5.1 Exercises

Name ______________________

Section ________ Date ________

Write each fraction in simplest form.

1. $\frac{16}{24}$

2. $\frac{56}{64}$

3. $\frac{80}{180}$

4. $\frac{18}{30}$

5. $\frac{4x^5}{6x^2}$

6. $\frac{10x^2}{15x^4}$

7. $\frac{9x^3}{27x^6}$

8. $\frac{25w^6}{20w^2}$

9. $\frac{10a^2b^5}{25ab^2}$

10. $\frac{18x^4y^3}{24x^2y^3}$

11. $\frac{42x^3y}{14xy^3}$

12. $\frac{18pq}{45p^2q^2}$

13. $\frac{2xyw^2}{6x^2y^3w^3}$

14. $\frac{3c^2d^2}{6bc^3d^3}$

15. $\frac{10x^5y^5}{2x^3y^4}$

16. $\frac{3bc^6d^3}{bc^3d}$

17. $\frac{-4m^3n}{6mn^2}$

18. $\frac{-15x^3y^3}{-20xy^4}$

19. $\frac{-8ab^3}{-16a^3b}$

20. $\frac{14x^2y}{-21xy^4}$

ANSWERS

1. $\frac{2}{3}$
2. $\frac{7}{8}$
3. $\frac{4}{9}$
4. $\frac{3}{5}$
5. $\frac{2x^3}{3}$
6. $\frac{2}{3x^2}$
7. $\frac{1}{3x^3}$
8. $\frac{5w^4}{4}$
9. $\frac{2ab^3}{5}$
10. $\frac{3x^2}{4}$
11. $\frac{3x^2}{y^2}$
12. $\frac{2}{5pq}$
13. $\frac{1}{3xy^2w}$
14. $\frac{1}{2bcd}$
15. $5x^2y$
16. $3c^3d^2$
17. $\frac{-2m^2}{3n}$
18. $\frac{3x^2}{4y}$
19. $\frac{b^2}{2a^2}$
20. $\frac{-2x}{3y^3}$

21. $\frac{-r}{2st^2}$

22. $\frac{-2a^2c^2}{3b^2}$

23. $\frac{3}{5}$

24. $\frac{4}{5}$

25. $\frac{3(x-2)}{5(x-3)}$

26. $\frac{x+5}{3}$

27. $\frac{6}{a+4}$

28. $\frac{5(x-1)}{(x+2)(x-2)}$

29. $\frac{x+1}{5}$

30. $\frac{4w}{w+3}$

31. $\frac{x+2}{x+8}$

32. $\frac{y+5}{y+4}$

33. $\frac{m-1}{m+3}$

34. $\frac{2x+1}{x-1}$

35. $\frac{p-3q}{p-5q}$

36. $\frac{2r+5s}{r+4s}$

37. $\frac{-2}{x+5}$

38. $\frac{-3}{a+4}$

39. $\frac{-a-5}{a+6}$

40. $\frac{-2x+1}{x+3}$

41. $\frac{-x-3y}{2y+x}$

42. $\frac{-w-4z}{2w+3z}$

21. $\frac{8r^2s^3t}{-16rs^4t^3}$

22. $\frac{-10a^3b^2c^3}{15ab^4c}$

23. $\frac{3x+18}{5x+30}$

24. $\frac{4x-28}{5x-35}$

25. $\frac{3x-6}{5x-15}$

26. $\frac{x^2-25}{3x-15}$

27. $\frac{6a-24}{a^2-16}$

28. $\frac{5x-5}{x^2-4}$

29. $\frac{x^2+3x+2}{5x+10}$

30. $\frac{4w^2-20w}{w^2-2w-15}$

31. $\frac{x^2-6x-16}{x^2-64}$

32. $\frac{y^2-25}{y^2-y-20}$

33. $\frac{2m^2+3m-5}{2m^2+11m+15}$

34. $\frac{6x^2-x-2}{3x^2-5x+2}$

35. $\frac{p^2+2pq-15q^2}{p^2-25q^2}$

36. $\frac{4r^2-25s^2}{2r^2+3rs-20s^2}$

37. $\frac{2x-10}{25-x^2}$

38. $\frac{3a-12}{16-a^2}$

39. $\frac{25-a^2}{a^2+a-30}$

40. $\frac{2x^2-7x+3}{9-x^2}$

41. $\frac{x^2+xy-6y^2}{4y^2-x^2}$

42. $\frac{16z^2-w^2}{2w^2-5wz-12z^2}$

43. $\dfrac{x^2 + 4x + 4}{x + 2}$

44. $\dfrac{4x^2 + 12x + 9}{2x + 3}$

45. $\dfrac{xy - 2y + 4x - 8}{2y + 6 - xy - 3x}$

46. $\dfrac{ab - 3a + 5b - 15}{15 + 3a^2 - 5b - a^2b}$

47. $\dfrac{y - 7}{7 - y}$

48. $\dfrac{5 - y}{y - 5}$

49. The area of the rectangle is represented by $6x^2 + 19x + 10$. What is the length?

50. The volume of the box is represented by $(x^2 + 5x + 6)(x + 5)$. Find the polynomial that represents the area of the bottom of the box.

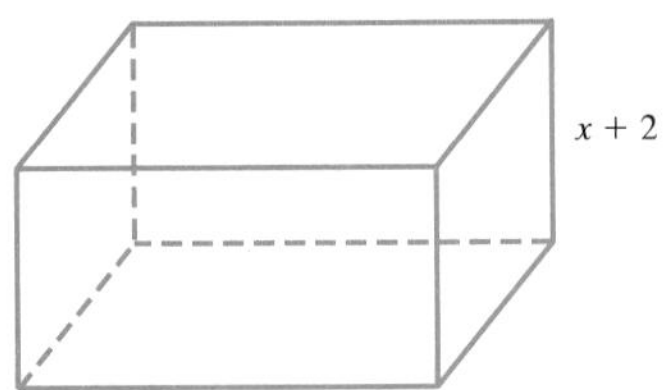

51. To work with algebraic fractions correctly, it is important to understand the difference between a *factor* and a *term* of an expression. In your own words, write difinitions for both, explaining the difference between the two.

52. Give some examples of terms and factors in algebraic fractions, and explain how both are affected when a fraction is reduced.

53. Show how the following algebraic fraction can be reduced:

$$\frac{x^2 - 9}{4x + 12}$$

Note that your reduced fraction is equivalent to the given fraction. Are there other algebraic fractions equivalent to this one? Write another algebraic fraction that you think is equivalent to this one. Exchange papers with another student. Do you agree that their fraction is equivalent to yours? Why or why not?

54. Explain the reasoning involved in each step of reducing the fraction $\frac{42}{56}$.

55. Describe why $\frac{3}{5}$ and $\frac{27}{45}$ are *equivalent fractions.*

ANSWERS

43. $x + 2$

44. $2x + 3$

45. $\dfrac{-(y + 4)}{y + 3}$

46. $\dfrac{-(a + 5)}{a^2 + 5}$

47. -1

48. -1

49. $2x + 5$

50. $(x + 3)(x + 5)$

51.

52.

53.

54.

55.

ANSWERS

a. $\frac{7}{10}$

b. $\frac{1}{8}$

c. $\frac{1}{3}$

d. $\frac{5}{8}$

e. $\frac{4}{5}$

f. 1

g. $\frac{3}{2}$

h. $\frac{4}{3}$

Getting Ready for Section 5.2 [Section 0.2]

Perform the indicated operations.

(a) $\frac{3}{10} + \frac{4}{10}$

(b) $\frac{5}{8} - \frac{4}{8}$

(c) $\frac{5}{12} - \frac{1}{12}$

(d) $\frac{7}{16} + \frac{3}{16}$

(e) $\frac{7}{20} + \frac{9}{20}$

(f) $\frac{13}{8} - \frac{5}{8}$

(g) $\frac{11}{6} - \frac{2}{6}$

(h) $\frac{5}{9} + \frac{7}{9}$

Answers

1. $\frac{2}{3}$ **3.** $\frac{4}{9}$ **5.** $\frac{2x^3}{3}$ **7.** $\frac{1}{3x^3}$ **9.** $\frac{2ab^3}{5}$ **11.** $\frac{3x^2}{y^2}$ **13.** $\frac{1}{3xy^2w}$

15. $5x^2y$ **17.** $\frac{-2m^2}{3n}$ **19.** $\frac{b^2}{2a^2}$ **21.** $\frac{-r}{2st^2}$ **23.** $\frac{3}{5}$ **25.** $\frac{3(x-2)}{5(x-3)}$

27. $\frac{6}{a+4}$ **29.** $\frac{x+1}{5}$ **31.** $\frac{x+2}{x+8}$ **33.** $\frac{m-1}{m+3}$ **35.** $\frac{p-3q}{p-5q}$

37. $\frac{-2}{x+5}$ **39.** $\frac{-a-5}{a+6}$ **41.** $\frac{-x-3y}{2y+x}$ **43.** $x+2$ **45.** $\frac{-(y+4)}{y+3}$

47. -1 **49.** $2x+5$ **51.**  **53.**  **55.** 

a. $\frac{7}{10}$ **b.** $\frac{1}{8}$ **c.** $\frac{1}{3}$ **d.** $\frac{5}{8}$ **e.** $\frac{4}{5}$ **f.** 1 **g.** $\frac{3}{2}$ **h.** $\frac{4}{3}$

Adding and Subtracting Like Fractions

5.2 OBJECTIVES

1. Write the sum of two like fractions in simplest form
2. Write the difference of two like fractions in simplest form

You probably remember from arithmetic that **like fractions** are fractions that have the same denominator. The same is true in algebra.

$\frac{2}{5}, \frac{12}{5}$, and $\frac{4}{5}$ are like fractions.

$\frac{x}{3}, \frac{y}{3}$, and $\frac{z-5}{3}$ are like fractions.

$\frac{3x}{2}, \frac{x}{4}$, and $\frac{3x}{8}$ are unlike fractions.

NOTE The fractions have different denominators.

$\frac{3}{x}, \frac{2}{x^2}$, and $\frac{x+1}{x^3}$ are unlike fractions.

In arithmetic, the sum or difference of like fractions was found by adding or subtracting the numerators and writing the result over the common denominator. For example,

$$\frac{3}{11} + \frac{5}{11} = \frac{3+5}{11} = \frac{8}{11}$$

In symbols, we have

Rules and Properties: To Add or Subtract Like Fractions

$$\frac{P}{R} + \frac{Q}{R} = \frac{P+Q}{R} \qquad R \neq 0$$

$$\frac{P}{R} - \frac{Q}{R} = \frac{P-Q}{R} \qquad R \neq 0$$

Adding or subtracting like fractions in algebra is just as straightforward. You can use the following steps.

Step by Step: To Add or Subtract Like Algebraic Fractions

Step 1 Add or subtract the numerators.
Step 2 Write the sum or difference over the common denominator.
Step 3 Write the resulting fraction in simplest form.

Example 1

Adding and Subtracting Algebraic Fractions

Add or subtract as indicated. Express your results in simplest form.

Add the numerators.

(a) $\frac{2x}{15} + \frac{x}{15} = \frac{2x + x}{15}$

$= \frac{3x}{15} = \frac{x}{5}$

Simplify.

Subtract the numerators.

(b) $\frac{5y}{6} - \frac{y}{6} = \frac{5y - y}{6}$

$= \frac{4y}{6} = \frac{2y}{3}$

Simplify.

(c) $\frac{3}{x} + \frac{5}{x} = \frac{3 + 5}{x} = \frac{8}{x}$

(d) $\frac{9}{a^2} - \frac{7}{a^2} = \frac{9 - 7}{a^2} = \frac{2}{a^2}$

(e) $\frac{7}{2ab} - \frac{5}{2ab} = \frac{7 - 5}{2ab}$

$= \frac{2}{2ab}$

$= \frac{1}{ab}$

CHECK YOURSELF 1

Add or subtract as indicated.

(a) $\frac{3a}{10} + \frac{2a}{10}$ (b) $\frac{7b}{8} - \frac{3b}{8}$ (c) $\frac{4}{x} + \frac{3}{x}$ (d) $\frac{5}{3xy} - \frac{2}{3xy}$

If polynomials are involved in the numerators or denominators, the process is exactly the same.

Example 2

Adding and Subtracting Algebraic Fractions

Add or subtract as indicated. Express your results in simplest form.

(a) $\frac{5}{x + 3} + \frac{2}{x + 3} = \frac{5 + 2}{x + 3} = \frac{7}{x + 3}$

(b) $$\frac{4x}{x-4} - \frac{16}{x-4} = \frac{4x-16}{x-4}$$

Factor and simplify.

$$= \frac{4\cancel{(x-4)}}{\cancel{x-4}} = 4$$

NOTE The final answer is always written in simplest form.

(c) $$\frac{a-b}{3} + \frac{2a+b}{3} = \frac{(a-b)+(2a+b)}{3}$$

$$= \frac{a-b+2a+b}{3}$$

$$= \frac{\cancel{3}a}{\cancel{3}} = a$$

Be sure to enclose the second numerator in parentheses!

(d) $$\frac{3x+y}{2x} - \frac{x-3y}{2x} = \frac{(3x+y)-(x-3y)}{2x}$$

Change both signs.

$$= \frac{3x+y-x+3y}{2x}$$

$$= \frac{2x+4y}{2x}$$

$$= \frac{\cancel{2}(x+2y)}{\cancel{2}x}$$

Factor and divide by the common factor of 2.

$$= \frac{x+2y}{x}$$

(e) $$\frac{3x-5}{x^2+x-2} - \frac{2x-4}{x^2+x-2} = \frac{(3x-5)-(2x-4)}{x^2+x-2}$$

Put the second numerator in parentheses.

Change both signs.

$$= \frac{3x-5-2x+4}{x^2+x-2}$$

$$= \frac{x-1}{x^2+x-2}$$

$$= \frac{\cancel{(x-1)}}{(x+2)\cancel{(x-1)}}$$

Factor and divide by the common factor of $x - 1$.

$$= \frac{1}{x+2}$$

(f) $\dfrac{2x + 7y}{x + 3y} - \dfrac{x + 4y}{x + 3y} = \dfrac{(2x + 7y) - (x + 4y)}{x + 3y}$

Change both signs.

$= \dfrac{2x + 7y - x - 4y}{x + 3y}$

$= \dfrac{x + 3y}{x + 3y} = 1$

CHECK YOURSELF 2

Add or subtract as indicated.

(a) $\dfrac{4}{x - 5} - \dfrac{2}{x - 5}$

(b) $\dfrac{3x}{x + 3} + \dfrac{9}{x + 3}$

(c) $\dfrac{5x - y}{3y} - \dfrac{2x - 4y}{3y}$

(d) $\dfrac{5x + 8}{x^2 - 2x - 15} - \dfrac{4x + 5}{x^2 - 2x - 15}$

CHECK YOURSELF ANSWERS

1. (a) $\dfrac{a}{2}$; (b) $\dfrac{b}{2}$; (c) $\dfrac{7}{x}$; (d) $\dfrac{1}{xy}$ **2.** (a) $\dfrac{2}{x - 5}$; (b) 3; (c) $\dfrac{x + y}{y}$; (d) $\dfrac{1}{x - 5}$

Name ____________

Section ______ Date ______

5.2 Exercises

Add or subtract as indicated. Express your results in simplest form.

1. $\frac{7}{18} + \frac{5}{18}$
2. $\frac{5}{18} - \frac{2}{18}$
3. $\frac{13}{16} - \frac{9}{16}$
4. $\frac{5}{12} + \frac{11}{12}$
5. $\frac{x}{8} + \frac{3x}{8}$
6. $\frac{5y}{16} + \frac{7y}{16}$
7. $\frac{7a}{10} - \frac{3a}{10}$
8. $\frac{5x}{12} - \frac{x}{12}$
9. $\frac{5}{x} + \frac{3}{x}$
10. $\frac{9}{y} - \frac{3}{y}$
11. $\frac{8}{w} - \frac{2}{w}$
12. $\frac{7}{z} + \frac{9}{z}$
13. $\frac{2}{xy} + \frac{3}{xy}$
14. $\frac{8}{ab} + \frac{4}{ab}$
15. $\frac{2}{3cd} + \frac{4}{3cd}$
16. $\frac{5}{4cd} + \frac{11}{4cd}$
17. $\frac{7}{x - 5} + \frac{9}{x - 5}$
18. $\frac{11}{x + 7} - \frac{4}{x + 7}$
19. $\frac{2x}{x - 2} - \frac{4}{x - 2}$
20. $\frac{7w}{w + 3} + \frac{21}{w + 3}$
21. $\frac{8p}{p + 4} + \frac{32}{p + 4}$
22. $\frac{5a}{a - 3} - \frac{15}{a - 3}$
23. $\frac{x^2}{x + 4} + \frac{3x - 4}{x + 4}$
24. $\frac{x^2}{x - 3} - \frac{9}{x - 3}$
25. $\frac{m^2}{m - 5} - \frac{25}{m - 5}$
26. $\frac{s^2}{s + 3} + \frac{2s - 3}{s + 3}$
27. $\frac{a - 1}{3} + \frac{2a - 5}{3}$
28. $\frac{y + 2}{5} + \frac{4y + 8}{5}$
29. $\frac{3x - 1}{4} - \frac{x + 7}{4}$
30. $\frac{4x + 2}{3} - \frac{x - 1}{3}$
31. $\frac{4m + 7}{6m} + \frac{2m + 5}{6m}$
32. $\frac{6x - y}{4y} - \frac{2x + 3y}{4y}$
33. $\frac{4w - 7}{w - 5} - \frac{2w + 3}{w - 5}$
34. $\frac{3b - 8}{b - 6} + \frac{b - 16}{b - 6}$
35. $\frac{x - 7}{x^2 - x - 6} + \frac{2x - 2}{x^2 - x - 6}$
36. $\frac{5a - 12}{a^2 - 8a + 15} - \frac{3a - 2}{a^2 - 8a + 15}$
37. $\frac{y^2}{2y + 8} + \frac{3y - 4}{2y + 8}$
38. $\frac{x^2}{4x - 12} - \frac{9}{4x - 12}$
39. $\frac{7w}{w + 3} + \frac{21}{w + 3}$
40. $\frac{2x}{x - 3} - \frac{6}{x - 3}$

ANSWERS

1. $\frac{2}{3}$
2. $\frac{1}{6}$
3. $\frac{1}{4}$
4. $\frac{4}{3}$
5. $\frac{x}{2}$
6. $\frac{3y}{4}$
7. $\frac{2a}{5}$
8. $\frac{x}{3}$
9. $\frac{8}{x}$
10. $\frac{6}{y}$
11. $\frac{6}{w}$
12. $\frac{16}{z}$
13. $\frac{5}{xy}$
14. $\frac{12}{ab}$
15. $\frac{2}{cd}$
16. $\frac{4}{cd}$
17. $\frac{16}{x - 5}$
18. $\frac{7}{x + 7}$
19. 2
20. 7
21. 8
22. 5
23. $x - 1$
24. $x + 3$
25. $m + 5$
26. $s - 1$
27. $a - 2$
28. $y + 2$
29. $\frac{x - 4}{2}$
30. $x + 1$
31. $\frac{m + 2}{m}$
32. $\frac{x - y}{y}$
33. 2
34. 4
35. $\frac{3}{x + 2}$
36. $\frac{2}{a - 3}$
37. $\frac{y - 1}{2}$
38. $\frac{x + 3}{4}$
39. 7
40. 2

41. 1

42. 1

43. 4

44. $\frac{2x + 16}{2x - 5}$

a. $\frac{5}{4}$

b. $\frac{1}{6}$

c. $\frac{1}{10}$

d. $\frac{11}{8}$

e. $\frac{29}{24}$

f. $\frac{11}{40}$

g. $\frac{23}{30}$

h. $\frac{29}{36}$

41. $\frac{x^2}{x^2 + x - 6} - \frac{6}{(x + 3)(x - 2)} + \frac{x}{(x^2 + x - 6)}$

42. $\frac{-12}{x^2 + x - 12} + \frac{x^2}{(x + 4)(x - 3)} + \frac{x}{x^2 + x - 12}$

43. Find the perimeter of the given figure.

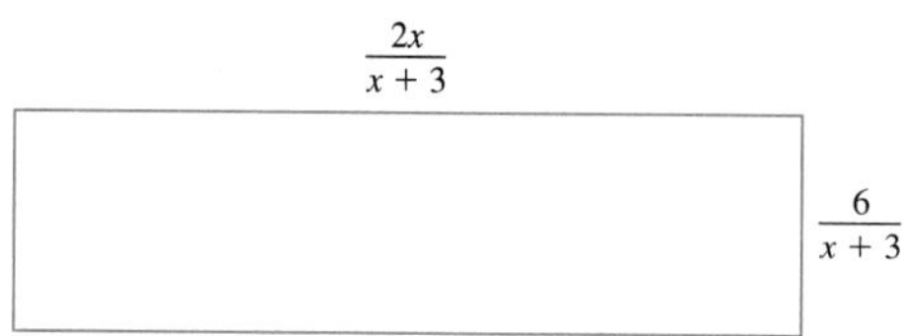

44. Find the perimeter of the following figure.

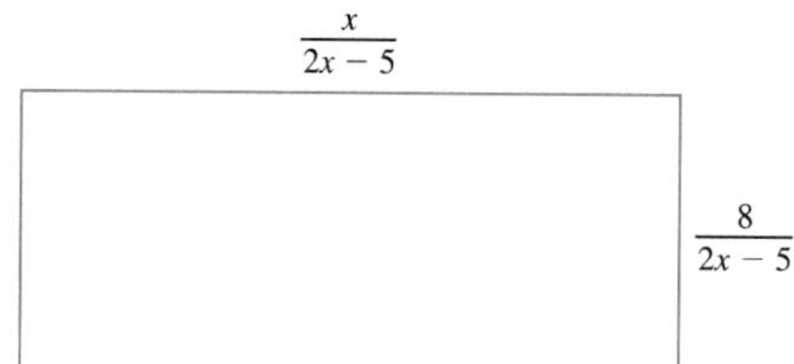

Getting Ready for Section 5.3 [Section 0.2]

(a) $\frac{3}{4} + \frac{1}{2}$

(b) $\frac{5}{6} - \frac{2}{3}$

(c) $\frac{7}{10} - \frac{3}{5}$

(d) $\frac{5}{8} + \frac{3}{4}$

(e) $\frac{5}{6} + \frac{3}{8}$

(f) $\frac{7}{8} - \frac{3}{5}$

(g) $\frac{9}{10} - \frac{2}{15}$

(h) $\frac{5}{12} + \frac{7}{18}$

Answers

1. $\frac{2}{3}$ **3.** $\frac{1}{4}$ **5.** $\frac{x}{2}$ **7.** $\frac{2a}{5}$ **9.** $\frac{8}{x}$ **11.** $\frac{6}{w}$ **13.** $\frac{5}{xy}$ **15.** $\frac{2}{cd}$ **17.** $\frac{16}{x - 5}$ **19.** 2 **21.** 8 **23.** $x - 1$ **25.** $m + 5$ **27.** $a - 2$ **29.** $\frac{x - 4}{2}$ **31.** $\frac{m + 2}{m}$ **33.** 2 **35.** $\frac{3}{x + 2}$ **37.** $\frac{y - 1}{2}$ **39.** 7 **41.** 1 **43.** 4 **a.** $\frac{5}{4}$ **b.** $\frac{1}{6}$ **c.** $\frac{1}{10}$ **d.** $\frac{11}{8}$ **e.** $\frac{29}{24}$ **f.** $\frac{11}{40}$ **g.** $\frac{23}{30}$ **h.** $\frac{29}{36}$

5.3 Adding and Subtracting Unlike Fractions

5.3 OBJECTIVES

1. Write the sum of two unlike fractions in simplest form
2. Write the difference of two unlike fractions in simplest form.

Adding or subtracting **unlike fractions** (fractions that do not have the same denominator) requires a bit more work than adding or subtracting the like fractions of the previous section. When the denominators are not the same, we must use the idea of the *lowest common denominator* (LCD). Each fraction is "built up" to an equivalent fraction having the LCD as a denominator. You can then add or subtract as before.

Let's review with an example from arithmetic.

Example 1

Finding the LCD

Add $\frac{5}{9} + \frac{1}{6}$.

Step 1 To find the LCD, factor each denominator.

$9 = 3 \cdot 3 \longleftarrow$ 3 appears twice.

$6 = 2 \cdot 3$

To form the LCD, include each factor the greatest number of times it appears in any single denominator. In this example, use one 2, because 2 appears only once in the factorization of 6. Use two 3s, because 3 appears twice in the factorization of 9. Thus the LCD for the fractions in $2 \cdot 3 \cdot 3 = 18$.

Step 2 "Build up" each fraction to an equivalent fraction with the LCD as the denominator. Do this by multiplying the numerator and denominator of the given fractions by the same number.

NOTE Do you see that this uses the fundamental principle in the following form?

$$\frac{P}{Q} = \frac{PR}{QR}$$

$$\frac{5}{9} = \frac{5 \cdot 2}{9 \cdot 2} = \frac{10}{18}$$

$$\frac{1}{6} = \frac{1 \cdot 3}{6 \cdot 3} = \frac{3}{18}$$

Step 3 Add the fractions.

$$\frac{5}{9} + \frac{1}{6} = \frac{10}{18} + \frac{3}{18} = \frac{13}{18}$$

$\frac{13}{18}$ is in simplest form, and so we are done!

CHECK YOURSELF 1

Add.

(a) $\frac{1}{6} + \frac{3}{8}$ **(b)** $\frac{3}{10} + \frac{4}{15}$

The process of finding the sum or difference is exactly the same in algebra as it is in arithmetic. We can summarize the steps with the following rule:

Step by Step: To Add or Subtract Unlike Fractions

Step 1 Find the lowest common denominator of all the fractions.
Step 2 Convert each fraction to an equivalent fraction with the LCD as a denominator.
Step 3 Add or subtract the like fractions formed in step 2.
Step 4 Write the sum or difference in simplest form.

Example 2

Adding Unlike Fractions

(a) Add $\frac{3}{2x} + \frac{4}{x^2}$.

Step 1 Factor the denominators.

$2x = 2 \cdot x$

$x^2 = x \cdot x$

NOTE Although the product of the denominators will be a common denominator, it is not necessarily the *lowest* common denominator (LCD).

The LCD must contain the factors 2 and x. The factor x must appear *twice* because it appears twice as a factor in the second denominator.

The LCD is $2 \cdot x \cdot x$, or $2x^2$.

Step 2

$$\frac{3}{2x} = \frac{3 \cdot x}{2x \cdot x} = \frac{3x}{2x^2}$$

$$\frac{4}{x^2} = \frac{4 \cdot 2}{x^2 \cdot 2} = \frac{8}{2x^2}$$

Step 3

$$\frac{3}{2x} + \frac{4}{x^2} = \frac{3x}{2x^2} + \frac{8}{2x^2} = \frac{3x + 8}{2x^2}$$

The sum is in simplest form.

(b) Subtract $\frac{4}{3x^2} - \frac{3}{2x^3}$.

Step 1 Factor the denominators.

$3x^2 = 3 \cdot x \cdot x$

$2x^3 = 2 \cdot x \cdot x \cdot x$

The LCD must contain the factors 2, 3, and x. The LCD is

$2 \cdot 3 \cdot x \cdot x \cdot x$ or $6x^3$

The factor x must appear 3 times. Do you see why?

Step 2

NOTE Both the numerator and the denominator must be multiplied by the same quantity.

$$\frac{4}{3x^2} = \frac{4 \cdot 2x}{3x^2 \cdot 2x} = \frac{8x}{6x^3}$$

$$\frac{3}{2x^3} = \frac{3 \cdot 3}{2x^3 \cdot 3} = \frac{9}{6x^3}$$

Step 3

$$\frac{4}{3x^2} - \frac{3}{2x^3} = \frac{8x}{6x^3} - \frac{9}{6x^3} = \frac{8x - 9}{6x^3}$$

The difference is in simplest form.

CHECK YOURSELF 2

Add or subtract as indicated.

(a) $\frac{5}{x^2} + \frac{3}{x^3}$ **(b)** $\frac{3}{5x} - \frac{1}{4x^2}$

We can also add fractions with more than one variable in the denominator. Example 3 shows this property.

Example 3

Adding Unlike Fractions

Add $\frac{2}{3x^2y} + \frac{3}{4x^3}$.

Step 1 Factor the denominators.

$3x^2y = 3 \cdot x \cdot x \cdot y$

$4x^3 = 2 \cdot 2 \cdot x \cdot x \cdot x$

The LCD is $12x^3y$. Do you see why?

Step 2

$$\frac{2}{3x^2y} = \frac{2 \cdot 4x}{3x^2y \cdot 4x} = \frac{8x}{12x^3y}$$

$$\frac{3}{4x^3} = \frac{3 \cdot 3y}{4x^3 \cdot 3y} = \frac{9y}{12x^3y}$$

Step 3

NOTE The y in the numerator and that in the denominator cannot be divided out because they are not factors.

$$\frac{2}{3x^2y} + \frac{3}{4x^3} = \frac{8x}{12x^3y} + \frac{9y}{12x^3y}$$

$$= \frac{8x + 9y}{12x^3y}$$

CHECK YOURSELF 3

Add.

$$\frac{2}{3x^2y} + \frac{1}{6xy^2}$$

Fractions with binomials in the denominator can also be added by taking the approach shown in Example 3. Example 4 illustrates this approach with binomials.

Example 4

Adding Unlike Fractions

(a) Add $\frac{5}{x} + \frac{2}{x - 1}$.

Step 1 The LCD must have factors of x and $x - 1$. The LCD is $x(x - 1)$.

Step 2

$$\frac{5}{x} = \frac{5(x - 1)}{x(x - 1)}$$

$$\frac{2}{x - 1} = \frac{2x}{x(x - 1)}$$

Step 3

$$\frac{5}{x} + \frac{2}{x - 1} = \frac{5(x - 1)}{x(x - 1)} + \frac{2x}{x(x - 1)}$$

$$= \frac{5x - 5 + 2x}{x(x - 1)}$$

$$= \frac{7x - 5}{x(x - 1)}$$

(b) Subtract $\dfrac{3}{x-2} - \dfrac{4}{x+2}$.

Step 1 The LCD must have factors of $x - 2$ and $x + 2$. The LCD is $(x - 2)(x + 2)$.

Step 2

NOTE Multiply numerator and denominator by $x + 2$.

$$\frac{3}{x-2} = \frac{3(x+2)}{(x-2)(x+2)}$$

NOTE Multiply numerator and denominator by $x - 2$.

$$\frac{4}{x+2} = \frac{4(x-2)}{(x+2)(x-2)}$$

Step 3

$$\frac{3}{x-2} - \frac{4}{x+2} = \frac{3(x+2) - 4(x-2)}{(x+2)(x-2)}$$

Note that the x term becomes negative and the constant term becomes positive.

$$= \frac{3x + 6 - 4x + 8}{(x+2)(x-2)}$$

$$= \frac{-x + 14}{(x+2)(x-2)}$$

CHECK YOURSELF 4

Add or subtract as indicated.

(a) $\dfrac{3}{x+2} + \dfrac{5}{x}$ **(b)** $\dfrac{4}{x+3} - \dfrac{2}{x-3}$

Example 5 will show how factoring must sometimes be used in forming the LCD.

Example 5

Adding Unlike Fractions

(a) Add $\dfrac{3}{2x-2} + \dfrac{5}{3x-3}$.

Step 1 Factor the denominators.

$2x - 2 = 2(x - 1)$

$3x - 3 = 3(x - 1)$

CAUTION

$x - 1$ is not used twice in forming the LCD.

The LCD must have factors of 2, 3, and $x - 1$. The LCD is $2 \cdot 3(x - 1)$, or $6(x - 1)$.

Step 2

$$\frac{3}{2x - 2} = \frac{3}{2(x - 1)} = \frac{3 \cdot 3}{2(x - 1) \cdot 3} = \frac{9}{6(x - 1)}$$

$$\frac{5}{3x - 3} = \frac{5}{3(x - 1)} = \frac{5 \cdot 2}{3(x - 1) \cdot 2} = \frac{10}{6(x - 1)}$$

Step 3

$$\frac{3}{2x - 2} + \frac{5}{3x - 3} = \frac{9}{6(x - 1)} + \frac{10}{6(x - 1)}$$

$$= \frac{9 + 10}{6(x - 1)}$$

$$= \frac{19}{6(x - 1)}$$

(b) Subtract $\frac{3}{2x - 4} - \frac{6}{x^2 - 4}$.

Step 1 Factor the denominators.

$2x - 4 = 2(x - 2)$

$x^2 - 4 = (x + 2)(x - 2)$

The LCD must have factors of 2, $x - 2$, and $x + 2$. The LCD is $2(x - 2)(x + 2)$.

Step 2

NOTE Multiply numerator and denominator by $x + 2$.

$$\frac{3}{2x - 4} = \frac{3}{2(x - 2)} = \frac{3(x + 2)}{2(x - 2)(x + 2)}$$

NOTE Multiply numerator and denominator by 2.

$$\frac{6}{x^2 - 4} = \frac{6}{(x + 2)(x - 2)} = \frac{6 \cdot 2}{2(x + 2)(x - 2)} = \frac{12}{2(x + 2)(x - 2)}$$

Step 3

$$\frac{3}{2x - 4} - \frac{6}{x^2 - 4} = \frac{3(x + 2) - 12}{2(x - 2)(x + 2)}$$

NOTE Remove the parentheses and combine like terms in the numerator.

$$= \frac{3x + 6 - 12}{2(x - 2)(x + 2)}$$

$$= \frac{3x - 6}{2(x - 2)(x + 2)}$$

Step 4 Simplify the difference.

NOTE Factor the numerator and divide by the common factor $x - 2$.

$$\frac{3x - 6}{2(x - 2)(x + 2)} = \frac{3\overset{1}{\cancel{(x - 2)}}}{2\underset{1}{\cancel{(x - 2)}}(x + 2)} = \frac{3}{2(x + 2)}$$

(c) Subtract $\dfrac{5}{x^2 - 1} - \dfrac{2}{x^2 + 2x + 1}$.

Step 1 Factor the denominators.

$$x^2 - 1 = (x - 1)(x + 1)$$

$$x^2 + 2x + 1 = (x + 1)(x + 1)$$

The LCD is $(x - 1)\underbrace{(x + 1)(x + 1)}$.

Two factors are needed.

Step 2

$$\frac{5}{(x - 1)(x + 1)} = \frac{5(x + 1)}{(x - 1)(x + 1)(x + 1)}$$

$$\frac{2}{(x + 1)(x + 1)} = \frac{2(x - 1)}{(x + 1)(x + 1)(x - 1)}$$

Step 3

$$\frac{5}{x^2 - 1} - \frac{2}{x^2 + 2x + 1} = \frac{5(x + 1) - 2(x - 1)}{(x - 1)(x + 1)(x + 1)}$$

NOTE Remove the parentheses and simplify in the numerator.

$$= \frac{5x + 5 - 2x + 2}{(x - 1)(x + 1)(x + 1)}$$

$$= \frac{3x + 7}{(x - 1)(x + 1)(x + 1)}$$

CHECK YOURSELF 5

Add or subtract as indicated.

(a) $\dfrac{5}{2x + 2} + \dfrac{1}{5x + 5}$

(b) $\dfrac{3}{x^2 - 9} - \dfrac{1}{2x - 6}$

(c) $\dfrac{4}{x^2 - x - 2} - \dfrac{3}{x^2 + 4x + 3}$

Recall from Section 5.1 that

$$a - b = -(b - a)$$

Let's see how this can be used in adding or subtracting algebraic fractions.

Example 6

Adding Unlike Fractions

Add $\dfrac{4}{x - 5} + \dfrac{2}{5 - x}$.

Rather than try a denominator of $(x - 5)(5 - x)$, let's simplify first.

NOTE Replace $5 - x$ with $-(x - 5)$. We now use the fact that

$$\frac{a}{-b} = \frac{-a}{b}$$

$$\frac{4}{x-5} + \frac{2}{5-x} = \frac{4}{x-5} + \frac{2}{-(x-5)}$$

$$= \frac{4}{x-5} + \frac{-2}{x-5}$$

The LCD is now $x - 5$, and we can combine the fractions as

$$= \frac{4-2}{x-5}$$

$$= \frac{2}{x-5}$$

CHECK YOURSELF 6

Subtract.

$$\frac{3}{x-3} - \frac{1}{3-x}$$

CHECK YOURSELF ANSWERS

1. **(a)** $\frac{13}{24}$; **(b)** $\frac{17}{30}$ **2.** **(a)** $\frac{5x+3}{x^3}$; **(b)** $\frac{12x-5}{20x^2}$ **3.** $\frac{4y+x}{6x^2y^2}$

4. **(a)** $\frac{8x+10}{x(x+2)}$; **(b)** $\frac{2x-18}{(x+3)(x-3)}$ **5.** **(a)** $\frac{27}{10(x+1)}$; **(b)** $\frac{-1}{2(x+3)}$;

(c) $\frac{x+18}{(x+1)(x-2)(x+3)}$ **6.** $\frac{4}{x-3}$

5.3 Exercises

Name ______

Section ______ Date ______

Add or subtract as indicated. Express your result in simplest form.

1. $\frac{3}{7} + \frac{5}{6}$

2. $\frac{7}{12} - \frac{4}{9}$

3. $\frac{13}{25} - \frac{7}{20}$

4. $\frac{3}{5} + \frac{7}{9}$

5. $\frac{y}{4} + \frac{3y}{5}$

6. $\frac{5x}{6} - \frac{2x}{3}$

7. $\frac{7a}{3} - \frac{a}{7}$

8. $\frac{3m}{4} + \frac{m}{9}$

9. $\frac{3}{x} - \frac{4}{5}$

10. $\frac{5}{x} + \frac{2}{3}$

11. $\frac{5}{a} + \frac{a}{5}$

12. $\frac{y}{3} - \frac{3}{y}$

13. $\frac{5}{m} + \frac{3}{m^2}$

14. $\frac{4}{x^2} - \frac{3}{x}$

15. $\frac{2}{x^2} - \frac{5}{7x}$

16. $\frac{7}{3w} + \frac{5}{w^3}$

17. $\frac{7}{9s} + \frac{5}{s^2}$

18. $\frac{11}{x^2} - \frac{5}{7x}$

19. $\frac{3}{4b^2} + \frac{5}{3b^3}$

20. $\frac{4}{5x^3} - \frac{3}{2x^2}$

21. $\frac{x}{x+2} + \frac{2}{5}$

22. $\frac{3}{4} - \frac{a}{a-1}$

23. $\frac{y}{y-4} - \frac{3}{4}$

24. $\frac{m}{m+3} + \frac{2}{3}$

25. $\frac{4}{x} + \frac{3}{x+1}$

26. $\frac{2}{x} - \frac{1}{x-2}$

ANSWERS

1. $\frac{53}{42}$
2. $\frac{5}{36}$
3. $\frac{17}{100}$
4. $\frac{62}{45}$
5. $\frac{17y}{20}$
6. $\frac{x}{6}$
7. $\frac{46a}{21}$
8. $\frac{31m}{36}$
9. $\frac{15 - 4x}{5x}$
10. $\frac{15 + 2x}{3x}$
11. $\frac{25 + a^2}{5a}$
12. $\frac{y^2 - 9}{3y}$
13. $\frac{5m + 3}{m^2}$
14. $\frac{4 - 3x}{x^2}$
15. $\frac{14 - 5x}{7x^2}$
16. $\frac{7w^2 + 15}{3w^3}$
17. $\frac{7s + 45}{9s^2}$
18. $\frac{77 - 5x}{7x^2}$
19. $\frac{9b + 20}{12b^3}$
20. $\frac{8 - 15x}{10x^3}$
21. $\frac{7x + 4}{5(x + 2)}$
22. $\frac{-a - 3}{4(a - 1)}$
23. $\frac{y + 12}{4(y - 4)}$
24. $\frac{5m + 6}{3(m + 3)}$
25. $\frac{7x + 4}{x(x + 1)}$
26. $\frac{x - 4}{x(x - 2)}$

ANSWERS

27. $\frac{3a + 2}{a(a - 1)}$
28. $\frac{7x + 6}{x(x + 2)}$
29. $\frac{2(8x - 3)}{3x(2x - 3)}$
30. $\frac{8y + 3}{2y(2y - 1)}$
31. $\frac{5x + 9}{(x + 1)(x + 3)}$
32. $\frac{7x + 8}{(x - 1)(x + 2)}$
33. $\frac{3(y + 2)}{(y - 2)(y + 1)}$
34. $\frac{2x - 17}{(x + 4)(x - 1)}$
35. $\frac{7}{2(b - 3)}$
36. $\frac{13}{4(a + 5)}$
37. $\frac{3x - 2}{3(x + 4)}$
38. $\frac{2x + 5}{2(x - 3)}$
39. $\frac{11}{6(m + 1)}$
40. $\frac{-1}{15(y - 1)}$
41. $\frac{7}{15(x - 2)}$
42. $\frac{19}{6(w + 1)}$
43. $\frac{49 - 6c}{21(c + 2)}$
44. $\frac{25 + 12c}{15(c - 4)}$
45. $\frac{2y - 3}{3(y + 1)}$
46. $\frac{2(x + 3)}{3(x - 2)}$
47. $\frac{2x - 1}{(x - 2)(x + 2)}$
48. $\frac{4x + 7}{(x - 2)(x + 1)}$

27. $\frac{5}{a - 1} - \frac{2}{a}$

28. $\frac{4}{x + 2} + \frac{3}{x}$

29. $\frac{4}{2x - 3} + \frac{2}{3x}$

30. $\frac{7}{2y - 1} - \frac{3}{2y}$

31. $\frac{2}{x + 1} + \frac{3}{x + 3}$

32. $\frac{5}{x - 1} + \frac{2}{x + 2}$

33. $\frac{4}{y - 2} - \frac{1}{y + 1}$

34. $\frac{5}{x + 4} - \frac{3}{x - 1}$

35. $\frac{2}{b - 3} + \frac{3}{2b - 6}$

36. $\frac{4}{a + 5} - \frac{3}{4a + 20}$

37. $\frac{x}{x + 4} - \frac{2}{3x + 12}$

38. $\frac{x}{x - 3} + \frac{5}{2x - 6}$

39. $\frac{4}{3m + 3} + \frac{1}{2m + 2}$

40. $\frac{3}{5y - 5} - \frac{2}{3y - 3}$

41. $\frac{4}{5x - 10} - \frac{1}{3x - 6}$

42. $\frac{2}{3w + 3} + \frac{5}{2w + 2}$

43. $\frac{7}{3c + 6} - \frac{2c}{7c + 14}$

44. $\frac{5}{3c - 12} + \frac{4c}{5c - 20}$

45. $\frac{y - 1}{y + 1} - \frac{y}{3y + 3}$

46. $\frac{x + 2}{x - 2} - \frac{x}{3x - 6}$

47. $\frac{3}{x^2 - 4} + \frac{2}{x + 2}$

48. $\frac{4}{x - 2} + \frac{3}{x^2 - x - 2}$

49. $\dfrac{3x}{x^2 - 3x + 2} - \dfrac{1}{x - 2}$

50. $\dfrac{a}{a^2 - 1} - \dfrac{4}{a + 1}$

51. $\dfrac{2x}{x^2 - 5x + 6} + \dfrac{4}{x - 2}$

52. $\dfrac{7a}{a^2 + a - 12} - \dfrac{4}{a + 4}$

53. $\dfrac{2}{3x - 3} - \dfrac{1}{4x + 4}$

54. $\dfrac{2}{5w + 10} - \dfrac{3}{2w - 4}$

55. $\dfrac{4}{3a - 9} - \dfrac{3}{2a + 4}$

56. $\dfrac{2}{3b - 6} + \dfrac{3}{4b + 8}$

57. $\dfrac{5}{x^2 - 16} - \dfrac{3}{x^2 - x - 12}$

58. $\dfrac{3}{x^2 + 4x + 3} - \dfrac{1}{x^2 - 9}$

59. $\dfrac{2}{y^2 + y - 6} + \dfrac{3y}{y^2 - 2y - 15}$

60. $\dfrac{2a}{a^2 - a - 12} - \dfrac{3}{a^2 - 2a - 8}$

61. $\dfrac{6x}{x^2 - 9} - \dfrac{5x}{x^2 + x - 6}$

62. $\dfrac{4y}{y^2 + 6y + 5} + \dfrac{2y}{y^2 - 1}$

63. $\dfrac{3}{a - 7} + \dfrac{2}{7 - a}$

64. $\dfrac{5}{x - 5} - \dfrac{3}{5 - x}$

65. $\dfrac{2x}{2x - 3} - \dfrac{1}{3 - 2x}$

66. $\dfrac{9m}{3m - 1} + \dfrac{3}{1 - 3m}$

Add or subtract, as indicated.

67. $\dfrac{1}{a - 3} - \dfrac{1}{a + 3} + \dfrac{2a}{a^2 - 9}$

68. $\dfrac{1}{p + 1} + \dfrac{1}{p - 3} - \dfrac{4}{p^2 - 2p - 3}$

69. $\dfrac{2x^2 + 3x}{x^2 - 2x - 63} + \dfrac{7 - x}{x^2 - 2x - 63} - \dfrac{x^2 - 3x + 21}{x^2 - 2x - 63}$

70. $\dfrac{3 - 2x^2}{x^2 - 9x + 20} - \dfrac{4x^2 + 2x + 1}{x^2 - 9x + 20} + \dfrac{2x^2 + 3x}{x^2 - 9x + 20}$

71. **Consecutive integers.** Use a rational expression to represent the sum of the reciprocals of two consecutive even integers.

72. **Integers.** One number is two less than another. Use a rational expression to represent the sum of the reciprocals of the two numbers.

73. Refer to the rectangle in the figure. Find an expression that represents its perimeter.

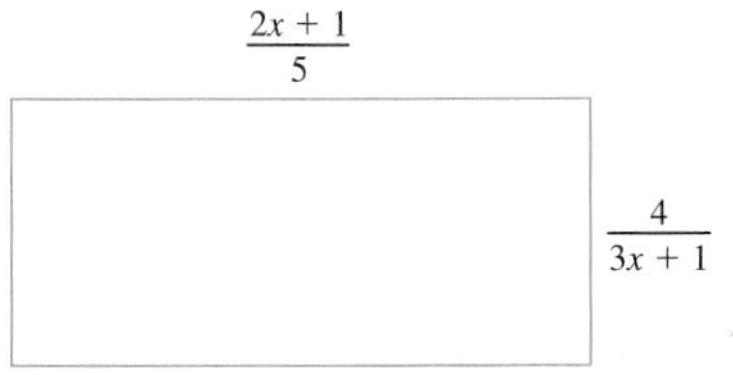

ANSWERS

49. $\dfrac{2x + 1}{(x - 1)(x - 2)}$

50. $\dfrac{-3a + 4}{(a - 1)(a + 1)}$

51. $\dfrac{6}{x - 3}$

52. $\dfrac{3}{a - 3}$

53. $\dfrac{5x + 11}{12(x - 1)(x + 1)}$

54. $\dfrac{-11w - 38}{10(w + 2)(w - 2)}$

55. $\dfrac{-a + 43}{6(a - 3)(a + 2)}$

56. $\dfrac{17b - 2}{12(b - 2)(b + 2)}$

57. $\dfrac{2x + 3}{(x + 4)(x - 4)(x + 3)}$

58. $\dfrac{2(x - 5)}{(x + 1)(x - 3)(x + 3)}$

59. $\dfrac{3y^2 - 4y - 10}{(y + 3)(y - 2)(y - 5)}$

60. $\dfrac{2a^2 + a - 9}{(a - 4)(a + 3)(a + 2)}$

61. $\dfrac{x}{(x - 3)(x - 2)}$

62. $\dfrac{6y}{(y + 5)(y - 1)}$

63. $\dfrac{1}{a - 7}$

64. $\dfrac{8}{x - 5}$

65. $\dfrac{2x + 1}{2x - 3}$

66. 3

67. $\dfrac{2}{a - 3}$

68. $\dfrac{2}{p + 1}$

69. $\dfrac{x - 2}{x - 9}$

70. $\dfrac{1}{x - 5}$

71. $\dfrac{2x + 2}{x(x + 2)}$

72. $\dfrac{2x - 2}{x(x - 2)}$

73. $\dfrac{2(6x^2 + 5x + 21)}{5(3x + 1)}$

ANSWERS

74. $\dfrac{3x + 24}{4x^2}$

a. $\dfrac{8}{15}$

b. $\dfrac{10}{33}$

c. $\dfrac{5}{14}$

d. $\dfrac{3}{14}$

e. $\dfrac{2}{3}$

f. $\dfrac{1}{2}$

g. $\dfrac{9}{5}$

h. $\dfrac{5}{3}$

74. Refer to the triangle in the figure. Find an expression that represents its perimeter.

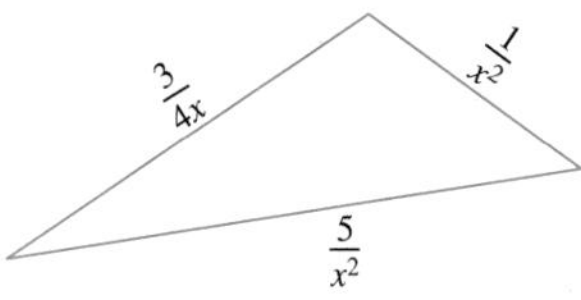

Getting Ready for Section 5.4 [Section 0.2]

Perform the indicated operations.

(a) $\dfrac{2}{3} \cdot \dfrac{4}{5}$ (b) $\dfrac{5}{6} \cdot \dfrac{4}{11}$

(c) $\dfrac{4}{7} \div \dfrac{8}{5}$ (d) $\dfrac{1}{6} \div \dfrac{7}{9}$

(e) $\dfrac{5}{8} \cdot \dfrac{16}{15}$ (f) $\dfrac{15}{21} \div \dfrac{10}{7}$

(g) $\dfrac{15}{8} \cdot \dfrac{24}{25}$ (h) $\dfrac{28}{16} \div \dfrac{21}{20}$

Answers

1. $\dfrac{53}{42}$ **3.** $\dfrac{17}{100}$ **5.** $\dfrac{17y}{20}$ **7.** $\dfrac{46a}{21}$ **9.** $\dfrac{15 - 4x}{5x}$ **11.** $\dfrac{25 + a^2}{5a}$ **13.** $\dfrac{5m + 3}{m^2}$ **15.** $\dfrac{14 - 5x}{7x^2}$ **17.** $\dfrac{7s + 45}{9s^2}$ **19.** $\dfrac{9b + 20}{12b^3}$ **21.** $\dfrac{7x + 4}{5(x + 2)}$ **23.** $\dfrac{y + 12}{4(y - 4)}$ **25.** $\dfrac{7x + 4}{x(x + 1)}$ **27.** $\dfrac{3a + 2}{a(a - 1)}$ **29.** $\dfrac{2(8x - 3)}{3x(2x - 3)}$ **31.** $\dfrac{5x + 9}{(x + 1)(x + 3)}$ **33.** $\dfrac{3(y + 2)}{(y - 2)(y + 1)}$ **35.** $\dfrac{7}{2(b - 3)}$ **37.** $\dfrac{3x - 2}{3(x + 4)}$ **39.** $\dfrac{11}{6(m + 1)}$ **41.** $\dfrac{7}{15(x - 2)}$ **43.** $\dfrac{49 - 6c}{21(c + 2)}$ **45.** $\dfrac{2y - 3}{3(y + 1)}$ **47.** $\dfrac{2x - 1}{(x - 2)(x + 2)}$ **49.** $\dfrac{2x + 1}{(x - 1)(x - 2)}$ **51.** $\dfrac{6}{x - 3}$ **53.** $\dfrac{5x + 11}{12(x - 1)(x + 1)}$ **55.** $\dfrac{-a + 43}{6(a - 3)(a + 2)}$ **57.** $\dfrac{2x + 3}{(x + 4)(x - 4)(x + 3)}$ **59.** $\dfrac{3y^2 - 4y - 10}{(y + 3)(y - 2)(y - 5)}$ **61.** $\dfrac{x}{(x - 3)(x - 2)}$ **63.** $\dfrac{1}{a - 7}$ **65.** $\dfrac{2x + 1}{2x - 3}$ **67.** $\dfrac{2}{a - 3}$ **69.** $\dfrac{x - 2}{x - 9}$ **71.** $\dfrac{2x + 2}{x(x + 2)}$ **73.** $\dfrac{2(6x^2 + 5x + 21)}{5(3x + 1)}$ **a.** $\dfrac{8}{15}$ **b.** $\dfrac{10}{33}$ **c.** $\dfrac{5}{14}$ **d.** $\dfrac{3}{14}$ **e.** $\dfrac{2}{3}$ **f.** $\dfrac{1}{2}$ **g.** $\dfrac{9}{5}$ **h.** $\dfrac{5}{3}$

Multiplying and Dividing Algebraic Fractions

5.4 OBJECTIVES

1. Write the product of two algebraic fractions in simplest form
2. Write the quotient of two algebraic fractions in simplest form
3. Simplify a complex fraction by the method of common denominators

In arithmetic, you found the product of two fractions by multiplying the numerators and the denominators. For example,

$$\frac{2}{5} \cdot \frac{3}{7} = \frac{2 \cdot 3}{5 \cdot 7} = \frac{6}{35}$$

In symbols, we have

Rules and Properties: Multiplying Algebraic Fractions

$$\frac{P}{Q} \cdot \frac{R}{S} = \frac{PR}{QS} \quad \text{when } Q \neq 0 \text{ and } S \neq 0$$

NOTE P, Q, R, and S again represent polynomials.

It is easier to divide the numerator and denominator by any common factors *before* multiplying. Consider the following.

$$\frac{3}{8} \cdot \frac{4}{9} = \frac{\overset{1}{\cancel{3}} \cdot \overset{1}{\cancel{4}}}{\underset{2}{\cancel{8}} \cdot \underset{3}{\cancel{9}}} = \frac{1}{6}$$

NOTE Divide by the common factors of 3 and 4. The alternative is to multiply *first:*

$$\frac{3}{8} \cdot \frac{4}{9} = \frac{12}{72}$$

and then use the GCF to reduce to lowest terms

$$\frac{12}{72} = \frac{1}{6}$$

In algebra, we multiply fractions in exactly the same way.

Step by Step: To Multiply Algebraic Fractions

Step 1 Factor the numerators and denominators.
Step 2 Divide the numerator and denominator by any common factors.
Step 3 Write the product of the remaining factors in the numerator over the product of the remaining factors in the denominator.

Example 1 illustrates this property.

Example 1

Multiplying Algebraic Fractions

Multiply the following fractions.

NOTE Divide by the common factors of 5, x^2, and y.

(a) $\frac{2x^3}{5y^2} \cdot \frac{10y}{3x^2} = \frac{2x^3 \cdot 10y}{5y^2 \cdot 3x^2} = \frac{4x}{3y}$

(b) $\dfrac{x}{x^2 - 3x} \cdot \dfrac{6x - 18}{9x} = \dfrac{x}{x(x - 3)} \cdot \dfrac{6(x - 3)}{9x}$ ← Factor

NOTE Divide by the common factors of 3, x, and $x - 3$.

$$= \frac{\overset{1}{\cancel{x}} \cdot \overset{2}{\cancel{6}}\overset{1}{\cancel{(x-3)}}}{\underset{1}{\cancel{x}}\underset{1}{\cancel{(x-3)}} \cdot \underset{3}{\cancel{9}}x}$$

$$= \frac{2}{3x}$$

(c) $\dfrac{4}{x^2 - 2x} \cdot \dfrac{10 - 5x}{8} = \dfrac{4}{x(x - 2)} \cdot \dfrac{5(2 - x)}{8}$

NOTE

$$\frac{2 - x}{x - 2} = \frac{-(x - 2)}{x - 2} = -1$$

$$= \frac{\overset{1}{\cancel{4}} \cdot 5\overset{-1}{\cancel{(2-x)}}}{x\underset{1}{\cancel{(x-2)}} \cdot \underset{2}{\cancel{8}}} = \frac{-5}{2x}$$

NOTE Divide by the common factors of $x - 4$, x, and 3.

(d) $\dfrac{x^2 - 2x - 8}{3x^2} \cdot \dfrac{6x}{3x - 12} = \dfrac{\overset{1}{\cancel{(x-4)}}(x + 2)}{\underset{x}{\cancel{3x^2}}} \cdot \dfrac{\overset{2}{\cancel{6x}}}{3\underset{1}{\cancel{(x-4)}}}$

$$= \frac{2(x + 2)}{3x}$$

(e) $\dfrac{x^2 - y^2}{5x - 5y} \cdot \dfrac{10xy}{x^2 + 2xy + y^2} = \dfrac{\overset{1}{\cancel{(x-y)}}\overset{1}{\cancel{(x+y)}}}{\underset{1}{\cancel{5}}\underset{1}{\cancel{(x-y)}}} \cdot \dfrac{\overset{2}{\cancel{10}}xy}{\underset{1}{\cancel{(x+y)}}(x + y)}$

$$= \frac{2xy}{x + y}$$

CHECK YOURSELF 1

Multiply.

(a) $\dfrac{3x^2}{5y^2} \cdot \dfrac{10y^5}{15x^3}$ **(b)** $\dfrac{5x + 15}{x} \cdot \dfrac{2x^2}{x^2 + 3x}$ **(c)** $\dfrac{x}{2x - 6} \cdot \dfrac{3x - x^2}{2}$

(d) $\dfrac{3x - 15}{6x^2} \cdot \dfrac{2x}{x^2 - 25}$ **(e)** $\dfrac{x^2 - 5x - 14}{4x^2} \cdot \dfrac{8x}{x^2 - 49}$

You can also use your experience from arithmetic in dividing fractions. Recall that, to divide fractions, we *invert the divisor* (the *second* fraction) and multiply. For example,

NOTE Recall, $\frac{6}{5}$ is the reciprocal of $\frac{5}{6}$.

$$\frac{2}{3} \div \frac{5}{6} = \frac{2}{3} \cdot \frac{6}{5} = \frac{2 \cdot 6}{3 \cdot 5} = \frac{4}{5}$$

In symbols, we have

Rules and Properties: Dividing Algebraic Fractions

NOTE Once more P, Q, R, and S are polynomials.

$$\frac{P}{Q} \div \frac{R}{S} = \frac{P}{Q} \cdot \frac{S}{R} = \frac{PS}{QR}$$

when $Q \neq 0$, $R \neq 0$, and $S \neq 0$.

Division of algebraic fractions is done in exactly the same way.

Step by Step: To Divide Algebraic Fractions

Step 1 Invert the divisor and change the operation to multiplication.
Step 2 Proceed, using the steps for multiplying algebraic fractions.

Example 2 illustrates this approach.

Example 2

Dividing Algebraic Fractions

Divide the following.

(a) $\frac{6}{x^2} \div \frac{9}{x^3} = \frac{6}{x^2} \cdot \frac{x^3}{9}$ Invert the divisor and multiply.

$= \frac{\overset{2}{\cancel{6}}\overset{x}{\cancel{x^3}}}{\underset{3}{\cancel{9}}\underset{1}{\cancel{x^2}}}$ No simplification can be done until the divisor is inverted. Then divide by the common factors of 3 and x^2.

$= \frac{2x}{3}$

(b) $\frac{3x^2y}{8xy^3} \div \frac{9x^3}{4y^4} = \frac{3x^2y}{8xy^3} \cdot \frac{4y^4}{9x^3}$

$= \frac{y^2}{6x^2}$

(c) $\frac{2x + 4y}{9x - 18y} \div \frac{4x + 8y}{3x - 6y} = \frac{2x + 4y}{9x - 18y} \cdot \frac{3x - 6y}{4x + 8y}$

$= \frac{\overset{1}{\cancel{2}}\overset{1}{\cancel{(x + 2y)}} \cdot \overset{1}{\cancel{3}}\overset{1}{\cancel{(x - 2y)}}}{\underset{3}{\cancel{9}}\underset{1}{\cancel{(x - 2y)}} \cdot \underset{2}{\cancel{4}}\underset{1}{\cancel{(x + 2y)}}}$

$= \frac{1}{6}$

NOTE Factor all numerators and denominators *before* dividing out any common factors.

(d) $\frac{x^2 - x - 6}{2x - 6} \div \frac{x^2 - 4}{4x^2} = \frac{x^2 - x - 6}{2x - 6} \cdot \frac{4x^2}{x^2 - 4}$

$= \frac{\overset{1}{\cancel{(x - 3)}}\overset{1}{\cancel{(x + 2)}} \cdot \overset{2}{\cancel{4}}x^2}{\underset{1}{\cancel{2}}\underset{1}{\cancel{(x - 3)}} \cdot \underset{1}{\cancel{(x + 2)}}(x - 2)}$

$= \frac{2x^2}{x - 2}$

CHECK YOURSELF 2

Divide.

(a) $\frac{4}{x^5} \div \frac{12}{x^3}$

(b) $\frac{5xy^2}{7x^3y} \div \frac{10y^2}{14x^3}$

(c) $\frac{3x - 9y}{2x + 10y} \div \frac{x^2 - 3xy}{4x + 20y}$

(d) $\frac{x^2 - 9}{4x} \div \frac{x^2 - 2x - 15}{2x - 10}$

Before we continue, let's review why the invert-and-multiply rule works for dividing fractions. We will use an example from arithmetic for the explanation. Suppose that we want to divide as follows:

$$\frac{3}{5} \div \frac{2}{3} \qquad (1)$$

We can write

$$\underbrace{\frac{3}{5} \div \frac{2}{3}}_{(1)} = \frac{\frac{3}{5}}{\frac{2}{3}} = \frac{\frac{3}{5} \cdot \frac{3}{2}}{\frac{2}{3} \cdot \frac{3}{2}}$$

We are multiplying by 1.

Interpret the division as a fraction.

$$= \frac{\frac{3}{5} \cdot \frac{3}{2}}{1}$$

$\frac{2}{3} \cdot \frac{3}{2} = 1$

$$= \frac{3}{5} \cdot \frac{3}{2} \qquad (2)$$

We then have

$$\overset{1}{\frac{3}{5} \div \frac{2}{3}} = \overset{2}{\frac{3}{5} \cdot \frac{3}{2}}$$

Comparing expressions (1) and (2), you should see the rule for dividing fractions. Invert the fraction that follows the division symbol and multiply.

A fraction that has a fraction in its numerator, in its denominator, or in both is called a **complex fraction.** For example, the following are complex fractions

$$\frac{\frac{5}{6}}{\frac{3}{4}} \quad \frac{\frac{4}{x}}{\frac{3}{x^2}} \quad \text{and} \quad \frac{\frac{a+2}{3}}{\frac{a-2}{5}}$$

Remember that we can always multiply the numerator and the denominator of a fraction by the same nonzero term.

NOTE This is the Fundamental Principle of Fractions.

$$\frac{P}{Q} = \frac{P \cdot R}{Q \cdot R} \qquad \text{in which } Q \neq 0 \text{ and } R \neq 0$$

To simplify a complex fraction, multiply the numerator and denominator by the LCD of all fractions that appear within the complex fraction.

Example 3

Simplifying Complex Fractions

Simplify $\dfrac{\frac{3}{4}}{\frac{5}{8}}$.

The LCD of $\frac{3}{4}$ and $\frac{5}{8}$ is 8. So multiply the numerator and denominator by 8.

$$\frac{\frac{3}{4}}{\frac{5}{8}} = \frac{\frac{3}{4} \cdot 8}{\frac{5}{8} \cdot 8} = \frac{3 \cdot 2}{5 \cdot 1} = \frac{6}{5}$$

CHECK YOURSELF 3

Simplify.

(a) $\frac{\frac{4}{7}}{\frac{3}{7}}$ (b) $\frac{\frac{3}{8}}{\frac{5}{6}}$

The same method can be used to simplify a complex fraction when variables are involved in the expression. Consider Example 4.

Example 4

Simplifying Complex Algebraic Fractions

Simplify $\frac{\frac{5}{x}}{\frac{10}{x^2}}$.

The LCD of $\frac{5}{x}$ and $\frac{10}{x^2}$ is x^2, so multiply the numerator and denominator by x^2.

NOTE Be sure to write the result in simplest form.

$$\frac{\frac{5}{x}}{\frac{10}{x^2}} = \frac{\left(\frac{5}{x}\right)x^2}{\left(\frac{10}{x^2}\right)x^2} = \frac{5x}{10} = \frac{x}{2}$$

CHECK YOURSELF 4

Simplify.

(a) $\frac{\frac{6}{x^3}}{\frac{9}{x^2}}$ (b) $\frac{\frac{m^4}{15}}{\frac{m^3}{20}}$

We may also have a sum or a difference in the numerator or denominator of a complex fraction. The simplification steps are exactly the same. Consider Example 5.

Example 5

Simplifying Complex Algebraic Fractions

Simplify $\dfrac{1 + \dfrac{x}{y}}{1 - \dfrac{x}{y}}$.

The LCD of 1, $\frac{x}{y}$, 1, and $\frac{x}{y}$ is y, so multiply the numerator and denominator by y.

NOTE Notice the use of the distributive property to multiply *each term* in the numerator and in the denominator by y.

$$\frac{1 + \frac{x}{y}}{1 - \frac{x}{y}} = \frac{\left(1 + \frac{x}{y}\right)y}{\left(1 - \frac{x}{y}\right)y} = \frac{1 \cdot y + \frac{x}{y} \cdot y}{1 \cdot y - \frac{x}{y} \cdot y}$$

$$= \frac{y + x}{y - x}$$

CHECK YOURSELF 5

Simplify.

$$\frac{\frac{x}{y} - 2}{\frac{x}{y} + 2}$$

A second method for simplifying complex fractions uses the fact that

NOTE To divide by a fraction, we invert the divisor (it *follows* the division sign) and multiply.

$$\frac{\frac{P}{Q}}{\frac{R}{S}} = \frac{P}{Q} \div \frac{R}{S} = \frac{P}{Q} \cdot \frac{S}{R}$$

To use this method, we must write the numerator and denominator of the complex fraction as single fractions. We can then divide the numerator by the denominator as before.

The following algorithm summarizes our work with simplifying complex fractions.

Step by Step: To Simplify Complex Fractions

Step 1 Multiply the numerator and denominator of the complex fraction by the LCD of all the fractions that appear within the complex fraction.

Step 2 Write the resulting fraction in simplest form.

CHECK YOURSELF ANSWERS

1. (a) $\frac{2y^3}{5x}$; (b) 10; (c) $\frac{-x^2}{4}$; (d) $\frac{1}{x(x + 5)}$; (e) $\frac{2(x + 2)}{x(x + 7)}$

2. (a) $\frac{1}{3x^2}$; (b) $\frac{x}{y}$; (c) $\frac{6}{x}$; (d) $\frac{x - 3}{2x}$

3. (a) $\frac{4}{3}$; (b) $\frac{9}{20}$ 4. (a) $\frac{2}{3x}$; (b) $\frac{4m}{3}$ 5. $\frac{x - 2y}{x + 2y}$

Name ______________________

Section ________ Date ________

5.4 Exercises

Multiply.

1. $\frac{3}{7} \cdot \frac{14}{27}$

2. $\frac{9}{20} \cdot \frac{5}{36}$

3. $\frac{x}{2} \cdot \frac{y}{6}$

4. $\frac{w}{2} \cdot \frac{5}{14}$

5. $\frac{3a}{2} \cdot \frac{4}{a^2}$

6. $\frac{5x^3}{3x} \cdot \frac{9}{20x}$

7. $\frac{3x^3y}{10xy^3} \cdot \frac{5xy^2}{9xy^2}$

8. $\frac{8xy^5}{5x^3y^2} \cdot \frac{15y^2}{16xy^3}$

9. $\frac{-4ab^2}{15a^3} \cdot \frac{25ab}{-16b^3}$

10. $\frac{-7xy^2}{12x^2y} \cdot \frac{24x^3y^5}{-21x^2y^7}$

11. $\frac{-3m^3n}{10mn^3} \cdot \frac{5mn^2}{-9mn^3}$

12. $\frac{3x}{2x - 6} \cdot \frac{x^2 - 3x}{6}$

13. $\frac{x^2 + 5x}{3x^2} \cdot \frac{10x}{5x + 25}$

14. $\frac{x^2 - 3x - 10}{5x} \cdot \frac{15x^2}{3x - 15}$

15. $\frac{p^2 - 8p}{4p} \cdot \frac{12p^2}{p^2 - 64}$

16. $\frac{a^2 - 81}{a^2 + 9a} \cdot \frac{5a^2}{a^2 - 7a - 18}$

17. $\frac{m^2 - 4m - 21}{3m^2} \cdot \frac{m^2 + 7m}{m^2 - 49}$

18. $\frac{2x^2 - x - 3}{3x^2 + 7x + 4} \cdot \frac{3x^2 - 11x - 20}{4x^2 - 9}$

19. $\frac{4r^2 - 1}{2r^2 - 9r - 5} \cdot \frac{3r^2 - 13r - 10}{9r^2 - 4}$

20. $\frac{a^2 + ab}{2a^2 - ab - 3b^2} \cdot \frac{4a^2 - 9b^2}{5a^2 - 4ab}$

ANSWERS

1. $\frac{2}{9}$
2. $\frac{1}{16}$
3. $\frac{xy}{12}$
4. $\frac{5w}{28}$
5. $\frac{6}{a}$
6. $\frac{3x}{4}$
7. $\frac{x^2}{6y^2}$
8. $\frac{3y^2}{2x^3}$
9. $\frac{5}{12a}$
10. $\frac{2}{3y}$
11. $\frac{m^2}{6n^3}$
12. $\frac{x^2}{4}$
13. $\frac{2}{3}$
14. $x^2 + 2x$
15. $\frac{3p^2}{p + 8}$
16. $\frac{5a}{a + 2}$
17. $\frac{m + 3}{3m}$
18. $\frac{x - 5}{2x + 3}$
19. $\frac{2r - 1}{3r - 2}$
20. $\frac{2a + 3b}{5a - 4b}$

CHECK YOURSELF 1

Solve and check.

$$\frac{x}{4} - \frac{1}{6} = \frac{4x - 5}{12}$$

Recall that, for any fraction, the denominator must not be equal to zero. When a fraction has a variable in the denominator, we must exclude any value for the variable that would result in division by zero.

Example 2

Finding Excluded Values for *x*

In the following algebraic fractions, what values for x must be excluded?

(a) $\frac{x}{5}$. Here x can have any value, so none need to be excluded.

(b) $\frac{3}{x}$. If $x = 0$, then $\frac{3}{x}$ is undefined; 0 is the excluded value.

(c) $\frac{5}{x - 2}$. If $x = 2$, then $\frac{5}{x - 2} = \frac{5}{2 - 2} = \frac{5}{0}$, which is undefined, so 2 is the excluded value.

CHECK YOURSELF 2

What values for x, if any, must be excluded?

(a) $\frac{x}{7}$ **(b)** $\frac{5}{x}$ **(c)** $\frac{7}{x - 5}$

If the denominator of an algebraic fraction contains a product of two or more variable factors, the zero-product principle must be used to determine the excluded values for the variable.

In some cases, you will have to factor the denominator to see the restrictions on the values for the variable.

Example 3

Finding Excluded Values for *x*

What values for x must be excluded in each fraction?

(a) $\frac{3}{x^2 - 6x - 16}$

Factoring the denominator, we have

$$\frac{3}{x^2 - 6x - 16} = \frac{3}{(x - 8)(x + 2)}$$

Letting $x - 8 = 0$ or $x + 2 = 0$, we see that 8 and -2 make the denominator 0 so both 8 and -2 must be excluded.

Name ________________

Section ________ Date ________

5.4 Exercises

Multiply.

1. $\frac{3}{7} \cdot \frac{14}{27}$

2. $\frac{9}{20} \cdot \frac{5}{36}$

3. $\frac{x}{2} \cdot \frac{y}{6}$

4. $\frac{w}{2} \cdot \frac{5}{14}$

5. $\frac{3a}{2} \cdot \frac{4}{a^2}$

6. $\frac{5x^3}{3x} \cdot \frac{9}{20x}$

7. $\frac{3x^3y}{10xy^3} \cdot \frac{5xy^2}{9xy^2}$

8. $\frac{8xy^5}{5x^3y^2} \cdot \frac{15y^2}{16xy^3}$

9. $\frac{-4ab^2}{15a^3} \cdot \frac{25ab}{-16b^3}$

10. $\frac{-7xy^2}{12x^2y} \cdot \frac{24x^3y^5}{-21x^2y^7}$

11. $\frac{-3m^3n}{10mn^3} \cdot \frac{5mn^2}{-9mn^3}$

12. $\frac{3x}{2x - 6} \cdot \frac{x^2 - 3x}{6}$

13. $\frac{x^2 + 5x}{3x^2} \cdot \frac{10x}{5x + 25}$

14. $\frac{x^2 - 3x - 10}{5x} \cdot \frac{15x^2}{3x - 15}$

15. $\frac{p^2 - 8p}{4p} \cdot \frac{12p^2}{p^2 - 64}$

16. $\frac{a^2 - 81}{a^2 + 9a} \cdot \frac{5a^2}{a^2 - 7a - 18}$

17. $\frac{m^2 - 4m - 21}{3m^2} \cdot \frac{m^2 + 7m}{m^2 - 49}$

18. $\frac{2x^2 - x - 3}{3x^2 + 7x + 4} \cdot \frac{3x^2 - 11x - 20}{4x^2 - 9}$

19. $\frac{4r^2 - 1}{2r^2 - 9r - 5} \cdot \frac{3r^2 - 13r - 10}{9r^2 - 4}$

20. $\frac{a^2 + ab}{2a^2 - ab - 3b^2} \cdot \frac{4a^2 - 9b^2}{5a^2 - 4ab}$

ANSWERS

1. $\frac{2}{9}$
2. $\frac{1}{16}$
3. $\frac{xy}{12}$
4. $\frac{5w}{28}$
5. $\frac{6}{a}$
6. $\frac{3x}{4}$
7. $\frac{x^2}{6y^2}$
8. $\frac{3y^2}{2x^3}$
9. $\frac{5}{12a}$
10. $\frac{2}{3y}$
11. $\frac{m^2}{6n^3}$
12. $\frac{x^2}{4}$
13. $\frac{2}{3}$
14. $x^2 + 2x$
15. $\frac{3p^2}{p + 8}$
16. $\frac{5a}{a + 2}$
17. $\frac{m + 3}{3m}$
18. $\frac{x - 5}{2x + 3}$
19. $\frac{2r - 1}{3r - 2}$
20. $\frac{2a + 3b}{5a - 4b}$

ANSWERS

21. $\frac{7x}{5}$
22. $\frac{6a}{7}$
23. $\frac{-6}{x+2}$
24. $\frac{-12}{x+3}$
25. $\frac{2}{3}$
26. $\frac{2}{3}$
27. $\frac{1}{2x}$
28. $3w$
29. $\frac{3y}{2}$
30. $\frac{5}{6xy^2}$
31. $\frac{9}{20}$
32. $\frac{1}{3}$
33. $\frac{a-3}{10a}$
34. $\frac{2(p+2)}{9}$
35. $\frac{x-2}{3x^2}$
36. $\frac{x}{x-2}$
37. $\frac{2x+1}{2x}$
38. $\frac{2m-7}{5(2m-3)}$
39. $\frac{3b}{a+2b}$
40. $\frac{5r}{r+3s}$

21. $\frac{x^2-4y^2}{x^2-xy-6y^2} \cdot \frac{7x^2-21xy}{5x-10y}$

22. $\frac{a^2-9b^2}{a^2+ab-6b^2} \cdot \frac{6a^2-12ab}{7a-21b}$

23. $\frac{2x-6}{x^2+2x} \cdot \frac{3x}{3-x}$

24. $\frac{3x-15}{x^2+3x} \cdot \frac{4x}{5-x}$

Divide.

25. $\frac{5}{8} \div \frac{15}{16}$

26. $\frac{4}{9} \div \frac{12}{18}$

27. $\frac{5}{x^2} \div \frac{10}{x}$

28. $\frac{w^2}{3} \div \frac{w}{9}$

29. $\frac{4x^2y^2}{9x^3} \div \frac{8y^2}{27xy}$

30. $\frac{8x^3y}{27xy^3} \div \frac{16x^3y}{45y}$

31. $\frac{3x+6}{8} \div \frac{5x+10}{6}$

32. $\frac{x^2-2x}{4x} \div \frac{6x-12}{8}$

33. $\frac{4a-12}{5a+15} \div \frac{8a^2}{a^2+3a}$

34. $\frac{6p-18}{9p} \div \frac{3p-9}{p^2+2p}$

35. $\frac{x^2+2x-8}{9x^2} \div \frac{x^2-16}{3x-12}$

36. $\frac{16x}{4x^2-16} \div \frac{4x-24}{x^2-4x-12}$

37. $\frac{x^2-9}{2x^2-6x} \div \frac{2x^2+5x-3}{4x^2-1}$

38. $\frac{2m^2-5m-7}{4m^2-9} \div \frac{5m^2+5m}{2m^2+3m}$

39. $\frac{a^2-9b^2}{4a^2+12ab} \div \frac{a^2-ab-6b^2}{12ab}$

40. $\frac{r^2+2rs-15s^2}{r^3+5r^2s} \div \frac{r^2-9s^2}{5r^3}$

41. $\dfrac{x^2 - 16y^2}{3x^2 - 12xy} \div (x^2 + 4xy)$

42. $\dfrac{p^2 - 4pq - 21q^2}{4p - 28q} \div (2p^2 + 6pq)$

43. $\dfrac{x - 7}{2x + 6} \div \dfrac{21 - 3x}{x^2 + 3x}$

44. $\dfrac{x - 4}{x^2 + 2x} \div \dfrac{16 - 4x}{3x + 6}$

Perform the indicated operations.

45. $\dfrac{x^2 - 2x - 8}{2x - 8} \cdot \dfrac{x^2 + 5x}{x^2 + 5x + 6} \div \dfrac{x^2 + 2x - 15}{x^2 - 9}$

46. $\dfrac{14x - 7}{x^2 + 3x - 4} \cdot \dfrac{x^2 + 6x + 8}{2x^2 + 5x - 3} \div \dfrac{x^2 + 2x}{x^2 + 2x - 3}$

47. $\dfrac{x^2 + 5x}{3x - 6} \cdot \dfrac{x^2 - 4}{3x^2 + 15x} \cdot \dfrac{6x}{x^2 + 6x + 8}$

48. $\dfrac{m^2 - n^2}{m^2 - mn} \cdot \dfrac{6m}{2m^2 + mn - n^2} \cdot \dfrac{8m - 4n}{12m^2 + 12mn}$

Simplify each complex fraction.

49. $\dfrac{\frac{2}{3}}{\frac{6}{8}}$

50. $\dfrac{\frac{5}{6}}{\frac{10}{15}}$

51. $\dfrac{1 + \frac{1}{2}}{2 + \frac{1}{4}}$

52. $\dfrac{1 + \frac{3}{4}}{2 - \frac{1}{8}}$

53. $\dfrac{\frac{x}{8}}{\frac{x^2}{4}}$

54. $\dfrac{\frac{m^2}{10}}{\frac{m^3}{15}}$

55. $\dfrac{\frac{3}{a}}{\frac{2}{a^2}}$

56. $\dfrac{\frac{6}{x^2}}{\frac{9}{x^3}}$

57. $\dfrac{\frac{y + 1}{y}}{\frac{y - 1}{2y}}$

58. $\dfrac{\frac{w + 3}{4w}}{\frac{w - 3}{2w}}$

ANSWERS

41. $\dfrac{1}{3x^2}$

42. $\dfrac{1}{8p}$

43. $\dfrac{-x}{6}$

44. $\dfrac{-3}{4x}$

45. $\dfrac{x}{2}$

46. $\dfrac{7}{x}$

47. $\dfrac{2x}{3(x + 4)}$

48. $\dfrac{2}{m(m + n)}$

49. $\dfrac{8}{9}$

50. $\dfrac{5}{4}$

51. $\dfrac{2}{3}$

52. $\dfrac{14}{15}$

53. $\dfrac{1}{2x}$

54. $\dfrac{3}{2m}$

55. $\dfrac{3a}{2}$

56. $\dfrac{2x}{3}$

57. $\dfrac{2(y + 1)}{y - 1}$

58. $\dfrac{w + 3}{2w - 6}$

ANSWERS

59. $\frac{2x-1}{2x+1}$

60. $\frac{3a+1}{3a-1}$

61. $\frac{3y-x}{6}$

62. $\frac{2y+x}{4}$

63. $\frac{x-y}{y}$

64. $\frac{b}{a-2b}$

65. $\frac{x+4}{x+3}$

66. $\frac{r-4}{r-3}$

67. $\frac{2y-1}{1+2x}$

68. $\frac{1+2y}{3x-1}$

69. $\frac{x+1}{x-4}$

70. $\frac{1-a}{a+4}$

71. $\frac{y+2}{(y-1)(y+4)}$

72. $\frac{x-3}{(x+2)(x-6)}$

73. $\frac{2x+1}{x+1}$

74. $\frac{2y-1}{y-1}$

75. $\frac{\frac{2}{3}}{\frac{1}{4}} = \frac{2}{3} \cdot \frac{4}{1} = \frac{8}{3}$

76. $\frac{\frac{1}{10}}{\frac{1}{4}} = \frac{1}{10} \cdot \frac{4}{1} = \frac{2}{5}$

59. $\dfrac{2 - \frac{1}{x}}{2 + \frac{1}{x}}$

60. $\dfrac{3 + \frac{1}{a}}{3 - \frac{1}{a}}$

61. $\dfrac{3 - \frac{x}{y}}{\frac{6}{y}}$

62. $\dfrac{2 + \frac{x}{y}}{\frac{4}{y}}$

63. $\dfrac{\frac{x^2}{y^2} - 1}{\frac{x}{y} + 1}$

64. $\dfrac{\frac{a}{b} + 2}{\frac{a^2}{b^2} - 4}$

65. $\dfrac{1 + \frac{3}{x} - \frac{4}{x^2}}{1 + \frac{2}{x} - \frac{3}{x^2}}$

66. $\dfrac{1 - \frac{2}{r} - \frac{8}{r^2}}{1 - \frac{1}{r} - \frac{6}{r^2}}$

67. $\dfrac{\frac{2}{x} - \frac{1}{xy}}{\frac{1}{xy} + \frac{2}{y}}$

68. $\dfrac{\frac{1}{xy} + \frac{2}{x}}{\frac{3}{y} - \frac{1}{xy}}$

69. $\dfrac{\frac{2}{x-1} + 1}{1 - \frac{3}{x-1}}$

70. $\dfrac{\frac{3}{a+2} - 1}{1 + \frac{2}{a+2}}$

71. $\dfrac{1 - \frac{1}{y-1}}{y - \frac{8}{y+2}}$

72. $\dfrac{1 + \frac{1}{x+2}}{x - \frac{18}{x-3}}$

73. $1 + \dfrac{1}{1 + \frac{1}{x}}$

74. $1 + \dfrac{1}{1 - \frac{1}{y}}$

75. **Ecology.** Herbicides constitute $\frac{2}{3}$ of all pesticides used in the United States. Insecticides are $\frac{1}{4}$ of all pesticides used in the United States. The ratio of herbicides to insecticides used in the United States can be written $\frac{2}{3} \div \frac{1}{4}$. Write this ratio in simplest form.

76. **Ecology.** Fungicides account for $\frac{1}{10}$ of the pesticides used in the United States. Insecticides account for $\frac{1}{4}$ of all the pesticides used in the United States. The ratio of fungicides to insecticides used in the United States can be written $\frac{1}{10} \div \frac{1}{4}$. Write this ratio in simplest form.

77. Ecology. The ratio of insecticides to herbicides applied to wheat, soybeans, corn, and cotton can be expressed as $\frac{7}{10} \div \frac{4}{5}$. Simplify this ratio.

78. Find the area of the rectangle shown.

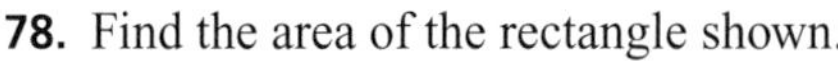

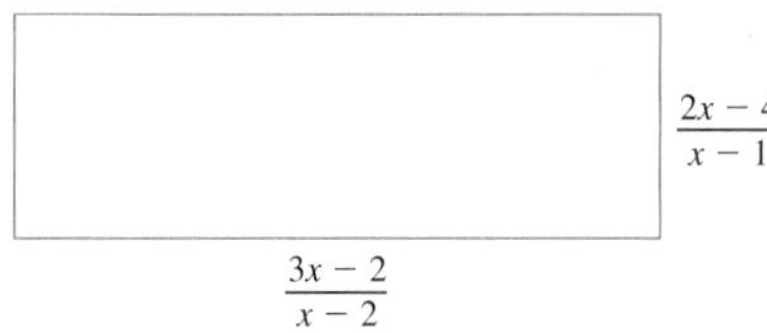

79. Find the area of the rectangle shown.

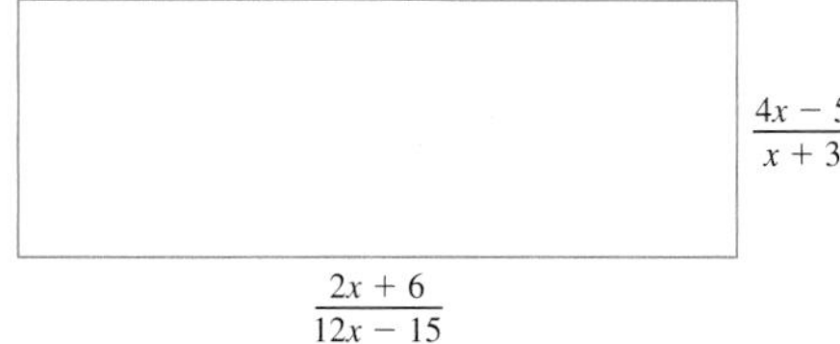

80. Electricity. The combined resistance of two resistors R_1 and R_2 in parallel is given by the formula

$$R_T = \frac{1}{\frac{1}{R_1} + \frac{1}{R_2}}$$

Simplify the formula.

81. Complex fractions have some interesting patterns. Work with a partner to evaluate each complex fraction in the sequence below. This is an interesting sequence of fractions because the numerators and denominators are a famous sequence of whole numbers, and the fractions get closer and closer to a number called "the golden mean."

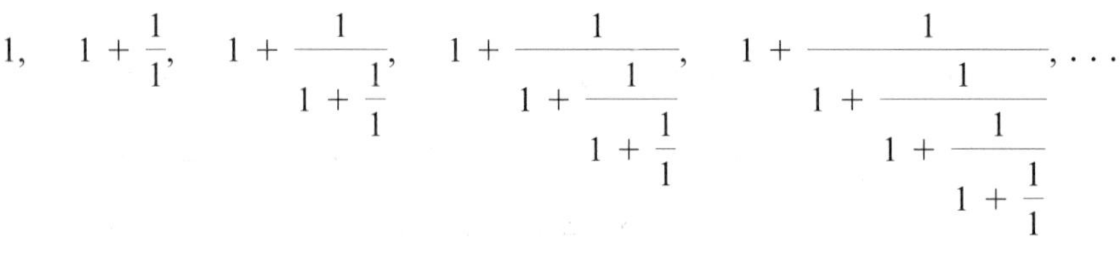

____, ____, ____, ____, ____, ____, ____, ____

After you have evaluated these first five, you no doubt will see a pattern in the resulting fractions that allows you to go on indefinitely without having to evaluate more complex fractions. Write each of these fractions as decimals. Write your observations about the sequence of fractions and about the sequence of decimal fractions.

ANSWERS

77. $\frac{\frac{7}{10}}{\frac{4}{5}} = \frac{7}{10} \cdot \frac{5}{4} = \frac{7}{8}$

78. $\frac{2(3x - 2)}{x - 1}$

79. $\frac{2}{3}$

80. $\frac{R_1R_2}{R_1 + R_2}$

81.

ANSWERS

a. 2

b. $\frac{6}{5}$

c. $-\frac{4}{3}$

d. 5

e. −3

f. −15

Getting Ready for Section 5.5 [Section 2.3]

Solve each of the following equations.

(a) $x + 8 = 10$

(b) $5x - 4 = 2$

(c) $3x + 8 = 4$

(d) $3(x - 2) - 4 = 5$

(e) $4(2x + 1) - 3 = -23$

(f) $4(2x - 5) - 3(3x + 1) = -8$

Answers

1. $\frac{2}{9}$ **3.** $\frac{xy}{12}$ **5.** $\frac{6}{a}$ **7.** $\frac{x^2}{6y^2}$ **9.** $\frac{5}{12a}$ **11.** $\frac{m^2}{6n^3}$ **13.** $\frac{2}{3}$ **15.** $\frac{3p^2}{p + 8}$ **17.** $\frac{m + 3}{3m}$ **19.** $\frac{2r - 1}{3r - 2}$ **21.** $\frac{7x}{5}$ **23.** $\frac{-6}{x + 2}$ **25.** $\frac{2}{3}$ **27.** $\frac{1}{2x}$ **29.** $\frac{3y}{2}$ **31.** $\frac{9}{20}$ **33.** $\frac{a - 3}{10a}$ **35.** $\frac{x - 2}{3x^2}$ **37.** $\frac{2x + 1}{2x}$ **39.** $\frac{3b}{a + 2b}$ **41.** $\frac{1}{3x^2}$ **43.** $\frac{-x}{6}$ **45.** $\frac{x}{2}$ **47.** $\frac{2x}{3(x + 4)}$ **49.** $\frac{8}{9}$ **51.** $\frac{2}{3}$ **53.** $\frac{1}{2x}$ **55.** $\frac{3a}{2}$ **57.** $\frac{2(y + 1)}{y - 1}$ **59.** $\frac{2x - 1}{2x + 1}$ **61.** $\frac{3y - x}{6}$ **63.** $\frac{x - y}{y}$ **65.** $\frac{x + 4}{x + 3}$ **67.** $\frac{2y - 1}{1 + 2x}$ **69.** $\frac{x + 1}{x - 4}$ **71.** $\frac{y + 2}{(y - 1)(y + 4)}$ **73.** $\frac{2x + 1}{x + 1}$ **75.** $\frac{8}{3}$ **77.** $\frac{7}{8}$ **79.** $\frac{2}{3}$ **81.** **a.** 2 **b.** $\frac{6}{5}$ **c.** $-\frac{4}{3}$ **d.** 5 **e.** −3 **f.** −15

5.5 Equations Involving Fractions

5.5 OBJECTIVES

1. Determine the excluded values for the variables of an algebraic fraction
2. Solve a fractional equation
3. Solve a proportion for an unknown

In Chapter 2, you learned how to solve a variety of equations. We now want to extend that work to the solution of **fractional equations,** which are equations that involve algebraic fractions as one or more of their terms.

NOTE The resulting equation *will* be equivalent unless a solution results that makes a denominator in the original equation 0. More about this later!

To solve a fractional equation, we multiply each term of the equation by the LCD of any fractions. The resulting equation should be equivalent to the original equation and be cleared of all fractions.

Example 1

Solving Fractional Equations

Solve

NOTE This equation has three terms: $\frac{x}{2}$, $-\frac{1}{3}$, and $\frac{2x+3}{6}$. The sign of the term is not used to find the LCD.

$$\frac{x}{2} - \frac{1}{3} = \frac{2x+3}{6} \qquad (1)$$

The LCD for $\frac{x}{2}, \frac{1}{3}$, and $\frac{2x+3}{6}$ is 6. Multiply both sides of the equation by 6. Using the distributive property, we multiply *each* term by 6.

NOTE By the multiplication property of equality, this equation is equivalent to the original equation, labeled (1).

$$6 \cdot \frac{x}{2} - 6 \cdot \frac{1}{3} = 6\left(\frac{2x+3}{6}\right) \qquad \text{or} \qquad 3x - 2 = 2x + 3 \qquad (2)$$

Solving as before, we have

$$3x - 2x = 3 + 2 \qquad \text{or} \qquad x = 5$$

To check, substitute 5 for x in the *original* equation.

$$\frac{5}{2} - \frac{1}{3} \stackrel{?}{=} \frac{2 \cdot 5 + 3}{6}$$

$$\frac{13}{6} = \frac{13}{6} \qquad \text{(True)}$$

Be Careful! Many students have difficulty because they don't distinguish between adding or subtracting *expressions* (as we did in Sections 5.2 and 5.3) and solving equations (illustrated in the above example). In the **expression**

$$\frac{x+1}{2} + \frac{x}{3}$$

we want to add the two fractions to form a single fraction. In the **equation**

$$\frac{x+1}{2} = \frac{x}{3} + 1$$

we want to solve for x.

CHECK YOURSELF 1

Solve and check.

$$\frac{x}{4} - \frac{1}{6} = \frac{4x - 5}{12}$$

Recall that, for any fraction, the denominator must not be equal to zero. When a fraction has a variable in the denominator, we must exclude any value for the variable that would result in division by zero.

Example 2

Finding Excluded Values for *x*

In the following algebraic fractions, what values for x must be excluded?

(a) $\frac{x}{5}$. Here x can have any value, so none need to be excluded.

(b) $\frac{3}{x}$. If $x = 0$, then $\frac{3}{x}$ is undefined; 0 is the excluded value.

(c) $\frac{5}{x - 2}$. If $x = 2$, then $\frac{5}{x - 2} = \frac{5}{2 - 2} = \frac{5}{0}$, which is undefined, so 2 is the excluded value.

CHECK YOURSELF 2

What values for x, if any, must be excluded?

(a) $\frac{x}{7}$ **(b)** $\frac{5}{x}$ **(c)** $\frac{7}{x - 5}$

If the denominator of an algebraic fraction contains a product of two or more variable factors, the zero-product principle must be used to determine the excluded values for the variable.

In some cases, you will have to factor the denominator to see the restrictions on the values for the variable.

Example 3

Finding Excluded Values for *x*

What values for x must be excluded in each fraction?

(a) $\frac{3}{x^2 - 6x - 16}$

Factoring the denominator, we have

$$\frac{3}{x^2 - 6x - 16} = \frac{3}{(x - 8)(x + 2)}$$

Letting $x - 8 = 0$ or $x + 2 = 0$, we see that 8 and -2 make the denominator 0 so both 8 and -2 must be excluded.

(b) $\dfrac{3}{x^2 + 2x - 48}$

The denominator is zero when

$x^2 + 2x - 48 = 0$

Factoring, we find

$(x - 6)(x + 8) = 0$

The denominator is zero when

$x = 6$ or $x = -8$

CHECK YOURSELF 3

What values for x must be excluded in the following fractions?

(a) $\dfrac{5}{x^2 - 3x - 10}$ (b) $\dfrac{7}{x^2 + 5x - 14}$

The steps for solving an equation involving fractions are summarized in the following rule.

Step by Step: To Solve a Fractional Equation

Step 1 Remove the fractions in the equation by multiplying each term by the LCD of all the fractions.

Step 2 Solve the equation resulting from step 1 as before.

Step 3 Check your solution in the *original equation.*

NOTE The equation that is formed in step 2 can be solved by the methods of Sections 2.3 and 4.4.

We can also solve fractional equations with variables in the denominator by using the above algorithm. Example 4 illustrates this approach.

Example 4

Solving Fractional Equations

Solve

$$\frac{7}{4x} - \frac{3}{x^2} = \frac{1}{2x^2}$$

NOTE The factor x appears twice in the LCD.

The LCD of the three terms in the equation is $4x^2$, and so we multiply both sides of the equation by $4x^2$.

$$4x^2 \cdot \frac{7}{4x} - 4x^2 \cdot \frac{3}{x^2} = 4x^2 \cdot \frac{1}{2x^2}$$

Simplifying, we have

$$7x - 12 = 2$$
$$7x = 14$$
$$x = 2$$

We'll leave the check to you. Be sure to return to the original equation.

CHECK YOURSELF 4

Solve and check.

$$\frac{5}{2x} - \frac{4}{x^2} = \frac{7}{2x^2}$$

The process of solving fractional equations is exactly the same when binomials are involved in the denominators.

Example 5

Solving Fractional Equations

(a) Solve

NOTE There are three terms.

$$\frac{x}{x-3} - 2 = \frac{1}{x-3}$$

The LCD is $x - 3$, and so we multiply each side (every term) by $x - 3$.

NOTE Each of the terms is multiplied by $x - 3$.

$$\overset{1}{\cancel{(x-3)}} \cdot \left(\frac{x}{\underset{1}{\cancel{x-3}}}\right) - 2(x-3) = \overset{1}{\cancel{(x-3)}} \cdot \left(\frac{1}{\underset{1}{\cancel{x-3}}}\right)$$

 **CAUTION**

Be careful of the signs!

Simplifying, we have

$$x - 2(x - 3) = 1$$
$$x - 2x + 6 = 1$$
$$-x = -5$$
$$x = 5$$

To check, substitute 5 for x in the original equation.

(b) Solve

NOTE Recall that $x^2 - 9 = (x - 3)(x + 3)$

$$\frac{3}{x-3} - \frac{7}{x+3} = \frac{2}{x^2-9}$$

In factored form, the three denominators are $x - 3$, $x + 3$, and $(x + 3)(x - 3)$. This means that the LCD is $(x + 3)(x - 3)$, and so we multiply:

$$\overset{1}{\cancel{(x-3)}}(x+3)\left(\frac{3}{\underset{1}{\cancel{x-3}}}\right) - \overset{1}{\cancel{(x+3)}}(x-3)\left(\frac{7}{\underset{1}{\cancel{x+3}}}\right) = \overset{1}{\cancel{(x+3)}}\overset{1}{\cancel{(x-3)}}\left(\frac{2}{\underset{1}{\cancel{x^2-9}}}\right)$$

Simplifying, we have

$$3(x + 3) - 7(x - 3) = 2$$
$$3x + 9 - 7x + 21 = 2$$
$$-4x + 30 = 2$$
$$-4x = -28$$
$$x = 7$$

CHECK YOURSELF 5

Solve and check.

(a) $\dfrac{x}{x-5} - 2 = \dfrac{2}{x-5}$ **(b)** $\dfrac{4}{x-4} - \dfrac{3}{x+1} = \dfrac{5}{x^2 - 3x - 4}$

You should be aware that some fractional equations have no solutions. Example 6 shows that possibility.

Example 6

Solving Fractional Equations

Solve

$$\frac{x}{x-2} - 7 = \frac{2}{x-2}$$

The LCD is $x - 2$, and so we multiply each side (every term) by $x - 2$.

$$(x-2)\left(\frac{x}{x-2}\right) - 7(x-2) = (x-2)\left(\frac{2}{x-2}\right)$$

Simplifying, we have

$$x - 7x + 14 = 2$$
$$-6x = -12$$
$$x = 2$$

Now, when we try to check our result, we have

NOTE 2 is substituted for x in the original equation.

$$\frac{2}{2-2} - 7 \stackrel{?}{=} \frac{2}{2-2} \quad \text{or} \quad \frac{2}{0} - 7 \stackrel{?}{=} \frac{2}{0}$$

These terms are undefined.

What went wrong? Remember that two of the terms in our original equation were $\dfrac{x}{x-2}$ and $\dfrac{2}{x-2}$. The variable x cannot have the value 2 because 2 is an excluded value (it makes the denominator 0). So our original equation has *no solution.*

CHECK YOURSELF 6

Solve, if possible.

$$\frac{x}{x+3} - 6 = \frac{-3}{x+3}$$

Equations involving fractions may also lead to quadratic equations, as Example 7 illustrates.

Example 7

Solving Fractional Equations

Solve

$$\frac{x}{x-4} = \frac{15}{x-3} - \frac{2x}{x^2 - 7x + 12}$$

The LCD is $(x-4)(x-3)$. Multiply each side (every term) by $(x-4)(x-3)$.

$$\frac{x}{\cancel{(x-4)}}\cancel{(x-4)}(x-3) = \frac{15}{\cancel{(x-3)}}(x-4)\cancel{(x-3)} - \frac{2x}{\cancel{(x-4)}\cancel{(x-3)}}\cancel{(x-4)}\cancel{(x-3)}$$

Simplifying, we have

$$x(x-3) = 15(x-4) - 2x$$

Multiply to clear of parentheses:

$$x^2 - 3x = 15x - 60 - 2x$$

NOTE Notice that this equation is *quadratic.* It can be solved by the methods of Section 4.4.

In standard form, the equation is

$$x^2 - 16x + 60 = 0 \qquad \text{or} \qquad (x-6)(x-10) = 0$$

Setting the factors to 0, we have

$$x - 6 = 0 \qquad \text{or} \qquad x - 10 = 0$$

$$x = 6 \qquad\qquad x = 10$$

So $x = 6$ and $x = 10$ are possible solutions. We will leave the check of *each* solution to you.

CHECK YOURSELF 7

Solve and check.

$$\frac{3x}{x+2} - \frac{2}{x+3} = \frac{36}{x^2 + 5x + 6}$$

The following equation is a special kind of equation involving fractions:

$$\frac{135}{t} = \frac{180}{t+1}$$

An equation of the form $\frac{a}{b} = \frac{c}{d}$ is said to be in **proportion form,** or more simply it is called a **proportion.** This type of equation occurs often enough in algebra that it is worth developing some special methods for its solution. First, we will need some definitions.

A **ratio** is a means of comparing two quantities. A ratio can be written as a fraction. For instance, the ratio of 2 to 3 can be written as $\frac{2}{3}$. A statement that two ratios are equal is called a *proportion.* A proportion has the form

$$\frac{a}{b} = \frac{c}{d}$$

In the proportion above, a and d are called the **extremes** of the proportion, and b and c are called the **means.**

A useful property of proportions is easily developed. If

$$\frac{a}{b} = \frac{c}{d}$$

NOTE *bd* is the LCD of the denominators.

and we multiply both sides by $b \cdot d$, then

$$\left(\frac{a}{b}\right)bd = \left(\frac{c}{d}\right)bd \qquad \text{or} \qquad ad = bc$$

Rules and Properties: Proportions

If $\frac{a}{b} = \frac{c}{d}$ then $ad = bc$

In words:

In any proportion, the product of the extremes (ad) is equal to the product of the means (bc).

Because a proportion is a special kind of fractional equation, this rule gives us an alternative approach to solving equations that are in the proportion form.

Example 8

Solving a Proportion

Solve the equations for x.

NOTE The extremes are x and 15. The means are 5 and 12.

(a) $\frac{x}{5} = \frac{12}{15}$

Set the product of the extremes equal to the product of the means.

$15x = 5 \cdot 12$

$15x = 60$

$x = 4$

Our solution is 4. You can check as before, by substituting in the original proportion.

(b) $\frac{x + 3}{10} = \frac{x}{7}$

Set the product of the extremes equal to the product of the means. Be certain to use parentheses with a numerator with more than one term.

$7(x + 3) = 10x$

$7x + 21 = 10x$

$21 = 3x$

$7 = x$

We will leave the checking of this result to the reader.

CHECK YOURSELF 8

Solve for x.

(a) $\frac{x}{8} = \frac{3}{4}$ **(b)** $\frac{x - 1}{9} = \frac{x + 1}{12}$

CHECK YOURSELF ANSWERS

1. $x = 3$ **2. (a)** none; **(b)** 0; **(c)** 5 **3. (a)** $-2, 5$; **(b)** $-7, 2$ **4.** $x = 3$
5. (a) $x = 8$; **(b)** $x = -11$ **6.** No solution **7.** $x = -5$ or $x = \frac{8}{3}$
8. (a) $x = 6$; **(b)** $x = 7$

As the examples of this section illustrated, *whenever* an equation involves algebraic fractions, the *first step* of the solution is to clear the equation of fractions by multiplication.

The following algorithm summarizes our work in solving equations that involve algebraic fractions.

Step by Step: To Solve an Equation Involving Fractions

Step 1 Remove the fractions appearing in the equation by multiplying each side (every term) by the LCD of all the fractions.

Step 2 Solve the equation resulting from step 1. If the equation is linear, use the methods of Section 2.3 for the solution. If the equation is quadratic, use the methods of Section 4.4.

Step 3 Check all solutions by substitution in the *original equation.* Be sure to discard any *extraneous* solutions, that is, solutions that would result in a zero denominator in the original equation.

5.5 Exercises

Name ______________________

Section ________ Date ________

What values for x, if any, must be excluded in each of the following algebraic fractions?

1. $\frac{x}{15}$

2. $\frac{8}{x}$

3. $\frac{17}{x}$

4. $\frac{x}{8}$

5. $\frac{3}{x - 2}$

6. $\frac{x - 1}{5}$

7. $\frac{-5}{x + 4}$

8. $\frac{4}{x + 3}$

9. $\frac{x - 5}{2}$

10. $\frac{x - 1}{x - 5}$

11. $\frac{3x}{(x + 1)(x - 2)}$

12. $\frac{5x}{(x - 3)(x + 7)}$

13. $\frac{x - 1}{(2x - 1)(x + 3)}$

14. $\frac{x + 3}{(3x + 1)(x - 2)}$

15. $\frac{7}{x^2 - 9}$

16. $\frac{5x}{x^2 + x - 2}$

17. $\frac{x + 3}{x^2 - 7x + 12}$

18. $\frac{3x - 4}{x^2 - 49}$

19. $\frac{2x - 1}{3x^2 + x - 2}$

20. $\frac{3x + 1}{4x^2 - 11x + 6}$

Solve each of the following equations for x.

21. $\frac{x}{2} + 3 = 6$

22. $\frac{x}{3} - 2 = 1$

ANSWERS

1. None
2. 0
3. 0
4. None
5. 2
6. None
7. −4
8. −3
9. None
10. 5
11. −1, 2
12. −7, 3
13. $-3, \frac{1}{2}$
14. $-\frac{1}{3}, 2$
15. −3, 3
16. −2, 1
17. 3, 4
18. −7, 7
19. $-1, \frac{2}{3}$
20. $\frac{3}{4}, 2$
21. 6
22. 9

CHECK YOURSELF 1

The sum of two-fifths of a number and one-half of that number is 18. Find the number.

Number problems that involve reciprocals can be solved by using fractional equations. Example 2 illustrates this approach.

Example 2

Solving a Numerical Application

One number is twice another number. If the sum of their reciprocals is $\frac{3}{10}$, what are the two numbers?

Step 1 You want to find the two numbers.

Step 2 Let x be one number. Then $2x$ is the other number.

Twice the first (points to $2x$)

Step 3

NOTE The reciprocal of a fraction is the fraction obtained by switching the numerator and denominator.

$$\frac{1}{x} + \frac{1}{2x} = \frac{3}{10}$$

The reciprocal of the first number, x (points to $\frac{1}{x}$)

The reciprocal of the second number, $2x$ (points to $\frac{1}{2x}$)

Step 4 The LCD of the fractions is $10x$, and so we multiply by $10x$.

$$10x\left(\frac{1}{x}\right) + 10x\left(\frac{1}{2x}\right) = 10x\left(\frac{3}{10}\right)$$

Simplifying, we have

$$10 + 5 = 3x$$

$$15 = 3x$$

$$5 = x$$

NOTE x was one number, and $2x$ was the other.

The numbers are 5 and 10.

Step 5

Again check the result by returning to the original problem. If the numbers are 5 and 10, we have

$$\frac{1}{5} + \frac{1}{10} = \frac{2+1}{10} = \frac{3}{10}$$

The sum of the reciprocals is $\frac{3}{10}$.

CHECK YOURSELF 2

One number is 3 times another. If the sum of their reciprocals is $\frac{2}{9}$, find the two numbers.

Name ______ Section ______ Date ______

5.5 Exercises

What values for x, if any, must be excluded in each of the following algebraic fractions?

1. $\frac{x}{15}$

2. $\frac{8}{x}$

3. $\frac{17}{x}$

4. $\frac{x}{8}$

5. $\frac{3}{x - 2}$

6. $\frac{x - 1}{5}$

7. $\frac{-5}{x + 4}$

8. $\frac{4}{x + 3}$

9. $\frac{x - 5}{2}$

10. $\frac{x - 1}{x - 5}$

11. $\frac{3x}{(x + 1)(x - 2)}$

12. $\frac{5x}{(x - 3)(x + 7)}$

13. $\frac{x - 1}{(2x - 1)(x + 3)}$

14. $\frac{x + 3}{(3x + 1)(x - 2)}$

15. $\frac{7}{x^2 - 9}$

16. $\frac{5x}{x^2 + x - 2}$

17. $\frac{x + 3}{x^2 - 7x + 12}$

18. $\frac{3x - 4}{x^2 - 49}$

19. $\frac{2x - 1}{3x^2 + x - 2}$

20. $\frac{3x + 1}{4x^2 - 11x + 6}$

Solve each of the following equations for x.

21. $\frac{x}{2} + 3 = 6$

22. $\frac{x}{3} - 2 = 1$

ANSWERS

1. None
2. 0
3. 0
4. None
5. 2
6. None
7. −4
8. −3
9. None
10. 5
11. −1, 2
12. −7, 3
13. $-3, \frac{1}{2}$
14. $-\frac{1}{3}, 2$
15. −3, 3
16. −2, 1
17. 3, 4
18. −7, 7
19. $-1, \frac{2}{3}$
20. $\frac{3}{4}, 2$
21. 6
22. 9

ANSWERS

23. 12
24. 24
25. 15
26. 12
27. 7
28. 5
29. 2
30. −4
31. 8
32. 6
33. 3
34. 2
35. 8
36. 4
37. 2
38. 2
39. 11
40. 4
41. −5
42. 11
43. −23
44. 5
45. No solution
46. No solution
47. 6
48. 6

23. $\frac{x}{2} - \frac{x}{3} = 2$

24. $\frac{x}{6} - \frac{x}{8} = 1$

25. $\frac{x}{5} - \frac{1}{3} = \frac{x - 7}{3}$

26. $\frac{x}{6} + \frac{3}{4} = \frac{x - 1}{4}$

27. $\frac{x}{4} - \frac{1}{5} = \frac{4x + 3}{20}$

28. $\frac{x}{12} - \frac{1}{6} = \frac{2x - 7}{12}$

29. $\frac{3}{x} + 2 = \frac{7}{x}$

30. $\frac{4}{x} - 3 = \frac{16}{x}$

31. $\frac{4}{x} + \frac{3}{4} = \frac{10}{x}$

32. $\frac{3}{x} = \frac{5}{3} - \frac{7}{x}$

33. $\frac{5}{2x} - \frac{1}{x} = \frac{9}{2x^2}$

34. $\frac{4}{3x} + \frac{1}{x} = \frac{14}{3x^2}$

35. $\frac{2}{x - 3} + 1 = \frac{7}{x - 3}$

36. $\frac{x}{x + 1} + 2 = \frac{14}{x + 1}$

37. $\frac{12}{x + 3} = \frac{x}{x + 3} + 2$

38. $\frac{5}{x - 3} + 3 = \frac{x}{x - 3}$

39. $\frac{3}{x - 5} + 4 = \frac{2x + 5}{x - 5}$

40. $\frac{24}{x + 5} - 2 = \frac{x + 2}{x + 5}$

41. $\frac{2}{x + 3} + \frac{1}{2} = \frac{x + 6}{x + 3}$

42. $\frac{6}{x - 5} - \frac{2}{3} = \frac{x - 9}{x - 5}$

43. $\frac{x}{3x + 12} + \frac{x - 1}{x + 4} = \frac{5}{3}$

44. $\frac{x}{4x - 12} - \frac{x - 4}{x - 3} = \frac{1}{8}$

45. $\frac{x}{x - 3} - 2 = \frac{3}{x - 3}$

46. $\frac{x}{x - 5} + 2 = \frac{5}{x - 5}$

47. $\frac{x - 1}{x + 3} - \frac{x - 3}{x} = \frac{3}{x^2 + 3x}$

48. $\frac{x}{x - 2} - \frac{x + 1}{x} = \frac{8}{x^2 - 2x}$

49. $\frac{1}{x-2} - \frac{2}{x+2} = \frac{2}{x^2-4}$

50. $\frac{1}{x+4} + \frac{1}{x-4} = \frac{12}{x^2-16}$

51. $\frac{5}{x-4} = \frac{1}{x+2} - \frac{2}{x^2-2x-8}$

52. $\frac{11}{x+2} = \frac{5}{x^2-x-6} + \frac{1}{x-3}$

53. $\frac{3}{x-1} - \frac{1}{x+9} = \frac{18}{x^2+8x-9}$

54. $\frac{2}{x+2} = \frac{3}{x+6} + \frac{9}{x^2+8x+12}$

55. $\frac{3}{x+3} + \frac{25}{x^2+x-6} = \frac{5}{x-2}$

56. $\frac{5}{x+6} + \frac{2}{x^2+7x+6} = \frac{3}{x+1}$

57. $\frac{7}{x-5} - \frac{3}{x+5} = \frac{40}{x^2-25}$

58. $\frac{3}{x-3} - \frac{18}{x^2-9} = \frac{5}{x+3}$

59. $\frac{2x}{x-3} + \frac{2}{x-5} = \frac{3x}{x^2-8x+15}$

60. $\frac{x}{x-4} = \frac{5x}{x^2-x-12} - \frac{3}{x+3}$

61. $\frac{2x}{x+2} = \frac{5}{x^2-x-6} - \frac{1}{x-3}$

62. $\frac{3x}{x-1} = \frac{2}{x-2} - \frac{2}{x^2-3x+2}$

63. $\frac{7}{x-2} + \frac{16}{x+3} = 3$

64. $\frac{5}{x-2} + \frac{6}{x+2} = 2$

65. $\frac{11}{x-3} - 1 = \frac{10}{x+3}$

66. $\frac{17}{x-4} - 2 = \frac{10}{x+2}$

Solve each of the following equations for x.

67. $\frac{x}{11} = \frac{12}{33}$

68. $\frac{4}{x} = \frac{16}{20}$

69. $\frac{5}{8} = \frac{20}{x}$

70. $\frac{x}{10} = \frac{9}{30}$

71. $\frac{x+1}{5} = \frac{20}{25}$

72. $\frac{2}{5} = \frac{x-2}{20}$

73. $\frac{3}{5} = \frac{x-1}{20}$

74. $\frac{5}{x-3} = \frac{15}{21}$

ANSWERS

49. 4

50. 6

51. −4

52. 4

53. −5

54. −3

55. No solution

56. $\frac{11}{2}$

57. $-\frac{5}{2}$

58. No solution

59. $-\frac{1}{2}, 6$

60. −4, 3

61. $-\frac{1}{2}$

62. $\frac{2}{3}$

63. $-\frac{1}{3}, 7$

64. $-\frac{1}{2}, 6$

65. −8, 9

66. $-\frac{9}{2}, 10$

67. 4

68. 5

69. 32

70. 3

71. 3

72. 10

73. 13

74. 10

ANSWERS

75. 3

76. 8

77. 10

78. 3

79. 6

80. −3

81. 10

82. 6

a. $\frac{1}{4}x + \frac{4}{5}x$

b. $6x - 12$

c. $\frac{x + 5}{6}$

d. $3x - 4$

e. $\frac{x}{5}$

f. $x - 5$

75. $\frac{x}{6} = \frac{x + 5}{16}$

76. $\frac{x - 2}{x + 2} = \frac{12}{20}$

77. $\frac{x}{x + 7} = \frac{10}{17}$

78. $\frac{x}{10} = \frac{x + 6}{30}$

79. $\frac{2}{x - 1} = \frac{6}{x + 9}$

80. $\frac{3}{x - 3} = \frac{4}{x - 5}$

81. $\frac{1}{x + 3} = \frac{7}{x^2 - 9}$

82. $\frac{1}{x + 5} = \frac{4}{x^2 + 3x - 10}$

Getting Ready for Section 5.6 [Section 1.1]

Write each of the following phrases using symbols. Use the variable x to represent the number in each case.

(a) One-fourth of a number added to four-fifths of the same number
(b) 6 times a number, decreased by 12
(c) The quotient when 5 more than a number is divided by 6
(d) Three times the length of a side of a rectangle decreased by 4
(e) A distance traveled divided by 5
(f) The speed of a truck that is 5 mi/h slower than a car

Answers

1. None **3.** 0 **5.** 2 **7.** −4 **9.** None **11.** −1, 2 **13.** $-3, \frac{1}{2}$ **15.** −3, 3 **17.** 3, 4 **19.** $-1, \frac{2}{3}$ **21.** 6 **23.** 12 **25.** 15 **27.** 7 **29.** 2 **31.** 8 **33.** 3 **35.** 8 **37.** 2 **39.** 11 **41.** −5 **43.** −23 **45.** No solution **47.** 6 **49.** 4 **51.** −4 **53.** −5 **55.** No solution **57.** $-\frac{5}{2}$ **59.** $-\frac{1}{2}, 6$ **61.** $-\frac{1}{2}$ **63.** $-\frac{1}{3}, 7$ **65.** −8, 9 **67.** 4 **69.** 32 **71.** 3 **73.** 13 **75.** 3 **77.** 10 **79.** 6 **81.** 10 **a.** $\frac{1}{4}x + \frac{4}{5}x$ **b.** $6x - 12$ **c.** $\frac{x + 5}{6}$ **d.** $3x - 4$ **e.** $\frac{x}{5}$ **f.** $x - 5$

Applications of Algebraic Fractions

5.6 OBJECTIVES

1. Solve a word problem that leads to a fractional equation
2. Apply proportions to the solution of a word problem

Many word problems will lead to fractional equations that must be solved by using the methods of the previous section. The five steps in solving word problems are, of course, the same as you saw earlier.

Example 1

Solving a Numerical Application

If one-third of a number is added to three-fourths of that same number, the sum is 26. Find the number.

Step 1 Read the problem carefully. You want to find the unknown number.

Step 2 Choose a letter to represent the unknown. Let x be the unknown number.

Step 3 Form an equation.

NOTE The equation expresses the relationship between the two parts of the number.

$$\frac{1}{3}x + \frac{3}{4}x = 26$$

One-third of number (arrow to $\frac{1}{3}x$); Three-fourths of number (arrow to $\frac{3}{4}x$)

Step 4 Solve the equation. Multiply each side (every term) of the equation by 12, the LCD.

$$12 \cdot \frac{1}{3}x + 12 \cdot \frac{3}{4}x = 12 \cdot 26$$

Simplifying yields

$$4x + 9x = 312$$

$$13x = 312$$

$$x = 24$$

NOTE Be sure to answer the question raised in the problem.

The number is 24.

Step 5 Check your solution by returning to the *original problem.* If the number is 24, we have

$$\frac{1}{3} \cdot 24 + \frac{3}{4} \cdot 24 = 8 + 18 = 26$$

and the solution is verified.

CHECK YOURSELF 1

The sum of two-fifths of a number and one-half of that number is 18. Find the number.

Number problems that involve reciprocals can be solved by using fractional equations. Example 2 illustrates this approach.

Example 2

Solving a Numerical Application

One number is twice another number. If the sum of their reciprocals is $\frac{3}{10}$, what are the two numbers?

Step 1 You want to find the two numbers.

Step 2 Let x be one number. Then $2x$ is the other number.

(Twice the first)

Step 3

NOTE The reciprocal of a fraction is the fraction obtained by switching the numerator and denominator.

$$\frac{1}{x} + \frac{1}{2x} = \frac{3}{10}$$

($\frac{1}{x}$: The reciprocal of the first number, x; $\frac{1}{2x}$: The reciprocal of the second number, $2x$)

Step 4 The LCD of the fractions is $10x$, and so we multiply by $10x$.

$$10x\left(\frac{1}{x}\right) + 10x\left(\frac{1}{2x}\right) = 10x\left(\frac{3}{10}\right)$$

Simplifying, we have

$$10 + 5 = 3x$$

$$15 = 3x$$

$$5 = x$$

NOTE x was one number, and $2x$ was the other.

The numbers are 5 and 10.

Step 5

Again check the result by returning to the original problem. If the numbers are 5 and 10, we have

$$\frac{1}{5} + \frac{1}{10} = \frac{2 + 1}{10} = \frac{3}{10}$$

The sum of the reciprocals is $\frac{3}{10}$.

CHECK YOURSELF 2

One number is 3 times another. If the sum of their reciprocals is $\frac{2}{9}$, find the two numbers.

The solution of many motion problems will also involve fractional equations. Remember that the key equation for solving all motion problems relates the distance traveled, the speed or rate, and the time:

Definitions: Motion Problem Relationships

$d = r \cdot t$

Often we will use this equation in different forms by solving for r or for t. So

$$r = \frac{d}{t} \qquad \text{or} \qquad t = \frac{d}{r}$$

Example 3

Solving an Application Involving $r = \frac{d}{t}$

Vince took 2 hours (h) longer to drive 225 miles (mi) than he did on a trip of 135 mi. If his speed was the same both times, how long did each trip take?

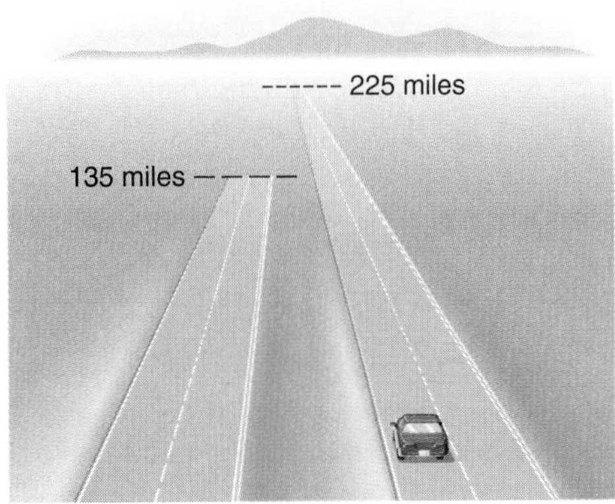

Step 1 You want to find the times taken for the 225-mi trip and for the 135-mi trip.

NOTE It is often helpful to choose your variable to "suggest" the unknown quantity—here t for time.

Step 2 Let t be the time for the 135-mi trip (in hours).

Then $\underbrace{t + 2}_{\text{2 h longer}}$ is the time for the 225-mi trip.

It is often helpful to arrange the information in tabular form such as that shown.

NOTE Remember that rate is distance divided by time. The rightmost column is formed by using that relationship.

	Distance	Time	Rate
135-mi trip	135	t	$\frac{135}{t}$
225-mi trip	225	$t + 2$	$\frac{225}{t+2}$

Step 3 In forming the equation, remember that the speed (or rate) for each trip was the same. That is the *key* idea. We can equate the rates for the two trips that were found in step 2.

The two rates are shown in the rightmost column of the table. Thus we can write

$$\frac{135}{t} = \frac{225}{t+2}$$

NOTE Notice that the equation is in proportion form. So we could solve by setting the product of the means equal to the product of the extremes.

Step 4 To solve the above equation, multiply each side by $t(t + 2)$, the LCD of the fractions.

$$\cancel{t}(t+2)\left(\frac{135}{\cancel{t}}\right) = t\cancel{(t+2)}\left(\frac{225}{\cancel{t+2}}\right)$$

Simplifying, we have

$$135(t + 2) = 225t$$

$$135t + 270 = 225t$$

$$270 = 90t$$

$$t = 3\text{ h}$$

The time for the 135-mi trip was 3 h, and the time for the 225-mi trip was 5 h. We'll leave the check to you.

CHECK YOURSELF 3

Cynthia took 2 h longer to bicycle 75 mi than she did on a trip of 45 mi. If her speed was the same each time, find the time for each trip.

Example 4 uses the $d = r \cdot t$ relationship to find the speed.

Example 4

Solving an Application Involving $d = r \cdot t$

A train makes a trip of 300 mi in the same time that a bus can travel 250 mi. If the speed of the train is 10 mi/h faster than the speed of the bus, find the speed of each.

Step 1 You want to find the speeds of the train and of the bus.

Step 2 Let r be the speed (or rate) of the bus (in miles per hour).

Then $\underbrace{r + 10}_{\text{10 mi/h faster}}$ is the rate of the train.

Again let's form a table of the information.

NOTE Remember that time is distance divided by rate. Here the rightmost column is found by using that relationship.

	Distance	Rate	Time
Train	300	$r + 10$	$\frac{300}{r+10}$
Bus	250	r	$\frac{250}{r}$

Step 3 To form an equation, remember that the times for the train and bus are the same. We can equate the expressions for time found in step 2. Again, working from the rightmost column, we have

$$\frac{250}{r} = \frac{300}{r + 10}$$

Step 4 We multiply each term by $r(r + 10)$, the LCD of the fractions.

$$\overset{1}{\cancel{r}}(r + 10)\left(\frac{250}{\underset{1}{\cancel{r}}}\right) = r\overset{1}{\cancel{(r + 10)}}\left(\frac{300}{\underset{1}{\cancel{r + 10}}}\right)$$

Simplifying, we have

$$\begin{aligned} 250(r + 10) &= 300r \\ 250r + 2500 &= 300r \\ 2500 &= 50r \\ r &= 50 \text{ mi/h} \end{aligned}$$

NOTE Remember to find the rates of both vehicles.

The rate of the bus is 50 mi/h, and the rate of the train is 60 mi/h. You can check this result.

CHECK YOURSELF 4

A car makes a trip of 280 mi in the same time that a truck travels 245 mi. If the speed of the truck is 5 mi/h slower than that of the car, find the speed of each.

The next example involves fractions in decimal form. Mixture problems often use percentages, and those percentages can be written as decimals. Example 5 illustrates this method.

Example 5

Solving an Application Involving Solutions

A solution of antifreeze is 20% alcohol. How much pure alcohol must be added to 12 quarts (qt) of the solution to make a 40% solution?

Step 1 You want to find the number of quarts of pure alcohol that must be added.

Step 2 Let x be the number of quarts of pure alcohol to be added.

Step 3 To form our equation, note that the amount of alcohol present before mixing *must be the same* as the amount in the combined solution.

A picture will help.

NOTE Express the percentages as decimals in the equation.

So

$$\underbrace{12(0.20)} + \underbrace{x(1.00)} = \underbrace{(12 + x)(0.40)}$$

The amount of alcohol in the first solution (20% is 0.20)

The amount of pure alcohol ("pure" is 100%, or 1.00)

The amount of alcohol in the mixture

Step 4 Most students prefer to clear the decimals at this stage. It's easy here—multiplying by 100 will move the decimal point *two places to the right.* We then have

$$12(20) + x(100) = (12 + x)(40)$$

$$240 + 100x = 480 + 40x$$

$$60x = 240$$

$$x = 4 \text{ qt}$$

CHECK YOURSELF 5

How much pure alcohol must be added to 500 cubic centimeters (cm^3) of a 40% alcohol mixture to make a solution that is 80% alcohol?

There are many types of applications that lead to proportions in their solution. Typically these applications will involve a common ratio, such as miles to gallons or miles to hours, and they can be solved with three basic steps.

Step by Step: To Solve an Application by Using Proportions

Step 1 Assign a variable to represent the unknown quantity.

Step 2 Write a proportion, using the known and unknown quantities. Be sure each ratio involves the same units.

Step 3 Solve the proportion written in step 2 for the unknown quantity.

Example 6 illustrates this approach.

Example 6

Solving an Application Using Proportions

A car uses 3 gallons (gal) of gas to travel 105 miles (mi). At that mileage rate, how many gallons will be used on a trip of 385 mi?

Step 1 Assign a variable to represent the unknown quantity. Let x be the number of gallons of gas that will be used on the 385-mi trip.

Step 2 Write a proportion. Note that the ratio of miles to gallons must stay the same.

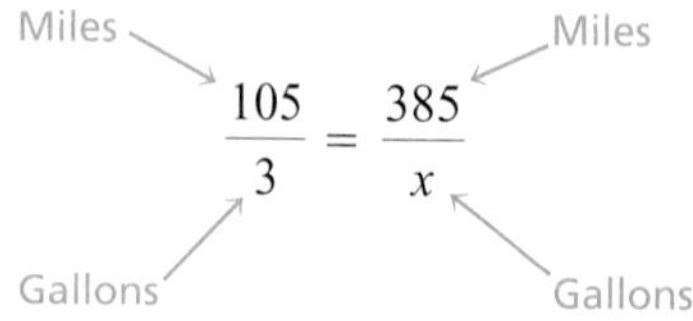

$$\frac{105}{3} = \frac{385}{x}$$

Step 3 Solve the proportion. The product of the extremes is equal to the product of the means.

$$105x = 3 \cdot 385$$

$$105x = 1155$$

$$\frac{105x}{105} = \frac{1155}{105}$$

$$x = 11 \text{ gal}$$

NOTE To verify your solution, return to the original problem and check that the two ratios are equivalent.

So 11 gal of gas will be used for the 385-mi trip.

CHECK YOURSELF 6

A car uses 8 liters (L) of gasoline in traveling 100 kilometers (km). At that rate, how many liters of gas will be used on a trip of 250 km?

Proportions can also be used to solve problems in which a quantity is divided by using a specific ratio. Example 7 shows how.

Example 7

Solving an Application Using Proportions

A piece of wire 60 inches (in.) long is to be cut into two pieces whose lengths have the ratio 5 to 7. Find the length of each piece.

Step 1 Let x represent the length of the shorter piece. Then $60 - x$ is the length of the longer piece.

NOTE A picture of the problem always helps.

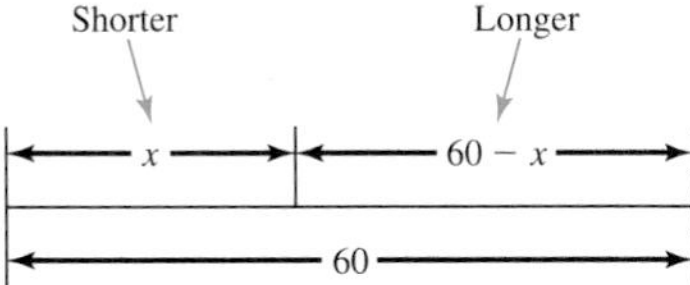

Step 2 The two pieces have the ratio $\frac{5}{7}$, so

$$\frac{x}{60 - x} = \frac{5}{7}$$

NOTE On the left and right, we have the ratio of the length of the shorter piece to that of the longer piece.

Step 3 Solving as before, we get

$$7x = (60 - x)5$$

$$7x = 300 - 5x$$

$$12x = 300$$

$$x = 25 \quad \text{(Shorter piece)}$$

$$60 - x = 35 \quad \text{(Longer piece)}$$

The pieces have lengths 25 in. and 35 in.

CHECK YOURSELF 7

A board 21 *feet (ft) long is to be cut into two pieces so that the ratio of their lengths is* 3 *to* 4. *Find the lengths of the two pieces.*

CHECK YOURSELF ANSWERS

1. The number is 20. **2.** The numbers are 6 and 18.
3. 75-mi trip: 5 h; 45-mi trip: 3 h **4.** Car: 40 mi/h; truck: 35 mi/h **5.** 1000 cm^3
6. 20 L **7.** 9 ft; 12 ft

Name ______________

Section ________ Date ________

5.6 Exercises

Solve the following word problems.

1. **Adding numbers.** If two-thirds of a number is added to one-half of that number, the sum is 35. Find the number.

2. **Subtracting numbers.** If one-third of a number is subtracted from three-fourths of that number, the difference is 15. What is the number?

3. **Subtracting numbers.** If one-fourth of a number is subtracted from two-fifths of a number, the difference is 3. Find the number.

4. **Adding numbers.** If five-sixths of a number is added to one-fifth of the number, the sum is 31. What is the number?

5. **Consecutive integers.** If one-third of an integer is added to one-half of the next consecutive integer, the sum is 13. What are the two integers?

6. **Consecutive integers.** If one-half of one integer is subtracted from three-fifths of the next consecutive integer, the difference is 3. What are the two integers?

7. **Reciprocals.** One number is twice another number. If the sum of their reciprocals is $\frac{1}{4}$, find the two numbers.

8. **Reciprocals.** One number is 3 times another. If the sum of their reciprocals is $\frac{1}{6}$, find the two numbers.

9. **Reciprocals.** One number is 4 times another. If the sum of their reciprocals is $\frac{5}{12}$, find the two numbers.

10. **Reciprocals.** One number is 3 times another. If the sum of their reciprocals is $\frac{4}{15}$, what are the two numbers?

11. **Reciprocals.** One number is 5 times another number. If the sum of their reciprocals is $\frac{6}{35}$, what are the two numbers?

12. **Reciprocals.** One number is 4 times another. The sum of their reciprocals is $\frac{5}{24}$. What are the two numbers?

13. **Reciprocals.** If the reciprocal of 5 times a number is subtracted from the reciprocal of that number, the result is $\frac{4}{25}$. What is the number?

14. **Reciprocals.** If the reciprocal of a number is added to 4 times the reciprocal of that number, the result is $\frac{5}{9}$. Find the number.

15. **Driving rate.** Lee can ride his bicycle 50 miles (mi) in the same time it takes him to drive 125 mi. If his driving rate is 30 mi/h faster than his rate bicycling, find each rate.

ANSWERS

1. 30
2. 36
3. 20
4. 30
5. 15, 16
6. 24, 25
7. 6, 12
8. 8, 24
9. 3, 12
10. 5, 15
11. 7, 35
12. 6, 24
13. 5
14. 9
15. 20 mi/h bicycling, 50 mi/h driving

ANSWERS

16. 4 mi/h running, 24 mi/h bicycling

17. Express 55 mi/h, local 45 mi/h

18. 3 h, 2 h

19. Freight 40 mi/h, passenger 65 mi/h

20. 3 h, 2 h

21. 5 h, 3 h

22. 8 h, 6 h

23. 5 h, 2 h

24. Train 40 mi/h, plane 140 mi/h

16. **Running rate.** Tina can run 12 mi in the same time it takes her to bicycle 72 mi. If her bicycling rate is 20 mi/h faster than her running rate, find each rate.

17. **Driving rate.** An express bus can travel 275 mi in the same time that it takes a local bus to travel 225 mi. If the rate of the express bus is 10 mi/h faster than that of the local bus, find the rate for each bus.

18. **Flying time.** A light plane took 1 hour (h) longer to travel 450 mi on the first portion of a trip than it took to fly 300 mi on the second. If the speed was the same for each portion, what was the flying time for each part of the trip?

19. **Train speed.** A passenger train can travel 325 mi in the same time a freight train takes to travel 200 mi. If the speed of the passenger train is 25 mi/h faster than the speed of the freight, find the speed of each.

20. **Flying time.** A small business jet took 1 h longer to fly 810 mi on the first part of a flight than to fly 540 mi on the second portion. If the jet's rate was the same for each leg of the flight, what was the flying time for each leg?

21. **Driving time.** Charles took 2 h longer to drive 240 mi on the first day of a vacation trip than to drive 144 mi on the second day. If his average driving rate was the same on both days, what was his driving time for each of the days?

22. **Driving time.** Ariana took 2 h longer to drive 360 mi on the first day of a trip than she took to drive 270 mi on the second day. If her speed was the same on both days, what was the driving time each day?

23. **Flying time.** An airplane took 3 h longer to fly 1200 mi than it took for a flight of 480 mi. If the plane's rate was the same on each trip, what was the time of each flight?

24. **Traveling time.** A train travels 80 mi in the same time that a light plane can travel 280 mi. If the speed of the plane is 100 mi/h faster than that of the train, find each of the rates.

25. **Canoeing time.** Jan and Tariq took a canoeing trip, traveling 6 mi upstream against a 2 mi/h current. They then returned to the same point downstream. If their entire trip took 4 h, how fast can they paddle in still water? [*Hint:* If r is their rate (in miles per hour) in still water, their rate upstream is $r - 2$ and their rate downstream is $r + 2$.].

26. **Flying speed.** A plane flies 720 mi against a steady 30 mi/h headwind and then returns to the same point with the wind. If the entire trip takes 10 h, what is the plane's speed in still air?

27. **Alcohol solution.** How much pure alcohol must be added to 40 ounces (oz) of a 25% solution to produce a mixture that is 40% alcohol?

28. **Mixtures.** How many centiliters (cL) of pure acid must be added to 200 cL of a 40% acid solution to produce a 50% solution?

29. **Speed conversion.** A speed of 60 miles per hour (mi/h) corresponds to 88 feet per second (ft/s). If a light plane's speed is 150 mi/h, what is its speed in feet per second?

30. **Cost.** If 342 cups of coffee can be made from 9 pounds (lb) of coffee, how many cups can be made from 6 lb of coffee?

31. **Fuel consumption.** A car uses 5 gallons (gal) of gasoline on a trip of 160 mi. At the same mileage rate, how much gasoline will a 384-mi trip require?

32. **Fuel consumption.** A car uses 12 liters (L) of gasoline in traveling 150 kilometers (km). At that rate, how many liters of gasoline will be used in a trip of 400 km?

33. **Yearly earnings.** Sveta earns $13,500 commission in 20 weeks in her new sales position. At that rate, how much will she earn in 1 year (52 weeks)?

34. **Investment earning.** Kevin earned $165 interest for 1 year on an investment of $1500. At the same rate, what amount of interest would be earned by an investment of $2500?

35. **Insect control.** A company is selling a natural insect control that mixes ladybug beetles and praying mantises in the ratio of 7 to 4. If there are a total of 110 insects per package, how many of each type of insect is in a package?

ANSWERS

25. 4 mi/h
26. 150 mi/h
27. 10 oz
28. 40 cL
29. 220 ft/s
30. 228 cups
31. 12 gal
32. 32 L
33. $35,100
34. $275
35. $x = 70$ ladybugs, $110 - x = 40$ praying mantises

ANSWERS

36. 6 ft

37. Brother $4800, Sister $7200

38. $3240.96

36. Individual height. A woman casts a shadow of 4 ft. At the same time, a 72-ft building casts a shadow of 48 ft. How tall is the woman?

37. Consumer affairs. A brother and sister are to divide an inheritance of $12,000 in the ratio of 2 to 3. What amount will each receive?

38. Taxes. In Bucks County, the property tax rate is $25.32 per $1000 of assessed value. If a house and property have a value of $128,000, find the tax the owner will have to pay.

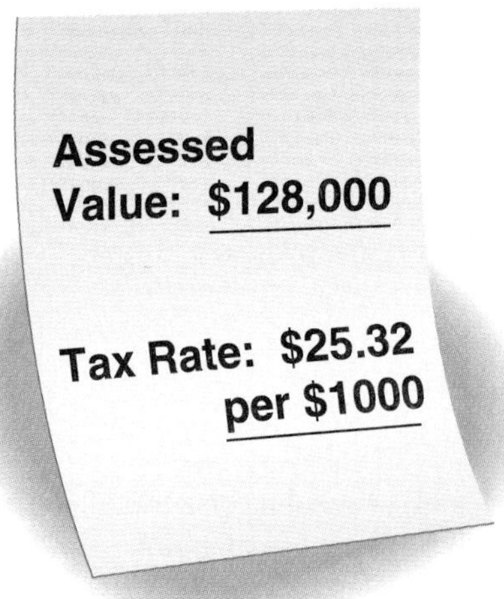

Answers

1. 30 **3.** 20 **5.** 15, 16 **7.** 6, 12 **9.** 3, 12 **11.** 7, 35 **13.** 5 **15.** 20 mi/h bicycling, 50 mi/h driving **17.** Express 55 mi/h, local 45 mi/h **19.** Freight 40 mi/h, passenger 65 mi/h **21.** 5 h, 3 h **23.** 5 h, 2 h **25.** 4 mi/h **27.** 10 oz **29.** 220 ft/s **31.** 12 gal **33.** $35,100 **35.** 70 ladybugs, 40 praying mantises **37.** Brother $4800, sister $7200

5 Summary

DEFINITION/PROCEDURE	EXAMPLE	REFERENCE
Simplifying Algebraic Fractions		**Section 5.1**
Algebraic Fractions These have the form Numerator → P, Fraction bar, Denominator → Q: $\dfrac{P}{Q}$ in which P and Q are polynomials and Q cannot have the value 0.	$\dfrac{x^2 - 3x}{x - 2}$ is an algebraic fraction. The variable x cannot have the value 2.	**p. 395**
Writing in Simplest Form A fraction is in simplest form if its numerator and denominator have no common factors other than 1. To write in simplest form: **1.** Factor the numerator and denominator. **2.** Divide the numerator and denominator by all common factors. The resulting fraction will be in simplest form.	$\dfrac{x + 2}{x - 1}$ is in simplest form. $\dfrac{x^2 - 4}{x^2 - 2x - 8}$ $= \dfrac{(x - 2)(x + 2)}{(x - 4)(x + 2)}$ $= \dfrac{(x - 2)\overset{1}{\cancel{(x + 2)}}}{(x - 4)\underset{1}{\cancel{(x + 2)}}}$ $= \dfrac{x - 2}{x - 4}$	**p. 395**
Adding and Subtracting Like Fractions		**Section 5.2**
Like Fractions **1.** Add or subtract the numerators. **2.** Write the sum or difference over the common denominator. **3.** Write the resulting fraction in simplest form.	$\dfrac{2x}{x^2 + 3x} + \dfrac{6}{x^2 + 3x}$ $= \dfrac{2x + 6}{x^2 + 3x}$ $= \dfrac{2\overset{1}{\cancel{(x + 3)}}}{x\underset{1}{\cancel{(x + 3)}}} = \dfrac{2}{x}$	**p. 405**
Adding and Subtracting Unlike Fractions		**Section 5.3**
The Lowest Common Denominator Finding the LCD: **1.** Factor each denominator. **2.** Write each factor the greatest number of times it appears in any single denominator. **3.** The LCD is the product of the factors found in step 2.	For $\dfrac{2}{x^2 + 2x + 1}$ and $\dfrac{3}{x^2 + x}$ Factor: $x^2 + 2x + 1 = (x + 1)(x + 1)$ $x^2 + x = x(x + 1)$ The LCD is $x(x + 1)(x + 1)$	**p. 411**

Continued

DEFINITION/PROCEDURE	EXAMPLE	REFERENCE
Unlike Fractions To add or subtract unlike fractions: 1. Find the LCD. 2. Convert each fraction to an equivalent fraction with the LCD as a common denominator. 3. Add or subtract the like fractions formed. 4. Write the sum or difference in simplest form.	$\frac{2}{x^2+2x+1} - \frac{3}{x^2+x}$ $= \frac{2x}{x(x+1)(x+1)}$ $- \frac{3(x+1)}{x(x+1)(x+1)}$ $= \frac{2x-3x-3}{x(x+1)(x+1)}$ $= \frac{-x-3}{x(x+1)(x+1)}$	p. 412
Multiplying and Dividing Algebraic Fractions		**Section 5.4**
Multiplying Fractions $\frac{P}{Q} \cdot \frac{R}{S} = \frac{PR}{QS}$ in which $Q \neq 0$ and $S \neq 0$.	$\frac{2}{3} \cdot \frac{4}{5} = \frac{2 \cdot 4}{3 \cdot 5} = \frac{8}{15}$	p. 423
Multiplying Algebraic Fractions 1. Factor the numerators and denominators. 2. Divide the numerator and denominator by any common factors. 3. Write the product of the remaining factors in the numerator over the product of the remaining factors in the denominator.	$\frac{2x-4}{x^2-4} \cdot \frac{x^2+2x}{6x+18}$ $= \frac{2(x-2)}{(x-2)(x+2)} \cdot \frac{x(x+2)}{6(x+3)}$ $= \frac{2\cancel{(x-2)}}{\cancel{(x-2)}\cancel{(x+2)}} \cdot \frac{x\cancel{(x+2)}}{6(x+3)}$ $= \frac{x}{3(x+3)}$	p. 423
Dividing Fractions $\frac{P}{Q} \div \frac{R}{S} = \frac{P}{Q} \cdot \frac{S}{R}$ in which $Q \neq 0$, $R \neq 0$, and $S \neq 0$. In words, invert the divisor (the second fraction) and multiply.	$\frac{4}{9} \div \frac{8}{12}$ $= \frac{4}{9} \cdot \frac{12}{8} = \frac{2}{3}$ $\frac{3x}{2x-6} \div \frac{9x^2}{x^2-9}$ $= \frac{3x}{2x-6} \cdot \frac{x^2-9}{9x^2}$ $= \frac{3x}{2\underset{1}{\cancel{(x-3)}}} \cdot \frac{(x+3)\overset{1}{\cancel{(x-3)}}}{9x^2}$ $= \frac{x+3}{6x}$	p. 424
Simplifying Complex Fractions $\frac{\frac{a}{b}}{\frac{c}{d}} = \frac{a}{b} \div \frac{c}{d} = \frac{a}{b} \cdot \frac{d}{c}$	$\frac{\frac{3}{8}}{\frac{5}{6}} = \frac{3}{8} \div \frac{5}{6}$ $= \frac{3}{\underset{4}{\cancel{8}}} \cdot \frac{\overset{3}{\cancel{6}}}{5}$ $= \frac{9}{20}$	p. 427

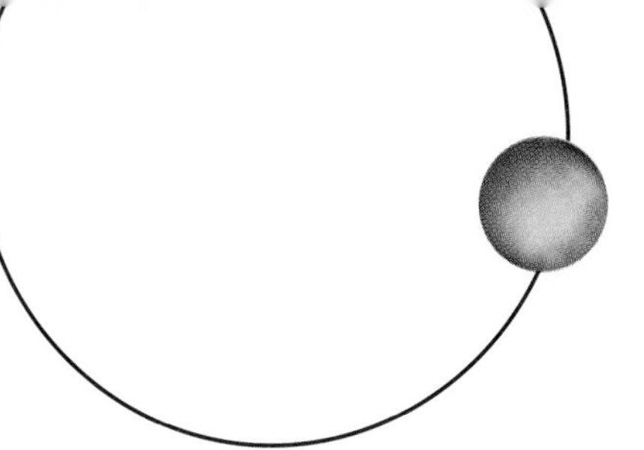

Summary Exercises

This summary exercise set is provided to give you practice with each of the objectives of the chapter. Each exercise is keyed to the appropriate chapter section. The answers are provided in the *Instructor's Manual.*

[5.1] Write each fraction in simplest form.

1. $\frac{6a^2}{9a^3}$ $\frac{2}{3a}$

2. $\frac{-12x^4y^3}{18x^2y^2}$ $\frac{-2x^2y}{3}$

3. $\frac{w^2 - 25}{2w - 8}$ $\frac{w^2 - 25}{2w - 8}$

4. $\frac{3x^2 + 11x - 4}{2x^2 + 11x + 12}$ $\frac{3x - 1}{2x + 3}$

5. $\frac{m^2 - 2m - 3}{9 - m^2}$ $\frac{-m - 1}{m + 3}$

6. $\frac{3c^2 - 2cd - d^2}{6c^2 + 2cd}$ $\frac{c - d}{2c}$

[5.2] Add or subtract as indicated.

7. $\frac{x}{9} + \frac{2x}{9}$ $\frac{x}{3}$

8. $\frac{7a}{15} - \frac{2a}{15}$ $\frac{a}{3}$

9. $\frac{8}{x + 2} + \frac{3}{x + 2}$ $\frac{11}{x + 2}$

10. $\frac{y - 2}{5} - \frac{2y + 3}{5}$ $\frac{-y - 5}{5}$

11. $\frac{7r - 3s}{4r} + \frac{r - s}{4r}$ $\frac{2r - s}{r}$

12. $\frac{x^2}{x - 4} - \frac{16}{x - 4}$ $x + 4$

13. $\frac{5w - 6}{w - 4} - \frac{3w + 2}{w - 4}$ 2

14. $\frac{x + 3}{x^2 - 2x - 8} + \frac{2x + 3}{x^2 - 2x - 8}$ $\frac{3}{x - 4}$

[5.3] Add or subtract as indicated.

15. $\frac{5x}{6} + \frac{x}{3}$ $\frac{7x}{6}$

16. $\frac{3y}{10} - \frac{2y}{5}$ $\frac{-y}{10}$

17. $\frac{5}{2m} - \frac{3}{m^2}$ $\frac{5m - 6}{2m^2}$

18. $\frac{x}{x - 3} - \frac{2}{3}$ $\frac{x + 6}{3(x - 3)}$

19. $\frac{4}{x - 3} - \frac{1}{x}$ $\frac{3x + 3}{x(x - 3)}$

20. $\frac{2}{s + 5} + \frac{3}{s + 1}$ $\frac{5s + 17}{(s + 5)(s + 1)}$

21. $\frac{5}{w-5} - \frac{2}{w-3}$ $\frac{3w-5}{(w-5)(w-3)}$

22. $\frac{4x}{2x-1} + \frac{2}{1-2x}$ 2

23. $\frac{2}{3x-3} - \frac{5}{2x-2}$ $\frac{-11}{6(x-1)}$

24. $\frac{4y}{y^2-8y+15} + \frac{6}{y-3}$ $\frac{10}{y-5}$

25. $\frac{3a}{a^2+5a+4} + \frac{2a}{a^2-1}$ $\frac{5a}{(a+4)(a-1)}$

26. $\frac{3x}{x^2+2x-8} - \frac{1}{x-2} + \frac{1}{x+4}$ $\frac{3}{x+4}$

[5.4] Multiply or divide as indicated.

27. $\frac{6x}{5} \cdot \frac{10}{18x^2}$ $\frac{2}{3x}$

28. $\frac{-2a^2}{ab^3} \cdot \frac{3ab^2}{-4ab}$ $\frac{3a}{2b^2}$

29. $\frac{2x+6}{x^2-9} \cdot \frac{x^2-3x}{4}$ $\frac{x}{2}$

30. $\frac{a^2+5a+4}{2a^2+2a} \cdot \frac{a^2-a-12}{a^2-16}$ $\frac{a+3}{2a}$

31. $\frac{3p}{5} \div \frac{9p^2}{10}$ $\frac{2}{3p}$

32. $\frac{8m^3}{5mn} \div \frac{12m^2n^2}{15mn^3}$ $2m$

33. $\frac{x^2+7x+10}{x^2+5x} \div \frac{x^2-4}{2x^2-7x+6}$ $\frac{2x-3}{x}$

34. $\frac{2w^2+11w-21}{w^2-49} \div (4w-6)$ $\frac{1}{2(w-7)}$

35. $\frac{a^2b+2ab^2}{a^2-4b^2} \div \frac{4a^2b}{a^2-ab-2b^2}$ $\frac{a+b}{4a}$

36. $\frac{2x^2+6x}{4x} \cdot \frac{6x+12}{x^2+2x-3} \div \frac{x^2-4}{x^2-3x+2}$ 3

[5.4] Simplify the complex fractions.

37. $\dfrac{\frac{x^2}{12}}{\frac{x^3}{8}}$ $\frac{2}{3x}$

38. $\dfrac{3+\frac{1}{a}}{3-\frac{1}{a}}$ $\frac{3a+1}{3a-1}$

39. $\dfrac{1+\frac{x}{y}}{1-\frac{x}{y}}$ $\frac{y+x}{y-x}$

40. $\dfrac{1+\frac{1}{p}}{p^2-1}$ $\frac{1}{p(p-1)}$

41. $\dfrac{\frac{1}{m}-\frac{1}{n}}{\frac{1}{m}+\frac{1}{n}}$ $\frac{n-m}{n+m}$

42. $\dfrac{2-\frac{x}{y}}{4-\frac{x^2}{y^2}}$ $\frac{y}{2y+x}$

43. $\dfrac{\frac{2}{a+1}+1}{1-\frac{4}{a+1}}$ $\frac{a+3}{a-3}$

44. $\dfrac{\frac{a}{b}-1-\frac{2b}{a}}{\frac{1}{b^2}-\frac{1}{a^2}}$ $\frac{ab(a-2b)}{a-b}$

[5.5] What values for x, if any, must be excluded in the following algebraic fractions?

45. $\frac{x}{5}$ None

46. $\frac{3}{x - 4}$ 4

47. $\frac{2}{(x + 1)(x - 2)}$ −1, 2

48. $\frac{7}{x^2 - 16}$ −4, 4

49. $\frac{x - 1}{x^2 + 3x + 2}$ −1, −2

50. $\frac{2x + 3}{3x^2 + x - 2}$ $-1, \frac{2}{3}$

[5.5] Solve the following proportions for x.

51. $\frac{x - 3}{8} = \frac{x - 2}{10}$ 7

52. $\frac{1}{x - 3} = \frac{7}{x^2 - x - 6}$ 5

[5.5] Solve the following equations for x.

53. $\frac{x}{4} - \frac{x}{5} = 2$ 40

54. $\frac{13}{4x} + \frac{3}{x^2} = \frac{5}{2x}$ −4

55. $\frac{x}{x - 2} + 1 = \frac{x + 4}{x - 2}$ 6

56. $\frac{x}{x - 4} - 3 = \frac{4}{x - 4}$ No solution

57. $\frac{x}{2x - 6} - \frac{x - 4}{x - 3} = \frac{1}{8}$ 7

58. $\frac{7}{x} - \frac{1}{x - 3} = \frac{9}{x^2 - 3x}$ 5

59. $\frac{x}{x - 5} = \frac{3x}{x^2 - 7x + 10} + \frac{8}{x - 2}$ 8

60. $\frac{6}{x + 5} + 1 = \frac{3}{x - 5}$ −10, 7

61. $\frac{24}{x + 2} - 2 = \frac{2}{x - 3}$ 4, 8

[5.6] Solve the following applications.

62. Number problem. If two-fifths of a number is added to one-half of that number, the sum is 27. Find the number. **30**

63. Number problem. One number is 3 times another. If the sum of their reciprocals is $\frac{1}{3}$, what are the two numbers? **4, 12**

64. Reciprocals. If the reciprocal of 4 times a number is subtracted from the reciprocal of that number, the result is $\frac{1}{8}$. What is the number? **6**

65. Driving speed. Robert made a trip of 240 miles (mi). Returning by a different route, he found that the distance was only 200 mi, but traffic slowed his speed by 8 miles per hour (mi/h). If the trip took the same time in both directions, what was Robert's rate each way? **48 mi/h, 40 mi/h**

66. Distance. On the first day of a vacation trip, Jovita drove 225 mi. On the second day it took her 1 h longer to drive 270 mi. If her average speed was the same both days, how long did she drive each day? **5 h, 6 h**

67. Plane speed. A light plane flies 700 mi against a steady 20 mi/h headwind and then returns, with the wind, to the same point. If the entire trip took 12 h, what was the speed of the plane in still air? **120 mi/h**

68. Solutions. How much pure alcohol should be added to 300 milliliters (mL) of a 30% solution to obtain a 40% solution? **50 mL**

69. Solutions. A chemist has a 10% acid solution and a 40% solution. How much of the 40% solution should be added to 300 mL of the 10% solution to produce a mixture with a concentration of 20%? **150 mL**

70. Investments. Melina wants to invest a total of $10,800 in two types of savings accounts. If she wants the ratio of the amounts deposited in the two accounts to be 4 to 5, what amount should she invest in each account? **$4800, $6000**

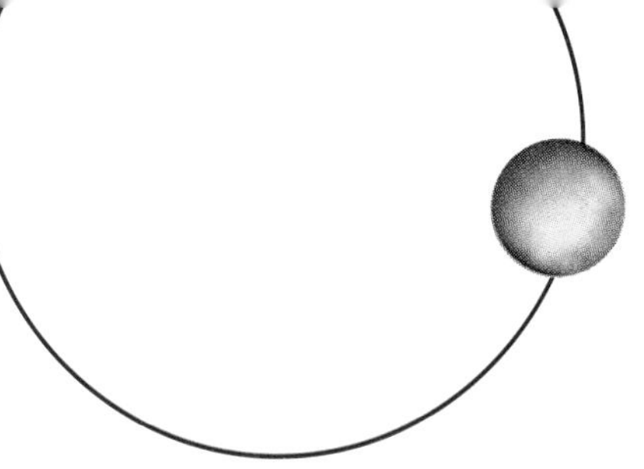

Self-Test for Chapter 5

Name ______________

Section ________ Date ________

The purpose of this self-test is to help you check your progress and to review for a chapter test in class. Allow yourself about an hour to take the test. When you are done, check your answers in the back of the book. If you missed any answers, be sure to go back and review the appropriate sections in the chapter and the exercises that are provided.

Write each fraction in simplest form.

1. $\dfrac{-21x^5y^3}{28xy^5}$ **2.** $\dfrac{4a - 24}{a^2 - 6a}$ **3.** $\dfrac{3x^2 + x - 2}{3x^2 - 8x + 4}$

Add or subtract as indicated.

4. $\dfrac{3a}{8} + \dfrac{5a}{8}$ **5.** $\dfrac{2x}{x + 3} + \dfrac{6}{x + 3}$ **6.** $\dfrac{7x - 3}{x - 2} - \dfrac{2x + 7}{x - 2}$

7. $\dfrac{x}{3} + \dfrac{4x}{5}$ **8.** $\dfrac{3}{s} - \dfrac{2}{s^2}$

9. $\dfrac{5}{x - 2} - \dfrac{1}{x + 3}$ **10.** $\dfrac{6}{w - 2} + \dfrac{9w}{w^2 - 7w + 10}$

Multiply or divide as indicated.

11. $\dfrac{3pq^2}{5pq^3} \cdot \dfrac{20p^2q}{21q}$ **12.** $\dfrac{x^2 - 3x}{5x^2} \cdot \dfrac{10x}{x^2 - 4x + 3}$

13. $\dfrac{2x^2}{3xy} \div \dfrac{8x^2y}{9xy}$ **14.** $\dfrac{3m - 9}{m^2 - 2m} \div \dfrac{m^2 - m - 6}{m^2 - 4}$

Simplify the complex fractions.

15. $\dfrac{\dfrac{x^2}{18}}{\dfrac{x^3}{12}}$ **16.** $\dfrac{2 - \dfrac{m}{n}}{4 - \dfrac{m^2}{n^2}}$

What values for x, if any, must be excluded in the following algebraic fractions?

17. $\dfrac{8}{x - 4}$ **18.** $\dfrac{3}{x^2 - 9}$

Solve the following equations for x.

19. $\dfrac{x}{3} - \dfrac{x}{4} = 3$ **20.** $\dfrac{5}{x} - \dfrac{x - 3}{x + 2} = \dfrac{22}{x^2 + 2x}$

ANSWERS

1. $\dfrac{-3x^4}{4y^2}$
2. $\dfrac{4}{a}$
3. $\dfrac{x + 1}{x - 2}$
4. a
5. 2
6. 5
7. $\dfrac{17x}{15}$
8. $\dfrac{3s - 2}{s^2}$
9. $\dfrac{4x + 17}{(x - 2)(x + 3)}$
10. $\dfrac{15}{w - 5}$
11. $\dfrac{4p^2}{7g}$
12. $\dfrac{2}{x - 1}$
13. $\dfrac{3}{4y}$
14. $\dfrac{3}{m}$
15. $\dfrac{2}{3x}$
16. $\dfrac{n}{2n + m}$
17. 4
18. −3, 3
19. 36
20. 2, 6

ANSWERS

21. 6

22. 4

23. 4, 12

24. 50 mi/h, 45 mi/h

25. 20 ft, 35 ft

Solve the following proportions.

21. $\frac{x - 1}{5} = \frac{x + 2}{8}$

22. $\frac{2x - 1}{7} = \frac{x}{4}$

Solve the following applications.

23. Number problem. One number is 3 times another. If the sum of their reciprocals is $\frac{1}{3}$, find the two numbers.

24. Driving rate. Mark drove 250 miles (mi) to visit Sandra. Returning by a shorter route, he found that the trip was only 225 mi, but traffic slowed his speed by 5 miles per hour (mi/h). If the two trips took exactly the same time, what was his rate each way?

25. Cable length. A cable that is 55 feet (ft) long is to be cut into two pieces whose lengths have the ratio 4 to 7. Find the lengths of the two pieces.

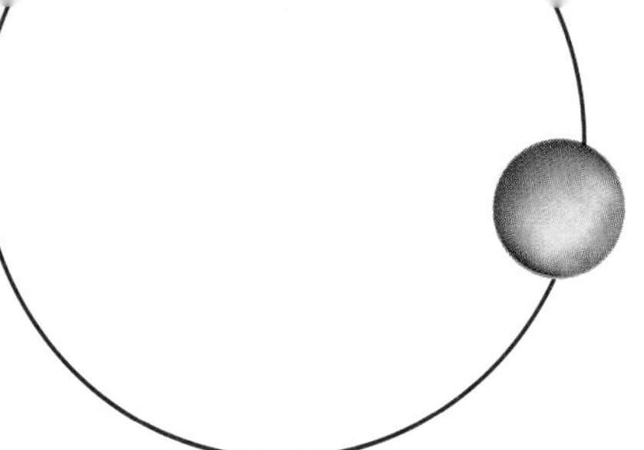

Cumulative Test for Chapters 0 to 5

Name ____________

Section ______ Date ______

The test covers selected topics from the first six chapters.

Perform the indicated operation.

1. $x^2y - 4xy - x^2y + 2xy$
2. $\dfrac{12a^3b}{9ab}$
3. $(5x^2 - 2x + 1) - (3x^2 + 3x - 5)$
4. $(5a^2 + 6a) - (2a^2 - 1)$

Evaluate the expression.

5. $4 + 3(7 - 4)^2$
6. $|3 - 5| - |-4 + 3|$

Multiply.

7. $(x - 2y)(2x + 3y)$
8. $(x + 7)(x + 4)$

Divide.

9. $(2x^2 + 3x - 1) \div (x + 2)$
10. $(x^2 - 5) \div (x - 1)$

Solve each equation and check your results.

11. $4x - 3 = 2x + 5$
12. $2 - 3(2x + 1) = 11$

Factor each polynomial completely.

13. $x^2 - 5x - 14$
14. $3m^2n - 6mn^2 + 9mn$
15. $a^2 - 9b^2$
16. $2x^3 - 28x^2 + 96x$

Solve the following word problems. Show the equation used for each solution.

17. **Number problem.** 2 more than 4 times a number is 30. Find the number.

18. **Number problem.** If the reciprocal of 4 times a number is subtracted from the reciprocal of that number, the result is $\frac{3}{16}$. What is the number?

19. **Speed.** A speed of 60 mi/h corresponds to 88 ft/s. If a race car is traveling at 180 mi/h, what is its speed in feet per second?

20. **Rectangle dimensions.** The length of a rectangle is 3 inches (in.) less than twice its width. If the area of the rectangle is 35 square inches (in.^{2}), find the dimensions of the rectangle.

ANSWERS

1. $-2xy$
2. $\dfrac{4a^2}{3}$
3. $2x^2 - 5x + 6$
4. $3a^2 + 6a + 1$
5. 31
6. 1
7. $2x^2 - xy - 6y^2$
8. $x^2 + 11x + 28$
9. $2x - 1 + \dfrac{1}{x + 2}$
10. $x + 1 - \dfrac{4}{x - 1}$
11. 4
12. -2
13. $(x - 7)(x + 2)$
14. $3mn(m - 2n + 3)$
15. $(a + 3b)(a - 3b)$
16. $2x(x - 6)(x - 8)$
17. 7
18. 4
19. 264 ft/s
20. 5 in. by 7 in.

ANSWERS

21. $\frac{m}{3}$

22. $\frac{a - 7}{3a + 1}$

23. $\frac{8r + 3}{6r^2}$

24. $\frac{x + 33}{3(x - 3)(x + 3)}$

25. $\frac{3}{x}$

26. $\frac{1}{3w}$

27. $\frac{x - 1}{2x + 1}$

28. $\frac{n}{3n + m}$

29. $\frac{6}{5}$

30. $\frac{-9}{2}, 7$

Write each fraction in simplest form.

21. $\dfrac{m^2 - 4m}{3m - 12}$

22. $\dfrac{a^2 - 49}{3a^2 + 22a + 7}$

Perform the indicated operations.

23. $\dfrac{4}{3r} + \dfrac{1}{2r^2}$

24. $\dfrac{2}{x - 3} - \dfrac{5}{3x + 9}$

25. $\dfrac{3x^2 + 9x}{x^2 - 9} \cdot \dfrac{2x^2 - 9x + 9}{2x^3 - 3x^2}$

26. $\dfrac{4w^2 - 25}{2w^2 - 5w} \div (6w + 15)$

Simplify the complex fractions.

27. $\dfrac{1 - \dfrac{1}{x}}{2 + \dfrac{1}{x}}$

28. $\dfrac{3 - \dfrac{m}{n}}{9 - \dfrac{m^2}{n^2}}$

Solve the following equations for x.

29. $\dfrac{5}{3x} + \dfrac{1}{x^2} = \dfrac{5}{2x}$

30. $\dfrac{10}{x - 3} - 2 = \dfrac{5}{x + 3}$

AN INTRODUCTION TO GRAPHING

6

INTRODUCTION

Graphs are used to discern patterns and trends that may be difficult to see when looking at a list of numbers or other kinds of data. The word *graph* comes from Latin and Greek roots and means "to draw a picture." This is just what a graph does in mathematics: It draws a picture of a relationship between two or more variables. But, as in art, these graphs can be difficult to interpret without a little practice and training. This chapter is the beginning of that training. And the training is important because graphs are used in every field in which numbers are used.

In the field of pediatric medicine, there has been controversy about the use of somatotropin (human growth hormone) to help children whose growth has been impeded by various health problems. The reason for the controversy is that many doctors are giving this expensive drug therapy to children who are simply shorter than average or shorter than their parents want them to be. The question of which children are not growing normally because of some serious health defect and need the therapy and which children are healthy and simply small of stature and thus should not be subjected to this treatment has been vigorously argued by professionals here and in Europe, where the therapy is being used.

Some of the measures used to distinguish between the two groups are blood tests and age and height measurements. The age and height measurements are graphed and monitored over several years of a child's life to monitor the rate of growth. If during a certain period the child's rate of growth slows to below 4.5 centimeters per year, this indicates that something may be seriously wrong. The graph can also indicate if the child's size fits within a range considered normal at each age of the child's life.

Name ____________

Section ________ Date ________

Pre-Test Chapter 6

ANSWERS

1. (15, 3); (18, 6)
2. (0, 3); (2, 0)
3. (1, 3); (0, 5); (−3, 11)
4. Answers vary
5. A(2, 3), B(0, −5), C(−3, 5)
6. See exercise
7. See exercise
8. See exercise
9. See exercise
10. 1
11. −1
12. 8
13. 4
14. $-\frac{2}{5}$
15. $1000

Determine which of the ordered pairs are solutions for the given equations.

1. $x - y = 12$ (15, 3), (9, 6), (18, 6)

2. $3x + 2y = 6$ (1, 2), (0, 3), (2, 0)

3. Complete the ordered pairs so that each is a solution for the given equation.

$2x + y = 5$ (1,), (0,), (, 11)

4. Find three solutions for each of the following equations.

$2x - 3y = 8$ $6x + y = 11$

5. Give the coordinates of the points graphed below

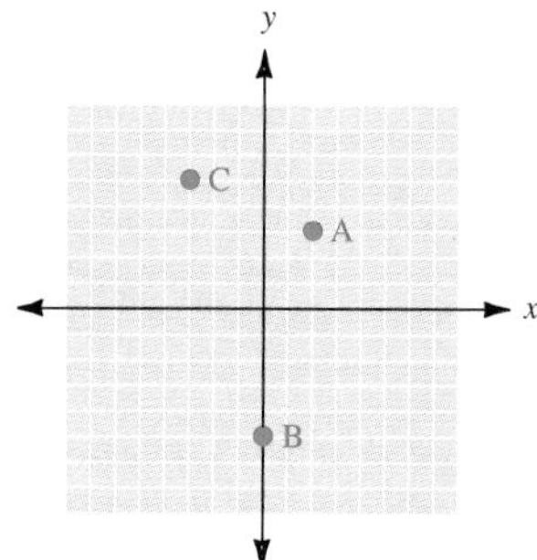

6. Plot the points with the given . coordinates. S(−1, 2), T(3, 0)

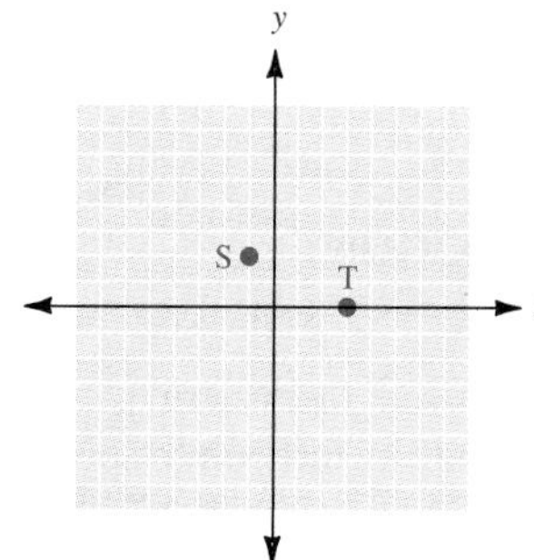

Graph each of the following equations.

7. $x + y = 5$

8. $y = \frac{1}{2}x - 1$

9. $y = -2$

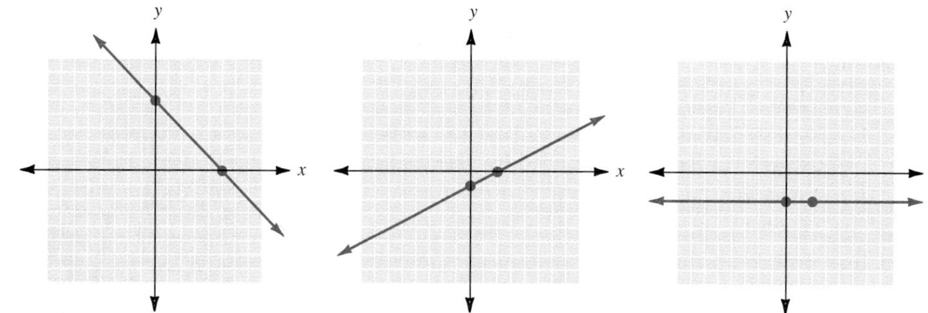

Find the slope of the line through the following pairs of points.

10. (−2,5) and (5,12)

11. (−1,−3) and (7,−11)

12. Find the constant of variation, k, if y varies directly with x and $y = 56$ when $x = 7$.

Find the slope of the line for the following equations.

13. $y = 4x - 5$

14. $y = -\frac{2}{5}x + 7$

15. Pete's commission varies directly as the number of appliances he sells. If his commission last month was $800 and he sold 20 appliances, what would his salary be if he sold 25 appliances?

6.1 Solutions of Equations in Two Variables

OBJECTIVES

1. Find solutions for an equation in two variables
2. Use ordered pair notation to write solutions for equations in two variables

We discussed finding solutions for equations in Chapter 2. Recall that a solution is a value for the variable that "satisfies" the equation, or makes the equation a true statement. For instance, we know that 4 is a solution of the equation

$$2x + 5 = 13$$

We know this is true because, when we replace x with 4, we have

$$2 \cdot 4 + 5 \stackrel{?}{=} 13$$

$$8 + 5 \stackrel{?}{=} 13$$

$$13 = 13 \quad \text{A true statement}$$

NOTE Recall that an equation is two expressions connected by an equal sign.

We now want to consider **equations in two variables.** An example is

$$x + y = 5$$

What will the solution look like? It is not going to be a single number, because there are two variables. Here the solution will be a pair of numbers—one value for each of the variables, x and y. Suppose that x has the value 3. In the equation $x + y = 5$, you can substitute 3 for x.

$$3 + y = 5$$

Solving for y gives

$$y = 2$$

NOTE An equation in two variables "pairs" two numbers, one for x and one for y.

So the pair of values $x = 3$ and $y = 2$ satisfies the equation because

$$3 + 2 = 5$$

That pair of numbers is then a *solution* for the equation in two variables.

How many such pairs are there? Choose any value for x (or for y). You can always find the other *paired* or *corresponding* value in an equation of this form. We say that there are an *infinite* number of pairs that will satisfy the equation. Each of these pairs is a solution. We will find some other solutions for the equation $x + y = 5$ in the following example.

Example 1

Solving for Corresponding Values

For the equation $x + y = 5$, find (a) y if $x = 5$ and (b) x if $y = 4$.

(a) If $x = 5$

$5 + y = 5 \qquad \text{or} \qquad y = 0$

(b) If $y = 4$,

$x + 4 = 5$ or $x = 1$

So the pairs $x = 5$, $y = 0$ and $x = 1$, $y = 4$ are both solutions.

CHECK YOURSELF 1

For the equation $2x + 3y = 26$,

(a) If $x = 4$, $y = ?$ **(b)** If $y = 0$, $x = ?$

To simplify writing the pairs that satisfy an equation, we use the **ordered-pair notation.** The numbers are written in parentheses and are separated by a comma. For example, we know that the values $x = 3$ and $y = 2$ satisfy the equation $x + y = 5$. So we write the pair as

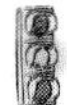

CAUTION

(3, 2) means $x = 3$ and $y = 2$.
(2, 3) means $x = 2$ and $y = 3$.
(3, 2) and (2, 3) are entirely different. That's why we call them *ordered pairs.*

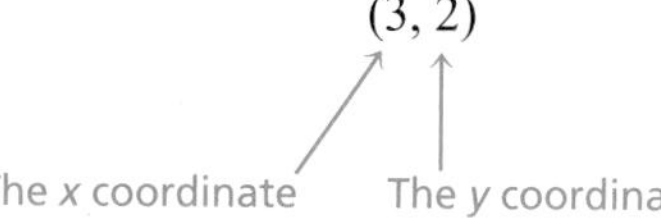

The first number of the pair is *always* the value for x and is called the ***x* coordinate.** The second number of the pair is *always* the value for y and is the ***y* coordinate.**

Using this ordered-pair notation, we can say that (3, 2), (5, 0), and (1, 4) are all *solutions* for the equation $x + y = 5$. Each pair gives values for x and y that will satisfy the equation.

Example 2

Identifying Solutions of Two-Variable Equations

Which of the ordered pairs (a) (2, 5), (b) (5, −1), and (c) (3, 4) are solutions for the equation $2x + y = 9$?

(a) To check whether (2, 5) is a solution, let $x = 2$ and $y = 5$ and see if the equation is satisfied.

$2x + y = 9$ The original equation.

$2 \cdot 2 + 5 \stackrel{?}{=} 9$ Substitute 2 for x and 5 for y.

$4 + 5 \stackrel{?}{=} 9$

$9 = 9$ A true statement

NOTE (2, 5) is a solution because a *true statement* results.

(2, 5) is a solution for the equation.

(b) For $(5, -1)$, let $x = 5$ and $y = -1$.

$$2 \cdot 5 - 1 \stackrel{?}{=} 9$$
$$10 - 1 \stackrel{?}{=} 9$$
$$9 = 9 \qquad \text{A true statement}$$

So $(5, -1)$ is a solution.

(c) For $(3, 4)$, let $x = 3$ and $y = 4$. Then

$$2 \cdot 3 + 4 \stackrel{?}{=} 9$$
$$6 + 4 \stackrel{?}{=} 9$$
$$10 = 9 \qquad \textit{Not} \text{ a true statement}$$

So $(3, 4)$ is *not* a solution for the equation.

CHECK YOURSELF 2

Which of the ordered pairs (3, 4), (4, 3), (1, −2), *and* (0, −5) *are solutions for the following equation?*

$$3x - y = 5$$

If the equation contains only one variable, then the missing variable can take on any value.

Example 3

Identifying Solutions of One-Variable Equations

Which of the ordered pairs, (2, 0), (0, 2), (5, 2), (2, 5), and (2, −1) are solutions for the equation $x = 2$?

A solution is any ordered pair in which the x coordinate is 2. That makes (2, 0), (2, 5), and (2, −1) solutions for the given equation.

CHECK YOURSELF 3

Which of the ordered pairs (3, 0), (0, 3), (3, 3), (−1, 3), *and* (3, −1) *are solutions for the equation* $y = 3$?

Remember that, when an ordered pair is presented, the first number is always the x coordinate and the second number is always the y coordinate.

Example 4

Completing Ordered Pair Solutions

Complete the ordered pairs (a) (9,), (b) (, −1), (c) (0,), and (d) (, 0) for the equation $x - 3y = 6$.

NOTE The x coordinate is sometimes called the **abscissa** and the y coordinate the **ordinate.**

(a) The first number, 9, appearing in (9,) represents the x value. To complete the pair (9,), substitute 9 for x and then solve for y.

$$9 - 3y = 6$$
$$-3y = -3$$
$$y = 1$$

(9, 1) is a solution.

(b) To complete the pair (, −1), let y be −1 and solve for x.

$$x - 3(-1) = 6$$
$$x + 3 = 6$$
$$x = 3$$

(3, −1) is a solution.

(c) To complete the pair (0,), let x be 0.

$$0 - 3y = 6$$
$$-3y = 6$$
$$y = -2$$

(0, −2) is a solution.

(d) To complete the pair (, 0), let y be 0.

$$x - 3 \cdot 0 = 6$$
$$x - 0 = 6$$
$$x = 6$$

(6, 0) is a solution.

CHECK YOURSELF 4

Complete the ordered pairs below so that each is a solution for the equation $2x + 5y = 10$.

(10,), (, 4), (0,), and (, 0)

Example 5

Finding Some Solutions of a Two-Variable Equation

Find four solutions for the equation

$2x + y = 8$

NOTE Generally, you'll want to pick values for *x* (or for *y*) so that the resulting equation in one variable is easy to solve.

In this case the values used to form the solutions are *up to you.* You can assign any value for x (or for y). We'll demonstrate with some possible choices.

Solution with $x = 2$:

$$2 \cdot 2 + y = 8$$
$$4 + y = 8$$
$$y = 4$$

(2, 4) is a solution.

Solution with $y = 6$:

$$2x + 6 = 8$$
$$2x = 2$$
$$x = 1$$

(1, 6) is a solution.

Solution with $x = 0$:

$$2 \cdot 0 + y = 8$$
$$y = 8$$

NOTE The solutions (0, 8) and (4, 0) will have special significance later in graphing. They are also easy to find!

(0, 8) is a solution.

Solution with $y = 0$:

$$2x + 0 = 8$$
$$2x = 8$$
$$x = 4$$

(4, 0) is a solution.

CHECK YOURSELF 5

Find four solutions for $x - 3y = 12$.

CHECK YOURSELF ANSWERS

1. (a) $y = 6$; **(b)** $x = 13$ **2.** (3, 4), (1, −2), and (0, −5) are solutions
3. (0, 3), (3, 3), and (−1, 3) are solutions **4.** (10, −2), (−5, 4), (0, 2), and (5, 0)
5. (6, −2), (3, −3), (0, −4), and (12, 0) are four possibilities

Name ______________

Section ________ Date ________

6.1 Exercises

Determine which of the ordered pairs are solutions for the given equation.

1. $x + y = 6$ (4, 2), (−2, 4), (0, 6), (−3, 9)

2. $x - y = 12$ (13, 1), (13, −1), (12, 0), (6, 6)

3. $2x - y = 8$ (5, 2), (4, 0), (0, 8), (6, 4)

4. $x + 5y = 20$ (10, −2), (10, 2), (20, 0), (25, −1)

5. $3x + y = 6$ (2, 0), (2, 3), (0, 2), (1, 3)

6. $x - 2y = 8$ (8, 0), (0, 4), (5, −1), (10, −1)

7. $2x - 3y = 6$ (0, 2), (3, 0), (6, 2), (0, −2)

8. $8x + 4y = 16$ (2, 0), (6, −8), (0, 4), (6, −6)

9. $3x - 2y = 12$ $(4, 0), \left(\frac{2}{3}, -5\right), (0, 6), \left(5, \frac{3}{2}\right)$

10. $3x + 4y = 12$ $(-4, 0), \left(\frac{2}{3}, \frac{5}{2}\right), (0, 3), \left(\frac{2}{3}, 2\right)$

11. $y = 4x$ (0, 0), (1, 3), (2, 8), (8, 2)

12. $y = 2x - 1$ $(0, -2), (0, -1), \left(\frac{1}{2}, 0\right), (3, -5)$

13. $x = 3$ (3, 5), (0, 3), (3, 0), (3, 7)

14. $y = 5$ (0, 5), (3, 5), (−2, −5), (5, 5)

Complete the ordered pairs so that each is a solution for the given equation.

15. $x + y = 12$ (4,), (, 5), (0,), (, 0)

16. $x - y = 7$ (, 4), (15,), (0,), (, 0)

17. $3x + y = 9$ (3,), (, 9), (, −3), (0,)

18. $x + 5y = 20$ (0,), (, 2), (10,), (, 0)

19. $5x - y = 15$ (, 0), (2,), (4,), (, −5)

20. $x - 3y = 9$ (0,), (12,), (, 0), (, −2)

21. $3x - 2y = 12$ (, 0), (, −6), (2,), (, 3)

ANSWERS

1. (4, 2), (0, 6), (−3, 9)
2. (13, 1), (12, 0)
3. (5, 2), (4, 0), (6, 4)
4. (10, 2), (20, 0), (25, −1)
5. (2, 0), (1, 3)
6. (8, 0)
7. (3, 0), (6, 2), (0, −2)
8. (2, 0), (6, −8), (0, 4)
9. $(4, 0), \left(\frac{2}{3}, -5\right), \left(5, \frac{3}{2}\right)$
10. $\left(\frac{2}{3}, \frac{5}{2}\right), (0, 3)$
11. (0, 0), (2, 8)
12. $(0, -1), \left(\frac{1}{2}, 0\right)$
13. (3, 5), (3, 0), (3, 7)
14. (0, 5), (3, 5), (5, 5)
15. 8, 7, 12, 12
16. 11, 8, −7, 7
17. 0, 0, 4, 9
18. 4, 10, 2, 20
19. 3, −5, 5, 2
20. −3, 1, 9, 3
21. 4, 0, −3, 6

52.

ANSWERS

a. See exercise

b. See exercise

c. See exercise

d. See exercise

e. See exercise

f. −7

g. −4

h. 4

i. 8

j. $\frac{19}{2}$

(c) The *amount of the* __________ in a restaurant is related to *the amount of the tip.*

(d) The *sales amount of a purchase in a store* determines __________.

(e) The *age of an automobile* is related to __________.

(f) The *amount of electricity you use in a month* determines __________.

(g) The *cost of food for a family of four* and __________.

Think of two more:

(h) __________.

(i) __________.

Getting Ready for Section 6.2 [Section 0.4]

Plot points with the following coordinates on the number line shown below.

(a) −3 (b) 7 (c) 0 (d) −8 (e) $\frac{3}{2}$

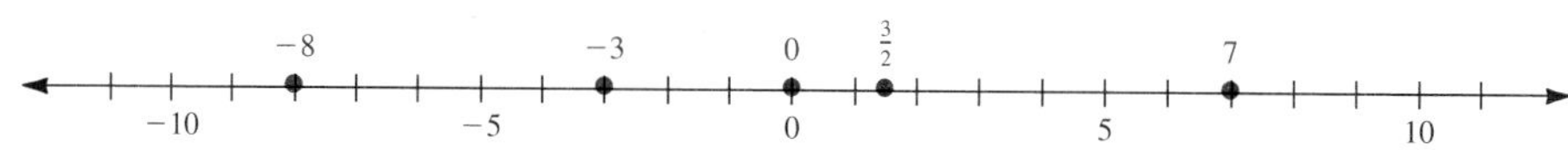

Give the coordinate of each of the following points.

(f) *A* (g) *B* (h) *C* (i) *D* (j) *E*

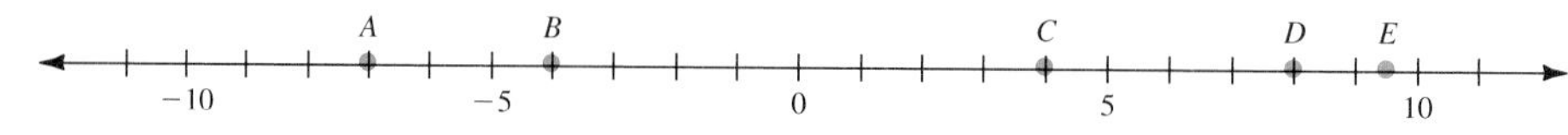

Answers

1. (4, 2), (0, 6), (−3, 9) **3.** (5, 2), (4, 0), (6, 4) **5.** (2, 0), (1, 3)
7. (3, 0), (6, 2), (0, −2) **9.** (4, 0), $\left(\frac{2}{3}, -5\right)$, $\left(5, \frac{3}{2}\right)$ **11.** (0, 0), (2, 8)
13. (3, 5), (3, 0), (3, 7) **15.** 8, 7, 12, 12 **17.** 0, 0, 4, 9 **19.** 3, −5, 5, 2
21. 4, 0, −3, 6 **23.** −3, 11, 9, 7 **25.** −4, 3, $\frac{4}{3}$, 1
27. (0, −7), (2, −5), (4, −3), (6, −1) **29.** (0, −6), (3, 0), (6, 6), (9, 12)
31. (8, 0), (−4, 3), (0, 2), (4, 1) **33.** (−5, −4), (0, −2), (5, 0), (10, 2)
35. (0, 3), (1, 5), (2, 7), (3, 9) **37.** (−5, 0), (−5, 1), (−5, 2), (−5, 3)
39. (2, −3, 1) **41.** (1, −6, 5) **43.** (−2, 5, 1)
45. \$9.50, \$11.75, \$15.50, \$19.25, \$23 **47.** 25 cm^2, 100 cm^2, 144 cm^2, 225 cm^2
49. 4527, 4689, 4851, 5013, 5337 **51.** 

a–e.

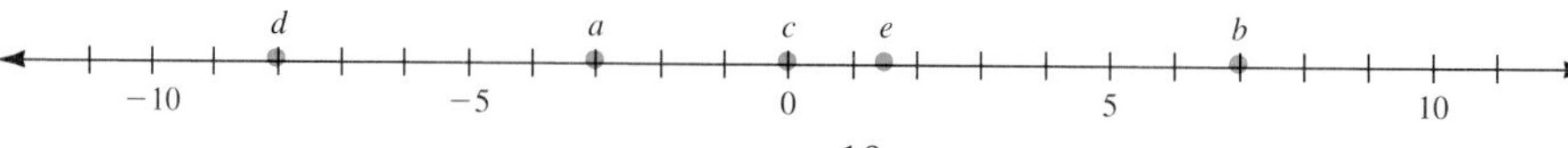

f. −7 **g.** −4 **h.** 4 **i.** 8 **j.** $\frac{19}{2}$

Name ______________________

6.1 Exercises

Section ________ Date ________

Determine which of the ordered pairs are solutions for the given equation.

1. $x + y = 6$ (4, 2), (−2, 4), (0, 6), (−3, 9)

2. $x - y = 12$ (13, 1), (13, −1), (12, 0), (6, 6)

3. $2x - y = 8$ (5, 2), (4, 0), (0, 8), (6, 4)

4. $x + 5y = 20$ (10, −2), (10, 2), (20, 0), (25, −1)

5. $3x + y = 6$ (2, 0), (2, 3), (0, 2), (1, 3)

6. $x - 2y = 8$ (8, 0), (0, 4), (5, −1), (10, −1)

7. $2x - 3y = 6$ (0, 2), (3, 0), (6, 2), (0, −2)

8. $8x + 4y = 16$ (2, 0), (6, −8), (0, 4), (6, −6)

9. $3x - 2y = 12$ $(4, 0), \left(\frac{2}{3}, -5\right), (0, 6), \left(5, \frac{3}{2}\right)$

10. $3x + 4y = 12$ $(-4, 0), \left(\frac{2}{3}, \frac{5}{2}\right), (0, 3), \left(\frac{2}{3}, 2\right)$

11. $y = 4x$ (0, 0), (1, 3), (2, 8), (8, 2)

12. $y = 2x - 1$ $(0, -2), (0, -1), \left(\frac{1}{2}, 0\right), (3, -5)$

13. $x = 3$ (3, 5), (0, 3), (3, 0), (3, 7)

14. $y = 5$ (0, 5), (3, 5), (−2, −5), (5, 5)

Complete the ordered pairs so that each is a solution for the given equation.

15. $x + y = 12$ (4,), (, 5), (0,), (, 0)

16. $x - y = 7$ (, 4), (15,), (0,), (, 0)

17. $3x + y = 9$ (3,), (, 9), (, −3), (0,)

18. $x + 5y = 20$ (0,), (, 2), (10,), (, 0)

19. $5x - y = 15$ (, 0), (2,), (4,), (, −5)

20. $x - 3y = 9$ (0,), (12,), (, 0), (, −2)

21. $3x - 2y = 12$ (, 0), (, −6), (2,), (, 3)

ANSWERS

1. (4, 2), (0, 6), (−3, 9)
2. (13, 1), (12, 0)
3. (5, 2), (4, 0), (6, 4)
4. (10, 2), (20, 0), (25, −1)
5. (2, 0), (1, 3)
6. (8, 0)
7. (3, 0), (6, 2), (0, −2)
8. (2, 0), (6, −8), (0, 4)
9. $(4, 0), \left(\frac{2}{3}, -5\right), \left(5, \frac{3}{2}\right)$
10. $\left(\frac{2}{3}, \frac{5}{2}\right), (0, 3)$
11. (0, 0), (2, 8)
12. $(0, -1), \left(\frac{1}{2}, 0\right)$
13. (3, 5), (3, 0), (3, 7)
14. (0, 5), (3, 5), (5, 5)
15. 8, 7, 12, 12
16. 11, 8, −7, 7
17. 0, 0, 4, 9
18. 4, 10, 2, 20
19. 3, −5, 5, 2
20. −3, 1, 9, 3
21. 4, 0, −3, 6

ANSWERS

22. 4, 2, 10, −5
23. −3, 11, 9, 7
24. 3, 3, 4, 1
25. −4, 3, $\frac{4}{3}$, 1
26. 5, 0, 2, 2
27. (0, −7), (2, −5), (4, −3), (6, −1)
28. (0, 18), (6, 12), (12, 6), (18, 0)
29. (0, −6), (3, 0), (6, 6), (9, 12)
30. (0, −12), (3, −3), (6, 6), (9, 15)
31. (8, 0), (−4, 3), (0, 2), (4, 1)
32. (0, 4), (3, 3), (6, 2), (9, 1)
33. (−5, −4), (0, −2), (5, 0), (10, 2)
34. (−7, 4), (0, 2), (7, 0), (14, −2)
35. (0, 3), (1, 5), (2, 7), (3, 9)
36. (0, −5), (1, 3), (2, 11), (3, 19)
37. (−5, 0), (−5, 1), (−5, 2), (−5, 3)
38. (0, 8), (1, 8), (2, 8), (3, 8)
39. (2, −3, 1)
40. (0, −1, 3)
41. (1, −6, 5)
42. (4, 0, 3)
43. (−2, 5, 1)
44. (−2, 1, −2)
45. \$9.50, \$11.75, \$15.50, \$19.25, \$23
46. 14°F, 32°F, 59°F, 212°F

22. $2x + 5y = 20$ $(0, \ ), (5, \ ), (\ , 0), (\ , 6)$

23. $y = 3x + 9$ $(\ , 0), \left(\frac{2}{3}, \ \right), (0, \ ), \left(-\frac{2}{3}, \ \right)$

24. $3x + 4y = 12$ $(0, \ ), \left(\ , \frac{3}{4}\right), (\ , 0), \left(\frac{8}{3}, \ \right)$

25. $y = 3x - 4$ $(0, \ ), (\ , 5), (\ , 0), \left(\frac{5}{3}, \ \right)$

26. $y = -2x + 5$ $(0, \ ), (\ , 5), \left(\frac{3}{2}, \ \right), (\ , 1)$

Find four solutions for each of the following equations. **Note:** Your answers may vary from those shown in the answer section.

27. $x - y = 7$

28. $x + y = 18$

29. $2x - y = 6$

30. $3x - y = 12$

31. $x + 4y = 8$

32. $x + 3y = 12$

33. $2x - 5y = 10$

34. $2x + 7y = 14$

35. $y = 2x + 3$

36. $y = 8x - 5$

37. $x = -5$

38. $y = 8$

An equation in three variables has an ordered triple as a solution. For example, (1, 2, 2) is a solution to the equation $x + 2y - z = 3$. Complete the ordered-triple solutions for each equation.

39. $x + y + z = 0$ $(2, -3, \ )$

40. $2x + y + z = 2$ $(\ , -1, 3)$

41. $x + y + z = 0$ $(1, \ , 5)$

42. $x + y - z = 1$ $(4, \ , 3)$

43. $2x + y + z = 2$ $(-2, \ , 1)$

44. $x + y - z = 1$ $(-2, 1, \ )$

45. Hourly wages. When an employee produces x units per hour, the hourly wage in dollars is given by $y = 0.75x + 8$. What are the hourly wages for the following number of units: 2, 5, 10, 15, and 20?

46. Temperature conversion. Celsius temperature readings can be converted to Fahrenheit readings using the formula $F = \frac{9}{5}C + 32$. What is the Fahrenheit temperature that corresponds to each of the following Celsius temperatures: −10, 0, 15, 100?

47. Area. The area of a square is given by $A = s^2$. What is the area of the squares whose sides are 5 centimeters (cm), 10 cm, 12 cm, 15 cm?

48. Unit pricing. When x number of units are sold, the price of each unit (in dollars) is given by $p = \frac{-x}{2} + 75$. Find the unit price when the following quantities are sold: 2, 7, 9, 11.

49. The number of programs for the disabled in the United States from 1993 to 1997 is approximated by the equation $y = 162x + 4365$ in which x is the number of years after 1993. Complete the following table

x	1	2	3	4	6
y	4527	4689	4851	5013	5337

50. Your monthly pay as a car salesman is determined using the equation $S = 200x + 1500$ in which x is the number of cars you can sell each month.

(a) Complete the following table.

x	12	15	17	18
S	3900	4500	4900	5100

(b) You are offered a job at a salary of $56,400 per year. How many cars would you have to sell per month to equal this salary?

51. You now have had practice solving equations with one variable and equations with two variables. Compare equations with one variable to equations with two variables. How are they alike? How are they different?

52. Each of the following sentences describes pairs of numbers that are related. After completing the sentences in parts (a) to (g), write two of your own sentences in (h) and (i).

(a) The *number of hours you work* determines the *amount you are* ______________.

(b) The *number of gallons of gasoline* you put in your car determines *the amount you* ______________.

ANSWERS

47. 25 cm², 100 cm², 144 cm², 225 cm²

48. $74, $71.50, $70.50, $69.50

49. See exercise

50. (a) See exercise

(b) 16 cars

51. ______________

52. 

ANSWERS

a. See exercise

b. See exercise

c. See exercise

d. See exercise

e. See exercise

f. −7

g. −4

h. 4

i. 8

j. $\frac{19}{2}$

(c) The *amount of the* ____________ in a restaurant is related to *the amount of the tip.*

(d) The *sales amount of a purchase in a store* determines ____________.

(e) The *age of an automobile* is related to ____________.

(f) The *amount of electricity you use in a month* determines ____________.

(g) The *cost of food for a family of four* and ____________.

Think of two more:

(h) ______________________________________.

(i) ______________________________________.

Getting Ready for Section 6.2 [Section 0.4]

Plot points with the following coordinates on the number line shown below.

(a) −3 (b) 7 (c) 0 (d) −8 (e) $\frac{3}{2}$

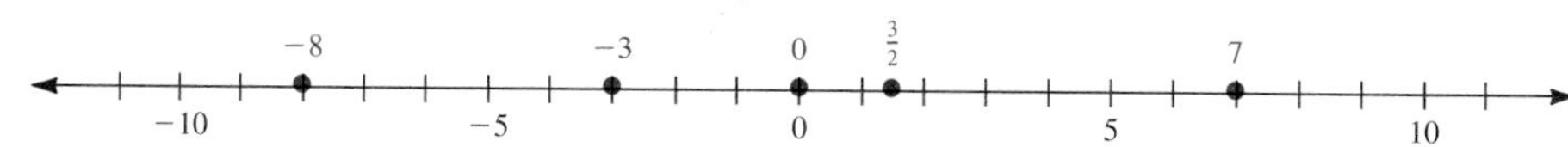

Give the coordinate of each of the following points.

(f) *A* (g) *B* (h) *C* (i) *D* (j) *E*

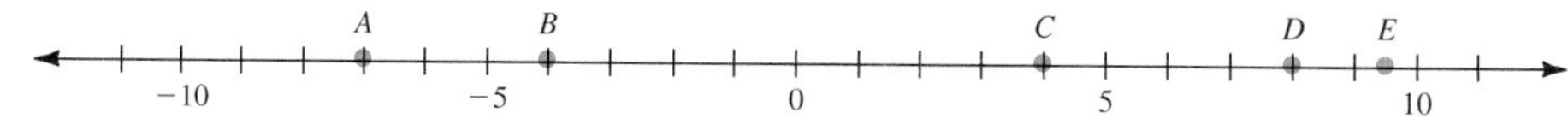

Answers

1. (4, 2), (0, 6), (−3, 9) **3.** (5, 2), (4, 0), (6, 4) **5.** (2, 0), (1, 3)
7. (3, 0), (6, 2), (0, −2) **9.** $(4, 0), \left(\frac{2}{3}, -5\right), \left(5, \frac{3}{2}\right)$ **11.** (0, 0), (2, 8)
13. (3, 5), (3, 0), (3, 7) **15.** 8, 7, 12, 12 **17.** 0, 0, 4, 9 **19.** 3, −5, 5, 2
21. 4, 0, −3, 6 **23.** −3, 11, 9, 7 **25.** $-4, 3, \frac{4}{3}, 1$
27. (0, −7), (2, −5), (4, −3), (6, −1) **29.** (0, −6), (3, 0), (6, 6), (9, 12)
31. (8, 0), (−4, 3), (0, 2), (4, 1) **33.** (−5, −4), (0, −2), (5, 0), (10, 2)
35. (0, 3), (1, 5), (2, 7), (3, 9) **37.** (−5, 0), (−5, 1), (−5, 2), (−5, 3)
39. (2, −3, 1) **41.** (1, −6, 5) **43.** (−2, 5, 1)
45. \$9.50, \$11.75, \$15.50, \$19.25, \$23 **47.** 25 cm^2, 100 cm^2, 144 cm^2, 225 cm^2
49. 4527, 4689, 4851, 5013, 5337 **51.** 

a–e.

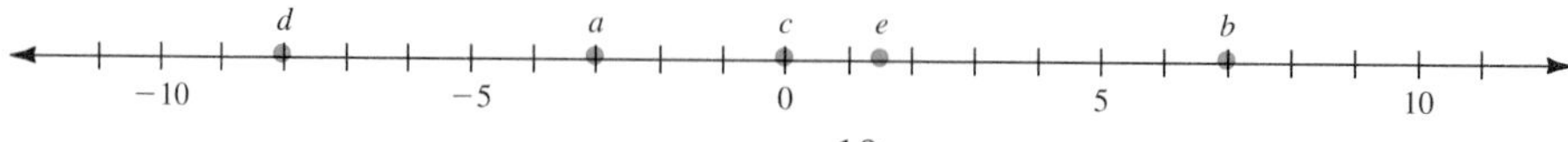

f. −7 **g.** −4 **h.** 4 **i.** 8 **j.** $\frac{19}{2}$

6.2 The Rectangular Coordinate System

6.2 OBJECTIVES

1. Give the coordinates of a set of points on the plane
2. Graph the points corresponding to a set of ordered pairs

In Section 6.1, we saw that ordered pairs could be used to write the solutions of equations in two variables. The next step is to graph those ordered pairs as points in a plane.

Because there are two numbers (one for x and one for y), we will need two number lines. One line is drawn horizontally, and the other is drawn vertically; their point of intersection (at their respective zero points) is called the *origin*. The horizontal line is called the **x axis,** and the vertical line is called the **y axis.** Together the lines form the **rectangular coordinate system.**

The axes divide the plane into four regions called **quadrants,** which are numbered (usually by Roman numerals) counterclockwise from the upper right.

NOTE This system is also called the **cartesian coordinate system,** named in honor of its inventor, René Descartes (1596–1650), a French mathematician and philosopher.

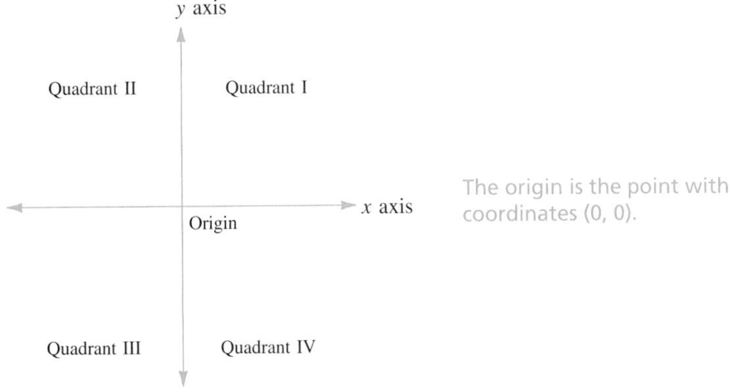

We now want to establish correspondences between ordered pairs of numbers (x, y) and points in the plane.

For any ordered pair

$$(x, y)$$

x coordinate   y coordinate

the following are true:

1. If the x coordinate is

Positive, the point corresponding to that pair is located x units to the *right* of the y axis.
Negative, the point is x units to the *left* of the y axis.
Zero, the point is on the y axis.

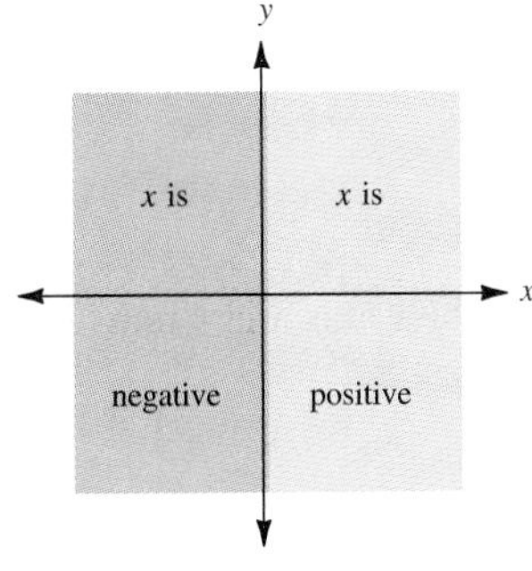

2. If the y coordinate is

   Positive, the point is y units *above* the x axis.
   Negative, the point is y units *below* the x axis.
   Zero, the point is on the x axis.

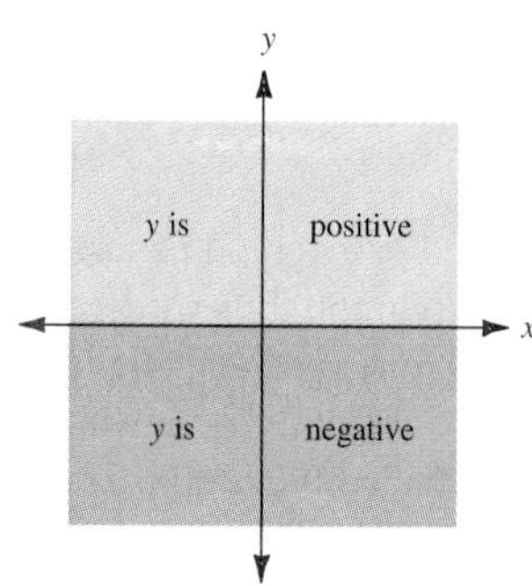

Example 1 illustrates how to use these guidelines to give coordinates to points in the plane.

REMEMBER: The x coordinate gives the *horizontal* distance from the y axis. The y coordinate gives the *vertical* distance from the x axis.

Example 1

Identifying the Coordinates for a Given Point

Give the coordinates for the given point

(a)

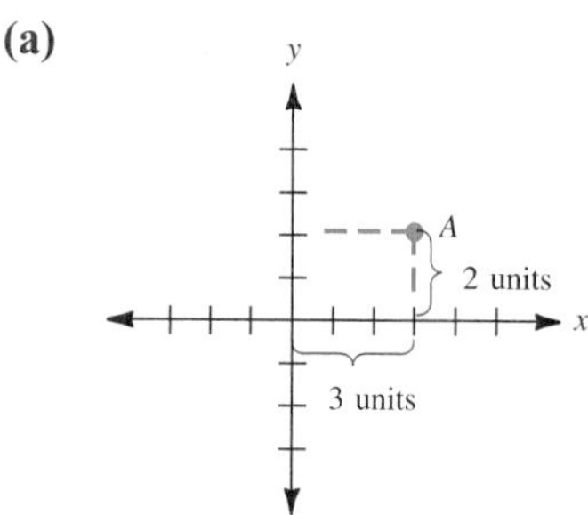

Point A is 3 units to the *right* of the y axis and 2 units *above* the x axis. Point A has coordinates (3, 2).

(b)

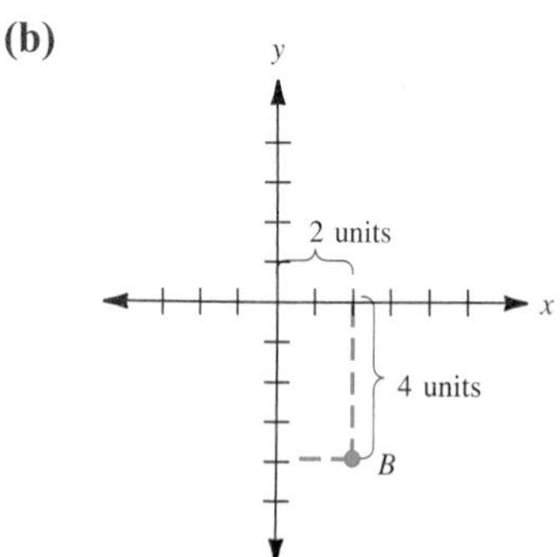

Point B is 2 units to the *right* of the y axis and 4 units *below* the x axis. Point B has coordinates (2, −4).

(c)

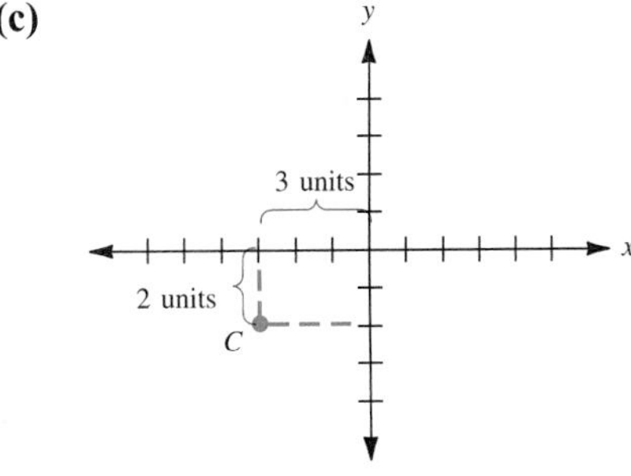

Point C is 3 units to the *left* of the y axis and 2 units *below* the x axis. C has coordinates $(-3, -2)$.

(d)

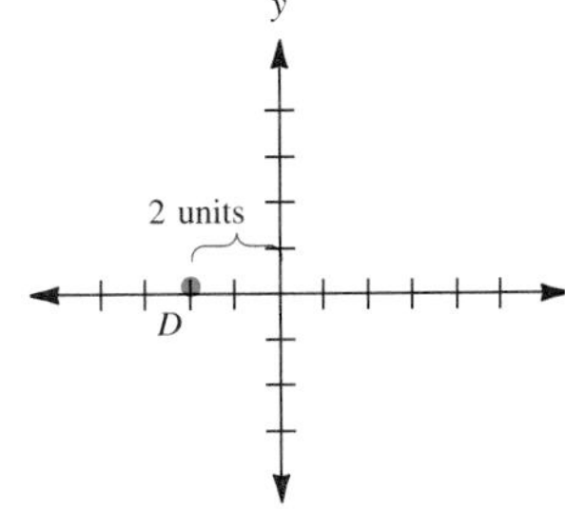

Point D is 2 units to the *left* of the y axis and *on* the x axis. Point D has coordinates $(-2, 0)$.

CHECK YOURSELF 1

Give the coordinates of points P, Q, R, and S.

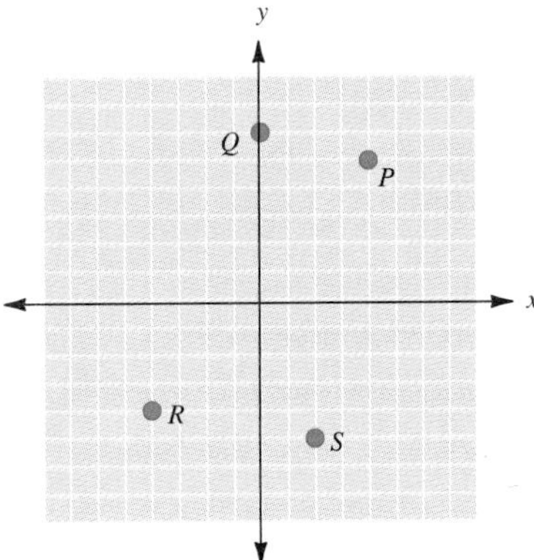

P ________

Q ________

R ________

S ________

Reversing the process above will allow us to graph (or plot) a point in the plane given the coordinates of the point. You can use the following steps.

NOTE The graphing of individual points is sometimes called **point plotting.**

Step by Step: To Graph a Point in the Plane

Step 1 Start at the origin.
Step 2 Move right or left according to the value of the x coordinate.
Step 3 Move up or down according to the value of the y coordinate.

Example 2

Graphing Points

(a) Graph the point corresponding to the ordered pair (4, 3).

Move 4 units to the right on the x axis. Then move 3 units up from the point you stopped at on the x axis. This locates the point corresponding to (4, 3).

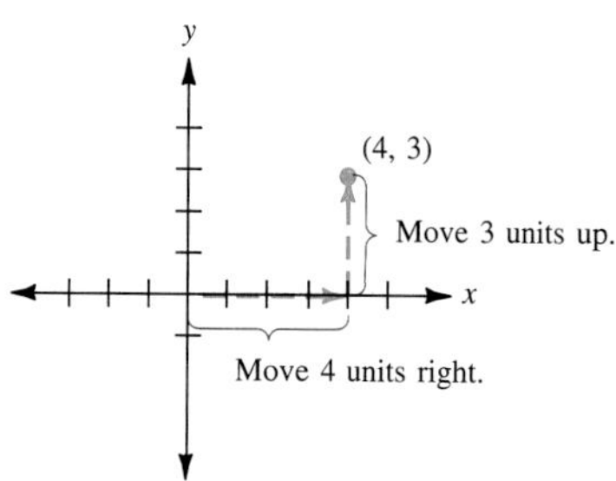

(b) Graph the point corresponding to the ordered pair (−5, 2).

In this case move 5 units *left* (because the x coordinate is negative) and then 2 units *up*.

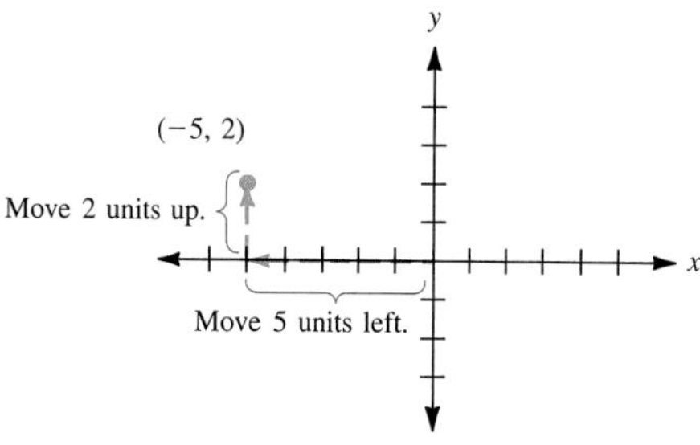

(c) Graph the point corresponding to (−4, −2).

Here move 4 units *left* and then 2 units *down* (the y coordinate is negative).

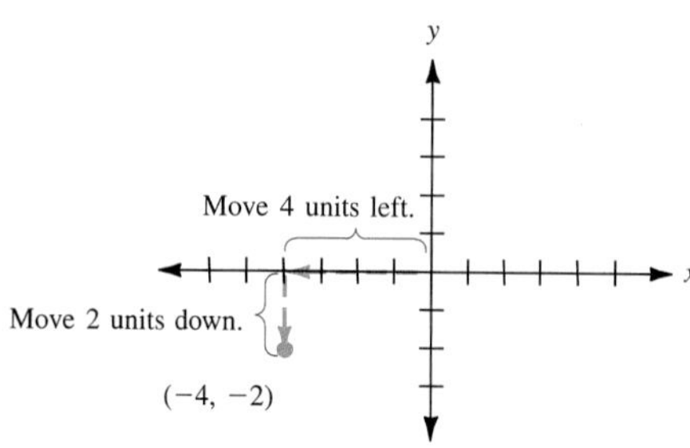

NOTE Any point on an axis will have 0 for one of its coordinates.

(d) Graph the point corresponding to $(0, -3)$.

There is *no* horizontal movement because the x coordinate is 0. Move 3 units *down*.

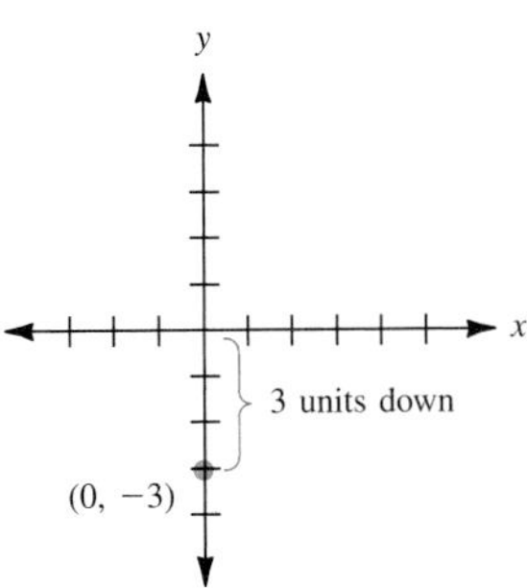

(e) Graph the point corresponding to $(5, 0)$.

Move 5 units *right*. The desired point is on the x axis because the y coordinate is 0.

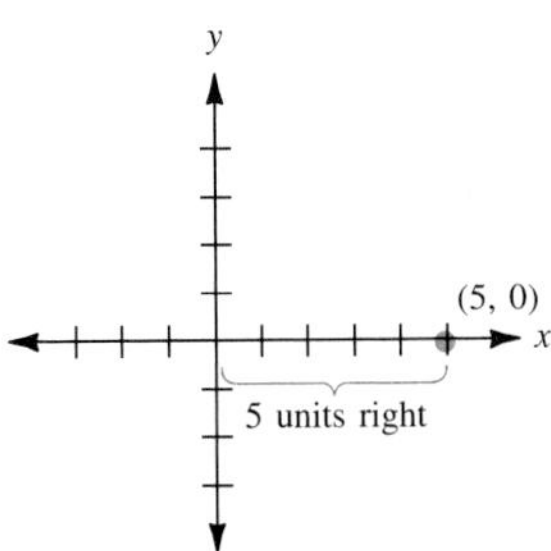

CHECK YOURSELF 2

Graph the points corresponding to $M(4, 3)$, $N(-2, 4)$, $P(-5, -3)$, and $Q(0, -3)$.

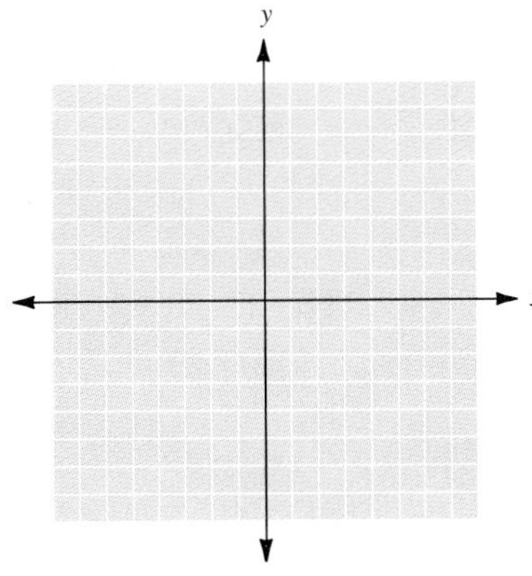

CHECK YOURSELF ANSWERS

1. $P(4, 5)$, $Q(0, 6)$, $R(-4, -4)$, and $S(2, -5)$
2.

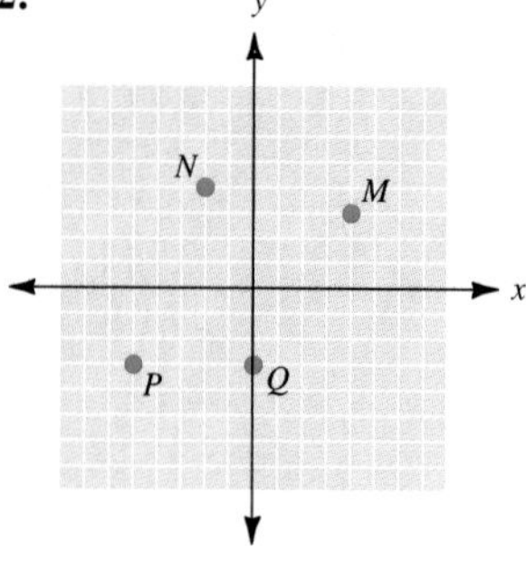

Name ____________________

Section ________ Date ________

6.2 Exercises

Give the coordinates of the points graphed below.

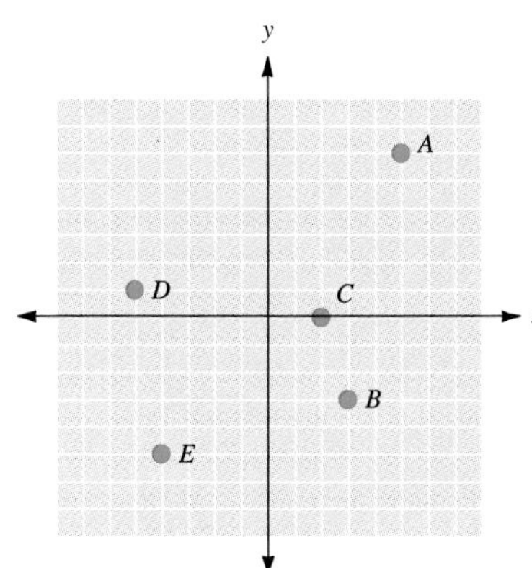

1. A
2. B
3. C
4. D
5. E

Give the coordinates of the points graphed below.

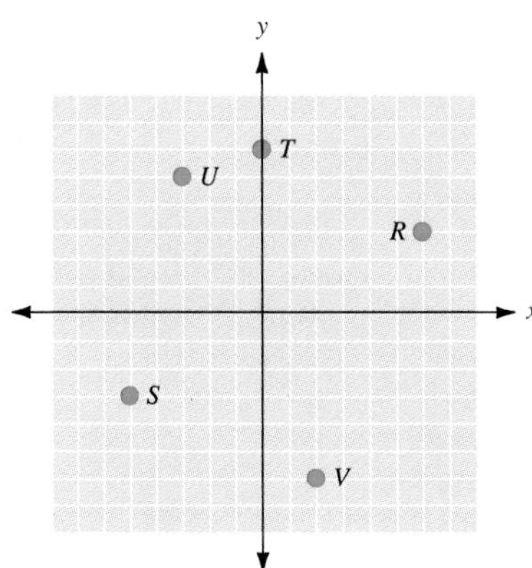

6. R
7. S
8. T
9. U
10. V

Plot points with the following coordinates on the graph below.

11. $M(5, 3)$
12. $N(0, -3)$
13. $P(-2, 6)$
14. $Q(5, 0)$
15. $R(-4, -6)$
16. $S(-3, -4)$

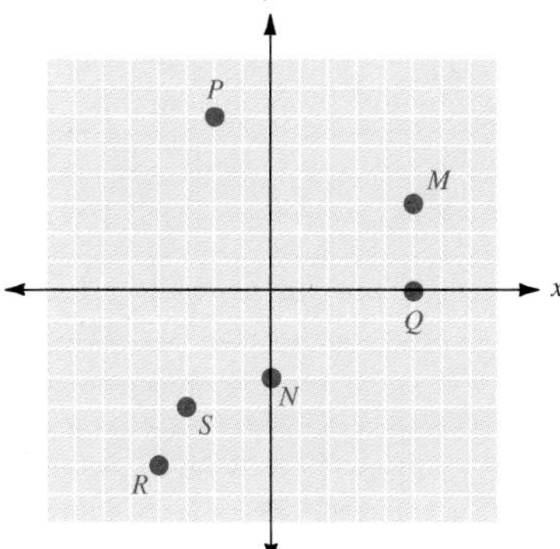

ANSWERS

1. (5, 6)
2. (3, −3)
3. (2, 0)
4. (−5, 1)
5. (−4, −5)
6. (6, 3)
7. (−5, −3)
8. (0, 6)
9. (−3, 5)
10. (2, −6)
11. See exercise
12. See exercise
13. See exercise
14. See exercise
15. See exercise
16. See exercise

ANSWERS

17. See exercise

18. See exercise

19. See exercise

20. See exercise

21. See exercise

22. See exercise

23. The points lie on a line; (1, 2)

24. The points lie on a line; (2, 1)

Plot points with the following coordinates on the graph below.

17. $F(-3, -1)$

18. $G(4, 3)$

19. $H(5, -2)$

20. $I(-3, 0)$

21. $J(-5, 3)$

22. $K(0, 6)$

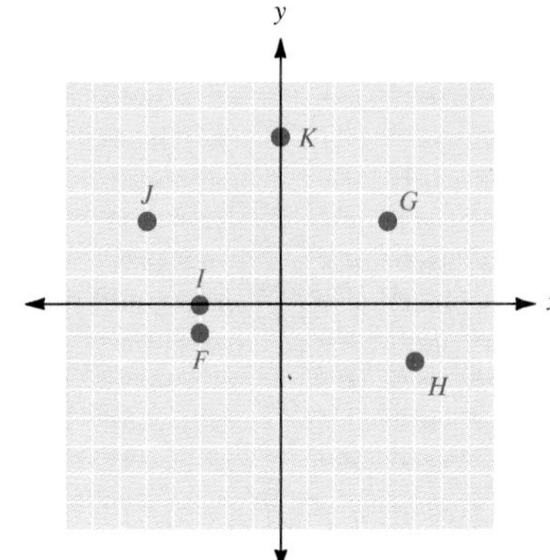

23. Graph points with coordinates (2, 3), (3, 4), and (4, 5) below. What do you observe? Can you give the coordinates of another point with the same property?

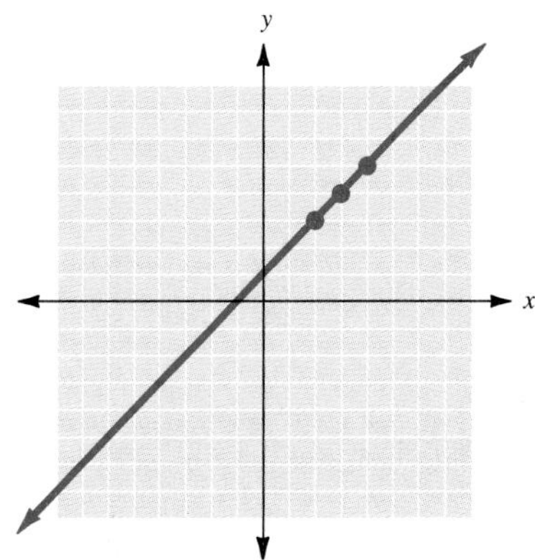

24. Graph points with coordinates (−1, 4), (0, 3), and (1, 2) below. What do you observe? Can you give the coordinates of another point with the same property?

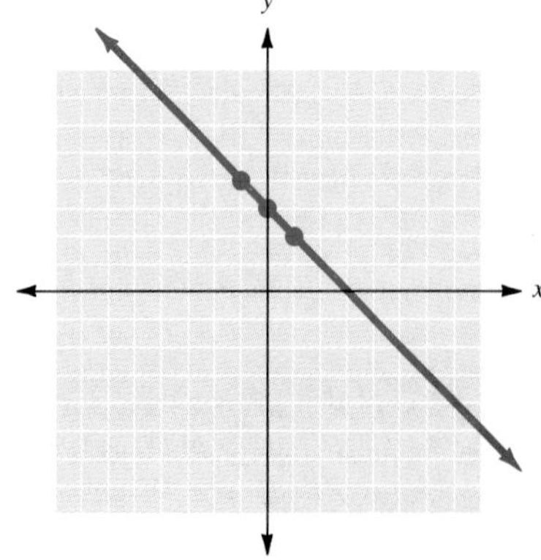

25. Graph points with coordinates $(-1, 3)$, $(0, 0)$, and $(1, -3)$ below. What do you observe? Can you give the coordinates of another point with the same property?

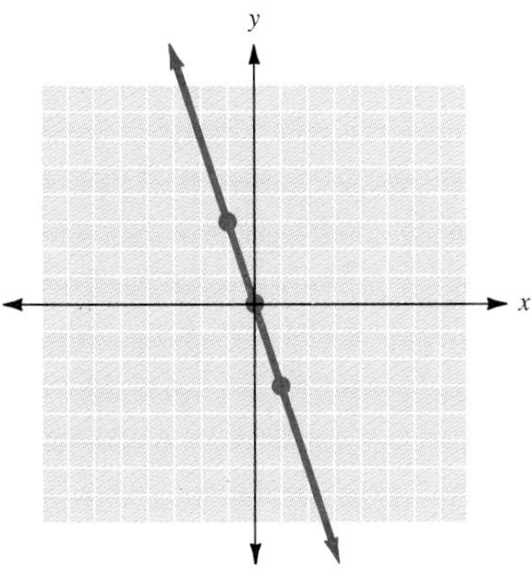

26. Graph points with coordinates $(1, 5)$, $(-1, 3)$, and $(-3, 1)$ below. What do you observe? Can you give the coordinates of another point with the same property?

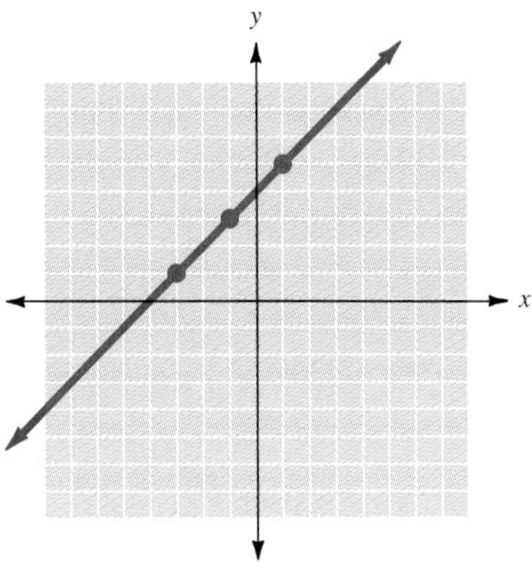

27. Environment. A local plastics company is sponsoring a plastics recycling contest for the local community. The focus of the contest is collecting plastic milk, juice, and water jugs. The company will award $200 plus the current market price of the jugs collected to the group that collects the most jugs in a single month. The number of jugs collected and the amount of money won can be represented as an ordered pair.

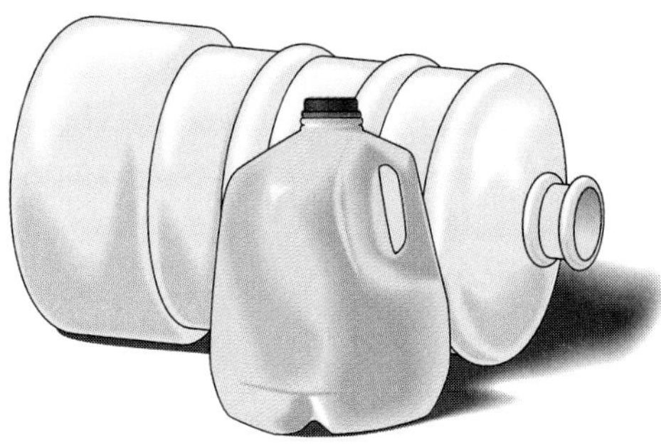

(a) In April, group A collected 1500 pounds (lb) of jugs to win first place. The prize for the month was $350. On the graph on the next page, x represents the pounds of jugs and y represents the amount of money that the group won. Graph the point that represents the winner for April.

ANSWERS

25. They lie on a line; $(2, -6)$

26. They lie on a line; $(3, 7)$

27. See exercise

28. See exercise

29. See exercise

(b) In May, group B collected 2300 lb of jugs to win first place. The prize for the month was \$430. Graph the point that represents the May winner on the same axis you used in part (a).

(c) In June, group C collected 1200 lb of jugs to win the contest. The prize for the month was \$320. Graph the point that represents the June winner on the same axis as used before.

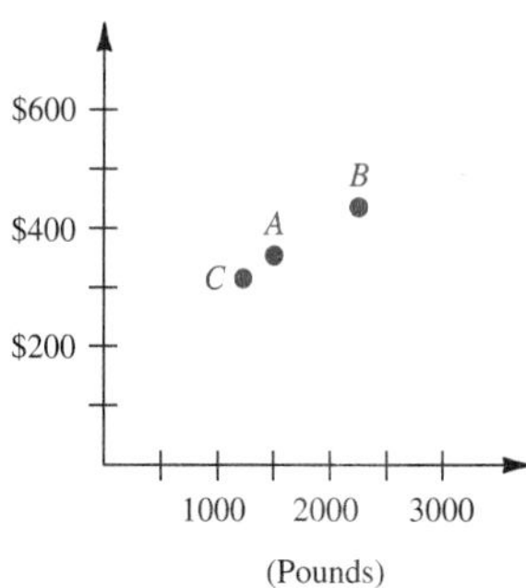

28. Education. The table gives the hours, x, that Damien studied for five different math exams and the resulting grades, y. Plot the data given in the table.

x	4	5	5	2	6
y	83	89	93	75	95

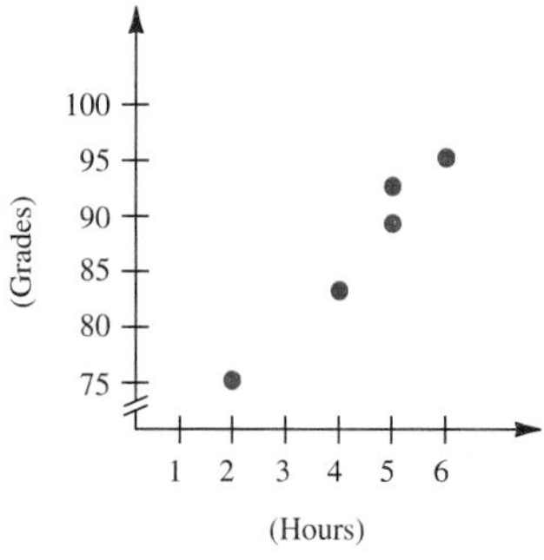

29. Science. The table gives the average temperature y (in degrees Fahrenheit) for the first 6 months of the year, x. The months are numbered 1 through 6, with 1 corresponding to January. Plot the data given in the table.

x	1	2	3	4	5	6
y	4	14	26	33	42	51

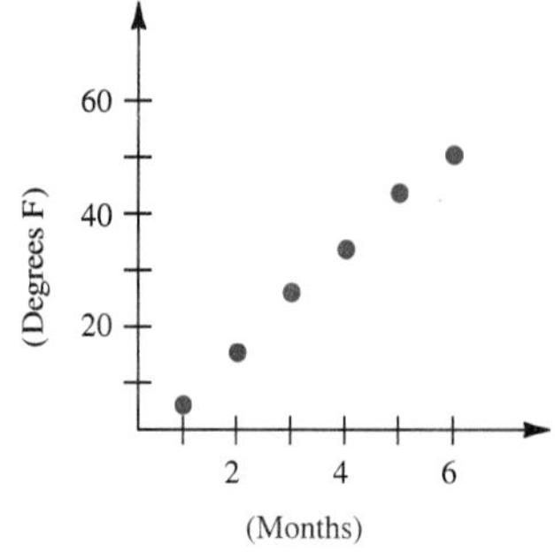

30. See exercise

31. See exercise

30. Business. The table gives the total salary of a salesperson, y, for each of the four quarters of the year, x. Plot the data given in the table.

x	1	2	3	4
y	\$6000	\$5000	\$8000	\$9000

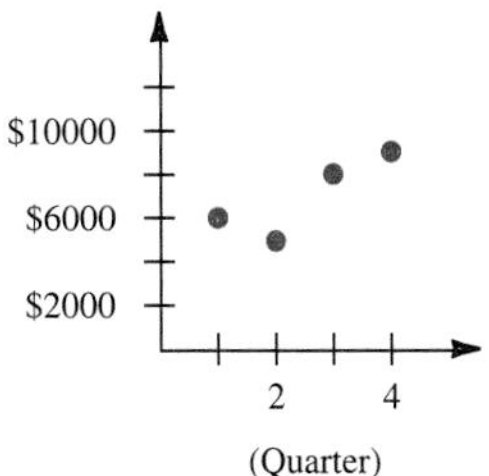

31. Sports. The table shows the number of runs scored by the New York Yankees in the 1999 World Series.

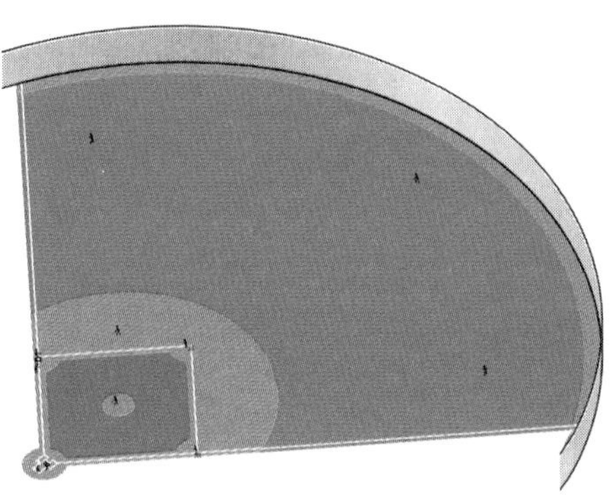

Game	1	2	3	4
Runs	4	7	6	4

Plot the data given in the table.

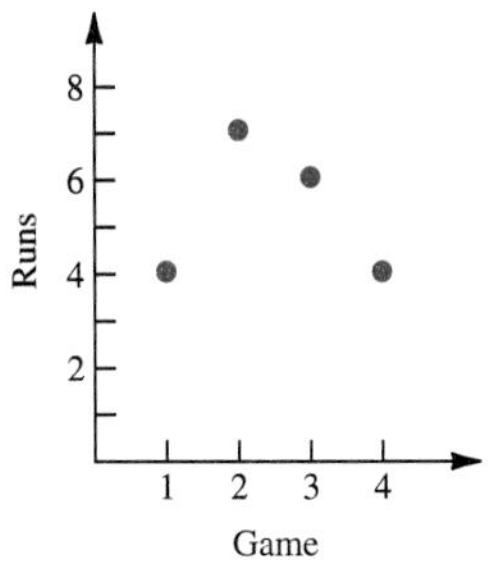

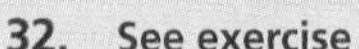

ANSWERS

32. See exercise

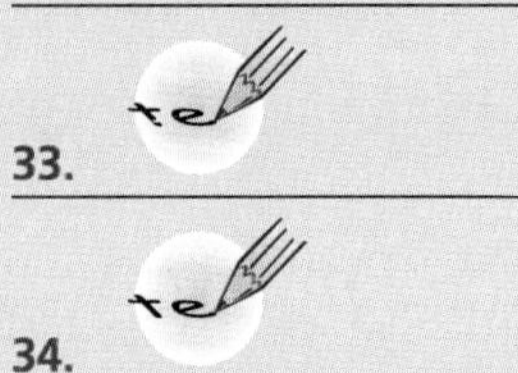

33.

34.

35. (a) A7, F2, C2

(b) Oysterville; Sweet Home; Mineral

32. **Sports.** The following table shows the number of wins and total points for the five teams in the Atlantic Division of the National Hockey League in the early part of the 1999–2000 season.

Team	Wins	Points
New Jersey Devils	5	12
Philadelphia Flyers	4	10
New York Rangers	4	9
Pittsburgh Penguins	2	6
New York Islanders	2	5

Plot the data given in the table.

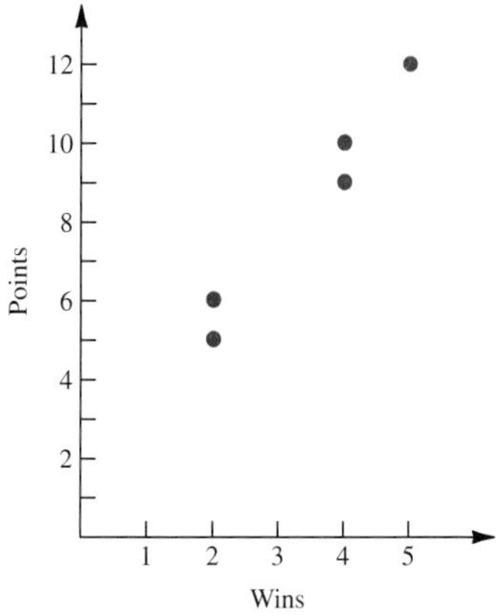

33. How would you describe a rectangular coordinate system? Explain what information is needed to locate a point in a coordinate system.

34. Some newspapers have a special day that they devote to automobile ads. Use this special section or the Sunday classified ads from your local newspaper to find all the want ads for a particular automobile model. Make a list of the model year and asking price for 10 ads, being sure to get a variety of ages for this model. After collecting the information, make a scatter plot of the age and the asking price for the car.

Describe your graph, including an explanation of how you decided which variable to put on the vertical axis and which on the horizontal axis. What trends or other information are given by the graph?

35. The map shown on the next page uses letters and numbers to label a grid that helps to locate a city. For instance, Salem is located at E-4.

(a) Find the coordinates for the following: White Swan, Newport, and Wheeler.

(b) What cities correspond to the following coordinates: A2, F4, and A5?

Getting Ready for Section 6.3 [Section 2.3]

Solve each of the following equations.

(a) $2x - 2 = 6$ (b) $2 - 5x = 12$

(c) $7y + 10 = -11$ (d) $-3 + 5x = 1$

(e) $6 - 3x = 8$ (f) $-4y + 6 = 3$

Answers

1. (5, 6) **3.** (2, 0) **5.** (−4, −5) **7.** (−5, −3) **9.** (−3, 5)

11–21.

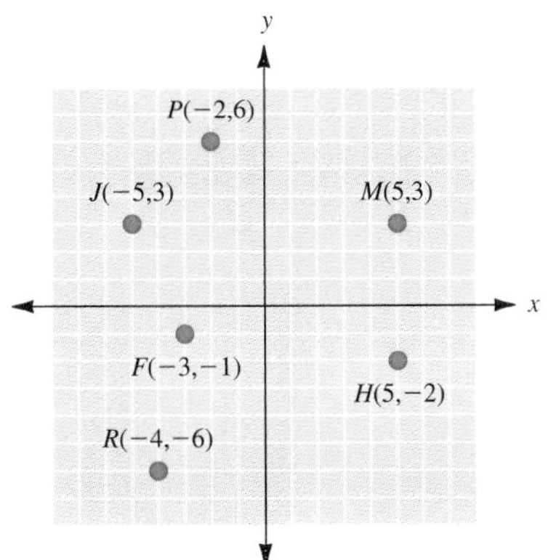

23. The points lie on a line; (1, 2)

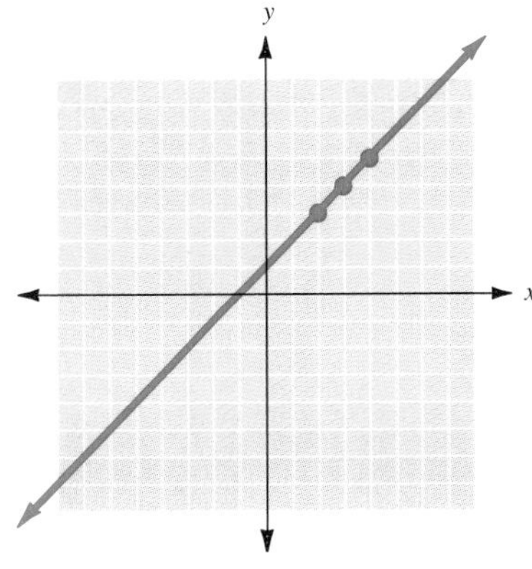

ANSWERS

a. 4

b. −2

c. −3

d. $\frac{4}{5}$

e. $-\frac{2}{3}$

f. $\frac{3}{4}$

25. The points lie on a line; (2, −6)

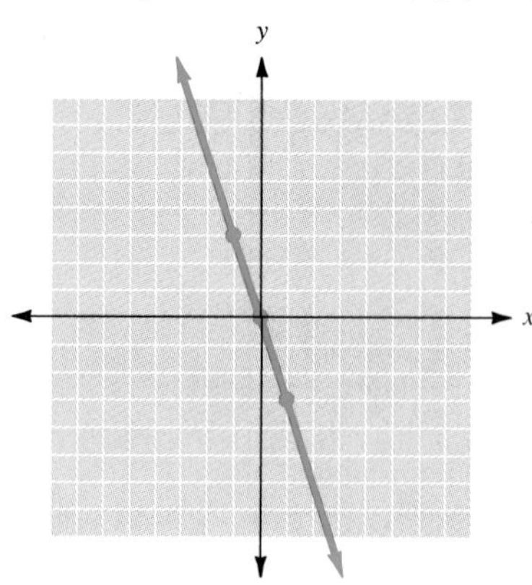

27. **(a)** (1500, 350); **(b)** (2300, 430); **(c)** (1200, 320)

29.

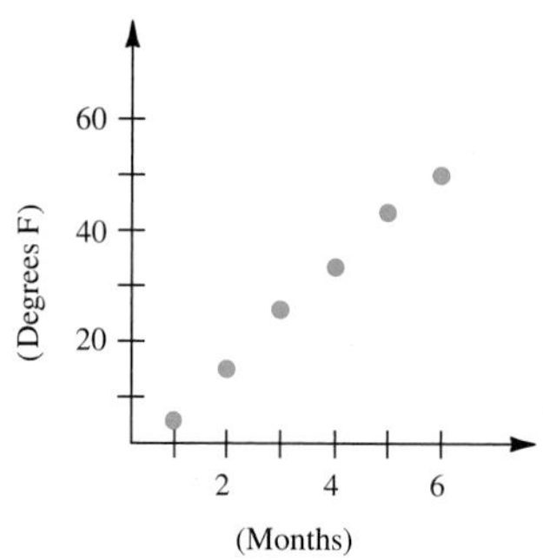

31.

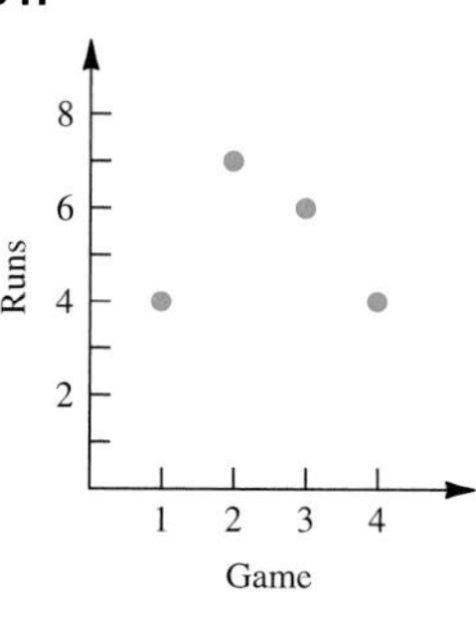

33.

35. **(a)** B7, F2, C2; **(b)** Oysterville, Sweet Home, Mineral

a. 4 **b.** −2 **c.** −3 **d.** $\frac{4}{5}$ **e.** $-\frac{2}{3}$ **f.** $\frac{3}{4}$

Graphing Linear Equations

OBJECTIVES

1. Graph a linear equation by plotting points
2. Graph a linear equation by the intercept method
3. Graph a linear equation by solving the equation for *y*

We are now ready to combine our work of the previous two sections. In Section 6.1 you learned to write the solutions of equations in two variables as ordered pairs. Then, in Section 6.2, these ordered pairs were graphed in the plane. Putting these ideas together will let us graph certain equations. Example 1 illustrates this approach.

Example 1

Graphing a Linear Equation

Graph $x + 2y = 4$.

NOTE We are going to find *three* solutions for the equation. We'll point out why shortly.

Step 1 Find some solutions for $x + 2y = 4$. To find solutions, we choose any convenient values for x, say $x = 0$, $x = 2$, and $x = 4$. Given these values for x, we can substitute and then solve for the corresponding value for y. So

If $x = 0$, then $y = 2$, so (0, 2) is a solution.
If $x = 2$, then $y = 1$, so (2, 1) is a solution.
If $x = 4$, then $y = 0$, so (4, 0) is a solution.

A handy way to show this information is in a table such as this:

NOTE The table is just a convenient way to display the information. It is the same as writing (0, 2), (2, 1), and (4, 0).

x	y
0	2
2	1
4	0

Step 2 We now graph the solutions found in step 1.

$x + 2y = 4$

x	y
0	2
2	1
4	0

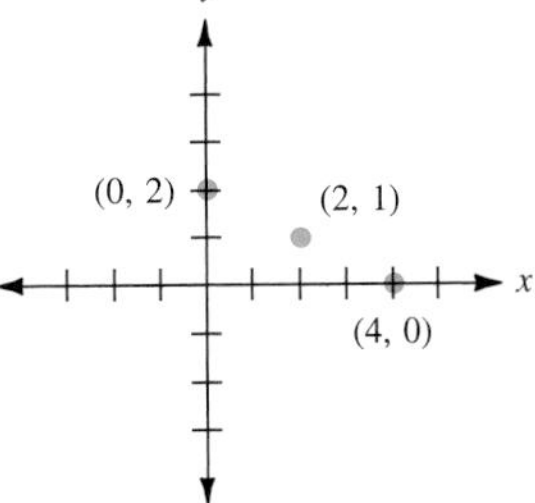

What pattern do you see? It appears that the three points lie on a straight line, and that is in fact the case.

Step 3 Draw a straight line through the three points graphed in step 2.

NOTE The arrows on the end of the line mean that the line extends indefinitely in either direction.

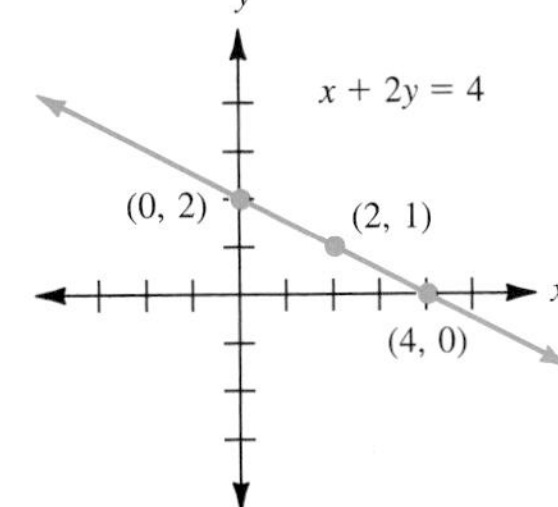

NOTE The graph is a "picture" of the solutions for the given equation.

The line shown is the **graph** of the equation $x + 2y = 4$. It represents *all* of the ordered pairs that are solutions (an infinite number) for that equation.

Every ordered pair that is a solution will have its graph on this line. Any point on the line will have coordinates that are a solution for the equation.

Note: Why did we suggest finding *three* solutions in step 1? Two points determine a line, so technically you need only two. The third point that we find is a check to catch any possible errors.

CHECK YOURSELF 1

Graph $2x - y = 6$, *using the steps shown in Example* 1.

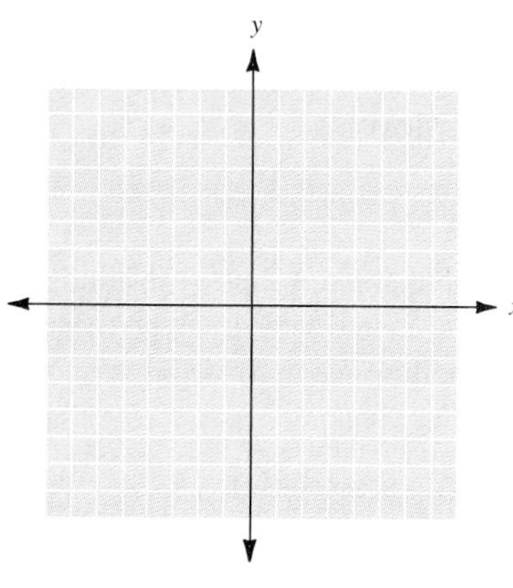

Let's summarize. An equation that can be written in the form

$$Ax + By = C$$

in which A, B, and C are real numbers and A and B cannot both be 0 is called a **linear equation in two variables.** The graph of this equation is a *straight line.*

The steps of graphing follow.

Step by Step: To Graph a Linear Equation

Step 1 Find at least three solutions for the equation, and put your results in tabular form.

Step 2 Graph the solutions found in step 1.

Step 3 Draw a straight line through the points determined in step 2 to form the graph of the equation.

Example 2

Graphing a Linear Equation

Graph $y = 3x$.

Step 1 Some solutions are

NOTE Let $x = 0$, 1, and 2, and substitute to determine the corresponding y values. Again the choices for x are simply convenient. Other values for x would serve the same purpose.

x	y
0	0
1	3
2	6

Step 2 Graph the points.

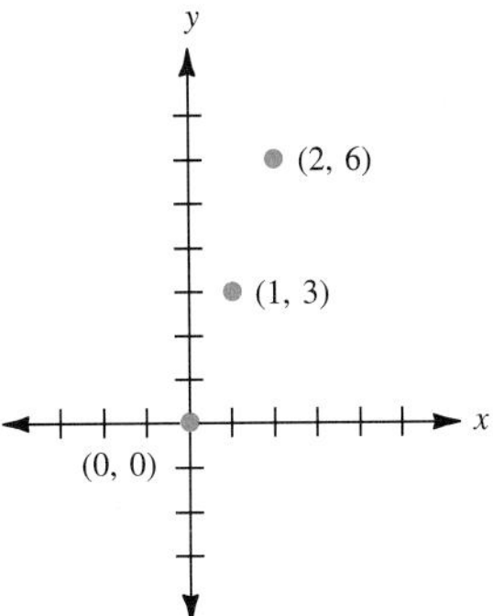

NOTE Notice that connecting any two of these points produces the same line.

Step 3 Draw a line through the points.

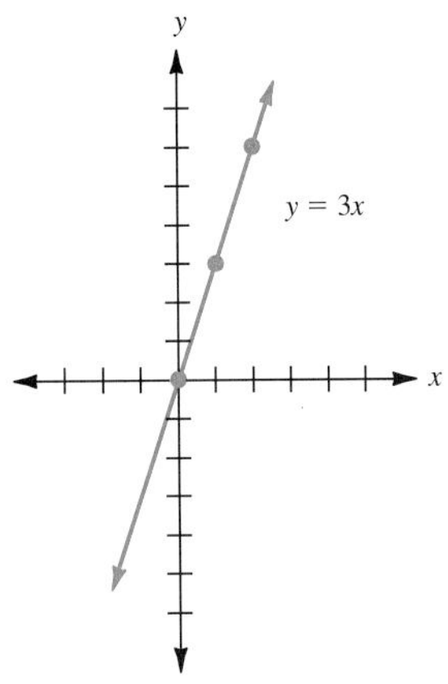

CHECK YOURSELF 2

Graph the equation $y = -2x$ after completing the table of values.

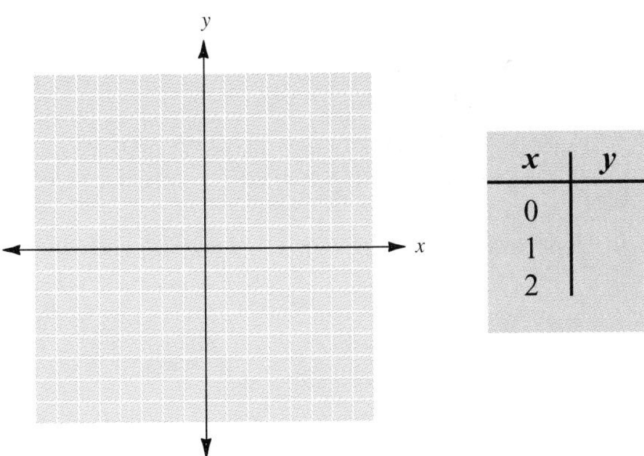

x	y
0	
1	
2	

Let's work through another example of graphing a line from its equation.

Example 3

Graphing a Linear Equation

Graph $y = 2x + 3$.

Step 1 Some solutions are

x	y
0	3
1	5
2	7

Step 2 Graph the points corresponding to these values.

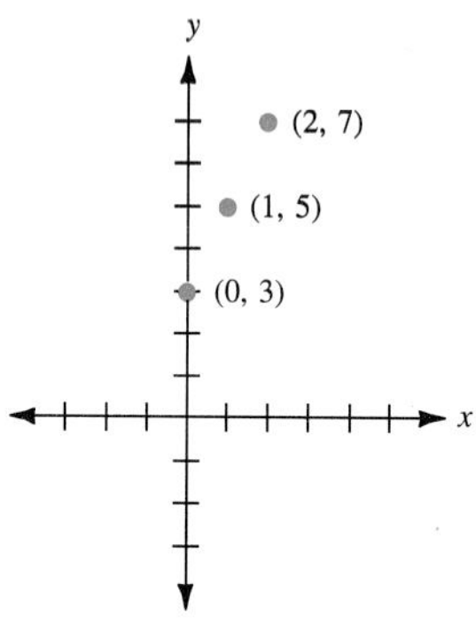

Step 3 Draw a line through the points.

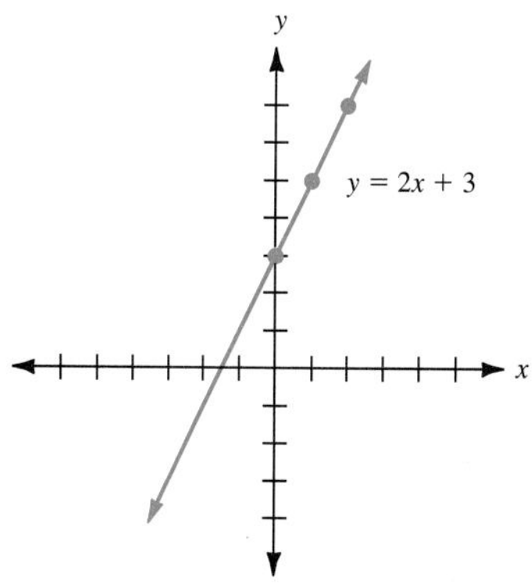

CHECK YOURSELF 3

Graph the equation $y = 3x - 2$ after completing the table of values.

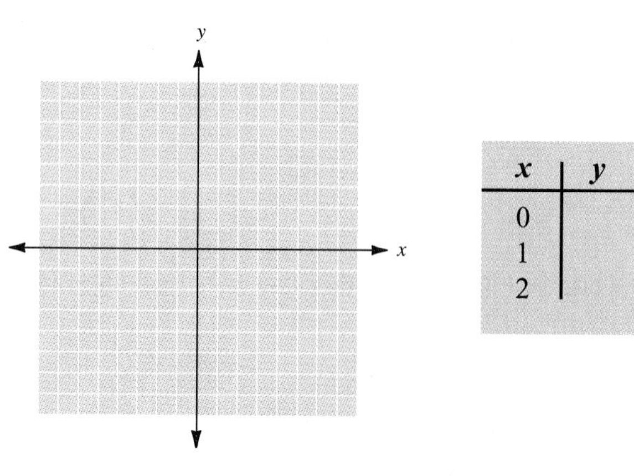

x	y
0	
1	
2	

In graphing equations, particularly when fractions are involved, a careful choice of values for x can simplify the process. Consider Example 4.

Example 4

Graphing a Linear Equation

Graph

$$y = \frac{3}{2}x - 2$$

As before, we want to find solutions for the given equation by picking convenient values for x. Note that in this case, choosing *multiples of 2* will avoid fractional values for y and make the plotting of those solutions much easier. For instance, here we might choose values of -2, 0, and 2 for x.

Step 1 If $x = -2$:

$$y = \frac{3}{2}(-2) - 2$$

$$= -3 - 2 = -5$$

If $x = 0$:

$$y = \frac{3}{2}(0) - 2$$

$$= 0 - 2 = -2$$

If $x = 2$:

$$y = \frac{3}{2}(2) - 2$$

$$= 3 - 2 = 1$$

NOTE Suppose we do *not* choose a multiple of 2, say, $x = 3$. Then

$$y = \frac{3}{2}(3) - 2$$

$$= \frac{9}{2} - 2$$

$$= \frac{5}{2}$$

$\left(3, \frac{5}{2}\right)$ is still a valid solution, but we must graph a point with fractional coordinates.

In tabular form, the solutions are

x	y
-2	-5
0	-2
2	1

Step 2 Graph the points determined above.

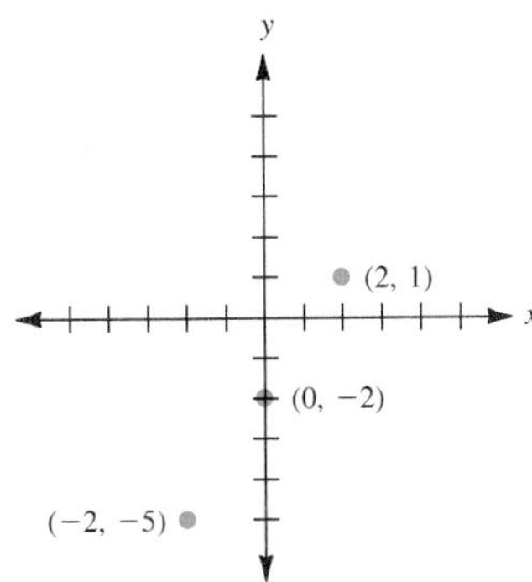

Step 3 Draw a line through the points.

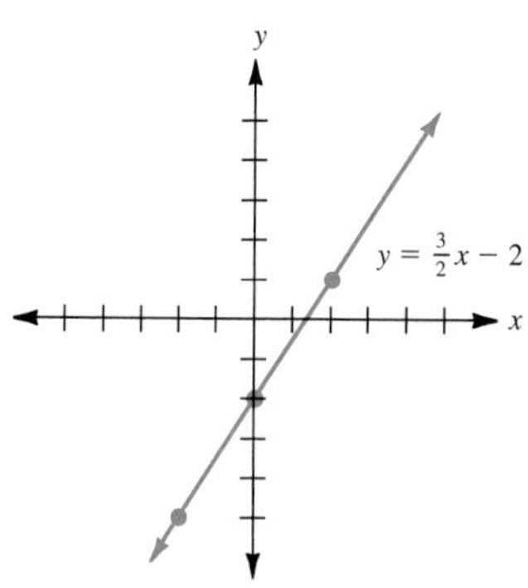

CHECK YOURSELF 4

Graph the equation $y = -\frac{1}{3}x + 3$ *after completing the table of values.*

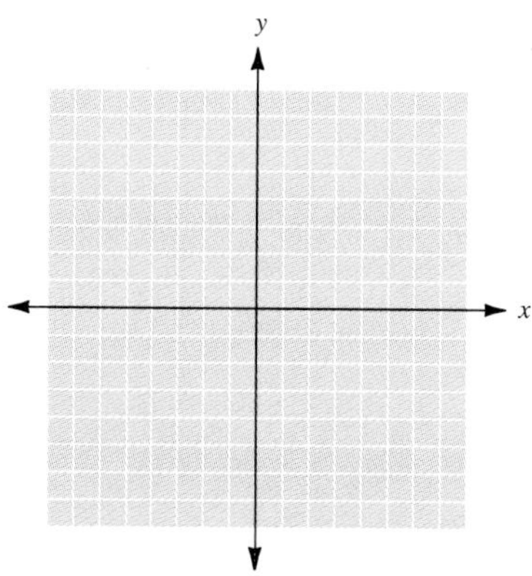

x	y
−3	
0	
3	

Some special cases of linear equations are illustrated in Examples 5 and 6.

Example 5

Graphing an Equation That Results in a Vertical Line

Graph $x = 3$.

The equation $x = 3$ is equivalent to $x + 0 \cdot y = 3$. Let's look at some solutions.

If $y = 1$:	If $y = 4$:	If $y = -2$:
$x + 0 \cdot 1 = 3$	$x + 0 \cdot 4 = 3$	$x + 0(-2) = 3$
$x = 3$	$x = 3$	$x = 3$

In tabular form,

x	y
3	1
3	4
3	−2

What do you observe? The variable x has the value 3, regardless of the value of y. Look at the graph on the following page.

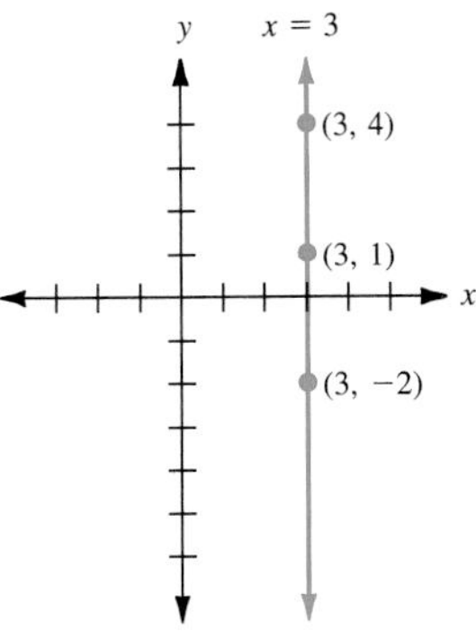

The graph of $x = 3$ is a vertical line crossing the x axis at (3, 0).

Note that graphing (or plotting) points in this case is not really necessary. Simply recognize that the graph of $x = 3$ *must* be a vertical line (parallel to the y axis) that intercepts the x axis at (3, 0).

CHECK YOURSELF 5

Graph the equation $x = -2$.

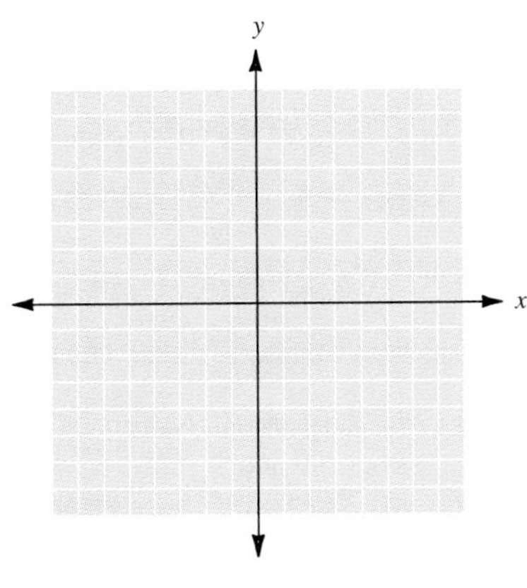

Example 6 is a related example involving a horizontal line.

Example 6

Graphing an Equation That Results in a Horizontal Line

Graph $y = 4$.

Because $y = 4$ is equivalent to $0 \cdot x + y = 4$, any value for x paired with 4 for y will form a solution. A table of values might be

x	y
−2	4
0	4
2	4

Here is the graph.

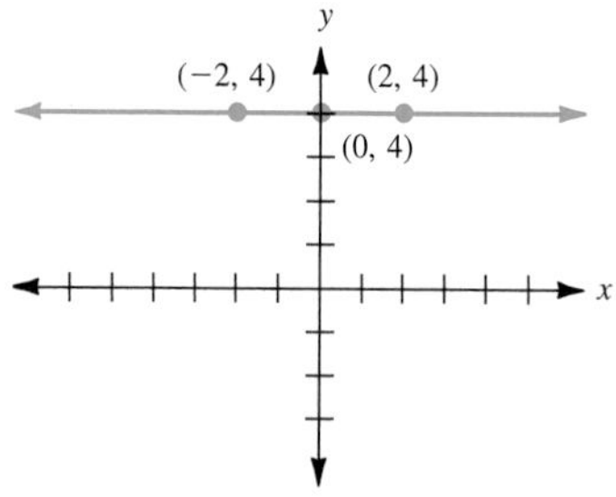

This time the graph is a horizontal line that crosses the y axis at (0, 4). Again the graphing of points is not required. The graph of $y = 4$ *must* be horizontal (parallel to the x axis) and intercepts the y axis at (0, 4).

CHECK YOURSELF 6

Graph the equation $y = -3$.

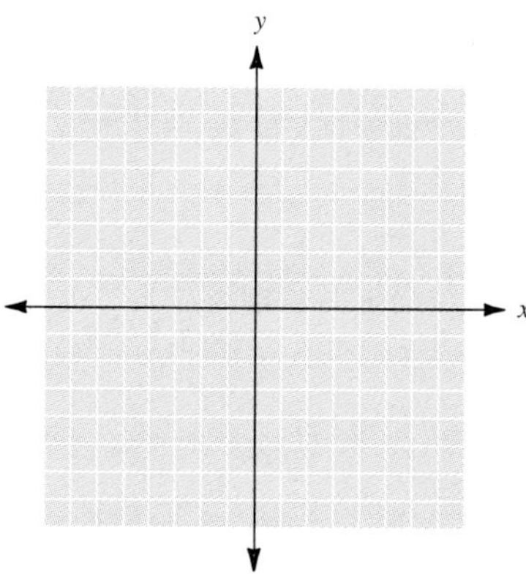

The following box summarizes our work in the previous two examples:

Definitions: Vertical and Horizontal Lines

1. The graph of $x = a$ is a *vertical line* crossing the x axis at $(a, 0)$.
2. The graph of $y = b$ is a *horizontal line* crossing the y axis at $(0, b)$.

To simplify the graphing of certain linear equations, some students prefer the **intercept method** of graphing. This method makes use of the fact that the solutions that are easiest to find are those with an x coordinate or a y coordinate of 0. For instance, let's graph the equation

$$4x + 3y = 12$$

NOTE With practice, all this can be done mentally, which is the big advantage of this method.

First, let $x = 0$ and solve for y.

$$4 \cdot 0 + 3y = 12$$

$$3y = 12$$

$$y = 4$$

So (0, 4) is one solution. Now we let $y = 0$ and solve for x.

$$4x + 3 \cdot 0 = 12$$
$$4x = 12$$
$$x = 3$$

A second solution is (3, 0).

The two points corresponding to these solutions can now be used to graph the equation.

NOTE Remember, only two points are needed to graph a line. A third point is used only as a check.

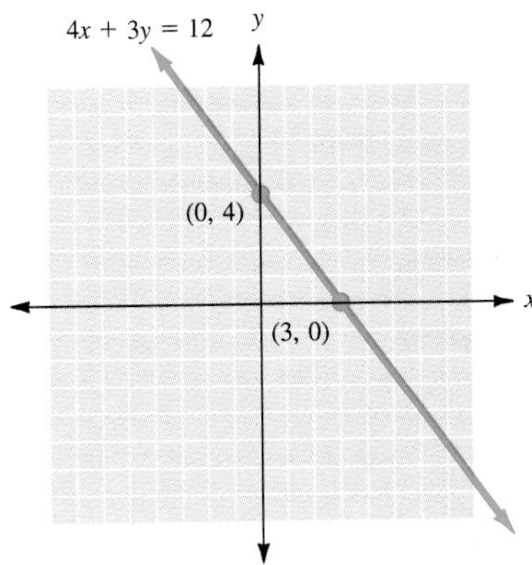

NOTE The intercepts are the points where the line cuts the x and y axes.

The ordered pair (3, 0) is called the ***x* intercept,** and the ordered pair (0, 4) is the ***y* intercept** of the graph. Using these points to draw the graph gives the name to this method. Let's look at a second example of graphing by the intercept method.

Example 7

Using the Intercept Method to Graph a Line

Graph $3x - 5y = 15$, using the intercept method.

To find the x intercept, let $y = 0$.

$$3x - 5 \cdot 0 = 15$$
$$x = 5$$

The x intercept is (5, 0)

To find the y intercept, let $x = 0$.

$$3 \cdot 0 - 5y = 15$$
$$y = -3$$

The y intercept is (0, −3)

So (5, 0) and (0, −3) are solutions for the equation, and we can use the corresponding points to graph the equation.

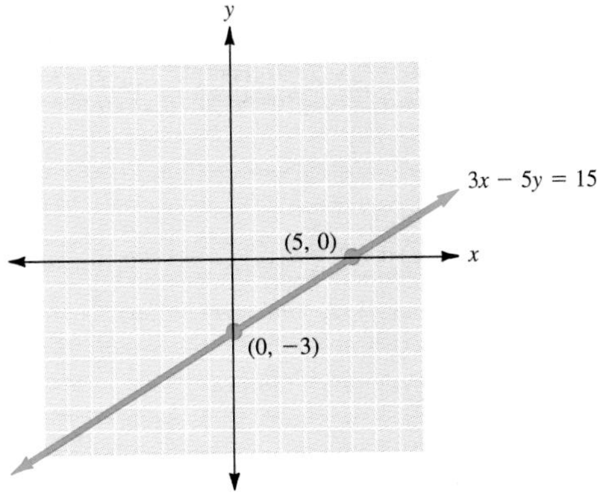

CHECK YOURSELF 7

Graph $4x + 5y = 20$, *using the intercept method.*

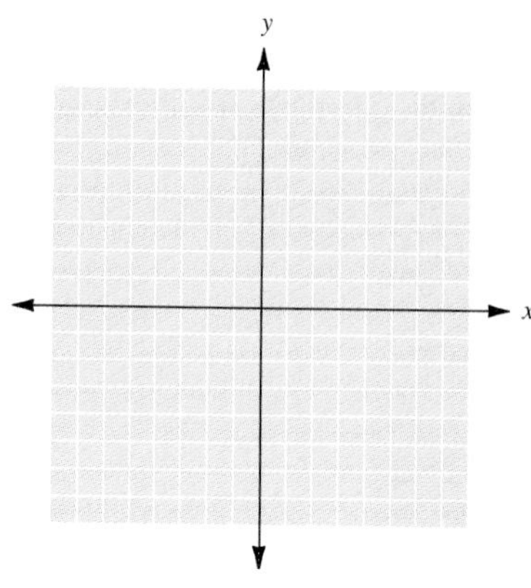

NOTE Finding the third "checkpoint" is always a good idea.

This all looks quite easy, and for many equations it is. What are the drawbacks? For one, you don't have a third checkpoint, and it is possible for errors to occur. You can, of course, still find a third point (other than the two intercepts) to be sure your graph is correct. A second difficulty arises when the x and y intercepts are very close to one another (or are actually the same point—the origin). For instance, if we have the equation

$$3x + 2y = 1$$

the intercepts are $\left(\frac{1}{3}, 0\right)$ and $\left(0, \frac{1}{2}\right)$. It is hard to draw a line accurately through these intercepts, so choose other solutions farther away from the origin for your points.

Let's summarize the steps of graphing by the intercept method for appropriate equations.

Step by Step: Graphing a Line by the Intercept Method

Step 1 To find the x intercept: Let $y = 0$, then solve for x.
Step 2 To find the y intercept: Let $x = 0$, then solve for y.
Step 3 Graph the x and y intercepts.
Step 4 Draw a straight line through the intercepts.

A third method of graphing linear equations involves **solving the equation for** y**.** The reason we use this extra step is that it often will make finding solutions for the equation much easier. Let's look at an example.

Example 8

Graphing a Linear Equation

Graph $2x + 3y = 6$.

Rather than finding solutions for the equation in this form, we solve for y.

NOTE Remember that solving for y means that we want to leave y isolated on the left.

$$2x + 3y = 6$$

$$3y = 6 - 2x \quad \text{Subtract } 2x.$$

$$y = \frac{6 - 2x}{3} \quad \text{Divide by 3.}$$

or $\quad y = 2 - \frac{2}{3}x$

Now find your solutions by picking convenient values for x.

NOTE Again, to pick convenient values for x, we suggest you look at the equation carefully. Here, for instance, picking multiples of 3 for x will make the work much easier.

If $x = -3$:

$$y = 2 - \frac{2}{3}(-3)$$

$$= 2 + 2 = 4$$

So $(-3, 4)$ is a solution.

If $x = 0$:

$$y = 2 - \frac{2}{3} \cdot 0$$

$$= 2$$

So $(0, 2)$ is a solution.

If $x = 3$:

$$y = 2 - \frac{2}{3} \cdot 3$$

$$= 2 - 2 = 0$$

So $(3, 0)$ is a solution.

We can now plot the points that correspond to these solutions and form the graph of the equation as before.

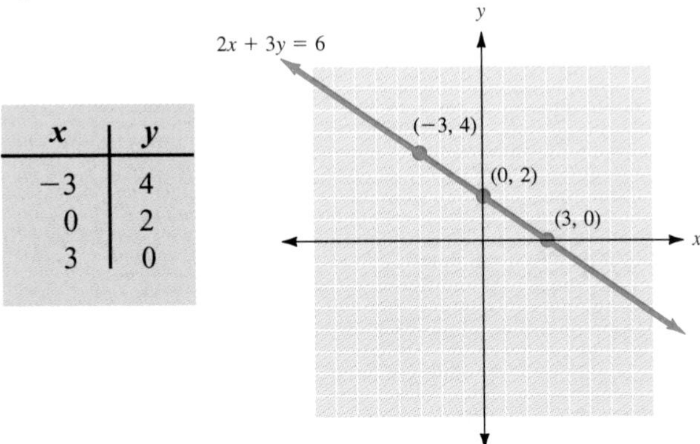

x	y
−3	4
0	2
3	0

CHECK YOURSELF 8

Graph the equation $5x + 2y = 10$. *Solve for y to determine solutions.*

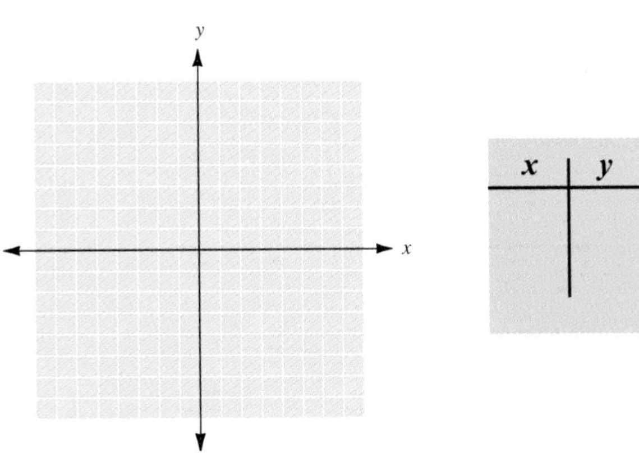

x	y

CHECK YOURSELF ANSWERS

1.

x	y
1	−4
2	−2
3	0

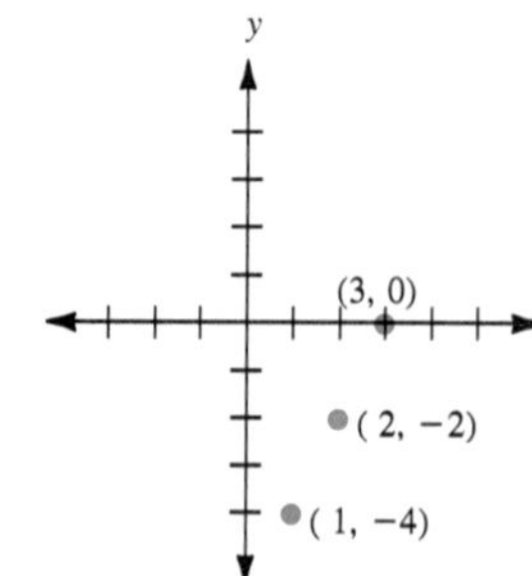

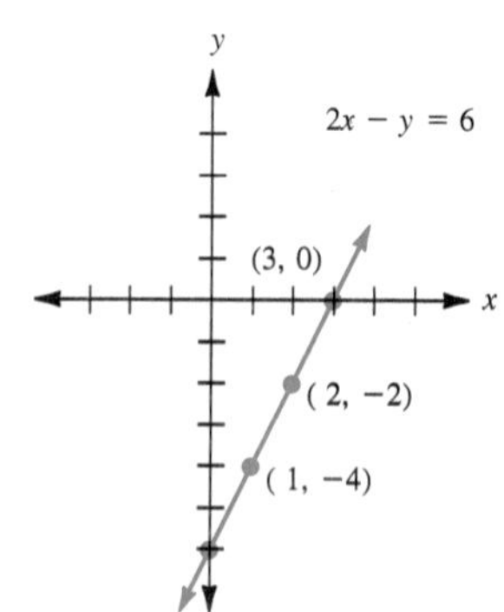

2.

x	y
0	0
1	−2
2	−4

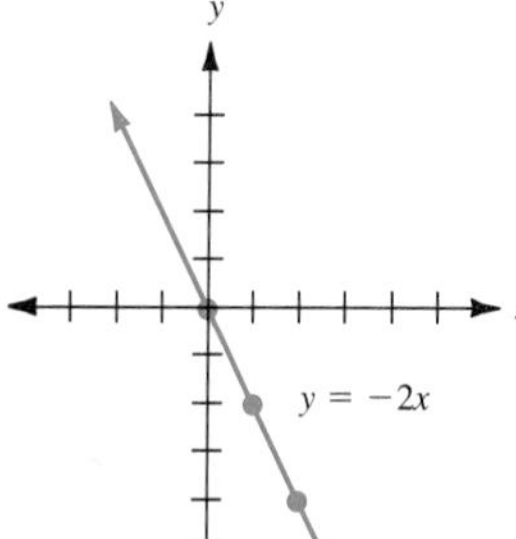

3.

x	y
0	−2
1	1
2	4

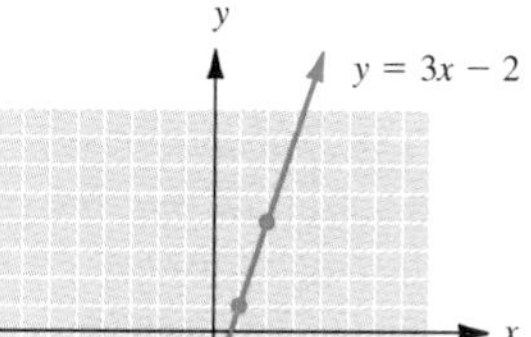

4.

x	y
−3	4
0	3
3	2

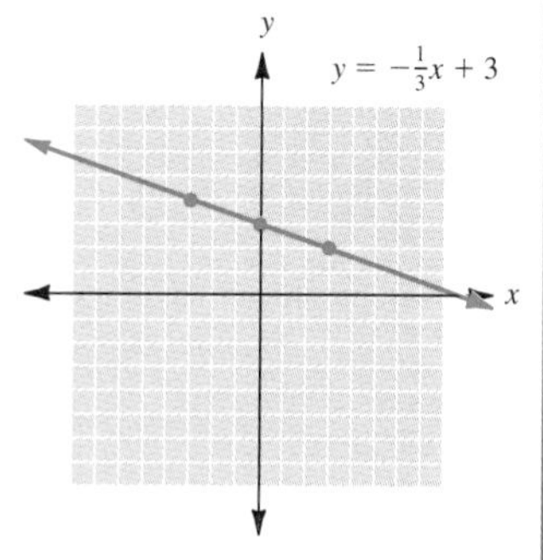

5.

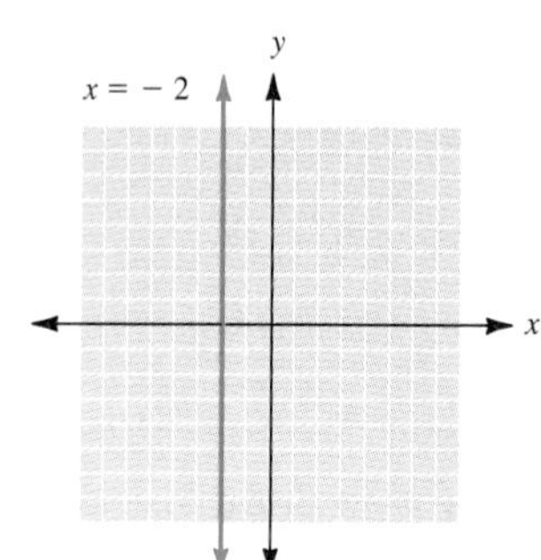

6.

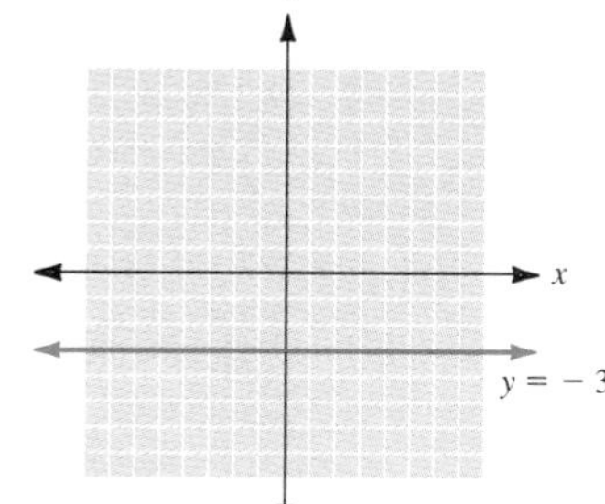

7.

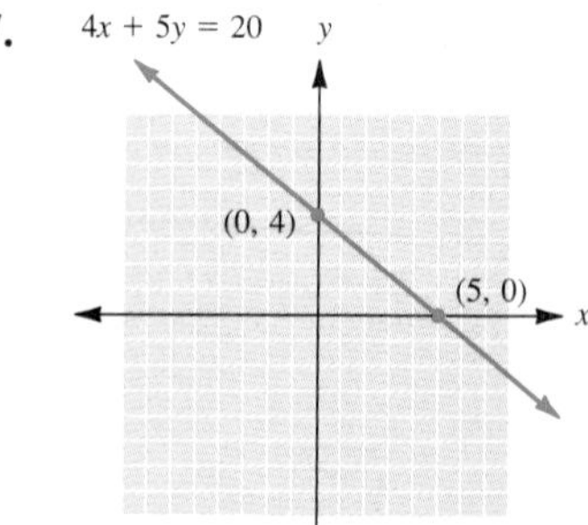

8.

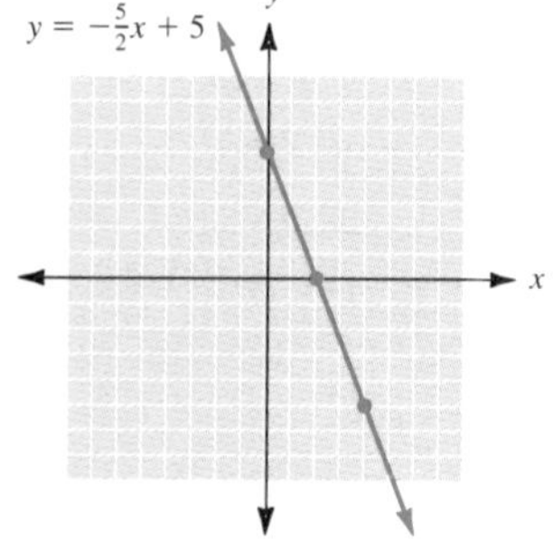

Name ____________

Section ________ Date ________

6.3 Exercises

Graph each of the following equations.

1. $x + y = 6$

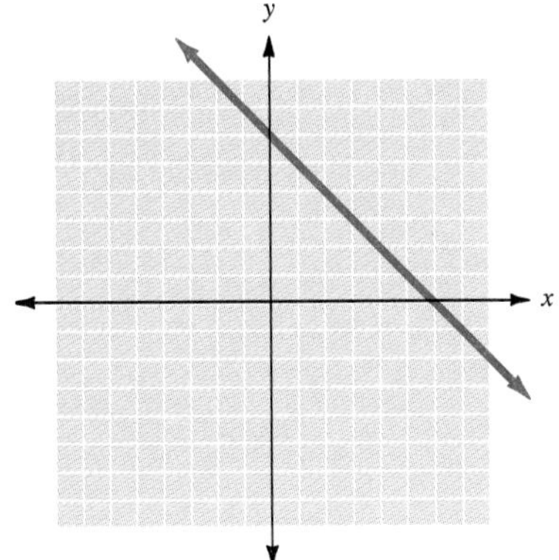

2. $x - y = 5$

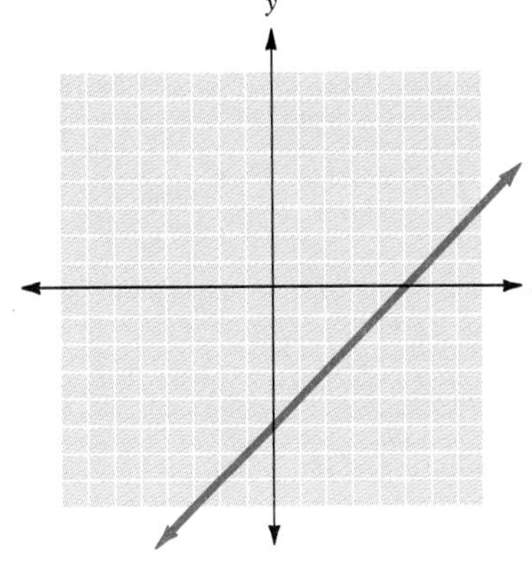

3. $x - y = -3$

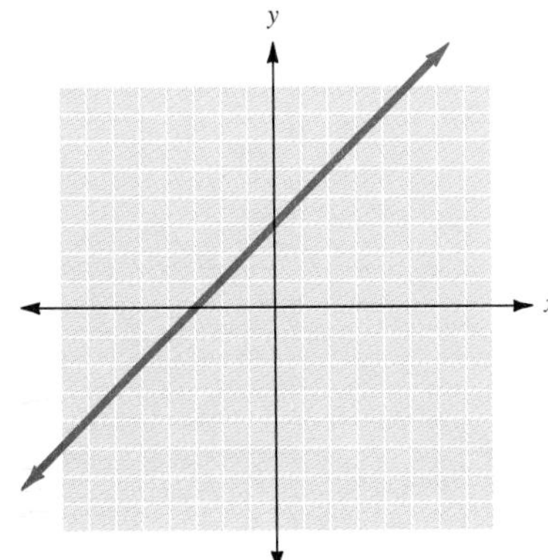

4. $x + y = -3$

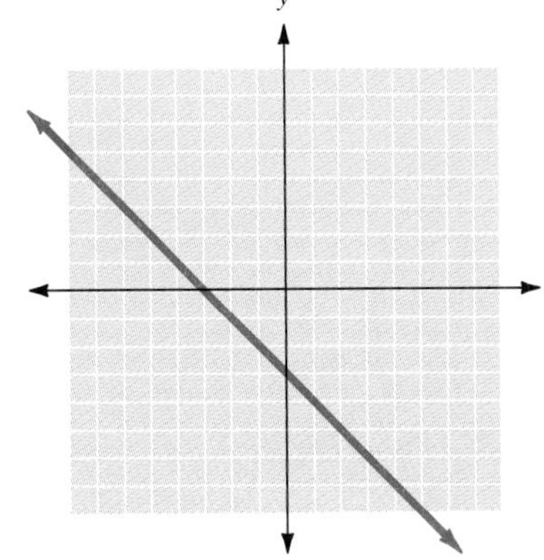

5. $2x + y = 2$

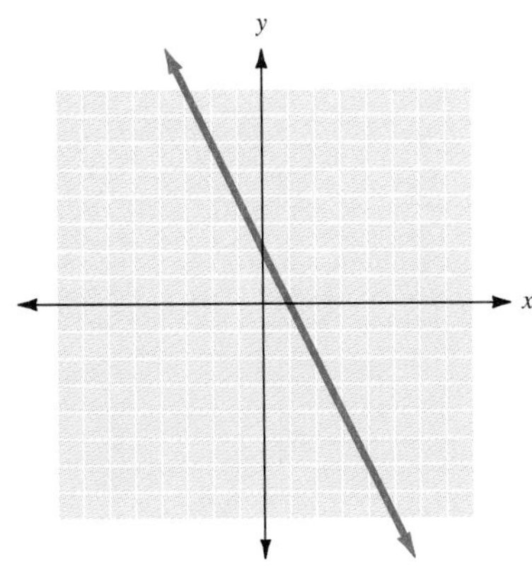

6. $x - 2y = 6$

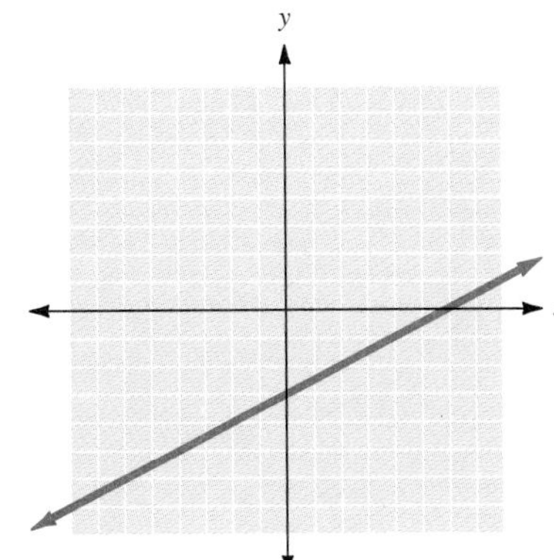

ANSWERS

1. See exercise
2. See exercise
3. See exercise
4. See exercise
5. See exercise
6. See exercise

ANSWERS

7. See exercise
8. See exercise
9. See exercise
10. See exercise
11. See exercise
12. See exercise

7. $3x + y = 0$

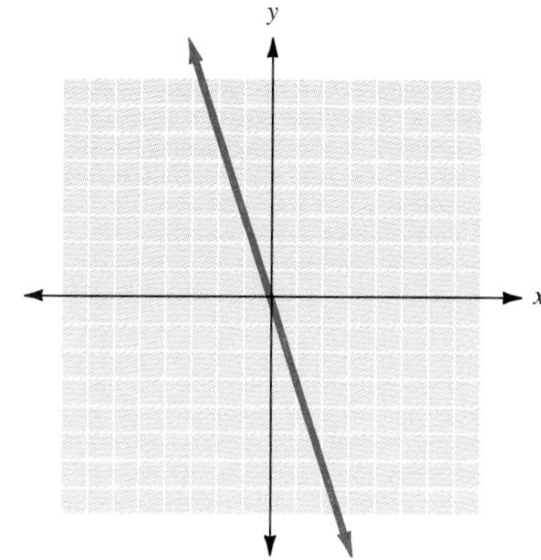

8. $3x - y = 6$

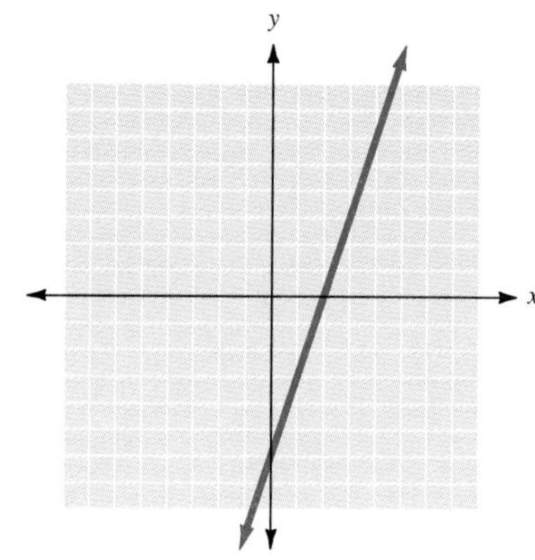

9. $x + 4y = 8$

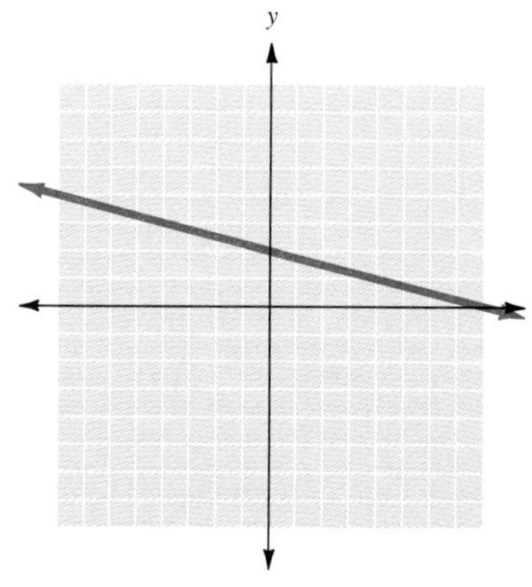

10. $2x - 3y = 6$

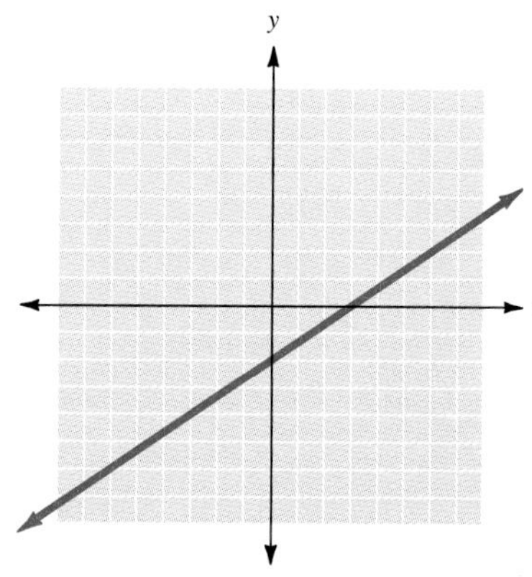

11. $y = 5x$

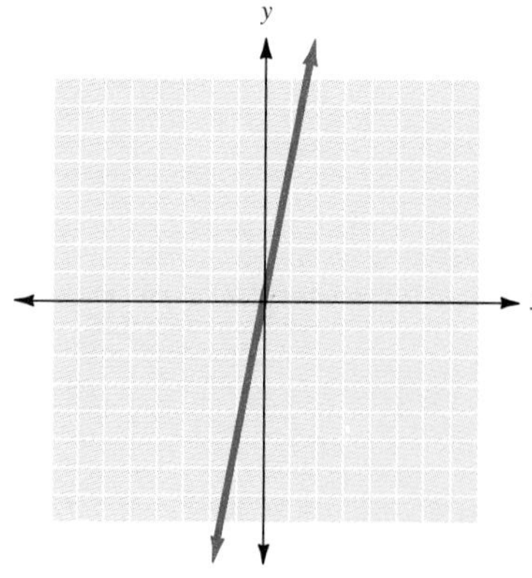

12. $y = -4x$

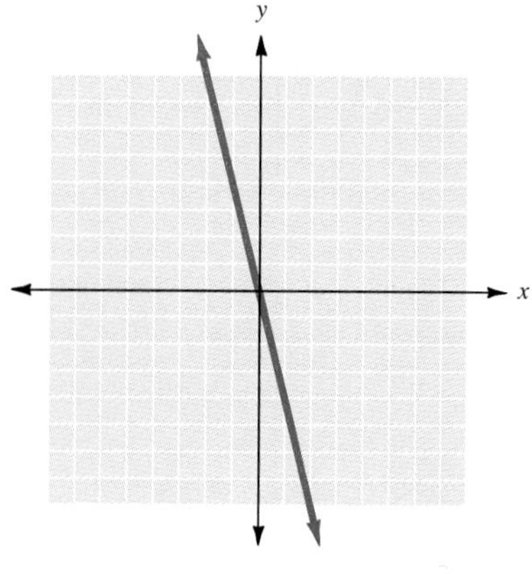

13. $y = 2x - 1$

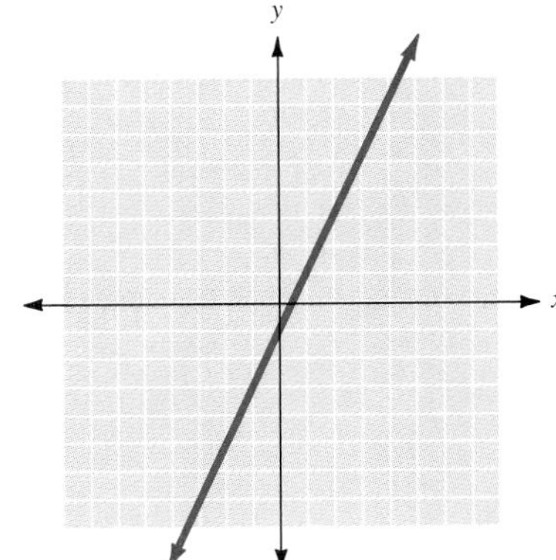

14. $y = 4x + 3$

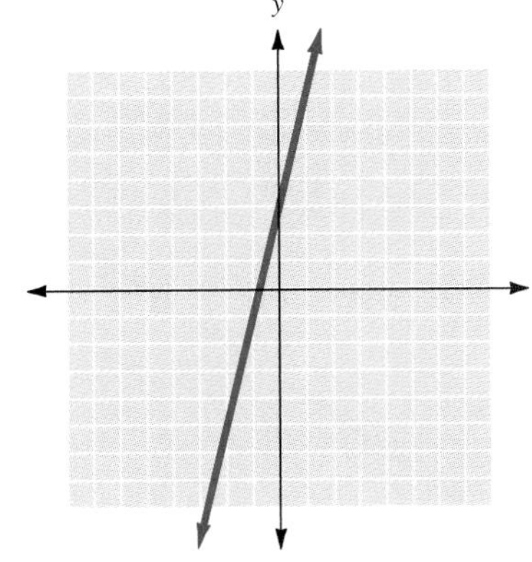

15. $y = -3x + 1$

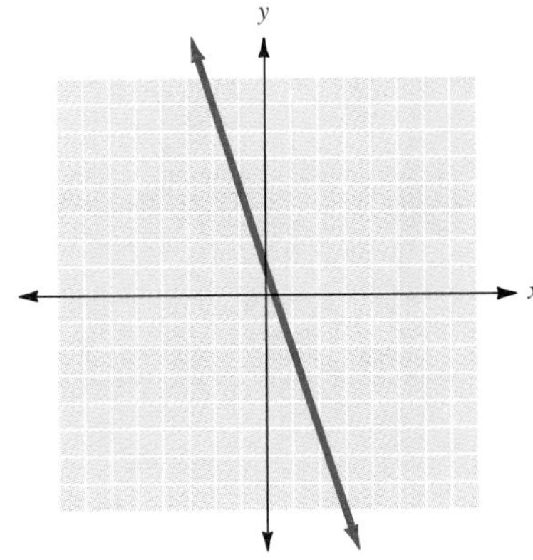

16. $y = -3x - 3$

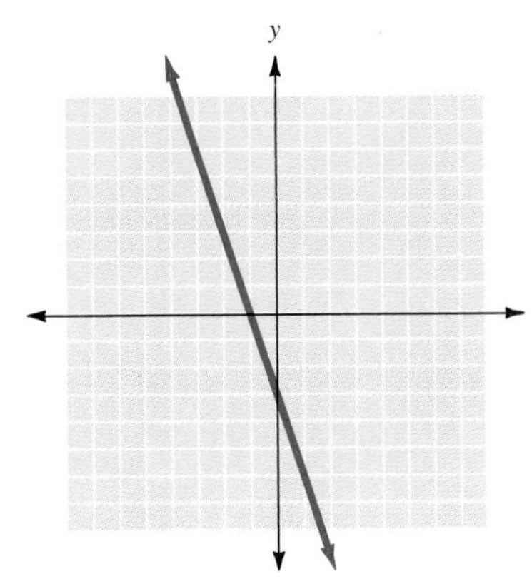

17. $y = \frac{1}{3}x$

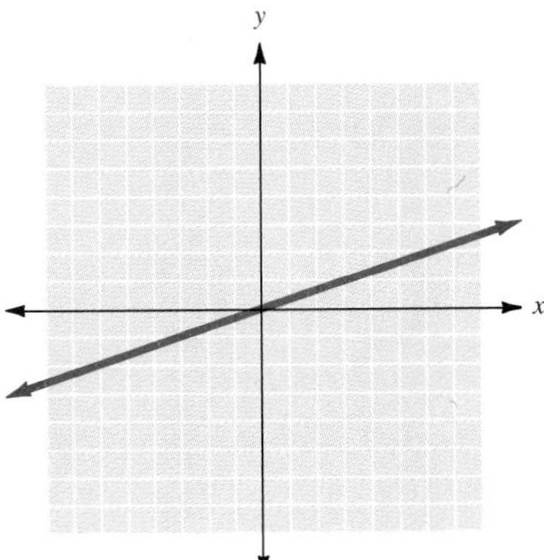

18. $y = -\frac{1}{4}x$

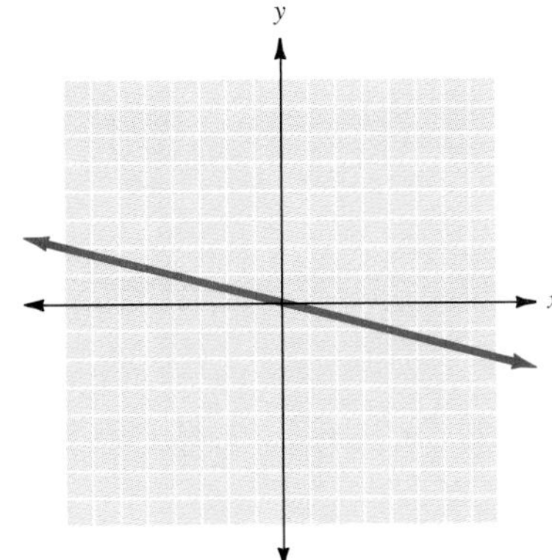

ANSWERS

13. See exercise

14. See exercise

15. See exercise

16. See exercise

17. See exercise

18. See exercise

ANSWERS

19. See exercise

20. See exercise

21. See exercise

22. See exercise

23. See exercise

24. See exercise

19. $y = \frac{2}{3}x - 3$

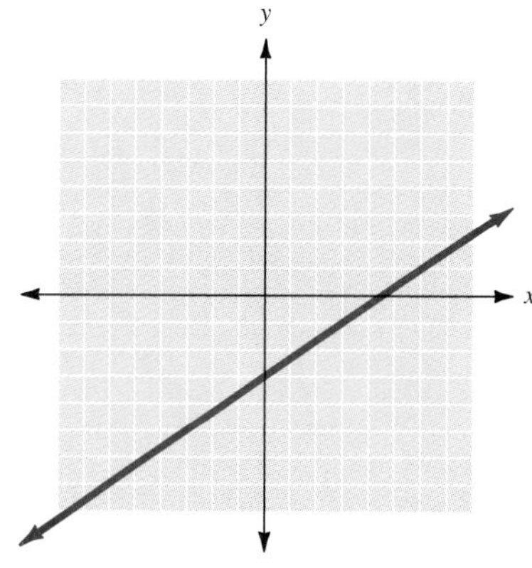

20. $y = \frac{3}{4}x + 2$

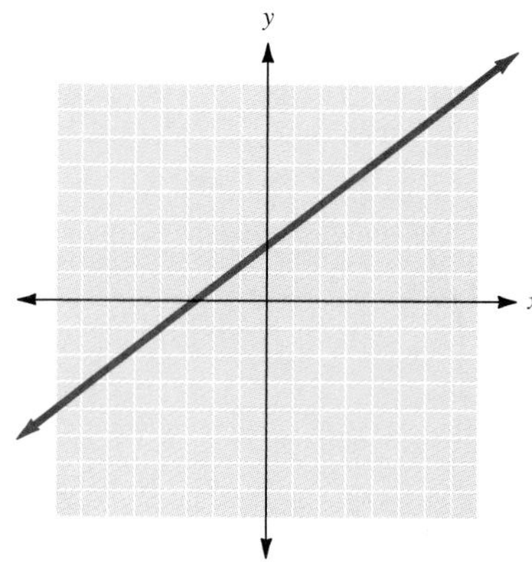

21. $x = 5$

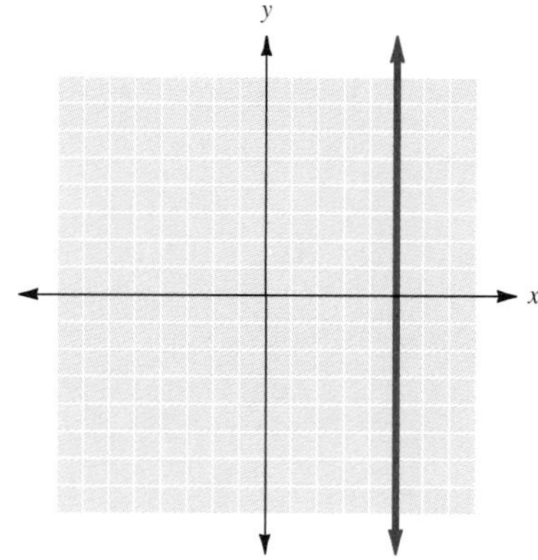

22. $y = -3$

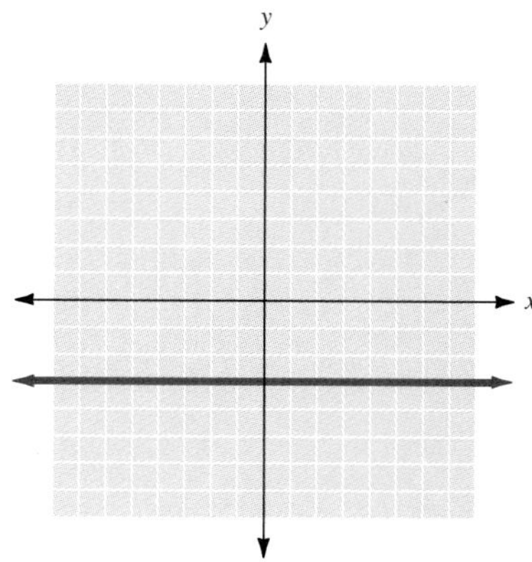

23. $y = 1$

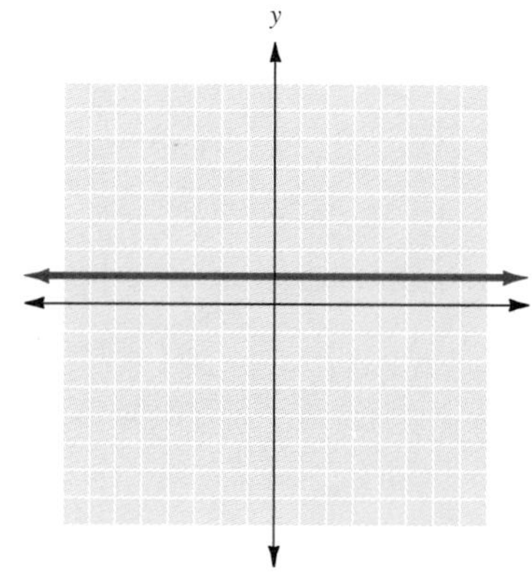

24. $x = -2$

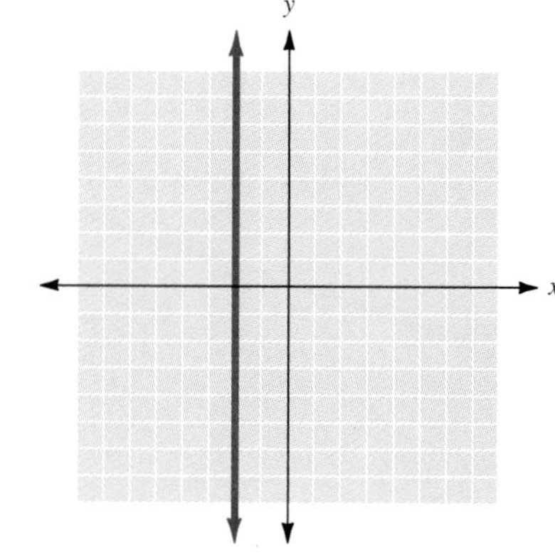

Graph each of the following equations, using the intercept method.

25. $x - 2y = 4$

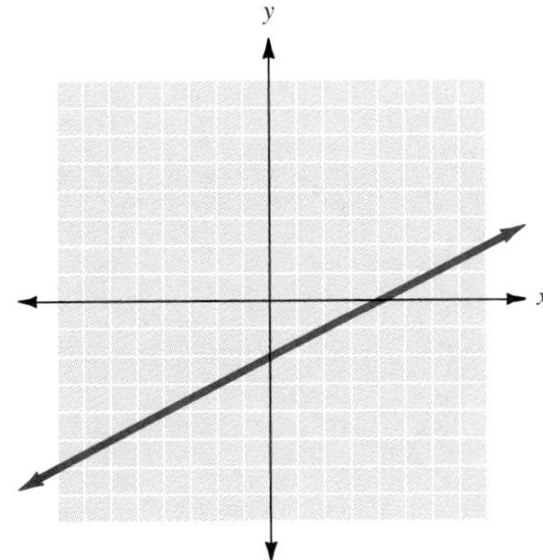

26. $6x + y = 6$

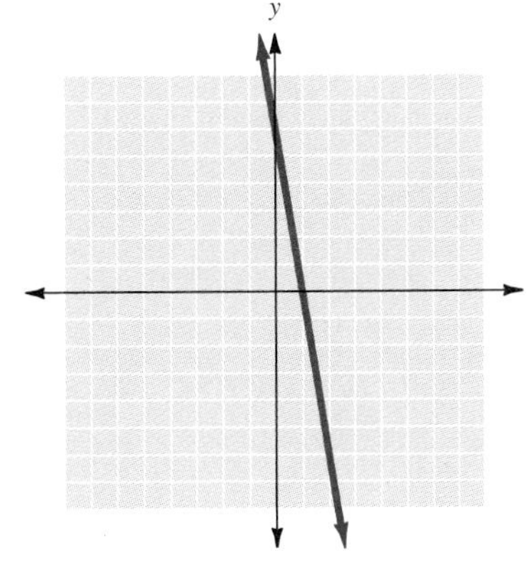

27. $5x + 2y = 10$

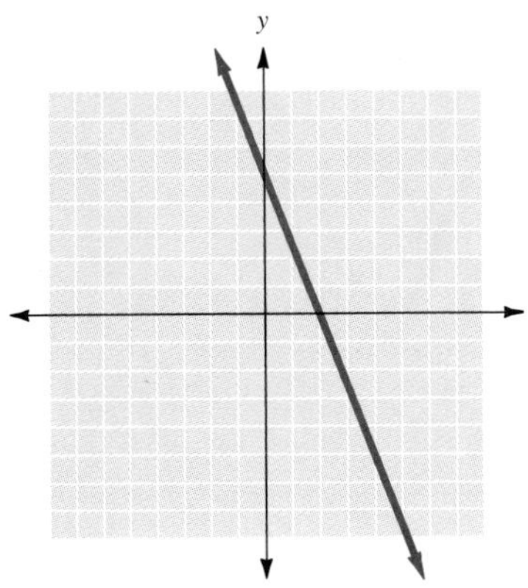

28. $2x + 3y = 6$

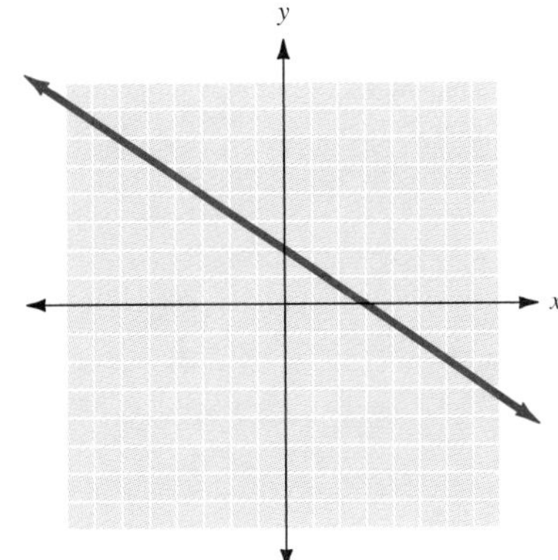

29. $3x + 5y = 15$

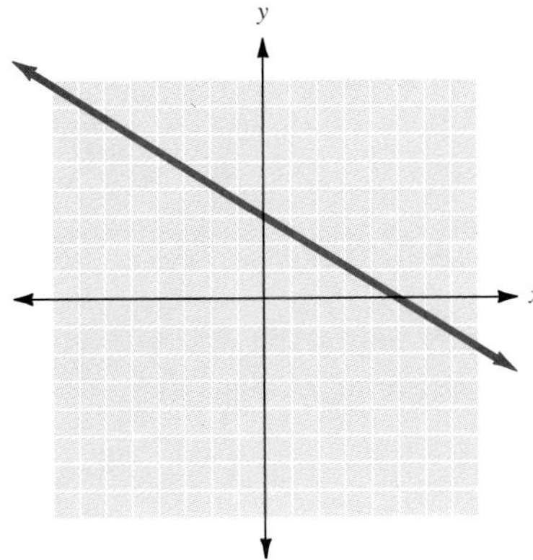

30. $4x + 3y = 12$

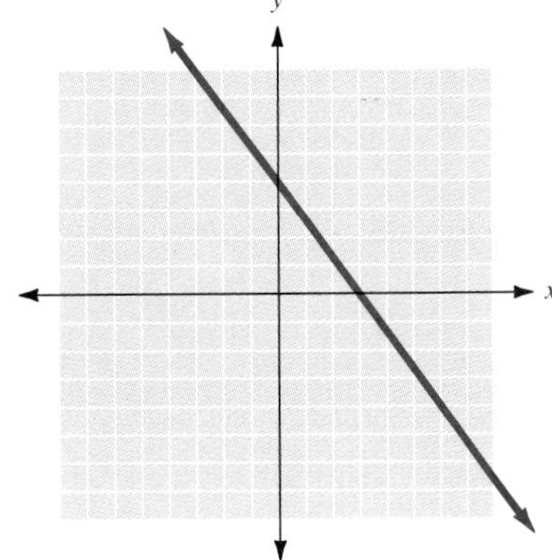

ANSWERS

25. See exercise

26. See exercise

27. See exercise

28. See exercise

29. See exercise

30. See exercise

ANSWERS

31. $y = 2 - \frac{x}{3}$

32. $y = -3 + \frac{x}{2}$

33. $y = 3 - \frac{3}{4}x$

34. $y = -4 + \frac{2}{3}x$

35. $y = -5 + \frac{5}{4}x$

36. $y = 7 - \frac{7}{3}x$

Graph each of the following equations by first solving for y.

31. $x + 3y = 6$

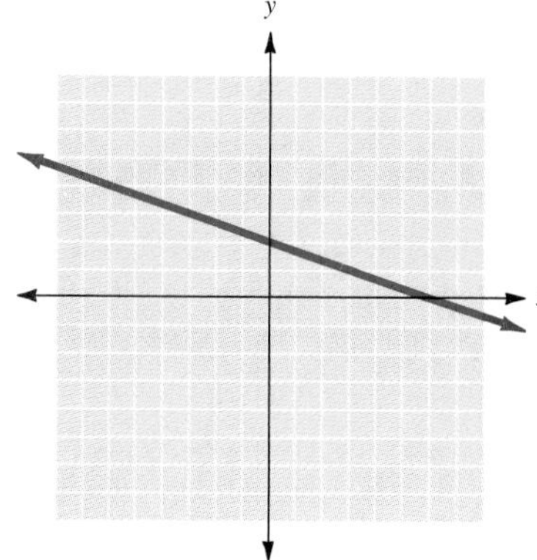

32. $x - 2y = 6$

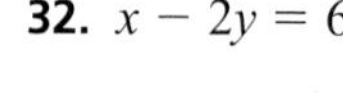

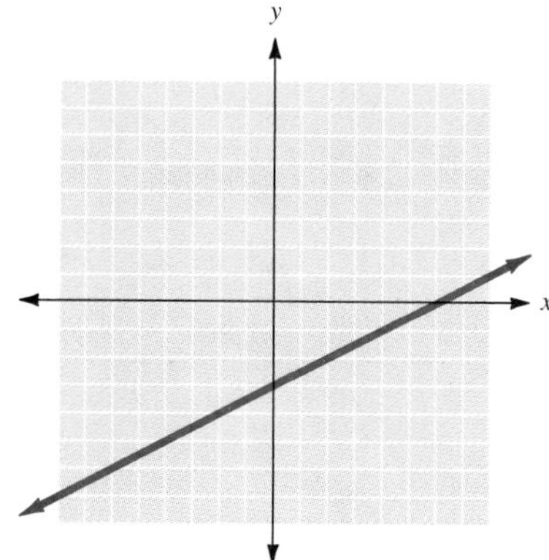

33. $3x + 4y = 12$

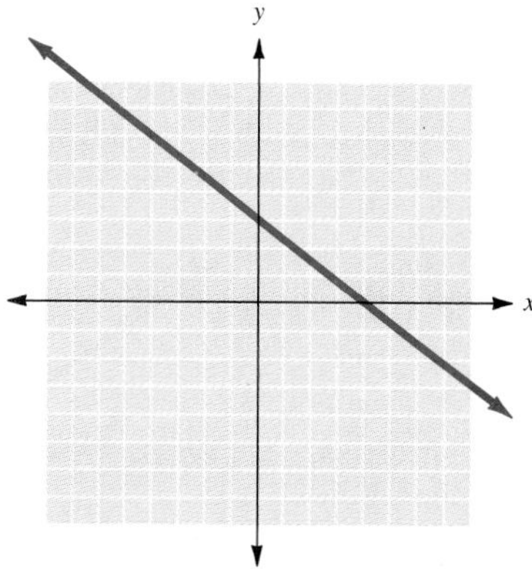

34. $2x - 3y = 12$

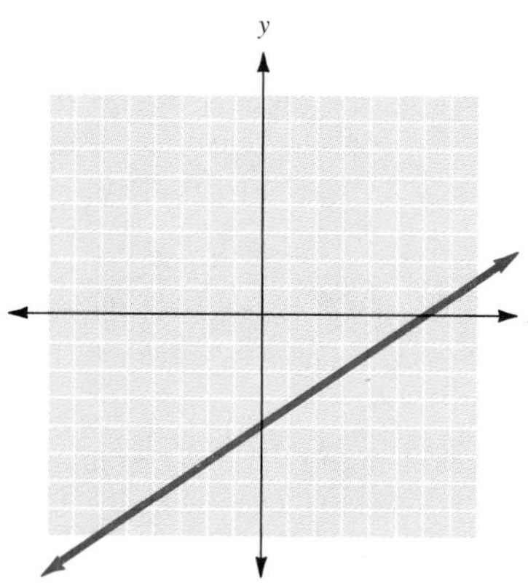

35. $5x - 4y = 20$

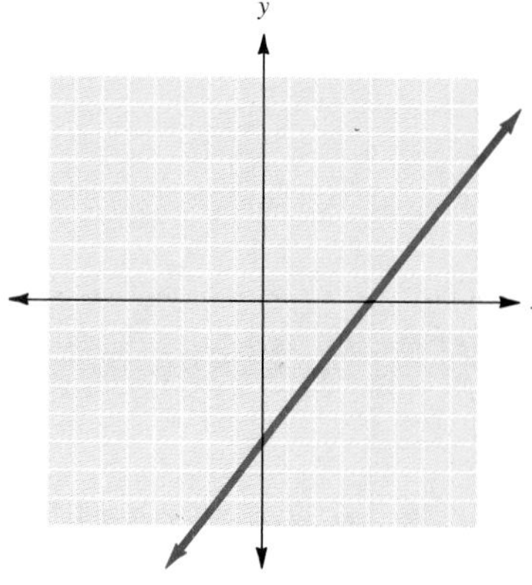

36. $7x + 3y = 21$

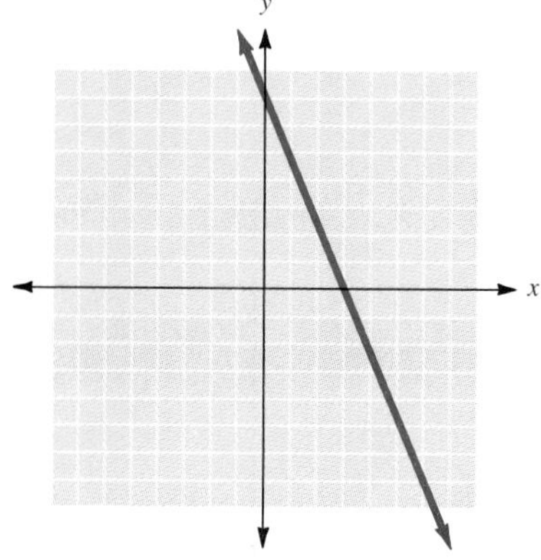

Write an equation that describes the following relationships between x and y. Then graph each relationship.

37. y is twice x.

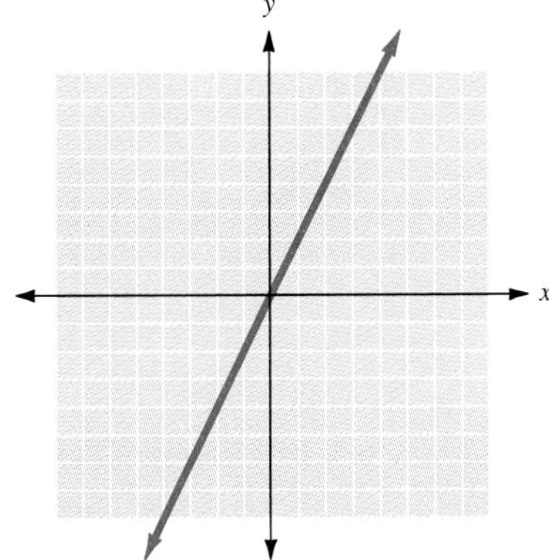

38. y is 3 times x.

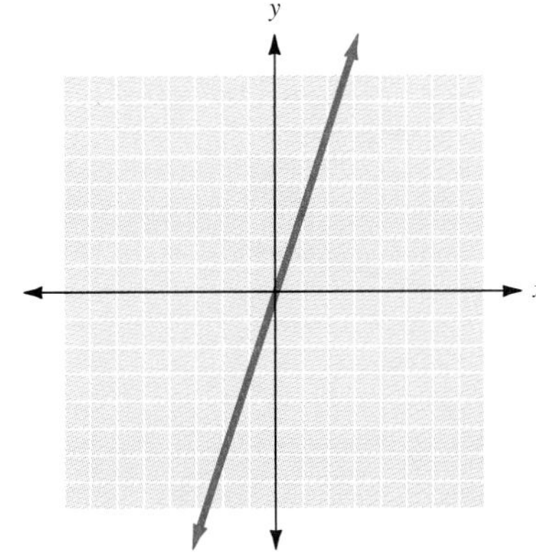

39. y is 3 more than x.

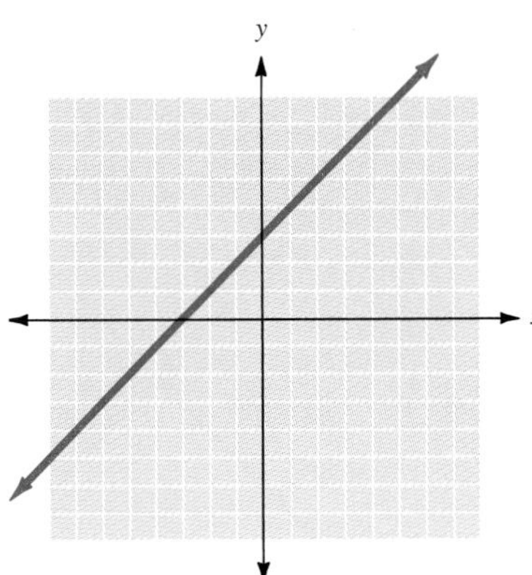

40. y is 2 less than x.

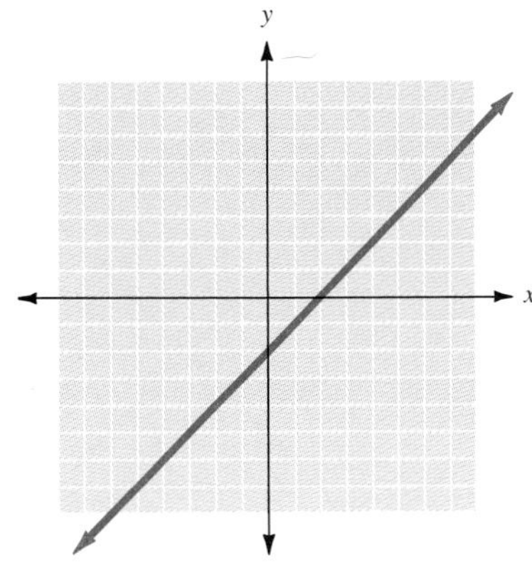

41. y is 3 less than 3 times x.

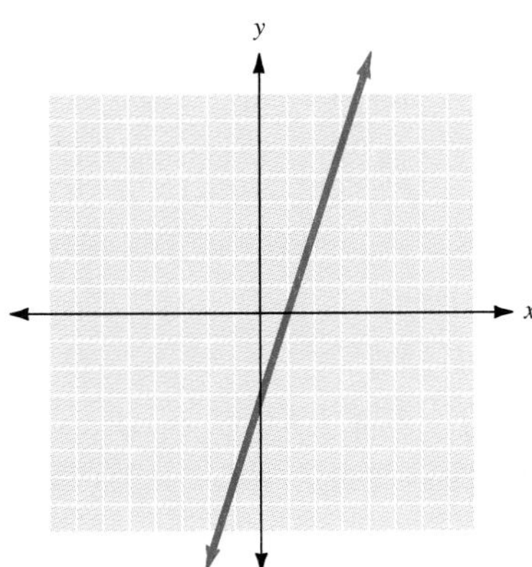

42. y is 4 more than twice x.

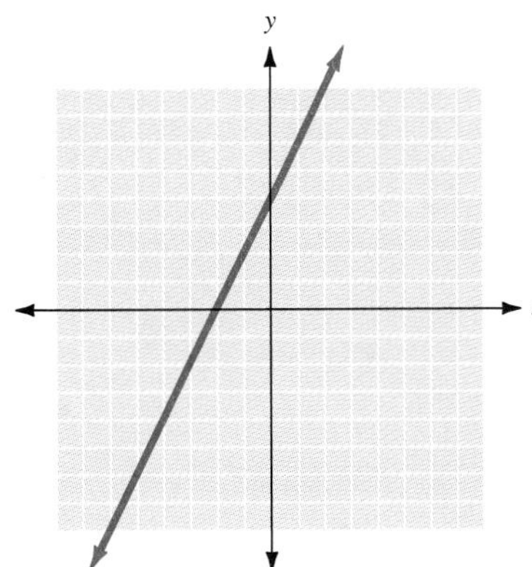

ANSWERS

37. $y = 2x$

38. $y = 3x$

39. $y = x + 3$

40. $y = x - 2$

41. $y = 3x - 3$

42. $y = 2x + 4$

ANSWERS

43. $x - 4y = 12$

44. $2x - y = 6$

45. (3, 1)

46. (4, 1)

47. See exercise

43. The difference of x and the product of 4 and y is 12.

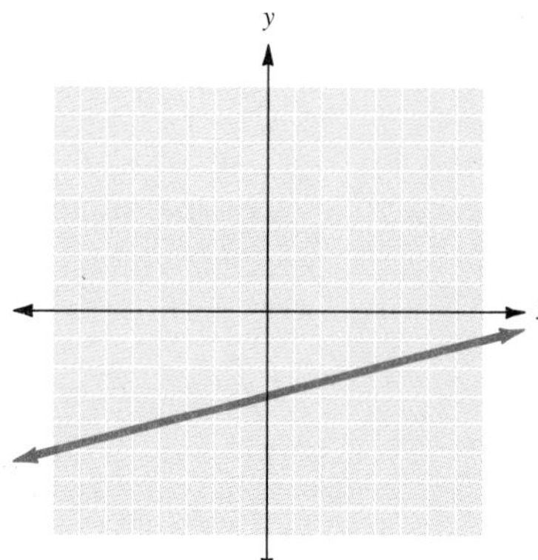

44. The difference of twice x and y is 6.

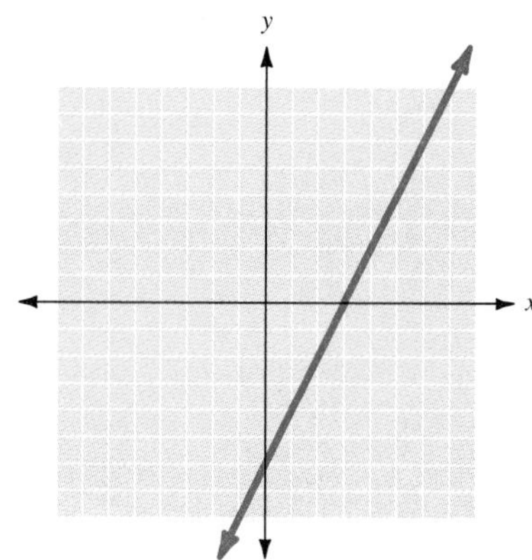

Graph each pair of equations on the same axes. Give the coordinates of the point where the lines intersect.

45. $x + y = 4$
$x - y = 2$

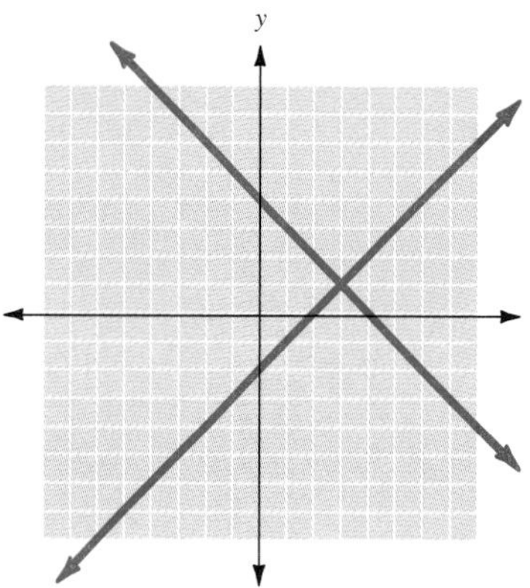

46. $x - y = 3$
$x + y = 5$

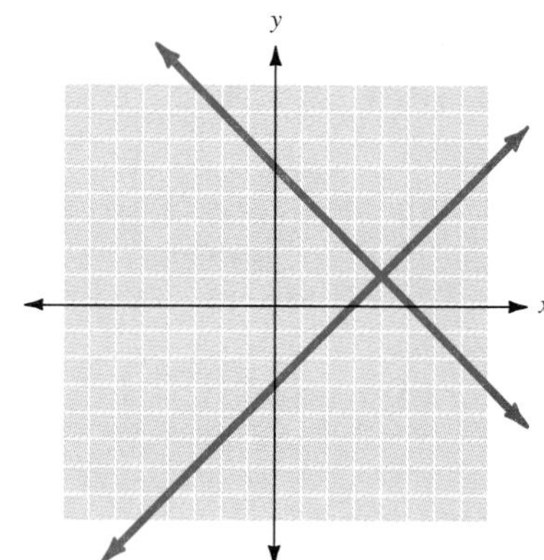

47. Graph of winnings. The equation $y = 0.10x + 200$ describes the amount of winnings a group earns for collecting plastic jugs in the recycling contest described in exercise 27 at the end of Section 6.2. Sketch the graph of the line on the coordinate system below.

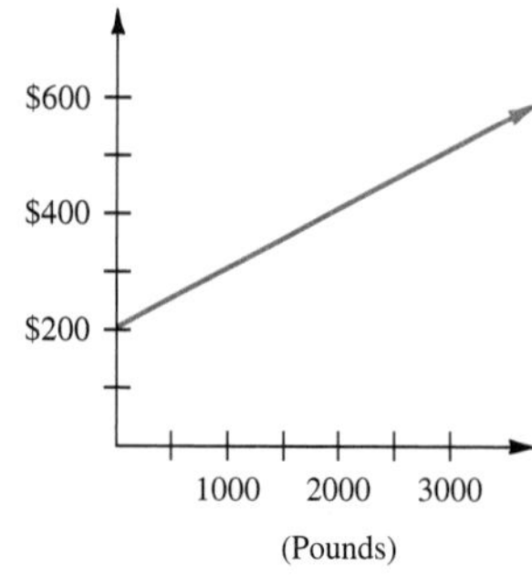

48. Minimum values. The contest sponsor will award a prize only if the winning group in the contest collects 100 lb of jugs or more. Use your graph in exercise 47 to determine the minimum prize possible.

49. Fundraising. A high school class wants to raise some money by recycling newspapers. They decide to rent a truck for a weekend and to collect the newspapers from homes in the neighborhood. The market price for recycled newsprint is currently \$11 per ton. The equation $y = 11x - 100$ describes the amount of money the class will make, in which y is the amount of money made in dollars, x is the number of tons of newsprint collected, and 100 is the cost in dollars to rent the truck.

(a) Using the axes below, draw a graph that represents the relationship between newsprint collected and money earned.

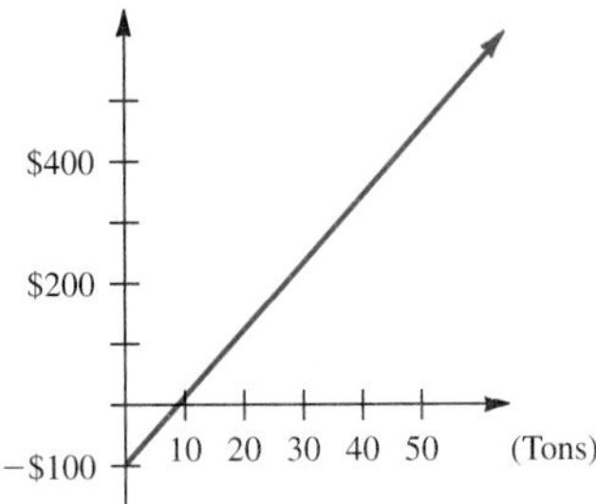

(b) The truck is costing the class \$100. How many tons of newspapers must the class collect to break even on this project?

(c) If the class members collect 16 tons of newsprint, how much money will they earn?

(d) Six months later the price of newsprint is \$17 dollars a ton, and the cost to rent the truck has risen to \$125. Write the equation that describes the amount of money the class might make at that time.

50. Production costs. The cost of producing a number of items x is given by $C = mx + b$, in which b is the fixed cost and m is the variable cost (the cost of producing one more item).

(a) If the fixed cost is \$40 and the variable cost is \$10, write the cost equation.

(b) Graph the cost equation.

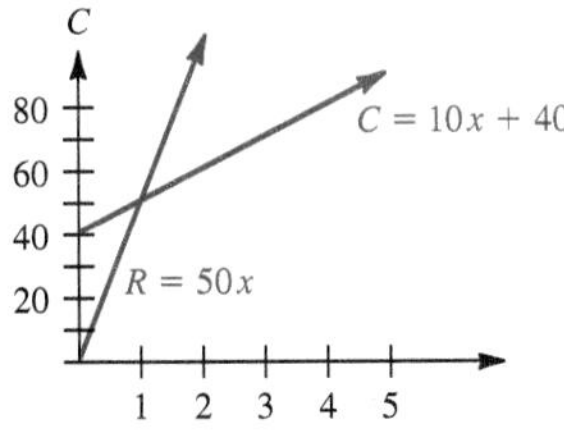

(c) The revenue generated from the sale of x items is given by $R = 50x$. Graph the revenue equation on the same set of axes as the cost equation.

(d) How many items must be produced for the revenue to equal the cost (the break-even point)?

ANSWERS

48. \$210

49. (a) See exercise

(b) $\frac{100}{11}$ or ≈9 tons

(c) \$76

(d) $y = 17x - 125$

50. (a) $C = 10x + 40$

(b) See exercise

(c) See exercise

(d) 1

51. The lines do not intersect. The y intercepts are (0,0), (0,4), and (0,−5).

52. The lines do not intersect. The y intercepts are (0,0), (0,3), and (0,−5).

a. 1

b. 2

c. $\frac{3}{2}$

d. 0

Graph each set of equations on the same coordinate system. Do the lines intersect? What are the y intercepts?

51. $y = 3x$
$y = 3x + 4$
$y = 3x - 5$

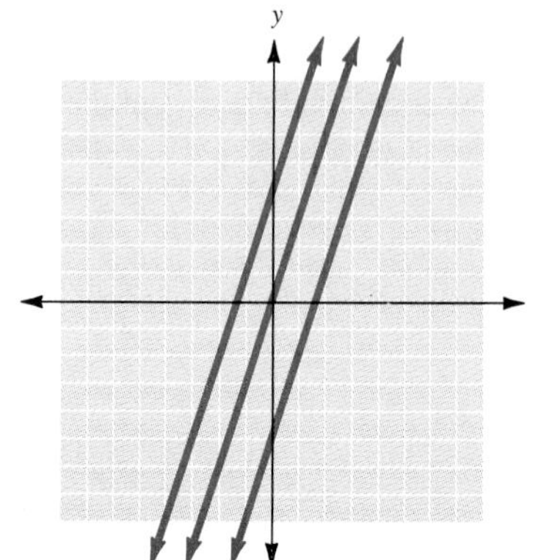

52. $y = -2x$
$y = -2x + 3$
$y = -2x - 5$

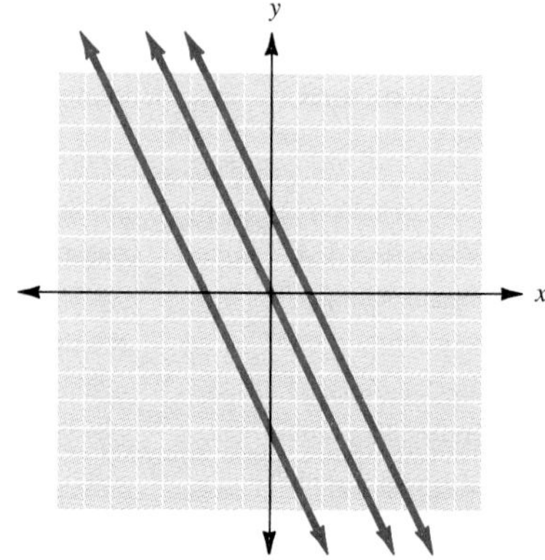

Getting Ready for Section 6.4 [Section 1.4]

Evaluate the following expressions.

(a) $\frac{7 - 3}{8 - 4}$ (b) $\frac{-9 - 5}{-4 - 3}$ (c) $\frac{4 - (-2)}{6 - 2}$ (d) $\frac{-4 - (-4)}{8 - 2}$

Answers

1. $x + y = 6$

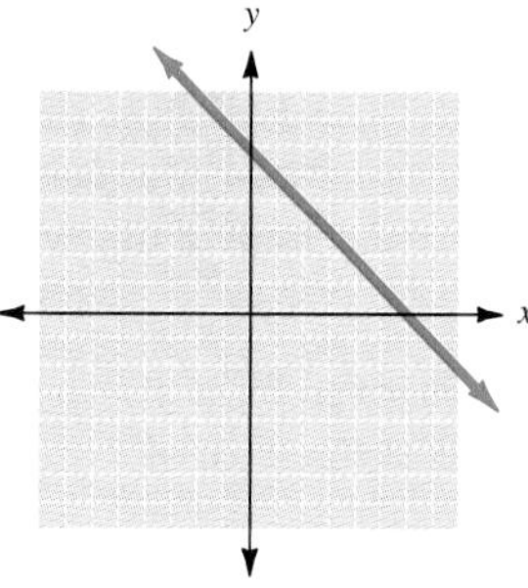

3. $x - y = -3$

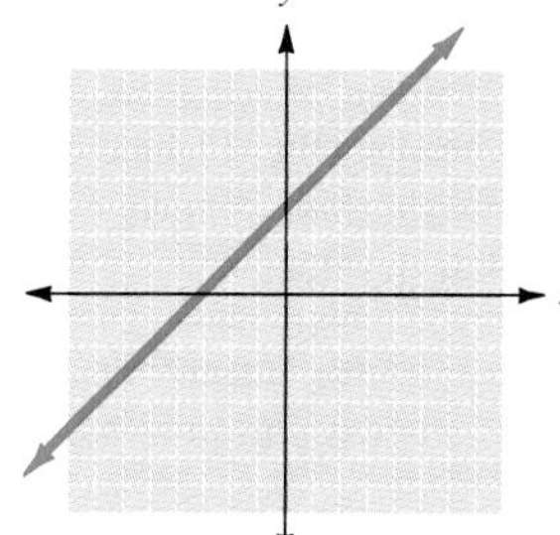

5. $2x + y = 2$

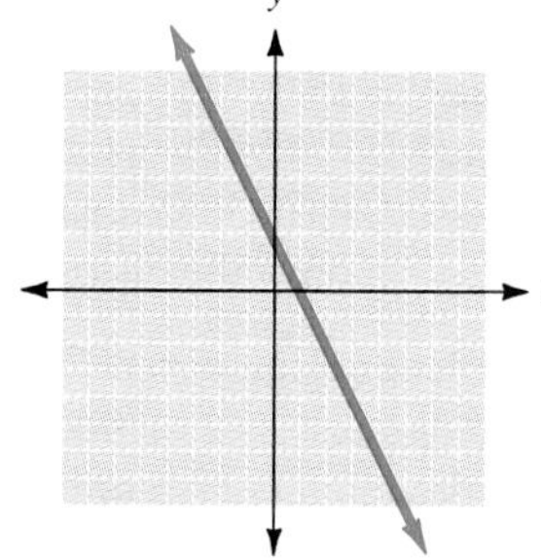

7. $3x + y = 0$

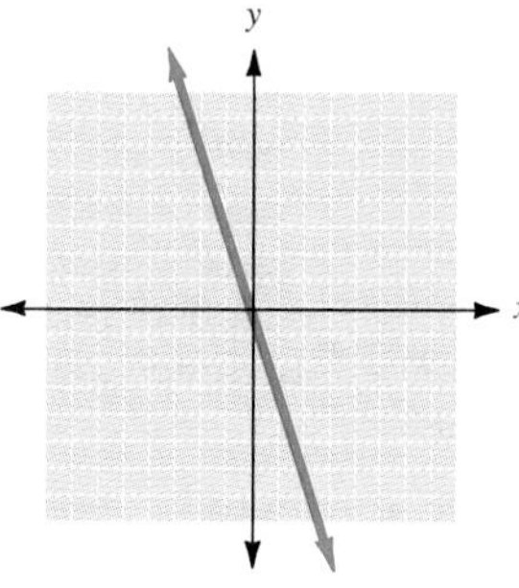

9. $x + 4y = 8$

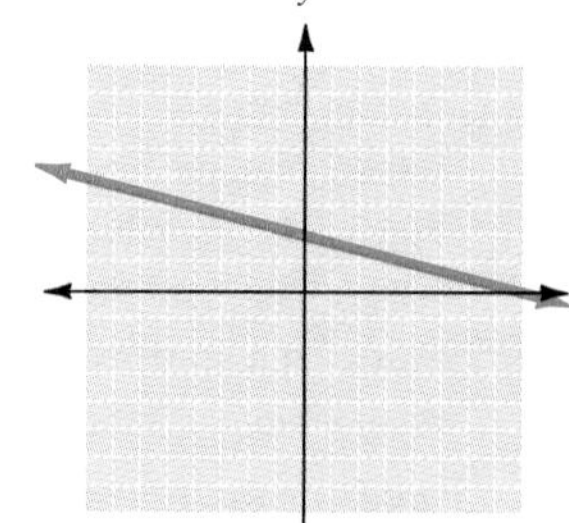

11. $y = 5x$

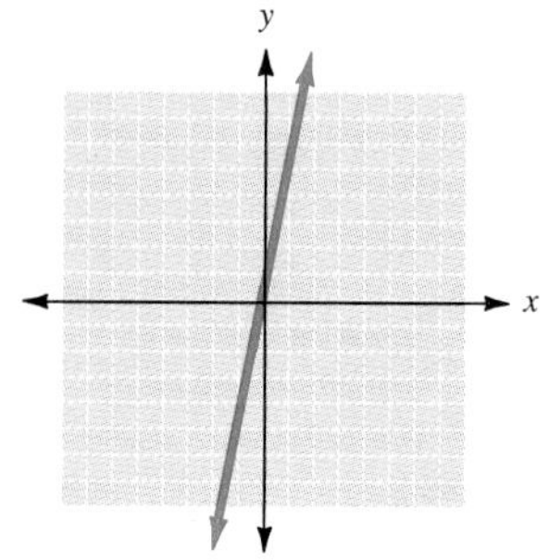

13. $y = 2x - 1$

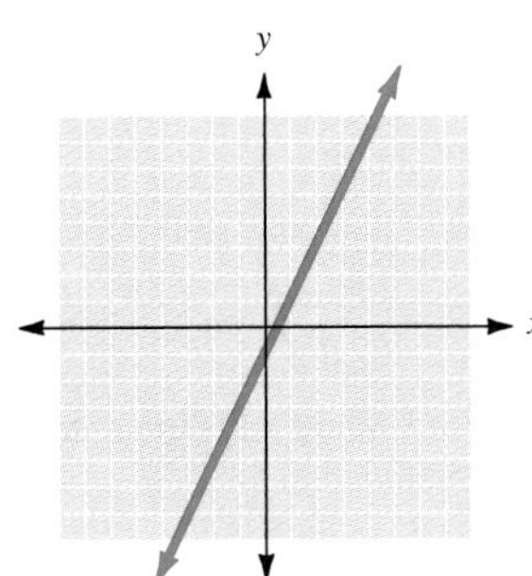

15. $y = -3x + 1$

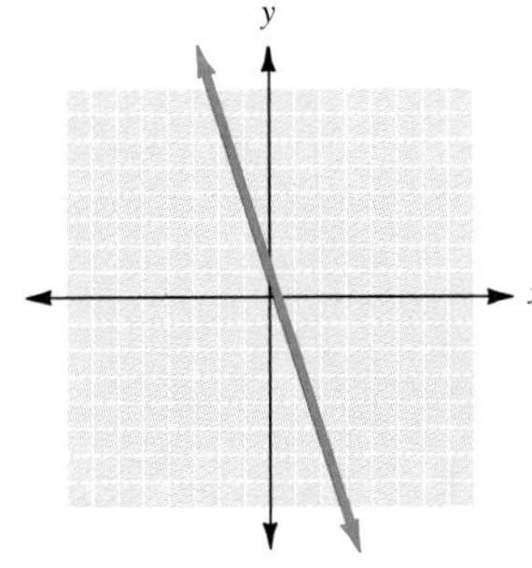

17. $y = \frac{1}{3}x$

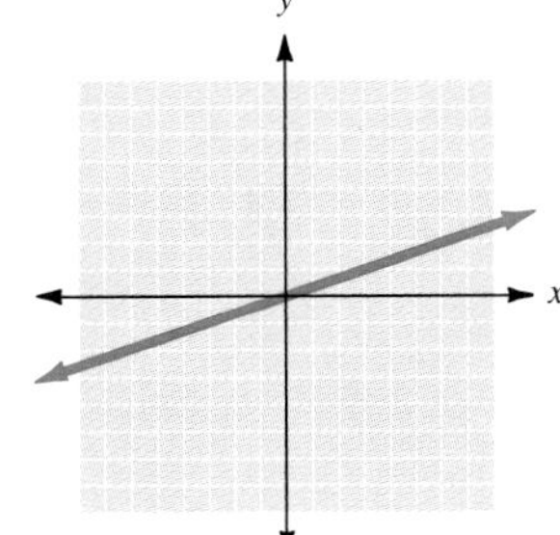

19. $y = \frac{2}{3}x - 3$

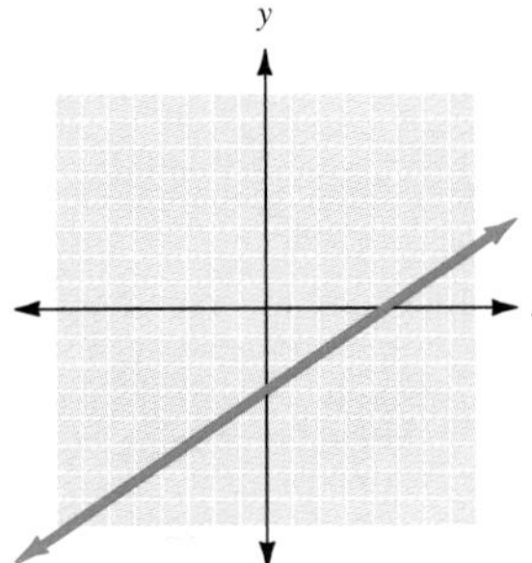

21. $x = 5$

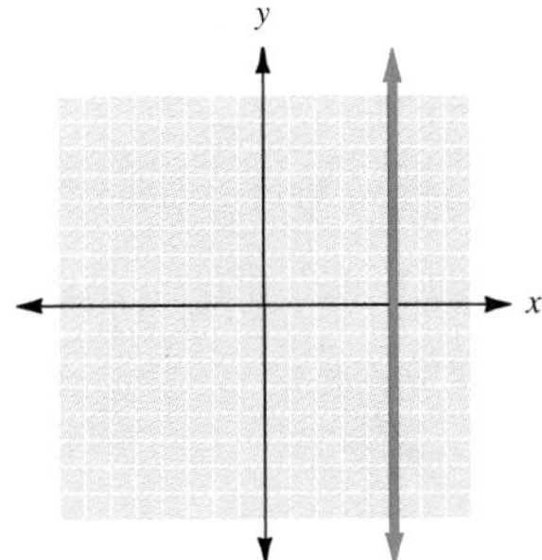

23. $y = 1$

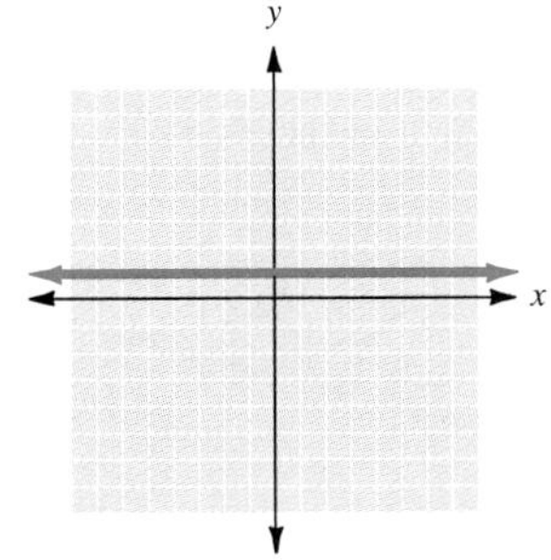

25. $x - 2y = 4$

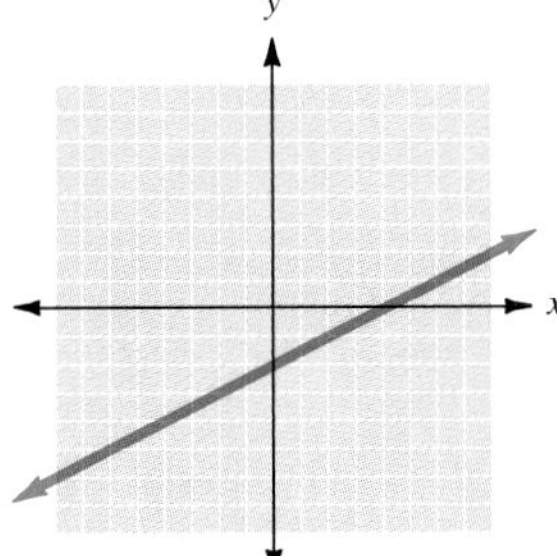

27. $5x + 2y = 10$

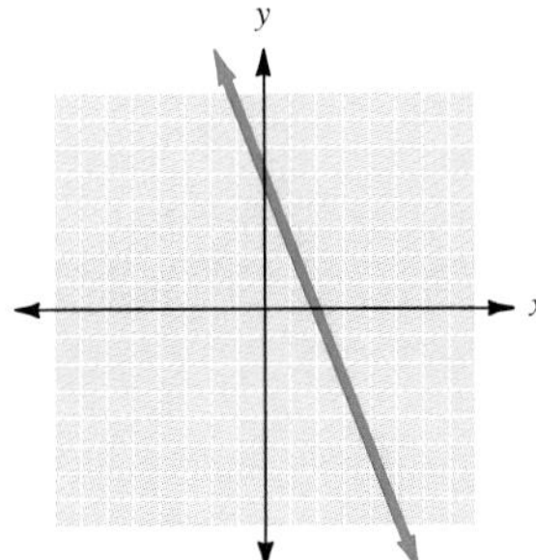

29. $3x + 5y = 15$

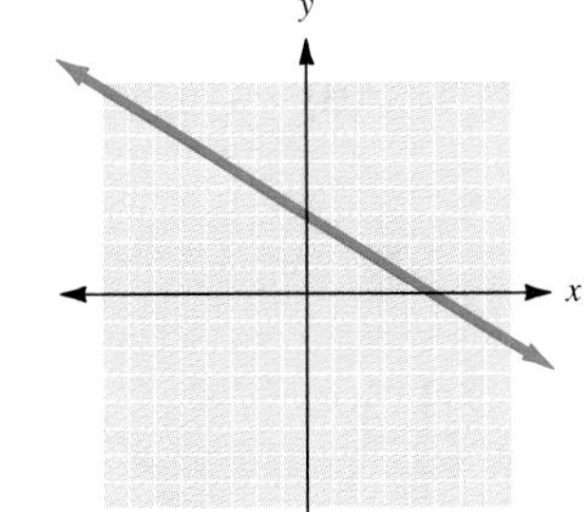

31. $y = 2 - \frac{x}{3}$

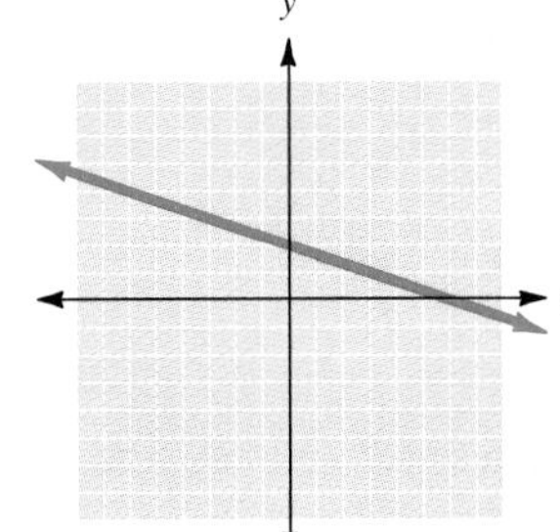

33. $y = 3 - \frac{3}{4}x$

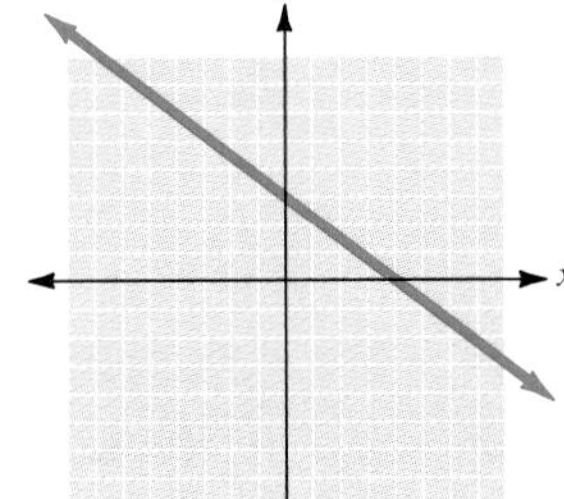

35. $y = -5 + \frac{5}{4}x$

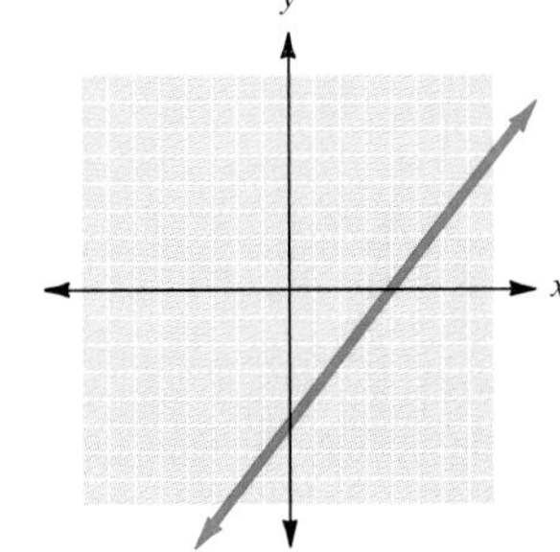

37. $y = 2x$

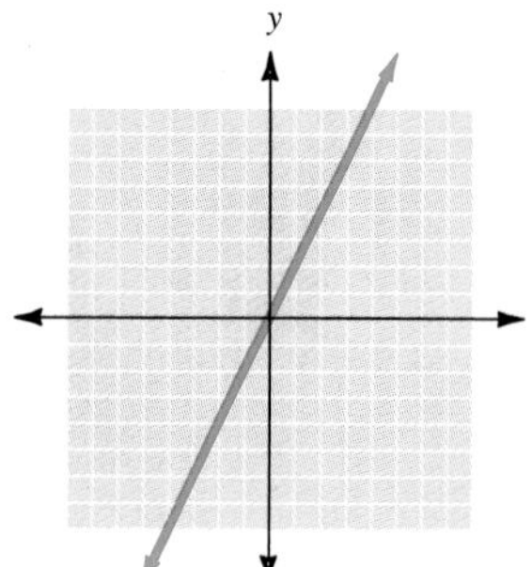

39. $y = x + 3$

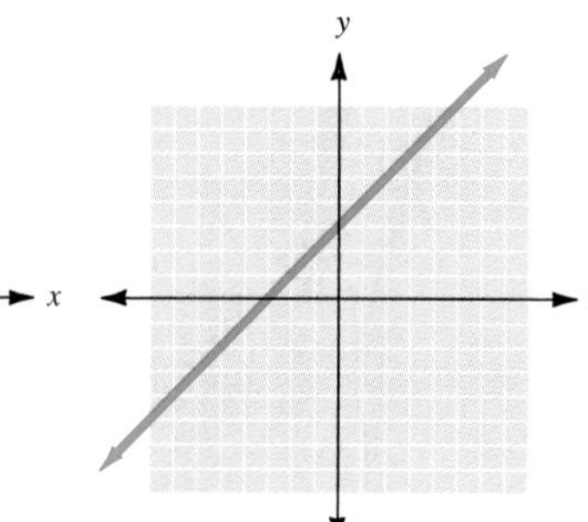

41. $y = 3x - 3$

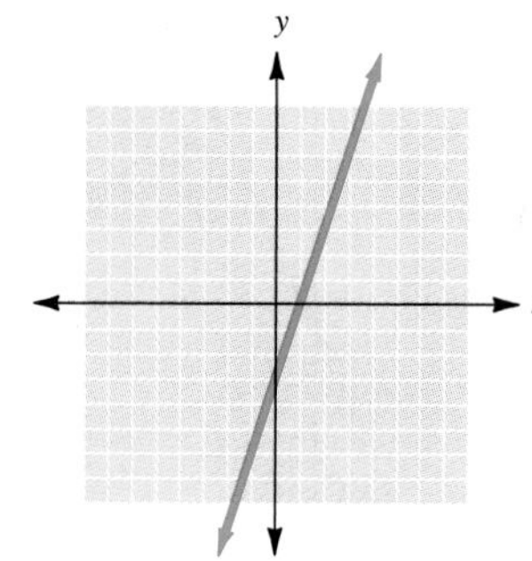

43. $x - 4y = 12$

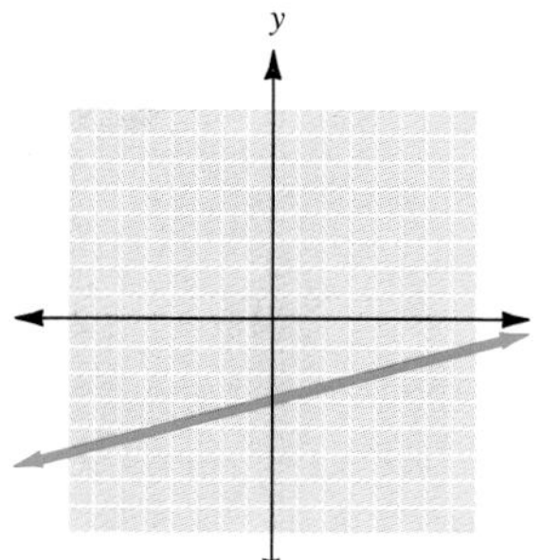

45. (3, 1)

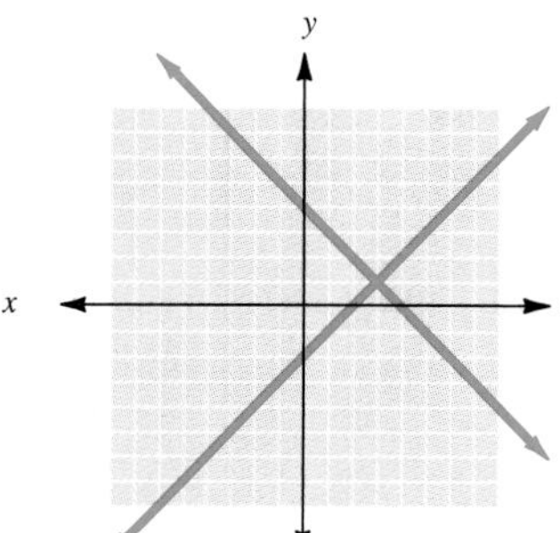

47. Graph

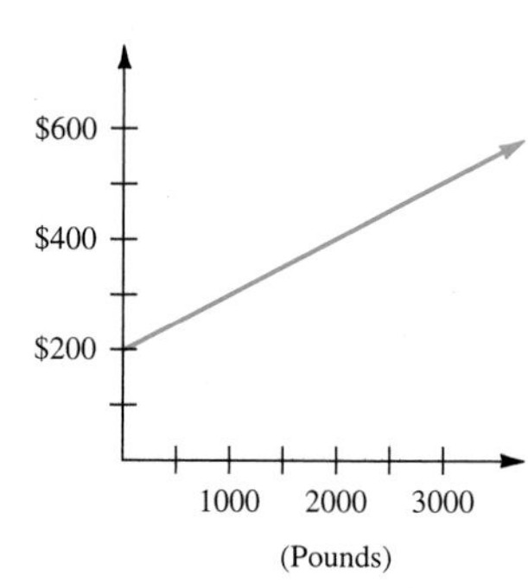

49. **(a)** Graph; **(b)** $\frac{100}{11}$ or $\approx$9 tons; **(c)** \$76; **(d)** $y = 17x - 125$

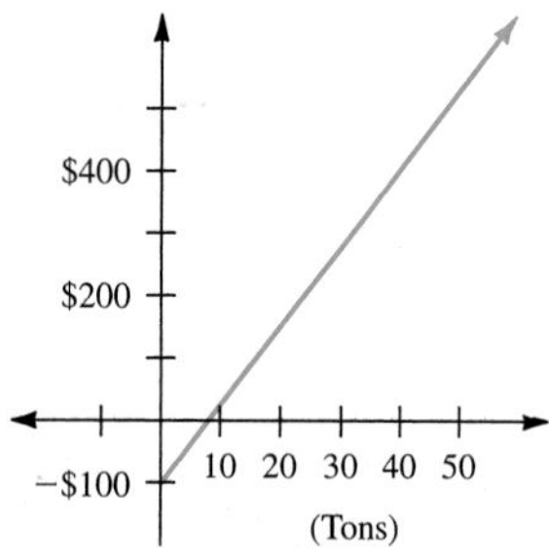

51. The lines do not intersect. The y intercepts are (0, 0), (0, 4), and (0, −5).

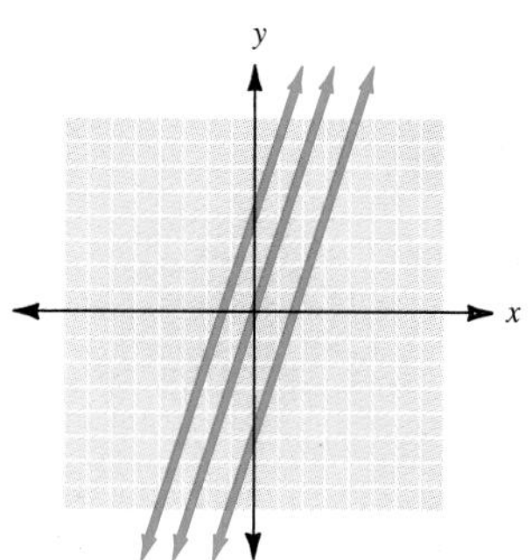

a. 1 **b.** 2 **c.** $\frac{3}{2}$ **d.** 0

6.4 The Slope of a Line

OBJECTIVES

1. Find the slope of a line through two given points
2. Find the slope of a line from its graph

We saw in Section 6.3 that the graph of an equation such as

$$y = 2x + 3$$

is a straight line. In this section we want to develop an important idea related to the equation of a line and its graph, called the **slope** of a line. Finding the slope of a line gives us a numerical measure of the "steepness" or inclination of that line.

NOTE Recall that an equation such as $y = 2x + 3$ is a *linear equation in two variables.* Its graph is always a straight line.

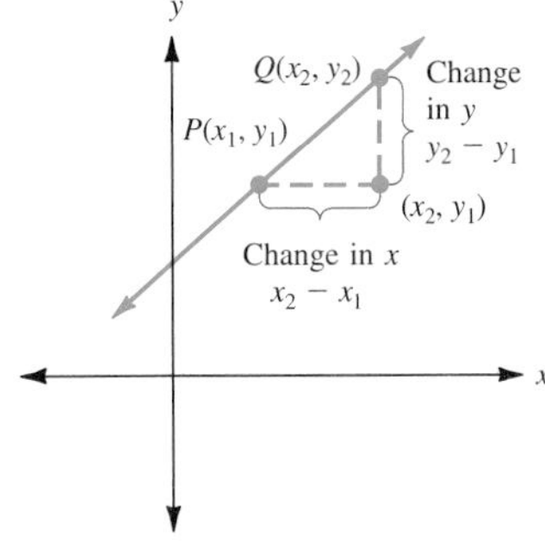

NOTE x_1 is read "x sub 1," x_2 is read "x sub 2," and so on. The 1 in x_1 and the 2 in x_2 are called **subscripts.**

To find the slope of a line, we first let $P(x_1, y_1)$ and $Q(x_2, y_2)$ be any two distinct points on that line. The **horizontal change** (or the change in x) between the points is $x_2 - x_1$. The **vertical change** (or the change in y) between the points is $y_2 - y_1$.

We call the ratio of the vertical change, $y_2 - y_1$, to the horizontal change, $x_2 - x_1$, the *slope* of the line as we move along the line from P to Q. That ratio is usually denoted by the letter m, and so we have the following formula:

NOTE The difference $x_2 - x_1$ is sometimes called the **run** between points P and Q. The difference $y_2 - y_1$ is called the **rise.** So the slope may be thought of as "rise over run."

Definitions: The Slope of a Line

If $P(x_1, y_1)$ and $Q(x_2, y_2)$ are any two points on a line, then m, the slope of the line, is given by

$$m = \frac{\text{vertical change}}{\text{horizontal change}} = \frac{y_2 - y_1}{x_2 - x_1} \qquad \text{when } x_2 \neq x_1$$

This definition provides exactly the numerical measure of "steepness" that we want. If a line "rises" as we move from left to right, the slope will be positive—the steeper the line, the larger the numerical value of the slope. If the line "falls" from left to right, the slope will be negative.

Let's proceed to some examples.

Example 1

Finding the Slope

Find the slope of the line containing points with coordinates (1, 2) and (5, 4).

Let $P(x_1, y_1) = (1, 2)$ and $Q(x_2, y_2) = (5, 4)$. By the definition of slope, we have

$$m = \frac{y_2 - y_1}{x_2 - x_1} = \frac{4 - 2}{5 - 1} = \frac{2}{4} = \frac{1}{2}$$

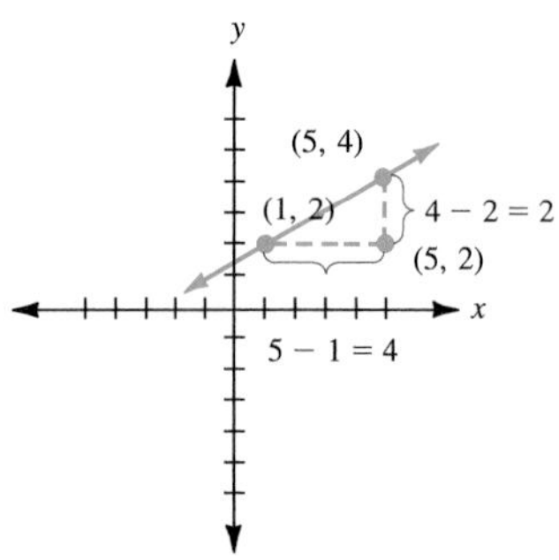

Note: We would have found the same slope if we had reversed P and Q and subtracted in the other order. In that case, $P(x_1, y_1) = (5, 4)$ and $Q(x_2, y_2) = (1, 2)$, so

$$m = \frac{2 - 4}{1 - 5} = \frac{-2}{-4} = \frac{1}{2}$$

It makes no difference which point is labeled (x_1, y_1) and which is (x_2, y_2). The resulting slope will be the same. You must simply stay with your choice once it is made and *not* reverse the order of the subtraction in your calculations.

CHECK YOURSELF 1

Find the slope of the line containing points with coordinates (2, 3) *and* (5, 5).

By now you should be comfortable subtracting negative numbers. Let's apply that skill to finding a slope.

Example 2

Finding the Slope

Find the slope of the line containing points with the coordinates (−1, −2) and (3, 6).

Again, applying the definition, we have

$$m = \frac{6 - (-2)}{3 - (-1)} = \frac{6 + 2}{3 + 1} = \frac{8}{4} = 2$$

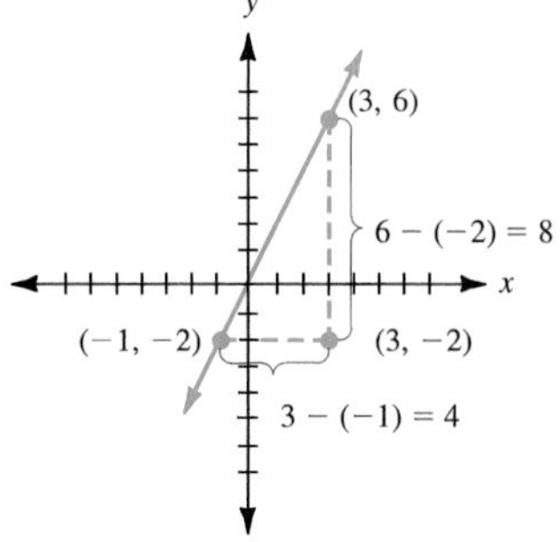

The figure below compares the slopes found in the two previous examples. Line l_1, from Example 1, had slope $\frac{1}{2}$. Line l_2, from Example 2, had slope 2. Do you see the idea of slope measuring steepness? The greater the slope, the more steeply the line is inclined upward.

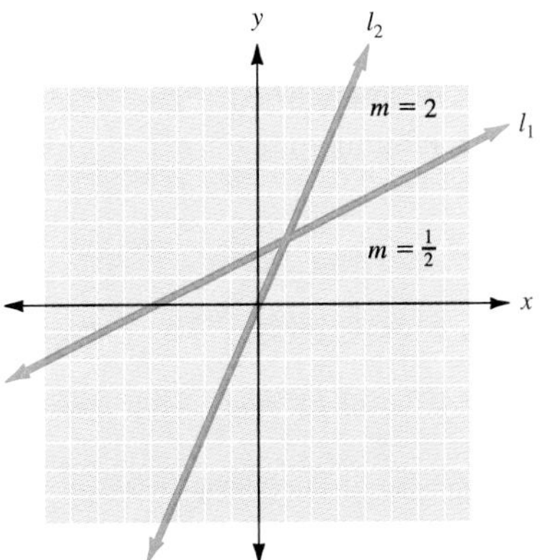

CHECK YOURSELF 2

Find the slope of the line containing points with coordinates (−1, 2) *and* (2, 7). *Draw a sketch of this line and the line of Check Yourself* 1. *Compare the lines and the two slopes.*

Let's look at lines with a negative slope.

Example 3

Finding the Slope

Find the slope of the line containing points with coordinates $(-2, 3)$ and $(1, -3)$.

By the definition,

$$m = \frac{-3 - 3}{1 - (-2)} = \frac{-6}{3} = -2$$

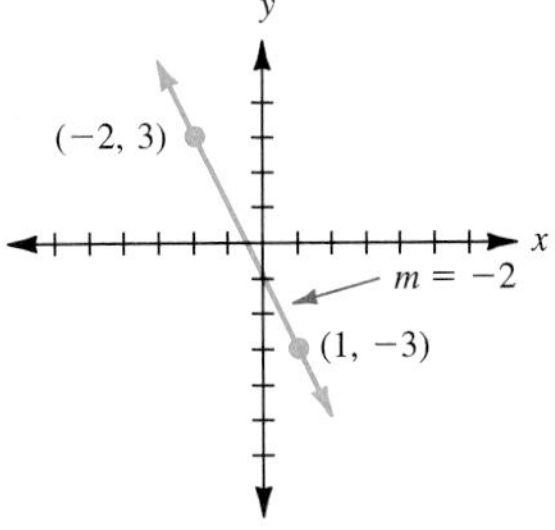

This line has a *negative* slope. The line *falls* as we move from left to right.

CHECK YOURSELF 3

Find the slope of the line containing points with coordinates (−1, 3) *and* (1, −3).

We have seen that lines with positive slope rise from left to right and lines with negative slope fall from left to right. What about lines with a slope of zero? A line with a slope of 0 is especially important in mathematics.

Example 4

Finding the Slope

Find the slope of the line containing points with coordinates (−5, 2) and (3, 2).

By the definition,

$$m = \frac{2 - 2}{3 - (-5)} = \frac{0}{8} = 0$$

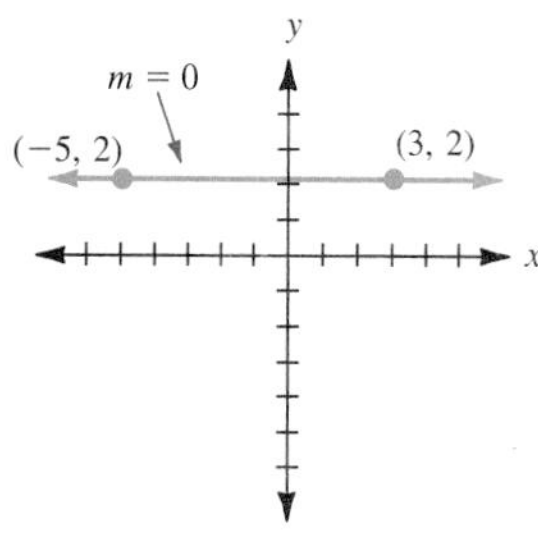

The slope of the line is 0. In fact, that will be the case for any horizontal line. Because any two points on the line have the same y coordinate, the vertical change $y_2 - y_1$ must always be 0, and so the resulting slope is 0.

CHECK YOURSELF 4

Find the slope of the line containing points with coordinates (−2, −4) *and* (3, −4).

Because division by 0 is undefined, it is possible to have a line with an undefined slope.

Example 5

Finding the Slope

Find the slope of the line containing points with coordinates (2, −5) and (2, 5).

By the definition,

$$m = \frac{5 - (-5)}{2 - 2} = \frac{10}{0}$$

Remember that division by zero is undefined.

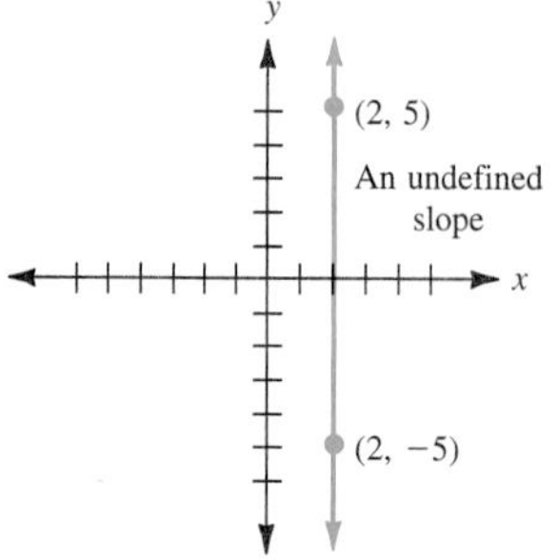

We say that the vertical line has an undefined slope. On a vertical line, any two points have the same x coordinate. This means that the horizontal change $x_2 - x_1$ must always be 0 and because division by 0 is undefined, the slope of a vertical line will always be undefined.

CHECK YOURSELF 5

Find the slope of the line containing points with the coordinates (−3, −5) and (−3, 2).

Given the graph of a line, we can find the slope of that line. Example 6 illustrates this.

Example 6

Finding the Slope from the Graph

Find the slope of the line graphed below.

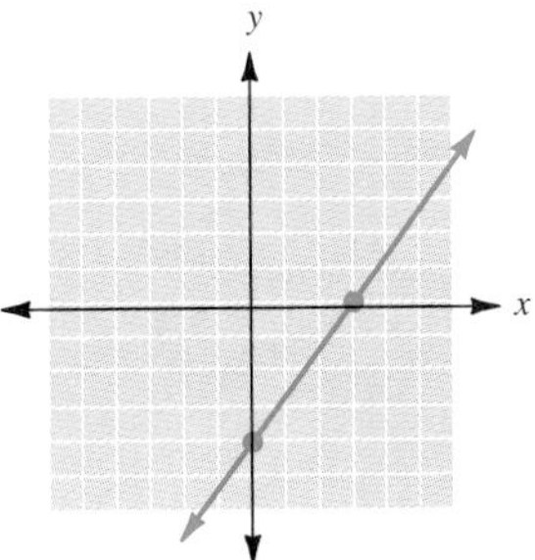

We can find the slope by identifying any two points. It is almost always easiest to use the x and y intercepts. In this case, those intercepts are (3, 0) and (0, −4).

Using the definition of slope, we find

$$m = \frac{0 - (-4)}{3 - 0} = \frac{4}{3}$$

The slope of the line is $\frac{4}{3}$.

CHECK YOURSELF 6

Find the slope of the line graphed below.

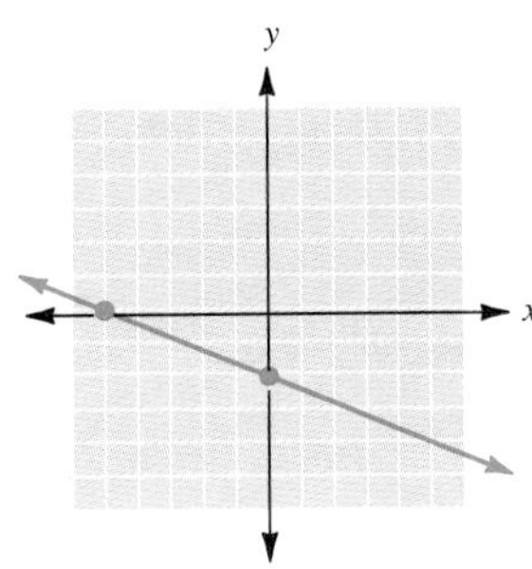

In Section 6.3, we saw that a line could be drawn from two ordered pairs. Given equations of the form $y = kx$, it is fairly easy to find two ordered pairs. In the next example, we will use those ordered pairs to find the graph of the equation.

Example 7

Graphing an Equation of the Form $y = kx$

(a) Find the graph of the equation $y = -2x$.

From the table to the right, we know that the ordered pairs (0, 0) and (1, −2) are solutions to the equation.

x	y
0	0
1	−2

The graph is displayed below

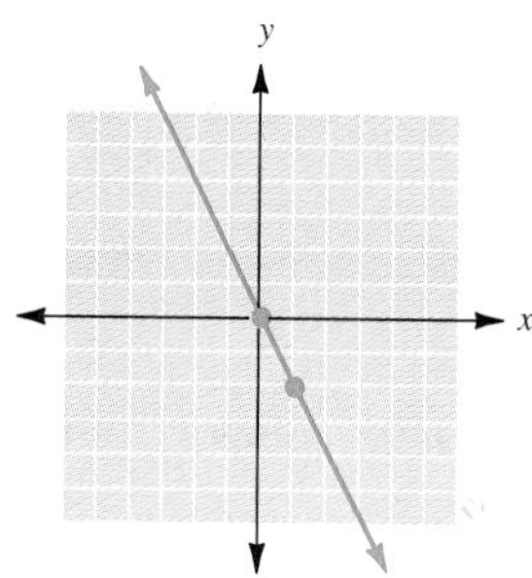

Note that the slope of the line that passes through the points (0, 0) and (1, −2) is

$$m = \frac{0 - (-2)}{0 - 1} = \frac{2}{-1} = -2$$

(b) Find the graph of the equation $y = \frac{1}{3}x$.

From the table to the right, we know that the ordered pairs (0, 0) and (3, 1) are solutions to the equation.

x	y
0	0
3	1

The graph is displayed below

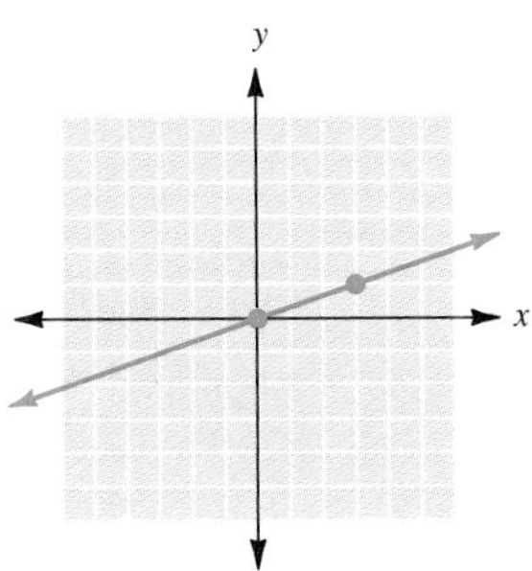

Note that the slope of the line that passes through the points (0, 0) and (3, 1) is

$$m = \frac{0 - 1}{0 - 3} = \frac{1}{3}$$

CHECK YOURSELF 7

Find the graph of the equation $y = -\frac{1}{2}x$.

In Example 7, we noted that the slope of the line for the equation $y = -2x$ is -2, and the slope of the line for the equation $y = \frac{1}{3}x$ is $\frac{1}{3}$. This leads us to the following observation.

The slope of a line for an equation of the form $y = kx$ will always be k. Because k is the slope, we generally write the form as

$$y = mx$$

Note that (0, 0) will be a solution for any equation of this form. As a result, the line for an equation of the form $y = mx$ will always pass through the origin.

The following sketch summarizes the results of our previous examples.

NOTE As the slope gets closer to 0, the line gets "flatter."

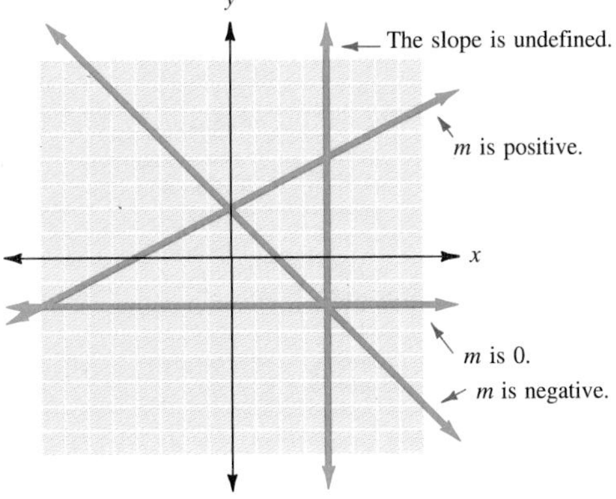

Four lines are illustrated in the figure. Note that

1. The slope of a line that rises from left to right is positive.
2. The slope of a line that falls from left to right is negative.
3. The slope of a horizontal line is 0.
4. A vertical line has an undefined slope.

CHECK YOURSELF ANSWERS

1. $m = \frac{2}{3}$
2. $m = \frac{5}{3}$

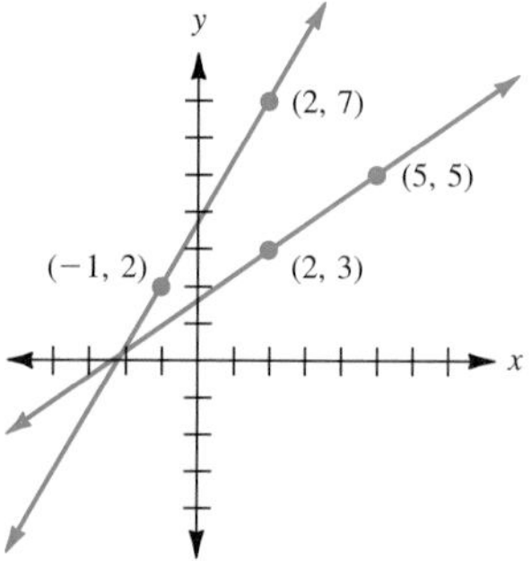

3. $m = -3$
4. $m = 0$
5. m is undefined
6. $m = -\frac{2}{5}$
7.

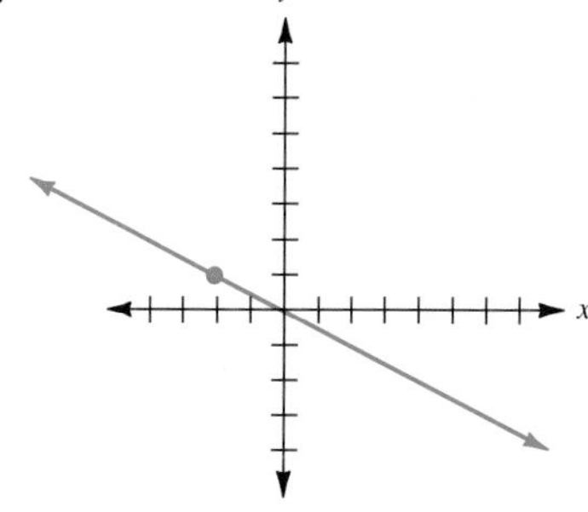

(Note: Your second point could have been $(-2, 1)$ or $(2, -1)$.)

6.4 Exercises

Find the slope of the line through the following pairs of points.

1. (5, 7) and (9, 11)

2. (4, 9) and (8, 17)

3. (−2, −5) and (2, 15)

4. (−3, 2) and (0, 17)

5. (−2, 3) and (3, 7)

6. (−3, −4) and (3, −2)

7. (−3, 2) and (2, −8)

8. (−6, 1) and (2, −7)

9. (3, 3) and (5, 0)

10. (−2, 4) and (3, 1)

11. (5, −4) and (5, 2)

12. (−5, 4) and (2, 4)

13. (−4, −2) and (3, 3)

14. (−5, −3) and (−5, 2)

15. (−3, −4) and (2, −4)

16. (−5, 7) and (2, −2)

17. (−1, 7) and (2, 3)

18. (−4, −2) and (6, 4)

In exercises 19 to 24, two points are shown. Find the slope of the line through the given points.

19.

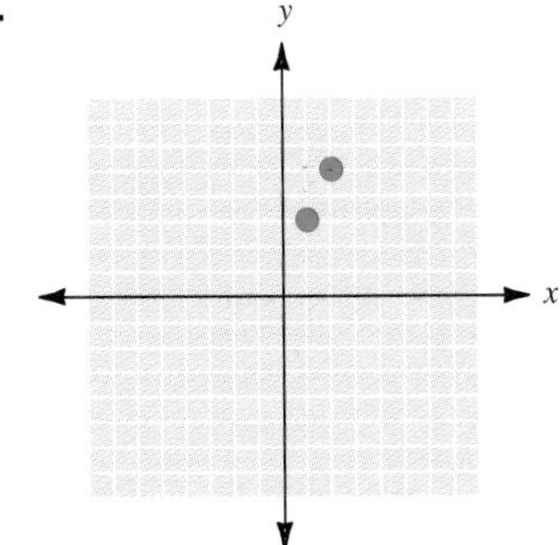

20.

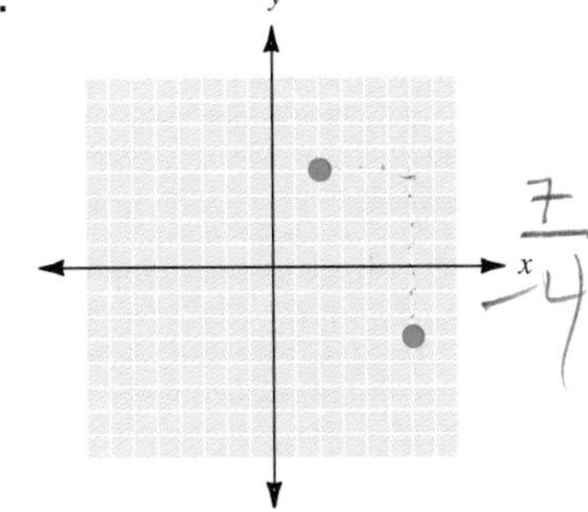

$\frac{7}{-4}$

ANSWERS

1. 1
2. 2
3. 5
4. 5
5. $\frac{4}{5}$
6. $\frac{1}{3}$
7. −2
8. −1
9. $-\frac{3}{2}$
10. $-\frac{3}{5}$
11. Undefined
12. 0
13. $\frac{5}{7}$
14. Undefined
15. 0
16. $-\frac{9}{7}$
17. $-\frac{4}{3}$
18. $\frac{3}{5}$
19. 2
20. $-\frac{7}{4}$

ANSWERS

21. −2
22. 3
23. 0
24. Undefined
25. 4
26. 2
27. −5
28. −3

21.

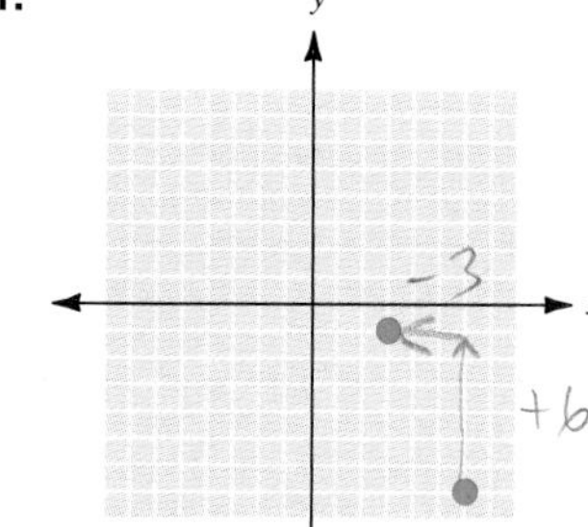

22.

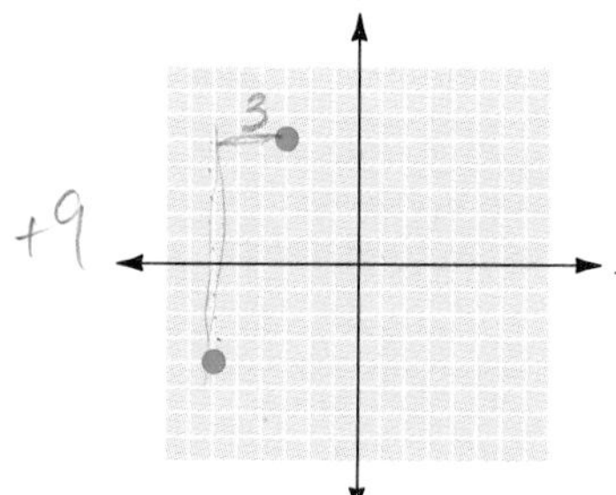

23.

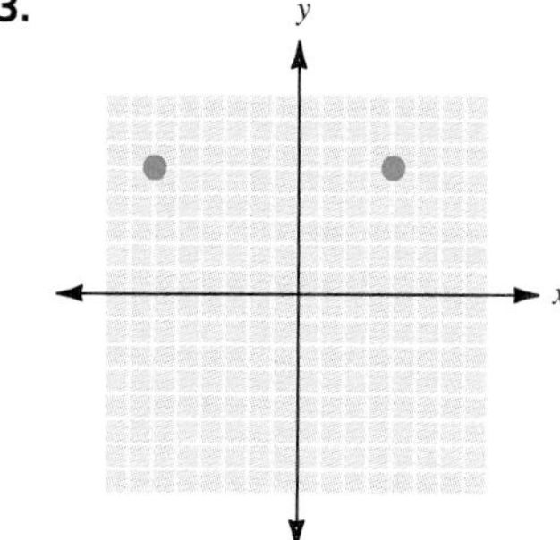

24.

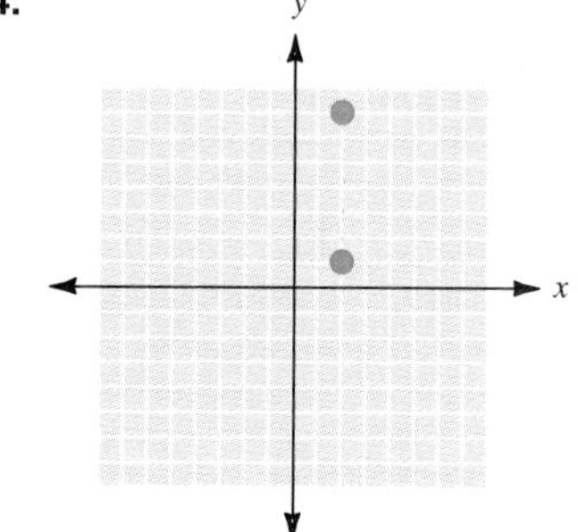

In exercises 25 to 30, find the slope of the lines graphed.

25.

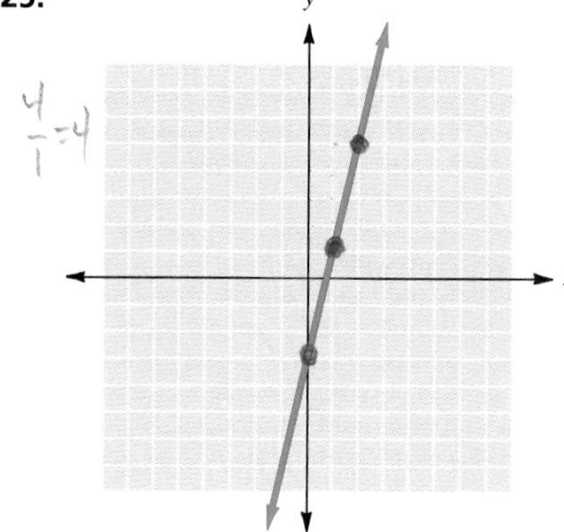

26.

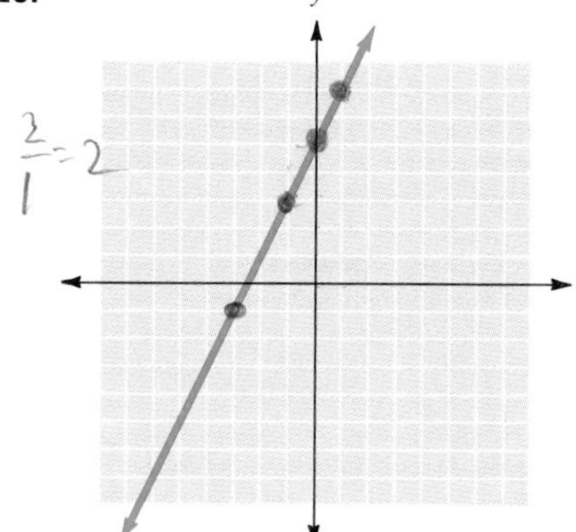

27.

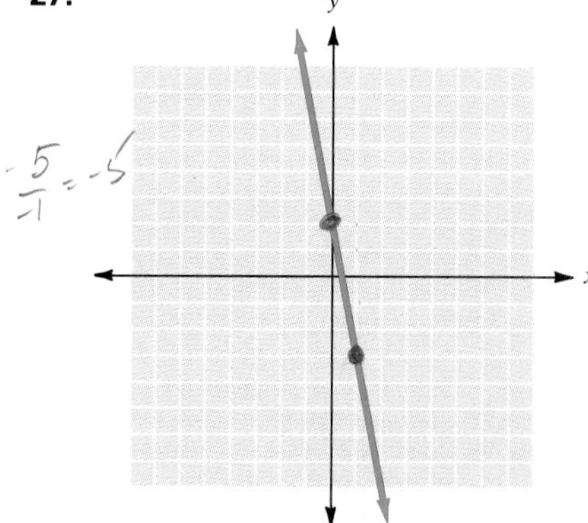

28.

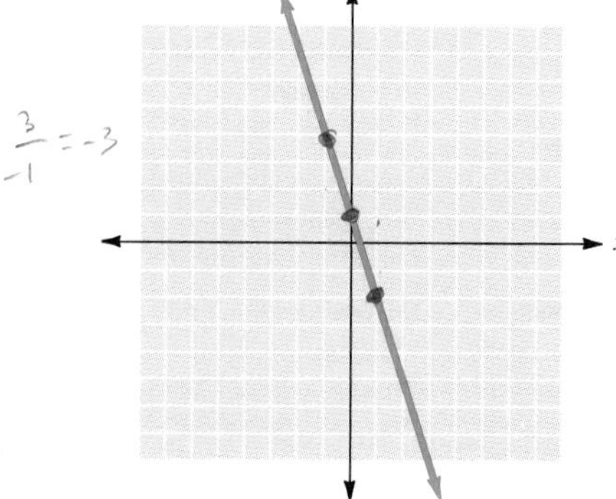

29.

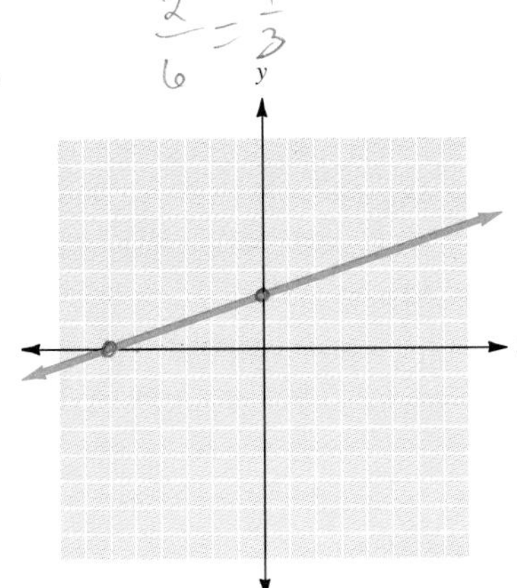

30.

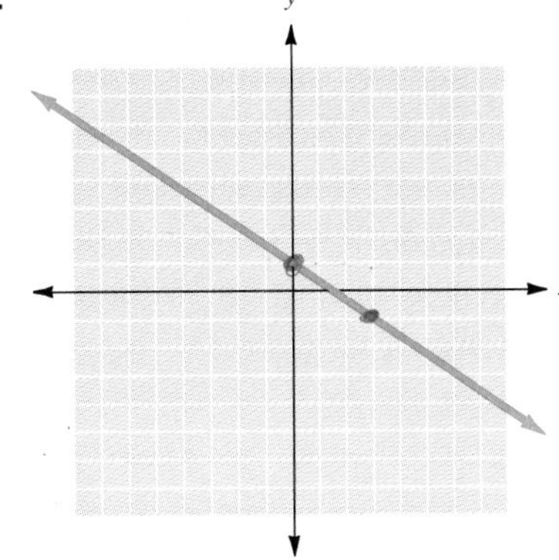

Find the graph of the following equations.

31. $y = -4x$

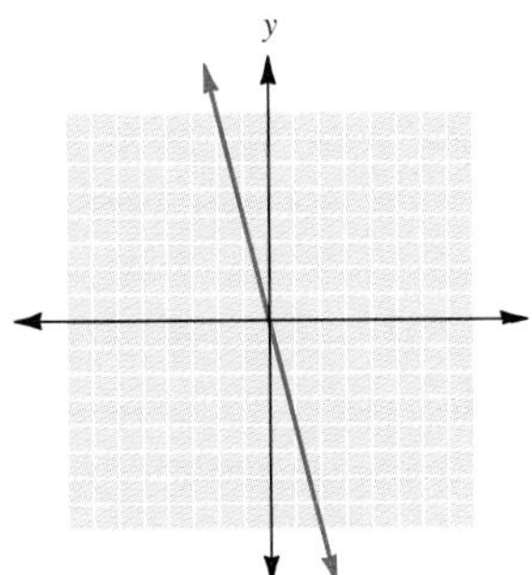

32. $y = 3x$

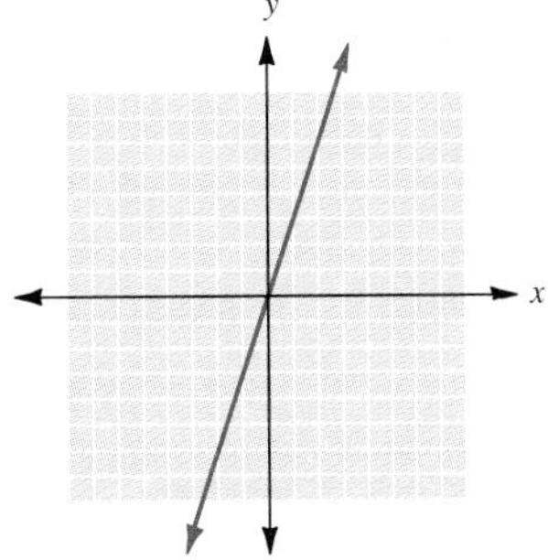

33. $y = \frac{2}{3}x$

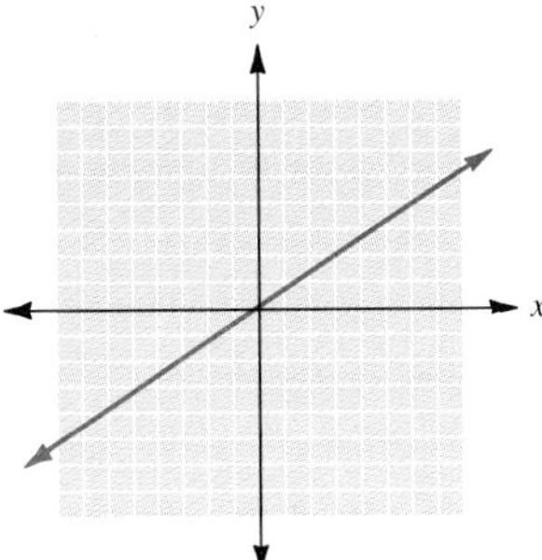

34. $y = -\frac{3}{4}x$

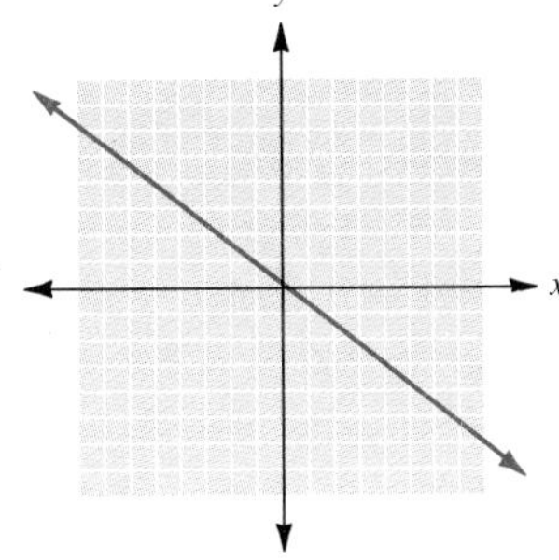

35. $y = \frac{5}{4}x$

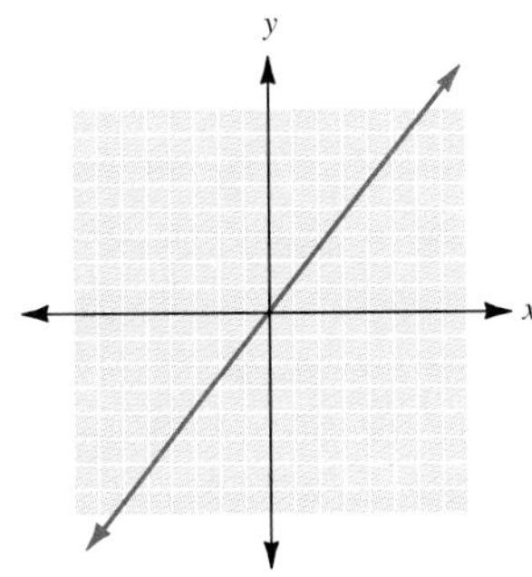

36. $y = -\frac{4}{5}x$

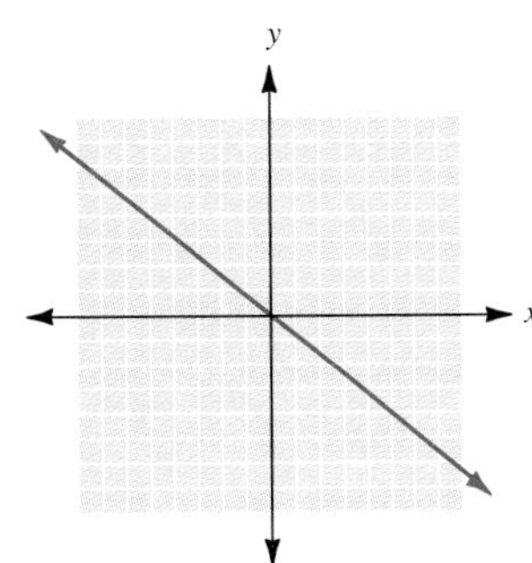

ANSWERS

29. $\frac{1}{3}$

30. $-\frac{2}{3}$

31. See exercise

32. See exercise

33. See exercise

34. See exercise

35. See exercise

36. See exercise

37. (a) (3, 1); (4, 3)
(b) 2
(c) slope equals coefficient of x

38. (a) (2, 8); (4, 11)
(b) $\frac{3}{2}$
(c) slope equals coefficient of x

39. (a) (3, 1); (6, 0)
(b) $-\frac{1}{3}$
(c) slope equals coefficient of x

40. (a) (−1, −2); (−3, 6)
(b) −4
(c) slope equals coefficient of x

41. (a) (5, 13); (6, 15); (7, 17); (8, 19); (9, 21)
(b) 2
(c) Yes
(d) increases by 2

42. (a) (5, 15); (6, 17); (7, 19); (8, 21); (9, 23)
(b) 2
(c) Yes
(d) increases by 2

43. (a) (5, 30); (6, 26); (7, 22); (8, 18); (9, 14)
(b) 4
(c) Yes
(d) decreases by 4

44. (a) (5, 20); (6, 16); (7, 12); (8, 8); (9, 4)
(b) 4
(c) Yes
(d) decreases by 4

37. Consider the equation $y = 2x - 5$.

(a) Complete the following table:

x	y
3	
4	

(b) Use the ordered pairs found in part (a) to calculate the slope of the line.

(c) What do you observe concerning the slope found in part (b) and the given equation?

38. Repeat exercise 37 for $y = \frac{3}{2}x + 5$ and

x	y
2	
4	

39. Repeat exercise 37 for $y = -\frac{1}{3}x + 2$ and

x	y
3	
6	

40. Repeat exercise 37 for $y = -4x - 6$ and

x	y
−1	
−3	

41. Consider the equation: $y = 2x + 3$

(a) Complete the following table of values, and plot the resulting points.

Point	x	y
A	5	
B	6	
C	7	
D	8	
E	9	

(b) As the x coordinate changes by 1 (for example, as you move from point A to point B), how much do the corresponding y coordinates change?

(c) Is your answer to part (b) the same if you move from B to C? from C to D? from D to E?

(d) Describe the "growth rate" of the line using these observations. Complete the following statement: When the x value grows by 1 unit, the y value __________.

42. Repeat exercise 41 using: $y = 2x + 5$

43. Repeat exercise 41 using: $y = -4x + 50$

44. Repeat exercise 41 using: $y = -4x + 40$

ANSWERS

45. See exercise

46. See exercise

47. See exercise

48. See exercise

a. 5

b. −3

c. 7

d. $\frac{14}{3}$

e. $-\frac{8}{3}$

f. −3

In the following exercises, (a) plot the given point; (b) using the given slope, move from the point plotted in (a) to plot a new point; (c) draw the line that passes through the points plotted in (a) and (b).

45. $(3, 1), m = 2$

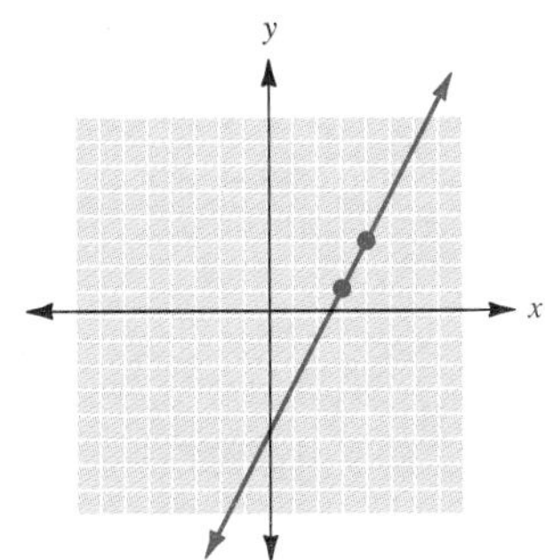

46. $(-1, 4), m = -2$

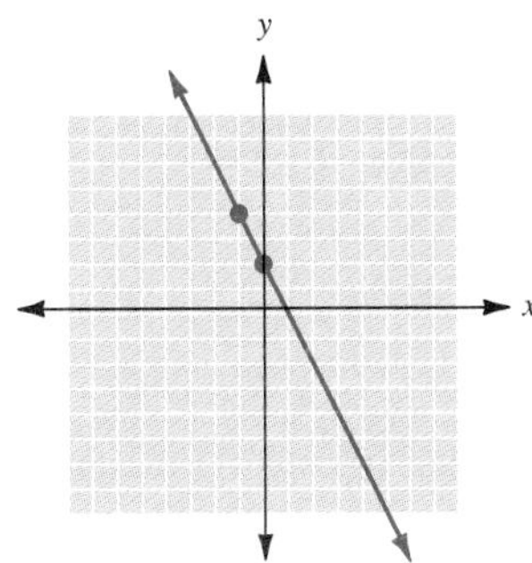

47. $(-2, -1), m = -4$

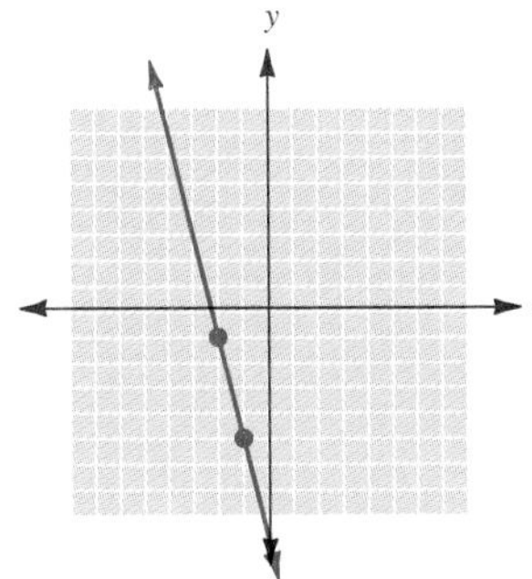

48. $(-3, 5), m = 2$

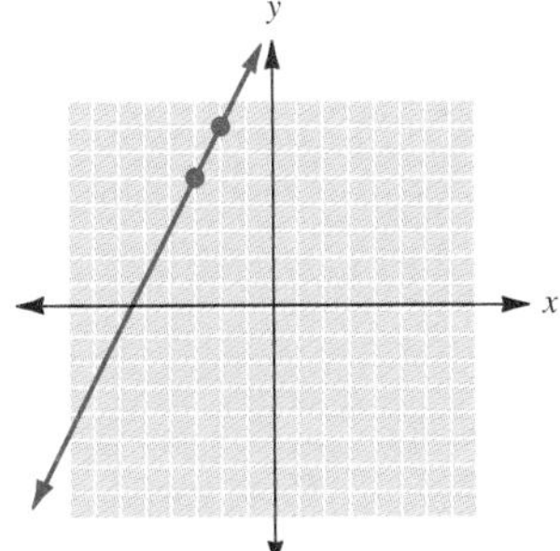

Getting Ready for Section 6.5 [Section 2.2]

Solve each equation for x.

(a) $25 = 5x$ (b) $36 = -12x$ (c) $-49 = -7x$

(d) $14 = 3x$ (e) $-24 = 9x$ (f) $72 = -24x$

Answers

1. 1 **3.** 5 **5.** $\frac{4}{5}$ **7.** −2 **9.** $-\frac{3}{2}$ **11.** Undefined **13.** $\frac{5}{7}$

15. 0 **17.** $-\frac{4}{3}$ **19.** 2 **21.** −2 **23.** 0 **25.** 4 **27.** −5 **29.** $\frac{1}{3}$

31. $y = -4x$

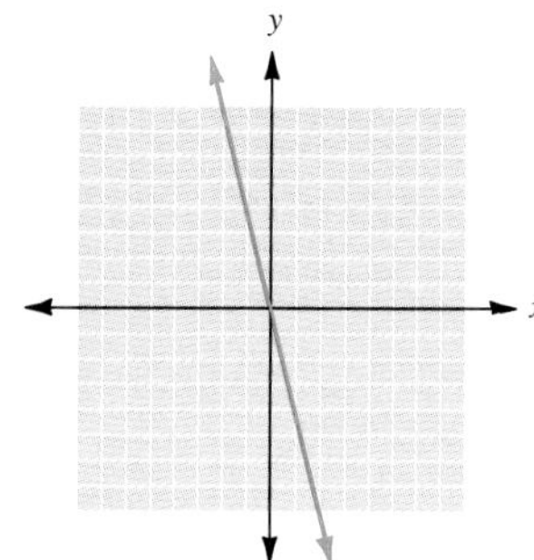

33. $y = \frac{2}{3}x$

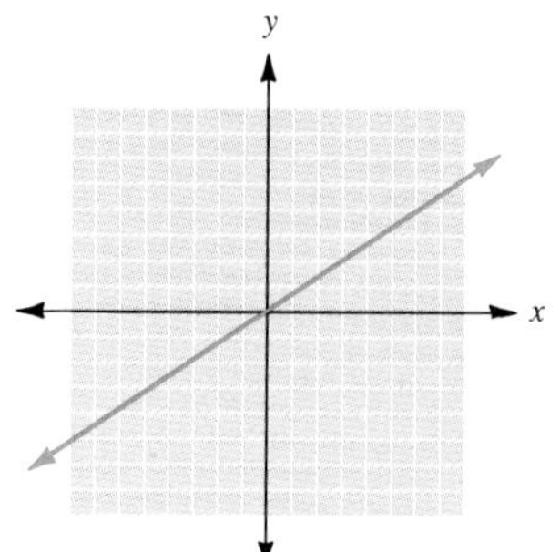

35. $y = \frac{5}{4}x$

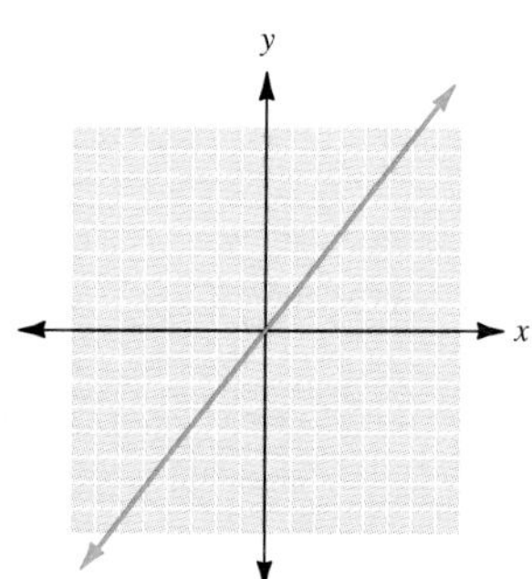

37. **(a)** (3, 1), (4, 3); **(b)** 2; **(c)** slope equals coefficient of x

39. **(a)** (3, 1), (6, 0); **(b)** $-\frac{1}{3}$; **(c)** slope equals coefficient of x

41. **(a)** (5, 13), (6, 15), (7, 17), (8, 19), (9, 21); **(b)** 2; **(c)** Yes; **(d)** increases by 2

43. **(a)** (5, 30), (6, 26), (7, 22), (8, 18), (9, 14); **(b)** 4; **(c)** Yes; **(d)** decreases by 4

45.

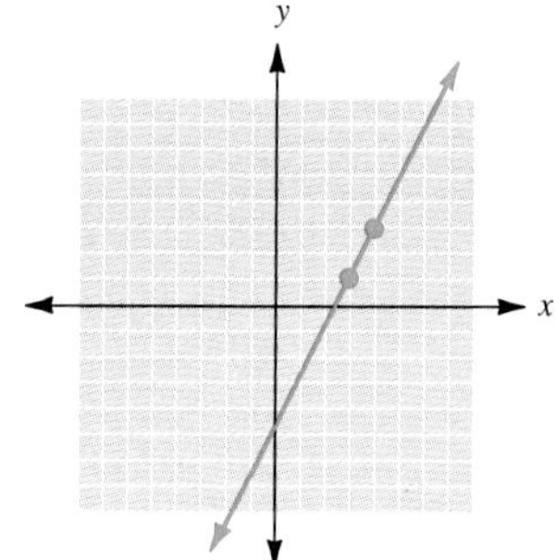

47.

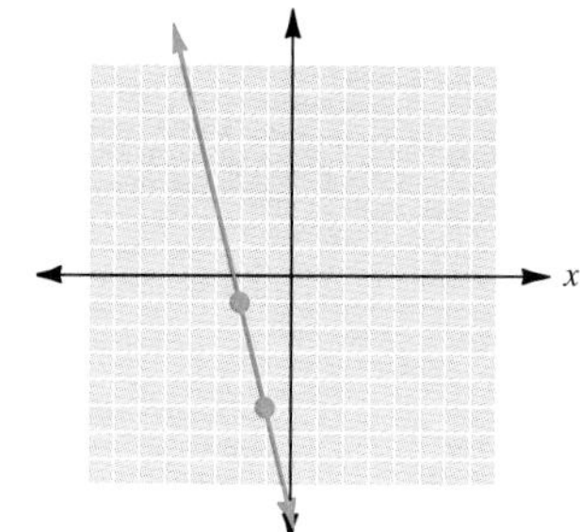

a. 5 **b.** -3 **c.** 7 **d.** $\frac{14}{3}$ **e.** $-\frac{8}{3}$ **f.** -3

6.5 Direct Variation

6.5 OBJECTIVES

1. Write an equation for a direct variation relationship
2. Graph the equation of a direct variation relationship

Pedro makes \$25 an hour as an electrician. If he works 1 hour, he makes \$25; if he works 2 hours, he makes \$50; and so on. We say his total pay **varies directly** with the number of hours worked.

Definitions: Direct Variation

If y is a constant multiple of x, we write

$y = kx$ in which k is a constant

We say that y *varies directly* as x, or that y is *directly proportional* to x. The constant k is called the **constant of variation.**

Example 1

Writing an Equation for Direct Variation

Marina earns \$9 an hour as a tutor. Write the equation that describes the relationship between the number of hours she works and her pay.

Her pay (P) is equal to the rate of pay (r) times the number of hours worked (h), so

$P = r \cdot h$ or $P = 9h$

CHECK YOURSELF 1

Sorina is driving at a constant rate of 50 m/h. Write the equation that shows the distance she travels (d) in h hours.

NOTE Remember that k is the constant of variation.

If two things vary directly and values are given for x and y, we can find k. This property is illustrated in Example 2.

Example 2

Finding the Constant of Variation

If y varies directly with x, and $y = 30$ when $x = 6$, find k.

Because y varies directly with x, we know from the definition that

$y = kx$

We need to find k. We do this by substituting 30 for y and 6 for x.

$30 = k(6)$ or $k = 5$

CHECK YOURSELF 2

If y varies directly with x and y = 100 when x = 25, find the constant of variation.

The graph for a linear equation of direct variation will always pass through the origin. The next example will illustrate.

Example 3

Graphing an Equation of Direct Variation

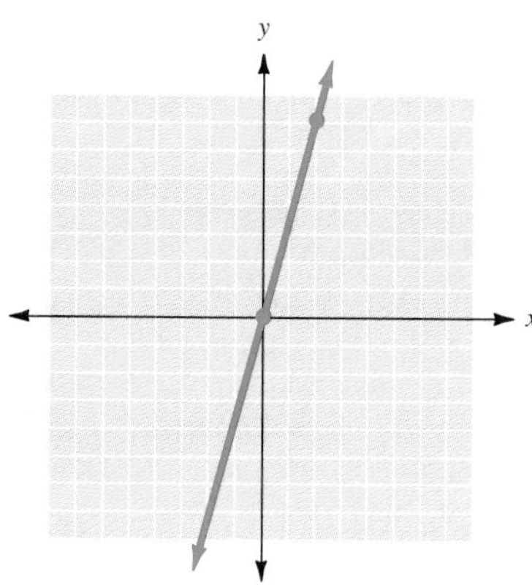

Let y vary directly as x, with a constant of variation $k = 3.5$. Graph the equation of variation.

The equation of variation is $y = 3.5x$, so the graph will have a slope of 3.5.

CHECK YOURSELF 3

Let y vary directly as x, with a constant of variation $k = \frac{7}{3}$. *Graph the equation of variation.*

With many applications it is necessary to adjust the scale on the x or y axis to present a reasonable graph. The next example will help prepare us for applications of this nature.

Example 4

Graphing an Equation of Direct Variation

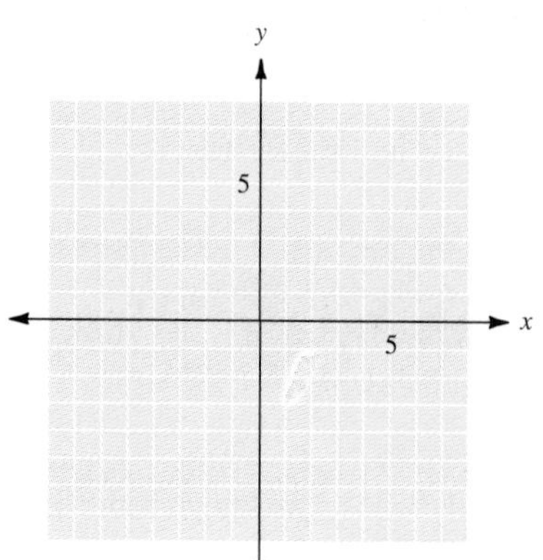

Let y vary directly as x, with a constant of variation $k = 200$. Graph the equation of variation.

The equation is $y = 200x$. What happens if we try to sketch this graph on a grid with the same scale on the x and y axes? The slope indicates that y increases by 200 when x increases by 1. Can you see that it would be impossible to produce a meaningful graph on this grid?

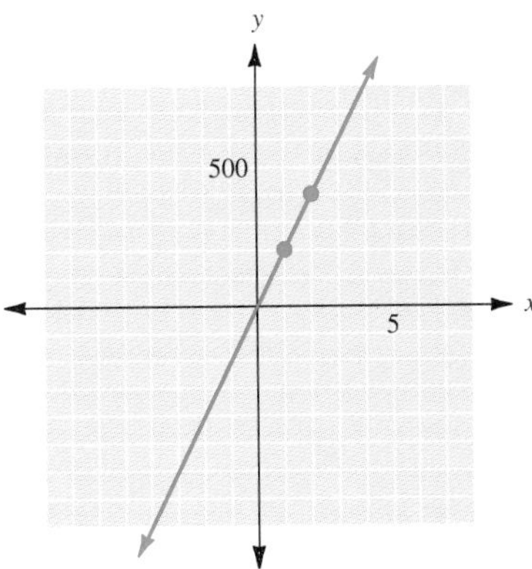

Instead, we create a grid that will make our graph readable. In this case, we will choose to have the y axis marked in units of 100. The x axis will retain the standard scale.

CHECK YOURSELF 4

Let y vary directly as x, with a constant of variation k = −1500. Graph the equation of variation.

Now we will examine an application that requires us to adjust the scale on the axes.

Example 5

Graphing a Direct Variation Equation

Tim Duncan earns approximately \$5000 per minute for playing basketball. Sketch the graph that represents the equation of direct variation between minutes played and money earned.

We know that the equation will be $y = 5000x$.

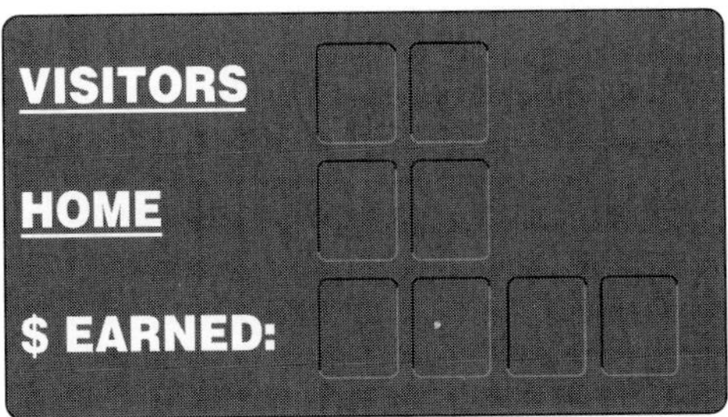

For this graph, we will use only the first quadrant. Do you see why? All of the other quadrants represent negative time, negative money, or both.

We will also use a different scale on each axis. The x axis, which represents minutes played, is marked every 200 minutes, to 2000 minutes. The y axis, representing money made, is marked every 1 million dollars for \$10,000,000.

As we do with any direct variation graph, we start at the origin. The slope of 5000 can be looked at in many ways. It is usually easiest to move one mark along the x axis and see how much change we have in the y direction.

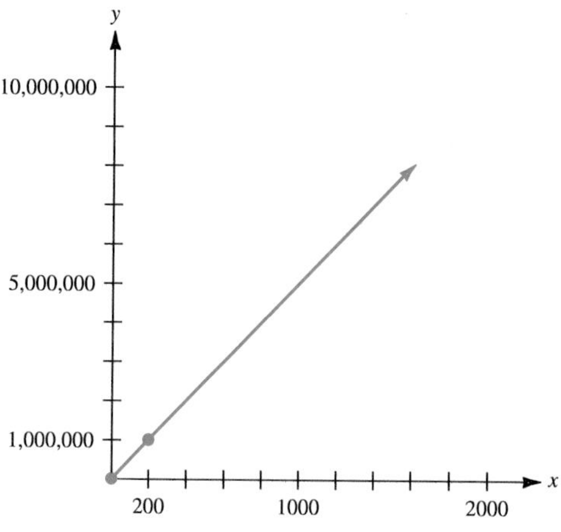

One mark in the x direction is 200 minutes. Because $y = 5000(200) = 1{,}000{,}000$, we find our second point at (200, \$1,000,000). Connecting that point to the origin, we get the graph shown.

CHECK YOURSELF 5

The average secretary makes about \$0.24 for each minute worked. Sketch the graph of the equation of direct variation.

CHECK YOURSELF ANSWERS

1. $d = 50h$ **2.** $k = 4$

3.

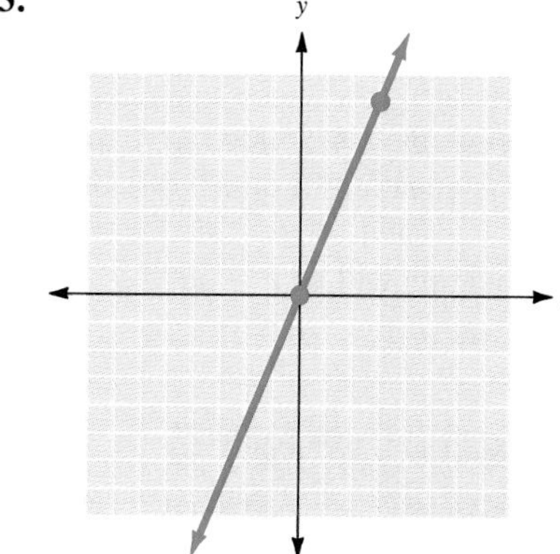

4.

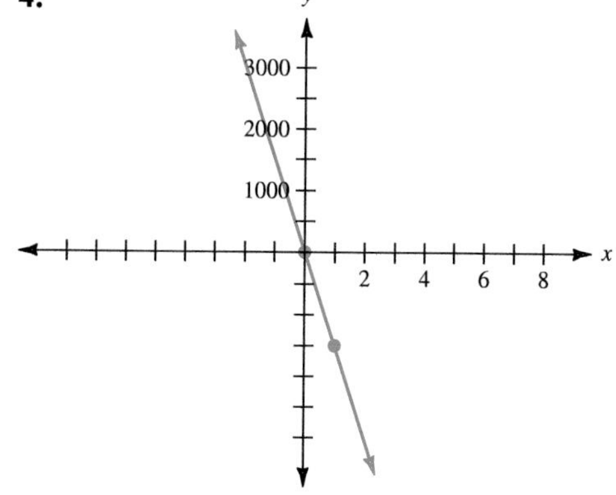

5.

y
250
125
25
Dollars
100
500
1000
x
Minutes

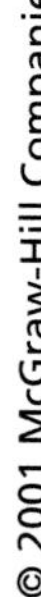

Name ______________________

Section __________ Date __________

6.5 Exercises

ANSWERS

1. $S = 12h$
2. $S = 11.5h$
3. $D = 55h$
4. $D = 450h$
5. 9
6. 18
7. 21
8. 50
9. 3.5
10. 0.4
11. See exercise
12. See exercise

1. **Salary.** Robin earns \$12 per hour. Write an equation that shows how much she makes (S) in h hours.

2. **Salary.** Kwang earns \$11.50 per hour. Write an equation that shows how much he earns (S) in h hours.

3. **Distance.** Lee is traveling at a constant rate of 55 miles per hour (mi/h). Write an equation that shows how far he travels (D) in h hours.

4. **Distance.** An airplane is traveling at a constant rate of 450 mi/h. Write an equation that shows how far the plane travels (D) in h hours.

In exercises 5 to 10, find the constant of variation k.

5. y varies directly with x; $y = 54$ when $x = 6$.

6. m varies directly with n; $m = 144$ when $n = 8$.

7. V varies directly with h; $V = 189$ when $h = 9$.

8. d varies directly with t; $d = 750$ when $t = 15$.

9. y varies directly with x; $y = 2100$ when $x = 600$.

10. y varies directly with x; $y = 400$ when $x = 1000$.

In exercises 11 to 18, y varies directly with x and the value of k is given. Graph the equation of variation.

11. $k = 2$

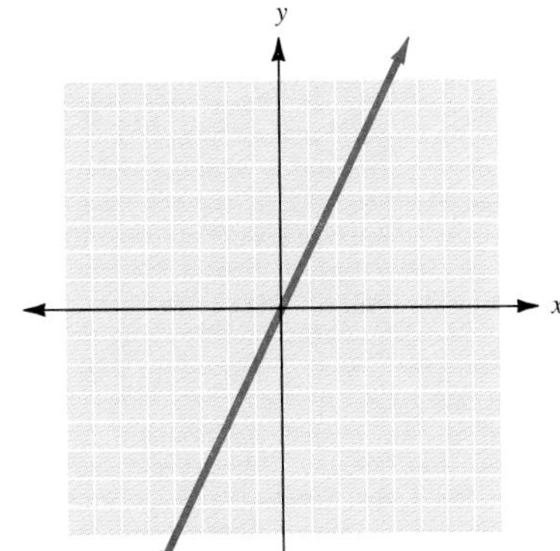

12. $k = 4$

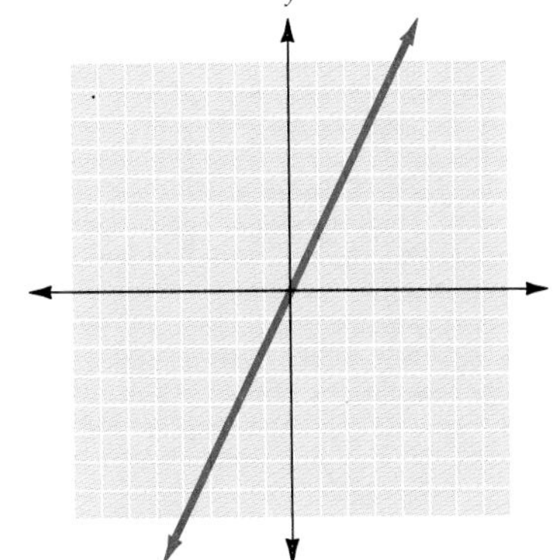

ANSWERS

13. See exercise

14. See exercise

15. See exercise

16. See exercise

17. See exercise

18. See exercise

13. $k = 2.5$

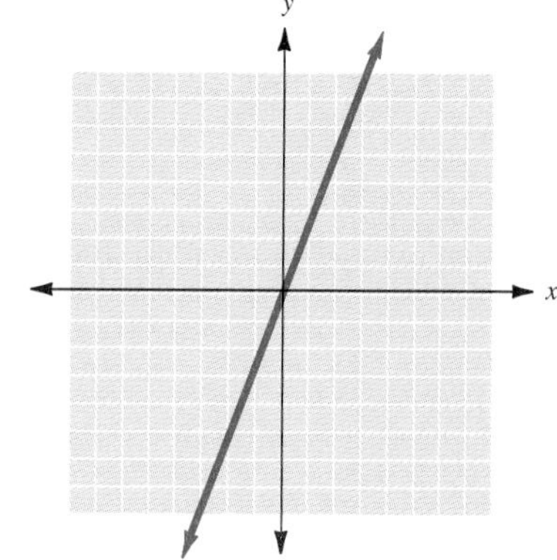

14. $k = \frac{11}{5}$

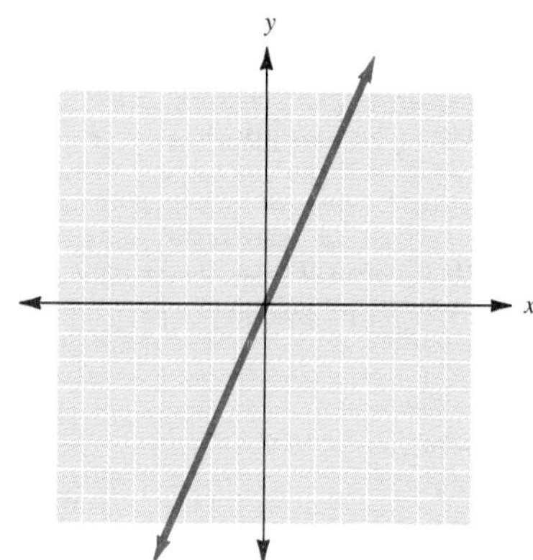

15. $k = 100$

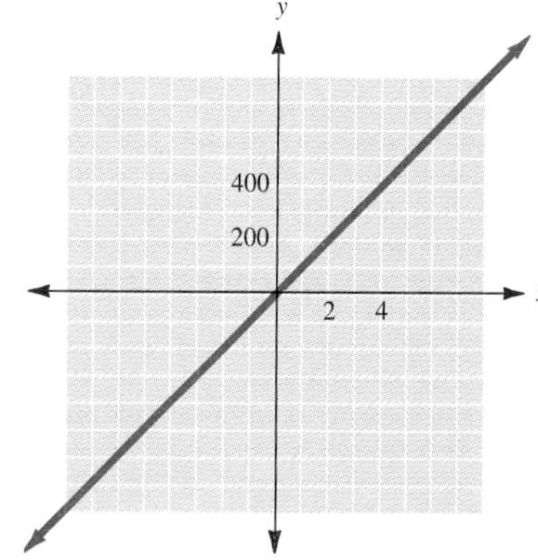

16. $k = 300$

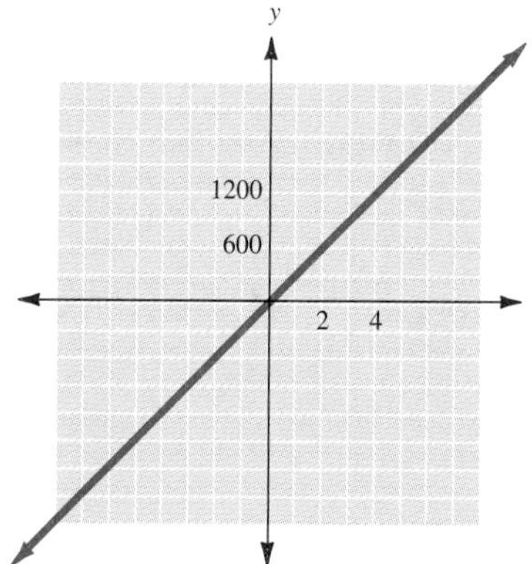

17. $k = 50$

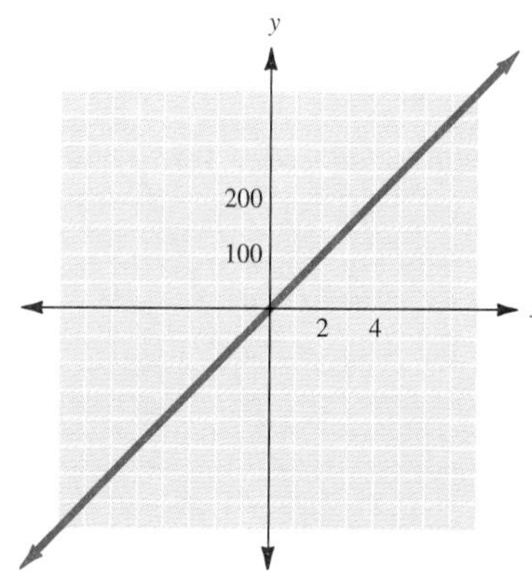

18. $k = 400$

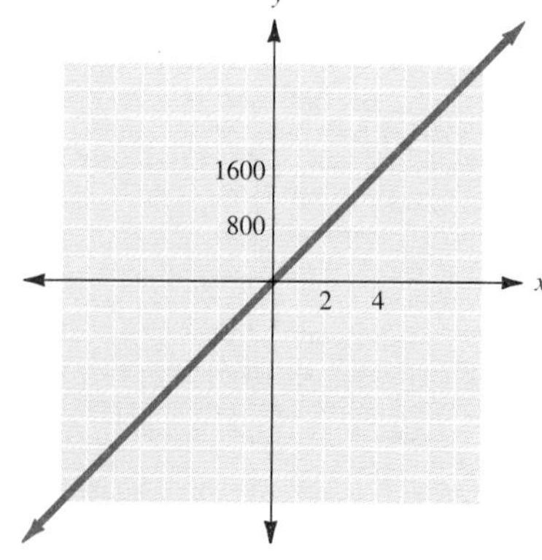

ANSWERS

19. See exercise

20. See exercise

21. See exercise

19. At a factory that makes grinding wheels, Kalila makes \$.20 for each wheel completed. Sketch the equation of direct variation.

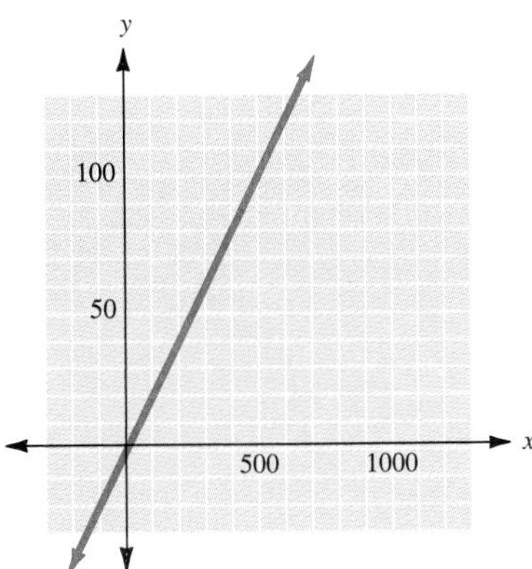

20. Palmer makes \$1.25 per page for each page that he types. Sketch the equation of direct variation.

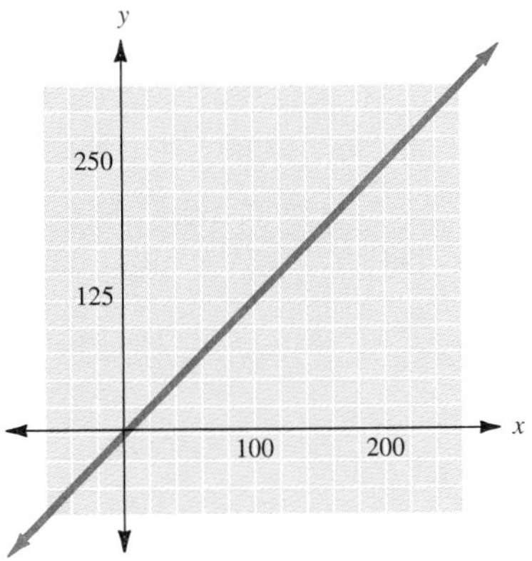

21. Cesar makes \$2.50 for each tire he details. Sketch the equation of direct variation.

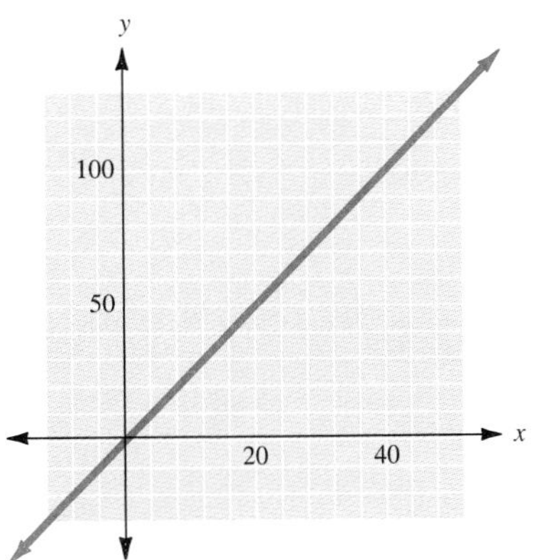

22. See exercise

23. 22 hours

24. 3000

22. Tanesha makes \$0.15 for each problem she checks in a math text. Sketch the equation of direct variation.

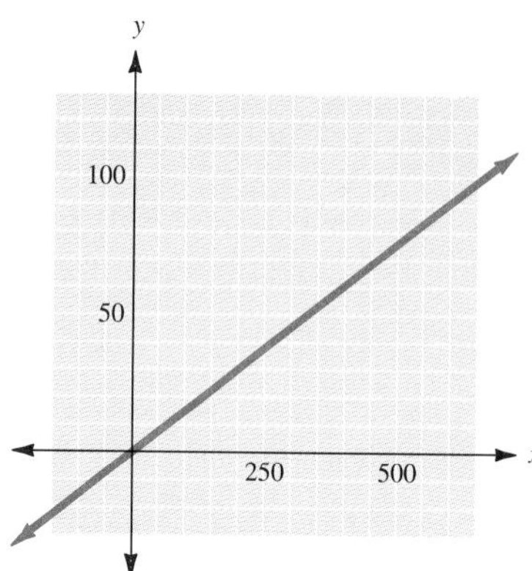

23. Salary. Josephine works part-time in a local video store. Her salary varies directly as the number of hours worked. Last week she earned \$43.20 for working 8 hours. This week she earned \$118.80. How many hours did she work this week?

24. Revenue. The revenue for a sandwich shop is directly proportional to its advertising budget. When the owner spent \$2000 a month on advertising, the revenue was \$120,000. If the revenue is now \$180,000, how much is the owner spending on advertising?

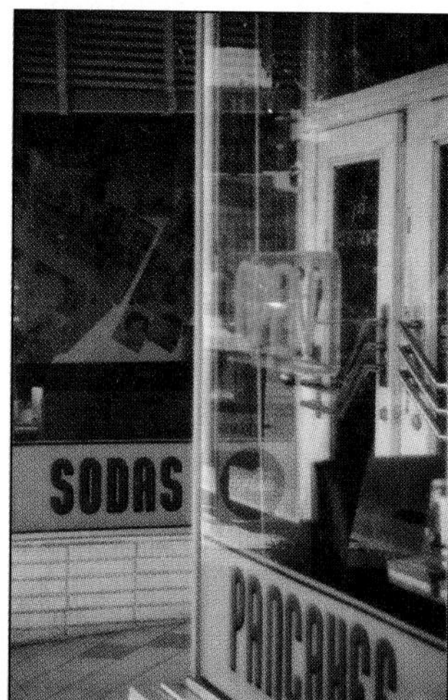

Answers

1. $S = 12h$ **3.** $D = 55h$ **5.** 9 **7.** 21 **9.** 3.5

11.

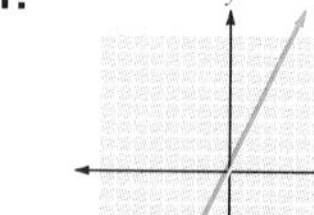
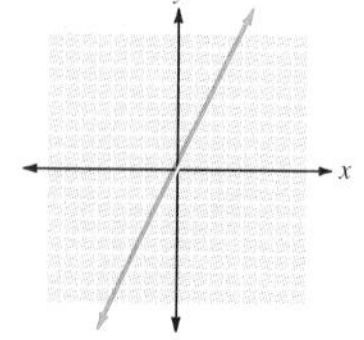

13.

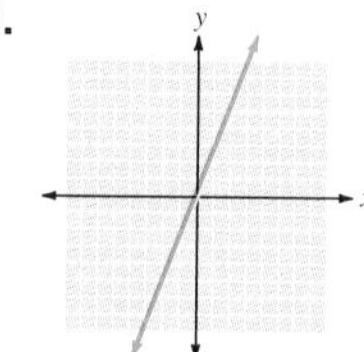

15.

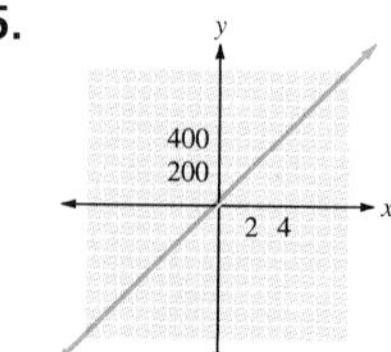

17.

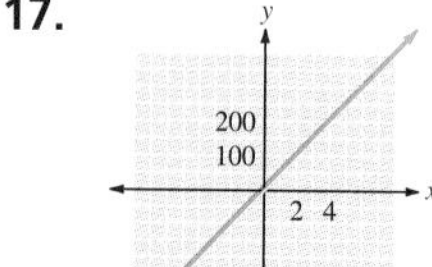

19.

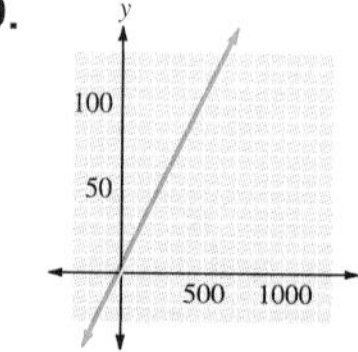

21.

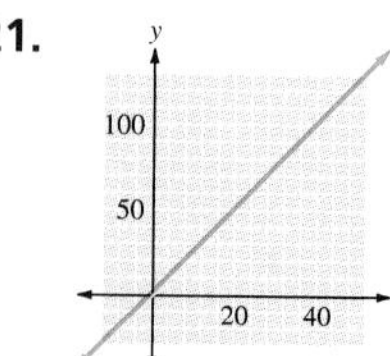

23. 22 hours

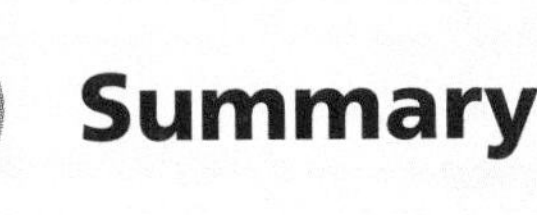

6 Summary

DEFINITION/PROCEDURE	**EXAMPLE**	**REFERENCE**
Solutions of Equations in Two Variables		**Section 6.1**
Solutions of Linear Equations A pair of values that satisfies the equation. Solutions for linear equations in two variables are written as *ordered pairs.* An ordered pair has the form (x, y) *x* coordinate *y* coordinate	If $2x - y = 10$, (6, 2) is a solution for the equation, because substituting 6 for x and 2 for y gives a true statement.	p. 472
The Rectangular Coordinate System		**Section 6.2**
The Rectangular Coordinate System A system formed by two perpendicular axes that intersect at a point called the **origin.** The horizontal line is called the ***x* axis.** The vertical line is called the ***y* axis.**	*y* *y* axis Origin *x* *x* axis	p. 481
Graphing Points from Ordered Pairs The coordinates of an ordered pair allow you to associate a point in the plane with every ordered pair. To graph a point in the plane, **1.** Start at the origin. **2.** Move right or left according to the value of the x coordinate: to the right if x is positive or to the left if x is negative. **3.** Then move up or down according to the value of the y coordinate: up if y is positive and down if y is negative.	To graph the point corresponding to (2, 3): *y* (2, 3) 3 units *x* 2 units	p. 484
Graphing Linear Equations		**Section 6.3**
Linear Equation An equation that can be written in the form $Ax + By = C$ in which A and B are not both 0.	$2x - 3y = 4$ is a linear equation.	p. 495

Continued

DEFINITION/PROCEDURE	EXAMPLE	REFERENCE
Graphing Linear Equations		**Section 6.3**
Graphing Linear Equations **1.** Find at least three solutions for the equation, and put your results in tabular form. **2.** Graph the solutions found in step 1. **3.** Draw a straight line through the points determined in step 2 to form the graph of the equation.	$x - y = 6$ (6, 0) (3, −3) (0, −6) x | y 0 | −6 3 | −3 6 | 0	p. 496
The Slope of a Line		**Section 6.4**
Slope The slope of a line gives a numerical measure of the steepness of the line. The slope m of a line containing the distinct points in the plane $P(x_1, y_1)$ and $Q(x_2, y_2)$ is given by $m = \frac{y_2 - y_1}{x_2 - x_1}$ when $x_2 \neq x_1$	To find the slope of the line through $(-2, -3)$ and $(4, 6)$, $m = \frac{6 - (-3)}{4 - (-2)}$ $= \frac{6 + 3}{4 + 2}$ $= \frac{9}{6} = \frac{3}{2}$	p. 519
The Graph of $y = mx$ A line passing through the origin with slope m.	To graph $y = -4x$ **1.** plot (0, 0) **2.** $\frac{\text{rise}}{\text{run}} = -4 = \frac{-4}{1}$	p. 524
Direct Variation		**Section 6.5**
If y is a constant multiple of x, we write $y = kx$ and say y varies directly as x	If $y = 20$ when $x = 5$, and y varies directly as x $20 = k(5)$ $k = 4$ $y = 4x$	p. 533

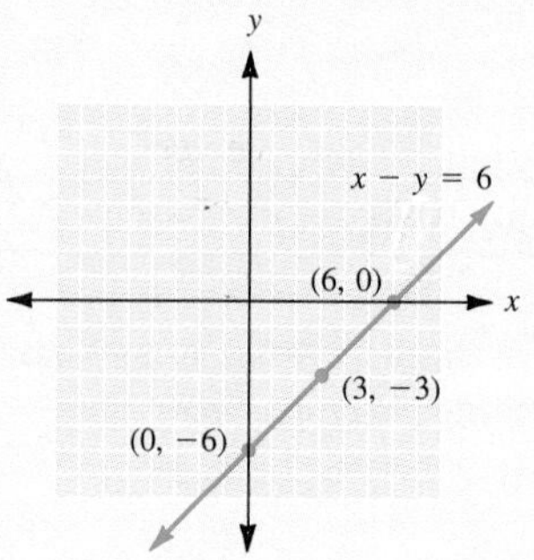

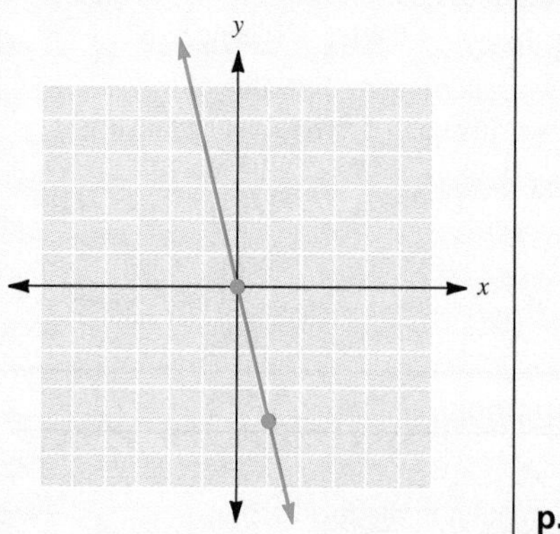

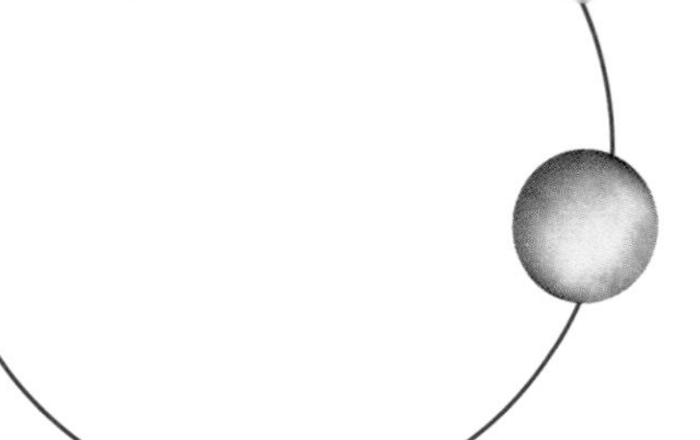

Summary Exercises

This summary exercise set is provided to give you practice with each of the objectives of the chapter. Each exercise is keyed to the appropriate chapter section. The answers are provided in the *Instructor's Manual.* Your instructor will give you guidelines on how to best use these exercises in your instructional setting.

[6.1] Tell whether the number shown in parentheses is a solution for the given equation.

1. $7x + 2 = 16$ (2) Yes

2. $5x - 8 = 3x + 2$ (4) No

3. $7x - 2 = 2x + 8$ (2) Yes

4. $4x + 3 = 2x - 11$ (−7) Yes

5. $x + 5 + 3x = 2 + x + 23$ (6) No

6. $\frac{2}{3}x - 2 = 10$ (21) No

[6.1] Determine which of the ordered pairs are solutions for the given equations.

7. $x - y = 6$ $(6, 0), (3, 3), (3, -3), (0, -6)$ (6, 0), (3, −3), (0, −6)

8. $2x + y = 8$ $(4, 0), (2, 2), (2, 4), (4, 2)$ (4, 0), (2, 4)

9. $2x + 3y = 6$ $(3, 0), (6, 2), (-3, 4), (0, 2)$ (3, 0), (−3, 4), (0, 2)

10. $2x - 5y = 10$ $(5, 0), \left(\frac{5}{2}, -1\right), \left(2, \frac{2}{5}\right), (0, -2)$ $(5, 0), \left(\frac{5}{2}, -1\right), (0, -2)$

[6.1] Complete the ordered pairs so that each is a solution for the given equation.

11. $x + y = 8$ $(4, \ ), (\ , 8), (8, \ ), (6, \ )$ (4, 4), (0, 8), (8, 0), (6, 2)

12. $x - 2y = 10$ $(0, \ ), (12, \ ), (\ , -2), (8, \ )$ (0, −5), (12, 1), (6, −2), (8, −1)

13. $2x + 3y = 6$ $(3, \ ), (6, \ ), (\ , -4), (-3, \ )$ (3, 0), (6, −2), (9, −4), (−3, 4)

14. $y = 3x + 4$ $(2, \ ), (\ , 7), \left(\frac{1}{3}, \ \right), \left(\frac{4}{3}, \ \right)$ $(2, 10), (1, 7), \left(\frac{1}{3}, 5\right), \left(\frac{4}{3}, 8\right)$

[6.1] Find four solutions for each of the following equations.

15. $x + y = 10$ (0, 10), (2, 8), (4, 6), (6, 4)

16. $2x + y = 8$ (0, 8), (2, 4), (4, 0), (6, −4)

17. $2x - 3y = 6$ (0, −2), (3, 0), (6, 2), (9, 4)

18. $y = -\frac{3}{2}x + 2$ (0, 2), (2, −1), (4, −4), (6, −7)

[6.2] Give the coordinates of the points graphed below.

19. A (4, 6)

20. B (0, 3)

21. E (− 1, −5)

22. F (5, −3)

[6.2] Plot points with the coordinates shown.

23. $P(6, 0)$

24. $Q(5, 4)$

25. $T(-2, 4)$

26. $U(4, -2)$

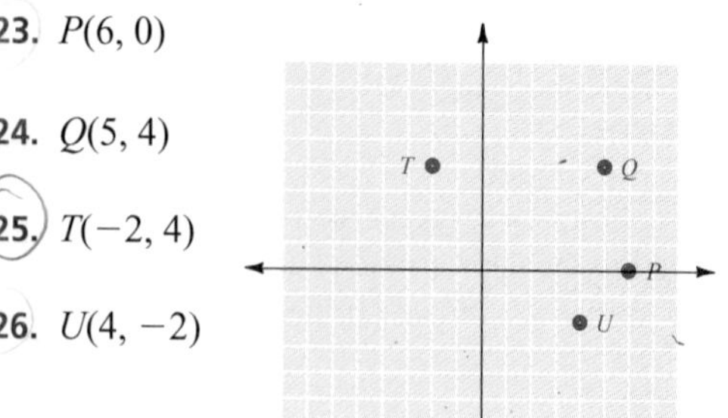

[6.3] Graph each of the following equations.

27. $x + y = 5$

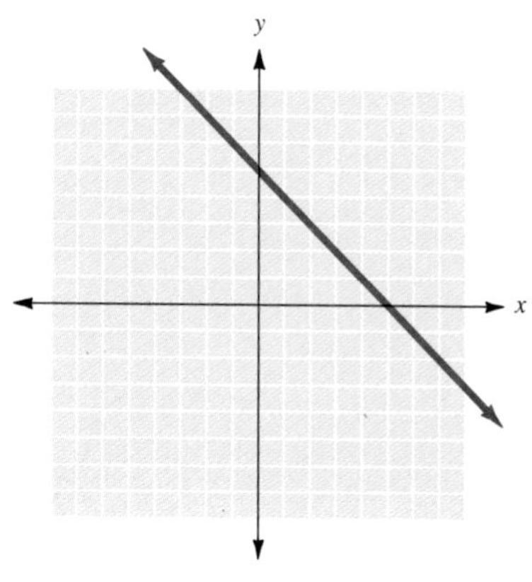

28. $x - y = 6$

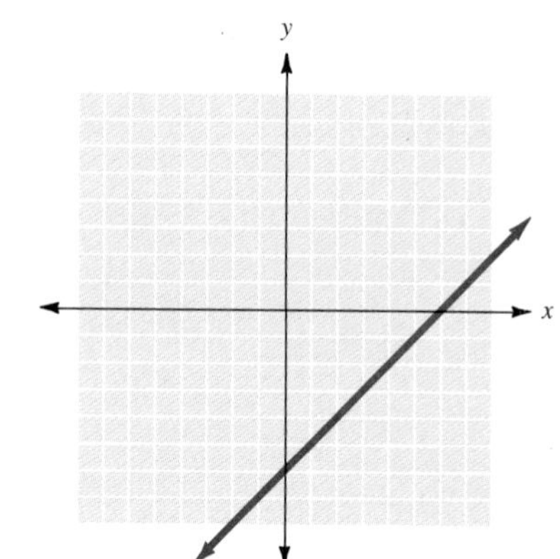

29. $y = 2x$

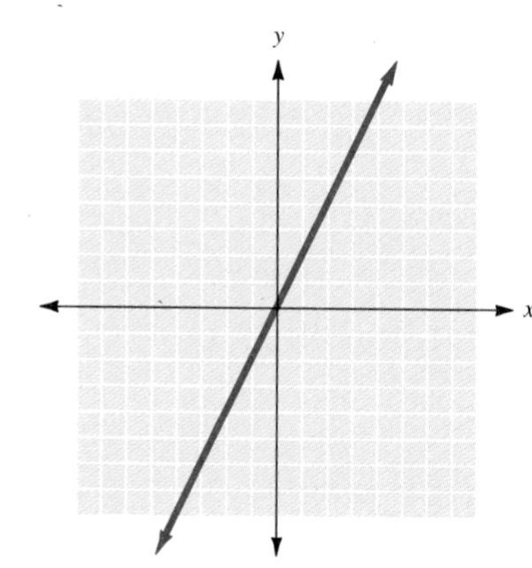

30. $y = -3x$

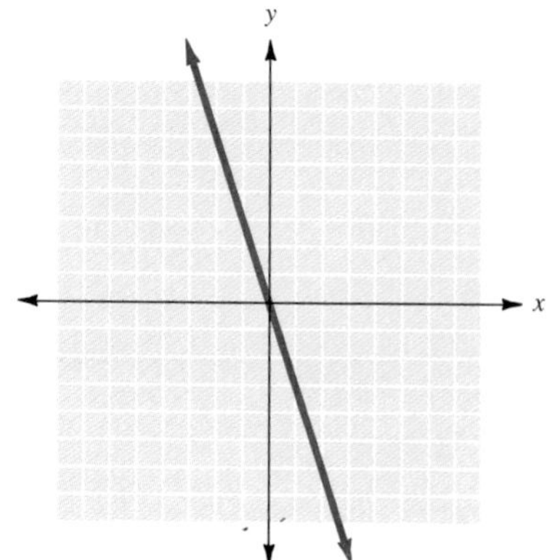

31. $y = \frac{3}{2}x$

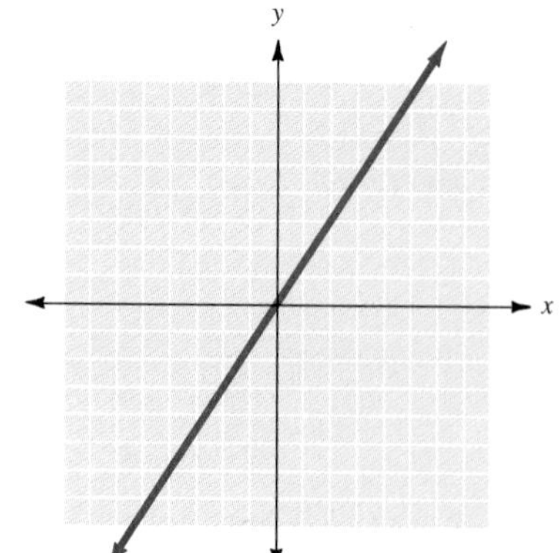

32. $y = 3x + 2$

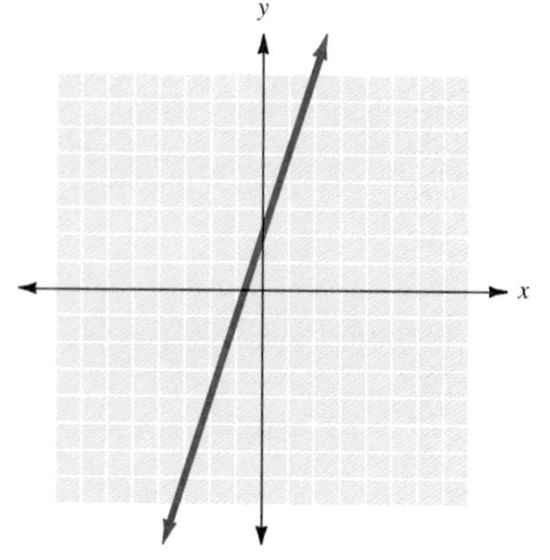

33. $y = 2x - 3$

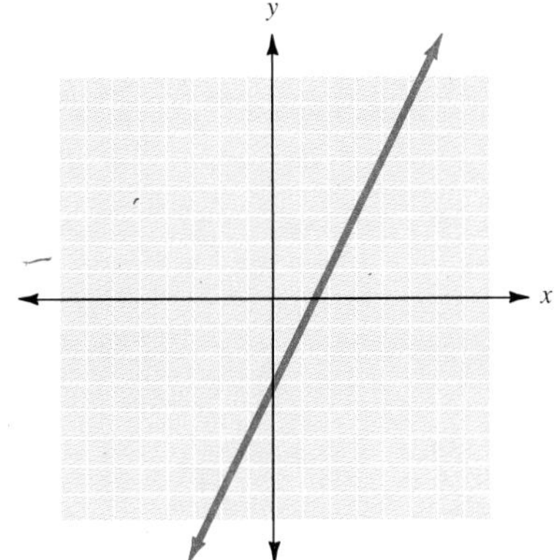

34. $y = -3x + 4$

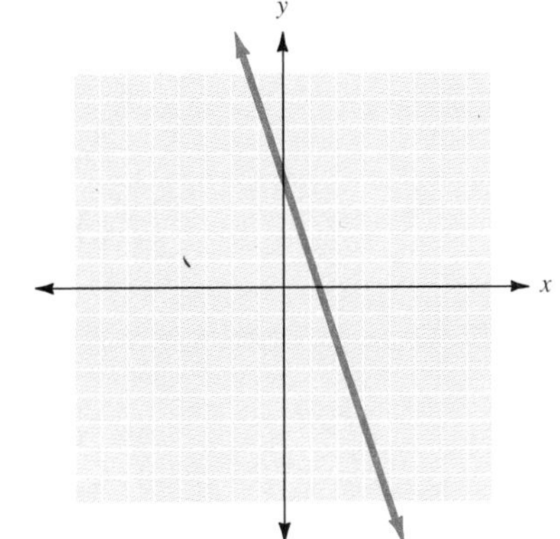

35. $y = \frac{2}{3}x + 2$

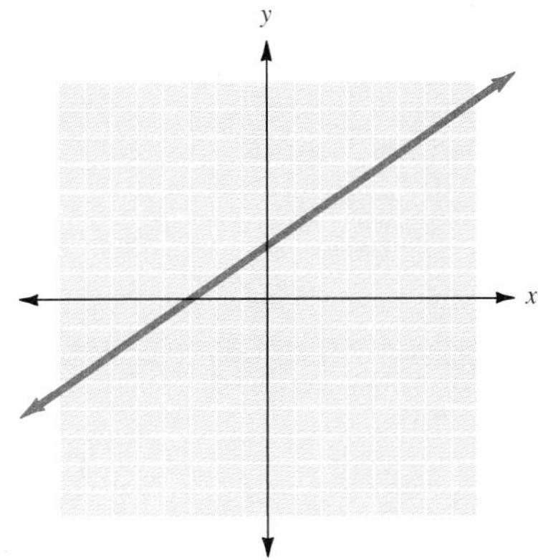

36. $3x - y = 3$

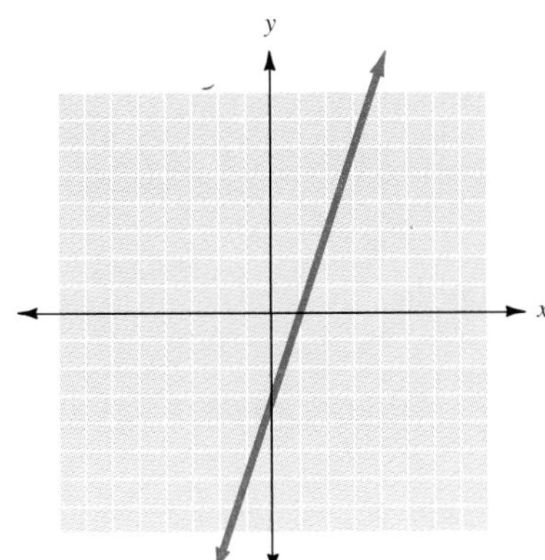

37. $2x + y = 6$

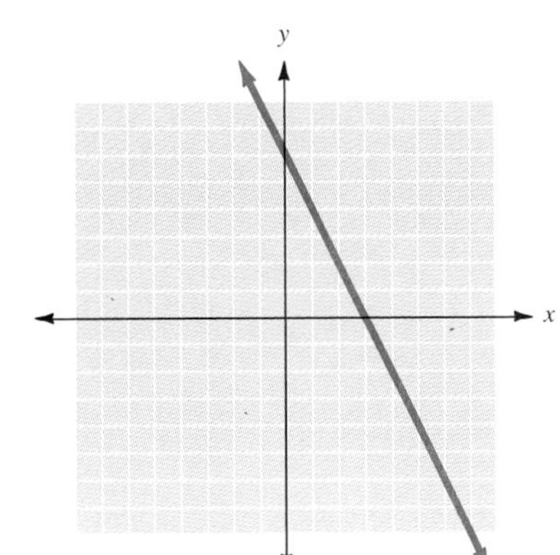

38. $3x + 2y = 12$

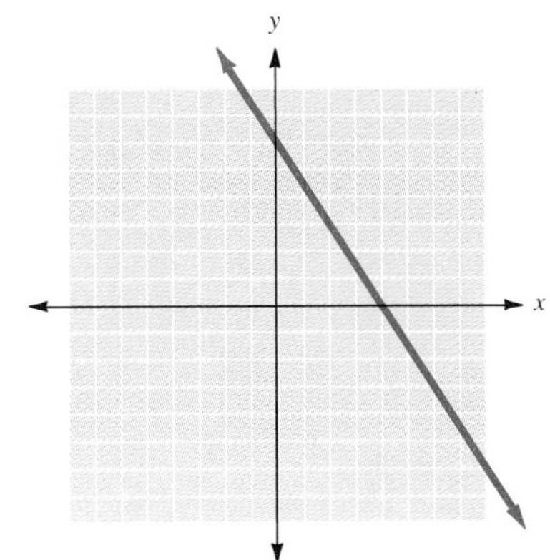

39. $3x - 4y = 12$

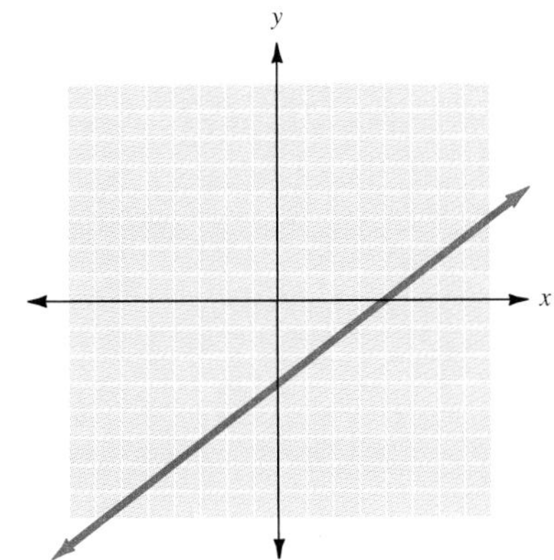

40. $x = 3$

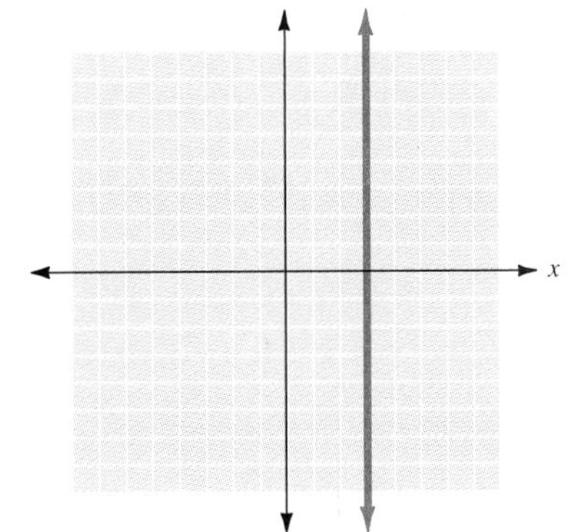

41. $y = -2$

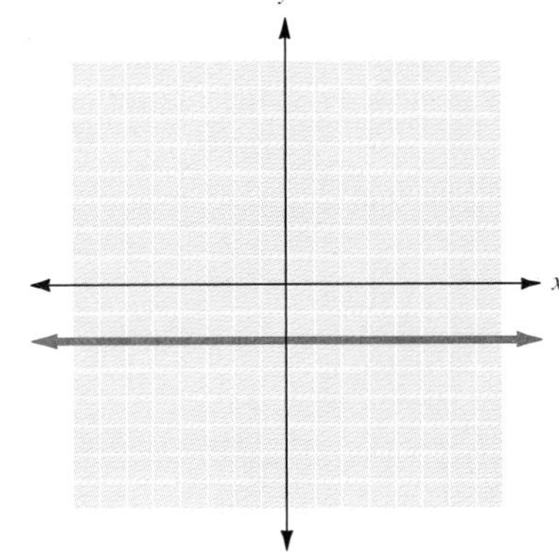

Graph each of the following equations.

42. $5x - 3y = 15$

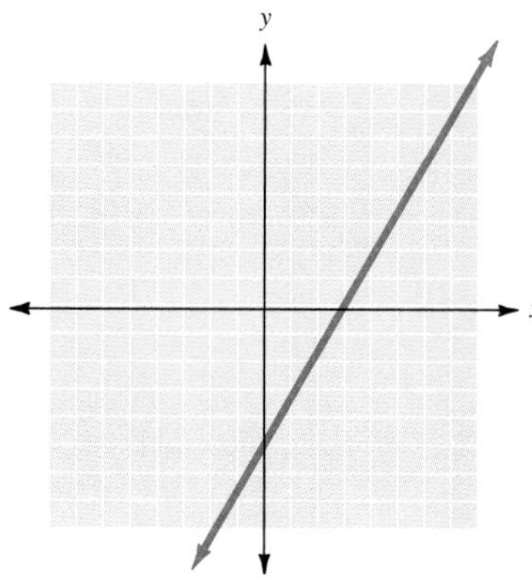

43. $4x + 3y = 12$

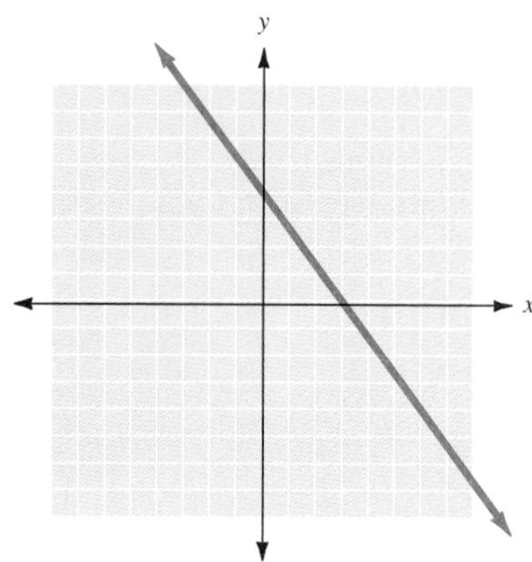

Graph each equation by first solving for y.

44. $2x + y = 6$ $y = -2x + 6$

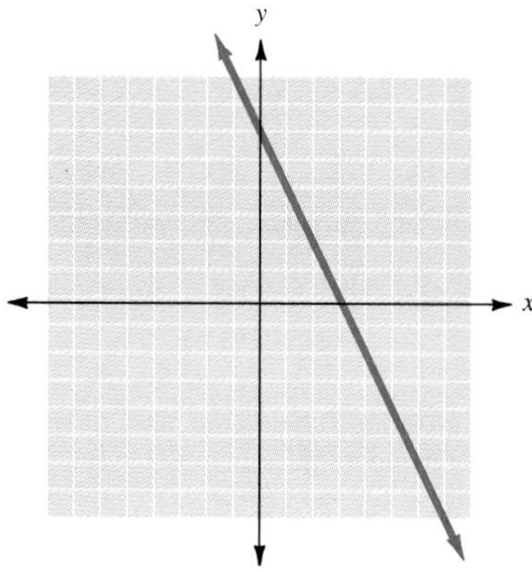

45. $3x + 2y = 6$ $y = -\frac{3}{2}x + 3$

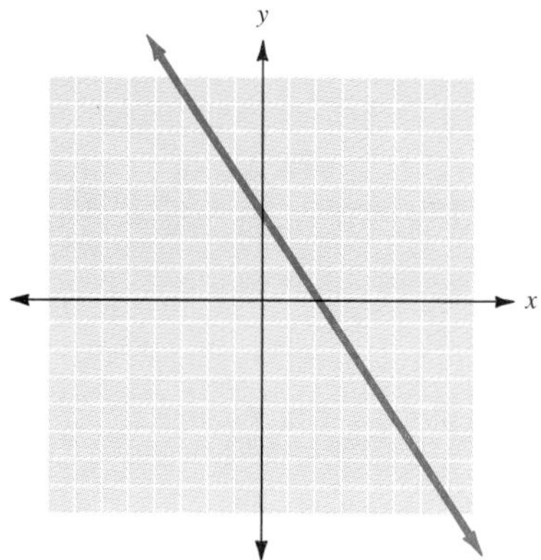

[6.4] Find the slope of the line through each of the following pairs of points.

46. $(3, 4)$ and $(5, 8)$ 2

47. $(-2, 3)$ and $(1, -6)$ -3

48. $(-2, 5)$ and $(2, 3)$ $-\frac{1}{2}$

49. $(-5, -2)$ and $(1, 2)$ $\frac{2}{3}$

50. $(-2, 6)$ and $(5, 6)$ 0

51. $(-3, 2)$ and $(-1, -3)$ $-\frac{5}{2}$

52. $(-3, -6)$ and $(5, -2)$ $\frac{1}{2}$

53. $(-6, -2)$ and $(-6, 3)$ Undefined

[6.4] In exercises 54 to 57, find the slope of the lines graphed.

54.

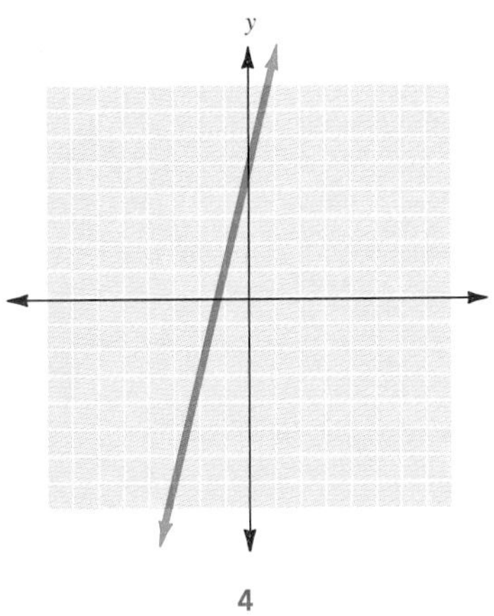

4

55.

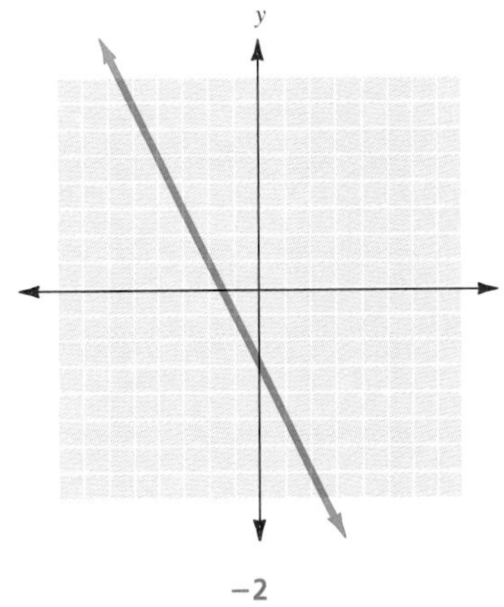

−2

56.

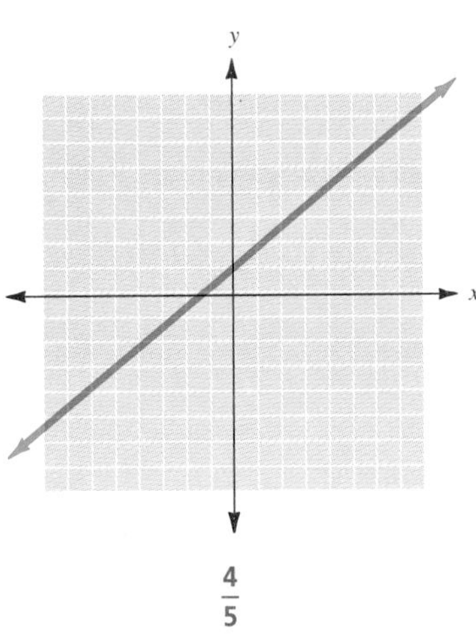

$\frac{4}{5}$

57.

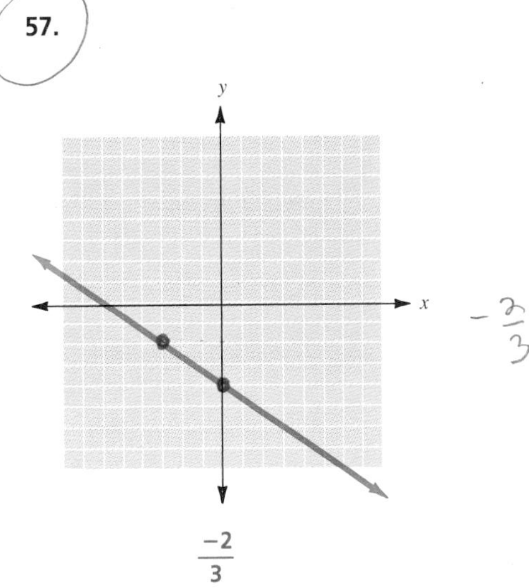

$\frac{-2}{3}$

[6.4] Graph each of the following equations.

58. $y = 6x$

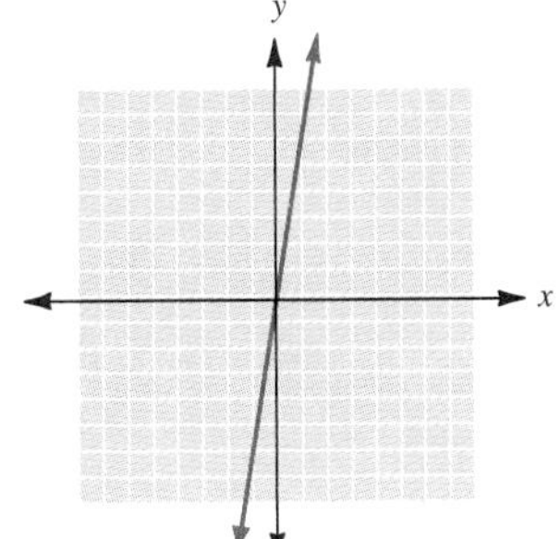

59. $y = -6x$

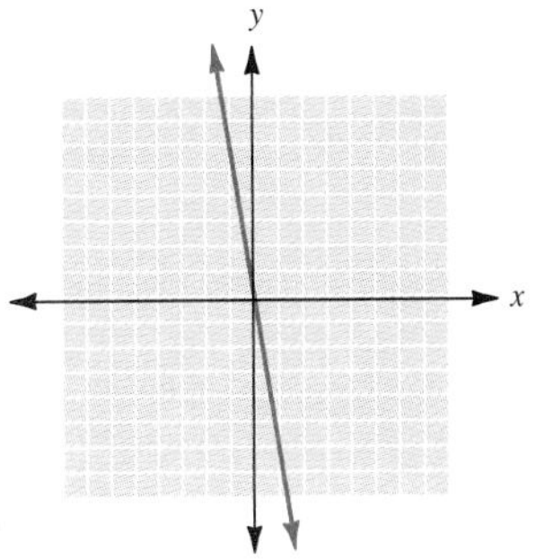

60. $y = \frac{2}{5}x$

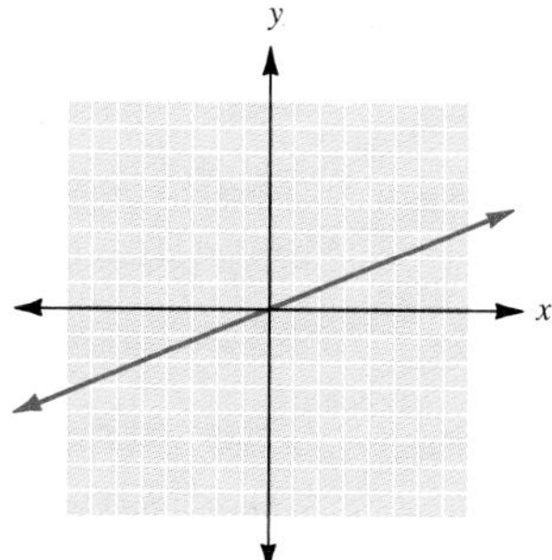

61. $y = -\frac{3}{4}x$

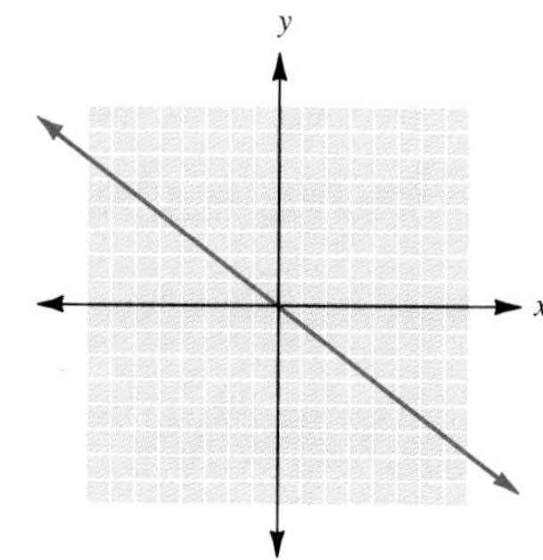

[6.5] Solve for k, the constant of variation.

62. y varies directly as x; $y = 20$ when $x = 40$ $\frac{1}{2}$

63. y varies directly as x; $y = 5$ when $x = 3$ $\frac{5}{3}$

[6.5] In exercises 64 and 65, y varies directly as x and the value of k is given. Graph the equation of variation.

64. $k = 4$

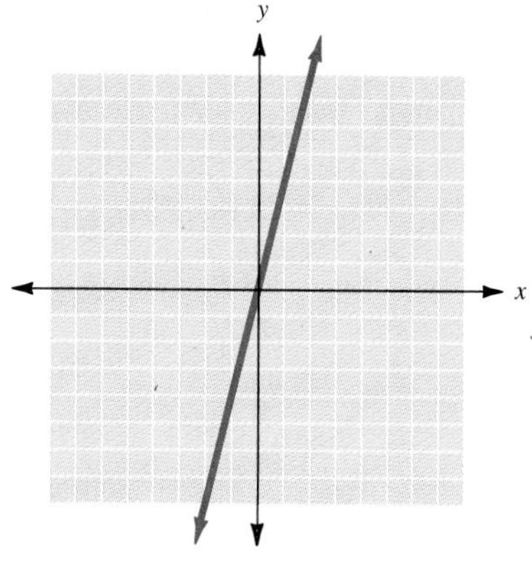

65. $k = -3.5$

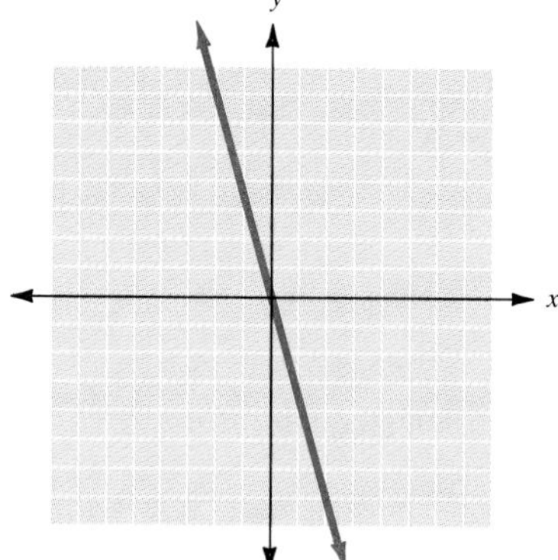

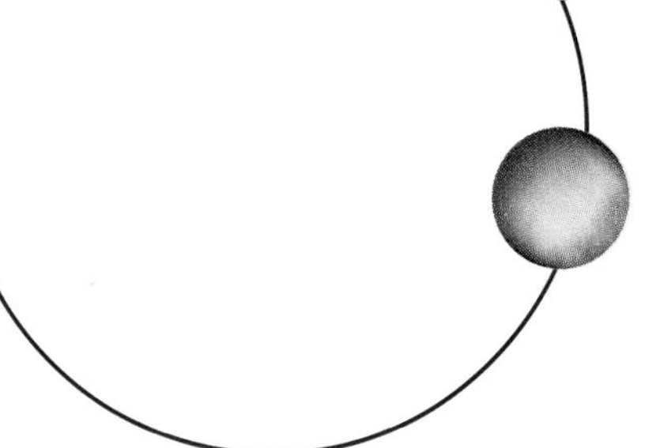

Self-Test for Chapter 6

Name ______________________

Section __________ Date __________

ANSWERS

1. (3, 6), (9, 0)
2. (4, 0), (5, 4)
3. (3, 3), (6, 2), (9, 1)
4. (3, 0), (0, 4), $\left(\frac{3}{4}, 3\right)$
5. Different answers are possible
6. Different answers are possible
7. (4, 2)
8. (−4, 6)
9. (0, −7)
10. See exercise
11. See exercise
12. See exercise
13. See exercise
14. See exercise
15. See exercise

The purpose of this test is to help you check your progress and to review for a chapter test in class. Allow yourself about an hour to take the test. When you are done, check your answers in the back of the book. If you missed any answers, be sure to go back and review the appropriate sections in the chapter.

Determine which of the ordered pairs are solutions for the given equations.

1. $x + y = 9$ (3, 6), (9, 0), (3, 2)

2. $4x - y = 16$ (4, 0), (3, −1), (5, 4)

Complete the ordered pairs so that each is a solution for the given equation.

3. $x + 3y = 12$ (3,), (, 2), (9,)

4. $4x + 3y = 12$ (3,), (, 4), (, 3)

Find four solutions for each of the following equations.

5. $x - y = 7$

6. $5x - 6y = 30$

Give the coordinates of the points graphed below.

7. A

8. B

9. C

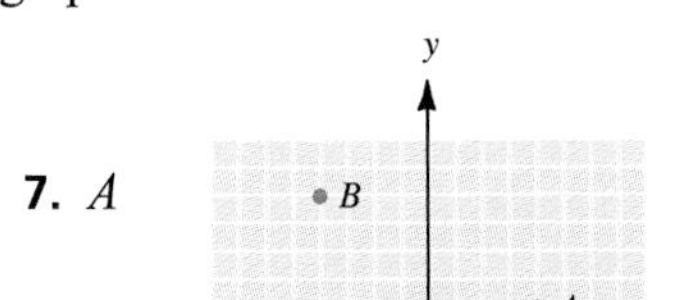

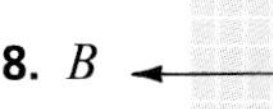

Plot points with the coordinates shown.

10. $S(1, -2)$

11. $T(0, 3)$

12. $U(-2, -3)$

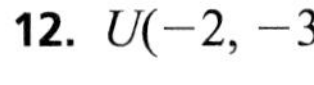

Graph each of the following equations.

13. $x + y = 4$

14. $y = 3x$

15. $y = \frac{3}{4}x - 4$

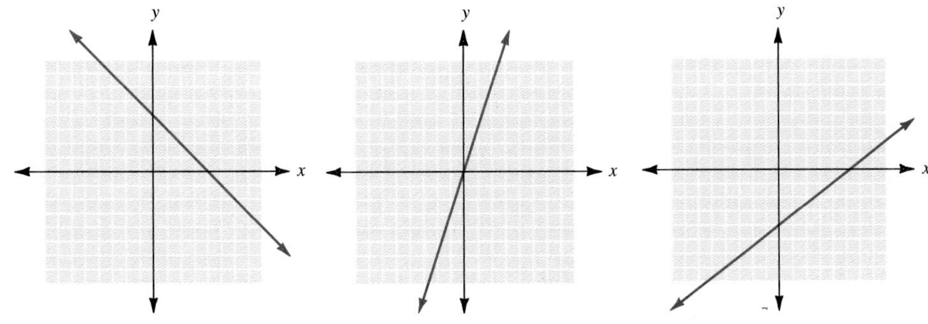

ANSWERS

16. See exercise
17. See exercise
18. See exercise
19. 1
20. $\frac{3}{4}$
21. Undefined
22. 0
23. −7
24. See exercise
25. 5

16. $x + 3y = 6$ **17.** $2x + 5y = 10$ **18.** $y = -4$

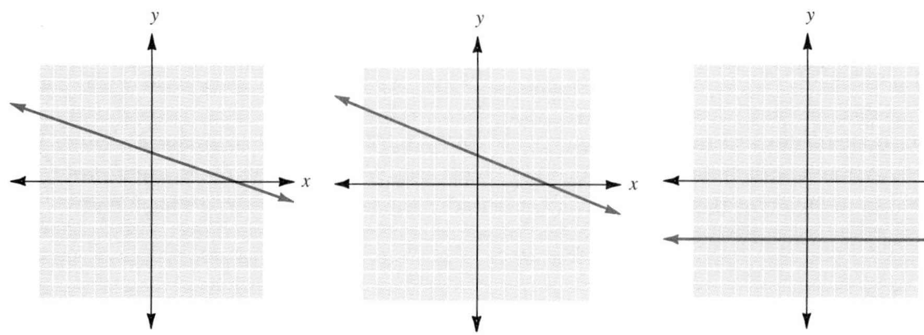

Find the slope of the line through the following pairs of points.

19. $(-3, 5)$ and $(2, 10)$

20. $(-2, 6)$ and $(2, 9)$

21. $(4, 6)$ and $(4, 8)$

22. $(7, 9)$ and $(3, 9)$

23. Find the slope of the line graphed.

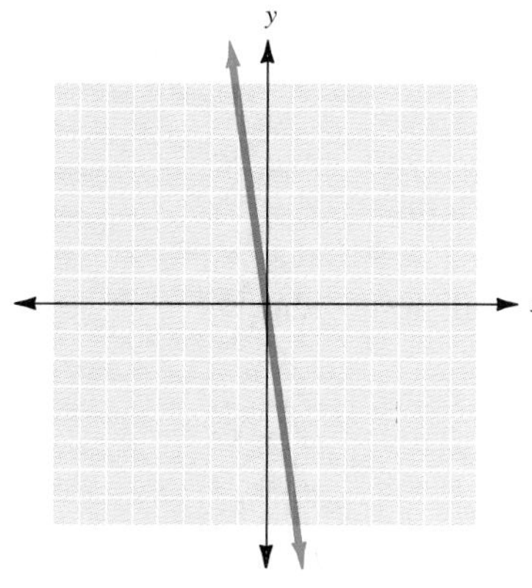

24. Graph the equation $y = 2x$

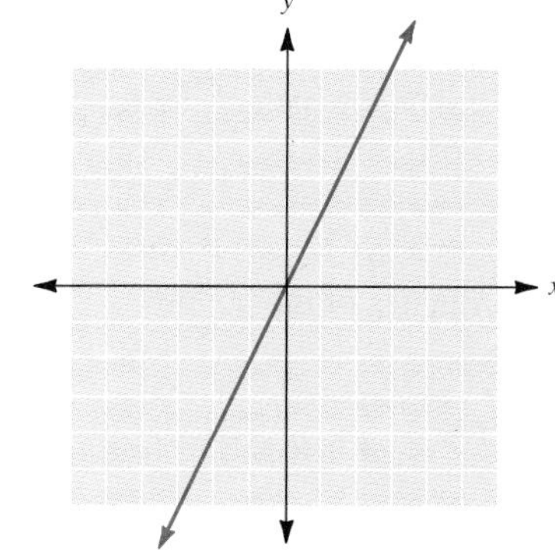

25. Solve for the constant of variation if y varies directly as x and $y = 35$ when $x = 7$.

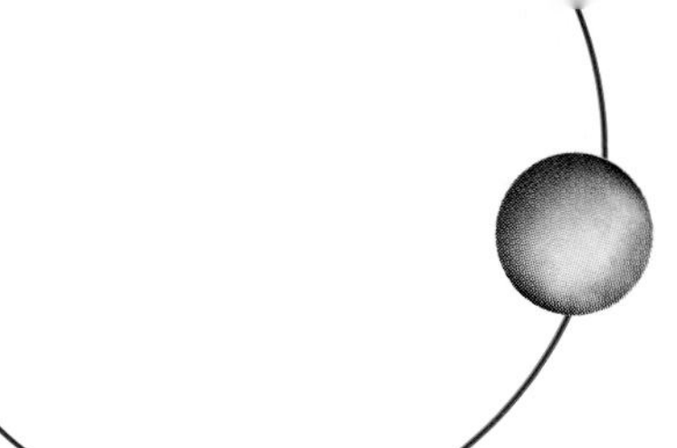

Cumulative Test Chapters 0 to 6

Name ____________

Section ________ Date ________

Perform the indicated operations.

1. $9 + (-6)$ **2.** $-4 - (-9)$ **3.** $25 - (-12)$ **4.** $-32 + (-21)$

5. $(-23)(-3)$ **6.** $(12)(-10)$ **7.** $30 \div (-6)$ **8.** $(-24) \div (-8)$

Evaluate the expressions if $x = -3$, $y = 4$, and $z = -5$.

9. $3x^2y$ **10.** $-3z - 3y$ **11.** $-3(-2y + 3z)$ **12.** $\dfrac{3y - 2x}{5y + 6x}$

Solve the following equations and check your results.

13. $5x - 2 = 2x - 6$ **14.** $3(x - 2) = 2(3x + 1)$

15. $\dfrac{5}{6}x - 3 = 2 + \dfrac{1}{3}x$ **16.** $4(2 - x) + 9 = 7 + 6x$

17. Solve the equation $F = \dfrac{9}{5}C + 32$ for C.

Solve the following inequalities.

18. $4x - 9 < 7$ **19.** $-5x + 15 \geq 2x - 6$

Use the properties of exponents to simplify the following expressions, and write the results with positive exponents.

20. $(x^2y^3)^{-2}$ **21.** $\dfrac{x^3y^2}{x^4y^{-3}}$ **22.** $(x^6y^{-3})^0$

Perform the indicated operation for each of the following polynomials.

23. Add $2x^2 + 4x - 6$ and $3x^2 - 4x - 4$.

24. Subtract $3a^2 - 2a + 5$ from the sum of $a^2 + 3a - 2$ and $-5a^2 + 2a + 9$.

Evaluate each polynomial for the indicated variable value.

25. $2x^2 - 5x + 7$ for $x = 4$ **26.** $x^3 + 3x^2 - 7x + 8$ for $x = -2$

Multiply the following polynomials.

27. $(3x - 5y)(2x + 4y)$ **28.** $3x(x - 3)(2x + 5)$ **29.** $(2a + 7b)(2a - 7b)$

Completely factor each polynomial.

30. $12p^2n^2 + 20pn^2 - 16pn^3$ **31.** $y^3 - 3y^2 - 5y + 15$ **32.** $9a^2b - 49b$

33. $6x^2 - 2x - 4$ **34.** $6a^2 + 7ab - 3b^2$

Solve each of the following by factoring.

35. $x^2 - 8x - 33 = 0$ **36.** $35x^2 - 38x = -8$

ANSWERS

1. 3 2. 5 3. 37
4. −53 5. 69
6. −120 7. −5
8. 3 9. 108 10. 3
11. 69 12. 9
13. $-\frac{4}{3}$ 14. $-\frac{8}{3}$
15. 10 16. 1
17. $C = \frac{5}{9}(F - 32)$
18. $x < 4$ 19. $x \leq 3$
20. $\frac{1}{x^4y^6}$ 21. $\frac{y^5}{x}$
22. 1 23. $5x^2 - 10$
24. $-7a^2 + 7a + 2$
25. 19 26. 26
27. $6x^2 + 2xy - 20y^2$
28. $6x^3 - 3x^2 - 45x$
29. $4a^2 - 49b^2$
30. $4pn^2(3p + 5 - 4n)$
31. $(y - 3)(y^2 - 5)$
32. $b(3a + 7)(3a - 7)$
33. $2(3x + 2)(x - 1)$
34. $(2a + 3b)(3a - b)$
35. −3, 11
36. $\frac{4}{5}, \frac{2}{7}$

ANSWERS

37. $\dfrac{-5a^3}{3b^2}$

38. $\dfrac{w-2}{w+3}$

39. $\dfrac{a+5}{a(a-5)}$

40. $\dfrac{10}{w-5}$

41. $\dfrac{3x^3}{4}$

42. $\dfrac{m-4}{4m}$

43. 6

44. 5

45. See exercise

46. See exercise

47. See exercise

48. 2

49. See exercise

50. 30

51. Width: 5 in. Length: 7 in.

52. 41, 43, 45

53. 200

Simplify each of the following rational expressions.

37. $\dfrac{-35a^4b^5}{21ab^7}$

38. $\dfrac{2w^2 - w - 6}{2w^2 + 9w + 9}$

Add or subtract as indicated. Simplify your answer.

39. $\dfrac{2}{a-5} - \dfrac{1}{a}$

40. $\dfrac{2w}{w^2 - 9w + 20} + \dfrac{8}{w-4}$

Multiply or divide as indicated.

41. $\dfrac{4xy^3}{5xy^2} \cdot \dfrac{15x^3y}{16y^2}$

42. $\dfrac{m^2 - 3m}{m^2 - 9} \div \dfrac{4m^2}{m^2 - m - 12}$

Solve each of the following equations.

43. $\dfrac{w}{w-2} + 1 = \dfrac{w+4}{w-2}$

44. $\dfrac{7}{x} - \dfrac{1}{x-3} = \dfrac{9}{x^2 - 3x}$

Graph each of the following equations.

45. $3x + 4y = 12$

46. $y = -7$

47. $x = 2y$

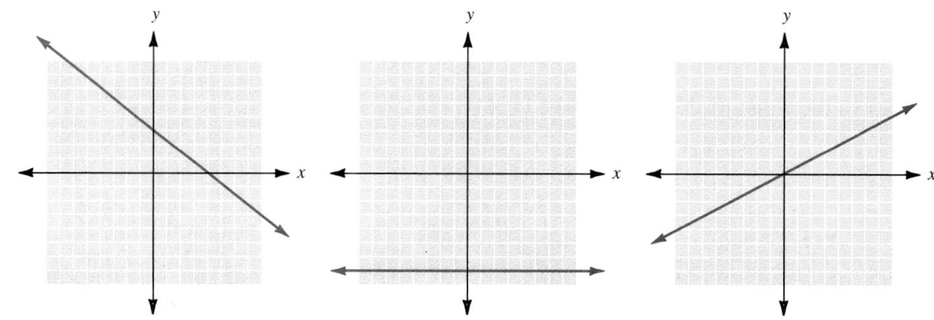

48. Find the slope of the line passing through the points $(-3, 5)$ and $(1, 13)$.

49. Graph the equation of variation if $k = -4$.

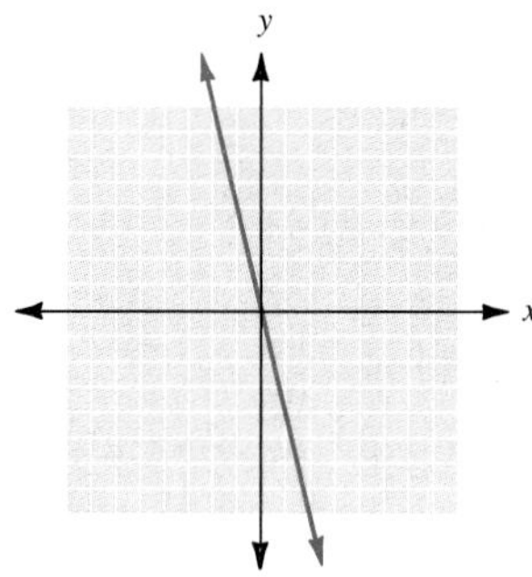

50. Find the constant of variation if y varies directly with x, and if $y = 450$ when $x = 15$.

Solve the following problems.

51. The length of a rectangle is 3 in. less than twice its width. If the perimeter is 24 in., find the dimensions of the rectangle.

52. The sum of three consecutive odd integers is 129. Find the three integers.

53. The carpet outlet is selling rug remnants at 25% off. If the sale price is \$150, what was the original price?

7 Graphing and Inequalities

INTRODUCTION

In the pharmaceutical-making process, great caution must be exercised to ensure that the medicines and drugs are pure and contain precisely what is indicated on the label. Guaranteeing such purity is a task the quality control division of the pharmaceutical company assumes.

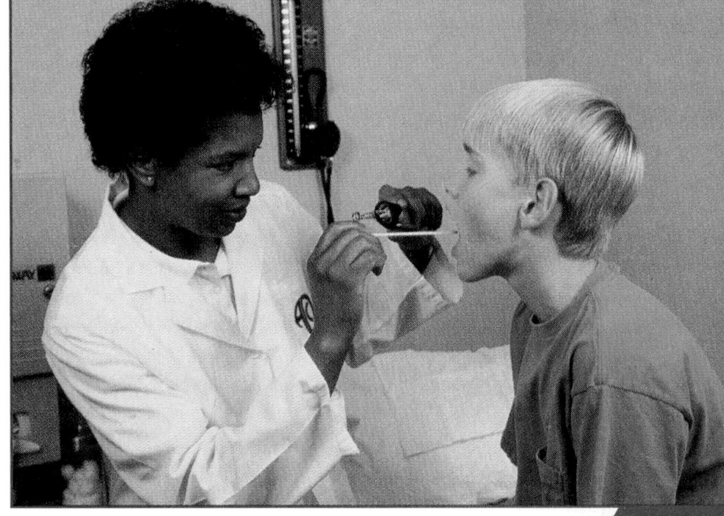

A lab technician working in quality control must run a series of tests on samples of every ingredient, even simple ingredients such as salt (NaCl). One such test is a measure of how much weight is lost as a sample is dried. The technician must set up a 3-hour procedure that involves cleaning and drying bottles and stoppers and then weighing them while they are empty and again when they contain samples of the substance to be heated and dried. At the end of the procedure, to compute the percentage of weight loss from drying, the technician uses the formula

$$L = \frac{W_g - W_f}{W_g - T} \cdot 100$$

in which L = percentage loss in drying

W_g = weight of container and sample

W_f = weight of container and sample after drying process completed

T = weight of empty container

The pharmaceutical company may have a standard of acceptability for this substance. For instance, the substance may not be acceptable if the loss of weight from drying is greater than 10%. The technician would then use the following inequality to calculate acceptable weight loss:

$$10 \geq \frac{W_g - W_f}{W_g - T} \cdot 100$$

We will further examine inequalities in this chapter.

Name ______________________

Section ________ Date ________

Pre-Test Chapter 7

ANSWERS

1. $\frac{10}{7}$

2. Slope: $\frac{5}{4}$ y intercept: (0, −5)

3. $y = 5x - 2$ see answers at the end of the text for graphs

4. $y = -4x + 6$ see answers at the end of the text for graphs

5. Parallel

6. Perpendicular

7. $y = -4x - 12$

8. $y = -\frac{5}{3}x + \frac{7}{3}$

9. $y = \frac{3}{5}x - 3$

10. See exercise

11. See exercise

12. 13; −3

1. Find the slope of the line through the pairs of points. $(-6, 2)$ and $(1, 12)$

2. Find the slope and y intercept from the equation. $5x - 4y = 20$

Write the equation of the line with the given slope and y intercept. Then graph each line.

3. Slope 5 and y intercept $(0, -2)$

4. Slope -4 and y intercept $(0, 6)$

Determine if the following pairs of lines are parallel, perpendicular, or neither.

5. L_1 through $(3, 7)$ and $(1, 11)$ $\quad$ L_2 through $(-5, -1)$ and $(6, -23)$

6. L_1 with equation $y = 3x + 9$ $\quad$ L_2 with equation $3y + x = 9$

7. Find the equation of the line through $(-5, 8)$ and parallel to the line $4x + y = 8$.

Write the equation of the line L satisfying each of the following sets of geometric conditions.

8. L passes through $(5, -6)$ and is perpendicular to $3x - 5y = 15$.

9. L has y intercept $(0, -3)$ and is parallel to $-3x + 5y = -15$.

Graph each of the following inequalities.

10. $2x + y < 5$

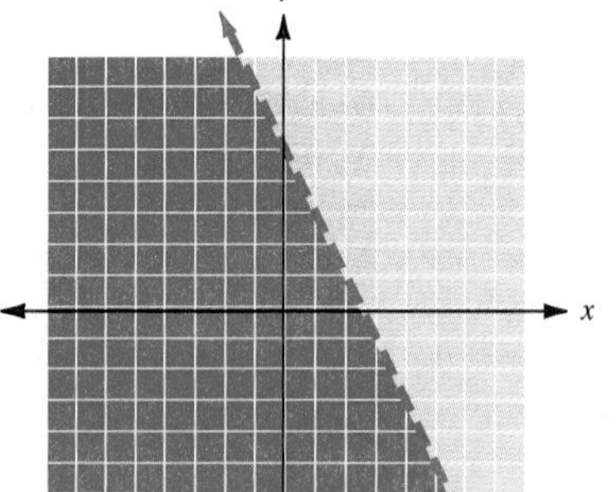

11. $-3x + 5y \geq 15$

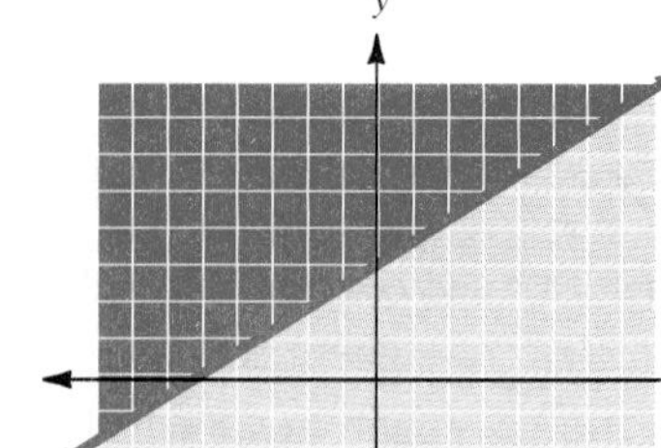

Evaluate $f(x)$ for the value given.

12. $f(x) = -2x^3 - 6x - 4x^2 + 9$; find $f(-1)$ and $f(1)$.

The Slope-Intercept Form

OBJECTIVES

1. Find the slope and y intercept from the equation of a line
2. Given the slope and y intercept, write the equation of a line
3. Use the slope and y intercept to graph a line

In Chapter 6, we used two points to find the slope of a line. In this chapter we will use the slope to find the graph of an equation.

First, we want to consider finding the equation of a line when its slope and y intercept are known.

Suppose that the y intercept of a line is $(0, b)$. Then the point at which the line crosses the y axis has coordinates $(0, b)$. Look at the sketch at left.

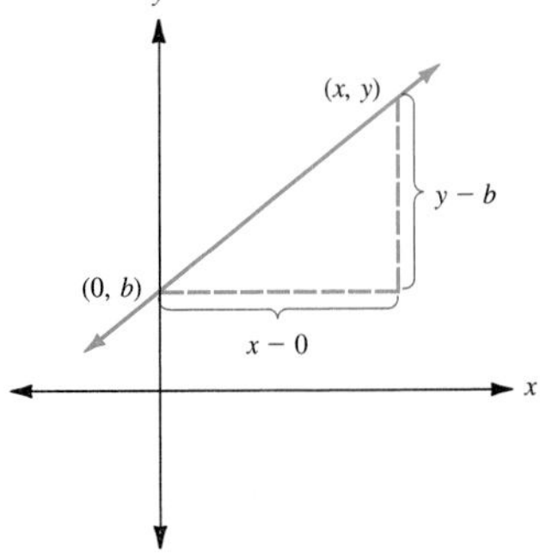

Now, using any other point (x, y) on the line and using our definition of slope, we can write

$$m = \frac{y - b}{x - 0} \qquad (1)$$

Change in y.

Change in x.

or

$$m = \frac{y - b}{x} \qquad (2)$$

Multiplying both sides of equation (2) by x, we have

$$mx = y - b \qquad (3)$$

Finally, adding b to both sides of equation (3) gives

$$mx + b = y$$

or

$$y = mx + b \qquad (4)$$

We can summarize the above discussion as follows:

NOTE In this form, the equation is *solved for y.* The coefficient of x will give you the slope of the line, and the constant term gives the y intercept.

Definitions: The Slope-Intercept Form for a Line

An equation of the line with slope m and y intercept $(0, b)$ is

$y = mx + b$

Example 1

Finding the Slope and y Intercept

Find the slope and y intercept for the graph of the equation

$$y = -\frac{2}{3}x - 5$$

($-\frac{2}{3}$ is m; -5 is b)

The slope of the line is $-\frac{2}{3}$; the y intercept is $(0, -5)$.

CHECK YOURSELF 1

Find the slope and y intercept for the graph of each of the following equations.

(a) $y = -3x - 7$ **(b)** $y = \frac{3}{4}x + 5$

As Example 2 illustrates, we may have to solve for y as the first step in determining the slope and the y intercept for the graph of an equation.

Example 2

Finding the Slope and y Intercept

Find the slope and y intercept for the graph of the equation

$3x + 2y = 6$

First, we must solve the equation for y.

NOTE If we write the equation as

$$y = \frac{-3x + 6}{2}$$

it is more difficult to identify the slope and the intercept.

$3x + 2y = 6$

$2y = -3x + 6$ Add $(-3x)$ to both sides.

$y = -\frac{3}{2}x + 3$ Divide each term by 2.

The equation is now in slope-intercept form. The slope is $-\frac{3}{2}$, and the y intercept is (0, 3).

CHECK YOURSELF 2

Find the slope and y intercept for the graph of the equation

$2x - 5y = 10$

As we mentioned earlier, knowing certain properties of a line (namely, its slope and y intercept) will also allow us to write the equation of the line by using the slope-intercept form. Example 3 illustrates this approach.

Example 3

Writing the Equation of a Line

Write the equation of a line with slope $-\frac{3}{4}$ and y intercept $(0, -3)$.

We know that $m = -\frac{3}{4}$ and $b = -3$. In this case,

$$y = \underset{m}{-\frac{3}{4}}x + \underset{b}{(-3)}$$

or

$$y = -\frac{3}{4}x - 3$$

which is the desired equation.

CHECK YOURSELF 3

Write the equation of a line with the following:

(a) slope -2 and y intercept $(0, 7)$ **(b)** slope $\frac{2}{3}$ and y intercept $(0, -3)$

We can also use the slope and y intercept of a line in drawing its graph. Consider Example 4.

Example 4

Graphing a Line

Graph the line with slope $\frac{2}{3}$ and y intercept $(0, 2)$.

Because the y intercept is $(0, 2)$, we begin by plotting the point $(0, 2)$. Because the horizontal change (or run) is 3, we move 3 units to the right *from that y intercept.* Then because the vertical change (or rise) is 2, we move 2 units up to locate another point on the desired graph. Note that we will have located that second point at $(3, 4)$. The final step is to simply draw a line through that point and the y intercept.

NOTE

$m = \frac{2}{3} = \frac{\text{rise}}{\text{run}}$

The line rises from left to right because the slope is positive.

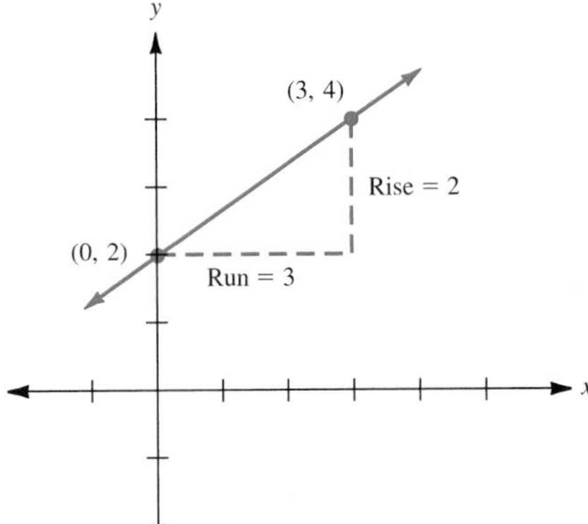

The equation of this line is $y = \frac{2}{3}x + 2$.

CHECK YOURSELF 4

Graph the equation of a line with slope $\frac{3}{5}$ *and y intercept* $(0, -2)$.

Step by Step: Graphing by Using the Slope-Intercept Form

Step 1 Write the original equation of the line in slope-intercept form.
Step 2 Determine the slope m and the y intercept $(0, b)$.
Step 3 Plot the y intercept at $(0, b)$.
Step 4 Use m (the change in y over the change in x) to determine a second point on the desired line.
Step 5 Draw a line through the two points determined above to complete the graph.

You have now seen two methods for graphing lines: the slope-intercept method (Section 7.1) and the intercept method (Section 6.3). When you graph a linear equation, you should first decide which is the appropriate method.

Example 5

Selecting an Appropriate Graphing Method

Decide which of the two methods for graphing lines—the intercept method or the slope-intercept method—is more appropriate for graphing equations (a), (b), and (c).

(a) $2x - 5y = 10$

Because both intercepts are easy to find, you should choose the intercept method to graph this equation.

(b) $2x + y = 6$

This equation can be quickly graphed by either method. As it is written, you might choose the intercept method. It can, however, be rewritten as $y = -2x + 6$. In that case the slope-intercept method is more appropriate.

(c) $y = \frac{1}{4}x - 4$

Because the equation is in slope-intercept form, that is the more appropriate method to choose.

CHECK YOURSELF 5

Which would be more appropriate for graphing each equation, the intercept method or the slope-intercept method?

(a) $x + y = -2$ **(b)** $3x - 2y = 12$ **(c)** $y = -\frac{1}{2}x - 6$

CHECK YOURSELF ANSWERS

1. **(a)** Slope is -3, y intercept is $(0, -7)$; **(b)** Slope is $\frac{3}{4}$, y intercept is $(0, 5)$
2. $y = \frac{2}{5}x - 2$; the slope is $\frac{2}{5}$; the y intercept is $(0, -2)$
3. **(a)** $y = -2x + 7$; **(b)** $y = \frac{2}{3}x - 3$
4. 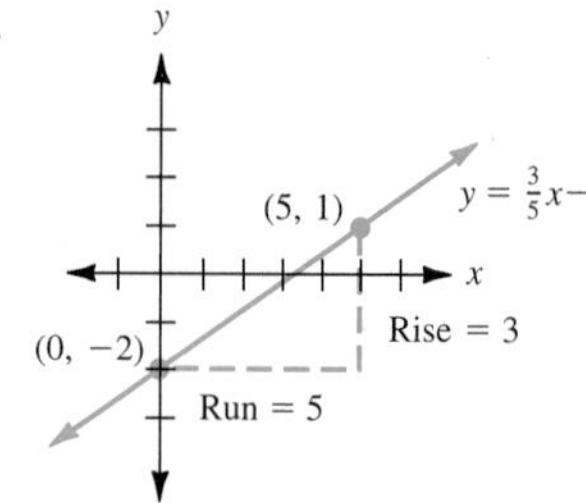

5. **(a)** Either; **(b)** intercept; **(c)** slope-intercept

Name ____________

Section ________ Date ________

7.1 Exercises

Find the slope and y intercept of the line represented by each of the following equations.

1. $y = 3x + 5$

2. $y = -7x + 3$

3. $y = -2x - 5$

4. $y = 5x - 2$

5. $y = \frac{3}{4}x + 1$

6. $y = -4x$

7. $y = \frac{2}{3}x$

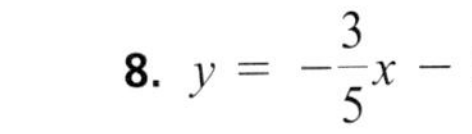

8. $y = -\frac{3}{5}x - 2$

9. $4x + 3y = 12$

10. $2x + 5y = 10$

11. $y = 9$

12. $2x - 3y = 6$

13. $3x - 2y = 8$

14. $x = 5$

Write the equation of the line with given slope and y intercept. Then graph each line, using the slope and y intercept.

15. Slope: 3; y intercept: (0, 5)

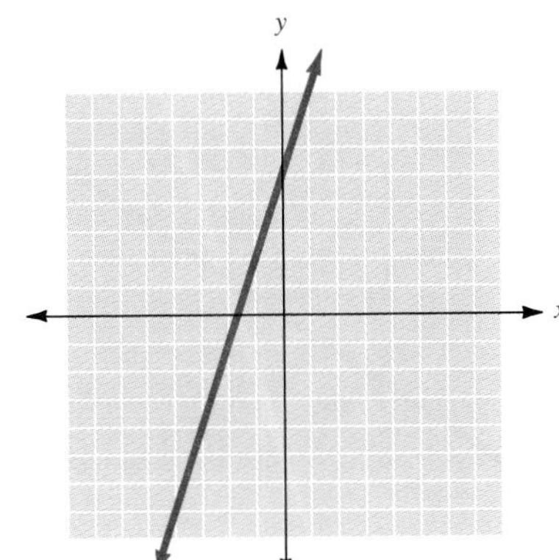

16. Slope: -2; y intercept: (0, 4)

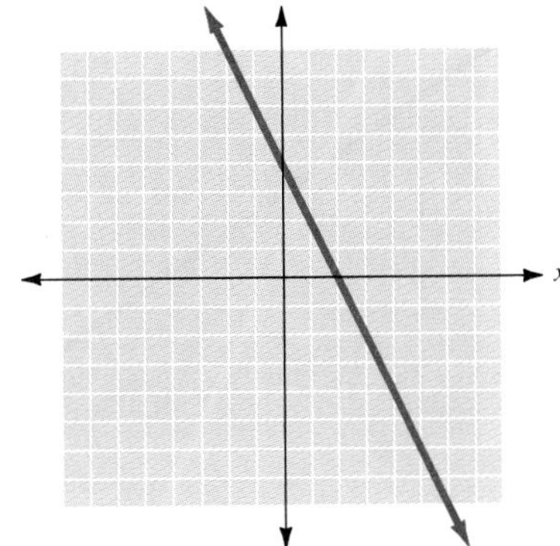

ANSWERS

1. Slope: 3; y intercept: (0, 5)
2. Slope: -7; y intercept: (0, 3)
3. Slope: -2; y intercept: (0, -5)
4. Slope: 5; y intercept: (0, -2)
5. Slope: $\frac{3}{4}$; y intercept: (0, 1)
6. Slope: -4; y intercept: (0, 0)
7. Slope: $\frac{2}{3}$; y intercept: (0, 0)
8. Slope: $-\frac{3}{5}$; y intercept: (0, -2)
9. Slope: $-\frac{4}{3}$; y intercept: (0, 4)
10. Slope: $-\frac{2}{5}$; y intercept: (0, 2)
11. Slope: 0; y intercept: (0, 9)
12. Slope: $\frac{2}{3}$; y intercept: (0, -2)
13. Slope: $\frac{3}{2}$; y intercept: (0, -4)
14. Slope is undefined; no y intercept
15. $y = 3x + 5$
16. $y = -2x + 4$

23. $y = \frac{3}{4}x + 3$

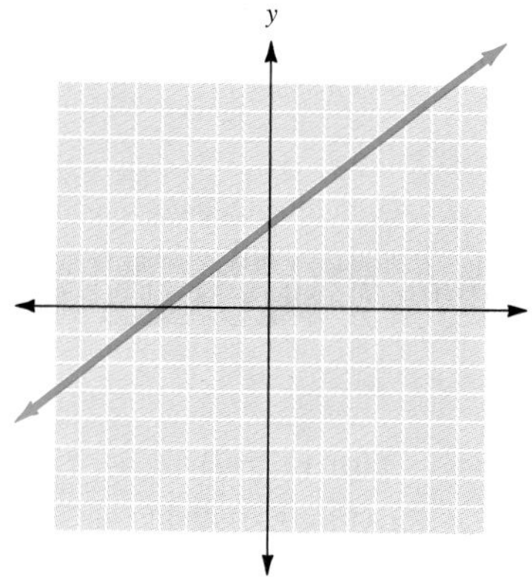

25. *g** **27.** *e* **29.** *h* **31.** *c* **33.** IV **35.** III **37.** I
39. III and IV **41.** Slope = 0.10; *y* intercept = (0, 200) **43.** 2°/hr

45. −0.30 **47.**  **49.**

51. Parallel lines; no

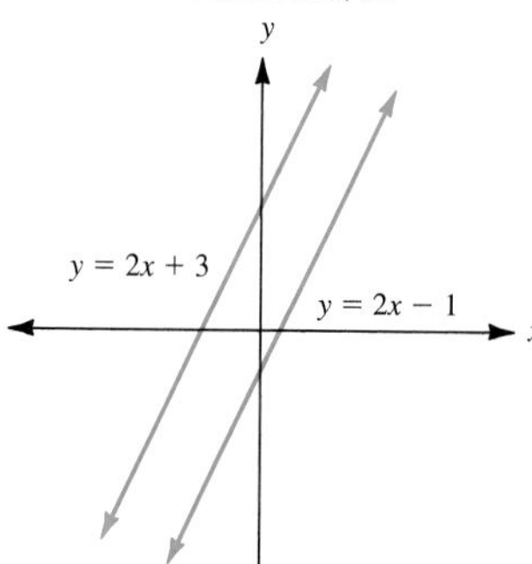

53. Perpendicular lines; −1

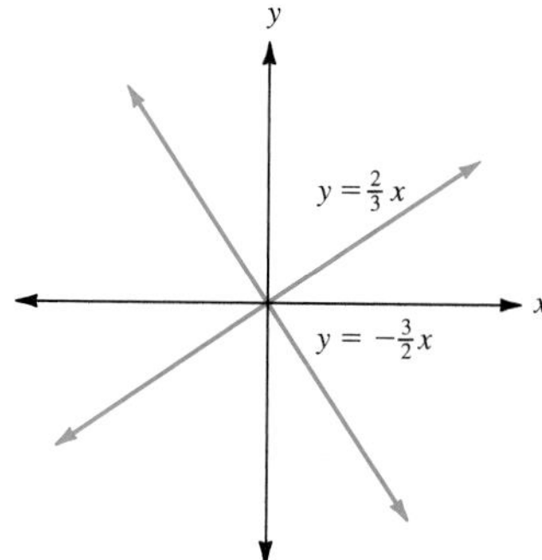

55. $y = -\frac{5}{3}x$ **a.** 2 **b.** 2 **c.** −1 **d.** Undefined **e.** 0 **f.** 1

Name ______________________

Section ________ Date ________

7.1 Exercises

Find the slope and y intercept of the line represented by each of the following equations.

1. $y = 3x + 5$

2. $y = -7x + 3$

3. $y = -2x - 5$

4. $y = 5x - 2$

5. $y = \frac{3}{4}x + 1$

6. $y = -4x$

7. $y = \frac{2}{3}x$

8. $y = -\frac{3}{5}x - 2$

9. $4x + 3y = 12$

10. $2x + 5y = 10$

11. $y = 9$

12. $2x - 3y = 6$

13. $3x - 2y = 8$

14. $x = 5$

Write the equation of the line with given slope and y intercept. Then graph each line, using the slope and y intercept.

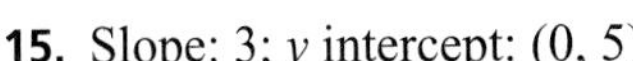

15. Slope: 3; y intercept: (0, 5)

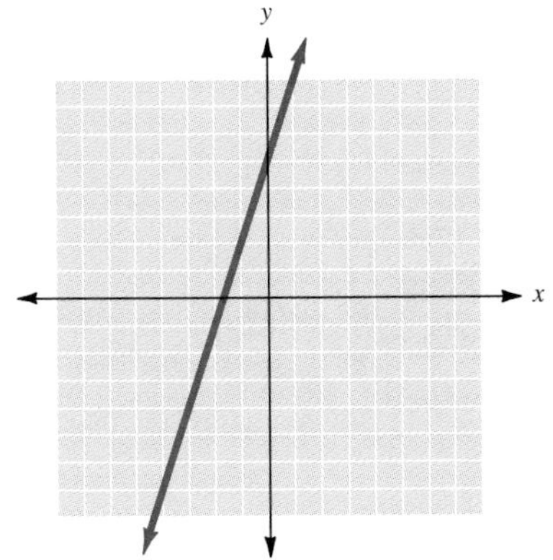

16. Slope: -2; y intercept: (0, 4)

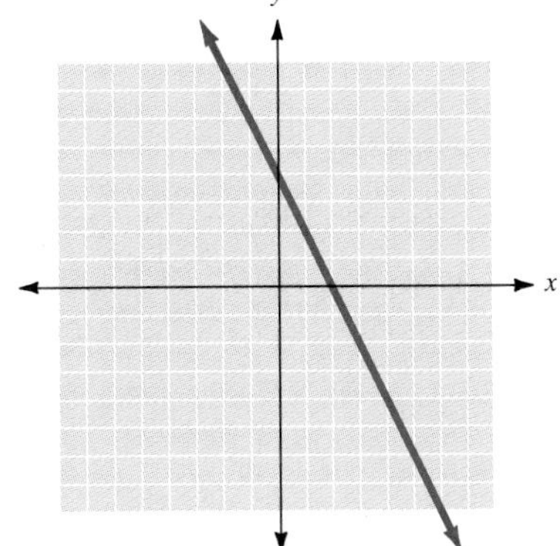

ANSWERS

1. Slope: 3; y intercept: (0, 5)
2. Slope: -7; y intercept: (0, 3)
3. Slope: -2; y intercept: (0, -5)
4. Slope: 5; y intercept: (0, -2)
5. Slope: $\frac{3}{4}$; y intercept: (0, 1)
6. Slope: -4; y intercept: (0, 0)
7. Slope: $\frac{2}{3}$; y intercept: (0, 0)
8. Slope: $-\frac{3}{5}$; y intercept: (0, -2)
9. Slope: $-\frac{4}{3}$; y intercept: (0, 4)
10. Slope: $-\frac{2}{5}$; y intercept: (0, 2)
11. Slope: 0; y intercept: (0, 9)
12. Slope: $\frac{2}{3}$; y intercept: (0, -2)
13. Slope: $\frac{3}{2}$; y intercept: (0, -4)
14. Slope is undefined; no y intercept
15. $y = 3x + 5$
16. $y = -2x + 4$

ANSWERS

17. $y = -3x + 4$

18. $y = 5x - 2$

19. $y = \frac{1}{2}x - 2$

20. $y = -\frac{3}{4}x + 8$

21. $y = -\frac{2}{3}x$

22. $y = \frac{2}{3}x - 2$

17. Slope: -3; y intercept: $(0, 4)$

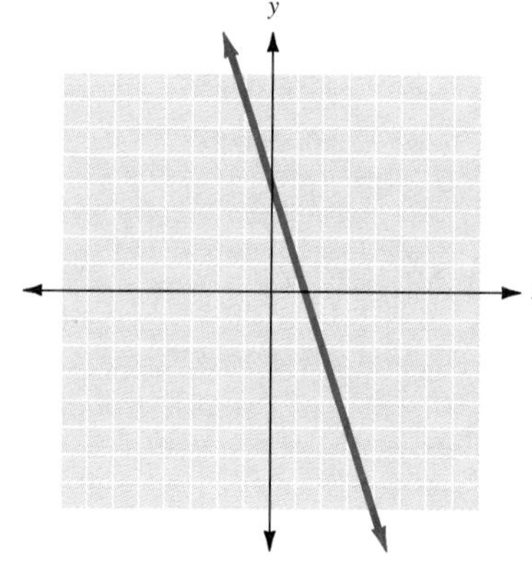

18. Slope: 5; y intercept: $(0, -2)$

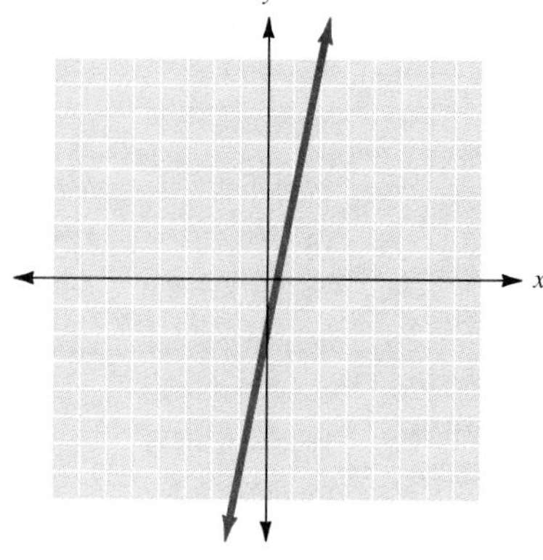

19. Slope: $\frac{1}{2}$; y intercept: $(0, -2)$

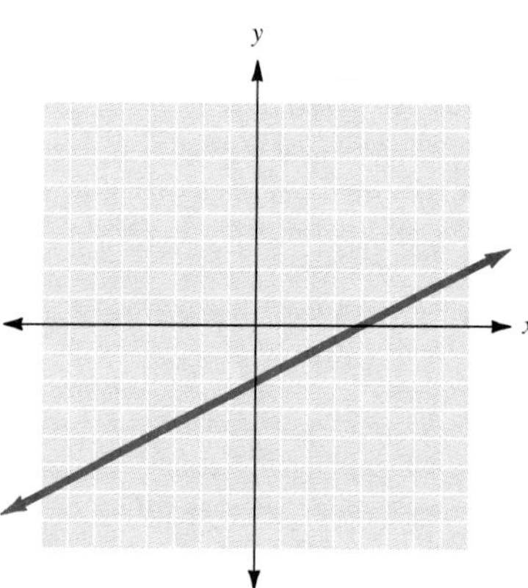

20. Slope: $-\frac{3}{4}$; y intercept: $(0, 8)$

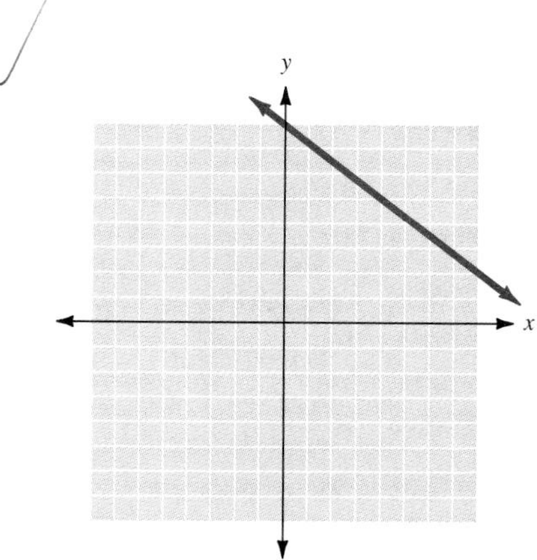

21. Slope: $-\frac{2}{3}$; y intercept: $(0, 0)$

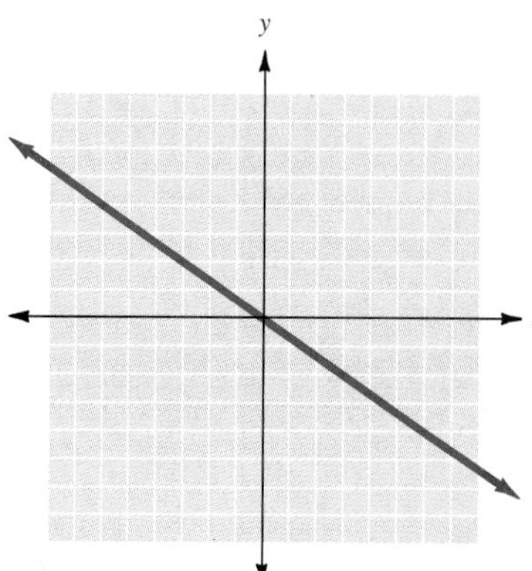

22. Slope: $\frac{2}{3}$; y intercept: $(0, -2)$

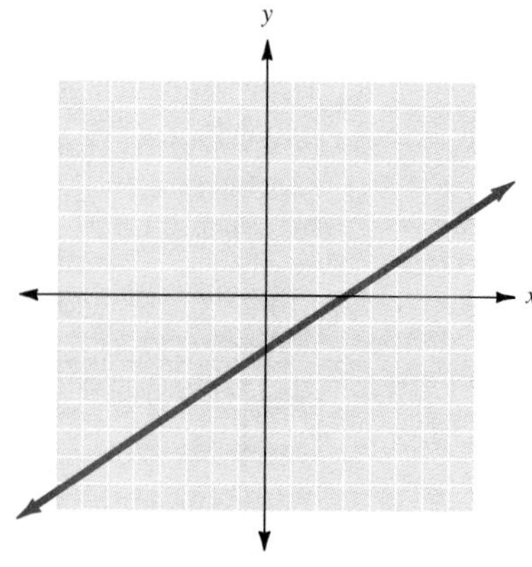

23. Slope: $\frac{3}{4}$; y intercept: (0, 3)

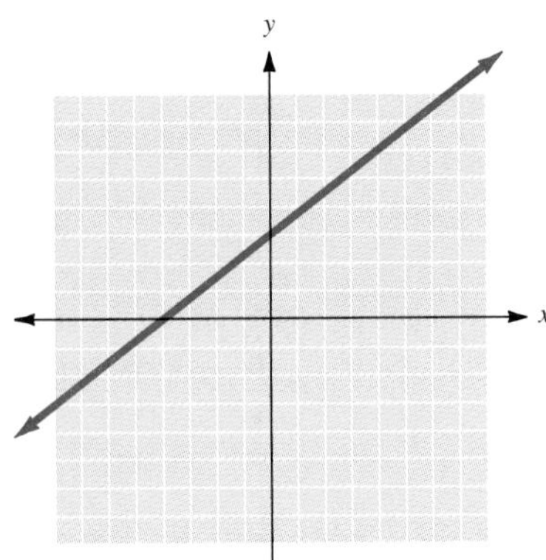

24. Slope: -3; y intercept: (0, 0)

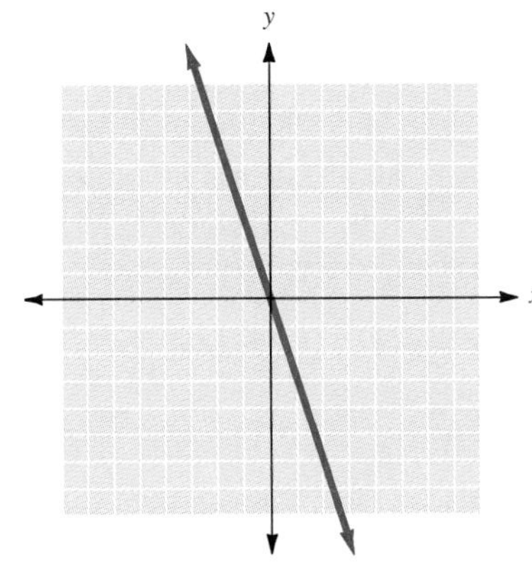

In exercises 25 to 32, match the graph with one of the equations below.

(a) $y = 2x$, **(b)** $y = x + 1$, **(c)** $y = -x + 3$, **(d)** $y = 2x + 1$,

(e) $y = -3x - 2$, **(f)** $y = \frac{2}{3}x + 1$, **(g)** $y = -\frac{3}{4}x + 1$, **(h)** $y = -4x$

25.

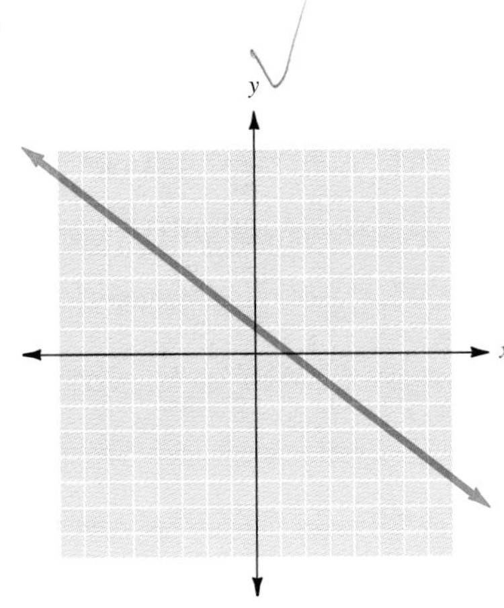

26.

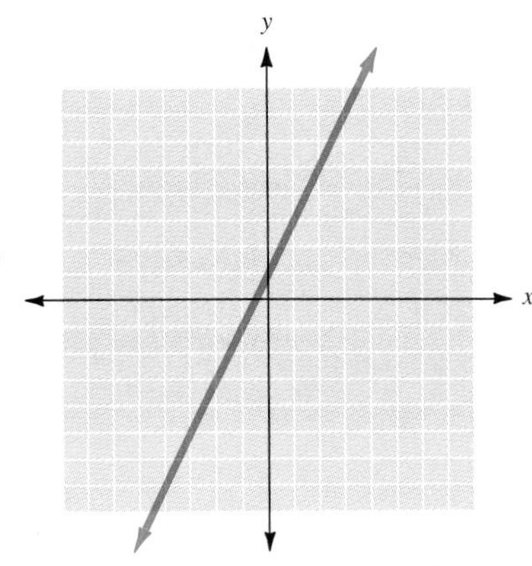

27.

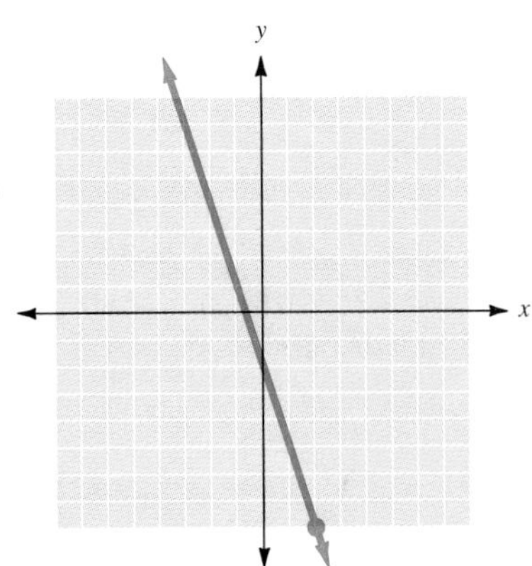

28.

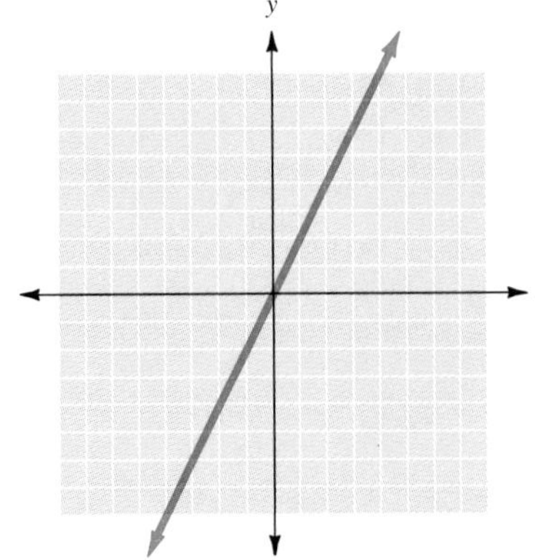

ANSWERS

23. $y = \frac{3}{4}x + 3$

24. $y = -3x$

25. g

26. d

27. e

28. a

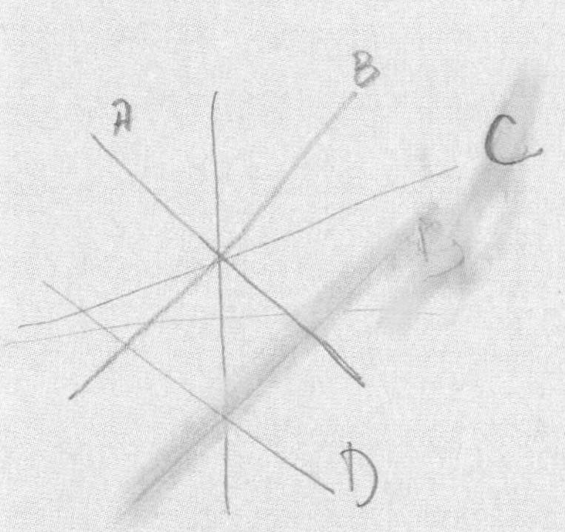

ANSWERS

29. h
30. f
31. c
32. b
33. IV
34. IV
35. III
36. III
37. I
38. I
39. III and IV
40. I and IV

29.

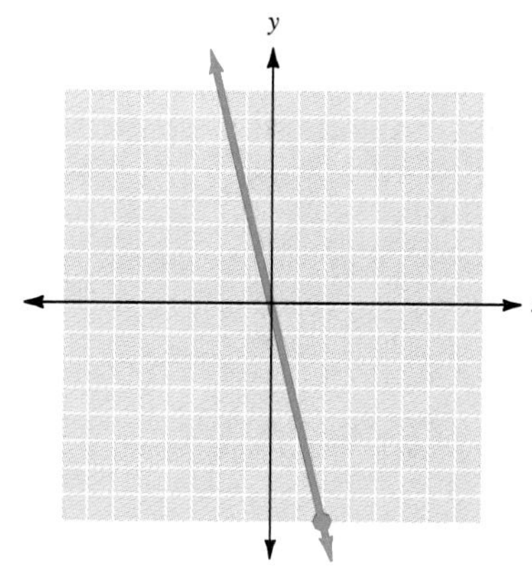

30.

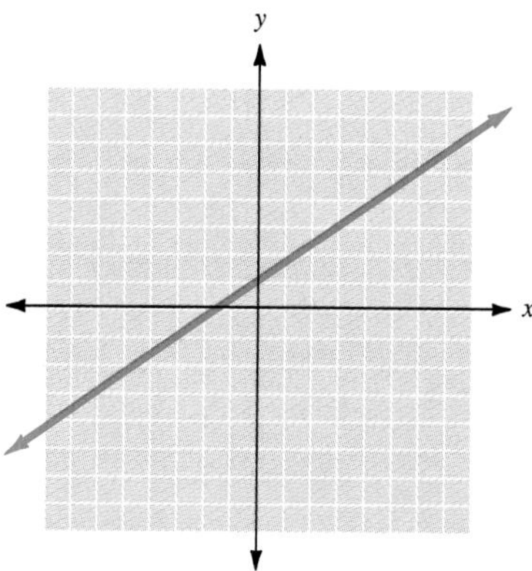

31.

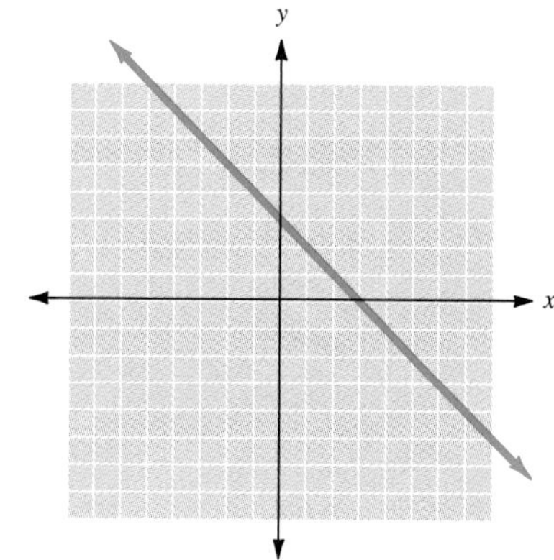

32.

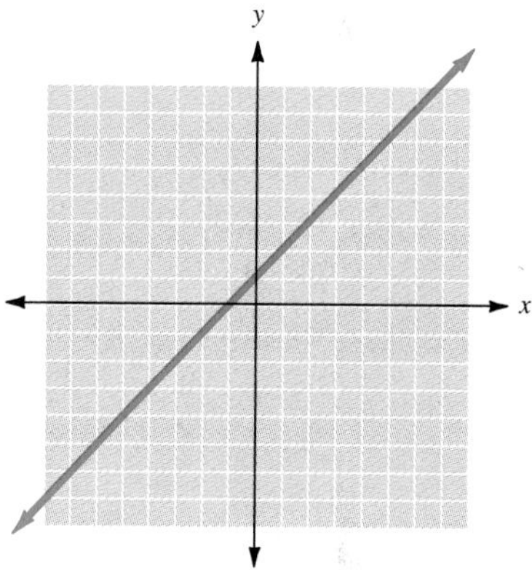

In which quadrant(s) are there no solutions for each line?

33. $y = 2x + 1$

34. $y = 3x + 2$

35. $y = -x + 1$

36. $y = -2x + 5$

37. $y = -2x - 5$

38. $y = -5x - 7$

39. $y = 3$

40. $x = -2$

41. **Recycling.** The equation $y = 0.10x + 200$ describes the award money in a recycling contest. What are the slope and the y intercept for this equation?

42. **Fundraising.** The equation $y = 15x - 100$ describes the amount of money a high school class might earn from a paper drive. What are the slope and y intercept for this equation?

43. **Science.** On a certain February day in Philadelphia, the temperature at 6:00 A.M. was 10°F. By 2:00 P.M. the temperature was up to 26°F. What was the hourly rate of temperature change?

44. **Slope of a roof.** A roof rises 8.75 feet (ft) in a horizontal distance of 15.09 ft. Find the slope of the roof to the nearest hundredth.

45. **Slope of airplane descent.** An airplane covered 15 miles (mi) of its route while decreasing its altitude by 24,000 ft. Find the slope of the line of descent that was followed. (1 mi = 5280 ft.) Round to the nearest hundredth.

46. **Slope of road descent.** Driving down a mountain, Tom finds that he has descended 1800 ft in elevation by the time he is 3.25 mi horizontally away from the top of the mountain. Find the slope of his descent to the nearest hundredth.

47. Complete the following statement: "The difference between undefined slope and zero slope is . . . ."

48. Complete the following: "The slope of a line tells you . . . ."

49. In a study on nutrition conducted in 1984, 18 normal adults aged 23 to 61 years old were measured for body fat, which is given as percentage of weight. The mean (average) body fat percentage for women 40 years old was 28.6 percent, and for women 53 years old was 38.4 percent. Work with a partner to decide how to show this information on a scatterplot. Try to find a linear equation that will tell you percentage of body fat based on a woman's age. What does your equation give for 20 years of age? For 60? Do you think a linear model works well for predicting body fat percentage in women as they age?

ANSWERS

41. $m = 0.10$; y intercept = (0, 200)

42. $m = 15$; y intercept = (0, −100)

43. 2°/h

44. 0.58

45. −0.30

46. −0.10

47.

48.

49.

50.

51.

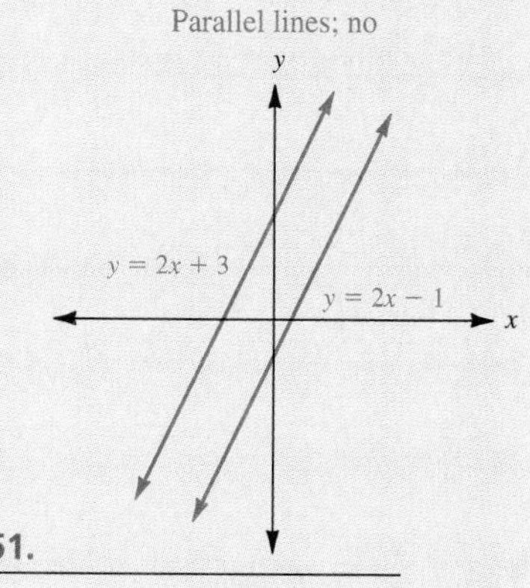

52.

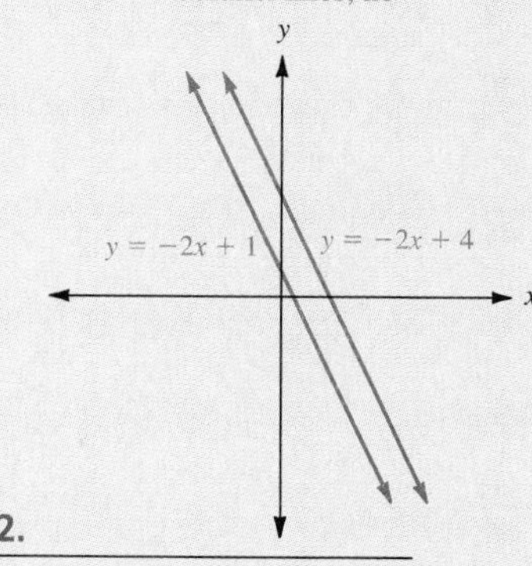

53.

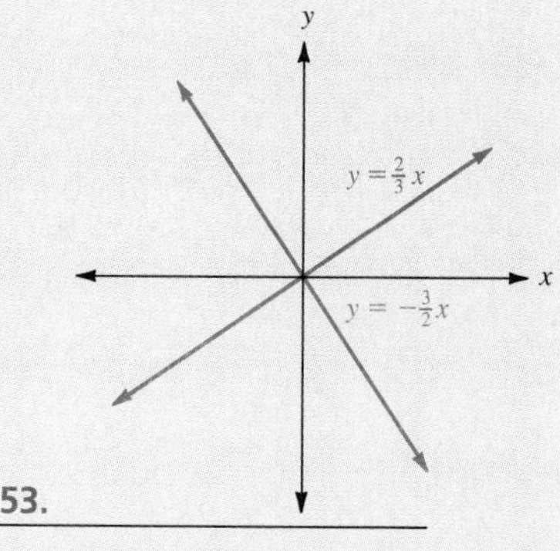

54.

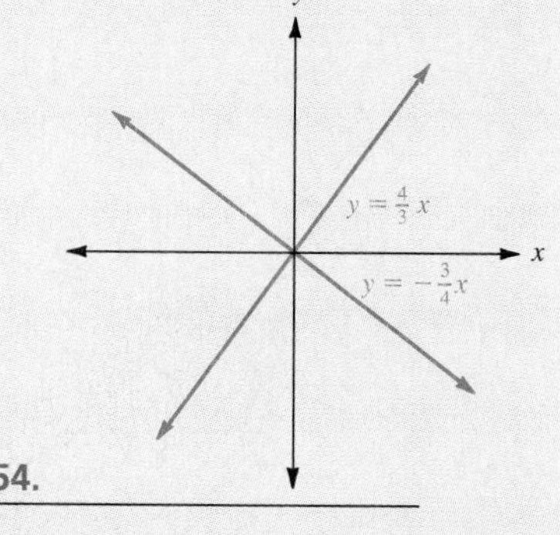

50. On two occasions last month, Sam Johnson rented a car on a business trip. Both times it was the same model from the same company, and both times it was in San Francisco. Sam now has to fill out an expense account form and needs to know how much he was charged per mile and the base rate. On both occasions he dropped the car at the airport booth and just got the total charge, not the details. All Sam knows is that he was charged \$210 for 625 miles on the first occasion and \$133.50 for 370 miles on the second trip. Sam has called accounting to ask for help. Plot these two points on a graph, and draw the line that goes through them. What question does the slope of the line answer for Sam? How does the y intercept help? Write a memo to Sam explaining the answers to his question and how a knowledge of algebra and graphing has helped you find the answers.

51. On the same graph, sketch the following lines:

$y = 2x - 1$ and $y = 2x + 3$

What do you observe about these graphs? Will the lines intersect?

52. Repeat exercise 51 using

$y = -2x + 4$ and $y = -2x + 1$

53. On the same graph, sketch the following lines:

$y = \frac{2}{3}x$ and $y = -\frac{3}{2}x$

What do you observe concerning these graphs? Find the product of the slopes of these two lines.

54. Repeat exercise 53 using

$y = \frac{4}{3}x$ and $y = -\frac{3}{4}x$

55. Based on exercises 53 and 54, write the equation of a line that is perpendicular to

$y = \frac{3}{5}x$

Getting Ready for Section 7.2 [Section 6.4]

Find the slope of the line connecting the given points.

(a) $(-4, 6)$ and $(3, 20)$ (b) $(2, 8)$ and $(-6, -8)$ (c) $(5, -7)$ and $(-5, 3)$
(d) $(2, 8)$ and $(2, 5)$ (e) $(6, 9)$ and $(3, 9)$ (f) $(4, 6)$ and $(-4, -2)$

Answers

1. Slope 3, y intercept $(0, 5)$ **3.** Slope -2, y intercept $(0, -5)$
5. Slope $\frac{3}{4}$, y intercept $(0, 1)$ **7.** Slope $\frac{2}{3}$, y intercept $(0, 0)$
9. Slope $-\frac{4}{3}$, y intercept $(0, 4)$ **11.** Slope 0, y intercept $(0, 9)$
13. Slope $\frac{3}{2}$, y intercept $(0, -4)$

15. $y = 3x + 5$

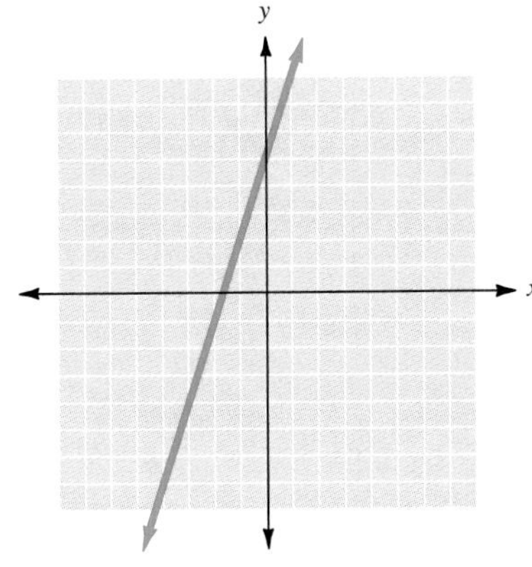

17. $y = -3x + 4$

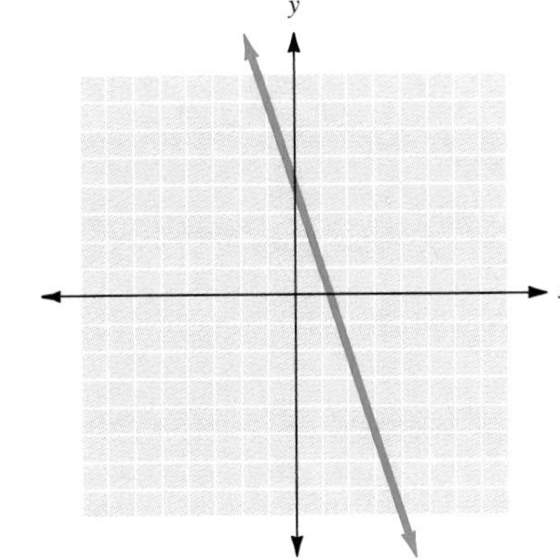

19. $y = \frac{1}{2}x - 2$

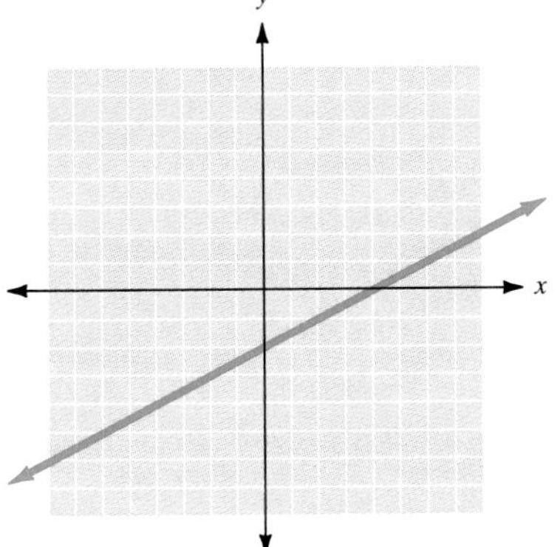

21. $y = -\frac{2}{3}x$

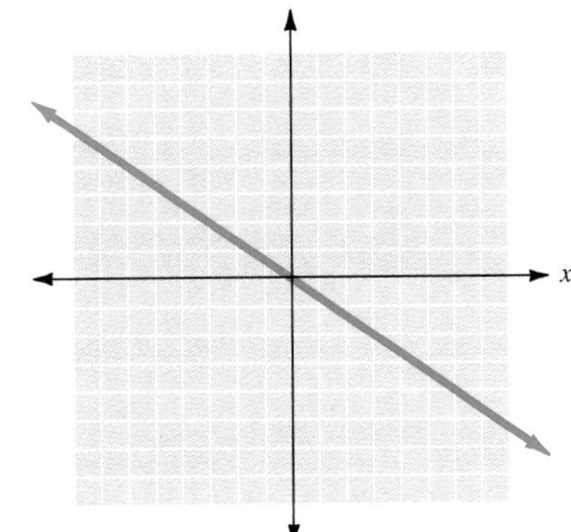

ANSWERS

55. $y = -\frac{5}{3}x$

a. 2

b. 2

c. −1

d. Undefined

e. 0

f. 1

23. $y = \frac{3}{4}x + 3$

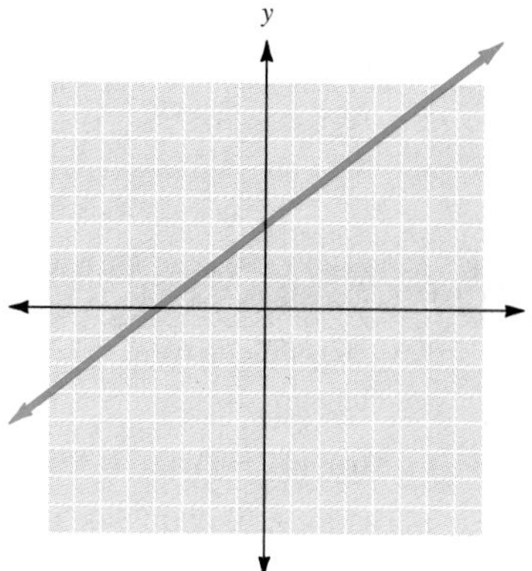

25. *g* 27. *e* 29. *h* 31. *c* 33. IV 35. III 37. I
39. III and IV 41. Slope = 0.10; *y* intercept = (0, 200) 43. 2°/hr

45. −0.30 47. 49.

51. Parallel lines; no

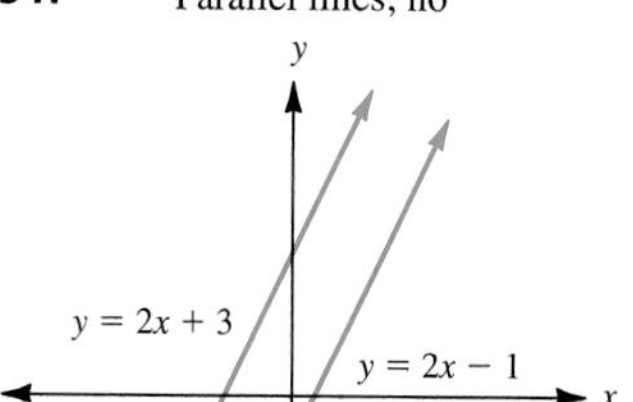

53. Perpendicular lines; −1

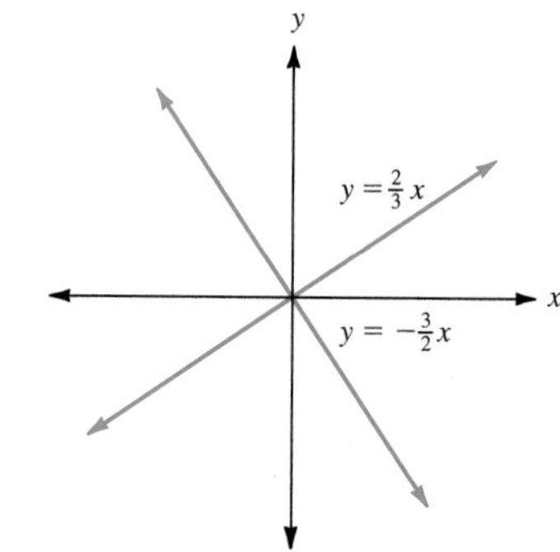

55. $y = -\frac{5}{3}x$ a. 2 b. 2 c. −1 d. Undefined e. 0 f. 1

Parallel and Perpendicular Lines

OBJECTIVES

1. Determine whether two lines are parallel
2. Determine whether two lines are perpendicular
3. Find the slope of a line perpendicular to a given line

For most inexperienced drivers, the most difficult driving maneuver to master is parallel parking. What is parallel parking? It is the act of backing into a curbside space so that the car's tires are parallel to the curb.

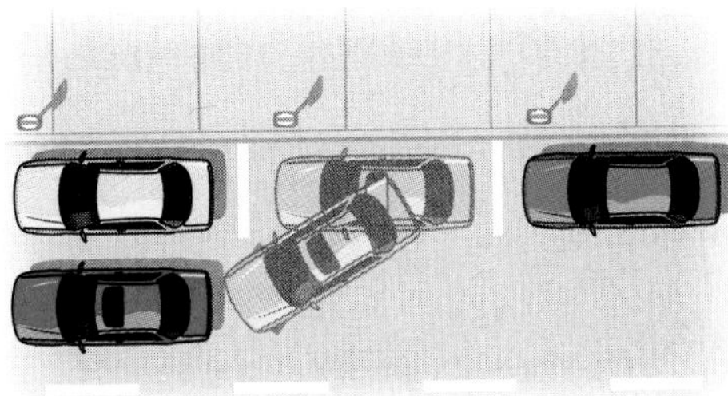

How can you tell that you've done a good job of parallel parking? Most people check to see that both the front tires and the back tires are the same distance (8 in. or so) from the curb. This is checking to be certain that the car is parallel to the curb.

How can we tell that two equations represent parallel lines? Look at the sketch below.

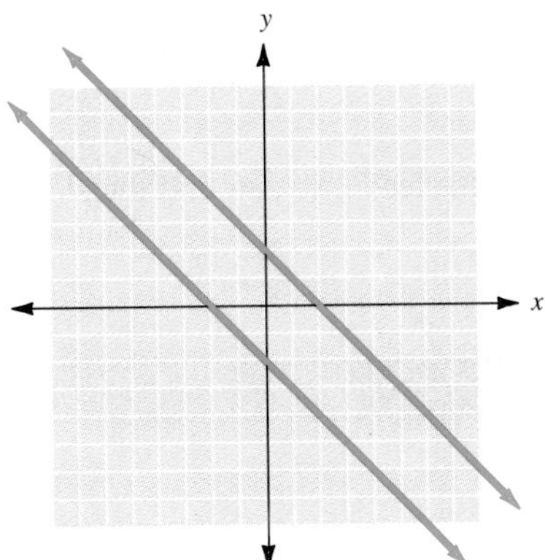

If two lines are parallel, they have the same slope. If their equations are in slope-intercept form, you simply compare the slopes.

Example 1

Determining That Two Lines Are Parallel

Which two equations represent parallel lines?

(a) $y = 2x + 3$

(b) $y = -\frac{1}{2}x - 5$

(c) $y = -2x + \frac{1}{2}$

(d) $y = -2x - 9$

Because (c) and (d) both have a slope of -2, the lines are parallel.

CHECK YOURSELF 1

Which two equations represent parallel lines?

(a) $y = -5x + 5$

(b) $y = \frac{1}{5}x - 5$

(c) $y = -5x + \frac{1}{2}$

(d) $y = 5x - 9$

More formally, we can state the following about parallel lines.

Definitions: Slope of Parallel Lines

For nonvertical lines L_1 and L_2, if line L_1 has slope m_1 and line L_2 has slope m_2, then

L_1 is parallel to L_2 if and only if $m_1 = m_2$

Note: All vertical lines are parallel to each other.

NOTE This means that if the lines are parallel, then their slopes are equal. Conversely, if the slopes are equal, then the lines are parallel.

As we discovered in Chapter 6, we can find the slope of a line from any two points on the line.

Example 2

Parallel Lines

Are lines L_1 through (2, 3) and (4, 6) and L_2 through $(-4, 2)$ and (0, 8) parallel, or do they intersect?

$$m_1 = \frac{6 - 3}{4 - 2} = \frac{3}{2}$$

$$m_2 = \frac{8 - 2}{0 - (-4)} = \frac{6}{4} = \frac{3}{2}$$

Because the slopes of the lines are equal, the lines are parallel. They do *not* intersect.

NOTE Unless, of course, L_1 and L_2 are actually the *same line.* In this case a quick sketch will show that the lines are distinct.

CHECK YOURSELF 2

Are lines L_1 through $(-2, -1)$ *and* $(1, 4)$ *and L_2 through* $(-3, 4)$ *and* $(0, 8)$ *parallel, or do they intersect?*

Many important characteristics of lines are evident from a city map.

Note that Fourth Street and Fifth Street are parallel. Just as these streets never meet, it is true that two parallel lines will never meet.

Recall that the point at which two lines meet is called their **intersection.** This is also true with two streets. We call the common area of the two streets the intersection.

In this case, the two streets meet at right angles. When two lines meet at right angles, we say that they are **perpendicular.**

Definitions: Slope of Perpendicular Lines

For nonvertical lines L_1 and L_2, if line L_1 has slope m_1 and line L_2 has slope m_2, then

L_1 is perpendicular to L_2 if and only if $m_1 = -\dfrac{1}{m_2}$

or equivalently

$m_1 \cdot m_2 = -1$

Note: Horizontal lines are perpendicular to vertical lines.

Example 3

Determining That Two Lines Are Perpendicular

Which two equations represent perpendicular lines?

(a) $y = 2x + 3$

(b) $y = -\frac{1}{2}x - 5$

(c) $y = -2x + \frac{1}{2}$

(d) $y = -2x - 9$

Because the product of the slopes for (a) and (b) is

$$2\left(-\frac{1}{2}\right) = -1$$

these two lines are perpendicular. Note that none of the other pairs of slopes have a product of -1.

CHECK YOURSELF 3

Which two equations represent perpendicular lines?

(a) $y = -5x + 5$

(b) $y = -\frac{1}{5}x - 5$

(c) $y = -5x + \frac{1}{2}$

(d) $y = 5x - 9$

Example 4

Perpendicular Lines

Are lines L_1 through points $(-2, 3)$ and $(1, 7)$ and L_2 through points $(2, 4)$ and $(6, 1)$ perpendicular?

$$m_1 = \frac{7 - 3}{1 - (-2)} = \frac{4}{3}$$

NOTE $\left(\frac{4}{3}\right)\left(-\frac{3}{4}\right) = -1$

$$m_2 = \frac{1 - 4}{6 - 2} = -\frac{3}{4}$$

Because the slopes are negative reciprocals, the lines are perpendicular.

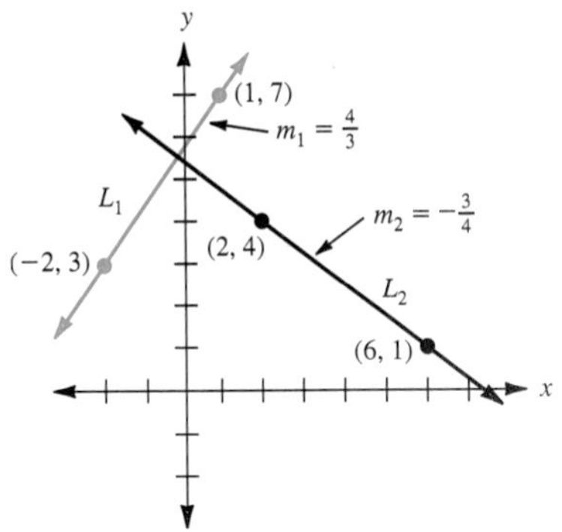

CHECK YOURSELF 4

Are lines L_1 through points (1, 3) and (4, 1) and L_2 through points (−2, 4) and (2, 10) perpendicular?

We can also use the slope-intercept form to determine whether the graphs of given equations will be parallel, intersecting, or perpendicular lines.

Example 5

Verifying That Two Lines Are Parallel

Show that the graphs of $3x + 2y = 4$ and $6x + 4y = 12$ are parallel lines.

First, we solve each equation for y:

$$3x + 2y = 4$$

$$2y = -3x + 4$$

$$y = -\frac{3}{2}x + 2 \qquad (1)$$

NOTE Notice that the slopes are the same, but the y intercepts are different. Therefore the lines are distinct.

$$6x + 4y = 12$$

$$4y = -6x + 12$$

$$y = -\frac{3}{2}x + 3 \qquad (2)$$

Because the two lines have the same slope, here $-\frac{3}{2}$, the lines are parallel.

CHECK YOURSELF 5

Show that the graphs of the equations

$-3x + 2y = 4$ *and* $2x + 3y = 9$

are perpendicular lines.

CHECK YOURSELF ANSWERS

1. (a) and (c) **2.** The lines intersect

3. (b) and (d) **4.** The lines are perpendicular

5.

$$y = \frac{3}{2}x + 2$$

$$y = -\frac{2}{3}x + 3$$

$$\left(\frac{3}{2}\right)\left(-\frac{2}{3}\right) = -1$$

7.2 Exercises

Name ______________

Section ________ Date ________

In exercises 1 to 4, determine which two equations represent parallel lines.

1. (a) $y = -4x + 5$ (b) $y = 4x + 5$ (c) $y = \frac{1}{4}x + 5$ (d) $y = -4x + 9$

2. (a) $y = 3x - 5$ (b) $y = -3x + 5$ (c) $y = 3x + 2$ (d) $y = -\frac{1}{3}x - 5$

3. (a) $y = \frac{2}{3}x + 3$ (b) $y = -\frac{3}{2}x - 6$ (c) $y = 4x + 12$ (d) $y = 4x - 3$

4. (a) $y = \frac{9}{4}x - 3$ (b) $y = \frac{4}{9}x + 7$ (c) $y = \frac{4}{9}x - 7$ (d) $y = -\frac{9}{4}x + 7$

In exercises 5 to 8, determine which two equations represent perpendicular lines.

5. (a) $y = 6x - 3$ (b) $y = \frac{1}{6}x + 3$ (c) $y = -\frac{1}{6}x + 3$ (d) $y = \frac{1}{6}x - 3$

6. (a) $y = \frac{2}{3}x - 8$ (b) $y = -\frac{2}{3}x - 6$ (c) $y = \frac{2}{3}x - 6$ (d) $y = \frac{3}{2}x - 6$

7. (a) $y = \frac{1}{3}x - 9$ (b) $y = 3x - 9$ (c) $y = \frac{1}{3}x + 9$ (d) $y = -\frac{1}{3}x + 9$

8. (a) $y = \frac{5}{9}x - 6$ (b) $y = 6x - \frac{5}{9}$ (c) $y = -\frac{1}{6}x + \frac{5}{9}$ (d) $y = \frac{1}{6}x - \frac{5}{9}$

Are the following pairs of lines parallel, perpendicular, or neither?

9. L_1 through $(-2, -3)$ and $(4, 3)$
   L_2 through $(3, 5)$ and $(5, 7)$

10. L_1 through $(-2, 4)$ and $(1, 8)$
    L_2 through $(-1, -1)$ and $(-5, 2)$

11. L_1 through $(8, 5)$ and $(3, -2)$
    L_2 through $(-2, 4)$ and $(4, -1)$

12. L_1 through $(-2, -3)$ and $(3, -1)$
    L_2 through $(-3, 1)$ and $(7, 5)$

13. L_1 with equation $x - 3y = 6$
    L_2 with equation $3x + y = 3$

14. L_1 with equation $x + 2y = 4$
    L_2 with equation $2x + 4y = 5$

15. Find the slope of any line parallel to the line through points $(-2, 3)$ and $(4, 5)$.

ANSWERS

1. a and d
2. a and c
3. c and d
4. b and c
5. a and c
6. b and d
7. b and d
8. b and c
9. Parallel
10. Perpendicular
11. Neither
12. Parallel
13. Perpendicular
14. Parallel
15. $\frac{1}{3}$

ANSWERS

16. $-\frac{1}{3}$

17. 12

18. −1

19. Parallelogram

20. Rectangle

21. Rectangle

22. No

23. Yes

16. Find the slope of any line perpendicular to the line through points (0, 5) and (−3, −4).

17. A line passing through (−1, 2) and (4, y) is parallel to a line with slope 2. What is the value of y?

18. A line passing through (2, 3) and (5, y) is perpendicular to a line with slope $\frac{3}{4}$. What is the value of y?

In exercises 19 to 21, use the concept of slope to determine if the given figure is a parallelogram or a rectangle.

19.

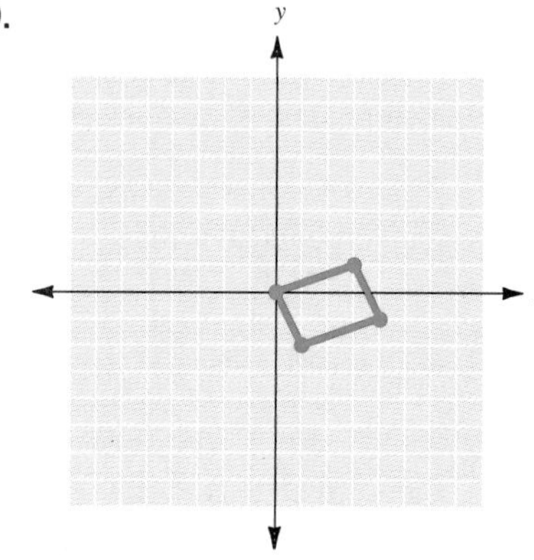

20.

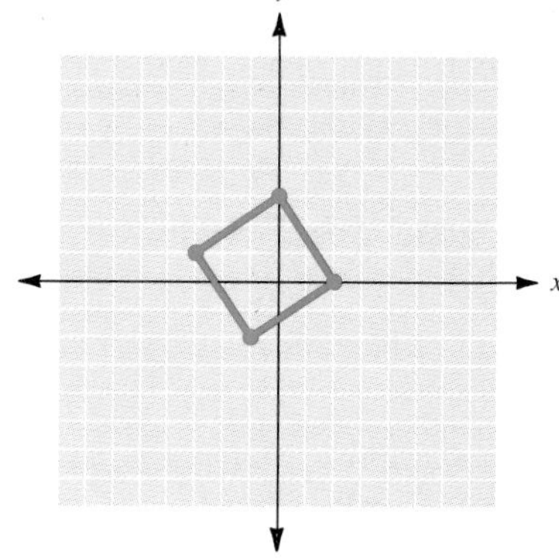

21.

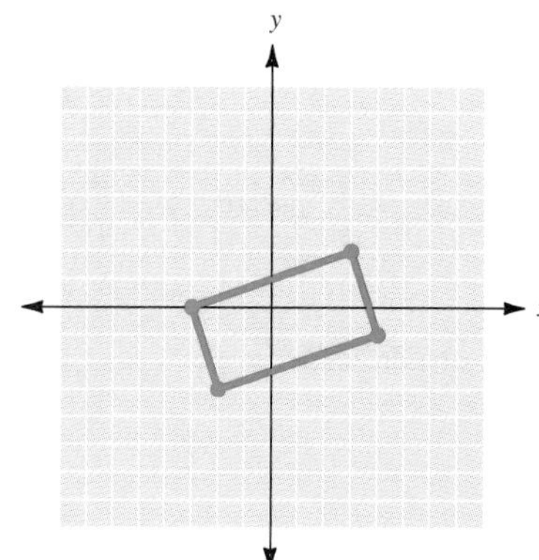

In exercises 22 to 24, use the concept of slope to determine whether the given figure is a right triangle (i.e., does the triangle contain a right angle?).

22.

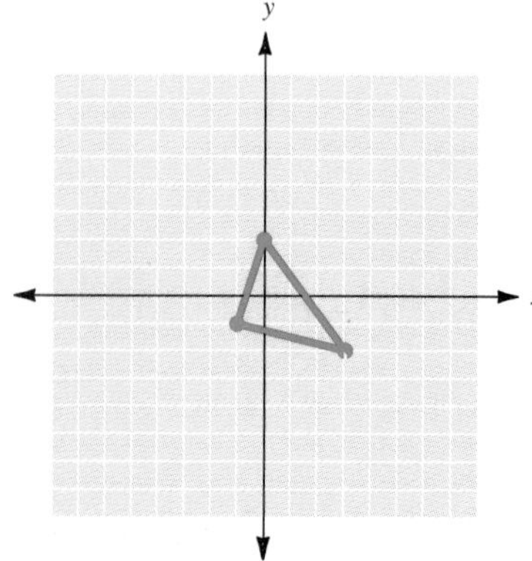

23.

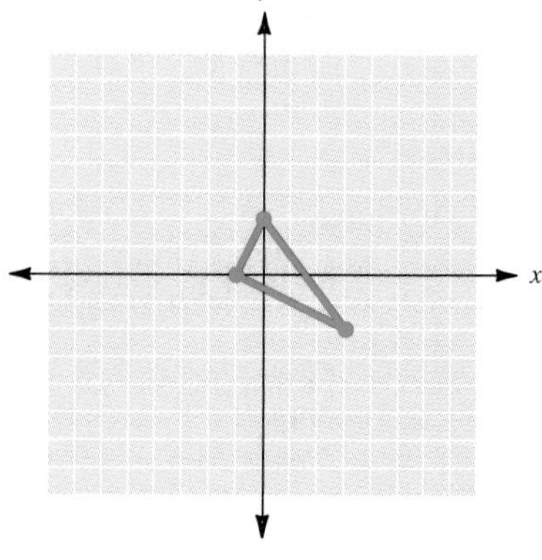

24.

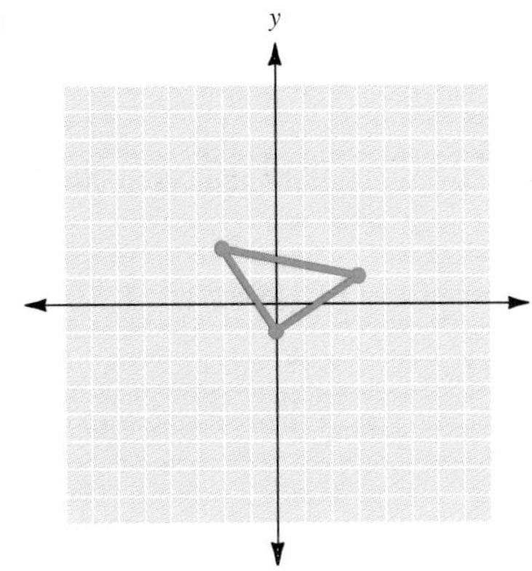

In exercises 25 to 27, use the concept of slope to draw a line perpendicular to the given line segment, passing through the marked point.

25.

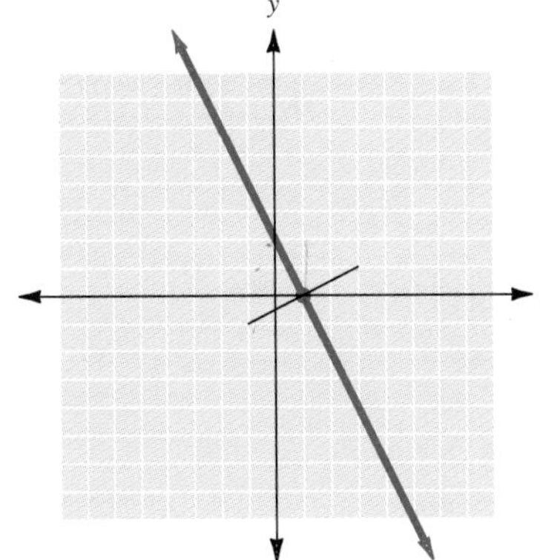

26.

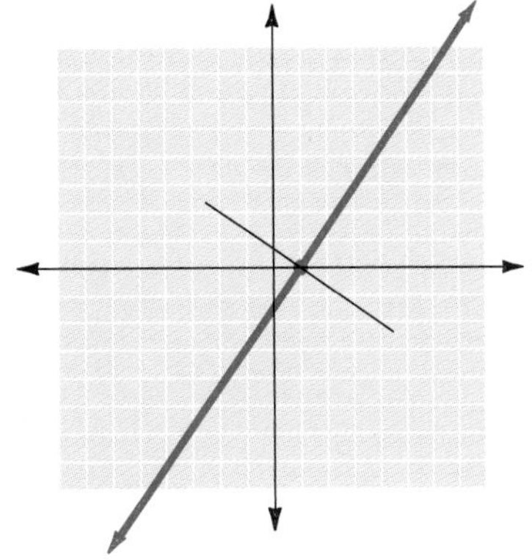

27.

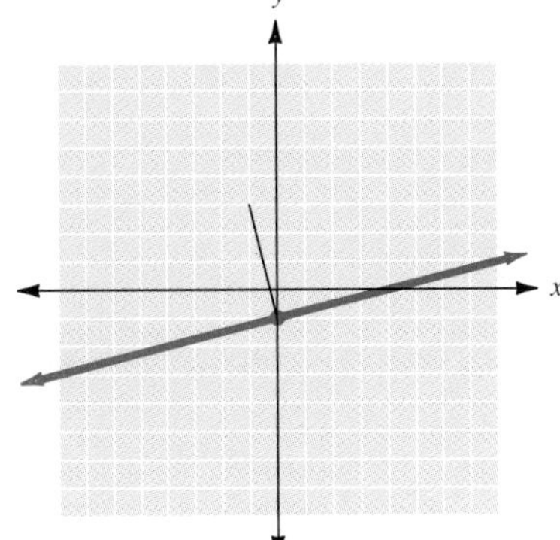

Getting Ready for Section 7.3 [Section 2.5]

Solve the following equations for y.

(a) $3x + 4y = 12$ (b) $x - y = -5$ (c) $2x + 6 = 4y$

(d) $y + 5 = x$ (e) $5x - 2y = 10$ (f) $-y + 3x = -5$

ANSWERS

24. Yes

25. See exercise

26. See exercise

27. See exercise

a. $y = -\frac{3}{4}x + 3$

b. $y = x + 5$

c. $y = \frac{1}{2}x + \frac{3}{2}$

d. $y = x - 5$

e. $y = \frac{5}{2}x - 5$

f. $y = 3x + 5$

Answers

1. a and d **3.** c and d **5.** a and c **7.** b and d **9.** Parallel

11. Neither **13.** Perpendicular **15.** $\frac{1}{3}$ **17.** 12 **19.** Parallelogram

21. Rectangle **23.** Yes

25.

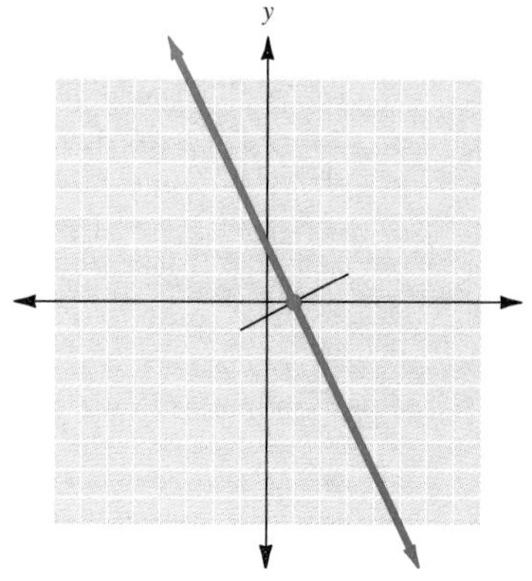

27.

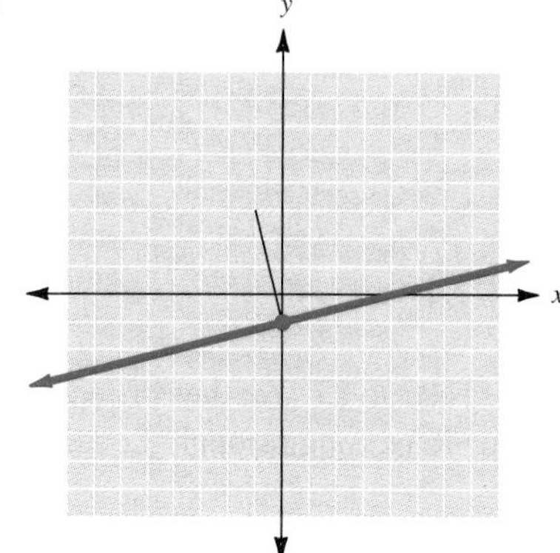

a. $y = -\frac{3}{4}x + 3$ **b.** $y = x + 5$ **c.** $y = \frac{1}{2}x + \frac{3}{2}$ **d.** $y = x - 5$

e. $y = \frac{5}{2}x - 5$ **f.** $y = 3x + 5$

7.3 The Point-Slope Form

OBJECTIVES

1. Given a point and a slope, find the graph of a line
2. Given a point and the slope, find the equation of a line
3. Given two points, find the equation of a line

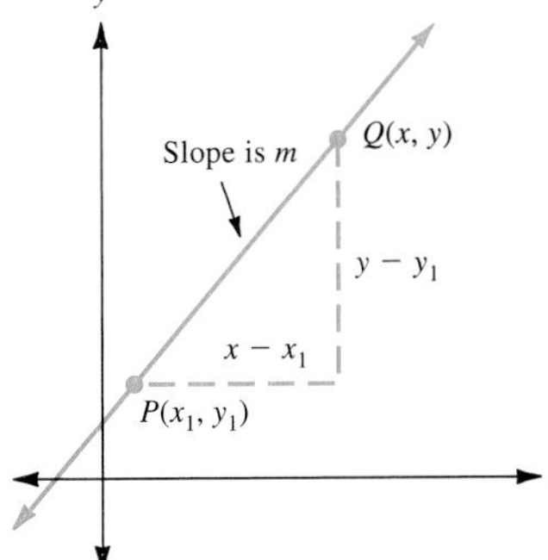

Often in mathematics it is useful to be able to write the equation of a line, given its slope and *any* point on the line. In this section, we will derive a third special form for a line for this purpose.

Suppose that a line has slope m and that it passes through the known point $P(x_1, y_1)$. Let $Q(x, y)$ be any other point on the line. Once again we can use the definition of slope and write

$$m = \frac{y - y_1}{x - x_1} \qquad (1)$$

Multiplying both sides of equation (1) by $x - x_1$, we have

$$m(x - x_1) = y - y_1$$

or

$$y - y_1 = m(x - x_1) \qquad (2)$$

Equation (2) is called the *point-slope form* for the equation of a line, and all points lying on the line [including (x_1, y_1)] will satisfy this equation. We can state the following general result.

NOTE The equation of a line with undefined slope passing through the point (x_1, y_1) is given by $x = x_1$.

Rules and Properties: Point-Slope Form for the Equation of a Line

The equation of a line with slope m that passes through point (x_1, y_1) is given by

$$y - y_1 = m(x - x_1)$$

Example 1

Finding the Equation of a Line

Write the equation for the line that passes through point $(3, -1)$ with a slope of 3.

Letting $(x_1, y_1) = (3, -1)$ and $m = 3$ in point-slope form, we have

$$y - (-1) = 3(x - 3)$$

or

$$y + 1 = 3x - 9$$

We can write the final result in slope-intercept form as

$$y = 3x - 10$$

CHECK YOURSELF 1

Write the equation of the line that passes through point $(-2, -4)$ with a slope of $\frac{3}{2}$. Write your result in slope-intercept form.

Because we know that two points determine a line, it is natural that we should be able to write the equation of a line passing through two given points. Using the point-slope form together with the slope formula will allow us to write such an equation.

Example 2

Finding the Equation of a Line

Write the equation of the line passing through (2, 4) and (4, 7).

First, we find m, the slope of the line. Here

$$m = \frac{7 - 4}{4 - 2} = \frac{3}{2}$$

NOTE We could just as well have chosen to let

$(x_1, y_1) = (4, 7)$

The resulting equation will be the same in either case. Take time to verify this for yourself.

Now we apply the point-slope form with $m = \frac{3}{2}$ and $(x_1, y_1) = (2, 4)$:

$$y - 4 = \frac{3}{2}(x - 2)$$

$$y - 4 = \frac{3}{2}x - 3$$

$$y = \frac{3}{2}x + 1$$

CHECK YOURSELF 2

Write the equation of the line passing through (−2, 5) *and* (1, 3). *Write your result in slope-intercept form.*

A line with slope zero is a horizontal line. A line with an undefined slope is vertical. The next example illustrates the equations of such lines.

Example 3

Finding the Equation of a Line

(a) Find the equation of a line passing through (7, −2) with a slope of zero.

We could find the equation by letting $m = 0$. Substituting the ordered pair (7, −2) into the slope-intercept form, we can solve for b.

$$y = mx + b$$

$$-2 = 0(7) + b$$

$$-2 = b$$

So,

$y = 0 \cdot x - 2$ or $y = -2$

It is far easier to remember that any line with a zero slope is a horizontal line and has the form

$$y = b$$

The value for b will always be the y coordinate for the given point.

(b) Find the equation of a line with undefined slope passing through $(4, -5)$.

A line with undefined slope is vertical. It will always be of the form $x = a$, in which a is the x coordinate for the given point. The equation is

$x = 4$

CHECK YOURSELF 3

(a) Find the equation of a line with zero slope that passes through point $(-3, 5)$.

(b) Find the equation of a line passing through $(-3, -6)$ with undefined slope.

Alternate methods for finding the equation of a line through two points exist and have particular significance in other fields of mathematics, such as statistics. The following example shows such an alternate approach.

Example 4

Finding the Equation of a Line

Write the equation of the line through points $(-2, 3)$ and $(4, 5)$.

First, we find m, as before:

$$m = \frac{5 - 3}{4 - (-2)} = \frac{2}{6} = \frac{1}{3}$$

We now make use of the slope-intercept equation, but in a slightly different form.

Because $y = mx + b$, we can write

$b = y - mx$

Now letting $x = -2$, $y = 3$, and $m = \frac{1}{3}$, we can calculate b:

NOTE We substitute these values because the line must pass through $(-2, 3)$.

$$b = 3 - \left(\frac{1}{3}\right)(-2)$$

$$= 3 + \frac{2}{3} = \frac{11}{3}$$

With $m = \frac{1}{3}$ and $b = \frac{11}{3}$, we can apply the slope-intercept form to write the equation of the desired line. We have

$$y = \frac{1}{3}x + \frac{11}{3}$$

CHECK YOURSELF 4

Repeat the Check Yourself 2 exercise, using the technique illustrated in Example 4.

We now know that we can write the equation of a line once we have been given appropriate geometric conditions, such as a point on the line and the slope of that line. In some applications the slope may be given not directly but through specified parallel or perpendicular lines.

Example 5

Finding the Equation of a Line

Find the equation of the line passing through $(-4, -3)$ and parallel to the line determined by $3x + 4y = 12$.

First, we find the slope of the given parallel line, as before:

$$3x + 4y = 12$$
$$4y = -3x + 12$$
$$y = -\frac{3}{4}x + 3$$

NOTE The slope of the given line is $-\frac{3}{4}$.

Now because the slope of the desired line must also be $-\frac{3}{4}$, we can use the point-slope form to write the required equation:

$$y - (-3) = -\frac{3}{4}[x - (-4)]$$

NOTE The line must pass through $(-4, -3)$, so let $(x_1, y_1) = (-4, -3)$.

This simplifies to

$$y = -\frac{3}{4}x - 6$$

and we have our equation in slope-intercept form.

CHECK YOURSELF 5

Find the equation of the line passing through $(5, 4)$ *and perpendicular to the line with equation* $2x - 5y = 10$.

Hint: Recall that the slopes of perpendicular lines are negative reciprocals of each other.

The following chart summarizes the various forms of the equation of a line.

Form	Equation for Line L	Conditions
Standard	$ax + by = c$	Constants a and b cannot both be zero.
Slope-intercept	$y = mx + b$	Line L has y intercept $(0, b)$ with slope m.
Point-slope	$y - y_1 = m(x - x_1)$	Line L passes through point (x_1, y_1) with slope m.
Horizontal	$y = k$	Slope is zero.
Vertical	$x = h$	Slope is undefined.

CHECK YOURSELF ANSWERS

1. $y = \frac{3}{2}x - 1$ **2.** $y = -\frac{2}{3}x + \frac{11}{3}$ **3.** **(a)** $y = 5$; **(b)** $x = -3$

4. $y = -\frac{2}{3}x + \frac{11}{3}$ **5.** $y = -\frac{5}{2}x + \frac{33}{2}$

7.3 Exercises

Write the equation of the line passing through each of the given points with the indicated slope. Give your results in slope-intercept form, where possible.

1. (0, 2), $m = 3$

2. (0, −4), $m = -2$

3. (0, 2), $m = \frac{3}{2}$

4. (0, −3), $m = -2$

5. (0, 4), $m = 0$

6. (0, 5), $m = -\frac{3}{5}$

7. (0, −5), $m = \frac{5}{4}$

8. (0, −4), $m = -\frac{3}{4}$

9. (1, 2), $m = 3$

10. (−1, 2), $m = 3$

11. (−2, −3), $m = -3$

12. (1, −4), $m = -4$

13. (5, −3), $m = \frac{2}{5}$

14. (4, 3), $m = 0$

15. (2, −3), m is undefined

16. (2, −5), $m = \frac{1}{4}$

Write the equation of the line passing through each of the given pairs of points. Write your result in slope-intercept form, where possible.

17. (2, 3) and (5, 6)

18. (3, −2) and (6, 4)

19. (−2, −3) and (2, 0)

20. (−1, 3) and (4, −2)

21. (−3, 2) and (4, 2)

22. (−5, 3) and (4, 1)

ANSWERS

1. $y = 3x + 2$
2. $y = -2x - 4$
3. $y = \frac{3}{2}x + 2$
4. $y = -2x - 3$
5. $y = 4$
6. $y = -\frac{3}{5}x + 5$
7. $y = \frac{5}{4}x - 5$
8. $y = -\frac{3}{4}x - 4$
9. $y = 3x - 1$
10. $y = 3x + 5$
11. $y = -3x - 9$
12. $y = -4x$
13. $y = \frac{2}{5}x - 5$
14. $y = 3$
15. $x = 2$
16. $y = \frac{1}{4}x - \frac{11}{2}$
17. $y = x + 1$
18. $y = 2x - 8$
19. $y = \frac{3}{4}x - \frac{3}{2}$
20. $y = -x + 2$
21. $y = 2$
22. $y = -\frac{2}{9}x + \frac{17}{9}$

ANSWERS

23. $y = \frac{3}{2}x - 3$

24. $x = 2$

25. $y = \frac{5}{2}x + 4$

26. $y = 1$

27. $y = 4x - 2$

28. $y = -\frac{2}{3}x + 4$

29. $y = -\frac{1}{2}x + 2$

30. $y = \frac{3}{4}x + \frac{3}{2}$

31. $y = 4$

32. $x = -2$

33. $y = 5x - 13$

34. $y = -\frac{3}{2}x - 7$

35. $y = 3x + 3$

36. $y = \frac{2}{3}x - 3$

37. $y = \frac{1}{2}x + 4$

38. $y = 2$

39. $y = 3$

40. $y = -\frac{3}{2}x + 2$

41. $y = 2x + 8$

42. $y = -2x - 5$

43. $y = \frac{4}{3}x - 2$

23. (2, 0) and (0, −3)

24. (2, −3) and (2, 4)

25. (0, 4) and (−2, −1)

26. (−4, 1) and (3, 1)

Write the equation of the line L satisfying the given geometric conditions.

27. L has slope 4 and y intercept (0, −2).

28. L has slope $-\frac{2}{3}$ and y intercept (0, 4).

29. L has x intercept (4, 0) and y intercept (0, 2).

30. L has x intercept (−2, 0) and slope $\frac{3}{4}$.

31. L has y intercept (0, 4) and a 0 slope.

32. L has x intercept (−2, 0) and an undefined slope.

33. L passes through point (3, 2) with a slope of 5.

34. L passes through point (−2, −4) with a slope of $-\frac{3}{2}$.

35. L has y intercept (0, 3) and is parallel to the line with equation $y = 3x - 5$.

36. L has y intercept (0, −3) and is parallel to the line with equation $y = \frac{2}{3}x + 1$.

37. L has y intercept (0, 4) and is perpendicular to the line with equation $y = -2x + 1$.

38. L has y intercept (0, 2) and is parallel to the line with equation $y = -1$.

39. L has y intercept (0, 3) and is parallel to the line with equation $y = 2$.

40. L has y intercept (0, 2) and is perpendicular to the line with equation $2x - 3y = 6$.

41. L passes through point (−3, 2) and is parallel to the line with equation $y = 2x - 3$.

42. L passes through point (−4, 3) and is parallel to the line with equation $y = -2x + 1$.

43. L passes through point (3, 2) and is parallel to the line with equation $y = \frac{4}{3}x + 4$.

44. L passes through point $(-2, -1)$ and is perpendicular to the line with equation $y = 3x + 1$.

45. L passes through point $(5, -2)$ and is perpendicular to the line with equation $y = -3x - 2$.

46. L passes through point $(3, 4)$ and is perpendicular to the line with equation $y = -\frac{3}{5}x + 2$.

47. L passes through $(-2, 1)$ and is parallel to the line with equation $x + 2y = 4$.

48. L passes through $(-3, 5)$ and is parallel to the x axis.

49. Describe the process for finding the equation of a line if you are given two points on the line.

50. How would you find the equation of a line if you were given the slope and the x *intercept?*

51. A temperature of 10°C corresponds to a temperature of 50°F. Also 40°C corresponds to 104°F. Find the linear equation relating F and C.

52. In planning for a new item, a manufacturer assumes that the number of items produced x and the cost in dollars C of producing these items are related by a linear equation. Projections are that 100 items will cost \$10,000 to produce and that 300 items will cost \$22,000 to produce. Find the equation that relates C and x.

53. A word processing station was purchased by a company for \$10,000. After 4 years it is estimated that the value of the station will be \$4000. If the value in dollars V and the time the station has been in use t are related by a linear equation, find the equation that relates V and t.

54. Two years after an expansion, a company had sales of \$42,000. Four years later the sales were \$102,000. Assuming that the sales in dollars S and the time in years t are related by a linear equation, find the equation relating S and t.

Getting Ready for Section 7.4 [Section 2.7]

Graph each of the following inequalities.

(a) $x < 3$

(b) $x \geq -2$

(c) $2x \leq 8$

(d) $3x \geq -9$

(e) $-3x < 12$

(f) $-2x \leq 10$

(g) $\frac{2}{3}x \leq 4$

(h) $-\frac{3}{4}x \geq 6$

ANSWERS

44. $y = -\frac{1}{3}x - \frac{5}{3}$

45. $y = \frac{1}{3}x - \frac{11}{3}$

46. $y = \frac{5}{3}x - 1$

47. $y = -\frac{1}{2}x$

48. $y = 5$

49.

50.

51. $F = \frac{9}{5}C + 32$

52. $C = 60x + 4000$

53. $V = -1500t + 10{,}000$

54. $S = 15{,}000t + 12{,}000$

a. See exercise

b. See exercise

c. See exercise

d. See exercise

e. See exercise

f. See exercise

g. See exercise

h. See exercise

Answers

1. $y = 3x + 2$ **3.** $y = \frac{3}{2}x + 2$ **5.** $y = 4$ **7.** $y = \frac{5}{4}x - 5$

9. $y = 3x - 1$ **11.** $y = -3x - 9$ **13.** $y = \frac{2}{5}x - 5$ **15.** $x = 2$

17. $y = x + 1$ **19.** $y = \frac{3}{4}x - \frac{3}{2}$ **21.** $y = 2$ **23.** $y = \frac{3}{2}x - 3$

25. $y = \frac{5}{2}x + 4$ **27.** $y = 4x - 2$ **29.** $y = -\frac{1}{2}x + 2$

31. $y = 4$ **33.** $y = 5x - 13$ **35.** $y = 3x + 3$ **37.** $y = \frac{1}{2}x + 4$

39. $y = 3$ **41.** $y = 2x + 8$ **43.** $y = \frac{4}{3}x - 2$ **45.** $y = \frac{1}{3}x - \frac{11}{3}$

47. $y = -\frac{1}{2}x$ **49.**  **51.** $F = \frac{9}{5}C + 32$

53. $V = -1500t + 10{,}000$

a. 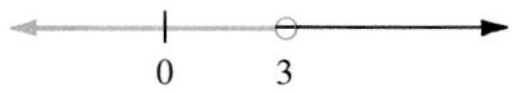 0 3

b. 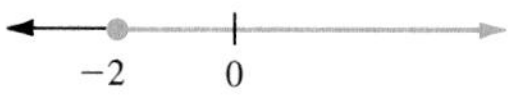 −2 0

c.  0 4

d. 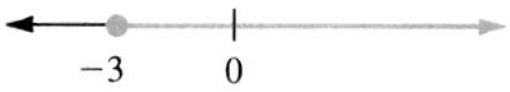 −3 0

e. 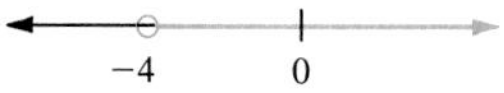 −4 0

f. 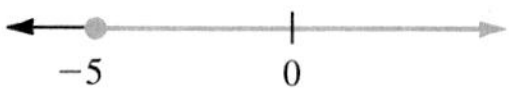 −5 0

g.  0 6

h.  −8 0

7.4 Graphing Linear Inequalities

7.4 OBJECTIVE

1. Graph a linear inequality in two variables

In Section 2.7 you learned to graph inequalities in one variable on a number line. We now want to extend our work with graphing to include linear inequalities in two variables. We begin with a definition.

Definitions: Linear Inequality in Two Variables

An inequality that can be written in the form

$Ax + By < C$

in which A and B are not both 0, is called a **linear inequality in two variables.**

NOTE The inequality symbols $\le$, $>$, and $\ge$ can also be used.

Some examples of linear inequalities in two variables are

$$x + 3y > 6 \qquad y \le 3x + 1 \qquad 2x - y \ge 3$$

The *graph* of a linear inequality is always a region (actually a half plane) of the plane whose boundary is a straight line. Let's look at an example of graphing such an inequality.

Example 1

Graphing a Linear Inequality

NOTE The dotted line indicates that the points on the line $2x + y = 4$ are *not* part of the solution to the inequality $2x + y < 4$.

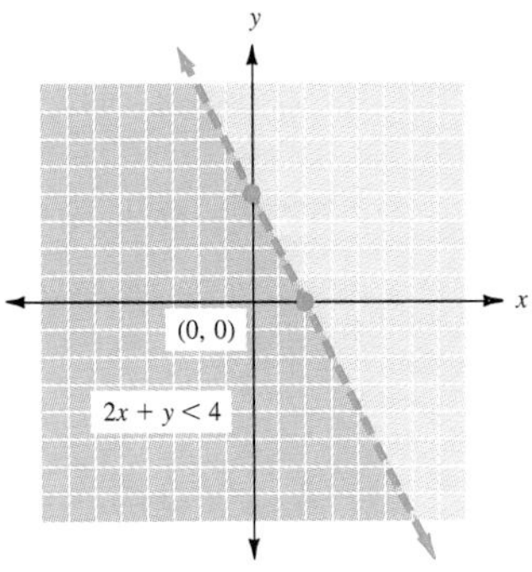

NOTE You can always use the origin for a test point unless the boundary line passes through the origin.

Graph $2x + y < 4$.

First, replace the inequality symbol ($<$) with an equals sign. We then have $2x + y = 4$. This equation forms the **boundary line** of the graph of the original inequality. You can graph the line by any of the methods discussed earlier.

The boundary line for our inequality is shown at left. We see that the boundary line separates the plane into two regions, each of which is called a **half plane.**

We now need to choose the correct half plane. Choose any convenient test point not on the boundary line. The origin (0, 0) is a good choice because it makes for easy calculation.

Substitute $x = 0$ and $y = 0$ into the inequality.

$$2 \cdot 0 + 0 < 4$$
$$0 + 0 < 4$$
$$0 < 4 \quad \text{A true statement}$$

Because the inequality is *true* for the test point, we shade the half plane containing that test point (here the origin). The origin and all other points *below* the boundary line then represent solutions for our original inequality.

CHECK YOURSELF 1

Graph the inequality $x + 3y < 3$.

The process is similar when the boundary line is included in the solution.

Example 2

Graphing a Linear Inequality

NOTE Again, we replace the inequality symbol (≥) with an equals sign to write the equation for our boundary line.

Graph $4x - 3y \geq 12$.

First, graph the boundary line, $4x - 3y = 12$.

Note: When equality *is included* (≤ or ≥), use a *solid line* for the graph of the boundary line. This means the line is included in the graph of the linear inequality.

The graph of our boundary line (a solid line here) is shown on the figure.

NOTE Although any of our graphing methods can be used here, the intercept method is probably the most efficient.

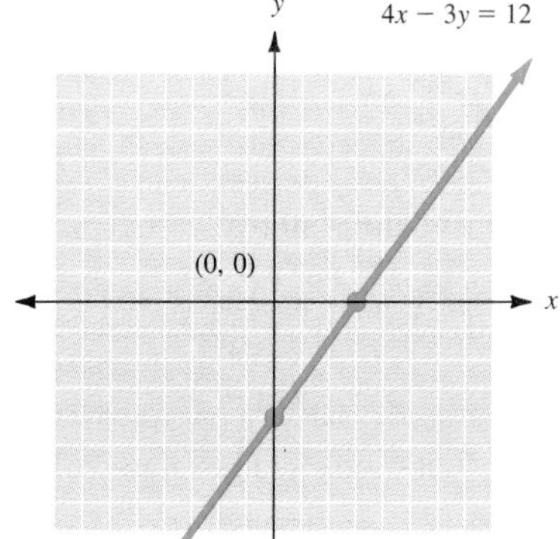

Again, we use (0, 0) as a convenient test point. Substituting 0 for x and for y in the original inequality, we have

$$4 \cdot 0 - 3 \cdot 0 \geq 12$$

$$0 \geq 12 \quad \text{A false statement}$$

Because the inequality is *false* for the test point, we shade the half plane that does *not* contain that test point, here (0, 0).

NOTE All points *on and below* the boundary line represent solutions for our original inequality.

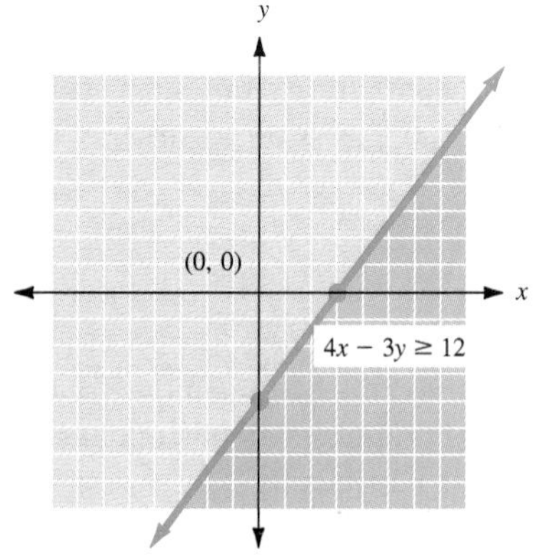

CHECK YOURSELF 2

Graph the inequality $3x + 2y \geq 6$.

Example 3

Graphing a Linear Inequality

Graph $x \le 5$.

The boundary line is $x = 5$. Its graph is a solid line because equality is included. Using (0, 0) as a test point, we substitute 0 for x with the result

$0 \le 5$ A true statement

Because the inequality is *true* for the test point, we shade the half plane containing the origin.

NOTE If the correct half plane is obvious, you may not need to use a test point. Did you know without testing which half plane to shade in this example?

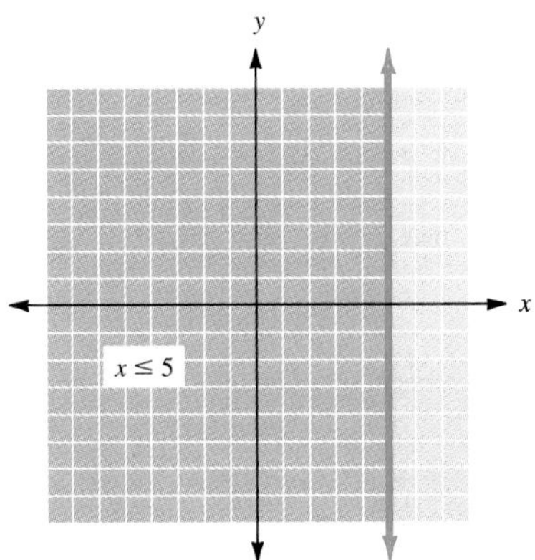

CHECK YOURSELF 3

Graph the inequality $y < 2$.

As we mentioned earlier, we may have to use a point other than the origin as our test point. Example 4 illustrates this approach.

Example 4

Graphing a Linear Inequality

Graph $2x + 3y < 0$.

The boundary line is $2x + 3y = 0$. Its graph is shown on the figure.

NOTE We use a dotted line for our boundary line because equality is not included.

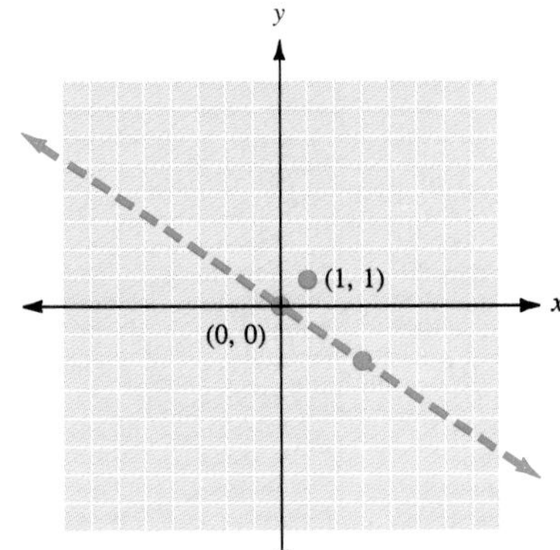

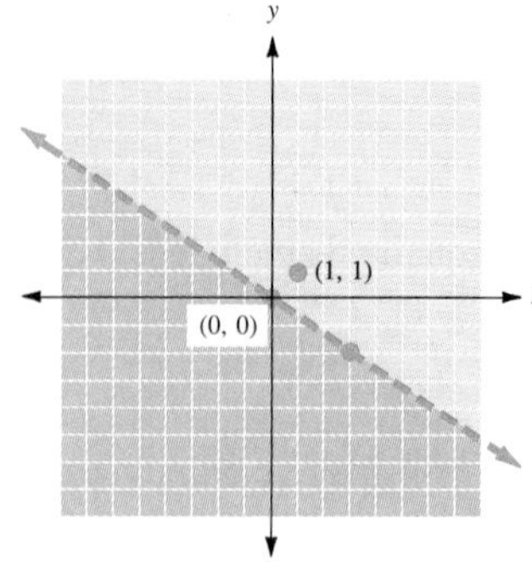

We cannot use (0, 0) as our test point in this case. Do you see why?

Choose any other point *not* on the line. For instance, we have picked (1, 1) as a test point. Substituting 1 for x and 1 for y gives

$$2 \cdot 1 + 3 \cdot 1 < 0$$
$$2 + 3 < 0$$
$$5 < 0 \quad \text{A false statement}$$

Because the inequality is *false* at our test point, we shade the half plane *not* containing (1, 1). This is shown in the graph in the margin.

CHECK YOURSELF 4

Graph the inequality $x - 2y < 0$.

The following steps summarize our work in graphing linear inequalities in two variables.

Step by Step: To Graph a Linear Inequality

Step 1 Replace the inequality symbol with an equals sign to form the equation of the boundary line of the graph.

Step 2 Graph the boundary line. Use a dotted line if equality is not included (< or >). Use a solid line if equality is included (≤ or ≥).

Step 3 Choose any convenient test point *not* on the line.

Step 4 If the inequality is *true* at the checkpoint, shade the half plane including the test point. If the inequality is *false* at the checkpoint, shade the half plane not including the test point.

CHECK YOURSELF ANSWERS

1.

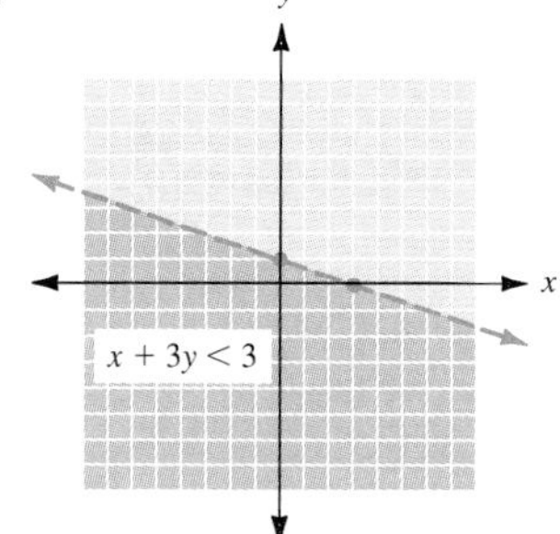

2.

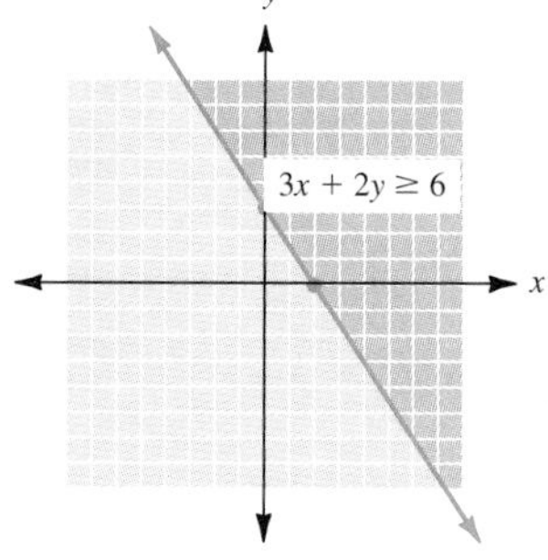

3.

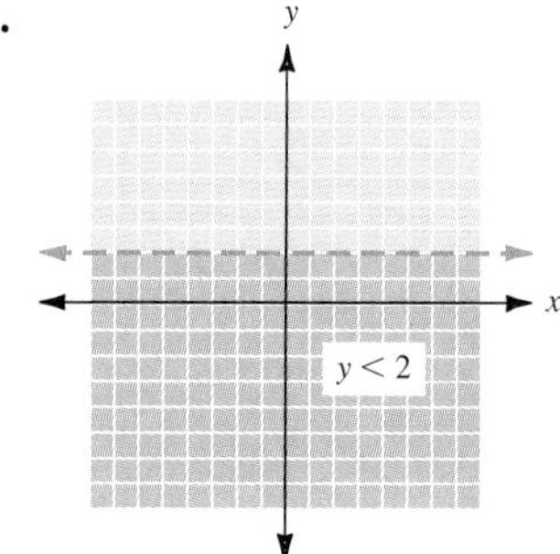

4.

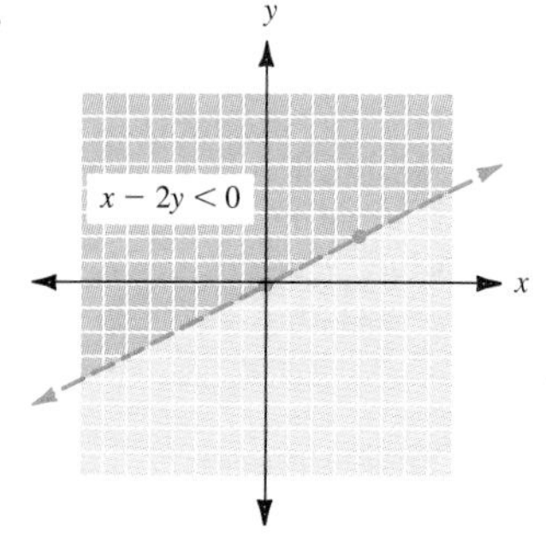

Name ____________________

Section ________ Date ________

7.4 Exercises

In exercises 1 to 8, we have graphed the boundary line for the linear inequality. Determine the correct half plane in each case, and complete the graph.

1. $x + y < 5$

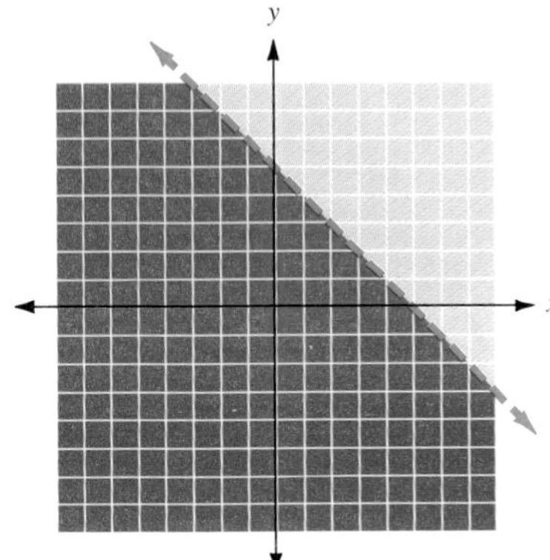

2. $x - y \geq 4$

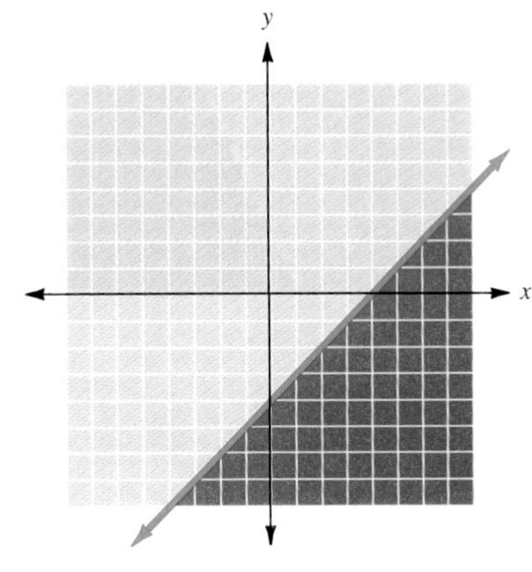

3. $x - 2y \geq 4$

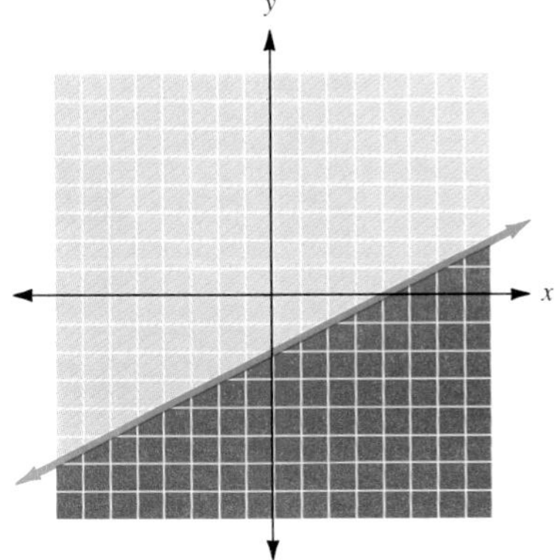

4. $2x + y < 6$

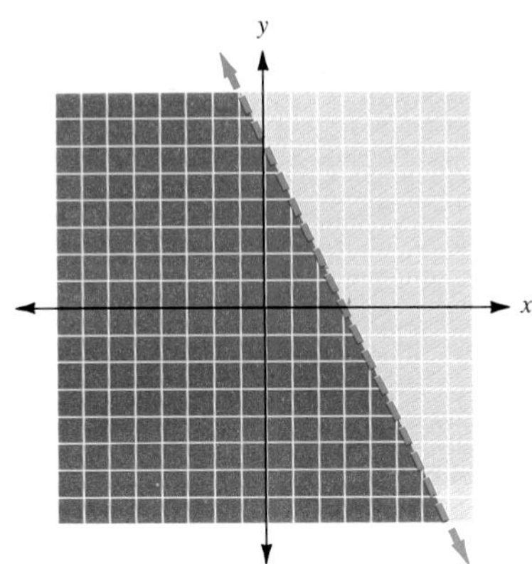

5. $x \leq -3$

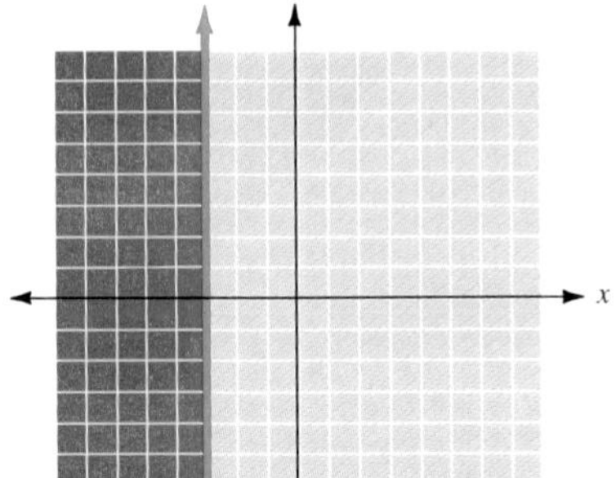

6. $y \geq 2x$

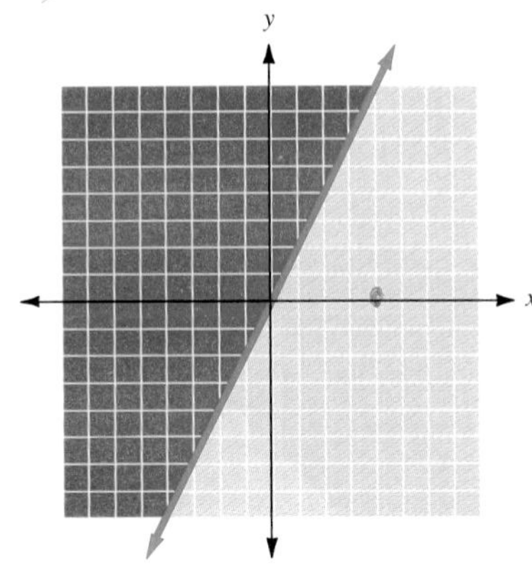

ANSWERS

1. See exercise
2. See exercise
3. See exercise
4. See exercise
5. See exercise
6. See exercise

CHECK YOURSELF 1

Evaluate the expression $2x^3 - 3x^2 + 3x + 1$ for the indicated value of x.

(a) $x = 0$ **(b)** $x = 1$ **(c)** $x = -2$

We could design a machine whose function would be to crank out the value of an expression for each given value of x. We could call this machine something simple such as f. Our *function* machine might look like this.

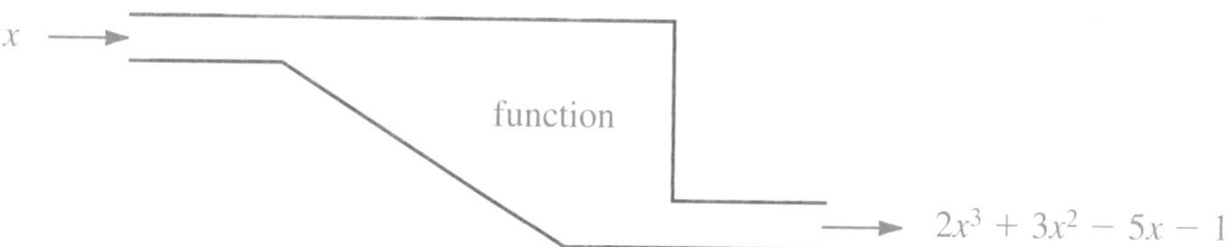

For example, when we put -1 into the machine, the machine would substitute -1 for x in the expression, and 5 would come out the other end because

$$2(-1)^3 + 3(-1)^2 - 5(-1) - 1 = -2 + 3 + 5 - 1 = 5$$

In fact, the idea of the function machine is very useful in mathematics. Your graphing calculator can be used as a function machine. You can enter the expression into the calculator as Y_1 and then evaluate Y_1 for different values of x.

Generally, in mathematics, we do not write $Y_1 = 2x^3 + 3x^2 - 5x - 1$. Instead, we write $f(x) = 2x^3 + 3x^2 - 5x - 1$, which is read as "$f$ of x is equal to . . . ." Instead of calling f a function machine, we say that f is a function of x. The greatest benefit of this notation is that it lets us easily note the input value of x along with the output of the function. Instead of "Evaluate Y_1 for $x = 4$" we say "Find $f(4)$."

Example 2

Evaluating Expressions with Function Notation

Given $f(x) = x^3 - 3x^2 + x + 5$, find the following:

(a) $f(0)$

Substituting 0 for x in the expression, we get

$$(0)^3 - 3(0)^2 + (0) + 5 = 5$$

(b) $f(-3)$

Substituting -3 for x in the expression, we get

$$(-3)^3 - 3(-3)^2 + (-3) + 5 = -27 - 27 - 3 + 5$$
$$= -52$$

(c) $f\left(\frac{1}{2}\right)$

7.5 An Introduction to Functions

7.5 OBJECTIVES

1. Evaluate expressions
2. Evaluate functions
3. Express the equation of a line as a linear function
4. Write an equation as a function
5. Graph a linear function

Variables can be used to represent unknown real numbers. Together with the operations of addition, subtraction, multiplication, division, and exponentiation, these numbers and variables form expressions such as

$3x + 5 \qquad 7x - 4 \qquad x^2 - 3x - 10 \qquad x^4 - 2x^2 + 3x + 4$

Four different actions can be taken with expressions. We can

1. Substitute values for the variable(s) and **evaluate the expression.**
2. Rewrite an expression as some simpler equivalent expression. This rewriting is called **simplifying the expression.**
3. Set two expressions equal to each other and **solve for the stated variable.**
4. Set two expressions equal to each other and **graph the equation.**

Throughout this book, everything we do will involve one of these four actions. We now return our focus to the first item, evaluating expressions. As we saw in Section 1.5, expressions can be evaluated for an indicated value of the variable(s). Example 1 illustrates.

Example 1

Evaluating Expressions

Evaluate the expression $x^4 - 2x^2 + 3x + 4$ for the indicated value of x.

(a) $x = 0$

Substituting 0 for x in the expression yields

$$\begin{aligned}(0)^4 - 2(0)^2 + 3(0) + 4 &= 0 - 0 + 0 + 4\\ &= 4\end{aligned}$$

(b) $x = 2$

Substituting 2 for x in the expression yields

$$\begin{aligned}(2)^4 - 2(2)^2 + 3(2) + 4 &= 16 - 8 + 6 + 4\\ &= 18\end{aligned}$$

(c) $x = -1$

Substituting -1 for x in the expression yields

$$\begin{aligned}(-1)^4 - 2(-1)^2 + 3(-1) + 4 &= 1 - 2 - 3 + 4\\ &= 0\end{aligned}$$

CHECK YOURSELF 1

Evaluate the expression $2x^3 - 3x^2 + 3x + 1$ *for the indicated value of x.*

(a) $x = 0$ **(b)** $x = 1$ **(c)** $x = -2$

We could design a machine whose function would be to crank out the value of an expression for each given value of x. We could call this machine something simple such as f. Our *function* machine might look like this.

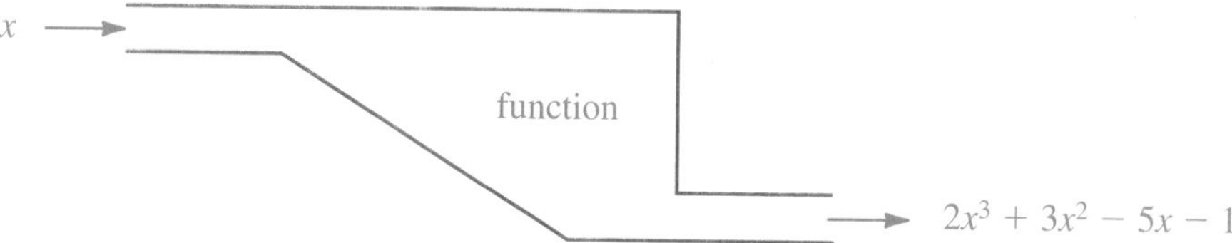

For example, when we put -1 into the machine, the machine would substitute -1 for x in the expression, and 5 would come out the other end because

$$2(-1)^3 + 3(-1)^2 - 5(-1) - 1 = -2 + 3 + 5 - 1 = 5$$

In fact, the idea of the function machine is very useful in mathematics. Your graphing calculator can be used as a function machine. You can enter the expression into the calculator as Y_1 and then evaluate Y_1 for different values of x.

Generally, in mathematics, we do not write $Y_1 = 2x^3 + 3x^2 - 5x - 1$. Instead, we write $f(x) = 2x^3 + 3x^2 - 5x - 1$, which is read as "$f$ of x is equal to . . . ." Instead of calling f a function machine, we say that f is a function of x. The greatest benefit of this notation is that it lets us easily note the input value of x along with the output of the function. Instead of "Evaluate Y_1 for $x = 4$" we say "Find $f(4)$."

Example 2

Evaluating Expressions with Function Notation

Given $f(x) = x^3 - 3x^2 + x + 5$, find the following:

(a) $f(0)$

Substituting 0 for x in the expression, we get

$$(0)^3 - 3(0)^2 + (0) + 5 = 5$$

(b) $f(-3)$

Substituting -3 for x in the expression, we get

$$\begin{aligned}(-3)^3 - 3(-3)^2 + (-3) + 5 &= -27 - 27 - 3 + 5 \\ &= -52\end{aligned}$$

(c) $f\left(\frac{1}{2}\right)$

13. $2x + y < 6$

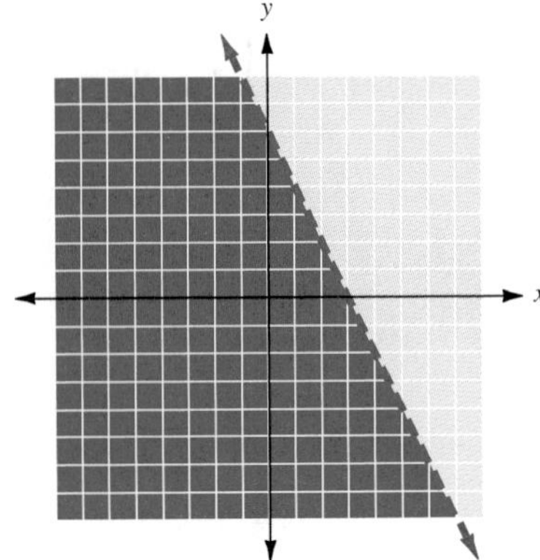

14. $3x + y \geq 6$

15. $x \leq 3$

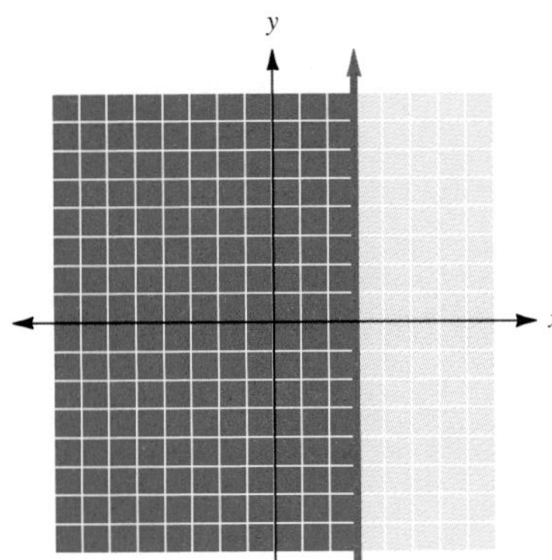

16. $4x + y \geq 4$

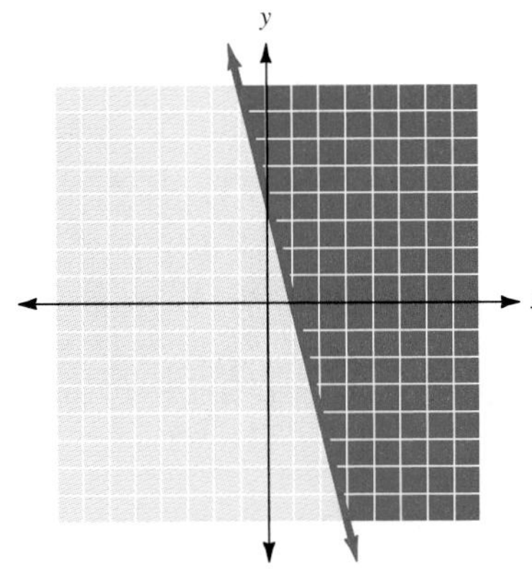

17. $x - 5y < 5$

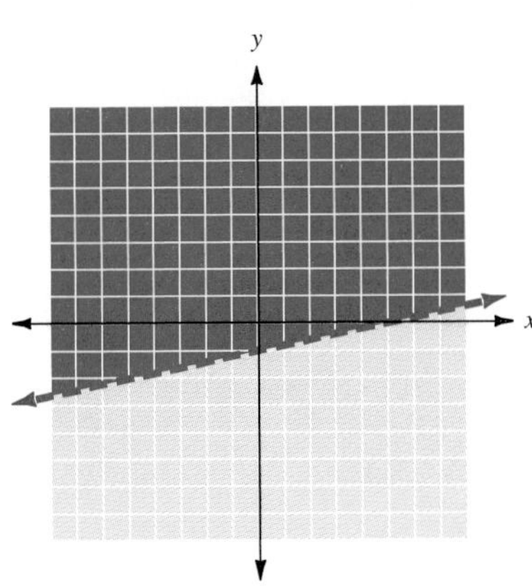

18. $y > 3$

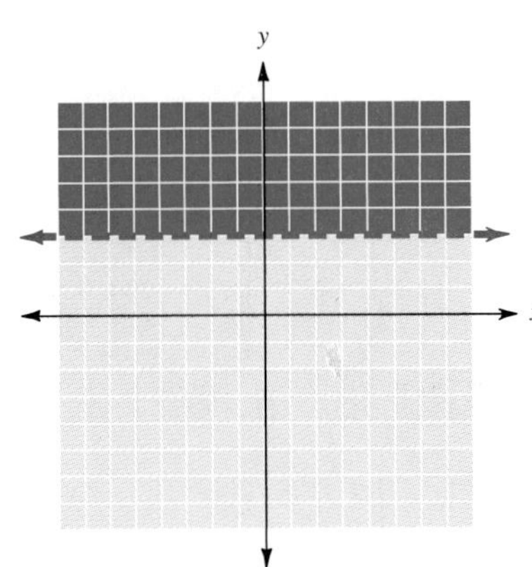

ANSWERS

13. See exercise

14. See exercise

15. See exercise

16. See exercise

17. See exercise

18. See exercise

ANSWERS

19. See exercise

20. See exercise

21. See exercise

22. See exercise

23. See exercise

24. See exercise

19. $y < -4$

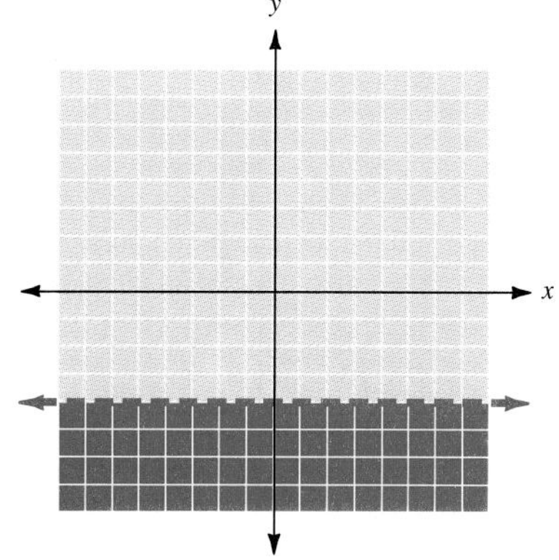

20. $4x + 3y > 12$

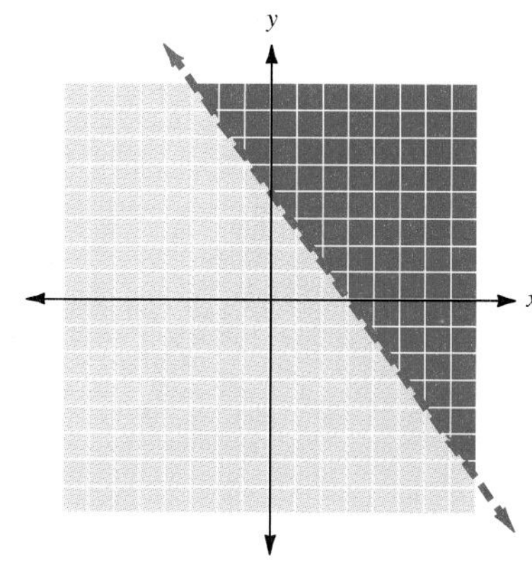

21. $2x - 3y \geq 6$

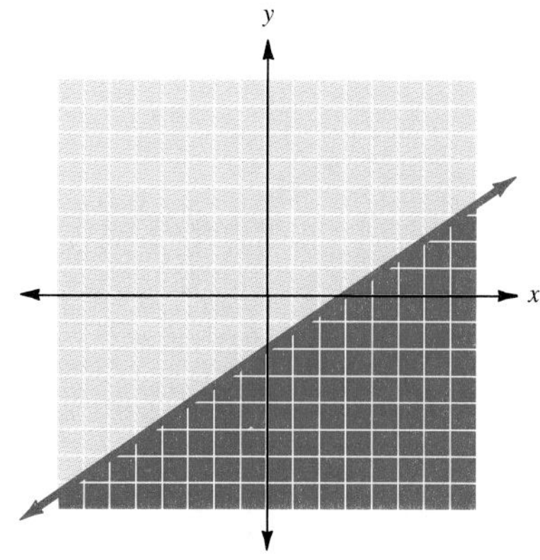

22. $x \geq -2$

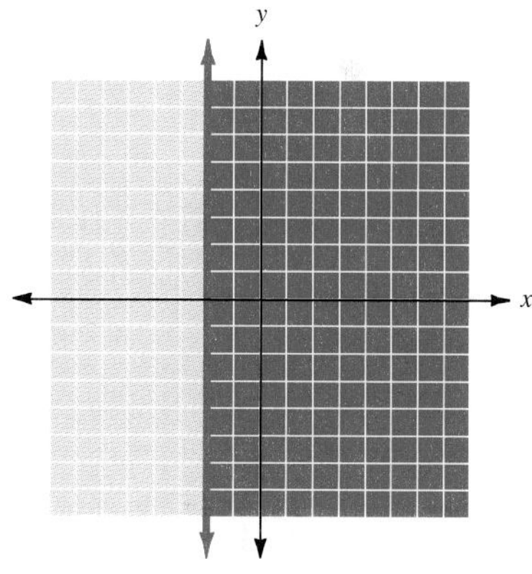

23. $3x + 2y \geq 0$

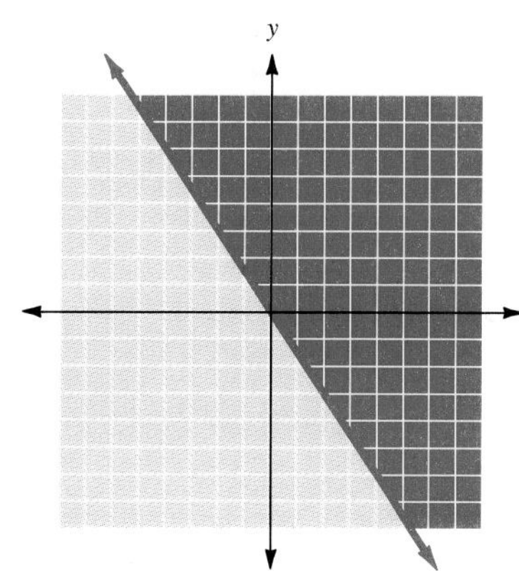

24. $3x + 5y < 15$

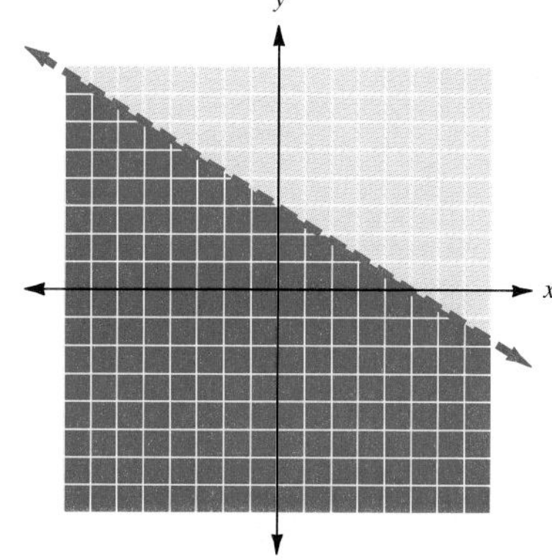

25. $5x + 2y > 10$

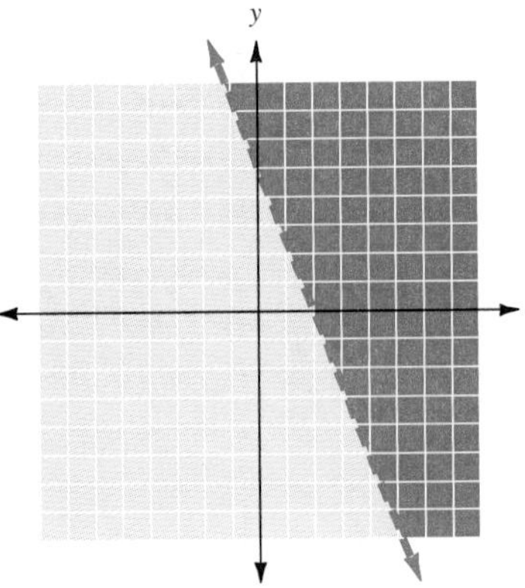

26. $x - 3y \geq 0$

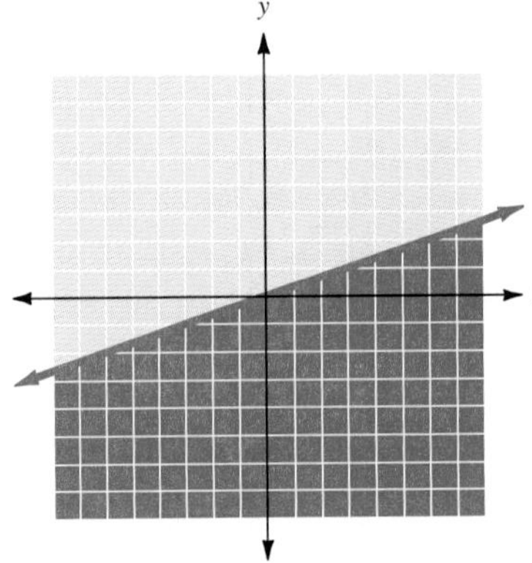

27. $y \leq 2x$

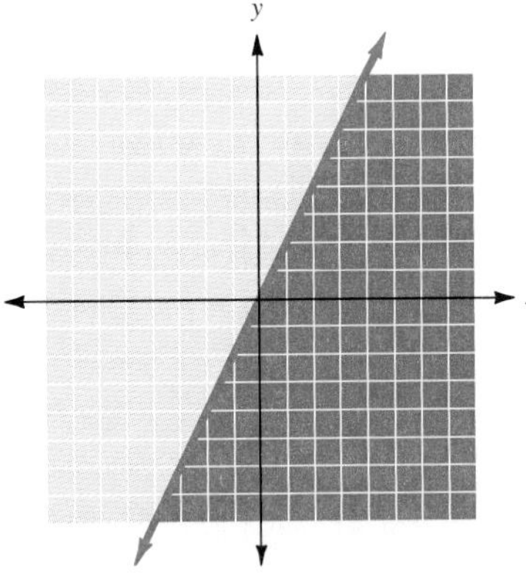

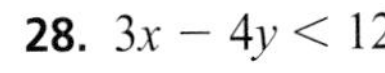

28. $3x - 4y < 12$

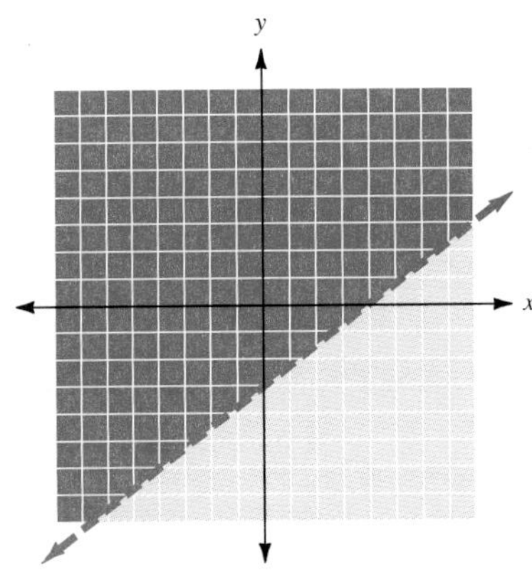

29. $y > 2x - 3$

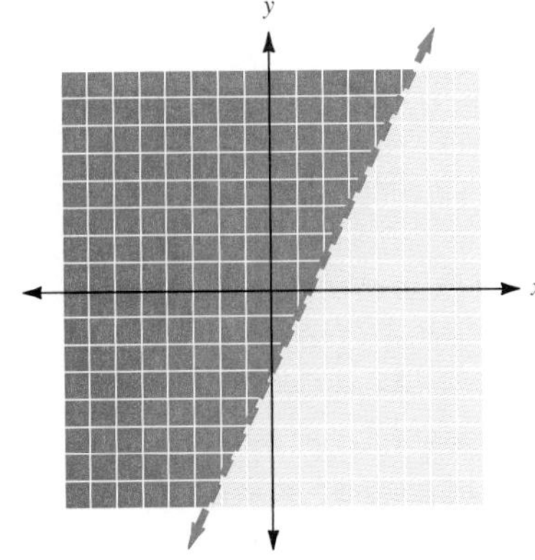

30. $y \geq -2x$

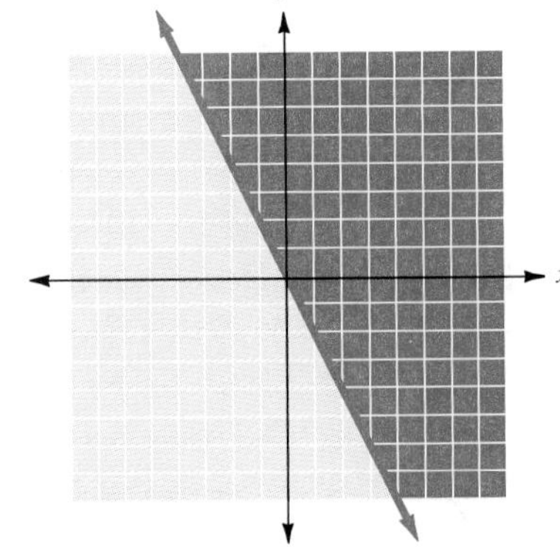

ANSWERS

25. See exercise

26. See exercise

27. See exercise

28. See exercise

29. See exercise

30. See exercise

ANSWERS

31. See exercise

32. See exercise

33. See exercise

34. See exercise

35. See exercise

36. See exercise

37. $9x + 8y \geq 240$

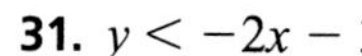

31. $y < -2x - 3$

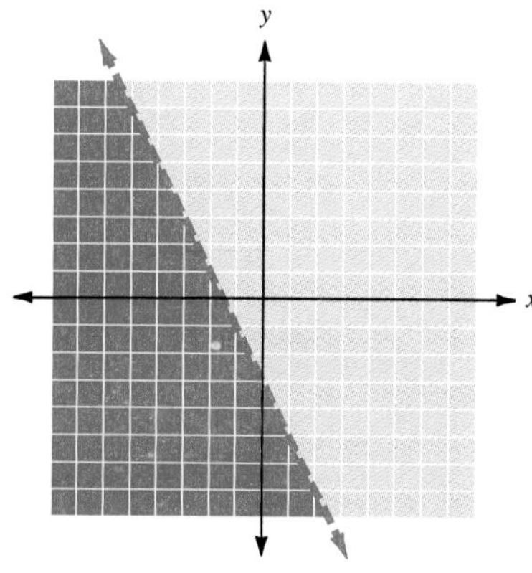

32. $y \leq 3x + 4$

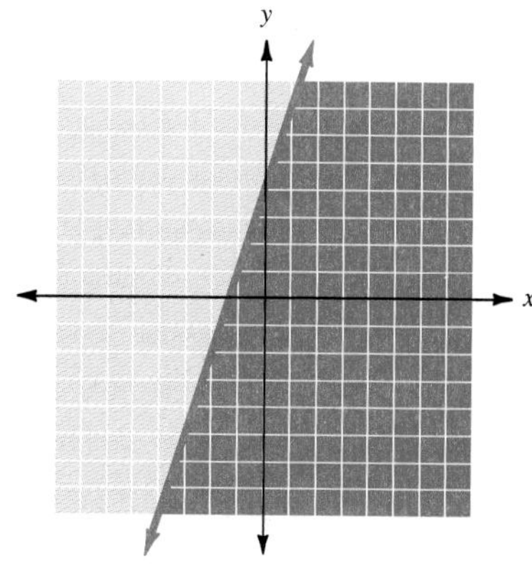

Graph each of the following inequalities.

33. $2(x + y) - x > 6$

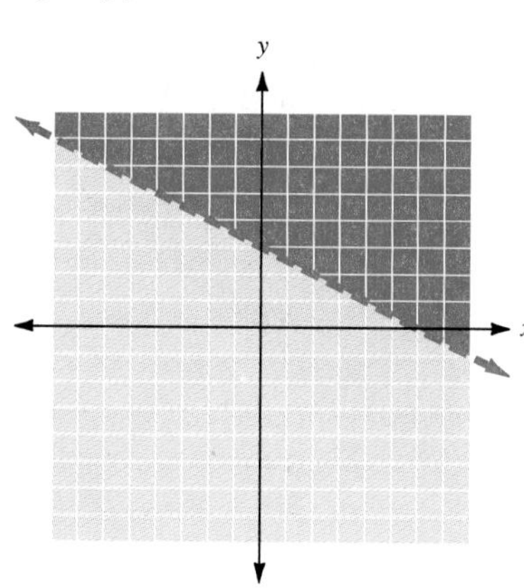

34. $3(x + y) - 2y < 3$

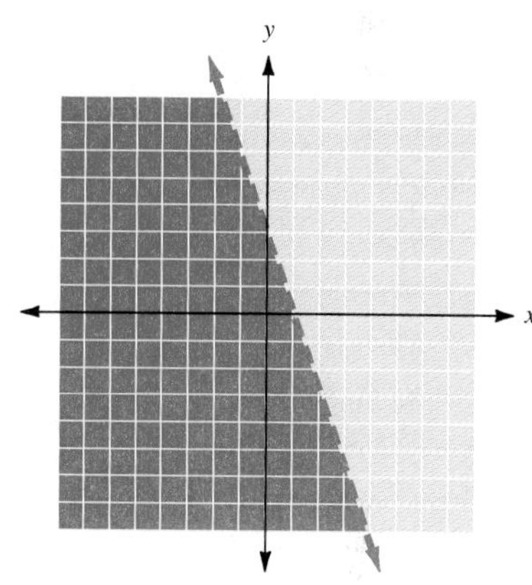

35. $4(x + y) - 3(x + y) \leq 5$

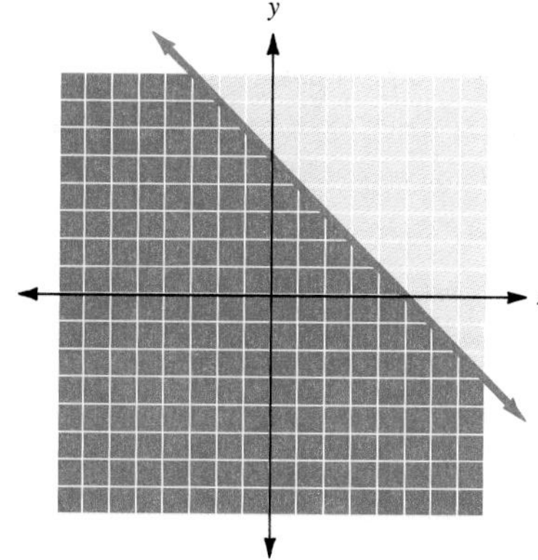

36. $5(2x + y) - 4(2x + y) \geq 4$

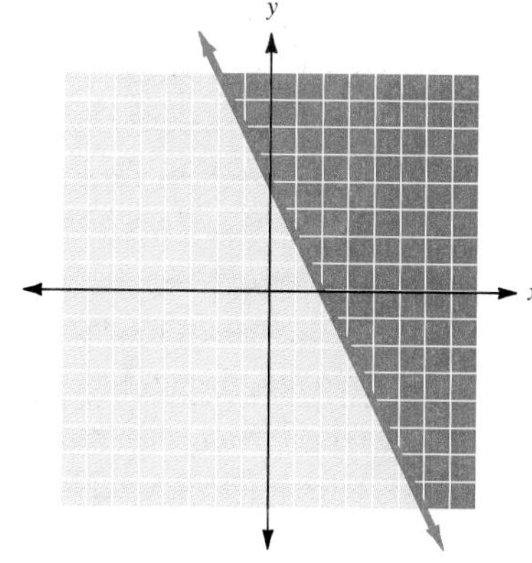

37. Hours worked. Suppose you have two part-time jobs. One is at a video store that pays \$9 per hour and the other is at a convenience store that pays \$8 per hour. Between the two jobs, you want to earn at least \$240 per week. Write an inequality that shows the various number of hours you can work at each job.

ANSWERS

38. $10x + 25y \geq 3000$

39.

40.

38. Money problem. You have at least \$30 in change in your drawer, consisting of dimes and quarters. Write an inequality that shows the different number of coins in your drawer.

39. Linda Williams has just begun a nursery business and seeks your advice. She has limited funds to spend and wants to stock two kinds of fruit-bearing plants. She lives in the northeastern part of Texas and thinks that blueberry bushes and peach trees would sell well there. Linda can buy blueberry bushes from a supplier for \$2.50 each and young peach trees for \$5.50 each. She wants to know what combination she should buy and keep her outlay to \$500 or less. Write an equation and draw a graph to depict what combinations of blueberry bushes and peach trees she can buy for the amount of money she has. Explain the graph and her options.

40. After reading an article on the front page of *The New York Times* titled "You Have to be Good at Algebra to Figure Out the Best Deal for Long Distance," Rafaella De La Cruz decided to apply her skills in algebra to try to decide between two competing long-distance companies. It was difficult at first to get the companies to explain their charge policies. They both kept repeating that they were 25% cheaper than their competition. Finally, Rafaella found someone who explained that the charge depended on when she called, where she called, how long she talked, and how often she called. "Too many variables!" she exclaimed. So she decided to ask one company what they charged as a base amount, just for using the service.

Company A said that they charged \$5 for the privilege of using their long-distance service whether or not she made any phone calls, and that because of this fee they were able to allow her to call anywhere in the United States after 6 P.M. for only \$0.15 a minute. Complete this table of charges based on this company's plan:

Total Minutes Long Distance in 1 Month (After 6 P.M.)	Total Charge
0 minutes	
10 minutes	
30 minutes	
60 minutes	
120 minutes	

Use this table to make a whole-page graph of the monthly charges from Company A based on the number of minutes of long distance.

ANSWERS

a. 5

b. −3

c. 1

d. 5

e. 2

f. 2

g. 6

h. 6

Rafaella wanted to compare this offer to Company B, which she was currently using. She looked at her phone bill and saw that one month she had been charged \$7.50 for 30 minutes and another month she had been charged \$11.25 for 45 minutes of long-distance calling. These calls were made after 6 P.M. to her relatives in Indiana and in Arizona. Draw a graph on the same set of axes you made for Company A's figures. Use your graph and what you know about linear inequalities to advise Rafaella about which company is best.

Getting Ready for Section 7.5 (Section 1.5)

Evaluate each expression for the given variable value.

(a) $2x + 1$ $(x = 2)$ (b) $2x + 1$ $(x = -2)$

(c) $3 - 2x$ $(x = 1)$ (d) $3 - 2x$ $(x = -1)$

(e) $x^2 - 2$ $(x = 2)$ (f) $x^2 - 2$ $(x = -2)$

(g) $x^2 + 5$ $(x = 1)$ (h) $x^2 + 5$ $(x = -1)$

Answers

1. $x + y < 5$

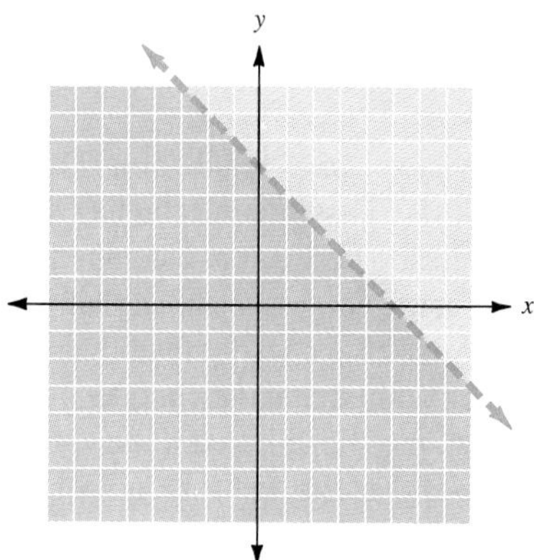

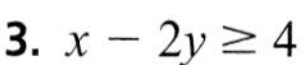

3. $x - 2y \geq 4$

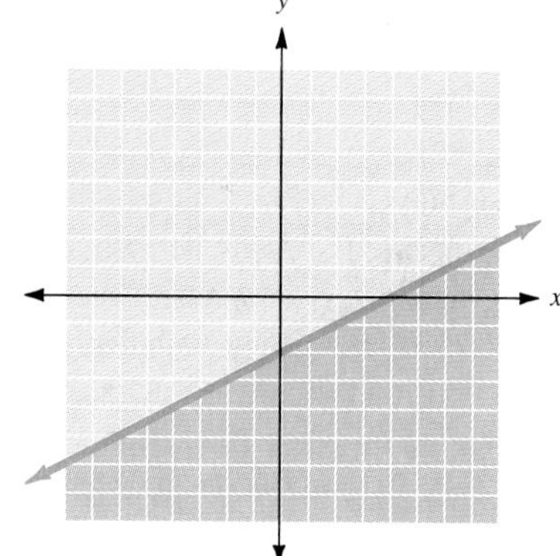

5. $x \leq -3$

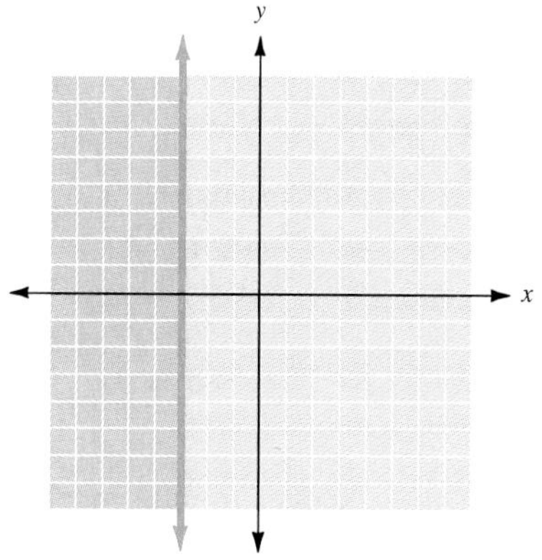

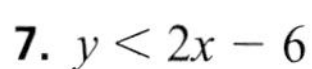

7. $y < 2x - 6$

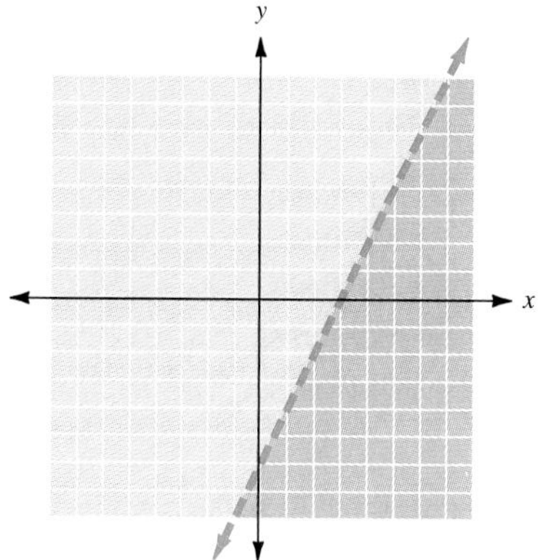

9. $x + y < 3$

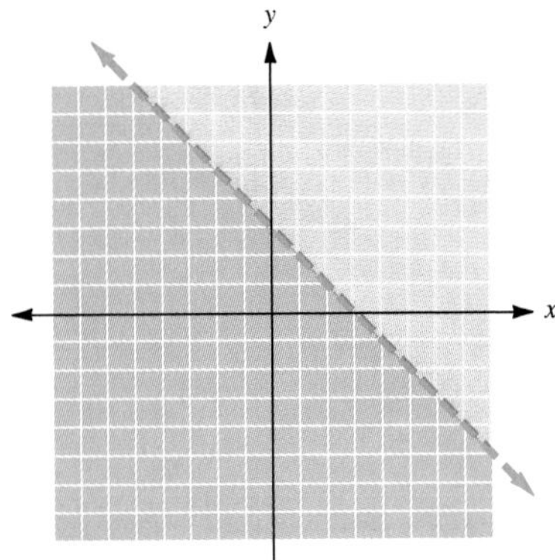

11. $x - y \le 5$

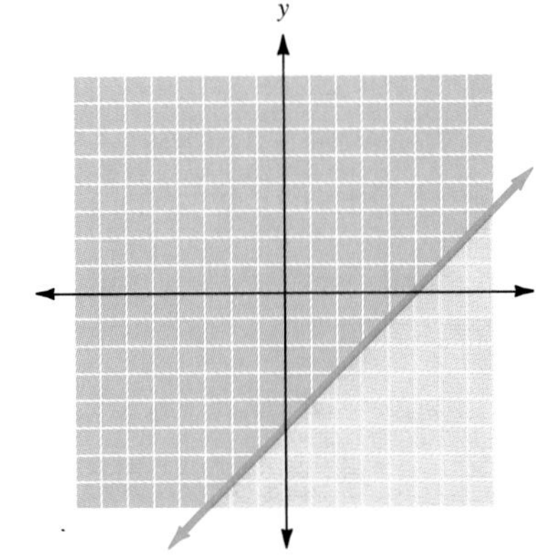

13. $2x + y < 6$

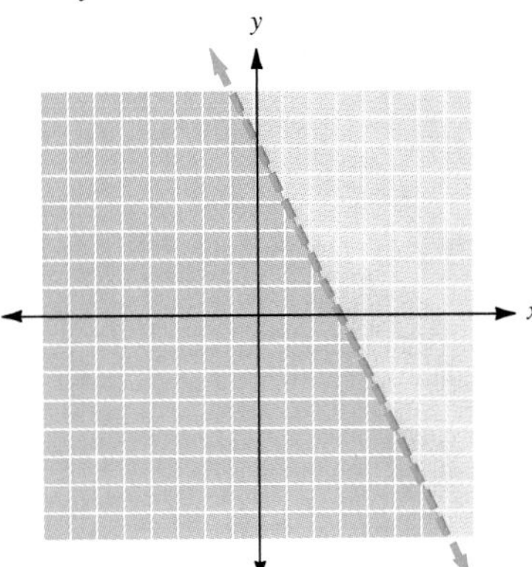

15. $x \le 3$

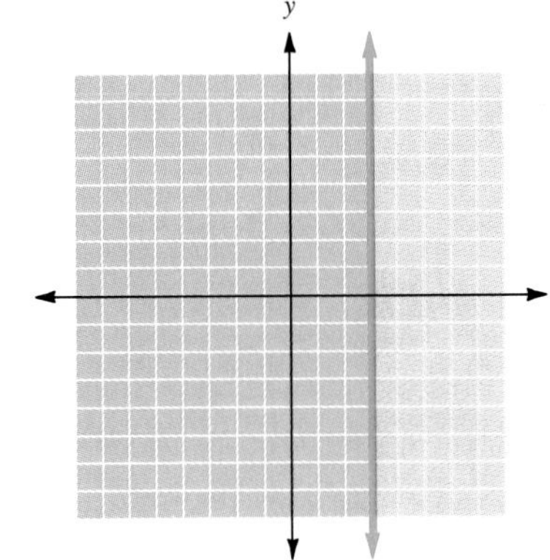

17. $x - 5y < 5$

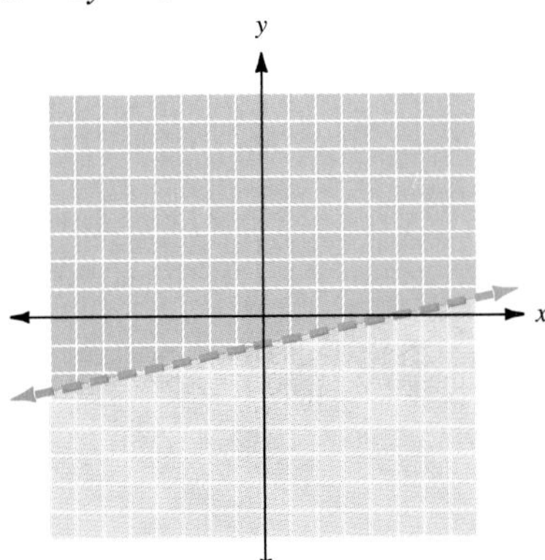

19. $y < -4$

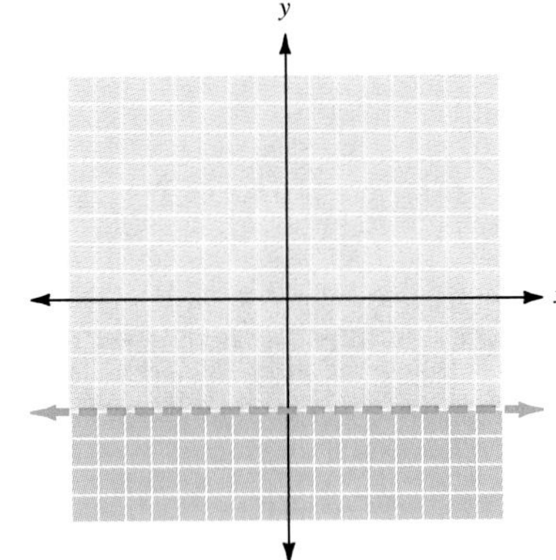

21. $2x - 3y \ge 6$

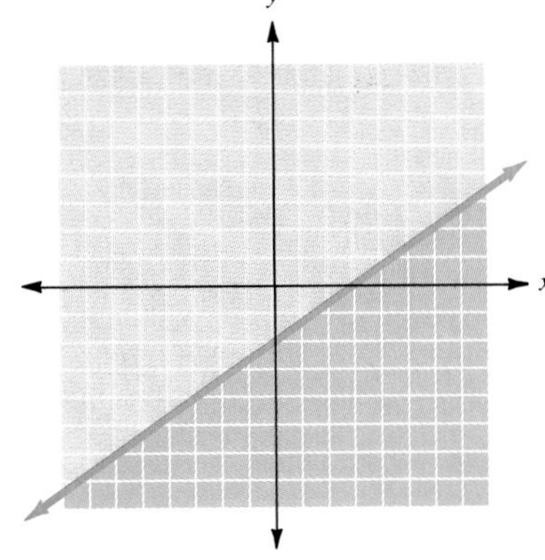

23. $3x + 2y \ge 0$

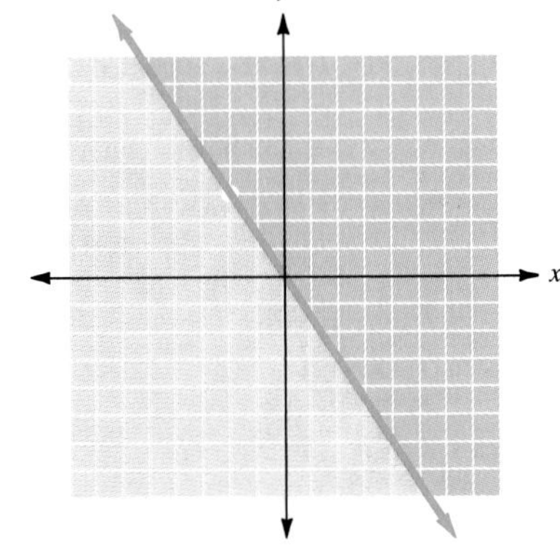

25. $5x + 2y > 10$

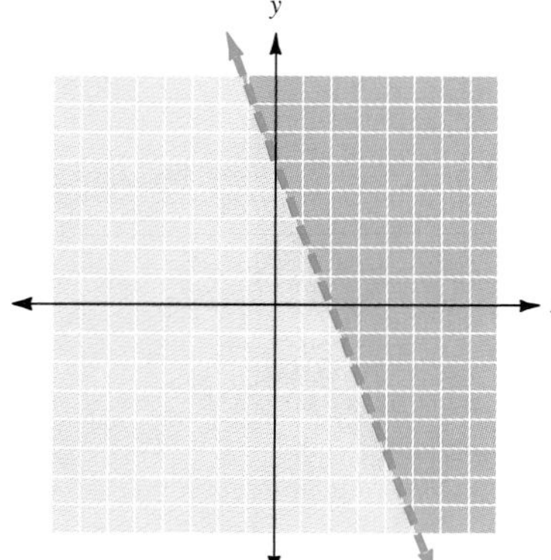

27. $y \le 2x$

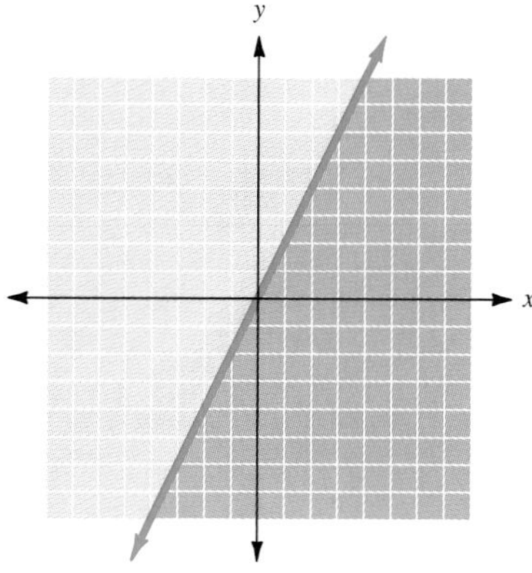

29. $y > 2x - 3$

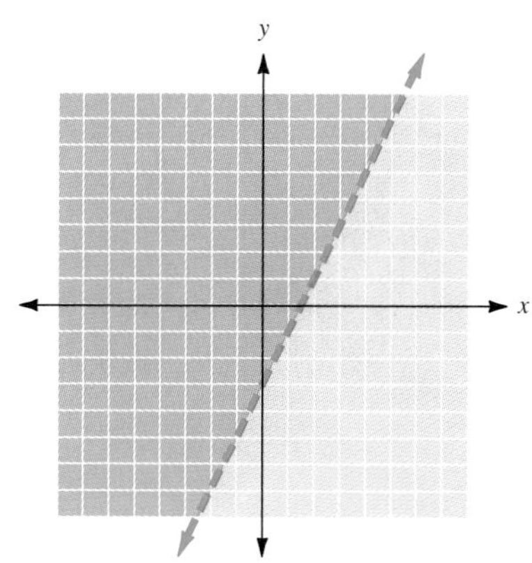

31. $y < -2x - 3$

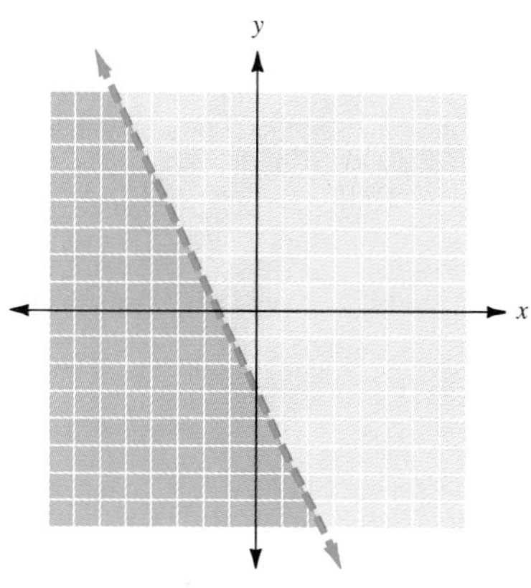

33. $x + 2y > 6$

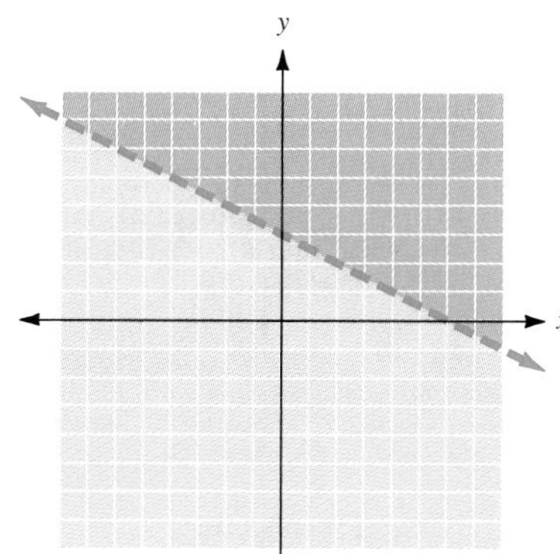

35. $x + y \le 5$

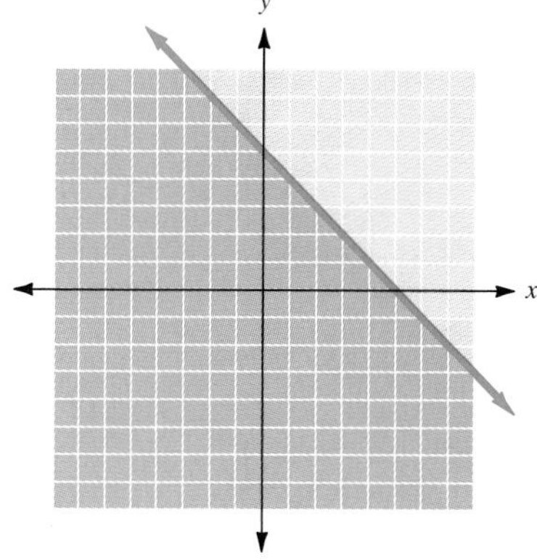

37. $9x + 8y \ge 240$ **39.**  **a.** 5 **b.** -3 **c.** 1 **d.** 5

e. 2 **f.** 2 **g.** 6 **h.** 6

Exercises

Name ______

Section ______ Date ______

In exercises 1 to 8, we have graphed the boundary line for the linear inequality. Determine the correct half plane in each case, and complete the graph.

They have line — which side is shaded!

1. $x + y < 5$

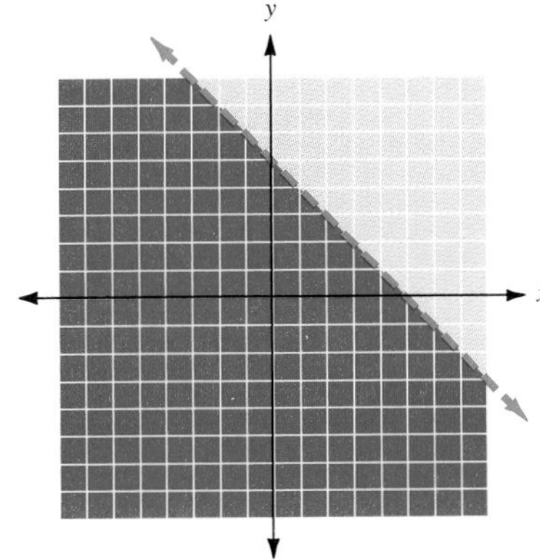

2. $x - y \geq 4$

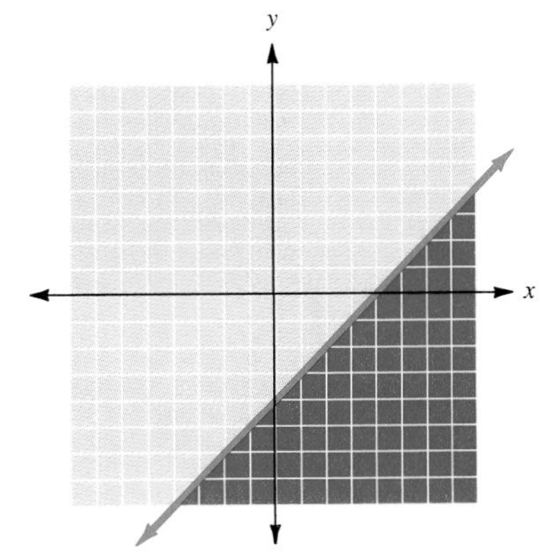

3. $x - 2y \geq 4$

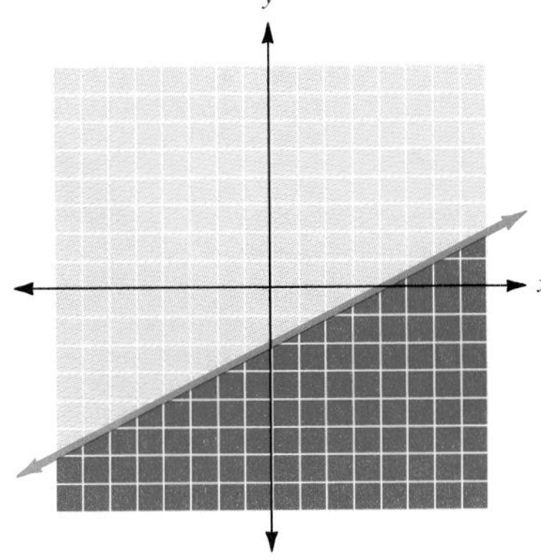

4. $2x + y < 6$

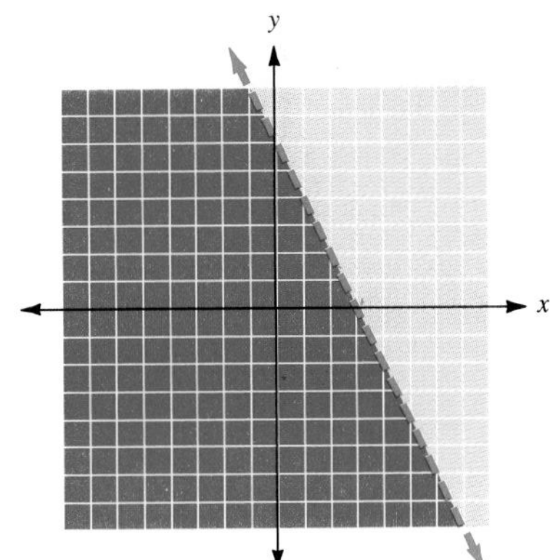

5. $x \leq -3$

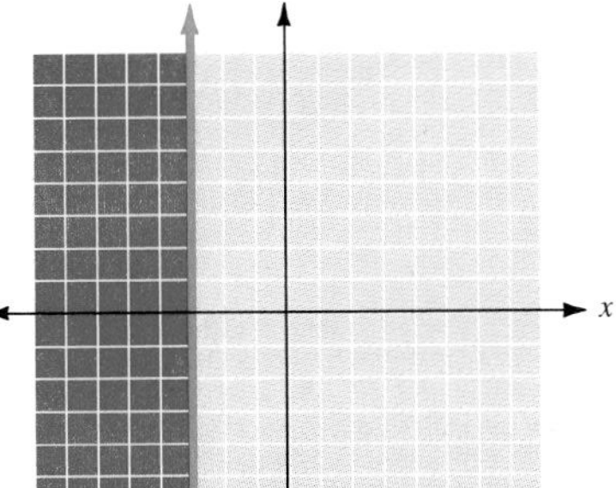

6. $y \geq 2x$

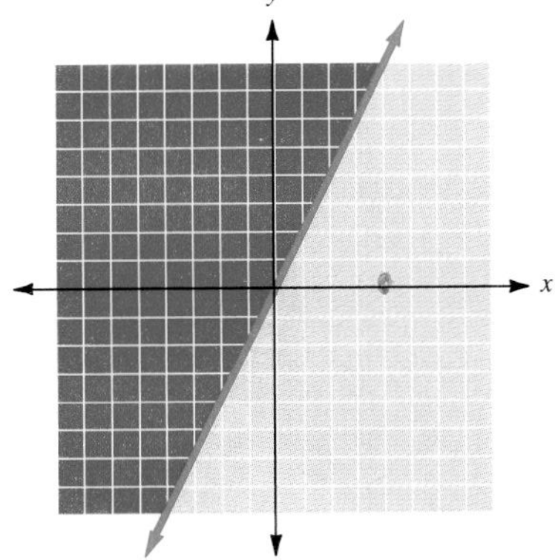

ANSWERS

1. See exercise
2. See exercise
3. See exercise
4. See exercise
5. See exercise
6. See exercise

ANSWERS

7. See exercise

8. See exercise

9. See exercise

10. See exercise

11. See exercise

12. See exercise

7. $y < 2x - 6$

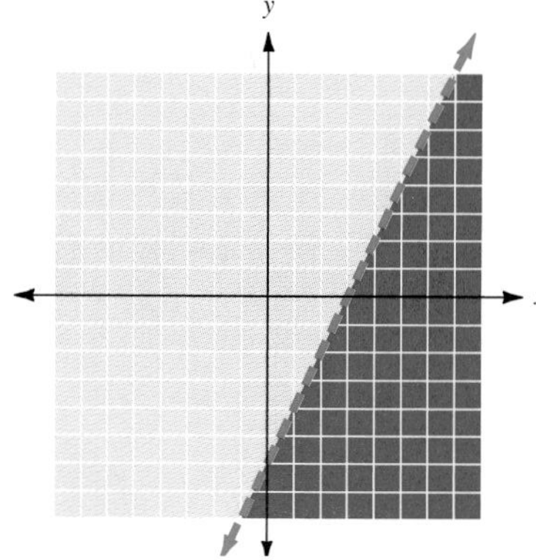

8. $y > 3$

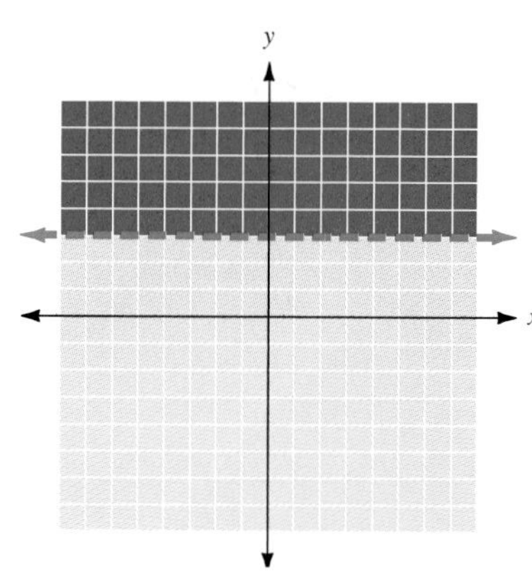

Graph each of the following inequalities.

9. $x + y < 3$

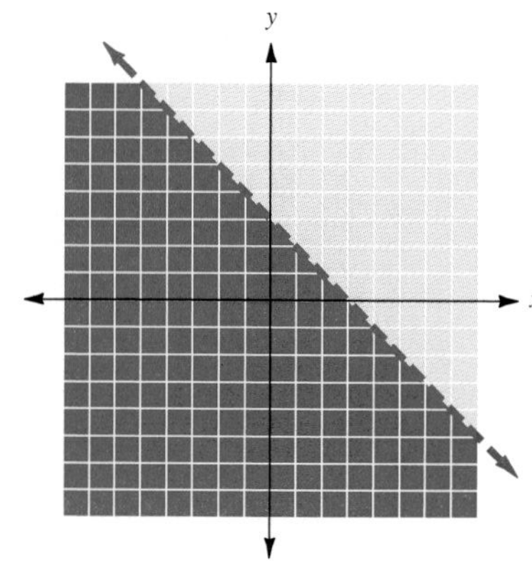

10. $x - y \geq 4$

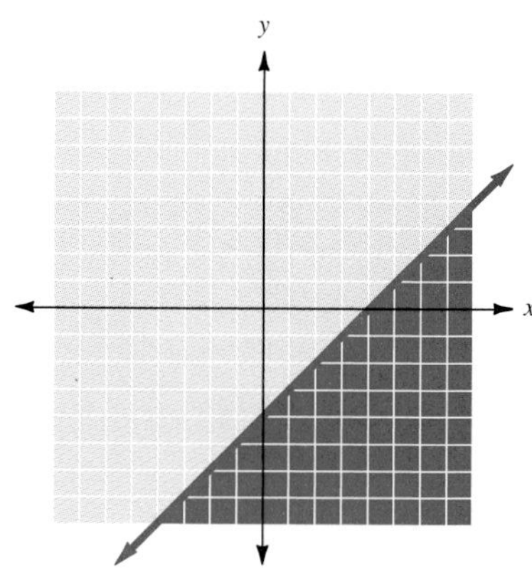

11. $x - y \leq 5$

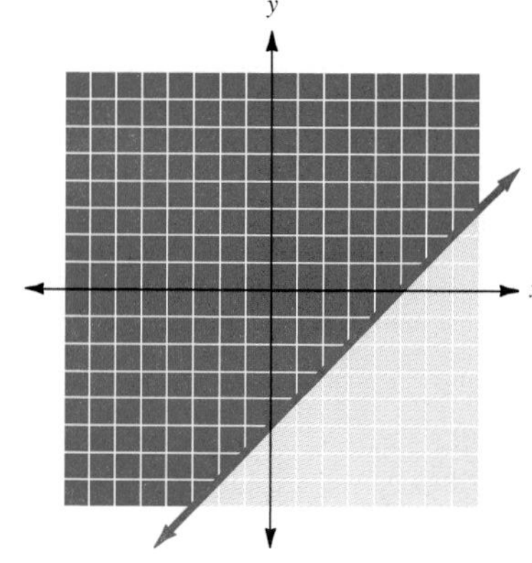

12. $x + y > 5$

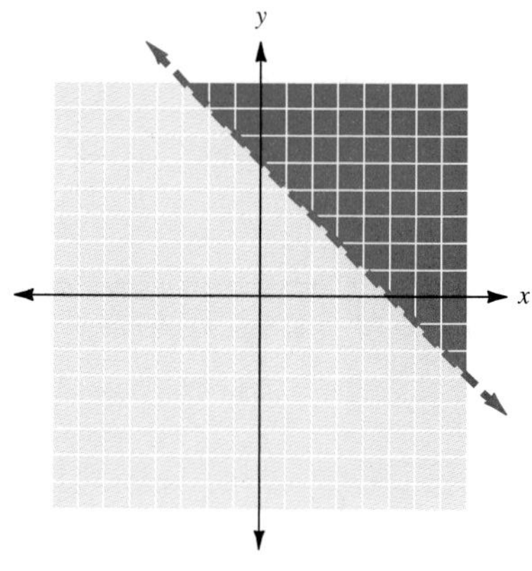

Substituting $\frac{1}{2}$ for x in the expression, we get

$$\left(\frac{1}{2}\right)^3 - 3\left(\frac{1}{2}\right)^2 + \left(\frac{1}{2}\right) + 5 = \frac{1}{8} - 3\left(\frac{1}{4}\right) + \frac{1}{2} + 5$$
$$= \frac{1}{8} - \frac{3}{4} + \frac{1}{2} + 5$$
$$= \frac{1}{8} - \frac{6}{8} + \frac{4}{8} + 5$$
$$= -\frac{1}{8} + 5$$
$$= 4\frac{7}{8} \text{ or } \frac{39}{8}$$

CHECK YOURSELF 2

Given $f(x) = 2x^3 - x^2 + 3x - 2$, find the following.

(a) $f(0)$ **(b)** $f(3)$ **(c)** $f\left(-\frac{1}{2}\right)$

Given a function f, the pair of numbers $(x, f(x))$ is very significant. We always write them in that order, hence the name *ordered pairs.* In Example 2, part a, we saw that given $f(x) = x^3 - 3x^2 + x + 5, f(0) = 5$, which meant that the ordered pair $(0, 5)$ was associated with the function. The ordered pair consists of the x value first and the function value at that x (the $f(x)$) second.

Example 3

Finding Ordered Pairs

Given the function $f(x) = 2x^2 - 3x + 5$, find the ordered pair $(x, f(x))$ associated with each given value for x.

(a) $x = 0$

$f(0) = 5$

so the ordered pair is $(0, 5)$.

(b) $x = -1$

$f(-1) = 2(-1)^2 - 3(-1) + 5 = 10$

The ordered pair is $(-1, 10)$.

(c) $x = \frac{1}{4}$

$$f\left(\frac{1}{4}\right) = 2\left(\frac{1}{16}\right) - 3\left(\frac{1}{4}\right) + 5 = \frac{35}{8}$$

The ordered pair is $\left(\frac{1}{4}, \frac{35}{8}\right)$.

CHECK YOURSELF 3

Given $f(x) = 2x^3 - x^2 + 3x - 2$, find the ordered pair associated with each given value of x.

(a) $x = 0$ **(b)** $x = 3$ **(c)** $x = -\frac{1}{2}$

In Chapters 6 and 7, we have discussed the graph of a linear equation. We saw that the graph for a vertical line had the form $x = a$. The equation for such a line cannot be rewritten as a function, but the equation for any nonvertical line can be written as a function.

Example 4

Writing Equations as Functions

Rewrite each linear equation as a function of x.

(a) $y = 3x - 4$

This can be rewritten as

$$f(x) = 3x - 4$$

(b) $2x - 3y = 6$

We must first solve the equation for y (recall that this will give us the slope-intercept form).

$$-3y = -2x + 6$$

$$y = \frac{2}{3}x - 2$$

This can be rewritten as

$$f(x) = \frac{2}{3}x - 2$$

CHECK YOURSELF 4

Rewrite each equation as a function of x.

(a) $y = -2x + 5$ **(b)** $3x + 5y = 15$

The process of finding the graph of a linear function is identical to the process of finding the graph of a linear equation.

Example 5

Graphing a Linear Function

Graph the function

$$f(x) = 3x - 5$$

We could use the slope and y intercept to graph the line, or we can find three points (the third is a checkpoint) and draw the line through them. We will do the latter.

$f(0) = -5 \qquad f(1) = -2 \qquad f(2) = 1$

We will use the three points $(0, -5)$, $(1, -2)$, and $(2, 1)$ to graph the line.

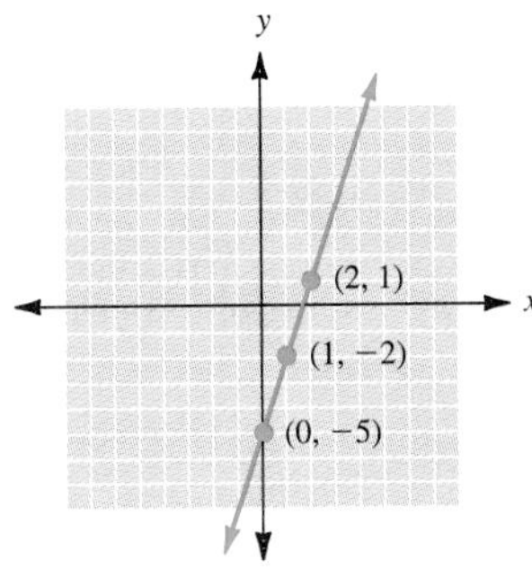

CHECK YOURSELF 5

Graph the function

$f(x) = 5x - 3$

One benefit of having a function written in $f(x)$ form is that it makes it fairly easy to substitute values for x. In Example 5, we substituted the values 0, 1, and 2. Sometimes it is useful to substitute nonnumeric values for x.

Example 6

Substituting Nonnumeric Values for x

Let $f(x) = 2x + 3$. Evaluate f as indicated.

(a) $f(a)$

Substituting a for x in our equation, we see that

$f(a) = 2a + 3$

(b) $f(2 + h)$

Substituting $2 + h$ for x in our equation, we get

$f(2 + h) = 2(2 + h) + 3$

Distributing the 2, then simplifying, we have

$$\begin{aligned} f(2 + h) &= 4 + 2h + 3 \\ &= 2h + 7 \end{aligned}$$

CHECK YOURSELF 6

Let f(x) = 4x − 2. Evaluate f as indicated.

(a) $f(b)$ **(b)** $f(4 + h)$

CHECK YOURSELF ANSWERS

1. **(a)** 1; **(b)** 3; **(c)** -33 **2.** **(a)** -2; **(b)** 52; **(c)** -4

3. **(a)** $(0, -2)$; **(b)** $(3, 52)$; **(c)** $\left(-\frac{1}{2}, -4\right)$ **4.** **(a)** $f(x) = -2x + 5$;

(b) $f(x) = -\frac{3}{5}x + 3$ **5.**

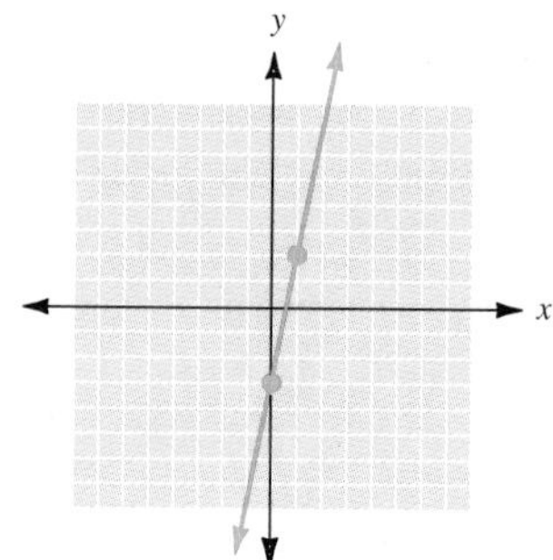

6. **(a)** $4b - 2$; **(b)** $4h + 14$

Name ____________

Section ______ Date ______

7.5 Exercises

In exercises 1 to 10, evaluate each function for the value specified.

1. $f(x) = x^2 - x - 2$; find (a) $f(0)$, (b) $f(-2)$, and (c) $f(1)$.

2. $f(x) = x^2 - 7x + 10$; find (a) $f(0)$, (b) $f(5)$, and (c) $f(-2)$.

3. $f(x) = 3x^2 + x - 1$; find (a) $f(-2)$, (b) $f(0)$, and (c) $f(1)$.

4. $f(x) = -x^2 - x - 2$; find (a) $f(-1)$, (b) $f(0)$, and (c) $f(2)$.

5. $f(x) = x^3 - 2x^2 + 5x - 2$; find (a) $f(-3)$, (b) $f(0)$, and (c) $f(1)$.

6. $f(x) = -2x^3 + 5x^2 - x - 1$; find (a) $f(-1)$, (b) $f(0)$, and (c) $f(2)$.

7. $f(x) = -3x^3 + 2x^2 - 5x + 3$; find (a) $f(-2)$, (b) $f(0)$, and (c) $f(3)$.

8. $f(x) = -x^3 + 5x^2 - 7x - 8$; find (a) $f(-3)$, (b) $f(0)$, and (c) $f(2)$.

9. $f(x) = 2x^3 + 4x^2 + 5x + 2$; find (a) $f(-1)$, (b) $f(0)$, and (c) $f(1)$.

10. $f(x) = -x^3 + 2x^2 - 7x + 9$; find (a) $f(-2)$, (b) $f(0)$, and (c) $f(2)$.

In exercises 11 to 20, rewrite each equation as a function of x.

11. $y = -3x + 2$
12. $y = 5x + 7$
13. $y = 4x - 8$
14. $y = -7x - 9$
15. $3x + 2y = 6$
16. $4x + 3y = 12$
17. $-2x + 6y = 9$
18. $-3x + 4y = 11$
19. $-5x - 8y = -9$
20. $4x - 7y = -10$

In exercises 21 to 26, graph the functions.

21. $f(x) = 3x + 7$

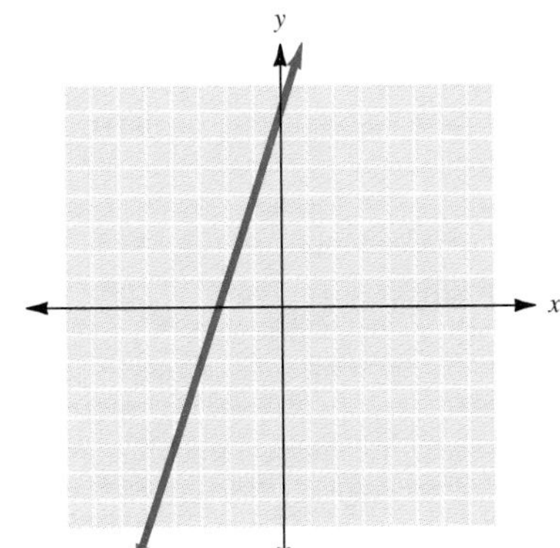

22. $f(x) = -2x - 5$

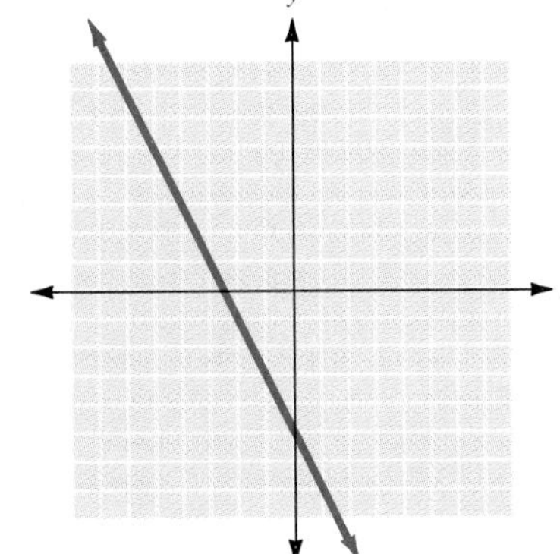

ANSWERS

1. (a) −2; (b) 4; (c) −2
2. (a) 10; (b) 0; (c) 28
3. (a) 9; (b) −1; (c) 3
4. (a) −2; (b) −2; (c) −8
5. (a) −62; (b) −2; (c) 2
6. (a) 7; (b) −1; (c) 1
7. (a) 45; (b) 3; (c) −75
8. (a) 85; (b) −8; (c) −10
9. (a) −1; (b) 2; (c) 13
10. (a) 39; (b) 9; (c) −5
11. $f(x) = -3x + 2$
12. $f(x) = 5x + 7$
13. $f(x) = 4x - 8$
14. $f(x) = -7x - 9$
15. $f(x) = -\frac{3}{2}x + 3$
16. $f(x) = -\frac{4}{3}x + 4$
17. $f(x) = \frac{1}{3}x + \frac{3}{2}$
18. $f(x) = \frac{3}{4}x + \frac{11}{4}$
19. $f(x) = -\frac{5}{8}x + \frac{9}{8}$
20. $f(x) = \frac{4}{7}x + \frac{10}{7}$
21. See exercise
22. See exercise

ANSWERS

23. See exercise
24. See exercise
25. See exercise
26. See exercise
27. 17
28. −3
29. 13
30. −7
31. −19
32. −1
33. $5a - 1$
34. $10r - 1$
35. $5x + 4$
36. $5a - 11$
37. $5x + 5h - 1$
38. 5

23. $f(x) = -2x + 7$

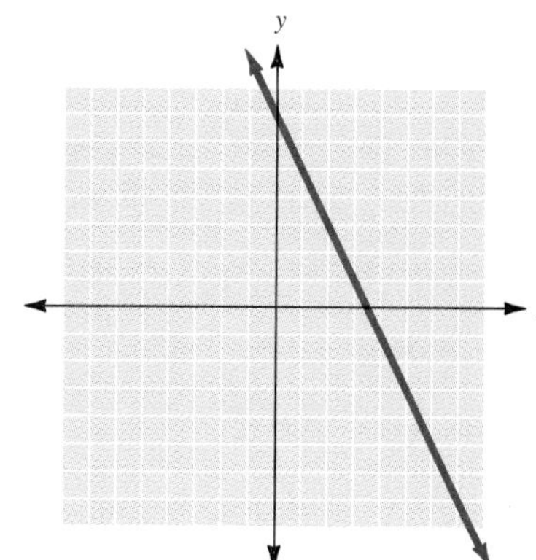

24. $f(x) = -3x + 8$

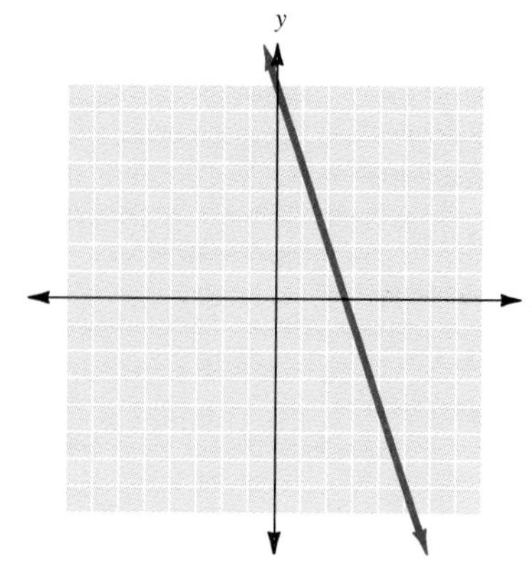

25. $f(x) = -x - 1$

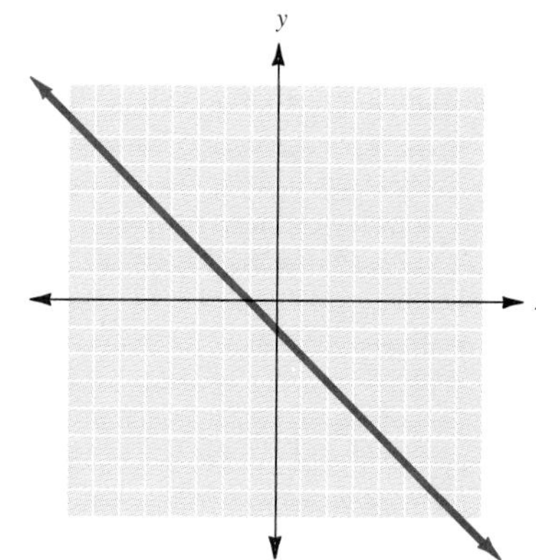

26. $f(x) = -2x - 5$

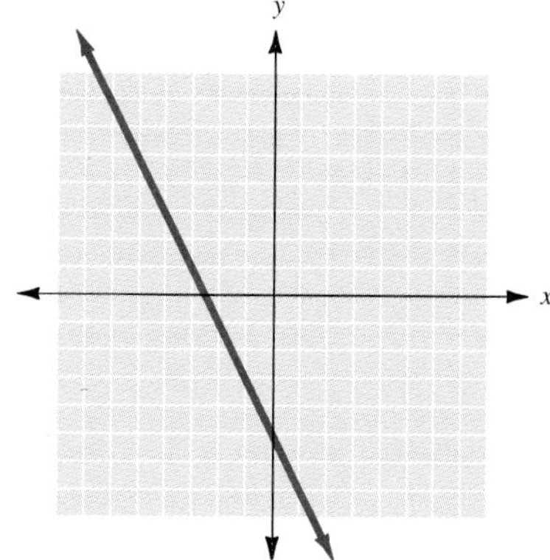

In exercises 27 to 32, if $f(x) = 4x - 3$, find the following:

27. $f(5)$ **28.** $f(0)$ **29.** $f(4)$

30. $f(-1)$ **31.** $f(-4)$ **32.** $f\left(\frac{1}{2}\right)$

In exercises 33 to 38, if $f(x) = 5x - 1$, find the following:

33. $f(a)$ **34.** $f(2r)$ **35.** $f(x + 1)$

36. $f(a - 2)$ **37.** $f(x + h)$ **38.** $\dfrac{f(x + h) - f(x)}{h}$

In exercises 39 to 42, if $g(x) = -3x + 2$, find the following:

39. $g(m)$ **40.** $g(5n)$ **41.** $g(x + 2)$ **42.** $g(s - 1)$

In exercises 43 to 46, let $f(x) = 2x + 3$.

43. Find $f(1)$. **44.** Find $f(3)$.

45. Form the ordered pairs $(1, f(1))$ and $(3, f(3))$.

46. Write the equation of the line passing through the points determined by the ordered pairs in exercise 45.

47. Let $f(x) = 5x - 2$. Find (a) $f(4) - f(3)$; (b) $f(9) - f(8)$; (c) $f(12) - f(11)$. (d) How do the results of (a) through (c) compare to the slope of the line that is the graph of f?

48. Repeat exercise 47 with $f(x) = 7x + 1$.

49. Repeat exercise 47 with $f(x) = mx + b$.

Answers

1. (a) -2, (b) 4, (c) -2 **3.** (a) 9, (b) -1, (c) 3
5. (a) -62, (b) -2, (c) 2 **7.** (a) 45, (b) 3, (c) -75
9. (a) -1, (b) 2, (c) 13 **11.** $f(x) = -3x + 2$ **13.** $f(x) = 4x - 8$
15. $f(x) = -\frac{3}{2}x + 3$ **17.** $f(x) = \frac{1}{3}x + \frac{3}{2}$ **19.** $f(x) = -\frac{5}{8}x + \frac{9}{8}$
21. $f(x) = 3x + 7$ **23.** $f(x) = -2x + 7$

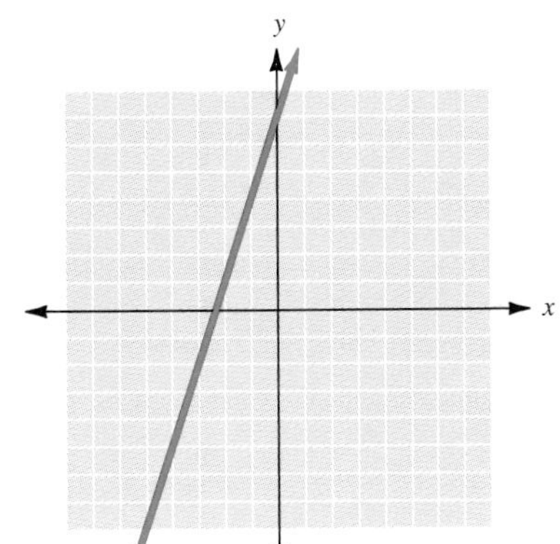

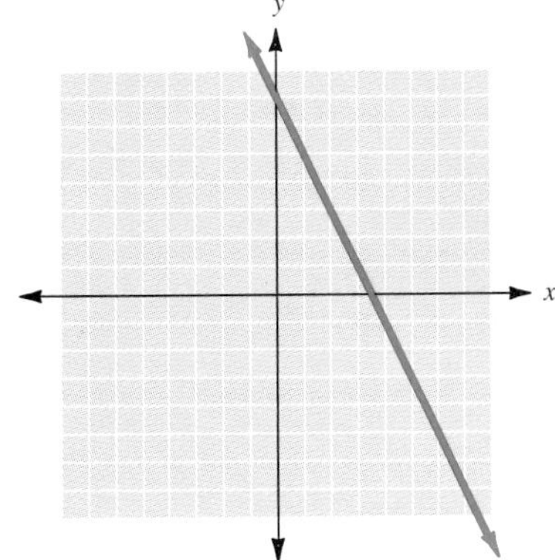

ANSWERS

39. $-3m + 2$
40. $-15n + 2$
41. $-3x - 4$
42. $-3s + 5$
43. 5
44. 9
45. (1, 5), (3, 9)
46. $y = 2x + 3$
47. (a) 5; (b) 5; (c) 5; (d) same
48. (a) 7; (d) 7; (c) 7; (d) same
49. (a) m; (b) m; (c) m; (d) same

25. $f(x) = -x - 1$

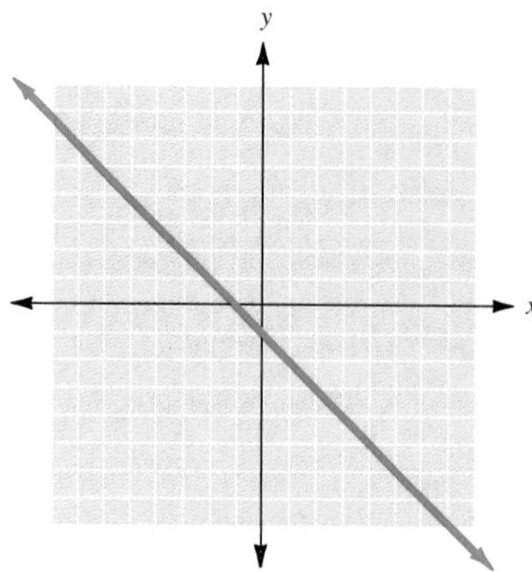

27. 17 **29.** 13 **31.** -19
33. $5a - 1$ **35.** $5x + 4$ **37.** $5x + 5h - 1$ **39.** $-3m + 2$
41. $-3x - 4$ **43.** 5 **45.** (1, 5), (3, 9) **47.** (a) 5, (b) 5, (c) 5, (d) same
49. (a) m, (b) m, (c) m, (d) same

Summary

Definition/Procedure	Example	Reference
The Slope-Intercept Form		**Section 7.1**
The slope-intercept form for the equation of a line is $$y = mx + b$$ in which the line has slope m and y intercept $(0, b)$.	For the equation $$y = \frac{2}{3}x - 3$$ the slope m is $\frac{2}{3}$ and b, which determines the y intercept, is -3.	**p. 555**
Parallel and Perpendicular Lines		**Section 7.2**
Two lines are parallel if they have the same slope, so $$m_1 = m_2$$	$y = 3x - 5$ and $y = 3x + 2$ are parallel	**p. 568**
Two lines are perpendicular if their slopes are negative reciprocals, i.e., when $$m_1 \cdot m_2 = -1$$	$y = 5x + 2$ and $y = -\frac{1}{5}x - 3$ are perpendicular	**p. 569**
The Point-Slope Form		**Section 7.3**
The equation of a line with slope m that passes through point (a, b) is $$y - b = m(x - a)$$	The line with slope $\frac{1}{3}$ passing through (4, 3) has the equation $$y - 3 = \frac{1}{3}(x - 4)$$	**p. 577**
Graphing Linear Inequalities		**Section 7.4**
The Graphing Steps **1.** Replace the inequality symbol with an equals sign to form the equation of the boundary line of the graph. **2.** Graph the boundary line. Use a dotted line if equality is not included ($<$ or $>$). Use a solid line if equality is included ($\le$ or $\ge$). **3.** Choose any convenient test point not on the line. **4.** If the inequality is *true* at the checkpoint, shade the half plane including the test point. If the inequality is *false* at the checkpoint, shade the half plane that does not include the checkpoint.	To graph $x - 2y < 4$: $x - 2y = 4$ is the boundary line. Using (0, 0) as the checkpoint, we have $0 - 2 \cdot 0 < 4$ $0 < 4$ (True) Shade the half plane that includes (0, 0).	**p. 588**

Continued

DEFINITION/PROCEDURE	EXAMPLE	REFERENCE
An Introduction to Functions		**Section 7.5**
Given a function f, $f(c)$ designates the value of the function when the variable is equal to c.	$f(x) = 2x^3 - x^2 + 1$ $f(-2) = 2(-2)^3 - (-2)^2 + 1$ $= 2(-8) - (4) + 1$ $= -19$	p. 600

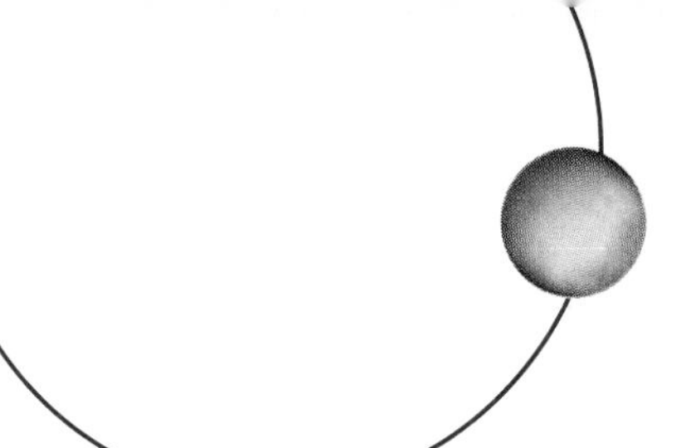

Summary Exercises

This summary exercise set is provided to give you practice with each of the objectives of the chapter. Each exercise is keyed to the appropriate chapter section. The answers are provided in the instructor's manual.

[7.1] Find the slope of the line through each of the following pairs of points.

1. (3, 4) and (5, 8) 2

2. (−2, 3) and (1, −6) −3

3. (−2, 5) and (2, 3) $-\frac{1}{2}$

4. (−5, −2) and (1, 2) $\frac{2}{3}$

5. (−2, 6) and (5, 6) 0

6. (−3, 2) and (−1, −3) $-\frac{5}{2}$

7. (−3, −6) and (5, −2) $\frac{1}{2}$

8. (−6, −2) and (−6, 3) Undefined

[7.1] Find the slope and *y* intercept of the line represented by each of the following equations.

9. $y = 2x + 5$ slope: 2; *y* intercept: (0, 5)

10. $y = -4x - 3$ slope: −4; *y* intercept: (0, −3)

11. $y = -\frac{3}{4}x$ slope: $-\frac{3}{4}$; *y* intercept: (0, 0)

12. $y = \frac{2}{3}x + 3$ slope: $\frac{2}{3}$; *y* intercept: (0, 3)

13. $2x + 3y = 6$ slope: $-\frac{2}{3}$; *y* intercept: (0, 2)

14. $5x - 2y = 10$ slope: $\frac{5}{2}$; *y* intercept: (0, −5)

15. $y = -3$ slope: 0; *y* intercept: (0, −3)

16. $x = 2$ Undefined slope, no *y* intercept

[7.1] Write the equation of the line with the given slope and *y* intercept. Then graph each line, *using* the slope and *y* intercept.

17. Slope = 2; *y* intercept: (0, 3) $y = 2x + 3$

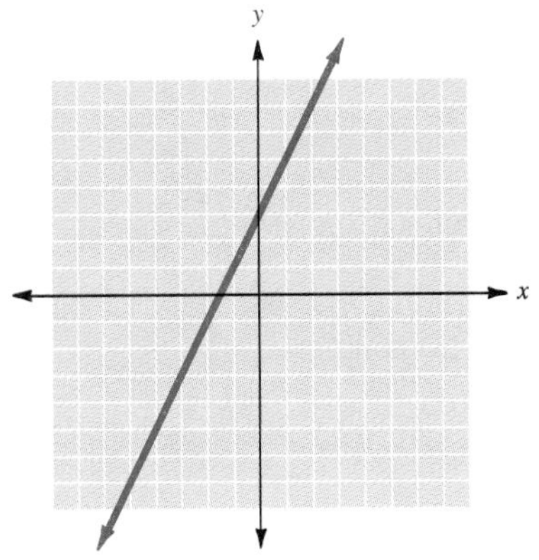

18. Slope = $\frac{3}{4}$; *y* intercept: (0, −2) $y = \frac{3}{4}x - 2$

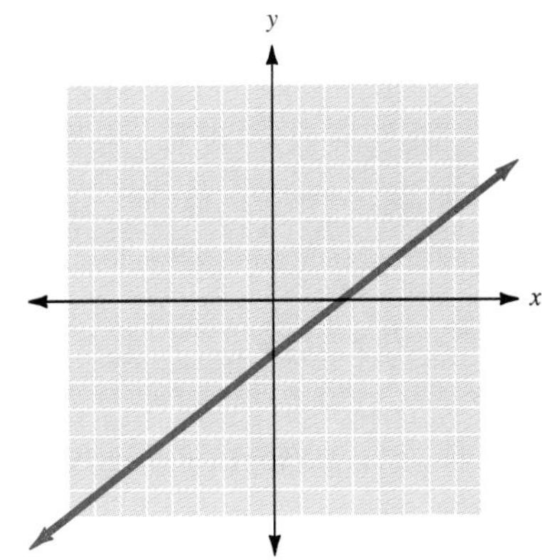

19. Slope $= -\frac{2}{3}$; y intercept $(0, 2)$ $y = -\frac{2}{3}x + 2$

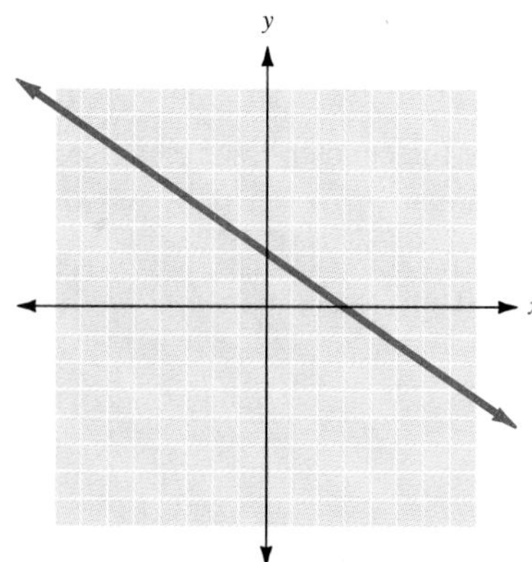

[7.2] Are the following pairs of lines parallel, perpendicular, or neither?

20. L_1 through $(-3, -2)$ and $(1, 3)$
L_2 through $(0, 3)$ and $(4, 8)$ parallel

21. L_1 through $(-4, 1)$ and $(2, -3)$
L_2 through $(0, -3)$ and $(2, 0)$ perpendicular

22. L_1 with equation $x + 2y = 6$
L_2 with equation $x + 3y = 9$ neither

23. L_1 with equation $4x - 6y = 18$
L_2 with equation $2x - 3y = 6$ parallel

[7.3] Write the equation of the line passing through each of the following points with the indicated slope. Give your results in slope-intercept form, where possible.

24. $(0, -5)$, $m = \frac{2}{3}$ $y = \frac{2}{3}x - 5$

25. $(0, -3)$, $m = 0$ $y = -3$

26. $(2, 3)$, $m = 3$ $y = 3x - 3$

27. $(4, 3)$, m undefined $x = 4$

28. $(3, -2)$, $m = \frac{5}{3}$ $y = \frac{5}{3}x - 7$

29. $(-2, -3)$, $m = 0$ $y = -3$

30. $(-2, -4)$, $m = -\frac{5}{2}$ $y = -\frac{5}{2}x - 9$

31. $(-3, 2)$, $m = -\frac{4}{3}$ $y = -\frac{4}{3}x - 2$

32. $\left(\frac{2}{3}, -5\right)$, $m = 0$ $y = -5$

33. $\left(-\frac{5}{2}, -1\right)$, m is undefined $x = -\frac{5}{2}$

[7.3] Write the equation of the line L satisfying each of the following sets of geometric conditions.

34. L passes through $(-3, -1)$ and $(3, 3)$. $y = \frac{2}{3}x + 1$

35. L passes through $(0, 4)$ and $(5, 3)$. $y = -\frac{1}{5}x + 4$

36. L has slope $\frac{3}{4}$ and y intercept $(0, 3)$. $y = \frac{3}{4}x + 3$

37. L passes through $(4, -3)$ with a slope of $-\frac{5}{4}$. $y = -\frac{5}{4}x + 2$

38. L has y intercept $(0, -4)$ and is parallel to the line with equation $3x - y = 6$. $y = 3x - 4$

39. L passes through $(3, -2)$ and is perpendicular to the line with equation $3x - 5y = 15$. $y = -\frac{5}{3}x + 3$

40. L passes through $(2, -1)$ and is perpendicular to the line with the equation $3x - 2y = 5$. $y = -\frac{2}{3}x + \frac{1}{3}$

41. L passes through the point $(-5, -2)$ and is parallel to the line with the equation $4x - 3y = 9$. $y = \frac{4}{3}x + \frac{14}{3}$

[7.4] Graph each of the following inequalities.

42. $x + y \le 4$

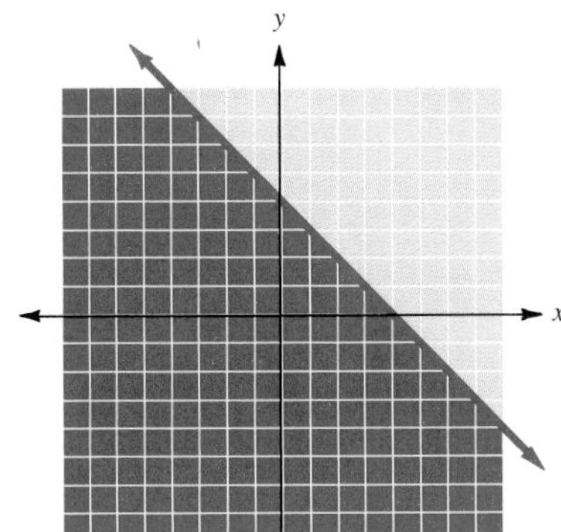

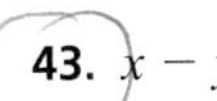

43. $x - y > 5$

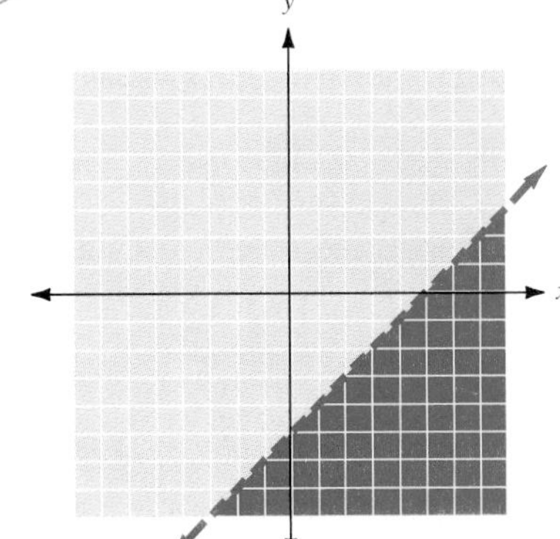

44. $2x + y < 6$

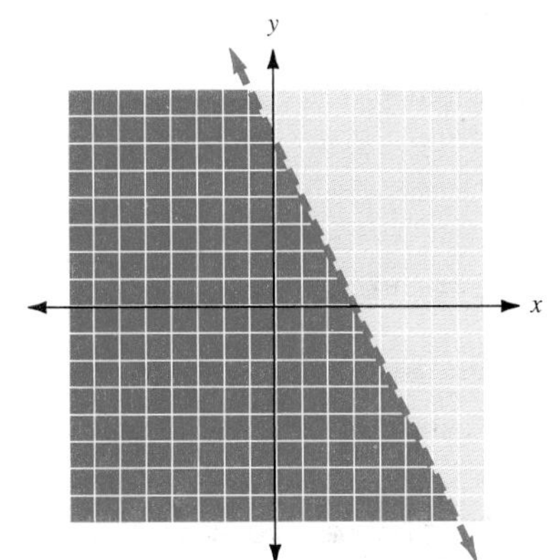

45. $2x - y \ge 6$

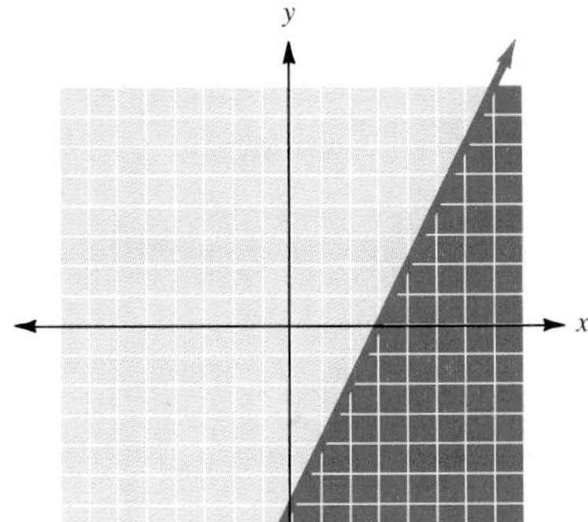

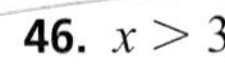

46. $x > 3$

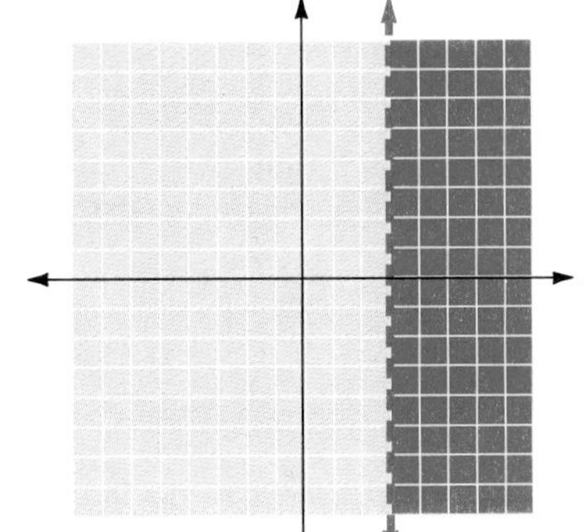

47. $y \le 2$

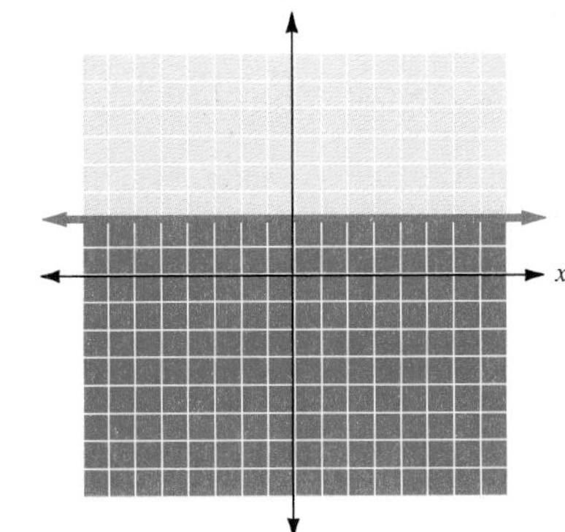

[7.5] In exercises 48 to 53, evaluate $f(x)$ for the value specified.

48. $f(x) = x^2 - 3x + 5$; find (a) $f(0)$, (b) $f(-1)$, and (c) $f(1)$. (a) 5 (b) 9 (c) 3

49. $f(x) = -2x^2 + x - 7$; find (a) $f(0)$, (b) $f(2)$, and (c) $f(-2)$. (a) −7 (b) −13 (c) −17

50. $f(x) = x^3 - x^2 - 2x + 5$; find (a) $f(-1)$, (b) $f(0)$, and (c) $f(2)$. (a) 5 (b) 5 (c) 5

51. $f(x) = -x^2 + 7x - 9$; find (a) $f(-3)$, (b) $f(0)$, and (c) $f(1)$. (a) −39 (b) −9 (c) −3

52. $f(x) = 3x^2 - 5x + 1$; find (a) $f(-1)$, (b) $f(0)$, and (c) $f(2)$. (a) 9 (b) 1 (c) 3

53. $f(x) = x^3 + 3x - 5$; find (a) $f(2)$, (b) $f(0)$, and (c) $f(1)$. (a) 9 (b) −5 (c) −1

[7.5] In exercises 54 to 57, rewrite each equation as a function of x.

54. $y = 4x + 7$ $f(x) = 4x + 7$

55. $y = -7x - 3$ $f(x) = -7x - 3$

56. $4x + 5y = 40$ $f(x) = -\frac{4}{5}x + 8$

57. $-3x - 2y = 12$ $f(x) = -\frac{3}{2}x - 6$

[7.5] In exercises 58 to 63, graph the function.

58. $f(x) = 2x + 3$

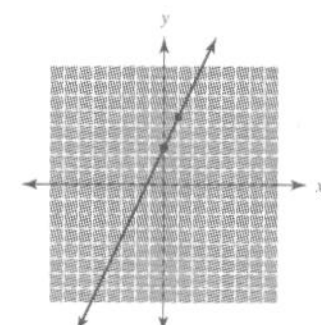

59. $f(x) = 3x - 6$

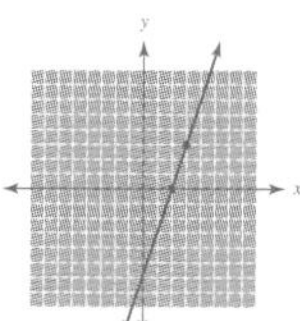

60. $f(x) = -5x + 6$

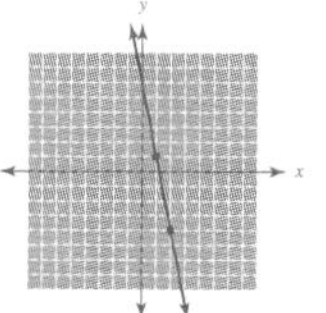

61. $f(x) = -x + 3$

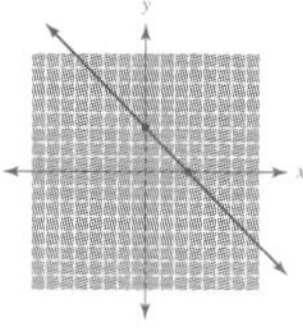

62. $f(x) = -3x - 2$

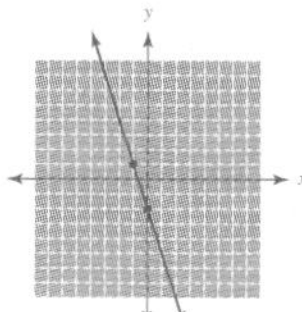

63. $f(x) = -2x + 6$

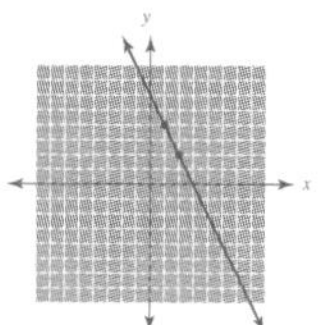

In exercises 64 to 69, evaluate each function as indicated.

64. $f(x) = 5x + 3$; find $f(2)$ and $f(0)$. 13, 3

65. $f(x) = -3x + 5$; find $f(0)$ and $f(1)$. 5, 2

66. $f(x) = 7x - 5$; find $f\left(\frac{5}{4}\right)$ and $f(-1)$. $\frac{15}{4}$, −12

67. $f(x) = -2x + 5$; find $f(0)$ and $f(-2)$. 5, 9

68. $f(x) = -5x + 3$; find $f(a)$, $f(2b)$, and $f(x + 2)$. $5a + 3, -10b + 3, -5x - 7$

69. $f(x) = 7x - 1$; find $f(a)$, $f(3b)$, and $f(x - 1)$. $7(a) - 1, 21b - 1, 7x - 8$

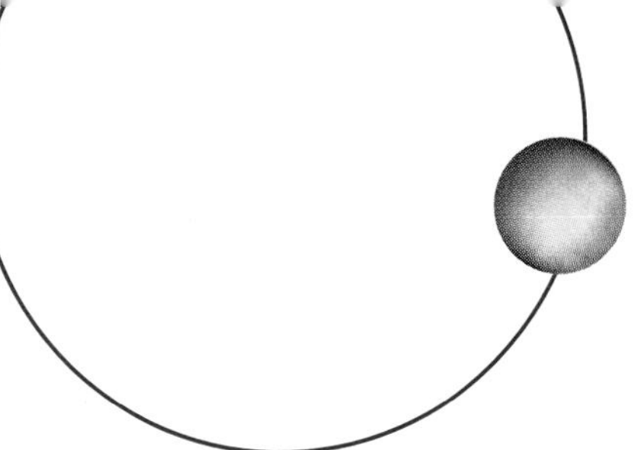

Self-Test for Chapter 7

Name ______________________

Section __________ Date __________

The purpose of this self-test is to help you check your progress and to review for a chapter test in class. Allow yourself about an hour to take the test. When you are done, check your answers in the back of the book. If you missed any answers, be sure to go back and review the appropriate sections in the chapter.

Find the slope of the line through the following pairs of points.

1. $(-3, 5)$ and $(2, 10)$

2. $(-2, 6)$ and $(2, 9)$

Find the slope and y intercept of the line represented by each of the following equations.

3. $4x - 5y = 10$

4. $y = -\frac{2}{3}x - 9$

Write the equation of the line with the given slope and y intercept. Then graph each line.

5. Slope -3 and y intercept $(0, 6)$

6. Slope 5 and y intercept $(0, -3)$

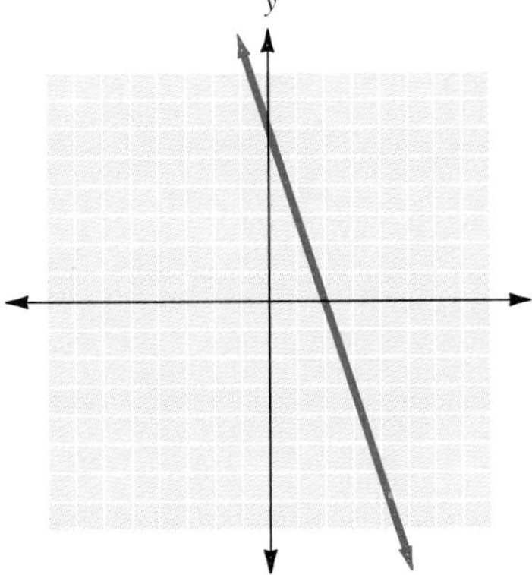

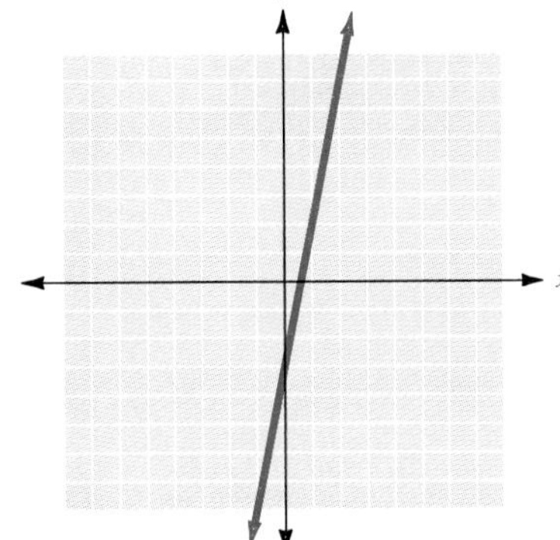

Determine if the following pairs of lines are parallel, perpendicular, or neither.

7. L_1 through $(2, 5)$ and $(4, 9)$ L_2 through $(-7, 1)$ and $(-2, 11)$

8. L_1 with equation $y = 5x - 8$ L_2 with equation $5y + x = 3$

9. Find the equation of the line through $(-5, 8)$ and perpendicular to the line $4x + 2y = 8$.

Write the equation of the line L satisfying each of the following sets of geometric conditions.

10. L passes through the points $(-4, 3)$ and $(-1, 7)$.

11. L passes through $(-3, 7)$ and is perpendicular to $2x - 3y = 7$.

12. L has y intercept $(0, 5)$ and is parallel to $4x + 6y = 12$.

13. L has y intercept $(0, -8)$ and is parallel to the x axis.

ANSWERS

1. 1
2. $\frac{3}{4}$
3. Slope: $\frac{4}{5}$; y intercept: $(0, -2)$
4. Slope: $-\frac{2}{3}$; y intercept: $(0, -9)$
5. $y = -3x + 6$
6. $y = 5x - 3$
7. Parallel
8. Perpendicular
9. $y = \frac{1}{2}x + \frac{21}{2}$
10. $y = \frac{4}{3}x + \frac{25}{3}$
11. $y = -\frac{3}{2}x + \frac{5}{2}$
12. $y = -\frac{2}{3}x + 5$
13. $y = -8$

ANSWERS

14. See exercise

15. See exercise

16. See exercise

17. −5; 7

18. −15, −7

19. −15, 6

20. 3a − 25; 3x − 28

Graph each of the following inequalities.

14. $x + y < 3$

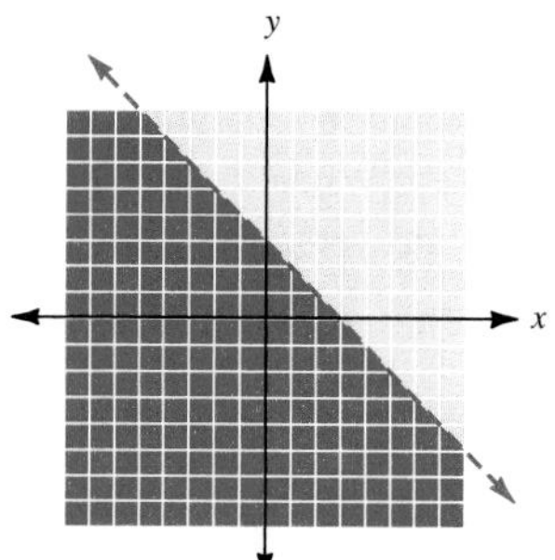

15. $3x + y \geq 9$

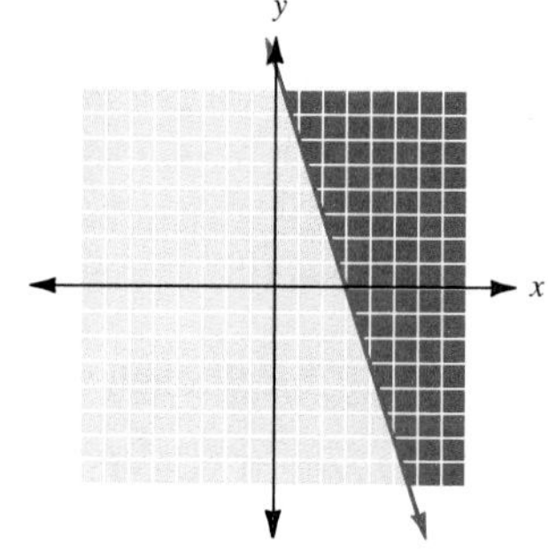

16. $x \leq 7$

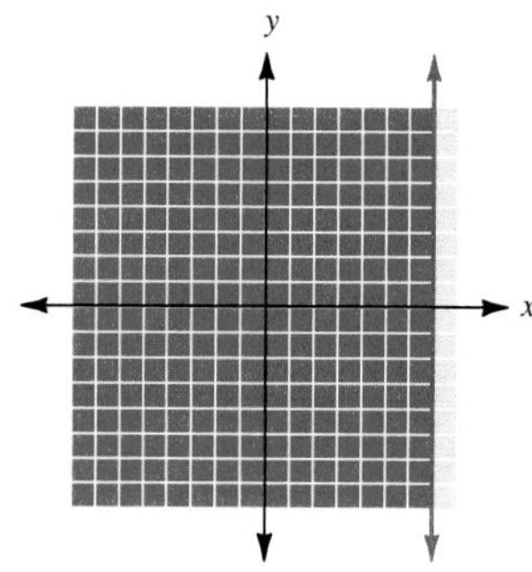

In each of the following, evaluate $f(x)$ for the value given.

17. $f(x) = x^2 - 4x - 5$; find $f(0)$ and $f(-2)$

18. $f(x) = -x^3 + 5x - 3x^2 - 8$; find $f(-1)$ and $f(1)$

19. $f(x) = -7x - 15$; find $f(0)$ and $f(-3)$

20. $f(x) = 3x - 25$; find $f(a)$ and $f(x - 1)$

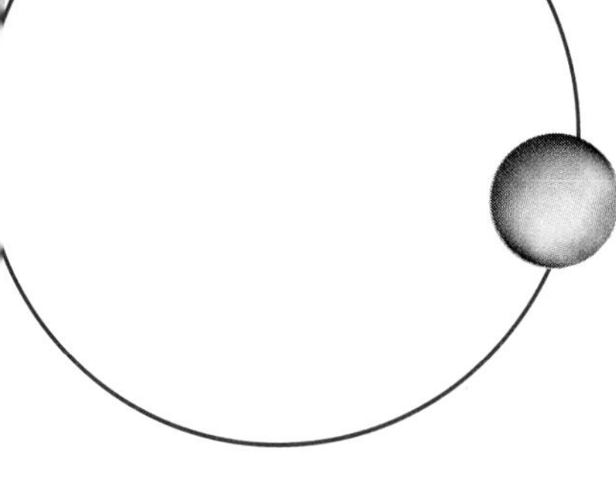

Cumulative Test for Chapters 0 to 7

Name ________________

Section ________ Date ________

This review covers selected topics from the first eight chapters.

Perform the indicated operation.

1. $3x^2y^2 - 5xy - 2x^2y^2 + 2xy$

2. $\dfrac{36m^5n^2}{27m^2n}$

3. $(x^2 - 3x + 5) - (x^2 - 2x - 4)$

4. $(5z^2 - 3z) - (2z^2 - 5)$

Multiply.

5. $(2x - 3)(x + 7)$

6. $(2a - 2b)(a + 4b)$

Divide

7. $(x^2 + 3x + 2) \div (x - 3)$

8. $(x^4 - 2x) \div (x + 2)$

Solve each equation and check your results.

9. $5x - 2 = 2x - 6$

10. $3(x - 2) = 2(3x + 1) - 2$

Factor each polynomial completely.

11. $x^2 - x - 56$

12. $4x^3y - 2x^2y^2 + 8x^4y$

13. $8a^3 - 18ab^2$

14. $15x^2 - 21xy + 6y^2$

Find the slope of the line through the following pairs of points.

15. $(2, -4)$ and $(-3, -9)$

16. $(-1, 7)$ and $(3, -2)$

Perform the indicated operations.

17. $\dfrac{x^2 + 7x + 10}{x^2 + 5x} \cdot \dfrac{2x^2 - 7x + 6}{x^2 - 4}$

18. $\dfrac{2a^2 + 11a - 21}{a^2 - 49} \div (2a - 3)$

19. $\dfrac{5}{2m} + \dfrac{3}{m^2}$

20. $\dfrac{4}{x - 3} - \dfrac{2}{x}$

21. $\dfrac{3y}{y^2 + 5y + 4} + \dfrac{2y}{y^2 - 1}$

ANSWERS

1. $x^2y^2 - 2xy$
2. $\dfrac{4m^3n}{3}$
3. $-x + 9$
4. $3z^2 - 3z + 5$
5. $2x^2 + 11x - 21$
6. $2a^2 + 6ab - 8b^2$
7. $x + 6 + \dfrac{20}{x - 3}$
8. $x^3 - 2x^2 + 4x - 10 + \dfrac{20}{x - 2}$
9. $-\dfrac{4}{3}$
10. -2
11. $(x - 8)(x + 7)$
12. $2x^2y(2x - y + 4x^2)$
13. $2a(2a + 3b)(2a - 3b)$
14. $3(5x - 2y)(x - y)$
15. 1
16. $-\dfrac{9}{4}$
17. $\dfrac{2x - 3}{x}$
18. $\dfrac{1}{a - 7}$
19. $\dfrac{5m + 6}{2m^2}$
20. $\dfrac{2x + 6}{x(x - 3)}$
21. $\dfrac{5y}{(y + 4)(y - 1)}$

ANSWERS

22. −4

23. 7, −10

24. 9

25. 49 mi/h going
42 mi/h returning

26. 126 min

27. $y = -\frac{1}{7}x + 2$

28. $y = -5x + 3$

29. See exercise

30. −2

Solve the following equations for x.

22. $\frac{13}{4x} + \frac{3}{x^2} = \frac{5}{2x}$

23. $\frac{6}{x + 5} + 1 = \frac{3}{x - 5}$

Solve the following applications.

24. If the reciprocal of 4 times a number is subtracted from the reciprocal of that number, the result is $\frac{1}{12}$. What is the number?

25. Kyoko drove 280 mi to attend a business conference. In returning from the conference along a different route, the trip was only 240 mi, but traffic slowed her speed by 7 mi/h. If her driving time was the same both ways, what was her speed each way?

26. A laser printer can print 400 form letters in 30 min. At that rate, how long will it take the printer to complete a job requiring 1680 letters?

27. Write the equation of the line perpendicular to the line $7x - y = 15$ with y intercept of (0, 2).

28. Write the equation of the line with slope of −5 and y intercept (0, 3).

29. Graph the inequality $4x - 2y \geq 8$.

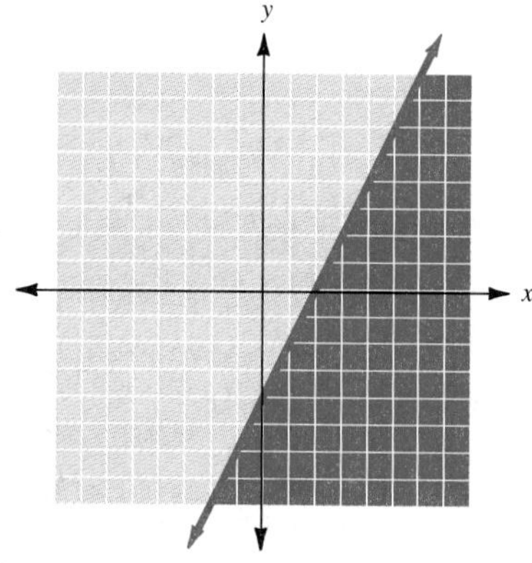

30. If $f(x) = x^2 + 3x$, find $f(-1)$.

SYSTEMS OF LINEAR EQUATIONS

8

INTRODUCTION

In the United States, almost all electricity is generated by the burning of the fossil fuels (coal, oil, or gas); by nuclear fission; or by water-powered turbines in hydroelectric dams. About 65% of the electric power we use comes from burning fossil fuels. Because of this dependence on a nonrenewable resource and concern over pollution caused by burning fossil fuels, there has been some urgency in developing ways to utilize other power sources. Some of the most promising projects have been in solar- and wind-generated energy.

Alternative sources of energy are expensive compared to the cost of the traditional methods of generating electricity described above. But alternative energy sources are looking more promising as the average price per kilowatt hour (kwh) of electric power sold to residential users has been going up—about \$0.0028 per kwh—since 1970. The costs of manufacturing and installing banks of wind turbines in windy locations have declined.

When will the cost of generating electricity for residential use using wind power be equal to or less than the cost of using the traditional energy mix? Economists use equations such as the following to make projections and then advise about the feasibility of investing in wind power plants for large cities:

$C_1 = \$0.054 + 0.0028t$

$C_2 = \$0.25 - 0.0035t$

in which C_1 and C_2 represent cost per kwh in 2000 and t is the time in years since 1970

C_1 = cost of present mix of energy sources*

C_2 = cost of wind-powered electricity

*Of course, the true cost of burning fossil fuels also includes the damage to the environment and people's health.

Name ______________________

Section ________ Date ________

ANSWERS

1. Inconsistent
2. (0, −2)
3. (6, −3)
4. Dependent system
5. (4, 1)
6. (3, −1)
7. 16, 24
8. 13 m, 17 m
9. 19 dimes, 26 quarters
10. canoe: 7.5 mi/h current: 1.5 mi/h
11. See exercise
12. See exercise

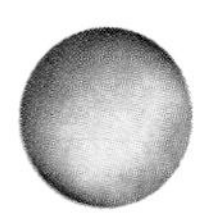

Pre-Test Chapter 8

Solve each of the following systems by graphing.

1. 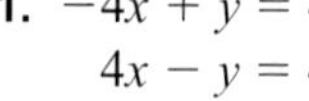
$-4x + y = 4$
$4x - y = 4$

2. 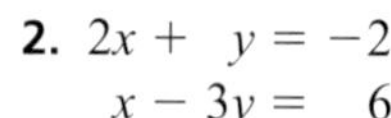
$2x + y = -2$
$x - 3y = 6$

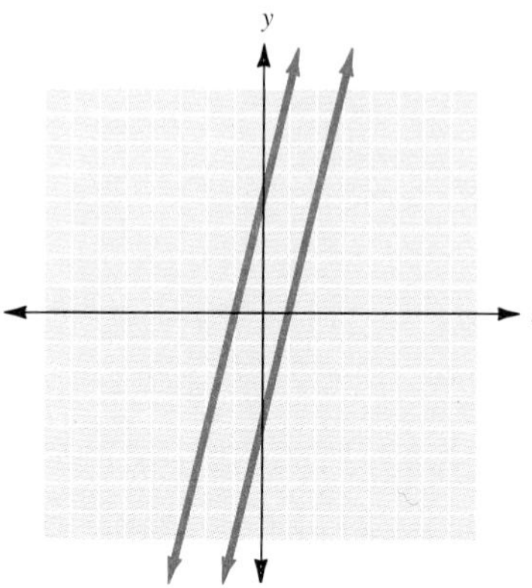

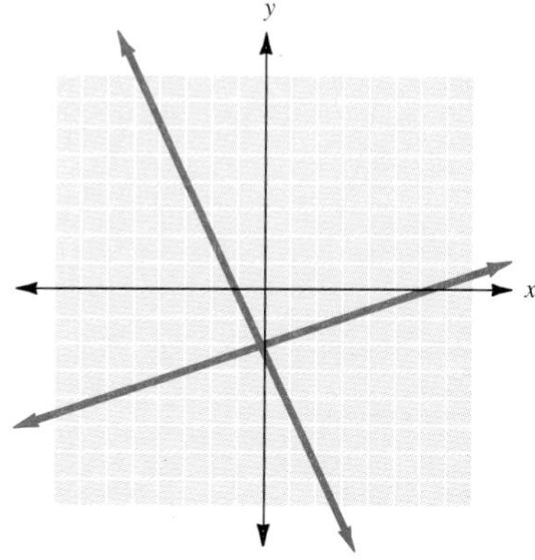

Solve each of the following systems.

3. $x - y = 9$
$x + y = 3$

4. $3x - 4y = 2$
$-6x + 8y = -4$

5. $x + y = 5$
$x = y + 3$

6. $x - 2y = 5$
$3x + y = 8$

Solve the following problems. Be sure to show the equations used.

7. The sum of two numbers is 40, and their difference is 8. Find the two numbers.

8. A rope 30 m long is cut into two pieces so that one piece is 4 m longer than the other. How long is each piece?

9. Nila has 45 coins with a value of $8.40. If the coins are all dimes and quarters, how many of each coin does he have?

10. Jackson was able to travel 36 mi downstream in 4 h. In returning upstream, it took 6 h to make the trip. How fast can his canoe travel in still water? What was the rate of the river current?

Solve each of the following linear inequalities graphically.

11. $x + 2y < 4$
$x - y < 5$

12. 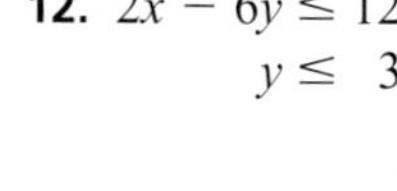
$2x - 6y \le 12$
$y \le 3$

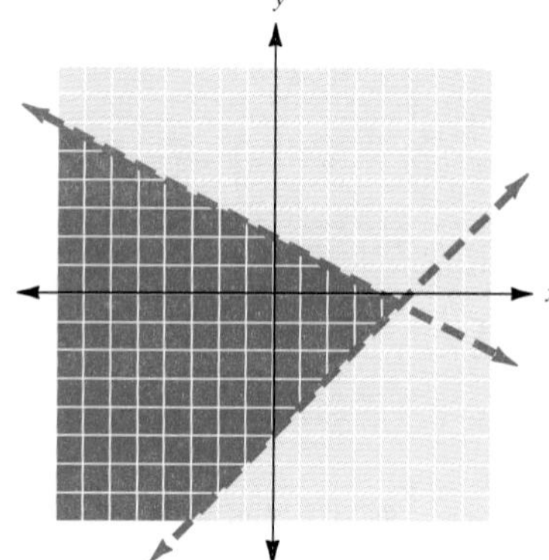

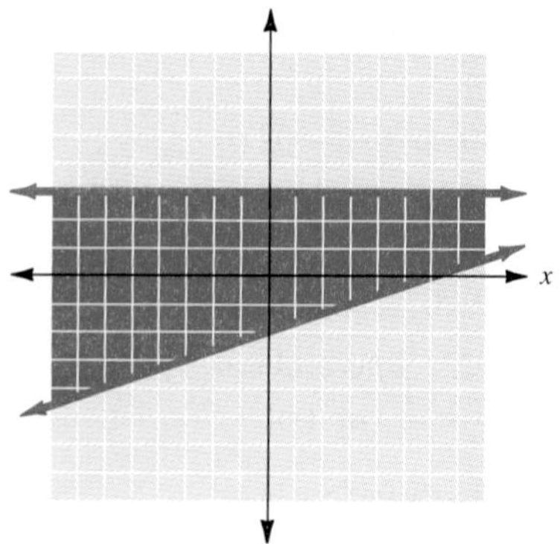

Systems of Linear Equations: Solving by Graphing

OBJECTIVE

1. Find the solution(s) for a set of linear equations by graphing

From our work in Section 6.1, we know that an equation of the form $x + y = 3$ is a linear equation. Remember that its graph is a straight line. Often we will want to consider two equations together. They then form a **system of linear equations.** An example of such a system is

$$x + y = 3$$
$$3x - y = 5$$

A solution for a linear equation in two variables is any ordered pair that satisfies the equation. Often there is just one ordered pair that satisfies both equations of a system. It is called the **solution for the system.** For instance, there is one solution for the system above, and it is (2, 1) because, replacing x with 2 and y with 1, we have

$x + y = 3$	$3x - y = 5$
$2 + 1 \stackrel{?}{=} 3$	$3 \cdot 2 - 1 \stackrel{?}{=} 5$
$3 = 3$	$6 - 1 \stackrel{?}{=} 5$
	$5 = 5$

NOTE There is no other ordered pair that satisfies both equations.

Because both statements are true, the ordered pair (2, 1) satisfies both equations.

One approach to finding the solution for a system of linear equations is the **graphical method.** To use this, we graph the two lines on the same coordinate system. The coordinates of the point where the lines intersect is the solution for the system.

Example 1

Solving by Graphing

Solve the system by graphing.

$$x + y = 6$$
$$x - y = 4$$

NOTE Use the intercept method to graph each equation.

First, we determine solutions for the equations of our system. For $x + y = 6$, two solutions are (6, 0) and (0, 6). For $x - y = 4$, two solutions are (4, 0) and (0, −4). Using these intercepts, we graph the two equations. The lines intersect at the point (5, 1).

NOTE By substituting 5 for x and 1 for y into the two original equations, we can check that (5, 1) is indeed the solution for our system.

$x + y = 6$	$x - y = 4$
$5 + 1 \stackrel{?}{=} 6$	$5 - 1 \stackrel{?}{=} 4$
$6 = 6$	$4 = 4$

Both statements must be true for (5, 1) to be a solution for the system.

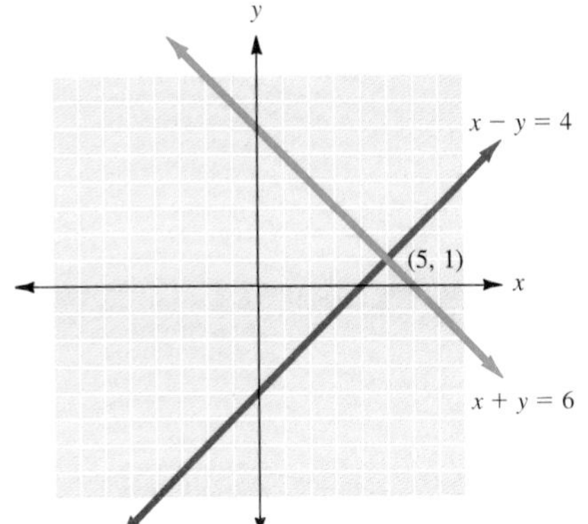

(5, 1) is the solution of the system. It is the only point that lies on both lines.

✓ CHECK YOURSELF 1

Solve the system by graphing.

$$2x - y = 4$$
$$x + y = 5$$

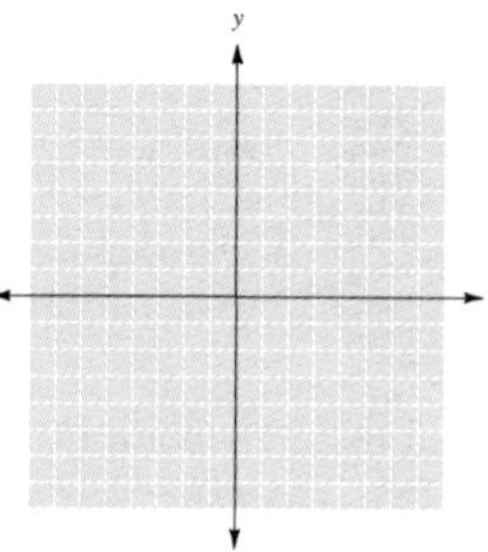

Example 2 shows how to graph a system when one of the equations represents a horizontal line.

Example 2

Solving by Graphing

Solve the system by graphing.

$$3x + 2y = 6$$
$$y = 6$$

For $3x + 2y = 6$, two solutions are (2, 0) and (0, 3). These represent the x and y intercepts of the graph of the equation. The equation $y = 6$ represents a horizontal line that crosses the y axis at the point (0, 6). Using these intercepts, we graph the two equations. The lines will intersect at the point $(-2, 6)$. So this is the solution to our system.

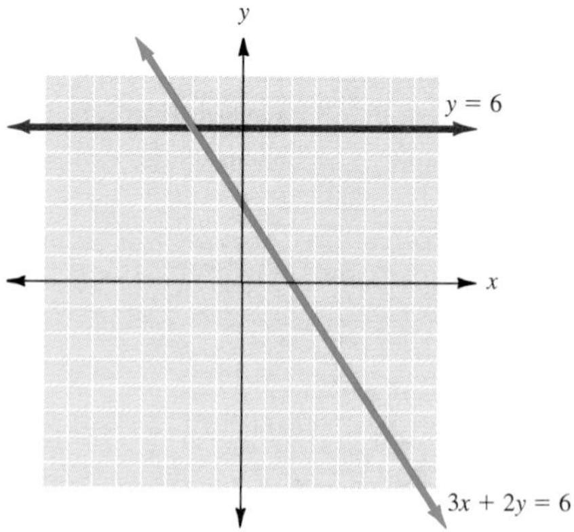

CHECK YOURSELF 2

Solve the system by graphing.

$4x + 5y = 20$

$y = 8$

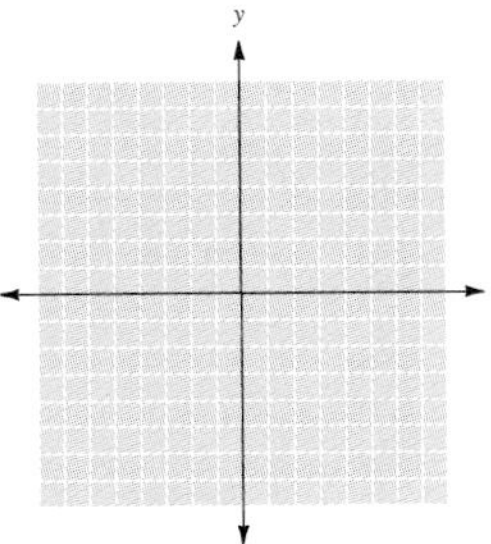

The systems in Examples 1 and 2 both had exactly one solution. A system with one solution is called a **consistent system.** It is possible that a system of equations will have no solution. Such a system is called an **inconsistent system.** We present such a system here.

Example 3

Solving an Inconsistent System

Solve by graphing.

$2x + y = 2$

$2x + y = 4$

We can graph the two lines as before. For $2x + y = 2$, two solutions are (0, 2) and (1, 0). For $2x + y = 4$, two solutions are (0, 4) and (2, 0). Using these intercepts, we graph the two equations.

NOTE In slope-intercept form, our equations are

$y = -2x + 2$

and

$y = -2x + 4$

Both lines have slope -2.

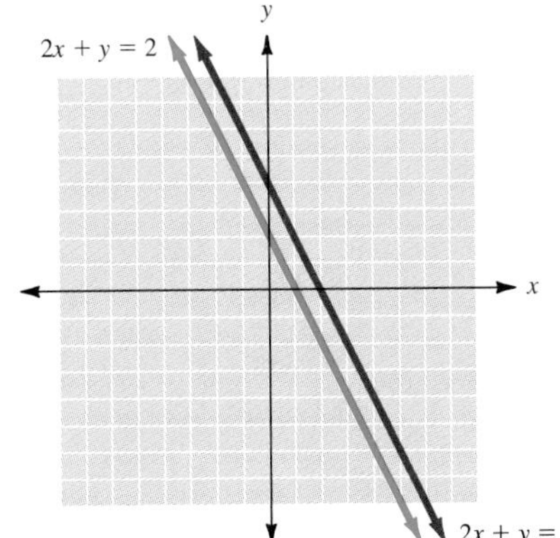

Notice that the slope for each of these lines is -2, but they have different y intercepts. This means that the lines are parallel (they will never intersect). Because the lines have no points in common, there is no ordered pair that will satisfy both equations. The system has no solution. It is *inconsistent.*

CHECK YOURSELF 3

Solve by graphing.

$x - 3y = 3$

$x - 3y = 6$

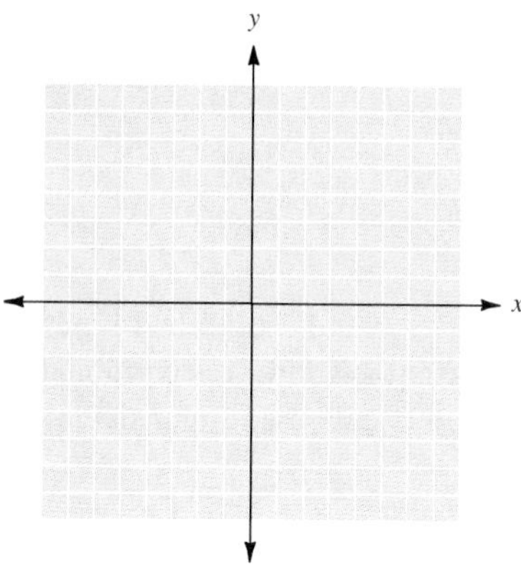

There is one more possibility for linear systems, as Example 4 illustrates.

Example 4

Solving a Dependent System

Solve by graphing.

$x - 2y = 4$

$2x - 4y = 8$

NOTE Notice that multiplying the first equation by 2 results in the second equation.

Graphing as before and using the intercept method, we find

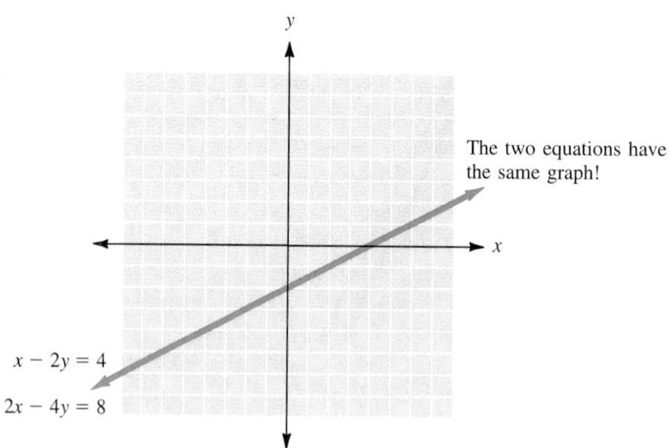

Because the graphs coincide, there are *infinitely many* solutions for this system. Every point on the graph of $x - 2y = 4$ is also on the graph of $2x - 4y = 8$, so any ordered pair satisfying $x - 2y = 4$ also satisfies $2x - 4y = 8$. This is called a *dependent* system, and any point on the line is a solution.

CHECK YOURSELF 4

Solve by graphing.

$$x + y = 4$$
$$2x + 2y = 8$$

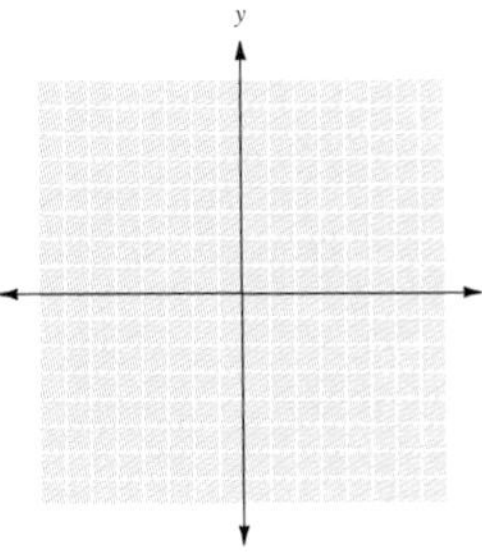

The following summarizes our work in this section.

Step by Step: To Solve a System of Equations by Graphing

Step 1 Graph both equations on the same coordinate system.

Step 2 Determine the solution to the system as follows.

a. If the lines intersect at one point, the solution is the ordered pair corresponding to that point. This is called a **consistent system.**

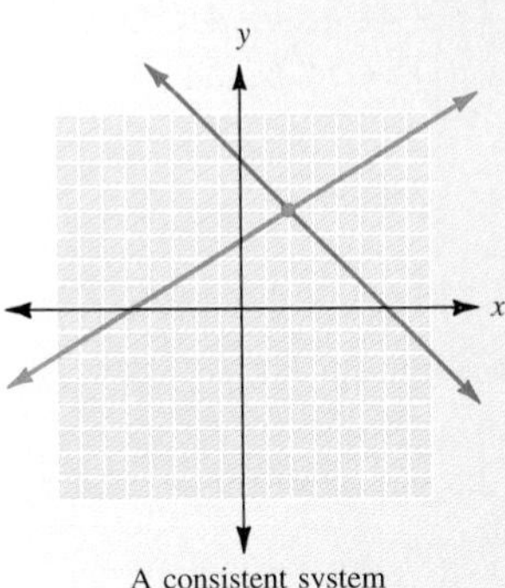

A consistent system

NOTE There is no ordered pair that lies on both lines.

b. If the lines are parallel, there are no solutions. This is called an **inconsistent system.**

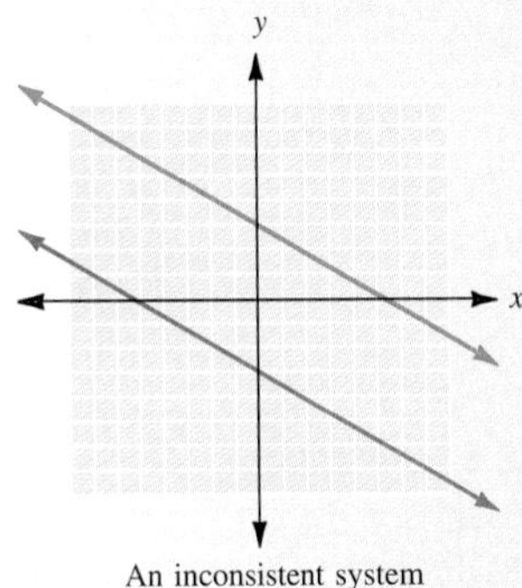

An inconsistent system

NOTE Any ordered pair that corresponds to a point on the line is a solution.

c. If the two equations have the same graph, then the system has infinitely many solutions. This is called a **dependent system.**

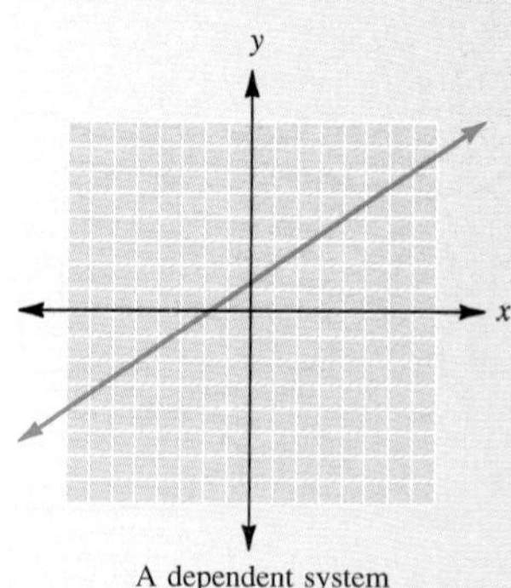

A dependent system

Step 3 Check the solution in both equations, if necessary.

CHECK YOURSELF ANSWERS

1.

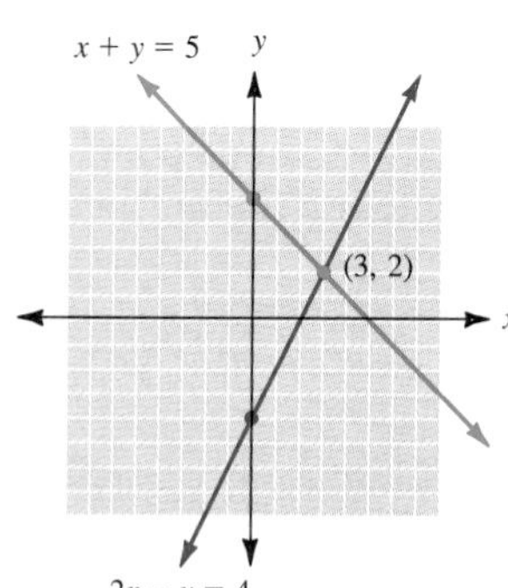

2.

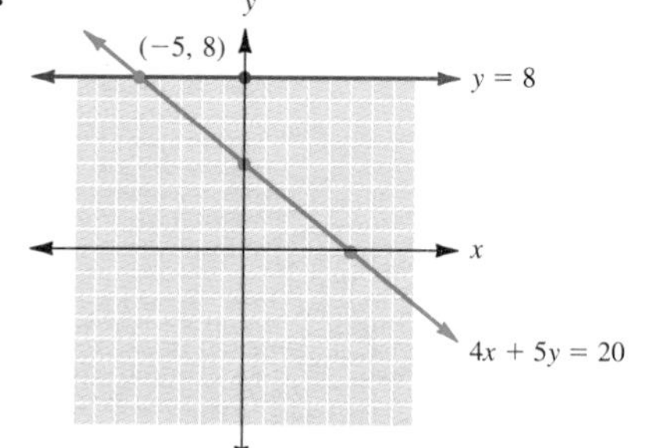

3. There is no solution. The lines are parallel, so the system is inconsistent.

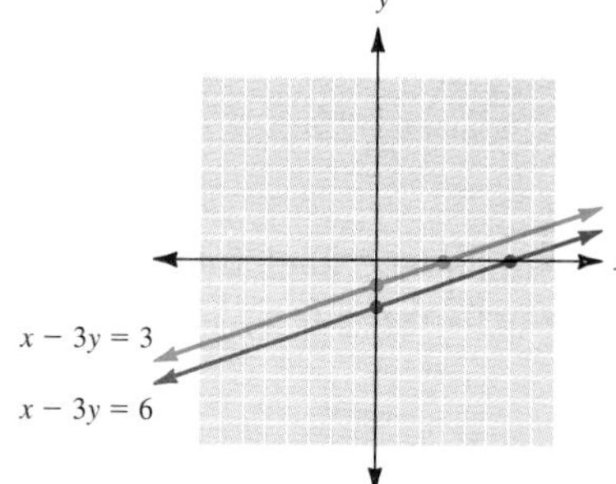

4.

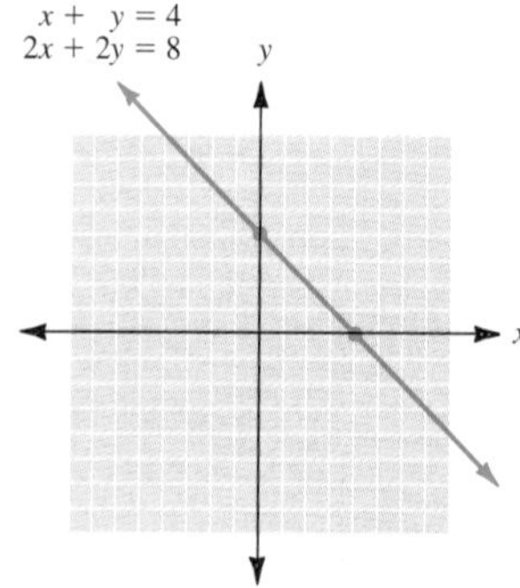

A dependent system

Name ____________

Section ________ Date ________

8.1 Exercises

Solve each of the following systems by graphing.

1. $x + y = 6$
$x - y = 4$

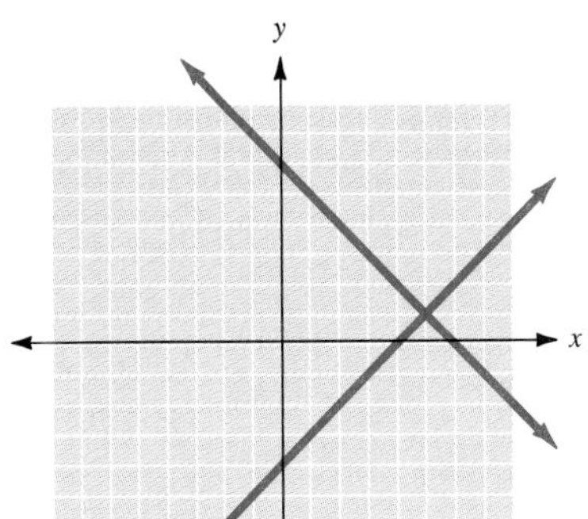

2. $x - y = 8$
$x + y = 2$

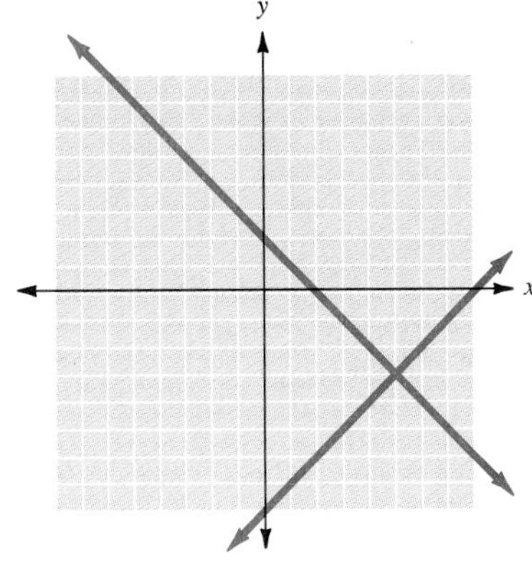

3. $-x + y = 3$
$x + y = 5$

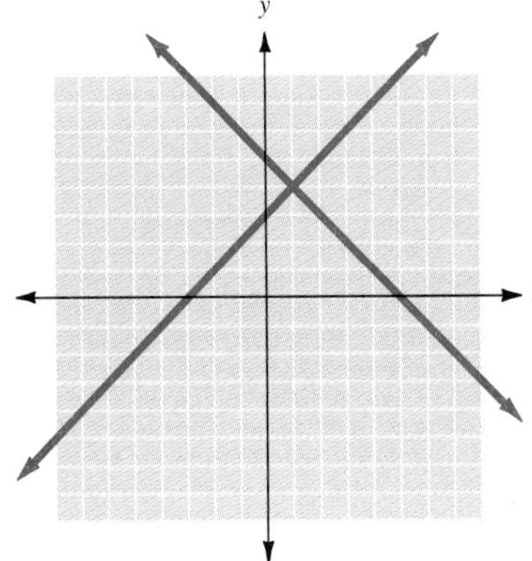

4. $x + y = 7$
$-x + y = 3$

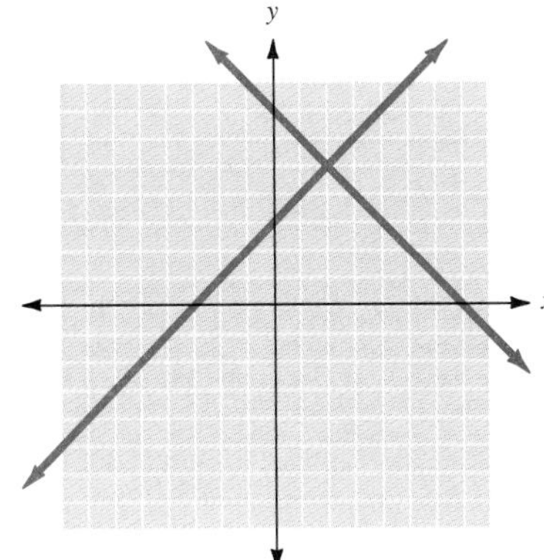

5. $x + 2y = 4$
$x - y = 1$

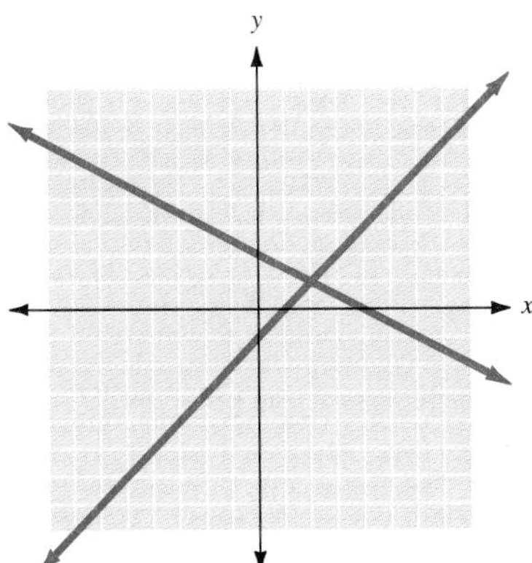

6. $3x + y = 6$
$x + y = 4$

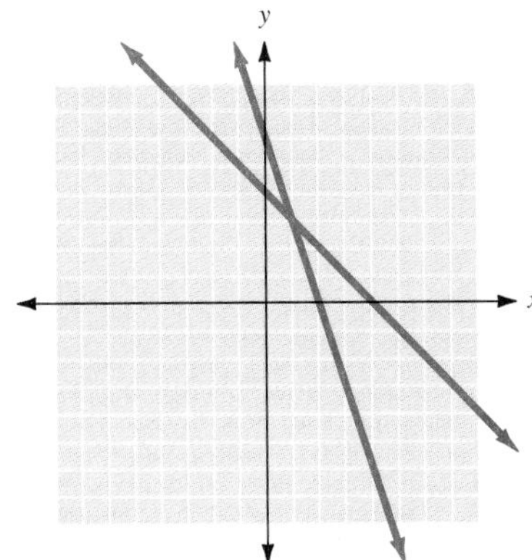

ANSWERS

1. (5, 1)
2. (5, −3)
3. (1, 4)
4. (2, 5)
5. (2, 1)
6. (1, 3)

ANSWERS

7. (2, 4)

8. (2, 2)

9. (6, 2)

10. No solution (Inconsistent system)

11. (2, 3)

12. (2, −3)

7. $2x + y = 8$
$2x - y = 0$

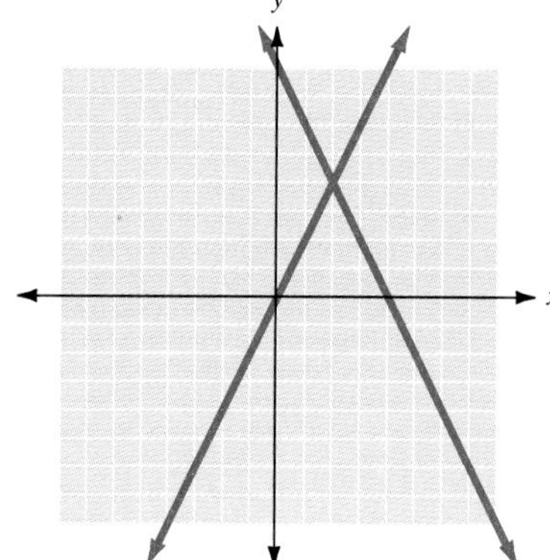

8. $x - 2y = -2$
$x + 2y = 6$

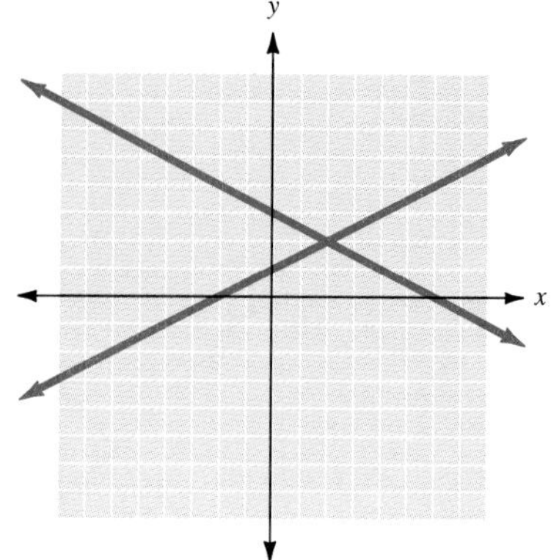

9. $x + 3y = 12$
$2x - 3y = 6$

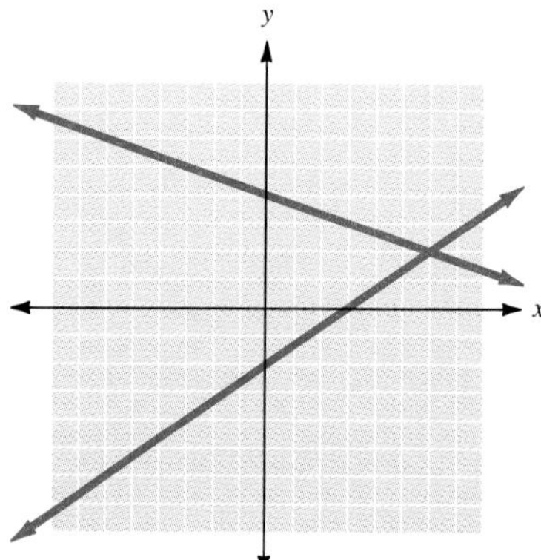

10. $2x - y = 4$
$2x - y = 6$

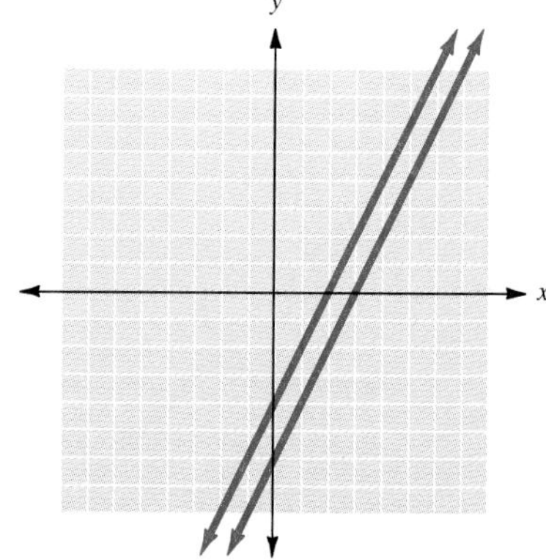

11. $3x + 2y = 12$
$y = 3$

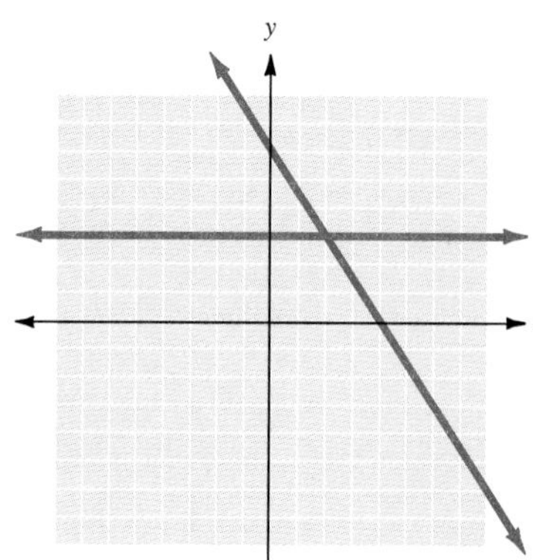

12. $x - 2y = 8$
$3x - 2y = 12$

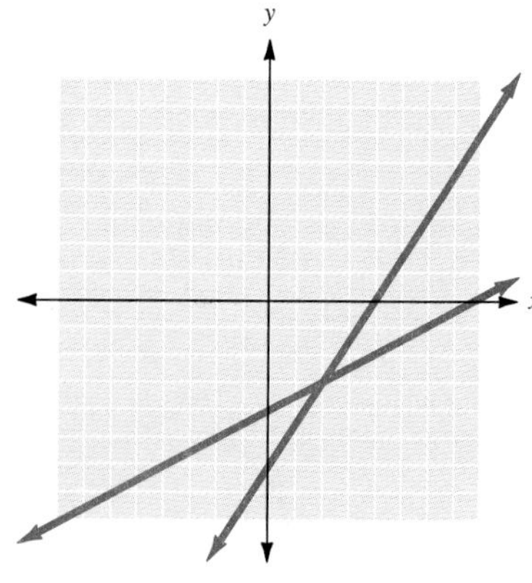

13. $x - y = 4$
$2x - 2y = 8$

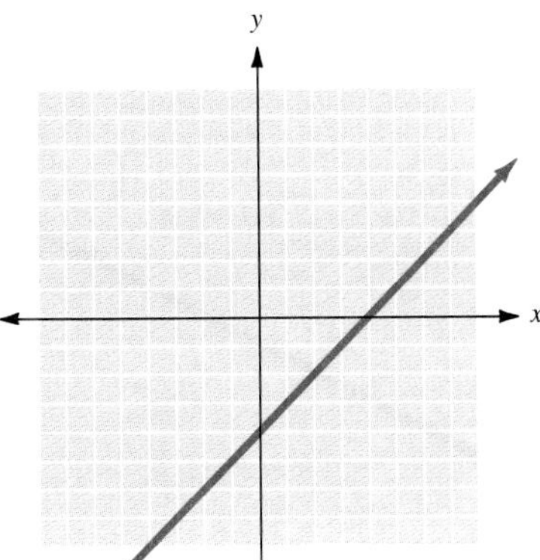

14. $2x - y = 8$
$x = 2$

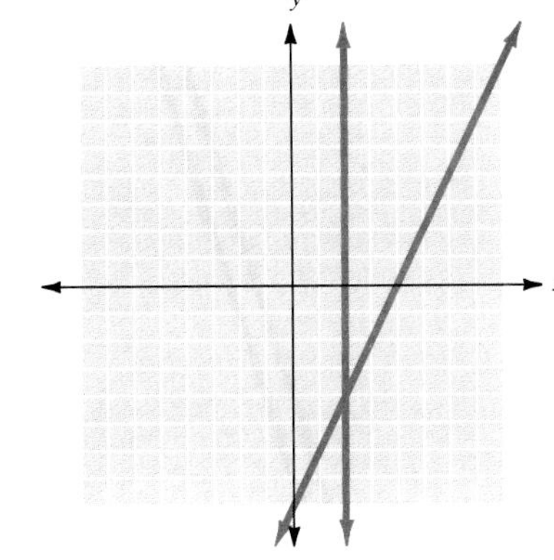

15. $x - 4y = -4$
$x + 2y = 8$

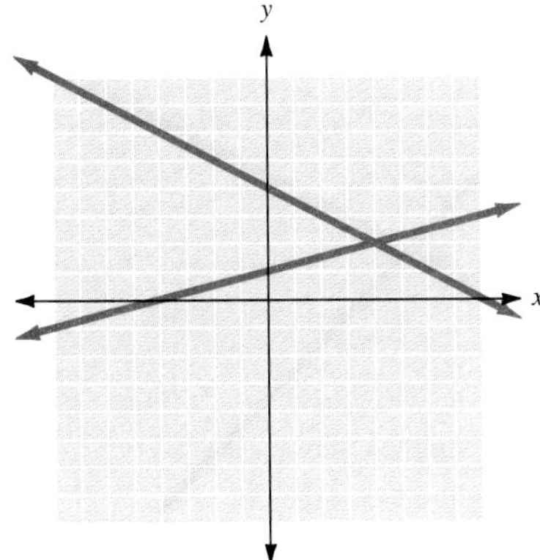

16. $x - 6y = 6$
$-x + y = 4$

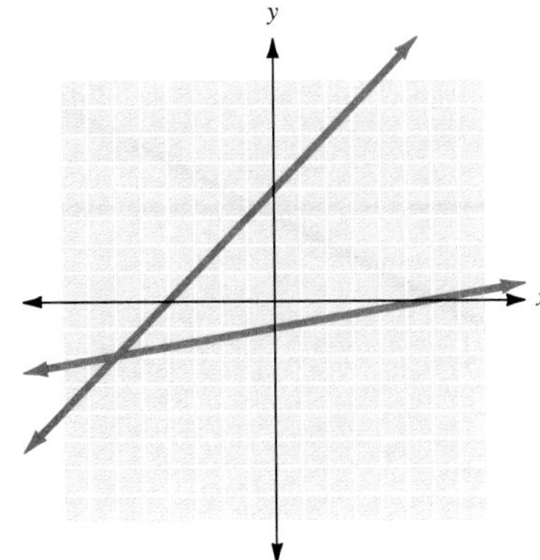

17. $3x - 2y = 6$
$2x - y = 5$

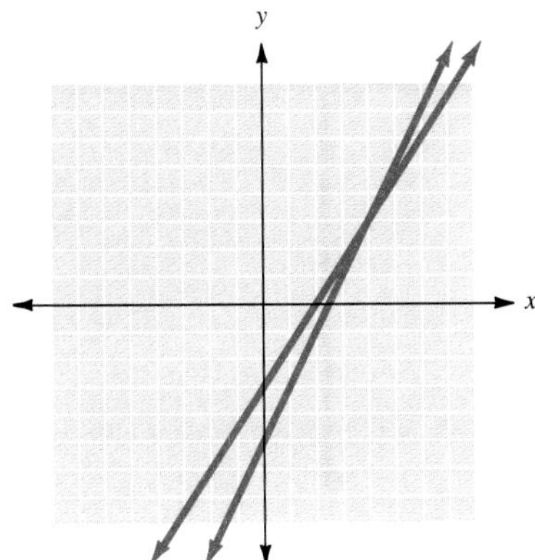

18. $4x + 3y = 12$
$x + y = 2$

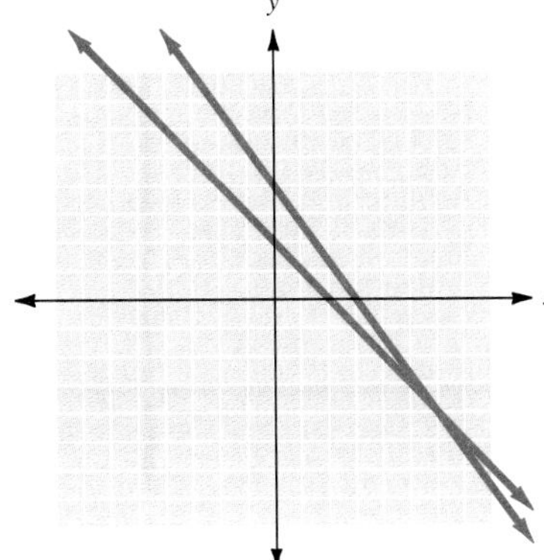

ANSWERS

13. Infinitely many solutions (Dependent system)

14. (2, −4)

15. (4, 2)

16. (−6, −2)

17. (4, 3)

18. (6, −4)

a. $3x$

b. $2y$

c. $-y$

d. $3x$

e. $5x$

f. $-5y$

g. $5y$

h. $10x$

Getting Ready for Section 8.2 [Section 1.6]

Simplify each of the following expressions.

(a) $(2x + y) + (x - y)$

(b) $(x + y) + (-x + y)$

(c) $(3x + 2y) + (-3x - 3y)$

(d) $(x - 5y) + (2x + 5y)$

(e) $2(x + y) + (3x - 2y)$

(f) $2(2x - y) + (-4x - 3y)$

(g) $3(2x + y) + 2(-3x + y)$

(h) $3(2x - 4y) + 4(x + 3y)$

Answers

1. $\left.\begin{aligned} x + y &= 6 \\ x - y &= 4 \end{aligned}\right\}$ (5, 1)

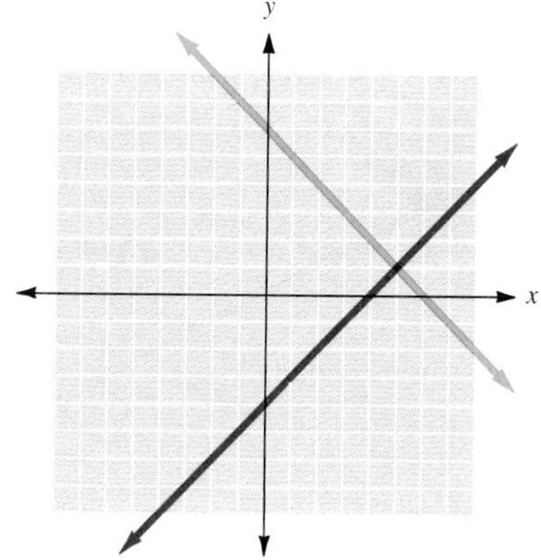

3. $\left.\begin{aligned} -x + y &= 3 \\ x + y &= 5 \end{aligned}\right\}$ (1, 4)

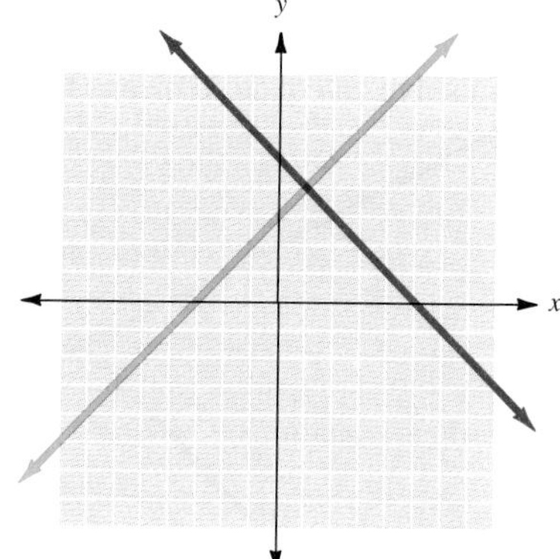

5. $\left.\begin{aligned} x + 2y &= 4 \\ x - y &= 1 \end{aligned}\right\}$ (2, 1)

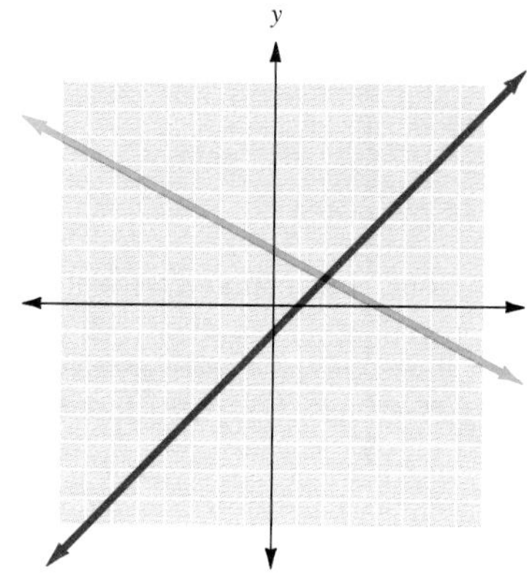

7. $\left.\begin{aligned} 2x + y &= 8 \\ 2x - y &= 0 \end{aligned}\right\}$ (2, 4)

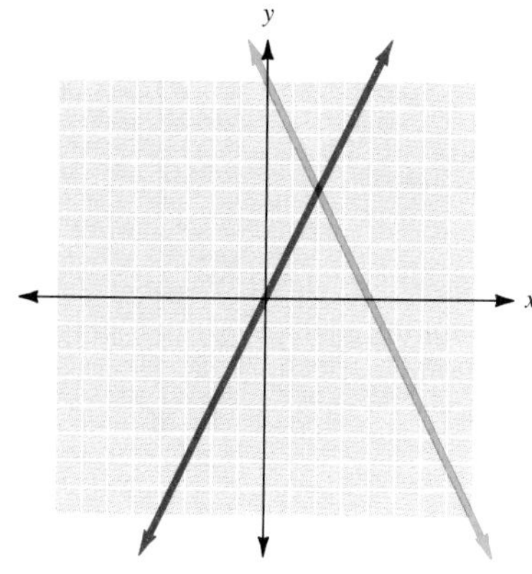

13. $x - y = 4$
$2x - 2y = 8$

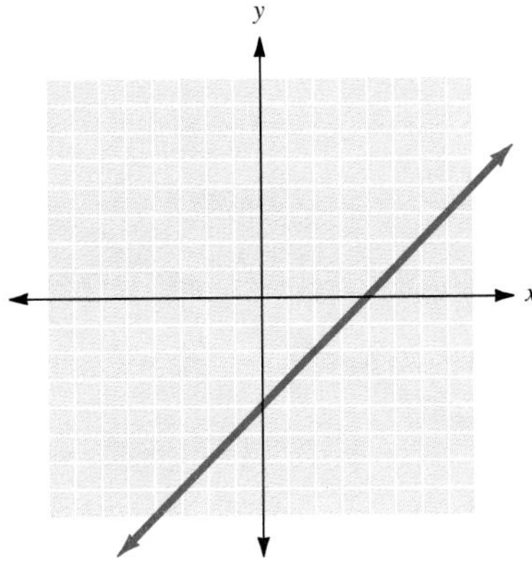

14. $2x - y = 8$
$x = 2$

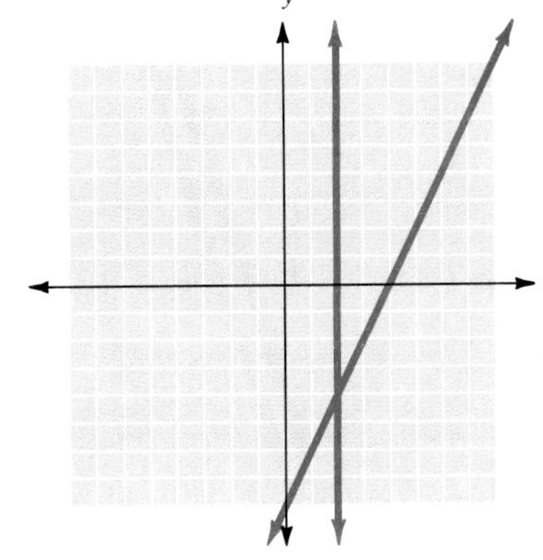

15. $x - 4y = -4$
$x + 2y = 8$

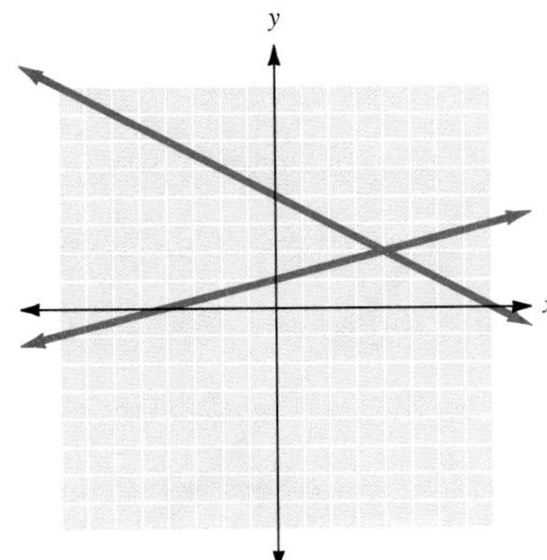

16. $x - 6y = 6$
$-x + y = 4$

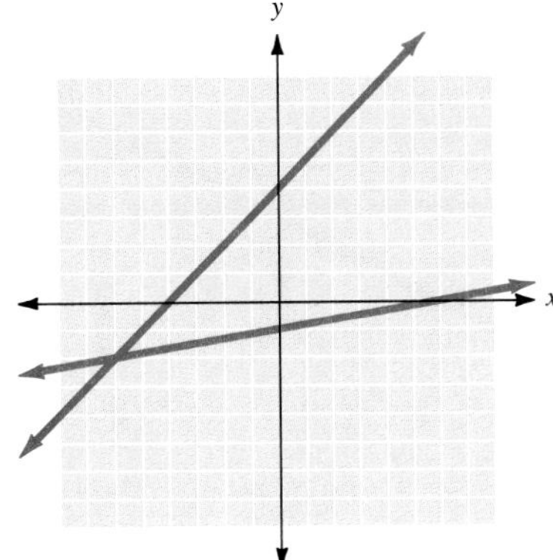

17. $3x - 2y = 6$
$2x - y = 5$

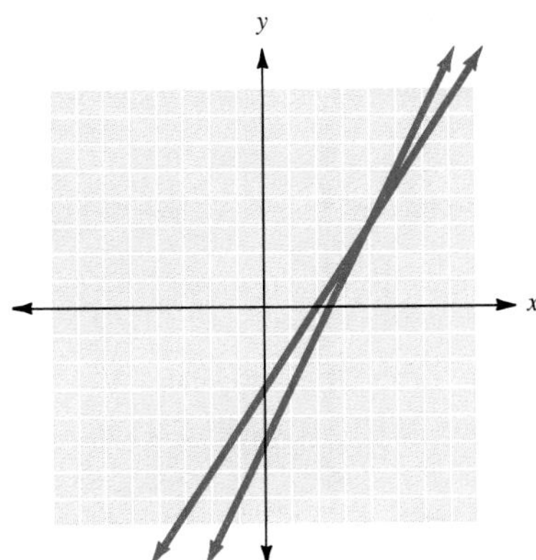

18. $4x + 3y = 12$
$x + y = 2$

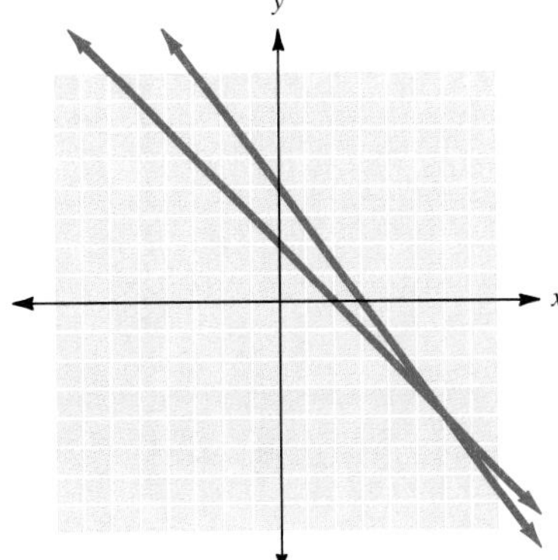

ANSWERS

13. Infinitely many solutions (Dependent system)
14. (2, −4)
15. (4, 2)
16. (−6, −2)
17. (4, 3)
18. (6, −4)

ANSWERS

19. Inconsistent system

20. Dependent system

21. $\left(0, \frac{3}{2}\right)$

22. (−6, 0)

23. (4, −6)

24. (−3, 5)

19. $3x - y = 3$
$3x - y = 6$

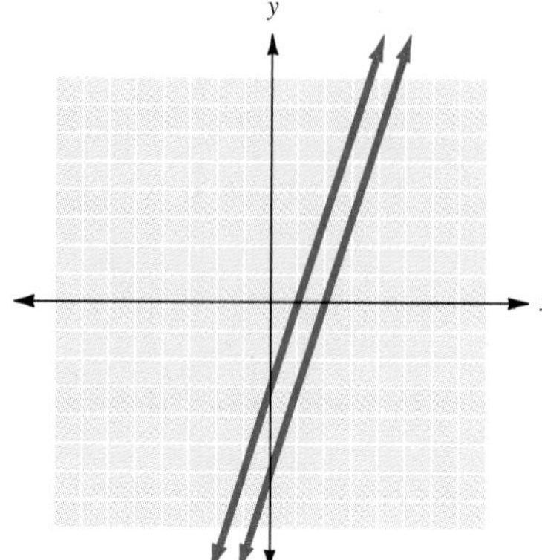

20. $3x - 6y = 9$
$x - 2y = 3$

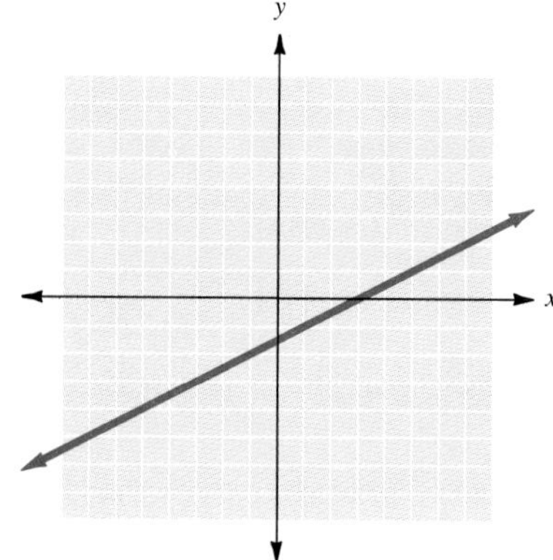

21. $2y = 3$
$x - 2y = -3$

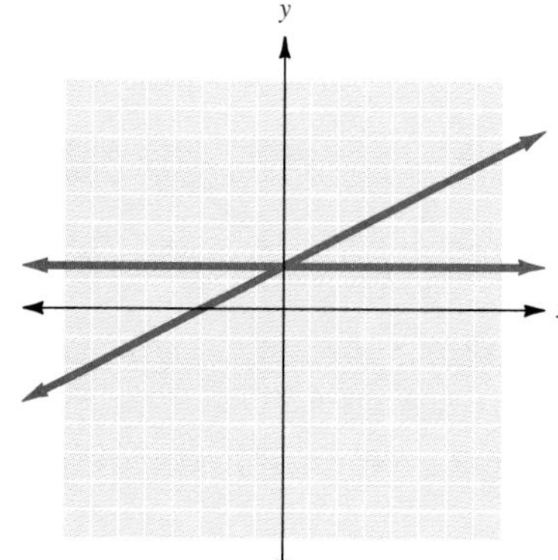

22. $x + y = -6$
$-x + 2y = 6$

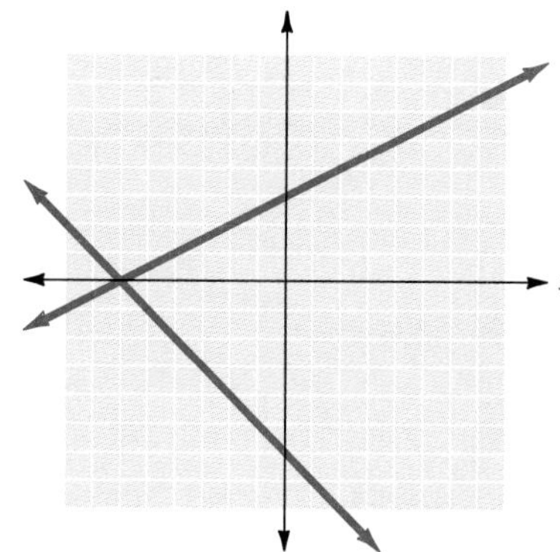

23. $x = 4$
$y = -6$

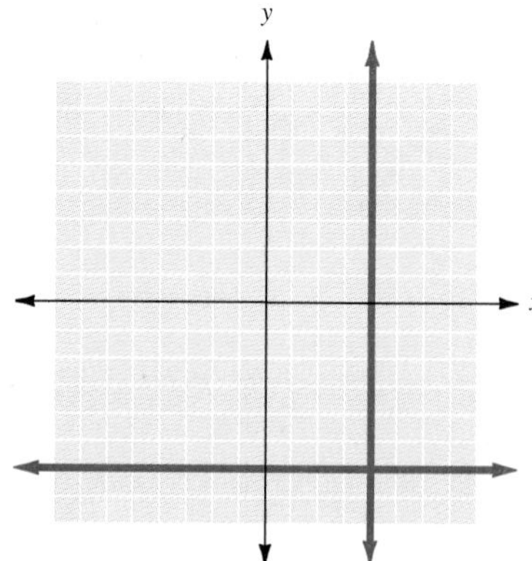

24. $x = -3$
$y = 5$

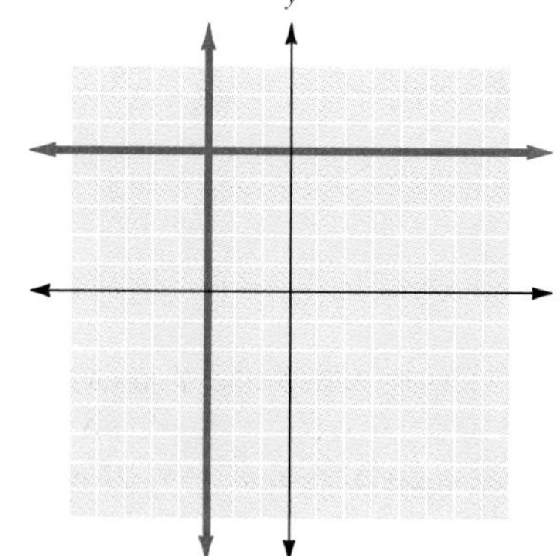

25. Find values for m and b in the following system so that the solution to the system is $(1, 2)$.

$$mx + 3y = 8$$
$$-3x + 4y = b$$

26. Find values for m and b in the following system so that the solution to the system is $(-3, 4)$.

$$5x + 7y = b$$
$$mx + y = 22$$

27. Complete the following statements in your own words:

"To solve an equation means to . . . ."

"To solve a system of equations means to . . . ."

28. A system of equations such as the one below is sometimes called a "2-by-2" system of linear equations."

$$3x + 4y = 1$$
$$x - 2y = 6$$

Explain this term.

29. Complete this statement in your own words: "All the points on the graph of the equation $2x + 3y = 6$ . . . ." Exchange statements with other students. Do you agree with other students' statements?

30. Does a system of linear equations always have a solution? How can you tell without graphing that a system of two equations graphs into two parallel lines? Give some examples to explain your reasoning.

ANSWERS

25. $m = 2, b = 5$

26. $m = -6, b = 13$

27.

28.

29.

30.

ANSWERS

a. $3x$

b. $2y$

c. $-y$

d. $3x$

e. $5x$

f. $-5y$

g. $5y$

h. $10x$

Getting Ready for Section 8.2 [Section 1.6]

Simplify each of the following expressions.

(a) $(2x + y) + (x - y)$

(b) $(x + y) + (-x + y)$

(c) $(3x + 2y) + (-3x - 3y)$

(d) $(x - 5y) + (2x + 5y)$

(e) $2(x + y) + (3x - 2y)$

(f) $2(2x - y) + (-4x - 3y)$

(g) $3(2x + y) + 2(-3x + y)$

(h) $3(2x - 4y) + 4(x + 3y)$

Answers

1. $\left.\begin{aligned} x + y &= 6 \\ x - y &= 4 \end{aligned}\right\} (5, 1)$

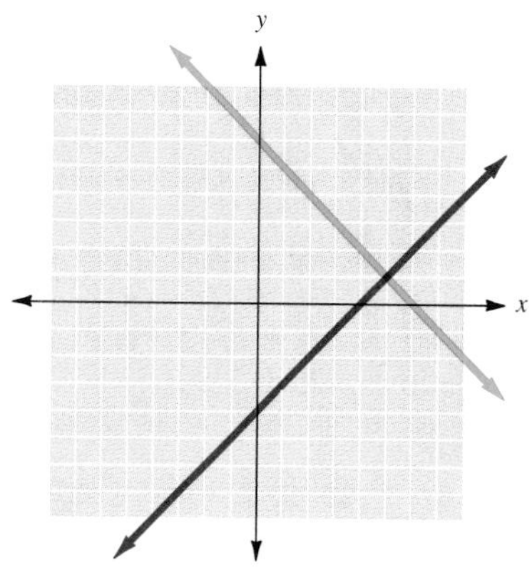

3. $\left.\begin{aligned} -x + y &= 3 \\ x + y &= 5 \end{aligned}\right\} (1, 4)$

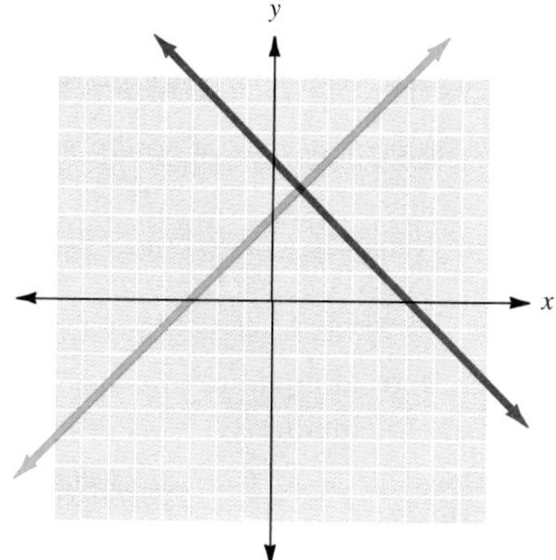

5. $\left.\begin{aligned} x + 2y &= 4 \\ x - y &= 1 \end{aligned}\right\} (2, 1)$

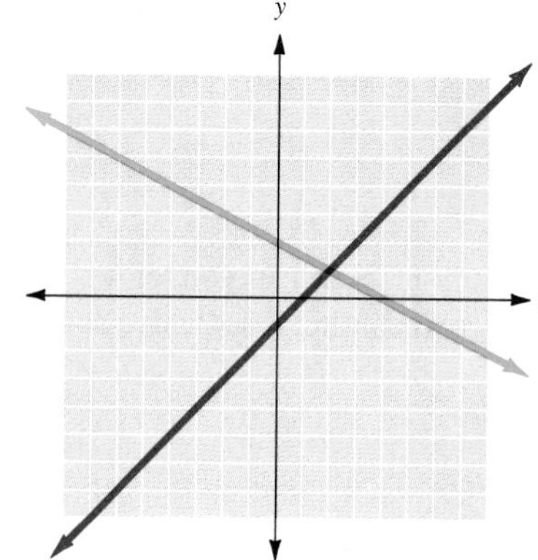

7. $\left.\begin{aligned} 2x + y &= 8 \\ 2x - y &= 0 \end{aligned}\right\} (2, 4)$

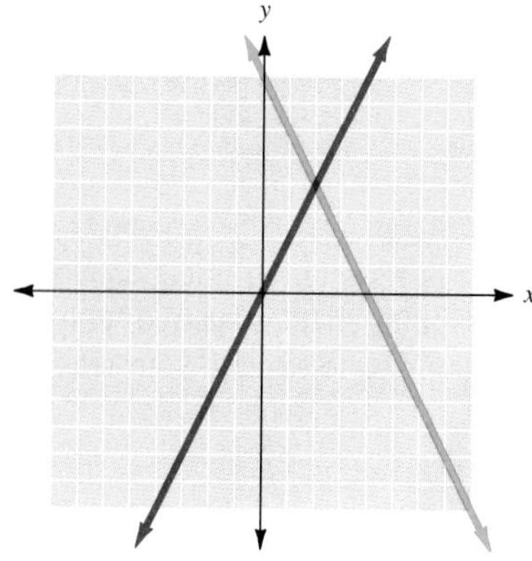

9. $\left.\begin{aligned} x + 3y &= 12 \\ 2x - 3y &= 6 \end{aligned}\right\}$ (6, 2)

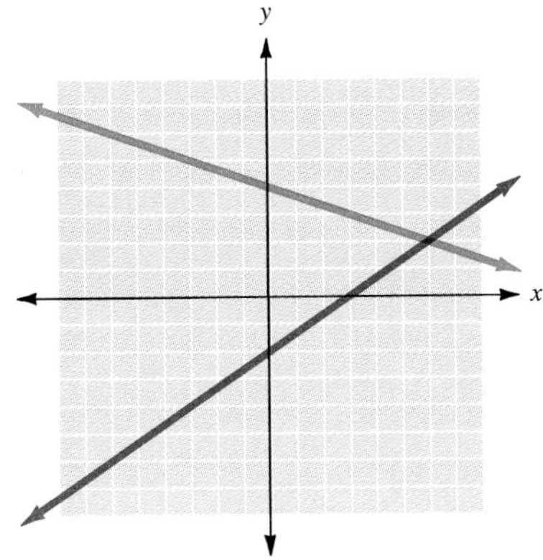

11. $\left.\begin{aligned} 3x + 2y &= 12 \\ y &= 3 \end{aligned}\right\}$ (2, 3)

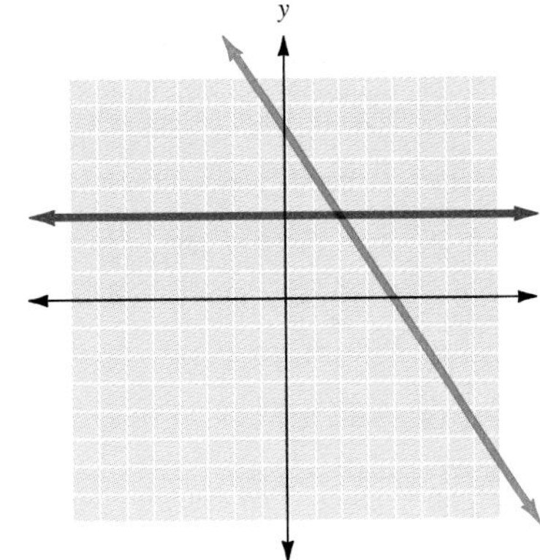

13. $\left.\begin{aligned} x - y &= 4 \\ 2x - 2y &= 8 \end{aligned}\right\}$ Dependent

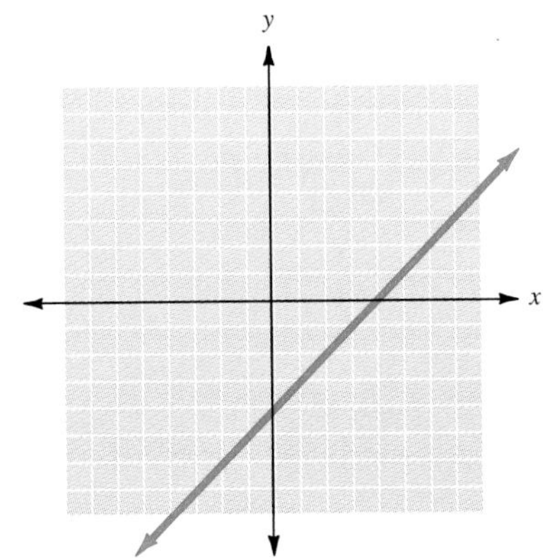

15. $\left.\begin{aligned} x - 4y &= -4 \\ x + 2y &= 8 \end{aligned}\right\}$ (4, 2)

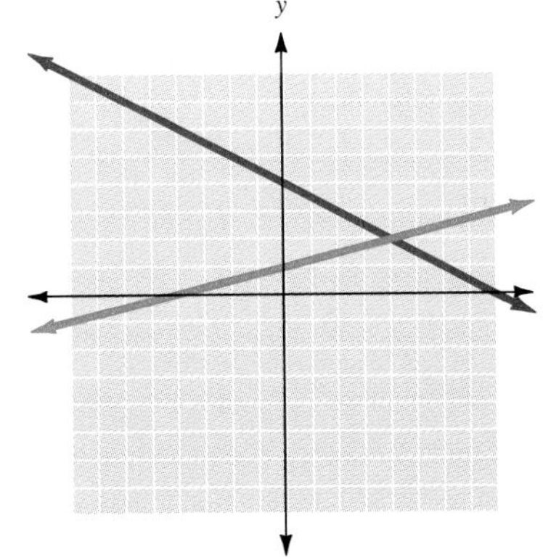

17. $\left.\begin{aligned} 3x - 2y &= 6 \\ 2x - y &= 5 \end{aligned}\right\}$ (4, 3)

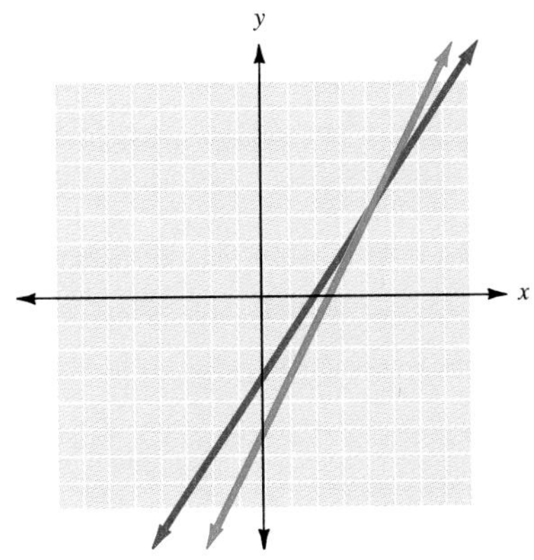

19. $\left.\begin{aligned} 3x - y &= 3 \\ 3x - y &= 6 \end{aligned}\right\}$ Inconsistent

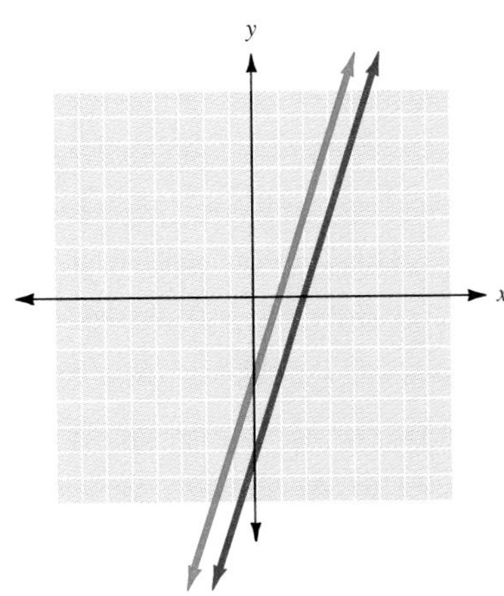

21. $\left.\begin{aligned} 2y &= 3 \\ x - 2y &= -3 \end{aligned}\right\} \left(0, \dfrac{3}{2}\right)$

23. 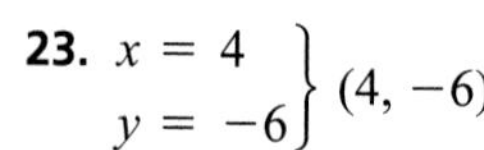
$\left.\begin{aligned} x &= 4 \\ y &= -6 \end{aligned}\right\} (4, -6)$

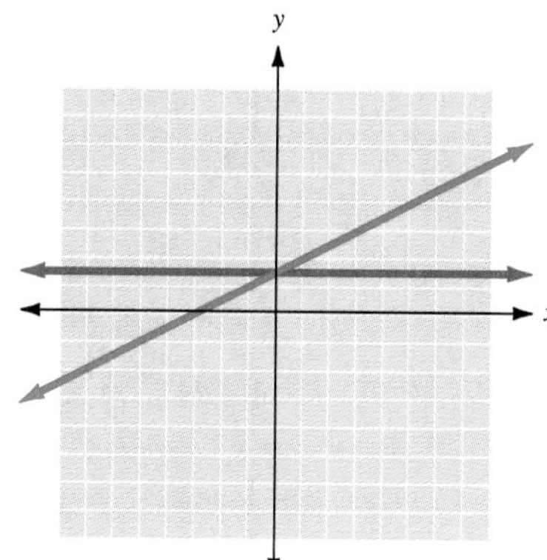

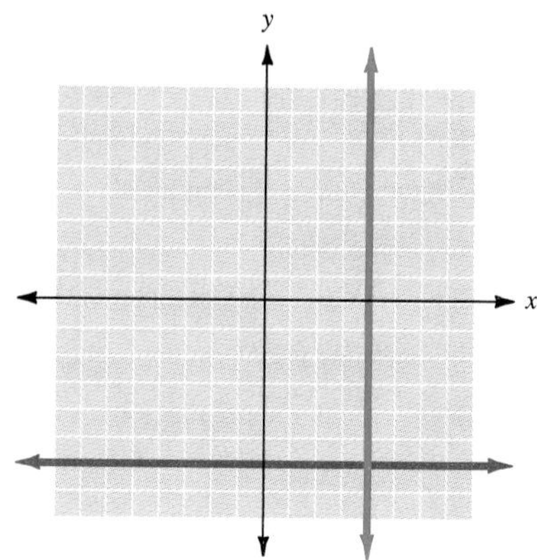

25. $m = 2, b = 5$ **27.** **29.** **a.** $3x$ **b.** $2y$

c. $-y$ **d.** $3x$ **e.** $5x$ **f.** $-5y$ **g.** $5y$ **h.** $10x$

Systems of Linear Equations: Solving by Adding

OBJECTIVES

1. Solve systems using the addition method
2. Solve applications of systems of equations

The graphical method of solving equations, shown in Section 8.1, has two definite disadvantages. First, it is time-consuming to graph each system that you want to solve. More importantly, the graphical method is not precise. For instance, look at the graph of the system

$$x - 2y = 4$$

$$3x + 2y = 6$$

which is shown below.

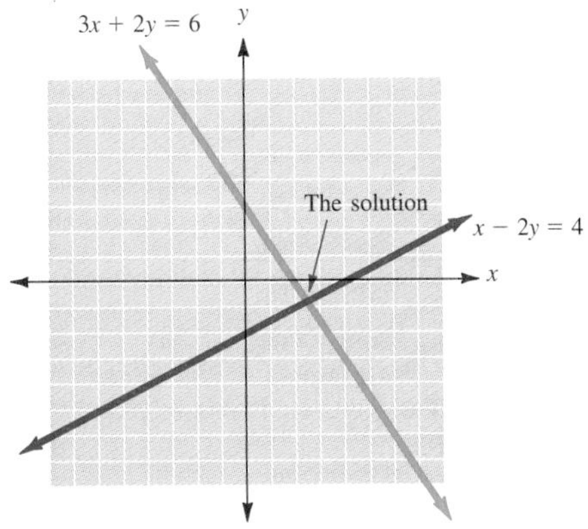

The exact solution for the system happens to be $\left(\frac{5}{2}, -\frac{3}{4}\right)$, but that would be difficult to read from the graph. Fortunately, there are algebraic methods that do not have this disadvantage and will allow you to find exact solutions for a system of equations.

Let's illustrate an algebraic method of finding a solution. It is called the **addition method.**

Example 1

Solving a System by the Addition Method

NOTE This method uses the fact that if

$a = b$ and $c = d$

then

$a + c = b + d$

This is the **additive property** of equality. Note that by the additive property, if equals are added to equals, the resulting sums are equal.

Solve the system.

$$x + y = 8$$

$$x - y = 2$$

Note that the coefficients of the y terms are the *additive inverses* of one another (1 and -1) and that adding the two equations will "eliminate" the variable y. That addition step is shown here.

NOTE This is also called **solution by elimination** for this reason.

$$\begin{aligned} x + y &= 8 \\ x - y &= 2 \\ \hline 2x \quad &= 10 \\ x &= 5 \end{aligned}$$

By adding, we eliminate the variable y. The resulting equation contains *only* the variable x.

We now know that 5 is the x coordinate of our solution. Substitute 5 for x into *either* of the original equations.

$$x + y = 8$$
$$5 + y = 8$$
$$y = 3$$

So (5, 3) is the solution.

To check, replace x and y with these values in *both* of the original equations.

$x + y = 8$		$x - y = 2$	
$5 + 3 = 8$		$5 - 3 = 2$	
$8 = 8$	(True)	$2 = 2$	(True)

Because (5, 3) satisfies both equations, it is the solution.

CHECK YOURSELF 1

Solve the system by adding.

$$x - y = -2$$
$$x + y = 6$$

Example 2

Solving a System by the Addition Method

Solve the system.

$$-3x + 2y = 12$$
$$3x - y = -9$$

In this case, adding will eliminate the x terms.

NOTE Notice that we don't care which variable is eliminated. Choose the one that requires the least work.

$$\begin{aligned} -3x + 2y &= 12 \\ 3x - y &= -9 \\ \hline y &= 3 \end{aligned}$$

Now substitute 3 for y in either equation. From the first equation

$$-3x + 2 \cdot 3 = 12$$
$$-3x = 6$$
$$x = -2$$

and $(-2, 3)$ is the solution.

Show that you get the same x coordinate by substituting 3 for y in the second equation rather than in the first. Then check the solution.

CHECK YOURSELF 2

Solve the system by adding.

$$5x - 2y = 9$$
$$-5x + 3y = -11$$

Note that in both Examples 1 and 2 we found an equation in a single variable by adding. We could do this because the coefficients of one of the variables were opposites. This gave 0 as a coefficient for one of the variables after we added the two equations. In some systems, you will not be able to directly eliminate either variable by adding. However, an equivalent system can always be written by multiplying one or both of the equations by a nonzero constant so that the coefficients of x (or of y) are opposites. Example 3 illustrates this approach.

Example 3

Solving a System by the Addition Method

Solve the system.

$$2x + y = 13 \qquad (1)$$

$$3x + y = 18 \qquad (2)$$

NOTE Remember that multiplying both sides of an equation by some nonzero number does not change the solutions. So even though we have "altered" the equations, they are equivalent and will have the same solutions.

Note that adding the equations in this form will not eliminate either variable. You will still have terms in x and in y. However, look at what happens if we multiply both sides of equation (2) by -1 as the first step.

$$2x + y = 13 \longrightarrow 2x + y = 13$$

$$3x + y = 18 \xrightarrow[\text{by } -1]{\text{Multiply}} -3x - y = -18$$

Now we can add.

$$\begin{aligned} 2x + y &= 13 \\ -3x - y &= -18 \\ \hline -x \qquad &= -5 \\ x &= 5 \end{aligned}$$

Substitute 5 for x in equation (1).

$$2 \cdot 5 + y = 13$$

$$y = 3$$

(5, 3) is the solution. We will leave it to the reader to check this solution.

CHECK YOURSELF 3

Solve the system by adding.

$$x - 2y = 9$$

$$x + 3y = -1$$

To summarize, multiplying both sides of one of the equations by a nonzero constant can yield an equivalent system in which the coefficients of the x terms or the y terms are opposites. This means that a variable can be eliminated by adding. Let's look at another example.

Example 4

Solving a System by the Addition Method

Solve the system.

$$x + 4y = 2 \quad (1)$$

$$3x - 2y = -22 \quad (2)$$

One approach is to multiply both sides of equation (2) by 2. Do you see that the coefficients of the y terms will then be opposites?

$$x + 4y = 2 \longrightarrow x + 4y = 2$$

$$3x - 2y = -22 \xrightarrow[\text{by 2}]{\text{Multiply}} 6x - 4y = -44$$

If we add the resulting equations, the variable y will be eliminated and we can solve for x.

NOTE Notice that the coefficients of the y terms are opposites.

$$\begin{aligned} x + 4y &= 2 \\ 6x - 4y &= -44 \\ \hline 7x \quad &= -42 \\ x &= -6 \end{aligned}$$

Now substitute -6 for x in equation (1) to find y.

NOTE Also, -6 could be substituted for x in equation (2) to find y.

$$-6 + 4y = 2$$

$$4y = 8$$

$$y = 2$$

So $(-6, 2)$ is the solution.

Again you should check this result. As is often the case, there are several ways to solve the system. For example, what if we multiply both sides of equation (1) by -3? The coefficients of the x terms will then be opposites, and adding will eliminate the variable x so that we can solve for y. Try that for yourself in the following Check Yourself exercise.

CHECK YOURSELF 4

Solve the system by eliminating x.

$$x + 4y = 2$$

$$3x - 2y = -22$$

It may be necessary to multiply each equation separately so that one of the variables will be eliminated when the equations are added. Example 5 illustrates this approach.

Example 5

Solving a System by the Addition Method

Solve the system.

$$4x + 3y = 11 \quad (1)$$

$$3x - 2y = 4 \quad (2)$$

Do you see that, if we want to have integer coefficients, multiplying in one equation will not help in this case? We will have to multiply in both equations.

NOTE The minus sign is used with the 4 so that the coefficients of the x term are opposites.

To eliminate x, we can multiply both sides of equation (1) by 3 and both sides of equation (2) by -4. The coefficients of the x terms will then be opposites.

$$4x + 3y = 11 \xrightarrow[\text{by } 3]{\text{Multiply}} 12x + 9y = 33$$

$$3x - 2y = 4 \xrightarrow[\text{by } -4]{\text{Multiply}} -12x + 8y = -16$$

Adding the resulting equations gives

$$17y = 17$$

$$y = 1$$

Now substituting 1 for y in equation (1), we have

$$4x + 3 \cdot 1 = 11$$

$$4x = 8$$

$$x = 2$$

NOTE Check (2, 1) in both equations of the original system.

and (2, 1) is the solution.

CHECK YOURSELF 5

Solve the system by eliminating y.

$$4x + 3y = 11$$

$$3x - 2y = 4$$

Let's summarize the solution steps that we have illustrated.

Step by Step: To Solve a System of Linear Equations by Adding

Step 1 If necessary, multiply both sides of one or both equations by nonzero numbers to form an equivalent system in which the coefficients of one of the variables are opposites.

Step 2 Add the equations of the new system.

Step 3 Solve the resulting equation for the remaining variable.

Step 4 Substitute the value found in step 3 into either of the original equations to find the value of the second variable.

Step 5 Check your solution in both of the original equations.

In Section 8.1 we saw that some systems had *infinitely* many solutions. Let's see how this is indicated when we are using the addition method of solving equations.

Example 6

Solving a Dependent System

Solve

$$x + 3y = -2 \qquad (1)$$

$$3x + 9y = -6 \qquad (2)$$

We multiply both sides of equation (1) by −3.

$$\begin{aligned} x + 3y &= -2 \xrightarrow[\text{by } -3]{\text{Multiply}} & -3x - 9y &= 6 \\ 3x + 9y &= -6 \longrightarrow & 3x + 9y &= -6 \\ & & 0 &= 0 \end{aligned}$$

Adding, we see that both variables have been eliminated, and we have the true statement $0 = 0$.

NOTE The lines coincide. That will be the case whenever *adding eliminates both variables* and a true statement results.

Look at the graph of the system.

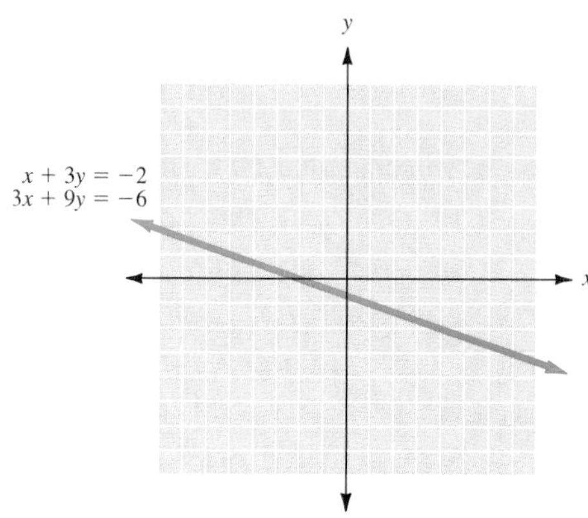

As we see, the two equations have the *same* graph. This means that the system is *dependent,* and there are *infinitely many solutions.* Any (x, y) that satisfies $x + 3y = -2$ will also satisfy $3x + 9y = -6$.

CHECK YOURSELF 6

Solve the system by adding.

$$\begin{aligned} x - 2y &= 3 \\ -2x + 4y &= -6 \end{aligned}$$

Earlier we encountered systems that had *no* solutions. Example 7 illustrates what happens when we try to solve such a system with the addition method.

Example 7

Solving an Inconsistent System

Solve the system.

$$3x - y = 4 \qquad (1)$$

$$-6x + 2y = -5 \qquad (2)$$

We multiply both sides of equation (1) by 2.

NOTE Be sure to multiply the 4 by 2.

$$\begin{aligned} 3x - y &= 4 \xrightarrow[\text{by } 2]{\text{Multiply}} & 6x - 2y &= 8 \\ -6x + 2y &= -5 \longrightarrow & -6x + 2y &= -5 \\ & & 0 &= 3 \end{aligned}$$

We now add the two equations

Again both variables have been eliminated by addition. But this time we have the *false* statement $0 = 3$ because we tried to solve a system whose graph consists of two parallel lines, as we see in the graph below. Because the two lines do not intersect, there is *no* solution for the system. It is *inconsistent.*

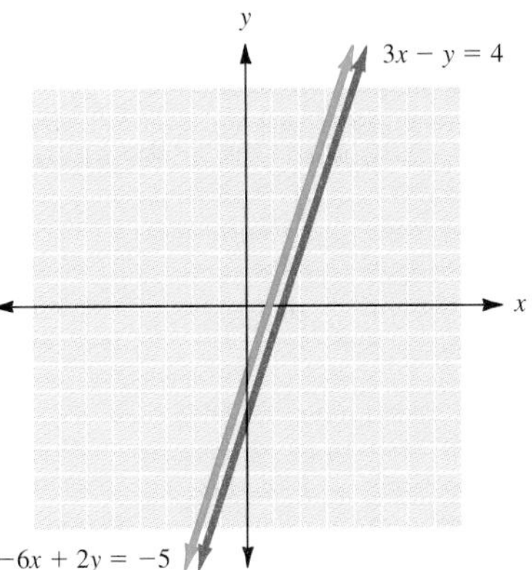

CHECK YOURSELF 7

Solve the system by adding.

$$5x + 15y = 20$$
$$x + 3y = 3$$

NOTE Remember that, in Chapter 2, all the unknowns in the problem had to be expressed in terms of that single variable.

In Chapter 2 we solved word problems by using equations in a single variable. Now that you have the background to use two equations in two variables to solve word problems, let's see how they can be applied. The five steps for solving word problems stay the same (in fact, we give them again for reference in our first application example). Many students find that using two equations and two variables makes writing the necessary equation much easier, as Example 8 illustrates.

Example 8

Solving an Application with Two Equations

Ryan bought 8 pens and 7 pencils and paid a total of \$14.80. Ashleigh purchased 2 pens and 10 pencils and paid \$7. Find the cost for a single pen and a single pencil.

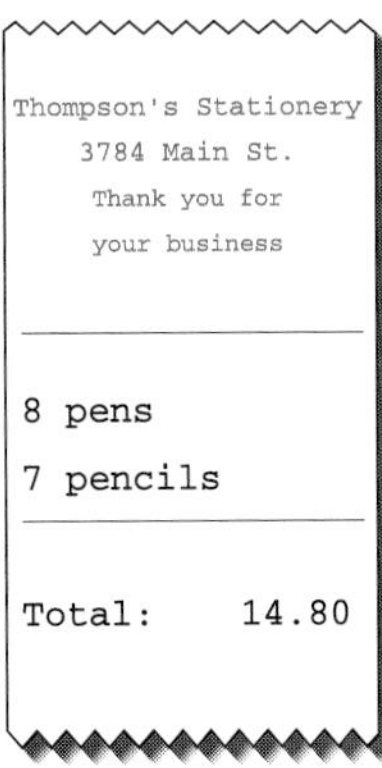

NOTE Here are the steps for using a single variable:

1. Read the problem carefully. What do you want to find?

Step 1 You want to find the cost of a single pen and the cost of a single pencil.

2. Assign variables to the unknown quantities.

Step 2 Let x be the cost of a pen and y be the cost of a pencil.

3. Write the equation for the solution.

Step 3 Write the two necessary equations.

$$8x + 7y = 14.80 \quad (1)$$

$$2x + 10y = 7.00 \quad (2)$$

In the first equation, $8x$ is the total cost of the pens Ryan bought and $7y$ is the total cost of the pencils Ryan bought. The second equation is formed in a similar fashion.

4. Solve the equation.

Step 4 Solve the system formed in step 3. We multiply equation (2) by -4. Adding will then eliminate the variable x.

$$8x + 7y = 14.80$$

$$-8x - 40y = -28.00$$

Now adding the equations, we have

$$-33y = -13.20$$

$$y = 0.40$$

Substituting 0.40 for y in equation (1), we have

$$8x + 7(0.40) = 14.80$$

$$8x + 2.80 = 14.80$$

$$8x = 12.00$$

$$x = 1.50$$

5. Verify your result by returning to the original problem.

Step 5 From the results of step 4 we see that the pens are \$1.50 each and that the pencils are 40¢ each.

To check these solutions, replace x with \$1.50 and y with 0.40 in equation (1).

$$8(1.50) + 7(0.40) = 14.80$$

$$12.00 + 2.80 = 14.80$$

$$14.80 = 14.80 \quad \text{(True)}$$

We leave it to you to check these values in equation (2).

CHECK YOURSELF 8

Alana bought three digital tapes and two compact disks on sale for \$66. At the same sale, Chen bought three digital tapes and four compact disks for \$96. Find the individual price for a tape and a disk.

Example 9 shows how sketches can be helpful in setting up a problem.

Example 9

Using a Sketch to Help Solve an Application

An 18-ft board is cut into two pieces, one of which is 4 ft longer than the other. How long is each piece?

NOTE You should always draw a sketch of the problem when it is appropriate.

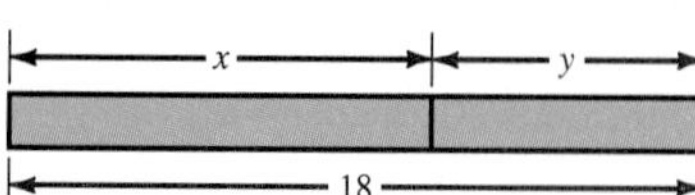

Step 1 You want to find the two lengths.

Step 2 Let x be the length of the longer piece and y the length of the shorter piece.

Step 3 Write the equations for the solution.

$x + y = 18 \longleftarrow$ The total length is 18.

$x - y = 4 \longleftarrow$ The difference in lengths is 4.

NOTE Our second equation could also be written as

$x = y + 4$

Step 4 To solve the system, add:

$$\begin{aligned} x + y &= 18 \qquad (1) \\ x - y &= 4 \qquad (2) \\ \hline 2x &= 22 \\ x &= 11 \end{aligned}$$

Replace x with 11 in equation (1).

$$11 + y = 18$$

$$y = 7$$

The longer piece has length 11 ft, the shorter piece 7 ft.

Step 5 We leave it to you to check this result in the original problem.

CHECK YOURSELF 9

A 20-ft board is cut into two pieces, one of which is 6 ft longer than the other. How long is each piece?

Using two equations in two variables also helps in solving **mixture problems.**

Example 10

Solving a Mixture Problem Involving Coins

Winnifred has collected \$4.50 in nickels and dimes. If she has 55 coins, how many of each kind of coin does she have?

Step 1 You want to find the number of nickels and the number of dimes.

Step 2 Let

NOTE Again we choose appropriate variables—n for nickels, d for dimes.

n = number of nickels

d = number of dimes

Step 3 Write the equations for the solution.

REMEMBER: The value of a number of coins is the value per coin times the number of coins: $5n$, $10d$, etc.

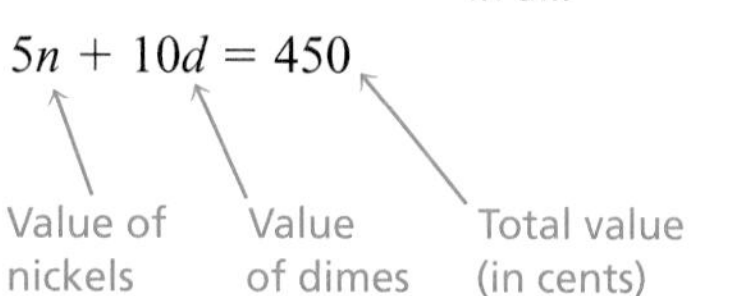

Step 4 We now have the system

$$n + d = 55 \quad (1)$$

$$5n + 10d = 450 \quad (2)$$

Let's solve this system by addition. Multiply equation (1) by -5. We then add the equation to eliminate the variable n.

$$-5n - 5d = -275$$
$$5n + 10d = 450$$
$$5d = 175$$
$$d = 35$$

We now substitute d for 35 in equation (1).

$$n + 35 = 55$$
$$n = 20$$

There are 20 nickels and 35 dimes.

Step 5 We leave it to you to check this result. Just verify that the value of these coins is \$4.50.

CHECK YOURSELF 10

Tickets for a play cost \$8 or \$6. If 350 tickets were sold in all and receipts were \$2500, how many of each price ticket were sold?

We can also solve mixture problems that involve percentages by using two equations in two unknowns. Look at Example 11.

Example 11

Solving a Mixture Problem Involving Chemicals

In a chemistry lab are two solutions: a 20% acid solution and a 60% acid solution. How many milliliters of each should be mixed to produce 200 milliliters (mL) of a 44% acid solution?

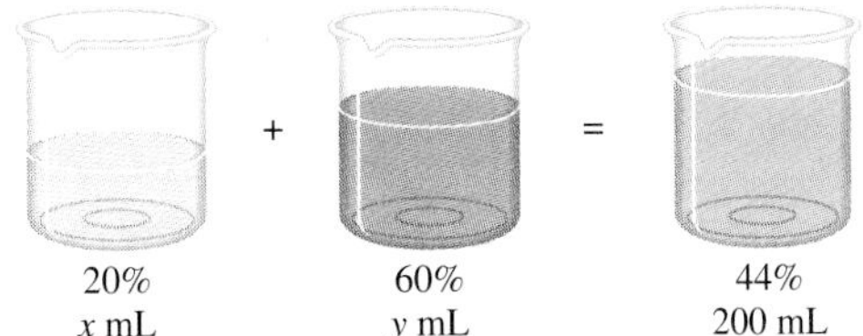

Step 1 You need to know the amount of each solution to use.

Step 2 Let

x = amount of 20% acid solution

y = amount of 60% acid solution

Step 3 A drawing will help. Note that a 20% acid solution is 20% acid and 80% water.

We can write equations from the total amount of the solution, here 200 mL, and from the amount of acid in that solution. Many students find a table helpful in organizing the information at this point. Here, for example, we might have

NOTE The amount of acid is the amount of solution times the percentage of acid (as a decimal). That is the key to forming the third column of our table.

	Amount of Solution	% Acid	Amount of Acid
	x	0.20	$0.20x$
	y	0.60	$0.60y$
Totals	200	0.44	(0.44)(200)

NOTE Equation (1) is the total amount of the solution from the first column of our table.

NOTE Equation (2) is the amount of acid from the third column of our table.

Now we are ready to form our system.

$$x + y = 200 \qquad (1)$$

$$0.20x + 0.60y = 0.44(200) \qquad (2)$$

Acid in 20% solution. Acid in 60% solution. Acid in mixture.

Step 4 If we multiply equation (2) by 100 to clear it of decimals, we have

$$x + y = 200 \xrightarrow[\text{by } -20]{\text{Multiply}} -20x - 20y = -4000$$

$$20x + 60y = 8800 \longrightarrow 20x + 60y = 8800$$

$$40y = 4800$$

$$y = 120$$

Substituting 120 for y in equation (1), we have

$$x + 120 = 200$$

$$x = 80$$

The amounts to be mixed are 80 mL (20% acid solution) and 120 mL (60% acid solution).

Step 5 You can check this solution by verifying that the amount of acid from the 20% solution added to the amount from the 60% solution is equal to the amount of acid in the mixture.

CHECK YOURSELF 11

You have a 30% alcohol solution and a 50% alcohol solution. How much of each solution should be combined to make 400 mL of a 45% alcohol solution?

A related kind of application involves interest. The key equation involves the *principal* (the amount invested), the annual *interest rate,* the *time* (in years) that the money is invested, and the amount of *interest* you receive.

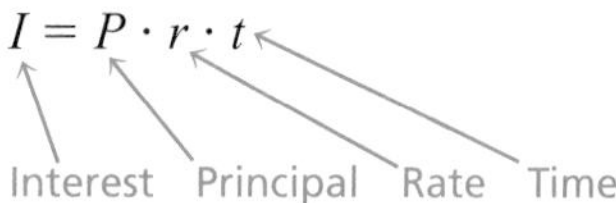

For 1 year we have

$I = P \cdot r$ because $t = 1$

Example 12

Solving an Investment Application

Jeremy inherits \$20,000 and invests part of the money in bonds with an interest rate of 11%. The remainder of the money is in savings at a 9% rate. What amount has he invested at each rate if he receives \$2040 in interest for 1 year?

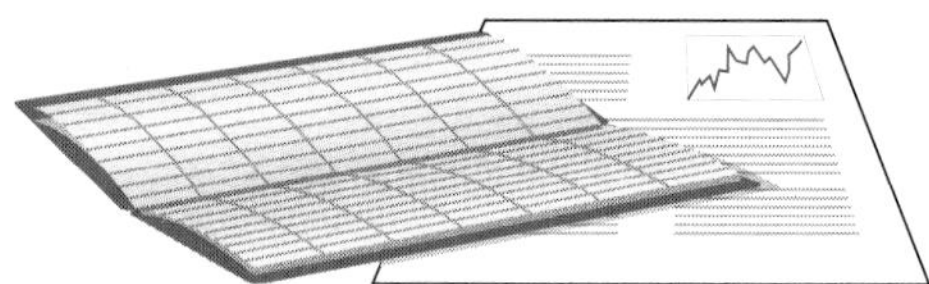

Step 1 You want to find the amounts invested at 11 and at 9%.

NOTE The amount invested at 11% could have been represented by y and the amount invested at 9% by x.

Step 2 Let $x =$ the amount invested at 11% and $y =$ the amount invested at 9%. Once again you may find a table helpful at this point.

NOTE The formula $I = P \cdot r$ (interest equals principal times rate) is the key to forming the third column of our table.

	Principal	Rate	Interest
	x	11%	$0.11x$
	y	9%	$0.09y$
Totals	20,000		2040

Step 3 Form the equations for the solution, using the first and third columns of the above table.

$x + y = 20{,}000$ ⟵ He has \$20,000 invested in all.

NOTE Notice the decimal form of 11% and 9% is used in the equation.

$0.11x + 0.09y = 2040$

The interest at 11% (rate · principal) — The interest at 9% — The total interest

Step 4 To solve the following system, use addition.

$$x + \quad y = 20{,}000 \qquad (1)$$

$$0.11x + 0.09y = 2{,}040 \qquad (2)$$

To do this, multiply both sides of equation (1) by -9. Multiplying both sides of equation (2) by 100 will clear decimals. Adding the resulting equations will eliminate y.

$$\begin{aligned} -9x - 9y &= -180{,}000 \\ 11x + 9y &= 204{,}000 \\ \hline 2x \qquad &= 24{,}000 \\ x &= 12{,}000 \end{aligned}$$

Now, substitute 12,000 for x in equation (1) and solve for y.

$$12{,}000 + y = 20{,}000$$

$$y = 8{,}000$$

NOTE Be sure to answer the question asked in the problem.

Jeremy has \$12,000 invested at 11% and \$8000 invested at 9%.

Step 5 To check, the interest at 11% is (\$12,000)(0.11), or \$1320. The interest at 9% is (\$8000)(0.09), or \$720. The total interest is \$2040, and the solution is verified.

CHECK YOURSELF 12

Jan has \$2000 *more invested in a stock that pays* 9% *interest than in a savings account paying* 8%. *If her total interest for* 1 *year is* \$860, *how much does she have invested at each rate?*

NOTE Distance, rate, and time of travel are related by the equation

Another group of applications is called **motion problems;** they involve a distance traveled, the rate, and the time of travel. Example 13 shows the use of $d = r \cdot t$ in forming a system of equations to solve a motion problem.

Example 13

Solving a Motion Problem

A boat can travel 36 miles (mi) downstream in 2 hours (h). Coming back upstream, the trip takes 3 h. Find the rate of the boat in still water and the rate of the current.

Step 1 You want to find the two rates (of the boat and the current).

Step 2 Let

x = rate of boat in still water

y = rate of current

Step 3 To write the equations, think about the following: What is the effect of the current? Suppose the boat's rate in still water is 10 mi/h and the current is 2 mi/h.

The current *increases* the rate *downstream* to 12 mi/h $(10 + 2)$. The current *decreases* the rate *upstream* to 8 mi/h $(10 - 2)$. So here the rate downstream will be $x + y$, and the rate upstream will be $x - y$. At this point a table of information is helpful.

	Distance	Rate	Time
Downstream	36	$x + y$	2
Upstream	36	$x - y$	3

From the relationship $d = r \cdot t$ we can now use our table to write the system

$36 = 2(x + y)$ (From line 1 of our table)

$36 = 3(x - y)$ (From line 2 of our table)

Step 4 Removing the parentheses in the equations of step 3, we have

$2x + 2y = 36$

$3x - 3y = 36$

By either of our earlier methods, this system gives values of 15 for x and 3 for y. The rate in still water is 15 mi/h, and the rate of the current is 3 mi/h. We leave the check to you.

CHECK YOURSELF 13

A plane flies 480 mi with the wind in 4 h. In returning against the wind, the trip takes 6 h. What is the rate of the plane in still air? What was the rate of the wind?

CHECK YOURSELF ANSWERS

1. (2, 4) **2.** (1, −2) **3.** (5, −2) **4.** (−6, 2) **5.** (2, 1)
6. There are infinitely many solutions. It is a dependent system.
7. There is no solution. The system is inconsistent. **8.** Tape \$12, disk \$15
9. 7 ft, 13 ft **10.** 150 \$6 tickets, 200 \$8 tickets
11. 100 mL (30%), 300 mL (50%) **12.** \$4000 at 8%, \$6000 at 9%
13. Plane's rate in still air, 100 mi/h; wind's rate, 20 mi/h

Name ______________

8.2 Exercises

Section ________ Date ________

Solve each of the following systems by addition. If a unique solution does not exist, state whether the system is inconsistent or dependent.

1. $x + y = 6$
   $x - y = 4$

2. $x - y = 8$
   $x + y = 2$

3. $-x + y = 3$
   $x + y = 5$

4. $x + y = 7$
   $-x + y = 3$

5. $2x - y = 1$
   $-2x + 3y = 5$

6. $x - 2y = 2$
   $x + 2y = -14$

7. $x + 3y = 12$
   $2x - 3y = 6$

8. $-3x + y = 8$
   $3x - 2y = -10$

9. $x + 2y = -2$
   $3x + 2y = -12$

10. $4x - 3y = 22$
    $4x + 5y = 6$

11. $4x - 3y = 6$
    $4x + 5y = 22$

12. $2x + 3y = 1$
    $5x + 3y = 16$

13. $2x + y = 8$
    $2x + y = 2$

14. $5x + 4y = 7$
    $5x - 2y = 19$

15. $3x - 5y = 2$
    $2x - 5y = -2$

16. $2x - y = 4$
    $2x - y = 6$

17. $x + y = 3$
    $3x - 2y = 4$

18. $x - y = -2$
    $2x + 3y = 21$

19. $-5x + 2y = -3$
    $x - 3y = -15$

20. $x + 5y = 10$
    $-2x - 10y = -20$

ANSWERS

1. (5, 1)
2. (5, −3)
3. (1, 4)
4. (2, 5)
5. (2, 3)
6. (−6, −4)
7. (6, 2)
8. (−2, 2)
9. $\left(-5, \frac{3}{2}\right)$
10. (4, −2)
11. (3, 2)
12. (5, −3)
13. Inconsistent system
14. (3, −2)
15. (4, 2)
16. Inconsistent system
17. (2, 1)
18. (3, 5)
19. (3, 6)
20. Dependent system

ANSWERS

69. *d*

70. *f*

71. *g*

72. *a*

73. *h*

74. *b*

Each of the following applications (Exercises 69 to 76) can be solved by the use of a system of linear equations. Match the application with the system on the right that could be used for its solution.

69. Number problem. One number is 4 less than 3 times another. If the sum of the numbers is 36, what are the two numbers?

(*a*) $\begin{aligned} 12x + 5y &= 116 \\ 8x + 12y &= 112 \end{aligned}$

70. Tickets sold. Suppose that a movie theater sold 300 adult and student tickets for a showing with a revenue of $1440. If the adult tickets were $6 and the student tickets $4, how many of each type of ticket were sold?

(*b*) $\begin{aligned} x + y &= 8000 \\ 0.06x + 0.09y &= 600 \end{aligned}$

71. Dimensions of a rectangle. The length of a rectangle is 3 centimeters (cm) more than twice its width. If the perimeter of the rectangle is 36 cm, find the dimensions of the rectangle.

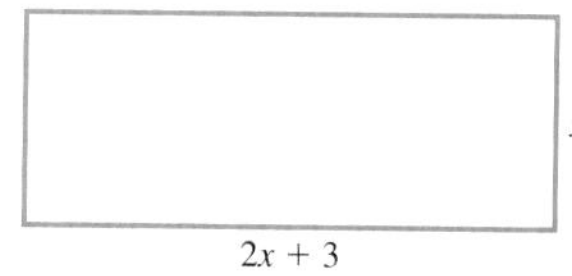

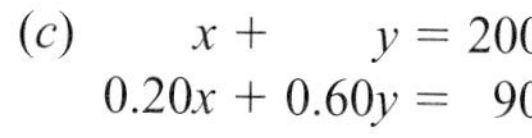
(*c*) $\begin{aligned} x + y &= 200 \\ 0.20x + 0.60y &= 90 \end{aligned}$

72. Cost of pens. An order of 12 dozen roller-ball pens and 5 dozen ballpoint pens cost $116. A later order for 8 dozen roller-ball pens and 12 dozen ballpoint pens cost $112. What was the cost of 1 dozen of each of the pens?

(*d*) $\begin{aligned} x + y &= 36 \\ y &= 3x - 4 \end{aligned}$

73. Nut mixture. A candy merchant wishes to mix peanuts selling at $2/lb with cashews selling for $5.50/lb to form 140 lb of a mixed-nut blend that will sell for $3/lb. What amount of each type of nut should be used?

(*e*) $\begin{aligned} 5(x - y) &= 80 \\ 4(x + y) &= 80 \end{aligned}$

74. Investments. Rolando has investments totaling $8000 in two accounts, one a savings account paying 6% interest and the other a bond paying 9%. If the annual interest from the two investments was $600, how much did he have invested at each rate?

(*f*) $\begin{aligned} x + y &= 300 \\ 6x + 4y &= 1440 \end{aligned}$

Name ______________________

Section ________ Date ________

8.2 Exercises

Solve each of the following systems by addition. If a unique solution does not exist, state whether the system is inconsistent or dependent.

1. $x + y = 6$
$x - y = 4$

2. $x - y = 8$
$x + y = 2$

3. $-x + y = 3$
$x + y = 5$

4. $x + y = 7$
$-x + y = 3$

5. $2x - y = 1$
$-2x + 3y = 5$

6. $x - 2y = 2$
$x + 2y = -14$

7. $x + 3y = 12$
$2x - 3y = 6$

8. $-3x + y = 8$
$3x - 2y = -10$

9. $x + 2y = -2$
$3x + 2y = -12$

10. $4x - 3y = 22$
$4x + 5y = 6$

11. $4x - 3y = 6$
$4x + 5y = 22$

12. $2x + 3y = 1$
$5x + 3y = 16$

13. $2x + y = 8$
$2x + y = 2$

14. $5x + 4y = 7$
$5x - 2y = 19$

15. $3x - 5y = 2$
$2x - 5y = -2$

16. $2x - y = 4$
$2x - y = 6$

17. $x + y = 3$
$3x - 2y = 4$

18. $x - y = -2$
$2x + 3y = 21$

19. $-5x + 2y = -3$
$x - 3y = -15$

20. $x + 5y = 10$
$-2x - 10y = -20$

ANSWERS

1. (5, 1)
2. (5, −3)
3. (1, 4)
4. (2, 5)
5. (2, 3)
6. (−6, −4)
7. (6, 2)
8. (−2, 2)
9. $\left(-5, \frac{3}{2}\right)$
10. (4, −2)
11. (3, 2)
12. (5, −3)
13. Inconsistent system
14. (3, −2)
15. (4, 2)
16. Inconsistent system
17. (2, 1)
18. (3, 5)
19. (3, 6)
20. Dependent system

ANSWERS

21. $(2, -4)$
22. $(6, 4)$
23. $\left(3, \frac{13}{2}\right)$
24. $(3, -2)$
25. Dependent system
26. $(-4, 3)$
27. $(5, -3)$
28. $(2, -9)$
29. $(6, 3)$
30. $(4, 2)$
31. $(-8, -2)$
32. $(-4, 3)$
33. $\left(2, \frac{3}{2}\right)$
34. $\left(3, -\frac{5}{2}\right)$
35. $(3, 0)$
36. $(-6, 0)$
37. $(4, 0)$
38. $(0, -5)$
39. $\left(\frac{1}{2}, -\frac{1}{2}\right)$
40. $\left(\frac{3}{4}, -\frac{1}{2}\right)$

21. $7x + y = 10$
$2x + 3y = -8$

22. $3x - 4y = 2$
$4x - y = 20$

23. $5x + 2y = 28$
$x - 4y = -23$

24. $7x + 2y = 17$
$x - 5y = 13$

25. $3x - 4y = 2$
$-6x + 8y = -4$

26. $-x + 5y = 19$
$4x + 3y = -7$

27. $5x - 2y = 31$
$4x + 3y = 11$

28. $7x + 3y = -13$
$5x + 2y = -8$

29. $3x - 2y = 12$
$5x - 3y = 21$

30. $-4x + 5y = -6$
$5x - 2y = 16$

31. $-2x + 7y = 2$
$3x - 5y = -14$

32. $3x + 4y = 0$
$5x - 3y = -29$

33. $7x + 4y = 20$
$5x + 6y = 19$

34. $5x + 4y = 5$
$7x - 6y = 36$

35. $2x - 7y = 6$
$-4x + 3y = -12$

36. $3x + 2y = -18$
$7x - 6y = -42$

37. $5x - y = 20$
$4x + 3y = 16$

38. $3x + y = -5$
$5x - 4y = 20$

39. $3x + y = 1$
$5x + y = 2$

40. $2x - y = 2$
$2x + 5y = -1$

41. $3x + 4y = 3$
$6x - 2y = 1$

42. $3x + 3y = 1$
$2x + 4y = 2$

43. $5x - 2y = \frac{9}{5}$
$3x + 4y = -1$

44. $2x + 3y = -\frac{1}{12}$
$5x + 4y = \frac{2}{3}$

Solve the following systems by adding. If a unique solution does not exist, state whether the system is inconsistent or dependent.

45. $\frac{x}{3} - \frac{y}{4} = -\frac{1}{2}$
$\frac{x}{2} - \frac{y}{5} = \frac{3}{10}$

46. $\frac{1}{3}x - \frac{1}{2}y = \frac{5}{6}$
$\frac{1}{2}x - \frac{2}{5}y = \frac{9}{10}$

47. $0.4x - 0.2y = 0.6$
$0.5x - 0.6y = 9.5$

48. $0.2x + 0.37y = 0.8$
$-0.6x + 1.4y = 2.62$

Solve each of the following problems. Be sure to show the equations used for the solution.

49. **Number problem.** The sum of two numbers is 40. Their difference is 8. Find the two numbers.

50. **Cost of stamps.** Eight eagle stamps and two raccoon stamps cost \$2.80. Three eagle stamps and four raccoon stamps cost \$2.35. Find the cost of each kind of stamp.

51. **Cost of food.** Robin bought four chocolate bars and a pack of gum and paid \$2.75. Meg bought two chocolate bars and three packs of gum and paid \$2.25. Find the cost of each.

ANSWERS

41. $\left(\frac{1}{3}, \frac{1}{2}\right)$
42. $\left(-\frac{1}{3}, \frac{2}{3}\right)$
43. $\left(\frac{1}{5}, -\frac{2}{5}\right)$
44. $\left(\frac{1}{3}, -\frac{1}{4}\right)$
45. (3, 6)
46. (1, −1)
47. (−11, −25)
48. (0.3, 2)
49. (24, 16)
50. Eagle 25¢, raccoon 40¢
51. Chocolate 60¢, gum 35¢

ANSWERS

52. Red delicious 49¢, Granny Smith 59¢

53. Disk $1.25, Zip disk $3.50

54. 18 m, 12 m

55. 12 ft, 6 ft

56. 30 nickels, 20 dimes

57. 10 dimes, 12 quarters

58. 160 general, 80 student

59. 250 at $7, 150 at $9

60. 25 lb at $9, 75 lb at $12

61. 15 lb peanuts, 5 lb cashews

52. Cost of apples. Xavier bought five red delicious apples and four Granny Smith apples at a cost of $4.81. Dean bought one of each of the two types at a cost of $1.08. Find the cost for each kind of apple.

53. Cost of disks. Eight disks and five zip disks cost a total of $27.50. Two disks and four zip disks cost $16.50. Find the unit cost for each.

54. Length. A 30-meter (m) rope is cut into two pieces so that one piece is 6 m longer than the other. How long is each piece?

55. Length. An 18-foot (ft) board is cut into two pieces, one of which is twice as long as the other. How long is each piece?

56. Number of coins. Jill has $3.50 in nickels and dimes. If she has 50 coins, how many of each type of coin does she have?

57. Number of coins. Richard has 22 coins with a total value of $4. If the coins are all quarters and dimes, how many of each type of coin does he have?

58. Tickets sold. Theater tickets are $8 for general admission and $5 for students. During one evening 240 tickets were sold, and the receipts were $1680. How many of each kind of ticket were sold?

59. Tickets sold. 400 tickets were sold for a concert. The receipts from ticket sales were $3100, and the ticket prices were $7 and $9. How many of each price ticket were sold?

60. Coffee mixture. A coffee merchant has coffee beans that sell for $9 per pound (lb) and $12 per pound. The two types are to be mixed to create 100 pounds of a mixture that will sell for $11.25 per pound. How much of each type of bean should be used in the mixture?

61. Nut mixture. Peanuts are selling for $2 per pound, and cashews are selling for $5 per pound. How much of each type of nut would be needed to create 20 lb of a mixture that would sell for $2.75 per pound?

62. Acid solution. A chemist has a 25% and a 50% acid solution. How much of each solution should be used to form 200 milliliters (mL) of a 35% acid solution?

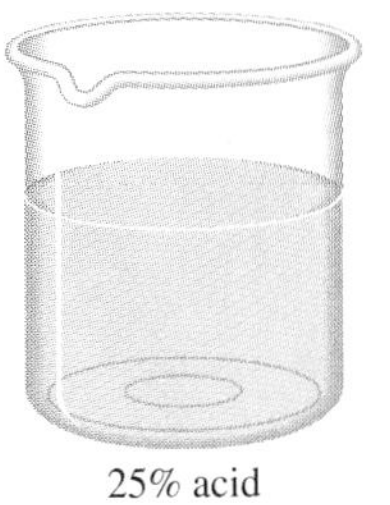
25% acid

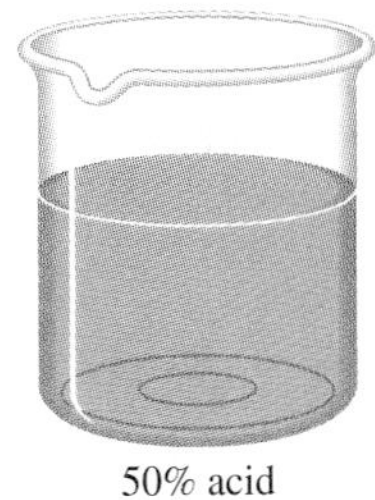
50% acid

63. Alcohol solution. A pharmacist wishes to prepare 150 mL of a 20% alcohol solution. She has a 30% solution and a 15% solution in her stock. How much of each should be used in forming the desired mixture?

64. Alcohol solution. You have two alcohol solutions, one a 15% solution and one a 45% solution. How much of each solution should be used to obtain 300 mL of a 25% solution?

65. Investment. Otis has a total of $12,000 invested in two accounts. One account pays 8% and the other 9%. If his interest for 1 year is $1010, how much does he have invested at each rate?

66. Investment. Amy invests a part of $8000 in bonds paying 12% interest. The remainder is in a savings account at 8%. If she receives $840 in interest for 1 year, how much does she have invested at each rate?

67. Motion problem. A plane flies 450 miles (mi) with the wind in 3 hours (h). Flying back against the wind, the plane takes 5 h to make the trip. What was the rate of the plane in still air? What was the rate of the wind?

68. Motion problem. An airliner made a trip of 1800 mi in 3 h, flying east across the country with the jetstream directly behind it. The return trip, against the jetstream, took 4 h. Find the speed of the plane in still air and the speed of the jetstream.

ANSWERS

62. 120 mL of 25%, 80 mL of 50%

63. 50 mL of 30%, 100 mL of 15%

64. 200 mL of 15%, 100 mL of 45%

65. $7000 at 8%, $5000 at 9%

66. $3000 at 8%, $5000 at 12%

67. 120 mi/h, 30 mi/h

68. 525 mi/h, 75 mi/h

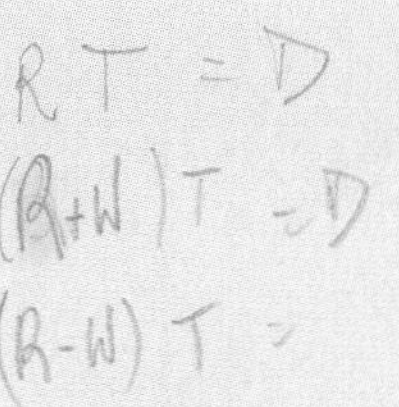

ANSWERS

69. *d*

70. *f*

71. *g*

72. *a*

73. *h*

74. *b*

Each of the following applications (Exercises 69 to 76) can be solved by the use of a system of linear equations. Match the application with the system on the right that could be used for its solution.

69. Number problem. One number is 4 less than 3 times another. If the sum of the numbers is 36, what are the two numbers?

70. Tickets sold. Suppose that a movie theater sold 300 adult and student tickets for a showing with a revenue of $1440. If the adult tickets were $6 and the student tickets $4, how many of each type of ticket were sold?

71. Dimensions of a rectangle. The length of a rectangle is 3 centimeters (cm) more than twice its width. If the perimeter of the rectangle is 36 cm, find the dimensions of the rectangle.

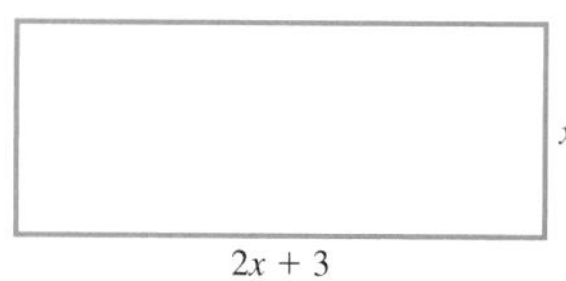

72. Cost of pens. An order of 12 dozen roller-ball pens and 5 dozen ballpoint pens cost $116. A later order for 8 dozen roller-ball pens and 12 dozen ballpoint pens cost $112. What was the cost of 1 dozen of each of the pens?

73. Nut mixture. A candy merchant wishes to mix peanuts selling at $2/lb with cashews selling for $5.50/lb to form 140 lb of a mixed-nut blend that will sell for $3/lb. What amount of each type of nut should be used?

74. Investments. Rolando has investments totaling $8000 in two accounts, one a savings account paying 6% interest and the other a bond paying 9%. If the annual interest from the two investments was $600, how much did he have invested at each rate?

(*a*) $12x + 5y = 116$
$8x + 12y = 112$

(*b*) $x + y = 8000$
$0.06x + 0.09y = 600$

(*c*) $x + y = 200$
$0.20x + 0.60y = 90$

(*d*) $x + y = 36$
$y = 3x - 4$

(*e*) $5(x - y) = 80$
$4(x + y) = 80$

(*f*) $x + y = 300$
$6x + 4y = 1440$

75. Alcohol solution. A chemist wants to combine a 20% alcohol solution with a 60% solution to form 200 milliliters (mL) of a 45% solution. How much of each of the solutions should be used to form the mixture?

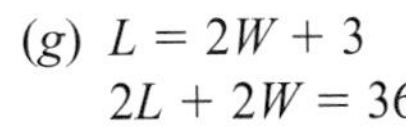

(g) $L = 2W + 3$
$2L + 2W = 36$

(h) $x + y = 140$
$2x + 5.5y = 420$

76. Motion Problem. A boat traveled 80 mi upstream in 5 h. Returning downstream with the current, the boat took 4 h to make the trip. What was the boat's speed in still water? What was the speed of the river's current?

77. Writing for reflection. Write in response to the questions below. 

Many people find word problems easier to do when two variables and two equations are used to model the problem. Compare the problems in this section with the problems in Section 2.5. How does this method help you in solving application problems? In general, how would you evaluate your ability to solve application or word problems? What can you do to improve?

78. Work with a partner to solve the following problems. 

Your friend, Valerie, has contacted you about going into business with her. She wants to start a small manufacturing business making and selling sweaters to specialty boutiques. She explains that the initial investment for each of you will be $1500 for a knitting machine. She has worked hard to come up with an estimate for expenses and thinks that they will be close to $1600 a month for overhead. She says that each sweater manufactured will cost $28 to produce and that the sweaters will sell for at least $70. She wants to know if you are willing to invest the money you have saved for college costs. You have faith in Valerie's ability to carry out her plan. But, you have worked hard to save this money. Use graphs and equations to help you decide if this is a good opportunity. Think about whether you need more information from Valerie. Write a letter summarizing your thoughts.

ANSWERS

75. c

76. e

77.

78.

ANSWERS

a. 2

b. 3

c. 5

d. −2

Getting Ready for Section 8.3 [Section 2.3]

Solve each of the following equations.

(a) $2x + 3(x + 1) = 13$

(b) $3(y - 1) + 4y = 18$

(c) $x + 2(3x - 5) = 25$

(d) $3x - 2(x - 7) = 12$

Answers

1. (5, 1) **3.** (1, 4) **5.** (2, 3) **7.** (6, 2) **9.** $\left(-5, \frac{3}{2}\right)$ **11.** (3, 2) **13.** Inconsistent system **15.** (4, 2) **17.** (2, 1) **19.** (3, 6) **21.** (2, −4) **23.** $\left(3, \frac{13}{2}\right)$ **25.** Dependent system **27.** (5, −3) **29.** (6, 3) **31.** (−8, −2) **33.** $\left(2, \frac{3}{2}\right)$ **35.** (3, 0) **37.** (4, 0) **39.** $\left(\frac{1}{2}, -\frac{1}{2}\right)$ **41.** $\left(\frac{1}{3}, \frac{1}{2}\right)$ **43.** $\left(\frac{1}{5}, -\frac{2}{5}\right)$ **45.** (3, 6) **47.** (−11, −25) **49.** (24, 16) **51.** Chocolate 60¢, gum 35¢ **53.** Disk \$1.25, zip disk \$3.50 **55.** 12 ft, 6 ft **57.** 10 dimes, 12 quarters **59.** 250 at \$7, 150 at \$9 **61.** 15 lb peanuts, 5 lb cashews **63.** 50 mL of 30%, 100 mL of 15% **65.** \$7000 at 8%, \$5000 at 9% **67.** 120 mi/h, 30 mi/h **69.** *d* **71.** *g* **73.** *h* **75.** *c* **77.**  **a.** 2 **b.** 3 **c.** 5 **d.** −2

8.3 Systems of Linear Equations: Solving by Substitution

8.3 OBJECTIVES

1. Solve systems using the substitution method
2. Solve applications of systems of equations

In Sections 8.1 and 8.2, we looked at graphing and addition as methods of solving linear systems. A third method is called **solution by substitution.**

Example 1

Solving a System by Substitution

Solve by substitution.

$$x + y = 12 \qquad (1)$$

$$y = 3x \qquad (2)$$

Notice that equation (2) says that y and $3x$ name the same quantity. So we may substitute $3x$ for y in equation (1). We then have

Replace y with $3x$ in equation (1).

↓

NOTE The resulting equation contains only the variable x, so substitution is just another way of eliminating one of the variables from our system.

$$x + 3x = 12$$

$$4x = 12$$

$$x = 3$$

We can now substitute 3 for x in equation (1) to find the corresponding y coordinate of the solution.

$$3 + y = 12$$

$$y = 9$$

NOTE The solution for a system is written as an ordered pair.

So (3, 9) is the solution.

This last step is identical to the one you saw in Section 8.2. As before, you can substitute the known coordinate value back into either of the original equations to find the value of the remaining variable. The check is also identical.

CHECK YOURSELF 1

Solve by substitution.

$$x - y = 9$$

$$y = 4x$$

The same technique can be readily used any time one of the equations is *already solved* for x or for y, as Example 2 illustrates.

Example 2

Solving a System by Substitution

Solve by substitution.

$$2x + 3y = 3 \quad (1)$$

$$y = 2x - 7 \quad (2)$$

Because equation (2) tells us that y is $2x - 7$, we can replace y with $2x - 7$ in equation (1). This gives

NOTE Now y is eliminated from the equation, and we can proceed to solve for x.

$$2x + 3\overbrace{(2x - 7)}^{y} = 3$$

$$2x + 6x - 21 = 3$$

$$8x = 24$$

$$x = 3$$

We now know that 3 is the x coordinate for the solution. So substituting 3 for x in equation (2), we have

$$y = 2 \cdot 3 - 7$$

$$= 6 - 7$$

$$= -1$$

And $(3, -1)$ is the solution. Once again you should verify this result by letting $x = 3$ and $y = -1$ in the original system.

CHECK YOURSELF 2

Solve by substitution.

$$2x - 3y = 6$$

$$x = 4y - 2$$

As we have seen, the substitution method works very well when one of the given equations is already solved for x or for y. It is also useful if you can readily solve for x or for y in one of the equations.

Example 3

Solving a System by Substitution

Solve by substitution.

$$x - 2y = 5 \quad (1)$$

$$3x + y = 8 \quad (2)$$

Neither equation is solved for a variable. That is easily handled in this case. Solving for x in equation (1), we have

$$x = 2y + 5$$

NOTE Equation (2) could have been solved for y with the result substituted into equation (1).

Now substitute $2y + 5$ for x in equation (2).

$$\overbrace{3(2y + 5)}^{x} + y = 8$$

$$6y + 15 + y = 8$$

$$7y = -7$$

$$y = -1$$

Substituting -1 for y in equation (2) yields

$$3x + (-1) = 8$$

$$3x = 9$$

$$x = 3$$

So $(3, -1)$ is the solution. You should check this result by substituting 3 for x and -1 for y in the equations of the original system.

CHECK YOURSELF 3

Solve by substitution.

$$3x - y = 5$$

$$x + 4y = 6$$

Inconsistent systems and dependent systems will show up in a fashion similar to that which we saw in Section 8.2. Example 4 illustrates this approach.

Example 4

Solving an Inconsistent or Dependent System

Solve the following systems by substitution.

(a) $4x - 2y = 6$ (1)

$y = 2x - 3$ (2)

From equation (2) we can substitute $2x - 3$ for y in equation (1).

NOTE Don't forget to change both signs in the parentheses.

$$4x - 2(2x - 3) = 6$$

$$4x - 4x + 6 = 6$$

$$6 = 6$$

Both variables have been eliminated, and we have the true statement $6 = 6$.

Recall from the last section that a true statement tells us that the lines coincide. We call this system dependent. There are an infinite number of solutions.

(b) $3x - 6y = 9$ (3)

$x = 2y + 2$ (4)

Substitute $2y + 2$ for x in equation (3).

$3(2y + 2) - 6y = 9$

$6y + 6 - 6y = 9$

$6 = 9$

This time we have a false statement.

This means that the system is *inconsistent* and that the graphs of the two equations are parallel lines. There is no solution.

CHECK YOURSELF 4

Indicate whether the systems are inconsistent (no solution) or dependent (an infinite number of solutions).

(a) $5x + 15y = 10$
$x = -3y + 1$

(b) $12x - 4y = 8$
$y = 3x - 2$

The following summarizes our work in this section.

Step by Step: To Solve a System of Linear Equations by Substitution

Step 1 Solve one of the given equations for *x* or *y*. If this is already done, go on to step 2.

Step 2 Substitute this expression for *x* or for *y* into the other equation.

Step 3 Solve the resulting equation for the remaining variable.

Step 4 Substitute the known value into either of the original equations to find the value of the second variable.

Step 5 Check your solution in both of the original equations.

You have now seen three different ways to solve systems of linear equations: by graphing, adding, and substitution. The natural question is, Which method should I use in a given situation?

Graphing is the least exact of the methods, and solutions may have to be estimated.

The algebraic methods—addition and substitution—give exact solutions, and both will work for any system of linear equations. In fact, you may have noticed that several examples in this section could just as easily have been solved by adding (Example 3, for instance).

The choice of which algebraic method (substitution or addition) to use is yours and depends largely on the given system. Here are some guidelines designed to help you choose an appropriate method for solving a linear system.

Rules and Properties: Choosing an Appropriate Method for Solving a System

1. If one of the equations is already solved for x (or for y), then substitution is the preferred method.
2. If the coefficients of x (or of y) are the same, or opposites, in the two equations, then addition is the preferred method.
3. If solving for x (or for y) in either of the given equations will result in fractional coefficients, then addition is the preferred method.

Example 5

Choosing an Appropriate Method for Solving a System

Select the most appropriate method for solving each of the following systems.

(a) $5x + 3y = 9$

$2x - 7y = 8$

Addition is the most appropriate method because solving for a variable will result in fractional coefficients.

(b) $7x + 26 = 8$

$x = 3y - 5$

Substitution is the most appropriate method because the second equation is already solved for x.

(c) $8x - 9y = 11$

$4x + 9y = 15$

Addition is the most appropriate method because the coefficients of y are opposites.

CHECK YOURSELF 5

Select the most appropriate method for solving each of the following systems.

(a) $2x + 5y = 3$
$8x - 5y = -13$

(b) $4x - 3y = 2$
$y = 3x - 4$

(c) $3x - 5y = 2$
$x = 3y - 2$

(d) $5x - 2y = 19$
$4x + 6y = 38$

Number problems, such as those presented in Chapter 2, are sometimes more easily solved by the methods presented in this section. Example 6 illustrates this approach.

Example 6

Solving a Number Problem by Substitution

The sum of two numbers is 25. If the second number is 5 less than twice the first number, what are the two numbers?

NOTE
1. What do you want to find?

Step 1 You want to find the two unknown numbers.

2. Assign variables. This time we use two letters, x and y.

Step 2 Let x = the first number and y = the second number.

3. Write equations for the solution. Here two equations are needed because we have introduced two variables.

Step 3

$$\underbrace{x + y}_{\text{The sum}} \underbrace{= 25}_{\text{is 25.}}$$

$$\underbrace{y}_{\text{The second number}} = \underbrace{2x - 5}_{\text{is 5 less than twice the first.}}$$

4. Solve the system of equations.

Step 4

$$x + y = 25 \qquad (1)$$

$$y = 2x - 5 \qquad (2)$$

NOTE We use the substitution method because equation (2) is already solved for y.

Substitute $2x - 5$ for y in equation (1).

$$x + (2x - 5) = 25$$

$$3x - 5 = 25$$

$$x = 10$$

From equation (1),

$$10 + y = 25$$

$$y = 15$$

The two numbers are 10 and 15.

5. Check the result.

Step 5 The sum of the numbers is 25. The second number, 15, is 5 less than twice the first number, 10. The solution checks.

CHECK YOURSELF 6

The sum of two numbers is 28. The second number is 4 more than twice the first number. What are the numbers?

Sketches are always helpful in solving applications from geometry. Let's look at such an example.

Example 7

Solving an Application from Geometry

The length of a rectangle is 3 meters (m) more than twice its width. If the perimeter of the rectangle is 42 m, find the dimensions of the rectangle.

Step 1 You want to find the dimensions (length and width) of the rectangle.

NOTE We used x and y as our two variables in the previous examples. Use whatever letters you want. The process is the same, and sometimes it helps you remember what letter stands for what. Here L = length and W = width.

Step 2 Let L be the length of the rectangle and W the width. Now draw a sketch of the problem.

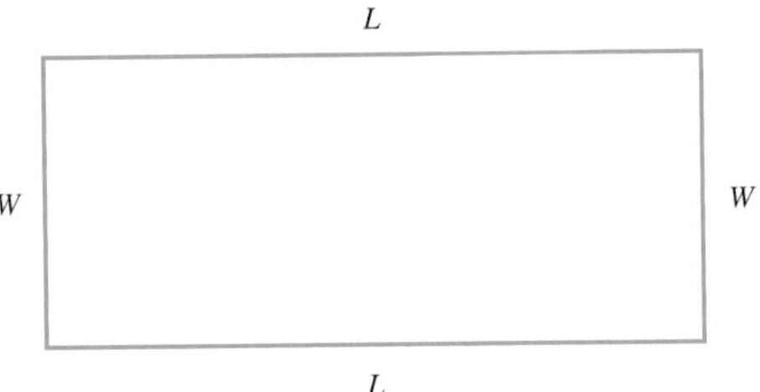

Step 3 Write the equations for the solution.

$L = \underbrace{2W + 3}$

3 more than twice the width

$\underbrace{2L + 2W} = 42$

The perimeter

Step 4 Solve the system.

$$L = 2W + 3 \quad (1)$$

$$2L + 2W = 42 \quad (2)$$

NOTE Substitution is used because one equation is already solved for a variable.

From equation (1) we can substitute $2W + 3$ for L in equation (2).

$$2(2W + 3) + 2W = 42$$

$$4W + 6 + 2W = 42$$

$$6W = 36$$

$$W = 6$$

Replace W with 6 in equation (1) to find L.

$$L = 2 \cdot 6 + 3$$

$$= 12 + 3$$

$$= 15$$

The length is 15 m, the width is 6 m.

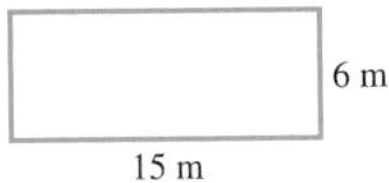

Step 5 Check these results. The perimeter is $2L + 2W$, which should give us 42 m.

$$2(15) + 2(6) \stackrel{?}{=} 42$$

$$30 + 12 \stackrel{\checkmark}{=} 42$$

CHECK YOURSELF 7

The length of each of the two equal legs of an isosceles triangle is 5 in. less than the length of the base. If the perimeter of the triangle is 50 in., find the lengths of the legs and the base.

CHECK YOURSELF ANSWERS

1. $(-3, -12)$ **2.** $(6, 2)$ **3.** $(2, 1)$

4. **(a)** Inconsistent system; **(b)** Dependent system

5. **(a)** Addition; **(b)** Substitution; **(c)** Substitution; **(d)** Addition

6. The numbers are 8 and 20. **7.** The legs have length 15 in.; the base is 20 in.

Name ____________________

Section ________ Date ________

8.3 Exercises

Solve each of the following systems by substitution.

1. $x + y = 10$
   $y = 4x$

2. $x - y = 4$
   $x = 3y$

3. $2x - y = 10$
   $x = -2y$

4. $x + 3y = 10$
   $3x = y$

5. $3x + 2y = 12$
   $y = 3x$

6. $4x - 3y = 24$
   $y = -4x$

7. $x + y = 5$
   $y = x - 3$

8. $x + y = 9$
   $x = y + 3$

9. $x - y = 4$
   $x = 2y - 2$

10. $x - y = 7$
    $y = 2x - 12$

11. $2x + y = 7$
    $y - x = -8$

12. $3x - y = -15$
    $x = y - 7$

13. $2x - 5y = 10$
    $x - y = 8$

14. $4x - 3y = 0$
    $y = x + 1$

15. $3x + 4y = 9$
    $y - 3x = 1$

16. $5x - 2y = -5$
    $y - 5x = 3$

17. $3x - 18y = 4$
    $x = 6y + 2$

18. $4x + 5y = 6$
    $y = 2x - 10$

19. $5x - 3y = 6$
    $y = 3x - 6$

20. $8x - 4y = 16$
    $y = 2x - 4$

21. $8x - 5y = 16$
    $y = 4x - 5$

22. $6x - 5y = 27$
    $x = 5y + 2$

ANSWERS

1. (2, 8)
2. (6, 2)
3. (4, −2)
4. (1, 3)
5. $\left(\frac{4}{3}, 4\right)$
6. $\left(\frac{3}{2}, -6\right)$
7. (4, 1)
8. (6, 3)
9. (10, 6)
10. (5, −2)
11. (5, −3)
12. (−4, 3)
13. (10, 2)
14. (3, 4)
15. $\left(\frac{1}{3}, 2\right)$
16. $\left(-\frac{1}{5}, 2\right)$
17. No solution
18. (4, −2)
19. (3, 3)
20. No solution
21. $\left(\frac{3}{4}, -2\right)$
22. $\left(5, \frac{3}{5}\right)$

ANSWERS

23. (4, 1)
24. (−3, −2)
25. Infinite number of solutions
26. (3, −1)
27. (10, 1)
28. No solution
29. (−2, −1)
30. (−3, −5)
31. (0, −2)
32. (4, 1)
33. (2, 3)
34. (6, 2)
35. Dependent system
36. (5, 0)
37. $\left(-5, \frac{3}{2}\right)$
38. Inconsistent system
39. $\left(\frac{5}{2}, -3\right)$
40. (5, −3)
41. $\left(\frac{3}{2}, 3\right)$
42. $\left(\frac{1}{4}, -\frac{3}{2}\right)$
43. (0, 10)
44. $\left(\frac{1}{2}, \frac{7}{20}\right)$

23. $\begin{aligned} x + 3y &= 7 \\ x - y &= 3 \end{aligned}$

24. $\begin{aligned} 2x - y &= -4 \\ x + y &= -5 \end{aligned}$

25. $\begin{aligned} 6x - 3y &= 9 \\ -2x + y &= -3 \end{aligned}$

26. $\begin{aligned} 5x - 6y &= 21 \\ x - 2y &= 5 \end{aligned}$

27. $\begin{aligned} x - 7y &= 3 \\ 2x - 5y &= 15 \end{aligned}$

28. $\begin{aligned} 4x - 12y &= 5 \\ -x + 3y &= -1 \end{aligned}$

29. $\begin{aligned} 4x + 3y &= -11 \\ 5x + y &= -11 \end{aligned}$

30. $\begin{aligned} 5x - 4y &= 5 \\ 4x - y &= -7 \end{aligned}$

Solve each of the following systems by using either addition or substitution. If a unique solution does not exist, state whether the system is dependent or inconsistent.

31. $\begin{aligned} 2x + 3y &= -6 \\ x &= 3y + 6 \end{aligned}$

32. $\begin{aligned} 7x + 3y &= 31 \\ y &= -2x + 9 \end{aligned}$

33. $\begin{aligned} 2x - y &= 1 \\ -2x + 3y &= 5 \end{aligned}$

34. $\begin{aligned} x + 3y &= 12 \\ 2x - 3y &= 6 \end{aligned}$

35. $\begin{aligned} 6x + 2y &= 4 \\ y &= -3x + 2 \end{aligned}$

36. $\begin{aligned} 3x - 2y &= 15 \\ -x + 5y &= -5 \end{aligned}$

37. $\begin{aligned} x + 2y &= -2 \\ 3x + 2y &= -12 \end{aligned}$

38. $\begin{aligned} 10x + 2y &= 7 \\ y &= -5x + 3 \end{aligned}$

39. $\begin{aligned} 2x - 3y &= 14 \\ 4x + 5y &= -5 \end{aligned}$

40. $\begin{aligned} 2x + 3y &= 1 \\ 5x + 3y &= 16 \end{aligned}$

41. $\begin{aligned} 4x - 2y &= 0 \\ x &= \frac{3}{2} \end{aligned}$

42. $\begin{aligned} 4x - 3y &= \frac{11}{2} \\ y &= -\frac{3}{2} \end{aligned}$

Solve each system.

43. $\begin{aligned} \frac{1}{3}x + \frac{1}{2}y &= 5 \\ \frac{x}{4} - \frac{y}{5} &= -2 \end{aligned}$

44. $\begin{aligned} \frac{5x}{2} - y &= \frac{9}{10} \\ \frac{3x}{4} + \frac{5y}{6} &= \frac{2}{3} \end{aligned}$

45. $0.4x - 0.2y = 0.6$
$2.5x - 0.3y = 4.7$

46. $0.4x - 0.1y = 5$
$6.4x + 0.4y = 60$

Solve each of the following problems. Be sure to show the equation used for the solution.

47. Number problem. The sum of two numbers is 100. The second is three times the first. Find the two numbers.

48. Number problem. The sum of two numbers is 70. The second is 10 more than 3 times the first. Find the numbers.

49. Number problem. The sum of two numbers is 56. The second is 4 less than twice the first. What are the two numbers?

50. Number problem. The difference of two numbers is 4. The larger is 8 less than twice the smaller. What are the two numbers?

51. Number problem. The difference of two numbers is 22. The larger is 2 more than 3 times the smaller. Find the two numbers.

52. Number problem. One number is 18 more than another, and the sum of the smaller number and twice the larger number is 45. Find the two numbers.

53. Number problem. One number is 5 times another. The larger number is 9 more than twice the smaller. Find the two numbers.

54. Package weight. Two packages together weigh 32 kilograms (kg). The smaller package weighs 6 kg less than the larger. How much does each package weigh?

55. Appliance costs. A washer-dryer combination costs $1200. If the washer costs $220 more than the dryer, what does each appliance cost separately?

56. Voting trends. In a town election, the winning candidate had 220 more votes than the loser. If 810 votes were cast in all, how many votes did each candidate receive?

ANSWERS

45. (2, 1)
46. (10, −10)
47. 25, 75
48. 15, 55
49. 20, 36
50. 16, 12
51. 32, 10
52. 3, 21
53. 3, 15
54. 13 kg, 19 kg
55. Washer $710 dryer $490
56. Winner 515, loser 295

ANSWERS

57. Desk $550, chair $300

58. Width 5 in., length 12 in.

59. 7 in., 15 in., 15 in.

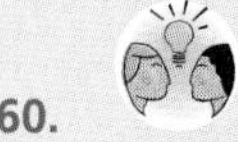

60.

a. See exercise

b. See exercise

57. Cost of furniture. An office desk and chair together cost $850. If the desk cost $50 less than twice as much as the chair, what did each cost?

58. Dimensions of a rectangle. The length of a rectangle is 2 inches (in.) more than twice its width. If the perimeter of the rectangle is 34 in., find the dimensions of the rectangle.

59. Perimeter. The perimeter of an isosceles triangle is 37 in. The lengths of the two equal legs are 6 in. less than 3 times the length of the base. Find the lengths of the three sides.

60. You have a part-time job writing the *Consumer Concerns* column for your local newspaper. Your topic for this week is clothes dryers, and you are planning to compare the Helpmate and the Whirlgarb dryers, both readily available in stores in your area. The information you have is that the Helpmate dryer is listed at $520, and it costs 22.5¢ to dry an average size load at the utility rates in your city. The Whirlgarb dryer is listed at $735, and it costs 15.8¢ to run for each normal load. The maintenance costs for both dryers are about the same. Working with a partner, write a short article giving your readers helpful advice about these appliances. What should they consider when buying one of these clothes dryers? 

Getting Ready for Section 8.4 [Section 2.7]

Graph the solution sets for the following linear inequalities.

(a) $x + y > 8$

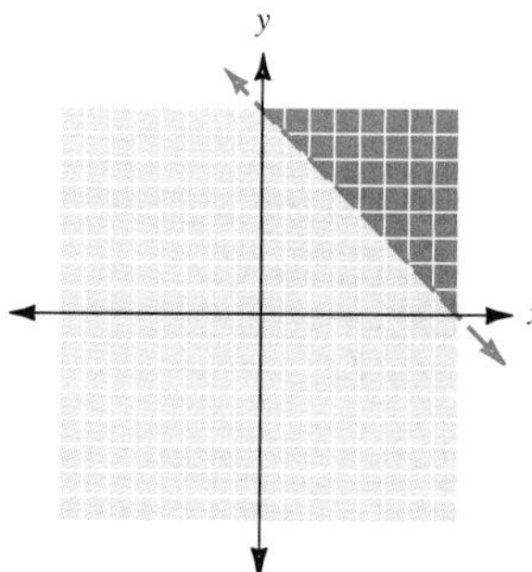

(b) $2x - y \leq 6$

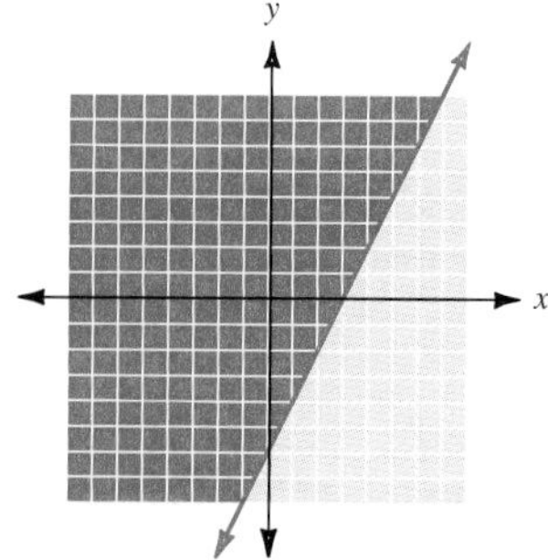

(c) $3x + 4y \geq 12$

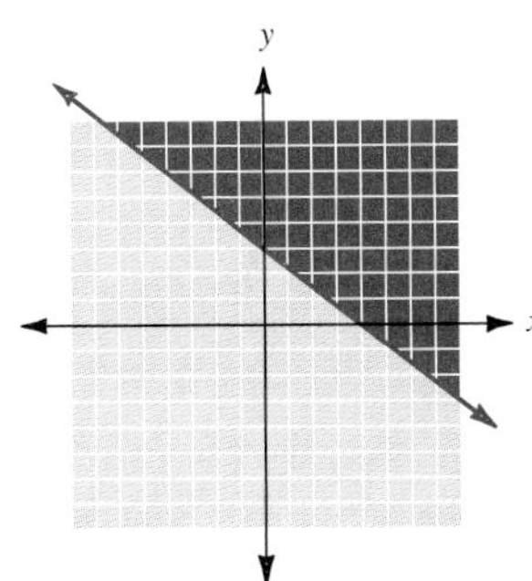

(d) $y > 2x$

(e) $y \leq -3$

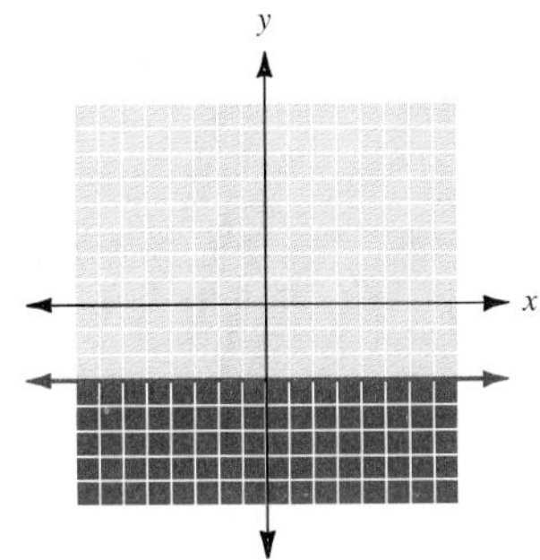

(f) $x > 5$

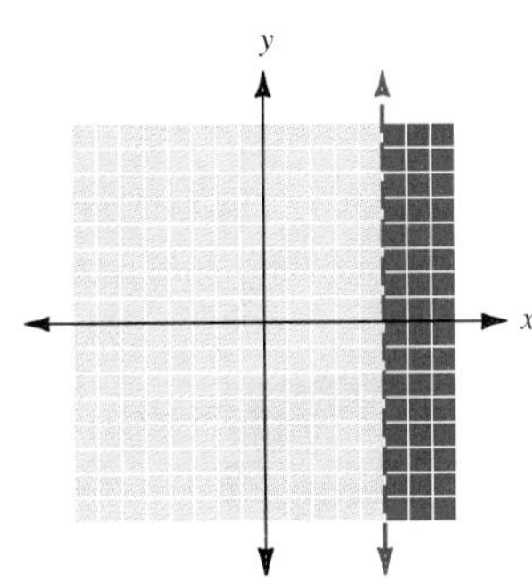

Answers

1. (2, 8) **3.** (4, −2) **5.** $\left(\frac{4}{3}, 4\right)$ **7.** (4, 1) **9.** (10, 6) **11.** (5, −3) **13.** (10, 2) **15.** $\left(\frac{1}{3}, 2\right)$ **17.** No solution **19.** (3, 3) **21.** $\left(\frac{3}{4}, -2\right)$ **23.** (4, 1) **25.** Infinite number of solutions **27.** (10, 1) **29.** (−2, −1) **31.** (0, −2) **33.** (2, 3) **35.** Dependent system **37.** $\left(-5, \frac{3}{2}\right)$ **39.** $\left(\frac{5}{2}, -3\right)$ **41.** $\left(\frac{3}{2}, 3\right)$ **43.** (0, 10) **45.** (2, 1) **47.** 25, 75 **49.** 20, 36 **51.** 32, 10 **53.** 3, 15 **55.** Washer \$710, dryer \$490 **57.** Desk \$550, chair \$300 **59.** 7 in., 15 in., 15 in.

a. $x + y > 8$

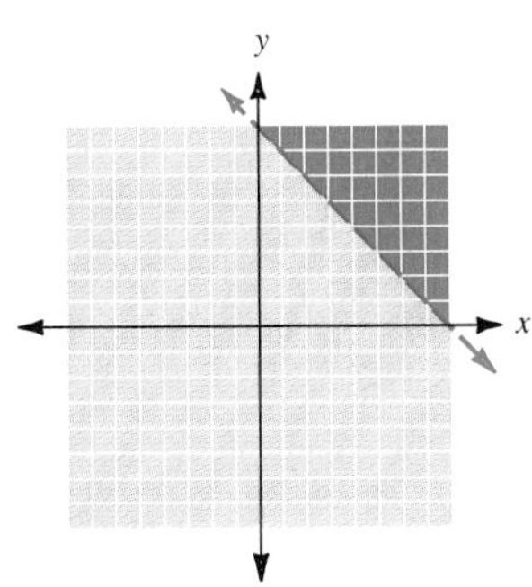

b. $2x - y \leq 6$

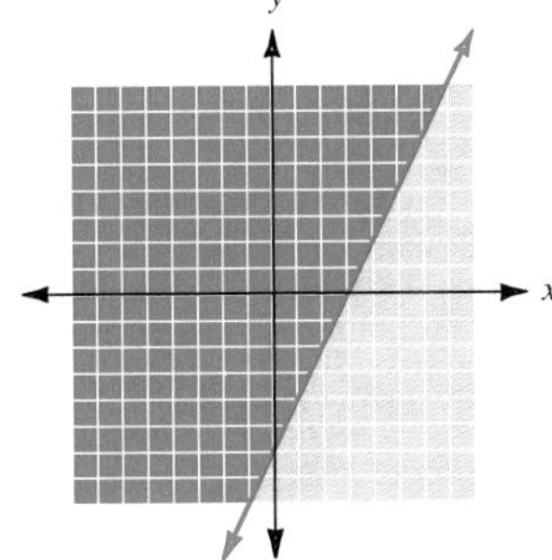

ANSWERS

c. See exercise

d. See exercise

e. See exercise

f. See exercise

c. $3x + 4y \geq 12$

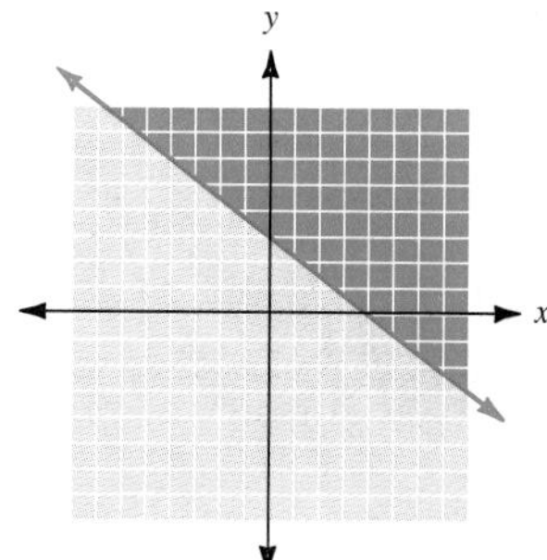

d. $y > 2x$

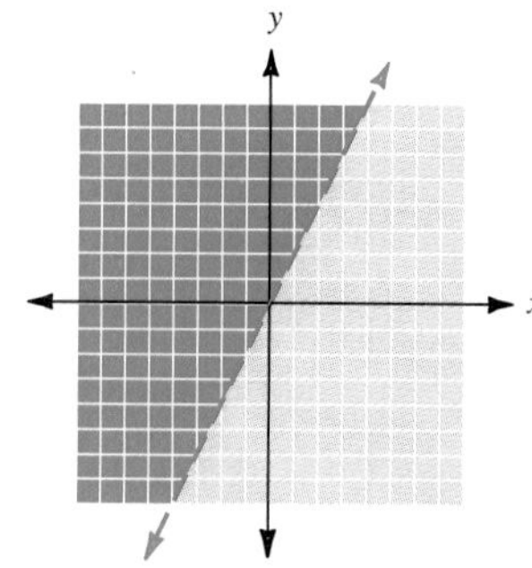

e. $y \leq -3$

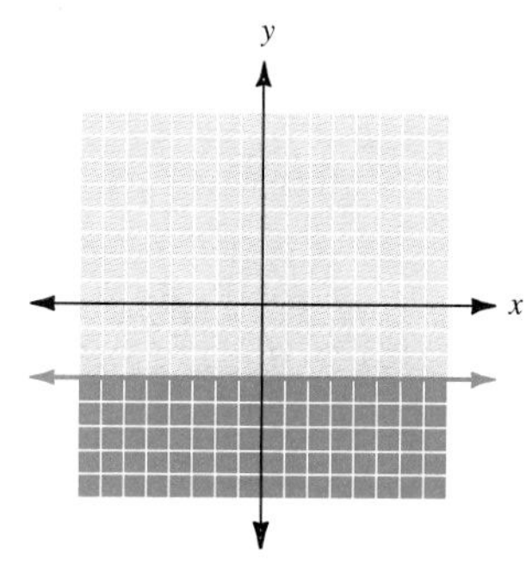

f. $x > 5$

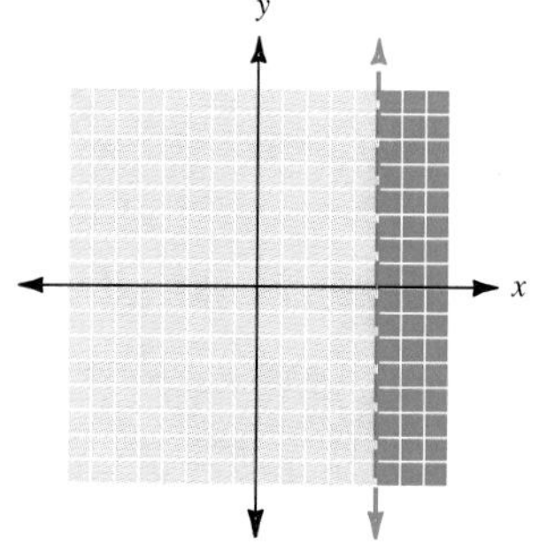

Systems of Linear Inequalities

OBJECTIVES

1. Graph a system of linear inequalities
2. Solve an application of linear inequalities

Our previous work in this chapter dealt with finding the solution set of a system of linear equations. That solution set represented the points of intersection of the graphs of the equations in the system. In this section, we extend that idea to include systems of linear inequalities.

NOTE You might want to review graphing linear inequalities in Section 7.4 at this point.

In this case, the solution set is all ordered pairs that satisfy each inequality. *The graph of the solution set of a system of linear inequalities* is then the intersection of the graphs of the individual inequalities. Let's look at an example.

Example 1

Solving a System by Graphing

Solve the following system of linear inequalities by graphing.

$x + y > 4$

$x - y < 2$

We start by graphing each inequality separately. The boundary line is drawn, and using (0, 0) as a test point, we see that we should shade the half plane above the line in both graphs.

NOTE Notice that the boundary line is dashed, to indicate it is *not* included in the graph.

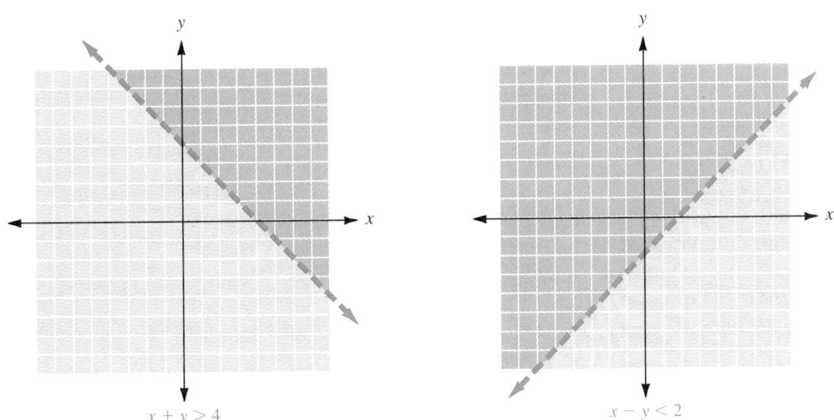

In practice, the graphs of the two inequalities are combined on the same set of axes, as is shown below. The graph of the solution set of the original system is the intersection of the graphs drawn above.

NOTE Points on the lines are not included in the solution.

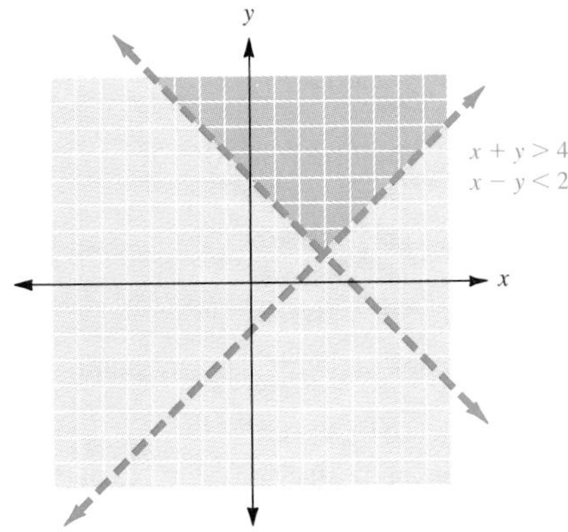

CHECK YOURSELF 1

Solve the following system of linear inequalities by graphing.

$$2x - y < 4$$
$$x + y < 3$$

Most applications of systems of linear inequalities lead to bounded regions. This requires a system of three or more inequalities, as shown in Example 2.

Example 2

Solving a System by Graphing

Solve the following system of linear inequalities by graphing.

$$x + 2y \le 6$$
$$x + y \le 5$$
$$x \ge 2$$
$$y \ge 0$$

On the same set of axes, we graph the boundary line of each of the inequalities. We then choose the appropriate half planes, indicating each with an arrow. The set of solutions is the intersection of those regions.

NOTE The vertices of the shaded region are given because they have particular significance in later applications of this concept. Can you see how the coordinates of the vertices were determined?

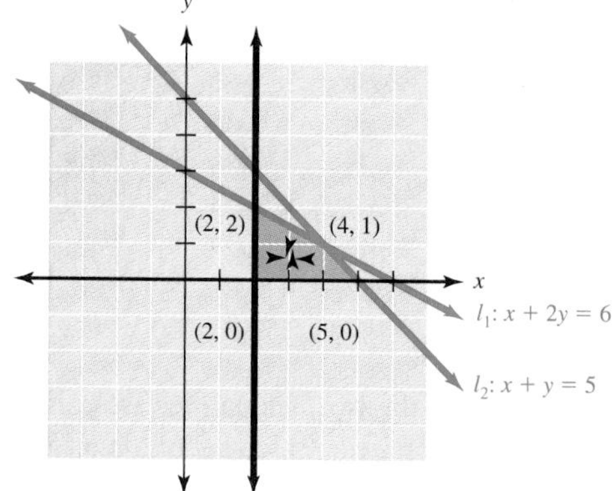

CHECK YOURSELF 2

Solve the following system of linear inequalities by graphing.

$2x - y \le 8 \qquad x \ge 0$

$x + y \le 7 \qquad y \ge 0$

Let's look at an application of our work with systems of linear inequalities. Consider the following example.

Example 3

Solving a Business-Based Application

A manufacturer produces a standard model and a deluxe model of a 13-in. television set. The standard model requires 12 h of labor to produce, whereas the deluxe model requires 18 h. The labor available is limited to 360 h per week. Also the plant capacity is limited to producing a total of 25 sets per week. Draw a graph of the region representing the number of sets that can be produced, given these conditions.

We let x represent the number of standard-model sets produced and y the number of deluxe-model sets. Because the labor is limited to 360 h, we have

NOTE The total labor is limited to (or less than or equal to) 360 h.

$$12x + 18y \leq 360 \quad (1)$$

(12 h per standard set; 18 h per deluxe set)

The total production, here $x + y$ sets, is limited to 25, so we can write

$$x + y \leq 25 \quad (2)$$

For convenience in graphing, we divide both members of inequality (1) by 6, to write the equivalent system:

NOTE We have $x \geq 0$ and $y \geq 0$ because the number of sets produced cannot be negative.

$$\begin{aligned} 2x + 3y &\leq 60 \\ x + y &\leq 25 \\ x &\geq 0 \\ y &\geq 0 \end{aligned}$$

We now graph the system of inequalities as before. The shaded area represents all possibilities in terms of the number of sets that can be produced.

NOTE The shaded area is called the *feasible region.* All points in the region meet the given conditions of the problem and represent possible production options.

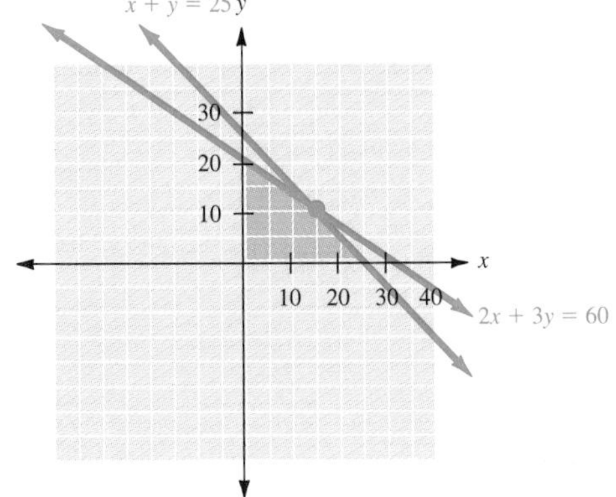

CHECK YOURSELF 3

A manufacturer produces DVD players and compact disk players. The DVD players require 10 h of labor to produce and the disk players require 20 h. The labor hours available are limited to 300 h per week. Existing orders require that at least 10 DVD players and at least 5 disk players be produced per week.

Draw a graph of the region representing the possible production options.

CHECK YOURSELF ANSWERS

1. $2x - y < 4$
$x + y < 3$

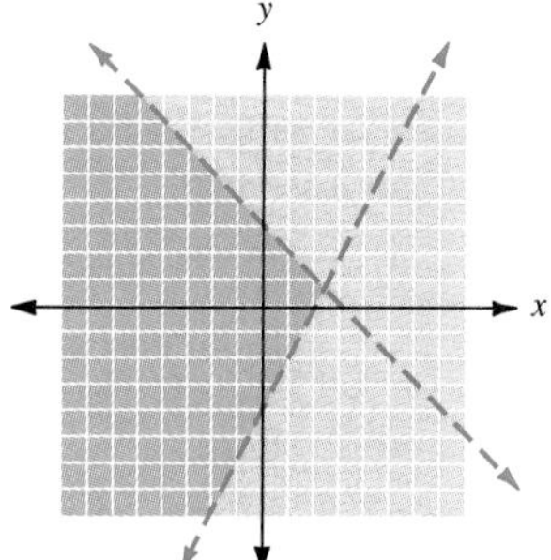

2. $2x - y \le 8$
$x + y \le 7$
$x \ge 0$
$y \ge 0$

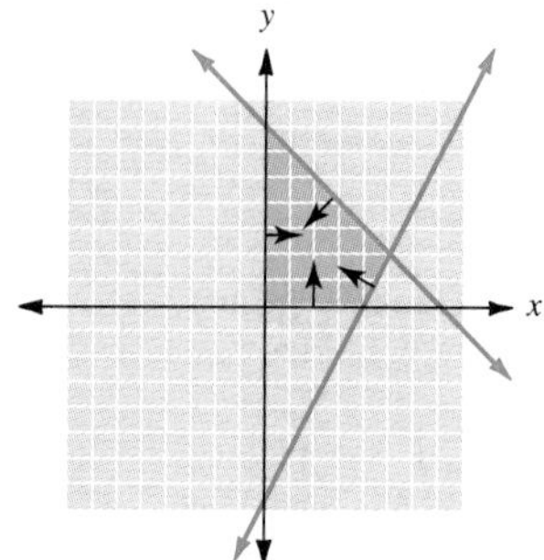

3. Let x be the number of DVD players and y be the number of CD players. The system is

$$10x + 20y \le 300$$
$$x \ge 10$$
$$y \ge 5$$

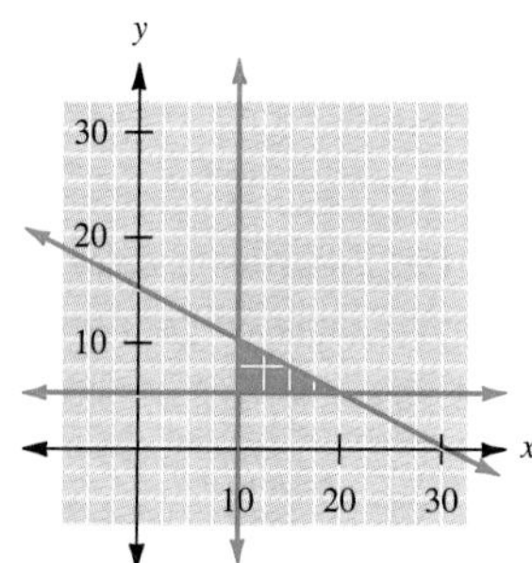

Name ______________________

Section ________ Date ________

8.4 Exercises

Solve each of the following systems of linear inequalities graphically.

1. $x + 2y \le 4$
$x - y \ge 1$

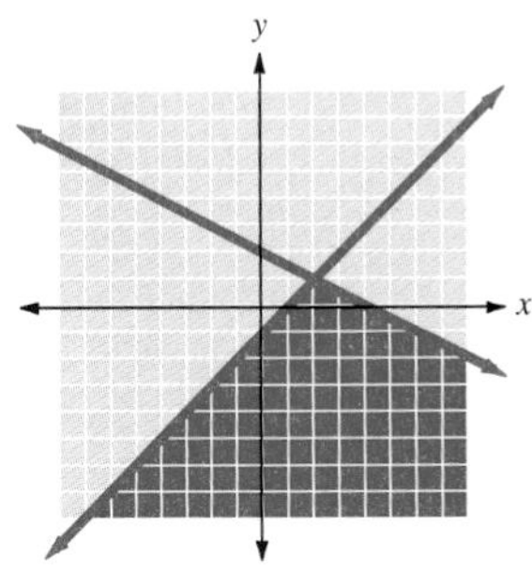

2. $3x - y > 6$
$x + y < 6$

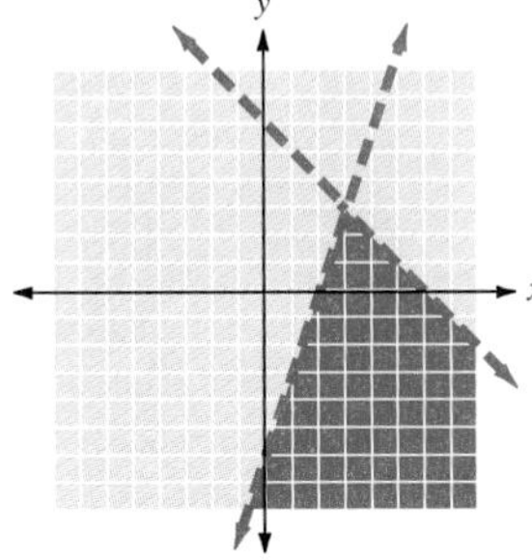

3. $3x + y < 6$
$x + y > 4$

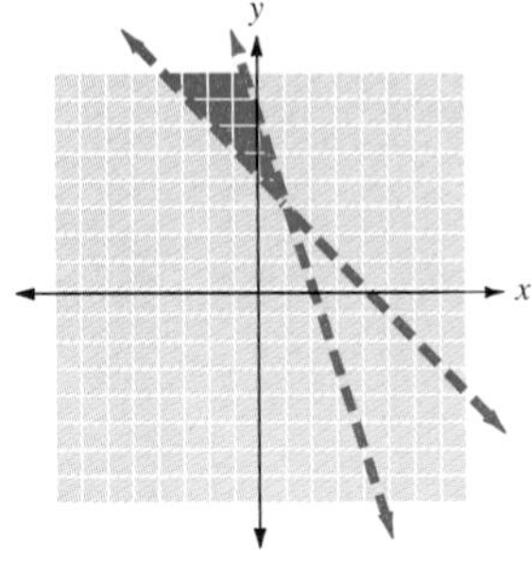

4. $2x + y \ge 8$
$x + y \ge 4$

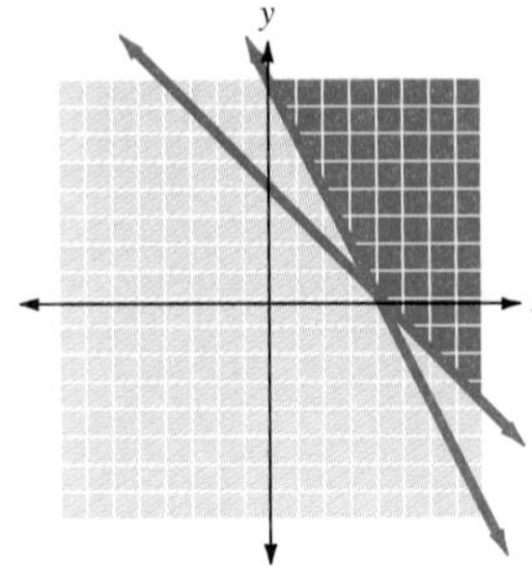

5. $x + 3y \le 12$
$2x - 3y \le 6$

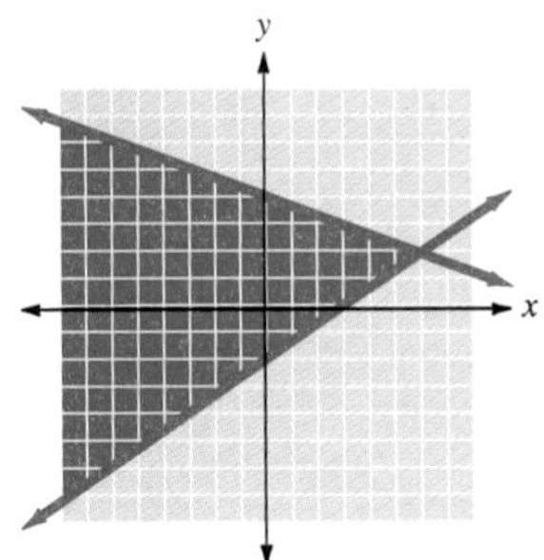

6. $x - 2y > 8$
$3x - 2y > 12$

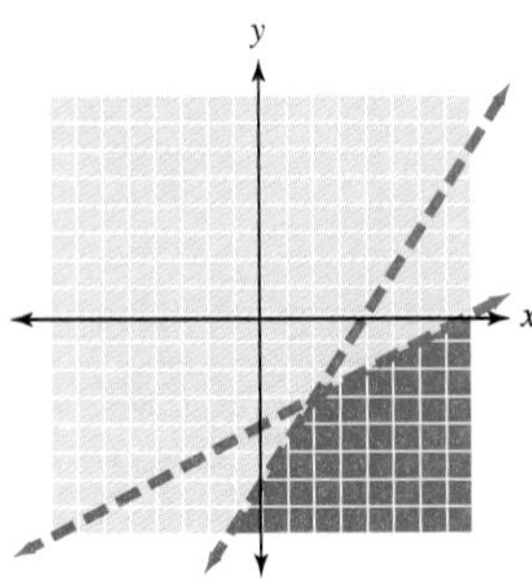

ANSWERS

1. See exercise
2. See exercise
3. See exercise
4. See exercise
5. See exercise
6. See exercise

7. See exercise

8. See exercise

9. See exercise

10. See exercise

11. See exercise

12. See exercise

7. $3x + 2y \le 12$
$x \ge 2$

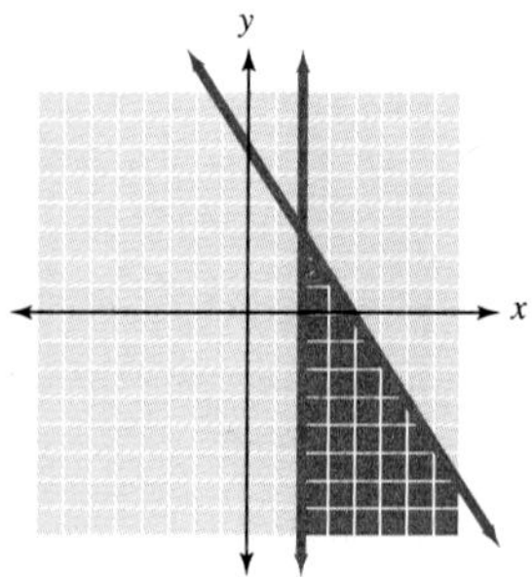

8. $2x + y \le 6$
$y \ge 1$

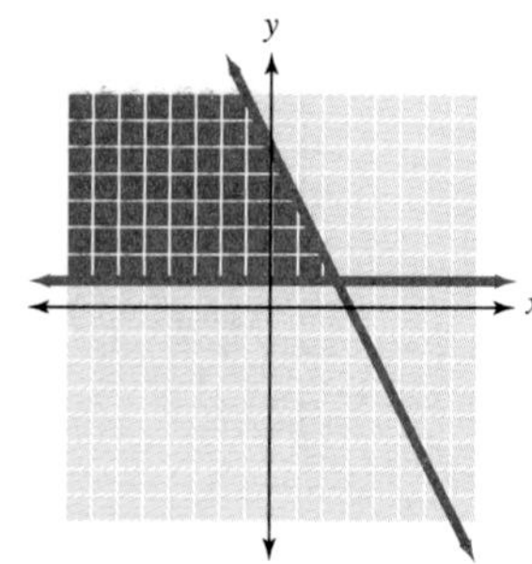

9. $2x + y \le 8$
$x > 1$
$y > 2$

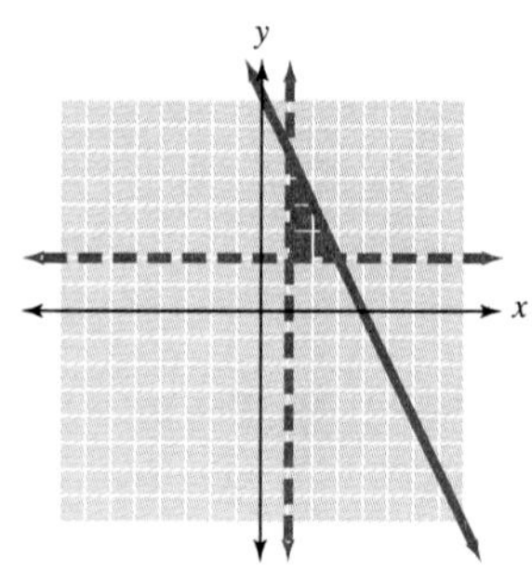

10. $3x - y \le 6$
$x \ge 1$
$y \le 3$

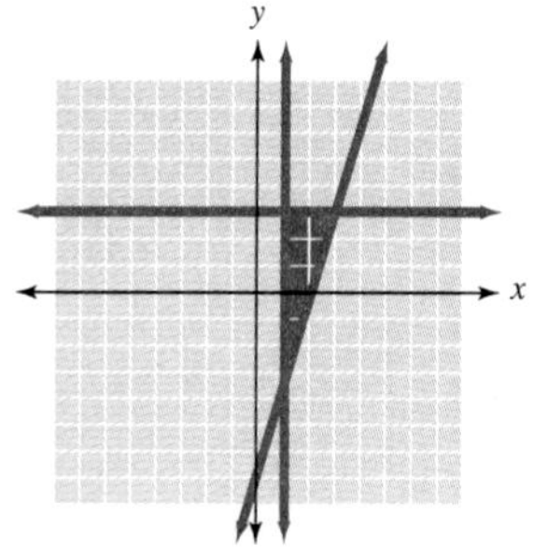

11. $x + 2y \le 8$
$2 \le x \le 6$
$y \ge 0$

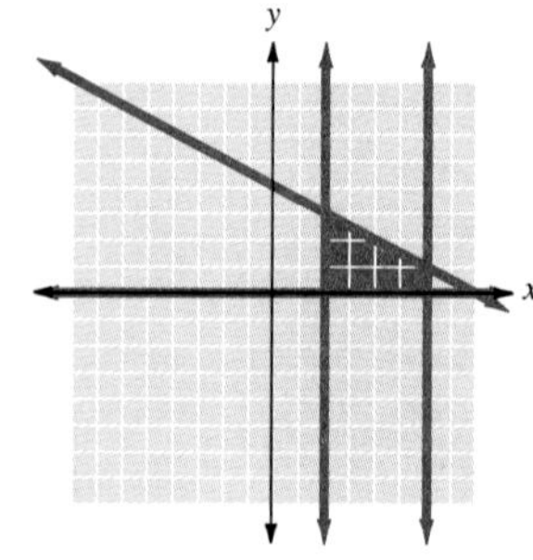

12. $x + y < 6$
$0 \le y \le 3$
$x \ge 1$

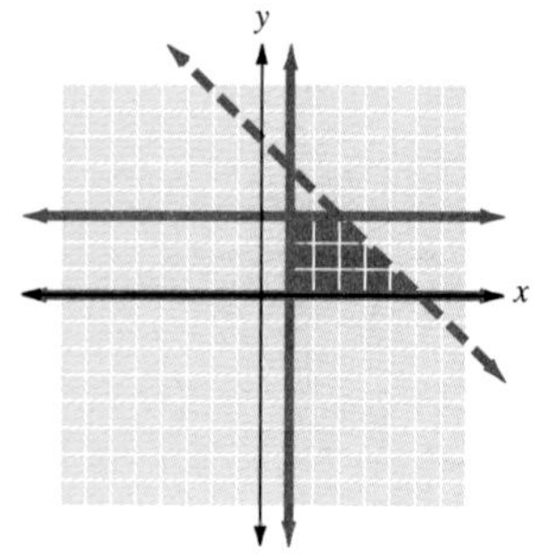

13. $3x + y \le 6$
$x + y \le 4$
$x \ge 0$
$y \ge 0$

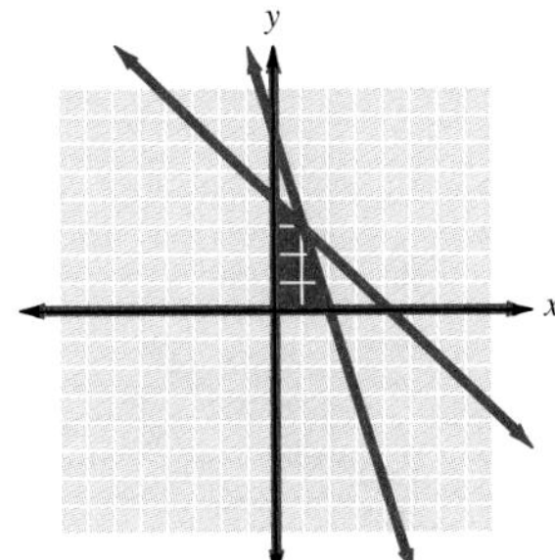

14. $x - 2y \ge -2$
$x + 2y \le 6$
$x \ge 0$
$y \ge 0$

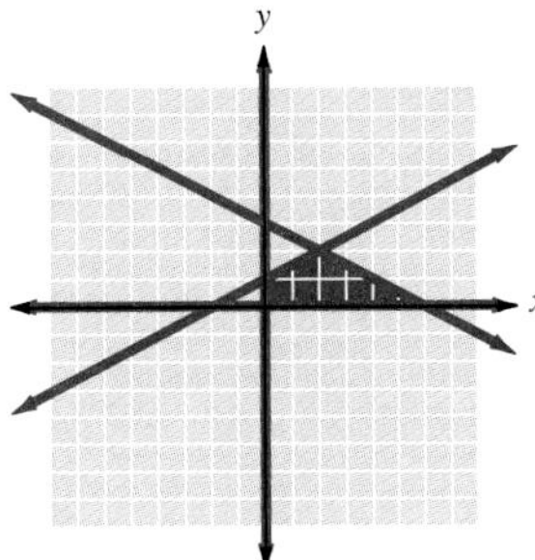

15. $4x + 3y \le 12$
$x + 4y \le 8$
$x \ge 0$
$y \ge 0$

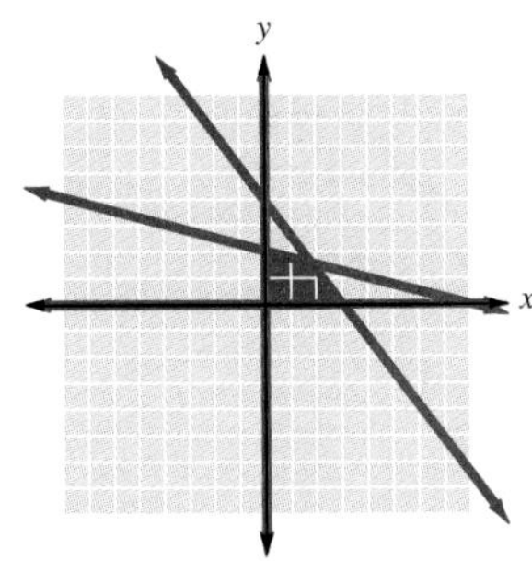

16. $2x + y \le 8$
$x + y \ge 3$
$x \ge 0$
$y \ge 0$

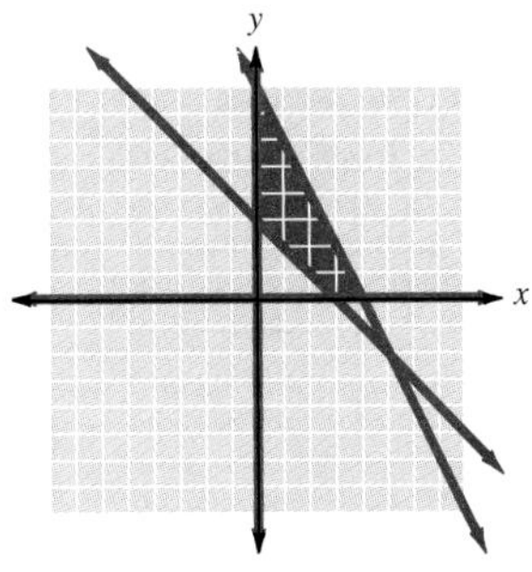

17. $x - 4y \le -4$
$x + 2y \le 8$
$x \ge 2$

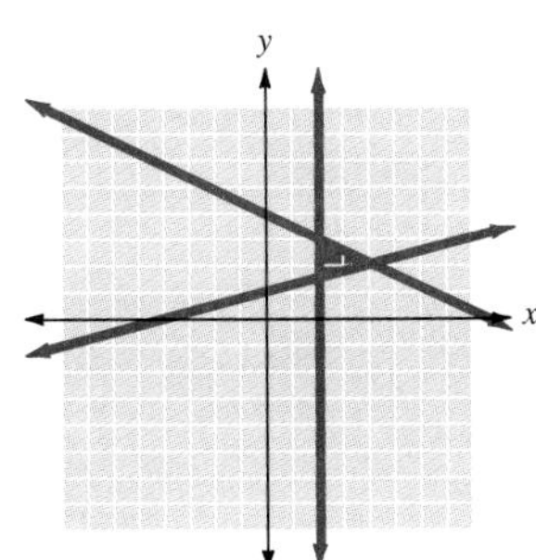

18. $x - 3y \ge -6$
$x + 2y \ge 4$
$x \le 4$

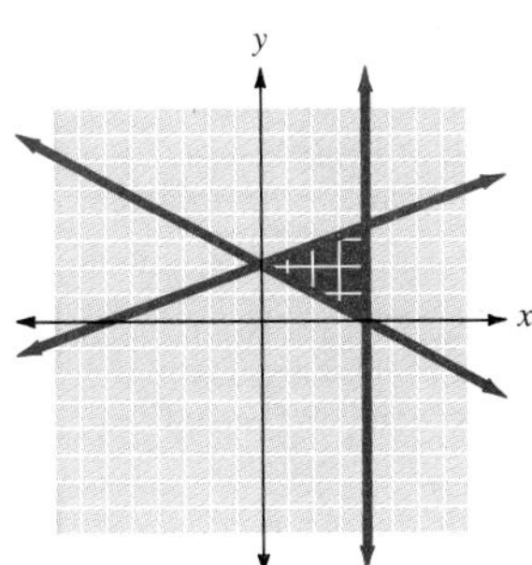

ANSWERS

13. See exercise
14. See exercise
15. See exercise
16. See exercise
17. See exercise
18. See exercise

ANSWERS

19. See exercise

20. See exercise

21.

22.

Draw the appropriate graphs in each of the following.

19. A manufacturer produces both two-slice and four-slice toasters. The two-slice toaster takes 6 h of labor to produce and the four-slice toaster 10 h. The labor available is limited to 300 h per week, and the total production capacity is 40 toasters per week. Draw a graph of the feasible region, given these conditions, in which x is the number of two-slice toasters and y is the number of four-slice toasters.

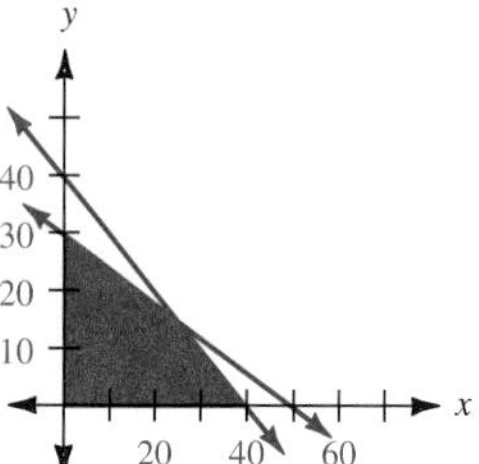

20. A small firm produces both AM and AM/FM car radios. The AM radios take 15 h to produce, and the AM/FM radios take 20 h. The number of production hours is limited to 300 h per week. The plant's capacity is limited to a total of 18 radios per week, and existing orders require that at least 4 AM radios and at least 3 AM/FM radios be produced per week. Draw a graph of the feasible region given these conditions, in which x is the number of AM radios and y the number of AM/FM radios.

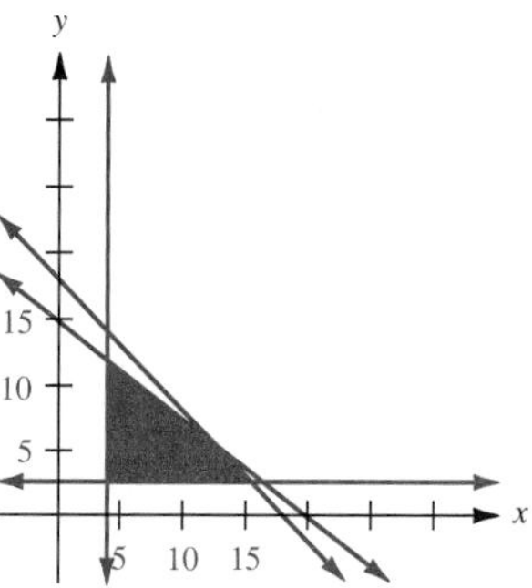

21. When you solve a system of linear inequalities, it's often easier to shade the region that is not part of the solution, rather than the region that is. Try this method, then describe its benefits.

22. Describe a system of linear inequalities for which there is no solution.

ANSWERS

$y \le 2x + 3$
$y \le -3x + 5$
23. $y \ge -x - 1$

$x \le 0$
$y < 3x + 4$
24. $y > -2x - 1$

23. Write the system of inequalities whose graph is the shaded region.

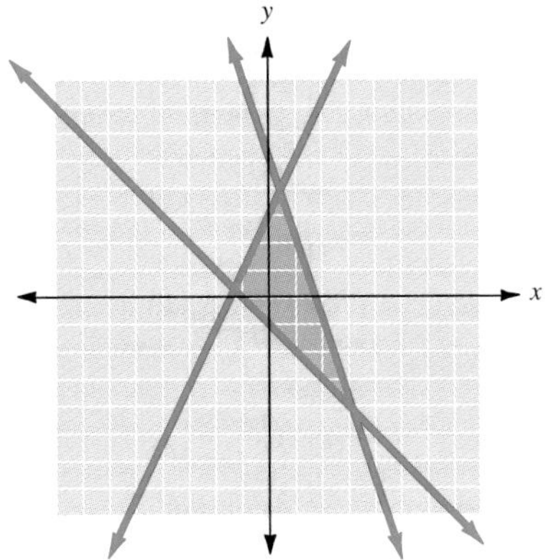

24. Write the system of inequalities whose graph is the shaded region.

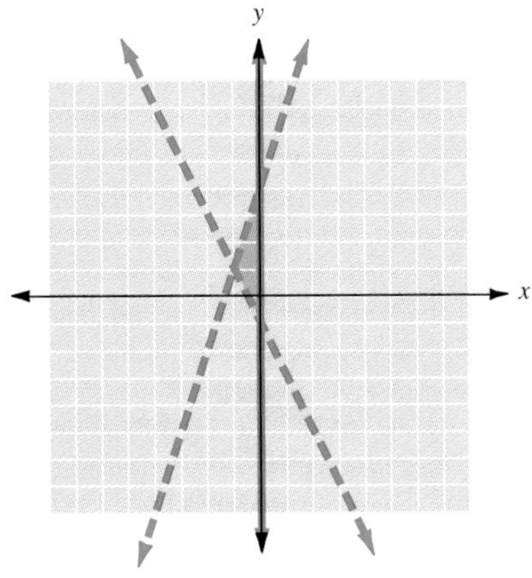

Answers

1. $x + 2y \le 4$
$x - y \ge 1$

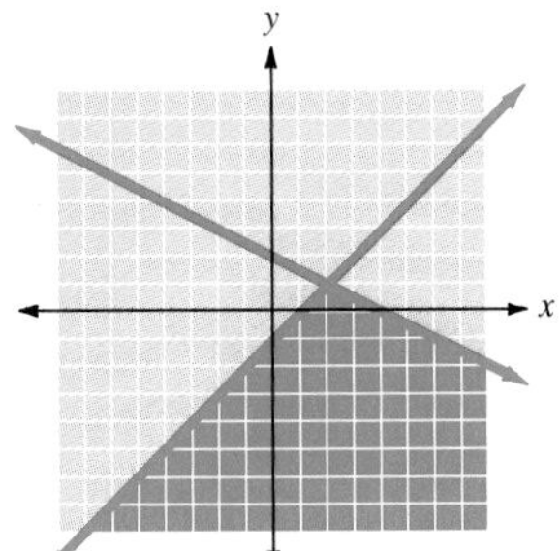

3. $3x + y < 6$
$x + y > 4$

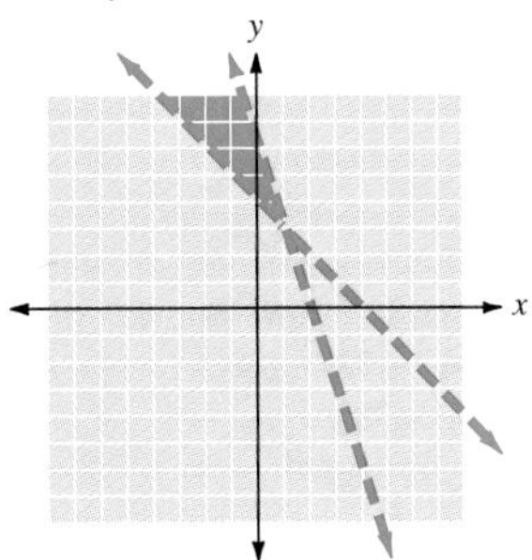

5. $x + 3y \le 12$
$2x - 3y \le 6$

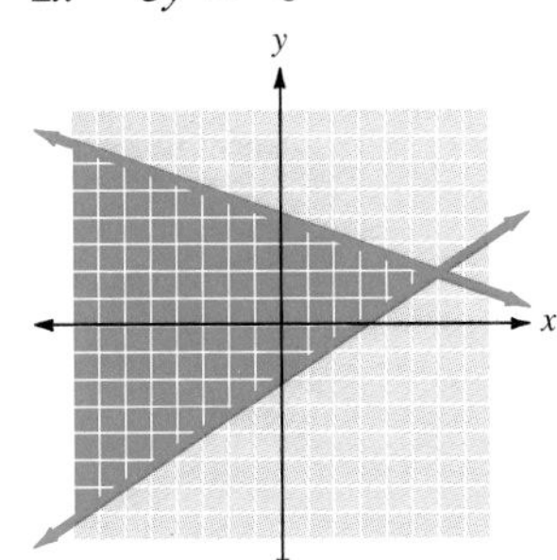

7. $3x + 2y \le 12$
$x \ge 2$

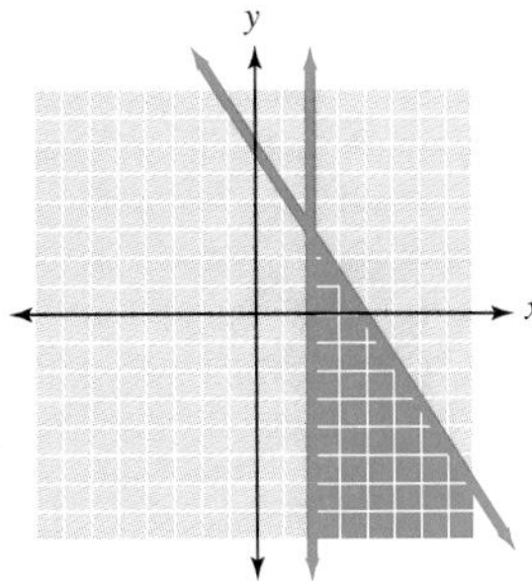

9. $2x + y \leq 8$
$x > 1$
$y > 2$

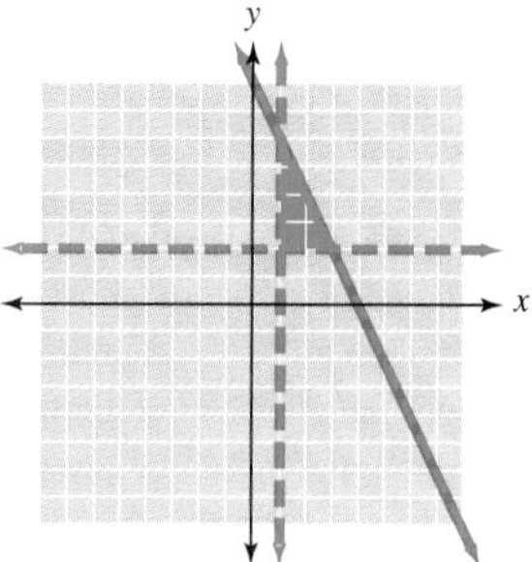

11. $x + 2y \leq 8$
$2 \leq x \leq 6$
$y \geq 0$

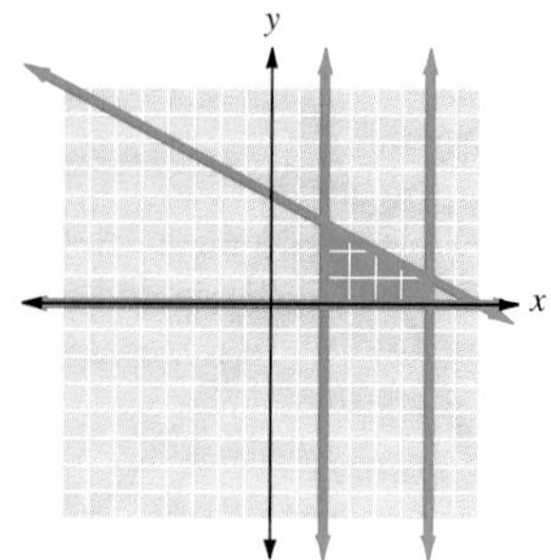

13. $3x + y \leq 6$
$x + y \leq 4$
$x \geq 0$
$y \geq 0$

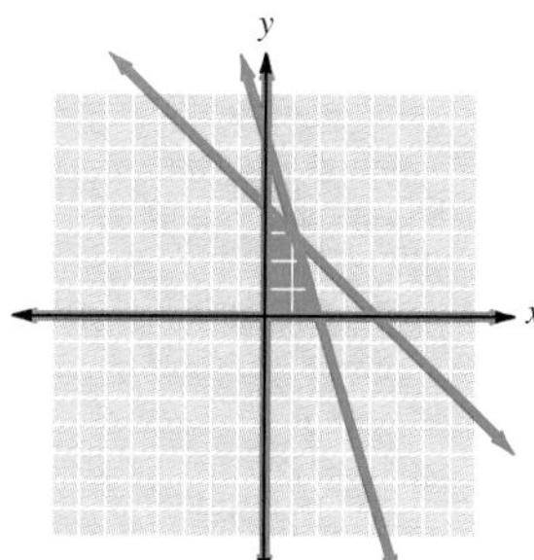

15. $4x + 3y \leq 12$
$x + 4y \leq 8$
$x \geq 0$
$y \geq 0$

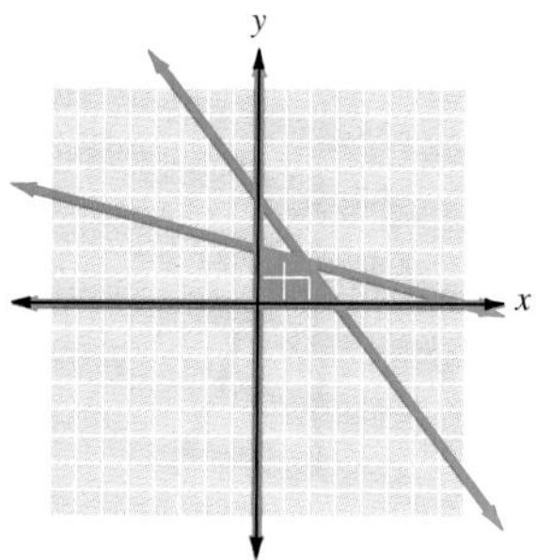

17. $x - 4y \leq -4$
$x + 2y \leq 8$
$x \geq 2$

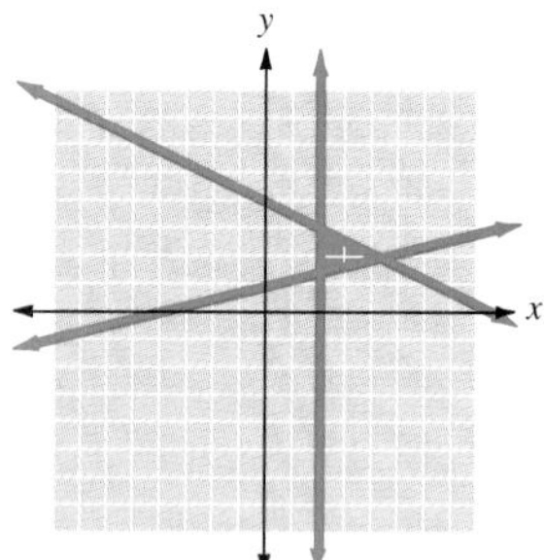

19. $6x + 10y \leq 300$
$x + y \leq 40$
$x \geq 0$
$y \geq 0$

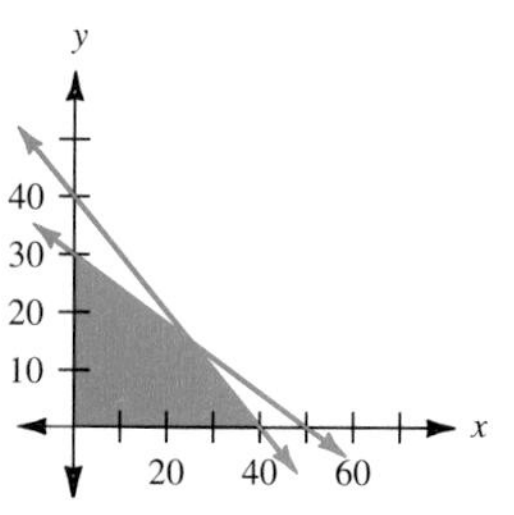

21. 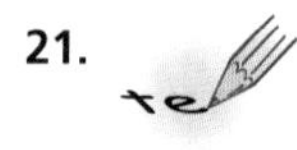

23. $y \leq 2x + 3$
$y \leq -3x + 5$
$y \geq -x - 1$

8 Summary

DEFINITION/PROCEDURE	EXAMPLE	REFERENCE
Systems of Linear Equations: Solving by Graphing		**Section 8.1**
A System of Equations Two or more equations considered together.	$x + y = 4$ $2x - y = 5$	**p. 621**
Solution The solution of a system of two equations in two unknowns is an ordered pair that satisfies each equation of the system.	$(x, y) = (3, 1)$	**p. 621**
Solving by Graphing **1.** Graph both equations on the same coordinate system. **2.** The system may have **a.** *One solution.* The lines intersect at one point (a consistent system). The solution is the ordered pair corresponding to that point. **b.** *No solution.* The lines are parallel (an inconsistent system). **c.** *Infinitely many solutions.* The two equations have the same graph (a dependent system). Any ordered pair corresponding to a point on the line is a solution.	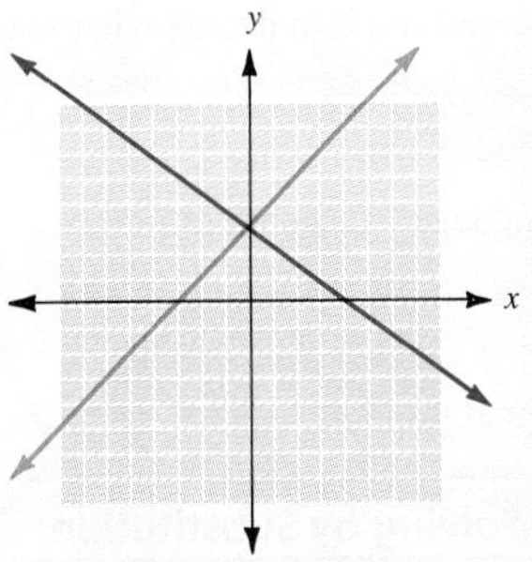 A consistent system 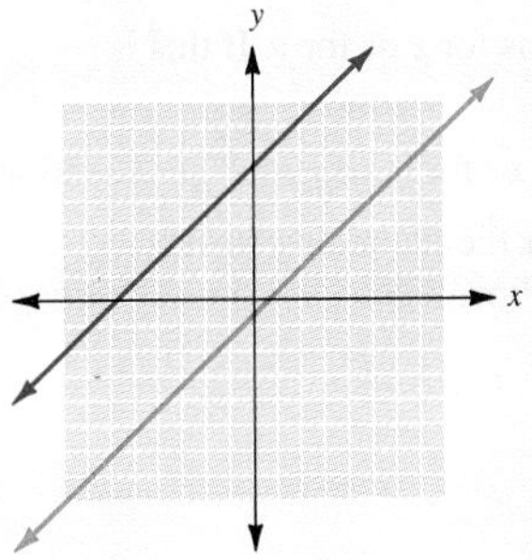 An inconsistent system 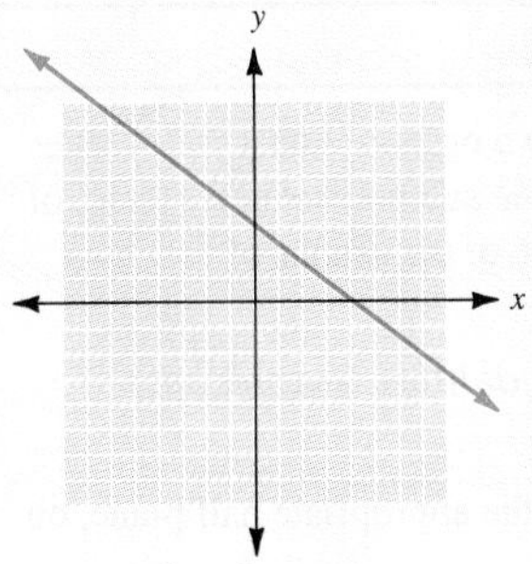 A dependent system	**p. 625**
Systems of Linear Equations: Solving by Adding		**Section 8.2**
Solving by Adding **1.** If necessary, multiply both sides of one or both equations by nonzero numbers to form an equivalent system in which the coefficients of one of the variables are opposites. **2.** Add the equations of the new system.	$2x - y = 4$ (1) $3x + 2y = 13$ (2) Multiply equation (1) by 2. $4x - 2y = 8$ $3x + 2y = 13$	

Continued

10. $2x + y = 7$
$3x - y = 3$
(2, 3)

11. $3x - 5y = 14$
$3x + 2y = 7$
(3, −1)

12. $2x - 4y = 8$
$x - 2y = 4$
Dependent system

13. $4x - 3y = -22$
$4x + 5y = -6$
(−4, 2)

14. $5x - 2y = 17$
$3x - 2y = 9$
$\left(4, \frac{3}{2}\right)$

15. $4x - 3y = 10$
$2x - 3y = 6$
$\left(2, -\frac{2}{3}\right)$

16. $2x + 3y = -10$
$-2x + 5y = 10$
(−5, 0)

17. $3x + 2y = 3$
$6x + 4y = 5$
Inconsistent system

18. $3x - 2y = 23$
$x + 5y = -15$
(5, −4)

19. $5x - 2y = -1$
$10x + 3y = 12$
$\left(\frac{3}{5}, 2\right)$

20. $x - 3y = 9$
$5x - 15y = 45$
Dependent system

21. $2x - 3y = 18$
$5x - 6y = 42$
(6, −2)

22. $3x + 7y = 1$
$4x - 5y = 30$
(5, −2)

23. $5x - 4y = 12$
$3x + 5y = 22$
(4, 2)

24. $6x + 5y = -6$
$9x - 2y = 10$
$\left(\frac{2}{3}, -2\right)$

25. $4x - 3y = 7$
$-8x + 6y = -10$
Inconsistent system

26. $3x + 2y = 8$
$-x - 5y = -20$
(0, 4)

27. $3x - 5y = -14$
$6x + 3y = -2$
$\left(-\frac{4}{3}, 2\right)$

[8.3] Solve each of the following systems by substitution. If a unique solution does not exist, state whether the system is inconsistent or dependent.

28. $x + 2y = 10$
$y = 2x$
(2, 4)

29. $x - y = 10$
$x = -4y$
(8, −2)

30. $2x - y = 10$
$x = 3y$
(6, 2)

31. $2x + 3y = 2$
$y = x - 6$
(4, −2)

32. $4x + 2y = 4$
$y = 2 - 2x$
Dependent system

33. $x + 5y = 20$
$x = y + 2$
(5, 3)

34. $6x + y = 2$
$y = 3x - 4$
$\left(\frac{2}{3}, -2\right)$

35. $2x + 6y = 10$
$x = 6 - 3y$
Inconsistent system

36. $2x + y = 9$
$x - 3y = 22$
(7, −5)

37. $x - 3y = 17$
$2x + y = 6$
(5, −4)

38. $2x + 3y = 4$
$y = 2$
(−1, 2)

39. $4x - 5y = -2$
$x = -3$
(−3, −2)

40. $-6x + 3y = -4$
$y = -\frac{2}{3}$
$\left(\frac{1}{3}, -\frac{2}{3}\right)$

41. $5x - 2y = -15$
$y = 2x + 6$
(−3, 0)

42. $3x + y = 15$
$x = 2y + 5$
(5, 0)

8 Summary

Definition/Procedure	Example	Reference
Systems of Linear Equations: Solving by Graphing		**Section 8.1**
A System of Equations Two or more equations considered together.	$x + y = 4$ $2x - y = 5$	p. 621
Solution The solution of a system of two equations in two unknowns is an ordered pair that satisfies each equation of the system.	$(x, y) = (3, 1)$	p. 621
Solving by Graphing **1.** Graph both equations on the same coordinate system. **2.** The system may have **a.** *One solution.* The lines intersect at one point (a consistent system). The solution is the ordered pair corresponding to that point. **b.** *No solution.* The lines are parallel (an inconsistent system). **c.** *Infinitely many solutions.* The two equations have the same graph (a dependent system). Any ordered pair corresponding to a point on the line is a solution.	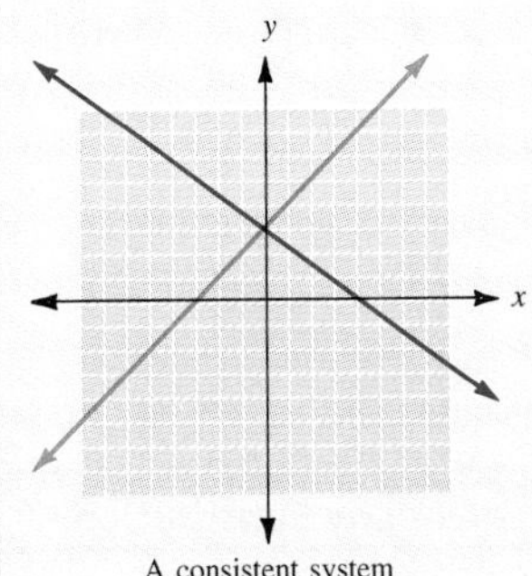 A consistent system An inconsistent system 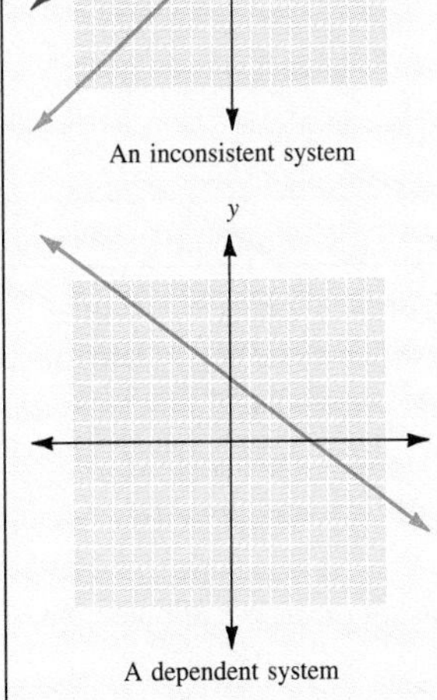 A dependent system	p. 625
Systems of Linear Equations: Solving by Adding		**Section 8.2**
Solving by Adding **1.** If necessary, multiply both sides of one or both equations by nonzero numbers to form an equivalent system in which the coefficients of one of the variables are opposites. **2.** Add the equations of the new system.	$2x - y = 4 \quad (1)$ $3x + 2y = 13 \quad (2)$ Multiply equation (1) by 2. $4x - 2y = 8$ $3x + 2y = 13$	

Continued

DEFINITION/PROCEDURE	EXAMPLE	REFERENCE
Systems of Linear Equations: Solving by Adding		**Section 8.2**
3. Solve the resulting equation for the remaining variable. **4.** Substitute the value found in step 3 into either of the original equations to find the value of the second variable. **5.** Check your solution in both of the original equations.	Add. $7x = 21$ $x = 3$ In equation (1), $2 \cdot 3 - y = 4$ $y = 2$ (3, 2) is the solution.	p. 639
Applying Systems of Equations Often word problems can be solved by using two variables and two equations to represent the unknowns and the given relationships in the problem. *The Solution Steps* **1.** Read the problem carefully. Then reread it to decide what you are asked to find. **2.** Choose letters to represent the unknowns. **3.** Translate the problem to the language of algebra to form a system of equations. **4.** Solve the system. **5.** Verify your solution in the original problem.		p. 641
Systems of Linear Equations: Solving by Substitution		**Section 8.3**
Solving by Substitution **1.** Solve one of the given equations for x or for y. If this is already done, go on to step 2. **2.** Substitute this expression for x or for y into the other equation. **3.** Solve the resulting equation for the remaining variable. Steps 4 and 5 are the same as above.	$x - 2y = 3$ (1) $2x + 3y = 13$ (2) From equation (1), $x = 2y + 3$ Substitute in equation (2): $2(2y + 3) + 3y = 13$ $4y + 6 + 3y = 13$ $7y + 6 = 13$ $7y = 7$ $y = 1$	p. 660
Systems of Linear Inequalities		**Section 8.4**
A *system of linear inequalities* is two or more linear inequalities considered together. The *graph of the solution set* of a system of linear inequalities is the intersection of the graphs of the individual inequalities. *Solving Systems of Linear Inequalities Graphically* **1.** Graph each inequality, shading the appropriate half plane, on the same set of coordinate axes. **2.** The graph of the system is the intersection of the regions shaded in step 1.	To solve $x + 2y \leq 8$ $x + y \leq 6$ $x \geq 0$ $y \geq 0$ graphically 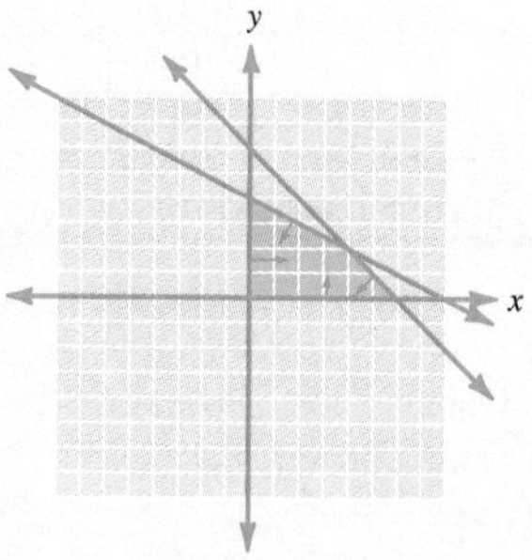 y x	p. 671

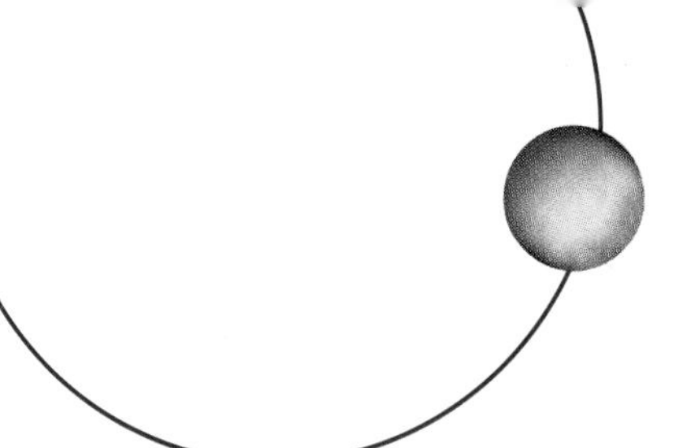

Summary Exercises

This summary exercise set is provided to give you practice with all the objectives of the chapter. Each exercise is keyed to the appropriate chapter section. The answers are provided in the *Instructor's Manual*.

[8.1] Solve each of the following systems by graphing.

1. $x + y = 6$
$x - y = 2$ **(4, 2)**

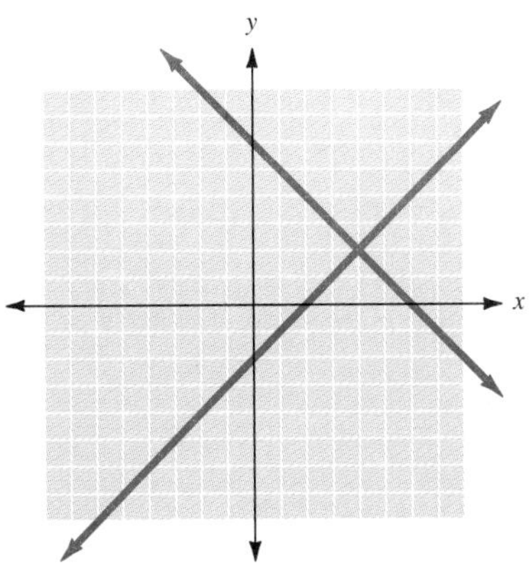

2. $x - y = 8$
$2x + y = 7$ **(5, −3)**

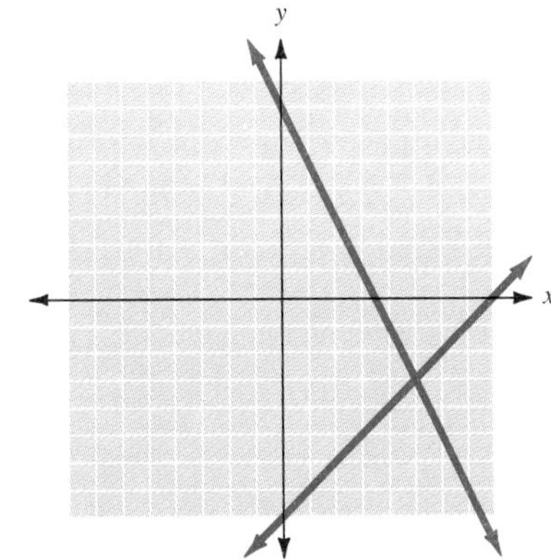

3. $x + 2y = 4$
$x + 2y = 6$ **No solution**

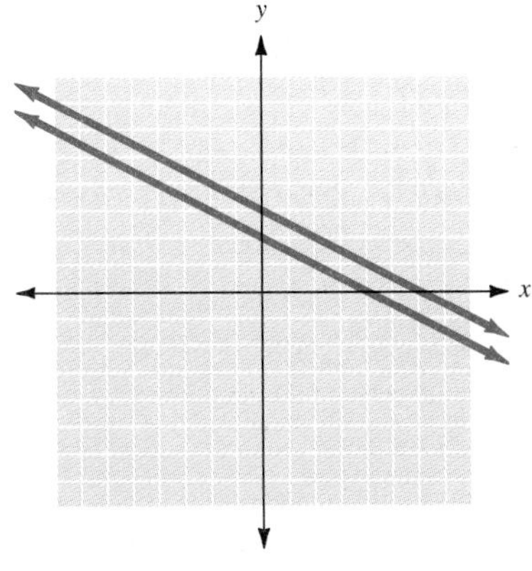

4. $2x - y = 8$
$y = 2$
(5, 2)

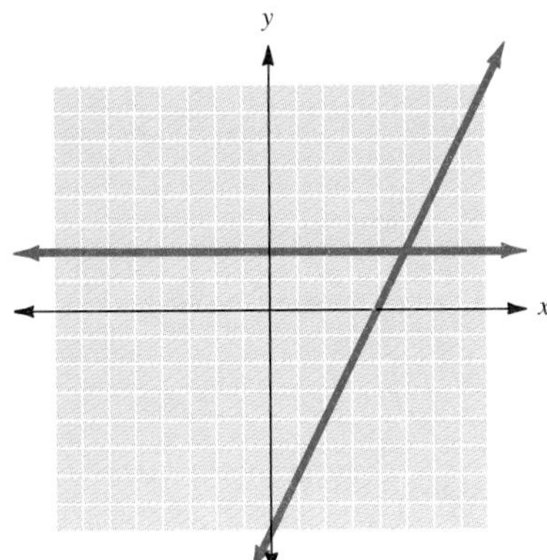

5. $2x - 4y = 8$
$x - 2y = 4$
Infinite number of solutions

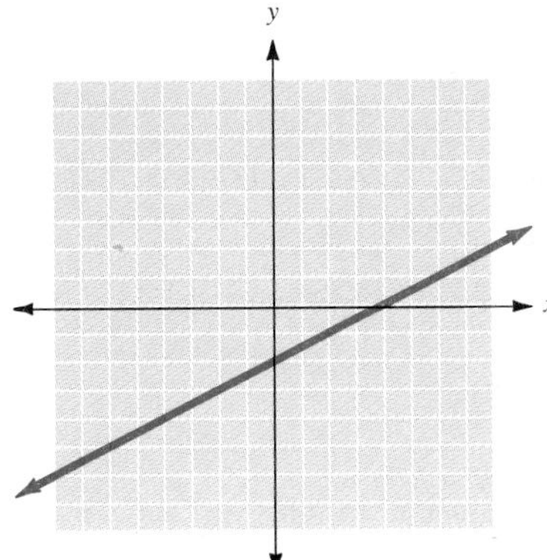

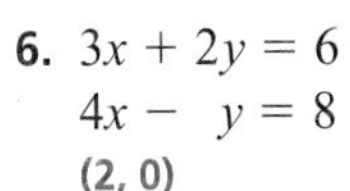

6. $3x + 2y = 6$
$4x - y = 8$
(2, 0)

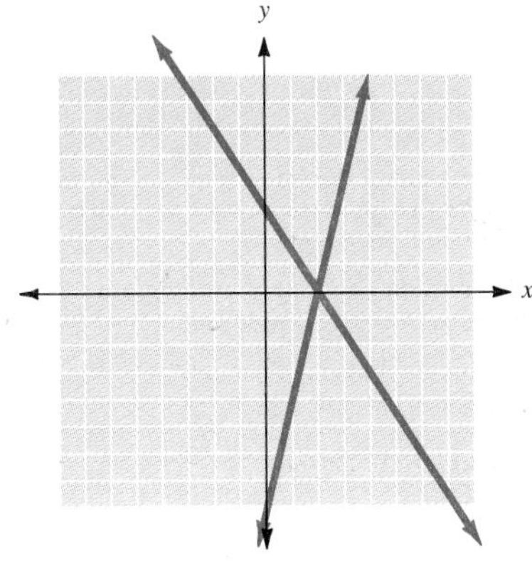

[8.2] Solve each of the following systems by addition. If a unique solution does not exist, state whether the system is inconsistent or dependent.

7. $x + y = 8$
$x - y = 2$ **(5, 3)**

8. $-x - y = 4$
$x - y = -8$ **(−6, 2)**

9. $2x - 3y = 16$
$5x + 3y = 19$ **(5, −2)**

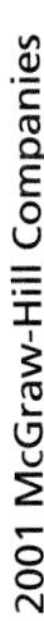

10. $2x + y = 7$
$3x - y = 3$
(2, 3)

11. $3x - 5y = 14$
$3x + 2y = 7$
(3, −1)

12. $2x - 4y = 8$
$x - 2y = 4$
Dependent system

13. $4x - 3y = -22$
$4x + 5y = -6$
(−4, 2)

14. $5x - 2y = 17$
$3x - 2y = 9$
$\left(4, \frac{3}{2}\right)$

15. $4x - 3y = 10$
$2x - 3y = 6$
$\left(2, -\frac{2}{3}\right)$

16. $2x + 3y = -10$
$-2x + 5y = 10$
(−5, 0)

17. $3x + 2y = 3$
$6x + 4y = 5$
Inconsistent system

18. $3x - 2y = 23$
$x + 5y = -15$
(5, −4)

19. $5x - 2y = -1$
$10x + 3y = 12$
$\left(\frac{3}{5}, 2\right)$

20. $x - 3y = 9$
$5x - 15y = 45$
Dependent system

21. $2x - 3y = 18$
$5x - 6y = 42$
(6, −2)

22. $3x + 7y = 1$
$4x - 5y = 30$
(5, −2)

23. $5x - 4y = 12$
$3x + 5y = 22$
(4, 2)

24. $6x + 5y = -6$
$9x - 2y = 10$
$\left(\frac{2}{3}, -2\right)$

25. $4x - 3y = 7$
$-8x + 6y = -10$
Inconsistent system

26. $3x + 2y = 8$
$-x - 5y = -20$
(0, 4)

27. $3x - 5y = -14$
$6x + 3y = -2$
$\left(-\frac{4}{3}, 2\right)$

[8.3] Solve each of the following systems by substitution. If a unique solution does not exist, state whether the system is inconsistent or dependent.

28. $x + 2y = 10$
$y = 2x$
(2, 4)

29. $x - y = 10$
$x = -4y$
(8, −2)

30. $2x - y = 10$
$x = 3y$
(6, 2)

31. $2x + 3y = 2$
$y = x - 6$
(4, −2)

32. $4x + 2y = 4$
$y = 2 - 2x$
Dependent system

33. $x + 5y = 20$
$x = y + 2$
(5, 3)

34. $6x + y = 2$
$y = 3x - 4$
$\left(\frac{2}{3}, -2\right)$

35. $2x + 6y = 10$
$x = 6 - 3y$
Inconsistent system

36. $2x + y = 9$
$x - 3y = 22$
(7, −5)

37. $x - 3y = 17$
$2x + y = 6$
(5, −4)

38. $2x + 3y = 4$
$y = 2$
(−1, 2)

39. $4x - 5y = -2$
$x = -3$
(−3, −2)

40. $-6x + 3y = -4$
$y = -\frac{2}{3}$
$\left(\frac{1}{3}, -\frac{2}{3}\right)$

41. $5x - 2y = -15$
$y = 2x + 6$
(−3, 0)

42. $3x + y = 15$
$x = 2y + 5$
(5, 0)

[8.3] Solve each of the following systems by either addition or substitution. If a unique solution does not exist, state whether the system is inconsistent or dependent.

43. $x - 4y = 0$
$4x + y = 34$
(8, 2)

44. $2x + y = 2$
$y = -x$
(2, −2)

45. $3x - 3y = 30$
$x = -2y - 8$
(4, −6)

46. $5x + 4y = 40$
$x + 2y = 11$
$\left(6, \frac{5}{2}\right)$

47. $x - 6y = -8$
$2x + 3y = 4$
$\left(0, \frac{4}{3}\right)$

48. $4x - 3y = 9$
$2x + y = 12$
$\left(\frac{9}{2}, 3\right)$

49. $9x + y = 9$
$x + 3y = 14$
$\left(\frac{1}{2}, \frac{9}{2}\right)$

50. $3x - 2y = 8$
$-6x + 4y = -16$
Dependent system

51. $3x - 2y = 8$
$2x - 3y = 7$
(2, −1)

[8.2–8.3] Solve the following problems. Be sure to show the equations used.

52. Number problem. The sum of two numbers is 40. If their difference is 10, find the two numbers. **25, 15**

53. Number problem. The sum of two numbers is 17. If the larger number is 1 more than 3 times the smaller, what are the two numbers? **4, 13**

54. Number problem. The difference of two numbers is 8. The larger number is 2 less than twice the smaller. Find the numbers. **10, 18**

55. Cost. Five writing tablets and three pencils cost \$8.25. Two tablets and two pencils cost \$3.50. Find the cost for each item. **Tablet \$1.50, pencil \$0.25**

56. Cable length. A cable 200 ft long is cut into two pieces so that one piece is 12 feet (ft) longer than the other. How long is each piece? **94 ft, 106 ft**

57. Cost. An amplifier and a pair of speakers cost \$925. If the amplifier costs \$75 more than the speakers, what does each cost? **Speakers \$425, amplifier \$500**

58. Cost. A sofa and chair cost \$850 as a set. If the sofa costs \$100 more than twice as much as the chair, what is the cost of each? **Chair \$250, sofa \$600**

59. Rectangular dimensions. The length of a rectangle is 4 centimeters (cm) more than its width. If the perimeter of the rectangle is 64 cm, find the dimensions of the rectangle. **Width 14 cm, length 18 cm**

60. Isosceles triangle. The perimeter of an isosceles triangle is 29 inches (in.). The lengths of the two equal legs are 2 in. more than twice the length of the base. Find the lengths of the three sides. **5 in., 12 in., 12 in.**

61. Coin problem. Darryl has 30 coins with a value of \$5.50. If they are all nickels and quarters, how many of each kind of coin does he have? **10 nickels, 20 quarters**

62. Ticket sales. Tickets for a concert sold for \$11 and \$8. If 600 tickets were sold for one evening and the receipts were \$5550, how many of each kind of ticket were sold? **250 at \$11, 350 at \$8**

63. Acid solution. A laboratory has a 20% acid solution and a 50% acid solution. How much of each should be used to produce 600 milliliters (mL) of a 40% acid solution? **200 mL of 20%, 400 mL of 50%**

64. Antifreeze mixture. A service station wishes to mix 40 L of a 78% antifreeze solution. How many liters of a 75% solution and a 90% solution should be used in forming the mixture? **32 L of 75%, 8 L of 90%**

65. Investment rates. Martha has \$18,000 invested. Part of the money is invested in a bond that yields 11% interest. The remainder is in her savings account, which pays 7%. If she earns \$1660 in interest for 1 year, how much does she have invested at each rate? **\$10,000 at 11%, \$8000 at 7%**

66. Motion problem. A boat travels 24 miles (mi) upstream in 3 hours (h). It then takes 3 h to go 36 mi downstream. Find the speed of the boat in still water and the speed of the current. **Boat 10 mi/h, current 2 mi/h**

67. Motion problem. A plane flying with the wind makes a trip of 2200 mi in 4 h. Returning against the wind, it can travel only 1800 mi in 4 h. What is the plane's rate in still air? What is the wind speed? **Plane 500 mi/h, wind 50 mi/h**

[8.4] Solve these systems of linear inequalities.

68. $x - y < 7$
$x + y > 3$

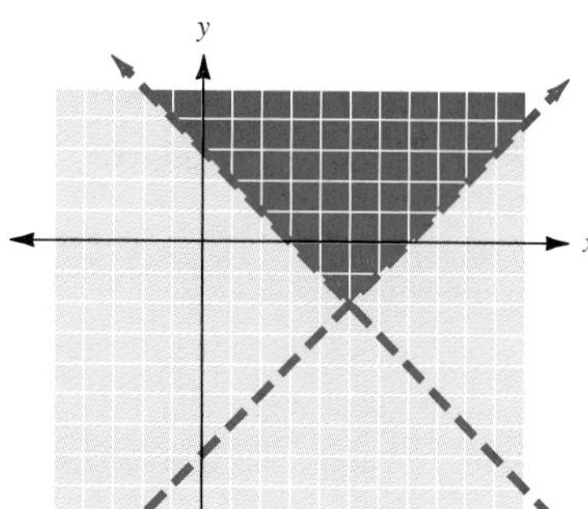

69. $x - 2y \leq -2$
$x + 2y \leq 6$

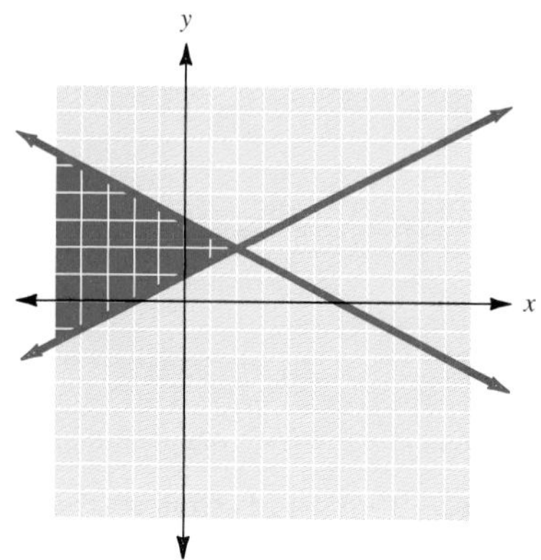

70. $x - 6y < 6$
$-x + y < 4$

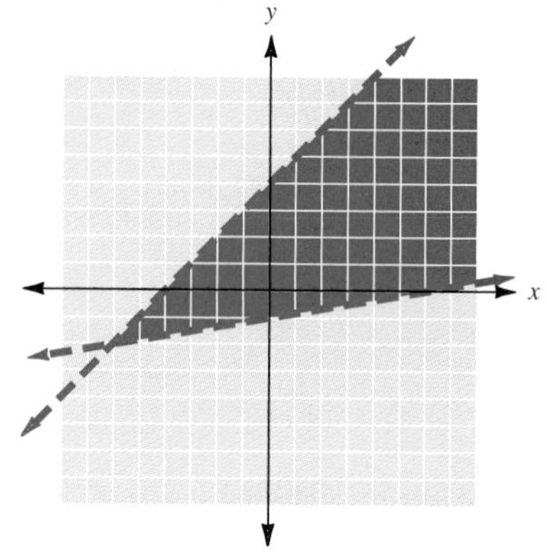

71. $2x + y \leq 8$
$x \geq 1$
$y \geq 0$

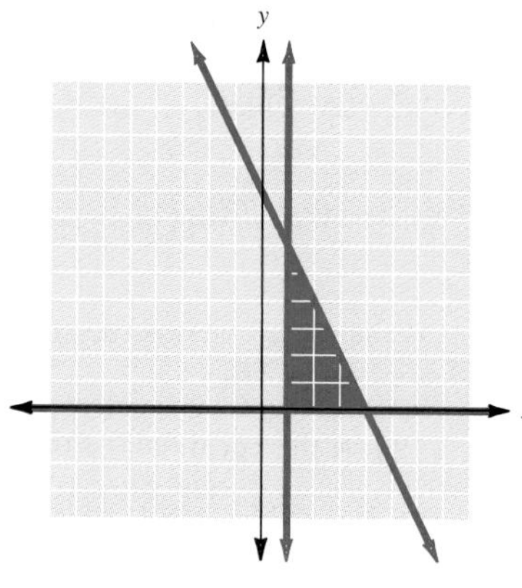

72. $2x + y \leq 6$
$x \geq 1$
$y \geq 0$

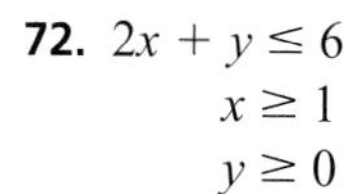

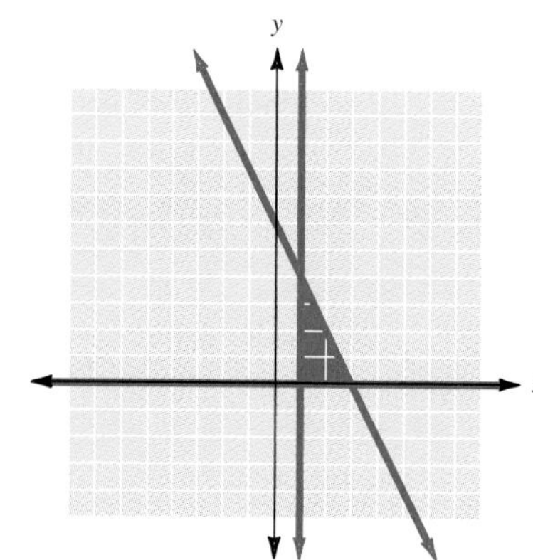

73. $4x + y \leq 8$
$x \geq 0$
$y \geq 2$

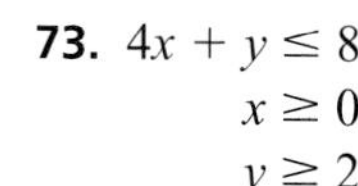

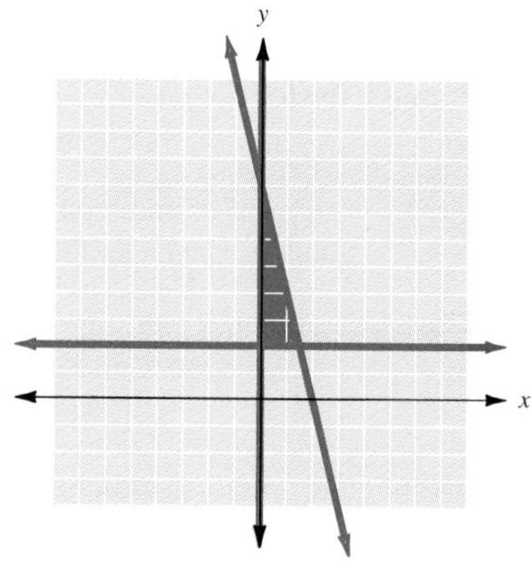

74. $4x + 2y \leq 8$
$x + y \leq 3$
$x \geq 0$
$y \geq 0$

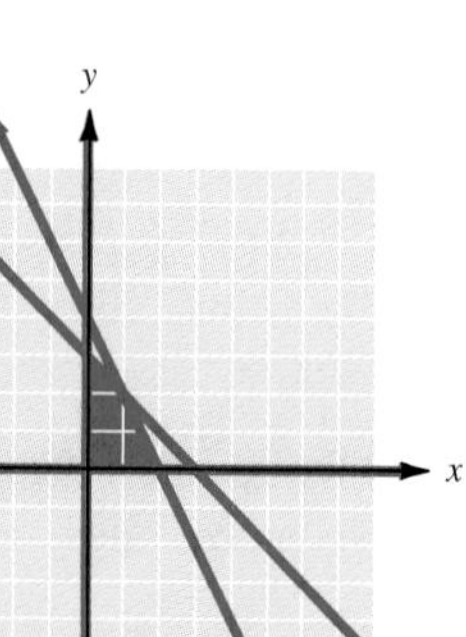

75. $3x + y \leq 6$
$x + y \leq 4$
$x \geq 0$
$y \geq 0$

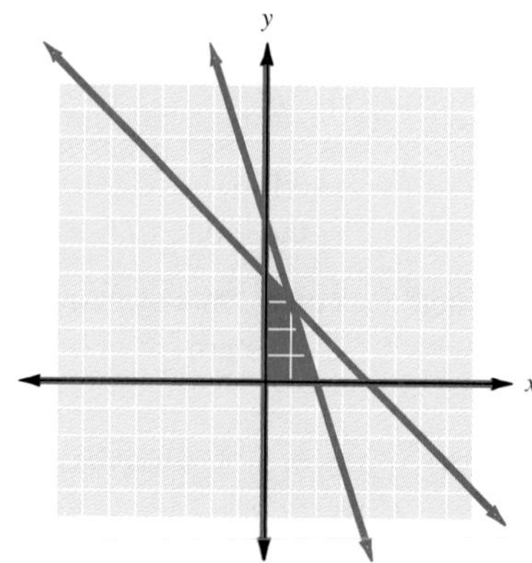

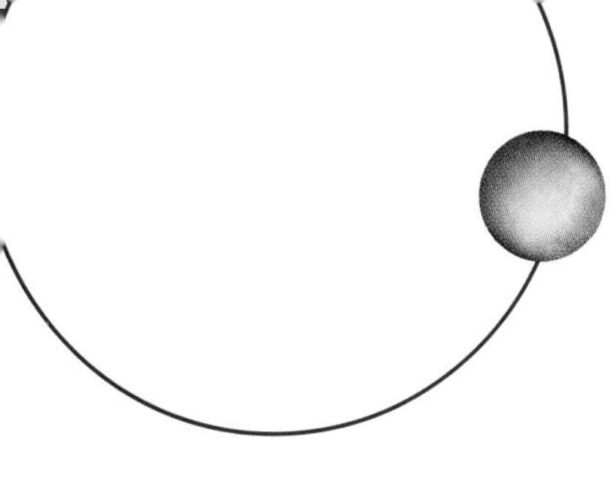

Self-Test for Chapter 8

Name ______________________

Section ________ Date ________

The purpose of this self-test is to help you check your progress and to review for a chapter test in class. Allow yourself about an hour to take the test. When you are done, check your answers in the back of the book. If you missed any answers, be sure to go back and review the appropriate sections in the chapter and the exercises that are provided.

Solve each of the following systems by graphing. If a unique solution does not exist, state whether the system is inconsistent or dependent.

1. $x + y = 5$
$x - y = 3$

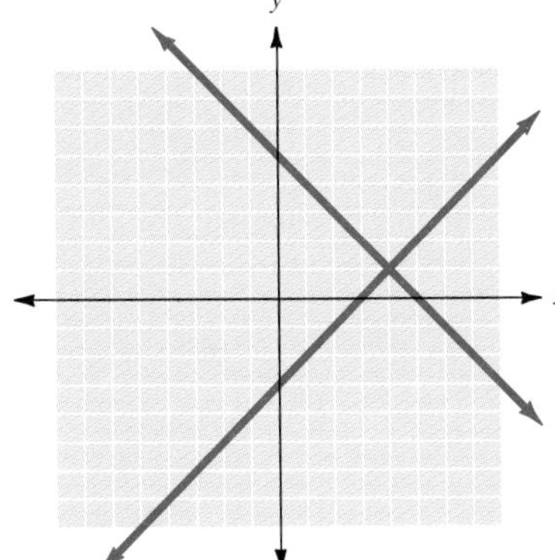

2. $x + 2y = 8$
$x - y = 2$

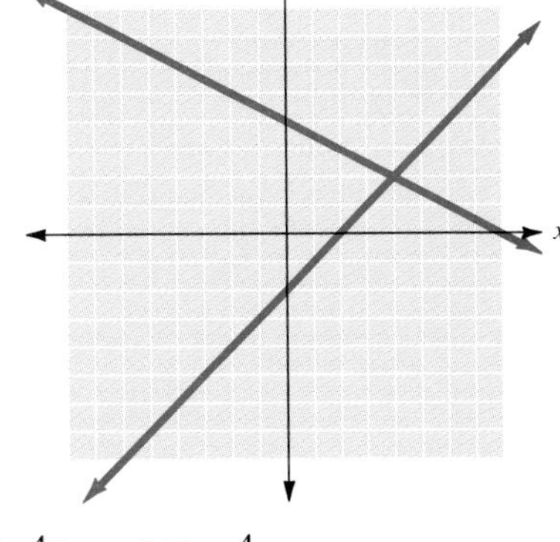

3. $x - 3y = 3$
$x - 3y = 6$

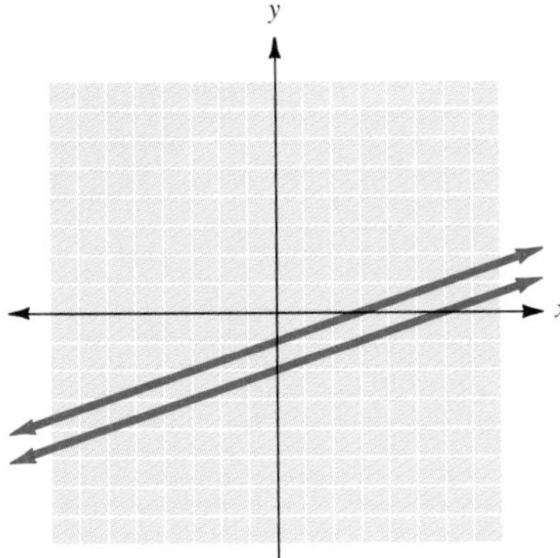

4. $4x - y = 4$
$x - 2y = -6$

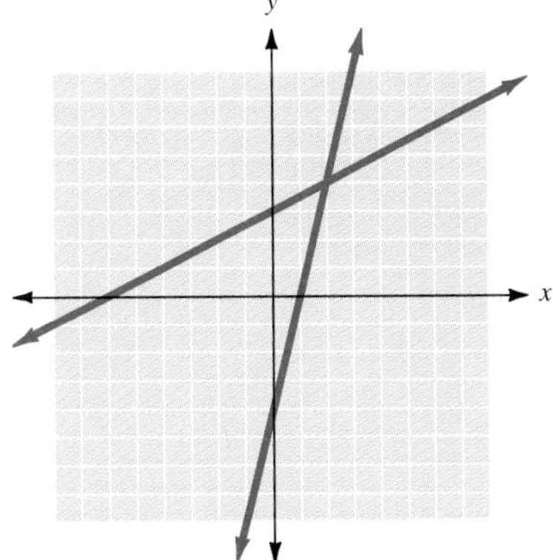

Solve each of the following systems by addition. If a unique solution does not exist, state whether the system is inconsistent or dependent.

5. $x + y = 5$
$x - y = 3$

6. $x + 2y = 8$
$x - y = 2$

7. $3x + y = 6$
$-3x + 2y = 3$

8. $3x + 2y = 11$
$5x + 2y = 15$

9. $3x - 6y = 12$
$x - 2y = 4$

10. $4x + y = 2$
$8x - 3y = 9$

11. $2x - 5y = 2$
$3x + 4y = 26$

12. $x + 3y = 6$
$3x + 9y = 9$

ANSWERS

1. (4, 1)
2. (4, 2)
3. Inconsistent system
4. (2, 4)
5. (4, 1)
6. (4, 2)
7. (1, 3)
8. $\left(2, \frac{5}{2}\right)$
9. Dependent system
10. $\left(\frac{3}{4}, -1\right)$
11. (6, 2)
12. Inconsistent system

ANSWERS

13. (2, 6)

14. (6, −3)

15. (6, 2)

16. (5, 4)

17. (−3, 3)

18. Inconsistent system

19. (3, −5)

20. (3, 2)

21. 12, 18

22. 21 m, 29 m

23. Width 12 in., length 20 in.

24. 12 dimes, 18 quarters

25. Boat 15 mi/h, current 3 mi/h

26. See exercise

27. See exercise

28. See exercise

Solve each of the following systems by substitution. If a unique solution does not exist, state whether the system is inconsistent or dependent.

13. $x + y = 8$
$y = 3x$

14. $x - y = 9$
$x = -2y$

15. $2x - y = 10$
$x = y + 4$

16. $x - 3y = -7$
$y = x - 1$

17. $3x + y = -6$
$y = 2x + 9$

18. $4x + 2y = 8$
$y = 3 - 2x$

19. $5x + y = 10$
$x + 2y = -7$

20. $3x - 2y = 5$
$2x + y = 8$

Solve each of the following problems. Be sure to show the equations used.

21. Number problem. The sum of two numbers is 30, and their difference is 6. Find the two numbers.

22. Rope length. A rope 50 meters (m) long is cut into two pieces so that one piece is 8 m longer than the other. How long is each piece?

23. Dimensions of a rectangle. The length of a rectangle is 4 inches (in.) less than twice its width. If the perimeter of the rectangle is 64 in., what are the dimensions of the rectangle?

24. Coin problem. Murray has 30 coins with a value of \$5.70. If the coins are all dimes and quarters, how many of each coin does he have?

25. Motion problem. Jackson was able to travel 36 miles (mi) downstream in 2 hours (h). In returning upstream, it took 3 h to make the trip. How fast can his boat travel in still water? What was the rate of the river current?

Solve the following systems of inequalities.

26. $x + y < 3$
$x - 2y < 6$

27. 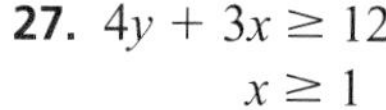
$4y + 3x \geq 12$
$x \geq 1$

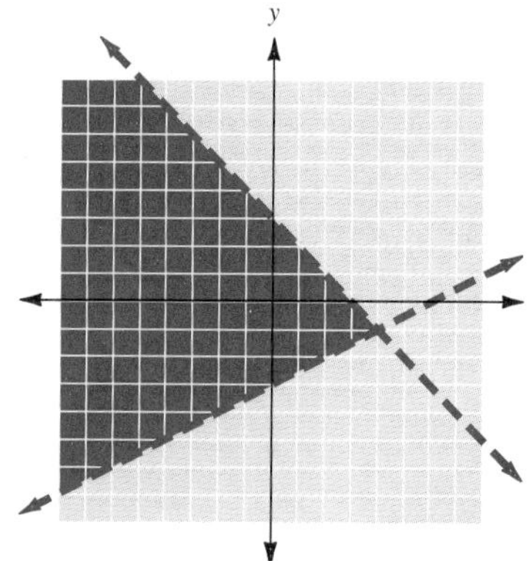

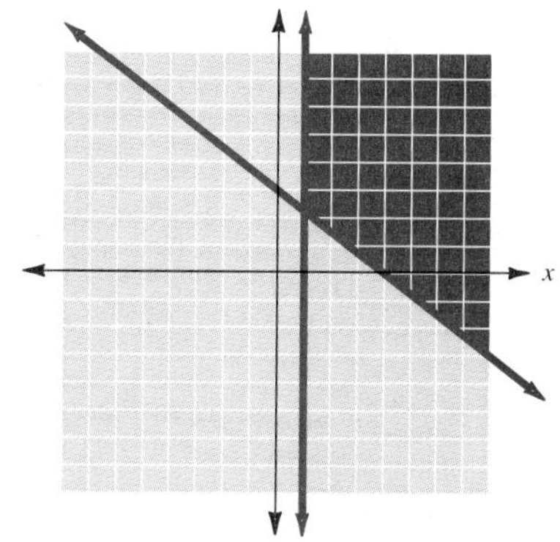

28. 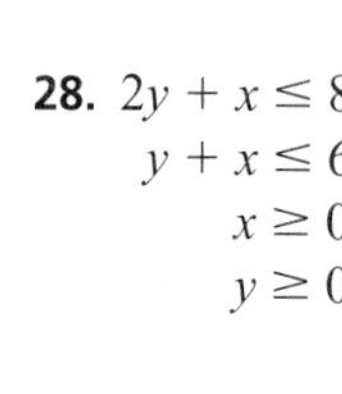
$2y + x \leq 8$
$y + x \leq 6$
$x \geq 0$
$y \geq 0$

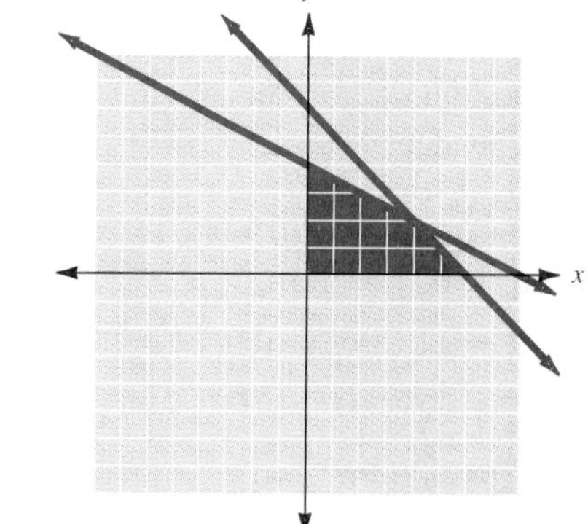

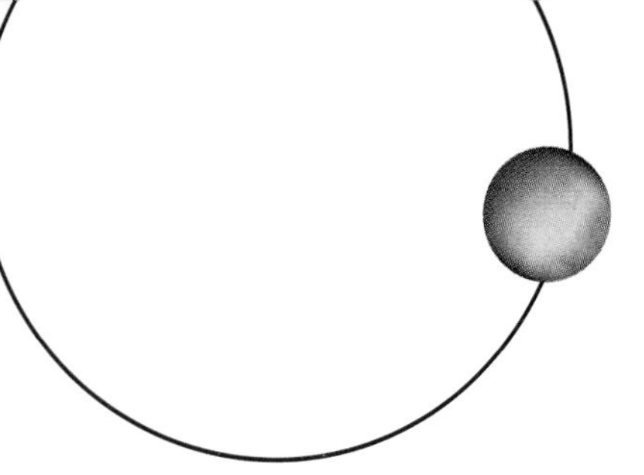

Cumulative Test for Chapters 0 to 8

Name ______________________

Section ________ Date ________

This test is provided to help you in the process of review of the previous chapters. Answers are provided in the back of the book. If you missed any answers, be sure to go back and review the appropriate chapter sections.

Perform each of the indicated operations.

1. $(5x^2 - 9x + 3) + (3x^2 + 2x - 7)$

2. Subtract $9w^2 + 5w$ from the sum of $8w^2 - 3w$ and $2w^2 - 4$.

3. $7xy(4x^2y - 2xy + 3xy^2)$

4. $(3s - 7)(5s + 4)$

5. $\dfrac{5x^3y - 10x^2y^2 + 15xy^2}{-5xy}$

6. $\dfrac{4x^2 + 6x - 4}{2x - 1}$

Solve the following equation for x.

7. $5 - 3(2x - 7) = 8 - 4x$

Factor each of the following polynomials completely.

8. $24a^3 - 16a^2$

9. $7m^2n - 21mn - 49mn^2$

10. $a^2 - 64b^2$

11. $5p^3 - 80pq^2$

12. $a^2 - 14a + 48$

13. $2w^3 - 8w^2 - 42w$

Solve each of the following equations.

14. $x^2 - 9x + 20 = 0$

15. $2x^2 - 32 = 0$

Solve the following applications.

16. Twice the square of a positive integer is 35 more than 9 times that integer. What is the integer?

17. The length of a rectangle is 2 inches (in.) more than 3 times its width. If the area of the rectangle is 85 square inches (in.^{2}), find the dimensions of the rectangle.

ANSWERS

1. $8x^2 - 7x - 4$
2. $w^2 - 8w - 4$
3. $28x^3y^2 - 14x^2y^2 + 21x^2y^3$
4. $15s^2 - 23s - 28$
5. $-x^2 + 2xy - 3y$
6. $2x + 4$
7. 9
8. $8a^2(3a - 2)$
9. $7mn(m - 3 - 7n)$
10. $(a + 8b)(a - 8b)$
11. $5p(p + 4q)(p - 4q)$
12. $(a - 6)(a - 8)$
13. $2w(w - 7)(w + 3)$
14. 4, 5
15. −4, 4
16. 7
17. 5 in. by 17 in.

ANSWERS

18. $\frac{m}{3}$

19. $\frac{a-7}{3a+1}$

20. $\frac{3}{x}$

21. $\frac{1}{3w}$

22. See exercise

23. See exercise

24. See exercise

25. See exercise

26. $\frac{10}{7}$

27. Slope: $\frac{5}{3}$; y intercept $(0, -5)$

Write each fraction in simplest form.

18. $\dfrac{m^2 - 4m}{3m - 12}$

19. $\dfrac{a^2 - 49}{3a^2 + 22a + 7}$

Perform the indicated operations.

20. $\dfrac{3x^2 + 9x}{x^2 - 9} \cdot \dfrac{2x^2 - 9x + 9}{2x^3 - 3x^2}$

21. $\dfrac{4w^2 - 25}{2w^2 - 5w} \div (6w + 15)$

Graph each of the following equations.

22. $x - y = 5$

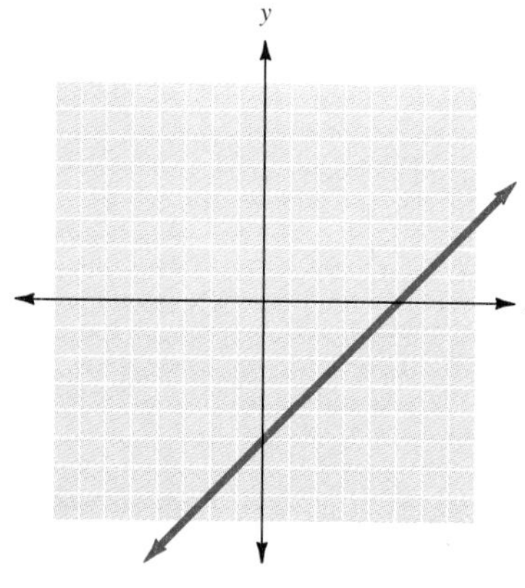

23. $y = \dfrac{2}{3}x + 3$

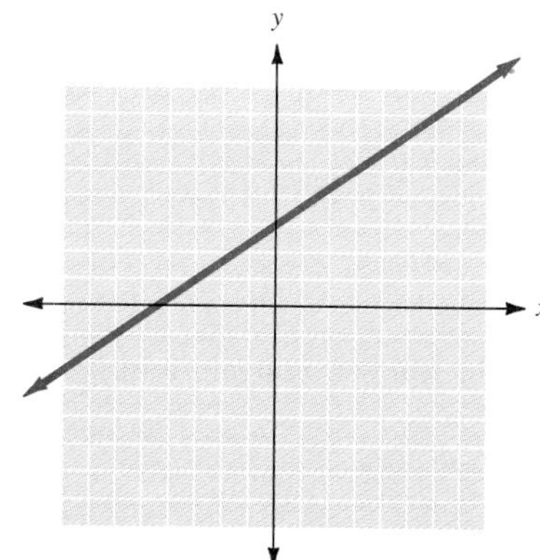

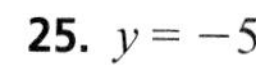

24. $2x - 5y = 10$

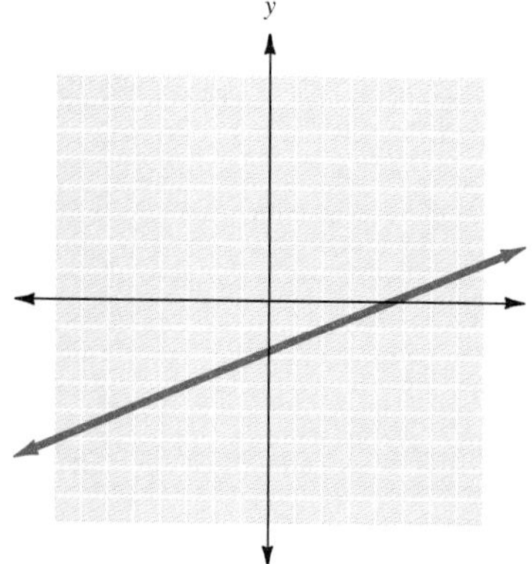

25. $y = -5$

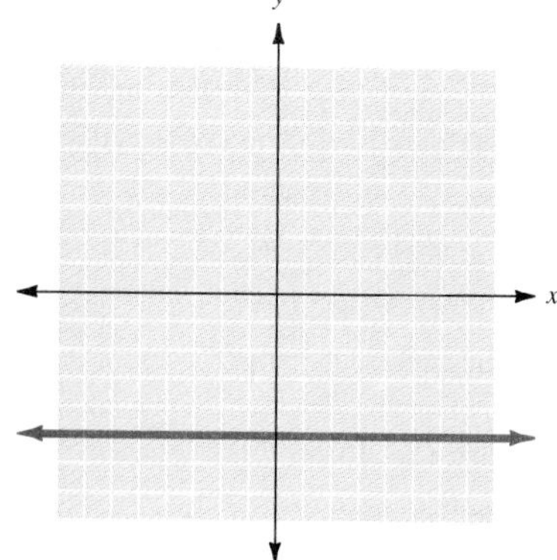

26. Find the slope of the line through the pair of points $(-2, -3)$ and $(5, 7)$.

27. Find the slope and y intercept of the line described by the equation $5x - 3y = 15$.

28. Given the slope and y intercept for the following line, write the equation of the line. Then graph the line.

Slope $= 2$; y intercept: $(0, -5)$

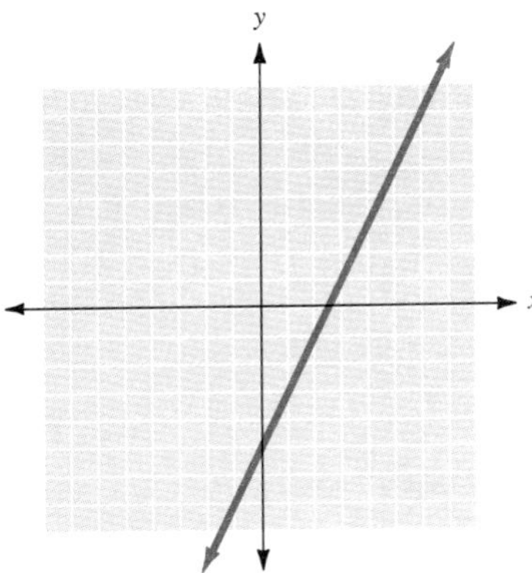

Graph each of the following inequalities.

29. $x + 2y < 6$

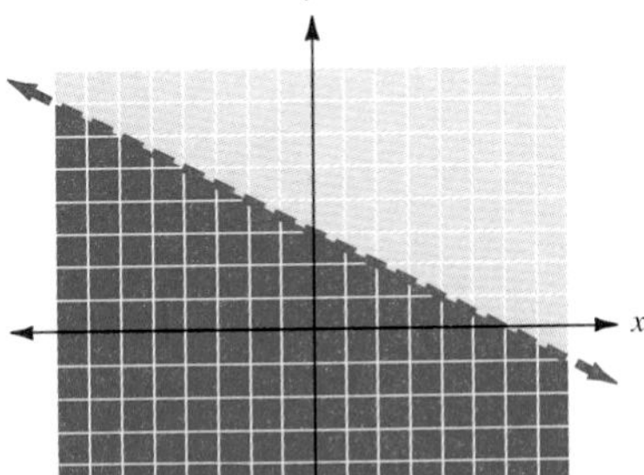

30. $3x - 4y \geq 12$

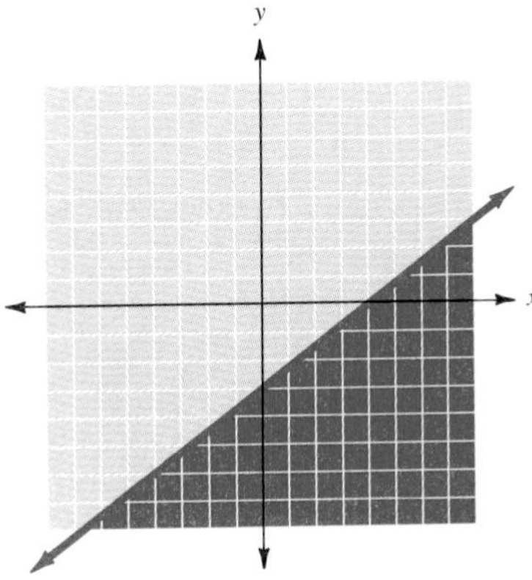

Solve the following system by graphing.

31. $3x + 2y = 6$
$x + 2y = -2$

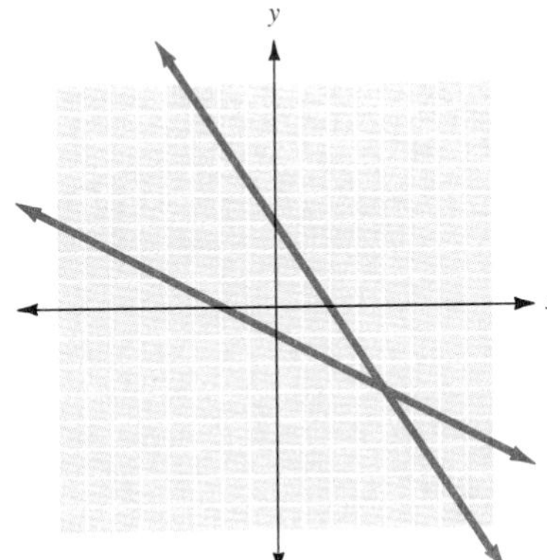

ANSWERS

28. $y = 2x - 5$

29. See exercise

30. See exercise

31. $(4, -3)$

ANSWERS

32. $\left(7, -\frac{5}{2}\right)$

33. Dependent system

34. (5, 0)

35. Inconsistent system

36. $\left(\frac{3}{2}, -\frac{1}{3}\right)$

37. 5, 21

38. VHS $4.50, cassette $1.50

39. 325 at $7, 125 at $4

40. $5000 at 6% $7000 at 9%

Solve each of the following systems. If a unique solution does not exist, state whether the system is inconsistent or dependent.

32. $5x + 2y = 30$
$x - 4y = 17$

33. $2x - 6y = 8$
$x = 3y + 4$

34. $4x - 5y = 20$
$2x + 3y = 10$

35. $4x + 2y = 11$
$2x + y = 5$

36. $4x - 3y = 7$
$6x + 6y = 7$

Solve each of the following applications.

37. One number is 4 less than 5 times another. If the sum of the numbers is 26, what are the two numbers?

38. Cynthia bought five blank VHS tapes and four cassette tapes for $28.50. Charlie bought four VHS tapes and two cassette tapes for $21.00. Find the cost of each type of tape.

39. Receipts for a concert, attended by 450 people, were $2775. If reserved-seat tickets were $7 and general-admission tickets were $4, how many of each type of ticket were sold?

40. Anthony invested part of his $12,000 inheritance in a bond paying 9% and the other part in a savings account paying 6%. If his interest from the two investments was $930 in the first year, how much did he have invested at each rate?

EXPONENTS AND RADICALS

9

INTRODUCTION

In designing a public building, an engineer or an architect must include a plan for safety. The Uniform Building Code states size and location requirements for exits: "If two exits are required in a building, they must be placed apart a distance not less than one-half the length of the maximum overall diagonal dimension of the building . . . ." Stated in algebraic terms, if the building is rectangular and if d is the distance between exits, l is the length of the building, and w is the width of the building, then

$$d \geq \frac{1}{2}\sqrt{l^2 + w^2}$$

So, for example, if a rectangular building is 50 by 40 ft, the diagonal dimension is $\sqrt{50^2 + 40^2}$, and the distance d between the exits must be equal to or more than half of this value. Thus,

$$d \geq \frac{1}{2}\sqrt{50^2 + 40^2}$$

In this case, the distance between the exits must be 32 ft or more.

The use of a radical sign often shows up in the measurement of distances and is based on the Pythagorean theorem, which describes the relationship between the sides of a right triangle: $a^2 + b^2 = c^2$. Using algebra to interpret statements such as the one in the building code quoted above is an example of how algebra can make complicated statements clearer and easier to understand.

Name ______________________

Section ________ Date ________

Pre-Test Chapter 9

ANSWERS

1. 12
2. 4
3. Not a real number
4. 5
5. $6\sqrt{2}$
6. $3x\sqrt{3x}$
7. $\frac{1}{2}$
8. $\frac{\sqrt{7}}{5}$
9. $7\sqrt{7}$
10. $3\sqrt{5}$
11. $3\sqrt{3}$
12. $\sqrt{2}$
13. $x\sqrt{10}$
14. $\sqrt{10} - 6$
15. $9 + 5\sqrt{3}$
16. $\frac{\sqrt{15}}{3}$
17. $4 - \sqrt{5}$
18. 15
19. 24
20. 8.49
21. 4.12
22. 19.21 cm
23. 14
24. 6.71
25. 15.26

Evaluate if possible.

1. $\sqrt{144}$

2. $\sqrt[3]{64}$

3. $\sqrt{-121}$

4. $-\sqrt[3]{-125}$

Simplify each of the following radical expressions.

5. $\sqrt{72}$

6. $\sqrt{27x^3}$

7. $\sqrt{\frac{9}{36}}$

8. $\sqrt{\frac{7}{25}}$

Simplify by combining like terms.

9. $3\sqrt{7} - 2\sqrt{7} + 6\sqrt{7}$

10. $\sqrt{5} + \sqrt{20}$

11. $3\sqrt{12} - \sqrt{27}$

12. $\sqrt{18} - \sqrt{98} + \sqrt{50}$

Simplify each of the following radical expressions.

13. $\sqrt{2x} \cdot \sqrt{5x}$

14. $\sqrt{2}(\sqrt{5} - 3\sqrt{2})$

15. $(\sqrt{3} + 2)(\sqrt{3} + 3)$

16. $\frac{\sqrt{5}}{\sqrt{3}}$

17. 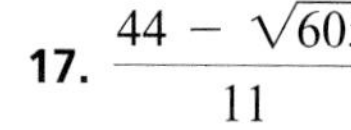 $\frac{44 - \sqrt{605}}{11}$

Find length x in each triangle. Express your answer to the nearest hundredth.

18.

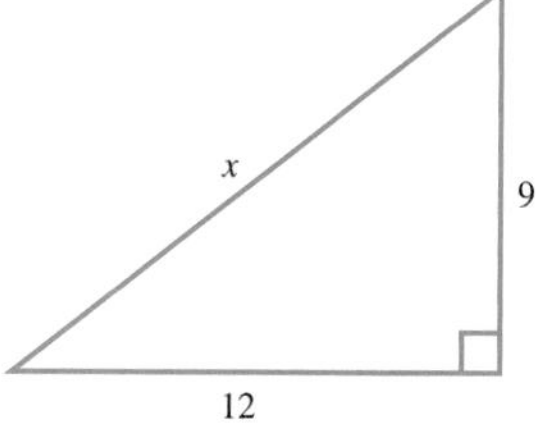

19.

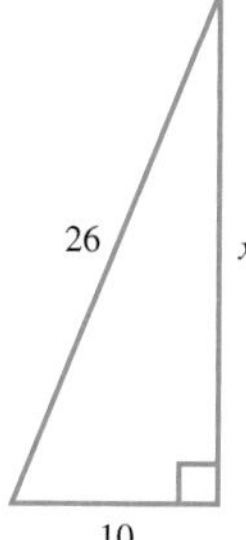

20.

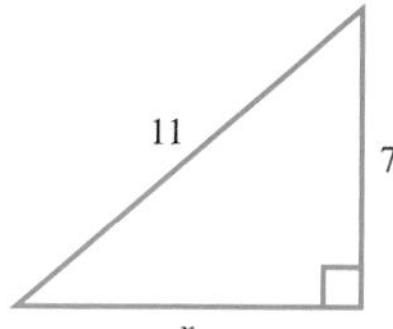

21.

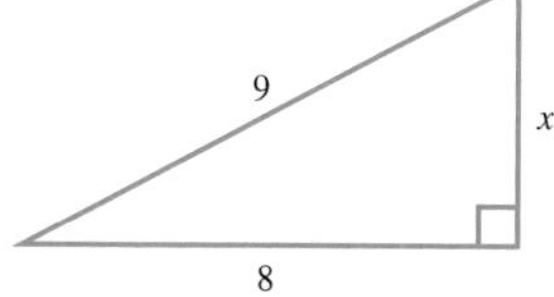

Solve the following word problem.

22. If the length of the diagonal of a rectangle is 25 cm and the width of the rectangle is 16 cm, what is the length of the rectangle? Express your answer to the nearest hundredth of a cm.

Find the distance between the given two points. Express your answer to the nearest hundredth.

23. $(-4, 9)$ and $(-18, 9)$ **24.** $(-3, 4)$ and $(-6, -2)$ **25.** $(5, 9)$ and $(-3, -4)$

9.1 Roots and Radicals

9.1 OBJECTIVES

1. Use the radical notation to represent roots
2. Distinguish between rational and irrational numbers

In Chapter 3, we discussed the properties of exponents. Over the next four sections, we will work with a new notation that "reverses" the process of raising to a power.

From our work in Chapter 0, we know that when we have a statement such as

$$x^2 = 9$$

it is read as "x squared equals 9."

Here we are concerned with the relationship between the variable x and the number 9. We call that relationship the **square root** and say, equivalently, that "x is the square root of 9."

We know from experience that x must be 3 (because $3^2 = 9$) or -3 [because $(-3)^2 = 9$]. We see that 9 has two square roots, 3 and -3. In fact, every positive number will have *two* square roots. In general, if $x^2 = a$, we call x a *square root* of a.

NOTE The symbol $\sqrt{}$ first appeared in print in 1525. In Latin, "radix" means **root,** and this was contracted to a small *r*. The present symbol may have evolved from the manuscript form of that small *r*.

We are now ready for our new notation. The symbol $\sqrt{}$ is called a **radical sign.** We saw above that 3 was the positive square root of 9. We also call 3 the **principal square root** of 9 and can write

$$\sqrt{9} = 3$$

to indicate that 3 is the principal square root of 9.

Definitions: Square Root

$\sqrt{a}$ is the *positive* (or *principal*) square root of *a*. It is the positive number whose square is *a*.

Example 1

Finding Principal Square Roots

Find the following square roots.

(a) $\sqrt{49} = 7$ Because 7 is the positive number we must square to get 49.

(b) $\sqrt{\frac{4}{9}} = \frac{2}{3}$ Because $\frac{2}{3}$ is the positive number we must square to get $\frac{4}{9}$.

CHECK YOURSELF 1

Find the following square roots.

(a) $\sqrt{64}$ (b) $\sqrt{144}$ (c) $\sqrt{\frac{16}{25}}$

NOTE When you use the radical sign, you are referring to the *positive square root:* $\sqrt{25} = 5$

Each positive number has two square roots. For instance, 25 has square roots of 5 and -5 because

$$5^2 = 25 \quad \text{and} \quad (-5)^2 = 25$$

If you want to indicate the negative square root, you must use a minus sign in front of the radical.

$-\sqrt{25} = -5$

Example 2

Finding Square Roots

Find the following square roots.

(a) $\sqrt{100} = 10$ The principal root

(b) $-\sqrt{100} = -10$ The negative square root

(c) $-\sqrt{\frac{9}{16}} = -\frac{3}{4}$

CHECK YOURSELF 2

Find the following square roots.

(a) $\sqrt{16}$ (b) $-\sqrt{16}$ (c) $-\sqrt{\frac{16}{25}}$

CAUTION

Be Careful! Do not confuse

$-\sqrt{9}$ with $\sqrt{-9}$

The expression $-\sqrt{9}$ is -3, whereas $\sqrt{-9}$ is not a real number.

Every number that we have encountered in this text is a **real number.** The square roots of negative numbers are *not* real numbers. For instance, $\sqrt{-9}$ is *not* a real number because there is *no* real number x such that

$x^2 = -9$

Example 3 summarizes our discussion thus far.

Example 3

Finding Square Roots

Evaluate each of the following square roots.

(a) $\sqrt{36} = 6$ (b) $\sqrt{121} = 11$

(c) $-\sqrt{64} = -8$ (d) $\sqrt{-64}$ is not a real number.

(e) $\sqrt{0} = 0$ (Because $0 \cdot 0 = 0$)

CHECK YOURSELF 3

Evaluate, if possible.

(a) $\sqrt{81}$ (b) $\sqrt{49}$ (c) $-\sqrt{49}$ (d) $\sqrt{-49}$

All calculators have square root keys, but the only integers for which the calculator gives the exact value of the square root are perfect square integers. For all other positive integers, *a calculator gives only an approximation of the correct answer.* In Example 4 you will use your calculator to approximate square roots.

Example 4

Approximating Square Roots

Use your calculator to approximate each square root to the nearest hundredth.

NOTE The ≈ sign means "is approximately equal to."

(a) $\sqrt{45} \approx 6.708203932 \approx 6.71$ **(b)** $\sqrt{8} \approx 2.83$

(c) $\sqrt{20} \approx 4.47$ **(d)** $\sqrt{273} \approx 16.52$

CHECK YOURSELF 4

Use your calculator to approximate each square root to the nearest hundredth.

(a) $\sqrt{3}$ **(b)** $\sqrt{14}$ **(c)** $\sqrt{91}$ **(d)** $\sqrt{756}$

As we mentioned earlier, finding the square root of a number is the reverse of squaring a number. We can extend that idea to work with other roots of numbers. For instance, the *cube root* of a number is the number we must cube (or raise to the third power) to get that number. For example, the cube root of 8 is 2 because $2^3 = 8$, and we write

NOTE $\sqrt[3]{8}$ is read "the cube root of 8."

$$\sqrt[3]{8} = 2$$

The parts of a radical expression are summarized as follows.

NOTE The index for $\sqrt[3]{a}$ is 3.

NOTE The index of 2 for square roots is generally not written. We understand that $\sqrt{a}$ is the principal square root of *a*.

Definitions: Parts of a Radical Expression

Every radical expression contains three parts as shown below. The principal *n*th root of *a* is written as

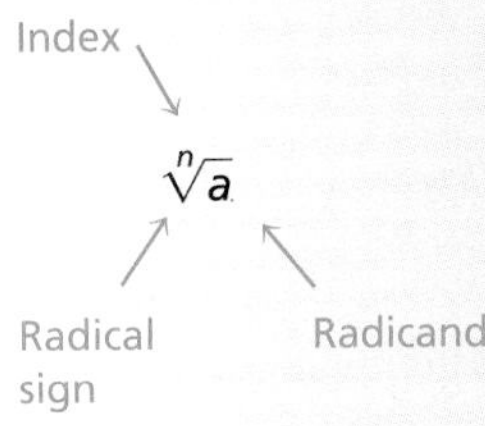

To illustrate, the *cube root* of 64 is written

Index of 3 ⟶ $\sqrt[3]{64} = 4$

because $4^3 = 64$. And

Index of 4 ⟶ $\sqrt[4]{81} = 3$

is the *fourth root* of 81 because $3^4 = 81$.

We can find roots of negative numbers as long as the index is *odd* (3, 5, etc.). For example,

$$\sqrt[3]{-64} = -4$$

because $(-4)^3 = -64$.

If the index is *even* (2, 4, etc.), roots of negative numbers are *not* real numbers. For example,

$$\sqrt[4]{-16}$$

NOTE The *even power* of a real number is always *positive* or *zero*.

is not a real number because there is no real number x such that $x^4 = -16$.

The following table shows the most common roots.

NOTE It would be helpful for your work here and in future mathematics classes to memorize these roots.

Square Roots		Cube Roots	Fourth Roots
$\sqrt{1} = 1$	$\sqrt{49} = 7$	$\sqrt[3]{1} = 1$	$\sqrt[4]{1} = 1$
$\sqrt{4} = 2$	$\sqrt{64} = 8$	$\sqrt[3]{8} = 2$	$\sqrt[4]{16} = 2$
$\sqrt{9} = 3$	$\sqrt{81} = 9$	$\sqrt[3]{27} = 3$	$\sqrt[4]{81} = 3$
$\sqrt{16} = 4$	$\sqrt{100} = 10$	$\sqrt[3]{64} = 4$	$\sqrt[4]{256} = 4$
$\sqrt{25} = 5$	$\sqrt{121} = 11$	$\sqrt[3]{125} = 5$	$\sqrt[4]{625} = 5$
$\sqrt{36} = 6$	$\sqrt{144} = 12$		

You can use the table in Example 5, which summarizes the discussion so far.

Example 5

Evaluating Cube Roots and Fourth Roots

Evaluate each of the following.

NOTE The cube root of a negative number will be negative.

NOTE The fourth root of a negative number is not a real number.

(a) $\sqrt[5]{32} = 2$ because $2^5 = 32$.

(b) $\sqrt[3]{-125} = -5$ because $(-5)^3 = -125$.

(c) $\sqrt[4]{-81}$ is not a real number.

CHECK YOURSELF 5

Evaluate, if possible.

(a) $\sqrt[3]{64}$ **(b)** $\sqrt[4]{16}$ **(c)** $\sqrt[4]{-256}$ **(d)** $\sqrt[3]{-8}$

The radical notation helps us to distinguish between two important types of numbers: rational numbers and irrational numbers.

A **rational number** can be represented by a fraction whose numerator and denominator are integers and whose denominator is nonzero. The form of a rational number is

$$\frac{a}{b} \qquad a \text{ and } b \text{ are integers, } b \neq 0$$

Certain square roots are rational numbers also. For example,

NOTE Notice that each radicand is a **perfect-square integer** (that is, an integer that is the square of another integer).

$$\sqrt{4} \qquad \sqrt{25} \qquad \text{and} \qquad \sqrt{64}$$

represent the rational numbers 2, 5, and 8, respectively.

NOTE The fact that the square root of 2 is irrational will be proved in later mathematics courses and was known to Greek mathematicians over 2000 years ago.

An **irrational number** is a number that *cannot* be written as the ratio of two integers. For example, the square root of any positive number that is not itself a perfect square is an irrational number. Because the radicands are *not* perfect squares, the expressions $\sqrt{2}$, $\sqrt{3}$, and $\sqrt{5}$ represent irrational numbers.

Example 6

Identifying Rational Numbers

Which of the following numbers are rational and which are irrational?

$$\sqrt{\frac{2}{3}} \qquad \sqrt{\frac{4}{9}} \qquad \sqrt{7} \qquad \sqrt{16} \qquad \sqrt{25}$$

Here $\sqrt{7}$ and $\sqrt{\frac{2}{3}}$ are irrational numbers. And $\sqrt{16}$ and $\sqrt{25}$ are rational numbers because 16 and 25 are perfect squares. Also $\sqrt{\frac{4}{9}}$ is rational because $\sqrt{\frac{4}{9}} = \frac{2}{3}$.

CHECK YOURSELF 6

Which of the following numbers are rational and which are irrational?

(a) $\sqrt{26}$ **(b)** $\sqrt{49}$ **(c)** $\sqrt{\frac{6}{7}}$ **(d)** $\sqrt{105}$ **(e)** $\sqrt{\frac{16}{9}}$

NOTE The decimal representation of a rational number always terminates or repeats. For instance,

$\frac{3}{8} = 0.375$

$\frac{5}{11} = 0.454545\ldots$

An important fact about the irrational numbers is that their decimal representations are always *nonterminating* and *nonrepeating*. We can therefore only approximate irrational numbers with a decimal that has been rounded off. A calculator can be used to find roots. However, note that the values found for the irrational roots are only approximations. For instance, $\sqrt{2}$ is approximately 1.414 (to three decimal places), and we can write

$$\sqrt{2} \approx 1.414$$

NOTE 1.414 is an approximation to the number whose square is 2.

With a calculator we find that

$$(1.414)^2 = 1.999396$$

The set of all rational numbers and the set of all irrational numbers together form the set of *real numbers.* The real numbers will represent every point that can be pictured on the number line. Some examples are shown below.

NOTE For this reason we refer to the number line as the **real number line.**

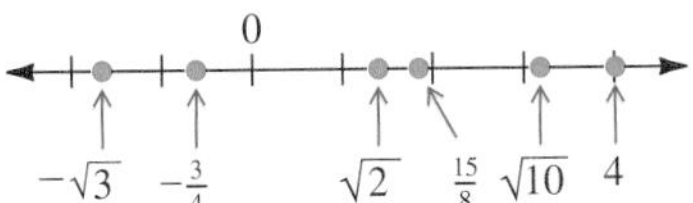

The following diagram summarizes the relationships among the various numeric sets.

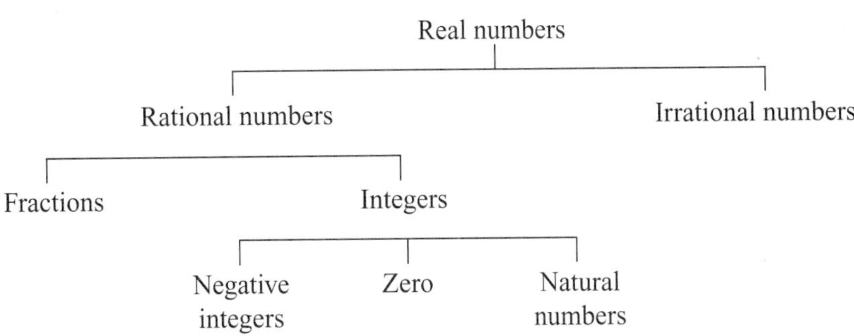

We conclude our work in this section by developing a general result that we will need later. Let's start by looking at two numerical examples.

$$\sqrt{2^2} = \sqrt{4} = 2 \qquad (1)$$

$$\sqrt{(-2)^2} = \sqrt{4} = 2 \qquad \text{because } (-2)^2 = 4 \qquad (2)$$

Consider the value of $\sqrt{x^2}$ when x is positive or negative.

NOTE This is because the principal square root of a number is always positive or zero.

In (1) when $x = 2$: In (2) when $x = -2$:

$$\sqrt{2^2} = 2 \qquad \sqrt{(-2)^2} \neq -2$$

$$\sqrt{(-2)^2} = -(-2) = 2$$

Comparing the results of (1) and (2), we see that $\sqrt{x^2}$ is x if x is positive (or 0) and $\sqrt{x^2}$ is $-x$ if x is negative. We can write

$$\sqrt{x^2} = \begin{cases} x & \text{when } x \geq 0 \\ -x & \text{when } x < 0 \end{cases}$$

From your earlier work with absolute values you will remember that

$$|x| = \begin{cases} x & \text{when } x \geq 0 \\ -x & \text{when } x < 0 \end{cases}$$

and we can summarize the discussion by writing

$$\sqrt{x^2} = |x| \qquad \text{for any real number } x$$

Example 7

Evaluating Radical Expressions

NOTE Alternatively in (b), we could write $\sqrt{(-4)^2} = \sqrt{16} = 4$

Evaluate each of the following.

(a) $\sqrt{5^2} = 5$ **(b)** $\sqrt{(-4)^2} = |-4| = 4$

CHECK YOURSELF 7

Evaluate.

(a) $\sqrt{6^2}$ **(b)** $\sqrt{(-6)^2}$

CHECK YOURSELF ANSWERS

1. (a) 8; **(b)** 12; **(c)** $\frac{4}{5}$ **2. (a)** 4; **(b)** −4; **(c)** $-\frac{4}{5}$ **3. (a)** 9; **(b)** 7; **(c)** −7; **(d)** not a real number **4. (a)** 1.73; **(b)** 3.74; **(c)** 9.54; **(d)** 27.50 **5. (a)** 4; **(b)** 2; **(c)** not a real number; **(d)** −2 **6. (a)** Irrational; **(b)** rational (because $\sqrt{49} = 7$); **(c)** irrational; **(d)** irrational **(e)** $\left(\text{because } \sqrt{\frac{16}{9}} = \frac{4}{3}\right)$ **7. (a)** 6; **(b)** 6

Name ____________

Section ________ Date ________

9.1 Exercises

Evaluate, if possible.

1. $\sqrt{16}$

2. $\sqrt{121}$

3. $\sqrt{400}$

4. $\sqrt{64}$

5. $-\sqrt{100}$

6. $\sqrt{-100}$

7. $\sqrt{-81}$

8. $-\sqrt{81}$

9. $\sqrt{\frac{16}{9}}$

10. $-\sqrt{\frac{1}{25}}$

11. $\sqrt{-\frac{4}{5}}$

12. $\sqrt{\frac{4}{25}}$

13. $\sqrt[3]{27}$

14. $\sqrt[4]{81}$

15. $\sqrt[3]{-27}$

16. $\sqrt[4]{-16}$

17. $\sqrt[4]{-81}$

18. $-\sqrt[3]{64}$

19. $-\sqrt[3]{27}$

20. $-\sqrt[3]{-8}$

21. $\sqrt[4]{625}$

22. $\sqrt[3]{1000}$

23. $\sqrt[3]{\frac{1}{27}}$

24. $\sqrt[3]{-\frac{8}{27}}$

ANSWERS

1. 4
2. 11
3. 20
4. 8
5. −10
6. Not a real number
7. Not a real number
8. −9
9. $\frac{4}{3}$
10. $-\frac{1}{5}$
11. Not a real number
12. $\frac{2}{5}$
13. 3
14. 3
15. −3
16. Not a real number
17. Not a real number
18. −4
19. −3
20. 2
21. 5
22. 10
23. $\frac{1}{3}$
24. $-\frac{2}{3}$

ANSWERS

25. Irrational
26. Rational
27. Rational
28. Irrational
29. Irrational
30. Rational
31. Rational
32. Rational
33. Irrational
34. Irrational
35. Rational
36. Rational
37. 3.32
38. 3.74
39. 2.65
40. 4.80
41. 6.78
42. 8.83
43. 0.63
44. 0.87
45. 0.94
46. 0.68

Which of the following roots are rational numbers and which are irrational numbers?

25. $\sqrt{19}$

26. $\sqrt{36}$

27. $\sqrt{100}$

28. $\sqrt{7}$

29. $\sqrt[3]{9}$

30. $\sqrt[3]{8}$

31. $\sqrt[4]{16}$

32. $\sqrt{\frac{4}{9}}$

33. $\sqrt{\frac{4}{7}}$

34. $\sqrt[3]{5}$

35. $\sqrt[3]{-27}$

36. $-\sqrt[4]{81}$

Use your calculator to approximate the square root to the nearest hundredth.

37. $\sqrt{11}$

38. $\sqrt{14}$

39. $\sqrt{7}$

40. $\sqrt{23}$

41. $\sqrt{46}$

42. $\sqrt{78}$

43. $\sqrt{\frac{2}{5}}$

44. $\sqrt{\frac{3}{4}}$

45. $\sqrt{\frac{8}{9}}$

46. $\sqrt{\frac{7}{15}}$

47. $-\sqrt{18}$

48. $-\sqrt{31}$

49. $-\sqrt{27}$

50. $-\sqrt{65}$

For exercises 51 to 56, find the two expressions that are equivalent.

51. $\sqrt{-16}, -\sqrt{16}, -4$

52. $-\sqrt{25}, -5, \sqrt{-25}$

53. $\sqrt[3]{-125}, -\sqrt[3]{125}, |-5|$

54. $\sqrt[5]{-32}, -\sqrt[5]{32}, |-2|$

55. $\sqrt[4]{10{,}000}, 100, \sqrt[3]{1000}$

56. $10^2, \sqrt{10{,}000}, \sqrt[3]{100{,}000}$

In exercises 57 to 62, label the statement as true or false.

57. $\sqrt{16x^{16}} = 4x^4$

58. $\sqrt{(x-4)^2} = x - 4$

59. $\sqrt{16x^{-4}y^{-4}}$ is a real number

60. $\sqrt{x^2 + y^2} = x + y$

61. $\dfrac{\sqrt{x^2 - 25}}{x - 5} = \sqrt{x + 5}$

62. $\sqrt{2} + \sqrt{6} = \sqrt{8}$

63. Dimensions of a square. The area of a square is 32 square feet (ft^2). Find the length of a side to the nearest hundredth.

64. Dimensions of a square. The area of a square is 83 ft^2. Find the length of the side to the nearest hundredth.

65. Radius of a circle. The area of a circle is 147 ft^2. Find the radius to the nearest hundredth.

66. Radius of a circle. If the area of a circle is 72 square centimeters (cm^2), find the radius to the nearest hundredth.

ANSWERS

47. -4.24

48. -5.57

49. -5.20

50. -8.06

51. $-\sqrt{16}, -4$

52. $-\sqrt{25}, -5$

53. $\sqrt[3]{-125}, -\sqrt[3]{125}$

54. $\sqrt[5]{-32}, -\sqrt[5]{32}$

55. $\sqrt[4]{10{,}000}, \sqrt[3]{1000}$

56. $10^2, \sqrt{10{,}000}$

57. False

58. False

59. True

60. False

61. False

62. False

63. 5.66 ft

64. 9.11 ft

65. 6.84 ft

66. 4.79 cm

ANSWERS

67. 7.07 s

68. 9.35 s

69. 3.16 ft

70. 3.61 ft

71. 4.12 ft

72. No

73. Conjugate: $4 + \sqrt{7}$
Opposite: $-4 + \sqrt{7}$

74. 0, 1

75. $\sqrt{3} + 2$

67. Freely falling objects. The time in seconds (s) that it takes for an object to fall from rest is given by $t = \frac{1}{4}\sqrt{s}$, in which s is the distance fallen. Find the time required for an object to fall to the ground from a building that is 800 ft high.

68. Freely falling objects. Find the time required for an object to fall to the ground from a building that is 1400 ft high. (Use the formula in exercise 67.)

In exercises 69 to 71, the area is given in square feet. Find the length of a side of the square. Round your answer to the nearest hundredth of a foot.

69. **70.** **71.**

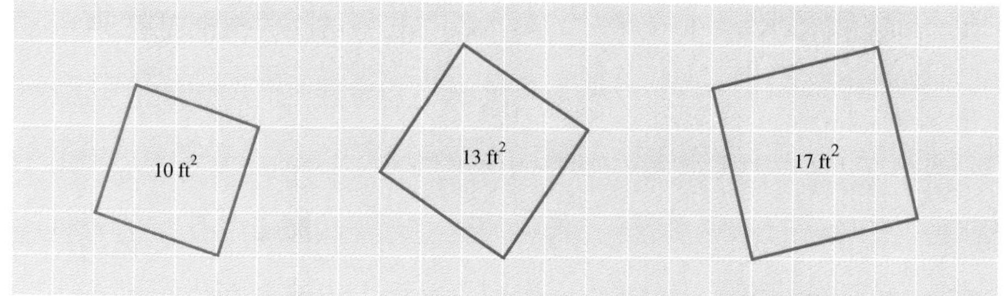

72. Is there any prime number whose square root is an integer? Explain your answer.

73. Explain the difference between the conjugate, in which the middle sign is changed, of a binomial and the opposite of a binomial. To illustrate, use $4 - \sqrt{7}$.

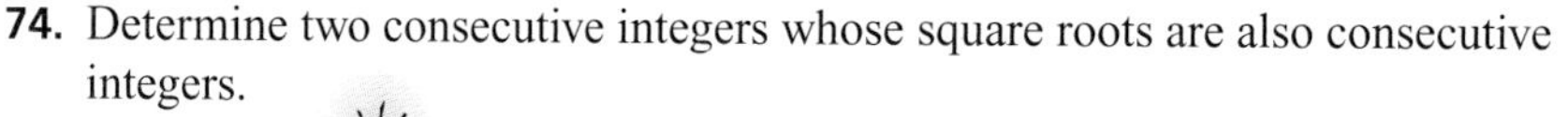

74. Determine two consecutive integers whose square roots are also consecutive integers.

75. Determine the missing binomial in the following: $(\sqrt{3} - 2)(\quad) = -1$.

76. Try the following using your calculator.

(a) Choose a number greater than 1 and find its square root. Then find the square root of the result and continue in this manner, observing the successive square roots. Do these numbers seem to be approaching a certain value? If so, what?

(b) Choose a number greater than 0 but less than 1 and find its square root. Then find the square root of the result, and continue in this manner, observing successive square roots. Do these numbers seem to be approaching a certain value? If so, what?

77. (a) Can a number be equal to its own square root?

(b) Other than the number(s) found in part a, is a number always greater than its square root? Investigate.

78. Let a and b be positive numbers. If a is greater than b, is it always true that the square root of a is greater than the square root of b? Investigate.

79. Suppose that a weight is attached to a string of length L, and the other end of the string is held fixed. If we pull the weight and then release it, allowing the weight to swing back and forth, we can observe the behavior of a simple pendulum. The period, T, is the time required for the weight to complete a full cycle, swinging forward and then back. The following formula may be used to describe the relationship between T and L.

$$T = 2\pi\sqrt{\frac{L}{g}}$$

If L is expressed in centimeters, then $g = 980 \text{ cm/s}^2$. For each of the following string lengths, calculate the corresponding period. Round to the nearest tenth of a second.

(a) 30 cm (b) 50 cm (c) 70 cm (d) 90 cm (e) 110 cm

Getting Ready for Section 9.2 [Section 1.7]

Find each of the following products.

(a) $(4x^2)(2x)$ (b) $(9a^4)(5a)$ (c) $(16m^2)(3m)$ (d) $(8b^3)(2b)$

(e) $(27p^6)(3p)$ (f) $(81s^4)(s^3)$ (g) $(100y^4)(2y)$ (h) $(49m^6)(2m)$

ANSWERS

76.

77.

78.

(a) 1.1 s
(b) 1.4 s
(c) 1.7 s
(d) 1.9 s
79. (e) 2.1 s

a. $8x^3$

b. $45a^5$

c. $48m^3$

d. $16b^4$

e. $81p^7$

f. $81s^7$

g. $200y^5$

h. $98m^7$

Answers

1. 4 **3.** 20 **5.** -10 **7.** Not a real number **9.** $\frac{4}{3}$
11. Not a real number **13.** 3 **15.** -3 **17.** Not a real number
19. -3 **21.** 5 **23.** $\frac{1}{3}$ **25.** Irrational **27.** Rational **29.** Irrational
31. Rational **33.** Irrational **35.** Rational **37.** 3.32 **39.** 2.65
41. 6.78 **43.** 0.63 **45.** 0.94 **47.** -4.24 **49.** -5.20
51. $-\sqrt{16}, -4$ **53.** $\sqrt[3]{-125}, -\sqrt[3]{125}$ **55.** $\sqrt[4]{10{,}000}, \sqrt[3]{1000}$
57. False **59.** True **61.** False **63.** 5.66 ft **65.** 6.84 ft
67. 7.07 s **69.** 3.16 ft **71.** 4.12 ft
73. Conjugate: $4 + \sqrt{7}$; opposite: $-4 + \sqrt{7}$ **75.** $\sqrt{3} + 2$ **77.**

79. **(a)** 1.1 s; **(b)** 1.4 s; **(c)** 1.7 s; **(d)** 1.9 s; **(e)** 2.1 s **a.** $8x^3$ **b.** $45a^5$
c. $48m^3$ **d.** $16b^4$ **e.** $81p^7$ **f.** $81s^7$ **g.** $200y^5$ **h.** $98m^7$

Simplifying Radical Expressions

OBJECTIVES

1. Simplify expressions involving numeric radicals
2. Simplify expressions involving algebraic radicals

In Section 9.1, we introduced the radical notation. For most applications, we will want to make sure that all radical expressions are in *simplest form.* To accomplish this, the following three conditions must be satisfied.

Rules and Properties: Square Root Expressions in Simplest Form

An expression involving square roots is in *simplest form* if

1. There are no perfect-square factors in a radical.
2. No fraction appears inside a radical.
3. No radical appears in the denominator.

For instance, considering condition 1,

$\sqrt{17}$ is in simplest form because 17 has *no* perfect-square factors

whereas

$\sqrt{12}$ is *not* in simplest form

because it does contain a perfect-square factor.

$$\sqrt{12} = \sqrt{4 \cdot 3}$$

A perfect square (pointing to 4)

To simplify radical expressions, we'll need to develop two important properties. First, look at the following expressions:

$$\sqrt{4 \cdot 9} = \sqrt{36} = 6$$

$$\sqrt{4} \cdot \sqrt{9} = 2 \cdot 3 = 6$$

Because this tells us that $\sqrt{4 \cdot 9} = \sqrt{4} \cdot \sqrt{9}$, the following general rule for radicals is suggested.

Rules and Properties: Property 1 of Radicals

For any positive real numbers a and b,

$$\sqrt{ab} = \sqrt{a} \cdot \sqrt{b}$$

In words, the square root of a product is the product of the square roots.

Let's see how this property is applied in simplifying expressions when radicals are involved.

Example 1

Simplifying Radical Expressions

NOTE Perfect-square factors are 1, 4, 9, 16, 25, 36, 49, 64, 81, 100, and so on.

Simplify each expression.

(a) $\sqrt{12} = \sqrt{4 \cdot 3}$

A perfect square

NOTE Apply Property 1.

$= \sqrt{4} \cdot \sqrt{3}$

$= 2\sqrt{3}$

NOTE Notice that we have removed the perfect-square factor from inside the radical, so the expression is in simplest form.

(b) $\sqrt{45} = \sqrt{9 \cdot 5}$

A perfect square

NOTE It would not have helped to write

$\sqrt{45} = \sqrt{15 \cdot 3}$

because neither factor is a perfect square.

$= \sqrt{9} \cdot \sqrt{5}$

$= 3\sqrt{5}$

NOTE We look for the *largest* perfect-square factor, here 36.

(c) $\sqrt{72} = \sqrt{36 \cdot 2}$

A perfect square

NOTE Then apply Property 1.

$= \sqrt{36} \cdot \sqrt{2}$

$= 6\sqrt{2}$

(d) $5\sqrt{18} = 5\sqrt{9 \cdot 2}$

A perfect square

$= 5 \cdot \sqrt{9} \cdot \sqrt{2} = 5 \cdot 3 \cdot \sqrt{2} = 15\sqrt{2}$

CAUTION

Be Careful! Even though

$\sqrt{a \cdot b} = \sqrt{a} \cdot \sqrt{b}$

$\sqrt{a + b}$ is *not the same* as $\sqrt{a} + \sqrt{b}$

Let $a = 4$ and $b = 9$, and substitute.

$\sqrt{a + b} = \sqrt{4 + 9} = \sqrt{13}$

$\sqrt{a} + \sqrt{b} = \sqrt{4} + \sqrt{9} = 2 + 3 = 5$

Because $\sqrt{13} \neq 5$, we see that the expressions $\sqrt{a + b}$ and $\sqrt{a} + \sqrt{b}$ are not in general the same.

CHECK YOURSELF 1

Simplify.

(a) $\sqrt{20}$ **(b)** $\sqrt{75}$ **(c)** $\sqrt{98}$ **(d)** $\sqrt{48}$

The process is the same if variables are involved in a radical expression. In our remaining work with radicals, we will assume that all variables represent positive real numbers.

Example 2

Simplifying Radical Expressions

Simplify each of the following radicals.

(a) $\sqrt{x^3} = \sqrt{x^2 \cdot x}$

A perfect square

$= \sqrt{x^2} \cdot \sqrt{x}$

$= x\sqrt{x}$

NOTE By our first rule for radicals.

NOTE $\sqrt{x^2} = x$ (as long as x is positive).

(b) $\sqrt{4b^3} = \sqrt{4 \cdot b^2 \cdot b}$

Perfect squares

$= \sqrt{4b^2} \cdot \sqrt{b}$

$= 2b\sqrt{b}$

(c) $\sqrt{18a^5} = \sqrt{9 \cdot a^4 \cdot 2a}$

Perfect squares

$= \sqrt{9a^4} \cdot \sqrt{2a}$

$= 3a^2\sqrt{2a}$

NOTE Notice that we want the perfect-square factor to have the largest possible even exponent, here 4. Keep in mind that

$a^2 \cdot a^2 = a^4$

CHECK YOURSELF 2

Simplify.

(a) $\sqrt{9x^3}$ **(b)** $\sqrt{27m^3}$ **(c)** $\sqrt{50b^5}$

To develop a second property for radicals, look at the following expressions:

$$\sqrt{\frac{16}{4}} = \sqrt{4} = 2$$

$$\frac{\sqrt{16}}{\sqrt{4}} = \frac{4}{2} = 2$$

Because $\sqrt{\frac{16}{4}} = \frac{\sqrt{16}}{\sqrt{4}}$, a second general rule for radicals is suggested.

Rules and Properties: Property 2 of Radicals

For any positive real numbers a and b,

$$\sqrt{\frac{a}{b}} = \frac{\sqrt{a}}{\sqrt{b}}$$

In words, the square root of a quotient is the quotient of the square roots.

This property is used in a fashion similar to Property 1 in simplifying radical expressions. Remember that our second condition for a radical expression to be in simplest form states that no fraction should appear inside a radical. Example 3 illustrates how expressions that violate that condition are simplified.

Example 3

Simplifying Radical Expressions

Write each expression in simplest form.

NOTE Apply Property 2 to write the numerator and denominator as separate radicals.

(a) $\sqrt{\frac{9}{4}} = \frac{\sqrt{9}}{\sqrt{4}}$ Remove any perfect squares from the radical.

$= \frac{3}{2}$

NOTE Apply Property 2.

(b) $\sqrt{\frac{2}{25}} = \frac{\sqrt{2}}{\sqrt{25}}$

$= \frac{\sqrt{2}}{5}$

NOTE Apply Property 2.

(c) $\sqrt{\frac{8x^2}{9}} = \frac{\sqrt{8x^2}}{\sqrt{9}}$

NOTE Factor $8x^2$ as $4x^2 \cdot 2$.

$= \frac{\sqrt{4x^2 \cdot 2}}{3}$

NOTE Apply Property 1 in the numerator.

$= \frac{\sqrt{4x^2} \cdot \sqrt{2}}{3}$

$= \frac{2x\sqrt{2}}{3}$

CHECK YOURSELF 3

Simplify.

(a) $\sqrt{\frac{25}{16}}$ (b) $\sqrt{\frac{7}{9}}$ (c) $\sqrt{\frac{12x^2}{49}}$

In our previous examples, the denominator of the fraction appearing in the radical was a perfect square, and we were able to write each expression in simplest radical form by removing that perfect square from the denominator.

If the denominator of the fraction in the radical is *not* a perfect square, we can still apply Property 2 of radicals. As we will see in Example 4, the third condition for a radical to be in simplest form is then violated, and a new technique is necessary.

Example 4

Simplifying Radical Expressions

Write each expression in simplest form.

NOTE We begin by applying Property 2.

(a) $\sqrt{\frac{1}{3}} = \frac{\sqrt{1}}{\sqrt{3}} = \frac{1}{\sqrt{3}}$

Do you see that $\frac{1}{\sqrt{3}}$ is still not in simplest form because of the radical in the denominator? To solve this problem, we multiply the numerator and denominator by $\sqrt{3}$. Note that the denominator will become

$$\sqrt{3} \cdot \sqrt{3} = \sqrt{9} = 3$$

We then have

NOTE We can do this because we are multiplying the fraction by $\frac{\sqrt{3}}{\sqrt{3}}$ or 1, which does not change its value.

$$\frac{1}{\sqrt{3}} = \frac{1 \cdot \sqrt{3}}{\sqrt{3} \cdot \sqrt{3}} = \frac{\sqrt{3}}{3}$$

The expression $\frac{\sqrt{3}}{3}$ is now in simplest form because all three of our conditions are satisfied.

(b) $\sqrt{\frac{2}{5}} = \frac{\sqrt{2}}{\sqrt{5}}$

$$= \frac{\sqrt{2} \cdot \sqrt{5}}{\sqrt{5} \cdot \sqrt{5}}$$

$$= \frac{\sqrt{10}}{5}$$

NOTE
$\sqrt{2} \cdot \sqrt{5} = \sqrt{2 \cdot 5} = \sqrt{10}$
$\sqrt{5} \cdot \sqrt{5} = 5$

and the expression is in simplest form because again our three conditions are satisfied.

(c) $\sqrt{\frac{3x}{7}} = \frac{\sqrt{3x}}{\sqrt{7}}$

$$= \frac{\sqrt{3x} \cdot \sqrt{7}}{\sqrt{7} \cdot \sqrt{7}}$$

$$= \frac{\sqrt{21x}}{7}$$

NOTE We multiply numerator and denominator by $\sqrt{7}$ to "clear" the denominator of the radical. This is also known as "rationalizing" the denominator.

The expression is in simplest form.

CHECK YOURSELF 4

Simplify.

(a) $\sqrt{\frac{1}{2}}$ **(b)** $\sqrt{\frac{2}{3}}$ **(c)** $\sqrt{\frac{2y}{5}}$

Both of the properties of radicals given in this section are true for cube roots, fourth roots, and so on. Here we have limited ourselves to simplifying expressions involving square roots.

CHECK YOURSELF ANSWERS

1. (a) $2\sqrt{5}$; **(b)** $5\sqrt{3}$; **(c)** $7\sqrt{2}$; **(d)** $4\sqrt{3}$ **2. (a)** $3x\sqrt{x}$; **(b)** $3m\sqrt{3m}$; **(c)** $5b^2\sqrt{2b}$ **3. (a)** $\frac{5}{4}$; **(b)** $\frac{\sqrt{7}}{3}$; **(c)** $\frac{2x\sqrt{3}}{7}$ **4. (a)** $\frac{\sqrt{2}}{2}$; **(b)** $\frac{\sqrt{6}}{3}$; **(c)** $\frac{\sqrt{10y}}{5}$

Name ____________________

Section ________ Date ________

9.2 Exercises

Use Property 1 to simplify each of the following radical expressions. Assume that all variables represent positive real numbers.

1. $\sqrt{18}$
2. $\sqrt{50}$
3. $\sqrt{28}$
4. $\sqrt{108}$
5. $\sqrt{45}$
6. $\sqrt{80}$
7. $\sqrt{48}$
8. $\sqrt{125}$
9. $\sqrt{200}$
10. $\sqrt{96}$
11. $\sqrt{147}$
12. $\sqrt{300}$
13. $3\sqrt{12}$
14. $5\sqrt{24}$
15. $\sqrt{5x^2}$
16. $\sqrt{7a^2}$
17. $\sqrt{3y^4}$
18. $\sqrt{10x^6}$
19. $\sqrt{2r^3}$
20. $\sqrt{5a^5}$
21. $\sqrt{27b^2}$
22. $\sqrt{98m^4}$
23. $\sqrt{24x^4}$
24. $\sqrt{72x^3}$

ANSWERS

1. $3\sqrt{2}$
2. $5\sqrt{2}$
3. $2\sqrt{7}$
4. $6\sqrt{3}$
5. $3\sqrt{5}$
6. $4\sqrt{5}$
7. $4\sqrt{3}$
8. $5\sqrt{5}$
9. $10\sqrt{2}$
10. $4\sqrt{6}$
11. $7\sqrt{3}$
12. $10\sqrt{3}$
13. $6\sqrt{3}$
14. $10\sqrt{6}$
15. $x\sqrt{5}$
16. $a\sqrt{7}$
17. $y^2\sqrt{3}$
18. $x^3\sqrt{10}$
19. $r\sqrt{2r}$
20. $a^2\sqrt{5a}$
21. $3b\sqrt{3}$
22. $7m^2\sqrt{2}$
23. $2x^2\sqrt{6}$
24. $6x\sqrt{2x}$

ANSWERS

25. $3a^2\sqrt{6a}$

26. $10y^3\sqrt{2}$

27. $xy\sqrt{x}$

28. $ab^2\sqrt{b}$

29. $\frac{2}{5}$

30. $\frac{8}{3}$

31. $\frac{3}{4}$

32. $\frac{7}{5}$

33. $\frac{\sqrt{3}}{2}$

34. $\frac{\sqrt{5}}{3}$

35. $\frac{\sqrt{5}}{6}$

36. $\frac{\sqrt{10}}{7}$

37. $\frac{2a\sqrt{2}}{5}$

38. $\frac{2y\sqrt{3}}{7}$

39. $\frac{\sqrt{5}}{5}$

40. $\frac{\sqrt{7}}{7}$

41. $\frac{\sqrt{6}}{2}$

42. $\frac{\sqrt{15}}{3}$

43. $\frac{\sqrt{15a}}{5}$

44. $\frac{\sqrt{14x}}{7}$

45. $\frac{x\sqrt{6}}{3}$

46. $\frac{m\sqrt{10}}{2}$

47. $\frac{2s\sqrt{14s}}{7}$

48. $\frac{2x\sqrt{15x}}{5}$

25. $\sqrt{54a^5}$

26. $\sqrt{200y^6}$

27. $\sqrt{x^3y^2}$

28. $\sqrt{a^2b^5}$

Use Property 2 to simplify each of the following radical expressions.

29. $\sqrt{\frac{4}{25}}$

30. $\sqrt{\frac{64}{9}}$

31. $\sqrt{\frac{9}{16}}$

32. $\sqrt{\frac{49}{25}}$

33. $\sqrt{\frac{3}{4}}$

34. $\sqrt{\frac{5}{9}}$

35. $\sqrt{\frac{5}{36}}$

36. $\sqrt{\frac{10}{49}}$

Use the properties for radicals to simplify each of the following expressions. Assume that all variables represent positive real numbers.

37. $\sqrt{\frac{8a^2}{25}}$

38. $\sqrt{\frac{12y^2}{49}}$

39. $\sqrt{\frac{1}{5}}$

40. $\sqrt{\frac{1}{7}}$

41. $\sqrt{\frac{3}{2}}$

42. $\sqrt{\frac{5}{3}}$

43. $\sqrt{\frac{3a}{5}}$

44. $\sqrt{\frac{2x}{7}}$

45. $\sqrt{\frac{2x^2}{3}}$

46. $\sqrt{\frac{5m^2}{2}}$

47. $\sqrt{\frac{8s^3}{7}}$

48. $\sqrt{\frac{12x^3}{5}}$

Decide whether each of the following is already written in simplest form. If it is not, explain what needs to be done.

49. $\sqrt{10mn}$

50. $\sqrt{18ab}$

51. $\sqrt{\dfrac{98x^2y}{7x}}$

52. $\dfrac{\sqrt{6xy}}{3x}$

53. Find the area and perimeter of this square:

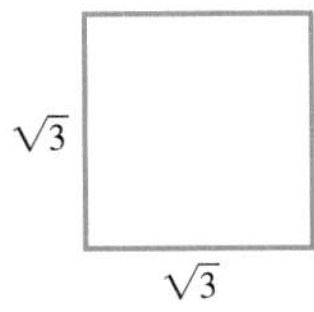

One of these measures, the area, is a rational number, and the other, the perimeter, is an irrational number. Explain how this happened. Will the area always be a rational number? Explain.

54. **(a)** Evaluate the three expressions $\frac{n^2 - 1}{2}$, n, $\frac{n^2 + 1}{2}$ using odd values of n: 1, 3, 5, 7, etc. Make a chart like the one below and complete it.

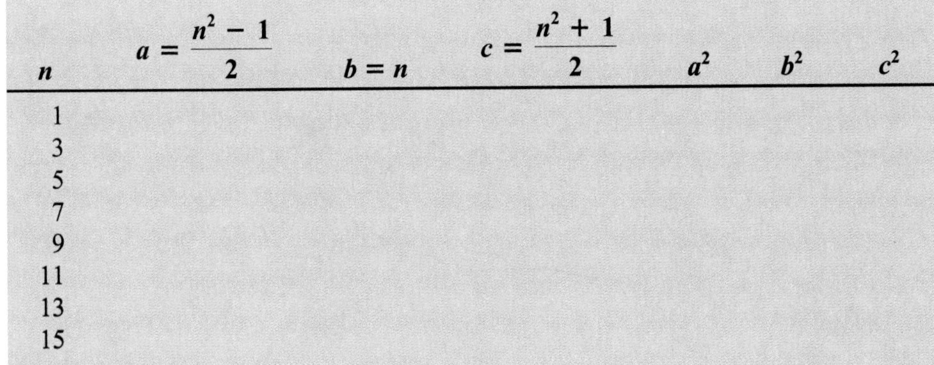

n	$a = \dfrac{n^2 - 1}{2}$	$b = n$	$c = \dfrac{n^2 + 1}{2}$	a^2	b^2	c^2
1						
3						
5						
7						
9						
11						
13						
15						

(b) Check for each of these sets of three numbers to see if this statement is true: $\sqrt{a^2 + b^2} = \sqrt{c^2}$. For how many of your sets of three did this work? Sets of three numbers for which this statement is true are called "Pythagorean triples" because $a^2 + b^2 = c^2$. Can the radical equation be written in this way: $\sqrt{a^2 + b^2} = a + b$? Explain your answer.

ANSWERS

49. Simplest form

50. Remove perfect-square factors from radical

51. Remove perfect-square factors from radical and simplify

52. Simplest form

53.

54.

ANSWERS

a. $11x$

b. $5a$

c. $-2y$

d. $17m$

e. $4a$

f. s

g. $6m + 3n$

h. $4x + 5y$

Getting Ready for Section 9.3 [Section 1.6]

Use the distributive property to combine the like terms in each of the following expressions.

(a) $5x + 6x$

(b) $8a - 3a$

(c) $10y - 12y$

(d) $7m + 10m$

(e) $9a + 7a - 12a$

(f) $5s - 8s + 4s$

(g) $12m + 3n - 6m$

(h) $8x + 5y - 4x$

Answers

1. $3\sqrt{2}$ **3.** $2\sqrt{7}$ **5.** $3\sqrt{5}$ **7.** $4\sqrt{3}$ **9.** $10\sqrt{2}$ **11.** $7\sqrt{3}$ **13.** $6\sqrt{3}$ **15.** $x\sqrt{5}$ **17.** $y^2\sqrt{3}$ **19.** $r\sqrt{2r}$ **21.** $3b\sqrt{3}$ **23.** $2x^2\sqrt{6}$ **25.** $3a^2\sqrt{6a}$ **27.** $xy\sqrt{x}$ **29.** $\frac{2}{5}$ **31.** $\frac{3}{4}$ **33.** $\frac{\sqrt{3}}{2}$ **35.** $\frac{\sqrt{5}}{6}$ **37.** $\frac{2a\sqrt{2}}{5}$ **39.** $\frac{\sqrt{5}}{5}$ **41.** $\frac{\sqrt{6}}{2}$ **43.** $\frac{\sqrt{15a}}{5}$ **45.** $\frac{x\sqrt{6}}{3}$ **47.** $\frac{2s\sqrt{14s}}{7}$ **49.** Simplest form **51.** Remove the perfect-square factors from the radical and simplify. **53.** 

a. $11x$ **b.** $5a$ **c.** $-2y$ **d.** $17m$ **e.** $4a$ **f.** s **g.** $6m + 3n$ **h.** $4x + 5y$

 Adding and Subtracting Radicals

9.3 OBJECTIVES

1. Add and subtract expressions involving numeric radicals
2. Add and subtract expressions involving algebraic radicals

Two radicals that have the same index and the same radicand (the expression inside the radical) are called **like radicals.** For example,

$2\sqrt{3}$ and $5\sqrt{3}$ are like radicals.

$\sqrt{2}$ and $\sqrt{5}$ are not like radicals—they have different radicands.

NOTE "Indices" is the plural of "index."

$\sqrt{2}$ and $\sqrt[3]{2}$ are not like radicals—they have different indices (2 and 3, representing a square root and a cube root).

Like radicals can be added (or subtracted) in the same way as like terms. We apply the distributive property and then combine the coefficients:

$2\sqrt{5} + 3\sqrt{5} = (2 + 3)\sqrt{5} = 5\sqrt{5}$

Example 1

Adding and Subtracting Like Radicals

Simplify each expression.

NOTE Apply the distributive property, then combine the coefficients.

(a) $5\sqrt{2} + 3\sqrt{2} = (5 + 3)\sqrt{2} = 8\sqrt{2}$

(b) $7\sqrt{5} - 2\sqrt{5} = (7 - 2)\sqrt{5} = 5\sqrt{5}$

(c) $8\sqrt{7} - \sqrt{7} + 2\sqrt{7} = (8 - 1 + 2)\sqrt{7} = 9\sqrt{7}$

CHECK YOURSELF 1

Simplify.

(a) $2\sqrt{5} + 7\sqrt{5}$ **(b)** $9\sqrt{7} - \sqrt{7}$

(c) $5\sqrt{3} - 2\sqrt{3} + \sqrt{3}$

If a sum or difference involves terms that are *not* like radicals, we may be able to combine terms after simplifying the radicals according to our earlier methods.

Example 2

Adding and Subtracting Radicals

Simplify each expression.

(a) $3\sqrt{2} + \sqrt{8}$

We do not have like radicals, but we can simplify $\sqrt{8}$. Remember that

$\sqrt{8} = \sqrt{4 \cdot 2} = 2\sqrt{2}$

so

$$3\sqrt{2} + \sqrt{8} = 3\sqrt{2} + \overset{\sqrt{8}}{\overbrace{2\sqrt{2}}}$$
$$= (3 + 2)\sqrt{2} = 5\sqrt{2}$$

NOTE Simplify $\sqrt{12}$.

NOTE The radicals can now be combined. Do you see why?

(b) $5\sqrt{3} - \sqrt{12} = 5\sqrt{3} - \sqrt{4 \cdot 3}$
$$= 5\sqrt{3} - \sqrt{4} \cdot \sqrt{3}$$
$$= 5\sqrt{3} - 2\sqrt{3}$$
$$= (5 - 2)\sqrt{3} = 3\sqrt{3}$$

CHECK YOURSELF 2

Simplify.

(a) $\sqrt{2} + \sqrt{18}$ **(b)** $5\sqrt{3} - \sqrt{27}$

If variables are involved in radical expressions, the process of combining terms proceeds in a fashion similar to that shown in previous examples. Consider Example 3. We again assume that all variables represent positive real numbers.

Example 3

Simplifying Expressions Involving Variables

Simplify each expression.

NOTE Because like radicals are involved, we apply the distributive property and combine terms as before.

(a) $5\sqrt{3x} - 2\sqrt{3x} = (5 - 2)\sqrt{3x} = 3\sqrt{3x}$

(b) $2\sqrt{3a^3} + 5a\sqrt{3a}$
$$= 2\sqrt{a^2 \cdot 3a} + 5a\sqrt{3a}$$

NOTE Simplify the first term.

$$= 2\sqrt{a^2} \cdot \sqrt{3a} + 5a\sqrt{3a}$$
$$= 2a\sqrt{3a} + 5a\sqrt{3a}$$

NOTE The radicals can now be combined.

$$= (2a + 5a)\sqrt{3a} = 7a\sqrt{3a}$$

CHECK YOURSELF 3

Simplify each expression.

(a) $2\sqrt{7y} + 3\sqrt{7y}$ **(b)** $\sqrt{20a^2} - a\sqrt{45}$

CHECK YOURSELF ANSWERS

1. **(a)** $9\sqrt{5}$; **(b)** $8\sqrt{7}$; **(c)** $4\sqrt{3}$ **2.** **(a)** $4\sqrt{2}$; **(b)** $2\sqrt{3}$
3. **(a)** $5\sqrt{7y}$; **(b)** $-a\sqrt{5}$

9.3 Exercises

Name ______________________

Section __________ Date __________

Simplify by combining like terms.

1. $2\sqrt{2} + 4\sqrt{2}$
2. $\sqrt{3} + 5\sqrt{3}$
3. $11\sqrt{7} - 4\sqrt{7}$
4. $5\sqrt{3} - 3\sqrt{2}$
5. $5\sqrt{7} + 3\sqrt{6}$
6. $3\sqrt{5} - 5\sqrt{5}$
7. $2\sqrt{3} - 5\sqrt{3}$
8. $2\sqrt{11} + 5\sqrt{11}$
9. $2\sqrt{3x} + 5\sqrt{3x}$
10. $7\sqrt{2a} - 3\sqrt{2a}$
11. $2\sqrt{3} + \sqrt{3} + 3\sqrt{3}$
12. $3\sqrt{5} + 2\sqrt{5} + \sqrt{5}$
13. $5\sqrt{7} - 2\sqrt{7} + \sqrt{7}$
14. $3\sqrt{10} - 2\sqrt{10} + \sqrt{10}$
15. $2\sqrt{5x} + 5\sqrt{5x} - 2\sqrt{5x}$
16. $5\sqrt{3b} - 2\sqrt{3b} + 4\sqrt{3b}$
17. $2\sqrt{3} + \sqrt{12}$
18. $5\sqrt{2} + \sqrt{18}$
19. $\sqrt{20} - \sqrt{5}$
20. $\sqrt{98} - 3\sqrt{2}$
21. $2\sqrt{6} - \sqrt{54}$
22. $2\sqrt{3} - \sqrt{27}$
23. $\sqrt{72} + \sqrt{50}$
24. $\sqrt{27} - \sqrt{12}$

ANSWERS

1. $6\sqrt{2}$
2. $6\sqrt{3}$
3. $7\sqrt{7}$
4. Cannot be simplified
5. Cannot be simplified
6. $-2\sqrt{5}$
7. $-3\sqrt{3}$
8. $-7\sqrt{11}$
9. $7\sqrt{3x}$
10. $4\sqrt{2a}$
11. $6\sqrt{3}$
12. $6\sqrt{5}$
13. $4\sqrt{7}$
14. $2\sqrt{10}$
15. $5\sqrt{5x}$
16. $7\sqrt{3b}$
17. $4\sqrt{3}$
18. $8\sqrt{2}$
19. $\sqrt{5}$
20. $4\sqrt{2}$
21. $-\sqrt{6}$
22. $-\sqrt{3}$
23. $11\sqrt{2}$
24. $\sqrt{3}$

ANSWERS

25. $2\sqrt{3}$
26. $16\sqrt{2}$
27. $2\sqrt{5}$
28. $2\sqrt{2}$
29. $4\sqrt{3}$
30. $7\sqrt{2}$
31. $4\sqrt{6}$
32. $4\sqrt{7}$
33. $15\sqrt{2}$
34. $7\sqrt{3}$
35. $a\sqrt{3}$
36. $-y\sqrt{2}$
37. $(5x + 6)\sqrt{3x}$
38. $(7a - 2)\sqrt{2a}$
39. 0.32
40. 5.96
41. 3.97
42. 0.52
43. −8.72
44. 19.25
45. 42.07
46. −3.37

25. $3\sqrt{12} - \sqrt{48}$

26. $5\sqrt{8} + 2\sqrt{18}$

27. $2\sqrt{45} - 2\sqrt{20}$

28. $2\sqrt{98} - 4\sqrt{18}$

29. $\sqrt{12} + \sqrt{27} - \sqrt{3}$

30. $\sqrt{50} + \sqrt{32} - \sqrt{8}$

31. $3\sqrt{24} - \sqrt{54} + \sqrt{6}$

32. $\sqrt{63} - 2\sqrt{28} + 5\sqrt{7}$

33. $2\sqrt{50} + 3\sqrt{18} - \sqrt{32}$

34. $3\sqrt{27} + 4\sqrt{12} - \sqrt{300}$

Simplify by combining like terms.

35. $a\sqrt{27} - 2\sqrt{3a^2}$

36. $5\sqrt{2y^2} - 3y\sqrt{8}$

37. $5\sqrt{3x^3} + 2\sqrt{27x}$

38. $7\sqrt{2a^3} - \sqrt{8a}$

Use a calculator to find a decimal approximation for each of the following. Round your answer to the nearest hundredth.

39. $\sqrt{3} - \sqrt{2}$

40. $\sqrt{7} + \sqrt{11}$

41. $\sqrt{5} + \sqrt{3}$

42. $\sqrt{17} - \sqrt{13}$

43. $4\sqrt{3} - 7\sqrt{5}$

44. $8\sqrt{2} + 3\sqrt{7}$

45. $5\sqrt{7} + 8\sqrt{13}$

46. $7\sqrt{2} - 4\sqrt{11}$

47. Perimeter of a rectangle. Find the perimeter of the rectangle shown in the figure.

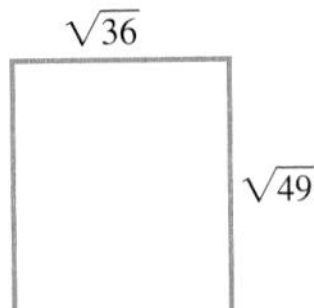

48. Perimeter of a rectangle. Find the perimeter of the rectangle shown in the figure. Write your answer in radical form.

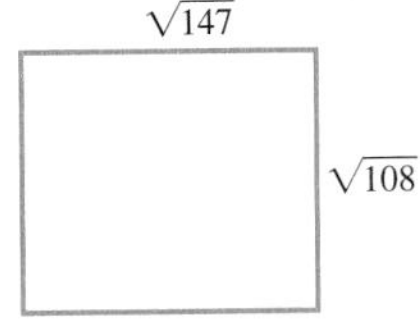

49. Perimeter of a triangle. Find the perimeter of the triangle shown in the figure.

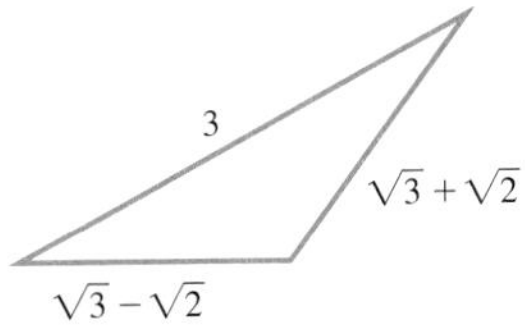

50. Perimeter of a triangle. Find the perimeter of the triangle shown in the figure.

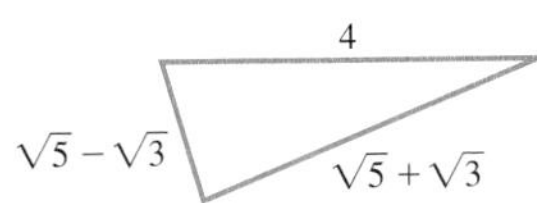

Getting Ready for Section 9.4 [Section 3.4]

Perform the indicated multiplication.

(a) $2(x + 5)$
(b) $3(a - 3)$
(c) $m(m - 8)$
(d) $y(y + 7)$
(e) $(w + 2)(w - 2)$
(f) $(x - 3)(x + 3)$
(g) $(x + y)(x + y)$
(h) $(b - 7)(b - 7)$

ANSWERS

47. 26
48. $26\sqrt{3}$
49. $2\sqrt{3} + 3$
50. $2\sqrt{5} + 4$
a. $2x + 10$
b. $3a - 9$
c. $m^2 - 8m$
d. $y^2 + 7y$
e. $w^2 - 4$
f. $x^2 - 9$
g. $x^2 + 2xy + y^2$
h. $b^2 - 14b + 49$

Answers

1. $6\sqrt{2}$ **3.** $7\sqrt{7}$ **5.** Cannot be simplified **7.** $-3\sqrt{3}$ **9.** $7\sqrt{3x}$
11. $6\sqrt{3}$ **13.** $4\sqrt{7}$ **15.** $5\sqrt{5x}$ **17.** $4\sqrt{3}$ **19.** $\sqrt{5}$ **21.** $-\sqrt{6}$
23. $11\sqrt{2}$ **25.** $2\sqrt{3}$ **27.** $2\sqrt{5}$ **29.** $4\sqrt{3}$ **31.** $4\sqrt{6}$ **33.** $15\sqrt{2}$
35. $a\sqrt{3}$ **37.** $(5x + 6)\sqrt{3x}$ **39.** 0.32 **41.** 3.97 **43.** -8.72
45. 42.07 **47.** 26 **49.** $2\sqrt{3} + 3$ **a.** $2x + 10$ **b.** $3a - 9$
c. $m^2 - 8m$ **d.** $y^2 + 7y$ **e.** $w^2 - 4$ **f.** $x^2 - 9$ **g.** $x^2 + 2xy + y^2$
h. $b^2 - 14b + 49$

Multiplying and Dividing Radicals

OBJECTIVES

1. Multiply and divide expressions involving numeric radicals
2. Multiply and divide expressions involving algebraic radicals

In Section 9.2 we stated the first property for radicals:

$\sqrt{ab} = \sqrt{a} \cdot \sqrt{b}$ when a and b are any positive real numbers

That property has been used to simplify radical expressions up to this point. Suppose now that we want to find a product, such as $\sqrt{3} \cdot \sqrt{5}$.

We can use our first radical rule in the opposite manner.

NOTE The product of square roots is equal to the square root of the product of the radicands.

$\sqrt{a} \cdot \sqrt{b} = \sqrt{ab}$

so

$\sqrt{3} \cdot \sqrt{5} = \sqrt{3 \cdot 5} = \sqrt{15}$

We may have to simplify after multiplying, as Example 1 illustrates.

Example 1

Simplifying Radical Expressions

Multiply then simplify each expression.

(a) $\sqrt{5} \cdot \sqrt{10} = \sqrt{5 \cdot 10} = \sqrt{50}$

$= \sqrt{25 \cdot 2} = 5\sqrt{2}$

(b) $\sqrt{12} \cdot \sqrt{6} = \sqrt{12 \cdot 6} = \sqrt{72}$

$= \sqrt{36 \cdot 2} = \sqrt{36} \cdot \sqrt{2} = 6\sqrt{2}$

An alternative approach would be to simplify $\sqrt{12}$ first.

$\sqrt{12} \cdot \sqrt{6} = 2\sqrt{3}\sqrt{6} = 2\sqrt{18}$

$= 2\sqrt{9 \cdot 2} = 2\sqrt{9}\,\sqrt{2}$

$= 2 \cdot 3\sqrt{2} = 6\sqrt{2}$

(c) $\sqrt{10x} \cdot \sqrt{2x} = \sqrt{20x^2} = \sqrt{4x^2 \cdot 5}$

$= \sqrt{4x^2} \cdot \sqrt{5} = 2x\sqrt{5}$

CHECK YOURSELF 1

Simplify.

(a) $\sqrt{3} \cdot \sqrt{6}$ **(b)** $\sqrt{3} \cdot \sqrt{18}$ **(c)** $\sqrt{8a} \cdot \sqrt{3a}$

If coefficients are involved in a product, we can use the commutative and associative properties to change the order and grouping of the factors. This is illustrated in Example 2.

Example 2

Multiplying Radical Expressions

Multiply.

NOTE In practice, it is not necessary to show the intermediate steps.

$$(2\sqrt{5})(3\sqrt{6}) = (2 \cdot 3)(\sqrt{5} \cdot \sqrt{6})$$
$$= 6\sqrt{5 \cdot 6}$$
$$= 6\sqrt{30}$$

CHECK YOURSELF 2

Multiply $(3\sqrt{7})(5\sqrt{3})$.

The distributive property can also be applied in multiplying radical expressions. Consider the following.

Example 3

Multiplying Radical Expressions

Multiply.

(a) $\sqrt{3}(\sqrt{2} + \sqrt{3})$

$= \sqrt{3} \cdot \sqrt{2} + \sqrt{3} \cdot \sqrt{3}$ The distributive property

$= \sqrt{6} + 3$ Multiply the radicals.

(b) $\sqrt{5}(2\sqrt{6} + 3\sqrt{3})$

$= \sqrt{5} \cdot 2\sqrt{6} + \sqrt{5} \cdot 3\sqrt{3}$ The distributive property

$= 2 \cdot \sqrt{5} \cdot \sqrt{6} + 3 \cdot \sqrt{5} \cdot \sqrt{3}$ The commutative property

$= 2\sqrt{30} + 3\sqrt{15}$

CHECK YOURSELF 3

Multiply.

(a) $\sqrt{5}(\sqrt{6} + \sqrt{5})$ **(b)** $\sqrt{3}(2\sqrt{5} + 3\sqrt{2})$

The FOIL pattern we used for multiplying binomials in Section 3.4 can also be applied in multiplying radical expressions. This is shown in Example 4.

Example 4

Multiplying Radical Expressions

Multiply.

(a) $(\sqrt{3} + 2)(\sqrt{3} + 5)$

$= \sqrt{3} \cdot \sqrt{3} + 5\sqrt{3} + 2\sqrt{3} + 2 \cdot 5$

$= 3 + 5\sqrt{3} + 2\sqrt{3} + 10$ Combine like terms.

$= 13 + 7\sqrt{3}$

NOTE You can use the pattern $(a + b)(a - b) = a^2 - b^2$, where $a = \sqrt{7}$ and $b = 2$, for the same result. $\sqrt{7} + 2$ and $\sqrt{7} - 2$ are called **conjugates** of each other. Note that their product is the rational number 3. The product of conjugates will *always be rational.*

Be Careful! This result *cannot* be further simplified: 13 and $7\sqrt{3}$ are *not* like terms.

(b) $(\sqrt{7} + 2)(\sqrt{7} - 2) = \sqrt{7} \cdot \sqrt{7} - 2\sqrt{7} + 2\sqrt{7} - 4$

$= 7 - 4 = 3$

(c) $(\sqrt{3} + 5)^2 = (\sqrt{3} + 5)(\sqrt{3} + 5)$

$= \sqrt{3} \cdot \sqrt{3} + 5\sqrt{3} + 5\sqrt{3} + 5 \cdot 5$

$= 3 + 5\sqrt{3} + 5\sqrt{3} + 25$

$= 28 + 10\sqrt{3}$

CHECK YOURSELF 4

Multiply.

(a) $(\sqrt{5} + 3)(\sqrt{5} - 2)$ **(b)** $(\sqrt{3} + 4)(\sqrt{3} - 4)$ **(c)** $(\sqrt{2} - 3)^2$

We can also use our second property for radicals in the opposite manner.

$$\frac{\sqrt{a}}{\sqrt{b}} = \sqrt{\frac{a}{b}}$$

NOTE The quotient of square roots is equal to the square root of the quotient of the radicands.

One use of this property to divide radical expressions is illustrated in Example 5.

Example 5

Simplifying Radical Expressions

Simplify.

NOTE The clue to recognizing when to use this approach is in noting that 48 is divisible by 3.

(a) $\frac{\sqrt{48}}{\sqrt{3}} = \sqrt{\frac{48}{3}} = \sqrt{16} = 4$

(b) $\frac{\sqrt{200}}{\sqrt{2}} = \sqrt{\frac{200}{2}} = \sqrt{100} = 10$

(c) $\frac{\sqrt{125x^2}}{\sqrt{5}} = \sqrt{\frac{125x^2}{5}} = \sqrt{25x^2} = 5x$

There is one final quotient form that you may encounter in simplifying expressions, and it will be extremely important in our work with quadratic equations in the next chapter. This form is shown in Example 6.

CHECK YOURSELF 5

Simplify.

(a) $\dfrac{\sqrt{75}}{\sqrt{3}}$ **(b)** $\dfrac{\sqrt{81s^2}}{\sqrt{9}}$

Example 6

Simplifying Radical Expressions

Simplify the expression

$$\frac{3 + \sqrt{72}}{3}$$

CAUTION

Be Careful! Students are sometimes tempted to write

$\dfrac{3 + 6\sqrt{2}}{3} = 1 + 6\sqrt{2}$

This is *not* correct. We must divide *both terms* of the numerator by the common factor.

First, we must simplify the radical in the numerator.

$$\frac{3 + \sqrt{72}}{3} = \frac{3 + \sqrt{36 \cdot 2}}{3}$$

Use Property 1 to simplify $\sqrt{72}$.

$$= \frac{3 + \sqrt{36} \cdot \sqrt{2}}{3} = \frac{3 + 6\sqrt{2}}{3}$$

$$= \frac{3(1 + 2\sqrt{2})}{3} = 1 + 2\sqrt{2}$$

Factor the numerator—then divide by the *common* factor 3.

CHECK YOURSELF 6

Simplify $\dfrac{15 + \sqrt{75}}{5}$.

CHECK YOURSELF ANSWERS

1. **(a)** $3\sqrt{2}$; **(b)** $3\sqrt{6}$; **(c)** $2a\sqrt{6}$ **2.** $15\sqrt{21}$ **3.** **(a)** $\sqrt{30} + 5$; **(b)** $2\sqrt{15} + 3\sqrt{6}$ **4.** **(a)** $-1 + \sqrt{5}$; **(b)** -13; **(c)** $11 - 6\sqrt{2}$ **5.** **(a)** 5; **(b)** $3s$ **6.** $3 + \sqrt{3}$

Name ______________________

Section ________ Date ________

9.4 Exercises

Perform the indicated multiplication. Then simplify each radical expression.

1. $\sqrt{7} \cdot \sqrt{5}$

2. $\sqrt{3} \cdot \sqrt{7}$

3. $\sqrt{5} \cdot \sqrt{11}$

4. $\sqrt{13} \cdot \sqrt{5}$

5. $\sqrt{3} \cdot \sqrt{10m}$

6. $\sqrt{7a} \cdot \sqrt{13}$

7. $\sqrt{2x} \cdot \sqrt{15}$

8. $\sqrt{17} \cdot \sqrt{2b}$

9. $\sqrt{3} \cdot \sqrt{7} \cdot \sqrt{2}$

10. $\sqrt{5} \cdot \sqrt{7} \cdot \sqrt{3}$

11. $\sqrt{3} \cdot \sqrt{12}$

12. $\sqrt{7} \cdot \sqrt{7}$

13. $\sqrt{10} \cdot \sqrt{10}$

14. $\sqrt{5} \cdot \sqrt{15}$

15. $\sqrt{18} \cdot \sqrt{6}$

16. $\sqrt{8} \cdot \sqrt{10}$

17. $\sqrt{2x} \cdot \sqrt{6x}$

18. $\sqrt{3a} \cdot \sqrt{15a}$

19. $2\sqrt{3} \cdot \sqrt{7}$

20. $3\sqrt{2} \cdot \sqrt{5}$

21. $(3\sqrt{3})(5\sqrt{7})$

22. $(2\sqrt{5})(3\sqrt{11})$

ANSWERS

1. $\sqrt{35}$
2. $\sqrt{21}$
3. $\sqrt{55}$
4. $\sqrt{65}$
5. $\sqrt{30m}$
6. $\sqrt{91a}$
7. $\sqrt{30x}$
8. $\sqrt{34b}$
9. $\sqrt{42}$
10. $\sqrt{105}$
11. 6
12. 7
13. 10
14. $5\sqrt{3}$
15. $6\sqrt{3}$
16. $4\sqrt{5}$
17. $2x\sqrt{3}$
18. $3a\sqrt{5}$
19. $2\sqrt{21}$
20. $3\sqrt{10}$
21. $15\sqrt{21}$
22. $6\sqrt{55}$

ANSWERS

23. $30\sqrt{2}$
24. $36\sqrt{2}$
25. $\sqrt{10} + 5$
26. $\sqrt{15} - 3$
27. $2\sqrt{15} - 9$
28. $2\sqrt{21} + 21$
29. $18 + 8\sqrt{3}$
30. $7 - 3\sqrt{5}$
31. $2 + 2\sqrt{5}$
32. $-19 - 4\sqrt{2}$
33. 1
34. -18
35. -15
36. 2
37. $x - 9$
38. $a - 16$
39. $7 + 4\sqrt{3}$
40. $14 - 6\sqrt{5}$
41. $y - 10\sqrt{y} + 25$
42. $x + 8\sqrt{x} + 16$
43. 7
44. 6
45. $6a$
46. $4m$

23. $(3\sqrt{5})(2\sqrt{10})$

24. $(4\sqrt{3})(3\sqrt{6})$

25. $\sqrt{5}(\sqrt{2} + \sqrt{5})$

26. $\sqrt{3}(\sqrt{5} - \sqrt{3})$

27. $\sqrt{3}(2\sqrt{5} - 3\sqrt{3})$

28. $\sqrt{7}(2\sqrt{3} + 3\sqrt{7})$

29. $(\sqrt{3} + 5)(\sqrt{3} + 3)$

30. $(\sqrt{5} - 2)(\sqrt{5} - 1)$

31. $(\sqrt{5} - 1)(\sqrt{5} + 3)$

32. $(\sqrt{2} + 3)(\sqrt{2} - 7)$

33. $(\sqrt{5} - 2)(\sqrt{5} + 2)$

34. $(\sqrt{7} + 5)(\sqrt{7} - 5)$

35. $(\sqrt{10} + 5)(\sqrt{10} - 5)$

36. $(\sqrt{11} - 3)(\sqrt{11} + 3)$

37. $(\sqrt{x} + 3)(\sqrt{x} - 3)$

38. $(\sqrt{a} - 4)(\sqrt{a} + 4)$

39. $(\sqrt{3} + 2)^2$

40. $(\sqrt{5} - 3)^2$

41. $(\sqrt{y} - 5)^2$

42. $(\sqrt{x} + 4)^2$

Perform the indicated division. Rationalize the denominator if necessary. Then simplify each radical expression.

43. $\dfrac{\sqrt{98}}{\sqrt{2}}$

44. $\dfrac{\sqrt{108}}{\sqrt{3}}$

45. $\dfrac{\sqrt{72a^2}}{\sqrt{2}}$

46. $\dfrac{\sqrt{48m^2}}{\sqrt{3}}$

47. $\dfrac{4 + \sqrt{48}}{4}$

48. $\dfrac{12 + \sqrt{108}}{6}$

49. $\dfrac{5 + \sqrt{175}}{5}$

50. $\dfrac{18 + \sqrt{567}}{9}$

51. $\dfrac{-8 - \sqrt{512}}{4}$

52. $\dfrac{-9 - \sqrt{108}}{3}$

53. $\dfrac{6 + \sqrt{18}}{3}$

53. $\dfrac{6 - \sqrt{20}}{2}$

55. $\dfrac{15 - \sqrt{75}}{5}$

56. $\dfrac{8 + \sqrt{48}}{4}$

57. **Area of a rectangle.** Find the area of the rectangle shown in the figure.

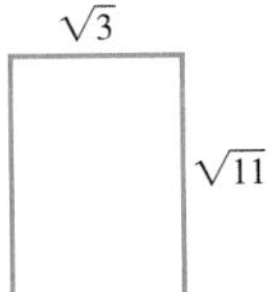

58. **Area of a rectangle.** Find the area of the rectangle shown in the figure.

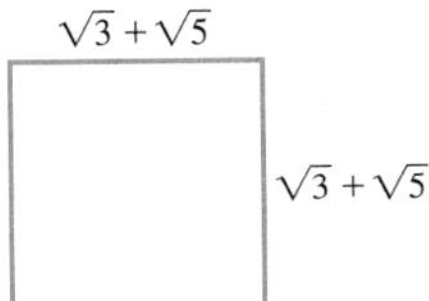

59. Complete this statement: "$\sqrt{2} \cdot \sqrt{5} = \sqrt{10}$ because . . . ."

60. Explain why $2\sqrt{3} + 5\sqrt{3} = 7\sqrt{3}$ but $7\sqrt{3} + 3\sqrt{5} \neq 10\sqrt{8}$.

61. When you look out over an unobstructed landscape or seascape, the distance to the visible horizon depends on your height above the ground. The equation

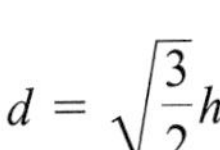

$$d = \sqrt{\frac{3}{2}h}$$

is a good estimate of this, in which d = distance to horizon in miles and h = height of viewer above the ground. Work with a partner to make a chart of distances to the horizon given different elevations. Use the actual heights of tall buildings or

ANSWERS

47. $1 + \sqrt{3}$
48. $2 + \sqrt{3}$
49. $1 + \sqrt{7}$
50. $2 + \sqrt{7}$
51. $-2 - 4\sqrt{2}$
52. $-3 - 2\sqrt{3}$
53. $2 + \sqrt{2}$
54. $3 - \sqrt{5}$
55. $3 - \sqrt{3}$
56. $2 + \sqrt{3}$
57. $\sqrt{33}$
58. $8 + 2\sqrt{15}$

59.

60.

61.

ANSWERS

a. 4

b. 7

c. 11

d. 3.464

e. 5.196

f. 9.899

prominent landmarks in your area. The local library should have a list of these. Be sure to consider the view to the horizon you get when flying in a plane. What would your elevation have to be to see from one side of your city or town to the other? From one side of your state or county to the other?

Getting Ready for Section 9.5 [Section 9.1]

Evaluate the following. Round your answer to the nearest thousandth.

(a) $\sqrt{16}$ (b) $\sqrt{49}$

(c) $\sqrt{121}$ (d) $\sqrt{12}$

(e) $\sqrt{27}$ (f) $\sqrt{98}$

Answers

1. $\sqrt{35}$ **3.** $\sqrt{55}$ **5.** $\sqrt{30m}$ **7.** $\sqrt{30x}$ **9.** $\sqrt{42}$ **11.** 6 **13.** 10 **15.** $6\sqrt{3}$ **17.** $2x\sqrt{3}$ **19.** $2\sqrt{21}$ **21.** $15\sqrt{21}$ **23.** $30\sqrt{2}$ **25.** $\sqrt{10} + 5$ **27.** $2\sqrt{15} - 9$ **29.** $18 + 8\sqrt{3}$ **31.** $2 + 2\sqrt{5}$ **33.** 1 **35.** -15 **37.** $x - 9$ **39.** $7 + 4\sqrt{3}$ **41.** $y - 10\sqrt{y} + 25$ **43.** 7 **45.** $6a$ **47.** $1 + \sqrt{3}$ **49.** $1 + \sqrt{7}$ **51.** $-2 - 4\sqrt{2}$ **53.** $2 + \sqrt{2}$ **55.** $3 - \sqrt{3}$ **57.** $\sqrt{33}$ **59.** **61.** **a.** 4 **b.** 7 **c.** 11 **d.** 3.464 **e.** 5.196 **f.** 9.899

Applications of the Pythagorean Theorem

9.5 OBJECTIVE

1. Apply the Pythagorean theorem in solving problems

Perhaps the most famous theorem in all of mathematics is the **Pythagorean theorem.** The theorem was named for the Greek mathematician Pythagoras, born in 572 B.C. Pythagoras was the founder of the Greek society the Pythagoreans. Although the theorem bears Pythagoras' name, his own work on this theorem is uncertain because the Pythagoreans credited new discoveries to their founder.

Rules and Properties: The Pythagorean Theorem

For every right triangle, the square of the length of the hypotenuse is equal to the sum of the squares of the lengths of the legs.

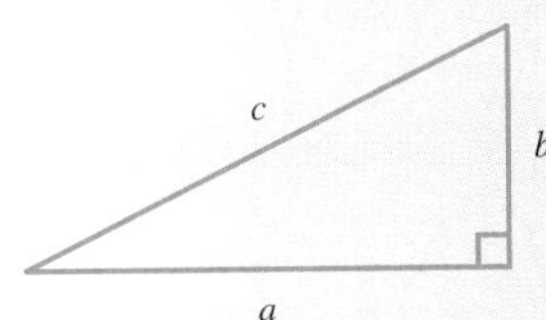

NOTE Here we use c to represent the length of the hypotenuse.

$c^2 = a^2 + b^2$

Example 1

Verifying the Pythagorean Theorem

Verify the Pythagorean theorem for the given triangles.

(a) For the right triangle below,

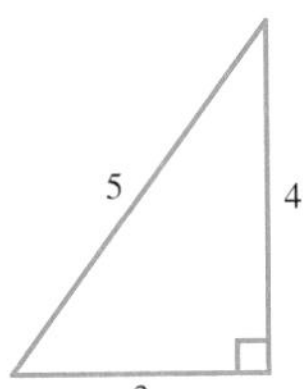

$5^2 \stackrel{?}{=} 3^2 + 4^2$

$25 \stackrel{?}{=} 9 + 16$

$25 = 25$

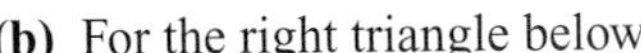

(b) For the right triangle below,

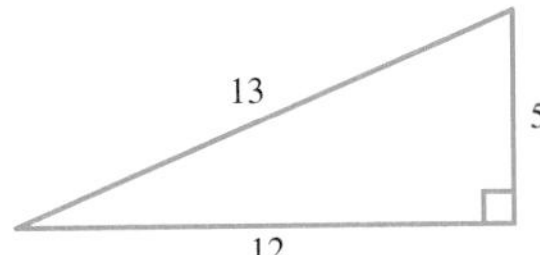

$13^2 \stackrel{?}{=} 12^2 + 5^2$

$169 \stackrel{?}{=} 144 + 25$

$169 = 169$

CHECK YOURSELF 1

Verify the Pythagorean theorem for the right triangle shown.

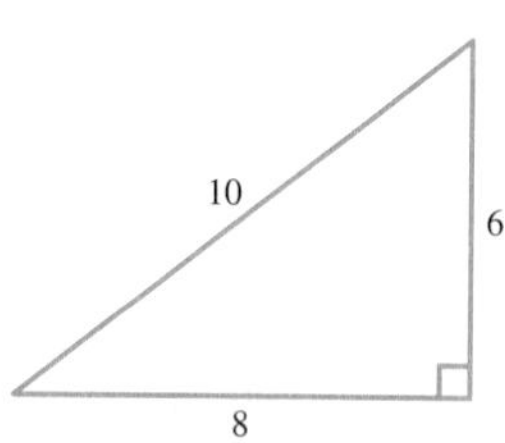

The Pythagorean theorem can be used to find the length of one side of a right triangle when the lengths of the two other sides are known.

Example 2

Solving for the Length of the Hypotenuse

Find length x.

NOTE Notice x will be longer than the given sides because it is the hypotenuse.

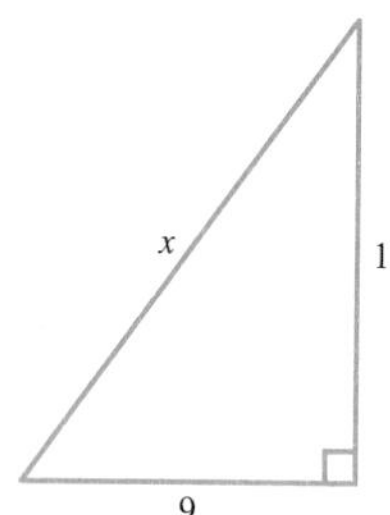

$$x^2 = 9^2 + 12^2$$
$$= 81 + 144$$
$$= 225$$

so

$x = 15$ or $x = -15$

We reject this solution because a length must be positive.

CHECK YOURSELF 2

Find length x.

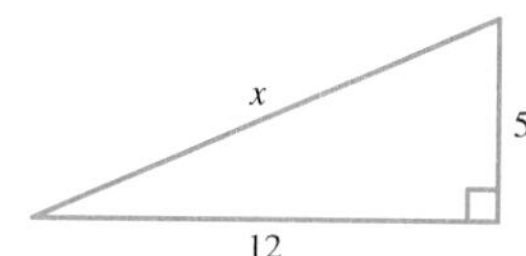

Sometimes, one or more of the lengths of the sides may be represented by an irrational number.

Example 3

Solving for the Length of the Leg

Find length x. Then use your calculator to give an approximation to the nearest tenth.

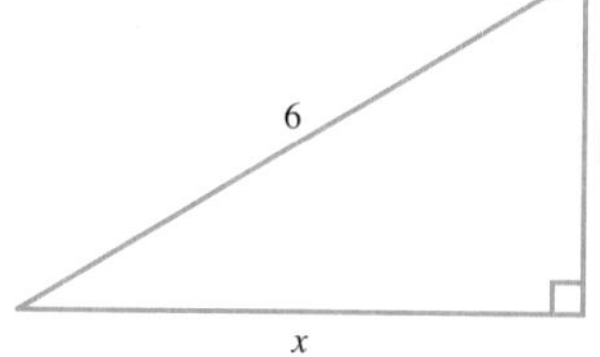

$$3^2 + x^2 = 6^2$$
$$9 + x^2 = 36$$
$$x^2 = 27$$
$$x = \pm\sqrt{27}$$

NOTE You can approximate $3\sqrt{3}$ (or $\sqrt{27}$) with the use of a calculator.

but distance cannot be negative, so

$x = \sqrt{27}$

So x is approximately 5.2.

CHECK YOURSELF 3

Find length x and approximate it to the nearest tenth.

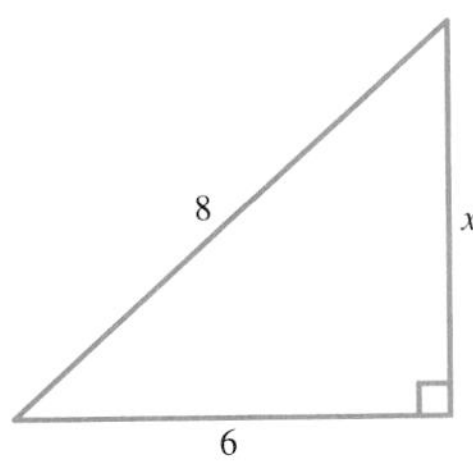

The Pythagorean theorem can be applied to solve a variety of geometric problems.

Example 4

Solving for the Length of the Diagonal

Find, to the nearest tenth, the length of the diagonal of a rectangle that is 8 centimeters (cm) long and 5 cm wide. Let x be the unknown length of the diagonal:

NOTE Always draw and label a sketch showing the information from a problem when geometric figures are involved.

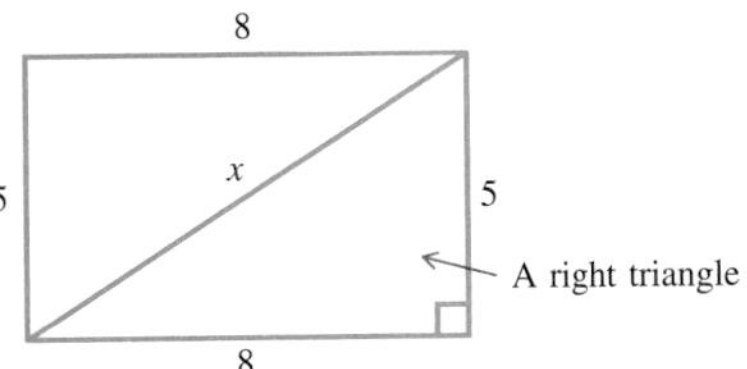

So

NOTE Again, distance cannot be negative, so we eliminate $x = -\sqrt{89}$.

$$x^2 = 5^2 + 8^2$$
$$= 25 + 64$$
$$= 89$$
$$x = \sqrt{89}$$

Thus

$x \approx 9.4$ cm

CHECK YOURSELF 4

The diagonal of a rectangle is 12 inches (in.) and its width is 6 in. Find its length to the nearest tenth.

The next application also makes use of the Pythagorean theorem.

Example 5

Solving an Application

NOTE Always check to see if your final answer is reasonable.

How long must a guywire be to reach from the top of a 30-foot (ft) pole to a point on the ground 20 ft from the base of the pole?

Again be sure to draw a sketch of the problem.

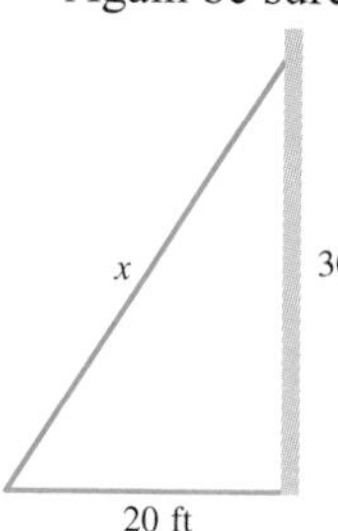

$$x^2 = 20^2 + 30^2$$
$$= 400 + 900$$
$$= 1300$$
$$x = \sqrt{1300}$$
$$\approx 36 \text{ ft}$$

CHECK YOURSELF 5

A 16-ft ladder leans against a wall with its base 4 ft from the wall. How far off the floor is the top of the ladder?

To find the distance between any two points in the plane, we use a formula derived from the Pythagorean theorem.

Definitions: Pythagorean Theorem

Given a right triangle in which c is the length of the hypotenuse, we have the equation

$c^2 = a^2 + b^2$

We can rewrite the formula as

$c = \sqrt{a^2 + b^2}$

NOTE A distance is always positive, so we use only the principal square root.

We use this form of the Pythagorean theorem in Example 6.

Example 6

Finding the Distance Between Two Points

Find the distance from (2, 3) to (5, 7).

The distance can be seen as the hypotenuse of a right triangle.

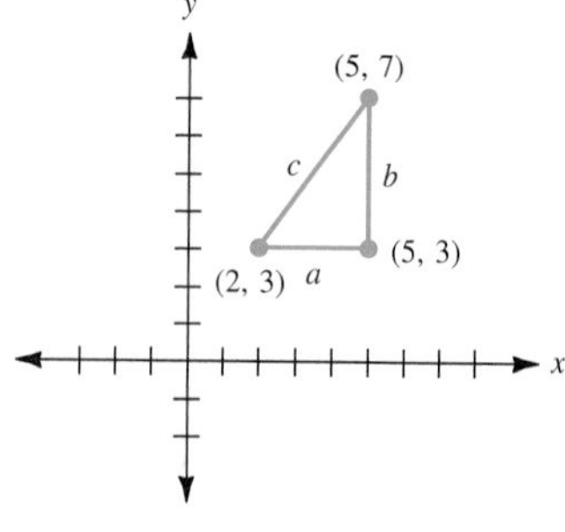

The lengths of the two legs can be found by finding the difference of the two x coordinates and the difference of the two y coordinates. So

$a = 5 - 2 = 3$ and $b = 7 - 3 = 4$

The distance, c, can then be found using the formula

$$c = \sqrt{a^2 + b^2}$$

or, in this case

$$c = \sqrt{3^2 + 4^2}$$
$$c = \sqrt{9 + 16}$$
$$= \sqrt{25}$$
$$= 5$$

The distance is 5 units.

CHECK YOURSELF 6

Find the distance between (0, 2) *and* (5, 14).

If we call our points (x_1, y_1) and (x_2, y_2), we can state the **distance formula.**

Definitions: Distance Formula

The distance between points (x_1, y_1) and (x_2, y_2) can be found using the formula

$$d = \sqrt{(x_2 - x_1)^2 + (y_2 - y_1)^2}$$

Example 7

Finding the Distance Between Two Points

Find the distance between (−2, 5) and (2, −3). Simplify the radical answer.

Using the formula,

NOTE $\sqrt{80} = \sqrt{16 \cdot 5} = 4\sqrt{5}$

$$d = \sqrt{[2 - (-2)]^2 + [(-3) - 5]^2}$$
$$= \sqrt{(4)^2 + (-8)^2}$$
$$= \sqrt{16 + 64}$$
$$= \sqrt{80}$$
$$= 4\sqrt{5}$$

CHECK YOURSELF 7

Find the distance between (2, 5) *and* (−5, 2). *Simplify the radical answer.*

In Example 7, you were asked to find the distance between (−2, 5) and (2, −3).

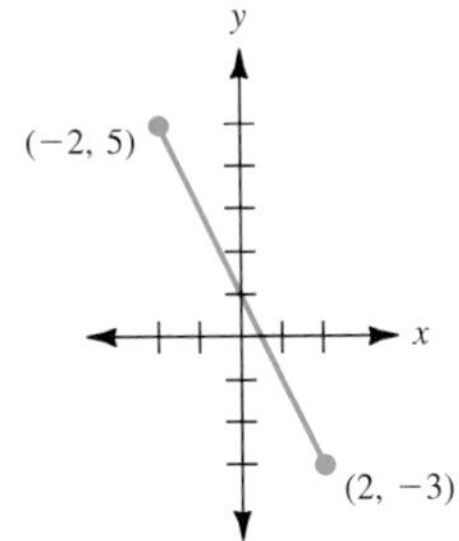

To form a right triangle, we include the point $(-2, -3)$

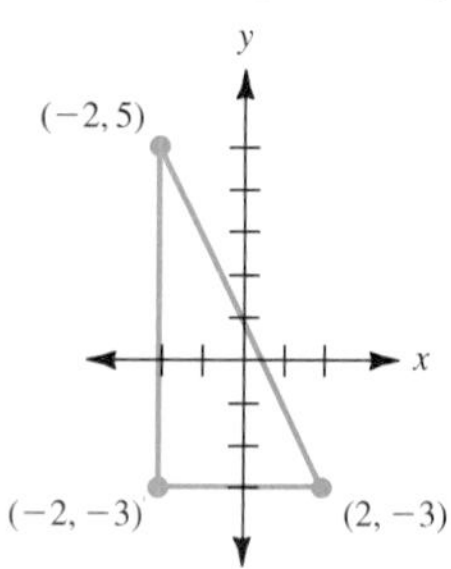

Note that the lengths of the two sides of the right triangle are 4 and 8. By the Pythagorean theorem, the hypotenuse must have length $\sqrt{4^2 + 8^2} = \sqrt{80} = 4\sqrt{5}$. The distance formula is an application of the Pythagorean theorem.

Using the square root key on a calculator, it is easy to approximate the length of a diagonal line. This is particularly useful in checking to see if an object is square or rectangular.

Example 8

Approximating Length with a Calculator

Approximate the length of the diagonal of a rectangle. The diagonal forms the hypotenuse of a triangle with legs 12.2 in. and 15.7 in. The length of the diagonal would be $\sqrt{12.2^2 + 15.7^2} = \sqrt{395.33} \approx 19.88$ in. Use your calculator to confirm the approximation.

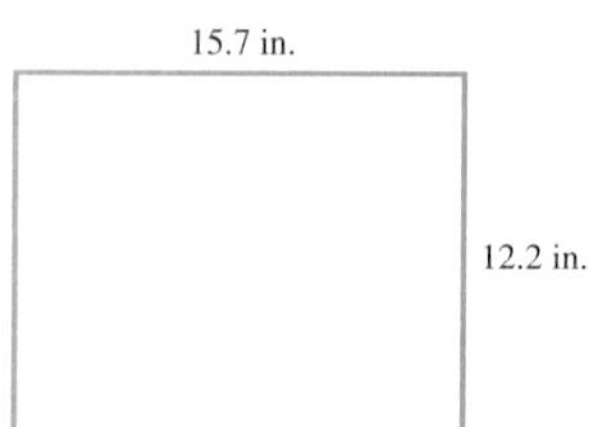

CHECK YOURSELF 8

Approximate the length of the diagonal of the rectangle to the nearest tenth.

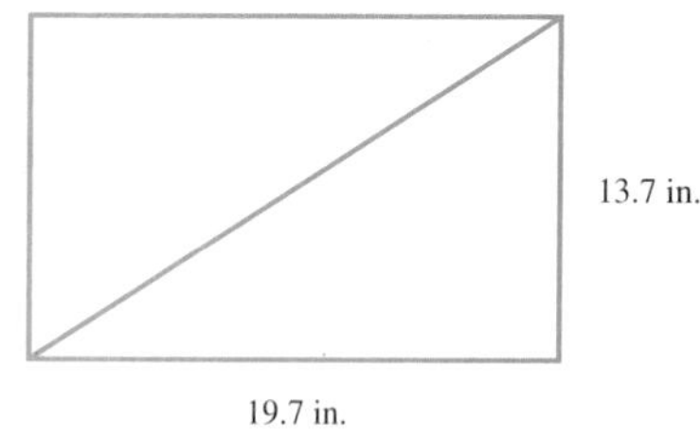

CHECK YOURSELF ANSWERS

1. $10^2 \stackrel{?}{=} 8^2 + 6^2$; $100 \stackrel{?}{=} 64 + 36$; $100 = 100$ **2.** 13 **3.** $2\sqrt{7}$; or approximately 5.3 **4.** Length is approximately 10.4 in. **5.** The height is approximately 15.5 ft. **6.** 13 **7.** $\sqrt{58}$ **8.** ≈ 24.0 in.

Name ______________________

Section ________ Date ________

9.5 Exercises

Find the length x in each triangle. Express your answer in simplified radical form.

1. 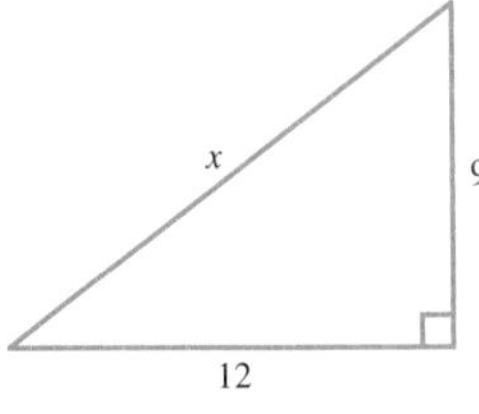

2. 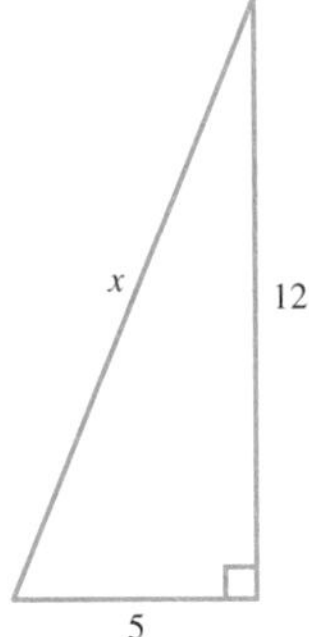

3. 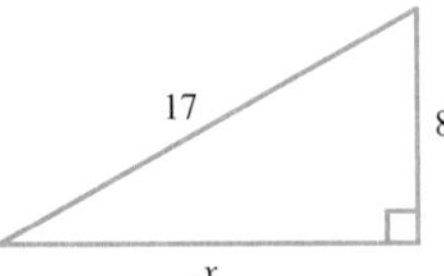

4. 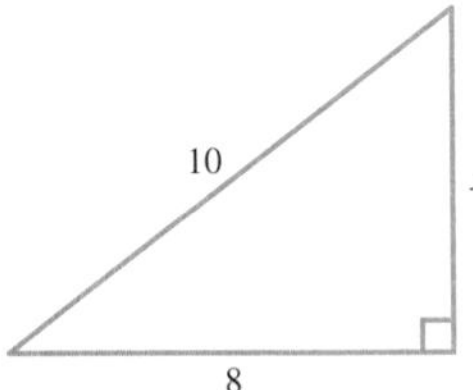

5. 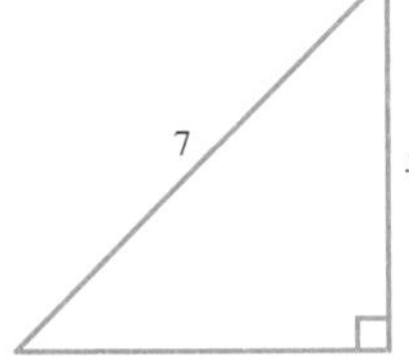

6. 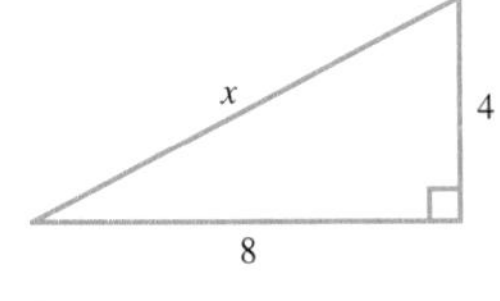

In exercises 7 to 12, express your answer to the nearest thousandth.

7. **Length of a diagonal.** Find the length of the diagonal of a rectangle with a length of 10 centimeters (cm) and a width of 7 cm.

8. **Length of a diagonal.** Find the length of the diagonal of a rectangle with 5 inches (in.) width and 7 in. length.

9. **Width of a rectangle.** Find the width of a rectangle whose diagonal is 12 feet (ft) and whose length is 10 ft.

10. **Length of a rectangle.** Find the length of a rectangle whose diagonal is 9 in. and whose width is 6 in.

11. **Length of a wire.** How long must a guywire be to run from the top of a 20-ft pole to a point on the ground 8 ft from the base of the pole?

ANSWERS

1. 15
2. 13
3. 15
4. 6
5. $2\sqrt{6}$
6. $4\sqrt{5}$
7. $\approx$12.207 cm
8. $\approx$8.602 in.
9. $\approx$6.633 ft
10. $\approx$6.708 in.
11. $\approx$21.541 ft

ANSWERS

12. ≈ 14.142 ft
13. 4
14. $\sqrt{56} = 2\sqrt{14}$
15. $4\sqrt{2}$ or ≈ 5.7 m
16. $2\sqrt{10}$ or ≈ 6.3 m
17. ≈ 3.6 m
18. ≈ 4.8 m

12. Height of a ladder. The base of a 15-ft ladder is 5 ft away from a wall. How high from the floor is the top of the ladder?

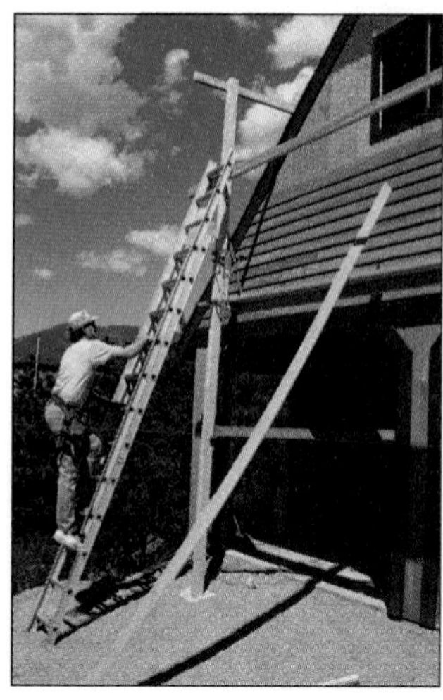

Find the altitude of each triangle.

13.

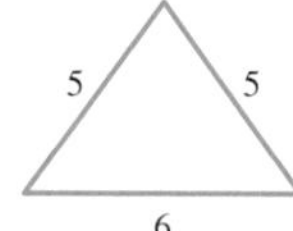

14.

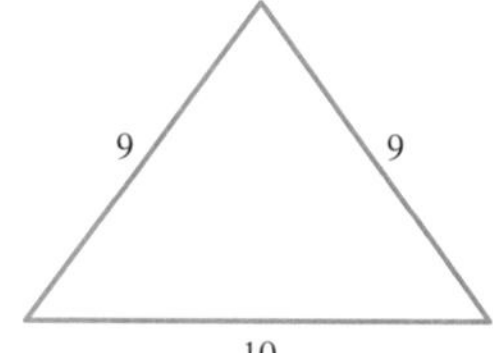

15. Length of insulation. A homeowner wishes to insulate her attic with fiberglass insulation to conserve energy. The insulation comes in 40-cm wide rolls that are cut to fit between the rafters in the attic. If the roof is 6 meters (m) from peak to eave and the attic space is 2 m high at the peak, how long does each of the pieces of insulation need to be? Round to the nearest tenth.

16. Length of insulation. For the home described in exercise 15, if the roof is 7 m from peak to eave and the attic space is 3 m high at the peak, how long does each of the pieces of insulation need to be? Round to the nearest tenth.

17. Base of a triangle. A solar collector and its stand are in the shape of a right triangle. The collector is 5.00 m long, the upright leg is 3.00 m long, and the base leg is 4.00 m long. Because of inefficiencies in the collector's position, it needs to be raised by 0.50 m on the upright leg. How long will the new base leg be? Round to the nearest tenth.

18. Base of a triangle. A solar collector and its stand are in the shape of a right triangle. The collector is 5.00 m long, the upright leg is 2.00 m long, and the base leg is 4.58 m long. Because of inefficiencies in the collector's position, it needs to be lowered by 0.50 m on the upright leg. How long will the new base leg be? Round to the nearest tenth.

Find the distance between each pair of points.

19. (2, 0) and (−4, 0)

20. (−3, 0) and (4, 0)

21. (0, −2) and (0, −9)

22. (0, 8) and (0, −4)

23. (2, 5) and (5, 2)

24. (3, 3) and (5, 7)

25. (5, 1) and (3, 8)

26. (2, 9) and (7, 4)

27. (−2, 8) and (1, 5)

28. (2, 6) and (−3, 4)

29. (6, −1) and (2, 2)

30. (2, −8) and (1, 0)

31. (−1, −1) and (2, 5)

32. (−2, −2) and (3, 3)

33. (−2, 9) and (−3, 3)

34. (4, −1) and (0, −5)

35. (−1, −4) and (−3, 5)

36. (−2, 3) and (−7, −1)

37. (−2, −4) and (−4, 1)

38. (−1, −1) and (4, −2)

39. (−4, −2) and (−1, −5)

40. (−2, −2) and (−4, −4)

41. (−2, 0) and (−4, −1)

42. (−5, −2) and (−7, −1)

Use the distance formula to show that each set of points describes an isosceles triangle (a triangle with two sides of equal length).

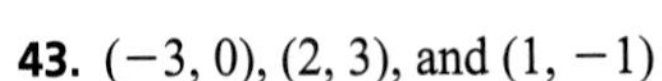

43. (−3, 0), (2, 3), and (1, −1)

44. (−2, 4), (2, 7), and (5, 3)

ANSWERS

19. 6

20. 7

21. 7

22. 12

23. $3\sqrt{2}$

24. $2\sqrt{5}$

25. $\sqrt{53}$

26. $5\sqrt{2}$

27. $3\sqrt{2}$

28. $\sqrt{29}$

29. 5

30. $\sqrt{65}$

31. $3\sqrt{5}$

32. $5\sqrt{2}$

33. $\sqrt{37}$

34. $4\sqrt{2}$

35. $\sqrt{85}$

36. $\sqrt{41}$

37. $\sqrt{29}$

38. $\sqrt{26}$

39. $3\sqrt{2}$

40. $2\sqrt{2}$

41. $\sqrt{5}$

42. $\sqrt{5}$

43. Sides are length $\sqrt{17}$, $\sqrt{17}$, $\sqrt{34}$

44. Sides are length 5, 5, $5\sqrt{2}$

ANSWERS

45. 9 in., 12 in.

46. 4 cm, 3 cm

47. 4.12

48. 3.61

49. 5

50. 10

51. 13

45. Dimensions of a triangle. The length of one leg of a right triangle is 3 in. more than the other. If the length of the hypotenuse is 15 in., what are the lengths of the two legs?

46. Dimensions of a rectangle. The length of a rectangle is 1 cm longer than its width. If the diagonal of the rectangle is 5 cm, what are the dimensions (the length and width) of the rectangle?

Use the Pythagorean theorem to determine the length of each line segment. Where appropriate, round to the nearest hundredth.

47.

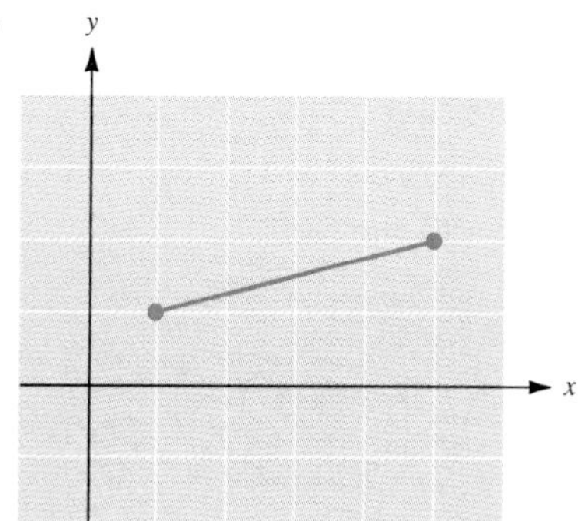

48.

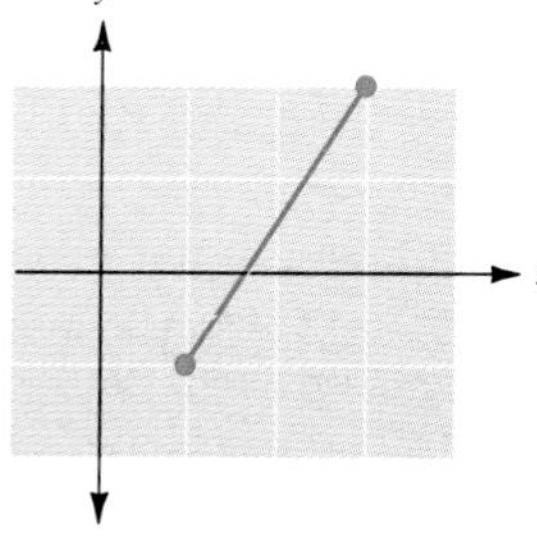

49.

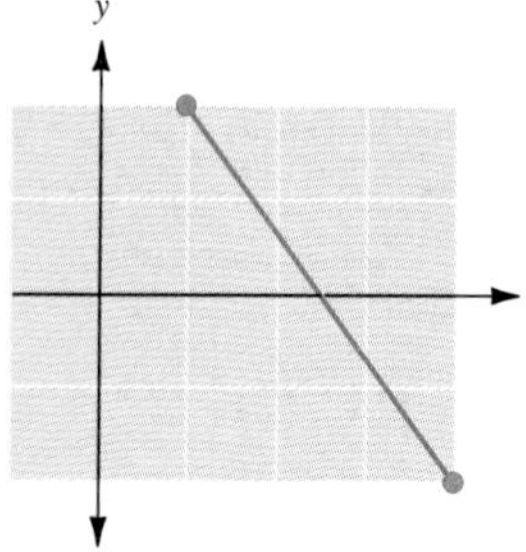

For each figure, use the slope concept and the Pythagorean theorem to show that the figure is a square. (Recall that a square must have four right angles and four equal sides.) Then give the area of the square to the nearest hundredth.

50.

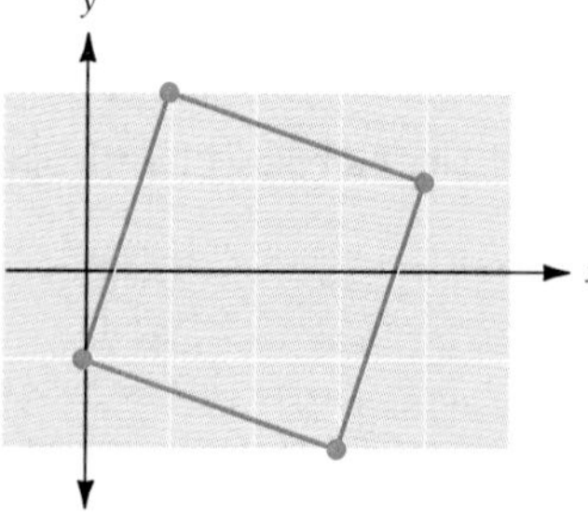

51.

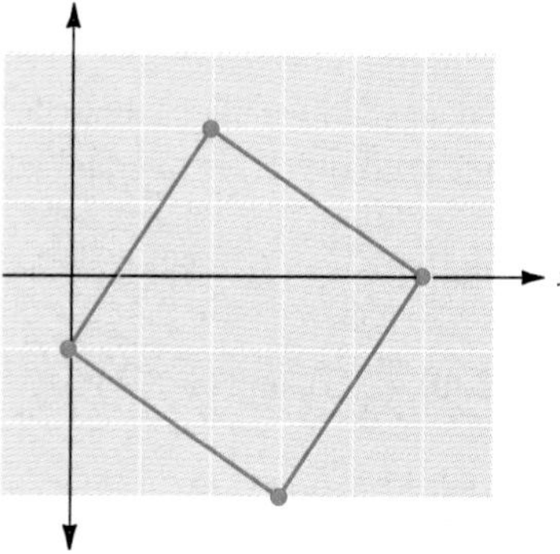

52.

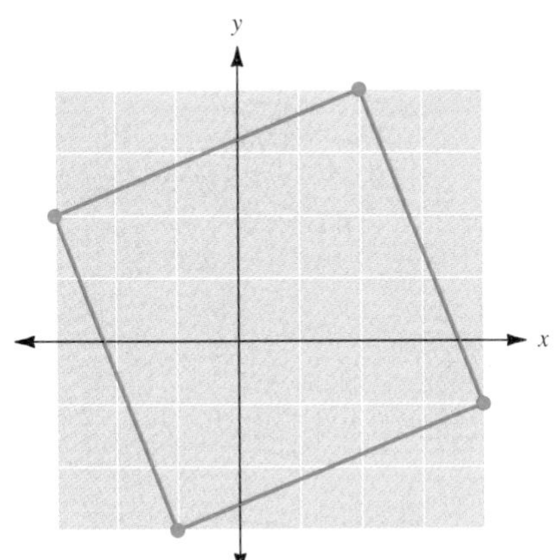

53. Your architecture firm just received this memo.

To: Algebra Expert Architecture, Inc.
From: Microbeans Coffee Company, Inc.
Re: Design for On-Site Day Care Facility
Date: Aug. 10, 2000

We are requesting that you submit a design for a nursery for preschool children. We are planning to provide free on-site day care for the workers at our corporate headquarters.

The nursery should be large enough to serve the needs of 20 preschoolers. There will be three child care workers in this facility. We want the nursery to be 3000 square feet in area. It needs a playroom, a small kitchen and eating space, and bathroom facilities. There should be some space to store toys and books, many of which should be accessible to children. The company plans to put this facility on the first floor on an outside wall so the children can go outside to play without disturbing workers. You are free to add to this design as you see fit.

Please send us your design drawn to a scale of 1 ft to 0.25 in., with precise measurements and descriptions. We would like to receive this design in 1 week from today. Please give us some estimate of the cost of this renovation to our building.

Submit a design, keeping in mind that the design has to conform to strict design specifications for buildings designated as nurseries, including:

1. Number of exits: Two exits for the first 7 people and one exit for every additional 7 people.
2. Width of exits: The total width of exits in inches shall not be less than the total occupant load served by an exit multiplied by 0.3 for stairs and 0.2 for other exits. No exit shall be less than 3 ft wide and 6 ft 8 in. high.
3. Arrangements of exits: If two exits are required, they shall be placed a distance apart equal to but not less than one-half the length of the maximum overall diagonal dimension of the building or area to be served measured in a straight line between exits. Where three or more exits are required, two shall be placed as above and the additional exits arranged a reasonable distance apart.
4. Distance to exits: Maximum distance to travel from any point to an exterior door shall not exceed 100 ft.

ANSWERS

52. 29

53.

Answers

1. 15 **3.** 15 **5.** $2\sqrt{6}$ **7.** $\approx$12.207 cm **9.** $\approx$6.633 ft
11. $\approx$21.541 ft **13.** 4 **15.** $4\sqrt{2} \approx 5.7$ m **17.** $\approx$3.6 m **19.** 6
21. 7 **23.** $3\sqrt{2}$ **25.** $\sqrt{53}$ **27.** $3\sqrt{2}$ **29.** 5 **31.** $3\sqrt{5}$
33. $\sqrt{37}$ **35.** $\sqrt{85}$ **37.** $\sqrt{29}$ **39.** $3\sqrt{2}$ **41.** $\sqrt{5}$
43. Sides have length $\sqrt{34}$, $\sqrt{17}$, and $\sqrt{17}$ **45.** 9 in., 12in. **47.** 4.12
49. 5 **51.** 13 **53.**

Summary

DEFINITION/PROCEDURE	EXAMPLE	REFERENCE
Roots and Radicals		**Section 9.1**
Square Roots $\sqrt{x}$ is the principal (or positive) square root of x. It is the positive number we must square to get x. $-\sqrt{x}$ is the negative square root of x. The square root of a negative number is not a real number.	$\sqrt{49} = 7$ $-\sqrt{49} = -7$ $\sqrt{-49}$ is not a real number.	**p. 695**
Other Roots $\sqrt[3]{x}$ is the cube root of x. $\sqrt[4]{x}$ is the fourth root of x.	$\sqrt[3]{64} = 4$ because $4^3 = 64$. $\sqrt[4]{81} = 3$ because $3^4 = 81$.	**p. 697**
Rational and Irrational Numbers Rational numbers can be expressed as the quotient of two integers with a nonzero denominator. Irrational numbers cannot be expressed as the quotient of two integers.	$\frac{2}{3}, \frac{-7}{12}, 5, \sqrt{36}$, and $\sqrt[3]{64}$ are rational numbers. $\sqrt{5}, \sqrt{37}$, and $\sqrt[3]{65}$ are irrational numbers.	**pp. 698, 699**
Real Numbers The real numbers are the set of rational numbers and the set of irrational numbers together.		**p. 699**
Definitions $\sqrt{x^2} = \lvert x \rvert$ for any real number x $\sqrt[3]{x^3} = x$ for any real number x	$\sqrt{5^2} = 5 \quad \sqrt{(-3)^2} = 3$ $\sqrt[3]{2^3} = 2 \quad \sqrt[3]{(-3)^3} = -3$	**p. 700**
Simplifying Radical Expressions		**Section 9.2**
An expression involving square roots is in *simplest form* if **1.** There are no perfect-square factors in a radical. **2.** No fraction appears inside a radical. **3.** No radical appears in the denominator.		**p. 707**
To simplify a radical expression, use one of the following properties. The square root of a product is the product of the square roots. $\sqrt{ab} = \sqrt{a} \cdot \sqrt{b}$	$\sqrt{40} = \sqrt{4 \cdot 10} = \sqrt{4} \cdot \sqrt{10} = 2\sqrt{10}$ $\sqrt{12x^3} = \sqrt{4x^2 \cdot 3x} = \sqrt{4x^2} \cdot \sqrt{3x} = 2x \cdot \sqrt{3x}$	**p. 707**
The square root of a quotient is the quotient of the square roots. $\sqrt{\frac{a}{b}} = \frac{\sqrt{a}}{\sqrt{b}}$	$\sqrt{\frac{5}{16}} = \frac{\sqrt{5}}{\sqrt{16}} = \frac{\sqrt{5}}{4}$ $\sqrt{\frac{2y}{3}} = \frac{\sqrt{2y}}{\sqrt{3}} = \frac{\sqrt{2y} \cdot \sqrt{3}}{\sqrt{3} \cdot \sqrt{3}} = \frac{\sqrt{6y}}{\sqrt{9}} = \frac{\sqrt{6y}}{3}$	**p. 710**

Continued

DEFINITION/PROCEDURE	EXAMPLE	REFERENCE
Adding and Subtracting Radicals		**Section 9.3**
Like radicals have the same index and the same radicand (the expression inside the radical). Like radicals can be added (or subtracted) in the same way as like terms. Apply the distributive law and combine the coefficients.	$3\sqrt{5}$ and $2\sqrt{5}$ are like radicals. $2\sqrt{3} + 3\sqrt{3} = (2 + 3)\sqrt{3}$ $= 5\sqrt{3}$ $5\sqrt{7} - 2\sqrt{7} = (5 - 2)\sqrt{7}$ $= 3\sqrt{7}$	**p. 717**
Certain expressions can be combined after one or more of the terms involving radicals are simplified.	$\sqrt{12} + \sqrt{3} = 2\sqrt{3} + \sqrt{3}$ $= (2 + 1)\sqrt{3}$	**p. 718**
Multiplying and Dividing Radicals		**Section 9.4**
Multiplying To multiply radical expressions, use the first property of radicals in the following way: $\sqrt{a} \cdot \sqrt{b} = \sqrt{ab}$	$\sqrt{6} \cdot \sqrt{15} = \sqrt{6 \cdot 15} = \sqrt{90}$ $= \sqrt{9 \cdot 10}$ $= 3\sqrt{10}$	**p. 723**
The distributive property can also be applied in multiplying radical expressions.	$\sqrt{5}(\sqrt{3} + 2\sqrt{5})$ $= \sqrt{5} \cdot \sqrt{3} + \sqrt{5} \cdot 2\sqrt{5}$ $= \sqrt{15} + 10$	**p. 724**
The FOIL pattern allows us to find the product of binomial radical expressions.	$(\sqrt{5} + 2)(\sqrt{5} - 1)$ $= \sqrt{5} \cdot \sqrt{5} - \sqrt{5} + 2\sqrt{5} - 2$ $= 3 + \sqrt{5}$ $(\sqrt{10} + 3)(\sqrt{10} - 3)$ $= 10 - 9 = 1$	**p. 725**
Dividing To divide radical expressions, use the second property of radicals in the following way: $\dfrac{\sqrt{a}}{\sqrt{b}} = \sqrt{\dfrac{a}{b}}$	$\dfrac{\sqrt{50}}{\sqrt{2}} = \sqrt{\dfrac{50}{2}}$ $= \sqrt{25}$ $= 5$	**p. 725**
Applications of the Pythagorean Theorem		**Section 9.5**
Hypotenuse c, Leg b, Leg a In words, given a right triangle, the square of the length of the hypotenuse is equal to the sum of the squares of the lengths of the legs. $c^2 = a^2 + b^2$	Find length x: (x, 6, 10) $x^2 = 10^2 + 6^2$ $= 100 + 36$ $= 136$ $x = \sqrt{136}$ or $2\sqrt{34}$	**p. 731**

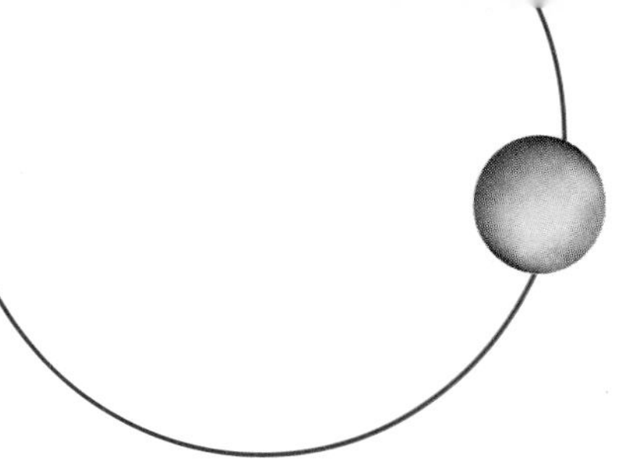

Summary Exercises

This summary exercise set is provided to give you practice with each of the objectives of the chapter. Each exercise is keyed to the appropriate chapter section. The answers are provided in the *Instructor's Manual*. Your instructor will give you guidelines on how to best use these exercises in your instructional setting.

[9.1] Evaluate if possible.

1. $\sqrt{81}$ 9

2. $-\sqrt{49}$ −7

3. $\sqrt{-49}$ Not a real number

4. $\sqrt[3]{64}$ 4

5. $\sqrt[3]{-64}$ −4

6. $\sqrt[4]{81}$ 3

7. $\sqrt[4]{-81}$ Not a real number

[9.2] Simplify each of the following radical expressions. Assume that all variables represent positive real numbers.

8. $\sqrt{50}$ $5\sqrt{2}$

9. $\sqrt{45}$ $3\sqrt{5}$

10. $\sqrt{7a^3}$ $a\sqrt{7a}$

11. $\sqrt{20x^4}$ $2x^2\sqrt{5}$

12. $\sqrt{49m^5}$ $7m^2\sqrt{m}$

13. $\sqrt{200b^3}$ $10b\sqrt{2b}$

14. $\sqrt{147r^3s^2}$ $7rs\sqrt{3r}$

15. $\sqrt{108a^2b^5}$ $6ab^2\sqrt{3b}$

16. $\sqrt{\dfrac{10}{81}}$ $\dfrac{\sqrt{10}}{9}$

17. $\sqrt{\dfrac{18x^2}{25}}$ $\dfrac{3x\sqrt{2}}{5}$

18. $\sqrt{\dfrac{12m^5}{49}}$ $\dfrac{2m^2\sqrt{3m}}{7}$

19. $\sqrt{\dfrac{3}{7}}$ $\dfrac{\sqrt{21}}{7}$

20. $\sqrt{\dfrac{3a}{2}}$ $\dfrac{\sqrt{6a}}{2}$

21. $\sqrt{\dfrac{8x^2}{7}}$ $\dfrac{2x\sqrt{14}}{7}$

[9.3] Simplify by combining like terms.

22. $\sqrt{3} + 4\sqrt{3}$ $5\sqrt{3}$

23. $9\sqrt{5} - 3\sqrt{5}$ $6\sqrt{5}$

24. $3\sqrt{2} + 2\sqrt{3}$ Cannot be simplified

25. $3\sqrt{3a} - \sqrt{3a}$ $2\sqrt{3a}$

26. $7\sqrt{6} - 2\sqrt{6} + \sqrt{6}$ $6\sqrt{6}$

27. $5\sqrt{3} + \sqrt{12}$ $7\sqrt{3}$

28. $3\sqrt{18} - 5\sqrt{2}$ $4\sqrt{2}$

29. $\sqrt{32} - \sqrt{18}$ $\sqrt{2}$

30. $\sqrt{27} - \sqrt{3} + 2\sqrt{12}$ $6\sqrt{3}$

31. $\sqrt{8} + 2\sqrt{27} - \sqrt{75}$ $2\sqrt{2} + \sqrt{3}$

32. $x\sqrt{18} - 3\sqrt{8x^2}$ $-3x\sqrt{2}$

[9.4] Simplify each radical expression.

33. $\sqrt{6} \cdot \sqrt{5}$ $\sqrt{30}$

34. $\sqrt{3} \cdot \sqrt{6}$ $3\sqrt{2}$

35. $\sqrt{3x} \cdot \sqrt{2}$ $\sqrt{6x}$

36. $\sqrt{2} \cdot \sqrt{8} \cdot \sqrt{3}$ $4\sqrt{3}$

37. $\sqrt{5a} \cdot \sqrt{10a}$ $5a\sqrt{2}$

38. $\sqrt{2}(\sqrt{3} + \sqrt{5})$ $\sqrt{6} + \sqrt{10}$

39. $\sqrt{7}(2\sqrt{3} - 3\sqrt{7})$ $2\sqrt{21} - 21$

40. $(\sqrt{3} + 5)(\sqrt{3} - 3)$ $-12 + 2\sqrt{3}$

41. $(\sqrt{15} - 3)(\sqrt{15} + 3)$ 6

42. $(\sqrt{2} + 3)^2$ $11 + 6\sqrt{2}$

43. $\dfrac{\sqrt{7x^3}}{\sqrt{3}}$ $\dfrac{x\sqrt{21x}}{3}$

44. $\dfrac{18 - \sqrt{20}}{2}$ $9 - \sqrt{5}$

[9.5] Find length x in each triangle. Express your answer in simplified radical form.

45. 10

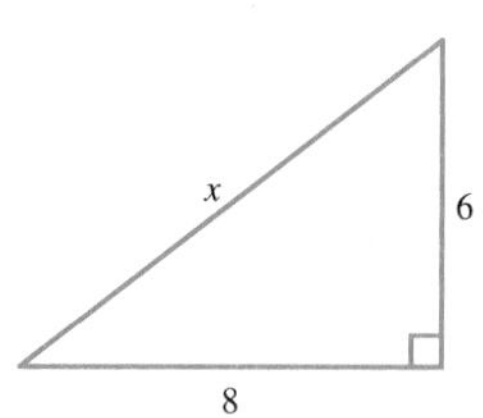

46. 13

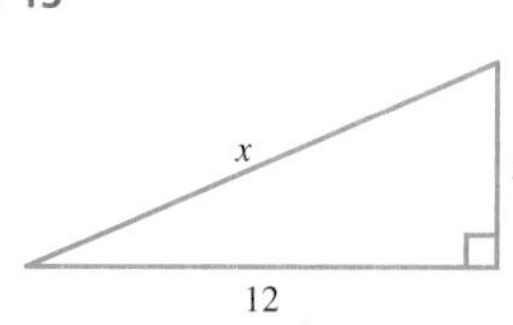

47. 15

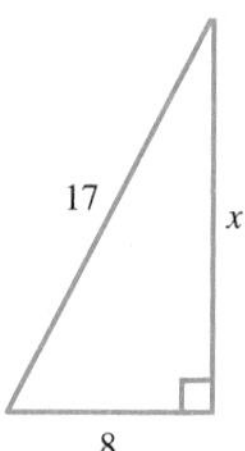

48. $\sqrt{89}$

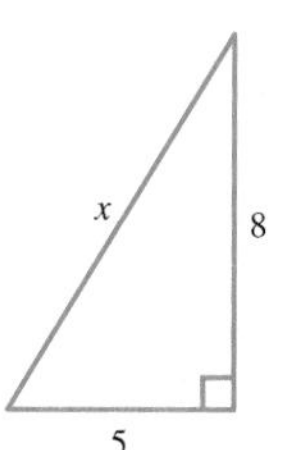

49. $5\sqrt{5}$

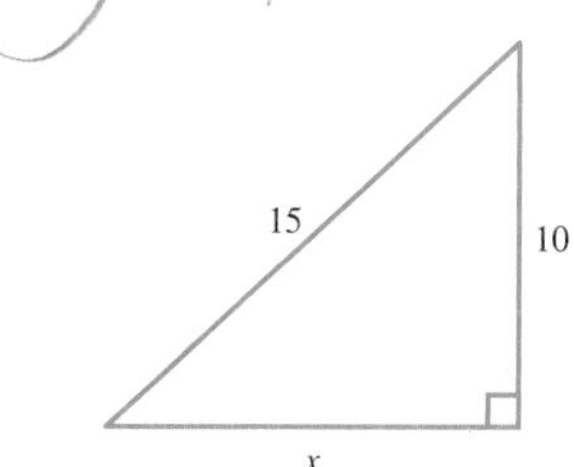

50. $4\sqrt{5}$

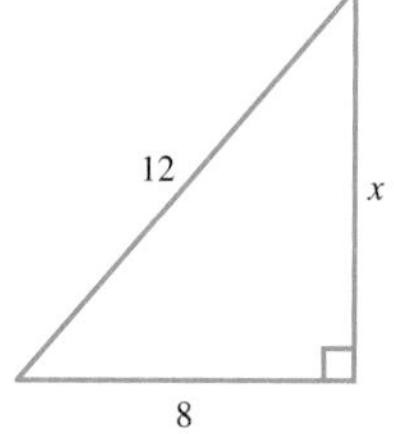

[9.5] Solve each of the following word problems. Approximate your answer to one decimal place where necessary.

51. Find the length of the diagonal of a rectangle whose length is 12 inches (in.) and whose width is 9 in. **15 in.**

52. Find the length of a rectangle whose diagonal has a length of 10 centimeters (cm) and whose width is 5 cm. **8.7 cm**

53. How long must a guywire be to run from the top of an 18-foot (ft) pole to a point on level ground 16 ft away from the base of the pole? **24.1 ft**

54. The length of one leg of a right triangle is 2 in. more than the length of the other. If the length of the hypotenuse of the triangle is 10 in., what are the lengths of the two legs? **6 in., 8 in.**

[9.5] Find the distance between each pair of points.

55. $(-3, 2)$ and $(-7, 2)$ 4

56. $(2, 0)$ and $(5, 9)$ $\sqrt{90}$

57. $(-2, 7)$ and $(-5, -1)$ $\sqrt{73}$

58. $(5, -1)$ and $(-2, 3)$ $\sqrt{65}$

59. $(-3, 4)$ and $(-2, -5)$ $\sqrt{82}$

60. $(6, 4)$ and $(-3, 5)$ $\sqrt{82}$

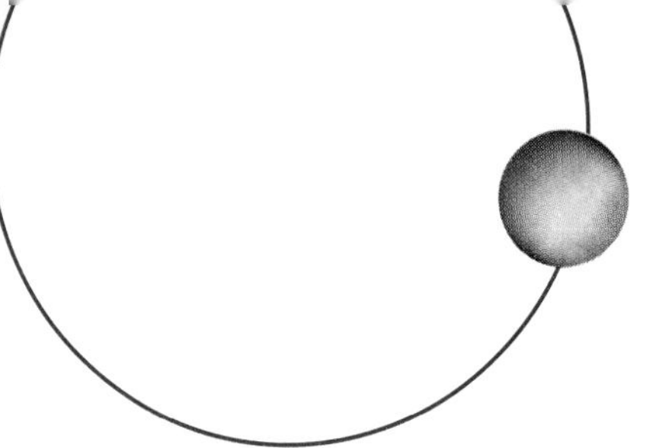

Self-Test for Chapter 9

Name ____________________

Section ________ Date ________

The purpose of this self-test is to help you check your progress and to review for a chapter test in class. Allow yourself about an hour to take the test. When you are done, check your answers in the back of the book. If you missed any answers, be sure to go back and review the appropriate sections in the chapter and the exercises that are provided.

Evaluate if possible.

1. $\sqrt{121}$

2. $\sqrt[3]{27}$

3. $\sqrt{-144}$

4. $-\sqrt[3]{-64}$

Simplify each of the following radical expressions.

5. $\sqrt{75}$

6. $\sqrt{24a^3}$

7. $\sqrt{\dfrac{16}{25}}$

8. $\sqrt{\dfrac{5}{9}}$

Simplify by combining like terms.

9. $2\sqrt{10} - 3\sqrt{10} + 5\sqrt{10}$

10. $3\sqrt{8} - \sqrt{18}$

11. $2\sqrt{50} - \sqrt{8} - \sqrt{50}$

12. $\sqrt{20} + \sqrt{45} - \sqrt{5}$

Simplify each of the following radical expressions.

13. $\sqrt{3x} \cdot \sqrt{6x}$

14. $(\sqrt{5} + 3)(\sqrt{5} + 2)$

15. $\dfrac{\sqrt{7}}{\sqrt{2}}$

16. $\dfrac{14 + 3\sqrt{98}}{7}$

17. $\dfrac{27 - \sqrt{243}}{9}$

Find length x in each triangle. Write the answer in simplified radical form.

18.

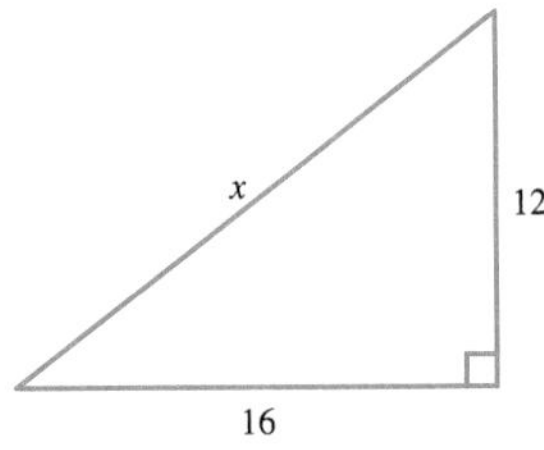

19.

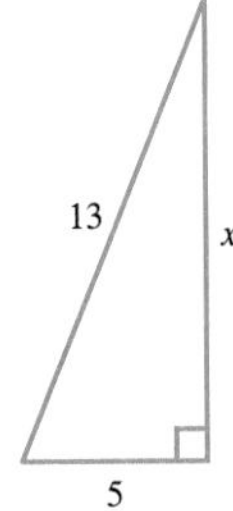

ANSWERS

1. 11
2. 3
3. Not a real number
4. 4
5. $5\sqrt{3}$
6. $2a\sqrt{6a}$
7. $\dfrac{4}{5}$
8. $\dfrac{\sqrt{5}}{3}$
9. $4\sqrt{10}$
10. $3\sqrt{2}$
11. $3\sqrt{2}$
12. $4\sqrt{5}$
13. $3x\sqrt{2}$
14. $11 + 5\sqrt{5}$
15. $\dfrac{\sqrt{14}}{2}$
16. $2 + 3\sqrt{2}$
17. $3 - \sqrt{3}$
18. 20
19. 12

ANSWERS

20. $3\sqrt{5}$

21. $\sqrt{15}$

22. Approximately 9.747 cm

23. 9

24. $\sqrt{85}$

25. $\sqrt{20} = 2\sqrt{5}$

20.

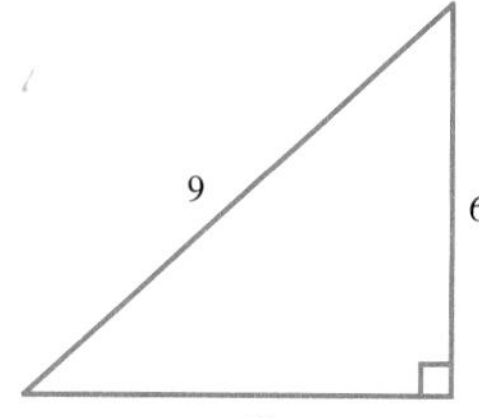

21.

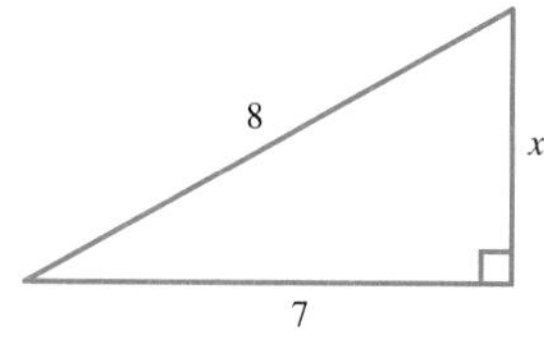

22. If the length of the diagonal of a rectangle is 12 centimeters (cm) and the width of the rectangle is 7 cm, what is the length of the rectangle? Round to the nearest thousandth.

Find the distance between the two points.

23. $(-3, 7)$ and $(-12, 7)$

24. $(-2, 5)$ and $(-9, -1)$

25. $(-3, -6)$ and $(-1, -2)$

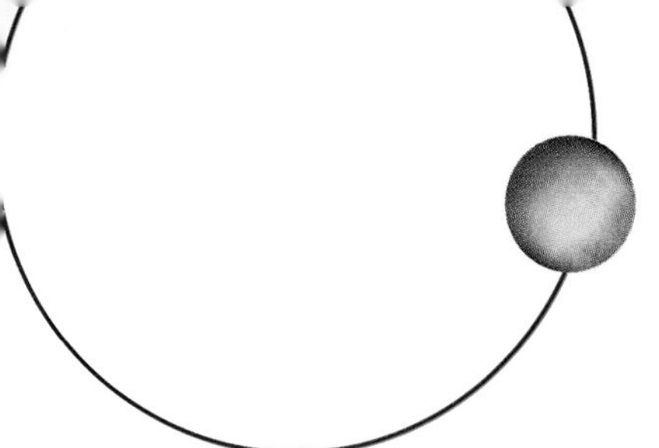

Cumulative Test for Chapters 0 to 9

Name ______________________

Section __________ Date __________

ANSWERS

1. $3x^2y^3 - 2x^3y$
2. $7x^2 - 6x + 12$
3. 0
4. -8
5. 2
6. 6
7. $x > 4$
8. $6x^4y - 10x^3y + 38x^2y$
9. $20x^2 - 23xy - 21y^2$
10. $9xy(4 - 3x^2y)$
11. $(4x - 3)(2x - 5)$
12. $\dfrac{1}{15(x + 7)}$
13. $\dfrac{x - 3}{x - 5}$
14. See exercise
15. See exercise

This test covers selected topics from the first 10 chapters.

Simplify the following expressions.

1. $8x^2y^3 - 5x^3y - 5x^2y^3 + 3x^3y$

2. $(4x^2 - 2x + 7) - (-3x^2 + 4x - 5)$

Evaluate each expression when $x = 2$, $y = -1$, and $z = -4$.

3. $2xyz^2 - 4x^2y^2z$

4. $-2xyz + 2x^2y^2$

Solve the following equations for x.

5. $-3x - 2(4 - 6x) = 10$

6. $5x - 3(4 - 2x) = 6(2x - 3)$

7. Solve the inequality $3x - 11 < 5x - 19$.

Perform the indicated operations.

8. $2x^2y(3x^2 - 5x + 19)$

9. $(5x + 3y)(4x - 7y)$

Factor each of the following completely.

10. $36xy - 27x^3y^2$

11. $8x^2 - 26x + 15$

Perform the indicated operations.

12. $\dfrac{2}{3x + 21} - \dfrac{3}{5x + 35}$

13. $\dfrac{x^2 - x - 6}{x^2 - x - 20} \div \dfrac{x^2 + x - 2}{x^2 + 3x - 4}$

Graph each of the following:

14. $4x + 5y = 20$

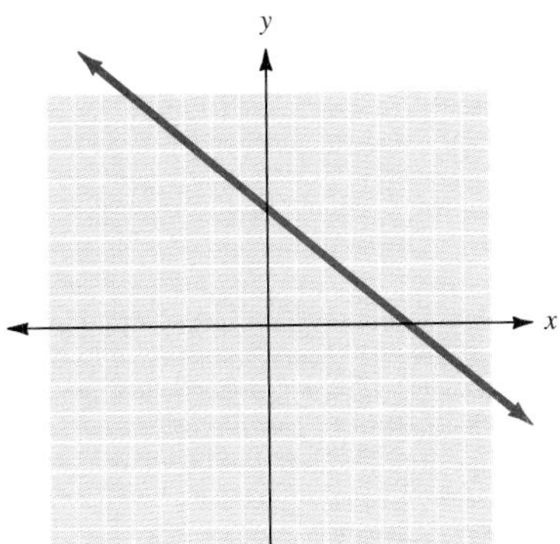

15. $5x - 4y \geq 20$

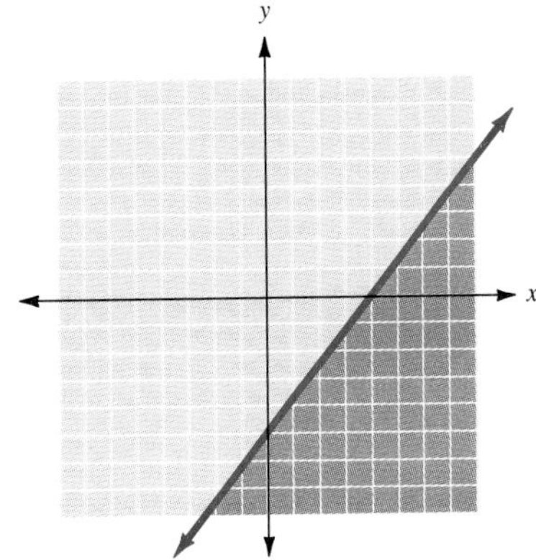

ANSWERS

16. 5

17. $y = -\frac{3}{2}x + 5$

18. (5, 0)

19. Inconsistent system

20. Boat: 13 mi/h, current: 3 mi/h

21. 12

22. −12

23. Not a real number

24. −3

25. $-4a\sqrt{5}$

26. $\frac{2x\sqrt{6x}}{3}$

27. $4 - 2\sqrt{2}$

28. $7x\sqrt{2}$

29. $5mn\sqrt{6m}$

30. $\frac{2a\sqrt{3}}{5}$

16. Find the slope of the line through the points (2, 9) and (−1, −6).

17. Given that the slope of a line is $-\frac{3}{2}$ and the y intercept is (0, 5), write the equation of the line.

Solve each of the following systems. If a unique solution does not exist, state whether the system is inconsistent or dependent.

18. $4x - 5y = 20$
$2x + 3y = 10$

19. $4x + 7y = 24$
$8x + 14y = 12$

Solve the following application. Be sure to show the system of equations used for your solution.

20. Amir was able to travel 80 miles (mi) downstream in 5 hours (h). Returning upstream, he took 8 h to make the trip. How fast can he travel in still water, and what was the rate of the current?

Evaluate each root, if possible.

21. $\sqrt{144}$

22. $-\sqrt{144}$

23. $\sqrt{-144}$

24. $\sqrt[3]{-27}$

Simplify each of the following radical expressions by combining like terms.

25. $a\sqrt{20} - 2\sqrt{45a^2}$

26. $\frac{\sqrt{8x^3}}{\sqrt{3}}$

27. $\frac{12 - \sqrt{72}}{3}$

28. $\sqrt{98x^2}$

29. $\sqrt{150m^3n^2}$

30. $\sqrt{\frac{12a^2}{25}}$

QUADRATIC EQUATIONS

10

INTRODUCTION

Large cities often commission fireworks artists to choreograph elaborate displays on holidays. Such displays look like beautiful paintings in the sky, in which the fireworks seem to dance to well-known popular and classical music. The displays are feats of engineering and very accurate timing. Suppose the designer wants a second set of rockets of a certain color and shape to be released after the first set of a different color and shape reaches a specific height and explodes. He must know the strength of the initial liftoff and use a quadratic equation to determine the proper time for setting off the second round.

The equation $h = -16t^2 + 100t$ gives the height in feet t seconds after the rockets are shot into the air if the initial velocity is 100 feet per second. Using this equation, the designer knows how high the rocket will ascend and when it will begin to fall. He can time the next round to achieve the effect he wishes. Displays that involve large banks of fireworks in shows that last up to an hour are programmed using computers, but quadratic equations are at the heart of the mechanism that creates the beautiful effects.

Name ______________________

Section __________ Date __________

ANSWERS

1. $\pm\sqrt{17}$
2. $\pm 2\sqrt{3}$
3. $1 \pm \sqrt{5}$
4. $\dfrac{\pm\sqrt{14}}{3}$
5. $-2, 5$
6. $\dfrac{5 \pm \sqrt{17}}{2}$
7. $2 \pm 2\sqrt{2}$
8. $\dfrac{2 \pm 3\sqrt{2}}{2}$
9. $-7, 2$
10. 5
11. $\dfrac{-3 \pm \sqrt{29}}{2}$
12. $\dfrac{3 \pm \sqrt{33}}{4}$
13. $\dfrac{2 \pm \sqrt{10}}{3}$
14. $\dfrac{-3 \pm \sqrt{37}}{2}$
15. see exercise
16. see exercise
17. see exercise
18. see exercise

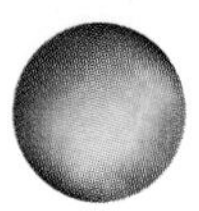

Pre-Test Chapter 10

Solve each of the equations for x.

1. $x^2 = 17$

2. $x^2 - 12 = 0$

3. $(x - 1)^2 = 5$

4. $9x^2 = 14$

Solve each of the equations by completing the square.

5. $x^2 - 3x - 10 = 0$

6. $x^2 - 5x + 2 = 0$

7. $x^2 - 4x - 4 = 0$

8. $2x^2 - 4x - 7 = 0$

Solve each of the equations by using the quadratic formula.

9. $x^2 + 5x - 14 = 0$

10. $x^2 - 10x + 25 = 0$

11. $x^2 + 3x = 5$

12. $2x^2 = 3x + 3$

13. $3x = 4 + \dfrac{2}{x}$

14. $(x - 1)(x + 4) = 3$

Graph each quadratic equation after completing the given table of values.

15. $y = x^2 + 2$

x	y
−2	6
−1	3
0	2
1	3
2	6

16. $y = x^2 + 4x$

x	y
−4	0
−3	−3
−2	−4
−1	−3
0	0

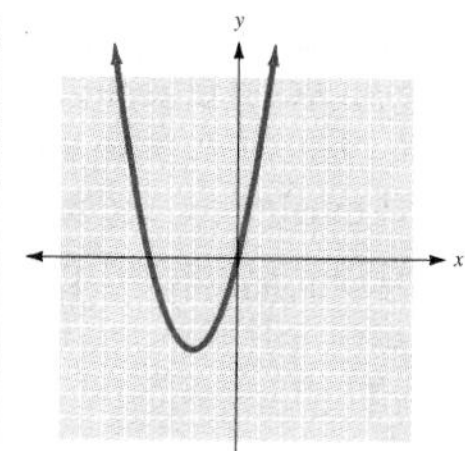

17. $y = x^2 + x - 6$

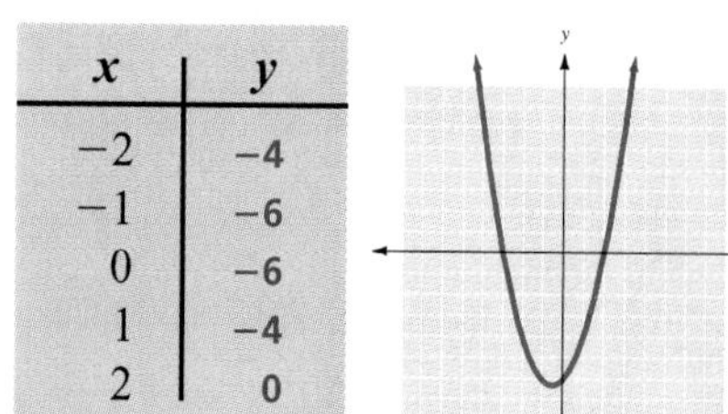

x	y
−2	−4
−1	−6
0	−6
1	−4
2	0

18. $y = -x^2 + 3$

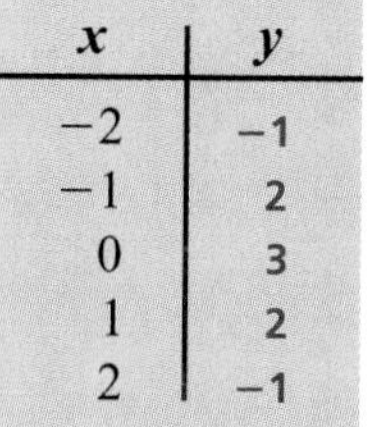

x	y
−2	−1
−1	2
0	3
1	2
2	−1

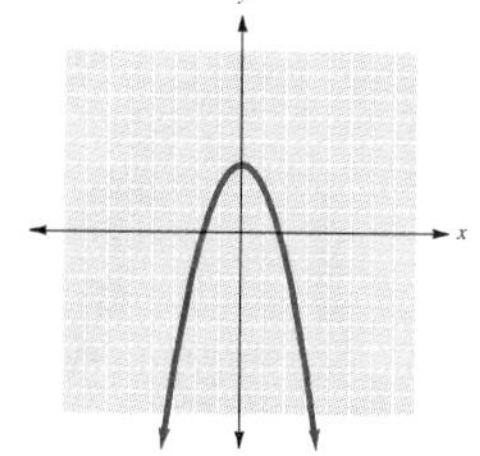

10.1 More on Quadratic Equations

10.1 OBJECTIVES

1. Solve equations of the form $ax^2 = k$
2. Solve equations of the form $(x - h)^2 = k$

We now have more tools for solving quadratic equations. In the next three sections we will be using the ideas of Chapter 9 to extend our solution techniques.

In Section 4.4 we identified all equations of the form

$$ax^2 + bx + c = 0$$

as quadratic equations in standard form. In that section, we discussed solving these equations whenever the quadratic expression was factorable. In this chapter, we want to extend our equation-solving techniques so that we can find solutions for all such quadratic equations.

Let's first review the factoring method of solution that we introduced in Chapter 4.

Example 1

Solving Quadratic Equations by Factoring

Solve each quadratic equation by factoring.

(a) $x^2 = -7x - 12$

First, we write the equation in standard form.

NOTE Add $7x$ and 12 to both sides of the equation. The quadratic expression must be *set equal to* 0.

$$x^2 + 7x + 12 = 0$$

Once the equation is in standard form, we can factor the quadratic member.

$$(x + 3)(x + 4) = 0$$

Finally, using the zero product rule, we solve the equations $x + 3 = 0$ and $x + 4 = 0$ as follows:

NOTE These solutions can be checked as before by substitution into the original equation.

$$x = -3 \quad \text{or} \quad x = -4$$

(b) $x^2 = 16$

Again, we write the equation in standard form.

NOTE Here we factor the quadratic member of the equation as a difference of squares.

$$x^2 - 16 = 0$$

Factoring, we have

$$(x + 4)(x - 4) = 0$$

Finally, the solutions are

$$x = -4 \quad \text{or} \quad x = 4$$

CHECK YOURSELF 1

Solve each of the following quadratic equations.

(a) $x^2 - 4x = 45$ **(b)** $w^2 = 25$

Certain quadratic equations can be solved by other methods, such as the square root method. Let's return to the equation of Example 1(*b*).

Beginning with

$$x^2 = 16$$

we can take the square root of each side, to write

$$\sqrt{x^2} = \sqrt{16}$$

From Section 9.1, we know that this is equivalent to

$$\sqrt{x^2} = 4 \qquad (1)$$

or

NOTE Recall that by definition $\sqrt{x^2} = |x|$

$$|x| = 4 \qquad (2)$$

Values for x of 4 or -4 will both satisfy equation (2), and so we have the two solutions

$$x = 4 \quad \text{or} \quad x = -4$$

We usually write the solutions as

NOTE $x = \pm 4$ is simply a convenient "shorthand" for indicating the two solutions, and we generally will go directly to this form.

$$x = \pm 4$$

Let's look at two more equations solved by this method in Example 2.

Example 2

Solving Equations by the Square Root Method

Solve each of the following equations by the square root method.

(a) $x^2 = 9$

By taking the square root of each side, we have

$$\sqrt{x^2} = \sqrt{9}$$
$$|x| = 3$$
$$x = \pm 3$$

(b) $x^2 = 5$

Again, we take the square root of each side to write our two solutions as

$$\sqrt{x^2} = \sqrt{5}$$
$$|x| = \sqrt{5}$$
$$x = \pm\sqrt{5}$$

CHECK YOURSELF 2

Solve.

(a) $x^2 = 100$ **(b)** $t^2 = 15$

You may have to add or subtract on both sides of the equation to write an equation in the form of those in the previous example, as Example 3 illustrates.

Example 3

Solving Equations by the Square Root Method

Solve $x^2 - 8 = 0$.

First, add 8 to both sides of the equation. We have

$$x^2 = 8$$

Now take the square root of both sides.

$$x = \pm\sqrt{8}$$

Normally, the solution should be written in the simplest form. In this case we have

$$x = \pm 2\sqrt{2}$$

NOTE Recall that

$$\sqrt{8} = \sqrt{4 \cdot 2} = \sqrt{4} \cdot \sqrt{2} = 2\sqrt{2}$$

CHECK YOURSELF 3

Solve.

(a) $x^2 - 18 = 0$ **(b)** $x^2 + 1 = 7$

To solve a quadratic equation of the form $ax^2 = k$, divide both sides of the equation by a as the first step. This is shown in Example 4.

NOTE In the form

$ax^2 = k$

a is the coefficient of x^2 and k is some number.

Example 4

Solving Equations by the Square Root Method

Solve $4x^2 = 3$.

Divide both sides of the equation by 4.

$$x^2 = \frac{3}{4}$$

Now take the square root of both sides.

$$x = \pm\sqrt{\frac{3}{4}}$$

Again write your result in the simplest form, so

$$x = \pm\frac{\sqrt{3}}{2}$$

NOTE Recall that

$$\sqrt{\frac{3}{4}} = \frac{\sqrt{3}}{\sqrt{4}}$$

$$= \frac{\sqrt{3}}{2}$$

CHECK YOURSELF 4

Solve $9x^2 = 5$.

Equations of the form $(x - h)^2 = k$ can also be solved by taking the square root of both sides. Consider Example 5.

Example 5

Solving Equations by the Square Root Method

Solve $(x - 1)^2 = 6$.

Again, take the square root of both sides of the equation.

$$x - 1 = \pm\sqrt{6}$$

Now add 1 to both sides of the equation to isolate x.

$$x = 1 \pm \sqrt{6}$$

CHECK YOURSELF 5

Solve $(x + 2)^2 = 12$.

Equations of the form $a(x - h)^2 = k$ can also be solved if each side of the equation is divided by a first, as shown in Example 6.

Example 6

Solving Equations by the Square Root Method

Solve $3(x - 2)^2 = 5$.

$$(x - 2)^2 = \frac{5}{3}$$

$$x - 2 = \pm\sqrt{\frac{5}{3}} = \frac{\pm\sqrt{15}}{3}$$

$$x = 2 \pm \frac{\sqrt{15}}{3}$$

$$x = \frac{6}{3} \pm \frac{\sqrt{15}}{3}$$

$$x = \frac{6 \pm \sqrt{15}}{3}$$

NOTE

$$\sqrt{\frac{5}{3}} = \frac{\sqrt{5}}{\sqrt{3}} \cdot \frac{\sqrt{3}}{\sqrt{3}} = \frac{\sqrt{15}}{3}$$

CHECK YOURSELF 6

Solve $5(x + 3)^2 = 2$.

What about an equation such as the following?

$$x^2 + 5 = 0$$

If we apply the above methods, we first subtract 5 from both sides, to write

$$x^2 = -5$$

Taking the square root of both sides gives

$$x = \pm\sqrt{-5}$$

But we know there are no square roots of -5 in the real numbers, so this equation has *no real number solutions*. You'll work with this type of equation in your next algebra course.

CHECK YOURSELF ANSWERS

1. (a) $-5, 9$; **(b)** $-5, 5$ **2. (a)** ± 10; **(b)** $\pm\sqrt{15}$ **3. (a)** $\pm 3\sqrt{2}$; **(b)** $\pm\sqrt{6}$
4. $\pm\frac{\sqrt{5}}{3}$ **5.** $-2 \pm 2\sqrt{3}$ **6.** $\frac{-15 \pm \sqrt{10}}{5}$

Name ______

Section ______ Date ______

10.1 Exercises

Solve each of the equations for x.

1. $x^2 = 5$

2. $x^2 = 15$

3. $x^2 = 33$

4. $x^2 = 43$

5. $x^2 - 7 = 0$

6. $x^2 - 13 = 0$

7. $x^2 - 20 = 0$

8. $x^2 = 28$

9. $x^2 = 40$

10. $x^2 - 54 = 0$

11. $x^2 + 3 = 12$

12. $x^2 - 7 = 18$

13. $x^2 + 5 = 8$

14. $x^2 - 4 = 17$

15. $x^2 - 2 = 16$

16. $x^2 + 6 = 30$

17. $9x^2 = 25$

18. $16x^2 = 9$

19. $49x^2 = 11$

20. $16x^2 = 3$

21. $4x^2 = 7$

22. $25x^2 = 13$

ANSWERS

1. $\pm\sqrt{5}$
2. $\pm\sqrt{15}$
3. $\pm\sqrt{33}$
4. $\pm\sqrt{43}$
5. $\pm\sqrt{7}$
6. $\pm\sqrt{13}$
7. $\pm 2\sqrt{5}$
8. $\pm 2\sqrt{7}$
9. $\pm 2\sqrt{10}$
10. $\pm 3\sqrt{6}$
11. ± 3
12. ± 5
13. $\pm\sqrt{3}$
14. $\pm\sqrt{21}$
15. $\pm 3\sqrt{2}$
16. $\pm 2\sqrt{6}$
17. $\pm\frac{5}{3}$
18. $\pm\frac{3}{4}$
19. $\pm\frac{\sqrt{11}}{7}$
20. $\frac{\pm\sqrt{3}}{4}$
21. $\frac{\pm\sqrt{7}}{2}$
22. $\frac{\pm\sqrt{13}}{5}$

Answers

1. $\pm\sqrt{5}$ **3.** $\pm\sqrt{33}$ **5.** $\pm\sqrt{7}$ **7.** $\pm2\sqrt{5}$ **9.** $\pm2\sqrt{10}$
11. ±3 **13.** $\pm\sqrt{3}$ **15.** $\pm3\sqrt{2}$ **17.** $\pm\frac{5}{3}$ **19.** $\frac{\pm\sqrt{11}}{7}$
21. $\frac{\pm\sqrt{7}}{2}$ **23.** $1 \pm \sqrt{5}$ **25.** $-1 \pm 2\sqrt{3}$ **27.** $3 \pm 2\sqrt{6}$
29. $-10, 0$ **31.** $\frac{15 \pm \sqrt{21}}{3}$ **33.** $-\frac{13}{2}, -\frac{7}{2}$ **35.** $-2 \pm \sqrt{3}$
37. $\frac{4 \pm \sqrt{5}}{2}$ **39.** $\frac{2 \pm 2\sqrt{2}}{5}$ **41.** $1 \pm \sqrt{7}$ **43.** No real number
45. $1, -2$ **47.** 30, 20 **49.** **(a)** 9, 3; **(b)** 25, 5; **(c)** $\frac{1}{4}, \frac{1}{2}$; **(d)** 36, 6;
(e) 100, 10; **(f)** 64, 8 **a.** $x^2 + 2x + 1$ **b.** $x^2 + 10x + 25$ **c.** $x^2 - 4x + 4$
d. $x^2 - 14x + 49$ **e.** $x^2 + 8x + 16$ **f.** $x^2 - 6x + 9$ **g.** $4x^2 + 20x + 25$
h. $4x^2 - 4x + 1$

Name ________

Section ________ Date ________

10.1 Exercises

Solve each of the equations for x.

1. $x^2 = 5$

2. $x^2 = 15$

3. $x^2 = 33$

4. $x^2 = 43$

5. $x^2 - 7 = 0$

6. $x^2 - 13 = 0$

7. $x^2 - 20 = 0$

8. $x^2 = 28$

9. $x^2 = 40$

10. $x^2 - 54 = 0$

11. $x^2 + 3 = 12$

12. $x^2 - 7 = 18$

13. $x^2 + 5 = 8$

14. $x^2 - 4 = 17$

15. $x^2 - 2 = 16$

16. $x^2 + 6 = 30$

17. $9x^2 = 25$

18. $16x^2 = 9$

19. $49x^2 = 11$

20. $16x^2 = 3$

21. $4x^2 = 7$

22. $25x^2 = 13$

ANSWERS

1. $\pm\sqrt{5}$
2. $\pm\sqrt{15}$
3. $\pm\sqrt{33}$
4. $\pm\sqrt{43}$
5. $\pm\sqrt{7}$
6. $\pm\sqrt{13}$
7. $\pm 2\sqrt{5}$
8. $\pm 2\sqrt{7}$
9. $\pm 2\sqrt{10}$
10. $\pm 3\sqrt{6}$
11. ± 3
12. ± 5
13. $\pm\sqrt{3}$
14. $\pm\sqrt{21}$
15. $\pm 3\sqrt{2}$
16. $\pm 2\sqrt{6}$
17. $\pm\frac{5}{3}$
18. $\pm\frac{3}{4}$
19. $\pm\frac{\sqrt{11}}{7}$
20. $\frac{\pm\sqrt{3}}{4}$
21. $\frac{\pm\sqrt{7}}{2}$
22. $\frac{\pm\sqrt{13}}{5}$

ANSWERS

23. $1 \pm \sqrt{5}$
24. $3 \pm \sqrt{10}$
25. $-1 \pm 2\sqrt{3}$
26. $-2 \pm 4\sqrt{2}$
27. $3 \pm 2\sqrt{6}$
28. $5 \pm 3\sqrt{3}$
29. $-10, 0$
30. $-6, 2$
31. $\frac{15 \pm \sqrt{21}}{3}$
32. $\frac{10 \pm \sqrt{6}}{2}$
33. $-\frac{13}{2}, -\frac{7}{2}$
34. $\frac{-13}{4}, \frac{-3}{4}$
35. $-2 \pm \sqrt{3}$
36. $-4 \pm \sqrt{2}$
37. $\frac{4 \pm \sqrt{5}}{2}$
38. $\frac{6 \pm \sqrt{11}}{3}$
39. $\frac{2 \pm 2\sqrt{2}}{5}$
40. $\frac{5 \pm \sqrt{14}}{3}$
41. $1 \pm \sqrt{7}$
42. $-2 \pm \sqrt{7}$
43. No real number
44. No real number
45. $1, -2$
46. $6, -10$
47. $30, 20$
48. $1, -11$

23. $(x - 1)^2 = 5$

24. $(x - 3)^2 = 10$

25. $(x + 1)^2 = 12$

26. $(x + 2)^2 = 32$

27. $(x - 3)^2 = 24$

28. $(x - 5)^2 = 27$

29. $(x + 5)^2 = 25$

30. $(x + 2)^2 = 16$

31. $3(x - 5)^2 = 7$

32. $2(x - 5)^2 = 3$

33. $4(x + 5)^2 = 9$

34. $16(x + 2)^2 = 25$

35. $-2(x + 2)^2 = -6$

36. $-5(x + 4)^2 = -10$

37. $-4(x - 2)^2 = -5$

38. $-9(x - 2)^2 = -11$

39. $(5x - 2)^2 = 8$

40. $(3x - 5)^2 = 14$

Solve each equation for x.

41. $x^2 - 2x + 1 = 7$
(*Hint:* Factor the left-hand side.)

42. $x^2 + 4x + 4 = 7$
(*Hint:* Factor the left-hand side.)

43. $(2x + 11)^2 + 9 = 0$

44. $(3x + 14)^2 + 25 = 0$

45. Number problem. The square of a number decreased by 2 is equal to the negative of the number. Find the number.

46. Number problem. The square of 2 more than a number is 64. Find the number.

47. Revenue. The revenue (in dollars) for selling x units of a product is given by

$$R = x\left(5 - \frac{1}{10}x\right) \qquad 0 < x < 50$$

Determine the number of units that must be sold if the revenue is to be \$60.

48. Number problem. The square of the sum of a number and 5 is 36. Find the number.

49. In this section, you solved quadratic equations by "extracting roots," taking the square root of both sides after writing one side as the square of a binomial. But what if the algebraic expression cannot be written this way? Work with another student to decide what needs to be added to each expression below to make it a "perfect square trinomial." Label the dimensions of the squares and the area of each section.

(a)

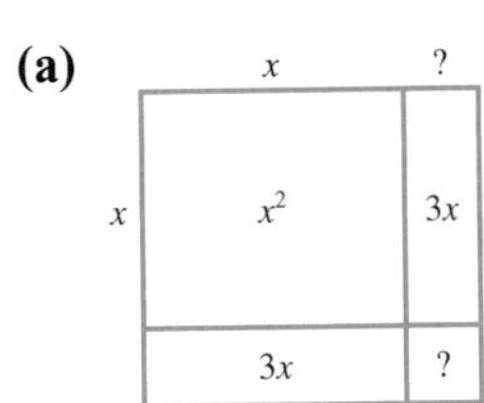

$x^2 + 6x + \_\_\_ = (x + ?)^2$

(b)

n
?
n
5n
?
5n
?

$n^2 + 10n + \_\_\_ = (n + ?)^2$

(c)

a
?
a
a^2
$\frac{a}{2}$
?
$\frac{a}{2}$
?

$a^2 + a + \_\_\_ = (a + ?)^2$

(d)

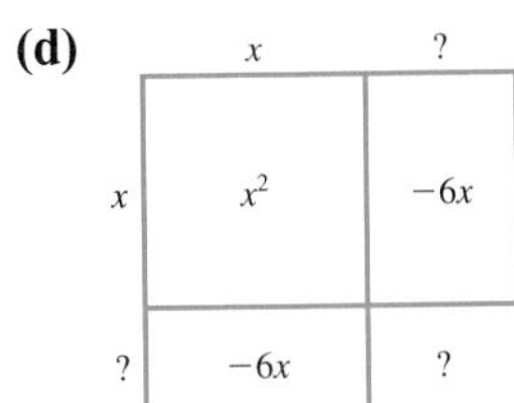

$x^2 - 12x + \_\_\_ = (x - ?)^2$

(e) $x^2 + 20x + \_\_\_ = (x + ?)^2$

(f) $n^2 - 16n + \_\_\_ = (n - ?)^2$

Getting Ready for Section 10.2 [Section 3.5]

Multiply each of the following expressions.

(a) $(x + 1)^2$
(b) $(x + 5)^2$
(c) $(x - 2)^2$
(d) $(x - 7)^2$
(e) $(x + 4)^2$
(f) $(x - 3)^2$
(g) $(2x + 5)^2$
(h) $(2x - 1)^2$

ANSWERS

49. (a) 9, 3
(b) 25, 5
(c) $\frac{1}{4}, \frac{1}{2}$
(d) 36, 6
(e) 100, 10
(f) 64, 8

a. $x^2 + 2x + 1$
b. $x^2 + 10x + 25$
c. $x^2 - 4x + 4$
d. $x^2 - 14x + 49$
e. $x^2 + 8x + 16$
f. $x^2 - 6x + 9$
g. $4x^2 + 20x + 25$
h. $4x^2 - 4x + 1$

Answers

1. $\pm\sqrt{5}$ **3.** $\pm\sqrt{33}$ **5.** $\pm\sqrt{7}$ **7.** $\pm 2\sqrt{5}$ **9.** $\pm 2\sqrt{10}$
11. ± 3 **13.** $\pm\sqrt{3}$ **15.** $\pm 3\sqrt{2}$ **17.** $\pm\frac{5}{3}$ **19.** $\frac{\pm\sqrt{11}}{7}$
21. $\frac{\pm\sqrt{7}}{2}$ **23.** $1 \pm \sqrt{5}$ **25.** $-1 \pm 2\sqrt{3}$ **27.** $3 \pm 2\sqrt{6}$
29. $-10, 0$ **31.** $\frac{15 \pm \sqrt{21}}{3}$ **33.** $-\frac{13}{2}, -\frac{7}{2}$ **35.** $-2 \pm \sqrt{3}$
37. $\frac{4 \pm \sqrt{5}}{2}$ **39.** $\frac{2 \pm 2\sqrt{2}}{5}$ **41.** $1 \pm \sqrt{7}$ **43.** No real number
45. $1, -2$ **47.** $30, 20$ **49.** **(a)** $9, 3$; **(b)** $25, 5$; **(c)** $\frac{1}{4}, \frac{1}{2}$; **(d)** $36, 6$;
(e) $100, 10$; **(f)** $64, 8$ **a.** $x^2 + 2x + 1$ **b.** $x^2 + 10x + 25$ **c.** $x^2 - 4x + 4$
d. $x^2 - 14x + 49$ **e.** $x^2 + 8x + 16$ **f.** $x^2 - 6x + 9$ **g.** $4x^2 + 20x + 25$
h. $4x^2 - 4x + 1$

Completing the Square

OBJECTIVES

1. Complete the square for a trinomial expression
2. Solve a quadratic equation by completing the square

We can solve a quadratic equation such as

$$x^2 - 2x + 1 = 5$$

very easily if we notice that the expression on the left is a perfect-square trinomial. Factoring, we have

$$(x - 1)^2 = 5$$

so

$$x - 1 = \pm\sqrt{5} \qquad \text{or} \qquad x = 1 \pm \sqrt{5}$$

The solutions for the original equation are then $1 + \sqrt{5}$ and $1 - \sqrt{5}$.

It is true that every quadratic equation can be written in the form above (with a perfect-square trinomial on the left). That is the basis for the **completing-the-square method** for solving quadratic equations.

First, let's look at two perfect-square trinomials.

$$x^2 + 6x + 9 = (x + 3)^2 \tag{1}$$

$$x^2 - 8x + 16 = (x - 4)^2 \tag{2}$$

There is an important relationship between the coefficient of the middle term (the x term) and the constant.

In equation (1),

$$\left(\frac{1}{2} \cdot 6\right)^2 = 3^2 = 9$$

The x coefficient   The constant

In equation (2),

$$\left[\frac{1}{2}(-8)\right]^2 = (-4)^2 = 16$$

The x coefficient   The constant

It is always true that, in a perfect-square trinomial with a coefficient of 1 for x^2, the square of one-half of the x coefficient is equal to the constant term.

Example 1

Completing the Square

(a) Find the term that should be added to $x^2 + 4x$ so that the expression is a perfect-square trinomial.

NOTE The coefficient of x^2 must be 1 before the added term is found.

To complete the square of $x^2 + 4x$, add the square of one-half of 4 (the x coefficient).

$$x^2 + 4x + \left(\frac{1}{2} \cdot 4\right)^2 \quad \text{or} \quad x^2 + 4x + 2^2 \quad \text{or} \quad x^2 + 4x + 4$$

The trinomial $x^2 + 4x + 4$ is a perfect square because

$$x^2 + 4x + 4 = (x + 2)^2$$

(b) Find the term that should be added to $x^2 - 10x$ so that the expression is a perfect-square trinomial.

To complete the square of $x^2 - 10x$, add the square of one-half of -10 (the x coefficient).

$$x^2 - 10x + \left[\frac{1}{2}(-10)\right]^2 \quad \text{or} \quad x^2 - 10x + (-5)^2 \quad \text{or} \quad x^2 - 10x + 25$$

Check for yourself, by factoring, that this is a perfect-square trinomial.

CHECK YOURSELF 1

Complete the square and factor.

(a) $x^2 + 2x$ **(b)** $x^2 - 12x$

We can now use the above process along with the solution methods of Section 10.1 to solve a quadratic equation.

Example 2

Solving a Quadratic Equation by Completing the Square

Solve $x^2 + 4x - 2 = 0$ by completing the square.

NOTE Add 2 to both sides to remove -2 from the left side.

$$x^2 + 4x = 2$$

We find the term needed to complete the square by squaring one-half of the x coefficient.

$$\left(\frac{1}{2} \cdot 4\right)^2 = 2^2 = 4$$

We now add 4 to both sides of the equation.

NOTE This *completes the square* on the left.

$$x^2 + 4x + 4 = 2 + 4$$

Now factor on the left and simplify on the right.

$$(x + 2)^2 = 6$$

Now solving as before, we have

$$x + 2 = \pm\sqrt{6}$$

$$x = -2 \pm \sqrt{6}$$

CHECK YOURSELF 2

Solve by completing the square.

$x^2 + 6x - 4 = 0$

For the completing-the-square method to work, the coefficient of x^2 must be 1. Example 3 illustrates the solution process when the coefficient of x^2 is not equal to 1.

Example 3

Solving a Quadratic Equation by Completing the Square

Solve $2x^2 - 4x - 5 = 0$ by completing the square.

$$2x^2 - 4x - 5 = 0$$ Add 5 to both sides.

$$2x^2 - 4x = 5$$

$$x^2 - 2x = \frac{5}{2}$$ Because the coefficient of x^2 is not 1 (here it is 2), divide every term by 2. This will make the new leading coefficient equal to 1.

$$x^2 - 2x + 1 = \frac{5}{2} + 1$$ Complete the square and solve as before.

$$(x - 1)^2 = \frac{7}{2}$$

$$x - 1 = \pm\sqrt{\frac{7}{2}}$$

$$x - 1 = \pm\frac{\sqrt{14}}{2}$$ Simplify the radical on the right.

$$x = 1 \pm \frac{\sqrt{14}}{2}$$

or

$$x = \frac{2 \pm \sqrt{14}}{2}$$

NOTE $\sqrt{\frac{7}{2}} = \sqrt{\frac{7}{2}} \cdot \sqrt{\frac{2}{2}} = \sqrt{\frac{14}{4}} = \frac{\sqrt{14}}{2}$

NOTE We have combined the terms on the right with the common denominator of 2.

CHECK YOURSELF 3

Solve by completing the square.

$3x^2 - 6x + 2 = 0$

Let's summarize by listing the steps to solve a quadratic equation by completing the square.

Step by Step: Solving a Quadratic Equation by Completing the Square

Step 1 Write the equation in the form

$$ax^2 + bx = k$$

so that the variable terms are on the left side and the constant is on the right side.

Step 2 If the coefficient of x^2 is not 1, divide both sides of the equation by that coefficient.

Step 3 Add the square of one-half the coefficient of x to both sides of the equation.

Step 4 The left side of the equation is now a perfect-square trinomial. Factor and solve as before.

CHECK YOURSELF ANSWERS

1. **(a)** $x^2 + 2x + 1 = (x + 1)^2$; **(b)** $x^2 - 12x + 36 = (x - 6)^2$

2. $-3 \pm \sqrt{13}$ **3.** $\dfrac{3 \pm \sqrt{3}}{3}$

10.2 Exercises

Name ______________

Section ________ Date ________

Determine whether each of the following trinomials is a perfect square.

1. $x^2 - 14x + 49$

2. $x^2 + 9x + 16$

3. $x^2 - 18x - 81$

4. $x^2 + 10x + 25$

5. $x^2 - 18x + 81$

6. $x^2 - 24x + 48$

Find the constant term that should be added to make each of the following expressions a perfect-square trinomial.

7. $x^2 + 6x$

8. $x^2 - 8x$

9. $x^2 - 10x$

10. $x^2 + 5x$

11. $x^2 + 9x$

12. $x^2 - 20x$

Solve each of the following quadratic equations by completing the square.

13. $x^2 + 4x - 12 = 0$

14. $x^2 - 6x + 8 = 0$

15. $x^2 - 2x - 5 = 0$

16. $x^2 + 4x - 7 = 0$

17. $x^2 + 3x - 27 = 0$

18. $x^2 + 5x - 3 = 0$

19. $x^2 + 6x - 1 = 0$

20. $x^2 + 4x - 4 = 0$

21. $x^2 - 5x + 6 = 0$

22. $x^2 - 6x - 3 = 0$

23. $x^2 + 6x - 5 = 0$

24. $x^2 - 2x = 1$

25. $x^2 = 9x + 5$

26. $x^2 = 4 - 7x$

27. $2x^2 - 6x + 1 = 0$

28. $2x^2 + 10x + 11 = 0$

29. $2x^2 - 4x + 1 = 0$

30. $2x^2 - 8x + 5 = 0$

31. $4x^2 - 2x - 1 = 0$

32. $3x^2 - x - 2 = 0$

ANSWERS

1. Yes 2. No 3. No

4. Yes 5. Yes 6. No

7. 9 8. 16 9. 25

10. $\frac{25}{4}$ 11. $\frac{81}{4}$

12. 100 13. $-6, 2$

14. 2, 4 15. $1 \pm \sqrt{6}$

16. $-2 \pm \sqrt{11}$

17. $\frac{-3 \pm 3\sqrt{13}}{2}$

18. $\frac{-5 \pm \sqrt{37}}{2}$

19. $-3 \pm \sqrt{10}$

20. $-2 \pm 2\sqrt{2}$

21. 2, 3 22. $3 \pm 2\sqrt{3}$

23. $-3 \pm \sqrt{14}$

24. $1 \pm \sqrt{2}$

25. $\frac{9 + \sqrt{101}}{2}$

26. $\frac{-7 + \sqrt{65}}{2}$

27. $\frac{3 \pm \sqrt{7}}{2}$

28. $\frac{-5 \pm \sqrt{3}}{2}$

29. $\frac{2 \pm \sqrt{2}}{2}$

30. $\frac{4 \pm \sqrt{6}}{2}$

31. $\frac{1 \pm \sqrt{5}}{4}$

32. $-\frac{2}{3}, 1$

ANSWERS

33. $1 \pm \sqrt{6}$

34. $3 \pm 2\sqrt{5}$

35. $-6, 0$

36. $-2, 6$

37. 19, 31

38. 6, 7

a. 13

b. 5

c. 76

d. 8

e. 0

f. -23

Solve each quadratic equation by completing the square.

33. $3x^2 - 4x + 7x - 9 = 2x^2 + 5x - 4$

34. $-4x^2 - 8x + 4x + 5 = -5x^2 + 2x + 16$

Solve the following problems.

35. **Number problem.** If the square of 3 more than a number is 9, find the number(s).

36. **Number problem.** If the square of 2 less than an integer is 16, find the number(s).

37. **Revenue.** The revenue for selling x units of a product is given by $R = x\left(25 - \frac{1}{2}x\right)$ Find the number of units sold if the revenue is \$294.50.

38. **Number problem.** Find two consecutive positive integers such that the sum of their squares is 85.

Getting Ready for Section 10.3 [Section 1.5]

Evaluate the expression $b^2 - 4ac$ for each set of values.

(a) $a = 1, b = 1, c = -3$
(b) $a = 1, b = -1, c = -1$
(c) $a = 1, b = -8, c = -3$
(d) $a = 1, b = -2, c = -1$
(e) $a = -2, b = 4, c = -2$
(f) $a = 2, b = -3, c = 4$

Answers

1. Yes **3.** No **5.** Yes **7.** 9 **9.** 25 **11.** $\frac{81}{4}$ **13.** $-6, 2$ **15.** $1 \pm \sqrt{6}$ **17.** $\frac{-3 \pm 3\sqrt{13}}{2}$ **19.** $-3 \pm \sqrt{10}$ **21.** 2, 3 **23.** $-3 \pm \sqrt{14}$ **25.** $\frac{9 \pm \sqrt{101}}{2}$ **27.** $\frac{3 \pm \sqrt{7}}{2}$ **29.** $\frac{2 \pm \sqrt{2}}{2}$ **31.** $\frac{1 \pm \sqrt{5}}{4}$ **33.** $1 \pm \sqrt{6}$ **35.** $-6, 0$ **37.** 19, 31 **a.** 13 **b.** 5 **c.** 76 **d.** 8 **e.** 0 **f.** -23

10.3 The Quadratic Formula

10.3 OBJECTIVES

1. Solve a quadratic equation by using the quadratic formula
2. Solve an application by using the quadratic formula

We are now ready to derive and use the **quadratic formula,** which will allow us to solve all quadratic equations. We derive the formula by using the method of completing the square.

To use the quadratic formula, the quadratic equation you want to solve must be in *standard form.* That form is

$ax^2 + bx + c = 0 \qquad \text{in which } a \neq 0$

Example 1

Writing Equations in Standard Form

Write each equation in standard form.

(a) $2x^2 - 5x + 3 = 0$

The equation is already in standard form.

$a = 2 \qquad b = -5 \qquad \text{and} \qquad c = 3$

(b) $5x^2 + 3x = 5$

The equation is *not* in standard form. Rewrite it by adding -5 to both sides.

$5x^2 + 3x - 5 = 0$ Standard form

$a = 5 \qquad b = 3 \qquad \text{and} \qquad c = -5$

CHECK YOURSELF 1

Rewrite each quadratic equation in standard form.

(a) $x^2 - 3x = 5$ **(b)** $3x^2 = 7 - 2x$

Once a quadratic equation is written in standard form, we will be able to find both solutions to the equation. Remember that a solution is a value for x that will make the equation true.

What follows is the derivation of the quadratic formula, which can be used to solve quadratic equations.

Step by Step: Deriving the Quadratic Formula

Let $ax^2 + bx + c = 0$, in which $a \neq 0$.

$ax^2 + bx = -c$	Subtract c from both sides.
$x^2 + \frac{b}{a}x = -\frac{c}{a}$	Divide both sides by a.
$x^2 + \frac{b}{a}x + \frac{b^2}{4a^2} = \frac{b^2}{4a^2} - \frac{c}{a}$	Add $\frac{b^2}{4a^2}$ to both sides.
$\left(x + \frac{b}{2a}\right)^2 = \frac{b^2 - 4ac}{4a^2}$	Factor on the left, and add the fractions on the right.
$x + \frac{b}{2a} = \pm\sqrt{\frac{b^2 - 4ac}{4a^2}}$	Take the square root of both sides.
$x + \frac{b}{2a} = \pm\frac{\sqrt{b^2 - 4ac}}{2a}$	Simplify the radical on the right.
$x = -\frac{b}{2a} \pm \frac{\sqrt{b^2 - 4ac}}{2a}$	Add $-\frac{b}{2a}$ to both sides.
$x = \frac{-b \pm \sqrt{b^2 - 4ac}}{2a}$	Use the common denominator, $2a$

NOTE This is the completing-the-square step that makes the left-hand side a perfect square.

Definitions: The Quadratic Formula

$$x = \frac{-b \pm \sqrt{b^2 - 4ac}}{2a}$$

Let's use the quadratic formula to solve some equations.

Example 2

Using the Quadratic Formula to Solve an Equation

Solve $x^2 - 5x + 4 = 0$ by formula.

The equation is in standard form, so first identify a, b, and c.

$$x^2 - 5x + 4 = 0$$

$a = 1 \quad b = -5 \quad c = 4$

NOTE The leading coefficient is 1, so $a = 1$.

We now substitute the values for a, b, and c into the formula.

$$x = \frac{-b \pm \sqrt{b^2 - 4ac}}{2a}$$

$$= \frac{-(-5) \pm \sqrt{(-5)^2 - 4(1)(4)}}{2(1)}$$

$$= \frac{5 \pm \sqrt{25 - 16}}{2}$$

NOTE Simplify the expression.

$$= \frac{5 \pm \sqrt{9}}{2}$$

$$= \frac{5 \pm 3}{2}$$

NOTE These results could also have been found by factoring the original equation. You should check that for yourself.

Now,

$$x = \frac{5 + 3}{2} \quad \text{or} \quad x = \frac{5 - 3}{2}$$

$$= 4 \qquad\qquad = 1$$

The solutions are 4 and 1.

CHECK YOURSELF 2

Solve $x^2 - 2x - 8 = 0$ by formula. Check your result by factoring.

The main use of the quadratic formula is to solve equations that *cannot* be factored.

Example 3

Using the Quadratic Formula to Solve an Equation

Solve $2x^2 = x + 4$ by formula.

First, the equation *must be written* in standard form to find a, b, and c.

$$2x^2 - x - 4 = 0$$

$a = 2$ $b = -1$ $c = -4$

$$x = \frac{-b \pm \sqrt{b^2 - 4ac}}{2a}$$

NOTE Substitute the values for a, b, and c into the formula.

$$= \frac{-(-1) \pm \sqrt{(-1)^2 - 4(2)(-4)}}{2(2)}$$

$$= \frac{1 \pm \sqrt{1 + 32}}{4}$$

$$= \frac{1 \pm \sqrt{33}}{4}$$

CHECK YOURSELF 3

Solve $3x^2 = 3x + 4$ by formula.

Example 4

Using the Quadratic Formula to Solve an Equation

Solve $x^2 - 2x = 4$ by formula.

In standard form, the equation is

$$x^2 - 2x - 4 = 0$$

$a = 1$ $b = -2$ $c = -4$

NOTE Again substitute the values into the quadratic formula.

$$x = \frac{-(-2) \pm \sqrt{(-2)^2 - 4(1)(-4)}}{2(1)}$$

$$= \frac{2 \pm \sqrt{20}}{2}$$

NOTE Because 20 has a perfect-square factor,

$$\sqrt{20} = \sqrt{4 \cdot 5} = 2\sqrt{5}$$

You should always write your solution in simplest form.

$$x = \frac{2 \pm 2\sqrt{5}}{2}$$

$$= \frac{2(1 \pm \sqrt{5})}{2}$$

$$= 1 \pm \sqrt{5}$$

NOTE Now factor the numerator and divide by the common factor 2.

CHECK YOURSELF 4

Solve $3x^2 = 2x + 4$ by formula.

Sometimes equations have common factors. Factoring first simplifies these equations, making them easier to solve. This is illustrated in Example 5.

Example 5

Using the Quadratic Formula to Solve an Equation

Solve $3x^2 - 6x - 3 = 0$ by formula.

Because the equation is in standard form, we could use

$a = 3$ $\quad b = -6$ $\quad$ and $\quad c = -3$

in the quadratic formula. There is, however, a better approach.

Note the common factor of 3 in the quadratic member of the original equation. Factoring, we have

$$3(x^2 - 2x - 1) = 0$$

and dividing both sides of the equation by 3 gives

$$x^2 - 2x - 1 = 0$$

Now let $a = 1$, $b = -2$, and $c = -1$. Then

NOTE The advantage to this approach is that these values will require much less simplification after we substitute into the quadratic formula.

$$x = \frac{-(-2) \pm \sqrt{(-2)^2 - 4(1)(-1)}}{2 \cdot 1}$$

$$= \frac{2 \pm \sqrt{8}}{2}$$

$$= \frac{2 \pm 2\sqrt{2}}{2}$$

$$= \frac{2(1 \pm \sqrt{2})}{2}$$

$$= 1 \pm \sqrt{2}$$

CHECK YOURSELF 5

Solve $4x^2 - 20x = 12$ by formula.

In applications that lead to quadratic equations, you may want to find approximate values for the solutions.

Example 6

Using the Quadratic Formula to Solve an Equation

Solve $x^2 - 5x + 5 = 0$ by formula, and write your solutions in approximate decimal form.

Substituting $a = 1$, $b = -5$, and $c = 5$ gives

$$x = \frac{-(-5) \pm \sqrt{(-5)^2 - 4(1)(5)}}{2(1)}$$

$$= \frac{5 \pm \sqrt{5}}{2}$$

Use your calculator to find $\sqrt{5} \approx 2.236$, so

$$x \approx \frac{5 + 2.236}{2} \quad \text{or} \quad x \approx \frac{5 - 2.236}{2}$$

$$= \frac{7.236}{2} \qquad\qquad = \frac{2.764}{2}$$

$$= 3.618 \qquad\qquad = 1.382$$

CHECK YOURSELF 6

Solve $x^2 - 3x - 5 = 0$ by formula, and approximate the solutions in decimal form to the thousandth.

You may be wondering whether the quadratic formula can be used to solve all quadratic equations. It can, but not all quadratic equations will have real number solutions, as Example 7 shows.

Example 7

Using the Quadratic Formula to Solve an Equation

NOTE Make sure the quadratic equation is in standard form. $x^2 - 3x = -5$ is equivalent to $x^2 - 3x + 5 = 0$.

Solve $x^2 - 3x = -5$ by formula.

Substituting $a = 1$, $b = -3$, and $c = 5$, we have

$$x = \frac{-(-3) \pm \sqrt{(-3)^2 - 4(1)(5)}}{2(1)}$$

$$= \frac{3 \pm \sqrt{-11}}{2}$$

In this case, there are no real number solutions because of the negative number in the radical.

CHECK YOURSELF 7

Solve $x^2 - 3x = -3$ by formula.

Let's review the steps used for solving equations by the use of the quadratic formula.

Step by Step: Solving Equations with the Quadratic Formula

Step 1 Rewrite the equation in standard form.

$ax^2 + bx + c = 0$

Step 2 If a common factor exists, divide both sides of the equation by that common factor.

Step 3 Identify the coefficients *a*, *b*, and *c*.

Step 4 Substitute values for *a*, *b*, and *c* into the formula

$$x = \frac{-b \pm \sqrt{b^2 - 4ac}}{2a}$$

Step 5 Simplify the right side of the expression formed in step 4 to write the solutions for the original equation.

Often, applied problems will lead to quadratic equations that must be solved by the methods of this or the previous section.

CHECK YOURSELF ANSWERS

1. **(a)** $x^2 - 3x - 5 = 0$; **(b)** $3x^2 + 2x - 7 = 0$

2. $x = 4, -2$ **3.** $x = \frac{3 \pm \sqrt{57}}{6}$ **4.** $x = \frac{1 \pm \sqrt{13}}{3}$

5. $x = \frac{5 \pm \sqrt{37}}{2}$ **6.** $x \approx 4.193$ or -1.193 **7.** $\frac{3 \pm \sqrt{-3}}{2}$, no real solutions

Name ____________

Section ________ Date ________

10.3 Exercises

Solve each of the following quadratic equations by formula.

1. $x^2 + 9x + 20 = 0$

2. $x^2 - 9x + 14 = 0$

3. $x^2 - 4x + 3 = 0$

4. $x^2 - 13x + 22 = 0$

5. $3x^2 + 2x - 1 = 0$

6. $x^2 - 8x + 16 = 0$

7. $x^2 + 5x = -4$

8. $4x^2 + 5x = 6$

9. $x^2 = 6x - 9$

10. $2x^2 - 5x = 3$

11. $2x^2 - 3x - 7 = 0$

12. $x^2 - 5x + 2 = 0$

13. $x^2 + 2x - 4 = 0$

14. $x^2 - 4x + 2 = 0$

15. $2x^2 - 3x = 3$

16. $3x^2 - 2x + 1 = 0$

17. $3x^2 - 2x = 6$

18. $4x^2 = 4x + 5$

19. $3x^2 + 3x + 2 = 0$

20. $2x^2 - 3x = 6$

21. $5x^2 = 8x - 2$

22. $5x^2 - 2 = 2x$

23. $2x^2 - 9 = 4x$

24. $3x^2 - 6x = 2$

ANSWERS

1. $-4, -5$
2. $2, 7$
3. $3, 1$
4. $11, 2$
5. $-1, \frac{1}{3}$
6. 4
7. $-4, -1$
8. $-2, \frac{3}{4}$
9. 3
10. $-\frac{1}{2}, 3$
11. $\frac{3 \pm \sqrt{65}}{4}$
12. $\frac{5 \pm \sqrt{17}}{2}$
13. $-1 \pm \sqrt{5}$
14. $2 \pm \sqrt{2}$
15. $\frac{3 \pm \sqrt{33}}{4}$
16. No real number solutions
17. $\frac{1 \pm \sqrt{19}}{3}$
18. $\frac{1 \pm \sqrt{6}}{2}$
19. No real number solutions
20. $\frac{3 \pm \sqrt{57}}{4}$
21. $\frac{4 \pm \sqrt{6}}{5}$
22. $\frac{1 \pm \sqrt{11}}{5}$
23. $\frac{2 \pm \sqrt{22}}{2}$
24. $\frac{3 \pm \sqrt{15}}{3}$

ANSWERS

25. $\frac{5 \pm \sqrt{37}}{6}$

26. $\frac{-3 \pm \sqrt{13}}{2}$

27. $\frac{1 \pm \sqrt{21}}{2}$

28. $\frac{1 \pm 3\sqrt{5}}{2}$

29. $1 \pm \sqrt{7}$

30. $\frac{-3 \pm \sqrt{5}}{2}$

31. $-2, 7$

32. $-1, \frac{1}{3}$

33. $\frac{1}{2}, \frac{10}{3}$

34. $-9, 2$

35. $\frac{4 \pm \sqrt{10}}{2}$

36. $-1 \pm \sqrt{2}$

37. $10, -1$

38. $-2, 0$

39. $\frac{2 \pm \sqrt{6}}{2}$

40. $3, -7$

41. $\frac{1 \pm \sqrt{21}}{6}$

42. $\frac{-4 \pm \sqrt{34}}{6}$

43. 5

44. 1, 10

45. 4.646

46. 0.268 or 3.732

47. 0.787 in.

---

25. $3x - 5 = \frac{1}{x}$

26. $x + 3 = \frac{1}{x}$

27. $(x - 2)(x + 1) = 3$

28. $(x - 3)(x + 2) = 5$

Solve the following quadratic equations by factoring or by any of the techniques of this chapter.

29. $(x - 1)^2 = 7$

30. $(2x + 3)^2 = 5$

31. $x^2 - 5x - 14 = 0$

32. $3x^2 + 2x - 1 = 0$

33. $6x^2 - 23x + 10 = 0$

34. $x^2 + 7x - 18 = 0$

35. $2x^2 - 8x + 3 = 0$

36. $x^2 + 2x - 1 = 0$

37. $x^2 - 9x - 4 = 6$

38. $5x^2 + 10x + 2 = 2$

39. $4x^2 - 8x + 3 = 5$

40. $x^2 + 4x = 21$

Solve the following equations.

41. $\frac{3}{x} + \frac{5}{x^2} = 9$

42. $\frac{8}{x} - \frac{3}{x^2} = -6$

43. $\frac{x}{x + 1} + \frac{10x}{x^2 + 4x + 3} = \frac{15}{x + 3}$

44. $x - \frac{9x}{x - 2} = \frac{-10}{x - 2}$

Use your calculator for the following exercises. Round your answer to the nearest thousandth.

45. Dimensions of a square. The perimeter of a square is numerically 3 less than its area. Find the length of one side.

46. Dimensions of a square. The perimeter of a square is numerically 1 more than its area. Find the length of one side.

47. Width of a picture frame. A picture frame is 15 inches (in.) by 12 in. The area of the picture that shows is 140 in^2. What is the width of the frame?

48. Width of a garden path. A garden area is 30 feet (ft) long by 20 ft wide. A path of uniform width is set around the edge. If the remaining garden area is 400 ft^2, what is the width of the path?

49. Solar frames. A solar collector is 2.5 meters (m) long by 2.0 m wide. It is held in place by a frame of uniform width around its outside edge. If the exposed collector area is 2.5 m^2, what is the width of the frame, to the nearest tenth of a centimeter?

50. Solar frames. A solar collector is 2.5 m long by 2.0 m wide. It is held in place by a frame of uniform width around its outside edge. If the exposed collector is 4 m^2, what is the width of the frame to the nearest tenth of a centimeter?

51. The part of the quadratic formula, $b^2 - 4ac$, that is under the radical is called the **discriminant.** Complete the following sentences to show how this value indicates whether there are *no* solutions, *one* solution, or *two* solutions for the quadratic equation.

(a) When $b^2 - 4ac$ is _______, there are no real number solutions because. . . .
(b) When $b^2 - 4ac$ is _______, there is one solution because. . . .
(c) When $b^2 - 4ac$ is _______, there are two solutions because. . . .
(d) When $b^2 - 4ac$ is _______, there are two *rational* solutions because. . . .
(e) When $b^2 - 4ac$ is _______, there are two *irrational* solutions because. . . .

52. Work with a partner to decide all values of b in the following equations that will give one or more real number solutions.

(a) $3x^2 + bx - 3 = 0$
(b) $5x^2 + bx + 1 = 0$
(c) $-3x^2 + bx - 3 = 0$
(d) Write a rule for judging if an equation has solutions by looking at it in standard form.

ANSWERS

48. 2.192 ft

49. ≈32.5 cm

50. ≈11.7 cm

51. _______

52. _______

53.

54.

a. 13

b. 5

c. −2

d. −49

e. 6

f. 27

53. Which method of solving a quadratic equation seems simplest to you? Which method do you try first?

54. Complete this statement: "You can tell an equation is quadratic and not linear by. . . ."

Getting Ready for Section 10.4 [Section 1.5]

Evaluate each of the given expressions for the value of the variable given.

(a) $x^2 + 3x - 5; x = 3$ (b) $x^2 - 3x - 5; x = -2$

(c) $3x^2 + 4x - 6; x = -2$ (d) $-2x^2 - 5x + 3; x = 4$

(e) $-5x^2 - 5x + 6; x = -1$ (f) $\frac{2}{3}x^2 - \frac{1}{3}x + 5; x = 6$

Answers

1. $-4, -5$ **3.** $3, 1$ **5.** $-1, \frac{1}{3}$ **7.** $-4, -1$ **9.** 3 **11.** $\frac{3 \pm \sqrt{65}}{4}$ **13.** $-1 \pm \sqrt{5}$ **15.** $\frac{3 \pm \sqrt{33}}{4}$ **17.** $\frac{1 \pm \sqrt{19}}{3}$ **19.** No real number solutions **21.** $\frac{4 \pm \sqrt{6}}{5}$ **23.** $\frac{2 \pm \sqrt{22}}{2}$ **25.** $\frac{5 \pm \sqrt{37}}{6}$ **27.** $\frac{1 \pm \sqrt{21}}{2}$ **29.** $1 \pm \sqrt{7}$ **31.** $-2, 7$ **33.** $\frac{1}{2}, \frac{10}{3}$ **35.** $\frac{4 \pm \sqrt{10}}{2}$ **37.** $10, -1$ **39.** $\frac{2 \pm \sqrt{6}}{2}$ **41.** $\frac{1 \pm \sqrt{21}}{6}$ **43.** 5 **45.** 4.646 **47.** 0.787 in. **49.** ≈32.5 cm **51.** **53.** 

a. 13 **b.** 5 **c.** −2 **d.** −49 **e.** 6 **f.** 27

Graphing Quadratic Equations

10.4 OBJECTIVE

1. Graph a quadratic equation by plotting points

In Section 6.3 you learned to graph first-degree equations. Similar methods will allow you to graph quadratic equations of the form

$$y = ax^2 + bx + c \qquad a \neq 0$$

The first thing you will notice is that the graph of an equation in this form is not a straight line. The graph is always the curve called a **parabola.**

Here are some examples:

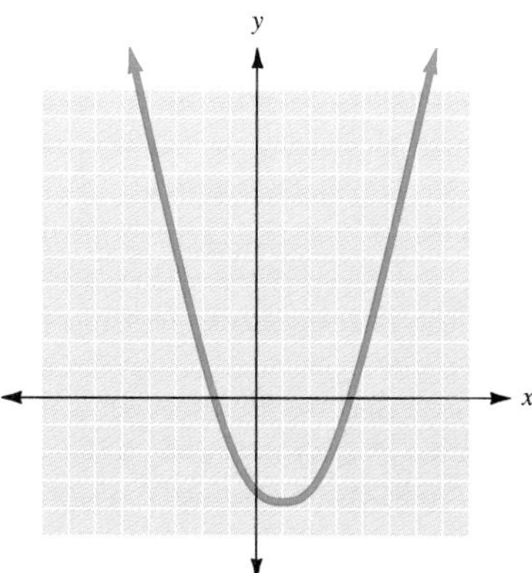

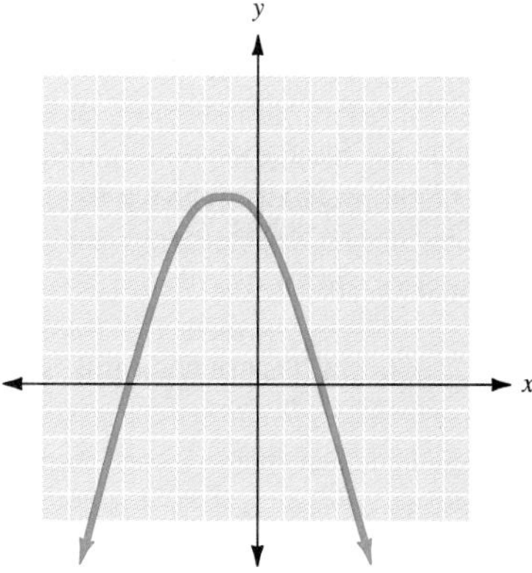

To graph quadratic equations, start by finding solutions for the equation. We begin by completing a table of values. This is done by choosing any convenient values for x. Then use the given equation to compute the corresponding values for y, as Example 1 illustrates.

Example 1

Completing a Table of Values

If $y = x^2$, complete the ordered pairs to form solutions. Then show these results in a table of values.

$(-2,\ ), (-1,\ ), (0,\ ), (1,\ ), (2,\ )$

For example, to complete the pair $(-2,\ )$, substitute -2 for x in the given equation.

$$y = (-2)^2 = 4$$

NOTE Remember that a solution is a pair of values that makes the equation a true statement.

So $(-2, 4)$ is a solution.

Substituting the other values for x in the same manner, we have the following table of values for $y = x^2$:

x	y
−2	4
−1	1
0	0
1	1
2	4

CHECK YOURSELF 1

If $y = x^2 + 2$, complete the ordered pairs to form solutions and form a table of values.

$(-2,\ ), (-1,\ ), (0,\ ), (1,\ ), (2,\ )$

We can now plot points in the cartesian coordinate system that correspond to solutions to the equation.

Example 2

Plotting Some Solution Points

Plot the points from the table of values corresponding to $y = x^2$ from Example 1.

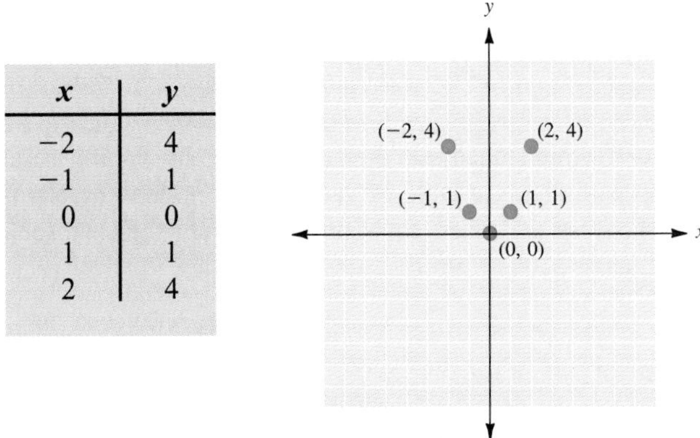

Notice that the y axis acts as a mirror. Do you see that any point graphed in quadrant I will be "reflected" in quadrant II?

CHECK YOURSELF 2

Plot the points from the table of values formed in Check Yourself 1.

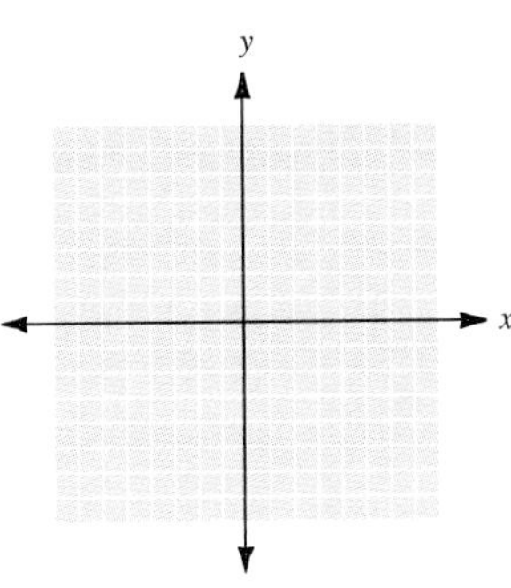

The graph of the equation can be drawn by joining the points with a smooth curve.

Example 3

Completing the Graph of the Solution Set

Draw the graph of $y = x^2$.

We can now draw a smooth curve between the points found in Example 2 to form the graph of $y = x^2$.

NOTE As we mentioned earlier, the graph must be the curve called a parabola.

NOTE Notice that a parabola *does* **not** come to a point.

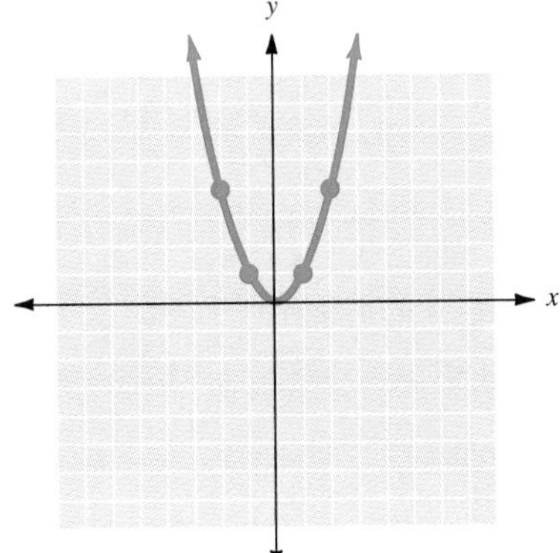

CHECK YOURSELF 3

Draw a smooth curve between the points plotted in the Check Yourself 2 exercise.

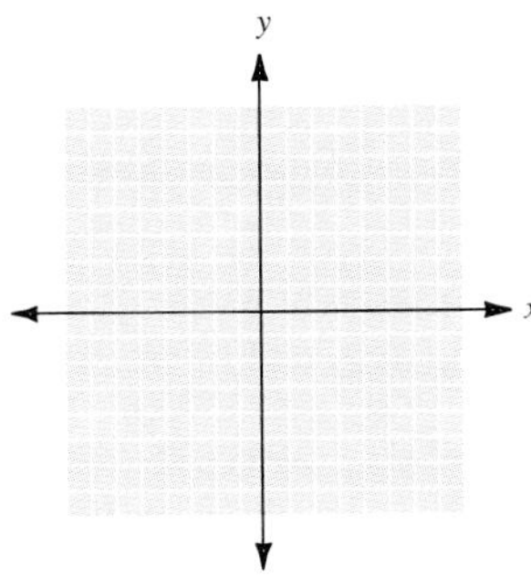

You can use any convenient values for x in forming your table of values. You should use as many pairs as are necessary to get the correct shape of the graph (a parabola).

Example 4

Graphing the Solution Set

Graph $y = x^2 - 2x$. Use values of x between -1 and 3.

First, determine solutions for the equation. For instance, if $x = -1$,

$$\begin{aligned} y &= (-1)^2 - 2(-1) \\ &= 1 + 2 \\ &= 3 \end{aligned}$$

then $(-1, 3)$ is a solution for the given equation.

Substituting the other values for x, we can form the table of values shown below. We then plot the corresponding points and draw a smooth curve to form our graph.

The graph of $y = x^2 - 2x$.

NOTE Any values can be substituted for x in the original equation.

x	y
-1	3
0	0
1	-1
2	0
3	3

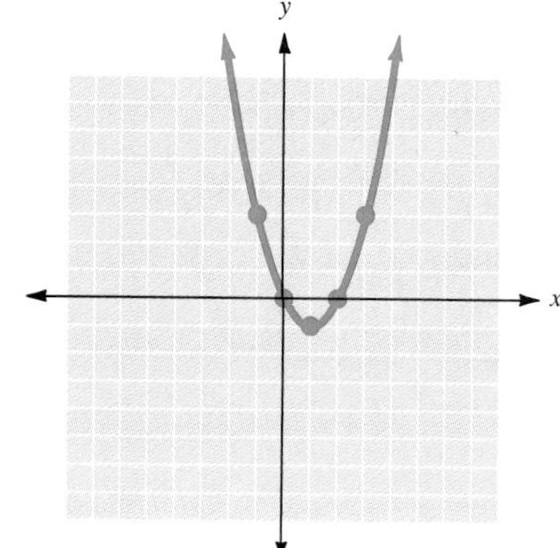

CHECK YOURSELF 4

Graph $y = x^2 + 4x$. Use values of x between -4 and 0.

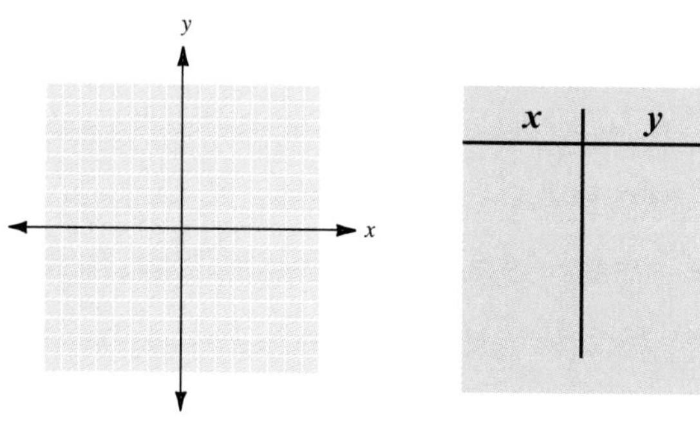

Choosing values for x is also a valid method of graphing a quadratic equation that contains a constant term.

Example 5

Graphing the Solution Set

Graph $y = x^2 - x - 2$. Use values of x between -2 and 3. We'll show the computation for two of the solutions.

If $x = -2$:

$$y = (-2)^2 - (-2) - 2$$
$$= 4 + 2 - 2$$
$$= 4$$

If $x = 3$:

$$y = 3^2 - 3 - 2$$
$$= 9 - 3 - 2$$
$$= 4$$

You should substitute the remaining values for x into the given equation to verify the other solutions shown in the table of values below.

The graph of $y = x^2 - x - 2$.

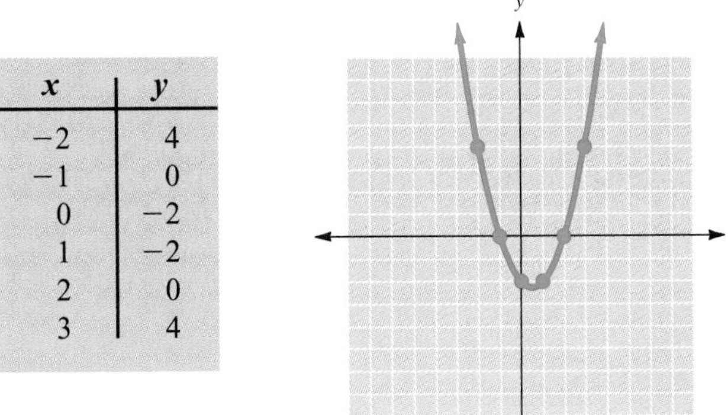

x	y
−2	4
−1	0
0	−2
1	−2
2	0
3	4

CHECK YOURSELF 5

Graph $y = x^2 - 4x + 3$. Use values of x between -1 and 4.

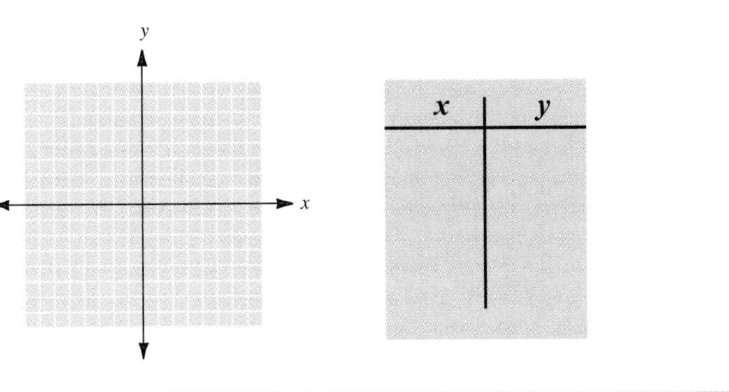

x	y

In Example 6, the graph looks significantly different from previous graphs.

Example 6

Graphing the Solution Set

Graph $y = -x^2 + 3$. Use x values between -2 and 2.

Again we'll show two computations.

NOTE $-(-2)^2 = -4$

If $x = -2$:

$$y = -(-2)^2 + 3$$
$$= -4 + 3$$
$$= -1$$

If $x = 1$:

$$y = -(1)^2 + 3$$
$$= -1 + 3$$
$$= 2$$

Verify the remainder of the solutions shown in the table of values below for yourself.

The graph of $y = -x^2 + 3$.

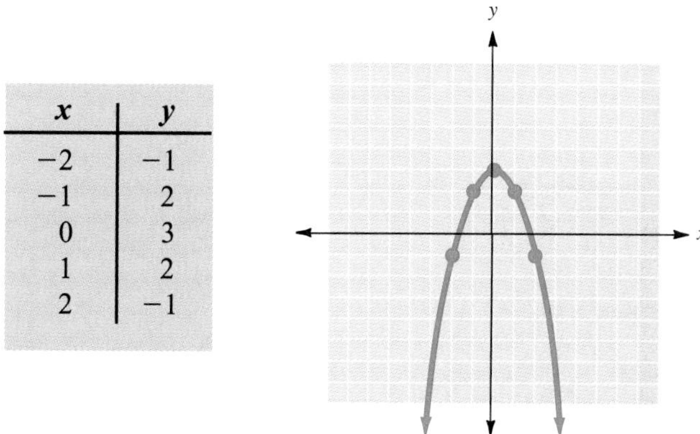

x	y
−2	−1
−1	2
0	3
1	2
2	−1

There is an important difference between this graph and the others we have seen. This time the parabola opens downward! Can you guess why? The answer is in the coefficient of the x^2 term.

If the coefficient of x^2 is *positive,* the parabola opens *upward.*

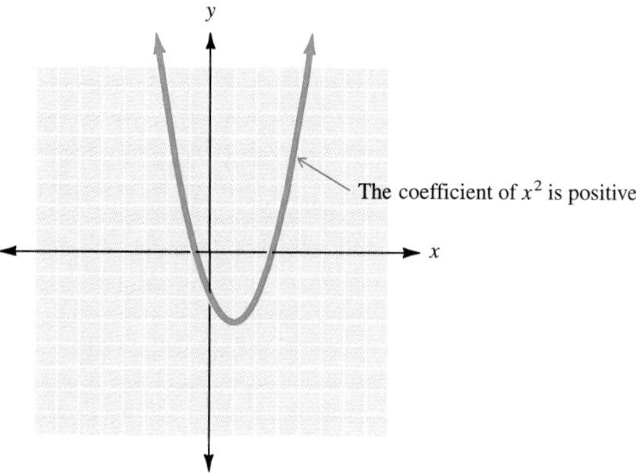

If the coefficient of x^2 is *negative,* the parabola opens *downward.*

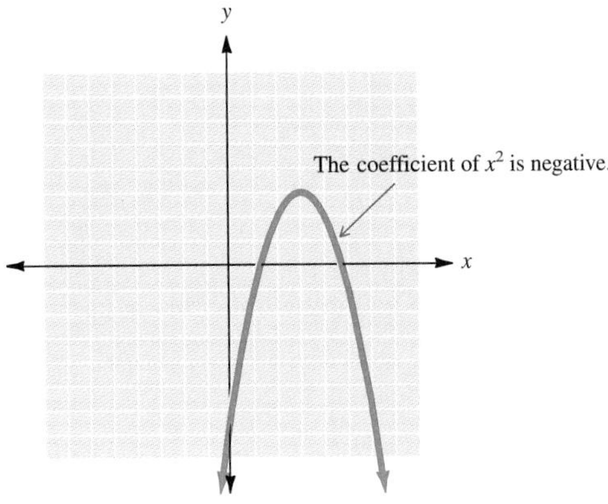

CHECK YOURSELF 6

Graph $y = -x^2 - 2x$. Use x values between −3 and 1.

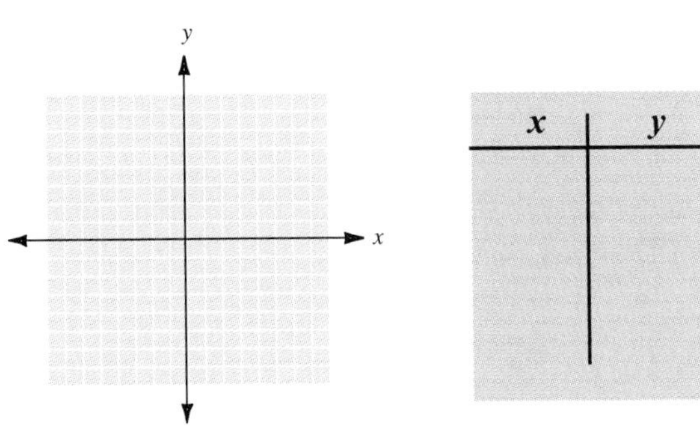

There are two other terms we would like to introduce before closing this section on graphing quadratic equations. As you may have noticed, all the parabolas that we graphed are symmetric about a vertical line. This is called the **axis of symmetry** for the parabola.

The point at which the parabola intersects that vertical line (this will be the lowest—or the highest—point on the parabola) is called the **vertex.** You'll learn more about finding the axis of symmetry and the vertex of a parabola in your next course in algebra.

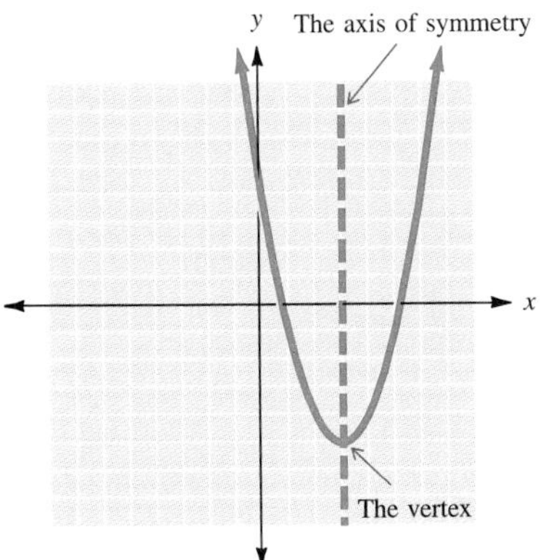

CHECK YOURSELF ANSWERS

1.

x	y
−2	6
−1	3
0	2
1	3
2	6

2.

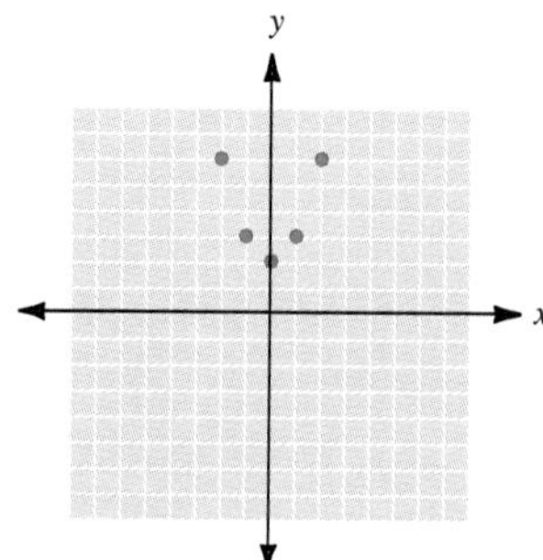

3. $y = x^2 + 2$

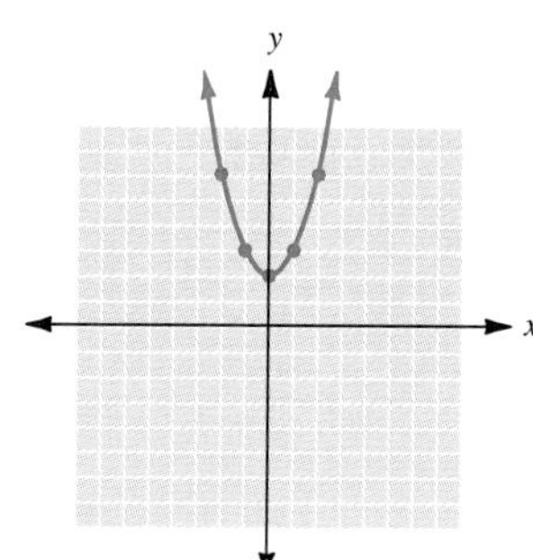

4. $y = x^2 + 4x$

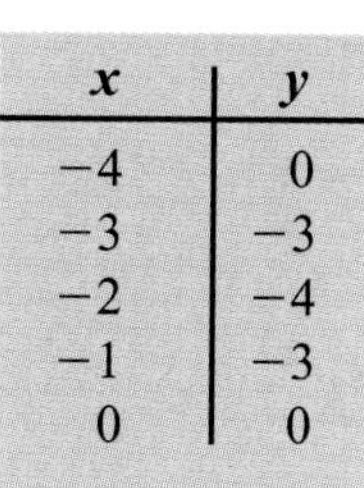

x	y
−4	0
−3	−3
−2	−4
−1	−3
0	0

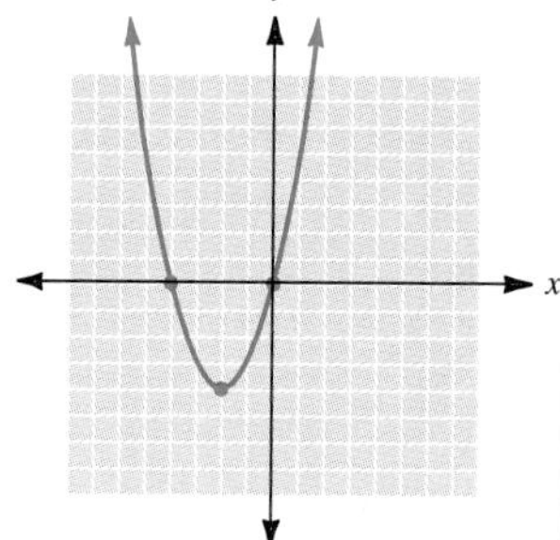

5. $y = x^2 - 4x + 3$

x	y
−1	8
0	3
1	0
2	−1
3	0
4	3

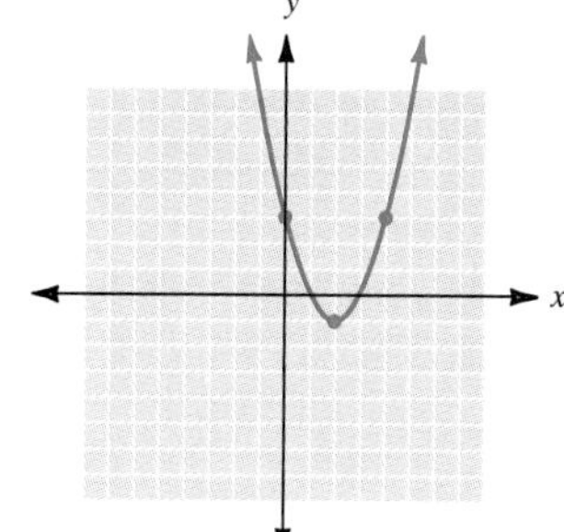

6. $y = -x^2 - 2x$

x	y
−3	−3
−2	0
−1	1
0	0
1	−3

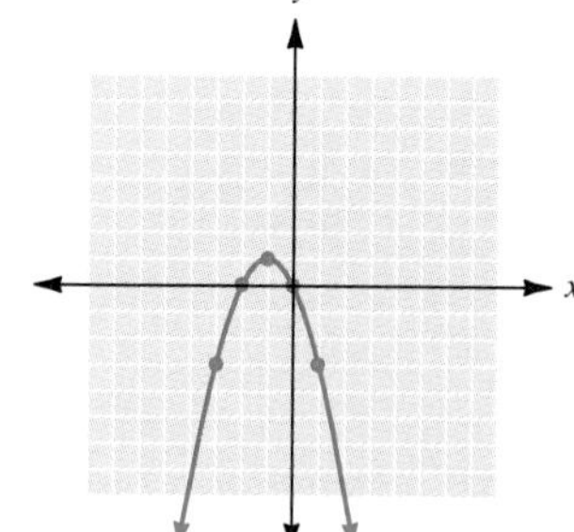

Name ______________

Section ________ Date ________

Exercises

Graph each of the following quadratic equations after completing the given table of values.

1. $y = x^2 + 1$

x	y
−2	5
−1	2
0	1
1	2
2	5

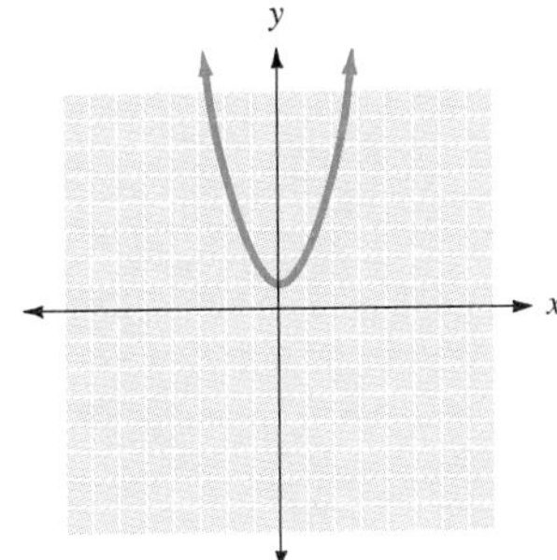

2. $y = x^2 - 2$

x	y
−2	2
−1	−1
0	−2
1	−1
2	2

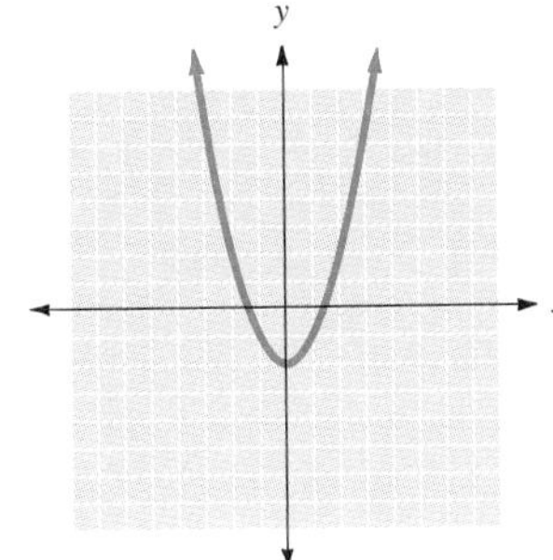

3. $y = x^2 - 4$

x	y
−2	0
−1	−3
0	−4
1	−3
2	0

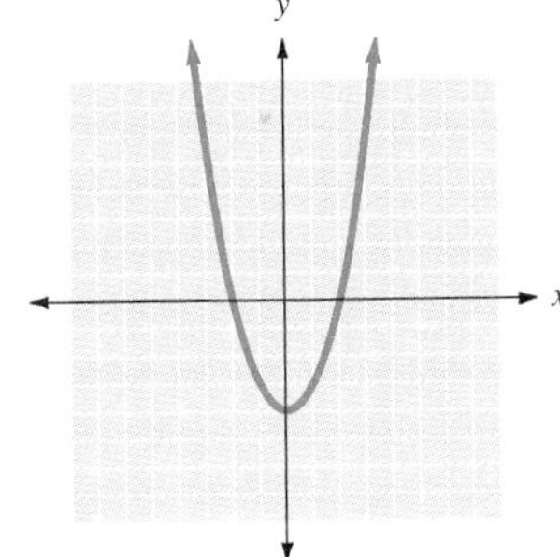

4. $y = x^2 + 3$

x	y
−2	7
−1	4
0	3
1	4
2	7

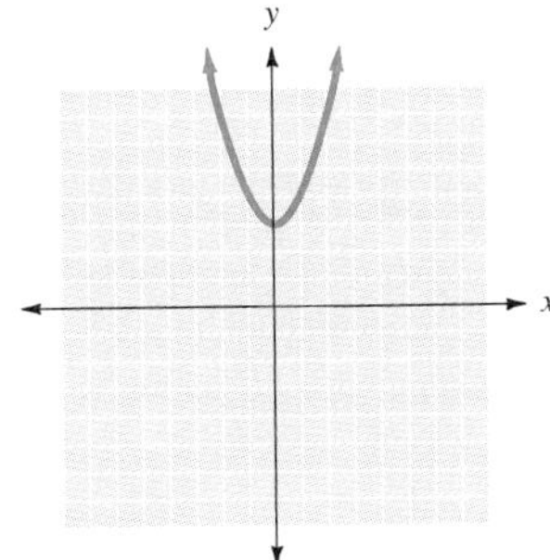

ANSWERS

1. See exercise
2. See exercise
3. See exercise
4. See exercise

ANSWERS

5. See exercise

6. See exercise

7. See exercise

8. See exercise

5. $y = x^2 - 4x$

x	y
0	0
1	−3
2	−4
3	−3
4	0

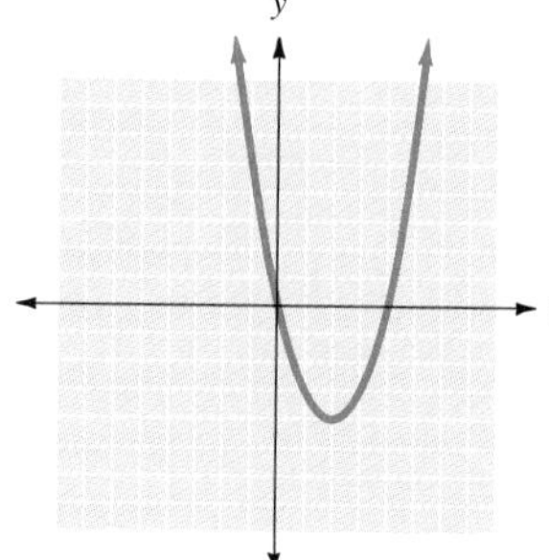

6. $y = x^2 + 2x$

x	y
−3	3
−2	0
−1	−1
0	0
1	3

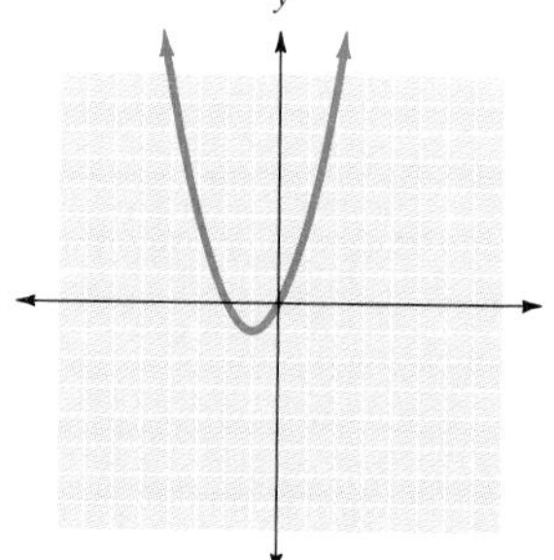

7. $y = x^2 + x$

x	y
−2	2
−1	0
0	0
1	2
2	6

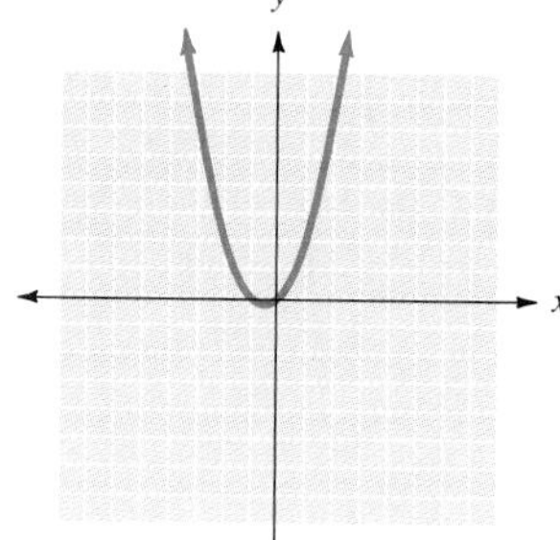

8. $y = x^2 - 3x$

x	y
−1	4
0	0
1	−2
2	−2
3	0

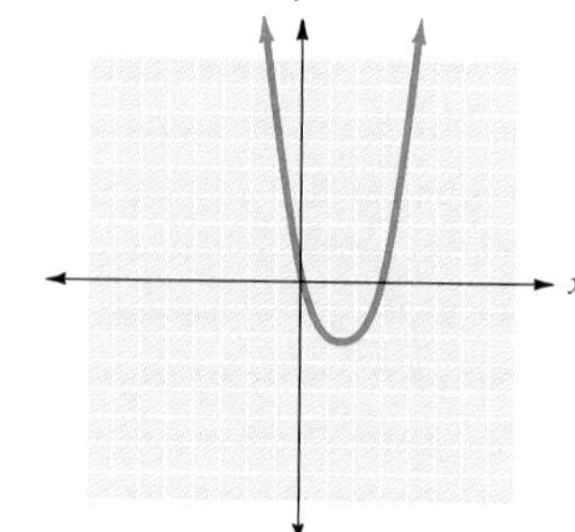

9. $y = x^2 - 2x - 3$

x	y
−1	0
0	−3
1	−4
2	−3
3	0

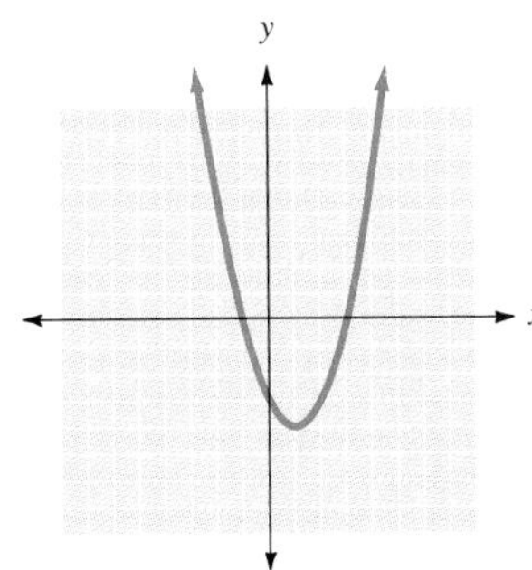

10. $y = x^2 - 5x + 6$

x	y
0	6
1	2
2	0
3	0
4	2

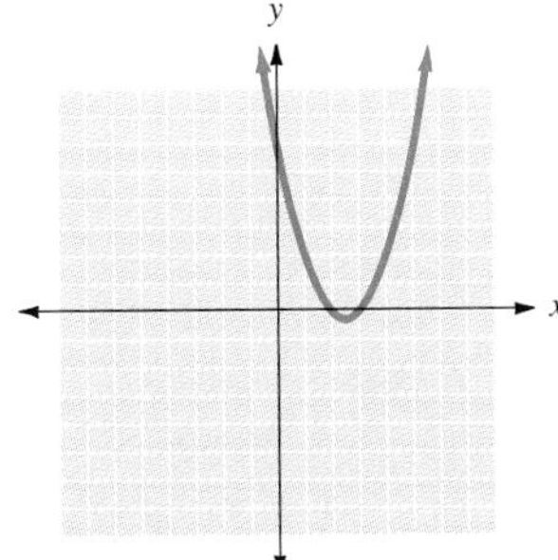

11. $y = x^2 - x - 6$

x	y
−1	−4
0	−6
1	−6
2	−4
3	0

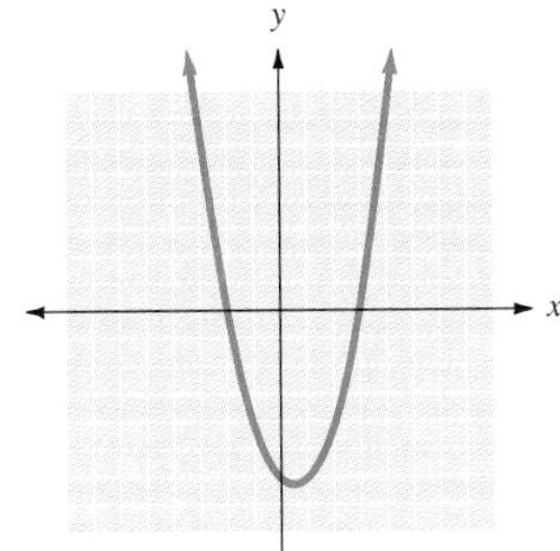

12. $y = x^2 + 3x - 4$

x	y
−4	0
−3	−4
−2	−6
−1	−6
0	−4

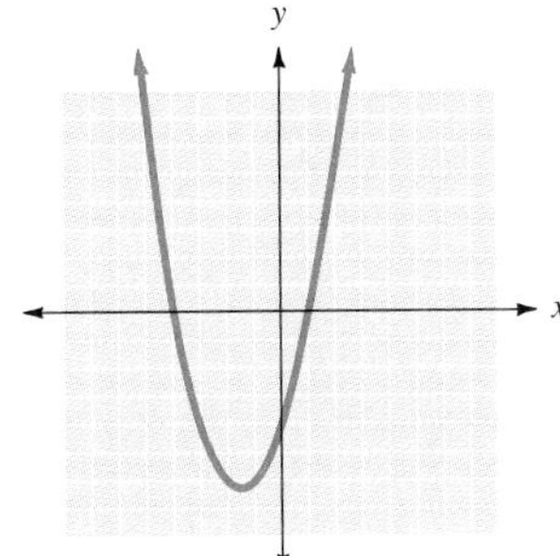

ANSWERS

9. See exercise

10. See exercise

11. See exercise

12. See exercise

13. See exercise

14. See exercise

15. See exercise

16. See exercise

13. $y = -x^2 + 2$

x	y
−2	−2
−1	1
0	2
1	1
2	−2

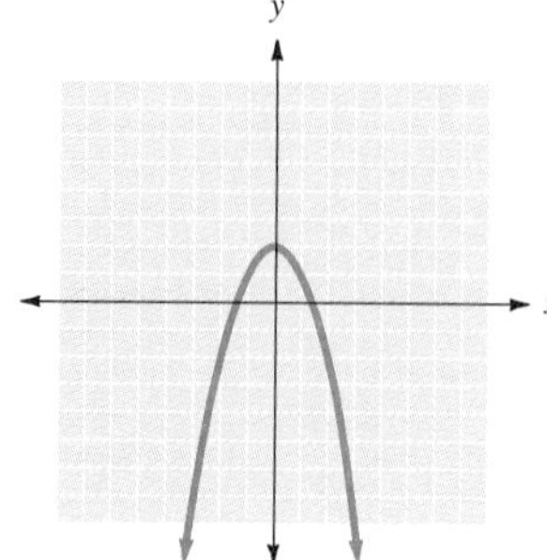

14. $y = -x^2 - 2$

x	y
−2	−6
−1	−3
0	−2
1	−3
2	−6

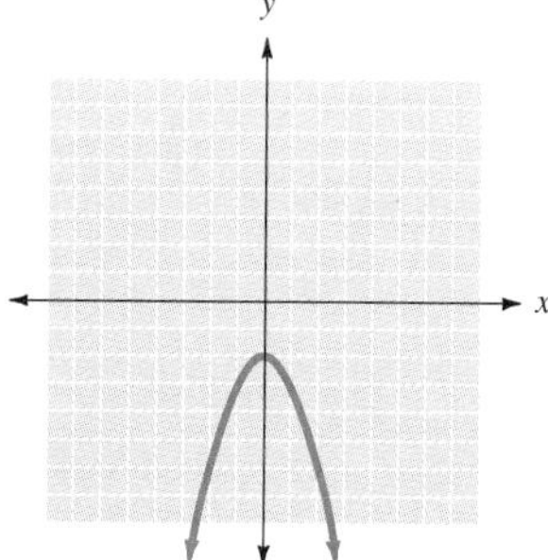

15. $y = -x^2 - 4x$

x	y
−4	0
−3	3
−2	4
−1	3
0	0

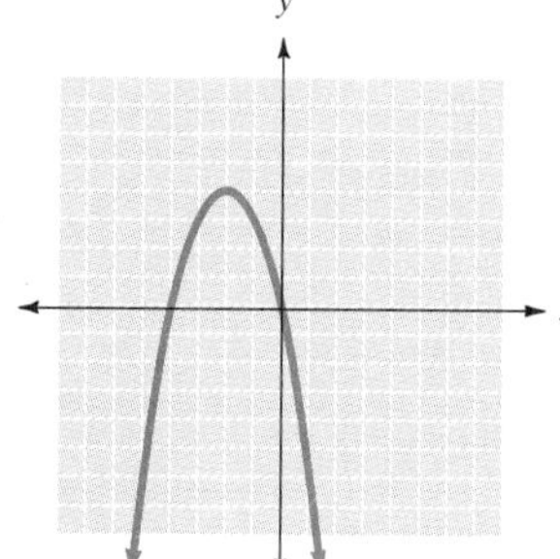

16. $y = -x^2 + 2x$

x	y
−1	−3
0	0
1	1
2	0
3	−3

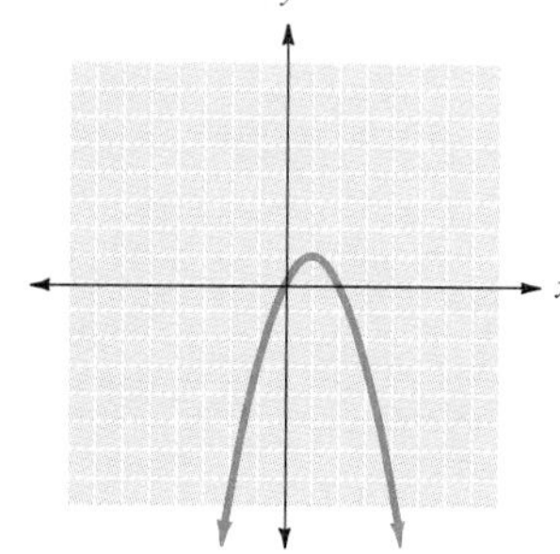

Match each graph with the correct equation on the right.

17.

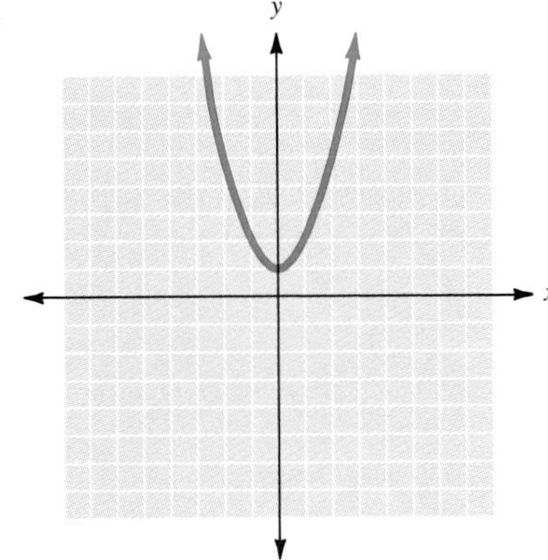

18.

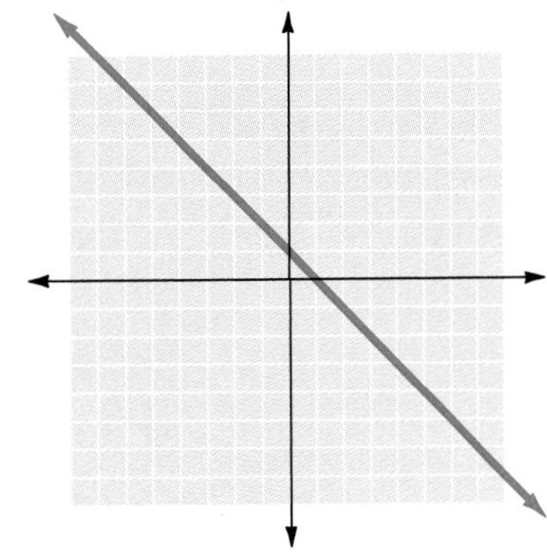

19.

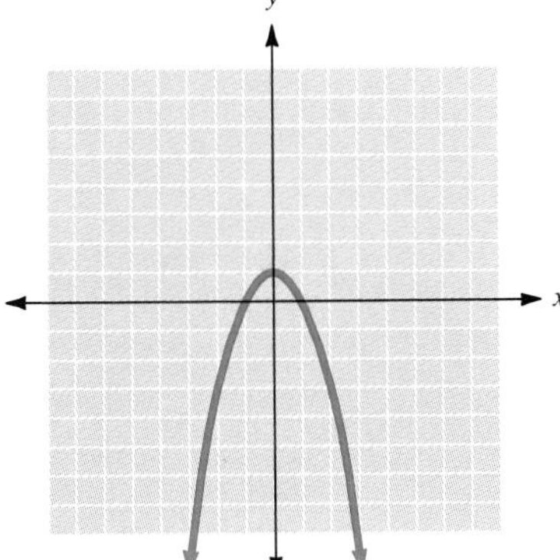

20.

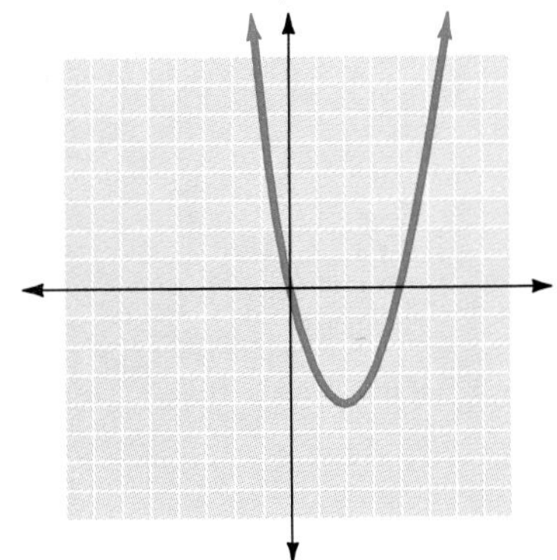

21.

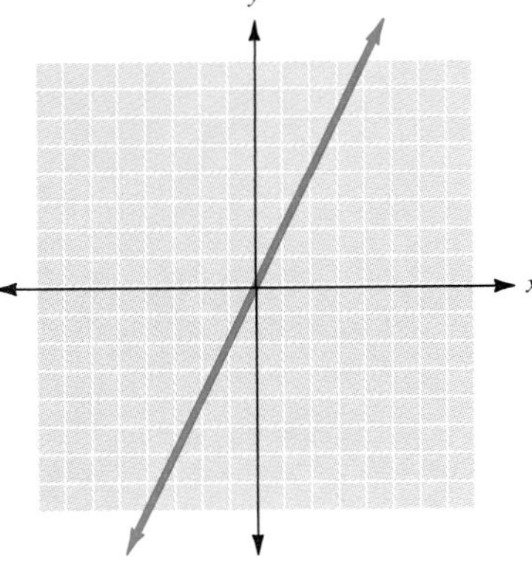

22.

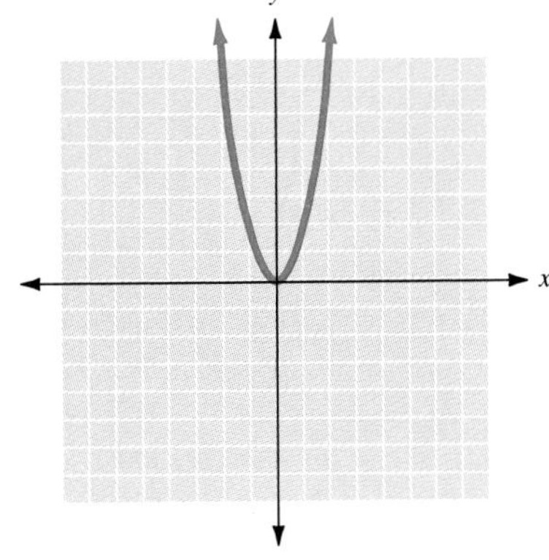

23.

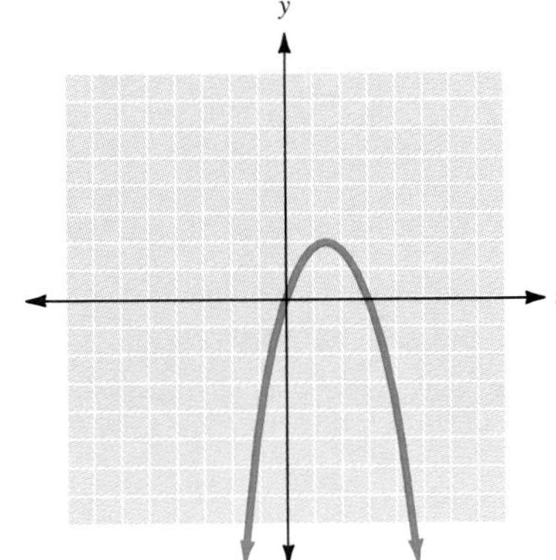

24.

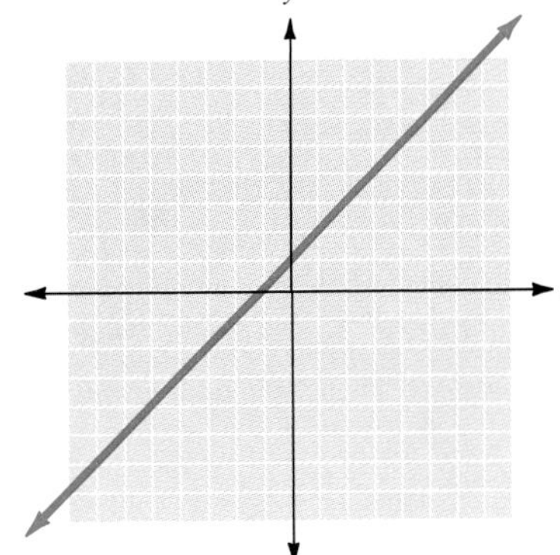

(a) $y = -x^2 + 1$

(b) $y = 2x$

(c) $y = x^2 - 4x$

(d) $y = -x + 1$

(e) $y = -x^2 + 3x$

(f) $y = x^2 + 1$

(g) $y = x + 1$

(h) $y = 2x^2$

ANSWERS

17. f
18. d
19. a
20. c
21. b
22. h
23. e
24. g

Answers

1. $y = x^2 + 1$

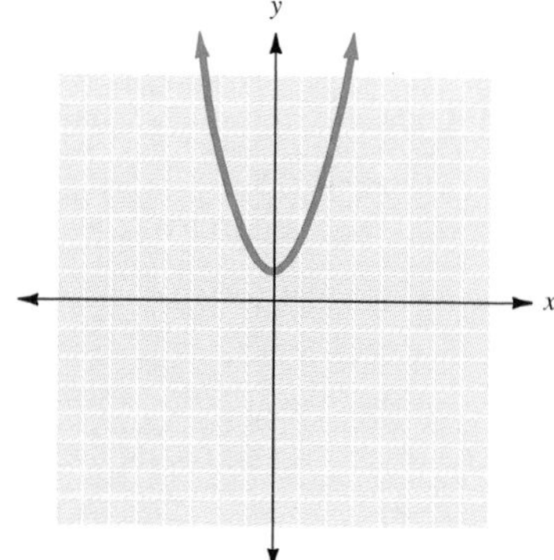

3. $y = x^2 - 4$

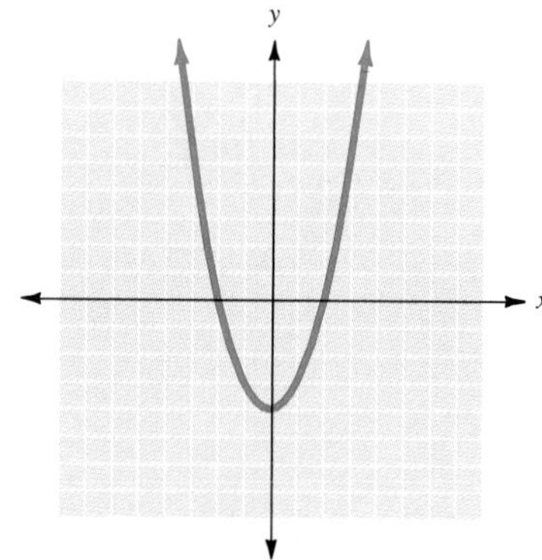

5. $y = x^2 - 4x$

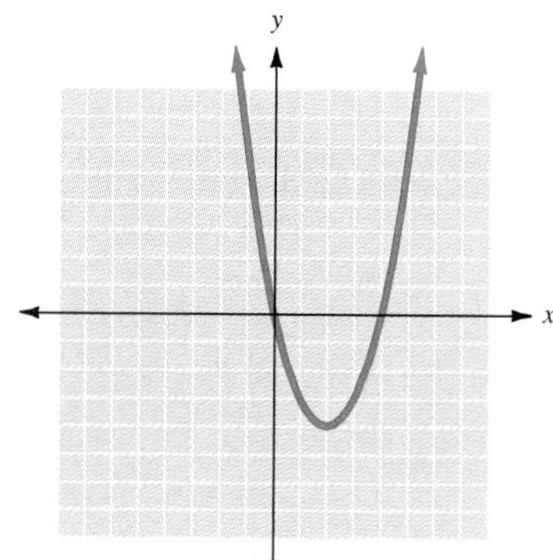

7. $y = x^2 + x$

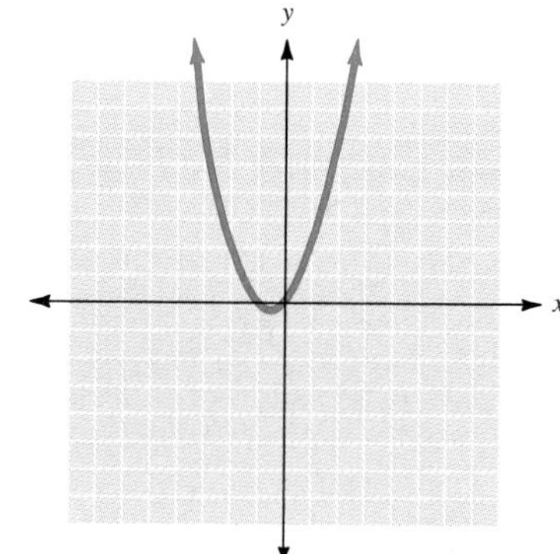

9. $y = x^2 - 2x - 3$

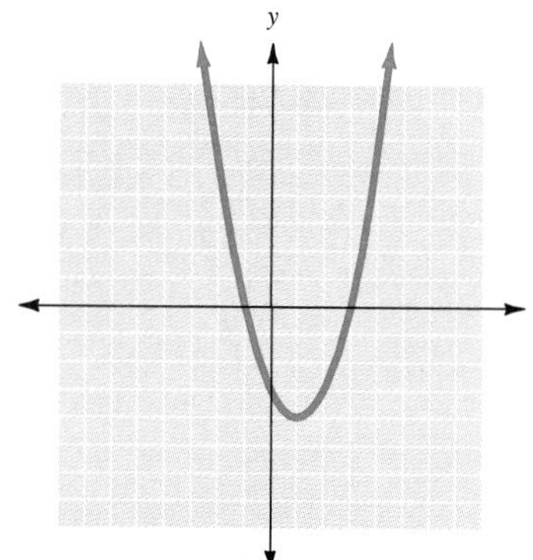

11. $y = x^2 - x - 6$

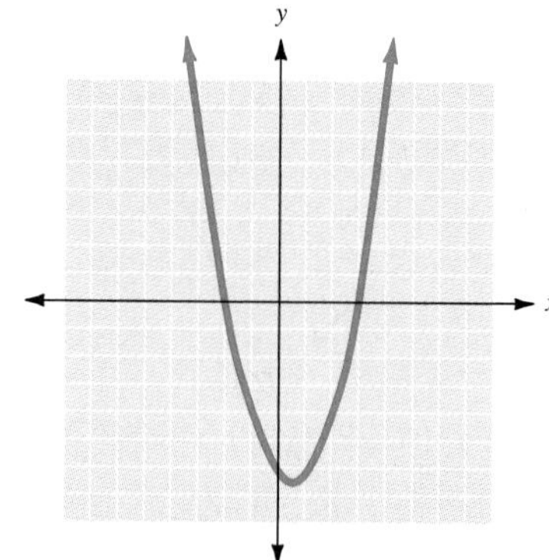

13. $y = -x^2 + 2$

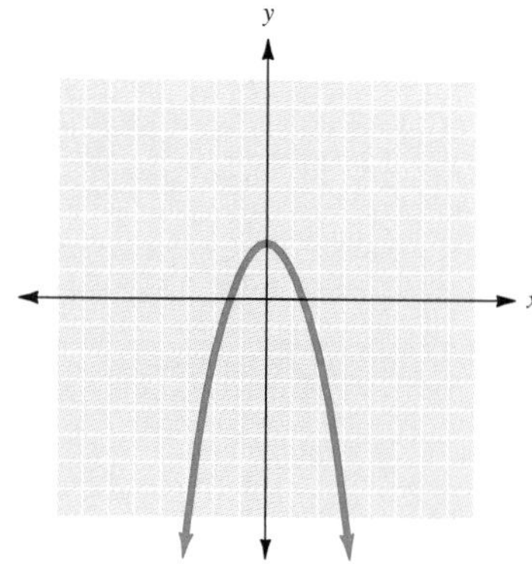

15. $y = -x^2 - 4x$

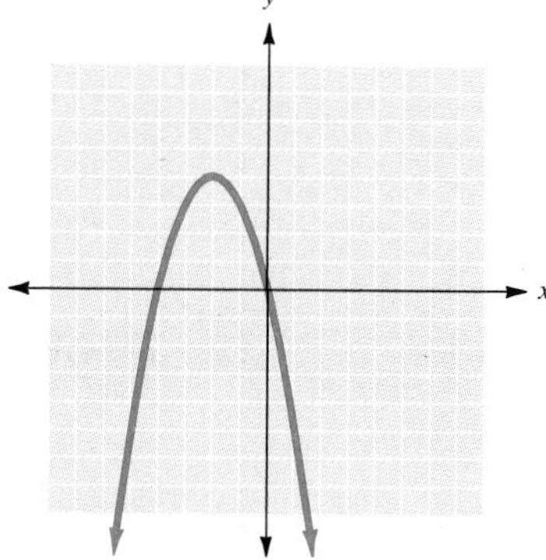

17. f **19.** a **21.** b **23.** e

10 Summary

Definition/Procedure	**Example**	**Reference**
More on Quadratic Equations		**Section 10.1**
Solving Quadratic Equations by Factoring **1.** Add or subtract the necessary terms on both sides of the equation so that the equation is in standard form (set equal to 0). **2.** Factor the quadratic expression. **3.** Set each factor equal to 0. **4.** Solve the resulting equations to find the solutions. **5.** Check each solution by substituting in the original equation.	To solve: $x^2 + 7x = 30$ $x^2 + 7x - 30 = 0$ $(x + 10)(x - 3) = 0$ $x + 10 = 0$ or $x - 3 = 0$ $x = -10$ and $x = 3$ are solutions.	**p. 753**
Completing the Square		**Section 10.2**
Completing the Square To solve a quadratic equation by completing the square: **1.** Write the equation in the form $ax^2 + bx = k$ so that the variable terms are on the left side and the constant is on the right side. **2.** If the leading coefficient (of x^2) is not 1, divide both sides by that coefficient. **3.** Add the square of one-half the middle (x) coefficient to both sides of the equation. **4.** The left side of the equation is now a perfect-square trinomial. Factor and solve as before.	To solve: $2x^2 + 2x - 1 = 0$ $2x^2 + 2x = 1$ $x^2 + x = \frac{1}{2}$ $x^2 + x + \left(\frac{1}{2}\right)^2 = \frac{1}{2} + \left(\frac{1}{2}\right)^2$ $\left(x + \frac{1}{2}\right)^2 = \frac{3}{4}$ $x + \frac{1}{2} = \pm\sqrt{\frac{3}{4}} = \pm\frac{\sqrt{3}}{2}$ $x = \frac{-1 \pm \sqrt{3}}{2}$	**p. 764**
The Quadratic Formula		**Section 10.3**
The Quadratic Formula To solve an equation by formula: **1.** Rewrite the equation in standard form. $ax^2 + bx + c = 0$ **2.** If a common factor exists, divide both sides of the equation by that common factor. **3.** Identify the coefficients a, b, and c. **4.** Substitute the values for a, b, and c into the quadratic formula. $x = \frac{-b \pm \sqrt{b^2 - 4ac}}{2a}$ **5.** Simplify the right side of the expression formed in step 4 to write the solutions for the original equation.	To solve: $x^2 - 2x = 4$ Write the equation as $x^2 - 2x - 4 = 0$ $a = 1 \quad b = -2 \quad c = -4$ $x = \frac{-(-2) \pm \sqrt{(-2)^2 - 4(1)(-4)}}{2(1)}$ $= \frac{2 \pm \sqrt{20}}{2}$ $= \frac{2 \pm 2\sqrt{5}}{2} = \frac{2(1 \pm \sqrt{5})}{2}$ $= 1 \pm \sqrt{5}$	**p. 772**

Continued

DEFINITION/PROCEDURE	EXAMPLE	REFERENCE
Graphing Quadratic Equations		**Section 10.4**
To graph equations of the form $$y = ax^2 + bx + c$$ **1.** Form a table of values by choosing convenient values for x and finding the corresponding values for y. **2.** Plot the points from the table of values. **3.** Draw a smooth curve between the points. The graph of a quadratic equation will always be a parabola. The parabola opens upward if a, the coefficient of the x^2 term, is positive.	$y = x^2 - 4x$ 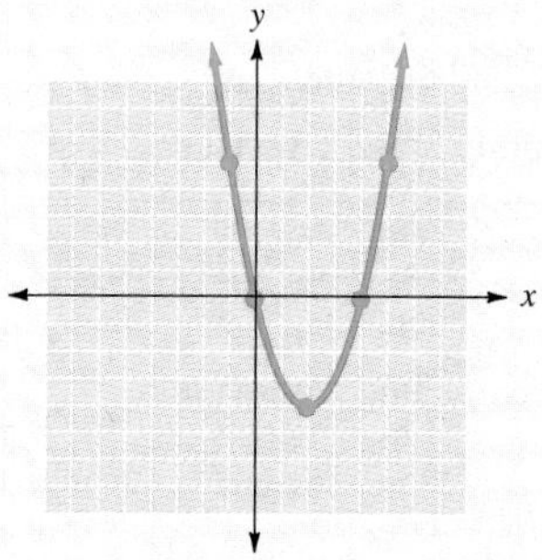	p. 777
The parabola opens downward if a, the coefficient of the x^2 term, is negative.	$y = -x^2 + 2x$ 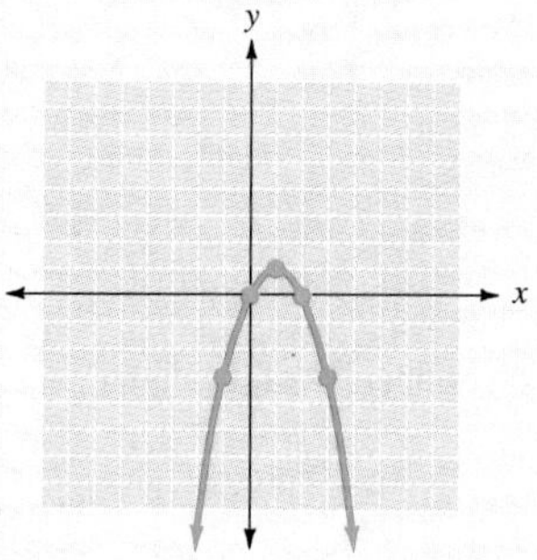	p. 782

$y = x^2 - 4x$

x	y
−1	5
0	0
1	−3
2	−4
3	−3
4	0
5	5

$y = -x^2 + 2x$

x	y
−1	−3
0	0
1	1
2	0
3	−3

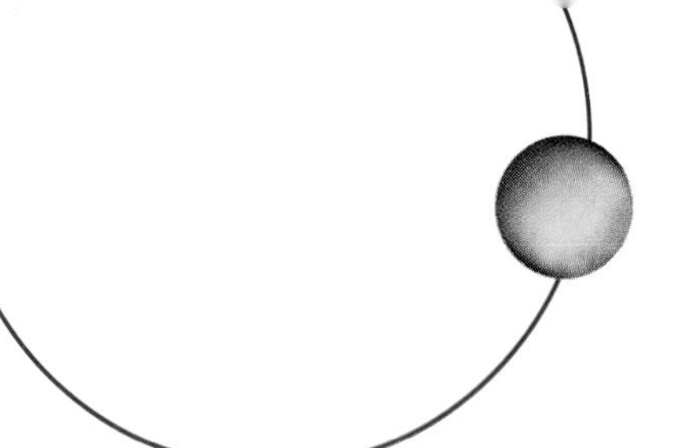

Summary Exercises

This summary exercise set is provided to give you practice with each of the objectives of the chapter. Each exercise is keyed to the appropriate chapter section. The answers are provided in the *Instructor's Manual*. Your instructor will give you guidelines on how to best use these exercises in your instructional setting.

[10.1] Solve each of the following equations for x by the square root method.

1. $x^2 = 10$
$\pm\sqrt{10}$

2. $x^2 = 48$
$\pm 4\sqrt{3}$

3. $x^2 - 20 = 0$
$\pm 2\sqrt{5}$

4. $x^2 + 2 = 8$
$\pm\sqrt{6}$

5. $(x - 1)^2 = 5$
$1 \pm \sqrt{5}$

6. $(x + 2)^2 = 8$
$-2 \pm 2\sqrt{2}$

7. $(x + 3)^2 = 5$
$-3 \pm \sqrt{5}$

8. $64x^2 = 25$
$\pm\frac{5}{8}$

9. $4x^2 = 27$
$\frac{\pm 3\sqrt{3}}{2}$

10. $9x^2 = 20$
$\frac{\pm 2\sqrt{5}}{3}$

11. $25x^2 = 7$
$\frac{\pm\sqrt{7}}{5}$

12. $7x^2 = 3$
$\frac{\pm\sqrt{21}}{7}$

[10.2] Solve each of the following equations by completing the square.

13. $x^2 - 3x - 10 = 0$
$-2, 5$

14. $x^2 - 8x + 15 = 0$
$3, 5$

15. $x^2 - 5x + 2 = 0$
$\frac{5 \pm \sqrt{17}}{2}$

16. $x^2 - 2x - 2 = 0$
$1 \pm \sqrt{3}$

17. $x^2 - 4x - 4 = 0$
$2 \pm 2\sqrt{2}$

18. $x^2 + 3x = 7$
$\frac{-3 \pm \sqrt{37}}{2}$

19. $x^2 - 4x = -2$
$2 \pm \sqrt{2}$

20. $x^2 + 3x = 5$
$\frac{-3 \pm \sqrt{29}}{2}$

21. $x^2 - x = 7$
$\frac{1 \pm \sqrt{29}}{2}$

22. $2x^2 + 6x = 12$
$\frac{-3 \pm \sqrt{33}}{2}$

23. $2x^2 - 4x - 7 = 0$
$\frac{2 \pm 3\sqrt{2}}{2}$

24. $3x^2 + 5x + 1 = 0$
$\frac{-5 \pm \sqrt{13}}{6}$

[10.3] Solve each of the following equations by using the quadratic formula.

25. $x^2 - 5x - 14 = 0$ **−2, 7**

26. $x^2 - 8x + 16 = 0$ **4**

27. $x^2 + 5x - 3 = 0$ $\frac{-5 \pm \sqrt{37}}{2}$

28. $x^2 - 7x - 1 = 0$ $\frac{7 \pm \sqrt{53}}{2}$

29. $x^2 - 6x + 1 = 0$ $3 \pm 2\sqrt{2}$

30. $x^2 - 3x + 5 = 0$ **No real number solution**

31. $3x^2 - 4x = 2$ $\frac{2 \pm \sqrt{10}}{3}$

32. $2x - 3 = \frac{3}{x}$ $\frac{3 \pm \sqrt{33}}{4}$

33. $(x - 1)(x + 4) = 3$ $\frac{-3 \pm \sqrt{37}}{2}$

34. $x^2 - 5x + 7 = 5$ $\frac{5 \pm \sqrt{17}}{2}$

35. $2x^2 - 8x = 12$ $2 \pm \sqrt{10}$

36. $5x^2 = 15 - 15x$ $\frac{-3 \pm \sqrt{21}}{2}$

Solve by factoring or by any of the methods of this chapter.

37. $5x^2 = 3x$ $0, \frac{3}{5}$

38. $(2x - 3)(x + 5) = -11$ $-4, \frac{1}{2}$

39. $(x - 1)^2 = 10$ $1 \pm \sqrt{10}$

40. $2x^2 = 7$ $\frac{\pm\sqrt{14}}{2}$

41. $2x^2 = 5x + 4$ $\frac{5 \pm \sqrt{57}}{4}$

42. $2x^2 - 4x = 30$ **5, −3**

43. $2x^2 = 5x + 7$ $\frac{7}{2}, -1$

44. $3x^2 - 4x = 2$ $\frac{2 \pm \sqrt{10}}{3}$

45. $3x^2 + 6x - 15 = 0$ $-1 \pm \sqrt{6}$

46. $x^2 - 3x = 2(x + 5)$ $\frac{5 \pm \sqrt{65}}{2}$

47. $x - 2 = \frac{2}{x}$ $1 \pm \sqrt{3}$

48. The perimeter of a square is numerically 2 less than its area. Find the length of one side. (Approximate your answer to three decimal places, using a calculator.) **4.449**

[10.4] Graph each quadratic equation after completing the table of values.

49. $y = x^2 + 3$

x	y
−2	7
−1	4
0	3
1	4
2	7

50. $y = x^2 - 2$

x	y
−2	2
−1	−1
0	−2
1	−1
2	2

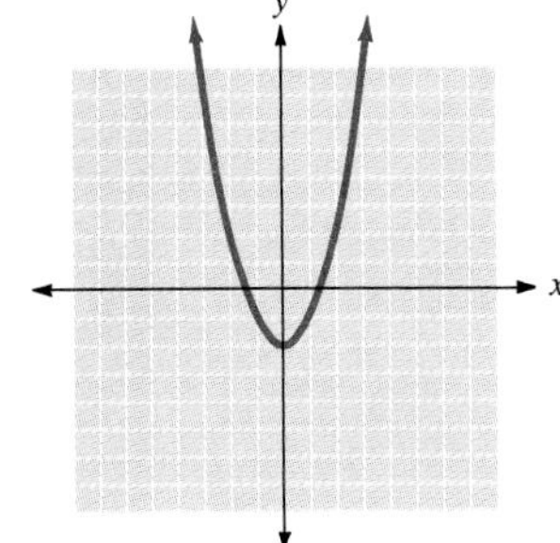

51. $y = x^2 - 3x$

x	y
−1	4
0	0
1	−2
2	−2
3	0

52. $y = x^2 + 4x$

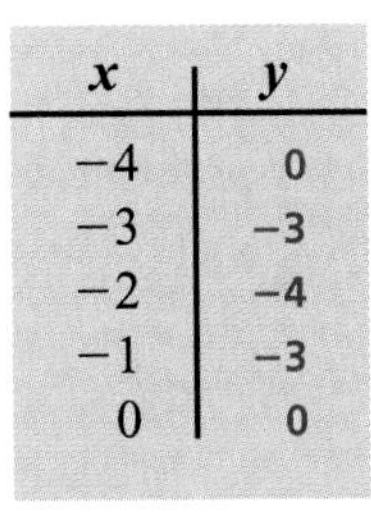

x	y
−4	0
−3	−3
−2	−4
−1	−3
0	0

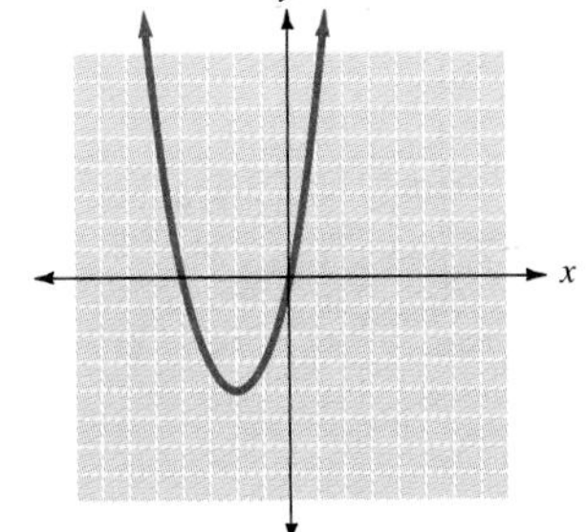

53. $y = x^2 - x - 2$

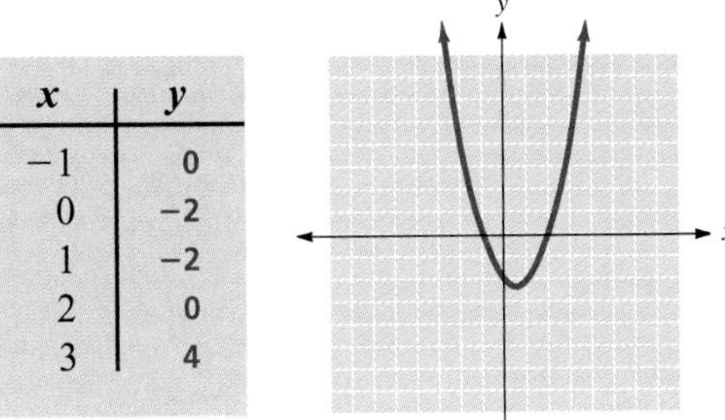

x	y
−1	0
0	−2
1	−2
2	0
3	4

54. $y = x^2 - 4x + 3$

x	y
0	3
1	0
2	−1
3	0
4	3

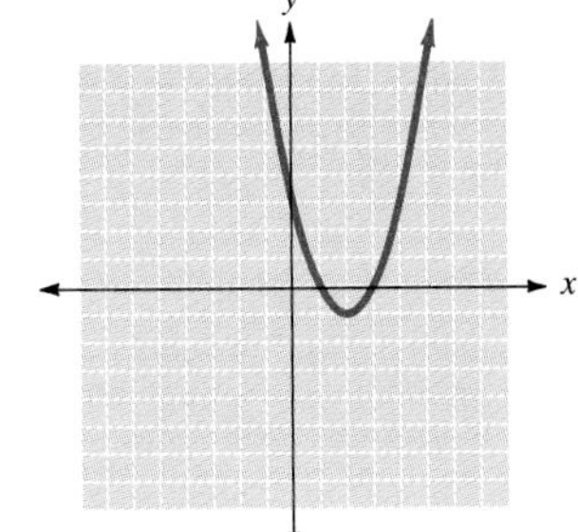

55. $y = x^2 + 2x - 3$

x	y
−3	0
−2	−3
−1	−4
0	−3
1	0

56. $y = 2x^2$

x	y
−2	8
−1	2
0	0
1	2
2	8

57. $y = 2x^2 - 3$

x	y
−2	5
−1	−1
0	−3
1	−1
2	5

58. $y = -x^2 + 3$

x	y
−2	−1
−1	2
0	3
1	2
2	−1

59. $y = -x^2 - 2$

x	y
−2	−6
−1	−3
0	−2
1	−3
2	−6

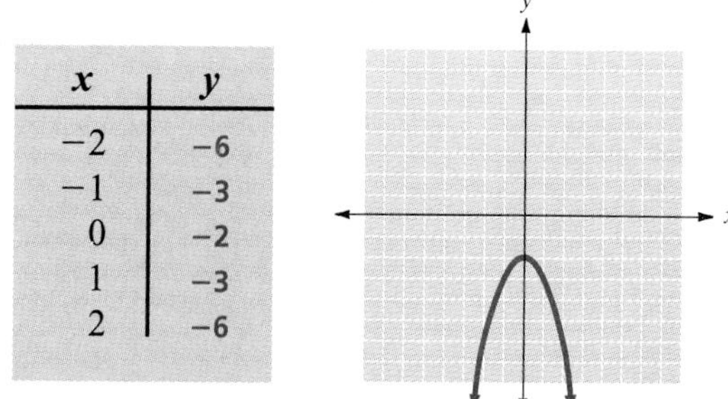

60. $y = -x^2 + 4x$

x	y
0	0
1	3
2	4
3	3
4	0

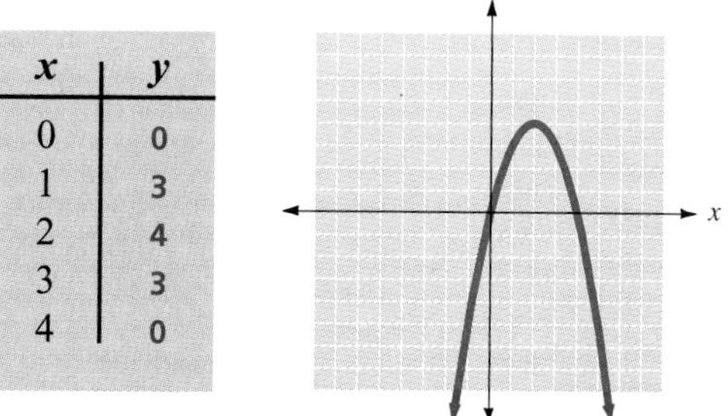

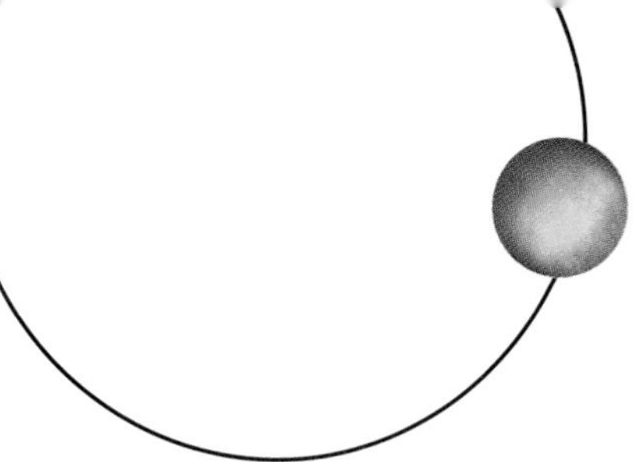

Self-Test for Chapter 10

Name ______________

Section ______ Date ______

The purpose of this self-test is to help you check your progress and to review for a chapter test in class. Allow yourself about an hour to take the test. When you are done, check your answers in the back of the book. If you missed any problems, be sure to go back and review the appropriate sections in the chapter and the exercises that are provided.

Solve each of the following equations for x.

1. $x^2 = 15$

2. $x^2 - 8 = 0$

3. $(x - 1)^2 = 7$

4. $9x^2 = 10$

Solve each of the following equations by completing the square.

5. $x^2 - 2x - 8 = 0$

6. $x^2 + 3x - 1 = 0$

7. $x^2 + 2x - 5 = 0$

8. $2x^2 - 5x + 1 = 0$

Solve each of the following equations by using the quadratic formula.

9. $x^2 - 2x - 3 = 0$

10. $x^2 - 6x + 9 = 0$

11. $x^2 - 5x = 2$

12. $2x^2 = 2x + 5$

13. $2x - 1 = \dfrac{4}{x}$

14. $(x - 1)(x + 3) = 2$

Graph each quadratic equation after completing the given table of values.

15. $y = x^2 + 4$

x	y
−2	8
−1	5
0	4
1	5
2	8

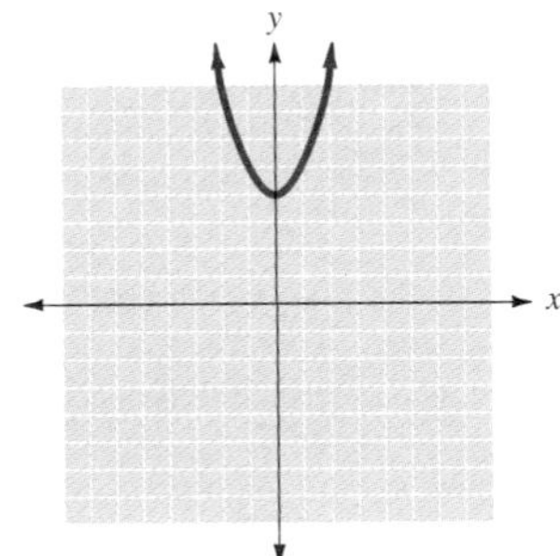

16. $y = x^2 - 2x$

x	y
−1	3
0	0
1	−1
2	0
3	3

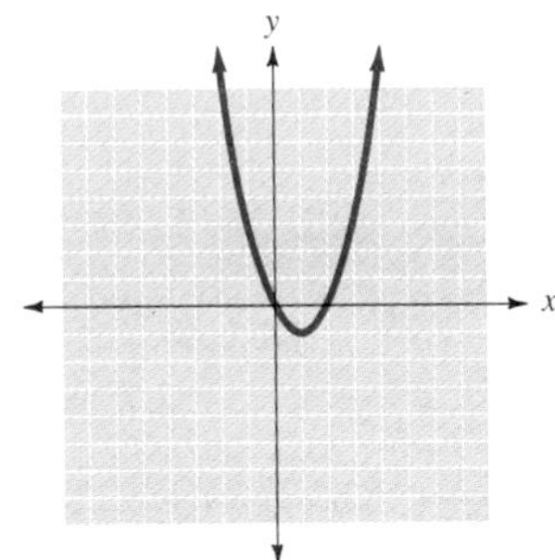

ANSWERS

1. $\pm\sqrt{15}$
2. $\pm 2\sqrt{2}$
3. $1 \pm \sqrt{7}$
4. $\dfrac{\pm\sqrt{10}}{3}$
5. 4, −2
6. $\dfrac{-3 \pm \sqrt{13}}{2}$
7. $-1 \pm \sqrt{6}$
8. $\dfrac{5 \pm \sqrt{17}}{4}$
9. −1, 3
10. 3
11. $\dfrac{5 \pm \sqrt{33}}{2}$
12. $\dfrac{1 \pm \sqrt{11}}{2}$
13. $\dfrac{1 \pm \sqrt{33}}{4}$
14. $-1 \pm \sqrt{6}$
15. See exercise
16. See exercise

ANSWERS

17. See exercise

18. See exercise

19. See exercise

20. See exercise

17. $y = x^2 - 3$

x	y
−2	1
−1	−2
0	−3
1	−2
2	1

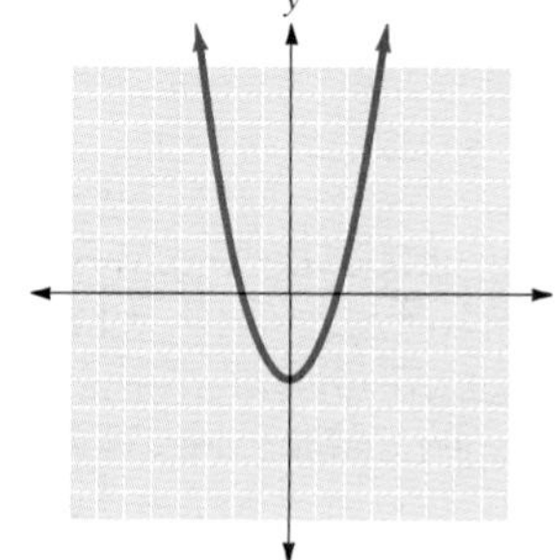

18. $y = x^2 + x - 2$

x	y
−2	0
−1	−2
0	−2
1	0
2	4

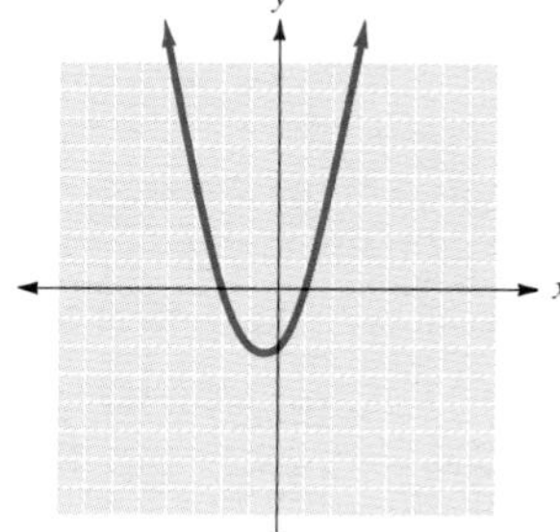

19. $y = -x^2 + 4$

x	y
−2	0
−1	3
0	4
1	3
2	0

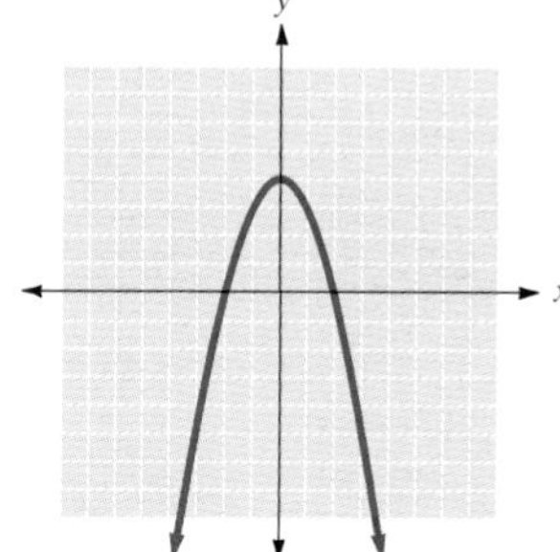

20. $y = -x^2 + 2x$

x	y
−1	−3
0	0
1	1
2	0
3	−3

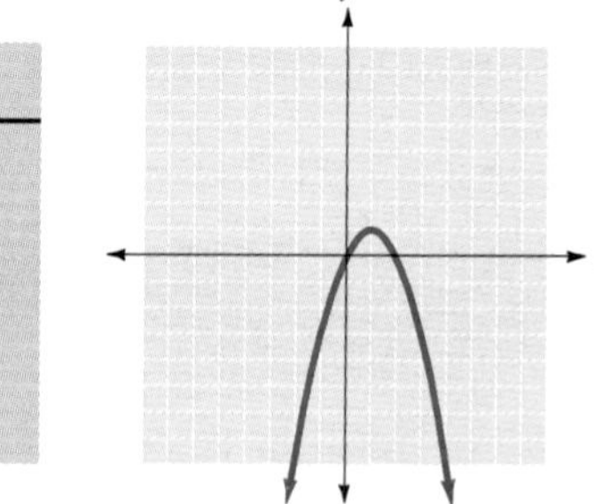

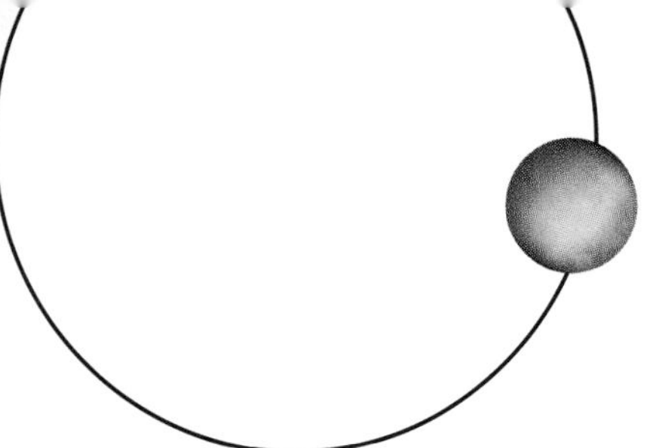

Cumulative Test for Chapters 0 to 10

Name ______________________

Section ________ Date ________

ANSWERS

1. $11x^2y - 6xy^2$
2. $x^2 + 5x - 7$
3. -480
4. -1680
5. 1
6. $x > 2$
7. $6x^3y - 3x^2y + 15xy$
8. $6x^2 + 11x - 10$
9. $9x^2 - 16y^2$
10. $8xy^2(2x - y)$
11. $(2x - 3)(4x + 5)$
12. $(5x - 4y)(5x + 4y)$
13. $\dfrac{29}{28(x + 2)}$
14. $\dfrac{5(x - 2)}{x - 1}$
15. $\dfrac{(3x - 1)(x + 6)}{15x^2}$

This test covers selected topics from all 11 chapters.

Simplify the following expressions.

1. $6x^2y - 4xy^2 + 5x^2y - 2xy^2$

2. $(3x^2 + 2x - 5) - (2x^2 - 3x + 2)$

Evaluate each expression when $x = 2, y = -3$, and $z = 4$.

3. $4x^2y - 3z^2y^2$

4. $-3x^2y^2z^2 - 2xyz$

5. Solve for x: $4x - 2(3x - 5) = 8$.

6. Solve the inequality $4x + 15 > 2x + 19$.

Perform the indicated operations.

7. $3xy(2x^2 - x + 5)$

8. $(2x + 5)(3x - 2)$

9. $(3x + 4y)(3x - 4y)$

Factor each of the following completely.

10. $16x^2y^2 - 8xy^3$

11. $8x^2 - 2x - 15$

12. $25x^2 - 16y^2$

Perform the indicated operations.

13. $\dfrac{7}{4x + 8} - \dfrac{5}{7x + 14}$

14. $\dfrac{5x + 5}{x - 2} \cdot \dfrac{x^2 - 4x + 4}{x^2 - 1}$

15. $\dfrac{3x^2 + 8x - 3}{15x^2} + \dfrac{3x - 1}{5x^2}$

ANSWERS

16. See exercise

17. See exercise

18. 5

19. $y = 2x - 5$

20. See exercise

Graph the following equations.

16. $3x - 2y = 6$

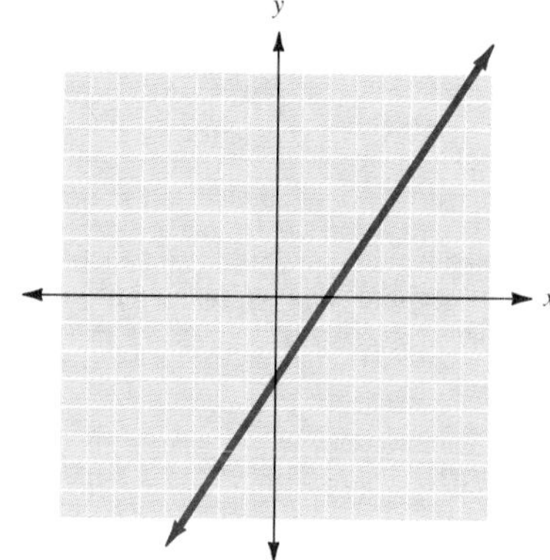

17. $y = 4x - 5$

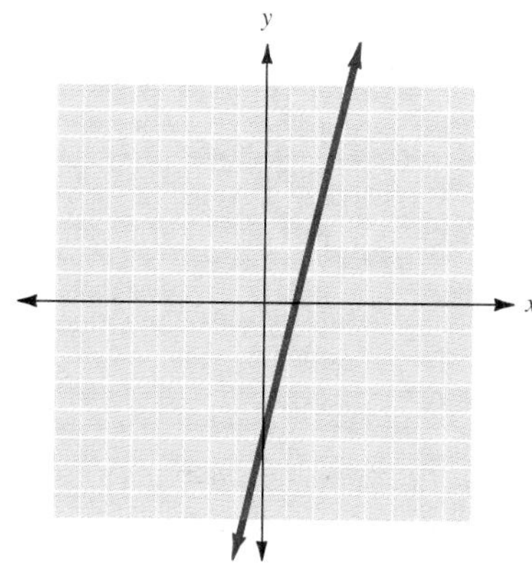

18. Find the slope of the line through the points (2, 9) and (−1, −6).

19. Given that the slope of a line is 2 and the y intercept is (0, −5), write the equation of the line.

20. Graph the inequality $x + 2y < 6$.

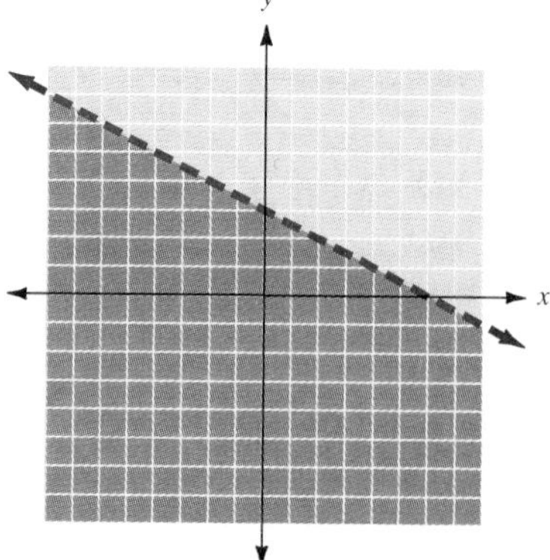

Solve each of the following systems. If a unique solution does not exist, state whether the system is inconsistent or dependent.

21. $2x - 3y = 6$
$x - 3y = 2$

22. $2x + y = 4$
$y = 2x - 8$

23. $5x + 2y = 8$
$x - 4y = 17$

24. $2x - 6y = 8$
$x = 3y + 4$

Solve each of the following applications. Be sure to show the system of equations used for your solution.

25. One number is 4 less than 5 times another. If the sum of the numbers is 26, what are the two numbers?

26. Receipts for a concert attended by 450 people were $2775. If reserved-seat tickets were $7 and general admission tickets were $4, how many of each type of ticket were sold?

27. A chemist has a 30% acid solution and a 60% solution already prepared. How much of each of the two solutions should be mixed to form 300 milliliters (mL) of a 50% solution?

Evaluate each root, if possible.

28. $\sqrt{169}$

29. $-\sqrt{169}$

30. $\sqrt{-169}$

31. $\sqrt[3]{-64}$

Simplify each of the following radical expressions.

32. $\sqrt{12} + 3\sqrt{27} - \sqrt{75}$

33. $3\sqrt{2a} \cdot 5\sqrt{6a}$

34. $(\sqrt{2} - 5)(\sqrt{2} + 3)$

35. $\dfrac{8 - \sqrt{32}}{4}$

Solve each of the following equations.

36. $x^2 - 72 = 0$

37. $x^2 + 6x - 3 = 0$

38. $2x^2 - 3x = 2(x + 1)$

ANSWERS

21. $\left(4, \frac{2}{3}\right)$
22. (3, −2)
23. $\left(3, \frac{-7}{2}\right)$
24. Dependent system
25. 5, 21
26. 325 reserved seat; 125 gen adm
27. 100 mL of 30%; 200 mL of 60%
28. 13
29. −13
30. Not a real number
31. −4
32. $6\sqrt{3}$
33. $30a\sqrt{3}$
34. $-13 - 2\sqrt{2}$
35. $2 - \sqrt{2}$
36. $\pm 6\sqrt{2}$
37. $-3 \pm 2\sqrt{3}$
38. $\dfrac{5 \pm \sqrt{41}}{4}$

39. See exercise

40. See exercise

Graph each of the following quadratic equations.

39. $y = x^2 - 2$

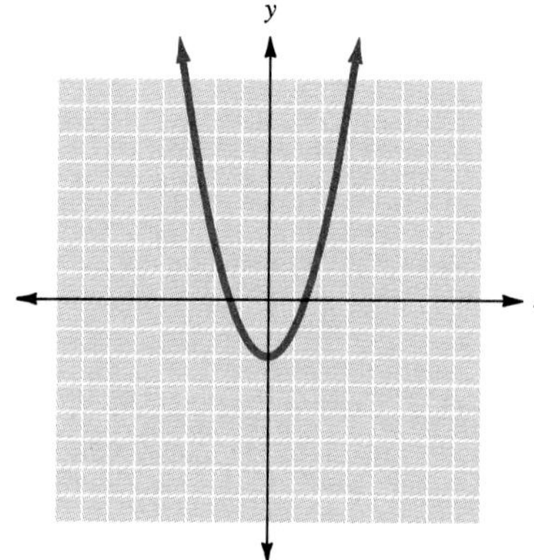

40. $y = x^2 - 4x$

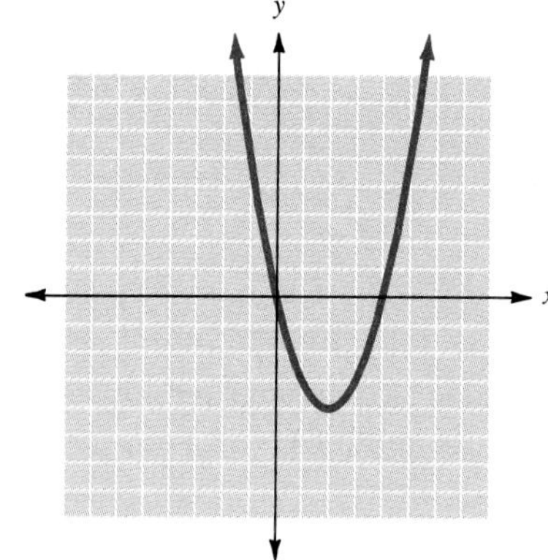

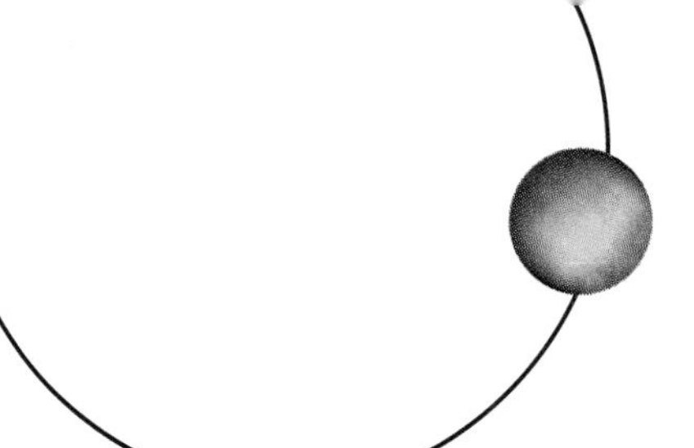

Final Examination

Name ______________________

Section __________ Date __________

In exercises 1 to 6, evaluate the given expressions.

1. $|-25| - |-11|$
2. $16 + (-22)$
3. $(-41) - (-15)$
4. $(-5)(-3)(-7)$
5. $\dfrac{3(-2) - 8}{-7 - (-4)(3)}$
6. $6 - 2^3 \cdot 5$

In exercises 7 and 8, evaluate the expressions for the given values of the variables.

7. $b^2 - 4ac$ for $a = -3$, $b = -4$, and $c = 2$
8. $-x^2 - 7x - 3$ for $x = -2$

9. Write the expression $9 \cdot p \cdot p \cdot p \cdot q \cdot q \cdot q \cdot q \cdot q$ in exponential form.

In exercises 10 to 12, simplify the expressions using the properties of exponents. Write all answers using positive exponents only.

10. $z^{-11} z^5$
11. $(5c^8d^7)^2$
12. $\dfrac{4x^8y^5z^3}{2x^6y^9z^7}$

In exercises 13 to 18, perform the indicated operations. Write each answer in simplified form.

13. $2x(x + 3) + 5$
14. $(6x^2 - 3x - 20) - 2(4x^2 - 16x + 11)$
15. $(7x - 9)(4x + 5)$
16. $(3x + 4y)(3x - 4y)$
17. $(5x - 2y)^2$
18. $x(x - 3y) - 2y(y + 6x)$

In exercises 19 to 21, solve the given equations.

19. $5x - 9 = -7x - 3$
20. $2x - 3(x - 2) = 8$
21. $\dfrac{5 - x}{-2} = 3x$
22. Solve the inequality $4x - 3 > 6x - 2$.

In exercises 23 to 26, factor each expression completely.

23. $10x^2 - 490$
24. $x^2 - 12x + 36$
25. $4p^2 - p - 18$
26. $3xy + 3xz - 5y - 5z$

In exercises 27 to 31, simplify the given expression.

27. $\dfrac{2y - 36}{3y^2 - 54y}$
28. $\dfrac{4z}{z + 8} + \dfrac{32}{z + 8}$
29. $\dfrac{6a}{a^2 - 9} - \dfrac{5a}{a^2 + a - 6}$
30. $\dfrac{y^2 + y - 2}{y + 5} \cdot \dfrac{3y + 3}{9y - 9}$
31. $\dfrac{x^2 - 3x - 10}{3x} \div \dfrac{5x - 25}{15x^2}$

ANSWERS

1. 14
2. −6
3. −26
4. −105
5. $-\dfrac{14}{5}$
6. −34
7. 40
8. 7
9. $9p^3q^5$
10. $\dfrac{1}{z^6}$
11. $25c^{16}d^{14}$
12. $\dfrac{2x^2}{y^4z^4}$
13. $2x^2 + 6x + 5$
14. $-2x^2 + 29x - 42$
15. $28x^2 - x - 45$
16. $9x^2 - 16y^2$
17. $25x^2 - 20xy + 4y^2$
18. $x^2 - 15xy - 2y^2$
19. $\dfrac{1}{2}$
20. −2
21. −1
22. $x < -\dfrac{1}{2}$
23. $10(x + 7)(x - 7)$
24. $(x - 6)(x - 6)$
25. $(4p - 9)(p + 2)$
26. $(y + z)(3x - 5)$
27. $\dfrac{2}{3y}$
28. 4
29. $\dfrac{a}{(a - 3)(a - 2)}$
30. $\dfrac{(y + 2)(y + 1)}{3(y + 5)}$
31. $x(x + 2)$

32. $\frac{1}{3}, -2$

33. $\frac{-1 \pm \sqrt{13}}{2}$

34. $-3 \pm \sqrt{14}$

35. -5 36. 4

37. $\frac{3}{4}$ 38. $y = -2x + 6$

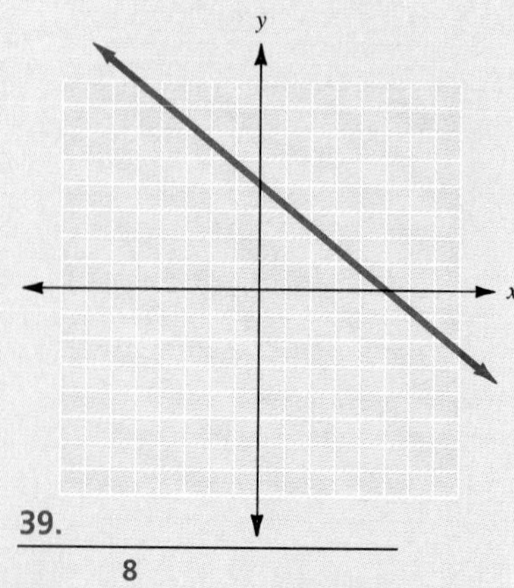

39.

40. $-\frac{8}{5}$ 41. $3x^2y^3\sqrt{2x}$

42. $-4\sqrt{5}$

43. $-11 - 6\sqrt{5}$

44. $(2, 0)$ 45. $y = x + 6$

46. width = 5 cm; length = 17 cm

47. $y = 3x$ 48. 5, 33

49. \$2.50

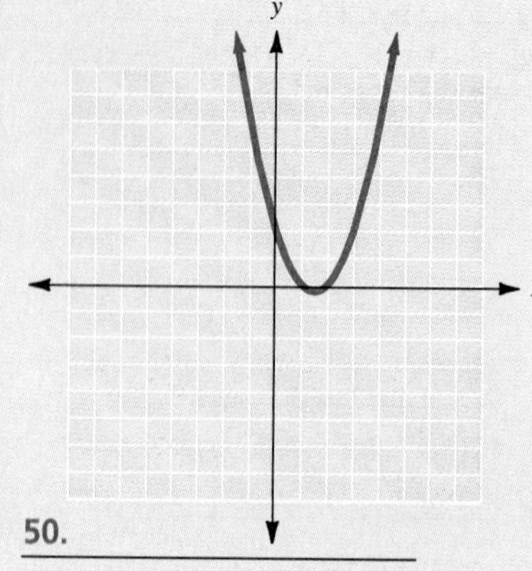

50.

In exercises 32 to 34, solve the given equation by the indicated method.

32. $3x^2 + 5x - 2 = 0$ by factoring

33. $x^2 + x - 3 = 0$ by using the quadratic formula

34. $x^2 + 6x = 5$ by completing the square

35. Solve the equation $\frac{1}{x - 1} + \frac{x + 1}{x^2 + 2x - 3} = \frac{1}{x + 3}$

36. Find the slope of the line through the points $(2, -3)$ and $(5, 9)$.

37. Find the slope of the line whose equation is $3x - 4y = 12$.

38. Find the equation of the line that passes through the point $(4, -2)$ and is parallel to the line $2x + y = 6$.

39. Graph the line whose equation is $4x + 5y = 20$.

In exercises 40 to 43, perform the indicated operations and simplify the result.

40. $-\sqrt{\frac{64}{25}}$

41. $\sqrt{18x^5y^6}$

42. $3\sqrt{20} - 2\sqrt{125}$

43. $(\sqrt{5} + 2)(\sqrt{5} - 8)$

44. Solve the system of equations

$$2x + 3y = 4$$
$$4x - 2y = 8$$

45. Determine the equation of the line in the following graph.

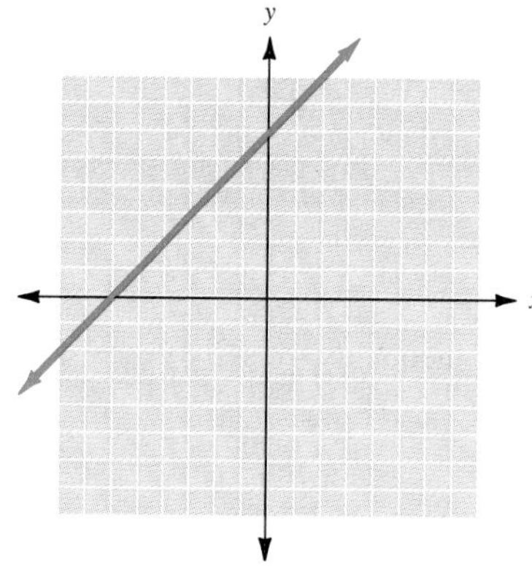

46. The length of a rectangle is 2 cm more than 3 times the width. The perimeter is 44 cm. Find the length.

47. Find the equation of the line that passes through the points $(-1, -3)$ and $(2, 6)$.

48. One number is 3 more than 6 times another. If the sum of the numbers is 38, find the two numbers.

49. A store marks up items to make a 30% profit. If an item sells for \$3.25, what does it cost before the mark up?

50. Graph the equation $y = x^2 - 3x + 2$.

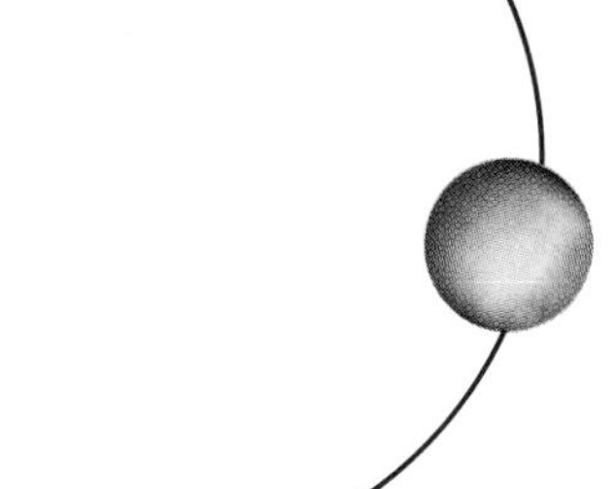

Answers to Pre-Tests, Self-Tests, and Cumulative Tests

Pre-Test for Chapter 0

1. 1, 2, 3, 6, 7, 14, 21, 42 **2.** Prime: 2, 3, 7, 17, 23; composite: 6, 9, 18, 21 **3.** $2 \times 2 \times 3 \times 5$ **4.** $2 \times 5 \times 5 \times 7$

5. 4 **6.** 6 **7.** $\frac{5}{4}$ **8.** $\frac{3}{2}$ **9.** $\frac{19}{12}$ **10.** $\frac{7}{18}$ **11.** 3.767 **12.** 22.8404

13. 6 **14.** 24 **15.** 4 **16.** 8 **17.** 13 **18.** 2

19.

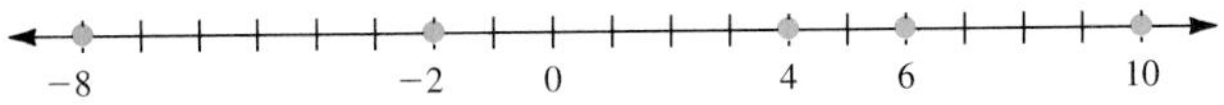

20. −4, −2, −1, 0, 1, 5 **21.** Max: 7; Min: −5 **22.** 5 **23.** 6 **24.** 6 **25.** 6 **26.** 6 **27.** 16 **28.** −23

Self-Test for Chapter 0

1. Prime: 5, 13, 17, 31; composite: 9, 22, 27, 45

2. $2 \times 2 \times 2 \times 3 \times 11$ **3.** 12 **4.** 8 **5.** $\frac{2}{7}$ **6.** $\frac{3}{4}$ **7.** $\frac{19}{12}$ **8.** $\frac{2}{21}$

9. 7.375 **10.** 3.884 **11.** 22.9635 **12.** 79.91 **13.** 4^4 **14.** 9^5 **15.** 3 **16.** 65 **17.** 144 **18.** 7 **19.** 8 **20.** 7

21.

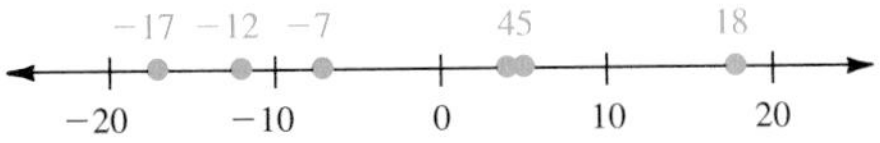

22. −6, −3, −2, 0, 2, 4, 5 **23.** Max: 6; Min: −5 **24.** 7 **25.** 7 **26.** 11 **27.** 11 **28.** −19 **29.** −40 **30.** 19

Pre-Test for Chapter 1

1. $x - 8$ **2.** $\frac{w}{17x}$ **3.** No **4.** Yes **5.** Commutative property of multiplication **6.** Distributive property **7.** Associative property of addition **8.** −10 **9.** −1 **10.** −5 **11.** −1 **12.** −3 **13.** −19

14. 12 **15.** 0 **16.** 21 **17.** 14 **18.** $\frac{1}{2}$ **19.** 7 **20.** −3 **21.** 55 **22.** −7 **23.** $8w^2t$ **24.** $-a^2 + 4a + 3$ **25.** $12xy^2$

Self-Test for Chapter 1

1. $a - 5$ **2.** $6m$ **3.** $4(m + n)$ **4.** $\frac{a + b}{3}$

5. Commutative property of multiplication **6.** Distributive property **7.** Associative property of addition **8.** 21 **9.** $20x + 12$ **10.** Not an expression **11.** Expression **12.** −13 **13.** −3 **14.** −21 **15.** 1 **16.** −6 **17.** −24 **18.** 9 **19.** 0 **20.** 3 **21.** 1 **22.** −40 **23.** 63 **24.** −27 **25.** −24 **26.** 14 **27.** −25 **28.** 3 **29.** −5 **30.** Undefined **31.** 3 **32.** 65 **33.** 144 **34.** −9 **35.** −4 **36.** $15a$ **37.** $19x + 5y$ **38.** $8a^2$ **39.** a^{14} **40.** $15x^3y^7$ **41.** $2x^3$ **42.** $4ab^3$ **43.** x^9 **44.** $2x - 8$ **45.** $2w + 4$

Pre-Test for Chapter 2

1. No **2.** Yes **3.** 8 **4.** −12 **5.** 7 **6.** 35 **7.** −2 **8.** 2

9. $\frac{P - 2L}{2}$ or $\frac{P}{2} - L$ **10.** $\frac{5x - 14}{3}$ or $\frac{5}{3}x - \frac{14}{3}$ **11.** $4x + 5 = 17$

12. $4(x + 6) = 10x + 6$ **13.** $x \le 15$ **14.** $x \le -1$ **15.** 6 **16.** 15, 17 **17.** 4 cm × 13 cm **18.** \$540 **19.** \$13,125 **20.** 15%

Self-Test for Chapter 2

1. No **2.** Yes **3.** 11 **4.** 12 **5.** 7 **6.** 7 **7.** −12 **8.** 25

9. 3 **10.** 4 **11.** $-\frac{2}{3}$ **12.** −5 **13.** $\frac{C}{2\pi}$ **14.** $\frac{3V}{B}$ **15.** $\frac{6 - 3x}{2}$

16. $x \le 14$ **17.** $x < -4$ **18.** $x \ge \frac{4}{3}$ **19.** $x > -1$ **20.** 7

21. 21, 22, 23 **22.** Juwan, 6; Jan, 12; Rick, 17 **23.** 10 in., 21 in. **24.** 5% **25.** \$35,000

Cumulative Test for Chapters 1 to 2

1. 4 **2.** −12 **3.** 8 **4.** 3 **5.** −18 **6.** 44 **7.** −5 **8.** 10 **9.** 0 **10.** Undefined **11.** 20 **12.** −11 **13.** 27 **14.** −28 **15.** −4 **16.** 2 **17.** $3x^2y$ **18.** $6x^4 - 10x^3y$ **19.** $x - 2y + 3$ **20.** $12x^2 + 3x$

21. 5 **22.** −24 **23.** $\frac{5}{4}$ **24.** $-\frac{2}{5}$ **25.** 5 **26.** $\frac{I}{Pt}$ **27.** $\frac{2A}{b}$

28. $\frac{c - ax}{b}$ **29.**

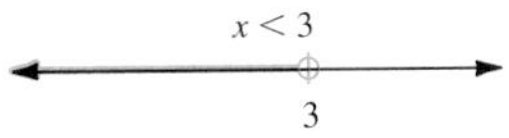

30.

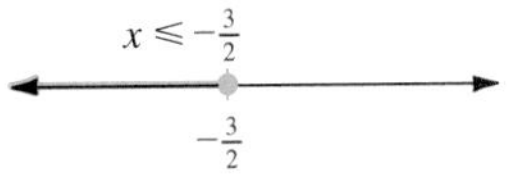

31.

32.

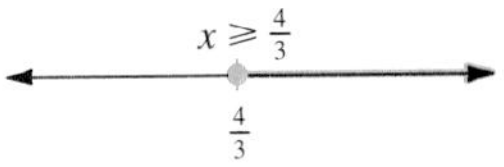

33. 13 **34.** 42, 43 **35.** 7

36. \$420 **37.** 5 cm, 17 cm **38.** 8 in., 13 in., 16 in. **39.** 2.5% **40.** 7.5%

Pre-Test for Chapter 3

1. x^{12} **2.** $8x^5y^7$ **3.** $3x^3y$ **4.** $4x^6y^8$ **5.** x^{16} **6.** $\frac{2y^3}{x^5}$ **7.** Binomial

8. Trinomial **9.** $2x^2 - 2x - 2$ **10.** $9x^2 - 11x + 6$ **11.** $12x^3y^3 - 6x^2y^2 + 21x^2y^4$ **12.** $6x^2 - 11x - 10$ **13.** $x^2 - 4y^2$ **14.** $16m^2 + 40m + 25$ **15.** $3x^3 - 14x^2y + 17xy^2 - 6y^3$ **16.** $9x^3 - 30x^2y + 25xy^2$ **17.** $4y - 5x^2y^3$ **18.** $x + 2$ **19.** $x - 3$

20. $3x - 5 - \frac{5}{x + 4}$

Self-Test for Chapter 3

1. a^{14} **2.** $15x^3y^7$ **3.** $2x^3$ **4.** $4ab^3$ **5.** $27x^6y^3$ **6.** $\frac{4w^4}{9t^6}$ **7.** $16x^{18}y^{17}$

8. 6 **9.** Binomial **10.** Trinomial **11.** $8x^4 - 3x^2 - 7$; 8, −3, −7; 4

12. 1 **13.** 6 **14.** $\frac{1}{y^5}$ **15.** $\frac{3}{b^7}$ **16.** $\frac{1}{y^4}$ **17.** $\frac{1}{p^{10}}$

18. $10x^2 - 12x - 7$ **19.** $7a^3 + 11a^2 - 3a$ **20.** $3x^2 + 11x - 12$ **21.** $b^2 - 7b - 5$ **22.** $7a^2 - 10a$ **23.** $4x^2 + 5x - 6$ **24.** $2x^2 - 7x + 5$ **25.** $15a^3b^2 - 10a^2b^2 + 20a^2b^3$ **26.** $3x^2 + x - 14$ **27.** $a^2 - 49b^2$ **28.** $8x^2 - 14xy - 15y^2$ **29.** $12x^3 + 11x^2y - 5xy^2$ **30.** $9m^2 + 12mn + 4n^2$

31. $2x^3 + 7x^2y - xy^2 - 2y^3$ **32.** $2x^2 - 3y$ **33.** $4c^2 - 6 + 9cd$
34. $x - 6$ **35.** $x + 2 + \frac{10}{2x - 3}$ **36.** $2x^2 - 3x + 2 + \frac{7}{3x + 1}$
37. $x^2 - 4x + 5 - \frac{4}{x - 1}$ **38.** 1.68×10^{20} **39.** 3.12×10^{-10}
40. 5.2×10^{19}

Cumulative Test for Chapters 1 to 3

1. 17 **2.** 6 **3.** 150 **4.** 4 **5.** 55 **6.** $-\frac{26}{21}$ **7.** $9x^{16}$ **8.** $\frac{x^{10}}{y^6}$
9. $8x^9y^3$ **10.** 7 **11.** 1 **12.** $\frac{1}{x^4}$ **13.** $\frac{3}{x^2}$ **14.** $\frac{1}{x^4}$ **15.** $\frac{1}{x^3y^3}$
16. $4x^5y$ **17.** $x^2 + 7x$ **18.** $4x^3 + 3x^2 + 2$ **19.** $x^2 - 2x - 15$
20. $x^2 + 2xy + y^2$ **21.** $9x^2 - 24xy + 16y^2$ **22.** $x + 4$ **23.** $x^3 - xy^2$
24. -2 **25.** -2 **26.** 84 **27.** 1 **28.** $2A - b$ **29.** $x \geq -2$
30. $x < -22$ **31.** Sam: \$510; Larry: \$250 **32.** 37, 39 **33.** \$2120
34. \$645

Pre-Test for Chapter 4

1. $5(3c + 7)$ **2.** $4q^3(2q - 5)$ **3.** $6(x^2 - 2x + 4)$
4. $7cd(c^2d - 3 + 2d^2)$ **5.** $(b - 3)(b + 5)$ **6.** $(x + 4)(x + 6)$
7. $(x - 9)(x - 5)$ **8.** $(a + 3b)(a + 4b)$ **9.** $(3y - 4)(y + 3)$
10. $(5w + 3)(w + 4)$ **11.** $(3x + 7y)(2x - 3y)$ **12.** $x(2x + 3)(x - 5)$
13. $(b + 7)(b - 7)$ **14.** $(6p + q)(6p - q)$ **15.** $(3x - 2y)^2$
16. $3x(3y - 4x)(3y + 4x)$ **17.** 4, 7 **18.** $-2, 7$ **19.** $\frac{3}{5}, -2$ **20.** 0, 2

Self-Test for Chapter 4

1. $6(2b + 3)$ **2.** $3p^2(3p - 4)$ **3.** $5(x^2 - 2x + 4)$
4. $6ab(a - 3 + 2b)$ **5.** $(a + 5)(a - 5)$ **6.** $(8m + n)(8m - n)$
7. $(7x + 4y)(7x - 4y)$ **8.** $2b(4a + 5b)(4a - 5b)$ **9.** $(a - 7)(a + 2)$
10. $(b + 3)(b + 5)$ **11.** $(x - 4)(x - 7)$ **12.** $(y + 10z)(y + 2z)$
13. $(x + 2)(x - 5)$ **14.** $(2x - 3)(3x + 1)$ **15.** $(2x - 1)(x + 8)$
16. $(3w + 7)(w + 1)$ **17.** $(4x - 3y)(2x + y)$ **18.** $3x(2x + 5)(x - 2)$
19. 3, 5 **20.** $-1, 4$ **21.** $-1, \frac{2}{3}$ **22.** 0, 3 **23.** 0, 4 **24.** $-3, 8$
25. 7

Cumulative Test for Chapters 1 to 4

1. 17 **2.** -2 **3.** $9x^2 - x - 5$ **4.** $-4a^2 - 2a - 5$
5. $6b^2 + 8b - 3$ **6.** $15r^3s^2 - 12r^2s^2 + 18r^2s^3$
7. $6a^3 - 5a^2b + 3ab^2 - b^3$ **8.** $-y^2 + 3xy - 2x^2$ **9.** $3a + 2$
10. $x^2 - 2x + \frac{5}{2x + 4}$ **11.** $x = -2$ **12.** $x \leq \frac{33}{5}$ **13.** $t = \frac{2S - na}{n}$
14. x^{17} **15.** $6x^5y^7$ **16.** $9x^4y^6$ **17.** $4xy^2$ **18.** $108x^8$
19. $12w^4(3w - 4)$ **20.** $5xy(x - 3 + 2y)$ **21.** $(5x + 3y)^2$
22. $4p(p + 6q)(p - 6q)$ **23.** $(a + 3)(a + 1)$ **24.** $2w(w^2 - 2w - 12)$
25. $(3x + 2y)(x + 3y)$ **26.** 3, 4 **27.** $-4, 4$ **28.** $\frac{2}{3}, -1$ **29.** 6
30. 5 in. by 21 in.

Pre-Test for Chapter 5

1. $\frac{-3b^6}{5a^2}$ **2.** $\frac{x + 4}{2}$ **3.** $\frac{x - 1}{2x}$ **4.** $\frac{13a}{6}$ **5.** 5 **6.** $x + 6$
7. $\frac{5w - 6}{2w^2}$ **8.** $\frac{3b + 3}{b(b - 3)}$ **9.** $\frac{-11}{6(x - 1)}$ **10.** $\frac{10}{x - 5}$ **11.** $\frac{a}{2b^2}$
12. $\frac{x + 3}{2x}$ **13.** $2b^2$ **14.** $\frac{x + y}{4x}$ **15.** $\frac{3x}{2}$ **16.** $\frac{y}{2y + x}$ **17.** 3
18. $-2, 5$ **19.** 40 **20.** No solution **21.** 5 **22.** 7 **23.** 5, 20
24. 48 mi/h going; 40 mi/h returning **25.** 15 ft, 40 ft

Self-Test for Chapter 5

1. $\frac{-3x^4}{4y^2}$ **2.** $\frac{4}{a}$ **3.** $\frac{x + 1}{x - 2}$ **4.** a **5.** 2 **6.** 5 **7.** $\frac{17x}{15}$
8. $\frac{3s - 2}{s^2}$ **9.** $\frac{4x + 17}{(x - 2)(x + 3)}$ **10.** $\frac{15}{w - 5}$ **11.** $\frac{4p^2}{7q}$ **12.** $\frac{2}{x - 1}$
13. $\frac{3}{4y}$ **14.** $\frac{3}{m}$ **15.** $\frac{2}{3x}$ **16.** $\frac{n}{2n + m}$ **17.** 4 **18.** $-3, 3$ **19.** 36
20. 2, 6 **21.** 6 **22.** 4 **23.** 4, 12
24. 50 mi/h going; 45 mi/h returning **25.** 20 ft, 35 ft

Cumulative Test for Chapters 1 to 5

1. $-2xy$ **2.** $\frac{4a^2}{3}$ **3.** $2x^2 - 5x + 6$ **4.** $3a^2 + 6a + 1$ **5.** 31
6. 1 **7.** $2x^2 - xy - 6y^2$ **8.** $x^2 + 11x + 28$ **9.** $2x - 1 + \frac{1}{x + 2}$
10. $x + 1 - \frac{4}{x - 1}$ **11.** 4 **12.** -2 **13.** $(x - 7)(x + 2)$
14. $3mn(m - 2n + 3)$ **15.** $(a + 3b)(a - 3b)$ **16.** $2x(x - 6)(x - 8)$
17. 7 **18.** 4 **19.** 264 ft/s **20.** 5 in. by 7 in. **21.** $\frac{m}{3}$ **22.** $\frac{a - 7}{3a + 1}$
23. $\frac{8r + 3}{6r^2}$ **24.** $\frac{x + 33}{3(x - 3)(x + 3)}$ **25.** $\frac{3}{x}$ **26.** $\frac{1}{3w}$ **27.** $\frac{x - 1}{2x + 1}$
28. $\frac{n}{3n + m}$ **29.** $\frac{6}{5}$ **30.** $\frac{-9}{2}, 7$

Pre-Test for Chapter 6

1. (15, 3); (18, 6) **2.** (0, 3); (2, 0) **3.** (1, 3); (0, 5); $(-3, 11)$
4. Answers vary **5.** $A(2, 3)$ $B(0, -5)$ $C(-3, 5)$
6.

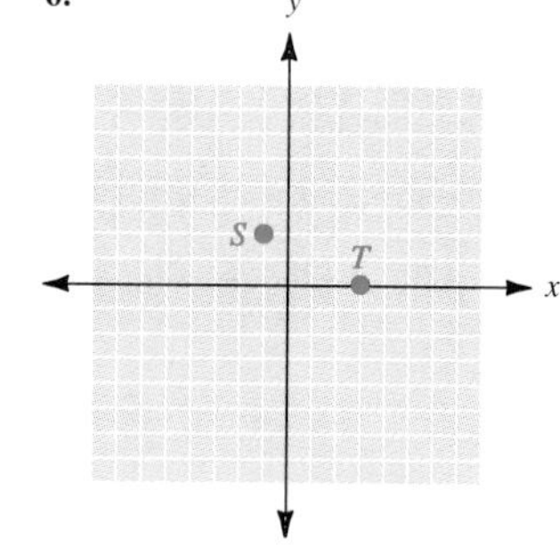

7.

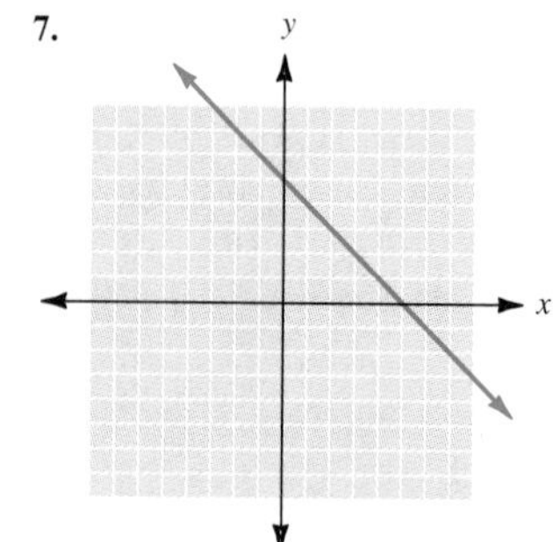

8.

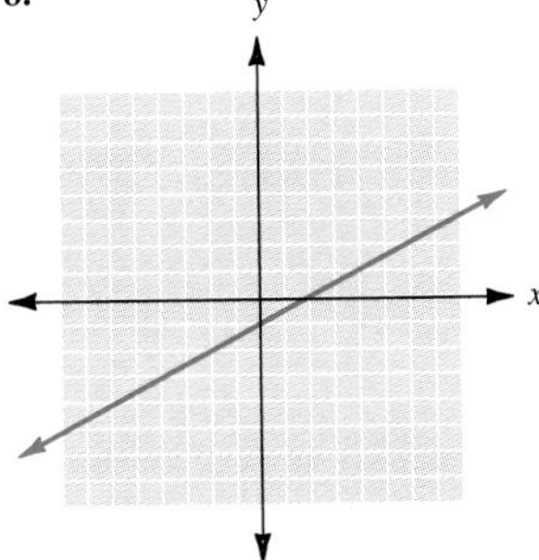

9.

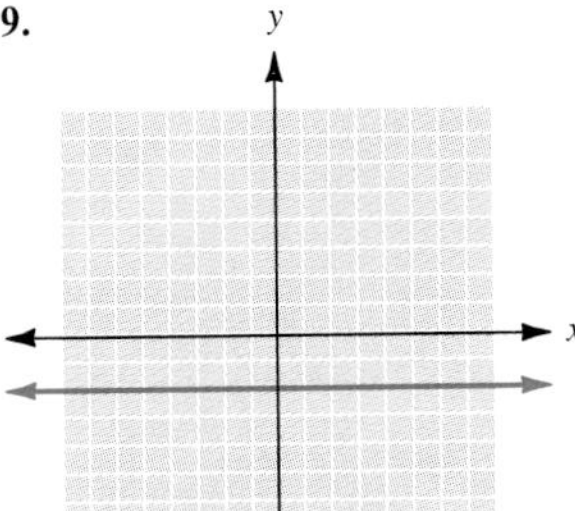

10. 1 11. −1 12. 8

13. 4 14. $-\frac{2}{5}$
15. $1000

Self-Test for Chapter 6

1. (3, 6), (9, 0) 2. (4, 0), (5, 4) 3. (3, 3), (6, 2), (9, 1)
4. (3, 0), (0, 4), $\left(\frac{3}{4}, 3\right)$ 5. Different answers are possible
6. Different answers are possible 7. (4, 2) 8. (−4, 6) 9. (0, −7)
10–12.

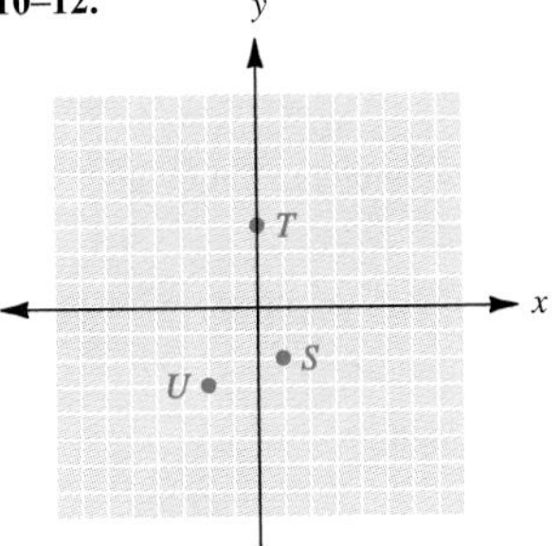

13. $x + y = 4$

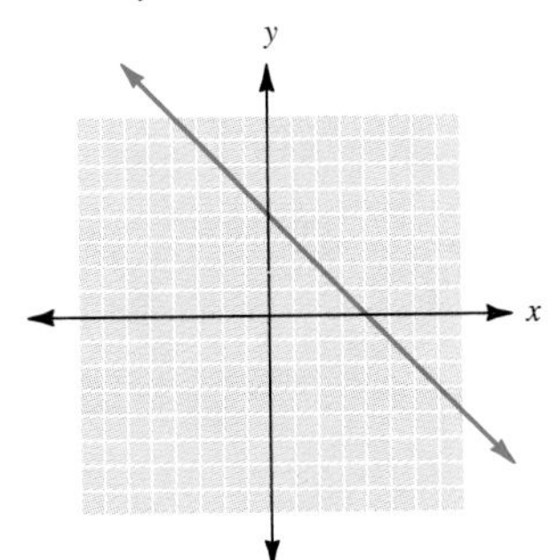

14. $y = 3x$

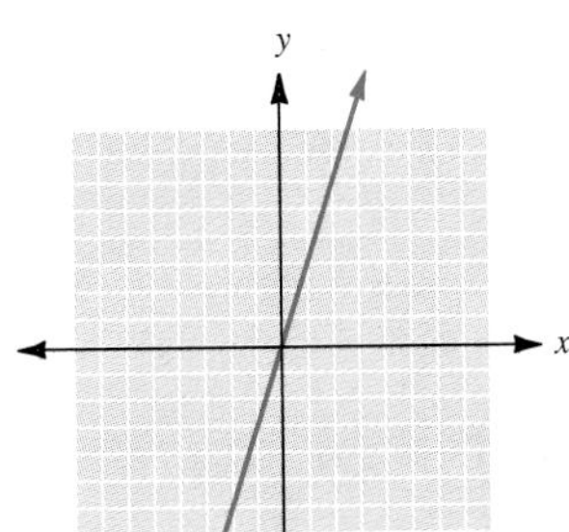

15. $y = \frac{3}{4}x - 4$

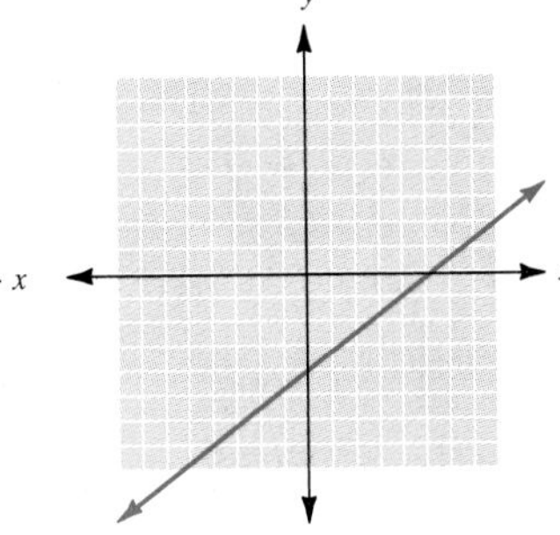

16. $x + 3y = 6$ 17. $2x + 5y = 10$

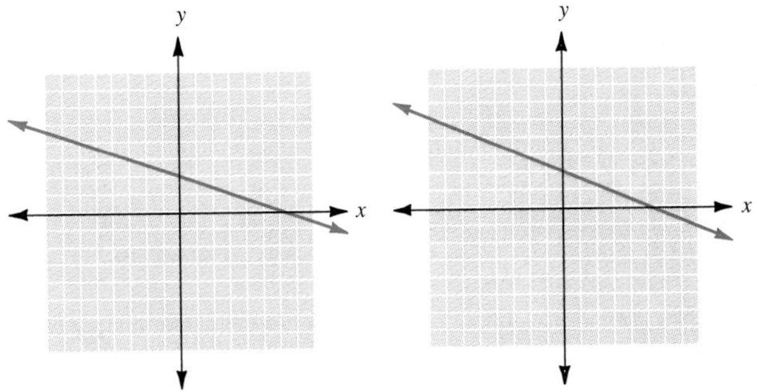

18. $y = -4$

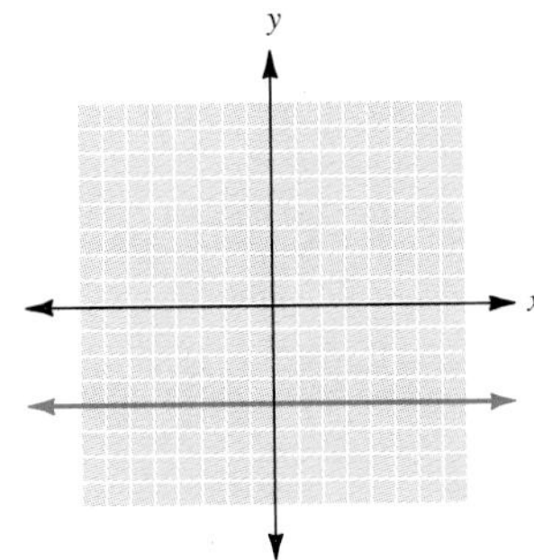

19. 1 20. $\frac{3}{4}$ 21. Undefined 22. 0 23. −7
24.

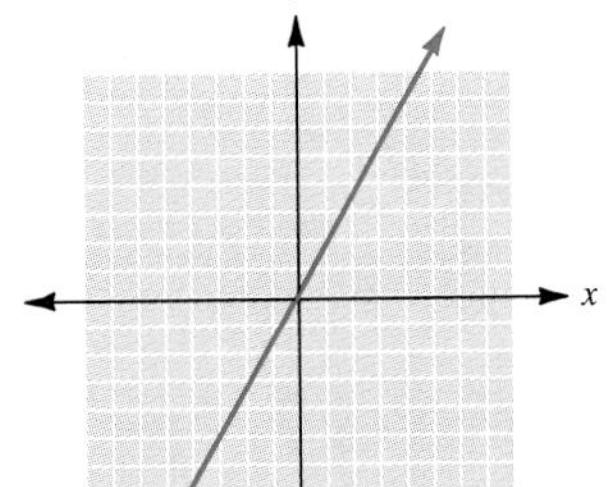

25. 5

Cumulative Test for Chapters 1 to 6

1. 3 2. 5 3. 37 4. −53 5. 69 6. −120 7. −5 8. 3
9. 108 10. 3 11. 69 12. 9 13. $-\frac{4}{3}$ 14. $-\frac{8}{3}$ 15. 10

16. 1 **17.** $C = \frac{5}{9}(F - 32)$ **18.** $x < 4$ **19.** $x \le 3$ **20.** $\frac{1}{x^4y^6}$

21. $\frac{y^5}{x}$ **22.** 1 **23.** $5x^2 - 10$ **24.** $-7a^2 + 7a + 2$ **25.** 19

26. 26 **27.** $6x^2 + 2xy - 20y^2$ **28.** $6x^3 - 3x^2 - 45x$

29. $4a^2 - 49b^2$ **30.** $4pn^2(3p + 5 - 4n)$

31. $(y - 3)(y^2 - 5)$ **32.** $b(3a + 7)(3a - 7)$

33. $2(3x + 2)(x - 1)$ **34.** $(2a + 3b)(3a - b)$

35. $-3, 11$ **36.** $\frac{4}{5}, \frac{2}{7}$ **37.** $\frac{-5a^3}{3b^2}$

38. $\frac{w-2}{w+3}$ **39.** $\frac{a+5}{a(a-5)}$ **40.** $\frac{10}{w-5}$

41. $\frac{3x^3}{4}$ **42.** $\frac{m-4}{4m}$ **43.** 6 **44.** 5

45. **46.**

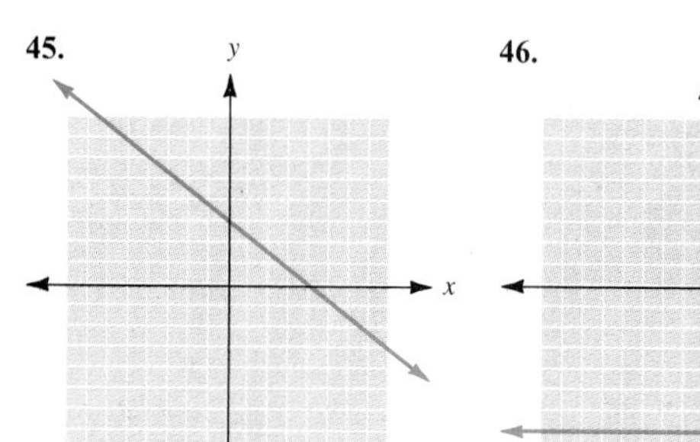

47.

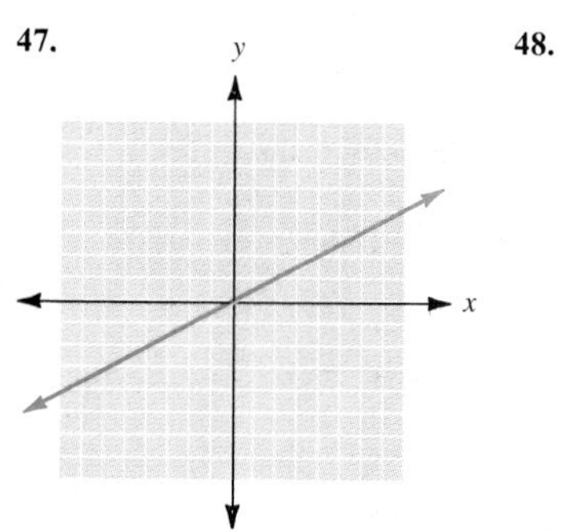

48. 2

49.

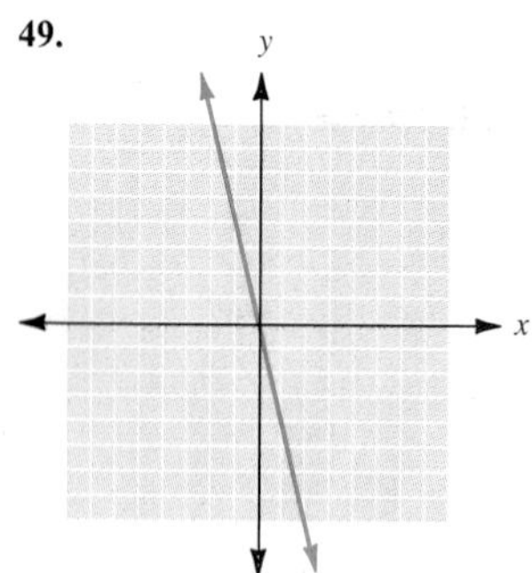

50. 30 **51.** Width: 5 in.; length: 7 in.

52. 41, 43, 45 **53.** $200

Pre-Test for Chapter 7

1. $\frac{10}{7}$ **2.** Slope: $\frac{5}{4}$; y intercept: $(0, -5)$

3. $y = 5x - 2$

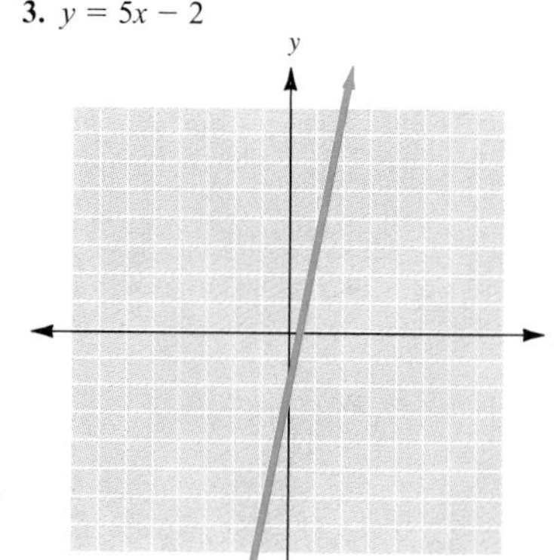

4. $y = -4x + 6$

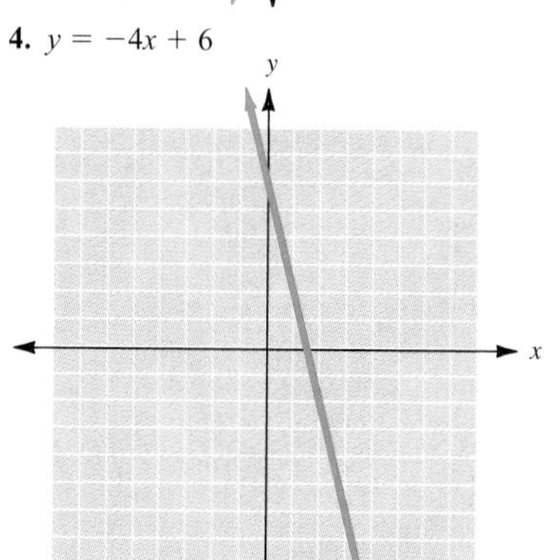

5. Parallel **6.** Perpendicular **7.** $y = -4x - 12$

8. $y = -\frac{5}{3}x + \frac{7}{3}$ **9.** $y = \frac{3}{5}x - 3$

10.

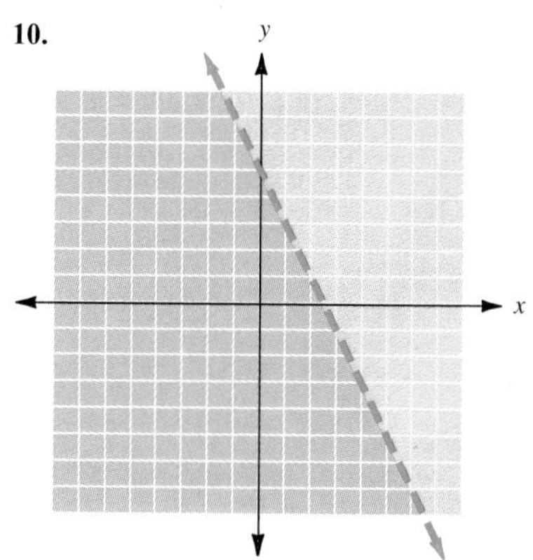

11.

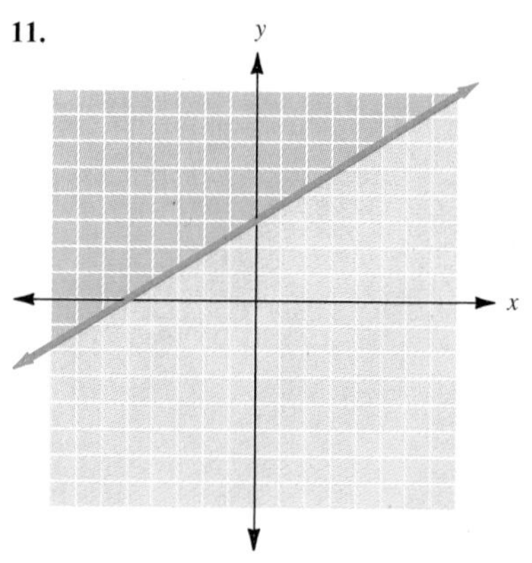

12. 13; -3

Self-Test for Chapter 7

1. 1 **2.** $\frac{3}{4}$ **3.** Slope: $\frac{4}{5}$; y intercept: $(0, -2)$

4. Slope: $-\frac{2}{3}$; y intercept: $(0, -9)$

5. $y = -3x + 6$

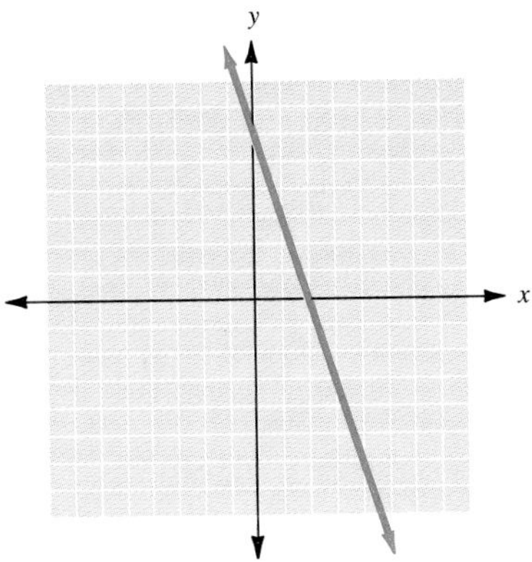

6. $y = 5x - 3$

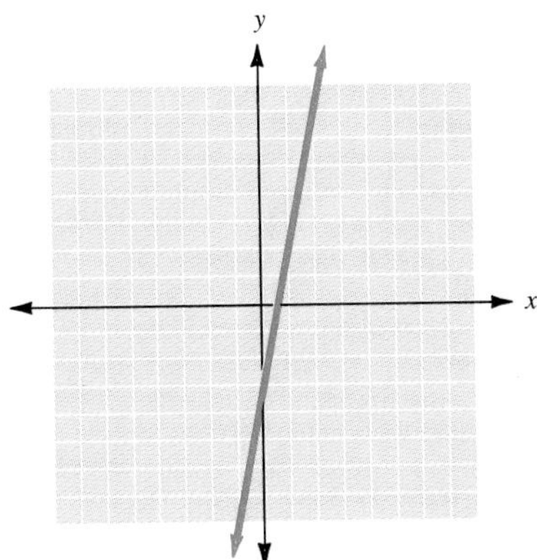

7. Parallel **8.** Perpendicular

9. $y = \frac{1}{2}x + \frac{21}{2}$

10. $y = \frac{4}{3}x + \frac{25}{3}$

11. $y = -\frac{3}{2}x + \frac{5}{2}$

12. $y = -\frac{2}{3}x + 5$

13. $y = -8$

14.

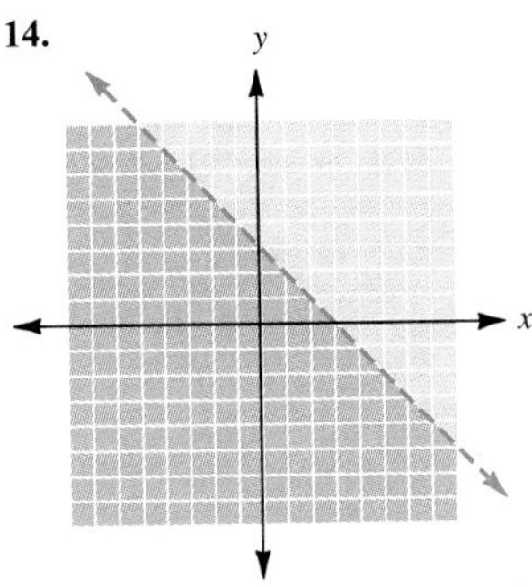

15.

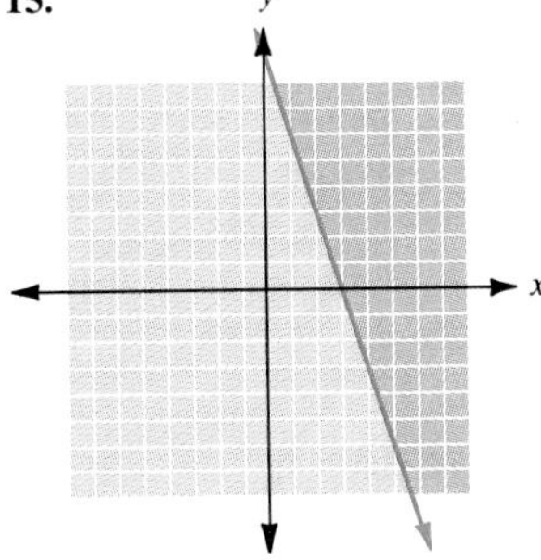

16.

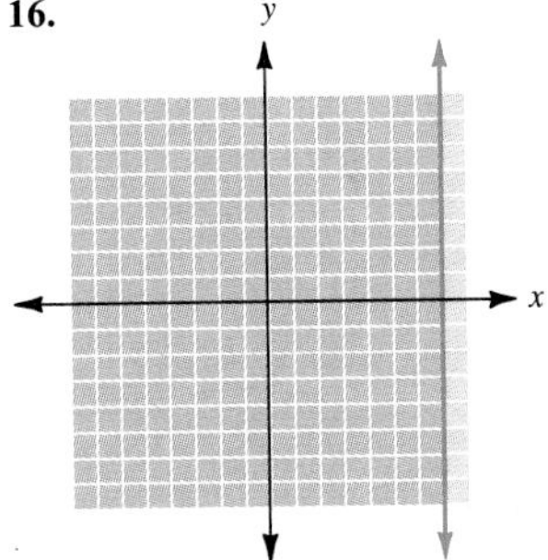

17. $-5; 7$ **18.** $-15; -7$ **19.** $-15; 6$ **20.** $3a - 25; 3x - 28$

Cumulative Test for Chapters 1 to 7

1. $x^2y^2 - 3xy$ **2.** $\frac{4m^3n}{3}$ **3.** $-x + 9$ **4.** $3z^2 - 3z + 5$

5. $2x^2 + 11x - 21$ **6.** $2a^2 + 6ab - 8b^2$ **7.** $x + 6 + \frac{20}{x - 3}$

8. $x^3 - 2x^2 + 4x - 10 + \frac{20}{x - 2}$ **9.** $-\frac{4}{3}$ **10.** -2

11. $(x - 8)(x + 7)$ **12.** $2x^2y(2x - y + 4x^2)$

13. $2a(2a + 3b)(2a - 3b)$ **14.** $3(5x - 2y)(x - y)$

15. 1 **16.** $-\frac{9}{4}$ **17.** $\frac{2x - 3}{x}$ **18.** $\frac{1}{a - 7}$

19. $\frac{5m + 6}{2m^2}$ **20.** $\frac{2x + 6}{x(x - 3)}$ **21.** $\frac{5y}{(y + 4)(y - 1)}$

22. -4 **23.** $7, -10$ **24.** 9

25. 49 mi/h going; 42 mi/h returning **26.** 126 min

27. $y = -\frac{1}{7}x + 2$ **28.** $y = -5x + 3$

29.

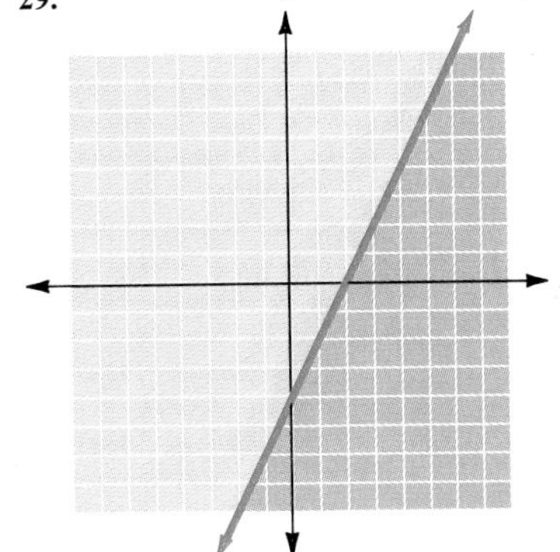

30. -2

Pre-Test for Chapter 8

1. Inconsistent system

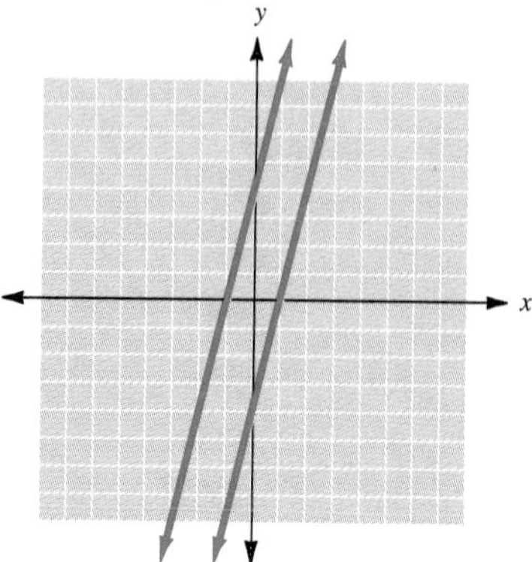

2. (0, −2)

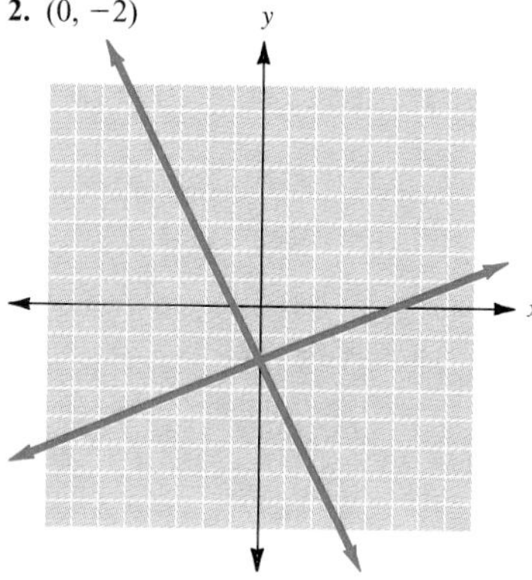

3. (6, −3) **4.** Dependent system **5.** (4, 1) **6.** (3, −1)
7. 16, 24 **8.** 13 m, 17 m **9.** 19 dimes, 26 quarters
10. canoe: 7.5 mi/h; current: 1.5 mi/h

11.

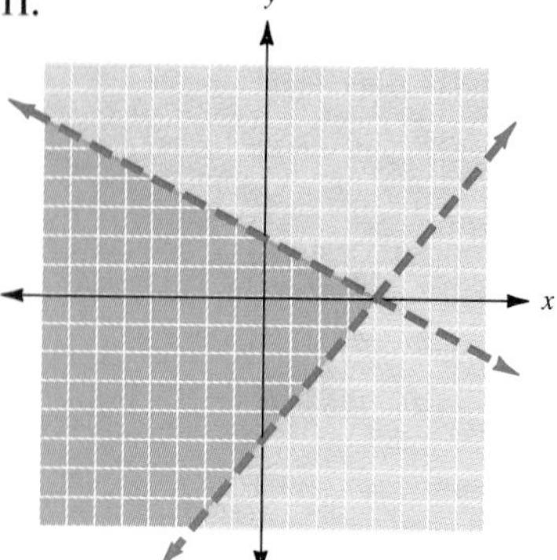

12.

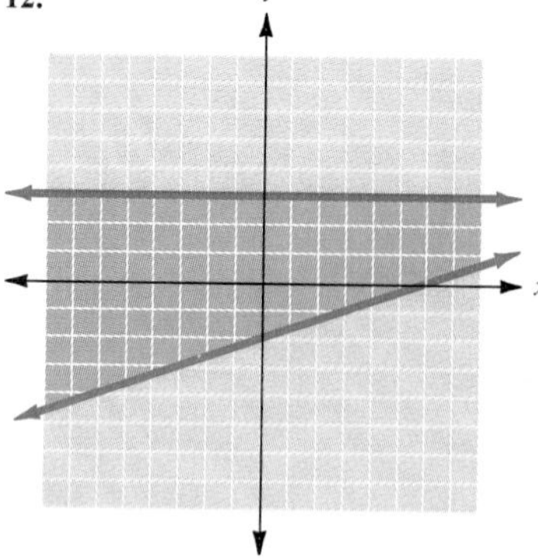

Self-Test for Chapter 8

1. (4, 1)

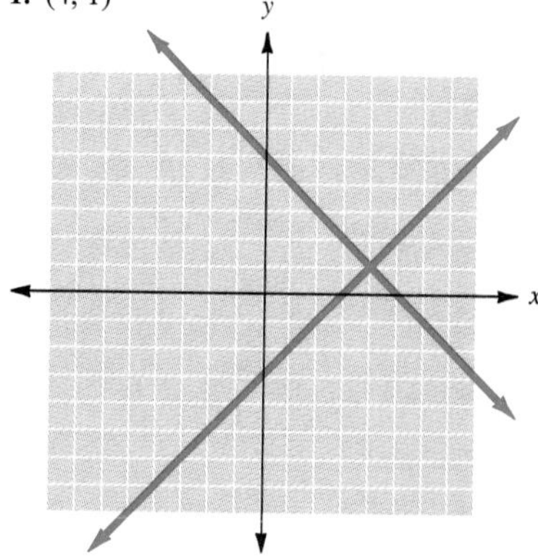

2. (4, 2)

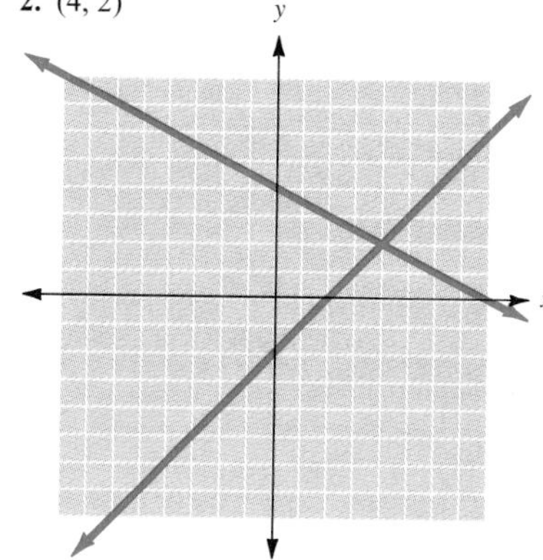

3. Inconsistent system

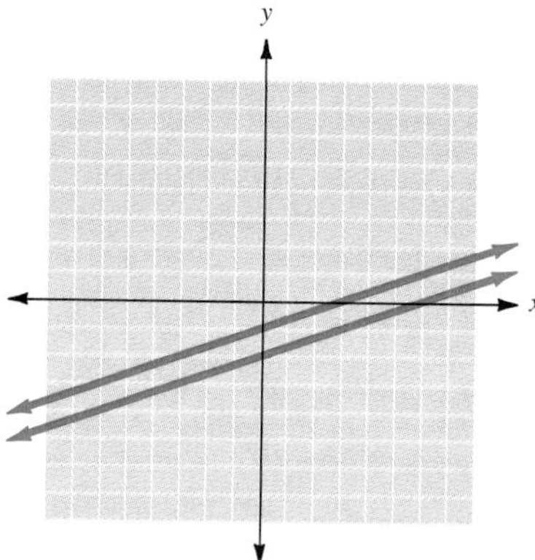

4. (2, 4)

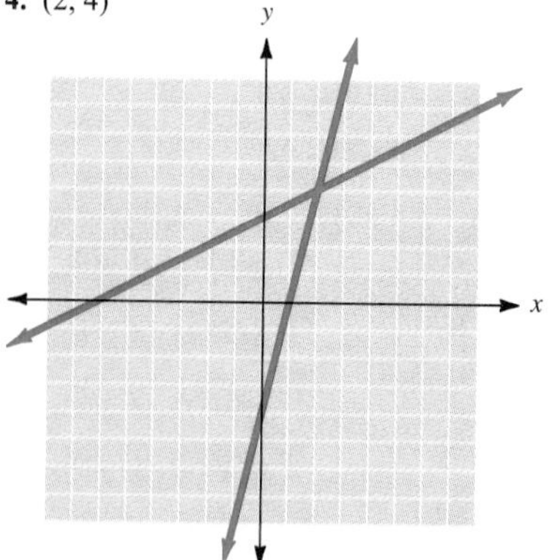

5. (4, 1) **6.** (4, 2) **7.** (1, 3) **8.** $\left(2, \frac{5}{2}\right)$

9. Dependent system **10.** $\left(\frac{3}{4}, -1\right)$ **11.** (6, 2)

12. Inconsistent system **13.** (2, 6) **14.** (6, −3)

15. (6, 2) **16.** (5, 4) **17.** (−3, 3) **18.** Inconsistent system

19. (3, −5) **20.** (3, 2) **21.** 12, 18 **22.** 21 m, 29 m

23. Width 12 in., length 20 in.

24. 12 dimes, 18 quarters **25.** Boat 15 mi/h, current 3 mi/h

26.

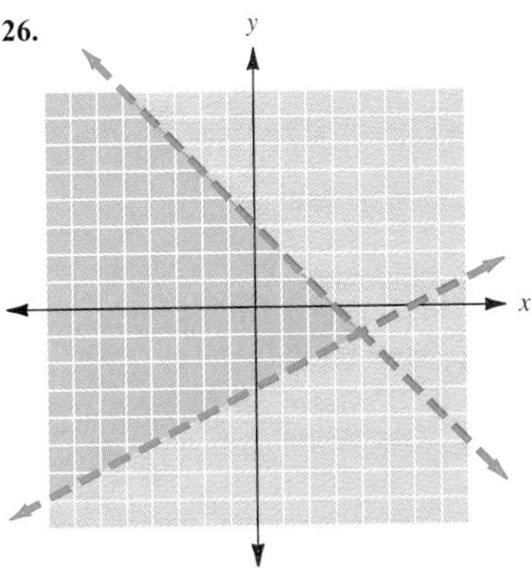

27.

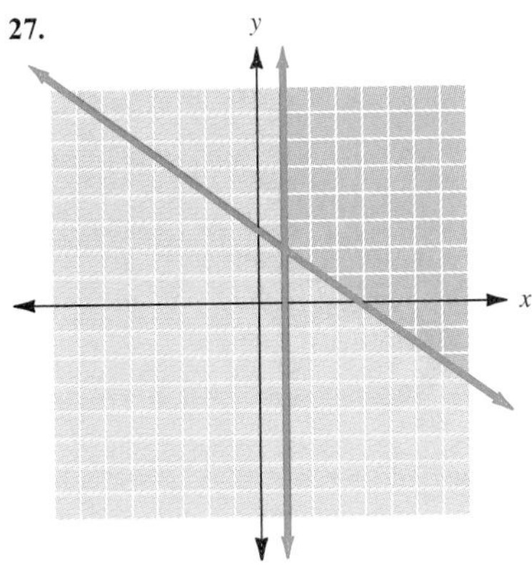

28.

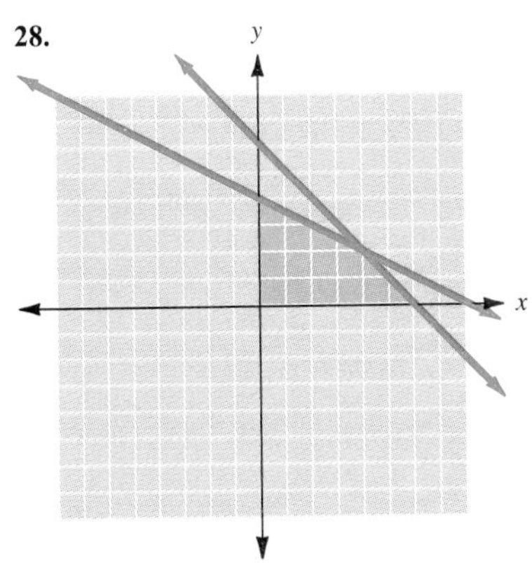

Cumulative Test for Chapters 1 to 8

1. $8x^2 - 7x - 4$ **2.** $w^2 - 8w - 4$

3. $28x^3y^2 - 14x^2y^2 + 21x^2y^3$ **4.** $15s^2 - 23s - 28$

5. $-x^2 + 2xy - 3y$ **6.** $2x + 4$ **7.** 9

8. $8a^2(3a - 2)$ **9.** $7mn(m - 3 - 7n)$

10. $(a + 8b)(a - 8b)$ **11.** $5p(p + 4q)(p - 4q)$

12. $(a - 6)(a - 8)$ **13.** $2w(w - 7)(w + 3)$

14. 4, 5 **15.** −4, 4 **16.** 7

17. 5 in. by 17 in. **18.** $\frac{m}{3}$

19. $\frac{a - 7}{3a + 1}$ **20.** $\frac{3}{x}$ **21.** $\frac{1}{3w}$

22.

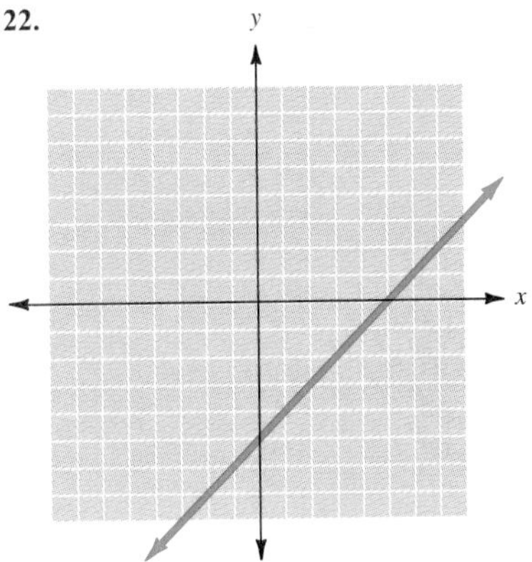

23.

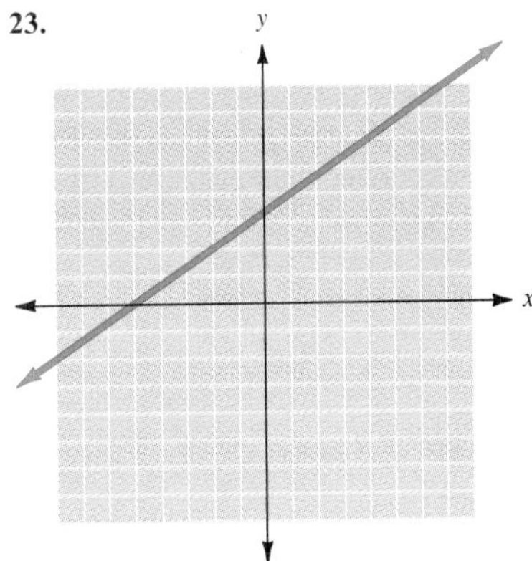

24.

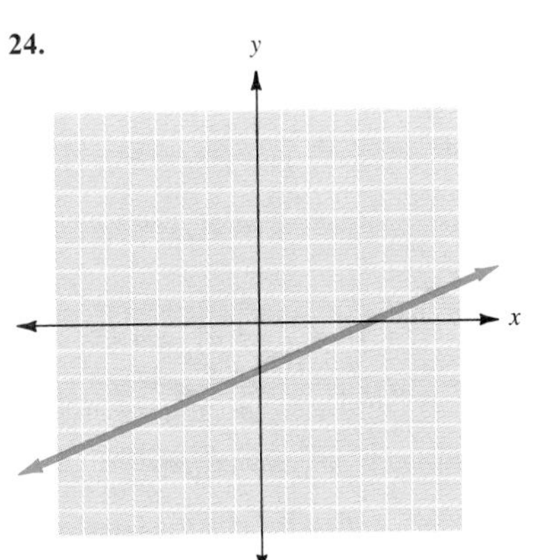

25.

26. $\frac{10}{7}$ 27. Slope: $\frac{5}{3}$; y intercept: $(0, -5)$ 28. $y = 2x - 5$

29.

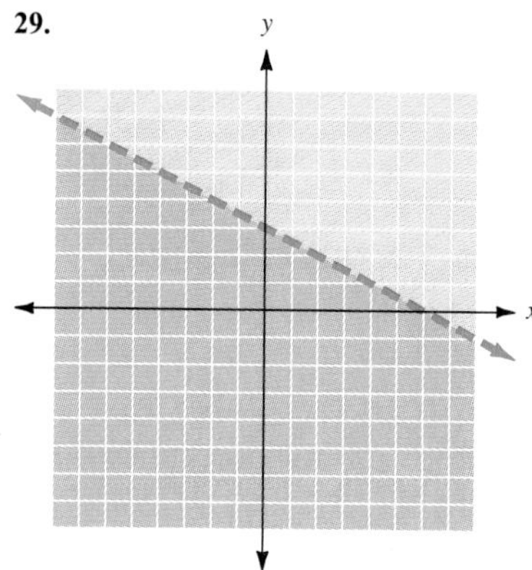

30.

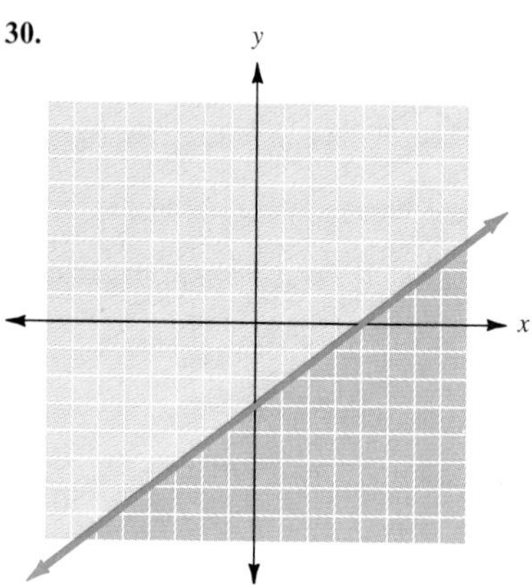

31. $(4, -3)$ 32. $\left(7, -\frac{5}{2}\right)$ 33. Dependent system 34. $(5, 0)$

35. Inconsistent system 36. $\left(\frac{3}{2}, -\frac{1}{3}\right)$ 37. 5, 21

38. VHS \$4.50, cassette \$1.50 39. 325 at \$7, 125 at \$4
40. \$5000 at 6%, \$7000 at 9%

Pre-Test for Chapter 9

1. 12 2. 4 3. Not a real number 4. 5 5. $6\sqrt{2}$ 6. $3x\sqrt{3x}$
7. $\frac{1}{2}$ 8. $\frac{\sqrt{7}}{5}$ 9. $7\sqrt{7}$ 10. $3\sqrt{5}$ 11. $3\sqrt{3}$ 12. $\sqrt{2}$ 13. $x\sqrt{10}$
14. $\sqrt{10} - 6$ 15. $9 + 5\sqrt{3}$ 16. $\frac{\sqrt{15}}{3}$ 17. $4 - \sqrt{5}$ 18. 15
19. 24 20. 8.49 21. 4.12 22. 19.21 cm 23. 14 24. 6.71
25. 15.26

Self-Test for Chapter 9

1. 11 2. 3 3. Not a real number 4. 4 5. $5\sqrt{3}$ 6. $2a\sqrt{6a}$
7. $\frac{4}{5}$ 8. $\frac{\sqrt{5}}{3}$ 9. $4\sqrt{10}$ 10. $3\sqrt{2}$ 11. $3\sqrt{2}$ 12. $4\sqrt{5}$
13. $3x\sqrt{2}$ 14. $11 + 5\sqrt{5}$ 15. $\frac{\sqrt{14}}{2}$ 16. $2 + 3\sqrt{2}$
17. $3 - \sqrt{3}$ 18. 20 19. 12 20. $3\sqrt{5}$ 21. $\sqrt{15}$
22. Approximately 9.747 cm 23. 9 24. $\sqrt{85}$ 25. $\sqrt{20} = 2\sqrt{5}$

Cumulative Test for Chapters 1 to 9

1. $3x^2y^3 - 2x^3y$ 2. $7x^2 - 6x + 12$ 3. 0 4. -8 5. 2 6. 6
7. $x > 4$ 8. $6x^4y - 10x^3y + 38x^2y$ 9. $20x^2 - 23xy - 21y^2$
10. $9xy(4 - 3x^2y)$ 11. $(4x - 3)(2x - 5)$ 12. $\frac{1}{15(x + 7)}$
13. $\frac{x - 3}{x - 5}$
14.

15.

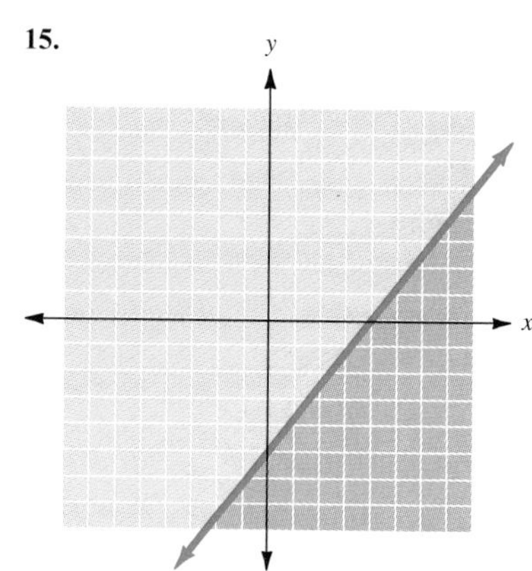

16. 5 17. $y = -\frac{3}{2}x + 5$ 18. $(5, 0)$ 19. Inconsistent system
20. Boat 13 mi/h, current 3 mi/h 21. 12 22. -12
23. Not a real number 24. -3 25. $-4a\sqrt{5}$ 26. $\frac{2x\sqrt{6x}}{3}$
27. $4 - 2\sqrt{2}$ 28. $7x\sqrt{2}$ 29. $5mn\sqrt{6m}$ 30. $\frac{2a\sqrt{3}}{5}$

Pre-Test for Chapter 10

1. $\pm\sqrt{17}$ 2. $\pm 2\sqrt{3}$ 3. $1 \pm \sqrt{5}$ 4. $\pm\frac{\sqrt{14}}{3}$
5. $-2, 5$ 6. $\frac{5 \pm \sqrt{17}}{2}$ 7. $2 \pm 2\sqrt{2}$ 8. $\frac{2 \pm 3\sqrt{2}}{2}$
9. $-7, 2$ 10. 5 11. $\frac{-3 \pm \sqrt{29}}{2}$ 12. $\frac{3 \pm \sqrt{33}}{4}$ 13. $\frac{2 \pm \sqrt{10}}{3}$
14. $\frac{-3 \pm \sqrt{37}}{2}$

15.

x	y
−2	6
−1	3
0	2
1	3
2	6

16.

x	y
−4	0
−3	−3
−2	−4
−1	−3
0	0

17.

x	y
−2	−4
−1	−6
0	−6
1	−4
2	0

18.

x	y
−2	−1
−1	2
0	3
1	2
2	−1

Self-Test for Chapter 10

1. $\pm\sqrt{15}$ **2.** $\pm 2\sqrt{2}$ **3.** $1 \pm \sqrt{7}$ **4.** $\frac{\pm\sqrt{10}}{3}$ **5.** 4, −2

6. $\frac{-3 \pm \sqrt{13}}{2}$ **7.** $-1 \pm \sqrt{6}$ **8.** $\frac{5 \pm \sqrt{17}}{4}$ **9.** −1, 3 **10.** 3

11. $\frac{5 \pm \sqrt{33}}{2}$ **12.** $\frac{1 \pm \sqrt{11}}{2}$ **13.** $\frac{1 \pm \sqrt{33}}{4}$ **14.** $-1 \pm \sqrt{6}$

15. $y = x^2 + 4$

x	y
−2	8
−1	5
0	4
1	5
2	8

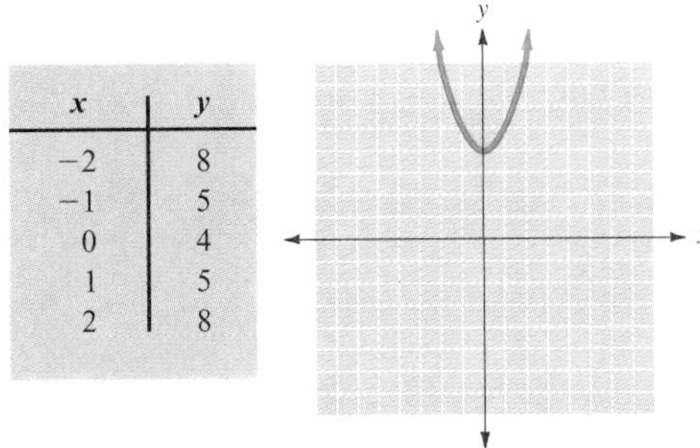

16. $y = x^2 - 2x$

x	y
−1	3
0	0
1	−1
2	0
3	3

17. $y = x^2 - 3$

x	y
−2	1
−1	−2
0	−3
1	−2
2	1

18. $y = x^2 + x - 2$

x	y
−2	0
−1	−2
0	−2
1	0
2	4

19. $y = -x^2 + 4$

x	y
−2	0
−1	3
0	4
1	3
2	0

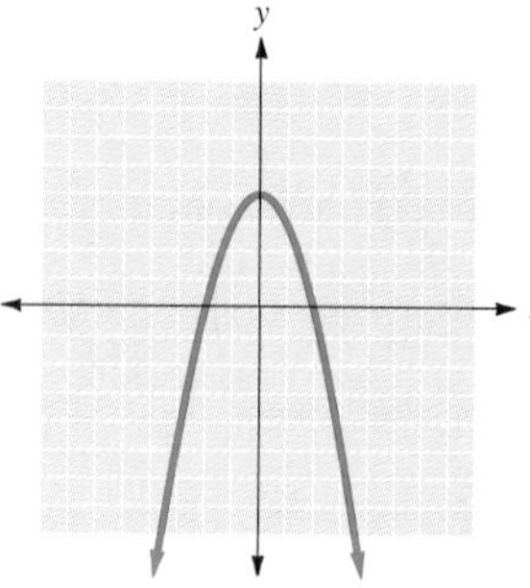

20. $y = -x^2 + 2x$

x	y
−1	−3
0	0
1	1
2	0
3	−3

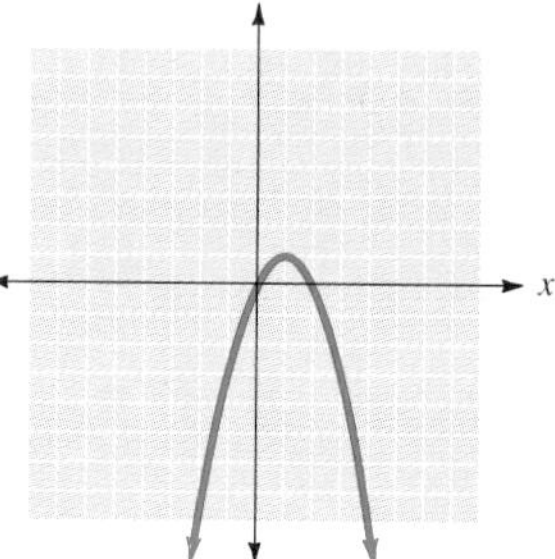

Cumulative Test for Chapters 0 to 10

1. $11x^2y - 6xy^2$ **2.** $x^2 + 5x - 7$ **3.** -480 **4.** -1680 **5.** 1 **6.** $x > 2$ **7.** $6x^3y - 3x^2y + 15xy$ **8.** $6x^2 + 11x - 10$ **9.** $9x^2 - 16y^2$ **10.** $8xy^2(2x - y)$ **11.** $(2x - 3)(4x + 5)$ **12.** $(5x - 4y)(5x + 4y)$ **13.** $\dfrac{29}{28(x + 2)}$ **14.** $\dfrac{5(x - 2)}{x - 1}$ **15.** $\dfrac{(3x - 1)(x + 6)}{15x^2}$

16. $3x - 2y = 6$

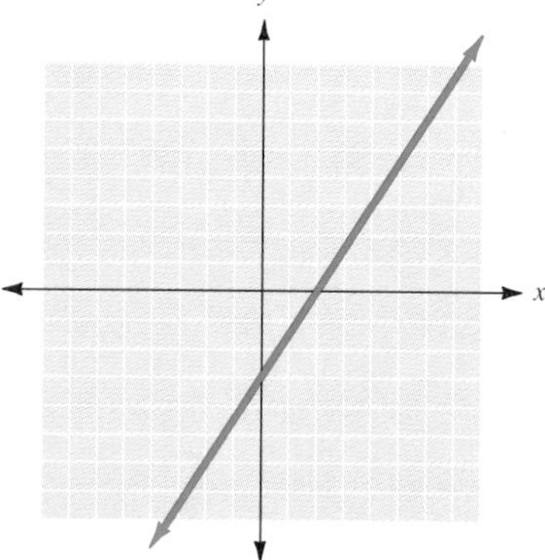

17. $y = 4x - 5$

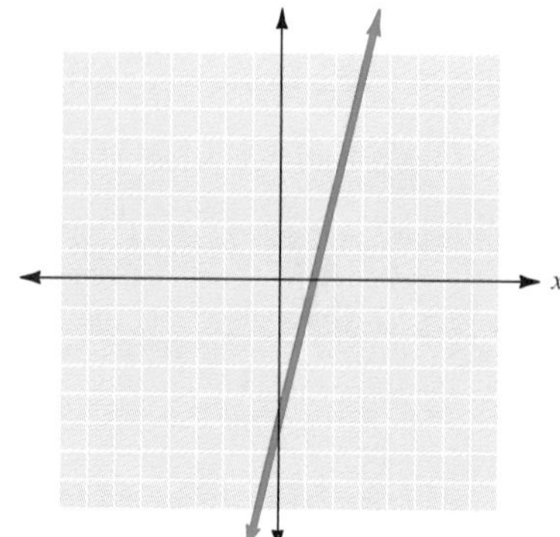

18. 5 **19.** $y = 2x - 5$

20.

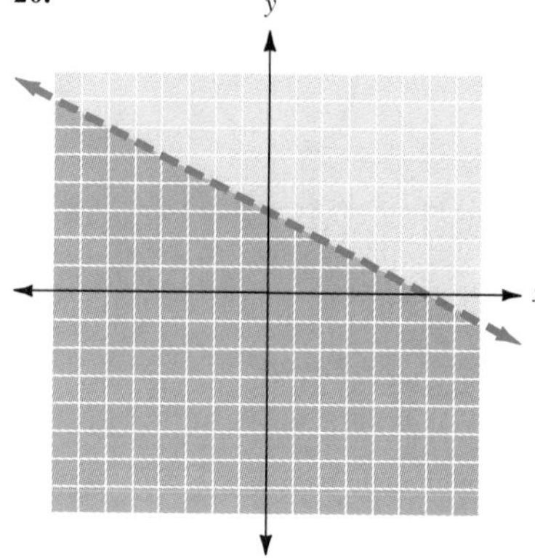

21. $\left(4, \dfrac{2}{3}\right)$ **22.** $(3, -2)$ **23.** $\left(3, \dfrac{-7}{2}\right)$

24. Dependent system **25.** 5, 21

26. 325 reserved seat, 125 gen adm

27. 100 mL of 30%, 200 mL of 60% **28.** 13 **29.** -13

30. Not a real number **31.** -4 **32.** $6\sqrt{3}$ **33.** $30a\sqrt{3}$

34. $-13 - 2\sqrt{2}$ **35.** $2 - \sqrt{2}$ **36.** $\pm 6\sqrt{2}$ **37.** $-3 \pm 2\sqrt{3}$

38. $\dfrac{5 \pm \sqrt{41}}{4}$

39. $y = x^2 - 2$

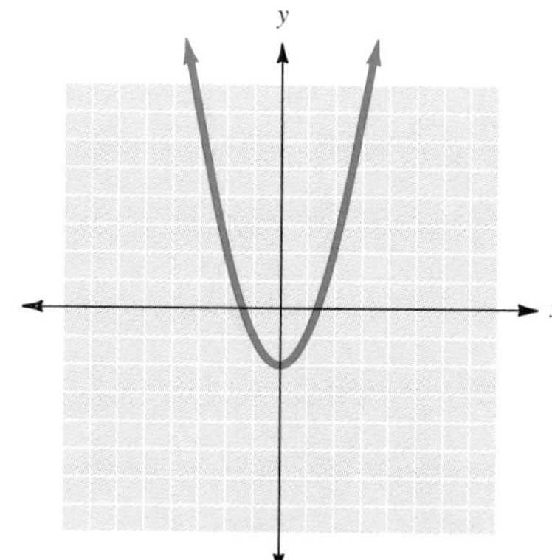

40. $y = x^2 - 4x$

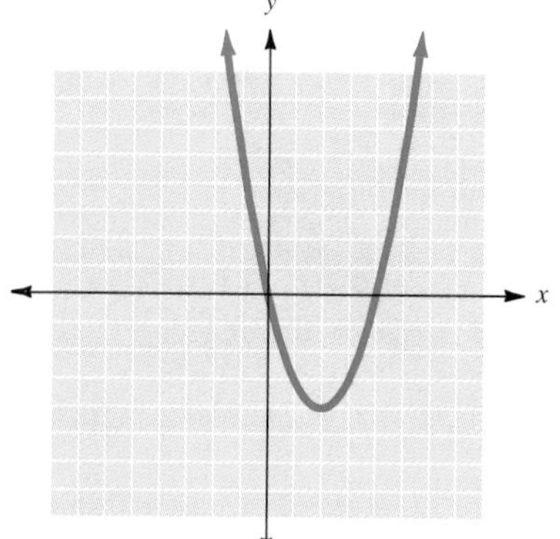

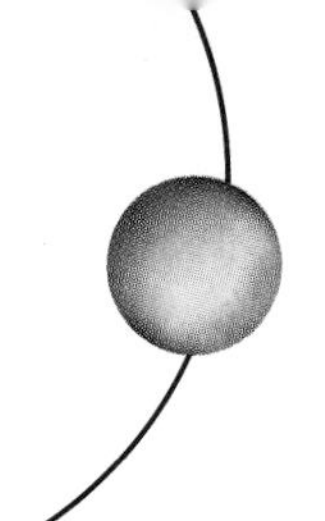

Answers to Final Examination

1. 14 **2.** -6 **3.** -26 **4.** -105 **5.** $-\frac{14}{5}$ **6.** -34 **7.** 40

8. 7 **9.** $9p^3q^5$ **10.** $\frac{1}{z^6}$ **11.** $25c^{16}d^{14}$ **12.** $\frac{2x^2}{y^4z^4}$ **13.** $2x^2 + 6x + 5$

14. $-2x^2 + 29x - 42$ **15.** $28x^2 - x - 45$ **16.** $9x^2 - 16y^2$

17. $25x^2 - 20xy + 4y^2$ **18.** $x^2 - 15xy - 2y^2$ **19.** $\frac{1}{2}$ **20.** -2

21. -1 **22.** $x < -\frac{1}{2}$ **23.** $10(x + 7)(x - 7)$ **24.** $(x - 6)(x - 6)$

25. $(4p - 9)(p + 2)$ **26.** $(y + z)(3x - 5)$ **27.** $\frac{2}{3y}$ **28.** 4

29. $\frac{a}{(a - 3)(a - 2)}$ **30.** $\frac{(y + 2)(y + 1)}{3(y + 5)}$ **31.** $x(x + 2)$

32. $\frac{1}{3}, -2$ **33.** $\frac{-1 \pm \sqrt{13}}{2}$ **34.** $-3 \pm \sqrt{14}$ **35.** -5

36. 4 **37.** $\frac{3}{4}$ **38.** $y = -2x + 6$

39.

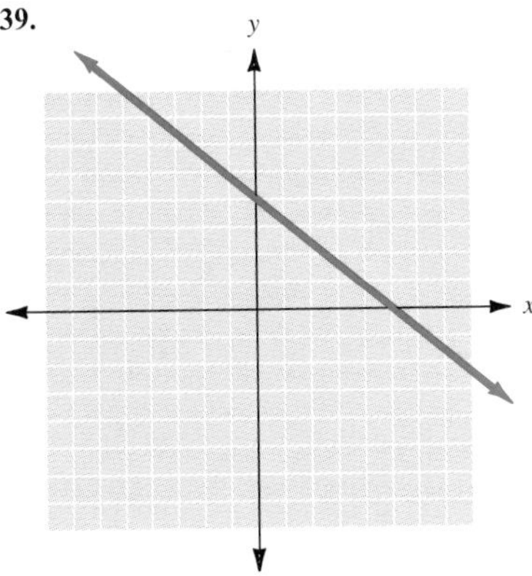

40. $-\frac{8}{5}$ **41.** $3x^2y^3\sqrt{2x}$ **42.** $-4\sqrt{5}$ **43.** $-11 - 6\sqrt{5}$

44. $(2, 0)$ **45.** $y = x + 6$ **46.** width = 5 cm; length = 17 cm

47. $y = 3x$ **48.** 5, 33 **49.** \$2.50

50.

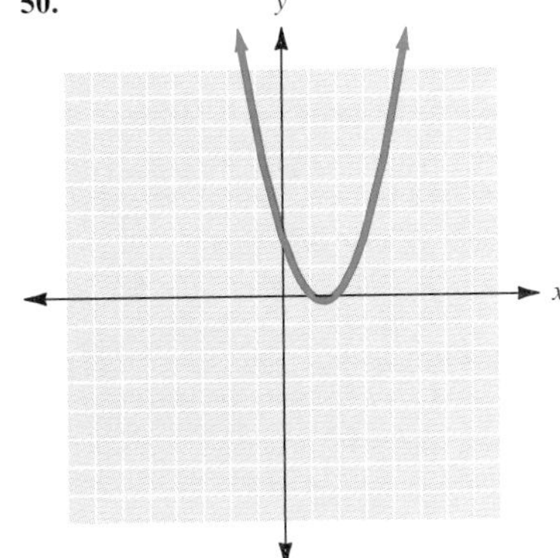

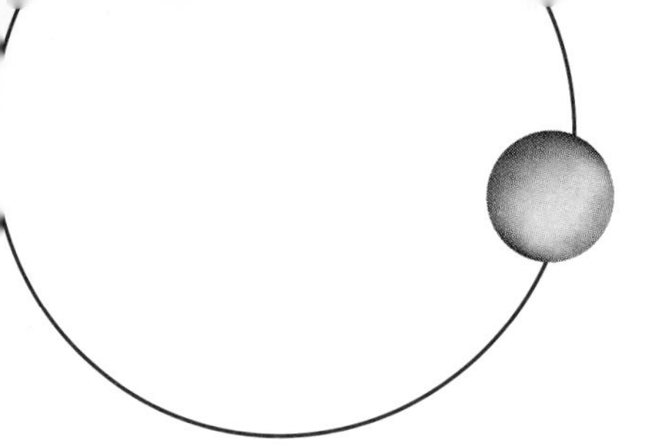

Index